1986

Experimental Microbial Ecology

Experimental Microbial Ecology

edited by Richard G. Burns *Biological Laboratory*
University of Kent, Canterbury
and J. Howard Slater *Department of Environmental Sciences*
University of Warwick, Coventry

Blackwell Scientific Publications *Oxford London*
Edinburgh Boston Melbourne

Editorial offices:
Osney Mead, Oxford, OX2 0EL
8 John Street, London, WC1N 2ES
9 Forest Road, Edinburgh, EH1 2QH
52 Beacon Street, Boston, Massachusetts 02108, USA
99 Barry Street, Carlton, Victoria 3053, Australia

First published 1982

Printed and bound in
Great Britain by
William Clowes (Beccles) Limited
Beccles and London

DISTRIBUTORS

USA
Blackwell Mosby Book Distributors
11830 Westline Industrial Drive
St Louis, Missouri 63141

Canada
Blackwell Mosby Book Distributors
120 Melford Drive, Scarborough
Ontario, M1B 2X4

Australia
Blackwell Scientific Book Distributors
214 Berkeley Street, Carlton
Victoria 3053

British Library
Cataloguing in Publication Data

Experimental microbial ecology.
1. Microbial ecology—Laboratory manuals
2. Biotic communities
I. Burns, Richard II. Slater, J. Howard
576'.15 QR100

ISBN 0 632 00765 6

Contents

List of Contributors

R. M. Atlas *Department of Biology, University of Louisville, Louisville, Kentucky 40208, USA*

H. J. Babich *Department of Biology, New York University, 952 Brown Building, Washington Square, New York, New York 10003, USA*

G. L. Barron *Department of Environmental Biology, University of Guelph, Guelph, Ontario, Canada N1G 2W1*

P. Berwick *Department of Microbiology, University College Cardiff, PO Box 97, Cardiff CF1 1XP, Wales*

C. M. Brown *Department of Brewing and Biological Sciences, Heriot-Watt University, Chambers Street, Edinburgh EH1 1HX, Scotland*

R. G. Burns *Biological Laboratory, University of Kent, Canterbury, Kent CT2 7NJ, England*

K-J. Cheng *Department of Biology, University of Calgary, 2500 University Drive NW, Calgary, Alberta, Canada T2N 1N4*

J. W. Costerton *Department of Biology, University of Calgary, 2500 University Drive NW, Calgary, Alberta, Canada T2N 1N4*

D. J. Cox *Department of Microbiology, Queen Elizabeth College, University of London, Campden Hill Road, London W8 7AH, England*

F. B. Dazzo *Department of Microbiology and Public Health, Michigan State University, East Lansing, Michigan 48824, USA*

C. H. Dickinson *Department of Plant Biology, Ridley Building, University of Newcastle upon Tyne, Newcastle upon Tyne NE1 7RU, England*

K-E. Eriksson *Biochemistry and Microbiology Research, Swedish Forest Products Research Laboratories, PO Box 5604, S-11486 Stockholm, Sweden*

B. W. Ferry *Department of Botany, Bedford College, University of London, Regent's Park, London NW1 4NS, England*

B. J. Finlay *Freshwater Biological Association, Windermere Laboratory, The Ferry House, Ambleside, Cumbria LA22 0LP, England*

K. P. Flint *Department of Environmental Sciences, University of Warwick, Coventry, West Midlands CV4 7AL, England*

D. D. Focht *Department of Soil and Environmental Sciences, University of California, Riverside, California 92521, USA*

M. P. Greaves *Agricultural Research Council, Weed Research Organisation, Begbroke Hill, Yarnton, Oxford, Oxfordshire OX5 1PF, England*

B. D-D. Grosovsky *Biological Science Center, Boston University, 2 Cummington Street, Boston, Massachusetts 02215, USA*

W. Harder *Department of Microbiology, University of Groningen, Kerklaan 30, Haren (GR), The Netherlands*

D. J. Hardman *Department of Environmental Sciences, University of Warwick, Coventry, West Midlands CV4 7AL, England*
Present address: *Department of Biochemistry, University of Manchester Institute of Science and Technology, PO Box 88, Manchester M60 1QD, England*

K. G. Hardy *Biogen S.A., Route de Troinex 3, 1227 Carouge/Geneva, Switzerland*

R. A. Herbert *Department of Biological Sciences, University of Dundee, Dundee DD1 4HN, Scotland*

D. E. Hughes *Department of Microbiology, University College Cardiff, PO Box 97, Cardiff CF1 1XP, Wales*

S. C. Johnsrud *Biochemistry and Microbiology Research, Swedish Forest Products Research Laboratories, PO Box 5604, S-11486 Stockholm, Sweden*

J. G. Kuenen *Department of Microbiology, University of Groningen, Kerklaan 30, Haren (GR), The Netherlands*
Present address: *Laboratory of Microbiology, Delft University of Technology, Julianalaan 67a, Delft 8, The Netherlands*

K. B. Logan *Department of Biological Sciences, University of Warwick, Coventry, West Midlands CV4 7AL, England*
Present address: *AMF CUNO Division Europe, Chemin du Contre Halage, Les Attaques, 62730 Marck, France*

J. M. Lynch *Agricultural Research Council, Letcombe Laboratory, Wantage, Oxfordshire OX12 9JT, England*

R. B. McKercher *Saskatchewan Institute of Pedology, University of Saskatchewan, Saskatoon, Canada S7N 0W0*

L. Margulis *Biological Science Center, Boston University, 2 Cummington Street, Boston, Massachusetts 02215, USA*

I. Morris *Bigelow Laboratory for Ocean Sciences, McKown Point, West Boothbay Harbor, Maine 04575, USA*
Present address: *Center for Environmental and Estuarine Studies, University of Maryland, Horn Point, PO Box 775, Cambridge, Maryland 21613, USA*

W. C. Noble *Department of Bacteriology, Institute of Dermatology, St John's Hospital for Diseases of the Skin, Hometon Grove, London E9 6BX, England*

C. G. Orpin *Agricultural Research Council, Institute of Animal Physiology, Babraham, Cambridge, Cambridgeshire CB2 4AT, England*

D. Parkinson *Department of Biology, University of Calgary, 2920 24 Avenue NW, Calgary, Canada T2N 1N4*

C. Ll. Powell *Ruakura Soil and Plant Research Station, Private Bag, Hamilton, New Zealand*

S. B. Primrose *Department of Biological Sciences, University of Warwick, Coventry, West Midlands CV4 7AL, England*
Present address: *G. D. Searle and Company Ltd, PO Box 53, Lane End Road, High Wycombe, Buckinghamshire HP12 4HL, England*

J. I. Prosser *Department of Microbiology, University of Aberdeen, Marischal College, Aberdeen AB9 1AS, Scotland*

S. C. Rittenberg *Department of Microbiology, University of California, Los Angeles, California 90024, USA*

F. E. Round *Department of Botany, University of Bristol, Woodland Road, Bristol, Avon BS8 1UG, England*

N. D. Seeley *Department of Biological Sciences, University of Warwick, Coventry, West Midlands CV4 7AL, England*

J. H. Slater *Department of Environmental Sciences, University of Warwick, Coventry, West Midlands CV4 7AL, England*

D. W. Smith *School of Life and Health Sciences, Ecology and Organismic Biology Section, University of Delaware, 117 Wolf Hall, Newark, Delaware 19711, USA*

D. A. Stafford *Department of Microbiology, University College Cardiff, PO Box 97, Cardiff CF1 1XP, Wales*

S. Stafford *Department of Microbiology, University College Cardiff, PO Box 97, Cardiff CF1 1XP, Wales*

J. W. B. Stewart *Saskatchewan Institute of Pedology, University of Saskatchewan, Saskatoon, Canada S7N 0W0*

G. Stotzky *Department of Biology, New York University, 952 Brown Building, Washington Square, New York, New York 10003, USA*

R. N. Strange *Department of Botany and Microbiology, University College, University of London, Gower Street, London WC1E 6BT, England*

M. J. Swift *Department of Plant Biology and Microbiology, Queen Mary College, University of London, Mile End Road, London E1 4NS, England*
Present address: *Department of Botany, University of Zimbabwe, PO Box MP 167, Mount Pleasant, Salisbury, Zimbabwe*

D. S. Weis *Department of Biology and Health Sciences, Cleveland State University, Cleveland, Ohio 44115, USA*

Preface

The subject of microbial ecology has undergone a quiet revolution in the last decade. It has become respectable: a legitimate field of research attractive to a large number of first-rate microbiologists. Explanations for this change of attitude are not difficult to find.

First, the concern voiced by the environmental lobby in the 1950s and 1960s was gradually translated into financial support for investigations into the effects of man-made chemicals on the microbiota (Chapters 34 to 36). This research has necessitated the study of microbial ecosystems in the *absence* of pollutants (Chapters 7 to 15) in order to establish a baseline against which any perturbations can be assessed. Second, exploration of the Moon and Mars has stimulated research into extreme terrestrial environments (Chapter 32) for the purpose of establishing the parameters for microbial growth and survival and thereby designing suitable microbiological experiments to be carried to these places. Third, the demands of an increasing World population for the efficient use of agricultural resources have served to advance our understanding of microbe–plant interactions (Chapters 7 to 14, 23 to 27, 29 and 31). There is also a fourth and rather more subtle factor which has persuaded microbiologists to study the ecology of protists. This has been the gradual realisation that the term mixed-culture microbiology need not be a euphemism for contaminated cultures. Indeed, the study of mixed cultures is now a branch of microbiology in its own right (Chapters 16 and 20) and, furthermore is likely to reflect more accurately the multi-species nature of the microbial environment. A parallel development in the study of microbe–eukaryote interactions (Chapters 23 to 31) has reinforced this trend.

Despite these stimuli the progress towards an understanding of microbes in their environment has been less than rapid. One basic problem is the difficulty in carrying out controlled, reproducible and informative experiments in the field. On the other hand, extrapolating observations made in the laboratory to the natural environment is a dangerous and controversial exercise. This apparent paradox, in the collection and interpretation of data in microbial ecology, is a continuing theme in this volume and is returned to again and again by the various authors. One way of resolving the problem is to design laboratory experiments which show a gradual increase in complexity, a step-by-step approach to a realistic microbial ecosystem: for example, a progression from monoaxenic cultures through co-cultures to mixed cultures containing three or more species; the use of simple defined media at optimum concentrations as well as complex media at realistic (often sub-optimal) concentrations; the choice of 'climatic' factors (e.g. pH, temperature, oxygen levels); that are likely to reflect conditions *in situ*; and encouraging interfacial phenomena characteristic of microbial activity in the environment (Chapters 7 and 17).

This volume is not intended as a recipe book for experiments in microbial ecology but rather an attempt to put research workers in touch with the advantages and shortcomings of the experimental techniques which are currently available. The experimental tools used at present to probe the complex microbial environment are barely adequate. However, given the recent surge of interest we feel that a collection of essays concerned with the plethora of techniques with which microbial ecologists are armed currently will be a spur to the development of novel techniques aimed at unravelling the mysteries of microbial ecology. Already some techniques are quite specific to microbial ecology because they have evolved from the study of highly integrated relationships (e.g. Chapters 18, 19, 22, 26, 30 and 31). Others have been borrowed from a more established microbial methodology such as that discussed in Chapters 1 to 6. Inevitably, therefore, a few authors have biased their chapters towards a review of their subject whilst others have concentrated upon a critical discussion of methodology.

The burgeoning interest in microbial ecology is

illustrated by the convening of three international meetings in the past ten years (Uppsala, Sweden 1972; Dunedin, New Zealand 1977; Warwick, England 1980) with a fourth planned for East Lansing, Michigan, USA in 1983. In addition, the major American and British Microbiology Societies (ASM and SGM) have thriving microbial ecology groups and publish a large number of ecology papers in their house journals (*Applied and Environmental Microbiology, Journal of General Microbiology*). Finally the research journal *Microbial Ecology* (Springer-Verlag, New York) is now in its 9th year and *Advances in Microbial Ecology* (Plenum Press, New York), established in 1977, has reached volume 5.

In conclusion, the editors would like to thank Anne Brown and Bob Campbell of Blackwell Scientific Publications for their encouragement, assistance and patience throughout the long gestation period of this volume.

Richard G. Burns, Canterbury
J. Howard Slater, Coventry

Part 1
Microbial Components of the Biosphere

Chapter 1 · Procedures for the Isolation, Cultivation and Identification of Bacteria

Rodney A. Herbert

1.1 Introduction

Natural environments are extremely diverse and the majority contain a wide range of microorganisms which reflect the nature of the habitat and the ability of individual members to compete successfully and coexist within that given ecosystem. In general terms the greater the heterogeneity of the environment, the more diverse and complex will be the microflora. For example, in an environment such as garden soil where numerous microenvironments exist, the microbial flora is extremely complex whereas in thermophilic or hypersaline environments where one physical or chemical characteristic dominates over all others only a few specialised species can grow under such extreme ecological conditions.

The microbial ecologist, seeking to determine what types of microorganisms are to be found in a particular environment, is, therefore, faced with the dilemma of trying to isolate them from a complex microbial community. In many instances the organism of interest is present in relatively low numbers and some form of enrichment is necessary before isolation can be attempted. The deliberate encouragement (selection) of one microorganism at the expense of others dates back to the work of Schloesing and Müntz (1877) who showed that the oxidation of ammoniacal liquors to nitrate was a biological process. This technique led Winogradsky and Beijerinck, early in this century, to lay the foundations of microbial ecology and physiology. Van Niel (1955) emphasised that, with the alteration of only a few words, Koch's postulates concerning the conditions which must be fulfilled before a microorganism may be considered responsible for causing a disease, can also be applied to enrichment cultures. The postulates can be stated as follows:

(1) the microorganism must always be present when the relevant chemical process is occurring;

(2) it must be possible to isolate and grow the microorganism in pure culture in the laboratory;

(3) the chemical process should occur when a suitable growth medium is inoculated with a pure culture of the microorganism and it should be possible to obtain the isolate from the growth medium at the end of the experiment.

Experimental

1.2 Types of enrichment systems

Two main types of enrichment cultures are used by microbiologists. The most commonly used enrichment procedure is the closed system in which the inoculum is added to a liquid or solid growth medium whose physico-chemical conditions are such that the desired microorganism is enriched relative to other components of the microflora. The desired bacterium forming the dominant population under these circumstances may be isolated by conventional techniques. A major attraction of the closed system is that it only requires simple equipment: bottles or flasks plus a few inexpensive chemicals. The particular microorganism which develops in enrichment culture clearly depends upon the chemical composition of the medium used, together with other factors such as temperature, Eh, pH, presence of selective inhibitors, light, gas phase and others. It is, therefore, apparent that to enrich successfully for a particular bacterium some knowledge of its physiology is a prerequisite yet, paradoxically, this can only be obtained from pure-culture studies. Many microbiologists have observed under the microscope microorganisms which never develop in any type of enrichment culture and our current knowledge does not yet allow us to overcome this hiatus. Traditionally in closed enrichment systems the initial nutrient levels are high in order to promote a large population of microorganisms. However, as growth proceeds there is a continuous change in the medium as the original nutrient levels are depleted and metabolic end-products accumulate. Since the chemical changes are due to the metabolic activities of the inoculum the chemical composition of the medium cannot be controlled and as a consequence it is important not to miss the stage when the desired organism becomes dominant.

The complexity of changes occurring during growth in closed enrichment systems are such that it is often difficult to predict with any certainty which species will become dominant. Another major disadvantage is that the nutrient levels in closed systems are often unnecessarily high particularly in comparison with many natural environments which are usually nutritionally poor, e.g. ocean waters. In an attempt to overcome some of the problems of the closed enrichment system Jannasch (1965, 1967) pioneered the use of continuous culture methods to enrich for microorganisms (see Chapters 16 and 20). In open enrichment systems, such as the chemostat, the exhaustion of nutrients and accumulation of end-products, which are inherent defects in the closed system, present no problems since fresh medium is continuously added and waste products are removed. Jannasch (1965, 1967) has argued that continuous culture enrichment methods have three particular advantages:

(1) no succession of species occurs and if there is no wall growth or interaction, the predominance of one species increases with time;

(2) the growth advantages of the successful competitor are not dependent upon substrate specificity but on the particular growth parameters of the organism and the cultural conditions provided: if these parameters are known and stable, then the enrichment is reproducible;

(3) enrichments may be carried out in the presence of extremely low concentrations of one particular growth-limiting nutrient and, therefore, at low population densities.

The theory and practice of the chemostat has been well reviewed (Herbert *et al.* 1956; Powell 1958; Kubitschek 1970; Tempest 1970; Veldkamp 1976; Slater 1979; see also Chapter 20) and therefore only those aspects relevant to competition will be discussed here.

The key to the chemostat lies in the way in which the specific growth rate (μ) of a microbial population depends upon the concentration of a growth-limiting substrate (s) in the culture medium. In its simplest form the relationship between μ and s can be described by the Monod equation (Monod 1942):

$$\mu = \mu_{\max} \cdot \frac{s}{K_s + s}$$

where μ is the specific growth rate; s the concentration of the growth-limiting nutrient; $\mu_{\max}$ the maximum specific growth rate at saturating values of s; and K_s the saturation constant which is equal to that concentration of s producing a growth rate which is half that of $\mu_{\max}$.

In a chemostat one nutrient in the incoming medium is usually maintained at a relatively low concentration and s in the growth vessel is fixed by the dilution rate. This, in turn, controls μ at some point on the μ/s curve. As a consequence μ is maintained at values below $\mu_{\max}$ and is fixed by the culture dilution rate.

When two bacteria are competing for the same growth-limiting substrate in a chemostat, assuming no interaction between the two populations, then the outcome will be determined by the μ/s relationships of the isolates involved (Veldkamp 1970). Fig. 1 shows the possible relationships between the two isolates. If bacterium A has higher $\mu_{\max}$ and lower K_s values than isolate B then it will outgrow B at all μ values imposed by the dilution

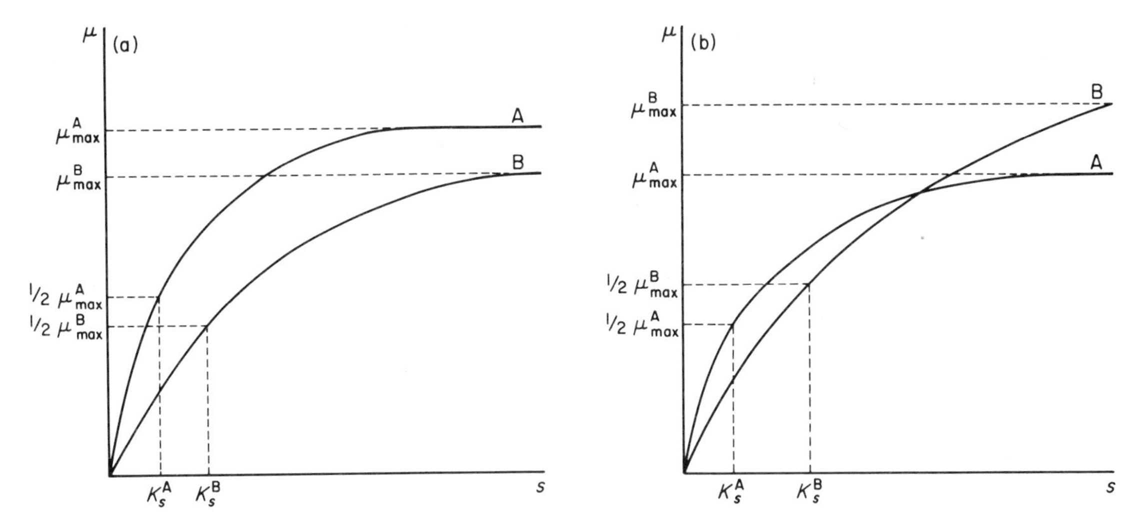

Figure 1. The μ/s relationship of two bacteria A and B (after Veldkamp 1970).

rate and will predominate in both batch and chemostat culture enrichments (Fig. 1(a)). If, however, the μ/s curves of isolates A and B intersect (Fig. 1(b)), then at high growth rates isolate B will outgrow isolate A whereas A will outgrow B at low growth rates. In a closed enrichment system with saturating values of s, selection is based only on μ_{max} and isolate B would always predominate. In an open flow system, such as a chemostat, the outcome depends upon the dilution rate and by using low flow rates, low K_s, low μ_{max} bacteria of type A would predominate, a situation which would never occur in closed-system enrichments. Several examples of μ/s curves which intersect have now been reported and these result in different bacteria predominating at different growth rates (Jannasch 1967; Meers 1971; Kuenen *et al.* 1976). Jannasch (1965, 1967) has shown that when samples of ocean water were inoculated into chemostats run at different dilution rates different bacteria became predominant. If these bacteria were isolated in pure culture, mixed and reinoculated into chemostats the enrichments were reproducible. The reason for the reproducibility was due to the fact that the growth conditions were closely defined. In the chemostat it is possible to select conditions favouring a particular bacterium of low substrate specificity. No other experimental system provides such well-defined conditions for the study of microbial processes in natural environments.

1.3 Selective agents used in enrichment cultures

The literature on enrichment methods is extensive. In this section the general principles underlying the use of particular selective agents will be discussed.

1.3.1 Temperature

Temperature is clearly an important selective agent in enrichment cultures as it is in natural environments. Most aquatic environments are at relatively low temperatures and it has been shown that microorganisms which commonly occur in these ecosystems are rapidly killed when exposed to temperatures above 20°C (Morita & Haight 1964; Stanley & Rose 1967). Thus, if these cold-loving (psychrophilic) microorganisms are to be obtained in enrichment cultures it is imperative that the inoculum be processed and incubated at low temperatures. Taking these precautions Sieburth (1967) was able to follow the seasonal variation in bacterial populations in Narragansett Bay, Rhode Island. In winter when the water temperature was low (minimum −2°C) psychrophiles predominated, whereas in summer (maximum temperature +23°C) mesophilic bacteria with temperature optima between 20 to 33°C were predominant. If samples taken during the winter had been enriched and incubated at 30°C, the dominant population of psychrophiles would have been missed.

Conversely thermophiles can be readily isolated by incubation at elevated temperatures (55 to 65°C). If *Bacillus stearothermophilus* is present in a sample, it can be isolated easily from other components of the microflora by incubation at 65°C (Wolfe & Barker 1968). Similarly, thermoactinomycetes, such as *Thermoactinomyces vulgaris*, can be readily isolated from mouldy hay on yeast extract agar by incubating at 55°C (Cross 1968).

Temperatures may also exert more subtle effects which need to be considered when isolating microorganisms in enrichment culture. At higher incubation temperatures the dissolved oxygen tensions are lower and this greatly affects the cell density of the enriched bacterial population (Sinclair & Stokes 1963). *Lactobacillus arabinosus* when grown in air at 39°C had a specific requirement for aspartate, whilst at 35°C there was no such requirement (Borek & Waelsch 1951). The effect of the increased temperature was to reduce the CO_2 concentration and the dependence on aspartate at 39°C was relieved by increasing the CO_2 concentration.

1.3.2 pH AS A SELECTIVE AGENT

The growth and reproduction of microorganisms is greatly influenced by the pH of the growth medium. Most bacteria can only grow within the range pH 4.0 to 9.0 (Thimann 1964) with optimal growth occurring between pH 6.5 and 8.5. Only a very few bacteria can grow at pH 3.0 or below, e.g. the acidophilic thiobacilli and lactobacilli. By using selective media which have a low buffering capacity, acid production allows the enrichment of acidophilic bacteria at the expense of the majority of the microflora of the original inoculum. In an analogous manner alkalophilic bacteria can be enriched by incubating at high pH (pH 10.0 to 11.0). Wiley and Stokes (1962) demonstrated that *Bacillus pasteurii* could be enriched and isolated from soil by inoculating onto a simple ammonium salts-yeast extract agar at pH 9.0. Under more extreme alkaline conditions (pH 10.0 to 11.0) *Bacillus* spp. still predominate (Gee *et al.* 1980) but alkalophilic strains of *Micrococcus* (Akiba & Horikoshi 1976), *Pseudomonas* (Hale 1977) and *Ectothiorhodospira* (Grant *et al.* 1979) have also been reported. Caldwell and Hirsch (1973) demonstrated the use of a two-dimensional steady-state diffusion system in which pH gradients can be developed to enrich for fastidious bacteria in a natural community.

When enriching for bacteria which are sensitive to pH changes, buffers are often incorporated into the medium. Usually they are of limited value in preventing pH changes since the buffering capacity is too low. Phosphates, whilst useful at low molarities, suffer from the disadvantages of giving rise to precipitates of insoluble phosphates. Tris buffers, since they are organic, provide a potential source of both carbon and nitrogen which may be undesirable in the enrichment medium. Calcium carbonate is frequently included in enrichment media when excessive acid production is anticipated but in stationary cultures it is not particularly effective and it makes microscopical examination of the enrichment difficult. To minimise pH changes occurring in closed system enrichments the concentration of the compound(s) responsible for the pH change should be reduced to a minimum. Alternatively frequent transfers of the developing enrichment into fresh media ensures that excessive pH changes do not occur.

1.3.3 LIGHT AS A SELECTIVE AGENT

The selective enrichment of phototrophic microorganisms is dependent not only on the light wavelengths supplied but the light irradiance levels used. Since the photosynthetic purple and green sulphur bacteria are obligately phototrophic they can be selectively enriched in mineral media in the presence of H_2S, anaerobic conditions and light. Purple non-sulphur bacteria can be similarly enriched except that the appropriate organic electron donor is substituted for H_2S. By using filters to exclude particular wavelengths of light, e.g. infrared filters that will only transmit wavelengths greater than 800 nm, purple sulphur bacteria can be selectively enriched (Pfennig 1967). Using infrared light with a wavelength greater than 900 nm as the selective agent, Eimhjellen *et al.* (1967) were able to isolate a hitherto unknown purple sulphur bacterium which contained a chlorophyll species with an in-vivo absorption maxima of 1017 to 1020 nm.

The light irradiance applied to enrichments for phototrophs may also significantly affect the enrichment and subsequent isolation of phototrophic bacteria. At high irradiance levels (500 to 2000 lux) and high sulphide levels (0.2% w/v) *Chlorobium* spp. usually predominate. In sharp contrast the enrichment of the green sulphur bacterium *Pelodictyon* sp. requires low light irradiance (50 to 100

Table 1. Optimal light irradiance levels for the enrichment of purple sulphur bacteria.

Light irradiance (lux)	Bacteria likely to predominate	Other conditions
100–300	*Chromatium okenii* *Chromatium weissii* *Thiospirillum jenense*	Alternating 16 h light and 8 h dark periods. Light from tungsten lamps
300–700	*Chromatium warmingii*	Continuous illumination from tungsten lamps
700–2000	Small *Chromatium* spp. *Thiocystis* sp. *Thiocapsa* sp. *Amoebobacter* sp.	Continuous illumination from tungsten lamps

lux). Pfennig (1965) has determined the optimum light irradiance for the enrichment of purple sulphur bacteria from fresh waters and the data are summarised in Table 1.

In order to exclude green sulphur bacteria the enrichment cultures were first exposed to infra-red light with a wavelength greater than 800 nm.

1.3.4 Aerobic or Anaerobic Conditions as Selective Agents

In closed enrichment systems aerobic conditions are usually achieved by dispensing the media as a shallow layer either in conical flasks or Petri dishes. However, even under these conditions the oxygen tension is high only at the surface of the medium and once pellicle growth develops the concentration declines even at the surface. Nevertheless the successful enrichment of species of *Azotobacter, Thiobacillus, Acetobacter, Nitrosomonas* and *Nitrobacter* occurs under these stationary conditions when the appropriate medium is used. If high oxygen tensions are required, these can be readily provided either by shaking the culture or by sparging with sterile air. However, even with strict aerobes the effect of oxygen tension on growth is subtle and in many instances enrichments are more successful in stationary culture, where oxygen is limiting, than when oxygen is provided in excess.

The incubation of enrichments in the absence of oxygen allows the development of facultative and obligate chemo-organotrophs and, if light is present, phototrophs. Terms such as micro-aerophiles, non-exacting anaerobes and strict anaerobes should be avoided since they are imprecise. A more reliable index is to express anaerobiosis in terms of the oxidation-reduction potential (Eh) which can be precisely measured. Numerous methods are available to grow anaerobic bacteria. These include the use of anaerobic jars (Willis 1969), roll-tube techniques (Hungate 1969), agar shakes (van Niel 1931) and Pankhurst tubes (Pankhurst 1966). The simplest form of anaerobic enrichment is to use a screw-topped bottle and it is preferable to use a high ratio of liquid to gas volume thereby minimising the ingress of oxygen. In primary enrichments the presence of slight traces of oxygen is not critical since it is rapidly removed by aerobes present in the inoculum and strict anaerobes, such as methanogens, which will only grow at low Eh values (−350 mV), will develop in the appropriate medium. However, where it is essential to maintain anoxic conditions in sample preparation prior to inoculation, an inflatable anaerobic glove bag of the type described by Leftley and Vance (1979) is eminently suitable. When working with anaerobes the media should always be freshly prepared since even when stored under allegedly anoxic conditions some ingress of oxygen often occurs.

Whilst absolutely anoxic conditions are not essential in primary enrichments, for reasons mentioned above, it is vital that in subsequent isolation procedures the medium be poised at the correct redox potential for growth. For example, Postgate (1966) showed that sulphate-reducing bacteria will not initiate growth unless the Eh is less than −100 mV and methanogens need an Eh in the range of −350 mV. To develop the desired Eh values reducing agents are frequently included in the growth medium: cysteine, mercapto-ethanol, mercapto-acetate, sulphide, dithionite, ascorbic acid and iron wire or nails are frequently used. To ensure that the medium has become sufficiently reduced, redox indicators are usually incorporated into the media. The most commonly used are resazurin (E_0' at pH 7.0, −30 to −40 mV), indigo-carmine (E_0' at pH 7.0, −123 mV) and benzyl viologen (E_0' at pH 7.0, −359 mV) at concentrations ranging from 0.0001 to 0.0005% (w/v). Reducing agents are often toxic and thus their concentrations in the media are critical. Cysteine is very useful at concentrations not exceeding 0.05% (w/v) but at higher concentrations it is toxic.

Sulphide is also extremely useful but the toxicity level is pH dependent.

Agar shakes (van Niel 1931) provide a convenient method for isolating anaerobes but for the more exacting forms the Hungate roll-tube method (Hungate 1969) is probably a more reliable technique. When working with strict anaerobes the inocula used should be about 10% (w/v) of the fresh medium used since too small an inoculum often fails to grow.

1.3.5 Trace Element and Co-factor Requirements

Under normal circumstances there are sufficient trace elements present in the reagents used to prepare the enrichment media. However, to avoid any trace element deficiency occurring they are often included in enrichment media.

One of the most useful trace element solutions is that of Hoagland as modified by Pfennig (1965). It comprises the following components in distilled water ($g\ l^{-1}$): $AlCl_3$, 1; KI, 0.5; KBr, 0.5; LiCl, 0.5; $MnCl_2 \cdot 4H_2O$, 7; H_3BO_3, 11; $ZnCl_2$, 1; $CuCl_2$, 1; $NiCl_2$, 1; $CoCl_2$, 5; $SnCl_2 \cdot 2H_2O$, 0.05; $BaCl_2$, 0.5; Na_2MoO_4, 0.5; $NaVO_3 \cdot H_2O$, 0.1; and selenium salt, 0.5. All the salts are dissolved separately in distilled water and mixed to give a final volume of 3.6 l at pH 3.0 to 4.0. Normally 5.0 to 6.0 ml of the trace element solution is added per litre of medium.

An important feature regarding trace elements is that not only should they be present but also available. A frequent problem when incorporating trace elements into enrichment media is that they co-precipitate out during autoclaving. To overcome this trace element solutions should be filter sterilised and added aseptically when the medium is cool.

In addition to trace element requirements some bacteria also need growth factors. For example, many strains of purple and green sulphur bacteria have a requirement for vitamin B_{12} whilst some purple non-sulphur bacteria, such as *Rhodopseudomonas sphaeroides*, require thiamine, biotin and nicotinic acid for growth (Pfennig & Trüper 1974). Many lactobacilli require a large number of growth factors, i.e. amino acids, purines, pyrimidines and vitamins. Growth factor requirements can often be satisfied by adding small quantities of yeast extract which contains a wide range of amino acids and vitamins of the B group. Vitamin B_{12}, however, is not present in yeast extract and must be added separately.

1.3.6 Inhibitors as Selective Agents

The addition of inhibitory compounds to enrichment media suppresses the development of the majority of the microflora enabling the development of the desired species to occur. A good example of this is the use of 0.5% (w/v) bile salts to inhibit non-intestinal bacteria allowing the growth of *Escherichia coli*. In an analogous manner 0.03% (w/v) cetrimide is a useful agent for the selective enrichment of *Pseudomonas aeruginosa* (Brown & Lowbury 1965). Enrichment media containing antibiotics have been used extensively. For example actidione agar (Difco Manual 1966; Oxoid Manual 1979) is invaluable for the isolation and enumeration of bacteria in samples containing large numbers of yeasts and fungi, e.g. bacterial contamination of pitching yeast. At a concentration of 10 $\mu g\ ml^{-1}$ actidione permits the growth of bacteria but inhibits the growth of most yeasts and fungi. Similarly, Cross (1968) used a combination of novobiocin (25 $\mu g\ ml^{-1}$) and actidione (50 $\mu g\ ml^{-1}$) to isolate selectively *Thermoactinomyces vulgaris* from thermophilic *Bacillus* spp. and fungi in mouldy hay. It is clearly impracticable to review the extensive range of chemical compounds that have been added to growth media for the selective isolation of microorganisms from natural environments, foodstuffs and clinical specimens and thus the reader is referred to the Difco (1966) and Oxoid manuals (1979) for further details.

1.4 Isolation methods

1.4.1 Winogradsky Column

The Winogradsky column (Winogradsky 1888) provides an extremely simple but effective way to simulate natural sediment environments in the laboratory and by providing the necessary selective pressures the enrichment of specific microorganisms from an extremely diverse initial microflora can be achieved. Aaronson (1970) gives a clear and concise account of the setting up and sampling of a Winogradsky column and this will not be discussed here. Within the glass or plastic column (Fig. 2) a gradient of nutrients and metabolic end-products develops as a result of fermentative

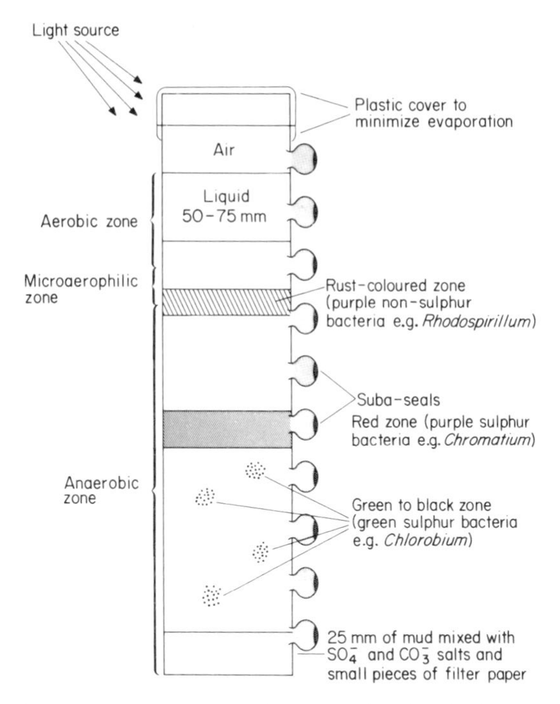

Figure 2. The Winogradsky column (modified after Aaronson 1970).

metabolism from the sediment upwards whilst oxygen diffuses in from the surface. These interacting gradients enable individual bacterial types, if present in the inoculum, to develop at specific points along the column where growth conditions are optimal. For example, in the upper regions of the column both oxygen and sulphide occur which allows the development of *Thiobacillus* spp. whilst in the depth of the sediment, where conditions are anoxic and sulphate levels are high, *Desulfovibrio* spp. predominate. Where light (artificial or natural), anaerobic conditions and hydrogen sulphide are present phototrophic bacteria develop. Green sulphur bacteria tolerate much higher sulphide levels than the purple sulphur bacteria (Pfennig 1967) and thus develop below them in the column. To facilitate the growth of purple non-sulphur bacteria which do not tolerate high sulphide levels the activities of sulphate-reducing bacteria must be suppressed. This may be achieved either by omitting exogenous sulphate from the column, that is, by replacing $CaSO_4$ with $CaCl_2$, or by using temperatures above 33°C whereupon the sulphate reduction rate is reduced (Schlegel & Pfennig 1961). As a consequence fermentation products accumulate and in the presence of light purple non-sulphur bacteria may predominate. From these few examples it can be seen that the Winogradsky column can be exploited to achieve the enrichment of a diverse range of metabolic types. In order to isolate the bacteria from the column a number of devices can be employed, for example, a spatula, wire-loop or hypodermic syringe. A more useful approach is to drill holes in the column (Fig. 2) and seal them with Suba seals or their equivalents. Samples can then be withdrawn with ease from the required sampling sites along the column.

1.4.2 Soil Perfusion Methods

Lees and Quastel (1946) developed the use of a soil reperfusion apparatus for enriching bacteria from soil and the percolation system shown in Fig. 3 is a typical design. After placing the soil sample or sediment within the main chamber of the apparatus, the enrichment medium is circulated through the soil column by applying a low vacuum pressure at the outlet. By incorporating the desired substrate, with or without inhibitors, into the percolation medium the required bacterium can often be successfully enriched by this technique. To minimise waterlogging of the soil column Jeffreys and Smith (1951) modified the soil perfusion apparatus of Lees and Quastel (1946) and incorporated a simple valve system to regulate the percolation rate. This method has an added advantage in that the percolation medium can be circulated using an inert gas, e.g. high-purity nitrogen or argon, and so anaerobes can be isolated.

1.4.3 Chemostat Enrichment Systems

The advantages of chemostat enrichments over conventional batch systems have already been discussed. A considerable number of chemostat designs are available. However, as a general principle, the simpler the design the more reliable the chemostat. The technical design and construction of chemostats has been well reviewed by Evans *et al.* (1970). In our own experience the simple one-stage all-glass chemostat of Baker (1968) has proved extremely useful in chemostat enrichment experiments (Dunn *et al.* 1978; 1980; see also Chapter 16). Jannasch (1967), in contrast, favoured a more complex arrangement involving three chemostat units and varied the dilution rate by using culture vessels of different volumes.

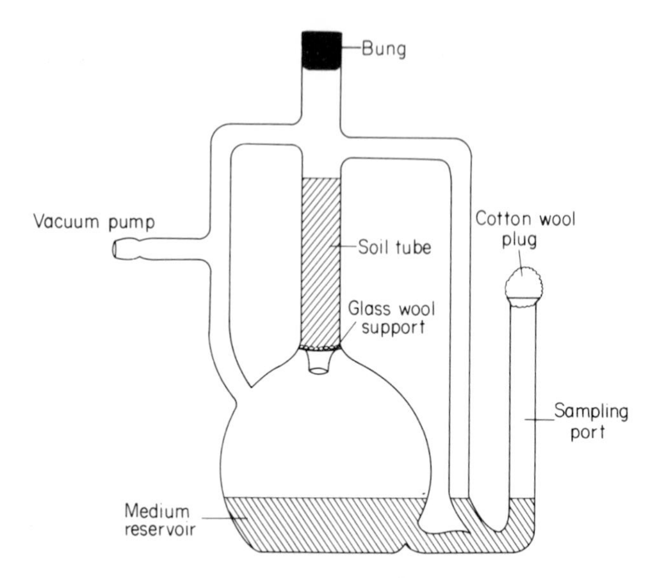

Figure 3. A typical soil perfusion apparatus.

After inoculation of the culture vessel(s) with either a water or sediment sample, small volumes of the culture are removed at daily intervals and streaked onto the corresponding agar medium. The dominant bacterial species can be classified to the genus level using conventional morphological and biochemical tests. In this manner chemostat enrichments for heterotrophic bacteria from aquatic environments have been made (Jannasch 1967; Veldkamp & Kuenen 1973; Brown *et al.* 1978). Dunn *et al.* (1980) extended this technique to isolate nitrate-dissimilating bacteria from estuarine sediments and R. A. Herbert (unpublished observations) has also successfully isolated *Desulfovibrio* spp. and *Clostridium* spp. from estuarine sediments by chemostat enrichment.

1.4.4 OTHER PROCEDURES

Enrichment cultures do not provide estimates of the abundance of a particular bacterium in a given habitat or whether or not it is physiologically active in that environment. They result in the isolation of those bacteria which are best adapted to the selective conditions and thus outgrow their competitors. Interpretation of the role played by particular bacteria in natural environments, when isolated by enrichment methods, should be treated with great caution.

Direct methods for the isolation of bacteria from natural environments involve the preparation of dilution series followed by either plating out onto solid media or distribution into tubes of media. To ensure maximum viability the choice of diluent is important and buffered diluents such as phosphate buffered saline or one-quarter-strength Ringer's solution (Cruickshank 1965) are preferable to 0.9% (w/v) saline. Indicators, such as 0.0001% (w/v) resazurin, should be incorporated into the diluent when handling anaerobes to ensure that the conditions are sufficiently reduced. Agar is usually used to gel the media but in certain circumstances, particularly with autotrophs, organic materials present in the agar are toxic and so silica gel should be used (Skerman 1967). To avoid spreading of the colonies over the agar surface the plates should be dried at 37°C for 2 d before use. Finally, spread-plates are preferable to pour-plates since many bacteria are killed by the thermal shock of being exposed to molten agar.

The isolation of anaerobic bacteria can be carried out either by using agar shake tubes (van Niel 1931) or by the roll-tube technique (Hungate 1969). The agar shake method has the great advantage of requiring simple equipment and yet it is extremely effective for the isolation of anaerobes. A series of tubes, each containing 9.0 ml volumes of the required agar medium, are kept molten at 42 to 44°C in a waterbath. The first tube in the series is inoculated with a sample of soil or sediment and, after mixing by inversion, a dilution series is prepared by pouring about 1.0 ml from one tube into the next, again mixing between each transfer. After mixing, the tubes are rapidly cooled in a bath of cold water and sealed with a plug of paraffin wax (1 part wax to 3 parts liquid paraffin). Finally the tubes are sealed with a rubber bung. As an additional safeguard, to prevent the ingress of oxygen, absorbent cotton wool plugs can be inserted into the tubes and a few crystals of pyrogallol plus 0.5 ml 2 M sodium carbonate added prior to sealing the tube with the rubber bung. Well-separated single colonies can be removed from the appropriate dilution tube by sucking into a fine capillary tip Pasteur pipette and transferring to fresh agar shakes. This procedure may be repeated until pure cultures are obtained.

The roll-tube method for the isolation and cultivation of strict anaerobes was pioneered by Hungate (1950) (see Chapter 6). In the roll-tube method exposure of the bacteria and culture medium to air is avoided by displacing residual air in the culture vessel with an oxygen-free gas, such as carbon dioxide, hydrogen, nitrogen or mixtures of these gases. The method is much more complicated than the agar shake system but for extremely oxygen-sensitive bacteria it is the preferred technique since the cultures are never exposed to oxygen. The equipment required and the procedures involved in the roll-tube method have been clearly described (Hungate 1969). Whilst the tubes can be rolled manually a much more satisfactory procedure is to use a tube roller which also incorporates a tube cooling system. Isolation of individual colonies from the roll tubes can be achieved in an analogous manner to that used for agar shakes except that the tubes are gassed continuously with oxygen-free gas during the process.

1.4.5 Membrane Filtration

This method is based on the use of highly porous cellulose acetate membranes which are manufactured in different pore sizes. The pore structure of these membranes enables large volumes of water to pass through under pressure but prevents the passage of any bacteria present in the sample. Bacteria are retained on the membrane surface which is brought into contact with an absorbent pad saturated with suitable nutrients. The growth substrates diffuse upwards through the pores permitting the bacteria to grow as surface colonies which can be isolated subsequently. The technique was first described by Windle-Taylor *et al.* (1953) and is used primarily for determining the bacteriological quality of water and sewage effluents (Department of Health and Social Security Report No. 71 1969). However, the technique has much more general applications as the following examples show. Watson (1965) successfully used membrane filtration to isolate *Nitrosomonas oceanus* from sea-water. Canale-Parola *et al.* (1967) were able to isolate *Spirillum* spp. from pond water using 0.45 μm Millipore filters. The membranes were inoculated with pond water and placed on a non-selective medium. After several days incubation the thin spirilla (0.25 μm diameter) migrated through the membrane pore and were growing on the agar medium from where they could be isolated in pure culture by standard methods. Membrane filtration has also been successfully used by Mouraret and Baldensperger (1977) to isolate autotrophic thiobacilli. For further details of the procedures involved in membrane filtration the reader should consult either the Oxoid or Millipore Data manuals.

1.5 Media for the isolation and cultivation of bacteria

A vast array of growth media has been formulated and described in the literature and it is clearly an impossible task to review all the types available. Manufacturers, such as Difco and Oxoid, produce a wide range of selective media for use in public health and food laboratories and for further details the relevant manuals should be consulted. In addition, Harrigan and McCance (1966), Skerman (1967), Norris and Ribbons (1970), Rodina (1972) and Collins and Lyne (1976) provide an extensive guide to media suitable for the isolation and cultivation of a wide range of bacterial species. To aid the reader a list of media and key references for the isolation and growth of chemo-organotrophs, chemolithotrophs and phototrophs from aquatic and terrestrial environments are given in Tables 2 to 8. The selective media listed are those which have been found by experience to be particularly useful for isolating specific bacterial groups from a wide range of habitats. However, when unusual habitats are being sampled the relevant references should be consulted to determine the most appropriate medium to be used. The choice of a particular isolation medium should be carefully considered in relation to the aims of the work. For example, where the requirement is merely to isolate a specific bacterial type for a subsequent laboratory study then a highly selective medium is appropriate,

Table 2. Media for isolating and cultivating chemo-organotrophs.

Bacterial genus	References	Bacterial genus	References
Acetobacter	Shimwell *et al.* (1960); Skerman (1967); Carr (1968); Asai (1973)	*Hyphomicrobium*	Hirsch & Conti (1964); Harder & Attwood (1978)
Actinomyces	Waksman (1959); Williams & Cross (1971); Lacey (1973)	*Lactobacilli*	Difco Manual (1966); Oxoid Manual (1979)
Azotobacter	Jensen (1954); Norris (1959); Skerman (1967)	Methane-oxidising bacteria	Skerman (1967); Whittenbury *et al.* (1970)
Bacillus	Smith *et al.* (1952); Claus (1965) Oxoid Manual (1979)	Methanogenic bacteria	Wolfe (1971); Edwards & McBride (1975)
Bacteroides	Beerens *et al.* (1963); Barnes & Impey (1971)	Nitrate-dissimilating bacteria	Skerman (1967); Dunn *et al.* (1980)
Bdellovibrio	Veldkamp (1970); Starr & Huang (1972)	*Pseudomonas*	Difco Manual (1966); Stanier *et al.* (1966); Oxoid Manual (1979)
Beggiatoa	Collins (1969); Joshi & Hollis (1976); Burton & Lee (1978)	*Photobacterium*	Spencer (1955); Hendry *et al.* (1970)
Caulobacter	Poindexter (1964); Hirsch & Conti (1965)	*Propionibacterium*	van Niel (1928); Veldkamp (1970)
Cellulolytic bacteria (aerobic)	Kadota (1956); Veldkamp (1970)	*Rhizobium*	Bergersen (1961); Skerman (1967)
Cellulolytic bacteria (anaerobic)	Hungate (1950); Skinner (1971)	Rumen bacteria	Hungate (1966, 1969); Hobson & Mann (1971)
Clostridia	Willis (1969); Shapton & Board (1971)	*Sphaerotilus* and *Leptothrix*	Collins (1963); Dondero (1975)
Cytophaga	Skerman (1967); Peterson (1969)	*Spirillum*	Jannasch (1967); Veldkamp (1970)
Desulfovibrio	Postgate (1966, 1969); Pankhurst (1971)	*Staphylococcus*	Baird-Parker (1962); Difco Manual (1966); Oxoid Manual (1979)
Enterobacteriaceae	Difco Manual (1966); Oxoid Manual (1979)	*Streptococcus*	Difco Manual (1966); Skinner & Quesnel (1978); Oxoid Manual (1979)
Halophilic bacteria	Eimhjellen (1965); Gibbons (1969)	*Vibrio*	Difco Manual (1966); Oxoid Manual (1979)
Hydrogen bacteria	Skerman (1967); Collins (1969); Veldkamp (1970)		

whereas if one is attempting to isolate bacteria carrying out a particular process such as nitrate dissimilation a highly selective medium may exclude some of the desired bacteria and lead to false conclusions.

1.6 Identification of bacteria

Identification is the process whereby an unknown bacterium is, on the basis of similarity, assigned to a previously described taxonomic group. Most microbiologists need to identify the bacteria they are handling at some stage and studies are often limited by the lack of good identification schemes. Before an unknown bacterium can be identified it is essential that it is a pure culture and this should be checked by plating on non-selective media. Another prerequisite for the successful identification of unknown bacteria is that all the tests must be carried out under well-defined conditions otherwise the results will be meaningless. Bacteria are primarily identified using a combination of morphological and biochemical reactions. In addition some bacterial groups are identified on the basis of serological tests and/or susceptibility to bacteriophage. More sophisticated methods, such as the determination of DNA base ratios (Mandel 1966; Hill 1968), gel electrophoresis of cell proteins (Lund 1965; Robinson 1968), cell wall analysis (Baird-Parker 1970), pyrolysis gas-liquid chromatography and DNA hybridisation, are also available but are not used routinely.

Due to the great diversity of bacteria the normal practice is to do a preliminary series of tests which direct the investigator to an appropriate bacterial group. Basic tests determine whether the bacteria

Table 3. Media for growth of chemolithotrophic *Thiobacillus* spp. Only *T. thiooxidans, T. ferrooxidans, T. thioparus, T. neapolitanus* and *T. denitrificans* are considered to be obligate chemolithotrophs (Kuenen 1975). (Key references: Vishniac & Santer 1957; Taylor *et al.* 1971; Kuenen & Veldkamp 1972; Kuenen 1975.)

	Composition (g l^{-1})				
				Aciduric medium	Non-aciduric medium
Component	Waksman & Starkey (1922)	Lieske (1912)	Starkey (1935)	Parker & Frisk (1953)	Parker & Frisk (1953)
$Na_2S_2O_3 \cdot 5H_2O$	—	5.0	5.0	5.0	10.0
NH_4Cl	—	—	—	0.1	—
$(NH_4)_2SO_4$	0.2	—	0.4	—	0.1
K_2HPO_4	—	0.2	4.0	—	4.0
KH_2PO_4	3.0	—	—	3.0	4.0
KNO_3	—	5.0	—	—	—
$CaCl_2 \cdot 2H_2O$	0.25	0.01	0.25	0.1	0.1
$MgCl_2 \cdot 6H_2O$	—	0.1	—	0.1	—
$MgSO_4 \cdot 7H_2O$	0.5	—	0.5	—	0.1
$FeSO_4$	0.01	—	0.01	—	—
$FeCl_3$	—	0.01	—	—	0.02
$NaHCO_3$	—	1.0	—	—	—
$MnSO_4 \cdot 4H_2O$	—	—	—	—	—
elemental sulphur	10.0	—	—	—	—

Table 4. Media for the growth and isolation of chemolithotrophic *Nitrobacter* spp. considered to be facultative chemolithotrophs (Delwiche & Finstein 1965). The Fe-EDTA stock solution contains in distilled water (mg in 50 ml): $FeSO_4 \cdot 7H_2O$, 77 and NaEDTA, 103. Trace element solution a contains in distilled water (mg in 100 ml): Mn^{2+}, 5.5; B^-, 5.2; Cu^{2+}, 4.2; Zn^{2+}, 4.0 and Co^{2+}, 3.0. Trace element solution b contains in distilled water (mg in 100 ml): $ZnSO_4 \cdot 2H_2O$, 2; $CuSO_4 \cdot 5H_2O$, 2 and $Na_2MoO_4 \cdot 2H_2O$, 2. (Key references: Pope & Skerman 1967; Carlucci & Strickland 1968; Sharma & Ahler 1977.)

	Composition (g l^{-1})	
Component	Skerman (1967)	Schmidt *et al.* (1973)
NaCl	0.3	—
$MgSO_4 \cdot 7H_2O$	0.14	0.02
Fe-EDTA	—	5.0 ml
Na_2HPO_4	—	5.1
KH_2PO_4	0.14	0.5
$FeSO_4 \cdot 7H_2O$	0.03	—
$NaNO_2$	0.5	1.4
trace element solution a or b	0.4 ml[a]	1.0 ml[b]
$CaCO_3$	10.0	—

are phototrophs, chemolithotrophs, heterotrophs or utilise carbohydrates oxidatively or fermentatively. These characters together with other features, such as cell shape and size, Gram reactions, mobility, oxygen requirement, endospore production, oxidase and catalase reaction, capsule formation and presence or absence of inclusion bodies, guide the investigator to the appropriate identification scheme. Some bacteria, e.g. the phototrophic purple and green sulphur bacteria, are identified principally on the basis of cell morphology and absorption spectra of their bacteriochlorophylls and carotenoids whereas most heterotrophs are routinely identified on the basis of biochemical tests. In selecting characters that are useful in bacterial identification a major consideration is to choose those that are constant within the taxa and give reproducible results. Unfortunately, the reliability of many microbiological tests is notoriously poor and obviously this has serious repercussions on the process of bacterial identification (Sneath 1974; Sneath & Collins 1974).

By definition an identification scheme depends upon the availability of data on existing bacteria with which the unknown isolates can be compared. Several different methods for identification have been devised. Skerman (1967) developed a dichotomous key for bacterial identification which depends upon individual characters being unequivocally positive or negative. However, bacteria have relatively few characters which can be considered to be entirely constant and some allowance must be made for variable reactions. Cowan and Steel (1974) favour a more flexible

Table 5. Media for the growth and isolation of chemolithotrophic *Nitrosomonas* spp. The Fe-EDTA solution, see Table 4. Trace element solution a, see Table 4. (Key references: Engel & Alexander 1958; Pope & Skerman 1967; Carlucci & Strickland 1968; Sharma & Ahler 1977.)

	Composition (g l^{-1})			
Component	Skerman (1967)	Skinner & Walker (1961)	Lewis & Pramer (1958)	Schmidt *et al.* (1973)
NaCl	0.3	—	—	—
$MgSO_4 \cdot 7H_2O$	0.14	0.04	0.01	0.04
Fe-EDTA	—	1.6 ml	—	1.6 ml
$FeCl_3$	—	—	0.014	—
$FeSO_4 \cdot 7H_2O$	0.03	—	—	—
KH_2PO_4	0.14	0.2	1.1	0.2
Na_2HPO_4	—	—	3.1	—
$CaCl_2$	—	0.04	0.018	0.04
Na_2CO_3	—	—	—	0.5
phenol red	—	0.0005	—	0.0005
$CaCO_3$	10.0	—	—	—
$(NH_4)_2SO_4$	0.66	0.5	0.5	0.5
trace element solution a	0.4 ml	—	—	—

Table 6. Media for purple non-sulphur bacteria. The basal medium (after Pfenning 1967, 1969a, b) contains in distilled water (g l^{-1}): KH_2PO_4, 1.0; NH_4Cl, 0.5; $MgSO_4 \cdot 7H_2O$, 0.4; NaCl, 0.4; $CaCl_2 \cdot 2H_2O$, 0.05; and 10.0 ml trace element solution SL6 (Pfennig & Lippert 1966). YE = Yeast Extract. (Key References: van Niel 1944; Drews & Giesebrecht 1966; Pfennig 1967; Whittenbury 1971.)

Species	Enrichment medium. The above basal medium supplemented with the following (g)	Special requirements (g)
Rhodospirillum		
R. rubrum	YE, 1.0; ethanol or methanol, 1.0; pH6.8	0.5 ascorbate as reducing agent
R. tenue } *R. fulvum* } *R. metricum* }	YE, 1.0; pelargonate or caprylate, 0.4; pH 6.8	0.5 ascorbate as reducing agent
Rhodopseudomonas		
R. palustris	YE, 1.0; ethanol, methanol or benzoate, 1.0 to 20.0; pH 6.8	none
R. viridis	YE, 1.0; succinate, 2.0; pH 6.8	Light through infra-red filter
R. acidophila	succinate, 2.0; pH 5.0 to 5.5.	none
R. capsulata	YE, 1.0; propionate, lactate or succinate, 2.0; pH 6.8	Grows with no organic carbon in presence of $H_2 + \cdot CO_2$ gas phase
R. sphaeroides	YE, 1.0; acetate or propionate, 1.0; pH 6.5	none
R. gelatinosa	YE, 1.0; citrate 2.0; pH 6.5	none
R. globiformis	YE, 1.0; fumarate, 1.0; pH 5.6	none
Rhodomicrobium		
R. vannielii	$Na_2S \cdot 9H_2O$, 0.3; acetate or ethanol, 2.0; pH 6.8	none

system in which primary, secondary and tertiary levels of identification are used, thereby enabling variable characters to be accommodated. A more wide-ranging and comprehensive system is Bergey's Manual of Determinative Bacteriology (Buchanan & Gibbons 1974) which gives extremely detailed descriptions of individual bacterial species with which the unknowns can be compared. Somewhat more sophisticated methods include the polyclave (multiple) key identification scheme in

Table 7. Media for purple and green sulphur bacteria. The basal medium contains in distilled water as follows:
solution 1: $CaCl_2 \cdot 2H_2O$, 1.5 g in 2.7 l distilled water;
solution 1b (for *Chromatiaceae*): $Na_2S \cdot 9H_2O$, 3.0 g in 400 ml water;
solution 1c (for *Chlorobiaceae*): $Na_2S \cdot 9H_2O$, 6.0 g in 400 ml water;
solution 2: $NaHCO_3$, 9.0 in 2.8 l distilled water;
solution 3: $MgSO_4 \cdot 7H_2O$, 3.0 g; NH_4Cl, 2.0 g; KCl, 2.0 g; K_2HPO_4, 2.0 g; vitamin B_{12}, 12.0 mg; 60 ml SL6 trace element solution; and 140 ml distilled water. The SL6 trace element solution contains in distilled water: $ZnSO_4 \cdot 7H_2O$, 10.0; $MnCl_2 \cdot 4H_2O$, 3.0; H_3BO_3, 30.0; $CoCl_2$, 20.0; $CuCl_2 \cdot 2H_2O$, 1.0; $NiCl_2 \cdot 6H_2O$, 2.0; and $Na_2MoO_4 \cdot 2H_2O$, 3.0. Solution 1 (46 ml) is autoclaved in 100 ml screw cap bottles. Solution 2 is flushed with carbon dioxide for 30 min, added to solution 3 and filter-sterilised. Portions (50 ml) are added aseptically to solution 1.
For purple sulphur bacteria, solution 1b is adjusted with 2 M Na_2CO_3 such that when 6.0 ml is added to each bottle the final pH is 7.3.
For green sulphur bacteria, 5.0 ml solution 1c is added to each bottle to give a final pH of 6.8.
(Key references: van Niel 1931; Larsen 1952; Pfennig 1961, 1965.)

Species	Enrichment medium	Special requirements
Chlorobium Species of other green sulphur bacteria (Chlorobiaceae)	Using solution 1c, high sulphide concentration. High light intensity (500 to 1000 lux). Temperature 25 to 30°C	—
Species of *Chromatium, Thiospirillum, Thiocystis, Thiocapsa* and other purple sulphur bacteria	Using solution 1b, low sulphide concentration. High light intensity (500 to 1000 lux). Temperature 25 to 30°C	Use infra-red filter
Species of *Amoebobacter, Thiodictyon, Lamprocystis, Pelodicyton* and *Clathrochloris*	Using solution 1b, low sulphide concentration. Low light intensity (100 to 250 lux). Temperature 4 to 15°C	Low temperature

which the data can either be stored in a computer or more easily on punched cards (Sneath 1978). The underlying principle of this method is to allocate a given position on each of a set of punched cards to every taxon. A separate card is used for every character, with a hole punched in a position corresponding to a particular taxon if that taxon is positive for that test. Computer cards are ideal for this method and can be punched manually. The system is used by picking up in any order the character cards appropriate to the features of the unknown isolate. Character cards are then added until only one punch hole remains through the pack. Since the holes must be either punched or not punched a polyclave scheme depends upon tests which give unequivocal yes-no responses although a limited number of variable characters can be accommodated.

An inherent weakness in all the identification schemes described so far is the fact that the identity of taxa are based upon single characters or a series of single characters such that the possession of all the properties is necessary before an unknown bacterium can be assigned to taxa so defined. There is a growing concern amongst microbiologists that such artificial monothetic classifications result in the mis-identification of strains aberrant in one or more of the diagnostic characters. Attempts have now been made to develop more satisfactory classification schemes in which the relationship of taxa are based upon several types of criteria, e.g. biochemical, chemical, genetical and serological data, and the success of such systems is a measure of the consistency found between the different types of information. Colwell (1970) has termed this type of classification scheme polyphasic and it is dependent upon the analysis of data according to Addansonian principles. In this system the similarity between bacteria is assessed numerically on the basis of the examination of a large number of equally weighted features and strains are grouped into phenons according to their affinity. The volume of data to be handled is large and this type of identification scheme is only practicable using a computer. The principles of data handling have been well reviewed (Lockhart & Liston 1970; Sneath 1972; Sneath & Sokal 1973). Taxa so delineated are considered to be natural and are

Table 8. Media for the isolation of cyanobacteria. Trace elements added (mg l^{-1});
(a) MoO_3, 0.02;$ZnSO_4 \cdot 7H_2O$, 0.2; and $CuSO_4 \cdot 5H_2O$; 0.08;
(b) MoO_3, 0.15; $ZnSO_4 \cdot 7H_2O$, 0.2; $CuSO_4 \cdot 5H_2O$, 0.08; $Co(NO_3)_2 \cdot 6H_2O$, 0.05; NH_4NO_3, 0.02; $NiSO_4$, 0.04; $Na_2NO_3 \cdot 2H_2O$, 0.02; and $TiO(C_2O_4)_2$, 0.01;
(c) $NaMoO_4 \cdot 2H_2O$, 0.39; $ZnSO_4 \cdot 7H_2O$, 0.2; $CuSO_4 \cdot 5H_2O$, 0.08; and $Co(NO_3)_2 \cdot 6H_2O$, 0.05.
* Marine medium made up with 750 ml seawater and 250 distilled water. (Key references: Pope & Skerman 1967; Carlucci & Strickland 1968; Sharma & Ahler 1977.)

	Composition (mg l^{-1})			
Component	Kratz & Myers (1955)	Allen & Arnon (1955)	Stanier *et al.* (1971)	Ripka *et al.** (1979)
$NaNO_3$	—	—	1500	750
KNO_3	1000	2000	—	—
NaCl	—	230	—	—
$MgSO_4 \cdot 7H_2O$	250	250	75	38
$CaCl_2 \cdot 2H_2O$	—	74	36	18
$Ca(NO_3)_2 \cdot 4H_2O$	25	—	—	—
$K_2HPO_4 \cdot 3H_2O$	1300	400	40	20
$Fe(SO_4)_3.6H_2O$	4	—	—	—
Fe-EDTA	—	26	—	—
Na_2-EDTA	—	—	1	0.5
Na citrate·$2H_2O$	165	—	—	—
Citric acid	—	—	6	3
Fe ammonium citrate	—	—	6	3
Na_2CO_3	—	—	20	20
H_2BO_3	2.86	2.86	2.86	2.86
$MnCl_2 \cdot 4H_2O$	1.81	—	1.81	1.81
$MgSO_4 \cdot 4H_2O$	—	2	—	—
trace elements	a	b	c	c
pH	6.9–9.0	not specified	7.1	7.5

termed phenetic. Phenetic groups contain bacteria that share a high degree of similarity indicating that no single character is either essential for group membership or is sufficient to make a bacterium a member of the group. The application of numerical taxonomy in identifying bacteria from natural environments has been limited at present by the lack of data on type strains. Where comparisons have been made between natural bacterial populations and type strains there is often little correlation (Goodfellow *et al.* 1976; Lee *et al.* 1977). This is understandable since many type strains are derived from clinical or industrial sources and not from natural environments. As more numerical taxonomy studies are performed this situation will be remedied.

Most microbiologists either do not have the facilities or the requirement to use numerical taxonomy techniques and standard texts for the identification of bacteria are more than satisfactory for their needs. In addition to Skerman (1967), Bergey's Manual (Buchanan & Gibbons 1974), Cowan and Steel (1974) and Wilson and Miles (1975), a range of more specialised texts on bacterial identification are available in the Society for Applied Bacteriology Technical Series (Gibbs & Skinner 1966; Gibbs & Shapton 1968; Goodfellow & Board 1980; Skinner & Lovelock 1980).

Irrespective of whether simple or sophisticated methods of identification are employed the ultimate requirement is for the intelligent interpretation of the data. As in all determinative bacteriology valid identification must be based on careful work using pure cultures and tests performed under standardised conditions otherwise the effort will be wasted.

Recommended reading

Aaronson S. (1970) *Experimental Microbial Ecology.* Academic Press, London.

Buchanan R. E. & Gibbons N. E. (1974) *Bergey's Manual of Determinative Bacteriology.* Williams & Wilkins, Baltimore.

Collins C. H. & Lyne P. M. (1976) *Microbiological*

Methods. Butterworths University Park Press, Baltimore.

COWAN S. T. & STEEL K. J. (1974) *Manual for the Identification of Medical Bacteria.* Cambridge University Press, Cambridge.

GIBBS B. M. & SHAPTON D. A. (1968) *Identification Methods for Microbiologists, part B.* Academic Press, London.

GIBBS B. M. & SKINNER F. A. (1966) *Identification Methods for Microbiologists, part A.* Academic Press, London.

GOODFELLOW M. & BOARD R. G. (1980) *Microbial Classification and Identification.* Academic Press, London.

NORRIS J. R. & RIBBONS D. W. (1970) *Methods in Microbiology,* vol. 3A. Academic Press, London.

NORRIS J. R & RIBBONS D. W. (1969) *Methods in Microbiology,* vol. 3B. Academic Press, London.

SHAPTON D. A. & BOARD R. G. (1971) *Isolation of Anaerobes.* Academic Press, London.

SKERMAN V. B. D. (1967) *A Guide to the Identification of the Genera of Bacteria.* Williams & Wilkins, Baltimore.

References

AARONSON S. (1970) *Experimental Microbial Ecology.* Academic Press, London.

AKIBA T. & HORIKOSHI K. (1976) Identification and growth of α galactosidase producing organisms. *Agricultural and Biological Chemistry* **40**, 1845–9.

ALLEN M. B. & ARNON D. I. (1955) Studies on nitrogen fixing blue-green algae. I. Growth and nitrogen fixation by *Anabaena cylindrica. Plant Physiology* **30**, 366–72.

ASAI T. (1973) *Acetic Acid Bacteria, Classification and Biochemical Activities.* Butterworths University Park Press, Baltimore.

BAIRD-PARKER A. C. (1962) An improved diagnostic and selective medium for isolating coagulase positive staphylococci. *Journal of Applied Bacteriology* **25**, 12–19.

BAIRD-PARKER A. C. (1970) The relationship of cell wall composition to the current classification of staphylococci and micrococci. *International Journal of Systematic Bacteriology* **20**, 483–90.

BAKER K. (1968) Low cost of continuous culture apparatus. *Laboratory Practice* **17**, 817–21.

BARNES E. M. & IMPEY C. S. (1971) The isolation of the anaerobic bacteria from chicken caeca with particular reference to members of the family Bacteroidaceae. In: *Isolation of Anaerobes* (Eds D. A. Shapton & R. G. Board), pp. 115–22. Academic Press, London.

BEERENS H., SCHAFFNER Y., GUILLAUME J. & CASTEL M. M. (1963). Les bacilles anaerobes non sporulé à Gramnégatif favorisés par la bile. *Annals Institut Pasteur* **14**, 5–11.

BERGERSEN F. J. (1961) The growth of rhizobia in synthetic media. *Australian Journal of Biological Sciences* **14**, 349–60.

BOREK E. & WAELSCH H. (1951) The effect of temperature on the nutritional requirements of microorganisms. *Journal of Biological Chemistry* **190**, 191–6.

BROWN C. M., ELLWOOD D. C. & HUNTER J. R. (1978) Enrichments in a chemostat. In: *Techniques for the Study of Mixed Populations* (Eds D. W. Lovelock & R. Davies), pp. 213–21. Academic Press, London.

BROWN V. I. & LOWBURY E. J. L. (1965) Use of an improved cetrimide agar medium and other culture methods for *Pseudomonas aeruginosa. Journal of Clinical Pathology* **18**, 752–6.

BUCHANAN R. E. & GIBBONS N. E. (1974) *Bergey's Manual of Determinative Bacteriology.* Williams & Wilkins, Baltimore.

BURTON S. D. & LEE J. D. (1978) Improved enrichment and isolation procedures for obtaining pure cultures of *Beggiatoa. Applied Environmental Microbiology* **35**, 614–17.

CALDWELL D. E. & HIRSCH P. (1973) Growth of microorganisms in two dimensional steady state diffusion gradients. *Canadian Journal of Microbiology* **19**, 53–58.

CANALE-PAROLA E., HOLT S. C. & UDRIS Z. (1967) Isolation of free living anaerobic spirochaetes. *Archiv für Mikrobiologie* **59**, 41–8.

CARLUCCI A. T. & STRICKLAND J. D. H. (1968) The isolation, purification and some kinetic studies of marine nitrifying bacteria. *Journal of Experimental Marine Biology and Ecology* **2**, 156–66.

CARR J. G. (1968) Methods for identifying acetic acid bacteria. In: *Identification Methods for Microbiologists* part B (Eds B. M. Gibbs & D. A. Shapton), pp. 1–8. Academic Press, London.

CARR N. G. (1969). Growth of phototrophic and blue-green algae. In: *Methods in Microbiology* (Eds J. R. Norris & D. W. Ribbons), vol. 3B, pp. 53–77. Academic Press, London.

CLAUS D. (1965). Isolation of *bacillus* species. In: *Anreicherungskultur und Mutants-auslee* (Ed. H. G. Schlegel), pp. 337–62. Gustav Fischer Verlag, Stuttgart.

COLLINS V. G. (1963) The distribution and ecology of bacteria in freshwater. *Proceedings of the Society for Water Treatment and Examination* **12**, 40–73.

COLLINS V. G. (1969) Isolation, cultivation and maintenance of autotrophs. In: *Methods in Microbiology* (Eds J. R. Norris & D. W. Ribbons), vol. 3B, pp. 1–52. Academic Press, London.

COLLINS C. H. & LYNE P. M. (1976) *Microbiological Methods.* Butterworths University Park Press, Baltimore.

COLWELL R. R. (1970) Polyphasic taxonomy of the genus *Vibrio. Journal of Bacteriology* **104**, 410–33.

COWAN S. T. & STEEL K. J. (1974) *Manual for the Identification of Medical Bacteria.* Cambridge University Press, Cambridge.

CROSS T. (1968) Thermophilic actinomycetes. *Journal of Applied Bacteriology* **31**, 36–53.

CRUICKSHANK B. (1965) *Medical Microbiology*. E. &. S. Livingstone, Edinburgh & London.

DELWICHE C. C. & FINSTEIN M. S. (1965) Carbon and energy sources for the nitrifying autotrophs *Nitrobacter*. *Journal of Bacteriology* **90**, 102–7.

Department of Health & Social Security (1969) *The Bacteriological Examination of Water Supplies*. Report No. 71. Her Majesty's Stationery Office, London.

Difco Manual (1966) Difco Laboratories, Detroit, Michigan.

DONDERO N. C. (1975) The *Sphaerotilus-Leptothrix* group. *Annual Review of Microbiology* **29**, 407–28.

DREWS G. & GIESEBRECHT P. (1966) *Rhodopseudomonas viridis* nov. spec., ein neu isoliertes, obligat phototrophes Bacterium. *Archiv für Mikrobiologie* **53**, 255–62.

DUNN G. M., HERBERT R. A. & BROWN C. M. (1978) Physiology of denitrifying bacteria from tidal mudflats in the River Tay. In: *Physiology and Behaviour of Marine Organisms* (Eds D. S. McLusky & A. J. Berry), pp. 135–40. Pergamon Press, Oxford.

DUNN G. M., WARDELL J. N., HERBERT R. A. & BROWN C. M. (1980) Enrichment, enumeration and characterisation of nitrate reducing bacteria present in sediments of the River Tay. *Proceedings of the Royal Edinburgh Society* **78**(B), 47–56.

EDWARDS T. & MCBRIDE B. C. (1975) New method for the isolation and identification of methanogenic bacteria. *Applied Microbiology* **29**, 540–5.

EIMHJELLEN K. (1965) Isolation of extremely halophilic bacteria. In: *Anreicherungskultur und Mutantsauslee* (Ed. H. G. Schlegel), pp. 126–38. Gustav Fischer Verlag, Stuttgart.

EIMHJELLEN K., STEENSLAND H. & TRAFFENBERG H. (1967) A *Thiococcus* sp. nov. gen., its pigments and internal membrane system. *Archiv für Mikrobiologie* **59**, 82–92.

ENGEL M. S. & ALEXANDER M. (1958) Growth and autotrophic metabolism of *Nitrosomonas europea*. *Journal of Bacteriology* **76**, 217–22.

EVANS G. G. T., HERBERT D. & TEMPEST D. W. (1970) The continuous cultivation of micro-organisms 2. Construction of a chemostat. In: *Methods in Microbiology* (Eds J. R. Norris & D. W. Ribbons), vol. 2, pp. 277–327. Academic Press, London.

FOGG G. E., STEWART W. D. P., FAY P. & WALSBY A. E. (1973) *The Blue-Green Algae*. Academic Press, London.

GEE J. M., LUND B. M., METCALF G. & PEEL J. L. (1980) Properties of a new group of alkalophilic bacteria. *Journal of General Microbiology* **117**, 9–17.

GIBBONS N. E. (1969) Isolation, growth and requirements of halophilic bacteria. In: *Methods in Microbiology* (Eds J. R. Norris & D. W. Ribbons), vol. 3B, pp. 169–83. Academic Press, London.

GIBBS B. M. & SHAPTON D. A. (1968) *Identification Methods for Microbiologists, part B*. Academic Press, London.

GIBBS B. M. & SKINNER F. A. (1966) *Identification Methods for Microbiologists, part A*. Academic Press, London.

GOODFELLOW M., AUSTIN B. & DAWSON D. (1976) Classification and identification of phylloplane bacteria using numerical taxonomy. In: *Microbiology of Aerial Plant Surfaces* (Eds C. H. Dickinson & A. Preece), pp. 275–92. Academic Press, London.

GOODFELLOW M. & BOARD R. G. (1980) *Microbial Classification and Identification*. Academic Press, London.

GRANT D. G., MILLS A. A. & SCHOFIELD A. K. (1979) An alkalophilic species of *Ectothiorhodospira* from a Kenyan salt lake. *Journal of General Microbiology* **110**, 137–42.

HALE E. M. (1977) Isolation of alkalophilic bacteria from an alkaline reservoir used in an industrial process. *Proceedings of the 77th Annual Meeting of the American Society for Microbiology*, abstract 150.

HARDER W. & ATTWOOD M. M. (1978) Biology, physiology and biochemistry of Hyphomicrobia. *Advances in Microbial Physiology* **17**, 303–59.

HARRIGAN W. F. & MCCANCE M. E. (1966) *Laboratory Methods in Microbiology*. Academic Press, London.

HENDRY M. S., HODGKISS W. & SHEWAN J. M. (1970) The identification and classification of luminous bacteria. *Journal of General Microbiology* **64**, 151–63.

HERBERT D., ELLSWORTH R. & TELLING R. C. (1956) The continuous culture of bacteria: a theoretical and experimental study. *Journal of General Microbiology* **14**, 601–22.

HILL L. R. (1968) The determination of deoxyribonucleic acid base compositions and its application to bacterial taxonomy. In: *Identification Methods for Microbiologists, part B* (Eds B. M. Gibbs & D. A. Shapton), pp. 177–86. Academic Press, London.

HIRSCH P. & CONTI S. F. (1964) Biology of budding bacteria. I. Enrichment, isolation and morphology of *Hyphomicrobium* spp. *Archiv für Mikrobiologie* **48**, 339–57.

HIRSCH P. & CONTI S. F. (1965) Enrichment and isolation of stalked and budding bacteria. In: *Anreicherungskultur und Mutantsauslee* (Ed. H. G. Schlegel), pp. 47–51. Gustav Fischer Verlag, Stuttgart.

HOBSON P. N. & MANN S. O. (1971) Isolation of celluloytic and lipolytic organisms from the rumen. In: *Isolation of Anaerobes* (Eds D. A. Shapton & R. G. Board), pp. 149–58. Academic Press, London.

HUNGATE R. E. (1950) The anaerobic mesophilic cellulolytic bacteria. *Bacteriological Reviews* **14**, 1–49.

HUNGATE R. E. (1966) *The Rumen and its Microbes*. Academic Press, London.

HUNGATE R. E. (1969) A roll tube method for cultivation of strict anaerobes. In: *Methods in Microbiology* (Eds

J. R. Norris & D. W. Ribbons), vol. 3B, pp. 117–32. Academic Press, London.

JANNASCH H. W. (1965) Continuous culture in microbial ecology. *Laboratory Practice* **14**, 1162–6.

JANNASCH H. W. (1967) Enrichments of aquatic bacteria in continuous culture. *Archiv für Mikrobiologie* **59**, 165–73.

JEFFREYS E. G. & SMITH W. K. (1951) A new type of soil percolator. *Proceedings of the Society for Applied Bacteriology* **14**, 169–71.

JENSEN H. L. (1954) The Azotobacteriaceae. *Bacteriological Reviews* **18**, 195–215.

JOSHI M. M. & HOLLIS J. P. (1976) Rapid enrichment of Beggiatoa from soils. *Journal of Applied Bacteriology* **40**, 223–4.

KADOTA K. (1956) A study on the marine aerobic cellulose decomposing bacteria. *Memoirs of the College of Agriculture No. 74 (Fisheries series 6)*, Kyoto University, Japan.

KRATZ W. A. & MYERS J. (1955) Nutrition and growth of several blue-green algae. *American Journal of Botany* **42**, 282–7.

KUBITSCHEK H. E. (1970) *Introduction to Research with Continuous Cultures*. Prentice-Hall, Englewood Cliffs, New Jersey.

KUENEN J. G. (1975) Colourless sulphur bacteria and their role in the sulphur cycle. *Plant and Soil* **43**, 49–76.

KUENEN J. G., BOONSTRA J., SCHRODER H. G. J. & VELDKAMP H. (1976) Competition for inorganic substrates among chemo-organotrophic and chemolithotrophic bacteria. *Microbial Ecology* **3**, 119–30.

KUENEN J. G. & VELDKAMP H. (1972) *Thiomicrospira pelophila*, nov. gen., nov. sp., a new obligately chemolithotrophic colourless sulfur bacterium. *Antonie van Leeuwenhoek* **38**, 241–56.

LACEY J. (1973) Actinomycetes in soils, composts and fodders. In: *Actinomycetales, Characteristics and Practical Importance* (Eds G. Sykes & F. A. Skinner), pp. 231–51. Academic Press, London.

LARSEN H. (1952) On the culture and general physiology of the green sulphur bacteria. *Journal of Bacteriology* **64**, 187–96.

LEE J. V., GIBSON D. M. & SHEWAN J. M. (1977) A numerical taxonomic study of some pseudomonas-like bacteria. *Journal of General Microbiology* **98**, 439–51.

LEES H. & QUASTEL J. H. (1946) Biochemistry of nitrification in soil. 1. Kinetics of, and effects of poisons on, soil nitrification as studied by a soil perfusion technique. *Biochemical Journal* **40**, 803–10.

LEFTLEY J. & VANCE I. (1979) An inflatable anaerobic glove bag. In: *Cold Tolerant Microbes in Spoilage and the Environment* (Eds A. D. Russell & R. Fuller), pp. 51–7. Academic Press, London.

LEWIS R. F. & PRAMER D. (1958) Isolation of *Nitrosomonas* in pure culture. *Journal of Bacteriology* **76**, 524–8.

LIESKE R. (1912) Untersuchungen über der Physiologie denitrifiziender Schwefelbakterien. *Deutsche Botanik Gesellschaft* **36**, 12–21.

LOCKHART W. R. & LISTON J. (1970) *Methods for Numerical Taxonomy*. American Society for Microbiology, Bethesda.

LUND B. M. (1965) A comparison by the use of gel electrophoresis of soluble protein components and esterase enzymes of some Group D Streptococci. *Journal of General Microbiology* **40**, 143–8.

MANDEL M. (1966) Deoxyribonucleic acid base composition in the genus *Pseudomonas*. *Journal of General Microbiology* **43**, 273–92.

MEERS J. L. (1971) Effect of dilution rate on the outcome of chemostat mixed culture experiments. *Journal of General Microbiology* **67**, 359–67.

MONOD J. (1942) *Recherches sur la Croissance des Cultures Bactériennes*. Hermann, Paris.

MORITA R. Y. & HAIGHT R. D. (1964) Temperature effects on the growth of an obligately psychrophilic marine bacterium. *Limnology and Oceanography* **9**, 103–6.

MOURARET M. & BALDENSPERGER J. (1977) Use of membrane filters for the enumeration of autotrophic thiobacilli. *Microbial Ecology* **3**, 345–59.

NORRIS J. R. (1959) The isolation and identification of Azotobacter. *Laboratory Practice* **8**, 239–43.

NORRIS J. R. & RIBBONS D. W. (1969) *Methods in Microbiology*, vol. 3B. Academic Press, London.

NORRIS J. R. & RIBBONS D. W. (1970) *Methods in Microbiology*, vol. 3A. Academic Press, London.

Oxoid Manual (1979) 4th edn. Oxoid Ltd., London.

PANKHURST E. S. (1966) Method and apparatus for growing anaerobic cultures. *British Patent* **1**, 041, 105.

PANKHURST E. S. (1971) The isolation and enumeration of sulphate-reducing bacteria. In: *Isolation of Anaerobes* (Eds D. A. Shapton & R. G. Board), pp. 223–40. Academic Press, London.

PETERSON J. E. (1969) Isolation, cultivation and maintenance of myxobacteria. In: *Methods in Microbiology* (Eds J. R. Norris & D. W. Ribbons), vol. 3B, pp. 185–210. Academic Press, London.

PFENNIG N. (1961) Eine vollsynthetische Nahrlösung zur selectiven Anreicherung einer Schwefelbakterien. *Naturwissenschaften* **48**, 136–7.

PFENNIG N. (1965) Anreicherungskulturen für rote und grüne Schwefelbakterien. In: *Anreicherungskultur und Mutantenauslee* (Ed. H. G. Schlegel), pp. 179–89. Gustav Fischer Verlag, Stuttgart.

PFENNIG N. (1967) Photosynthetic bacteria. *Annual Review of Microbiology* **21**, 285–324.

PFENNIG N. (1969a) *Rhodopseudomonas acidophila* sp. n. a new species of the budding purple non-sulfur bacteria. *Journal of Bacteriology* **99**, 597–602.

PFENNIG N. (1969b) *Rhodospirillum tenue* sp. n. a new species of the budding purple non-sulfur bacteria. *Journal of Bacteriology* **99**, 619–20.

PFENNIG N. & LIPPERT K. D. (1966) Über das Vitamin

B_{12} Bedürfnis phototropher Schwefelbakterien. *Archiv für Mikrobiologie* **55**, 245–56.

PFENNIG N. & TRÜPER H. G. (1974) The Rhodospirillales. In: *Handbook of Microbiology* (Eds A. I. Laskin & H. A. Lechevalies), pp. 14–24. CRC Press, Cleveland, Ohio.

POINDEXTER J. S. (1964) Biological properties and classification of the Caulobacter group. *Bacteriological Reviews* **28**, 231–95.

POPE A. & SKERMAN V. B. D. (1967). Mineral salts media for the cultivation of non-exacting autotrophs and heterotrophs. In: *A Guide to the Identification of the Genera of Bacteria* (Ed. V. B. D. Sherman), pp. 213–16. Williams & Wilkins, Baltimore.

POSTGATE J. R. (1966) Media for sulphur bacteria. *Laboratory Practice* **15**, 1239–41.

POSTGATE J. R. (1969) Media for sulphur bacteria: some amendments. *Laboratory Practice* **18**, 286–9.

POWELL E. O. (1958) Criteria for the growth of contaminants and mutants in continuous culture. *Journal of General Microbiology* **18**, 259–67.

RIPKA R., DERUELLES J., WATERBURY J. B., HERDMAN M. & STANIER R. Y. (1979) Generic assignments, strain histories and properties of pure cultures of cyanobacteria. *Journal of General Microbiology* **111**, 1–61.

ROBINSON K. (1968) The use of cell wall analysis and gel electrophoresis for the identification of coryneform bacteria. In: *Identification Methods for Microbiologists part B* (Eds B. M. Gibbs & D. A. Shapton), pp. 85–92. Academic Press, London.

RODINA A. G. (1972) *Methods in Aquatic Microbiology*. Butterworths University Park Press, Baltimore.

SCHLEGEL H. G. & PFENNIG N. (1961) Die Anreicherungskultur einiger Schwefelpurpurbakterien. *Archiv für Mikrobiologie* **38**, 1–39.

SCHLOESING T. & MÜNTZ A. (1877) Sur la nitrification par les ferments organisés. *Comptes Rendus Acadamie Science Paris* **84**, 30–9.

SCHMIDT E. L., MOLINA J. A. E. & CHIANG C. (1973) Isolation of chemoautotrophic nitrifiers from Moroccan soils. *Bulletin Ecological Research Communications (Stockholm)* **17**, 166–7.

SHAPTON D. A. & BOARD R. G. (1971) *Isolation of Anaerobes*. Academic Press, London.

SHARMA B. & AHLER R. C. (1977) Nitrification and nitrogen removal. *Water Research* **11**, 897–925.

SHIMWELL J. L., CARR J. G. & RHODES M. E. J. (1960) Differentiation of *Acetomonas* and *Pseudomonas*. *Journal of General Microbiology* **23**, 283–6.

SIEBURTH J. McN. (1967) Seasonal selection of estuarine bacteria by water temperature. *Journal of Experimental Marine Biology and Ecology* **1**, 98–121.

SINCLAIR N. A. & STOKES J. L. (1963) Role of oxygen in the high cell yields of psychrophiles and mesophiles at low temperatures. *Journal of Bacteriology* **85**, 164–7.

SKERMAN V. B. D. (1967) *A Guide to the Identification of the Genera of Bacteria*. 2nd edn. Williams & Wilkins, Baltimore.

SKINNER F. A. (1971) The isolation of soil clostridia. In: *Isolation of Anaerobes* (Eds D. A. Shapton & R. G. Board), pp. 57–80. Academic Press, London.

SKINNER F. A. & LOVELOCK D. W. (1980) *Identification Methods for Microbiologists*. Academic Press, London.

SKINNER F. A. & QUESNEL L. B. (1978) Streptococci. *Society for Applied Bacteriology Symposium*. Academic Press, London.

SKINNER F. A. & WALKER N. (1961) Growth of *Nitrosomonas europea* in batch and continuous culture. *Archiv für Mikrobiologie* **38**, 339–49.

SLATER J. H. (1979) Microbial population and community dynamics. In: *Microbial Ecology: a Conceptual Approach* (Eds. J. M. Lynch & N. J. Poole), pp. 45–63. Blackwell Scientific Publications, Oxford.

SMITH N. R., GORDON R. E. & CLARK F. E. (1952) Aerobic Spore-forming Bacteria. *U.S. Department Agricultural Monograph* **16**, Washington.

SNEATH P. H. A. (1972) Computer taxonomy. In: *Methods in Microbiology* (Eds J. R. Norris & D. W. Ribbons), vol. 7A, pp. 29–98. Academic Press, London.

SNEATH P. H. A. (1974) Test reproducibility in relation to identification. *International Journal of Systematic Bacteriology* **24**, 508–23.

SNEATH P. H. A. (1978) Identification of Microorganisms In: *Essays in Microbiology* (Eds J. R. Norris & M. H. Richmond), pp. 10/1–10/32. John Wiley, Chichester.

SNEATH P. H. A. & COLLINS V. G. (1974) A study in test reproducibility between laboratories: Report of a Pseudomonas working party. *Antonie van Leeuwenhoek* **40**, 481–527.

SNEATH P. H. A. & SOKAL R. R. (1973) *Numerical Taxonomy*. W. H. Freeman, San Francisco.

SPENCER R. (1955) The taxonomy of certain luminous bacteria. *Journal of General Microbiology* **13**, 111–18.

STANIER R. Y., KUNISAWA R., MANDEL M. & COHEN-BAZIRE G. (1971) Purification and properties of unicellular blue-green algae. *Bacteriological Reviews* **35**, 171–205.

STANIER R. Y., PALLERONI N. J. & DOUDOROFF M. (1966) The aerobic pseudomonads: a taxonomic study. *Journal of General Microbiology* **43**, 159–271.

STANLEY S. O. & ROSE A. H. (1967) Bacteria and yeasts from lakes on Deception Island. *Proceedings of the Royal Society Series B* **252**, 199–207.

STARKEY R. L. (1935) Isolation of some bacteria which oxidise thiosulphate. *Journal of Soil Science* **39**, 197–206.

STARR M. P. & HUANG J. C. C. (1972) Physiology of the Bdellovibrios. *Advances in Microbial Physiology* **8**, 215–61.

TAYLOR B. F., HOARE D. S. & HOARE S. L. (1971) *Thiobacillus denitrificans* as an obligate chemolithotroph. Isolation and growth studies. *Archiv für Mikrobiologie* **78**, 193–204.

TEMPEST D. W. (1970) The continuous cultivation of micro-organisms. 1. Theory of the chemostat. In: *Methods in Microbiology* (Eds J. R. Norris & D. W. Ribbons), vol. 2, pp. 259–76. Academic Press, London.

THIMANN K. V. (1964) *Das Leben der Bakterien*. Fischer, Jena.

VAN NIEL C. B. (1928) *The Propionic Acid Bacteria*. Uitgeverszaak & Boissevain, Haarlem.

VAN NIEL C. B. (1931) On the morphology and physiology of the purple and green sulphur bacteria. *Archiv für Mikrobiologie* **3**, 1–112.

VAN NIEL C. B. (1944) The culture, general physiology, morphology and classification of the non-sulphur purple and brown bacteria. *Bacteriological Reviews* **8**, 1–118.

VAN NIEL C. B. (1955) Natural selection in the microbial world. *Journal of General Microbiology* **13**, 201–17.

VELDKAMP H. (1970) Enrichment cultures of prokaryotic microorganisms. In: *Methods in Microbiology* (Eds J. R. Norris & D. W. Ribbons), vol. 3A, pp. 305–61. Academic Press, London.

VELDKAMP H. (1976) Ecological studies with the chemostat. *Advances in Microbial Ecology* **1**, 59–94.

VELDKAMP H. & KUENEN J. G. (1973) The chemostat as a model system for ecological studies. *Bulletin Ecological Research Commission (Stockholm)* **17**, 247–8.

VISHNIAC W. & SANTER M. (1957) The thiobacilli. *Bacteriological Reviews* **21**, 195–213.

WAKSMAN S. A. (1959) *The Actinomycetes*, vol. 1. Williams & Wilkins, Baltimore.

WAKSMAN S. A. & STARKEY R. L. (1922) On the growth and respiration of sulphur-oxidising bacteria. *Journal of General Physiology* **5**, 285–92.

WATSON S. W. (1965) Characteristics of a marine nitrifying bacterium, *Nitrocystis oceanus* sp. n. *Limnology and Oceanography* **10**, 274–89.

WHITTENBURY R. (1971) Enrichment and isolation of photosynthetic bacteria. In: *Isolation of Anaerobes* (Eds D. A. Shapton & R. G. Board), pp. 241–9. Academic Press, London.

WHITTENBURY R., PHILLIPS K. C. & WILKINSON J. F. (1970) Enrichment, isolation and some properties of methane utilising bacteria. *Journal of General Microbiology* **61**, 205–16.

WILEY W. R. & STOKES J. L. (1962) Requirement of an alkaline pH and ammonia for substrate oxidation by *Bacillus pasteurii*. *Journal of Bacteriology* **84**, 730–4.

WILLIAMS S. T & CROSS T. (1971) Actinomycetes. In: *Methods in Microbiology* (Ed. C. Booth), vol. 4, pp. 295–334. Academic Press, London.

WILLIS A. T. (1969) Techniques for the study of anaerobic, spore forming bacteria. *Methods in Microbiology* (Eds J. R. Norris & D. W. Ribbons), vol. 3B. Academic Press, London.

WILSON G. S. & MILES A. A. (1975) *Topley and Wilson's Principles of Bacteriology, Virology and Immunity*. Edward Arnold, London.

WINDLE-TAYLOR E., BURMAN N. P. & OLIVER C. W. (1953) Use of the membrane filter in the bacteriological examination of water. *Journal of Applied Chemistry* **3**, 233–9.

WINOGRADSKY S. (1888) Beiträge zur Morphologie und Physiologie der Bakterium. *Pflanzenforschung* **1**, 1–120.

WOLFE J. & BARKER A. N. (1968) The genus Bacillus: aids to the identification of its species. In: *Identification Methods for Microbiologists, part B* (Eds B. M. Gibbs & D. A. Shapton), pp. 93–109. Academic Press, London.

WOLFE R. S. (1971) Microbial fermentation of methane. *Advances in Microbial Physiology* **6**, 107–46.

Chapter 2 · Procedures for the Isolation, Cultivation and Identification of Fungi

Dennis Parkinson

2.1 Introduction

As heterotrophic organisms the fungi must live as saprophytes or parasites. Thus they play important roles in terrestrial and aquatic ecosystems as decomposers of organic matter (with attendant roles in nutrient cycling), as pathogens, or as symbionts with terrestrial plants. This chapter deals only with saprophytic fungi (for mycorrhizal and pathogenic fungi see Chapters 26 and 27).

The ecologically important characteristics of terrestrial fungi were summarised (Harley 1971) as:

(1) their ability to alter their environment via the production of exo-enzymes and the release of end-products of metabolism (e.g. nutrients, inhibitors and antibiotics);

(2) their large surface to volume ratio plus a hyphal structure which allows spreading growth and enables penetration of substrates, such as leaf litter and wood;

(3) the ability of septate fungi to form hyphal fusions which allow the production of hyphal networks and heterokaryosis (with the attendant possibilities for increased genetic variability);

(4) their ability to accumulate nutrients in their thalli;

(5) the fact that they act as sources of nutrients for other organisms, either indirectly, through the production of soluble metabolic products, or directly where fungal hyphae are eaten by microarthropods.

Most of these characters are also applicable to fungi in aquatic environments.

2.2 Terrestrial fungi

Detailed reviews on the nature and activities of soil fungi have been given by Warcup (1967), Griffin (1972) and Pugh (1974) and many of the contributions in Parkinson and Waid (1960) still provide valuable ideas. Swift (1976) has reviewed ecological concepts regarding microbial communities in terrestrial habitats and Parkinson (1982) has attempted to trace the development of soil fungal ecology.

In view of this substantial background literature only a brief restatement of major points follows.

(1) A large number of species of fungi can be isolated from soil which represent all the major groups of the fungi. The pattern of fungal species isolated from a given soil is affected by the quality of the organic matter entering the soil (i.e. the nature and quality of the vegetation) plus the environmental conditions of moisture availability and temperature.

(2) Soil fungi are not uniformly distributed but are associated with microhabitats, such as decomposing organic matter and living roots.

(3) The majority of soil fungi can exist in a variety of morphological forms (and physiological states), such as spores, other resting structures and hyphae. Fungal hyphae may range from short-lived

structures frequently utilising soluble, simple carbon compounds to long-lived structures utilising insoluble, complex carbon compounds. The soil fungi also exhibit various growth patterns in soil (Burges 1960) ranging from dense colonisation of organic substrates with no growth into surrounding soil to the growth of isolated hyphae through soil with no obvious association with fragments of organic matter.

(4) The methods which are available for the isolation of fungi from soil are all selective. Therefore the species lists obtained in studies of the qualitative nature of soil mycofloras must be incomplete. Nevertheless, such studies are valuable for their intrinsic ecological interest and because interactions between species of fungi and the patterns of distribution of such fungi may be important factors in determining the rates and patterns of organic matter decomposition (Swift 1976). In addition, it may be possible to deduce from such species lists and data on the physiological properties of individual species (Domsch & Gams 1972) information concerning the activities of fungi in soil.

Qualitative studies on soil fungi are either described as synecological when the objective is to assess the total number of species in the fungal community, or as autecological when detailed information on an individual species or group of related species is required.

2.3 Aquatic fungi

Fungi in fresh water and marine ecosystems have not been studied to the same extent as the soil fungi. Nevertheless, just as in terrestrial systems, they play important roles as decomposers of organic matter and as pathogens, and serve as a food source for aquatic invertebrates. Several major works on the taxonomy and general biology of aquatic fungi are available (Sparrow 1960; Johnson & Sparrow 1961; Jones 1976; Kohlmeyer & Kohlmeyer 1979). Of the numerous shorter reviews on aquatic fungi, reference should be made to the many individual contributions in Sparrow (1968) and Jones (1974, 1976).

Representatives of each major group of the fungi (Phycomycetes, Ascomycetes, Basidiomycetes and Fungi Imperfecti) have been isolated from water (marine and fresh). However only a small number of Basidiomycetes appear to occur in aquatic habitats (Jones 1974).

Fungi isolated from water fall into one of two categories: the obligate aquatic fungi which require the presence of water to complete their life cycles, and the facultative aquatic fungi which are typically soil fungi which can survive and grow in/on particular substrates in the aquatic environment. An ecological scheme for classification of aquatic fungi has been given by Park (1972).

Experimental

2.4 Terrestrial fungi

2.4.1 Synecological Studies

Given that the methods available for the isolation of soil fungi are all selective, then it is obvious that the major difficulty in assessing the ecological roles of such fungi is that of determining what species occur in a particular environment together with the state of their vegetative activity (Harley 1971). It must be emphasised that, in ecological investigations, methods must be chosen in order to answer specific questions and these methods must be tested prior to their regular use.

2.4.2 Choice of Medium

Qualitative studies of soil fungi require plating out of soil, organic matter, or fungal hyphae and spores on to a nutrient agar medium. Thus the choice of an appropriate isolation medium is of great importance. The old idea that a non-selective medium could be developed for such purposes has given way to the concept of using a nutrient agar medium which will allow the isolation of the maximum number of fungal species from the substrata under study. In some studies, plain water agar can be used for the primary isolation of fungi from soil or organic matter in an attempt to eliminate the selectivity of nutrient media. The isolates obtained must be transferred to nutrient agar as soon as they are observed on the primary isolation plates.

One of the initial questions to be answered prior to beginning mass isolations of fungi from soil is, which nutrient medium should be used? In such studies it would be desirable to use several selective media, but time constraints rarely allow this to be done. The choice of an isolation medium for use in studying a particular soil or group of substrates should follow preliminary comparative tests using a range of media. Details on media which have been commonly used for studies on soil fungi are given in Booth (1971b), Parkinson *et al.* (1971) and Johnson and Curl (1972).

If the investigation is restricted to that of studying an individual physiological group of the soil fungi, such as cellulolytic or lignolytic species, then selective media should be used. Examples of such media have been given (Parkinson *et al.* 1971; Johnson & Curl 1972), but frequently modification must be made to known media in order to make them suitable for specific problems.

When isolating fungi from soils containing high numbers of bacteria it is frequently necessary to reduce competition from the bacteria by adding anti-bacterial substances to the medium. Simple acidification of the medium to *c.* pH 5.0 can be used effectively in some studies and the addition of crystal violet, rose bengal, sodium deoxycholate or sodium propionate restricts bacterial development (Papavizas & Davey 1959). Currently the addition of antibacterial antibiotics (e.g. aureomycin at 30 mg l^{-1} or streptomycin at 30 mg l^{-1}) to isolation media is the most frequently used method for restricting bacterial growth. Whenever these compounds are added to isolation media tests need to be carried out in order to show that they do not have any selective effects on the fungi.

2.4.3 ENUMERATION

The soil dilution plate method was, until recently, used predominantly for studying the nature of soil fungal communities and involves the following: the preparation of an initial soil suspension; the preparation from the soil suspension of a dilution series; the plating of suitable dilutions on/in an appropriate nutrient medium. It is described in detail elsewhere in this volume (see Chapter 6, p. 87).

If this method is adopted, the investigator must determine such details as: the character of the initial soil suspension (weight of soil, volume and type of the suspending fluid); the type and time of agitation to produce the initial soil suspension; the type of diluent for the dilution series; the nutrient isolation medium chosen; and the plating method to be used (surface spread or pour plate).

The soil dilution plate method is inappropriate for studies on fungi which occur as hyphae in soil as it is selective for fungi present as spores (Warcup 1957). Therefore it can be valuable for assessing the spore content of soil, even though it is impossible to gain information on the microhabitat relations of the fungi which are isolated.

Warcup (1950) developed a simple variant of the soil dilution plate method in which small amounts of soil are dispersed in known volumes of nutrient agar. Using sterile needles with flattened tips, a 5 to 15 mg sub-sample of soil is placed in a sterile Petri dish, a drop of sterile water added and the soil thoroughly broken up and dispersed over the base of the dish. A known volume (10 ml) of molten, cool (*c.* 45°C) nutrient agar is added to each Petri dish, the dishes are rotated to disperse the soil into the agar medium and, after setting, the dishes are incubated. Fungi developing on the isolation plates are isolated into pure culture for subsequent identification. The simplicity of this method and its speed of execution make it useful for preliminary studies on soil mycofloras, but it must be remembered that it suffers from the same defects as the soil dilution plate method in that it is selective for fungi present as spores.

2.4.4 ISOLATION

In an attempt to isolate fungi which occur in soil as actively growing hyphae, various immersion methods have been developed. The original technique, designed by Chesters (1940), has been modified in a number of ways (Thornton 1952; Mueller & Durrell 1957; Parkinson 1957; Wood & Wilcoxson 1960; Anderson & Huber 1965; Luttrell 1967) but all have the same basic principle, i.e. the isolating medium is placed in soil in such a manner that it is separated from the soil by an air gap. The medium is left in the soil for a period (usually 5 to 7 d) after which it is brought into the laboratory where small samples are plated on to a selective medium. Fungi growing from the inoculum are isolated and purified. It is assumed that these fungi were present as active hyphae in the soil since they must have grown into the isolating medium across the separating air gap.

A simple immersion method is described by Mueller and Durrell (1957) in which autoclavable plastic centrifuge tubes are used. Holes (*c.* 5.0 cm diameter) are bored at desired positions in the walls of each centrifuge tube, the tubes are wrapped with plastic tape, filled to within 4 cm of the top with the isolating medium, plugged and autoclaved. In the field the plastic tape is pierced opposite the holes in the tube using sterile needles and the tube placed in the soil. After a predetermined time, the tubes are retrieved and taken to the laboratory where the fungi are isolated by removing the agar core from each tube, cutting the core into small pieces, and plating each piece on to nutrient agar.

Immersion methods have certain disadvantages. The placement of immersion apparatus (tubes, plates, etc.) into soil appears to induce the germination of fungal spores in the vicinity of the apparatus, and thus the fungi isolated may originate from spores and not from previously active hyphae. In addition, small soil animals may enter the isolating medium held in the immersion apparatus bringing fungal propagules with them. There is also the possibility of competition between fungi for entry into the immersion apparatus and for subsequent survival in the enclosed medium.

In an attempt to achieve a more precise isolation of fungi present in a hyphal state in soil, Warcup (1955) applied hyphal isolation. Small volumes of soil (even individual soil crumbs) were saturated with sterile water and dispersed with a fine jet of sterile water. The coarser soil particles were allowed to sediment and the finer particles decanted off. The supernatant was resuspended and the procedure repeated several times but no details were given as to sedimentation time or particle size. The coarser particles were spread in a film of water and, working under a dissecting microscope, hyphal fragments picked out using sterile fine forceps or a mounted needle. The hyphae were carefully drawn through semisolid agar to remove bacteria, spores and organic debris. The cleansed hyphae were then plated on to nutrient agar plates, the position of each hypha being marked on the base of the plate. The plates were incubated and observed every 24 h in order to ensure that any growth was the result of growth from the plated hyphae and not from spores which survived the cleaning process.

Data obtained by the use of this method for fungal isolation from soil are very different from those obtained by soil dilution plating or direct soil plating. Thus Warcup (1957) showed that, in a wheat field soil, the most abundant taxa isolated by direct hyphal isolation were sterile forms, Basidiomycetes, *Rhizoctonia solani, Gaeumannomyces graminis*, and species of *Pythium, Rhizopus, Mortierella* and *Fusarium*. The most abundant taxa isolated from the same soil using the dilution plate method were species of *Pencillium, Mucor, Cladosporium, Fusarium* and *Rhizopus*.

This method would appear to be the ideal for the isolation of active fungal hyphae from soil. However, it must be appreciated that it is time-consuming and therefore impractical for studies which require regular isolations from many soil samples. Furthermore the method requires considerable technical skill and is open to unintentional bias because robust, pigmented hyphae of large diameter tend to be selectively picked from soil whilst delicate, fine, hyaline hyphae tend to be ignored. In addition hyphae which are closely associated with organic matter fragments are extremely difficult to isolate by this method.

Harley and Waid (1955) showed that efficient washing of root material could, by removal of detachable fungal propagules, allow the isolation of fungi in a hyphal state. They also reviewed washing methods for isolation of fungi from different types of organic material. As a result of this demonstration a number of washing methods have been developed for studies of soil fungi. These methods range from a simple washing-decanting method (Watson 1960) to the use of automated machines which can deal with several soil samples at one time (Hering 1966; Bissett & Widden 1972).

Certain general features are evident in the soil washing methods. Soil samples are placed in sterile boxes which contain a number of sieves of graded size—the sieve sizes being determined by the type (inorganic or organic) and size of soil particle in which the investigator is interested. The soil samples are then vigorously washed (by aeration or mechanical agitation) with several changes of sterile water, the number of washings required to remove the maximum number of fungal spores being determined by preliminary experimentation (Harley & Waid 1955). Finally the washed soil particles and/or organic fragments are removed from the sieves, dried on sterile filter paper, and plated on to nutrient agar plates (the number of particles per plate depends on the degree of fungal colonisation of the particles, but is usually two to four). Williams *et al.* (1965) examined the efficiency of this method when applied to soils of different textures and showed that the method was more efficient when applied to sandy soils than when applied to soils with a high clay content.

Soil washing methods have the advantage of being simple in execution and, when the automatic, multi-box apparatus is used, allow the handling of numerous soil samples in a short time. Although it is unlikely that washing removes all the spores from soil samples, it is highly likely that fungi isolated from the plating of washed soil particles have arisen from hyphae rather than spores.

Serial washing, followed by plating of small pieces of washed material has proved to be a valuable method for studying fungi associated with discrete organic matter in soil. Indeed many of the data on succession of fungi on decomposing leaf litter (Hayes 1979) and on fungi associated with root surfaces (Parkinson 1967) have been obtained using this method. Detailed information on methods for studying fungi associated with plant roots is given elsewhere (see Chapters 23 and 27).

During the foregoing discussion reference has been made to the incubation of isolation plates. The choice of temperature for such incubation must be based on the aims of the study. For most general studies of mesophilic soil fungi, the common incubation temperatures range between 15 and 25°C. Clearly, if the interests centre on psychophilic or thermophilic forms, then the incubation temperatures are adjusted accordingly.

2.4.5 IDENTIFICATION

Following the primary isolation of fungi, individual colonies must be subcultured as soon as possible into pure culture for subsequent identification. In most studies it is necessary to observe the isolation plates under the dissecting microscope to ensure that overgrown, slow-growing species are also isolated. In attempts to isolate individual fungi, initial transfer on to fresh nutrient agar plates is preferable to transfer on to nutrient agar slopes. After the purity of the isolates is assured, the cultures may be maintained on slopes.

A wide range of media have been used for the maintenence of pure cultures of soil fungi. Many of these are natural complex media (Gams *et al.* 1975), such as malt extract agar, potato extract agar, potato dextrose agar, potato carrot agar, cherry agar and V8 vegetable juice agar. Eventually, when identification of specific taxa is being attempted, defined nutrient media and growth conditions must be used. Descriptions of these can be obtained from the taxonomic monographs used for the identification of the isolates. General keys for the identification of soil fungi are notoriously inadequate. The publication *The Genera of Hyphomycetes from Soil* (Baron 1968) provided enormous help for Soil ecologists, but normally more specialised monographs are required for identifying specific orders and genera, e.g. *Mortierella* (Linneman 1941), Mucorales (Zycha *et al.* 1969), *Fusarium* (Booth 1971a), *Cephalosporium* (Gams 1971), *Aspergillus* (Raper & Fennell 1965), *Penicillium* (Raper & Thom 1949) and *Trichoderma* (Rifai 1969). Gams *et al.* (1975) give a detailed list of the literature available for identification of soil fungi.

2.4.6 AUTECOLOGICAL STUDIES

Most autecological investigations of soil fungi have dealt with soil-borne pathogens, where data on occurrence, inoculum potential and competitive abilities are required. To this end methods of isolation selective for different taxa have been developed. An exhaustive description of such methods is not possible in this chapter because of their diversity and number. Detailed information on such methods may be obtained from Domsch and Schwinn (1965), Tsao (1970) and Johnson and Curl (1972).

Mention should be made here of one of the major problems in qualitative studies of soil fungi: the infrequency of isolation of basidiomycetes from soil samples. This may be the result of the specialised nutrient requirements of many of these fungi and/or their susceptibility to competition (at least on isolation plates) from faster growing species. Hale and Savory (1976) have reviewed selective media available for studying this group of the fungi.

2.5 Isolation of aquatic fungi

Details of the major methods for isolation of fungi from water have been given by Jones (1971, 1976) and the media used are described by Jones (1971) and Booth (1971b).

The major methods for isolating aquatic fungi are these: baiting, dilution plating, particle plating and concentration using centrifugation and filtration.

2.5.1 BAITING METHODS

Baiting methods involve placing pieces of sterile organic matter (bait) into the water under study to provide a substrate for fungal development and can be carried out either in the laboratory or in the field.

In laboratory studies water samples (or samples

of submerged soil or organic debris) are collected in sterile bottles and returned to the laboratory in cool conditions. The isolation of fungi from the samples should be begun within 24 h of collection.

A few drops of the collected water (or small amounts of soil or organic matter) are added to several Petri dishes which are half full of sterile distilled water. Baits (a single type or several types) are added to each dish and incubated and observed daily for two weeks. When mycelia and sporangia are observed growing on the bait they are dissected out, washed in several changes of sterile water and plated on to antibiotic-amended cornmeal agar. When hyphal growth occurs on these agar plates, the fungi are subcultured. This process is continued until pure cultures are obtained.

Another method of obtaining pure cultures of individual aquatic Phycomycetes is the harvesting of individual zoospores (using sterile capillary tubes). The harvested spores are transferred to an appropriate nutrient medium (e.g. water agar, agar with 0.15% (w/v) maltose and/or 0.004% (w/v) peptone, or cornmeal agar. In studies of this type, the majority of interest has centred on the aquatic Phycomycetes. A large variety of sterile baits have been used including conifer pollen, cellulose (wettable cellophane), boiled hemp seed, boiled grass leaves and dead insects. Some of these baits (e.g. hemp seed, grass leaves) attract the motile spores of a wide range of Mastigomycotina, while others are more specific in the range of fungi attracted (e.g. pollen grains attract Chytridiales).

In field studies bait is placed in small mesh (usually 2 to 4 mm) boxes made of plastic-covered or galvanised wire. The boxes are suspended in water for an appropriate time, usually from one to five weeks. Subsequently the baits are retrieved, returned (under cool conditions, *c.* 15 to 20°C) to the laboratory where, after thorough washing with sterile water, the bait is placed in sterile water along with fresh sterile bait. The fungi subsequently growing on the fresh bait are isolated and grown in pure culture in the manner described for laboratory studies.

For studies on the Mastigomycotina, the types of bait commonly used include rosaceous fruits, tomatoes, grapes, cellophane, twigs of various tree species and human hair. For studies on marine yeasts, Fell *et al.* (1960) used cores of banana stalks as bait.

Among the Ascomycetes, the lignolytic (lignicolous) forms have received the most attention. In such studies, blocks of wood are submerged in water (Jones 1971). The period that the wood blocks are submerged depends on the environmental conditions (e.g. temperature) in the water and the type of experiment.

The sampled blocks are placed in sterile containers and taken to the laboratory in conditions which are cool (*c.* 15 to 20°C) and which prevent the drying of the wood blocks. In the laboratory the blocks are placed in sterile plastic boxes which contain sterile tissue and are incubated for long periods (e.g. 15 weeks). Sterile water may be added to prevent the complete drying of the wood blocks. Fruit bodies develop on the wood and ascospores may be removed from these and streaked on to nutrient agar medium. The media used for these isolations are usually low in nutrient content (Jones 1971). Fungi from the wood blocks can be directly plated out by cutting off surface slivers of wood and placing these on nutrient agar. Jones (1971) has given a detailed account of the variety of methods available for isolating lignolytic Ascomycetes from marine and freshwater environments.

2.5.2 Dilution Plating

Park (1972) described a method for isolating fungi from organic detritus in water, in which the collected detritus was macerated (1 g wet weight detritus in 20 ml sterile water). From this initial suspension a dilution series was made and one ml aliquots of appropriate dilutions were spread on the surface of dried nutrient agar containing rose bengal or aureomycin. These isolation plates were incubated at 20°C and examined regularly over a 5-week period. Fungi were isolated and purified for subsequent identification.

While conventional dilution plating of water samples has been used for studies on aquatic bacteria, it is rarely appropriate for studies of aquatic fungi because of their low propagule density. However, direct plating of undiluted water samples has been found valuable in some investigations. Thus Vishniac (1956) used surface plating of water samples to isolate marine Phycomycetes. In this method seawater agar (supplemented with glucose and a range of growth factors) was used as the isolation medium. Poured plates of the medium were dried, flooded with an antibiotic mixture (penicillin G and streptomycin sulphate), and each plate inoculated with 0.2 ml sea-water sample using the standard spread plate method.

For the isolation of aquatic hyphomycetes, scum or foam can be collected from the water surface and must be dealt with quickly following sampling. Usually the samples are diluted with sterile water and 1 ml aliquots are inoculated (spread plate method) on to cornmeal or cellulose agar plates to which antibiotics have been added. Germinating spores can be subcultured on to fresh medium.

2.5.3 Particle Plating

The particle plating method requires that organic matter (e.g. algae, stems, leaves) is collected from the aquatic environment and transported in cool sterile containers to the laboratory where the samples can be dealt with in several ways.

(1) Small pieces of the organic material are plated on to the surface of dried nutrient agar plates. The plates are incubated (*c.* 20°C), and fungi growing from the plated particles subcultured for subsequent identification in pure culture.

Presumably, in studying organic matter decomposition in aquatic environments, the application of serial washing of organic matter with sterile water followed by plating of small pieces of washed organic matter would remove loosely adhering propagules and bacteria, and would allow the isolation of fungi growing as hyphae on (and in) the organic matter.

(2) After washing with sterile water the organic matter can be placed in Petri dishes containing sterile water and observed for development of fungal structures. This type of plating out has been used successfully for studies on aquatic hyphomycetes in the following way. Decaying leaves from well aerated streams are collected, returned to the laboratory, washed in sterile water and placed in sterile water. If the leaves are colonised by aquatic hyphomycetes, spores will be produced and these can be picked off and transferred to an appropriate, antibiotic-amended agar medium.

(3) The organic matter can be placed in damp chambers (e.g. sterile Petri dishes containing damp tissue or filter paper). The organic matter is observed regularly for the development of fungal fruiting structures from which isolation and purification can be attempted. This method is particularly useful for studying fungi associated with stems and leaves in water.

2.5.4 Concentration Methods

As stated earlier the distribution of fungal propagules (in particular those of the Mastigomycotina) in free water is very sparse. In order to quantify fungi, methods must first be developed to extract fungi from large volumes of water. To this end continuous-flow centrifugation and filtration have been applied. Fuller and Poynton (1964) used continuous-flow centrifugation to study monocentric Mastigomycotina; the concentrate obtained being plated on to antibiotic-amended nutrient agar.

Miller (1967) used Millipore filtration (0.8 μm, 1.2 μm or 3.0 μm) to remove zoospores from large water samples of known volumes. The filtrate obtained on the Millipore filter was suspended in 0.5 ml sterile water and was spread on to antibiotic-amended low nutrient agar (g l^{-1} distilled water: agar, 30; glucose, 0.05; peptone, 0.05; yeast extract, 0.05; streptomycin sulphate, 0.5; penicillin G, 0.5).

Concentration techniques are time consuming but offer the means of adding a quantitative aspect to the study of aquatic Phycomycetes (Jones 1971).

The major references quoted at the beginning of this section on aquatic fungi either provide keys for the identification of aquatic fungi or give detailed reference lists of monographs available for study of specific taxa. Further references are given by Gams *et al.* (1975), and Ingold (1975) has provided a guide to the aquatic hyphomycetes.

Recommended reading

Barron G. L. (1968) *The Genera of Hyphomycetes from Soil.* Williams and Wilkins Co., Baltimore.

Booth C. (1971b). Fungal culture media. In: *Methods in Microbiology* (Ed. C. Booth), vol 4, pp. 49–94. Academic Press, London.

Domsch K. H. & Gams W. (1972) *Fungi in Agricultural Soils.* Longmans Group Ltd., London.

Griffin D. M. (1972) *Ecology of Soil Fungi.* Chapman & Hall, London.

Harley J. L. (1971) Fungi in ecosystems. *Journal of Ecology*, **59**, 653–68.

Johnson T. W. & Sparrow F. K. (1961) *Fungi in Oceans and Estuaries.* J. Cramer, Weinheim.

Jones E. B. G. (1971) Aquatic fungi. In: *Methods in Microbiology* (Ed. C. Booth), vol 4, pp. 335–65. Academic Press, London.

Jones E. B. G. (1974) Aquatic fungi: freshwater and marine. In: *Biology of Plant Litter Decomposition* (Eds C. H. Dickinson & G. J. F. Pugh), vol. 2, pp. 337–83. Academic Press, London.

PARKINSON D. (1982) Filamentous fungi. In: *Methods of Soil Analysis* (Ed. R. Miller). American Society Agronomy Publication, Madison, Wisconsin.

PARKINSON D. & WAID J. S. (1960) *The Ecology of Soil Fungi.* Liverpool University Press, Liverpool.

PUGH G. J. F. (1974) Terrestrial fungi. In: *Biology of Plant Litter Decomposition* (Eds C. H. Dickinson & G. J. F. Pugh), vol. 2, pp. 303–36. Academic Press, London.

WARCUP J. H. (1967) Fungi in soil. In: *Soil Biology* (Eds N. A. Burges & F. Raw), pp. 51–110. Academic Press, London.

References

ANDERSON A. L. & HUBER D. M. (1965) The plate-profile technique for isolating soil fungi and studying their activity in the vicinity of roots. *Phytopathology* **55**, 592–4.

BARRON G. L. (1968) *The Genera of Hyphomycetes from Soil.* Williams & Wilkins Co., Baltimore.

BISSETT J. & WIDDEN P. (1972) An automatic, multichamber soil-washing apparatus for removing fungal spores from soil. *Canadian Journal of Microbiology* **18**, 1399–409.

BOOTH C. (1971a) *The Genus Fusarium.* Commonwealth Mycological Institute, Kew.

BOOTH C. (1971b) Fungal culture media. In: *Methods in Microbiology* (Ed. C. Booth), vol. 4, pp. 49–94. Academic Press, London.

BRIERLEY W. B., JEWSON S. T. & BRIERLEY M. (1928) The quantitative study of soil fungi. *Proceedings of the 1st International Congress of Soil Science* **3**, 48–71.

BURGESS N. A. (1960) Dynamic equilibria in the soil. In: *The Ecology of Soil Fungi* (Eds D. Parkinson & J. S. Waid), pp. 185–91. Liverpool University Press, Liverpool.

CHESTERS C. G. C. (1940) A method for isolating soil fungi. *Transactions of the British Mycological Society* **24**, 352–5.

DOMSCH K. H. & GAMS W. (1972) *Fungi in Agricultural Soils.* Longmans Group Ltd., London.

DOMSCH K. H. & SCHWINN F. J. (1965) Nachweis und Isolierung von pflanzenpathogenen Bodenpilze mit Selektiven Verfahren. *Zentralblatt für Bakteriologie, Parasitenkunde, Infektionskrankheiten und Hygiene,* Abt. **1**, (Suppl. 1), 461–85.

FELL J. W., AHEARN D. G., MEYERS S. P. & ROTH F. J. (1960) Isolation of yeasts from Biscayne Bay, Florida and adjacent benthic areas. *Limnology and Oceanography* **5**, 366–71.

FULLER M. S. & POYNTON R. O. (1964) A new technique for the isolation of aquatic fungi. *Bioscience* **14**, 45–6.

GAMS W. (1971) *Cephalosporium-artige Schimmelpilze.* Gustav Fischer Verlag, Stuttgart.

GAMS W., VAN DER AA H. A., VAN DER PLAATS-NITERINK A. J., SAMSON R. A. & STALPERS J. S. (1975) *CBS Course of Mycology.* Centraalbureau voor Schimmelcultures, Baarn.

GRIFFIN D. M. (1972) *Ecology of Soil Fungi.* Chapman & Hall, London.

HALE M. D. C. & SAVORY J. G. (1976) Selective agar media for the isolation of basidiomycetes from wood—a review. *International Biodeterioration Bulletin* **12**, 112–15.

HARLEY J. L. (1971) Fungi in ecosystems. *Journal of Ecology* **59**, 653–68.

HARLEY J. L. & WAID J. S. (1955) A method for studying active mycelia on living roots and other surfaces in the soil. *Transactions of the British Mycological Society* **38**, 104–18.

HAYES A. J. (1979) The microbiology of plant litter decomposition. *Science Progress (Oxford)* **66**, 25–42.

HERING T. F. (1966) An automatic soil washing apparatus for fungal isolation. *Plant and Soil* **25**, 195–200.

INGOLD C. T. (1975) An illustrated guide to aquatic and water borne hyphomycetes with notes on their biology. *Scientific Publication of the Freshwater Biological Association* **30**, 1–96.

JOHNSON L. F. & CURL E. A. (1972) *Methods for Research on the Ecology of Soil-borne Plant Pathogens.* Burgess Publishing Co., Minnesota.

JOHNSON T. W. & SPARROW F. K. (1961) *Fungi in Oceans and Estuaries.* J. Cramer, Weinheim.

JONES E. B. G. (1971) Aquatic fungi. In: *Methods in Microbiology* (Ed. C. Booth), vol. 4, pp. 335–65. Academic Press, London.

JONES E. B. G. (1974) Aquatic fungi: freshwater and marine. In: *Biology of Plant Litter Decomposition* (Eds C. H. Dickinson & G. J. F. Pugh), vol 2, pp. 337–83. Academic Press, London.

JONES E. B. G. (1976) *Recent Advances in Aquatic Mycology.* Elek Science, London.

KOHLMEYER J. & KOHLMEYER E. (1979) *Marine Mycology, The Higher Fungi.* Academic Press, London.

LINNEMANN G. (1941) Die Mucorineen-Gattung *Mortierella* Coemans. *Pflanzenforschung (Jena)* **23**, 64.

LUTTRELL E. S. (1967) A strip bait for studying the growth of fungi in soil and aerial habitats. *Phytopathology* **57**, 1266–7.

MILLER C. E. (1967) Isolation and pure culture of aquatic phycomycetes by membrane filtration. *Mycologia* **59**, 524–7.

MUELLER K. E. & DURRELL L. W. (1957) Sampling tubes for soil fungi. *Phytopathology* **47**, 243.

PAPAVIZAS G. C. & DAVEY C. B. (1959) Evaluation of various media and anti-microbial agents for isolation of soil fungi. *Soil Science* **88**, 112–17.

PARK D. (1972) Methods of detecting fungi in organic detritus in water. *Transactions of the British Mycological Society* **58**, 281–90.

PARKINSON D. (1957) New methods for qualitative and quantitative study of fungi in the rhizosphere. *Pedologie Gand* **7**, 146–54.

PARKINSON D. (1967) Soil microorganisms and plant roots. In: *Soil Biology* (Eds N. A. Burges & F. Raw), pp. 449–78. Academic Press, London.

PARKINSON D. (1982) Filamentous fungi. In: *Methods of Soil Analysis* (Ed. R. Miller), American Society Agronomy Publication, Madison, Wisconsin.

PARKINSON D., GRAY T. R. G., HOLDING J. & NAGEL-DE-BOOIS H. M. (1970) Heterotrophic microflora. In: *Methods in the Study of Quantitative Soil Ecology* (Ed. J. Phillipson), pp. 34–50. Blackwell Scientific Publications, Oxford.

PARKINSON D., GRAY T. R. G. & WILLIAMS S. T. (1971) *Methods for Studying the Ecology of Soil Microorganisms*. Blackwell Scientific Publications. Oxford.

PARKINSON D. & WAID J. S. (1960) *The Ecology of Soil Fungi*. Liverpool University Press, Liverpool.

PUGH G. J. F. (1974) Terrestrial fungi. In: *Biology of Plant Litter Decomposition* (Eds C. H. Dickinson & G. J. F. Pugh), vol. 2, pp. 303–36. Academic Press, London.

RAPER K. B. & FENNELL D. I. (1965) *The Genus Aspergillus*. Williams & Wilkins, Baltimore.

RAPER K. B. & THOM C. (1949) *A Manual of the Penicillia*. Williams & Wilkins, Baltimore.

RIFAI M. A. (1969) *A Revision of the Genus Trichoderma*. Mycological Paper 116, Commonwealth Mycological Institute, Kew.

SPARROW F. K. (1960) *Aquatic Phycomycetes*. 2nd Ed. University of Michigan Press, Ann Arbor.

SPARROW F. K. (1968) Ecology of freshwater fungi. In: *The Fungi* (Eds G. C. Ainsworth & A. S. Sussman), vol. 3, pp. 41–93. Academic Press, London.

SWIFT M. J. (1976) Species diversity and the structure of microbial communities in terrestrial habitats. In: *The Role of Terrestrial and Aquatic Organisms in Decomposition Processes* (Eds J. M. Anderson & A. Macfadyen), pp. 185–222. Blackwell Scientific Publications, Oxford.

THORNTON R. H. (1952) The screened immersion plate. A method for isolating soil microorganisms. *Research, London* **5**, 190–1.

TSAO P. H. (1970) Selective media for isolation of pathogenic fungi. *Annual Review of Phytopathology* **8**, 157–86.

VISHNIAC H. S. (1956) On the ecology of lower marine fungi. *Biological Bulletin of the Marine Biological Laboratory, Woods Hole, Massachussetts* **111**, 410–14.

WARCUP J. H. (1950) The soil plate method for isolation of fungi from soil. *Nature, London* **116**, 117.

WARCUP J. H. (1955) Isolation of fungi from hyphae present in soil. *Nature, London* **175**, 953–4.

WARCUP J. H. (1957) Studies on the occurrence and activity of fungi in a wheat field soil. *Transactions of the British Mycological Society* **40**, 237–62.

WARCUP J. H. (1967) Fungi in soil. In: *Soil Biology* (Eds N. A. Burgess & F. Raw), pp. 51–110. Academic Press, London.

WATSON R. D. (1960) Soil washing improves the value of the soil dilution and plate count method of estimating populations of soil fungi. *Phytopathology* **50**, 792–4.

WILLIAMS S. T., PARKINSON D. & BURGES N. A. (1965) An examination of the soil washing technique by its application to several soils. *Plant and Soil* **22**, 167–86.

WOOD F. A. & WILCOXSON R. D. (1960) Another screened immersion plate for isolating soil fungi. *Plant Disease Reporter* **44**, 594.

ZYCHA H., SIEPMANN R. & LINNEMANN G. (1969) *Mucorales*. J. Cramer, Lehre.

Chapter 3 · Procedures for the Isolation, Cultivation and Identification of Algae

Frank E. Round

3.1 The freshwater environment

Fresh waters are generally defined as those which have less than 0.5 g l^{-1} of dissolved solids and are geologically transient features of the landscape. They are not synonymous with inland waters since in many parts of the world the standing waters have a very high salt content (e.g. 300 g l^{-1} in the Dead Sea and 203 g l^{-1} in the Great Salt Lake) making them hypersaline waters relative to the oceans at 35 g l^{-1}. The habitats may be divided into those where unidirectional flow is a dominant factor and those where water movements are more random. This division can become somewhat blurred. In other words, through-flow is important in lake studies although the time scale may be in the order of months compared with fractions of a day in a river. Some lakes behave rather like slow-flowing rivers and vice versa. The development of a river phytoplankton population is dependent upon a slow enough flow rate and the species composition differs from that of lakes, possibly due to the flow rate but also because of the methods of perennation and recruitment. In both habitats algal growth is linked with rates of nutrient supply, depletion of key nutrients and the seasonal interplay of water supply rates, temperature, light intensity and grazing. Habitats can be divided initially into those which support planktonic forms and those supporting benthic forms of life. Planktonic habitats are those associated with the water mass (standing or flowing) and benthic habitats those with the bottom surface and other surfaces, excluding the air/water interface.

Phytoplanktonic species are mainly free-living and, since they are heavier than water, they sink unless they have some mechanism for maintaining their position in the water column, e.g. buoyancy by gas vacuoles, by active swimming upwards along a light gradient or rarely, and disputedly, by accumulation of oil. Most species, however, are suspended by turbulent water motion. A small number of species (neuston) live associated with the air/water interface (attached to the water with their bodies either submerged or protruding into the atmosphere) and some species have other algae attached to them. The prime benthic habitat is the surface of sediments in the illuminated zone, i.e. down to depths of 5 to 10 m or perhaps 50 m in very clear lakes. Two communities live in this region: an attached assemblage on sand grains (epipsammon), and a motile assemblage moving over the surface sediment which is mixed into the upper few mm (epipelon). This latter community is particularly well developed in most waters and comprises a large number of species (in the order of hundreds), whereas the epipsammic community which is more widespread than generally appreciated, comprises only a small number of species of mainly diatoms, some green algae and cyanobacteria. The epipelic species maintain their surface position by rapid phototactic movement whereas the epipsammic species move slowly from grain to grain and many can withstand lengthy periods of burial. These habitats are exceedingly rich in bacteria, protozoa and microfauna and form a major source of food for many animals up to and including fish (especially in the tropics). Submerged rocks and macroscopic plant surfaces act as favourable anchors for many epilithic and epiphytic species respectively. Some species can live on both surfaces whilst others are confined to one or the other. The algae form a layer of adnate species and an upwardly growing felt of stalked (pedunculate) species (Round 1981a) and on some surfaces the brown diatoms or green algal layers may be several mm thick. Within this felt another assemblage (metaphyton) of unattached species can be found, especially in acid waters. These habitats, also, are rich in bacteria and associated microfauna and they extend onto artificial substrata placed in the water. Blue-green algae (cyanobacteria) live in hot springs and at a somewhat lower temperature diatoms enter the association.

3.2 The marine environment

Three-quarters of the planet's surface is covered with saline water and clearly this is the most extensive environment for microorganisms. Unlike fresh waters, the salinity of the oceans is fairly constant at around 35 g l^{-1} and only where evaporation is excessive is this exceeded (up to 45 g l^{-1} but never to the salinities of inland saline waters), or reduced in regions of fresh water or ice melt. However, unlike fresh waters, which have an extremely variable, but dilute, ionic composition, the seas have a constant proportion of ions present. Thus the range of communities in the seas is much more limited than those encountered in fresh waters and the temperature and light regimes are more important than the ionic ratios in determining the species composition. The marine phytoplankton constitute the largest plant assemblage on the planet and can occur regularly down to a depth of 35 to 40 m or occasionally 100 to 150 m. The top 100 m or so of the central regions of tropical gyres may be depleted of organisms but richly populated just above the thermocline (Venrick *et al.* 1973). Growth at the bottom of the illuminated zone is now known to be much more common than believed even a decade ago, e.g. it occurs between 10 to 40 m in European seas (Holligan & Harbour 1977).

Similar life forms and factors operate in the marine plankton as in fresh waters and some of the parallels are quite remarkable. It is often not appreciated that the number of species involved in the phytoplankton communities can exceed 1000 over a year's cycle. The phytoplankton does not maintain a constant flora but changes from week to week: very few species in either marine or freshwater environments have growth periods exceeding a few weeks (Round 1981a). The marine benthic habitats are similar to those in fresh water but the epipsammon makes a greater contribution. However, the marine benthic communities are more complex since the communities either live in the fluctuating intertidal region or the more stable, shallow subtidal region. Whilst diatoms are important components of the flora of fresh waters they are dominants in the marine habitats and the flagellates (mainly dinoflagellates) are the subdominants. Microscopic algae coat many wave-exposed surfaces in the intertidal zone but this is a neglected field of study. They are also important in coral reefs, especially on and in the dead coral debris.

3.3 The soil environment

The soil is the most neglected of all the algal habitats, yet few soil samples can be collected which do not contain numerous diatoms, green algae and cyanobacteria. In some rich soils, the particles are actually coated with algae. The flora is more restricted compared with the aquatic habitats but still contains 200 or so species and more continue to be described (Broady 1979a, b). The forms are very similar to those found in the aquatic epipelon, but the soil algae tend to have a much greater capacity for forming resistant stages. Numerous studies have attempted to estimate algal cells within soil profiles to compare with the changes noted in fungal and bacterial assemblages but most of the active algae live in the surface millimetre and those studies are merely reactivating resting spores. Soil cyanobacteria are extremely common and add greatly to the nitrogen fixation and fertility of the soil. Most soil algal studies have been carried out by Soviet scientists and their very extensive literature should be consulted (Gollerbakh & Shtina 1969; Shtina & Gollerbakh 1976). Algae are important early colonisers of virgin soil, e.g. in Antarctica and in some other regions they form an algal crust which is important in stabilising the soil.

Subaerial algae occur not only on soils but on plant surfaces: the green powdery algae on tree bark is familiar enough, but many other algae are involved on bark and leaves in the tropics. Rock surfaces, especially in mountain regions but also apparently in the dry and hot tropics, support a rich flora of micro-algae, the study of which has been neglected. Micro-algae can be found growing on and in crevices of the underside of stones in desert habitats (Friedmann 1971).

Experimental

3.4 Sampling and enumeration methods

The study of all algal communities ought to be a two-stage process:

(1) the micro-algae must be obtained in a representative manner from their habitats;

(2) the algal biomass (crop) must be estimated and characterised.

A problem at both stages is that populations have to be quantitatively estimated yet different organisms vary in size from a few μm up to 2 mm for some of the phytoplankton (e.g. the marine tropical diatom, *Ethmodiscus* sp.). Furthermore for the benthic habitats, the various forms have to be sampled and usually no single sampling technique will suffice. The heterogeneous, spatial distribution of every alga in the community adds greatly to the problem. It is not necessary to separate this section into marine and freshwater methods since similar techniques are applicable to both habitats.

3.4.1 PHYTOPLANKTON

To obtain a sample for the study of individual species and as a rough estimate of dominance, a plankton net (meshes vary from 100 to 5 μm) drawn through the water is adequate. In poor waters it may have to be towed slowly for some time but a single passage in rich waters may be sufficient. The smaller phytoplankton species can be completely missed when using coarse nets, and towing for long periods may lead to compaction and damage of the cells. The fine nets block more rapidly and mesh size is no indication of the size of the cells trapped. Such samples cannot be quantified easily. To obtain integrated samples from the whole water column, the net is weighted, lowered to a point below the expected limit of living cells and raised to the surface. Alternatively nets can be towed at set depths in order to analyse the algal communities at specific depths.

At sea the Hardy Continuous Plankton Recorder (Colebrook 1960) has been widely used and this consists of a net which filters water onto a rotating screen which is rolled into a chamber containing preservative. It is extremely useful for synoptic surveys across oceans. Other techniques involve sampling the cooling water intake of ships, or deck-pumps are used to sample known volumes from known depths.

Samples of known volume are required for strict quantitative work. In shallow fresh waters a length of hosepipe can be used to obtain an integrated sample (Lund & Talling 1957) and, if specific depth samples are required, then water bottles closing at set depths are used. This latter technique is the commonest method used at sea. There is, however, often great vertical variation in the biomass with depth, and isolated water bottle sampling is subject to considerable error. Samples obtained from depths are susceptible to changes in light intensity and temperature, especially if brought onto the deck of a ship in the tropics. This may not affect enumeration of rigid species but could greatly affect subsequent experimental techniques (see below). The neuston can be sampled only by simple devices consisting either of steel or nylon screens or simply a glass plate immersed vertically in the water then drawn out, drained and the film clinging to it drained or scraped into a collection bottle.

An estimate of the biomass in the sample can be obtained by either identification and calculation of the volume of the component species or measurement of some biochemical component of the sample. Whenever possible the samples should be examined live as soon after sampling as possible and this is especially important if shape and position of the cytoplasmic organelles is to be studied. Concentrated samples rapidly decay and these should be preserved using a few drops of Lugol's iodine or 20% (w/v) formaldehyde neutralised with hexamethylene-tetramine or acidified with acetic acid and added to give a final concentration of about 0.5% (v/v). It is important to remember that preservatives can distort and remove flagella; acid preservatives dissolve coccoliths; and alkali dissolves siliceous phytoplankton.

Identification is the first essential step and it is usually necessary to have a fairly large sample in order to find the rare species and note the variation of the populations. A net sample may be the most suitable.

The commonest method for counting mixed populations is to use an inverted microscope. Samples of known volumes of unconcentrated plankton are sedimented by the addition of a known volume of Lugol's iodine in special counting chambers. After settling overnight the cells can be identified and counted. A normal microscope can be used to count algae using a counting slide (such as Sedgwick-Rafter, Palmer-Maloney or Haemocytometers). This method has the advantage that counts can be performed immediately and higher magnifications can be used but problems arise when large species are present. Other techniques involve concentrating the sample by centrifugation, reverse osmosis or membrane filtration, but considerable controversy exists about the accuracy of these techniques.

Natural populations contain organisms which are variable in size and shape and this usually precludes the use of electronic counters, though these have been used. There is a problem inherent in all counting techniques and that is recognition of live cells and exclusion of dead cells.

Counting methods do not give any indication of the size differences between organisms which for length can vary by a factor of 1×10^3 (exceptionally 2×10^3) and for cell volume by a factor of 1×10^6, for example *Asterionella formosa* has a volume of 700 μm^3 whereas *Ceratium hirundinella* may be 70 000 μm^3. Thus in order to obtain accurate biomass values from cell count data various transformations must be made to allow for differences in cell volume, cell area and plasma volume. Cell volumes are calculated for algae with a regular shape from standard relationships. For the algae with more complex and irregular shapes, models can be made and a displacement method used to measure the volume. It should be noted that the volumes of the same species from different sites often vary and so measurements must be made on each sample, precluding the use of standard tables. All data of this nature must be critically assessed since 1000 small cells may have the same volume as one large alga. Volume data is converted into approximate cell carbon or nitrogen values by applying appropriate equations (Smayda 1978).

Almost all these procedures underestimate the phytoplankton cell number and biomass since the small nanoplankton are difficult to estimate accurately. In most situations the nanoplankton form an appreciable amount of the total biomass (up to 70% of total primary production is attributed to this group of algae in some habitats). Some are small flagellates which are very difficult to detect, but other flagellate and coccoid groups can be estimated by plating techniques using nutrient-enriched medium (Happey 1970). However, this technique has the standard problem in that it provides selective conditions which may exclude the growth of some of the nanoplankton. Nevertheless, it is extremely valuable and should be more widely employed.

Estimation of cell constituents is used when a total component estimate is acceptable irrespective of the species composition. The component most commonly measured is chlorophyll *a* since this is confined to the algal component and can be extracted simply with either acetone or methanol. Phaeophytin and phaeophorbide, both chlorophyll *a* degradation products, interfere with chlorophyll *a* determinations and so these pigments are also measured. This is particularly important in senescent populations where these pigments can account for significant amounts of material. Other pigments are sometimes measured using standard extraction techniques for spectrophotometry or fluorometric measurements for estimations *in vivo* (Vollenweider 1974). Carbon (Strickland & Parsons 1968; Ganf & Milburn 1971) and nitrogen determinations are sometimes made but not on a routine basis. ATP measurements are also used (Holm-Hansen & Paerl 1972) but this is not specific to the algae. Specific component composition is usually expressed as an amount per unit dry weight. The dry weight is obtained normally by drying at 105°C but this leads to the loss of some volatile components. Thus drying over dessicating agents or freeze-drying may be preferred.

A sound statistical treatment is desirable in all these studies, although it is not always easy to apply for reasons such as limitations of sampling time at sea. Venrick (1978) has summarised the major considerations.

3.4.2 BENTHOS

Initial sampling of the benthos is difficult since the sites need to be sampled on an area basis. Sediments can be removed by coring devices or suction systems although sandy sediments are particularly difficult to sample (Round & Hickman 1971). Some samples are best obtained by scuba diving. Another important point is to realise that sediment samples contain a mixture of the epipelon and epipsammon communities. The latter is very difficult to separate and analyse quantitatively: the best approach is to remove the loose sediment and epipelon by agitating the sample in a bottle and allowing the particles with the attached epipsammon to settle. This may need to be repeated several times but yields a sample in which the epipsammon can be studied. There are no completely satisfactory means of estimating the epipsammon since direct observation is necessary but only the tops of the sand grains can be studied and the cells are often very small and difficult to identify. Sonication removes the epipsammon species which can be counted subsequently (Hickman & Round 1970). Chlorophyll extractions can be made directly on sand or indirectly on cells removed by sonication.

The epipelon is easier to quantify. The sample cannot be observed directly since the cells are obscured by the non-living particles. Instead the sample should be allowed to settle, the supernatant liquid poured off and the sediment spread in a Petri dish—for the best quantitative results, the area sampled should equal that of the dish. The sediment is allowed to settle in the Petri dish, the excess water removed and cover glasses or squares of lens-cleaning tissue placed on the surface (Eaton & Moss 1966). The dish is left overnight and the cover glasses or tissues removed the following day. The sediment must not be too liquid, which causes the cover glasses to float, or too dry, in which case they stick to and pick up sediment. Since most of the epipelic species are positively phototactic they move to the surface and attach to the glass or the fibres of the tissue and can be removed. The efficiency of removal varies with the sediment, but is can be as high as 70%. Cells can be identified, counted or extracted for chlorophyll determination. This technique is particularly valuable when used on intertidal sediments (Round & Palmer 1966; Palmer & Round 1967).

3.4.3 EPIPHYTON

Sampling of the epiphyton is also difficult. Known areas of plant material can be agitated, boiled in dilute acid or brushed to remove the algal populations which may be estimated by counting on an inverted microscope or in a haemocytometer. Cylindrical aquatic plant material (e.g. *Equisetum* sp.) can be rotated and a fine spray of water played over the surface to remove the cells. However, the firm adnate species will not be removed by this technique.

3.4.4 EPILITHON

The epilithon on stones can be removed by brushing or sonicating. A method of quantifying the epilithon used by Round (unpublished observations) involves drying the stone and gluing a small glass ring onto it. When the ring is securely in place water is added to the ring and the area inside the ring scraped with a fine brush (a dentist's tooth-cleaning brush is suitable). The washings are transferred into a graduated centrifuge tube. After several washings the sample is made up to a standard volume and this sampled to determine the species composition and the cell numbers estimated using a haemocytometer. If there is much sediment or there are many filamentous species on the surface, the method is less successful. An alternative, less popular method involves using a sheet of cellulose acetate formed on the surface. The sheet is stripped off together with any attached algae which may then be examined and quantified.

3.4.5 SOIL

Soil algae can be treated like the epipelon but in the author's experience the recovery is not good, probably because the species are not so actively motile. All the benthic associations can also be treated like the phytoplankton and subsamples plated out on nutrient agar. A surprising number of flagellate and coccoid species are usually recovered. If bacterial and fungal contamination becomes a significant problem, antibiotics can be incorporated into the medium although some may affect the growth of the algae.

3.5 Observation

Algae growing in liquid culture and on solid media give suspensions and colonies with characteristic colours and colonies with specific forms. Colonies on agar plates can be observed and photographed whilst on the plate (Happey-Wood 1978). Light microscopy, using standard and inverted microscopes can be employed and samples can be prepared for observation by scanning electron microscopy. Sometimes material dried onto stubs is satisfactory, e.g. diatoms on other algae, but the samples can also be freeze-dried or transferred via a cryostage attached to the scanning microscope. Many delicate forms are difficult to observe since they tend to collapse in the vacuum. Transmission electron microscopy can be employed on flagellate populations. The sample is placed on a grid, fixed by suspension over osmic acid, dried, coated (often by metal shadowing) and observed directly in the electron microscope. This technique is particularly valuable with freshwater flagellates, such as *Mallomonas* spp., and marine nanoplankton, such as *Chrysochromulina* spp., and indeed it is often necessary in order to reveal the arrangement of scales on these species before complete identification can be made.

Various staining techniques or the fluorescence of chlorophyll have been used to determine the live cells in a sample. Live cells also can be determined if the sample is stained with acridine orange which binds to the cellular DNA.

3.6 Cultivation

Cultivation in order to estimate cell numbers is normally only used for small flagellate or coccoid forms (see section 3.5, above). This technique results in the isolation of specific algae which can be maintained as permanent cultures: many algae survive indefinitely by subculturing on agar media. For species isolation or the study of the cellular contents of individual species, the natural sample can be enriched even though this inevitably involves the selection of certain forms. Culturing is often necessary before detailed studies of small species, especially flagellates, can be made. Delicate species which do not preserve well must be cultured if type collections are to be maintained. However, it is also valuable to increase the numbers available for isolation. At the isolation stage individual algae can be picked out using micropipettes, working with a low-power microscope or with an inverted microscope or by dilution culture techniques.

For many cultures of algae, the Erd-Schreiber growth medium can be used (Pringsheim 1946). This medium utilises a soil extract which is added to sea-water, fresh water or distilled water. For some algae this is sufficient for growth but usually a vitamin mixture, trace metals and EDTA are added. No general recipe can be given and each alga must be tried in media with various concentrations of nutrients and, indeed, some may require additional nutrients. Although the algae are predominantly autotrophic many have special organic and inorganic requirements which must be satisfied. It is usually necessary to add a number of the same algae to any new culture tube or flask as individual cells often have difficulty in surviving.

Another useful technique is the soil–water culture in which a small amount of sterilised soil is placed at the bottom of a test-tube, covered with some sterile sand and then culture medium added.

Some algae are extremely difficult to culture, indeed a few species have never been grown in the laboratory. Diatoms are particularly difficult to maintain for long periods since auxospore formation may not occur. Many algae are homothallic or heterothallic, e.g. volvocalean flagellates, desmids and filamentous zygnemaphyceae, but the individual strains can be maintained.

Axenic cultures of most algae can be maintained but not easily in soil–water combination. In a few instances bacteria have to be present for full morphological development or reproduction of the algae (Provasoli 1971). At some stages it may be desirable to prevent the growth of certain algal groups, e.g. during isolation enrichment. For instance, diatoms and Cyanophyta often successfully compete with other algae and the former can be prevented from growing by adding germanium oxide to the growth medium and the latter by adding streptomycin.

Preservation of cultures by cryotechniques has been developed and some algae can be frozen and preserved at liquid nitrogen temperatures or freeze-dried and stored in a vacuum (Holm–Hansen 1973). However, these techniques only maintain a small proportion of cells in a viable state.

3.7 Ecological energetics

The estimation of biomass is dealt with in section 3.4.1 (p. 33), and this section will deal with the measurement of rates of carbon fixation (productivity), gross and net contribution to the community (production) and energy relationships. The primary source of external energy is solar radiation but a smaller, and much less investigated contribution, is chemical.

Continuous rate measurements, such as those which can be made on populations grown in the laboratory, are not easily obtained from natural populations. Instead short-term determinations of photosynthesis are made and extrapolated to give average rates over days or seasons (see Chapter 15, p. 243). Only two common techniques are in use: namely the measurement of oxygen production by the Winkler method and measurement of [^{14}C]carbon uptake from labelled sodium bicarbonate. In general, the oxygen method is more reliable and does not involve such complex apparatus and expense but it cannot be used in extreme oligotrophic waters. The [^{14}C]carbon method, on the other hand, is more sensitive but errors can occur if photoassimilation of organic compounds and/or if soluble products of photosynthesis are released from the algal cells in any quantity. There are, of course, techniques to measure these and correct the data accordingly but this is probably not feasible for routine ecological studies. Obviously a representative sample of the community must be obtained and all the sampling techniques and statistic controls referred to previously need to be evaluated for each community before significant results can be obtained. Additionally, for sampling prior to rate studies, care must be taken to ensure that a large enough sample fills all the reaction bottles with a homogenous mixture. Furthermore, metal containers should be avoided since the metal may either inhibit or stimulate growth, the sample should be protected from full daylight especially if the sample is taken from a depth, and temperature changes should be avoided. In short the sample must be undamaged. Enclosure of the sample in a glass bottle adversely affects the nature of the light which reaches the sample. Quartz vessels are better but much more expensive and vessels made from plexiglass are almost as good. The volume should be large enough to minimise the effect of wall growth (100 to 200 ml).

Some experiments have involved enclosing large volumes of water in plastic spheres, experiments which involve teams of workers. In lakes, a technique, developed by Lund (1975), of enclosing large volumes of water in plastic tubes is very valuable since the effects of water inflow and outflow can be eliminated and a captive population studied. A simpler approach involves the study of cultures of single algal species in nature but extrapolation from these populations to mixed communities is hardly feasible. Nevertheless, these studies have great value in determining the constraints of the environment on single species, factors which are much more difficult to determine from studies of community fixation (see autoradiographic techniques below).

Whatever the source of the samples, they are enclosed in bottles and the bottles either suspended in the water or positioned in an appropriate region of the habitat in an attempt to reproduce as near as possible the conditions which the natural population would normally meet. This is known as the in-situ method. Alternatively the bottles are placed in an incubator in which an attempt is made to simulate the natural conditions. The latter method is often used at sea where it is expensive in terms of ship's time to stay in one place and suspend the bottles down a profile. Whatever method is used, it is clear that conditions are in many cases artificial, since enclosure of the sample prevents the normal circulation up and down the water column and replenishment of nutrients, light climate and grazing are all different from those in nature. Nevertheless, this is the most commonly used technique. A much less common approach is to attempt to measure the overall metabolism of the community in the natural environment by measuring changes in oxygen, pH and conductivity, and relating these changes in the rate of carbon dioxide fixation.

The oxygen method (Winkler titration or electrochemical determination) is precise and a sensitivity of ± 0.02 mg l^{-1} can be obtained. The amount of oxygen produced in the bottle incubated in the light gives the net photosynthetic activity, and the amount of oxygen consumed in the dark bottle gives the rate of loss of oxygen due to respiration which is, of course, not simply due to phytoplankton but also due to heterotrophic activity. The sum of these two measurements gives the gross rate of photosynthesis which, in general, is a more reliable measure than the net rate. Under some conditions photorespiration affects these measurements and then inhibitors have to be used to block various pathways. This introduces all the problems of inhibition and stimulation inherent in such techniques. The time of exposure is important and, on the whole, long exposure times should be avoided, since the unnatural effects of enclosure are exaggerated. In rich waters even a short exposure may lead to oxygen supersaturation and the raising of pH values. Short exposures and summation of data for the light period are to be preferred.

It must be remembered that there is a distinct diurnal variation in the photosynthetic rate and 3 h morning exposures may give rise to quite different rates of photosynthesis compared to 3 h afternoon exposures. Unless sampling occurs just before dawn, it should be remembered that the population will have completed some of its daily quota of photosynthesis prior to the experiment and this affects the rate determined.

In order to estimate integrated production, individual estimations have to be made at several depths with material collected from each depth. The depth of the euphotic zone and the depth of maximal photosynthesis have to be determined and then the production data from the depth profile integrated to give the activity below a unit area of the surface. Considerable experimentation is still needed to improve the technique, the aim always being to establish as near as possible the natural rate of carbon fixation.

The [^{14}C]carbon technique involves similar sampling, bottling and incubation *in situ* with [^{14}C]-sodium bicarbonate being added to the bottles. The total alkalinity of the initial water must be known and also the pH value. The complexity of the carbon dioxide, carbonate, bicarbonate and carbonic acid ion system in water is a problem (Talling 1976) and the methods for determining the carbon dioxide concentration are given in Talling (1976) and Vollenweider (1974).

The algal material from the bottles is filtered in semi-darkness onto membrane filters which are placed onto planchets for counting in Geiger Müller counters of the thin window or windowless type, or the filters are placed in counting vials and a mixture of solvent and scintillator added for liquid scintillation counting (Vollenweider 1974). Activity is obtained as counts per second and the carbon assimilated calculated after application of various correction factors. Some recent studies (Gieskes *et al.* 1979) have shown that, contrary to earlier experience, long incubation times (12 h) in large bottles give a more accurate picture of the rate of carbon dioxide fixation: they suggest rates 5 to 15 times greater than usual in tropical waters and indicated that fixation rates in these situations were just as high as elsewhere.

Techniques developed for laboratory studies of algal photosynthesis using polarographic measurements of oxygen exchange may provide valuable insights into ecological problems but the need for dense cell suspensions is a drawback (Jassby 1978). They may, however, be valuable in studying photosynthesis during bloom or red tide conditions.

3.8 Identification

Algae of bacterial dimensions (< 10 μm in length) may be identified directly using the light microscope if the morphological features of the vegetative cells alone are distinctive. However, many small green unicellular algae require culturing (see section 3.6, p. 36) in order to obtain enough material with which to make detailed measurements. This applies particularly to the flagellate and coccoid Chlorophyta, Chrysophyta and Prymnesiophyta. Some have to be examined by transmission electron microscopy, using characters of the scales coating the cells and this can only be done from fairly dense natural populations or from laboratory cultures. For example, the common chrysophyte *Synura* and the prymnesiophyte *Chrysochromulina* can rarely be identified in any other way. Filamentous genera such as *Oedogonium, Mougeotia, Spirogyra* are difficult, if not impossible to identify without the reproductive stages (oogonia or zygospores). On the other hand Cyanophyta have to be identified on morphological features alone; indeed, it is such features on which the descriptions in floras are based as culturing of members in this group often modifies the very distinctive natural morphological forms. However, there is much work, particularly of a statistical nature, still to be undertaken in the study of natural populations of Cyanophyta.

Reference to the genus only is rarely of any value in ecological studies and identifications in many publications cannot be checked since they are not accompanied by illustrations (preferably made with the help of a drawing attachment or camera lucida on the microscope) or photographs (although these can be valueless unless taken with skill and appreciation of the characters necessary for identification). Permanent slides can be made of many groups of algae, e.g. desmids, diatoms (before or after cleaning in acid) and dinoflagellates (especially the armoured forms) and material of some species can be preserved (see section 3.5, p. 35) for subsequent positive identification. New genera and species must be published according to the rules of botanical nomenclature and it is important to emphasise that failure to do this merely adds confusion. Ecological studies are particularly difficult owing to the very large number of species involved and they should only be undertaken if the worker is prepared to go to the necessary lengths required for accurate identification—anything less is merely an accumulation of spurious data.

A more reliable approach is an autecological one where a single species is studied in detail and the identification problem reduced to manageable proportions. Certain communities are particularly problematical owing to the preponderance of large numbers of coccoid or flagellate groups. One such is the algal community which is present on the surface of all soils, is often extremely important, yet is rarely studied and is complicated by the difficulties of identification—but see the excellent studies of Texas soils (Bold 1970) and of Icelandic and Antarctic soils (Broady 1979a, b). Communities in which diverse populations of naviculoid diatoms occur are also difficult, e.g. shallow marine sediments and salt marshes.

The following are some key references for the various phyla of algae. A discussion of the latest taxonomic systems of algae is given in Round (1981a). For freshwater genera the volumes by Bourelly (1966, 1968, 1970, 1972) are essential for tracking down many genera and equally important are the volumes of the Süsswasserflora (Hustedt 1930a; Ettl 1978), Rabenhorsts Kryptogamen Flora (Hustedt 1930b, 1959, 1962; Geitler 1932; Schiller 1933, 1937; Kolkwitz & Krieger 1937) and Die Binnengewässer (Huber-Pestalozzi 1941, 1950, 1955).

3.8.1 Cyanochloronta

Cyanochloronta is the phylum name used by Bold and Wynne (1978) for the blue-green algae (Cyanophyta) and is the preferred name since these organisms have both algal and bacterial characteristics. Terms implying a purely bacterial nature are as misleading as those implying a purely algal nature. The classic work, still of immense value, is Geitler (1932) whilst other useful publications are Desikachary (1959, 1972). Attempts to reduce the taxa to small numbers have been made but these are unusable for ecological studies.

3.8.2 Chlorophyta

Much taxonomic revision is taking place in the Chlorophyta. A marine British flora will soon appear but until then the group is difficult and specialist literature on individual genera must be consulted. The freshwater genera are covered in the volumes of Süsswasserflora (1914 onwards) although these are somewhat dated. For identification to the generic level Bourelly (1972) is indispensable. Volume 16 Part 5 (1961) of Die Binnengewässer deals with the flagellate volvocales. Desmids are dealt with in the volumes by West and West (1904–1912) and parts are being reprinted in the Kryptogamen Flora (Kolkwitz & Krieger 1937, 1941–1944). The filamentous Ulotrichales, Chaetophorales, Coleochaetales and Trentepohliales, need re-examination, although Printz (1964) is useful for the Chaetophorales.

The Chlorococcales have been covered best by the Prague school of phycologists and numerous papers published since 1945 by Fott and co-workers can be found in the special volumes of *Archiv für Hydrobiologie* entitled Algologische Studien. Many of the coccoid, flagellate and chlorosarcinoid soil genera have been dealt with by Bold and co-workers (Bold 1970).

Detailed monographic studies of most genera need to be prepared. Cultural studies of some, especially the Chlorococcales, may reveal that some genera are merely life-cycle forms or pleomorphic stages of others.

3.8.3 Prasinophyta

There is no up-to-date survey of the Prasinophyta but flagellate, coccoid and dendroid forms occur in both marine and freshwater habitats. Some authorities retain this group in the chlorophyta but their morphology is distinct and it is desirable to maintain them as a separate group, but see also Stewart and Mattox (1975).

3.8.4 Euglenophyta

The pigmented genera of the Euglenophyta are clearly functional algae and this is in no way diminished by the recent suggestions that the chloroplasts are possibly derived from a symbiotic event with an unknown chlorophyte (Gibbs 1978; Ragan & Chapman 1978). For comment on apochlorotic forms and general discussion on derivative apochlorotic genera of other groups see Round (1980). Euglenophyta are well described in Leedale (1967) and classification is covered by Pringsheim

(1956), Godjics (1953) and Huber–Pestallozzi (1955). Euglenophyta are an extremely important group of microbial algae, being particularly common in waters of high organic matter content.

3.8.5 Charophyta

Charophyta are not of microbial size but are often dealt with in algal ecological studies and they have interesting effects on other microbial groups when they decay Groves and Bullock–Webster (1917–1925) provide an old taxonomic account and Wood and Imahori (1964–1965) a more modern view.

3.8.6 Xanthophyta

Xanthophyta are green algae which belong in the chromophyte series both on biochemical and ultrastructural grounds and are much more common than generally supposed. The filamentous *Tribonema* spp., the siphonal *Vaucheria* spp., and *Botrydium* spp., are especially widespread. There is an excellent new monograph by Ettl (1978) although the older volume in the *Rabenhorst Kryptogamen Flora* (1939) is adequate. The coccoid genera require further study. Round (1981b) presents a brief discussion of the relationships between this group and the Chrysophyta and Bacillariophyta.

3.8.7 Chrysophyta

The phylum Chrysophyta is frequently represented in fresh waters, in planktonic and more especially in epiphytic situations. It is one of the most varied groups with both flagellate and non-motile, mainly coccoid genera. Bourelly (1968) treats the genera as a whole, and various other monographs of these genera possessing siliceous scales exist, the latest being Takahashi (1978). There has been a recent development to use these scales in paleolimnological studies. Huber–Pestalozzi (1941) gave a fairly detailed taxaonomic survey and Hibberd (1976) provided details of their ultrastructure.

3.8.8 Bacillariophyta

Bacillariophyta are probably the most widespread and common of the microbial-size algae. Identification is based entirely upon the morphology of the siliceous wall components though recent workers are beginning to appreciate the value of the cytoplasmic details. The classification at the light microscope level is partially described by the volumes of *Rabenhorst Kryptogamen Flora* (Hustedt 1930b–1962), and freshwater genera by Hustedt (1930a) and Patrick and Reimer (1966, 1975). Over the last decade the group has been studied again using transmission and scanning electron microscope techniques and this is leading to a considerable refinement of the classification (Helmke & Krieger 1962–1977, see also volumes reporting meetings of Fossil & Marine Diatom Symposia issued as Beihefte of *Nova Hedwiga* and published by J. Kramer–latest volume 1979). The classification of many genera is now under revision but the old floras still have value.

3.8.9 Prymnesiophyta (= Haptophyta)

The Prymnesiophyta is a most important phylum of microbial-sized flagellates especially common in marine habitats and fixing a considerable proportion of the total carbon dioxide. It is divided into two groups—those with only mineralised scales (e.g. *Chrysochromulina*) and the forms with mineral scales *and* calcium carbonate deposits (coccoliths), which form a most important element of the marine plankton; the latter are surveyed by Hay (1977). Ultrastructural characterisation of the prymnesiophytes can be found in Hibberd (1976) who also refers to the earlier papers by Parke, Manton and co-workers.

3.8.10 Chloromonadophyceae

The family Chloromonadophyceae is a small group not yet assigned to the phylum. It contains only the genera *Vacuolaria* and *Gonyostomun*.

3.8.11 Dinophyta

The Dinophyta comprise a widespread freshwater and marine group of flagellates. There is a good old flora in the *Rabenhorst Kryptogamen Flora* series (Schiller 1933, 1937) but the phylum is being considerably revised (Dodge & Crawford 1970) on the basis of electron microscope study of the cell wall components. Freshwater species are described by Huber-Pestalozzi (1950). The group is large and relatively unstudied when compared with some other algal groups. A very useful volume on marine forms has been published (Drebes 1974) and this also includes planktonic diatoms.

Recommended reading

BOLD H. C. & WYNNE M. J. (1978) *Introduction to the Algae, Structure and Reproduction.* Prentice-Hall, New Jersey, USA.

DREBES G. (1974) *Marine Phytoplankton.* Georg Thieme Verlag, Stuttgart.

PRINGSHEIM Ė. G. (1946) *Pure Cultures of Algae, Their Preparation and Maintenance.* University Press, Cambridge.

ROUND F. E. (1981a) *The Ecology of Algae.* Cambridge University Press, New York & Melbourne.

VOLLENWEIDER R. A. (1974) *A Manual on Methods for Measuring Primary Production in Aquatic Environments.* IBP Handbook, No. 12, Blackwell Scientific Publications, Oxford.

References

BOLD H. C. (1970) Some aspects of the taxonomy of soil algae. *Annals of the New York Academy of Sciences* **175**, 607–16.

BOLD H. C. & WYNNE M. J. (1978) *Introduction to the Algae, Structure and Reproduction.* Prentice-Hall, New Jersey, USA.

BOURELLY P. (1966) *Les algues d'eau douce. Initiation à la systématique.* Vol. I. *Les algues vertes.* Boubee et Cie, Paris.

BOURELLY P. (1968) *Les algues d'eau douce. Initiation à la systématique.* Vol. II. *Les algues jaunes et brunes.* Boubee et Cie, Paris.

BOURELLY P. (1970) *Les algues d'eau douce. Initiation à la systématique.* Vol. III. *Les algues bleues et rouges les Eugleniens, peridiniens et Cryptomonadines.* Boubee et Cie, Paris.

BOURELLY P. (1972) *Les algues d'eau douce. Initiation à la systématique.* Vol. I. *Les algues vertes* (revised). Boubee et Cie, Paris.

BROADY P. A. (1979a) The terrestrial algae of Glerardalur, Akureyiri, Iceland. *Acta Botanica Islandica* **5**, 3–60.

BROADY P. A. (1979b) The terrestrial algae of Signy Island, South Orkney Islands. *British Antarctic Survey Scientific Reports* No. 98.

COLEBROOK J. M. (1960) Continuous plankton records: methods of analysis, 1950–1959. *Bulletin of Marine Ecology* **5**, 51–64.

DESIKACHARY T. V. (1959) *Cyanophyta.* Indian Council for Agricultural Research Series, New Delhi.

DESIKACHARY T. V. (1972) *Taxonomy and Biology of Blue-green Algae.* University of Madras Centre for Advanced Study on Botany.

DIXON P. S. & IRVINE L. M. (1977) *Seaweeds of the British Isles*, vol. 1, Rhodophyta Part 1. Introduction, Namaliales, Gigartinales. British Museum (Natural History), London.

DODGE J. D. & CRAWFORD R. R. (1970) A survey of thecal fine structure in the Dinophyceae. *Botanical Journal of the Linnean Society* **63**, 53–67.

DREBES G. (1974) *Marine Phytoplankton.* Georg Thieme Verlag, Stuttgart.

EATON J. W. & MOSS B. (1966) The estimation of numbers and pigment content in epipelic algal populations. *Limnology and Oceanography* **11**, 585–95.

ETTL A. (1978) Xanthophyceae. *Süsswasserflora von Mitteleuropa* Bd. 3, Teil 1. Gustav Fischer Verlag, Stuttgart.

FRIEDMANN E. I. (1971) Light and scanning electron microscopy of the endolithic desert habitat. *Phycologia* **10**, 411–78.

GANF G. C. & MILBURN T. R. (1971) A conductometric method for the determination of total inorganic and particulate organic fractions in freshwater. *Archiv für Hydrobiologie* **69**, 1–13.

GEITLER L. (1932) Cyanophyceae. In: *Kryptogamenflora von Deutschland, Österreich und der Schweiz* vol. 14. Akademie Verlagsgesellschaft, Leipzig.

GIBBS S. P. (1978) The chloroplasts of *Euglena* may have evolved from symbiotic green algae. *Canadian Journal of Botany* **56**, 2883–9.

GEISKES W. W. G., KRAAY G. W. & BAARS M. A. (1979) Current ^{14}C methods for measuring primary production: Gross underestimations in oceanic waters. *Netherlands Journal of Sea Research* **13**, 58–78.

GODJICS M. (1953) *The Genus Euglena.* University of Wisconsin Press, Madison, Wisconsin.

GOLLERBAKH M. M. & SHTINA E. A. (1969) *Soil Algae.* Nauka, Leningrad.

GROVES J. & BULLOCK-WEBSTER G. R. (1917–1924) *The British Charophyta* Vols 1 & 2. Ray Society, London.

HAPPEY C. M. (1970) The estimation of cell numbers of flagellate and coccoid Chlorophyta in natural populations. *British Phycological Journal* **5**, 71–8.

HAPPEY-WOOD C. M. (1978) The application of culture methods in studies of the ecology of small green algae. *Mitteilungen der Internationalen Vereinigung für Limnologie* **21**, 385–97.

HAY W. W. (1977) Calcareous nannofossils. In: *Oceanic Micropalaentology* (Ed. A. B. Ramsay), vol. 2, pp. 1055–200. Academic Press, London & New York.

HELMKE J. G. & KRIEGER W. (1962–1977) *Diatomanschalen im elektronenmikroskopischen Bild.* [in 10 parts]. J. Cramer, Weinheim.

HIBBERD D. J. (1976) The ultrastructure and taxonomy of the Chrysophyceae and Prymnesiophyceae (Haptophyceae); a survey with some new observations on the ultrastructure of the Chrysophyceae. *Botanical Journal of the Linnean Society* **72**, 55–80.

HICKMAN M. & ROUND F. E. (1970) Primary production

and standing crops of epipsammic and epipelic algae. *British Phycological Journal* **5**, 247–55.

HOLLIGAN P. M. & HARBOUR D. S. (1977) The vertical distribution and succession of phytoplankton in the western English Channel in 1975 and 1976. *Journal of the Marine Biological Association of the United Kingdom* **57**, 1075–93.

HOLM-HANSEN O. (1973) Preservation by freezing and freeze drying. In: *Phycological Methods* (Ed. J. R. Stein), pp. 195–205. Cambridge University Press, London.

HOLM-HANSEN O. & PAERL H. W. (1972) The applicability of ATP determination for estimation of microbial biomass and metabolic activity. In: *Detritus and its Role in Aquatic Ecosystems*, Proceedings of the IBP-UNESCO Symposium, pp. 149–68. Pallanza, Italy.

HUBER-PESTALOZZI G. (1941) Das Phytoplankton des Süsswassers. Teil 2. Chrysophyceen, Farblose Flagellaten, Heterokonten. Bd. 16 *Die Binnengewässer*. E. Schweizerbartsche, Verlagsbuchhandlung, Stuttgart.

HUBER-PESTALOZZI G. (1950) Das Phytoplankton des Süsswassers. Teil 3. Cryptophyceen, Chloromonadinen, Peridineen. Bd. 16 *Die Binnengewässer*. E. Schweizerbartsche, Verlagsbuchhandlung, Stuttgart.

HUBER-PESTALOZZI G. (1955) Das Phytoplankton des Süsswassers. Teil 4. Euglenophyceen. Bd. 16 *Die Binnengewässer*. E. Schweizerbartsche, Verlagsbuchhandlung, Stuttgart.

HUSTEDT F. (1930a) Bacillariophyta (Diatomeae) Heft 10, *Die Süsswasserflora Mitteleuropas*.

HUSTEDT F. (1930b) Die Kieselalgen Deutschlands. Österreichs und der Schweiz. Part 1 of vol. 7 of *Kryptogamen Flora von Deutschland, Österreich und der Schweiz*. Akademie Verlagsgesellschaft, Leipzig.

HUSTEDT F. (1959–1962) Die Kieselalgen Deutschlands, Österreichs und der Schweiz. Part 2 of vol. 7 of *Kryptogamen Flora von Deutschland, Österreich und der Schweiz*. Akademie Verlagsgesellschaft, Leipzig.

JASSBY A. D. (1978) Polarographic measurements of photosynthesis and respiration. In: *Handbook of Phycological Methods*. Vol. 2. *Phycological and Biochemical Methods* (Eds J. A. Hellebust & J. S. Craigie), pp. 285–96. Cambridge University Press, London.

KOLKWITZ R. & KRIEGER H. (1937) Conjugatae. Part I of vol. 13 of *Rabenhorst Kryptogamen Flora von Deutschlands, Österreich und der Schweiz*.

KOLKWITZ R. & KRIEGER H. (1941–1944) Zygnemales. Part II of vol. 13 of *Rabenhorst Kryptogamen Flora von Deutschlands, Österreich und der Schweiz*.

LEEDALE G. F. (1967) *Euglenoid Flagellates*. Prentice-Hall Inc., Englewood Cliffs, New Jersey.

LUND J. W. G. (1975) The uses of large experimental tubes in lakes. In: *The Effects of Storage on Water Quality*, 12. Water Research Centre Symposium.

LUND J. W. G. & TALLING J. F. (1957) Botanical limnological methods with special reference to the algae. *Botanical Review* **23**, 489–583.

PALMER J. D. & ROUND F. E. (1967) Persistent, vertical-migration rhythms in benthic microflora. VI. The tidal and diurnal nature of the rhythm in the diatom *Hantzschia virgata*. *Biological Bulletin (Woods Hole)* **132**, 44–55.

PATRICK R. & REIMER C. W. (1960) *The diatoms of the United States exclusive of Alaska and Hawaii*. Vol. 1 Fragilariaceae, Eunotiaceae, Achnanthaceae, Naviculaceae. Monographs of the Academy of Natural Sciences of Philadelphia.

PATRICK R. & REIMER C. W. (1975) *The diatoms of the United States exclusive of Alaska and Hawaii*. Vol. 2 part 1, Entomoneidaceae, Cymbellaceae, Gomphonemaceae, Epithemiaceae. Monograms of the Academy of Natural Sciences of Philadelphia.

PRINGSHEIM E. G. (1946) *Pure Cultures of Algae, Their Preparation and Maintenance*. University Press, Cambridge.

PRINGSHEIM E. G. (1956) Contributions towards a monograph of the genus *Euglena*. *Nova Acta Leopoldina* **18**, 125.

PRINTZ H. (1964) Die Chaetophoralen der Binnengewässer—eine systematische Übersicht. *Hydrobiologia* **24**, 1–376.

PROVASOLI L. (1971) Nutritional relationships in marine organisms. In: *Fertility of the Sea*, (ed. J. D. Costlow), vol. 2, pp. 369–82. Gordon Breach.

RAGAN M. A. & CHAPMAN D. J. (1978) *A Biochemical Phylogeny of Protists*. Academic Press, New York.

ROUND F. E. (1980) The evolution of pigmented and unpigmented unicells—a reconsideration of the Protista. *Biosystems* **12**, 61–9.

ROUND F. E. (1981a) *The Ecology of Algae*. Cambridge University Press, New York and Melbourne.

ROUND F. E. (1981b) The phyletic relationships of the silicified algae and the archetypal diatom—monophyly and polyphyly. In: *Silicon and Siliceous Structures in Biological Systems*. (Eds B. E. Volcani & T. L. Simpson). Springer Verlag. (In press.)

ROUND F. E. & HICKMAN M. (1971) Phytobenthos sampling and estimation of primary production. In: *Methods for the Study of Marine Benthos* (Eds N. A. Holme & A. D. McIntyre), pp. 169–96. Blackwell Scientific Publications, Oxford.

ROUND F. E. & PALMER J. D. (1966) Persistent-vertical-migration rhythms in benthic microflora. II Field and laboratory studies on diatoms from the banks of the River Avon. *Journal of the Marine Biological Association of the United Kingdom* **46**, 191–214.

SCHILLER J. (1933) Dinoflagellatae (Peridineae) in monographischer Behandlung. Part 1. In vol. 10 of *Rabenhorst's Kryptogamen Flora von Deutschland, Österreichs und der Schweiz*. 2nd edn. Akademie Verlagsgesellschaft, Leipzig.

SCHILLER J. (1937) Dinoflagellatae (Peridineae) in monographischer Behandlung. Part II. In vol. 10 of *Rabenhorst's Kryptogamen Flora von Deutschland, Österreichs und der Schweiz*. 2nd edn. Akademie Verlagsgesellschaft, Leipzig.

SHTINA E. A. & GOLLERBAKH M. M. (1976) *Ecology of Soil Algae*. Nauka, Moscow.

SMAYDA T. J. (1978) From phytoplanktons to biomass. In: *Phytoplankton Manual* (Ed. A. Sournia), pp. 273–9. UNESCO, Paris.

STEWART K. D. & MATTOX K. R. (1975) Comparative cytology, evolution and classification of the green algae with some consideration of the origin of other organisms with chlorophylls *a* and *b*. *Botanical Review* **41**, 104–35.

STRICKLAND J. D. H. & PARSONS T. R. (1968) A practical handbook of seawater analysis. *Bulletin of the Fisheries Research Board of Canada* **167**, 1–311.

TAKAHASHI E. (1978) *Electron Microscopical Studies of the Synuraceae (Chrysophyceae) in Japan: Taxonomy and Ecology*. Tokai University Press, Japan.

TALLING J. F. (1976) The depletion of carbon dioxide from lake water by phytoplankton. *Journal of Ecology* **64**, 79–121.

VENRICK E. L. (1978) Statistical considerations. In: *Phytoplankton Manual* (Ed. A. Sournia), pp. 238–50. UNESCO, Paris.

VENRICK E. L., MCGOWAN J. A. & MANTYLA A. W. (1973) Deep maxima of photosynthetic chlorophyll in the Pacific Ocean. *Fisheries Bulletin* **71**, 41–52.

VOLLENWEIDER R. A. (1974) *A Manual on Methods for Measuring Primary Production in Aquatic Environments*. IPB Handbook, No. 12, Blackwell Scientific Publications, Oxford.

WEST W. & WEST G. S. (1904–1912) *A Monograph of the British Desmidiaceae*, vols 1–4. Ray Society, London.

WOOD R. D. & IMAHORI K. (1964–1965) *A Revision of the Characeae—Iconograph of the Characeae. Monograph of the Characeae*, vol. 2. J. Cramer, Weinheim.

Chapter 4 · Procedures for the Isolation, Cultivation and Identification of Protozoa

Bland J. Finlay

4.1 Distribution

Active, free-living protozoa can be found in most parts of the biosphere where free water persists. They are common and often abundant in both salt and fresh water, in damp terrestrial habitats and, in the case of a few species, in extreme environments from hot springs to the abyssal benthos.

Many species are cosmopolitan, a result of their small size, the ease with which cysts are transported over comparatively large distances (Maguire 1963; Corliss & Esser 1974) and, possibly, genetic stability in some species (Stout & Heal 1967). Many protozoa are also ubiquitous e.g. the ciliate genus *Euplotes*, which is common in freshwater, marine and terrestrial environments from the tropics to polar regions.

Individual protozoan species can often tolerate relatively broad ranges of many environmental factors (Finley 1930; Fenchel 1969; Bick 1972; Stössel 1979) but some factors are more important than others in determining the composition and distribution of protozoan communities. For example, although there is little doubt that some species can tolerate and even adapt to large changes in the salt content of water (Noland & Gojdics 1967), whole orders have apparently evolved almost exclusively in either saline or fresh waters. Testate amoebae are only rarely found in salt water but the Foraminifera and Radiolaria are almost all marine. Table 1 includes all the major groups of free-living protozoa together with an indication of their relative abundance and diversity in fresh and salt water.

The presence of oxygen and the level of reducing conditions are also known to be important in determining the size and composition of protozoan communities (Fenchel 1969; Goulder 1971a; Wefer 1976; Finlay 1980). Many protozoa found in anaerobic environments are probably micro-aero-tolerant (Hartwig 1977) but free-living, obligate anaerobes undoubtedly exist (Fenchel *et al.* 1977). It is interesting to note that the obligate anaerobes are not evenly distributed throughout the protozoan kingdom but confined to a few orders, probably only three orders in the case of the ciliates (Table 1).

In common with all poikilotherms, the rates of physiological processes in protozoa are usually directly related to the ambient temperature but clear evidence for the importance of temperature in influencing distribution is available only for a few groups, especially the Foraminifera (Cushman 1948; Boltovskoy & Wright 1976; Williams & Healy-Williams 1980). The living benthic Foraminifera may be separated primarily on the basis of temperature into cold-water and warm-water faunas with representatives of the former being common to both the Arctic and the Antarctic.

4.2 Function

Because protozoa are common in most aquatic environments, investigators usually have little difficulty in finding a suitable habitat for studying protozoa *in situ*. Results of recent research in a wide variety of habitats indicate similarities in the function of protozoan communities in different ecosystems. For example, the protozoa of soil, sediment, detritus and water column are all known to be intimately connected with the rate of decomposition of plants and plant products (Beers & Stewart 1971; Stout 1974; Beers *et al.* 1975; Harrison & Mann 1975; Fenchel & Harrison 1976).

The functional importance of protozoa in the biosphere is largely a result of their abundance and diversity in many habitats together with their characteristically high reproductive rates. Numbers of ciliates at the sediment–water interface often reach several thousand per cubic centimetre (Goulder 1971b; Finlay 1980); terrestrial testate amoebae may be even more abundant (Heal 1964a; Bamforth 1971), whilst in marine microzooplankton communities, protozoa are often the dominant component (Beers & Stewart 1971; Beers *et al.* 1975). The number of protozoan species recorded from individual habitats is also often large, usually more than 50 (Wang 1928; Schafer & Pelletier 1976; Hartwig & Parker 1977; Finlay *et al.* 1979a). More extensive surveys or long-term studies may reveal well over 100 species (Cairns 1965; Bonnet 1974). Laboratory-determined generation times may overestimate the reproductive capacity of protozoa *in situ* (Fenchel 1968; Finlay 1977; Taylor 1978a) but if estimates of reproductive rates in dinoflagellates using in-situ methods (Heller 1977) are representative of protozoa generally, then protozoa probably divide, on average, every few days, which makes them highly productive.

Protozoa are major consumers of other microorganisms, especially bacteria, though there is now considerable evidence that not all bacteria serve equally well as food sources for protozoa (Taylor & Berger 1976; Curds 1977; Fenchel 1980a). When the available bacteria are readily consumed, however, predation may significantly reduce the bacterial number (Curds *et al* 1968). Grazing by protozoa has received considerable attention in recent years (Laybourn 1975; Heinbokel 1978a, b; Taylor 1978b; Heinbokel & Beers 1979; Fenchel 1980a, b, c, d) partly because of pure interest in protozoan feeding and partly because of the apparent ability of protozoa to stimulate growth and activity in the microbial populations they graze (Fenchel & Harrison 1976; Curds 1977). Although the mechanisms involved are incompletely understood, it is clear that protozoa, through their modification of bacterial activity, may be influential in determining rates of phosphorus regeneration (Barsdate *et al.* 1974; Cole *et al.* 1978), nitrogen fixation (Nikoljuk 1969; Darbyshire 1972a, b) and in modifying the structure of microbial communities (Curds 1974; Güde 1979).

Although much data has been collected regarding the predatory activity of protozoa (Curds 1977), little is known of the importance of protozoa in the diet of other organisms. Indeed they have often been considered unimportant in the past, perhaps because they often leave no recognisable remains in the digestive tracts of predators. This applies especially to 'naked' protozoa which may easily be included in the 'unidentified detritus' category of ingesta. However, there is evidence that ciliates may be important components in the diet of some earthworms (Miles 1963; Piearce & Phillips 1980) and marine crustacea (Hedin 1975; Berk *et al.* 1977; Heinle *et al.* 1977). Foraminifera are also probably digested by a wide range of marine invertebrates (Mageau & Walker 1976).

Protozoa in general, and ciliates in particular, have a potential as indicators of water quality (Cairns 1974, 1978), especially of the organic content of water. Stössel (1979) has recently found good agreement between the occurrence of certain species and the dissolved organic carbon content of water. Other workers (Šrámek-Hušek 1958; Curds 1969; Bick & Kunze 1971; Bick 1972) have classified protozoa in terms of the saprobic category (Noland & Gojdics 1967; Curds 1969) in which they tend to occur. Protozoa undoubtedly have a role to play as biological indicators of water quality and specific organic toxins (Schultz & Dumont 1977) but the results of recent research on the phenomenon of acquired tolerance to heavy metals (Berk *et al.* 1978) casts suspicion on the usefulness of protozoa as indicators of some pollutants.

These are some of the reasons for studying the ecology of protozoa, their abundance and diversity, their many interactions with other organisms, their importance in decomposition and mineralisation and their potential as valuable indicators of pollution.

Table 1. The diversity of Protozoa and the habitats they occupy. All major groups containing free-living representatives are included. The revised classification of the sub-kingdom Protozoa (Levine *et al.* 1980) includes seven phyla of which only two, the Sarcomastigophora and the Ciliophora are composed largely of free-living forms.

Group	Habitats occupied		Examples of genera
Phylum Sarcomastigophora Sub-phylum Sarcodina (Amoeboid protozoa)			
Order Amoebida (naked amoebae)	Fresh water; salt water; soil; some parasitic	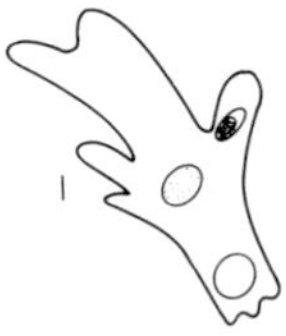	*Amoeba* (1), *Naegleria*, *Pelomyxa*, *Acanthamoeba*
Sub-class Testacealobosia (testate amoebae)	Mainly fresh water; few in salt water; soil (especially in mosses)	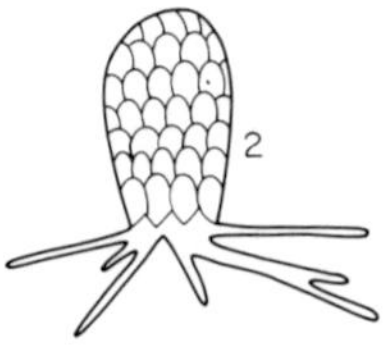	*Difflugia* (2), *Arcella*, *Nebela*
Order Foraminiferida ('forams')	Probably exclusively marine		*Elphidium* (3), *Globigerina*, *Allogromia*
Super-class Actinopoda			
Class Heliozoea (sun animalcules)	Mainly fresh water; some marine	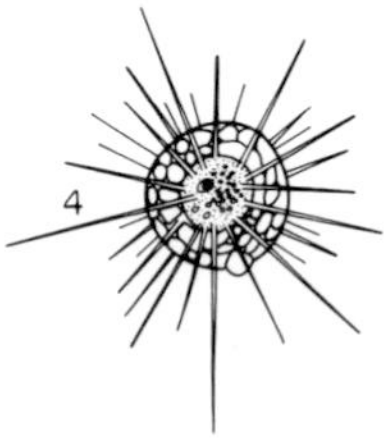	*Acanthocystis*, *Actinophaerium* (4)
Class Polycystinea Class Phaeodarea (both classes radiolarians)	Probably exclusively marine (pelagic)	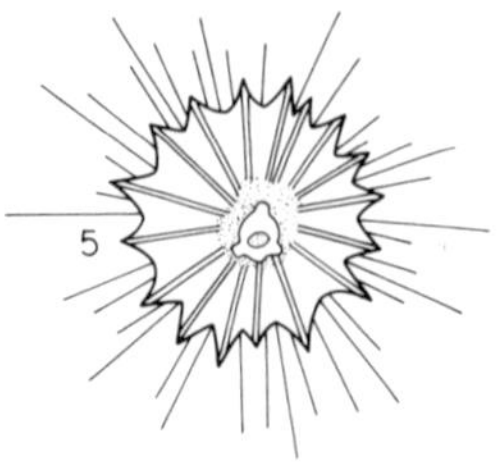	*Heliosphaera*, *Aulocantha* (5)
Class Acantharea	Probably exclusively marine		*Acanthometron*

Group	Habitats occupied	Examples of genera
Sub-phylum Mastigophora (flagellate protozoa)		
Order Chrysomonadida ('phytoflagellates')	Sea-water; fresh water; few in soil	*Ochromonas*, *Paraphysomonas* (syn. *Monas*, colourless)
Order Prymnesiida (includes coccolithophorids) ('phytoflagellates')	Mainly sea-water	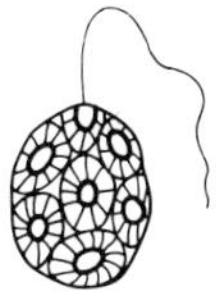 *Coccolithophora* (1)
Order Silicoflagellida ('phytoflagellates')	Sea-water	*Distephanus*
Order Cryptomonadida ('phytoflagellates')	Fresh water; some in sea-water	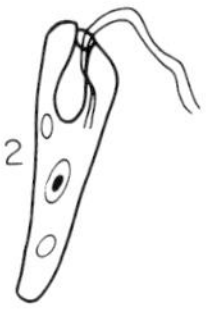 *Cryptomonas*, *Chilomonas* (2) (colourless)
Order Volvocida ('phytoflagellates')	Mainly fresh water; some in sea-water; few in soil	*Chlamydomonas*, *Volvox*
Order Euglenida ('phytoflagellates')	Mainly fresh water; some in sea-water; few in soil; few parasitic	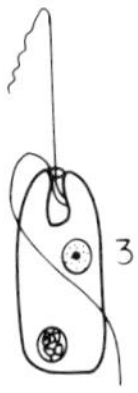 *Euglena*, *Phacus*, *Klebsiella*, *Astasia*, *Peranema* (3) (the last two are colourless
Order Dinoflagellida ('phytoflagellates')	Fresh water; sea-water; some parasitic	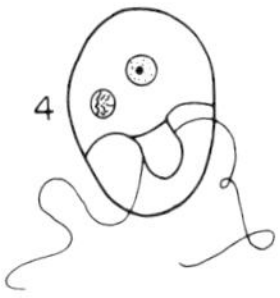 *Ceratium*, *Gymnodinium*, *Noctiluca*, *Oxyrrhis* (4) (the last two are colourless)
Sub-order Bodonina ('zooflagellates')	Fresh water; sea-water; few in soil; many parasitic	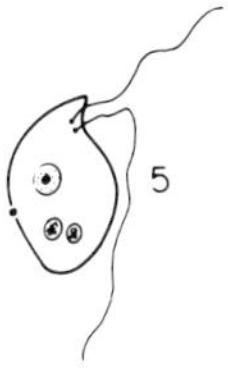 *Bodo* (5)
Order Choanoflagellida ('zooflagellates')	Fresh water; sea-water; many ectocommensals	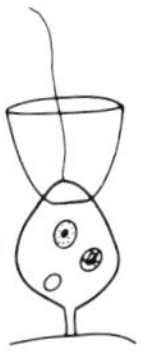 *Monosiga* (6), *Codosiga*, *Salpingoeca* (almost exclusively colourless)
Order Diplomonadida ('zooflagellates')	Possibly in stagnant fresh water; many parasitic	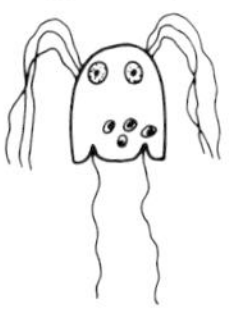 *Hexamita* (7) (all colourless)

Table 1. (cont.)

Group	Habitats occupied	Examples of genera
Phylum Ciliophora (ciliate protozoa)		
Sub-class Gymnostomatia	Fresh water; salt water; some endocommensals	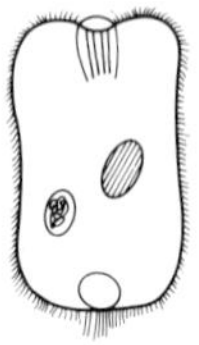 *Loxodes, Tracheloraphis, Prorodon* (1), *Didinium, Litonotus*
Sub-class Vestibuliferia		
Order Trichostomatida	Most are parasites and endocommensals of invertebrates. Those in fresh water and salt water are largely polysaprobic and anaerobic forms	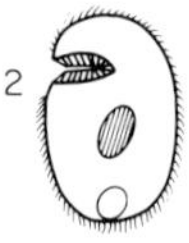 *Plagiopyla* (2)
Order Colpodida	Mainly fresh water and soil; few in salt water	*Colpoda*
Order Bursariomorphida	Predominantly in fresh water	*Bursaria*
Sub-class Hypostomatia	Fresh water; salt water (especially as commensals or parasites of invertebrates)	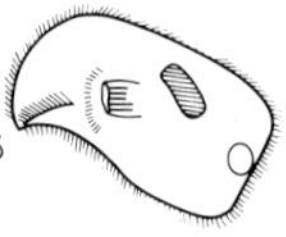 *Nassulopsis, Nassula, Chilodonella* (3), *Spirochona, Hypocomella, Gymnodinioidos*
Sub-class Suctoria	Fresh water; salt water; many ectocommensals, some parasites	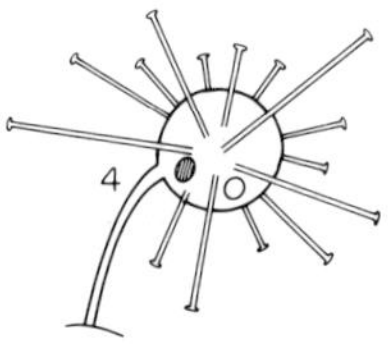 *Podophrya* (4), *Acineta*
Sub-class Hymenostomatia Order Hymenostomatida	Mainly fresh water; some in salt water; some terrestrial; some endocommensals and parasites of invertebrates and fish	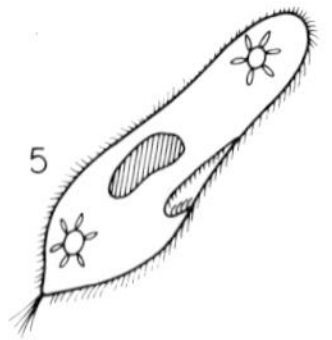 *Paramecium* (5) *Colpidium, Tetrahymena, Frontonia, Lembadion*
Order Scuticociliatida	Fresh water; salt water; some ecto/endocommensals of invertebrates	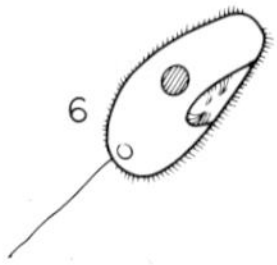 *Uronema* (6), *Loxocephalus, Cyclidium*
Sub-class Peritrichia	Fresh water; salt water; many ectocommensals, some parasitic	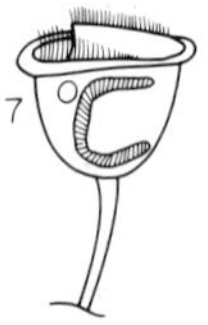 *Vorticella* (7), *Epistylis, Vaginicola*

Group	Habitats occupied		Examples of genera
Sub-class Spirotrichia Order Heterotrichida	Fresh water (including many anaerobes); salt water; some in soil; some parasitic	8	*Spirostomum* (8) *Metopus, Stentor, Folliculina, Caenomorpha*
Order Odontostomatida	Almost exclusively anaerobes in fresh water	9	*Epalxella, Saprodinium* (9), *Discomorphella, Mylestoma*
Order Oligotrichida	Fresh water; salt water (the tintinnids are predominantly marine)	10	*Halteria, Tintinnidium, Tintinnopsis* (10) *Codonella, Metacylis*
Order Hypotrichida	Fresh water; salt water; some in soil	11	*Stylonychia, Aspidisca, Euplotes* (11)

Experimental

4.3 Sampling, isolation and enumeration

4.3.1 Planktonic Protozoa

Planktonic protozoa have been collected with most types of traditional water-sampling apparatus. These include nets hauled vertically, horizontally or obliquely, usually with a mesh size of 20 to 35 μm, bottle samplers (e.g. Niskin, Ruttner, van Dorn, Friedinger), submersible pumps, continuous plankton recorders, syringe samplers and simple glass jars filled by SCUBA divers.

The principle advantage in using nets is that the sample is concentrated during the collection process but this advantage is offset by drawbacks, such as the clogging of the fine net making its operation less quantitative (Beers 1978a) and possible distortion of fragile specimens caused by friction and high pressures within the nets (Graham *et al.* 1976).

Samples collected using bottle samplers (Leadbetter 1972; Rigler *et al.* 1974; Beers *et al.* 1975) are usually more easily handled after concentration or sedimentation. Rapid methods of concentration, such as centrifugation and membrane filtration, probably destroy many planktonic protozoa, although Sawyer (1971) and Davis *et al.* (1978) successfully collected naked amoebae on Millipore and Nucleopore filters respectively. Gentler methods, such as those described by Dodson and Thomas (1964), are probably less selective. The most commonly used substance for sedimenting protozoa is Lugol's iodine (see section 4.4, p. 54). Adding a few drops to a water sample in a measuring cylinder or tube for enumeration by the inverted microscope technique (Utermöhl 1958) causes most protozoa to settle in minutes. Most species remain recognisable although those forms with a shell or rigid pellicle are most easily identified.

When a submersible pumping system is linked to a deck-mounted plankton collection and sorting unit (Beers *et al.* 1967; Beers 1978b) the large amounts of water removed (*c.* 150 l min^{-1}) provide a representative microzooplankton sample (zooplankton passing through a 202 μm mesh) when allowances are made for the more fragile forms. A similar picture is obtained using continuous plankton recorders (CPR) (Fornshell 1979) although the necessary trapping of protozoa on a mesh and fixing of the record as soon as it is obtained makes the sampling techniques more selective for larger, less fragile protozoa.

When the protozoa to be sampled are present in relatively high numbers, the pneumatically operated multiple syringe (10 to 20 ml) sampler described by Heaney (1974) is particularly useful, especially if the objective is to map vertical distribution over transition zones (e.g. oxycline, thermocline) in the water column (Goulder 1972).

Planktonic Foraminifera usually occur in the size range 0.2 to 1.0 mm and they are easily collected with plankton nets (Boltovskoy & Wright 1976). Several techniques have been described for the subsequent separation and concentration of Foraminifera from mixed plankton samples. These are based either on their high specific gravity (*c.* 1.4) and differential settling through concentrated sodium chloride solutions (Bé 1959) or on resistance of the test to procedures for digestion and ignition of interfering organic matter (the 'ignition method'—Smith 1967).

The planktonic sarcodina are notoriously difficult to collect without disturbance while still retaining their capacity to grow in the laboratory. In the case of Foraminifera, one solution has been to use SCUBA divers to collect individual specimens in pint-sized (475 ml) glass jars (Alldredge & Jones 1973). If the light intensity is high enough, sunlight is reflected from their spines making them visible, even from several metres distance (Bé *et al.* 1977). Another sampling device apparently capable of collecting Foraminifera and Acantharia in excellent condition is that described by Graham *et al.* (1976). The sampler consists of a 1.0 l polyethylene bag which gently fills with water at the required depth. The authors believe that the higher numbers of Acantharia collected with this device are due to elimination of the 'advance warning' characteristic of net sampling.

A novel, semi-quantitative method of collecting planktonic protozoa involves the use of polyurethane foam units ('PF units' or 'artificial sponges') which are left suspended in the water column of lakes, usually for several weeks (Cairns *et al.* 1976). During the period of immersion they are colonised by a protozoan community which can be released by squeezing the foam over a sampling bottle.

4.3.2 Freshwater Benthos

Freshwater sediments generally differ from intertidal and shallow-water marine sediments with respect to particle size distribution and the nature and amount of detritus present. The greater average particle size in marine sediments creates a larger interstitial volume. This, together with the action of periodic flushing and the characteristic absence of large accumulations of detrital aggregates, alleviates the physico-chemical barriers which restrict depth penetration of microorganisms. Thus, while large populations of protozoa are often found at depths of more than 10 cm in shallow, marine sediments (Fenchel 1969), they are seldom found deeper than 5 cm in cores from the freshwater benthos where almost all occur in the top centimetre. This means that the depth of sediment which must be examined to include all protozoa occurring in the freshwater benthos is usually much lower than that in most marine sediments. Nevertheless, the types of sampling equipment used in freshwater sediments are similar if not identical to those used in the marine benthos. Most quantitative studies involve either the use of corers whose principle of operation is based on that originally described by Moore and Neill (1930) (Moore 1939; Bryant & Laybourn 1972/1973; Finlay 1978, 1980) or more sophisticated devices such as the Jenkin corer (Webb 1961; Goulder 1971b, 1974).

Qualitative samples from shallow-water sites have been obtained by a variety of means including medicine droppers (Taylor 1978a), hose samplers (Arlt 1973) and ooze suckers (Moore 1939; Cole 1955; Grabacka 1971) often analogous to the 'pooter' used for collecting insects (Noland 1925).

There is no single method of extracting all protozoa from freshwater sediments which approaches the efficiency of Uhlig's method for marine sediments. The high mud content of most freshwater sediments makes use of the latter method unsuitable (Arlt 1973). Electromigration techniques (Hairston & Kellermann 1964; Borkott 1975) often work for ciliates, such as *Paramecium* spp., but they are of no use with slow-moving or non-motile protozoa. Other techniques, such as sucrose gradient flotation or centrifugation (Berk *et al.* 1976), may be useful for isolating some of the less fragile forms.

Direct observation is the most efficient if the most laborious method of examining and counting protozoa in freshwater sediments. The sediment is either examined undiluted (Goulder 1974) or it is diluted with filtered lake water before being spotted out on a slide, the spots being of even (Finlay *et al.* 1979b) or uneven volume (Goulder 1971b). In a few cases the use of chambers which restrain the movement of protozoa might be suitable (Tseeb 1937).

One form of natural migratory pattern in protozoa which can be exploited to separate them from sediment (especially from productive lakes) is their response to anoxia and migration to a source of oxygen. If core samples complete with the original overlying water are kept airtight for about 24 h, oxygen depletion at the sediment surface forces many protozoa into the overlying water. They can be concentrated by slow-speed centrifugation, filtering through nylon mesh (only the larger forms) or with more gentle methods such as that described by Dodson and Thomas (1964).

It is often particularly time-consuming to examine freshwater sediments if they contain much organic matter which has formed aggregates with the inorganic particles. These aggregates are invariably inhabited by the smaller protozoa and must be split open with a fine needle during microscopic examination.

Techniques of adding diluted sediment to culture media as a means of enhancing the growth of protozoa are well-suited to the isolation and enumeration of specific groups (e.g. the naked amoebae–O'Dell 1979) but rarely are they suitable for the complete protozoan community.

4.3.3 Marine and Intertidal Benthos

There exists in the marine benthos a great diversity of both sediment types and protozoa. The requirement to sample both has led to the development of a wide range of apparatus and sampling techniques. Extracting and enumerating thigmotactic ciliates from estuarine mud flats requires quite different techniques to those involved in isolating amoebae from the abyssal depths of the open ocean.

The method selected for obtaining the initial sediment sample depends on the depth of the overlying water and the nature of the sediment. Depths greater than about 1 m require the use of grabs, corers, wedges and trawls (see Uhlig (1968) and Holme & McIntyre (1971) for descriptions of the more common benthos samplers). Such sampling devices usually have no peculiar adaptations which make them suitable for sampling protozoa, although when the internal diameter is variable, as with different corers, the smaller sizes tend to be those used for protozoa. Bearing in mind the comparatively small size of protozoa, some of these devices are less suitable than others; grabs, for example, allow the percolation and loss of interstitial water containing the protozoa.

In shallower waters the superficial layer of sediment is often scraped off (Hartwig & Parker 1977) but the most commonly used sampling vessel is a tube made of PVC or glass which can be pushed or dropped into the sediment. This may or may not have devices incorporated for preventing loss of the sample (Fenchel 1967).

Two important factors are related to the internal diameter of such tubes. First, the narrower the tube the more compacted the sample but the more likely that it will remain enclosed as the tube is retracted from the sediment. Second, the bore used is usually a compromise, tending towards the larger when the study is concerned with the nature of vertical distribution. Typical internal diameters of the tubes and the materials used are: 9 mm, glass (Mare 1942); 13 mm, PVC (Fenchel & Jansson 1966); 17.8 mm, PVC (Hartwig 1977); 21 mm, PVC (Fenchel 1967), and 26 mm, glass (Elliott & Bamforth 1975). Borror (1963) used a glass tube tapering from 7 to 3 mm at the tip and bent through 45°. This was filled with sediment by quickly removing and then replacing a thumb over one end.

Many methods are also available for identifying and counting the protozoa contained in a sediment sample. The sediment can be observed with the aid of a microscope directly or after dilution with filtered sea-water. Alternatively, the sediment can be shaken with filtered water in a Petri dish. This allows the larger particles to settle more quickly from the supernatant containing the protozoa (Fauré-Fremiet 1950; Fjeld 1955). Only the more sophisticated methods of elutriation (Uhlig *et al.* 1973) are suitable for the less fragile protozoa. The sediment sample can be diluted and added to nutrients in which the protozoa can grow (Lighthart 1969). Such a technique is usually selective for those protozoa capable of growing in the conditions provided. Nevertheless, for certain groups (notably bacterial feeders) it may allow enumeration using a most-probable-number (MPN) technique (see Chapter 6. p. 88).

One technique that has been widely used is the sea-water–ice method of Uhlig (1964, 1968). The changing salinity gradients and the flow of water created by melting ice, force the protozoa (especially ciliates) down through the sediment sample, across a stretched nylon gauze and into a dish of filtered sea-water. The technique is most effective (up to 90%—Fenchel 1967) when extracting communities of predominantly motile ciliates from inshore sandy sediments in temperate or colder waters. In extracting ciliates from sandy sediments in Bermuda, Hartwig (1977) found the temperature drop in the process so great that it killed or damaged most of the ciliates, and Burnett (1973) considered the process too harsh for protozoa from the deep benthos. Some of the problems of sampling protozoa from the abyssal benthos are discussed by Burnett (1977).

4.3.4 Foraminifera

Living benthic Foraminifera have been observed in most sediment types from tidal marshes (Scott 1975) to deep-sea trenches (>9000 m, Thiel 1975). They are often common in the intertidal benthos and in sediments just below the low water mark, particularly if the sediment is in a sheltered habitat and composed predominantly of finer particles. The sampling devices used include augers, simple tubes and corers in shallow waters (often operated by SCUBA-divers), to grabs and dredges in the deepest waters. The latter are usually subsampled by inserting smaller-diameter coring tubes which can be fractioned vertically. Investigations are often confined to the top centimetre of sediment but, as Thiel (1975) points out, living Foraminifera can often be recorded at depths of 10 cm or more, especially in inshore, sandy sediments.

The two simplest methods of isolating Foraminifera from sediment, each making use of their characteristic negative geotropism, are:

(1) allowing them to climb up the sides of a shallow tray; and

(2) concentration by migration and attachment to the underside of a filter paper placed on the sediment surface (Murray 1979).

Because Foraminifera are relatively resilient, the procedures for isolating them from sediment need not be as gentle as those required for naked protozoa. The sediment is first passed through a coarse sieve (e.g. 0.5 mm) to remove the larger particles and detritus. Most Foraminifera can then be retained on a 0.063 mm sieve. If much organic material is still present, the sample can be suspended in filtered sea-water and the excess decanted off. If there remains an excessive amount of sediment in relation to the number of Foraminifera, the latter can be floated out by adding carbon tetrachloride (specific gravity = 1.58) or preferably a bromoform-acetone (specific gravity = 2.2) or bromoform-alcohol (specific gravity = 2.0) mixture.

For qualitative investigation of Foraminifera attached to marine plants, Arnold (1954, 1974) gives a comprehensive description of methods used in separation and concentration.

The identification of living Foraminifera is aided by fixation and staining. Isopropanol, methanol, formaldehyde and glutaraldehyde are the most commonly used fixatives, with Rose Bengal and Sudan Black B being the most common stains. There is some controversy concerning the reliability of Rose Bengal and the pre-treatment necessary to give unambiguous results. Marshall (1975) and Walker *et al.* (1974) reported favourably on the use of Rose Bengal and Sudan Black B respectively with the latter authors including a critical analysis of the adequacy of various fixatives and stains. The procedure they recommended involves fixing with formaldehyde and calcium chloride, followed by staining with heated acetylated Sudan Black B and washing in ethanol. Scott (1975) had no success with this technique, preferring instead to use Rose Bengal with the Foraminifera suspended in isopropanol. Clearly the most suitable technique depends on the species concerned, whether the preparation is wet or dry and whether or not it is obtained from culture (Walker *et al.* 1974).

4.3.5 Soil

Living terrestrial protozoa are largely confined to the upper horizons of soil, especially if these are rich in decaying organic matter. Samples are usually obtained with soil corers or soil augers, depending on the importance of retaining the original integrity of the vertical profile. The soil sample is often passed through a sieve (mesh *c.* 3 mm) to remove the larger inorganic particles before subsequent analysis. In many soils, a large proportion of the soil protozoan community is often inactive or encysted and it is important that samples are not subjected to uncontrolled variations (especially temperature and humidity) that might significantly alter that proportion before treatment or examination.

Direct examination of soil often yields little more than testate amoebae so the 'direct' methods tend to refer specifically to this group (see Stout & Heal (1967) and Heal (1970) for references relating to the variety of methods). Only a few of these are described below.

The method of stained smears (variations of the Jones and Mollison (1948) technique) involves dispersion or maceration of the soil sample, mixing it with 0.5 to 1.5% (w/v) agar, followed by staining (erythrosin in carbolic acid or phenolic aniline blue) to locate living protoplasmic contents (Heal 1964a; Korganova & Geltser 1977), and examination on a slide. The technique of dispersion depends on the soil type but may include grinding, homogenisation, shaking or mixing by feeding a gas (carbon dioxide or air) through the suspension. By counting the number of tests in a specific area of the slide, an estimate can be obtained for the number of tests per unit weight of the original sample. The usual low density of Testacea on slides makes use of this technique very time-consuming (Heal 1964a). The method of preparing soil sections (Burges & Nicholas 1961) is also applicable to testate amoebae and from a knowledge of the section thickness it is possible to calculate the number of tests in the original sample. Like the agar film technique, soil sections are time-consuming to prepare, making them impracticable in quantitative studies involving many samples. However, they do have the advantage of retaining the structural integrity of the soil and may be of some use in providing information on the more qualitative aspects of microbial interactions, especially the spatial relationships of Testacea in their microenvironment.

Direct techniques are probably simplest when used for observation and enumeration of protozoa in mosses, especially *Sphagnum* spp. The material can be macerated and homogenised before being examined in a grid cell (Heal 1964b) or the protozoa can simply be squeezed out of the plant (Grolière 1977).

The method of Coûteaux (1967, 1975) involves enumeration on a membrane filter. Fresh soil is fixed, stained (with scarlet xylidine), diluted and dispersed before being filtered. Recently, in a rare comparison of techniques, Lousier and Parkinson (1981) quantified the differences in results obtained between the Millipore (membrane) filtering method and the agar film technique. These results showed that higher numbers of testate species and greater numbers of active, encysted and empty tests were obtained following 10 s maceration and filtration and that these figures were much higher than those obtained with the agar film technique. Another important advantage of the membrane technique is the higher density of Testacea on the filter, making the process of enumeration less time-consuming.

Testate amoebae may also be separated from soil by flotation. Gas is fed in beneath the diluted soil sample causing the lighter, gas-filled materials, such as the testate amoebae, to rise. At its simplest the technique depends on the air supply delivered from an aquarium pump (Chardez 1959) but carbon dioxide (Bonnet & Thomas 1958; Décloitre 1966) and hydrogen (Chardez 1964) have also been used. The technique becomes more efficient and practicable as the amount of extraneous organic matter in the sediment decreases. The efficiency of extraction by this method has not been quantified.

Extensive use has been made of culture techniques with all groups of soil protozoa. Whether the investigation is quantitative or qualitative, the prime criticism of culture techniques is their selectivity. The only protozoa examined are those that are amenable to cultivation. However, Heal (1964a) concluded that, at least in the case of the Testacea, liquid and agar soil extract media together supported the growth of almost all species recorded for the same soil sample by direct observation, and Stout (1963) identified over 170 species of protozoa from soil cultures. The capacity of protozoa to appear in culture through proliferation is also the basis of a well-established method for enumerating soil protozoa, namely, the dilution culture method. A dilution series is produced, the protozoa in each dilution are encouraged to grow and after a period of time the number of individuals of each species in the original sample is derived from the number of higher-dilution cultures showing no growth in the species concerned. Treatment of a replicate soil sample with HCl is reputed to kill only trophic (non-encysted) forms. The treated sample is cultured and the size of the trophic population calculated by difference. The technique is fully described by Singh (1955), Darbyshire (1973) and Elliot and Coleman (1977) and is discussed critically by Heal (1970).

Viability and capacity for reproduction of single cells is also the basis of the modified dilution culture method of enumerating soil amoebae described by Menapace *et al.* (1975). The technique is essentially a double-layer assay method with amoebae dividing to produce colonies which consume the surrounding layer of bacteria, producing clear, observable plaques. The authors found no sure method of differentiating between trophic and encysted forms.

4.4 Observation

Both living and dead protozoa are usually observed on glass slides although Foraminifera are best observed on a dark background using incident light. Most protozoa remain active and alive until the sample dries but the more fragile, heat- and light-sensitive forms are easily killed. The use of precooled slides and low light intensities often alleviates such problems.

There is little doubt that observation and identification of all naked protozoa is best carried out without the aid of immobilising agents or fixatives but often their use is inevitable. Different species react in varying ways to different immobilising agents and the protozoan community in a water, soil or sediment sample will usually contain some species that are resistant and some that are susceptible to the substance used. This variation in response is, in part, reflected in the variety of techniques described including the use of physical restraint, deciliation, paralysis and immobilisation with antibodies or magnetic fields (Patterson 1980). In examining the protozoa contained in water and sediment samples, the author has had some success in immobilising ciliates with polyethylene oxide (Spoon *et al.* 1977) and either nickel sulphate or formaldehyde applied on the end of a needle.

Identification and observation of protozoa is often facilitated by using so-called 'vital' stains to emphasise cellular features, such as the nucleus and food vacuoles. These stains, such as neutral red and methyl blue or green, are applied before fixation and may or may not eventually kill the protozoa (Mackinnon & Hawes 1961; Kudo 1966). When it is not possible to examine samples immediately or within a reasonably short period of time, fixation is required. The best general fixatives for protozoa are probably dilute (<2% to 4% (w/v)) buffered formaldehyde, glutaraldehyde (<5% (w/v)), osmium tetroxide (<2% (w/v)) and Lugol's iodine (i.e. Lugol's solution or Lugol's fixative). Lugol's iodine is prepared by dissolving 10 g iodine and 20 g potassium iodine in 200 ml distilled water with optional acidification by adding 20 ml glacial acetic acid. The solution is stored in the dark. The fixative is effective in low concentrations which should not exceed 1 ml per 100 ml sample. If necessary, the dark staining of cell contents can be removed by adding sodium thiosulphate. Taylor (1976) described the use of combined formaldehyde and Lugol's iodine for the fixation of flagellates. The problem of decalcification and dissolution of formaldehyde-fixed calcareous marine protozoa is discussed by Bé and Anderson (1976) and a recommended fixation procedure given. It should be emphasised that a single fixative rarely acts equally well on protozoa in a community and that some losses and disfigured specimens are inevitably produced.

The more sophisticated staining techniques are outside the scope of this chapter. For protargol fixation see Tuffrau (1967) or Uhlig (1972) and for silver impregnation of ciliates see Curds (1969) or the extensive bibliography given by Corliss (1979).

4.5 Cultivation

The general principles of isolation and cultivation of the major groups of protozoa are considered below. The reader is referred to the comprehensive review by Kinne (1977) and Provasoli (1977) for methods concerning individual species, especially marine forms.

4.5.1 Amoeboid Protozoa

The culture of naked amoebae has perhaps received more attention than any other group of amoeboid protozoa with the result that a wide range of standard liquid and solid media have been devised. Page (1976) included preparation instructions for media supporting the growth of most freshwater and soil forms. The slight variations of some of these required for a variety of marine forms are described in Page (1970, 1973). Recently, O'Dell (1979) evaluated the suitability of several new media and the overlay plaque technique of Menapace *et al.* (1975) for culturing amoebae from the lake benthos.

The medium selected for an amoebae must also be suitable for the food organism, while at the same time preventing the food organism from outgrowing the amoebae and swamping it (hence the occasional necessity for non-nutrient media). Although the bacteria carried over in the subculturing procedure can provide the sole and adequate source of food for amoebae, additional organisms are often necessary. Many of the more common supplementary protozoan food organisms are listed by Page (1976) who also includes details of isolation of amoebae (both trophic forms and cysts) from solid and liquid media. The technique of isolating soil amoebae from enrichment cultures on solid media is described by Dawid (1975). Most amoebae are apparently dependent on the presence of living or dead bacteria (Willaert 1971) in the culture medium, even if they are not used as the principal food source. This dependence has hindered the establishment of naked amoebae in axenic culture with a few exceptions, e.g. *Acanthamoeba* sp. Goldstein and Ko (1976) provided details of their simple technique for growing large numbers of amoebae in batch culture.

4.5.2 Foraminifera

There is considerable variation in the ease with which different Foraminifera can be cultured, the least difficult being the near-shore benthic forms that do not have an alternation of generations. Thus, some species dividing by multiple fission and utilising as a food source the products of intracellular symbiotic algae, can be maintained for many months in clonal culture in the laboratory while the maintenance of others is terminated with the onset of gametogenesis (Bé *et al.* 1977). Generally, the smaller species with reproductive cycles lasting a few weeks are more easily cultured than slower-growing species. Arnold (1974) included a list of the more easily cultured species and those with a potential for being cultured. Planktonic Foraminifera from the open ocean have never been successfully cultured.

Various media have been used to support Foraminifera including sea-water, either natural (Röttger & Berger 1972), membrane-filtered (Anderson *et al.* 1979), artificial (Lee *et al.* 1961; Lee & Pierce 1963) or enriched- 'Erdschreiber' (Arnold 1954; Muller 1974; Lee & Muller 1973). Foraminifera can be cleaned and isolated by picking off attached microflora with fine glass needles drawn from Pasteur pipettes (Lee *et al.* 1969; see description of 'algal choking' in Arnold 1954), serial washing through sterile sea-water followed by bathing in a mild antibiotic solution (Lee *et al.* 1961). Those Foraminifera that feed on algae are usually cultured in a light–dark cycle in the presence of apparently indispensable bacteria (Muller & Lee 1969). Although a wide variety of algae serve as food for Foraminifera, the more commonly used species are listed by Boltovskoy and Wright (1976).

4.5.3 Testate Amoebae

Testate amoebae have been successfully cultured in both soil extract medium (Heal 1964a, b, 1965; Bamforth 1971) and Chalkley's *Amoeba* medium (Heal 1964b) with the addition of algae providing an adequate food source. Periodic transfer to fresh medium obviates the inhibitory effects of excessive fungal and bacterial growth. The addition of some extra particulate matter (e.g. fine soil particles) as a source of test-building materials may also be necessary. Heal (1964b) has shown that testate amoebae can adapt to utilising materials other than those typically employed in the natural environment.

4.5.4 Ciliate Protozoa

Ciliates are cultured most easily in liquid media, either axenically or in association with living food organisms. Plant infusions, especially those of dried hay and lettuce (Sonneborn 1950), have been widely used, as has soil extract medium (George 1976) and cerophyl (a commercially available mixture of dried cereal leaves usually prepared in strengths of $<0.25\%$ (w/v). Such infusions are produced by boiling the plant material in water, filtering the suspension and adjusting the pH, usually close to pH 7.0, followed by sterilisation (steaming or autoclaving) of the medium. The medium is inoculated with a suitable food source or the latter is carried over in the isolation procedure from a mixed culture. Most ciliates that grow well in culture utilise one or several species of bacteria from a relatively limited range (Curds &

Vandyke 1966; Taylor & Berger 1976). Infusions have also been used as basal media for marine ciliates (Soldo & Merlin 1972) although both sea-water itself, aged and filtered (Parker 1976) and Erdschreiber medium (Andrews & Floodgate 1974) may also be suitable. Culture methods for the specifically marine tintinnids and folliculinids are given by Gold (1973) and Uhlig (1965), respectively. The latter author also describes the incorporation of invertebrate 'culture partners' with the ciliates as a means of keeping cultures free of excessive detrital films.

It is relatively easy to produce large numbers of ciliates from most samples of soil, sediment and water by adding the nutrients required for a luxurious bacterial growth. For example, by adding a few drops of casein-peptone-starch medium (Collins & Willoughby 1962) or a medium which supports the growth of *Tetrahymena* spp. (1% (w/v) proteose peptone and 0.25% (w/v) yeast extract) with a few grains of rice or wheat to a flask containing the diluted sample, a community of fast-growing bacterial-feeding ciliates invariably develops. Stimulating the growth of mixed communities of algal-feeding ciliates is more difficult but a combination of low light intensities and low levels of additional nutrients (if any) often selects for these types of ciliates. The required species can be isolated using finely drawn-out pipettes.

Although greater densities of ciliates may be obtained with peptone-based or other highly enriched media, the number of species developing is rarely as great (Stout 1956). Those that thrive are usually the faster-growing bacterial-feeders and osmotrophic forms. Thus, the less highly enriched infusion-based media are generally more suitable for supporting growth of the diversity of species in the original sample while maintaining lower population densities of each.

There is no single suitable vessel used for containing ciliate cultures, although most workers prefer ones of glass for ease of acid cleaning and steam sterilisation. Narrow-necked vessels are certainly more convenient if the risk of introducing microbial contaminants during subculturing is to be avoided. An air space must be left over the medium if the ciliates are aerobic although other ciliates (e.g. *Spirostomum* spp.) are apparently dependent on low levels of dissolved oxygen and they prefer greater liquid:air volume ratios. The internal surface area available is also usually important especially for the crawling and attached forms and as a site for the attachment and accumulation of food organisms.

The axenic culture of ciliates is now possible for many species. The technique usually involves the transfer of individuals by serial washing in antibiotic solutions or migration from a culture containing microbial contaminants to a defined or semi-defined sterile growth medium (Soldo & Merlin 1972, 1977; Allen & Nerad 1978; Fok & Allen 1979).

Although the advantages of using axenic cultures in experimental work are obvious, it should perhaps be remembered that the ciliates concerned rarely, if ever, exist in the field dissociated from a variety of other microorganisms. Conclusions reached regarding nutrition and growth in axenic culture may have limited applicability to ciliates in the natural environment. For references concerning the more commonly cultured ciliates, such as species of *Tetrahymena, Paramecium* and *Euplotes*, the reader is referred to Corliss (1979).

4.5.5 Flagellate Protozoa

This section is restricted to a consideration of the colourless, free-living flagellates. Typical representatives in the various flagellate orders are indicated in Table 1. Such flagellates may be osmotrophic and/or phagotrophic, in the latter case ingesting such microorganisms as bacteria, cyanobacteria and other flagellates.

The basal media described above for ciliates (especially soil-extract medium—Leedale 1967) are often also suitable for flagellates. Indeed, *Chilomonas* spp. and *Oxyrrhis* spp. are common contaminants of freshwater and marine infusions respectively. Other commonly used media include the *Polytomella* medium (Starr 1964), Musgrave and Clegg's medium and the *Oxyrrhis* medium (both these last two are described in Mackinnon & Hawes 1961).

Marine colourless flagellates have been successfully cultured in aged sea-water and Erdschreiber medium. Enrichment of the latter with peptone (80 $\mu g\ l^{-1}$) and yeast extract (6 $\mu g\ l^{-1}$) encourages the growth of bacteria, providing suitable conditions for the growth of choanoflagellates especially (Leadbetter & Morton 1974; Leadbetter 1977). Pure cultures of choanoflagellates are produced by growing them in the dark and repeatedly subculturing with the aid of fine drawn-out pipettes to remove photosynthesising contaminants. In a few cases (Gold *et al.* 1970) axenic cultivation on semi-defined media has been shown to be possible.

The ease with which a colourless flagellate can be cultured using a particular method is probably a function of:

(1) its dependence on a bacterial component in the diet;

(2) its ability to utilise dissolved organic substances as carbon sources (Haas & Webb 1979).

The comprehensive list of defined and semi-defined media for algae (including colourless flagellates) given by Droop (1969) includes formulations based on both soluble and particulate carbon sources.

4.6 Calculation of biomass and production

4.6.1 Biomass

The biomass of protozoa is not usually measured directly because of their small size, but calculated from cell volume estimates and an assumption regarding the specific gravity of cell protoplasm. The population biomass is calculated by multiplication with the number of protozoa per unit area or volume of the sample concerned.

Cell volume is easily estimated by assuming some resemblance between the cell shape of the species concerned and a standard geometric shape (Heal 1965; Beers & Stewart 1971; Murray 1973; Finlay 1977). Some protozoa, especially the naked amoebae, can be encouraged to become spherical if they are disturbed. Hopkins (1946) pipetted amoebae back and forth until they had rounded up into spheres whose volumes were then easily calculated. Rogerson (personal communication) has refined the technique by fixing rounded amoebae in Carnoy's fluid (90% (v/v) ethanol, 10% (v/v) acetic acid) whilst simultaneously compressing them to 20 μg in a Hawksley bacterial counting chamber. The area of an image projected by camera-lucida is measured with a planimeter. When the cell shape differs markedly from a standard geometric shape it is often easier to construct a series of plastic or plasticine models in a range of sizes (Dillon & Hobbs 1973; Wefer & Lutze 1977). Their displacement in water can then be measured.

Michiels (1974) described a graphical method which is accurate but time-consuming and Taylor (1978a) reduced the error attached to volume estimates derived from only two linear measurements by measuring the area of a projected, fixed specimen together with its length. When the depth of a living cell is also known (Cameron & Prescott 1961), the accuracy is increased further at the expense of time lost in preparation.

Electronic particle counters such as the Coulter Counter can be used to size protozoa although such instruments are better established for the counting and sizing of monospecific populations of the more robust protozoa (Dive 1975). It may also be necessary to make a volume correction for the change resulting from suspending the cells in the electrolyte.

Fenchel (1968) produced length–weight distributions for 14 species of marine ciliates, the volumes being calculated from the area and depth of cells retarded in a rotocompressor. The interspecific relationship is linear in a double-log plot.

The conversion from cell volume to cell weight is usually made assuming a protoplasm specific gravity of 1.0 or 1.027 for Foraminifera (Wefer & Lutze 1977). Other commonly used rough conversion factors are that dry weight is approximately 20% of wet weight and organic carbon is 40 to 50% of the total dry weight. However, Curds and Cockburn (1971) showed that the dry weight value was only 7.2% of the wet weight, whilst Kimball *et al.* (1959) found a value of 6.5%.

4.6.2 Production

There are few published estimates of protozoan production in the field, the lack of data resulting from difficulties in accurately measuring population or community growth rates *in situ*. Using laboratory-determined growth rates, Finlay (1978) and Burkovskii (1978) have estimated production in freshwater benthic and marine sand-dwelling ciliates respectively. There is probably a difference between laboratory and field growth rates but the magnitude of that difference is as yet unknown.

The error attached to calculating production in shell-bearing protozoa is probably smaller when:

(1) the reproductive rate is known to be low and size variation curves can be constructed as in the Foraminifera (Wefer & Lutze 1977); or

(2) changes in numbers of empty and occupied tests and loricae in production chambers can be monitored as with the Testacea and loricate ciliates (Schönborn 1977).

However in the case of naked protozoa, the most

Table 2. Some of the texts commonly used in identifying representatives of the major groups of free-living protozoa.

	Ciliated protozoa	'Zooflagellates'	Radiolaria/Acantharia	Heliozoa	Foraminifera	Testate amoebae	Naked amoebae
Collin 1912	●						
Penard 1922	●						
Fauré-Fremiet 1924	●						
Kahl 1930–35	●						
Kahl 1934	●						
Noland 1959	●						
Dragesco 1960	●						
Borror 1963	●						
Curds 1969	●						
Bick 1972	●						
Patsch 1974	●						
Corliss 1979	●						
Curds 1981	●						
Kent 1880–2	●	●					
Stokes 1888	●	●					
Grassé 1952		●					
Lackey 1959		●					
Calaway & Lackey 1962		●					
Schewiakoff 1926			●				
Hollande & Enjumet 1960			●				
Nigrini 1967			●				
Renz 1976			●				
Penard 1904				●			
Cash, *et al.* 1921				●			
Rainer 1968				●			
Cushman 1948					●		
Murray 1973					●		
Murray 1979					●		
Davis 1955	●	●	●		●		
Chardez 1964						●	
Ogden & Hedley 1980						●	
Singh & Das 1970							●
Page 1976							●
Penard 1902						●	●
Cash & Hopkinson 1905, 1909						●	●
Cash *et al.* 1915, 1919						●	●
Deflandre 1959				●		●	●
Grassé 1953			●	●	●	●	●
Pennak 1978	●	●		●		●	●
Kudo 1966	●	●	●	●	●	●	●
Jahn & Jahn 1979	●	●	●	●	●	●	●

accurate techniques are probably those based on employing the frequency of dividing cells in the population. Calculation of generation times *in situ* using such frequencies has so far been confined to the dinoflagellates and other microorganisms (Weiler & Chisholm 1976; Heller 1977; Hagström *et al.* 1979).

4.7 Identification

Protozoa are often difficult to identify. The small, motile forms are especially difficult and often there is no substitute for patient observation and the judicious use of stains, fixatives and narcotising agents (see section 4.4, p. 54). The slower-moving Foraminifera and testate amoebae may be easier to identify from shell characteristics but in ecological studies additional tests may be necessary such as staining for living cytoplasmic inclusions (see section 4.3.4, p. 52). Some of the more relevant texts available for the identification of protozoa have been listed in Table 2. These references include both keys and supporting literature with descriptions down to species level. For further references the reader should consult Kerrich *et al.* (1978) and Sims (1980).

Acknowledgements

I am grateful to Dr C. Curds and Dr O. W. Heal for useful discussions and to Dr G. George for reading and criticising parts of the manuscript. I am also indebted to the library staff of the Freshwater Biological Association, especially Miss D. L. Powell for providing computer-assisted literature searches and Miss E. M. Evans for typing the manuscript.

Recommended reading

BÉ A. W. H. & TOLDERLUND D. S. (1971) Distribution and ecology of living planktonic Foraminifera in surface waters of the Atlantic and Indian Oceans. In: *The Micropalaeontology of Oceans* (Eds B. M. Funnel & W. R. Riedel), pp. 105–49. Cambridge University Press, Cambridge.

BOLTOVSKOY E. & WRIGHT R. (1976) *Recent Foraminifera*. W. Junk, The Hague.

CURDS C. R. (1977) Microbial interactions involving protozoa. In: *Aquatic Microbiology* (Eds F. A. Skinner & J. M. Shewan), pp. 69–105. Academic Press, London.

FENCHEL T. (1969) The ecology of marine microbenthos. IV. Structure and function of the benthic ecosystem, its chemical and physical factors and the microfauna communities with special reference to the ciliated protozoa. *Ophelia* **6**, 1–182.

HEDLEY R. H. & ADAMS C. G. (1974) *Foraminifera*. Academic Press, London.

NOLAND L. E. & GOLDICS M. (1967) Ecology of free-living protozoa. In: *Research in Protozoology* (Ed. T-T. Chen), vol. 2, pp. 215–66. Pergamon Press, New York.

SIEBURTH J. McN. (1979) *Sea Microbes*. Oxford University Press, New York.

STOUT J. D. (1981) The role of protozoa in nutrient cycling and energy flow. In: *Advances in Microbial Ecology* (Ed. M. Alexander), vol. 4, pp. 1–50. Plenum Press, New York and London.

References

ALLDREDGE A. L. & JONES B. M. (1973) *Hastigerina pelagica*: Foraminiferal habitat for planktonic dinoflagellates. *Marine Biology* **22**, 131–5.

ALLEN S. L. & NERAD T. A. (1978) Method for the simultaneous establishment of many axenic cultures of *Paramecium*. *Journal of Protozoology* **25**, 134–9.

ANDERSON O. R., SPINDLER M., BÉ A. W. H. & HEMLEBEN CH. (1979) Trophic activity of planktonic foraminifera. *Journal of the Marine Biological Association U.K.* **59**, 791–9.

ANDREWS A. R. & FLOODGATE G. D. (1974) Some observations on the interactions of marine protozoa and crude oil residues. *Marine Biology* **25**, 7–12.

ARLT G. (1973) Vertical and horizontal distribution of the meiofauna in the Greifswälder Boden. *Oikos* **15**, 105–11.

ARNOLD Z. M. (1954) Culture methods in the study of living foraminifera. *Journal of Paleontology* **28**, 404–16.

ARNOLD Z. M. (1974) Field and laboratory techniques for the study of living foraminifera. In: *Foraminifera* (Eds R. H. Hedley & C. G. Adams), pp. 153–206. Academic Press, London.

BAMFORTH S. S. (1971) The numbers and proportions of testacea and ciliates in litters and soils. *Journal of Protozoology* **18**, 24–8.

BARSDATE R. J., PRENTKI R. T. & FENCHEL T. (1974) Phosphorus cycle of model ecosystems: significance for decomposer food chains and effect of bacterial grazers. *Oikos* **25**, 239–51.

BÉ A. W. H. (1959) A method for rapid sorting of foraminifera from marine plankton samples. *Journal of Paleontology* **33**, 846–8.

BÉ A. W. H. & ANDERSON O. R. (1976) Gametogenesis in planktonic foraminifera. *Science* **192**, 890–2.

BÉ A. W. H., HEMLEBEN C., ANDERSON O. R., SPINDLER M., HACUNDA J. & TUNTIVATE-CHOY S. (1977) Laboratory and field observations of living planktonic foraminifera. *Micropaleontology* **23**, 155–79.

BÉ A. W. H. & TOLDERLUND D. S. (1971) Distribution and ecology of living planktonic Foraminifera in surface waters of the Atlantic and Indian Oceans. In: *The Micropalaeontology of Oceans* (Eds B. M. Funnel & W. R. Riedel), pp. 105–49. Cambridge University Press, Cambridge.

BEERS J. R. (1978a) About microzooplankton. In: *Phytoplankton Manual* (Ed. A. Sournia), pp. 288–96. UNESCO, Paris.

BEERS J. R. (1978b) Pump sampling. In: *Phytoplankton Manual* (Ed. A. Sournia), pp. 41–9. UNESCO, Paris.

BEERS J. R., REID F. M. H. & STEWART G. L. (1975) Microplankton of the North Pacific Central Gyre. Population structure and abundance, June 1973. *Internationale Revue der gasamten Hydrobiologie u. Hydrographie* **60**, 607–38.

BEERS J. R. & STEWART G. L. (1971) Micro-zooplankters in the plankton communities of the upper waters of the eastern tropical Pacific. *Deep-Sea Research* **18**, 861–83.

BEERS J. R., STEWART G. L. & STRICKLAND J. D. H. (1967) A pumping system for sampling small plankton. *Journal of the Fisheries Research Board of Canada* **24**, 1811–18.

BERK S. G., BROWNLEE D. C., HEINLE D. R., KLING H. J. & COLWELL R. R. (1977) Ciliates as a food source for marine planktonic copepods. *Microbial Ecology* **4**, 27–40.

BERK S. G., GUERRY P. & COLWELL R. R. (1976) Separation of small ciliate protozoa from bacteria by sucrose gradient centrifugation. *Applied and Environmental Microbiology* **31**, 450–2.

BERK S. G., MILLS A. L., HENDRICKS D. K. & COLWELL R. R. (1978) Effects of ingesting mercury-containing bacteria on mercury tolerance and growth rates of ciliates. *Microbial Ecology* **4**, 319–30.

BICK H. (1972) *Ciliated Protozoa*. WHO, Geneva.

BICK H. & KUNZE S. (1971) Eine Zusammenstellung von autokologischen und saprobiologischen Befunden an Susswasserciliaten. *Internationale Revue der gesamten Hydrobiologie u. Hydrographie* **56**, 337–84.

BOLTOVSKOY E. & WRIGHT R. (1976) *Recent Foraminifera*. W. Junk, The Hague.

BONNET L. (1974) Quelques aspects du peuplement thécamoebiens des sols de la province de Québec (Canada). *Canadian Journal of Zoology* **52**, 29–41.

BONNET L. & THOMAS R. (1958) Une technique d'isolement des Thécamoebiens (Rhizopoda, testacea) du sol et ses resultats. *Comptes Rendus de l'Académie des Sciences* **247**, 1901–3.

BORKOTT H. (1975) A method for quantitative isolation and preparation of particle-free suspensions of bacteriophagous ciliates from different substrates for electronic counting. *Archiv für Protistenkunde* **117**, 261–8.

BORROR A. C. (1963) Morphology and ecology of the benthic ciliated protozoa of Alligator Harbor, Florida. *Archiv für Protistenkunde* **106**, 465–534.

BRYANT V. M. T. & LAYBOURN J. E. M. (1972/3) The vertical distribution of ciliophora and nematoda in the sediments of Loch Leven, Kinross. *Proceedings of the Royal Society of Edinburgh (B)* **74**, 265–73.

BURGES A. & NICHOLAS D. P. (1961) Use of soil sections

in studying amount of fungal hyphae in soil. *Soil Science* **92**, 25–9.

BURKOVSKII I. V. (1978) Structure dynamics and production of a community of marine psammophilous infusorians. *Zoolohichnyĭ zhurnal Ukrayiny̆* **57**, 325–37.

BURNETT B. R. (1973) Observation of the microfauna of the deep-sea benthos using light and scanning electron microscopy. *Deep-Sea Research* **20**, 413–17.

BURNETT B. R. (1977) Quantitative sampling of microbiota of the deep-sea benthos. I. Sampling techniques and some data from the abyssal Central North Pacific. *Deep-Sea Research* **24**, 781–9.

CAIRNS J. (1965) The protozoa of the Conestoga Basin. *Notulae Naturae* **375**, 1–14.

CAIRNS J. (1974) Protozoans (Protozoa). In: *Pollution Ecology of Freshwater Invertebrates* (Eds C. W. Hart & S. L. H. Fuller), pp. 1–28. Academic Press, London.

CAIRNS J. (1978) Zooperiphyton (especially Protozoa) as indicators of water quality. *Transactions of the American Microscopical Society* **97**, 44–9.

CAIRNS J., PLAFKIN J. L., YONGUE W. H. & KAESLER R. L. (1976) Colonization of artificial substrates by protozoa: replicated samples. *Archiv für Protistenkunde* **118**, 259–67.

CALAWAY W. T. & LACKEY J. B. (1962) *Waste Treatment Protozoa: Flagellata*. Florida Engineering Series No. 3, Gainesville, Florida.

CAMERON I. L. & PRESCOTT D. M. (1961) Relations between cell growth and cell division. V. Cell and macronuclear volumes of *Tetrahymena pyriformis* HSM during the cell life cycle. *Experimental Cell Research* **23**, 354–60.

CASH J. & HOPKINSON J. (1905) *The British Freshwater Rhizopoda and Heliozoa*, vol. 1. The Ray Society, London.

CASH J. & HOPKINSON J. (1909) *The British Freshwater Rhizopoda and Heliozoa*, vol. 2. The Ray Society, London.

CASH J., WAILES G. H. & HOPKINSON J. (1915) *The British Freshwater Rhizopoda and Heliozoa*, vol. 3. The Ray Society, London.

CASH J., WAILES G. H. & HOPKINSON J. (1919) *The British Freshwater Rhizopoda and Heliozoa*, vol. 4. The Ray Society, London.

CASH J., WAILES G. H. & HOPKINSON J. (1921) *The British Freshwater Rhizopoda and Heliozoa*, vol. 5. The Ray Society, London.

CHARDEZ D. (1959) Thécamoebiens des terres de Belgique, I. *Hydrobiologia* **14**, 72–8.

CHARDEZ D. (1964) Sur la séparation verticale des Thécamoebiens endogés. *Bulletin de l'Institut agronomique et des Stations de recherches de Gembloux* **32**, 26–32.

COLE C. V., ELLIOTT E. T., HUNT H. W. & COLEMAN D. C. (1978) Trophic interactions in soils as they affect energy and nutrient dynamics. V. Phosphorus transformations. *Microbial Ecology* **4**, 381–7.

COLE G. A. (1955) An ecological study of the microbenthic fauna of two Minnesota lakes. *American Midland Naturalist* **53**, 213–30.

COLLIN B. (1912) Étude monographique sur les Acinétiens. II. Morphologie, Physiologie, Systématique. *Archives de Zoologie Expérimentale et Générale* **51**, 1–457.

COLLINS V. G. & WILLOUGHBY L. G. (1962) The distribution of bacteria and fungal spores in Blelham Tarn with particular reference to an experimental overturn. *Archiv für Mikrobiologie* **43**, 294–307.

CORLISS J. O. (1979 *Ciliated Protozoa*. Pergamon Press, Oxford.

CORLISS J. O. & ESSER S. C. (1974) Comments on the role of the cyst in the life cycle and survival of free-living protozoa. *Transactions of the American Microscopical Society* **93**, 578–93.

COÛTEAUX M-M. (1967) Une technique d'observation des Thécamoebiens du sol pour l'estimation de leur densité absolue. *Revue d'Écologie et de Biologie du Sol* **4**, 593–6.

COÛTEAUX M-M. (1975) Estimation quantitative des Thécamoebiens édaphiques par rapport à la surface du sol. *Comptes Rendus de l'Académie des Sciences* **281**, 739–41.

CURDS C. R. (1969) *An Illustrated Key to the British Freshwater Ciliated Protozoa Commonly Found in Activated Sludge*. HMSO, London.

CURDS C. R. (1974) Computer simulations of some complex microbial food chains. *Water Research* **8**, 769–80.

CURDS C. R. (1977) Microbial interactions involving protozoa. In: *Aquatic Microbiology* (Eds F. A. Skinner & J. M. Shewan), pp. 69–105. Academic Press, London.

CURDS C. R. (1981) *British Freshwater Ciliated Protozoa, vol. 1: Kinetofragminophora* (Ed. D. M. Kermack), Academic Press, London.

CURDS C. R. & COCKBURN A. (1971) Continuous monoxenic culture of *Tetrahymena pyriformis*. *Journal of General Microbiology* **66**, 95–108.

CURDS C. R., COCKBURN A. & VANDYKE J. M. (1968) An experimental study of the role of the ciliated protozoa in the activated-sludge process. *Water Pollution Control* **67**, 312–29.

CURDS C. R. & VANDYKE J. M. (1966) The feeding habits and growth rates of some freshwater ciliates found in activated-sludge. *Journal of Applied Ecology* **3**, 127–37.

CUSHMAN J. A. (1948) *Foraminifera, their Classification and Economic Use*. Harvard University Press, Cambridge, Massachusetts.

DARBYSHIRE J. F. (1972a) Nitrogen fixation by *Azotobacter chroococcum* in the presence of *Colpoda steini*. I. The influence of temperature. *Soil Biology and Biochemistry* **4**, 359–69.

DARBYSHIRE J. F. (1972b) Nitrogen fixation by *Azotobacter chroococcum* in the presence of *Colpoda steini*, II. The influence of agitation. *Soil Biology and Biochemistry* **4**, 371–6.

DARBYSHIRE J. F. (1973) The estimation of soil protozoan populations. In: *Sampling, Microbiological Monitoring of Environments* (Eds R. G. Board & D. W. Lovelock), pp. 175–88. Academic Press, London.

DAVIS C. C. (1955) *The Marine and Fresh-Water Plankton.* Constable & Co., London.

DAVIS P. G., CARON D. A. & SIEBURTH J. McN. (1978) Oceanic amoebae from the North Atlantic: culture, distribution and taxonomy. *Transactions of the American Microscopical Society* **97**, 73–88.

DAWID W. (1975) Kultur von Erdamöben. *Mikrokosmos* **64**, 195–200.

DÉCLOITRE L. (1966) Comment compter le nombre de Thécamoebiens dans une récolte. *Limnologica* **4**, 489–92.

DEFLANDRE G. (1959) Rhizopoda and Actinopoda. In: *Freshwater Biology* (Ed. W. T. Edmondson), pp. 232–64. John Wiley, New York.

DILLON R. D. & HOBBS J. T. (1973) Estimating quantity and quality of the biomass of benthic protozoa. *Proceedings of the South Dakota Academy of Sciences* **52**, 47–9.

DIVE D. (1975) Influence de la concentration bactérienne sur la croissance de *Colpidium campylum. Journal of Protozoology* **22**, 545–50.

DODSON A. N. & THOMAS W. H. (1964) Concentrating plankton in a gentle fashion. *Limnology and Oceanography* **9**, 455–6.

DRAGESCO J. (1960) Ciliés mésopsammiques littoraux. *Travaux de la Station Biologique de Roscoff* **12**.

DROOP M. R. (1969) Algae. In: *Methods in Microbiology* (Eds J. R. Norris & D. W. Ribbons), vol. 3B, pp. 269–313. Academic Press, London.

ELLIOTT P. B. & BAMFORTH S. S. (1975) Interstitial protozoa and algae of Louisiana salt marshes. *Journal of Protozoology* **22**, 514–19.

ELLIOTT P. B. & COLEMAN D. C. (1977) Soil protozoan dynamics in a shortgrass prairie. *Soil Biology and Biochemistry* **9**, 113–18.

FAURÉ-FREMIET E. (1924) Contribution à la connaissance des infusoires planktoniques. *Bulletin Biologique de la France et de la Belgique* **6**, 1–171.

FAURÉ-FREMIET E. (1950) Ecology of ciliate infusoria. *Endeavour* **9**, 183–87.

FENCHEL T. (1967) The ecology of marine microbenthos. I. The quantitative importance of ciliates as compared with metazoans in various types of sediments. *Ophelia* **4**, 121–37.

FENCHEL T. (1968) The ecology of marine microbenthos. III. The reproductive potential of ciliates. *Ophelia* **5**, 123–36.

FENCHEL T. (1969) The ecology of marine microbenthos. IV. Structure and function of the benthic ecosystem, its chemical and physical factors and the microfauna communities with special reference to the ciliated protozoa. *Ophelia* **6**, 1–182.

FENCHEL T. (1980a) Suspension feeding in ciliated protozoa: functional response and particle size selection. *Microbial Ecology* **6**, 1–11.

FENCHEL T. (1980b) Suspension feeding in ciliated protozoa: feeding rates and their ecological significance. *Microbial Ecology* **6**, 13–25.

FENCHEL T. (1980c) Relation between particle size selection and clearance in suspension-feeding ciliates. *Limnology and Oceanography* **25**, 733–8.

FENCHEL T. (1980d) Suspension feeding in ciliated protozoa: structure and function of feeding organelles. *Archiv für Protistenkunde* **123**, 239–60.

FENCHEL T. & HARRISON P. (1976) The significance of bacterial grazing and mineral cycling for the decomposition of particulate detritus. In: *The Role of Terrestrial and Aquatic Organisms in Decomposition Processes* (Ed. J. M. Anderson), pp. 285–99. Blackwell Scientific Publications, Oxford.

FENCHEL T. & JANSSON B-O. (1966) On the vertical distribution of the microfauna in the sediments of a brackish-water beach. *Ophelia* **3**, 161–77.

FENCHEL T., PERRY T. & THANE A. (1977) Anaerobiosis and symbiosis with bacteria in free-living ciliates. *Journal of Protozoology* **24**, 154–63.

FINLAY B. J. (1977) The dependence of reproductive rate on cell size and temperature in freshwater ciliated protozoa. *Oecologia* **30**, 75–81.

FINLAY B. J. (1978) Community production and respiration by ciliated protozoa in the benthos of a small eutrophic loch. *Freshwater Biology* **8**, 327–41.

FINLAY B. J. (1980) Temporal and vertical distribution of ciliophoran communities in the benthos of a small eutrophic loch with particular reference to the redox profile. *Freshwater Biology* **10**, 15–34.

FINLAY B., BANNISTER P. & STEWART J. (1979a) Temporal variation in benthic ciliates and the application of association analysis. *Freshwater Biology* **9**, 45–53.

FINLAY B. J., LAYBOURN J. & STRACHAN I. (1979b) A technique for the enumeration of benthic ciliated protozoa. *Oecologia* **39**, 375–7.

FINLEY H. E. (1930) Tolerance of fresh water Protozoa to increased salinity. *Ecology* **11**, 337–46.

FJELD P. (1955) On some marine psammobiotic ciliates from Drøbak (Norway). *Nytt Magasin for Zoologi* **3**, 5–64.

FOK A. K. & ALLEN R. D. (1979) Axenic *Paramecium caudataum.* I. Mass culture and structure. *Journal of Protozoology* **26**, 463–70.

FORNSHELL J. A. (1979) Microplankton patchiness in the Northwest Atlantic Ocean. *Journal of Protozoology* **26**, 270–72.

GEORGE E. A. (1976) *List of Strains.* NERC, Culture Centre for Algae and Protozoa, Cambridge.

GOLD K. (1973) Methods for growing tintinnida in continuous culture. *American Zoologist* **13**, 203–8.

GOLD K., PFISTER R. M. & LIGUORI V. R. (1970) Axenic cultivation and electron microscopy of two species of choanoflagellida. *Journal of Protozoology* **17**, 210–12.

GOLDSTEIN L. & KO C. (1976) A method for the mass culturing of large free-living amebas. In: *Methods in Cell Biology* (Ed. D. M. Prescott), vol. 13, pp. 239–46. Academic Press, London.

GOULDER R. (1971a) The effects of saprobic conditions on some ciliated Protozoa in the benthos and hypolimnion of a eutrophic pond. *Freshwater Biology* **1**, 307–18.

GOULDER R. (1971b) Vertical distribution of some ciliated protozoa in two freshwater sediments. *Oikos* **22**, 199–203.

GOULDER R. (1972) The vertical distribution of some ciliated protozoa in the plankton of a eutrophic pond during summer stratification. *Freshwater Biology* **2**, 163–76.

GOULDER R. (1974) The seasonal and spatial distribution of some benthic ciliated protozoa in Esthwaite Water. *Freshwater Biology* **4**, 127–47.

GRABACKA E. (1971) Ciliata in bottom sediments of fingerling ponds. *Polskie Archiwum Hydrobiologii* **18**, 225–33.

GRAHAM L. B., COLBURN A. D. & BURKE J. C. (1976) A new, simple method for gently collecting planktonic Protozoa. *Limnology and Oceanography* **21**, 336–41.

GRASSÉ P-P. (1952) *Traité de Zoologie*, vol. I (i). Masson & Cie, Paris.

GRASSÉ P-P. (1953) *Traité de Zoologie*, vol. I (ii). Masson & Cie, Paris.

GRELL K. G. (1973) *Protozoology*. Springer-Verlag, Heidelberg.

GROLIÈRE C-A. (1977) Contribution à l'étude des ciliés des sphnaignes. II. Dynamique des populations. *Protistologica* **13**, 335–52.

GÜDE H. (1979) Grazing by protozoa as selection factor for activated sludge bacteria. *Microbial Ecology* **5**, 225–37.

HAAS L. W. & WEBB K. L. (1979) Nutritional mode of several non-pigmented microflagellates from the York River estuary. *Journal of Experimental Marine Biology and Ecology* **39**, 125–34.

HAGSTRÖM A., LARSSON U., HÖRSTEDT P. & NORMARK S. (1979) Frequency of dividing cells, a new approach to the determination of bacterial growth rates in aquatic environments. *Applied and Environmental Microbiology* **37**, 805–12.

HAIRSTON N. G. & KELLERMANN S. L. (1964) *Paramecium* ecology: Electromigration for field samples and observations on density. *Ecology* **45**, 373–6.

HARRISON P. G. & MANN K. H. (1975) Detritus formation from eelgrass (*Zostera marina* L.): the relative effects of fragmentation, leaching and decay. *Limnology and Oceanography* **20**, 924–34.

HARTWIG E. (1977) Investigations on the ecophysiology of *Geleia nigriceps* Kahl (Ciliophora, Gymnostomata) inhabiting a sandy beach in Bermuda. *Oecologia* **31**, 159–75.

HARTWIG E. & PARKER J. G. (1977) On the systematics and ecology of interstitial ciliates of sandy beaches in North Yorkshire. *Journal of the Marine Biological Association U.K.* **57**, 735–60.

HEAL O. W. (1964a) The use of cultures for studying Testacea (Protozoa: Rhizopoda) in soil. *Pedobiologia* **4**, 1–7.

HEAL O. W. (1964b) Observations on the seasonal and spatial distribution of Testacea (Protozoa: Rhizopoda) in *Sphagnum*. *Journal of Animal Ecology* **33**, 395–412.

HEAL O. W. (1965) Observations on testate amoebae (Protozoa: Rhizopoda) from Signy Island, South Orkney Islands. *British Antarctic Survey Bulletin* **6**, 43–7.

HEAL O. W. (1970) Methods of study of soil protozoa. In: *Methods of Study in Soil Ecology* (Ed. J. Phillipson), pp. 119–26. UNESCO, New York.

HEANEY S. I. (1974) A pneumatically operated water sampler for close intervals of depth. *Freshwater Biology* **4**, 103–6.

HEDIN H. (1975) On the ecology of tintinnids on the Swedish west coast. *Zoon* **3**, 125–40.

HEDLEY R. H. & ADAMS C. G. (1974) *Foraminifera*. Academic Press, London.

HEINBOKEL J. F. (1978a) Studies on the functional role of tintinnids in the Southern California Bight. I. Grazing and growth rates in laboratory cultures. *Marine Biology* **47**, 177–89.

HEINBOKEL J. F. (1978b) Studies on the functional role of tintinnids in the Southern California Bight. II. Grazing rates of field populations. *Marine Biology* **47**, 191–7.

HEINBOKEL J. F & BEERS J. R. (1979) Studies on the functional role of tintinnids in the Southern California Bight. III. Grazing impact of natural assemblages. *Marine Biology* **52**, 23–32.

HEINLE D. R., HARRIS R. P., USTACH J. F. & FLEMER D. A. (1977) Detritus as food for marine copopods. *Marine Biology* **40**, 341–53.

HELLER M. D. (1977) The phased division of the freshwater dinoflagellate *Ceratium hirundinella* and its use as a method of assessing growth in natural populations. *Freshwater Biology* **7**, 527–33.

HOLLANDE A. & ENJUMET M. (1960) Cytologie, évolution et systématique des sphaeroïdes (Radiolaires). *Archives du Muséum National d'Histoire Naturelle* **7**, 1–134.

HOLME N. A. & MCINTYRE A. D. (1971) *Methods for the Study of Marine Benthos*. Blackwell Scientific Publications, Oxford.

HOPKINS D. W. (1946) The contractile vacuole and the adjustment to changing concentrations in freshwater amoebae. *Biological Bulletin. Marine Biological Laboratory, Woods Hole, Massachusetts* **90**, 158–76.

JAHN T. L. & JAHN F. L. (1979) *How to Know the Protozoa*. Wm. C. Brown, Dubuque, Iowa.

JONES P. C. T. & MOLLISON T. (1948) A technique for the quantitative estimation of soil micro-organisms. *Journal of General Microbiology* **2**, 54–69.

KAHL (1934) Suctoria. In: *Tierwelt der Nord- und Ostee* (Eds G. Grimpe & E. Wagler), vol. 26, pp. 184–226. Leipzig.

KAHL A. (1930–1935) Urtiere oder Protozoa. I. Wimpertiere oder Ciliata (Infusoria). *Die Tierwelt Deutschlands*, vols. 18, 21, 25, 30. Fischer, Jena.

KENT W. S. (1880–1882) *A Manual of the Infusoria*, vols I–III. David Bogue, London.

KERRICH G. J. HAWKSWORTH D. L. & SIMS R. W. (1978) *Key Works to the Fauna and Flora of the British Isles and Northwestern Europe*. Academic Press, London.

KIMBALL R. F., CASPERSSON T. O., SVENSSON G. & CARLSON L. (1959) Quantitative cytochemical studies on *Paramecium aurelia*. *Experimental Cell Research* **17**, 160–172.

KINNE O. (1977) Cultivation of animals—research cultivation. In: *Marine Ecology* (Ed. O. Kinne), vol. 3, part 2, pp. 584–627. John Wiley, New York.

KORGANOVA G. A. & GELTSER J. G. (1977) Stained smears for the study of soil Testacida (Protozoa, Rhizopoda). *Pedobiologia* **17**, 222–5.

KUDO R. (1966) *Protozoology*, Charles C. Thomas, Springfield, Illinois.

LACKEY J. B. (1959) Zooflagellates. In: *Freshwater Biology* (Ed. W. T. Edmondson), pp. 190–231. John Wiley, New York.

LAYBOURN J. E. M. (1975) An investigation of the factors influencing mean cell volume in populations of the ciliate *Colpidium campylum*. *Journal of Zoology* **177**, 171–7.

LEADBETTER B. S. C. (1972) Fine structural observations on some marine choanoflagellates from the coast of Norway. *Journal of the Marine Biological Association U.K.* **52**, 67–79.

LEADBETTER B. S. C. (1977) Observations on the life-history and ultrastructure of the marine choanoflagellate *Choanoeca perplexa* Ellis. *Journal of the Marine Biological Association U.K.* **57**, 285–301.

LEADBETTER B. S. C. & MORTON C. (1974) A microscopical study of a marine species of *Codosiga* James-Clark (Choanoflagellata) with special reference to the ingestion of bacteria. *Biological Journal of the Linnean Society* **6**, 337–47.

LEE J. J., MCENERY M. E. & RUBIN H. (1969) Quantitative studies on the growth of *Allogromia laticollaris* (Foraminifera). *Journal of Protozoology* **16**, 377–95.

LEE J. J. & MULLER W. A. (1973) Trophic dynamics and niches of salt marsh Foraminifera. *American Zoologist* **13**, 215–23.

LEE J. J. & PIERCE S. (1963) Growth and physiology of Foraminifera in the laboratory. IV. Monoxenic culture of an allogromiid with notes on its morphology. *Journal of Protozoology* **10**, 404–11.

LEE J. J., PIERCE S., TENTCHOFF M. & MCLAUGHLIN J. J. A. (1961) Growth and physiology of foraminifera in the laboratory. I. Collection and maintenance. *Micropaleontology* **7**, 461–6.

LEEDALE G. F. (1967) *Euglenoid Flagellates*. Prentice-Hall Inc. New Jersey.

LEVINE N. D., CORLISS J. O., COX F. E. G., DEROUX G., GRAIN J., HONIGBERG B. M., LEEDALE G. F., LOEBLICH A. R., LOM J., LYNN D., MERINFELD E. G., PAGE F. C., POLJANSKY G., SPRAGUE V., VAVRA J. & WALLACE F. G. (1980) A newly revised classification of the Protozoa. *Journal of Protozoology* **27**, 37–58.

LIGHTHART B. (1969) Planktonic and benthic bacterivorous protozoa at eleven stations in Puget Sound and adjacent Pacific Ocean. *Journal of the Fisheries Research Board of Canada* **26**, 299–304.

LOUSIER J. D. (1976) Testate amoebae (Rhizopoda, Testacea) in some Canadian Rocky Mountain soils. *Archiv für Protistenkunde* **118**, 191–201.

LOUSIER J. D. & PARKINSON D. (1981) Evaluation of a membrane filter technique to count soil and litter Testacea. *Soil Biology and Biochemistry* **13**, 209–14.

MACKINNON D. L. & HAWES R. S. J. (1961) *An Introduction to the Study of Protozoa*. Oxford University Press, London.

MAGEAU N. C. & WALKER D. A. (1976) Effects of ingestion of Foraminifera by larger invertebrates. In: *Maritime Sediments* (Eds C. T. Schafer & B. R. Pelletier), pp. 89–105. Halifax, Canada.

MAGUIRE B. (1963) The passive dispersal of small aquatic organisms and their colonization of isolated bodies of water. *Ecological Monographs* **33**, 161–85.

MARE M. F. (1942) A study of a marine benthic community with special reference to the micro-organisms. *Journal of the Marine Biological Association U.K.* **25**, 517–54.

MARSHALL P. R. (1975) Some relationships between living and total foraminiferal faunas on Pedro Bank, Jamaica. In: *Maritime Sediments Special Publication No. 1* (Eds C. T. Schafer & B. R. Pelletier), pp. 61–70. Halifax, Canada.

MENAPACE D., KLEIN D. A., MCCLELLAN J. F. & MAYEUX J. V. (1975) A simplified overlay plaque technique for evaluating responses of small free-living amoebae in grassland soils. *Journal of Protozoology* **22**, 405–10.

MICHIELS M. (1974) Biomass determination of some freshwater ciliates. *Biologisch jaarboek Dodonaea* **42**, 132–6.

MILES H. B. (1963) Soil protozoa and earthworm nutrition. *Soil Science* **95**, 407–9.

MOORE G. M. (1939) A limnological investigation of the microscopic benthic fauna of Douglas Lake, Michigan. *Ecological Monographs* **9**, 537–82.

MOORE H. B. & NEILL R. G. (1930) An instrument for sampling marine muds. *Journal of the Marine Biological Association U.K.* **16**, 589–94.

MULLER P. H. (1974) Sediment production and population biology of the benthic foraminifer *Amphistegina madagascariensis*. *Limnology and Oceanography* **19**, 802–9.

MULLER W. A. & LEE J. J. (1969) Apparent indispensability of bacteria in foraminiferan nutrition. *Journal of Protozoology* **16**, 471–8.

MURRAY J. W. (1973) *Distribution and Ecology of Living Benthic Foraminiferids*. Heinemann, London.

MURRAY J. W. (1979) *British Nearshore Foraminiferids. Key and Notes for the Identification of Species* Academic Press, London.

NIGRINI C. (1967) Radiolaria in pelagic sediments from the Indian and Atlantic Oceans. *Bulletin of the Scripps Institute of Oceanography* **11**, 1–106.

NIKOLJUK V. F. (1969) Some aspects of the study of soil protozoa. *Acta Protozoologica* **7**, 99–109.

NOLAND L. E. (1925) Factors influencing the distribution of fresh-water ciliates. *Ecology* **6**, 437–52.

NOLAND L. E. (1959) Ciliophora. In: *Freshwater Biology* (Ed. W. T. Edmondson), pp. 265–97. John Wiley, New York.

NOLAND L. E. & GOJDICS M. (1967) Ecology of free-living protozoa. In: *Research in Protozoology* (Ed. T-T. Chen), vol. 2, pp. 215–66. Pergamon Press, New York.

O'DELL W. D. (1979) Isolation, enumeration and identification of amoebae from a Nebraska Lake. *Journal of Protozoology* **26**, 265–9.

OGDEN C. G., & HEDLEY R. H. (1980) *Freshwater Testate Amoebae*. British Museum (Natural History), London.

PAGE F. C. (1970) Two new species of *Paramoeba* from Maine. *Journal of Protozoology* **17**, 421–7.

PAGE F. C. (1973) *Paramoeba*: A common marine genus. *Hydrobiologia* **41**, 183–8.

PAGE F. C. (1976) *An Illustrated Key to Freshwater and Soil Amoebae*. Scientific Publications Freshwater Biological Association, no. 34.

PARKER J. G. (1976) Cultural characteristics of the marine ciliated protozoan *Uronema marinum* Dujardin. *Journal of Experimental Marine Biology and Ecology* **24**, 213–26.

PATSCH B. (1974) Die Aufwuchsciliaten des Naturlehrparks Haus Wildenrath, Dissertation, Institut für Landwirtschaftliche Zoologie und Bienenkunde, University of Bonn.

PATTERSON D. J. (1980) Contractile vacuoles and associated structures: their organisation and function. *Biological Reviews* **55**, 1–46.

PENARD E. (1902) *Faune Rhizopodique du Bassin du Léman*. H. Kündig, Geneva.

PENARD E. (1904) *Les Héliozoaires d'Eau Douce*. H. Kündig, Geneva.

PENARD E. (1922) *Études sur les Infusoires d'Eau Douce*. Georg et Cie, Geneva.

PENARD E. (1922) *Études sur les Infusoires d'Eau Douce*. Georg et Cie, Geneva.

PENNAK R. W. (1978) Protozoa. In: *Fresh-Water Invertebrates of the United States*, pp. 19–79. John Wiley, New York.

PIEARCE T. G. & PHILLIPS M. J. (1980) The fate of ciliates in the earthworm gut: an *in vitro* study. *Microbial Ecology* **5**, 313–19.

PROVASOLI L. (1977) Cultivation of animals—axenic cultivation. In: *Marine Ecology* (Ed. O. Kinne), vol. 3, pp. 1301–12. John Wiley, New York.

RAINER H. (1968) Heliozoa. Systematik und Taxonomie, Biologie, Verbreitung und Ökologie der Arten der Erde. *Die Tierwelt Deutschlands* **56**, 1–176.

RENZ G. W. (1976) The distribution and ecology of Radiolaria in the Central Pacific: plankton and surface sediments. *Bulletin of the Scripps Institute of Oceanography* **22**, 1–267.

RIGLER, F. H., MACCALLUM M. E. & ROFF J. C. (1974) Production of zooplankton in Char Lake. *Journal of the Fisheries Research Board of Canada* **31**, 637–46.

RÖTTGER R. & BERGER W. H. (1972) Benthic Foraminifera: morphology and growth in clone cultures of *Heterostegina depressa*. *Marine Biology* **15**, 89–94.

SAWYER T. K. (1971) Isolation and identification of free-living marine amoebae from Upper Chesapeake Bay, Maryland. *Transactions of the American Microscopical Society* **90**, 43–51.

SCHAFER C. T. & PELLETIER B. R. (1976) *Maritime Sediments*. Halifax, Canada.

SCHEWIAKOFF W. (1926) Die Acantharia des Golfes von Neapel. *Fauna Flora Golfo Napoli* **37**, 1–755.

SCHÖNBORN W. (1977) Production studies on protozoa. *Oecologia* **27**, 171–84.

SCHULTZ T. W. & DUMONT J. N. (1977) Cytotoxicity of synthetic fuel products on *Tetrahymena pyriformis*. *Journal of Protozoology* **24**, 104–72.

SCOTT D. R. (1975) Quantitative studies on marsh foraminifera patterns in Southern California and their application to holocene stratigraphic problems. In: *Maritime Sediments* (Eds C. T. Schafer & B. R. Pelletier), pp. 153–70. Halifax, Canada.

SIEBURTH J. MCN. (1979) *Sea Microbes*. Oxford University Press, New York.

SIMS R. W. (1980) *Animal Identification, a Reference Guide*, pp. 1–6. British Museum (Natural History) and John Wiley, Chichester.

SINGH B. N. (1955) Culturing soil protozoa and estimating their numbers in soil. In: *Soil Zoology* (Ed. D. K. McE. Kevan), pp. 403–11. Butterworths, London.

SINGH B. N. & DAS S. R. (1970) Studies on pathogenic and non-pathogenic small free-living amoebae and the bearing of nuclear division on the classification of the order Amoebida. *Philosophical Transactions of the Royal Society (B)* **259**, 435–76.

SMITH R. K. (1967) Ignition and filter methods of concentrating shelled organisms. *Journal of Paleontology* **41**, 1288–91.

SOLDO A. T. & MERLIN E. J. (1972) The cultivation of symbiote-free marine ciliates in axenic medium. *Journal of Protozoology* **19**, 519–24.

SOLDO A. T. & MERLIN E. J. (1977) The nutrition of *Parauronema acutum*. *Journal of Protozoology* **24**, 556–62.

SONNEBORN T. M. (1950) Methods in the general biology

and genetics of *Paramecium aurelia*. *Journal of Experimental Zoology* **113**, 87–148.

SPOON D. M., FEISE C. O. & YOUN R. S. (1977) Poly-(ethylene)oxide, a new slowing agent for protozoa. *Journal of Protozoology* **24**, 471–4.

ŠRÁMEK-HUŠEK R. (1958) Die Rolle der Ciliatenanalyse bei der biologischen Kontrolle von Flussverunreinigungen. *Verhandlungen der Internationalen Vereinigung für theoretische und angewandte Limnologie* **13**, 636–45.

STARR R. C. (1964) The culture collection of algae at Indiana University. *American Journal of Botany* **51**, 1013–44.

STOKES A. C. (1888) A preliminary contribution towards a history of the fresh-water infusoria of the United States. *Journal of the Trenton Natural History Society* **1**, 71–344.

STÖSSEL F. (1979) Autökologische Analyse der in schweizerischen Fliessengewässern häufig vorkommenden Ciliatenarten und ihre Eignung als Bioindikatoren. *Schweizerische Zeitschrift für Hydrobiologie* **41**, 113–40.

STOUT J. (1956) Reaction of ciliates to environmental factors. *Ecology* **37**, 179–91.

STOUT J. D. (1963) The distribution of rhizopod and ciliate protozoa in the soils, forest litters and peats of the New Zealand area. In: *Progress in Protozoology* (Eds J. Ludvík, J. Lom & J. Vávra), p. 334. Academic Press, London.

STOUT J. D. (1974) Protozoa. In: *Biology of Plant Litter Decomposition* (Eds C. H. Dickinson & G. J. F. Pugh), pp. 385–420. Academic Press, London.

STOUT J. D. (1981) The role of protozoa in nutrient cycling and energy flow. In: *Advances in Microbial Ecology* (Ed. M. Alexander), vol. 4, pp. 1–50. Plenum Press, New York and London.

STOUT J. F. & HEAL O. W. (1967) Protozoa. In: *Soil Biology* (Eds A. Burges & F. Raw), pp. 149–95. Academic Press, London.

TAYLOR F. J. R. (1976) Flagellates. In: *Zooplankton Fixation and Preservation* (Ed. H. F. Steedman), pp. 259–64. UNESCO, Paris.

TAYLOR W. D. (1978a) Maximum growth rate, size and commonness in a community of bactivorous ciliates. *Oecologia* **36**, 263–72.

TAYLOR W. D. (1978b) Growth responses of ciliate protozoa to the abundance of their bacterial prey. *Microbial Ecology* **4**, 207–14.

TAYLOR W. D. & BERGER H. (1976) Growth responses of cohabiting ciliate protozoa to various prey bacteria. *Canadian Journal of Zoology* **54**, 1111–14.

THIEL H. (1975) The size structure of the deep-sea benthos. *Internationale Revue der gesamten Hydrobiologie* **60**, 575–606.

TSEEB Y. Y. (1937) A method for quantitatively estimated microfauna for purposes of its use in salt lakes of the Crimea. *Zoologicheskii zhurnal SSSR* **16**, 499–509.

TUFFRAU M. (1967) Perfectionnements et pratique de la technique d'impregnation au protargol des infusoires cilies. *Protistologica* **3**, 91–8.

UHLIG G. (1964) Eine einfache Methode zur Extraction der vagilen, mesopsammon Mikrofauna. *Helgoländer wissenschaftliche Meeresuntersuchungen* **11**, 178–85.

UHLIG G. (1965) Die mehrgliedrige Kultur litoraler Folliculiniden. *Helgoländer wissenschaftliche Meeresuntersuchungen* **12**, 52–60.

UHLIG G. (1968) Quantitative methods in the study of interstitial fauna. *Transactions of the American Microscopical Society* **87**, 226–32.

UHLIG G. (1972) Protozoa. In: *Research Methods in Marine Biology* (Ed. C. Schlieper), pp. 129–41. University of Washington Press, Seattle.

UHLIG G., THIEL H. & GRAY J. S. (1973) The quantitative separation of microfauna. *Helgoländer wissenschaftliche Meeresuntersuchungen* **25**, 175–95.

UTERMÖHL H. (1958) Zur Vervollkommnung der quantitativen Phytoplankton—Methodik. *Mitteilungen der internationalen Vereinigung für theoretische und angewandte Limnologie* **13**, 236–51.

WALKER S. A., LINTON A. E. & SCHAFER C. T. (1974) Sudan Black B: a superior stain to Rose Bengal for distinguishing living from non-living Foraminifera. *Journal of Foraminiferal Research* **4**, 205–15.

WANG C. C. (1928) Ecological studies of the seasonal distribution of protozoa in a fresh-water pond. *Journal of Morphology* **46**, 431–78.

WEBB M. G. (1961) The effects of thermal stratification on the distribution of benthic protozoa in Esthwaite Water. *Journal of Animal Ecology* **30**, 137–51.

WEFER G. (1976) Environmental effects on growth rates of benthic Foraminifera (shallow water, Baltic Sea). In: *Maritime Sediments* (Eds C. T. Schafer & B. R. Pelletier), pp. 39–50. Halifax, Canada.

WEFER G. & LUTZE G. F. (1977) Benthic Foraminifera biomass production in the Western Baltic. *Kieler Meeresforschung Sonderheft* **3**, 76–81.

WEILER C. S. & CHISHOLM S. W. (1976) Phased cell division in natural populations of marine dinoflagellates from shipboard cultures. *Journal of Experimental Marine Biology and Ecology* **25**, 239–47.

WILLAERT E. (1971) Isolement et culture *in vitro* des amibes du genre Naegleria. *Annales de la Société belge de médecine tropicale* **51**, 701–8.

WILLIAMS D. F. & HEALY-WILLIAMS N. (1980) Oxygen isotope—hydrographic relationships among recent planktonic Foraminifera from the Indian Ocean. *Nature, London* **283**, 848–52.

Chapter 5 • Methods for the Study of Virus Ecology

Sandy B. Primrose, Nigel D. Seeley and Kelvin B. Logan

5.1 Introduction

All viruses are obligate intracellular parasites and so it is impossible to discuss their ecology without considering their hosts as well. Over the last few decades a considerable amount of information has been gathered concerning the epidemiology and ecology of viruses infecting man or his domestic animals and plants (Andrewes 1967). This review will be concerned mainly with:

(1) what is known at present about the ecology of viruses infecting other organisms, e.g. bacteria, fungi, algae and others;

(2) what methods are available for the isolation of viruses from the natural environment and for their identification.

Virus ecology may be defined as the interaction of viruses with their host cells in the environment and the effects that such interactions have on the composition and productivity of particular habitats. Thus, to study virus ecology we must know about the distribution of viruses in the environment and the effect of environmental factors on their ability to infect cells and subsequently reproduce. At present little is known about the effects of viruses on their host populations in the natural environment. However, it is known that viruses have the potential to either reduce the vigour of their host or to kill it. If the vigour of the host is reduced, its competitiveness may be reduced or lost. If the host is killed, not only does it no longer contribute to the biological productivity of its habitat, but its place may be taken by another, unrelated organism. Thus viruses may upset the balance between the growth of different hosts in a particular habitat. In addition, some viruses confer new phenotypic traits on the cells they infect and this may give them a selective advantage, or disadvantage, in certain environments.

To study virus ecology requires methods for the isolation and identification of viruses from the environment. It must be stressed that absolute proof that a virus is present requires successful transmission of purified particles to virus-free host cells and their subsequent multiplication therein. This may not be easy and sources of confusion abound; for example, there are numerous reports of virus-like particles in algae (Brown 1972). The particles most commonly found are large angular polyhedra about 150 to 250 nm in diameter, although smaller (40 nm), DNA-containing particles have been described. While similar to viruses these particles have not been transmitted to particle-free algal isolates. Furthermore, they bear a striking resemblance to the carboxysomes of prokaryotes which, like some viruses, contain DNA (Westphal *et al.* 1979). By contrast, virus-like particles found in the green alga *Chara corallina* have been purified, transmitted by injection to particle-free cells which later become chlorotic and have been shown to multiply (Gibbs *et al.* 1975). This confirms that they are virus particles and at the time of writing this was the only authentic report of a virus infecting a eukaryotic alga. Based on similar considerations there are relatively few reports of authentic viruses infecting protozoa, fungi and mycoplasmas. It is for this reason that most of the ensuing discussion is centred on the ecology of bacteriophages.

5.2 The distribution of phages in the environment

Reliable dates for phage distribution are not available and this reflects the lack of suitable methodology. Viruses are obligate parasites whose assay usually involves living cells. This is the source of the problem, for most phages only infect a limited range of bacteria. Even within a single species, e.g. *Escherichia coli*, there are many different strains all of which differ in their sensitivity toward different phages. In the absence of a universal phage assay host it is not possible to determine the total number of bacteriophage particles present in any sample. What is possible is the quantification of a few selected bacteriophages; e.g. it is possible to assay the numbers of male-specific (F-specific) bacteriophages in a water sample. It should be noted that since each bacterial strain can serve as a host for more than one bacteriophage there are probably more different bacteriophage types than bacterial species in the environment.

5.3 The importance of host numbers

Being obligate parasites phages have little, if any, impact on their environment in the absence of host cells. The concentration of host cells is more important than might at first be imagined. For example, the time (T) for a given proportion of phage (P_o/P_T) to adsorb to a bacterium is given by:

$$T = \frac{2.3}{Bk}\left[\log\left(\frac{P_o}{P_T}\right)\right]$$

where k = adsorption rate constant (ml min^{-1}); P_o = free phage at time 0; P_T = free phage at time T and B is the bacterial concentration (organisms ml^{-1}). Measured values of k are in the range 1×10^{-9} to 1×10^{-11} and these values have been used to construct a plot of T versus B for adsorption of 1% and 50% of the added phages (Fig. 1). From the graph it can be seen that if a value of $k = 1 \times 10^{-11}$ is chosen then, at a bacterial density of 1×10^9 organisms ml^{-1}, 1% of added phage will adsorb in 1 min. However, if the bacterial concentration is only 1×10^5 organisms ml^{-1} then 1% phage adsorption requires 1×10^4 min. Bearing in mind that sewage usually contains only 1×10^5 to 1×10^6 *Escherichia coli* ml^{-1} (Scarpino 1978) and that only a proportion of these will be susceptible to a given phage, it would appear that phage infection occurs very infrequently in aquatic habitats. However, the large number of phages detected in such habitats suggest that the rate of infection is much higher than theoretically possible. The most likely solution to this paradox is that phage infection largely occurs at interfaces and on surfaces where the numbers of hosts are much greater due to concentration effects.

5.4 Population studies

A number of workers have used chemostat cultures to study the effect of prolonged incubation of phage with bacteria. Paynter and Bungay (1969) observed that following addition of phage T2 to chemostat populations of *Escherichia coli* B the numbers of T2-resistant cells increased dramatically. However, T2-sensitive cells persisted as a significant proportion of the total number of organisms although they underwent periodic fluctuations in numbers which were followed by similar changes in phage titre. Horne (1971) studied the interaction of phages T3 and T4 with *E. coli* B in chemostat culture. He observed that infection was followed by a period in which phage and bacterial numbers fluctuated erratically. Eventually the numbers of bacteria stabilised at the same level as those found in comparable phage-free chemostats. Fluctuations continued in the numbers of phage but they, too,

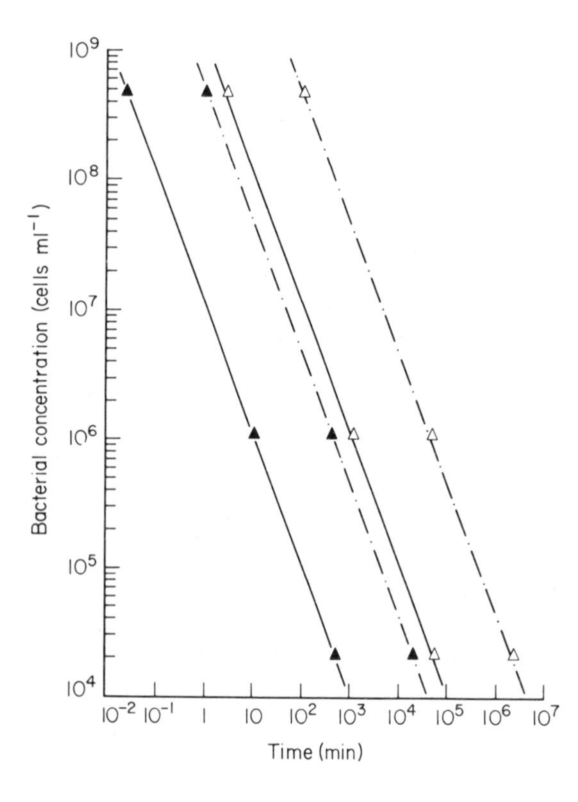

Figure 1. The effect of host cell concentration on the time required for adsorption of a given proportion of a phage population. The lines were constructed according to the equation given in the text. They show the time required for adsorption of 1% (——) and 50% (·—·—·) of the phage using arbitrary k values of 1×10^{-9} ml min^{-1} (▲) and 1×10^{-11} ml min^{-1} (△).

eventually stabilised although the phage had mutated such that they exhibited an increased latent period and a decreased burst size.

Levin and his co-workers (Chao *et al.* 1977; Levin *et al.* 1977) have modelled the population dynamics of phage-bacteria interactions and experimentally have tested these models by monitoring the events which followed infection of chemostat cultures of *Escherichia coli* with phages T2 and T7. In each case they observed a steady state in which phage-sensitive organisms and phage coexisted. Much more interesting, however, was the observation that the final number of host organisms was only 1% of that which an identical phage-free medium could support.

One criticism of all these population studies is that they were done with levels of bacteria and/or phage which were several orders of magnitude greater than those normally encountered in the environment. As already pointed out, the proportion of phages that attach to susceptible host organisms in a given time period is strongly influenced by the concentration of host organisms. For this reason it would be a useful exercise to use the equations for population dynamics derived by Levin and his colleagues to calculate the outcome of phage infections when there are only low levels ($< 1 \times 10^6$ organisms ml^{-1}) of bacteria and phage.

5.5 Factors affecting virus multiplication

Any factor which influences phage multiplication will concomitantly affect the rate at which new phage particles are produced. The most obvious environmental parameters are the specific growth rate of the host, temperature, pH, the degree of anaerobiosis and salt concentration.

5.5.1 Growth Rate of the Host

Since a phage is dependent on its host cell for the provision of all precursor molecules it is obvious that the growth rate of the host could influence the length of the eclipse and rise periods and the burst size. However, at the present moment no representative data are available. This is unfortunate because most studies of the kinetics of phage multiplication have been done with cells growing near their maximum specific growth rate: for example with *Escherichia coli* growing with a specific growth rate of 2.0 to 1.0 h^{-1}. In nature it is unlikely that such growth rates are ever achieved, for even in the human gut, enteric bacteria have a specific growth rate of 0.09 to 0.07 h^{-1} (Meynell 1959).

The situation with temperate phages may be more complex. On infection of a cell a choice has to be made between the lysogenic and lytic responses. The factor that governs this choice is the cellular level of cyclic AMP which is controlled by a number of physiological factors including growth rate (Hong *et al.* 1971). Furthermore, in chemostat culture certain nutrient limitations can result in induction of lysogens (Taylor & Sleytr 1977).

5.5.2 Temperature

As well as its effect on the growth rate of the host, temperature can affect phage infection and multiplication *per se*. In considering the effects of temperature it is necessary to take into account the temperature of the environment and the upper and lower temperatures for growth of the host. For example, *Escherichia coli* can grow in the range 10°C to 46°C but the temperature of aquatic habitats in Britain seldom rises above 20°C. Seeley and Primrose (1980) have shown that coliphages fall into three physiological classes (Table 1) depending upon the effect of temperature on their efficiency of plating. These three classes are called low temperature (LT), mid-temperature (MT) and high temperature (HT). Only LT and MT phages are capable of multiplication at environmental temperatures and thus HT phages cannot be active components of the ecosystem. However, the ratio of plaque-forming units (p.f.u.) at 20°C and 30°C may provide a measure of faecal pollution, for the origin of HT phages is the gut of warm-blooded animals.

Table 1. The three physiological classes of coliphage isolated from aquatic habitats.

Class	Minimum plating temperature °C	Maximum plating temperature °C
Low temperature (LT)	< 15	30
Mid-temperature (MT)	< 15	45
High-temperature (HT)	25	45

Relatively little is known about the effects of temperature on phage multiplication. Temperature could affect one or more of the following stages: adsorption, eclipse, penetration and multiplication. Bacteriophages λ, ε^{15} and ϕX174 are typical HT phages but low temperatures affect their host cell interaction in different ways. With phages λ and ε^{15}, decreasing temperatures decrease the proportion of phage particles which inject their DNA (Mackay & Bode 1976; McConnell *et al.* 1979), whereas with phage ϕX174 low temperatures tures inhibit eclipse and the late stages of replication (Newbold & Sinsheimer 1970; Espejo & Sinsheimer 1976). Seeley and Primrose (1980) tested a number of HT and LT phages and found with some that adsorption was blocked at the restrictive temperature whereas with others it was some stage in replication. It should be noted that if phage DNA enters a cell at a restrictive temperature it may well kill the cell even though a productive infection does not ensue. Generally, within the range of permissive temperatures for phage multiplication, temperature influences the length of the latent period but, surprisingly, does not always affect the burst size (Ellis & Delbruck 1939; Seeley & Primrose 1980).

5.5.3 Salt Effects

Most bacteriophages require divalent cations for adsorption and this requirement has been well documented. However, there are no data on the relationship between levels of divalent cations in a particular habitat and the divalent cation requirements of phages isolated from it.

Numerous studies have established that bacteriophages are present in marine and estuarine waters (Zachary 1974). However, few attempts have been made to estimate the distribution, abundance and significance of bacteriophages in this ecosystem. Zachary (1974) has shown that bacteriophages active against the marine organisms *Beneckea natriegens* are widely distributed in coastal salt marshes. These phages are limited to marine waters and in estuaries their distribution appears to be salinity-dependent. Laboratory studies have shown that these *Beneckea* sp. phages require sodium ions for multiplication (Zachary 1976) and this explains, in part, their distribution in estuaries.

Torsvik and Dundas (1980) found that cultures of *Halobacterium salinarium* are persistently infected with a virulent and halophilic phage. The nature of this phage infection depends on the salt concentration in the medium, changing from lytic to persistent as the salt concentration increased from 17.5% to 30% (w/v) NaCl.

5.5.4 Anaerobiosis

McConnell and Wright (1975) showed that anaerobic conditions do not affect the ability of phages to plate on facultative anaerobes but no information was presented on the effects of anaerobiosis on burst size or the length of the latent period. Zachary (1978) studied two phages of *Beneckea natriegens* and found that replication under anaerobic conditions resulted in longer latent periods. One of the phages also had a reduced burst size but the other had an increased burst size resulting in a rate of phage production nearly equal to that observed under aerobic conditions.

5.5.5 pH

Virtually nothing is known about the effect of pH on the multiplication of bacteriophages despite the fact that different natural habitats have widely different pH values. Of particular significance is the fact that the pH of many rivers and lakes fluctuates during the year from pH 5.0 to pH 9.0; the higher values occurring in the summer months when the temperature is highest.

One effect of pH on intact virus particles is to alter the charge on the protein capsid and this could influence the rate of adsorption. However, the pH of the environment can have other effects; Fujimura and Kaesberg (1962) showed that eclipse of bacteriophage ϕX174 does not occur below pH 6.0.

5.5.6 Light

Light is not known, nor is it expected, to affect phage multiplication in chemoheterotrophs. However, it can affect cyanophage multiplication in cyanobacteria. The available evidence (Sherman & Brown 1979) shows that the dependence of phage growth on photosynthesis strongly parallels the host metabolic capabilities. Facultative heterotrophs allow phage development under conditions where photosynthesis is severely impaired while the obligate photoautotrophs must have a functioning photosynthetic apparatus for phage to replicate.

5.5.7 Persistence

It should be borne in mind that in laboratory studies it is usual to select the optimal conditions for phage multiplication, whereas such optimal conditions may never be encountered in the natural environment. Thus the classical observation of lysis of liquid cultures may be a laboratory artefact whereas persistent infections may be the rule in nature. This would prevent the bacterial populations from being eliminated but would ensure perpetuation of the phages. To an animal virologist the term 'persistent infection' has a very precise meaning and the only phage analogy would be the filamentous male specific phages which are released from infected cells without cell lysis. Here we are using the term in a different way. Laboratory studies have indicated a number of reasons why phage may establish a persistent infection and these are detailed below.

(1) There exist only a few receptor sites per bacterium and therefore a fixed but low probability of infection per cell. Thus some phages are made but at the expense of only a fraction of the cells. This is the case for phage T7 infection of *Shigella dysenteriae* (Li *et al.* 1961). A number of bacteria, such as *Caulobacter* spp., have an obligate dimorphic life cycle. Many of the phages which infect these species only attach to a cell structure formed at one point in the life cycle and could give rise to persistent infections. Even with bacteria which do not exhibit a characteristic morphogenesis it is conceivable that environmental conditions could limit the number of phage receptors.

(2) A phage-sensitive bacterium gives rise to a resistant mutant at a high frequency (1×10^{-3} to 1×10^{-4} per bacterium per generation). Baker *et al.* (1949) have cited an example in which *Shigella sonnei* II mutates to *S. sonnei* I with a concomitant change in sensitivity to phages T3, T4 and T7. A similar situation exists with *Hyphomicrobium* and *Rhodomicrobium* spp. These two organisms have two vegetative life cycles and the proportion of the organisms in each cycle depends on the environmental conditions (Dow & France 1980; C. S. Dow personal communication). Phages isolated so far infect organisms from only one of the cycles (C. S. Dow personal communication).

(3) Early infected cells produce enough receptor-destroying enzyme to render much of the population resistant so that the population is always a mixture of sensitive (phage-producing) and pheno-

typically resistant cells. Li *et al.* (1961) have described such a situation following phage T7 infection of *Shigella dysenteriae*.

(4) A bacterial culture is lysogenic for a temperate phage and therefore immune to superinfection. However, the prophage mutates at a low frequency to a virulent form which can lyse lysogens. The replication of the virulent phage in the lysogens can give rise to a persistent infection as described by Coetzee and Hawtrey (1962) and Baess (1971).

5.6 Genetic interactions

Many bacteriophages are temperate, i.e. they can lysogenise their host. Lysogeny can be advantageous to both the phage and the host. The phage has a means of survival when conditions are not favourable for its replication and the host may benefit by the acquisition of a new phenotype. This latter phenomenon is known as lysogenic conversion and examples of phenotypes acquired include new surface antigens and toxin production. A more complete list is given by Barksdale and Arden (1974). In addition, many temperate phages can acquire transposons, e.g. P1Cm (Kondo & Mitsuhashi 1964) and this extends the range of new phenotypes to include drug resistance, heavy metal resistance and carbohydrate utilisation (Bukhari *et al.* 1977). Lysogeny does not always result in the acquisition of a directly selectable phenotype. Thus λ, Mu, P2 and P1 lysogens all reproduce more rapidly than the corresponding non-lysogens during aerobic growth in glucose-limited chemostats (Edlin *et al.* 1977). However, λ lysogens are less fit than non-lysogens when growth becomes anaerobic (Lin *et al.* 1977).

As well as lysogenic conversion, which is nothing more than specialised transduction by a plaque-forming phage (Primrose 1976), many temperate phages have been shown to mediate generalised transduction. Although the frequency of transduction may be as low as 1×10^{-6} per phage particle, it is still of considerable evolutionary and ecological importance. It has been shown that some virulent phages, e.g. phages T1 and T4 (Drexler 1970; Wilson *et al.* 1979), can also mediate transduction and it is possible that this is a general property of phages with a terminally redundant genome. The problem with detecting transduction with virulent phages is that transduced cells are lysed by wild-type phage released after infection of neighbouring cells. To detect transduction by such virulent phages it is essential to use conditional phage mutants whose replication in the recipient cell is blocked prior to any host-killing step. In the natural environment two factors may favour transduction by virulent phages.

(1) Environmental conditions may favour persistent rather than wide-scale lytic infections.

(2) Many wild-type phages are naturally temperature- or cold-sensitive (see section 5.5.2, p. 69). It is worth noting at this stage that persistent phage infection can also lead to phenotypic changes in populations (Li *et al.* 1961; Coetzee & Hawtrey 1962).

5.7 Virus inactivation

An important aspect of the ecology of viruses is their destruction or inactivation in the natural environment. Loss in this way may be the result of either physical or biological factors or a combination of both. Unfortunately most of the studies have been laboratory based and the results cannot always be extrapolated to a natural situation. It should be noted that virus aggregation (Young & Sharp 1977) and association with suspended solids lead to a reduced apparent infectivity. But such behaviour is generally reversible: phages adsorbed to particles have been shown to retain their infectivity (Moore *et al.* 1975) and changes in the composition of the suspending medium can result in the elution of phages from solids or their disaggregation (Schaub *et al.* 1974).

Physical parameters which may directly affect virus titres are temperature (Niemi 1976), sunlight, exposure to air/water interfaces (Trouwborst *et al.* 1974), the chemical nature of the water (Mitchell & Jannasch 1969) and virus particle concentration (Tyler & Beswick 1976). Of these, it appears that water type and temperature have the greatest influence on virus survival times and that increasing temperatures result in decreased survival.

Mitchell and Jannasch (1969) found that following phage ϕX174 addition to sea-water, a specific group of antagonistic microorganisms developed which inactivated the phage. By contrast, killed microbial cells protected the virus from chemical inactivation. Similar protective effects were exerted on phage MS2 when it was associated with inorganic clays (Stagg *et al.* 1977) indicating that viruses temporarily inactivated by association with solids may be protected from other forms of inactivation. Such results demonstrate the complex interactions between the different factors which control virus survival and explain the conflicting results obtained by different groups of workers (Katzenelson 1978).

5.8 Summary

From the foregoing account it should be clear that we still know very little about the ecology of viruses. The reason for this dearth of knowledge is not any difficulty associated with its collection but the fact that until recently phage were only playthings for molecular biologists. Hopefully this situation will right itself in the near future and we hope that the methodology provided in the following section will facilitate matters.

Experimental

5.9 Recovery of viruses from water

An essential component of any study of virus ecology is virus enumeration and this demands efficient recovery of viruses from the habitat being sampled. The simplest method is to directly plate water samples but usually this is not practicable because the viruses are present in low numbers. For example, raw sewage may contain 1×10^5 p.f.u. ml^{-1} of coliphages (on any one *Escherichia coli* indicator) and effluent from a sewage treatment plant about 10 times less (Scarpino 1978). However, in aquatic habitats which are not polluted with sewage effluent, the numbers of coliphages are much less. This, of course, does not imply either that phages for other hosts are present in lower numbers than coliphages or that high levels of phages do not occur, e.g. in the vicinity of submerged surfaces or in sediments.

As a consequence of the low numbers of viruses present in most aquatic habitats it is necessary to concentrate viruses from large volumes of water (10 to 500 l) before they can be detected. A number of concentration methods are available and most of them have exploited various physico-chemical properties of the viruses, e.g. physical adsorption, precipitation, phase partitioning, sedimentation or filtration. Virus recovery by these methods depends greatly on water quality and is influenced by such factors as particulates, organics and salts. Excellent reviews of virus concentration methodology have been presented by Hill *et al.* (1971), Shuval and Katzenelson (1972), Sobsey (1976) and Gerba *et al.* (1978).

Ideally, the method used for virus concentration should be rapid, cheap, efficient (i.e. give high recoveries), capable of handling large water volumes and be suitable for field use. Only filter adsorption/elution methodology fulfils all these requirements and this is discussed in more detail below. Where it is intended that only small volumes of water (< 2 l) are to be processed then the hydroxylapatite method (Primrose & Day 1977) can be used. In this method water is passed through a bed of hydroxylapatite held in a Buchner funnel and the virus eluted in a small volume of 0.8M sodium phosphate buffer. Recoveries with this method are very good but it is not suitable for large-scale use because of the low flow rate of water through the hydroxylapatite bed.

All methods in current use for concentrating viruses from large volumes of water rely on membrane chromatography. The development of these methods is almost entirely due to Melnick and his co-workers and has been reviewed by Wallis *et al.* (1979). The basic method involves adsorption of the viruses on to a microporous membrane filter and their elution in a small volume. However, Melnick's group were interested in the recovery of human enteroviruses, primarily from drinking water and their methods are not applicable to ecological studies on all virus types. There are two basic problems, both of which stem from the fact that enteroviruses are much more resistant to inactivation than most other viruses.

(1) The eluent used with enteroviruses is glycine-buffer pH 11.5 but some bacteriophages are inactivated at pH 9.5 (Seeley & Primrose 1979).

(2) Until recently the only suitable adsorbing filters which were available were negatively charged at neutral pH as are virus particles. Thus it was essential to reduce the pH of the test water to give the virus particles a net positive charge to permit binding to the filters which still retained a net negative charge. This pH alteration can also inactivate viruses: for example, the lipid-containing phages PM2 and $\phi6$ (K. B. Logan unpublished observations). Positively charged filters (Zeta-plus; Flowtech Fluid Handling Ltd; Deacon Way, Reading, Berks.) are now available (Sobsey & Jones 1979) and can be used to adsorb viruses at near neutral pH values (Logan *et al.* 1980, 1981).

The method we use to concentrate viruses (Logan *et al.* 1981) consists of:

(1) pre-filtration of water through a series of non-absorbing filters;

(2) adjustment of the water to pH 5.5 to 6.0;

(3) passage through a Zeta-plus filter;

(4) elution of the bound virus with 1% (w/v) beef extract, pH 9.0.

Natural waters may contain large amounts of negatively charged organic material which competes with viruses for sites on the filter. Since the iso-electric point of Zeta-plus filters is between pH 5.0 and 6.0 (Sobsey & Jones 1979) maximum capacity of the filters is achieved by adjusting the water to this pH range. Using a 500 mm diameter Zeta-plus filter 300 l of river or lake water can be processed and the bound virus eluted in less than 2 h. The eluent is 1 l of beef extract and can be further concentrated to 50 to 100 ml in 60 mins using a hollow fibre ultrafiltration device (Logan *et al.* 1981). With this method, phage recoveries greater than 60% are obtained routinely.

Since Zeta-plus filters are particularly useful for concentrating viruses from rivers and lakes it is unfortunate that they cannot be used for the concentration of viruses from sea-water. The reason for this is that in the presence of high concentrations of cationic salts viruses become less electronegative and thus do not adsorb to the positively charged filter. Two alternatives are available:

(1) to concentrate the viruses by adsorption on to negatively charged filters at low pH (Farrah *et al.* 1977) although there is always the attendant risk of phage inactivation (see p. 72);

(2) the viruses might be concentrated in a large-scale, hollow fibre ultrafiltration device. These are expensive, require careful handling and still remain to be tested for this purpose.

One problem associated with the concentration of viruses from water samples is the loss of viruses adsorbed to particulate matter suspended in the water. In most virus concentration systems this particulate matter is filtered out before the adsorption step. However, in most instances it is not too difficult to devise methods of recovering this particulate material and then the problem lies in eluting the bound virus particles (see next section).

5.10 Recovery of viruses from solids

The lack of good methods for the extraction of viruses from solids represents a major obstacle to the study of phage ecology in terrestrial habitats. The water industry has developed methods for the recovery of animal viruses from sludges and sediments but, by and large, these are not suitable for the recovery of phages. S. Lanning (personal communication) has made an intensive study of the factors affecting phage recovery from soils and has devised a suitable method. She recommends that the soil is extracted with an equal volume of nutrient broth, pH 9.0 containing 1.0 g egg albumin l^{-1} by shaking at minimum speed for 30 min on a wrist-action flask shaker. Use of orbital shakers or magnetic stirrers results in a reduced recovery. After the extraction step the solids are left overnight at 4°C to settle and the supernate is removed and titred. If the solids are removed by centrifugation or filtration this can reduce the titre by up to 90%.

5.11 Enumeration

Bacteriophages and cyanophages are enumerated by the plaque assay method but this technique has not been adapted for use with fungi, protozoa or eukaryotic algae for which a suitable assay method is lacking. Although the plaque assay technique has been described many times before, most accounts fail to describe the numerous pitfalls which can be encountered.

5.11.1 The Standard Plaque Assay

The usual procedure is to melt some nutrient medium which has been solidified with 0.8% (w/v) agar (top agar) and to dispense it in 2.0 to 2.5 ml portions into tubes held at 45°C. When the top agar has cooled to this holding temperature 0.1 to 0.3 ml of a suspension of host cells (indicator) containing 1×10^8 cells ml^{-1} are added to each tube. Finally 0.1 ml of the phage-containing sample is added to the tube. Immediately the contents are mixed (either on a vortex mixer or by vigorously rolling the tube between the palms of the hands) and poured over the surface of a Petri dish of nutrient medium solidified with 1.2% (w/v) agar (bottom agar). After the top agar has set the plates are incubated until plaques are clearly visible. This can take 4 h to 7 d depending on the growth rate of the host.

There are a number of common errors which are made with the plaque assay technique. For example, if the host cells are held at 45°C for more than 10 min, many may die resulting in the formation of poor lawns. Poor lawns or an unequal distribution of phages can result from inadequate mixing of the tube contents. Another reason for an uneven distribution of the bacterial lawn and the plaques is the preparation of the bottom agar plates on a surface which is not level. When the top agar is added to plates prepared in this way the agar runs to one half of the plate. Perhaps the commonest error is to waste time between removing the tube of seeded agar from the 45°C bath and pouring the contents over the surface of the bottom agar. Some of the other problems which can be encountered are more subtle and these are outlined below.

Age of the bottom agar plates

Freshly prepared plates have to be dried otherwise the surface of the solidified top agar becomes wet during incubation and the plaques run together. However, with old bottom agar plates, the top agar dries out during incubation and the increased viscosity of the top agar results in decreased plaque size or lack of plaque formation altogether. For this reason it is best to use plates within 2 d of preparation. Some phages always give small plaques, even under optimum conditions, and this can make their enumeration difficult. If the host cells are facultative anaerobes, the plaque size can be increased by incubating the plates anaerobically (McConnell & Wright 1975).

Incubation temperature

The incubation temperature can be critical and some examples should make this clear. As mentioned in section 5.2 (p. 67) three classes of coliphage have been identified based on the effect of temperature on their efficiency of plating. Thus when concentrates from clean water, i.e. little or no faecal pollution, are plated the number of plaques obtained is 200 to 2000 times greater at 15°C than at 37°C. The converse is true with concentrates from faecally polluted waters (Seeley & Primrose 1980).

It should always be borne in mind that the temperature range of a phage may not mirror that of its host. For this reason it is always best to incubate assay plates at a temperature which is normally found in the environment being sampled. For example, in laboratory culture *Rhodopseudomonas blastica* grows rapidly at 30°C but many of the phages for this organism do not plaque at this temperature. Instead the assay has to be performed at 20°C, a temperature at which the host cells grow much more slowly with the consequence that the bacterial lawns take up to 5 days to develop (K. Eckersley, personal communication).

The indicator organism

With many of the common laboratory bacteria, e.g. *Escherichia coli* and *Pseudomonas* sp., the condition of the host cells used for the assay is not critical. However, with organisms which grow much more slowly, e.g. *Rhizobium* sp., it is essential to use a fresh culture of the correct density.

As mentioned earlier, there is no bacterium which is a universal phage indicator. Consequently the indicators used are selected arbitrarily. Even within one species some strains may consistently give 10 to 100 times more phages with a virus concentrate but the only way to detect them is to do a blind screening of all available strains. There are at least two reasons for this variability between strains.

(1) Although some phages adsorb to surface appendages such as pili or flagella, the majority attach to wall components, principally lipopolysaccharide (Lindbergh 1977). The efficiency of plating of virus concentrates can be significantly affected by the presence of excess capsular material or the degree of 'capping' of lipopolysaccharide cores with O repeat units.

(2) Many bacteria possess host-restriction and modification systems which act as primitive defence systems (Primrose & Dimmock 1980). When the phage DNA enters the cell it may be degraded by a restriction endonuclease unless it is first modified, usually by methylation. Thus, use of a cell which lacks any restriction system could result in increased titres.

An important factor to be considered is the morphology of the host cells. For example, some genera, such as *Rhodomicrobium* and *Hyphomicrobium* have two vegetative life cycles (Dow & France 1980; C. S. Dow, personal communication) and the life cycle that predominates depends on environmental conditions, e.g. CO_2 tension. It is conceivable that the cycle selected during laboratory cultivation may differ from that operating in the natural environment. Since phages for these organisms may plate only on cells in one stage of the life cycle (C. S. Dow, personal communication), the choice of incubation conditions can be critical.

Again, walls of *Bacillus* sp. contain teichoic acids and these can act as receptors for bacteriophages (Archibald 1976). However, under phosphorus limitation the cell synthesises teichuronic acids instead of teichoic acids and these modified walls may no longer be susceptible to the same phages. Since many aquatic habitats are phosphorus-limited this is an important consideration.

Handling virus concentrates

Care needs to be taken when handling virus concentrates. Most methods for concentrating phages from water will also concentrate animal viruses. Thus, when the source material is faecally polluted water there is always a chance that the concentrates contain human pathogenic viruses, e.g. hepatitis viruses. All concentrates should be treated carefully and autoclaved before disposal.

Most concentrates contain significant numbers of bacteria and these can interfere with the plaque assay, particularly where the assay takes more than one day. Although it is possible to reduce the bacterial numbers by the addition of chloroform, this is not recommended. Not only are the filamentous phages and some lipid-containing phages sensitive to chloroform, but so also are a number of tailed phages. An alternative solution is to use indicator organisms which are resistant to antibiotics which can be included in the plating medium. The best antibiotics to use are rifampicin or nalidixic acid since resistance to them is not plasmid borne. Where a large number of different host organisms are being used it may be inconvenient to make them all rifampicin or nalidixic-acid-resistant. In this instance it may be worthwhile to include an antibiotic which is selective against Gram-negative bacteria (e.g. polymyxin B) or against Gram-positive bacteria (e.g. ionophores). Yet another approach is to filter the concentrate through a 0.45 μm membrane filter. This does not remove all the bacteria but the use of a smaller porosity filter is not practicable because of the viscosity of the concentrates.

5.12 Isolation and enumeration of phages

A number of methods can be used to isolate and enumerate selected groups of phages and a few examples of each are outlined below. In all cases, however, it is assumed that a virus concentrate is available. Full details of all these techniques are given by Primrose *et al.* (1982).

5.12.1 Use of Mixed Indicators

The classic method of Hershey and Rotman (1949) employs mixed indicators for genetic analysis of phage T4 mutants. This method can be adapted for use in an ecological context, e.g. for the isolation and enumeration of plasmid-specific phages and broad-host-range phages. When a concentrate is assayed on a male (F^+ or Hfr) strain of *Escherichia coli*, a certain proportion of the phage which form plaques will be male-specific, i.e. they will only infect cells carrying the sex factor F. This proportion can be enumerated by using a mixed indicator comprised of equal proportions of male *E. coli* and male *Salmonella* sp. cells. Any clear plaques which develop are either due to male specific phages or to phages which plate on both *E. coli* and *Salmonella* sp. The latter type can be enumerated by plating the same concentrate on a mixture of female strains of the same two hosts.

It is even easier to isolate phages specific for the P group plasmids, such as R68.45. These plasmids can be transferred to a wide range of Gram-negative bacteria (Datta *et al.* 1971) and the mixed indicator in this case is plasmid-containing *Salmonella* sp. and *Pseudomonas aeruginosa*. Any clear plaques which develop are due to plasmid-specific phages, for no other kind of phage can plate on both hosts.

5.12.2 Pre-adsorption

To enumerate or isolate phages which adsorb to a particular cell surface component, the concentrate must first be mixed with a dense suspension of an isogenic bacterium which lacks the component in question. In this way all the phage which adsorb to other components are removed from the concentrate. For example, using this technique Verhoef *et al.* (1977) were able to isolate a phage which adsorbed to protein C, the major outer membrane protein of *Escherichia coli.*

5.12.3 Isopycnic Gradient Centrifugation

Most phages have a characteristic buoyant density which is a direct measure of their protein to nucleic acid ratio, and which can be used to facilitate isolation and enumeration. Thus the filamentous coliphages have about 10% nucleic acid and a buoyant density in caesium chloride of 1.3 g ml^{-1}. The ϕX174-like phages have 25% nucleic acid and a buoyant density of 1.41 g ml^{-1}. Finally, the majority of tailed phages have about 50% nucleic acid and a buoyant density 1.5 g ml^{-1}. Exceptions are provided by the lipid-containing phages whose buoyant density is usually less than 1.3 g ml^{-1} because of the lipid component. In addition, the buoyant density of these phages may be variable (Seeley 1980). Figure 2 shows the result obtained when a virus concentrate was mixed with caesium chloride and centrifuged to equilibrium. The fractions at the top of the gradient contain mostly lipid-containing phages and probably a few filamentous male-specific phages. The large peak of phages at 1.5 g ml^{-1} contains the tailed phages.

5.12.4 Rate-Zonal Centrifugation

Macromolecules and viruses sediment in a centrifugal field at a rate which is proportional to their size and shape. Rate-zonal centrifugation can be used to isolate specific groups of viruses providing their sedimentation coefficient (*S* value) is known. An example of the use of this technique is given by Godson (1974) who isolated a whole series of new ϕX174-like phages.

5.12.5 Combination of Techniques

A number of techniques can be combined in order to make them more selective. For example, the filamentous male-specific and the spherical male-specific phages can be distinguished from each other by the greater heat-resistance of the former

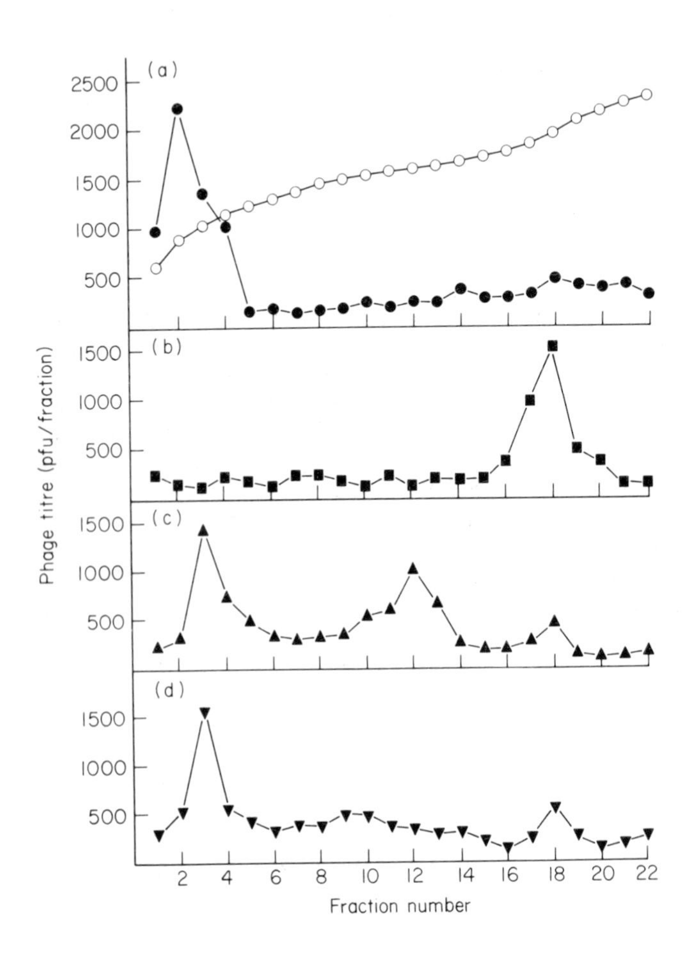

Figure 2. Separation of the different structural types of bacteriophage on the basis of differing buoyant densities. Phage in 200 l natural river water were concentrated to 80 ml by the method of Logan *et al.* (1980, 1981). The concentrate was mixed with solid caesium chloride to a final density of 1.39 g ml^{-1} and centrifuged for 40 h at 4°C at 110 000 g in an MSE 6 × 14 ml swing-out rotor. Fractions were collected by upward displacement and assayed on: (a) mixed indicator of plasmid RP1-containing *Salmonella* sp. and *Pseudomonas* sp.; (b) *Escherichia coli* K12; (c) mixed indicator of male derivatives of *E. coli* and *Salmonella typhimurium*; (d) as (c) but assayed in the presence of 100 μg ml^{-1} RNAase. Note that in (b) the majority of the *E. coli* K12 specific phages have the buoyant density characteristic of tailed phages, in (a) the RP1-specific phages have a low buoyant density due to their lipid content, in (c) the filamentous DNA and spherical RNA male-specific phages are readily separable, and in (d) RNAase reduces the height of the peak at 1.4 g ml^{-1} indicating the presence of RNA-containing phages.

and the RNAase sensitivity of the latter. A combination of isopycnic centrifugation with mixed indicators is a particularly powerful technique (Fig. 2).

The lipid-containing phages which are easily identified in caesium chloride gradients can be further subdivided into those with an outer lipid coat and those with an inner lipid coat. The former are inactivated by monoglycerides, such as monopalmitolein, whereas the latter are not (Primrose *et al.* 1982).

5.13 Identification and characterisation of viruses

Although animal and plant viruses fall into reasonably well-defined groups this is not true of the bacteriophages—despite what publications of the International Committee for the Nomenclature of Viruses (ICNV) may suggest. The most common morphological group of bacteriophages are those with a head and tail but examination of published electron micrographs shows that almost every conceivable variation on this structure can be encountered (Ackermann 1973). At the present moment it is not very difficult to assign a filamentous or icoschedral bacteriophage to its correct family but this is not true if it is a tailed phage. Although detailed physico-chemical characterisation of a phage might assist in its identification, such characterisation is a lengthy process and unless there is good reason for doing it, it should not be attempted. References to the appropriate methods are given by Ackermann *et al.* (1978).

From an ecological point of view the most interesting vital properties are morphology (to assist recognition), host range and the effect of environmental factors on biological parameters, such as the latent period and burst size (to determine adaptation of the phages to the environment from which they were obtained). If phages of similar morphology are isolated from different habitats, comparison, either by serology or restriction endonuclease mapping of phage DNA, may be worthwhile. Suitable descriptions of serological techniques are given by Adams (1959) and Eisenstark (1967) and some necessary precautions are listed by Ackermann *et al.* (1978). A good example of the use of restriction endonuclease mapping is that of Godson and Roberts (1976).

5.13.1 Host Range

Determining the host range of a bacteriophage is a relatively simple matter but frequently it is done incorrectly. The commonest method is to apply drops of phage suspension to a lawn of bacteria and to look for zones of lysis after incubation. This method is not recommended for it takes no account of the effects of host restriction and modification systems or of the selection of host range mutants. In addition, many phages can kill cells without multiplying in them: for example, phage T7 kills male cells of *Escherichia coli* without replicating in them but replicates normally in female *E. coli* (Morrison & Malamy 1971). Again, many phages have capsule depolymerases associated with their outer surface. In spot tests such phages can cause zones of turbid lysis which are due to depolymerisation of capsular material but which can be mistaken for lysogeny.

The correct way to determine the host range of a phage is to start with a stock phage preparation with a titre $> 1 \times 10^9$ p.f.u. ml^{-1}. Dilutions of this stock are assayed on all the hosts to be tested for sensitivity. Following incubation the efficiency of plating (e.o.p.) is calculated as follows:

$$\text{e.o.p.} = \frac{\text{titre on test host}}{\text{titre on usual plating host}}$$

E.o.p. values from 1×10^1 to 1×10^{-1} are not uncommon and simply reflect differences in the chemical structure, number and extent of masking of cell wall receptors (Lindbergh 1977). Values of the e.o.p. from 1×10^{-1} to 1×10^{-5} indicate that the test cell is sensitive to the phage but possesses restriction and modification systems. Absolute proof that a cell carries a host restriction system requires a considerable amount of work and the reader is referred to Clowes and Hayes (1968) for a detailed example of the methodology. When the e.o.p. is less than 1×10^{-6} then the cell is resistant and any plaques which are observed contain spontaneous host range mutants. Such mutants on further purification should plaque with similar efficiency on the original two host bacteria.

Where a large number of bacteria have to be tested for susceptibility to a range of phages it is not logistically feasible to perform the large number of plaque assays which the above method requires. A suitable alternative is to make phage dilutions as before and to put small drops of each dilution on to a single lawn of each bacterium. In effect this is the virological equivalent of the Miles and Misra method for counting bacteria. When using this method it is essential that the plates of bottom agar have been well-dried otherwise the spots run together during handling of the plates.

When determining host ranges it should be borne in mind that the absence of plaque formation or clearing in a spot test, does not necessarily indicate resistance. For example, the promiscuous plasmids RP1, RP4 and R68.45 can be transferred to a wide range of Gram-negative bacteria. In conventional phage sensitivity tests, such as those described above, not all plasmid-containing cells appear sensitive to the plasmid-specific phage PRR1 (Olsen & Thomas 1973). However, after the phage is added to plasmid-containing cells an increase in titre can be detected indicating that the cells are phage-sensitive.

5.13.2 Electron Microscopy

Electron microscopy can give a considerable amount of information about the phage being studied. However, the use of inappropriate methods of sample preparation can give results which are misleading. Before carrying out a detailed analysis of phage morphology the reader is advised to consult Ackermann *et al.* (1978). High resolution electron microscopy demands highly purified phage preparations but for preliminary analysis it is possible to examine phages isolated directly from a plaque. Suitable methods for electron microscopy of phage from plaques are not easy to find in the published literature and the method used in this laboratory is detailed below.

A small amount of agar from the centre of a plaque is removed with a Pasteur pipette and transferred to the bottom of a conical microcentrifuge tube. Sterile distilled water (10 μl) is added to the tube which is stored at 4°C for 3 to 4 h to allow the phage to elute. The contents of the tube are vortexed and the agar pelleted by centrifugation. Five μl of the supernate are spotted onto a Formvar-coated grid and after one min are removed by gently touching with filter paper. Immediately thereafter 5 μl of 1% (w/v) phosphotungstic acid or 1% (w/v) ammonium molybdate is added and immediately removed with filter paper. After air-drying, the grids can be examined in the electron microscope. Note that if phosphotungstic acid is used as the stain it must be freshly prepared and adjusted to pH 7.0 with freshly prepared KOH (not NaOH).

5.14 Factors affecting phage productivity

The most important facet of any study on phage ecology is undoubtedly the influence of environmental factors on phage productivity. However, before beginning any studies it should be borne in mind that the magnitude of, and changes in, these factors should approximate to those found in the natural environment. The simplest, but least informative, measure of phage productivity is the e.o.p. (see p. 77). For example, it is possible to measure the e.o.p. at different temperatures with the same medium or at the same temperature with media containing different concentrations of divalent cations or buffered at different pH values. The effect of growth rate on e.o.p. can be determined by assaying the phage on cells growing on the same basal medium but containing different carbon sources, e.g. glucose, succinate or acetate. Using these three carbon sources it is possible to vary the generation time of *Escherichia coli* from 40 to 160 min (Lark 1966).

A more informative way of assessing the influence of environmental factors is to determine their effect on the latent period and average burst size of the phage. In order to do this it is essential to:

(1) use a multiplicity of infection (m.o.i.) less than 0.5;

(2) have a synchronous infection;

(3) have good adsorption of the phage to the host cells.

A low m.o.i. minimises the number of cells which are multiply infected and this number can be calculated from the Poisson distribution. Thus, when an m.o.i. of 0.5 is used, 10% of the phage-infected cells receive more than one particle but this drops to 1% when the m.o.i. is reduced to 0.1. An additional advantage of using low m.o.i. values is that it is easier to detect any changes in the phage population which occur at the end of the latent period.

There are a number of ways of synchronising phage infections but the commonest is to temporarily poison the cells with cyanide. The usual procedure is to add 1 M sodium or potassium cyanide to the cell suspension to give a final concentration of 0.01 M. After a few minutes the phage are added at low multiplicity. The cyanide is removed 10 to 30 min after by diluting the suspension 100 times into fresh, prewarmed medium. The major drawback to the use of cyanide is its toxicity to humans, particularly in acid solution when it is rapidly converted to volatile prussic acid. Sometimes it is possible to use chloramphenicol instead of cyanide but the precise concentration required is not always easy to determine. For example, low concentrations of chloramphenicol (30 $\mu g\ ml^{-1}$) block only the later steps of phage ϕX multiplication whereas higher levels (150 $\mu g\ ml^{-1}$) block the earliest steps (Sinsheimer *et al.* 1967). The alternative to the use of chemical inhibitors is to starve the cells in buffer for 60 to 90 min prior to adsorption.

No assessment of the latent period and burst size can be made unless the phage adsorb efficiently to the host cells under the test conditions. For this reason it is essential that the extent of adsorption be monitored. Although several methods exist the most reliable method is to measure the number of p.f.u. which survive chloroform treatment. The rationale of this method is that phage which have irreversibly adsorbed to cells cannot form a plaque if the viability of the host cell is destroyed by chloroform. Thus any chloroform-resistant plaque-forming units remaining at the end of the adsorption are due to unadsorbed phage. In practice, 0.1 ml samples are removed from the adsorption mixture, added to 10 ml broth containing 0.5 ml chloroform, vortexed for 30 s and diluted with broth before assaying. Obviously this method cannot be used with those phages which are chloroform-sensitive.

Environmental factors not only affect phage multiplication but also phage infection. In fact, two steps in infection can be recognised:

(1) the conversion of the phage to a form which cannot be eluted from the cell in a viable form (irreversible adsorption);

(2) nucleic acid penetration.

Either of these steps can be affected by environmental conditions but in practice it is not necessary to determine which one.

5.15 Genetic interactions

As outlined earlier (section 5.6, p. 71) both lysogeny and persistent infections can bring about real or apparent changes in the characteristic properties of the bacterial population. However, it is clear from the existing literature that persistent bacteriophage infections have sometimes been mistakenly equated with lysogeny and that changes found in such phage-infected populations have been called lysogenic conversions. For this reason we include here a summary of the characteristics which separate lysogeny from non-lysogenic perpetuation of phage.

Lysogeny is the perpetuation of prophages as part of the bacterial replicating system. In the change from the non-lysogenic to the lysogenic state every bacterium becomes converted to the capacity to liberate phage and to an immunity to that liberated phage. The acquisition of these two properties is essential for lysogeny. The lysogenic state is stable to repeated subculture and this provides the most useful practical feature. If a suspected lysogen is streaked on solid media for individual colonies, each colony should be a lysogen. Since the majority of single colonies arise from single bacteria deposited on the plate, most of them would be expected to liberate phage spontaneously and to be immune to that phage. If the cells were persistently infected, rather than lysogenised, many of the colonies would be expected to be phage-free and phage-sensitive. This kind of test is conclusive when persistence is due to a reduced number of phage receptors on the bacterial cell but does not suffice for those infections which do not result in host cell lysis, e.g. filamentous phage infections of male *Escherichia coli* (Hoffman-Berling & Maze 1964).

If anti-serum to the presumed temperate phage is available then it is easy to distinguish between lysogeny and persistence. Lysogenic cells are never cured of phage by growth in the presence of antiserum. Phage-free clones can be isolated after repeated subculture of persistently infected cells in antiserum-containing media (Li *et al.* 1961). Another useful test is inducibility. Some lysogens can be induced to produce phage by mitomycin C treatment (Otsuji *et al.* 1959) or irradiation with UV light (Lwoff *et al.* 1950). Inducibility is, therefore, absolute proof of lysogeny. However, two cautions are necessary:

(1) the phage which is induced must be identical to the presumed temperate phage for it is conceivable for a cell lysogenised with one phage to be persistently infected with a second phage;

(2) lack of inducibility is not proof of persistence for many genuine temperate phages, such as phage P2, are not inducible.

The detection of transduction is not always easy. Specialised transducing phages only transduce those genetic markers close to their attachment site on the host cell chromosome. If this site is not known, as will be the case for most new temperate phage isolates, it is really a matter of luck whether transduction is detected. Generalised transduction is much easier to detect. The important features here are:

(1) transducing particles for any given marker are usually present in lysates at a frequency of 1×10^{-6} to 1×10^{-8} per p.f.u.;

(2) transducing particles are only found after lytic infection of cells (Ikeda & Tomizawa 1965);

(3) it is important that the transduced cells are not killed by the wild-type particles which are present in excess.

In practice, the potential transducing phage is grown on the donor cell and a high titre lysate ($> 1 \times 10^{10}$ p.f.u. ml^{-1}) prepared. The recipient cells are infected at m.o.i. values of 1.0, 0.1 and 0.01 and plated on selective media. Where the phage is known to have an absolute requirement for divalent cations for adsorption, killing of transductants by progeny wild-type phage can be prevented by inclusion of citrate or EDTA in the medium. The choice of marker to be transduced is an important practical consideration. The marker must be stable (i.e. no revertants per 1×10^9 to 1×10^{10} cells). Either the recipient cell has to be made auxotrophic or else the donor cell is made antibiotic resistant. The latter possibility is often the easiest and a good marker is streptomycin resistance. With many bacteria streptomycin-resistant mutants are difficult to obtain ($< 1 \times 10^{-10}$ $cell^{-1}$) and thus they are good, stable mutations for transduction.

The filamentous, male-specific (F-specific) phages have been shown to transduce small bacterial plasmids (Ohsumi *et al.* 1978). Whether or not these phages can transduce any bacterial gene or just small plasmids remains to be seen. Nor is it known whether the other filamentous phages (Matthews 1979) can transduce genes. It should be noted, however, that F-specific filamentous phages are relatively common in the environment (Logan *et al.* 1980; Schluederberg *et al.* 1980).

Recommended reading

ANDERSON E. S. (1959) The relations of bacteriophages to bacterial ecology. In: *Microbial Ecology* (Eds R. E. O. Williams & C. C. Spicer), pp. 189–217. Cambridge University Press, Cambridge.

BARKSDALE L. & ARDEN S. B. (1974) Persisting bacteriophage infections, lysogeny and phage conversions. *Annual Review of Microbiology* **28**, 265–99.

BERG G., BODILY H. L., LENNETTE E. H., MELNICK J. L. & METCALF T. G. (1976) *Viruses in Water*. American Public Health Association, Washington D.C.

CAMPBELL A. (1961) Conditions for the existence of bacteriophage. *Evolution* **15**, 153–65.

HORNE M. T. (1970) Coevolution of *Escherichia coli* and bacteriophages in chemostat culture. *Science* **168**, 992–3.

PRIMROSE S. B., LOGAN K. B. & SEELEY N. D. (1981) Recovery of viruses from water: methods and applications. In: *Viruses and Waste Water Treatment* (Eds M. Butler & M. Goddard), pp. 211–34. Pergamon Press, Oxford.

REANNEY D. C. (1976) Extrachromosomal elements as possible agents of adaptation and development. *Bacteriological Reviews* **40**, 552–90.

References

ACKERMANN H-W. (1973) The morphology of bacteriophages. In: *Handbook of Microbiology* (Eds A. I. Laskin & H. A. Lechevalier), vol. 1, pp. 573–8. Chemical Rubber Company, Cleveland, Ohio.

ACKERMANN H-W., AUDURIER A., BERTHIAUME L., JONES L. A., MAYO J. A. & VIDAVER A. K. (1978) Guidelines for bacteriophage characterisation. *Advances in Virus Research* **23**, 1–24.

ADAMS M. H. (1959) *Bacteriophages*. Interscience, New York.

ANDREWES C. H. (1967) *The Natural History of Viruses*. Norton & Company, New York.

ANDERSON E. S. (1959) The relation of bacteriophages to bacterial ecology. In: *Microbial Ecology* (Eds R. E. O. Williams & C. C. Spicer), pp. 189–217. Cambridge University Press, Cambridge.

ARCHIBALD A. R. (1976) The use of bacteriophages to detect alterations in the cell surface of *Bacillus subtilis*. In: *Continuous Culture 6: Applications and New Fields* (Eds A. C. R. Dean, D. C. Ellwood, C. G. T. Evans & J. Melling), pp. 262–9. Ellis Horwood, Chichester.

BAESS I. (1971) Report on a pseudolysogenic *Mycobacterium* and a review of the literature concerning pseudolysogeny. *Acta Pathologica et Microbiologica Scandinavica* **79**, 428–34.

BAKER E. E., GOEBEL W. F. & PERLMAN E. (1949) Specific antigens of variants of *Shigella sonnei*. *Journal of Experimental Medicine* **89**, 325–38.

BARKSDALE L. & ARDEN S. B. (1974) Persisting bacteriophage infections, lysogeny and phage conversions. *Annual Review of Microbiology* **28**, 265–99.

BERG G., BODILY H. L., LENNETTE E. H., MELNICK J. L. & METCALF T. G. (1976) *Viruses in Water*. American Public Health Association, Washington D.C.

BROWN R. M. (1972) Algal viruses. *Advances in Virus Research* **17**, 243–77.

BUKHARI A. I., SHAPIRO J. A. & ADHYA S. L. (1977) *DNA Insertion Elements, Plasmids and Episomes*. Cold Spring Harbor Laboratory, Cold Spring Harbor.

CAMPBELL A. (1961) Conditions for the existence of bacteriophage. *Evolution* **15**, 153–65.

CHAO L., LEVIN B. R. & STEWART F. M. (1977) A complex community in a simple habitat: an experimental study with bacteria and phage. *Ecology* **58**, 369–78.

CLOWES R. C. & HAYES W. (1968) *Experiments in Microbial Genetics*. Blackwell Scientific Publications, Oxford.

COETZEE J. N. & HAWTREE A. O. (1962) A change in phenotype associated with the bacteriophage carrier state in a strain of *Proteus mirabilis*. *Nature, London* **194**, 1196–7.

DATTA N., HEDGES R. W., SHAW E. J., SYKES R. B. & RICHMOND M. H. (1971) Properties of an R factor from *Pseudomonas aeruginosa*. *Journal of Bacteriology* **108**, 1244–9.

DOW C. S. & FRANCE A. D. (1980) Simplified vegetative cell cycle of *Rhodomicrobium vannielii*. *Journal of General Microbiology* **117**, 47–55.

DREXLER H. (1970) Transduction by bacteriophage T1. *Proceedings of the National Academy of Sciences U.S.A.* **66**, 1083–8.

EDLIN G., LIN L. & BITNER R. (1977) Reproductive fitness of P1, P2 and Mu lysogens of *Escherichia coli*. *Journal of Virology* **21**, 560–4.

EISENSTARK A. (1967) Bacteriophage techniques. In: *Methods in Virology* (Eds K. Maramorosch & H. Koprowski), vol. 1, pp. 449–524. Academic Press, London.

ELLIS E. E. & DELBRUCK M. (1939) The growth of bacteriophage. *Journal of General Physiology* **22**, 365–84.

ESPEJO R. T. & SINSHEIMER R. L. (1976) Process of infection with bacteriophage φX174.XLI. Synthesis of defective particles at 15°C. *Journal of Virology* **19**, 732–42.

FARRAH S. R., GOYAL S. M., GERBA C. P., WALLIS C. & MELNICK J. M. (1977) Concentration of enteroviruses from estuarine water. *Applied and Environmental Microbiology* **33**, 1192–6.

FUJIMURA R. & KAESBERG P. (1962) Adsorption of bacteriophage φX174 to its host. *Biophysical Journal* **2**, 433–49.

GERBA C. P., FARRAH S. R., GOYAL S. M., WALLIS C. & MELNICK J. L. (1978) Concentration of enteroviruses from large volumes of tap water, treated sewage and seawater. *Applied and Environmental Microbiology* **35**, 540–8.

GIBBS A., SKOTNICKI A. H., GARDINER J. E. & WALKER E. S. (1975) A tobamovirus of a green alga. *Virology* **64**, 571–4.

GODSON G. N. (1974) Evolution of φX174: isolation of four new φX-like phages and comparison with φX174. *Virology* **58**, 272–89.

GODSON G. N. & ROBERTS R. J. (1976) A catalogue of cleavages of φX174, S13, G4 and St1 by 25 different restriction endonucleases. *Virology* **73**, 561–7.

HERSHEY A. D. & ROTMAN R. (1949) Genetic recombination between host range and plaque-type mutants of bacteriophage in single bacterial cells. *Genetics* **34**, 44–71.

HILL W. F., AKIN E. W. & BENTON W. H. (1971) Detection of viruses in water: a review of methods and application. *Water Research* **5**, 967–95.

HOFFMANN-BERLING H. & MAZE R. (1964) Release of male specific phages from surviving host bacteria. *Virology* **22**, 305–13.

HONG J., SMITH G. R. & AMES B. N. (1971) Adenosine 3′,5′-cylic monophosphate concentration in the bacterial host regulates the viral decision between lysogeny and lysis. *Proceedings of the National Academy of Sciences U.S.A.* **68**, 2258–62.

HORNE M. T. (1970) Coevolution of *Escherichia coli* and bacteriophage in chemostat culture. *Science* **168**, 992–3.

IKEDA H. & TOMIZAWA J. (1965) Transducing fragments in generalised transduction by phage P1. I. Molecular origin of the fragments. *Journal of Molecular Biology* **14**, 85–109.

KATZENELSON E. (1978) Survival of viruses. In: *Indicators of Viruses in Water and Food* (Ed. G. Berg), pp. 39–50. John Wiley, Chichester.

KONDO E. & MITZUHASHI S. (1964) Drug resistance of enteric bacteria. IV. Active transducing bacteriophage P1Cm produced by the combination of R factor with bacteriophage P1. *Journal of Bacteriology* **88**, 1266–76.

LARK K. G. (1966) Regulation of chromosome replication and segregation in bacteria. *Bacteriological Reviews* **30**, 3–32.

LEVIN B. R., STEWART F. M. & CHAO L. (1977) Resource-limited growth, competition and predation: a model and experimental studies with bacteria and bacteriophage. *American Naturalist* **111**, 3–24.

LI K., BARKSDALE L. & GARMISE L. (1961) Phenotypic alterations associated with the bacteriophage carrier state of *Shigella dysenteriae*. *Journal of General Microbiology* **24**, 355–67.

LIN L., BITNER R. & EDLIN G. (1977) Increased reproductive fitness of *Escherichia coli* Lambda lysogens. *Journal of Virology* **21**, 554–9.

LINDBERGH A. A. (1977) Bacterial surface carbohydrates and bacteriophage adsorption. In: *Surface Carbohydrates of the Prokaryotic Cell* (Ed. I. W. Sutherland), pp. 289–356. Academic Press, London.

LOGAN K. B., REES G. E., SEELEY N. D. & PRIMROSE S. B. (1980) Rapid concentration of bacteriophages from large volumes of freshwater: evaluation of positively charged, microporous filters. *Journal of Virological Methods* **1**, 87–97.

LOGAN K., SCOTT G. E., SEELEY N. D. & PRIMROSE S. B. (1981) A portable device for the rapid concentration of viruses from large volumes of natural fresh water. *Journal of Virological Methods.* (In press.)

LWOFF A., SIMINOVITCH L. & KJELGAARD N. (1950) Induction de la production de bactériophages chez une bactérie lysogène. *Annals de l'Institut Pasteur* **79**, 815–59.

MACKAY D. J. & BODE V. C. (1976) Events in Lambda injection between phage adsorption and DNA entry *Virology* **72**, 154–66.

MCCONNELL M. & WRIGHT A. (1975) An anaerobic technique for increasing bacteriophage plaque size. *Virology* **35**, 588–90.

MCCONNELL M., REZNICK A. & WRIGHT A. (1979) Studies on the initial interactions of bacteriophage ε^{15} with its host cell *Salmonella anatum. Virology* **94**, 10–23.

MATTHEWS R. E. F. (1979) Classification and nomenclature of viruses. *Intervirology* **12**, 139–296.

MEYNELL G. G. (1959) Use of superinfecting phage for estimating the division rate of lysogenic bacteria in infected animals. *Journal of General Microbiology* **21**, 421–37.

MITCHELL R. & JANNASCH H. W. (1969) Processes controlling virus inactivation in seawater. *Environmental Science and Technology* **3**, 941–3.

MOORE B. E. SAGIK B. P. & MALINA J. F. (1975) Viral association with suspended solids. *Water Research* **9**, 197–203.

MORRISON T. G. & MALAMY M. H. (1971) T7 translational control mechanisms and their inhibition by F factors. *Nature, London* **231**, 37–41.

NEWBOLD J. E. & SINSHEIMER R. L. (1970) Process of infection with bacteriophage ϕX174. XXXIV. Kinetics of the attachment and eclipse steps of the infection. *Journal of Virology* **5**, 427–31.

NIEMI M. (1976) Survival of *Escherichia coli* phage T7 in different water types. *Water Research* **10**, 751–5.

OHSUMI M., VOUIS G. F. & ZINDER N. D. (1978) Isolation and characterisation of an *in vivo* recombinant between filamentous bacteriophage f1 and plasmid pSC101. *Virology* **89**, 438–49.

OLSEN R. H. & THOMAS D. D. (1973) Characteristics and purification of PRR1, an RNA phage specific for the broad host range *Pseudomonas* R1822 drug resistance plasmid. *Journal of Virology* **12**, 1560–7.

OTSUJI N., SEKIGUCHI M., IIJIMA T. & TAKAGI Y. (1959) Induction of phage formation in the lysogenic *Escherichia coli* K-12 by mitomycin C. *Nature, London* **184**, 1079–80.

PAYNTER M. J. B. & BUNGAY H. R. (1969) Dynamics of coliphage infections. In: *Fermentation Advances* (Ed. D. Perlman), pp. 323–6. Academic Press, New York.

PRIMROSE S. B. (1976) *Bacterial Transduction.* Meadowfield Press, Durham.

PRIMROSE S. B. & DAY M. (1977) Rapid concentration of bacteriophages from aquatic habitats. *Journal of Applied Bacteriology* **42**, 417–21.

PRIMROSE S. B. & DIMMOCK D. J. (1980) *An Introduction to Modern Virology.* 2nd ed. Blackwell Scientific Publications, Oxford.

PRIMROSE S. B., LOGAN K. B. & SEELEY N. D. (1981) Recovery of viruses from water: methods and applications. In: *Viruses and Waste Water Treatment* (Eds M. Butler & M. Goddard), pp. 211–34. Pergamon Press, Oxford.

PRIMROSE S. B., SEELEY N. D., LOGAN K. B. & NICOLSON J. W. (1982) Methods for studying the ecology of aquatic bacteriophages. *Applied and Environmental Microbiology.* (In press.)

REANNEY D. C. (1976) Extrachromosomal elements as possible agents of adaptation and development. *Bacteriological Reviews* **40**, 552–90.

SCARPINO P. V. (1978) Bacteriophage indicators. In: *Indicators of Viruses in Water and Food* (Ed. G. Berg), pp. 201–37. John Wiley, Chichester.

SCHAUB S. A., SORBER C. A. & TAYLOR G. W. (1974) The association of enteric viruses with natural turbidity in the aquatic environment. In: *Virus Survival in Water and Wastewater Systems* (Eds J. F. Malina & B. P. Sagik), pp. 71–83. Center for Research in Water Resources, University of Texas at Austin, Austin.

SCHLUEDERBERG S. A., MARSHALL B., TACHIBANA C. & LEVY S. B. (1980) Recovery frequency of phages λ and M-13 from human and animal faeces. *Nature, London* **283**, 792–4.

SEELEY N. D. (1980) Ph.D. thesis, University of Warwick.

SEELEY N. D. & PRIMROSE S. B. (1979) Concentration of bacteriophages from natural waters. *Journal of Applied Bacteriology* **46**, 103–16.

SEELEY N. D. & PRIMROSE S. B. (1980) The effect of temperature on the ecology of aquatic bacteriophages. *Journal of General Virology* **46**, 87–95.

SHERMAN L. A. & BROWN R. M. (1979) Cyanophages and viruses of eukaryotic algae. *Comprehensive Virology* **12**, 145–234.

SHUVAL H. I. & KATZENELSON E. (1972) The detection of enteric viruses in the water environment. In: *Water Pollution Microbiology* (Ed. R. Mitchell), pp. 347–61. Interscience, New York.

SINSHEIMER R. L., HUTCHISON C. & LINDQUIST B. (1967) Bacteriophage ϕX174: viral functions. In: *Molecular Biology of Viruses* (Eds J. S. Colter & W. Paranchych), pp. 175–92. Academic Press, London.

SOBSEY M. D. (1976) Methods for detecting enteric viruses in water and wastewater. In: *Viruses in Water* (Eds G. Berg, H. L. Bodily, E. H. Lennette, J. L.

Melnick & T. G. Metcalf), pp. 89–127. American Public Health Association, Washington D.C.

SOBSEY M. D. & JONES B. L. (1979) Concentration of poliovirus from tap water using positively charged microporous filters. *Applied and Environmental Microbiology* **37**, 588–95.

STAGG C. H., WALLIS C. & WARD C. H. (1977) Inactivation of clay-associated bacteriophage MS2 by chlorine. *Applied and Environmental Microbiology* **33**, 385–91.

TAYLOR P. W. & SLEYTR U. B. (1977) Release of a lysogenic bacteriophage from a smooth urinary *Escherichia coli* strain following magnesium-limitation in the chemostat. *FEMS Microbiology Letters* **2**, 189–92.

TORSVIK T. & DUNDAS I. D. (1980) Persisting phage infection in *Halobacterium salinarium* str 1. *Journal of General Virology* **47**, 29–36.

TROUWBORST T., KUYPER S., DE JONG J. C. & PLANTINGA A. D. (1974) Inactivation of some bacterial and animal viruses by exposure to liquid air interfaces. *Journal of General Virology* **24**, 155–65.

TYLER J. M. & BESWICK F. M. (1976) The effect of virus particle concentration on the inactivation of bacteriophage MS2 in seawater. *Journal of Applied Bacteriology* **41** (3), viii–ix.

VERHOEF C., DE GRAAFF P. J. & LUGTENBERG E. J. J. (1977) Mapping of a gene for a major outer membrane protein of *Escherichia coli* K12 with the aid of a newly isolated bacteriophage. *Molecular and General Genetics* **150**, 103–5.

WALLIS C., MELNICK J. L. & GERBA C. P. (1979) Concentration of viruses from water by membrane chromatography. *Annual Review of Microbiology* **33**, 413–37.

WESTPHAL K., BOCK E., CANNON G. & SHIVELY J. M. (1979) Deoxyribonucleic acid in *Nitrobacter* carboxysomes. *Journal of Bacteriology* **140**, 285–8.

WILSON G. C., YOUNG K. K. Y., EDLIN G. J. & KONIGSBERG W. (1979) High-frequency generalised transduction by bacteriophage T4. *Nature, London* **280**, 80–2.

YOUNG D. C. & SHARP D. G. (1977) Poliovirus aggregates and their survival in water. *Applied and Environmental Microbiology* **33**, 168–77.

ZACHARY A. (1974) Isolation of bacteriophages of the marine bacterium *Beneckea natriegens* from coastal salt marshes. *Applied Microbiology* **27**, 980–2.

ZACHARY A. (1976) Physiology and ecology of bacteriophages of the marine bacterium *Beneckea natriegens*: salinity. *Applied and Environmental Microbiology* **31**, 415–22.

ZACHARY A. (1978) An ecological study of bacteriophages of *Vibrio natriegens*. *Canadian Journal of Microbiology* **24**, 321–4.

Chapter 6 • Enumeration and Estimation of Microbial Biomass

Ronald M. Atlas

6.1 Introduction

Enumerating microorganisms and estimating microbial biomass in the biosphere is complex. The diversity of the microorganisms and the habitats in which they occur, requires the development and use of a variety of enumeration procedures (Black 1965; Parkinson *et al.* 1971; American Public Health Association 1975; Costerton & Colwell 1979; Jones 1979). There are no universal methods which can be applied to all microorganisms and all habitats. The successful enumeration of microbial components in the biosphere requires the ability to define carefully which populations are to be counted and from which habitats.

Sample collection and handling is an integral part of the enumeration procedure. The sampling and handling procedures must attempt to ensure that what is enumerated is what existed in the natural habitat. Frequently, however, the numbers of viable microorganisms are altered during sample collection and processing, sometimes due to the growth and sometimes due to the death of microbial populations. Different approaches are required for enumerating different components of the total microbial population and these approaches are themselves determined by the habitat. Remote habitats, such as deep ocean trenches, present particular problems in ensuring that the sample collection procedure does not radically change the numbers of viable microorganisms. Contamination with microorganisms is of serious concern when enumerating microorganisms from environmental samples.

A fundamental difficulty in enumerating microbial populations is that of differentiating living from dead microorganisms. Enumeration procedures which discriminate between living and dead microorganisms often severely underestimate the numbers of microorganisms. Most often the viability of microorganisms is equated with the ability to reproduce successfully. The conditions of the assay will have a marked effect on which of the living organisms actually reproduce and are thus counted as viable. Other procedures, such as direct counts using the microscope, generally lack the capability of differentiating living and dead microorganisms.

The diversity of microbial populations necessitates a variety of methodological approaches and thus different procedures are required for enumerating viral, bacterial, fungal, algal and protozoan populations. For most bacterial populations, numbers of individual cells can be used as a measure of enumeration, since individual organisms generally occur as single cells. However, similar procedures cannot be applied to multicellular organisms, such as filamentous fungi and algae.

There are differences between procedures aimed at enumerating microbial populations and those aimed at estimating microbial biomass. Enumeration involves determining how many; biomass estimation involves determining how much. In this context, biomass refers to the living weight of organisms. Determination of microbial biomass yields information on some of the energy stored within an ecosystem at the microbial level. Units of biomass may be given as a weight measure, e.g. grams, or as an energy measure, e.g. calories. Direct measures of microbial biomass are rarely applicable to environmental samples. It is normally impossible to separate the microbial biomass from their abiotic and biotic surroundings with sufficient efficiency to permit accurate weighing of the recovered microorganisms in order to determine the biomass directly. Rather, it is normally necessary to estimate biomass from an enumeration determination or other indirect procedures. Direct weight measurements, cell volume measurements and microcalorimetry cannot normally be used for determining microbial biomass in environmental samples.

Experimental

6.2 Sample collection and handling

The enumeration procedure begins at the time of sample collection (Board & Lovelock 1973). Sampling and handling procedures must ensure that the numbers determined accurately reflect the numbers actually present in the biosphere. In certain habitats (e.g. within the water column of some aquatic habitats) microbial populations are free-living and relatively evenly dispersed (Parkinson *et al.* 1971; Costerton & Geesey 1979). Such homogeneous distributions are relatively easy to sample, especially if the habitat is easily accessible. In other habitats microbial populations are heterogeneously distributed since microorganisms can exist in microhabitats. Many microorganisms exist in their natural habitats in tight association with surfaces (Marshall 1976). For example, in soils many microorganisms form microcolonies which are absorbed onto the soil particles (Burns 1980). Animal and plant surfaces are similarly colonised by microbial populations. Sometimes there are also close associations between microbial populations, e.g. algae are often colonised by epiphytic bacteria. In some enumeration procedures it is essential that the association between the microorganisms and their surroundings be preserved, whilst in others it is important that the association is disrupted even though the microorganisms must be recovered without being damaged.

In soil habitats, grab surface samples or cores are most often collected for microbial enumeration procedures. The natural abundance of microorganisms in most soils lessens concern about sample contamination and in many studies aseptic procedures are not rigorously adhered to. Preparing soil and sediment samples for viable count procedures normally involves attempting to separate the microorganisms from the particles and establishing a homogeneous suspension. Addition of a diluent and agitation are normally required to distribute microorganisms evenly. Dispersants may also be added to enhance separation of the microorganisms. Efficiency of recovery of microorganisms has been found to be highly dependent upon the chemical composition and osmotic strength of the diluent, the time of mixing, temperature and the degree of agitation (Jensen 1968; Litchfield *et al.* 1975). Not surprisingly, conditions which are incompatible with the physiological tolerance ranges of the microbial populations being enumerated lead to a loss of viability and an underestimation of the microorganisms. When the enumeration procedure employed does not discriminate between living and dead microorganisms, it is sometimes possible to preserve the microorganisms by the addition of formaldehyde or glutaraldehyde. Subsequently the microorganisms can be counted by direct microscopic observations using techniques such as fluorescence-staining light microscopy or scanning electron microscopy. These techniques will also permit observation of microorganisms in association with surfaces, e.g. with soil particles. However, when specific biochemicals, such as ATP, chlorophyll, muramic acid or DNA, are used to estimate microbial biomass, the microorganisms must be desorbed from particles. The efficiency of recovery of these biochemicals must be determined and optimised for the given sample. In some enumeration procedures the sampling involves baiting or trapping microorganisms using glass slides or capillary tubes which are implanted in the habitat, e.g. soil. The microorganisms which become associated with the implanted tubes or slides are recovered for enumeration (Cholodny 1930; Rossi *et al.* 1936; Tribe 1957; Aristovskaya & Parinkina 1961; Perfilev & Gabe 1969; Hirsch & Pankratz 1970; Aristovskaya 1973).

Sampling problems are normally greater in aquatic habitats than in soils. Microorganisms in aquatic habitats may exist in the surface neuston layer, within the water column—either freely or in association with particles—or within the sediment (Costerton & Geesey 1979). Sampling of the thin surface neuston layer requires special collection techniques such as absorption onto teflon or nitrocellulose filters (Crow *et al.* 1975). Sampling within the water column and from sediment generally requires remote sampling devices. The relatively low numbers of microorganisms present in oligotrophic lakes and in offshore marine waters normally requires that special caution be taken to avoid contamination during sampling. Collecting samples at a given depth requires that the sampling devices be designed to preclude contamination from overlying waters during deployment and recovery of the sampling apparatus. Sampling devices for collecting water samples generally consist of an evacuated sterile chamber which can be lowered on a cable, opened and filled at depth upon a signal from a messenger (Bordner & Winter 1978). Some devices have specialised inlet systems to prevent possible contamination from the supporting hydrographic cable (Jannasch & Maddux 1967). More elaborate devices are used for collecting deep-sea samples where pressure must be maintained during recovery to prevent decompression and loss of viability of the collected microorganisms (Jannasch & Wirsen 1977). Sediment samples are normally collected with grab samplers or coring devices (Ross 1970), although many sampling devices greatly disturb the sample and mix surface and subsurface microorganisms. In some sediments the disturbance results in mixing of aerobic and anaerobic zones. The processing of samples from aquatic habitats may involve extracting specific biochemicals for biomass estimation, preserving samples for determination of numbers of microorganisms without differentiating living from dead or attempting to maintain viability for enumeration of living microorganisms. Specific concern must be given to maintaining viability of microorganisms with limited physiological tolerance ranges. For example, pressure must be regulated carefully during sampling and processing to maintain viability of barophilic microorganisms (ZoBell 1964); even brief exposure to oxygen can cause loss of viability of obligate anaerobes (Shapton & Board 1971); cold temperatures must be maintained throughout sampling and processing to maintain viability of psychrophilic microorganisms (Morita 1975). Processing of some aquatic samples involves dilution; in other cases it is necessary to concentrate the microorganisms by collection on filters in order to recover a sufficient number of microorganisms for subsequent counting. The chemical composition and the pore size of filters used for concentrating microorganisms has been found to have a pronounced effect on the results of enumeration procedures (Lin 1976).

Sampling from the atmosphere may be achieved by passage of a known volume of air through a 0.2 to 0.45 μm filter (Gregory 1973). For viable counting of microorganisms in air samples it is also possible to collect the microorganisms in a sampling device containing a graded series of grids of decreasing pore size (Andersen 1958). An agar plate is placed under each grid to collect microorganisms which impact on the surface of the agar. The size of the grid opening determines the air velocity and thus the size of the microorganisms impacting the agar through their inertia.

Counting microorganisms associated with plants or animals can be approached in several ways. Microorganisms may be recovered from surfaces by swabbing or washing the surface with a sterile solution (Dickinson & Preece 1976). For example, microorganisms can be recovered from human epithelial tissues by swabbing (Rosebury 1962) and can be washed or scraped from tooth surfaces (Gibbons & van Houte 1975). Similarly, leaves or stems of plants can be washed in order to recover microorganisms. Such non-destructive sampling requires that the microorganisms are not tightly bonded to the plant or animal tissues. In some cases vital fluids, such as tree sap, or excreted products, such as faecal material, can be collected and used to enumerate microbial populations. In other cases dissection procedures can be utilised to recover plant or animal tissues and their associated microorganisms although such methods are of necessity destructive. The cellular integrity of the tissues may be maintained for direct observation and enumeration of microbial populations or the tissues may be macerated and homogenised to produce a homogeneous suspension of microorganisms (Waid 1957; Clarke & Parkinson 1960). In summary, sample collection and processing should maintain the numbers of microorganisms actually present in the sample in the desired form. For some enumeration procedures viability must be maintained; for others it is critical to preserve the samples. In some instances the microorganisms must be separated from attached surfaces; in others the integrity of the attachment must be maintained. Often it is necessary to process samples rapidly in order to minimise changes in the numbers of microorganisms.

6.3 Counting procedures for enumeration and biomass determinations

6.3.1 Viable Count

Viable count procedures for enumerating microorganisms employ two basic approaches:
(1) plate count techniques;
(2) most-probable-number (MPN) techniques.

Both procedures require that microorganisms grow and divide during the assay in order to establish viability. Both plate count and MPN techniques require that the microorganisms are separated into individual reproductive units in a homogeneous suspension in order to obtain accurate counts. In cases where the numbers of microorganisms require concentration, the filters may be placed directly on supporting media and the colonies that develop on the filters counted following incubation. Some consideration must be given to possible toxicity associated with some membrane filters (Green *et al.* 1975; Lin 1976).

The use of the agar plate count in microbial ecology has been severely criticised (Postgate 1969; Buck 1979). The reason lies in the misuse of the techniques and/or misinterpretation of the data generated by viable plate counts. The problem is mainly with the use of the term *total* viable microorganisms. The viable plate count procedure, by its very nature, is selective and therefore cannot yield accurate data on total numbers of microorganisms in environmental samples. Plate count procedures employ a variety of media and incubation conditions for the enumeration of different microorganisms (Jannasch & Jones 1959; Clark 1965a, b, c, d, e; Jensen 1968; Jones 1970; Shapton & Board 1971; Pratt & Reynolds 1972; Simidu 1972). The conditions of the plate-count procedure should meet the nutritional and physiological requirements of the microorganisms being enumerated. Meeting the requirements of one physiological type of microorganism, e.g. aerobes, by necessity excludes other physiological types, e.g. anaerobes. Once the selectivity of the plate count procedure is recognised, this technique can be used effectively to yield valuable data.

The principle of the plate-count procedure is to inoculate a homogeneously dispersed microbial suspension into or onto a solid medium containing the nutrients necessary to support the growth of the microorganisms. Following incubation for a given period, under controlled environmental conditions, the numbers of colonies which have formed on or in the medium are counted. The assumption of the technique is that each viable microorganism, for which the particular enumeration procedure is designed, will form a colony during the incubation period and that each colony will arise from a single microorganism. Therefore, it is necessary to ensure that the microorganisms are well dispersed in or on the solid medium because overcrowding of colonies gives rise to erroneous estimations of microbial numbers.

In most plate-count procedures agar is used although other solidifying agents, such as silica gel, are sometimes employed. Agar is relatively inert, in that few microorganisms are agarolytic. However, agar may contain a variety of organic contaminants, limiting its usefulness for enumeration of specific, nutritionally defined groups of microorganisms, such as obligate chemolithotrophs. In most agar plate-count procedures high concentrations of nutrients are employed. Such high concentrations of nutrients are appropriate in some cases, but clearly discriminate against slow-growing, oligotrophic microorganisms.

Plate-count procedures are most frequently used for the enumeration of bacteria but may also be used for other groups of microorganisms, including algae and fungi. However, the use of viable plate-count procedures for counting fungi is strongly biased (Menzies 1965). Spores and single-celled vegetative forms (e.g. yeast and yeast-like fungi) yield high numbers by plate-count procedures, while filamentous forms yield low numbers. Filamentous fungi also tend to overgrow agar plates, often making differentiation of single colonies impossible.

Two basic approaches utilised in plate-count procedures are the surface-spread-plate techniques and the pour-plate technique (Buck 1979). In the surface-spread-plate technique, the suspension of microorganisms is uniformly spread over the surface of the solidified medium. In the pour-plate technique the agar medium is melted and held at approximately 40°C. The suspension of microorganisms is added to the liquid agar which is poured into a Petri dish or other suitable container.

Alternatively, the suspension is added to the plate, the molten agar poured and the mixture homogenised by gentle swirling. The pour-plate technique has the disadvantage that the microorganisms are exposed to a temperature of 40°C, which may be above the tolerance limit for some microbial populations (e.g. for the psychrophiles which may be killed by exposure at 25°C for even a few seconds). The pour-plate technique, however, generally ensures better distribution of microorganisms, i.e. there is less problem with overlapping colonies, and protects some microorganisms against exposure to oxygen. The roll-tube method, which is used for enumeration of strict anaerobes, is an extension of the pour-plate method (Hungate 1969; Hungate & Macy 1973; see also Chapter 1, p. 8).

In some plating procedures the method is designed to be differential and/or selective. In other words it should permit the growth and thus the counting of only a selected population and/or growth of multiple populations, but permit differentiation of a population of interest from the others. For example, it is possible to add antibacterial substances to a medium, e.g. streptomycin, to preclude the growth of bacteria and to selectively enumerate fungi (Menzies 1965). Dyes may be added to agar media to permit the differentiation of bacteria carrying out a particular metabolic activity, e.g. acid production from lactose utilisation by *Escherichia coli* (Wolf 1972).

6.3.2 Most Probable Number

The most-probable-number (MPN) technique, for determining viable numbers of microorganisms, employs a statistical approach in which successive dilutions are performed to reach an extinction point (Cochran 1950; Alexander 1965a; Melchiorri-Santolini 1972; Colwell 1979). In the MPN technique, replicates, usually three to ten, of each dilution are made and the pattern of positive and negative scores are recorded. A statistical table based on a Poisson distribution is used to determine the most probable number of viable microorganisms present in the original sample. Different MPN procedures employ different criteria for establishing positive and negative scores. In many procedures, a tube is scored positive when there is visible growth, i.e. an increase in turbidity. Other procedures employ more quantitative criteria, such as an increase in protein concentration or differential criteria, such as the production of acids from carbohydrates or the production of [^{14}C]carbon dioxide from radiolabelled substrates (Alexander 1965b; Alexander & Clark 1965; Olson 1978; Atlas 1979; Colwell 1979; Lehmicke *et al.* 1979). Production of chlorophyll can be used to establish a positive score in algal enumeration procedures using MPN techniques (Clark & Durrell 1965).

The MPN techniques have several advantages and disadvantages compared to viable plate-count procedures (Colwell 1979). The use of liquid culture in most MPN techniques eliminates the need for a solidifying agent, such as agar, and thus the problem of contamination with organic compounds. This is of particular advantage when enumerating viable numbers of microorganisms capable of utilising a particular organic compound as sole carbon source, e.g. in the enumeration of hydrocarbon-utilising microorganisms. The MPN technique is subject, however, to most of the same inherent problems as the plate-count procedure; for example it is necessary for the microorganisms to reproduce in order to establish viability and be counted as positive scores and it is impossible to establish a medium and set up incubation conditions which will allow for simultaneous growth of all viable microorganisms in the sample. Microorganisms must be separated initially into individual reproductive units in order to satisfy the statistical assumptions used in this technique, aggregates of microorganisms will lead to an underestimation of numbers of viable microorganisms. The MPN technique is less accurate than the plate-count procedure in the sense that it establishes an MPN and confidence limit rather than an actual number of reproductive units. In comparison to plate-count techniques, MPN techniques generally are more laborious, employing many tubes of liquid media in order to establish the necessary dilutions to reach a point of extinction. Large numbers of replicates are needed to increase accuracy. However, the use of microtitre plates, coupled with automated spectrophotometric readings to establish positive and negative scores, has greatly simplified the use of MPN techniques in some procedures (Rowe *et al.* 1977).

Most-probable-number procedures can be employed for enumerating viable numbers of viruses, bacteria, fungi, algae and protozoa since they can be designed to be selective or differential. Various combinations of incubation conditions and media can be used for enumerating specific groups of microorganisms (Wolf 1972). In the case of enteric viruses, suitable host cells in tissue culture are employed as the medium. Tubes can be examined for evidence of cytopathic effects, i.e. death of infected cells (Farrah *et al.* 1977), and the number of viruses in the sample can be calculated. For enumeration of protozoa, suspensions of bacteria growing on agar surfaces are employed as a suitable medium (Singh 1946; Clark & Beard 1965) and the growth of protozoa on the bacteria produces zones of clearing. The dilutions in which clearing zones are observed can be used to calculate the number of protozoa. For determinations of MPN of algae, it is necessary to carry out the incubations in the light employing a suitable mineral salts medium supplemented with necessary growth factors. Production of chlorophyll is generally used as the criterion for establishing a positive score for the presence of algae.

6.3.3 Direct Count

Microorganisms can be counted and their biomass estimated using direct microscopic observation (Frederick 1965; Gray *et al.* 1968). Relatively large microorganisms, such as protozoa and algae, can be enumerated using a counting chamber when they are present in relatively high numbers (Finlay *et al.* 1979). A variety of counting chambers, e.g. a haemocytometer or Petroff-Hauser chamber, may be used in these procedures. These chambers are designed to hold a known volume of liquid which permits the determination of concentrations of microorganisms from counts within a measured area (Parkinson *et al.* 1971). Estimates of numbers of microorganisms using counting chambers are generally inaccurate when concentrations are low. Small and viable motile microorganisms are difficult to count in such chambers. The use of phase contrast microscopy is advantageous as it eliminates the need for staining to increase contrast.

The enumeration of filamentous fungi generally is achieved by determining the length of hyphae rather than the numbers of individual cells. This can be accomplished using direct microscopic observation with the aid of a micrometer. Various modifications of the agar film technique of Jones and Mollison (1948) have been applied for the direct measurement of lengths of mycelia (Skinner *et al.* 1952; Thomas *et al.* 1965; Parkinson 1973). The method involves mixing a sample with agar and pipetting onto a glass slide forming a film which can be stained. Difficulties have been noted in estimating lengths of hyphae by this technique when there is background interference. For example, it has been found when estimating lengths of fungal mycelia in soil that higher estimates are obtained at higher dilutions, due to the fact that at low dilutions soil particles obscure some of the fungal mycelia (Skinner *et al.* 1952).

When microorganisms are present in low concentrations they can be concentrated and counted on filters of a suitable pore size. Prior to counting, the microorganisms collected on the filters may be stained to enhance contrast. The use of fluorescence microscopy and fluorescence stains, such as acridine orange (AODC method), permit accurate counting of even small bacteria (Strugger 1948; Trolldenier 1973; Zimmerman & Meyer-Reil 1974; Daley & Hobbie 1975; Daley 1979; Geesey & Costerton 1979). For the best results it has been found that polycarbonate nuclepore filters are superior to cellulose filters, since the former retain microorganisms on a flat surface while in the latter microorganisms are trapped in different planes of the filter making their observation difficult (Hobbie *et al.* 1977). For increased contrast, dark (black-stained) filters are used (Jones & Simon 1975a). When acridine orange is used, microorganisms fluoresce green or orange. Direct epifluorescent counts of bacteria are typically several orders of magnitude higher than estimates of numbers of microorganisms obtained by viable count procedures. Obtaining higher numbers is generally equated with superiority in enumeration techniques based on the presumption that a higher proportion of the microorganisms actually present in the sample are being counted; lower numbers are taken as evidence of inefficiency in enumeration procedures. Direct epifluorescent counts are applicable to a variety of habitats. Microorganisms can be counted even in the presence of background material such as soil or sediment particles. It is sometimes necessary to decolourise the background in order to decrease non-specific fluorescence (Zimmerman & Meyer-Reil 1974). If coupled with size measurements, direct counts can be converted to estimates of microbial biomass.

The problem with direct counts by methods such as the AODC method, is that it is not possible to differentiate living from dead microorganisms. Generally, direct epifluorescent counts of microorganisms can be used as a good estimate of total numbers of microorganisms but cannot be used to determine if the microorganisms are living, dead or dormant. As a consequence, several procedures have been developed which attempt to separate living from dead microorganisms.

Fluorescein diacetate has been used to estimate the biomass of living fungi and bacteria (Babuik & Paul 1970; Soderstrom 1977). Only metabolically active hyphae and cells are stained in this procedure since fluorescence occurs only after enzymatic cleavage of the stain molecule. Use of the fluorescein diacetate stain in conjunction with the use of a vital stain (which does not distinguish living from dead hyphae) allows for the estimation of both active and total biomass.

The presence of active microbial dehydrogenase enzymes can be detected and similarly used to separate living from dead microorganisms (Patriquin & Dobereiner 1978; Zimmerman *et al.* 1978). One method uses 2-(*p*-iodophenyl) 3-(*p*-nitrophenyl) 5-phenyl tetrazolium chloride (INT) as an indicator of dehydrogenase activity. Microbial electron transport systems which are active in living or respiring microorganisms reduce the tetrazolium chloride to the corresponding formazan: the accumulation of INT-formazan can be seen as intracellular dark red spots. When coupled with the AODC microscopy method, the use of INT allows the separation of metabolically active cells from dormant or dead microorganisms.

Living microorganisms can also be detected by the use of autoradiography combined with direct microscopic observation (Brock & Brock 1968; Waid *et al.* 1973; Ramsay 1974; Fliermans & Schmidt 1975; Hoppe 1976; Faust & Correll 1977; Meyer-Reil 1978). In this approach microorganisms are incubated with a radiolabelled substrate. Labelled substrates have been employed in these procedures. Following incubation of natural microbial populations under appropriate conditions, the microorganisms are collected on filters. The filters are placed on glass slides and coated with a thin film emulsion and following a second period of incubation, the film developed. Actively metabolising microorganisms are associated with dark silver grains which are deposited in close proximity to the cell. Fluorescent stains and epifluorescent microscopy can be used to visualise the cells and determine the total numbers of microorganisms present in the sample. Often the incorporation of [^{3}H]- or [^{32}P]-labelled nucleotides into DNA is employed in these procedures.

Another approach used for the separation of living from dormant or dead microorganisms is based on inhibition of cell division (Kogure *et al.* 1979). Samples are incubated with nalidyxic acid which prevents cell division but not growth and results in the formation of elongated cells. Fluorescence staining and epifluorescent microscopy permits the determination of the numbers of elongated cells (actively growing) and the numbers of normal cells (dormant or dead microorganisms).

The use of immunofluorescence permits the determination of numbers of a specific microorganism even when other microorganisms are present in the sample (Schmidt & Bankole 1963; Hill & Gray 1967; Bohlool & Schmidt 1968; Schmidt *et al.* 1968; Gray 1973; Schmidt 1973; Pugsley & Evison 1975; Dazzo & Brill 1977; Strayer & Tiedje 1978; Fliermans *et al.* 1979; Stanley *et al.* 1979). In this method a specific fluorescent antibody is produced by conjugating an antibody with a fluorochrome, e.g. fluorescein isothiocyanate (FITC). The method relies on the specificity of antibody–antigen reactions. The specificity is both an advantage and a disadvantage: it permits the identification and enumeration of specific microorganisms within a sample but generally precludes the estimation of numbers of even closely related microorganisms using the same conjugated antibody.

An electron microscope can be used instead of a light microscope for the direct observation and enumeration of microorganisms (Gray 1967; Harris *et al.* 1972; Nikitin 1973; Paerl & Shimp 1973; Todd *et al.* 1973; Bowden 1977; Larsson *et al.* 1978). The scanning electron microscope is particularly useful in this regard as it permits enumeration within the context of a three-dimensional matrix. Thus, it is not necessary to separate microorganisms from particles or tissues to which they are adhering prior to observation by scanning electron microscopy. Results of enumerations with scanning electron microscopy have been found to be comparable to those obtained by epifluorescent microscopy. When using electron microscopy, caution must always be given to the possibility that artefacts may occur due to the use of high vacuum and/or metal coating. Electron microscopic observation can be coupled with the nalidyxic acid inhibition technique or autoradiographic procedures in order to separate active from inactive microorganisms.

As with the viable count procedure the determination of numbers of active microorganisms requires that activity occurs during the incubation period. In the INT method activity does not require growth; in most autoradiographic methods activity requires DNA replication but not necessarily cell division; in the nalidyxic acid method activity requires growth and an attempt to divide. Many of the criticisms of the viable plating and MPN procedures apply here as well, in particular the problem that virtually all incubation conditions used with a collected (disturbed) sample will be selective.

Another method for the direct counting of microorganisms employs the use of a Coulter counter (Kubitschek 1969). This type of instrument permits discrimination between different particle sizes allowing one to set a range of particle sizes and to count all the particles in that size. It is possible to set the ranges so that particles in the size range of one microbial group, e.g. bacteria, are counted separately from those within the normal size range of another group, e.g. protozoa. Using counts of particles within these ranges it is possible to estimate total numbers of microorganisms. There are severe problems with the accuracy of such determinations, however, when dealing with environmental samples. Microorganisms in nature can vary greatly in size depending on species and state of growth. Aggregation of microorganisms which is common in environmental samples causes problems both with errors in estimating the number of particles and in the particle size ranges which should be counted. Soil, sediment or other non-microbial particles present in the sample will interfere with this analytical procedure. Furthermore, the method does not discriminate between living and dead microorganisms. These problems generally have precluded the widespread use of particle size determinations for estimating numbers of microorganisms.

6.4 Biochemical approaches to enumeration and biomass determinations

In contrast to the previously described methods, all of which depend on maintaining the integrity of the microbial cell, the biochemical approaches for enumerating and estimating biomass rely on quantitative estimates of biochemical constituents of the cell. Various biochemicals have been used for estimating microbial biomass but, to be useful, the biochemical used should be correlated with the biomass of the microorganisms being enumerated. Ideally the biochemical assayed should be present in the microorganisms being enumerated and absent in other populations in the sample. Furthermore all microorganisms being counted should have the same amount of the biochemical assayed. In addition, it should be possible to quantify accurately the biochemical in question in samples from various habitats without interference from other components present in the sample. It is difficult to establish the validity of the above requirements in many situations and appropriate controls and caution are required for estimating microbial biomass from biochemical measurements.

6.4.1. Energy Components

ATP is the universal biochemical form of energy in biological systems. ATP is normally present in living cells in constant proportions relative to total cellular carbon. ATP rapidly degenerates following the death of a cell and thus can be used to estimate living biomass (Holm-Hansen 1969; Paerl & Williams 1976; Deming *et al.* 1979; Stevenson *et al* 1979). Concentrations of ATP can be determined using the luciferin-luciferase assay. Light is produced when reduced luciferin reacts with oxygen to form oxidised luciferin in the presence of luciferase enzyme, magnesium ions and ATP. The light emitted can be detected by a photomultiplier and quantified. When ATP is the limiting reactant the quantity of light emitted is proportional to the concentration of ATP. There are a variety of commercial instruments available for the measurement of ATP. Some of the instruments are designed to measure the peak intensity of light emission: others are designed to integrate a portion of the emitted light and calculate total ATP. The sensitivity, repeatability and reliability of these instruments vary. Commercial preparations of luciferin-luciferase are available for determining ATP concentrations. Differences in purity between different lots of reactants make frequent calibration a requirement for accurate quantification of ATP concentrations. Deming *et al.* (1979) concluded that the instantaneous peak height of the light response of the sample was a more valid measure of the original ATP concentration of the sample than integration of the area under the light reaction curve and that purified luciferase enzymes provided the sensitivity required for measuring low levels of ATP.

The methodology employed in the extraction of ATP is important in determining the sensitivity and reliability of the method (Deming *et al.* 1979; Stevenson *et al.* 1979). Different approaches have been used for extracting ATP from microorganisms in soil sediment and aquatic habitats. ATP determinations have been used to estimate microbial biomass in aquatic and terrestrial habitats (Hamilton & Holm-Hansen 1965; Holm-Hansen & Booth 1966; Holm-Hansen 1969, 1973; Ausmus 1973; Bancroft *et al.* 1976; Sutcliffe *et al.* 1976; Jones & Simon 1977; Paul & Johnson 1977; Jenkinson & Oades 1979; Oades & Jenkinson 1979).

The ATP must be extracted rapidly and in a manner that maintains the integrity of the nucleotide molecule. ATP-degrading enzymes must be inactivated. Conversions of ATP determinations to estimates of cell biomass are extremely sensitive to the efficiency of the extraction procedure. A variety of organic solvents, e.g. chloroform, inorganic acids (such as sulphuric acid), and boiling buffers (such as Tris), have been employed to extract ATP from microbial cells in various samples (Deming *et al.* 1979). The efficiency of a particular extraction procedure depends on the nature of the sample and the microbial community. Adsorption onto inorganic particles can greatly reduce the efficiency of extraction (Ausmus 1973; Bancroft *et al.* 1976; Jenkinson & Oades 1979).

A possible complication of the ATP assay involves reactions of GTP (Karl 1978). GTP does not produce light emissions when highly purified luciferin-luciferase reagents are used. However, with crude preparations of these reagents, GTP can react to produce ATP within the assay chamber increasing the amount of light emitted. Due to the pattern of light emission, the reaction with GTP does not result in a serious overestimation of ATP when the peak height is used for quantification but, when an integrated measure of light emission is used for the quantification, contamination with GTP can create a serious problem. The problem with using ATP concentrations to estimate microbial biomass lies with the fact that ATP concentrations within a cell depend on the physiological state of an organism. ATP levels are higher in actively growing microorganisms than in viable but dormant or slowly growing microorganisms. When used in conjunction with the determinations of the total adenylate pool the ATP concentrations can be used to calculate the energy charge which is a measure of the physiological state of the organisms within a sample (Wiebe & Bancroft 1975; Karl & Holm-Hansen 1978). The total adenylate pool can be used to estimate total microbial biomass but will not effectively separate living from dead biomass.

The presence of ATP in a sample from plant or animal cells can interfere with the assay procedure, preventing the determination of ATP associated with microorganisms in the sample and thus precluding the accurate estimation of microbial biomass. Several digest treatments can be employed in an attempt to degrade ATP from plant and animal cells without reducing the concentrations of ATP within microbial cells (Chappelle *et al.* 1978).

The following points summarise the advantages and disadvantages of the use of ATP measurements to estimate microbial biomass:

(1) the assay is highly sensitive and very low concentrations of microorganisms (less than 100 per ml) can be detected;

(2) the assay exclusively measures viable microorganisms;

(3) the assay quantifies ATP from bacteria, fungi, algae and protozoa within the sample;

(4) assays can be performed rapidly and with great precision.

The disadvantages of the assay lie with the sensitivity of the assay to the physiological state of the organisms in the sample and to the extraction and assay methodological procedures. As a result, conversion of ATP measurements to estimates of microbial biomass can be difficult and it is necessary to use different conversion factors for samples from different habitats. Failure to take into consideration the potential pitfalls of converting ATP determinations to estimates of biomass can lead to serious errors.

6.4.2 Cell Wall Constituents

Cell wall components of bacteria and fungi have been proposed as biomass indicators. With few exceptions, bacterial cell walls contain a unique biochemical, murein, which is not found in other organisms. The specific relationship between murein and bacteria makes quantification of this biochemical potentially quite useful for estimating bacterial biomass (Millar & Casida 1970; Moriarty 1975, 1977, 1978; King & White 1977; White *et al.* 1979). The method involves quantifying the muramic acid moiety of the murein or peptidoglycan layer of the cell wall.

In this approach the bacterial cell wall component, muramic acid is assayed by its conversion to lactate, followed by enzymatic (Morarity 1977) or chemical analysis (King & White 1977) to determine the concentration of lactate. The lactate is released from the cells by acid and base hydrolysis. In the enzymatic analysis the lactate is converted to pyruvate, using lactic acid dehydrogenase, and then to alanine, using glutamate pyruvate transaminase. The NADH produced in this assay is reacted with a phosphate-luciferase preparation and the light emission quantified. In the chemical procedure for determining muramic acid absorbance at 560 nm is used to quantify the amount of lactate and thus muramic acid. Differences in accuracy and sensitivity have been reported between the chemical and enzymatic analysis methods.

The conversion of muramic acid determinations to estimates of bacterial biomass is difficult when dealing with mixed microbial communities. Gram-positive bacteria contain approximately three times the concentrations of muramic acid per gram cell dry weight as Gram-negative bacteria. Endospores contain higher concentrations of muramic acid, and cyanobacteria may contain 500 times the concentration of muramic acid per gram cell compared with Gram-negative heterotrophic bacteria. Obviously this disparity in the concentrations of muramic acid can lead to serious errors in biomass estimation unless the bacterial community is well defined. In some cases, e.g. in marine waters where Gram-negative bacteria are dominant, it is possible to apply a single conversion factor and estimate bacterial biomass from muramic acid determinations with relatively good precision.

For the estimation of fungal biomass, chitin, which is a component of the cell wall of some fungi, has been used (Swift 1973a, b; Sharma *et al.* 1977; Willoughby 1978). Not all fungi, however, contain chitin in their cell wall, limiting the application of this method largely to comparative studies where similar groups of fungi are believed to be present. A further problem with this assay is that different fungi contain different amounts of chitin relative to their biomass and the relative concentrations of chitin vary with the age and physiological state of the fungus. Chitin is relatively resistant to degradation and may remain intact following death of fungi, making the use of this biochemical for the estimation of living fungal biomass questionable.

Chitin can be quantified by the release of N-acetylglucosamine, which can be accomplished enzymatically using chitinase or by acid or base hydrolysis. The N-acetylglucosamine is assayed with Ehrlich's reagent (*p*-dimethylaminobenzaldehyde) which is quantified spectrophotometrically using a wavelength of approximately 530 nm. The hydrolysis of chitin can take a long period of time (days to weeks) and should be allowed to proceed until a constant determination value for N-acetylglucosamine is reached for an accurate estimation of fungal biomass.

6.4.3 Cell Envelope Components

The lipopolysaccharide complex associated with the cell envelope of Gram-negative bacteria can be quantified and used to estimate bacterial biomass: this is known as the LPS method (Coates 1977; Watson *et al.* 1977; Watson & Hobbie 1978). The assay utilizes an aqueous extract from the blood cells of the horseshoe crab (*Limulus amoebocyte* lysate) which forms a turbid solution when it reacts with the lipopolysaccharide complex. The turbidity is proportional to the concentration of endotoxin and can be quantified spectrophotometrically. The reaction of the *Limulus amoebocyte* lysate with lipopolysaccharide is specific. In aquatic habitats, where Gram-negative bacterial populations dominate, this assay can effectively be used to estimate bacterial biomass. Good correlation has been obtained between estimates of bacterial biomass based on direct epifluorescent microscopic counts and those obtained by the LPS method. The assay is very sensitive and can detect less than 1000 bacteria per 1 ml sample.

6.4.4 Nucleic Acids

Determination of concentrations of nucleic acids may be used as an estimator of microbial biomass (Holm-Hansen *et al* 1968; Hobbie *et al.* 1972). Concentrations of RNA vary greatly depending on the physiological state of an organism and thus RNA determinations are not reliable estimations of microbial biomass. Concentrations of DNA, however, are maintained in relatively constant proportions within microorganisms. DNA can be recovered from microbial cells, purified and quantified (Herbert *et al.* 1971). Concentrations of DNA can be estimated spectrophotometrically but the sensitivity required for an accurate DNA estimation in microbial biomass generally requires reaction with a chemical reagent, such as diaminobenzoic acid or ethidium bromide followed by spectrofluorometric analysis to determine the quantity of DNA (Udenfriend 1969; Cattolico & Gibbs 1975; Jones & Simond 1975b; Setaro & Morely 1976; Kapuscinski & Skoczylas 1977). Attention must be paid to interference with this assay from chemicals in the sample, e.g. the quenching of fluorescence by humic materials in soil samples. Careful purification of the DNA before analysis can overcome some problems. Furthermore, nonmicrobial sources of DNA can interfere with determination of microbial biomass from DNA determinations and DNA can exist in degraded and dead cells. Thus estimates of DNA may not accurately reflect amounts of living microbial biomass.

6.4.5. Photosynthetic Pigments

Determination of concentrations of photosynthetic pigments can be utilised to estimate the biomass of photosynthetic microorganisms (Edmondson 1974; Cohen *et al.* 1977; Holm-Hansen & Riemann 1978). Although the evidence suggests that there may not be a constant relationship between biomass and chlorophyll content, chlorophyll is still a useful estimate of the biomass of photosynthetic populations (Banse 1977). Estimates of biomass based on chlorophyll determinations have been found to correlate well with those based on other biochemicals, e.g. ATP (Paerl *et al.* 1976). Chlorophyll *a* is the most abundant photosynthetic pigment in algae and cyanobacteria. Chlorophyll *a* can be extracted from cells using solvents, most commonly hot methanol or cold acetone (Heaney 1978; Holm-Hansen & Rieman 1978). With some algae and cyanobacteria it may be necessary to disrupt the cells initially, e.g. by lysozyme treatment, sonication or freezing and thawing, prior to extraction of the chlorophyll. The concentrations of chlorophyll can be determined spectrophotometrically using wavelengths corresponding to the absorption maxima. Generally, for chlorophyll *a* determinations, absorbance is measured at 665 nm and 750 nm and the absorbance at 750 nm is subtracted from that at 665 nm to correct for background turbidity. The wavelengths may be altered to detect and quantify other photosynthetic pigments (Stanier & Smith 1960; Loftus & Carpenter 1971; Caldwell & Tiedje 1975). For example, bacterial chlorophylls from purple photosynthetic bacteria have absorption maxima at 850 nm and 1000 nm. It is possible to estimate the biomass of different photosynthetic microbial populations in the same sample by estimating various chlorophylls which absorb at different wavelengths.

It is also possible to quantify photosynthetic pigments spectrofluorometrically (Lorenzen 1966; Trüper & Yentsch 1967; Loftus & Carpenter 1971; Kieffer 1973; Sharabi & Pramer 1973; Caldwell 1977). Chlorophyll *a* excited at a wavelength of 436 nm emits light at 685 nm. The spectrofluorometric determination is more sensitive than the spectrophotometric method. The increased sensitivity allows determination of chlorophyll quantities *in vivo*, i.e. without solvent extraction. However, fluorescence is subject to interference within the sample and up to a tenfold variation has been shown in estimates of natural phytoplankton biomass based on fluorescence *in vivo* and determination of extractable chlorophyll *a*. Adaptation of the photosynthetic populations, i.e. establishment of a steady physiological state, appears to eliminate some of this variability. Fluorescence determinations *in vivo* are applied more readily to planktonic populations, as light scattering and quenching are serious problems associated with these measurements when soil or sediment particles are present. Using different excitation and emission wavelengths, bacterial chlorophylls can be measured spectrofluorometrically and used to estimate the biomass of photosynthetic bacteria.

6.4.6 Protein

All microorganisms contain protein which can readily be quantified by methods such as that of Lowry *et al.* (1951). Protein determinations are used for estimating numbers of microorganisms and microbial biomass when dealing with pure cultures, but there are extreme difficulties in applying protein determinations for estimating microbial biomass in environmental samples. The relative concentrations of protein vary greatly depending on the physiological state of the microorganism and the particular species. This virtually precludes accurate estimation of biomass in environmental samples containing mixed microbial populations which may be in different states of growth.

Some procedures for estimating numbers of microorganisms involve the estimation of specific classes of proteins. For example, titres of immunoglobulin proteins can be used to estimate the extent of a microbial infection (i.e. the numbers of infecting microorganisms) within an animal system (Dubos & Hirsch 1965). Bacterial haeme proteins can be detected by chemiluminescence (Oleniacz *et al.* 1968). The reaction involves the stimulation of peroxide by haeme proteins present in bacteria which is reacted with luminol resulting in light emission. The amount of light emitted can be detected spectrophotometrically. A number of factors can influence the amount of light emitted, e.g. acids and solutes such as iron induce chemiluminescence. It is possible, using membrane filtered controls and alkaline assay conditions, to overcome these problems. Detection limits of 1×10^3 to 1×10^6 bacteria ml^{-1} have been reported for this technique. Miller and Vogelhut (1978) reported though that the limit of sensitivity of the chemiluminescent assay determined, both experimentally and theoretically, was no lower than 1×10^5 to 1×10^6 viable bacteria ml^{-1}. The relatively high concentrations of bacteria necessary for detection by luminol chemiluminescence and the possible sources of interference in environmental samples probably limit the applicability of this process.

6.5 Physiological approaches to enumeration and biomass determinations

Physiological approaches for estimating microbial biomass have been developed and applied to soil habitats. One method involves sterilising a sample, i.e. killing the microbial cells in the sample, reinoculating the samples and quantifying the mineralisation (carbon dioxide production) from the microbial biomass killed in the sterilisation procedure (Jenkinson 1976; Jenkinson & Powlson 1976; Anderson & Domsch 1978a). Chloroform fumigation has been used to sterilise samples in these procedures. The difference between the amount of carbon dioxide evolved from fumigated and non-fumigated replicate samples is used to estimate the biomass of microorganisms in the sample. A high correlation has been shown between biomass estimates based on ATP determinations and those derived from the soil fumigation method.

Another physiological approach for estimating microbial biomass involves measurement of respiration rates following substrate addition (Anderson & Domsch 1973, 1974, 1975, 1978b). The peak respiration measured during a short period is assumed to be proportional to the numbers of viable microorganisms present in the sample. Microbial inhibitors can be added to separate estimates of bacterial and fungal biomass in this method. A high correlation has been found between estimates of microbial biomass based on initial respiratory response of microbial populations to amendment with an excess of carbon and energy and the chloroform fumigation methods. These methods have not yet been applied to the variety of soils necessary to evaluate the advantages and limitations in interpretation of estimating microbial numbers and biomass by physiological approaches. The soil fumigation approach appears to have the advantage of directly assessing the amount of carbon which is particularly useful for estimating biomass.

Recommended reading

AMERICAN PUBLIC HEALTH ASSOCIATION (1975) *Standard Methods for the Examination of Water and Waste Water* 14th edn. American Public Health Association, Washington D.C.

ATLAS R. M. & BARTHA R. (1981) *Microbial Ecology: Fundamentals and Applications.* Addison-Wesley Publishing Co., Reading, Massachusets.

BLACK C. A. (1965) *Methods of Soil Analysis. Part 2, Chemical and Microbiological Properties.* American Society of Agronomy, Madison, Wisconsin.

COSTERTON J. W. & COLWELL R. R. (1979) *Native Aquatic Bacteria: Enumeration, Activity and Ecology.* ASTM Special Technical Publication No. 695, American Society for Testing Materials, Philadelphia.

JONES J. G. (1979) *A Guide to Methods for Estimating Microbial Numbers and Biomass in Fresh Water.* Scientific Publication No. 39. Freshwater Biological Association, Cumbria.

NORRIS J. R. & RIBBONS D. W. (1971) *Methods in Microbiology.* Academic Press, London.

PARKINSON D., GRAY T. R. G. & WILLIAMS S. T. (1971) *Methods for Studying the Ecology of Soil Microorganisms.* IBP Handbook No. 19. Blackwell Scientific Publications, Oxford.

ROSSWALL T. (1973) *Modern Methods in the Study of Microbial Ecology.* Ecological Research Committee of NFR, The Swedish National Research Council, Stockholm.

References

ALEXANDER M. (1965a) Most-probable-number method for microbial populations. In: *Methods of Soil Analysis. Part 2, Chemical and Microbiological Properties* (Ed. C. A. Black), pp. 1467–72. American Society of Agronomy, Madison, Wisconsin.

ALEXANDER M. (1965b) Nitrifying bacteria. In: *Methods of Soil Analysis. Part 2, Chemical and Microbiological Properties* (Ed. C. A. Black), pp. 1484–6. American Society of Agronomy, Madison, Wisconsin.

ALEXANDER M. & CLARK F. E. (1965) Nitrifying bacteria. In: *Methods of Soil Analysis. Part 2, Chemical and Microbiological Properties* (Ed. C. A. Black), pp. 1477–83. American Society of Agronomy, Madison, Wisconsin.

AMERICAN PUBLIC HEALTH ASSOCIATION (1975) *Standard Methods for the Examination of Water and Waste Water* 14th edn. American Public Health Association, Washington D.C.

ANDERSEN A. A. (1958) A new sampler for the collection, sizing and enumeration of the viable air-borne bacteria. *Journal of Bacteriology* **76**, 471–84.

ANDERSON J. P. E. & DOMSCH K. H. (1973) Quantification of bacterial and fungal contributions to soil respiration. *Archiv für Mikrobiologie* **93**, 113–27.

ANDERSON J. P. E. & DOMSCH K. H. (1974) Use of selective inhibitors in the study of respiratory activities and shifts in bacterial and fungal populations in soil. *Annales de Microbiologie* **124**, 189–94.

ANDERSON J. P. E. & DOMSCH K. H. (1975) Measurement of bacterial and fungal contributions to respiration of selected agricultural and forest soils. *Canadian Journal of Microbiology* **21**, 314–22.

ANDERSON J. P. E. & DOMSCH K. H. (1978a) Mineralization of bacteria and fungi in chloroform-fumigated soils. *Soil Biology and Biochemistry* **10**, 207–13.

ANDERSON J. P. E. & DOMSCH K. H. (1978b) A physiological method for the quantitative measurement of microbial biomass in soils. *Soil Biology and Biochemistry* **10**, 215–21.

ARISTOVSKAYA T. V. (1973) The use of capillary techniques in ecological studies of microorganisms. *Bulletins from the Ecological Research Committee (Stockholm)* **17**, 47–52.

ARISTOVSKAYA T. V. & PARINKINA O. M. (1961) New methods of studying soil microorganism associations. *Soviet Soil Science* **1**, 12–20.

ATLAS R. M. (1979) Measurement of hydrocarbon biodegradation potentials and enumeration of hydrocarbon-utilizing microorganisms using carbon-14 hydrocarbon-spiked crude oil. In: *Native Aquatic Bacteria: Enumeration, Activity and Ecology* (Eds J. W. Costerton & R. R. Colwell), pp. 196–204. ASTM Special Technical Publication No. 695, American Society for Testing and Materials, Philadelphia.

ATLAS R. M. & BARTHA R. (1981) *Microbial Ecology: Fundamentals and Applications*. Addison-Wesley Publishing Co., Reading, Massachusetts.

AUSMUS B. S. (1973) The use of the ATP assay in terrestrial decomposition studies. *Bulletins from the Ecological Research Committee (Stockholm)* **17**, 223–4.

BABUIK L. A. & PAUL E. A. (1970) The use of fluorescein isothiocyanate in the determination of the bacterial biomass in grassland soil. *Canadian Journal of Microbiology* **16**, 57–62.

BANCROFT K., PAUL E. A. & WIEBE W. J. (1976) The extraction and measurement of adenosine triphosphate from marine sediments. *Limnology and Oceanography* **21**, 473–80.

BANSE K. (1977) Determining the carbon to chlorophyll ratio of natural phytoplankton. *Marine Biology* **41**, 199–212.

BLACK C. A. (1965) *Methods of Soil Analysis. Part 2, Chemical and Microbiological Properties*. American Society of Agronomy, Madison, Wisconsin.

BOARD R. G. & LOVELOCK D. W. (1973) *Sampling—Microbiological Monitoring of Environments*. Academic Press, London.

BOHLOOL B. B. & SCHMIDT E. L. (1968) Non-specific staining: its control in immunofluorescence examination of soil. *Science* **162**, 1012–14.

BOHLOOL B. B. & SCHMIDT E. L. (1973) A fluorescent antibody technique for determination of growth rates of bacteria in soil. *Bulletins from the Ecological Research Committee (Stockholm)* **17**, 336–8.

BORDNER R. & WINTER J. (1978) *Microbiological Methods for Monitoring the Environment. I. Water and Wastes*. Environmental Protection Agency, Cincinnati, Ohio.

BOWDEN W. B. (1977) Comparison of two direct-count techniques for enumerating aquatic bacteria. *Applied and Environmental Microbiology* **33**, 1229–32.

BROCK M. L. & BROCK T. D. (1968) The application of micro-autoradiographic techniques to ecological studies. *Mitteilungen Internationale Vereinigung für theoretische und angewandte Limnologie* **15**, 1–29.

BUCK J. D. (1979) The plate count in aquatic microbiology. In: *Native Aquatic Bacteria: Enumeration, Activity and Ecology* (Eds J. W. Costerton & R. R. Colwell), pp. 19–28. ASTM Special Technical Publication No. 695, American Society for Testing and Materials, Philadelphia.

BURNS R. G. (1980) Microbial adhesion to Soil Surfaces: consquences for growth and enzyme activities. In: *Microbial Adhesion to Surfaces* (Eds R. C. W. Berkeley, J. M. Lynch, J. Melling, P. R. Rutter & B. Vincint), pp. 249–62. Ellis Harwood, Chichester.

CALDWELL D. E. (1977) Accessory pigment fluorescence for quantitation of photosynthetic microbial populations. *Canadian Journal of Microbiology* **23**, 1594–7.

CALDWELL D. E. & TIEDJE J. M. (1975) The structure of anaerobic bacterial communities in the hypolimnia of several Michigan lakes. *Canadian Journal of Microbiology* **21**, 377–85.

CATTOLICO R. A. & GIBBS S. P. (1975) Rapid filter method for the microfluorimetric analysis of DNA. *Analytical Biochemistry* **69**, 572–82.

CHAPPELLE E. W., PICCIOLO G. L. & DEMING J. W. (1978) Determination of bacterial content in fluids. In: *Methods in Enzymology. Bioluminescence and Chemilu-*

minescence (Ed. M. A. DeLuca), vol. 57, pp. 65–72. Academic Press, New York.

CHOLODNY N. (1930) Über eine neue Methode zur Untersuchung der Bodenmikroflora. *Archiv für Mikrobiologie* **1**, 650–2.

CLARK F. E. (1965a) Agar-plate method for total microbial count. In: *Methods of Soil Analysis. Part 2, Chemical and Microbiological Properties* (Ed. C. A. Black), pp. 1460–6. American Society of Agronomy, Inc., Madison, Wisconsin.

CLARK F. E. (1965b) Aerobic spore-forming bacteria. In: *Methods of Soil Analysis. Part 2, Chemical and Microbiological Properties* (Ed. C. A. Black), pp. 1473–6. American Society of Agronomy, Madison, Wisconsin.

CLARK F. E. (1965c) Azotobacter. In: *Methods of Soil Analysis. Part 2, Chemical and Microbiological Properties* (Ed. C. A. Black), pp. 1493–7. American Society of Agronomy, Madison, Wisconsin.

CLARK F. E. (1965d) Rhizobia. In: *Methods of Soil Analysis. Part 2, Chemical and Microbiological Properties* (Ed. C. A. Black), pp. 1487–92. American Society of Agronomy, Madison, Wisconsin.

CLARK F. E. (1965e) Actinomycetes. In: *Methods of Soil Analysis. Part 2, Chemical and Microbiological Properties* (Ed. C. A. Black), pp. 1498–501. American Society of Agronomy, Madison, Wisconsin.

CLARK F. E. & BEARD W. E. (1965) Protozoa. In: *Methods of Soil Analysis. Part 2, Chemical and Microbiological Properties* (Ed. C. A. Black), pp. 1513–16. American Society of Agronomy, Madison, Wisconsin.

CLARK F. E. & DURRELL L. W. (1965) Algae. In: *Methods of Soil Analysis. Part 2, Chemical and Microbiological Properties* (Ed. C. A. Black), pp. 1506–12. American Society of Agronomy, Madison, Wisconsin.

CLARKE J. H. & PARKINSON D. (1960) A comparison of three methods for the assessment of fungal colonization of seedling roots of leek and broad beans. *Nature, London* **188**, 166–7.

COATES D. A. (1977) Enhancement of the sensitivity of the *Limulus* assay for the detection of Gram-negative bacteria. *Journal of Applied Bacteriology* **42**, 445–9.

COCHRAN W. G. (1950) Estimation of bacterial densities by means of the "most probable number". *Biometrics* **2**, 105–16.

COHEN Y., KRUMBEIN W. & SHILO M. (1977) Solar Lake (Sinai). 3. Bacterial distribution and production. *Limnology and Oceanography* **22**, 621–34.

COLWELL R. R. (1979) Enumeration of specific populations by the most-probable-number (MPN) method. In: *Native Aquatic Bacteria: Enumeration, Activity and Ecology* (Eds J. W. Costerton & R. R. Colwell), pp. 56–64. ASTM Special Technical Publication No. 695, American Society for Testing and Materials, Philadelphia.

COSTERTON J. W. & COLWELL R. R. (1979) *Native Aquatic Bacteria: Enumeration, Activity and Ecology.* ASTM Special Technical Publication No. 695, American Society for Testing and Materials, Philadelphia.

COSTERTON J. W. & GEESEY G. G. (1979) Which populations of aquatic bacteria should we enumerate? In: *Native Aquatic Bacteria: Enumeration, Activity and Ecology* (Eds J. W. Costerton & R. R. Colwell), pp. 7–18. ASTM Special Technical Publication No. 695, American Society for Testing and Materials, Philadelphia.

CROW S. A., AHEARN D. G., COOK W. L. & BOURQUIN A. W. (1975) Densities of bacteria and fungi in coastal surface films as determined by a membrane adsorption procedure. *Limnology and Oceanography* **20**, 644–55.

DALEY R. J. (1979) Direct epifluorescence enumeration of native aquatic bacteria: uses, limitations and comparative accuracy. In: *Native Aquatic Bacteria: Enumeration, Activity and Ecology* (Eds J. W. Costerton & R. R. Colwell), pp. 29–45. ASTM Special Technical Publication No. 695, American Society for Testing and Materials, Philadelphia.

DALEY R. J. & HOBBIE J. E. (1975) Direct counts of aquatic bacteria by a modified epifluorescence technique. *Limnology and Oceanography* **20**, 875–82.

DAZZO F. B. & BRILL W. J. (1977) Receptor sites on clover and alfalfa roots for *Rhizobium. Applied and Environmental Microbiology* **33**, 132–6.

DEMING J. W., PICCIOLO G. L. & CHAPPELLE E. W. (1979) Important factors in adenosine triphosphate determinations using firefly luciferase: applicability of the assay to studies of native aquatic bacteria. In: *Native Aquatic Bacteria: Enumeration, Activity and Ecology* (Eds J. W. Costerton & R. R. Colwell), pp. 89–98. ASTM Special Technical Publication No. 695, American Society for Testing and Materials, Philadelphia.

DICKINSON C. H. & PREECE T. F. (1976) *Microbiology of Aerial Plant Surfaces.* Academic Press, London.

DUBOS R. J. & HIRSCH J. G. (1965) *Bacterial and Mycotic Infections of Man.* J. B. Lippincott, Philadelphia.

EDMONDSON W. T. (1974) A simplified method for counting phytoplankton. In: *A Manual on Methods for Measuring Primary Production in Aquatic Environments* (Ed. R. A. Vollenweider), pp. 14–16. IBP Handbook No. 12, 2nd edn. Blackwell Scientific Publications, Oxford.

FARRAH S. R., GOYAL S. M., GERBA C. P., WALLIS C. & MELNICK J. L. (1977) Concentration of enteroviruses from estuarine water. *Applied and Environmental Microbiology* **33**, 1192–6.

FAUST M. A. & CORRELL D. L. (1977) Autoradiographic study to detect metabolically active phytoplankton and bacteria in Rhode River Estuary. *Marine Biology* **41**, 293–305.

FINLAY B. J., LAYBOURN J. & STRACHAN I. (1979) A technique for the enumeration of benthic ciliated protozoa. *Oecologia* **39**, 375–7.

FLIERMANS C. B., CHERRY W. D., ORRISON L. H. &

THACKER L. (1979) Isolation of *Legionella pneumophila* from nonepidemic-related aquatic habitats. *Applied and Environmental Microbiology* **37**, 1239–42.

FLIERMANS C. B. & SCHMIDT E. L. (1975) Autoradiography and immunofluorescence combined for autecological study of single cell activity with *Nitrobacter* as a model system. *Applied Microbiology* **30**, 674–7.

FREDERICK L. R. (1965) Microbial populations by direct microscopy. In: *Methods of Soil Analysis. Part 2, Chemical and Microbiological Properties* (Ed. C. A. Black), pp. 1452–9. American Society of Agronomy, Madison, Wisconsin.

GEESEY G. G. & COSTERTON J. W. (1979) Bacterial biomass determinations in a silt-laden river: comparison of direct count epifluorescence microscopy and extractable adenosine triphosphate techniques. In: *Native Aquatic Bacteria: Enumeration, Activity and Ecology* (Eds J. W. Costerton & R. R. Colwell), pp. 117–30. ASTM Special Technical Publication No. 695, American Society for Testing and Materials, Philadelphia.

GIBBONS R. J. & VAN HOUTE J. (1975) Bacterial adherence in oral microbiology. *Annual Reviews of Microbiology*, **29**, 19–44.

GRAY T. R. G. (1967) Stereoscan electron microscopy of soil microorganisms. *Science* **155**, 1668–70.

GRAY T. R. G. (1973) The use of the fluorescent-antibody technique to study the ecology of *Bacillus subtilis* in soil. *Bulletins from the Ecological Research Committee (Stockholm)* **17**, 119–22.

GRAY T. R. G., BAXBY P., HALL I. R. & GOODFELLOW M. (1968) Direct observation of bacteria in soil. In: *The Ecology of Soil Bacteria* (Eds T. R. G. Gray & D. Parkinson), pp. 171–97. University of Toronto Press, Toronto.

GREEN B. L., CLAUSEN E. & LITSKY W. (1975) Comparison of the new Millipore HC with conventional membrane filters for the enumeration of faecal coliform bacteria. *Applied Microbiology* **30**, 697–9.

GREGORY P. H. (1973) *The Microbiology of the Atmosphere*. John Wiley & Sons, New York.

HAMILTON R. D. & HOLM-HANSEN O. (1955) Adenosine triphosphate content of marine bacteria. *Limnology and Oceanography* **12**, 319–24.

HARRIS J. E., MCKEE T. R., WILSON R. C. & WHITEHOUSE U. G. (1972) Preparation of membrane filter samples for direct examination with an electron microscope. *Limnology and Oceanography* **17**, 784–7.

HEANEY S. I. (1978) Some observations on the use of *in vivo* fluorescence technique to determine chlorophyll a in natural populations and cultures of freshwater phytoplankton. *Freshwater Biology* **8**, 115–26.

HERBERT D., PHIPPS P. J. & STRANGE R. E. (1971) Chemical analysis of microbial cells. In: *Methods in Microbiology* (Eds J. R. Norris & D. W. Ribbons), vol. 5B, pp. 209–344. Academic Press, London.

HILL I. R. & GRAY T. R. G. (1967) Application of the fluorescent antibody technique to an ecological study of bacteria in soil. *Journal of Bacteriology* **93**, 1888–96.

HIRSCH P. & PANKRATZ ST. H. (1970) Study of bacterial populations in natural environments by use of submerged electron microscope grids. *Zeitschrift für allgemeine Mikrobiologie* **10**, 589–605.

HOBBIE J. E., DALEY R. J. & JASPER S. (1977) Use of Nuclepore filters for counting bacteria by fluorescence microscopy. *Applied and Environmental Microbiology* **33**, 1225–8.

HOBBIE J. E., HOLM-HANSEN O., PACKARD T. T., POMEROY L. R., SHELDON R. W., THOMAS J. P. & WIEBE W. J. (1972) A study of the distribution and activity of micro-organisms in ocean water. *Limnology and Oceanography* **17**, 544–55.

HOLM-HANSEN O. (1969) Determination of microbial biomass in ocean profiles. *Limnology and Oceanography* **19**, 31–4.

HOLM-HANSEN O. (1973) The use of ATP determinations in ecological studies. *Bulletins from the Ecological Research Committee (Stockholm)* **17**, 215–22.

HOLM-HANSEN O. & BOOTH C. R. (1966) The measurement of adenosine triphosphate in the ocean and its ecological significance. *Limnology and Oceanography* **11**, 510–19.

HOLM-HANSEN O. & RIEMAN B. (1978) Chlorophyll a determination: improvement in methodology. *Oikos* **30**, 438–47.

HOLM-HANSEN O., SUTCLIFFE W. H. & SHARPE J. (1968) Measurement of deoxyribonucleic acid in the ocean and its ecological significance. *Limnology and Oceanography* **13**, 507–14.

HOPPE H. G. (1976) Determination and properties of actively metabolising heterotrophic bacteria in the sea, investigated by means of microautoradiography. *Marine Biology* **36**, 291–302.

HUNGATE R. E. (1969) A roll tube method for cultivation of strict anaerobes. In: *Methods in Microbiology* (Eds J. R. Norris & D. W. Ribbons), vol. 3B, pp. 117–32. Academic Press, London.

HUNGATE R. E. & MACY J. (1973) The roll-tube method for cultivation of strict anaerobes. *Bulletins from the Ecological Research Committee (Stockholm)* **17**, 123–6.

JANNASCH H. W. & JONES G. E. (1959) Bacterial populations in sea water as determined by different methods of enumeration. *Limnology and Oceanography* **4**, 128–39.

JANNASCH H. W. & MADDUX W. S. (1967) A note on bacteriological sampling of seawater. *Journal of Marine Research* **25**, 185–9.

JANNASCH H. W. & WIRSEN C. O. (1977) Retrieval of concentrated and undecompressed microbial populations from the deep sea. *Applied and Environmental Microbiology* **33**, 642–6.

JENKINSON D. S. (1976) The effects of biocidal treatments on metabolism in soil. 4. The decomposition of fumigated organisms in soil. *Soil Biology and Biochemistry* **8**, 203–8.

JENKINSON D. S. & OADES J. M. (1979) A method for measuring adenosine triphosphate in soil. *Soil Biology and Biochemistry* **11**, 193–9.

JENKINSON D. S. & POWLSON D. S. (1976) The effects of biocidal treatments on metabolism in soil. 5. A method for measuring soil biomass. *Soil Biology and Biochemistry* **8**, 209–13.

JENSEN V. (1968) The plate count technique. In: *The Ecology of Soil Bacteria* (Eds T. R. G. Gray & D. Parkinson), pp. 158–70. Liverpool University Press, Liverpool.

JONES J. G. (1970) Studies on freshwater bacteria: effect of medium composition and method on estimates of bacterial population. *Journal of Applied Bacteriology* **33**, 679–87.

JONES J. G. (1979) *A Guide to Methods for Estimating Microbial Numbers and Biomass in Fresh Water*. Scientific Publication No. 39, Freshwater Biological Association, Cumbria.

JONES J. G. & SIMON B. M. (1975a) An investigation of errors in direct counts of aquatic bacteria by epifluorescence microscopy, with reference to a new method for dyeing membrane filters. *Journal of Applied Bacteriology* **39**, 317–29.

JONES J. G. & SIMON B. M. (1975b) Some observations on the fluorometric determination of glucose in fresh water. *Limnology and Oceanography* **20**, 882–7.

JONES J. G. & SIMON B. M. (1977) Increased sensitivity in the measurement of ATP in freshwater samples with a comment on the adverse effect of membrane filtration. *Freshwater Biology* **7**, 253–60.

JONES P. C. T. & MOLLISON J. E. (1948) A technique for the quantitative estimation of soil micro-organisms. *Journal of General Microbiology* **2**, 54–69.

KAPUSCINSKI J. & SKOCZYLAS B. (1977) Simple and rapid fluorometric method for DNA microassay. *Analytical Biochemistry* **83**, 252–7.

KARL D. M. (1978) Occurrence and ecological significance of GTP in the ocean and in microbial cells. *Applied and Environmental Microbiology* **36**, 349–55.

KARL D. M. & HOLM-HANSEN O. (1978) Methodology and measurement of adenylate energy charge ratios in environmental samples. *Marine Biology* **48**, 185–97.

KIEFFER D. A. (1973) Fluorescence properties of natural phytoplankton populations. *Marine Biology* **22**, 263–9.

KING J. D. & WHITE D. C. (1977) Muramic acid as a measure of microbial biomass in estuarine and marine samples. *Applied and Environmental Microbiology* **33**, 777–83.

KOGURE K., SIMIDU U. & TAGA N. (1979) A tentative direct microscopic method for counting living marine bacteria. *Canadian Journal of Microbiology* **25**, 415–20.

KUBITSCHEK H. E. (1969) Counting and sizing microorganisms with the Coulter counter. In: *Methods in Microbiology* (Eds J. R. Norris & D. W. Ribbons), vol. 1, pp. 593–610. Academic Press, London.

LARSSON K., WENBULL C. & CRONBERG G. (1978) Comparison of light and electron microscopic determinations of the number of bacteria and algae in lake water. *Applied and Environmental Microbiology* **35**, 397–404.

LEHMICKE L. G., WILLIAMS R. I. & CRAWFORD R. L. (1979) ^{14}C-most-probable-number method for enumeration of active heterotrophic microorganisms in natural waters. *Applied and Environmental Microbiology* **38**, 644–9.

LIN S. D. (1976) Evaluation of Millipore HA and HC membrane filters for the enumeration of indicator bacteria. *Applied and Environmental Microbiology* **32**, 300–2.

LITCHFIELD C. D., RAKER J. B., ZINDULIS J., WATANABE R. T. & STEIN D. J. (1975) Optimization of procedures for the recovery of heterotrophic bacteria from marine sediments. *Microbial Ecology* **1**, 219–233.

LOFTUS M. E. & CARPENTER J. H. (1971) A fluorometric method for determining chlorophylls a, b and c. *Journal of Marine Research* **29**, 319–38.

LORENZEN C. J. (1966) A method for the continuous measurement of *in vivo* chlorophyll concentration. *Deep-Sea Research* **13**, 223–7.

LOWRY O. H., ROSEBROUGH N. J., FARR A. L. & RANDALL R. J. (1951) Protein measurement with the Folin phenol reagent. *Journal of Biological Chemistry* **193**, 265–75.

MARSHALL K. C. (1976) *Interfaces in Microbial Ecology*. Harvard University Press, Cambridge, Massachusetts.

MELCHIORRI-SANTOLINI U. (1972) Enumeration of microbial concentration of dilution series (MPN). In: *Techniques for the Assessment of Microbial Production and Decomposition in Fresh Waters* (Eds Y. I. Sorokin & H. Kadota), pp. 64–70. IBP Handbook No. 23, Blackwell Scientific Publications, Oxford.

MENZIES J. D. (1965) Fungi. In: *Methods of Soil Analysis. Part 2, Chemical and Microbiological Properties* (Ed. C. A. Black), pp. 1502–5. American Society of Agronomy, Madison, Wisconsin.

MEYER-REIL L. A. (1978) Autoradiography and epifluorescence microscopy combined for the determination of number and spectrum of actively metabolizing bacteria in natural waters. *Applied and Environmental Microbiology* **36**, 506–12.

MILLAR W. N. & CASIDA L. E. JR. (1970) Evidence foı muramic acid in soil. *Canadian Journal of Microbiology* **16**, 299–304.

MILLER C. A. & VOGELHUT P. O. (1978) Chemiluminescent detection of bacteria: experimental and theoretical limits. *Applied and Environmental Microbiology* **35**, 813–16.

MORARITY D. J. W. (1975) A method for estimating the

biomass of bacteria in aquatic sediments and its application in trophic studies. *Oecologia* **20**, 219–29.

MORARITY D. J. W. (1977) Improved method using muramic acid to estimate biomass of bacteria in sediments. *Oceologia* **26**, 317–23.

MORARITY D. J. W. (1978) Estimation of microbial biomass in water and sediments using muramic acid. In: *Microbial Ecology* (Eds M. W. Loutit & J. A. R. Miles), pp. 31–3. Springer Verlag, Berlin.

MORITA R. Y. (1975) Psychrophilic bacteria. *Bacteriological Reviews* **39**, 144–67.

NIKITIN D. I. (1973) Direct electron microscopic techniques for the obervation of microorganisms in soil. *Bulletins from the Ecological Research Committee (Stockholm)* **17**, 85–92.

NORRIS J. R. & RIBBONS D. W. (1971) *Methods in Microbiology*. Academic Press, London.

OADES J. M. & JENKINSON D. S. (1979) Adenosine triphosphate content of the soil microbial biomass. *Soil Biology and Biochemistry* **11**, 201–4.

OLENIACZ W. S., PISANO M. A., ROSENFELD M. H. & ELGART R. L. (1968) Chemiluminescent method for detecting micro-organisms in water. *Environmental Science and Technology* **2**, 1030–3.

OLSON B. H. (1978) Enhanced accuracy of coliform testing in seawater by a modification of the most probable number method. *Applied and Environmental Microbiology* **36**, 438–44.

PAERL H. W. & SHIMP S. L. (1973) Preparation of filtered plankton and detritus for study with scanning electron microscopy. *Limnology and Oceanography* **18**, 802–5.

PAERL H. W., TILZER M. M. & GOLDMAN C. R. (1976) Chlorophyll a versus adenosine triphosphate as algal biomass indicators in lakes. *Journal of Phycology* **12**, 242–6.

PAERL H. W. & WILLIAMS N. J. (1976) The relation between adenosine triphosphate and microbial biomass in diverse aquatic ecosystems. *Internationale Revue der gesamten Hydrobiologie* **61**, 659–64.

PARKINSON D. (1973) Techniques for the study of soil fungi. *Bulletins from the Ecological Research Committee (Stockholm)* **17**, 29–36.

PARKINSON D., GRAY T. R. G. & WILLIAMS S. T. (1971) *Methods for Studying the Ecology of Soil Micro-Organisms*. IBP Handbook No. 19, Blackwell Scientific Publications, Oxford.

PATRIQUIN D. G. & DOBEREINER J. (1978) Light microscopy obervations of tetrazolium-reducing bacteria in the endorhizosphere of maize and other grasses in Brazil. *Canadian Journal of Microbiology* **24**, 734–42.

PAUL E. A. & JOHNSON R. L. (1977) Microscope counting and adenosine 5′- triphosphate measurement in determining microbial growth in soils. *Applied and Environmental Microbiology* **34**, 263–9.

PERFILEV B. V. & GABE D. R. (1969) *Capillary Methods of Investigating Microorganisms*. Oliver & Boyd, Edinburgh.

POSTGATE J. R. (1969) Viable counts and viability. In: *Methods in Microbiology* (Eds J. R. Norris & D. W. Ribbons), vol. 1, pp. 611–28. Academic Press, London.

PRATT D. & REYNOLDS J. (1972) Selective media for characterizing marine bacterial populations. In: *Effects of the Ocean Environment on Microbial Activities* (Eds R. R. Colwell & R. Y. Morita), pp. 258–67. University Park Press, Baltimore.

PUGSLEY A. P. & EVISON L. M. (1975) A fluorescent antibody technique for the enumeration of faecal streptococci in water. *Journal of Applied Bacteriology* **38**, 63–5.

RAMSAY A. J. (1974) The use of autoradiography to determine the proportion of bacteria metabolising in an aquatic habitat. *Journal of General Microbiology* **80**, 363–73.

ROSEBURY I. (1962) *Microorganisms Indigenous to Man*. McGraw-Hill, New York.

ROSS D. A. (1970) *Introduction to Oceanography*. Appleton-Century-Crofts, New York.

ROSSI G., RICCARDO S., GESUE G., STANGANELLI M. & WANG T. K. (1936) Direct microscopic and bacteriological investigations of the soil. *Soil Science* **41**, 53–66.

ROSSWALL T. (1973) *Modern Methods in the Study of Microbial Ecology*. Ecological Research Committee of NFR, The Swedish National Research Council, Stockholm.

ROWE R., TODD R. & WAIDE J. (1977) Microtechnique for most-probable-number analysis. *Applied and Environmental Microbiology* **33**, 675–80.

SCHMIDT E. L. (1973) Fluorescent antibody techniques for the study of microbial ecology. *Bulletins from the Ecological Research Committee (Stockholm)* **17**, 67–76.

SCHMIDT E. L. & BANKOLE R. O. (1963) The use of fluorescent antibody with the buried slide technique. In: *Soil Organisms* (Eds J. Doeksen & J. van der Drift), pp. 197–203. North Holland Publishing Co., Amsterdam.

SCHMIDT E. L., BANKOLE R. O. & BOHLOOL B. B. (1968) Fluorescent antibody approach to study of *Rhizobia* in soil. *Journal of Bacteriology* **95**, 1987–92.

SETARO F. & MORELY C. G. D. (1976) A modified fluorometric method for the determination of microgram quantities of DNA for cell or tissue cultures. *Analytical Biochemistry* **71**, 313–17.

SHAPTON D. A. & BOARD R. G. (1971) *Isolation of Anaerobes*. The Society for Applied Bacteriology Technical Series No. 5. Academic Press, London.

SHARABI N. EL.-D. & PRAMER D. (1973) A spectrophotofluorometric method for studying algae in soil. *Bulletins from the Ecological Research Committee (Stockholm)*, **17**, 77–84.

SHARMA P. D., FISHER P. J. & WEBSTER J. (1977) Critique of the chitin assay technique for estimation of fungal biomass. *Transactions of the British Mycological Society* **69**, 479–83.

SIMIDU U. (1972) Improvement of media for enumeration

and isolation of heterotrophic bacteria in seawater. In: *Effect of the Ocean Environment on Microbial Activities* (Eds R. R. Colwell & R. Y. Morita), pp. 249–57. University Park Press, Baltimore.

SINGH B. N. (1946) A method of estimating the numbers of soil protozoan, especially amoebae, based on their differential feeding on bacteria. *Annals of Applied Biology* **33**, 112–19.

SKINNER F. A., JONES P. C. T. & MOLLISON J. E. (1952) A comparison of a direct- and a plate counting technique for quantitative estimation of soil micro-organisms. *Journal of General Microbiology* **6**, 261–71.

SODERSTROM B. E. (1977) Vital staining of fungi in pure cultures and in soil with fluorescein diacetate. *Soil Biology and Biochemistry* **9**, 59–63.

STANIER R. Y. & SMITH J. H. C. (1960) The chlorophylls of green bacteria. *Biochimica et Biophysica Acta* **41**, 478–84.

STANLEY P. M., GAGE M. A. & SCHMIDT E. L. (1979) Enumeration of specific populations by immunufluorescence. In: *Native Aquatic Bacteria: Enumeration, Activity and Ecology* (Eds J. W. Costerton & R. R. Colwell), pp. 46–55. ASTM Special Technical Publication No. 695, American Society for Testing and Materials, Philadelphia.

STEVENSON L. H., CHRAZANOWSKI T. H. & ERKENBRECHER C. W. (1979) The adenosine triphosphate assay; conceptions and misconceptions. In: *Native Aquatic Bacteria: Enumeration, Activity and Ecology* (Eds J. W. Costerton & R. R. Colwell), pp. 99–116. ASTM Special Technical Publication No. 695, American Society for Testing and Materials, Philadelphia.

STRAYER R. F. & TIEDJE J. M. (1978) Application of fluorescent-antibody technique to the study of a methanogenic bacterium in lake sediments. *Applied and Environmental Microbiology* **35**, 192—8.

STRUGGER S. (1948) Fluorescence microscope examination of bacteria. *Canadian Journal of Research, Series C* **26**, 188–93.

SUTCLIFFE W. H. JR., ORR E. A. & HOLM-HANSEN O. (1976) Difficulties with ATP measurements in inshore waters. *Limnology and Oceanography* **21**, 145–9.

SWIFT M. J. (1973a) The estimation of mycelial biomass by determination of the hexosamine content of wood tissue decayed by fungi. *Soil Biology and Biochemistry* **5**, 321–2.

SWIFT M. J. (1973b) Estimation of mycelial growth during decomposition of plant litter. *Bulletins from the Ecological Research Committee (Stockholm)* **17**, 323–8.

THOMAS A., NICHOLAS D. P. & PARKINSON D. (1965) Modification of the agar film technique for assaying lengths of mycelium in soil. *Nature, London* **205**, 105.

TODD R. L., CROMACK K. JR. & KNUTSON R. M. (1973) Scanning electron microscopy in the study of terrestrial microbial ecology. *Bulletins from the Ecological Research Committee (Stockholm)* **17**, 109–18.

TRIBE H. I. (1957) Ecology of microorganisms in soils as observed during their development upon buried cellulose film. In: *Microbial Ecology* (Eds C. C. Spicer & R. E. O. Williams), pp. 287–98. Cambridge University Press, Cambridge.

TROLLDENIER G. (1973) The use of fluorescence microscopy for counting soil microorganisms. *Bulletins from the Ecological Research Committee (Stockholm)* **17**, 53–9.

TRÜPER H. G. & YENTSCH C. S. (1967) Use of glass fiber filters for the rapid preparation of *in vivo* absorption. Spectra of phososynthetic bacteria. *Journal of Bacteriology* **94**, 1255–6.

UDENFRIEND S. (1969) *Fluorescence Assay in Biology and Medicine, vol. 2.* Academic Press, New York.

WAID J. S. (1957) Distribution of fungi within the decomposing tissues of rye-grass roots. *Transactions of the British Mycological Society* **40**, 391–406.

WAID J. S., PRESTON K. J. & HARRIS P. J. (1973) Autoradiographic techniques to detect active microbial cells in natural habitats. *Bulletins from the Ecological Research Committee (Stockholm)* **17**, 317–22.

WATSON S. W. & HOBBIE J. E. (1979) Measurement of bacterial biomass as lipopolysaccharide. In: *Native Aquatic Bacteria: Enumeration, Activity and Ecology* (Eds J. W. Costerton & R. R. Colwell), pp. 82–8. ASTM Special Technical Publication No. 695, American Society for Testing and Materials, Philadelphia.

WATSON S. W., NOVITSKY T. J., QUINBY H. L. & VALOIS F. W. (1977) Determination of bacterial number and biomass in the marine environment. *Applied and Environmental Microbiology* **33**, 940–6.

WHITE D. C., BOBBIE R. J., HERRON J. S., KING J. D. & MORRISON S. J. (1979) Biochemical measurements of microbial mass and activity from environmental samples: in: *Native Aquatic Bacteria: Enumeration, Activity and Ecology* (Eds J. W. Costerton & R. R. Colwell), pp. 69–81. ASTM Special Technical Publication No. 695, American Society for Testing and Materials, Philadelphia.

WIEBE W. J. & BANCROFT K. (1975) Use of adenylate energy charge ratio to measure growth state of natural microbial communities. *Proceedings of the National Academy of Sciences, USA*, **72**, 2112–5.

WILLOUGHBY L. G. (1978) Methods for studying micro-organisms on decaying leaves and wood in fresh water. In: *Techniques for the Study of Mixed Populations* (Eds D. W. Lovelock & R. Davies), pp. 31–50. Academic Press, London.

WOLF H. W. (1972) The coliform count. In: *Water Pollution Microbiology* (Ed. R. Mitchell), pp. 333–45. John Wiley & Sons, New York.

ZIMMERMANN R., ITURRIAGA R. & BECKER-BIRCK J. (1978) Simultaneous determination of the total number of aquatic bacteria and the number thereof involved in respiration. *Applied and Environmental Microbiology* **36**, 926–35.

ZIMMERMANN R. & MEYER-REIL L. A. (1974) A new

method for fluorescence staining of bacterial populations on membrane filters. *Kieler Meeresforschungen Wissenshaftliche* **30**, 24–7.

ZoBell C. E. (1964) Hydrostatic measure as a factor affecting the activity of marine microbes. In: *Recent Researches in the Field of Hydrosphere, Atmosphere and Nuclear Geochemistry* (Eds M. Miyahe & T. Koyama), pp. 83–116. Marazen, Tokyo.

Part 2
The Biosphere

Chapter 7 • The Soil Environment: Clay–Humus–Microbe Interactions

Guenther Stotzky and Richard G. Burns

7.1 Introduction

Any study of the microbial ecology of soils must consider the spatial and temporal interrelations of the potential reactants. A conclusion about the soil environment based solely on gross events in a large volume of soil and at a particular moment in time is incomplete and often misleading; for example, measuring carbon dioxide evolution from, or dehydrogenase activity in, a cm^3 or even a mm^3 volume of soil reveals little about the numerous and diverse events that occur in microhabitats and that contribute to the data obtained. Similarly, determinations of the bulk soil pH (pH_b) yield an average value and may conceal variations of two or more pH units from one microsite to another (McLaren & Skujins 1968). These variations in pH can be of profound importance in the ecology of the local microbial population. In an axenic culture that contains an abundant, homogenised and easily assimilable source of nutrients and is maintained at the optimum pH and temperature (e.g. in a chemostat), all viable microbes may be performing the same transformations at approximately the same rate within the same environment. In soils, however, the very opposite applies, as a genetically diverse population is competing for a small quantity of unevenly distributed substrate and is subject to fluctuations in pH, redox potential, ionic concentration and level of hydration. In addition, the physical barriers erected by inorganic particles, either individually or as components of aggregates, and various microbe–microbe, microbe–plant and microbe–animal interactions, create unique locales.

In the past decade, significant progress has been made in understanding the micro-environments in which microbes reside in soil (Stotzky 1972, 1973, 1980; Hattori & Hattori 1976; Marshall 1976; Burns 1979, 1980; Marshall 1980). What emerges from these studies is that the physical and chemical properties of clay and humic materials impart an important influence on the microorganisms in their vicinity and that this influence is the result principally of four characteristics of colloidal clay and organic matter;

(1) their high surface to volume ratios;
(2) their cation exchange capacity;
(3) their ability to retain water;
(4) their ionic properties.

Consequently, microorganisms, viruses, substrates, enzymes, products, inorganic ions and water molecules tend to be concentrated and, therefore, to interact at the soil colloid/liquid interface. The geometry of micro-environments in sediments is assumed to be similar.

7.2. Nature and properties of soil colloids

The principal colloids in soil are clay minerals, humic substances, viruses and the cells of bacteria. The clay minerals are primarily crystalline hydrous aluminosilicates, the component silicon oxide tetrahedral and aluminium hydroxide octahedral sheets of which are associated either in a 2:1 ratio (Si–Al–Si) or a 1:1 ratio (Si–Al). These associations, the unit layers, are held together tenaciously by hydrogen bonds (e.g. kaolinite–serpentine group) or are only weakly associated through van der Waals forces (e.g. smectite–vermiculite group). Intermediate and amorphous forms also exist. As a result, 2:1 layer clays, especially smectites, tend to expand upon wetting, exposing a significant internal surface area between adjacent silicon tetrahedral layers. 1:1 layer clays do not normally expand upon wetting. 2:1 layer clays are also particularly subject during morphogenesis to isomorphous substitution: a process wherein structural cations (e.g. Si^{4+}, Al^{3+}) are replaced by those of a lower valency (e.g. Al^{3+}, Fe^{2+}) and which imparts a permanent net negative charge to the clays. This electronegativity is compensated by exchangeable cations (e.g. Ca^{2+}, Mg^{2+}, K^{+}, Na^{+}, H^{+}, NH_4^{+}) adsorbed from the soil solution and is expressed as the cation exchange capacity (CEC) of the clay. Clays also have a small anion exchange capacity (AEC). The charged surface and its associated ions are collectively termed the electrical or diffuse double layer (DDL), a zone that will vary from 0.5 to 100 nm in thickness depending upon the valency and concentration of electrolyte (Santoro & Stotzky 1967a, b; 1968). Detailed descriptions of clay mineralogy can be found elsewhere (Grim 1968; Gieseking 1975; Brown *et al.* 1978).

Soil organic matter can be considered as being composed of three discernible fractions:

(1) a macroscopic component represented by particulate plant, animal and microbial debris in the early stages of disintegration and decay;

(2) a biochemically well-defined and largely soluble component (e.g. carbohydrates, proteins) arising from the degradation of the material in (1);

(3) a complex dark-coloured component, largely aromatic and polymeric, arising during lignin breakdown as well as being synthesised *de novo* by the soil microbiota. This humic material may be further differentiated, on the basis of its solubility, into humic acids, fulvic acids and humins.

The third type of soil organic matter ranks with the expanding lattice clays in terms of importance to microbial activities. This is because humic substances are polydisperse materials that expand upon wetting and therefore, like the 2:1 clays, have an extensive internal surface area that is potentially available for associated biological activity. Soil organic matter is predominantly anionic, with the principal functional groups being carboxyl, phenolic and alcoholic hydroxyl, methyl, amino and carbonyl, and the sign and density of charge will vary according to the pH_b and the isoelectric point (pI) of the constituent moieties. Soil organic colloids are described in detail by Schnitzer and Khan (1972), Flaig *et al.* (1975) and Hayes and Swift (1978).

Some of the properties of soil clay and organic colloids are listed in Table 1. However, it should be emphasised that clays and humic materials (as well as other organic molecules) are often intimately associated to form organo-mineral complexes (Theng 1979), and these associations can have profound effects on the ionic properties and exposed surface areas of the individual components. Furthermore, amorphous and crystalline oxides of iron, aluminium and manganese interact with clays and reduce the density of their negative and positive charges.

Microorganisms and viruses have a net negative charge at the pH of most microbial habitats (i.e. the pH_b is above the pI or the dissociation constant

Table 1. Properties of soil clay and organic colloids.

Colloid type	Layering	Swelling	Surface area ($m^2\ g^{-1}$)	CEC ($mEq\ g^{-1}$)	Basal spacing (nm)
Kaolinite	1:1	non-expanding	25–50	0·02–0·1	0.7
Vermiculite	2:1	expanding	500–700	1.2–2.0	0.93–1.57*
Smectite	2:1	expanding	700–750	0.6–1.3	0.95–2.2*
Organic matter	—	expanding	500–800	2.0–5.0	—

*Varies according to level of hydration and species of interlayer cation.

(p*K*) of constituent molecules). The cell walls of Gram-positive bacteria have teichoic acids, which are anionic polyelectrolytes whose charge is derived from phosphate ester groupings, concentrated towards their outer surface. The magnitude of the electronegativity of these cells is regulated by pH, alanine ester residues and, possibly, conformational changes in wall components (Archibald *et al.*·1973). Depending on the pH, the proteins of the envelopes of Gram-negative bacteria can have either a net negative or a net positive charge that arises from the ionisation of carboxyl or amino groups, respectively. At the pH of the pI, the net surface charge of these bacteria, due to their envelope protein constituents, will be zero. The structure of bacterial cell walls has been reviewed by Ward and Berkeley (1980), and Marshall (1976) has summarised the variable ionogenic properties of the Gram-negative bacterial genus, *Rhizobium*. The presence of aminopolysaccharides, polyuronides and proteins in fungal cell walls suggests that these microorganisms have a range of ionic properties similar to those of the bacteria.

It is obvious from the foregoing that a number of mechanisms exist that can bring about the association of soil colloids. In addition to cation and anion exchange, other forces (e.g. van der Waals attraction, ligand exchange, hydrogen bonding, hydrophobic and lipophilic interactions, polyvalent bridging) permit the intimate association of like-, oppositely- and non-charged soil constituents (Rutter & Vincent 1980). Microbes may also adhere tenaciously to surfaces by way of exopolymers, pili and prostheca (Rogers 1979). Initial 'long-distance' attraction of a microbe to a surface is still something of a mystery but may be brought about by hydrophobicity (i.e. rejection of a microbe from the aqueous phase), electrostatic repulsion from a previous location, chemotaxis, motility and chance contact caused by Brownian motion. The electrokinetic properties of microbial cells and soil particles can be studied by micro-electrophoresis (Stotzky 1972; Marshall 1976; James 1979; see also p. 122).

7.3 Effect of soil colloids on microorganisms

7.3.1 Indirect Effects of Clay Minerals on Microbial Ecology

Clay minerals (and humic substances) affect the activity, ecology and population dynamics of microorganisms in soils and sediments. Of the various clay minerals studied, montmorillonite (smectite) has the most influence, primarily because of its high cation exchange capacity (CEC), although its large surface area and ability to swell (which enhances access to interlayer cation exchange sites and increases viscosity) also appear to be involved in some microbial responses. These effects of clays are primarily indirect, that is the clays modify the physico-chemical environment of the microbes and this either enhances or attenuates the growth of individual microbial populations, which, in turn, influences the growth of other populations within that habitat (Stotzky 1972). Some of the indirect effects of clays, especially of montmorillonite, are summarised in Table 2. Although the mechanisms involved in these effects are fairly well understood, relatively little is known about direct interactions between clay minerals and microorganisms or between clays and organic substances that are either substrates for, products of, or toxic to microbial growth. Not only is information on these interactions scant, but essentially nothing is known about the importance of such interactions to microbial events, to survival of viruses, and to migration of organic and inorganic substances in soil.

Studies of the effects of clays on microbial ecology must be conducted on various levels of

Table 2. Some effects of montmorillonite (M) on microbes at different levels of experimental complexity (after Stotzky 1980).

Level of experimental complexity	Effect	References
Field observations	*Fusarium* wilt of bananas: faster spread in soils without M	Stotzky *et al.* (1961); Stotzky & Martin (1963)
	Histoplasma capsulatum: isolated only from soils without M	Stotzky & Post (1967); Stotzky (1971, 1972)
	Coccidioides immitis: isolated from both soils with and without M	Stotzky (1972)
	Other mycotic pathogens of humans: isolated only from soils without M	Stotzky (1972)
	Enzootic leptospirosis: found only where soils and sediments do not contain M	Stotzky (1972)
Pure culture studies	Fungi	
	respiration, radial growth and spore germination decreased by concentrations greater than 2% (w/v) (related to increases in viscosity which, in turn, reduces access to O_2)	Stotzky & Rem (1967); Santoro *et al.* (1967)
	Bacteria	
	respiration stimulated	Stotzky & Rem (1966)
	stimulation due to 'buffering effect' (pH)	Stotzky (1966a)
	stimulation related to CEC and not to particle size or specific surface of clays	Stotzky (1966b)
	lag phase of growth decreased	Stotzky & Rem (1966)
	protected against hypertonic osmotic pressures (related to CEC)	Stotzky & Rem (1966); Nanfara & Stotzky (1979)
	Protects both bacteria and fungi against heavy metal toxicities (related to CEC)	Babich & Stotzky (1977a, 1978a, b, 1979, 1980a)
	Sorbs organic volatiles from germinating seeds and bacteria	Moore–Landecker & Stotzky (1974); Schenck & Stotzky (1976); Stotzky & Schenck (1976)
Spread through soil (soil replica plate studies)	Fungi slower in presence of M; bacteria faster in presence of M	Stotzky (1965, 1972, 1973)
	Competitive effects of bacteria against fungi greater in presence of M	Stotzky (1965, 1972)
	related to CEC, 'buffering effect', and more rapid utilisation of nutrients by bacteria (competition)	Rosenzweig & Stotzky (1979, 1980a, b)
	Conjugation of bacteria stimulated by presence of M	Weinberg & Stotzky (1972)
	Protects bacteria and fungi against heavy metal toxicities (related to CEC)	Babich & Stotzky (1977b, 1978a, b, 1980a)
Metabolism in soil (autotrophic and heterotrophic)	Nitrification rate enhanced by M	Macura & Stotzky (1980); Kunc & Stotzky (1980)
	Decomposition of aldehydes enhanced by M	Kunc & Stotzky (1977)
	Decomposition of other organics either enhanced, decreased or unaffected by M	Kunc & Stotzky (1974)
	Protects against inhibition of nitrification by sulphur dioxide	Bozian & Stotzky (1976)
	Protects against inhibition by heavy metals of organic matter decomposition	Babich & Stotzky (1980a); Debosz & Stotzky (unpublished observations)
Greenhouse studies	Spread of fusarial wilts decreased by presence of M in soil	Stotzky (1972)
Field studies	Incorporation of M into soils: effect on persistence and spread of plant and animal pathogens	Experiments to be conducted

experimental and system complexity, so that the veracity of results and observations from these different levels can be compared and, most important, can be related to real events that occur *in situ*. This approach is illustrated in Fig. 1.

The emphasis in this chapter will be on the surface interaction studies shown in the lower right-hand corner of the flow diagram, but, as the arrows indicate, these studies are intimately related to those on the indirect effects of clays. Furthermore, it must be emphasised that the physicochemical mechanisms indicated in Fig. 1 and in Table 1 and discussed herein have no importance unless they can be related to how they actually impact on microbes *in situ*. As will be illustrated, this has not always occurred, and many unanswered questions and apparent paradoxes exist.

7.3.2 Surface Interactions Between Clay Minerals, Microorganisms and Organic Substances

The prelude to any surface interactions between clays and microorganisms or organic substances must be a sufficient reduction in the electrokinetic potentials (EKP) of the components for them to come close enough for any attractive forces, either physical or chemical, to overcome electrostatic repulsion. These surface interactions are restricted to physical and ionic mechanisms as clays are not normally capable of participating in covalent bonding (Theng 1979). Clay minerals are rigid, and their charge distribution is relatively stable, when compared to the fluidity of the lipid bilayer membranes of living cells. Most cells have membranes containing polymers which can rapidly alter their conformation and that of their associated integral and peripheral proteins in response to changes in their ambient environment. This fluidity, the apparent specificity of some receptor sites (e.g. for hormones, lectins, certain drugs, and viruses; antigen–antibody reactions), the ability to form covalent bonds, and the pH-dependence of the surface charge of cells in contrast to the predominantly pH-independent charge of most clays distinguished living cells from clay minerals as adsorbents.

From the point of view of the microbial ecologist, definition of the mechanisms involved in surface interactions between clays, microorganisms and organic substances is secondary to knowing the degree of stability of the resultant complexes. For example, the microbial ecologist wants to know whether a complexed organic molecule is more or less available as a nutrient for or active as a toxicant to microbes than when not complexed with clays; whether and how the physiology of a microbe complexed with clays differs from that of a noncomplexed microbe; and whether viruses associated with clays are more or less resistant to inactivation by biotic and abiotic factors than are free viruses. However, the microbial ecologist must be able to distinguish clearly between equilibrium adsorption and binding of the biological component to the clays, as the activity, availability and toxicity of the complexed biological component cannot be evaluated if the complex is readily disrupted and the adsorbate released into the aqueous phase. Consequently, equilibrium adsorption and binding isotherms must be constructed so that both the quantity of adsorbate stably complexed with the clay is known and experiments can be conducted with confidence that the complex, rather than the noncomplexed biological component in the presence of clay, is being studied (Harter & Stotzky 1971).

For convenience, clays will be considered here as adsorbents, and microbes, viruses, and organic substances as adsorbates, even though clays are usually smaller than the microbes studied. Because of this difference in size, because the direction of adsorption is not known (e.g. does clay adsorb to a bacterium or vice versa?), and because some soluble organics may intercalate clays, the term 'sorption' will be primarily used when discussing surface interactions between clays and various biological components.

Equilibrium adsorption and binding: general concepts

Equilibrium adsorption is reversible, exhibits low specificity between the adsorbent and the adsorbate, may result in multilayer sorption, and involves primarily physical forces, such as various London–van der Waals interactions, simple hydrogen bonds, protonation, coordination bonds and water bridging (Theng 1979; Rutter & Vincent 1980). Equilibrium is attained when the rate of adsorption equals the rate of desorption. As coverage of the adsorbent surface increases, it becomes increasingly difficult for adsorption to proceed and mutual repulsion of the adsorbate molecules may result in a reduction in the bond strength and in lower values for the energy of adsorption.

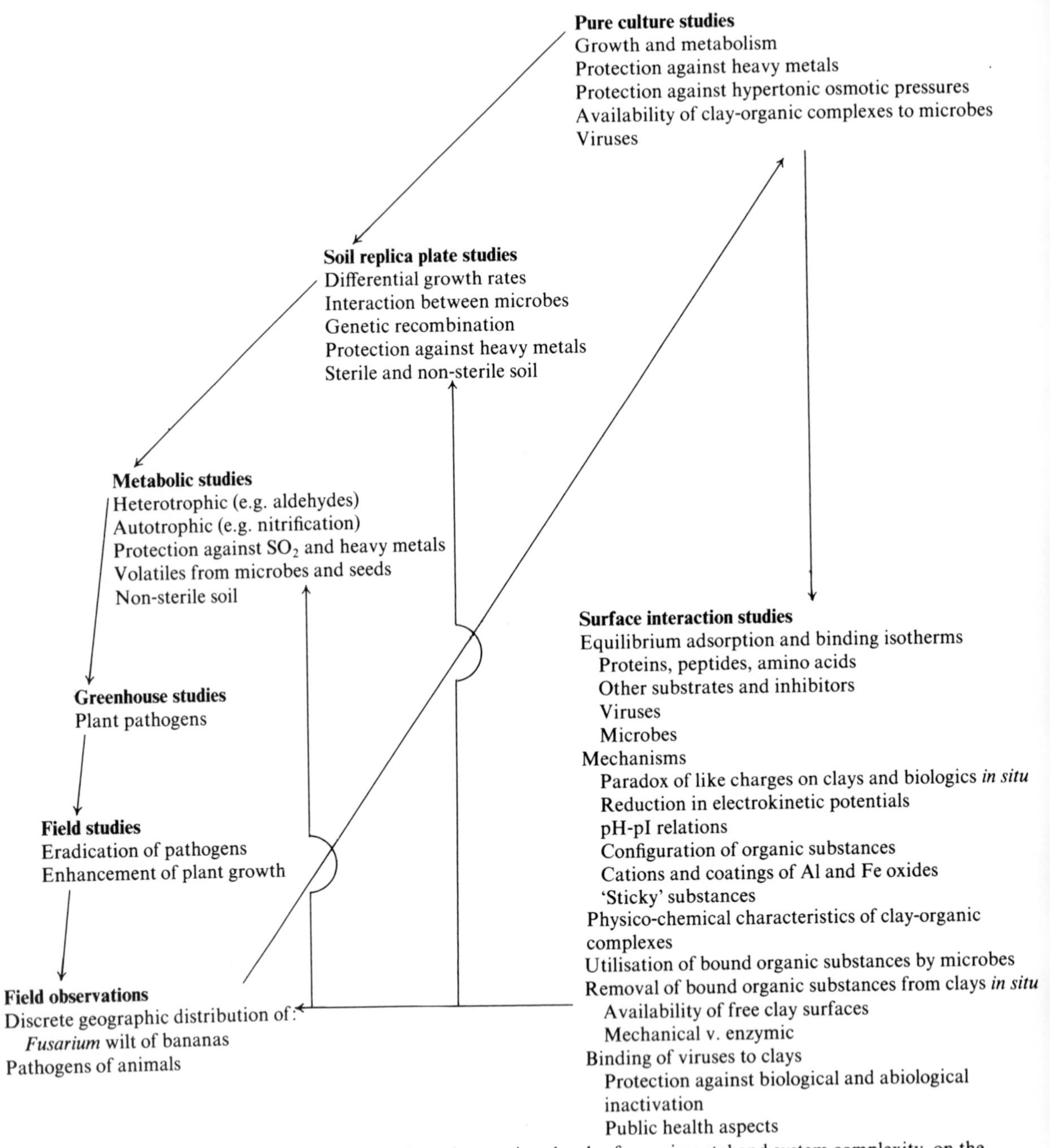

Figure 1. Flow diagram of studies being conducted, at various levels of experimental and system complexity, on the effects of clay minerals on the activity, ecology and population dynamics of microbes and viruses (after Stotzky 1980).

Binding, often referred to as chemical adsorption or chemisorption, can be either reversible or irreversible; may exhibit high specificity of the adsorbate for binding sites on the adsorbent; usually results in only single-layer adsorption, as only the reactive groups of the adsorbate and the adsorbent enter into the reaction; and usually involves the formation of a chemical bond, such as an ionic or covalent bond. However, there appear to be significant differences in chemisorption between systems containing a free surface of a solid and those not involving a solid. Reactive groups in a liquid or gas phase are usually unrestrained and can form conventional chemical bonds, whereas the reactive groups of a surface are components of surface molecules which, in turn, are part of the solid and, therefore, these molecules are restrained (Clark 1974).

These restraints, in the form of bonds in three directions to neighbouring molecules, may influence the properties of the reactive groups of the adsorbent and, consequently, their ability to form bonds with the adsorbate. If the adsorbate is also a solid, such as a bacterium or a virus, whose surface molecules are also restrained in three directions, the effect of these restraints on surface interactions between clays and such microorganisms and viruses may be exacerbated. Consequently, experimental models using relatively unrestrained (flexible) adsorbates (e.g. proteins, peptides, amino acids) and restrained (rigid) adsorbents (e.g. clays) may not be predictive of sorptive reactions between clays and solid biological adsorbates, even if the reactive surface groups of the restrained adsorbates (e.g. the proteins of the surfaces of cells and viruses) are similar to the model of unrestrained adsorbates.

Although binding is often the result of a chemical bond, it may also result from the attachment of an adsorbate by numerous physical forces, the sum of which exceeds the tendency of the adsorbate to detach from the adsorbent (e.g. the 'zipper effect'—Stotzky 1980). If these physical forces are sufficiently dense, they can, especially in conjunction with hydrophobic interactions, result in binding that is of greater strength than that resulting solely from chemical bonds. Because both chemical and physical forces are probably involved in binding, it would be incorrect to refer to the binding between biological moieties and clays as chemisorption. Consequently, adsorption will be used here to describe equilibrium adsorption (which is reversible and probably the result primarily of physical forces), and binding will be used to describe the condition in which the biological component cannot be separated from the clay after extensive washing with appropriate solutions.

Electrokinetic potentials of clay minerals and bacteria

To determine the importance of reducing the EKP of particulates before surface interactions between them can occur, sorption between bacterial cells and either montmorillonite or kaolinite homoionic to different cations was studied by measuring changes in particle size distribution and by electron microscopy (Santoro & Stotzky 1967a, 1968; Stotzky 1972). Sorption between these populations of particulates (and flocculation of clays by microbial metabolites—Santoro & Stotzky 1967b), occurred only when the EKP of the clays and cells were sufficiently decreased by polyvalent cations, and the amount of sorption increased as the valency of the cations increased. Sorption also occurred at pH values below the pI of the cells, when the net charge on the cells became positive but that of the clays remained negative, resulting in ionic interactions.

These changes in EKP that resulted in sorption were verified by direct measurement of the electrophoretic mobility of clays and bacteria in the presence of different cations (Santoro & Stotzky 1967c; Stotzky 1972). The electronegativity of bacteria decreased, at pH levels above their pI, as the valency of the cations in the ambient medium increased and also as the concentration of the cations, even of monovalent ones, increased. These observations were consistent with a presumed reduction in the thickness of the diffuse electrical double-layer surrounding the cells. At pH values below their pI, the net charge of the bacteria was positive. The measured pI of some bacterial species was dependent on the type of cations present; for example, the pI in the presence of low concentrations (ionic strength = 3×10^{-4}) of chloride salts of mono- or divalent cations was between pH 2.5 and 3.5, whereas in solutions of the same ionic strength but containing lanthanum chloride or chromium chloride the pI was approximately at pH 5.0, and in the presence of ferric chloride or aluminium chloride the pI was shifted to approximately pH 7.0, probably reflecting the effect of pH on the hydroxylated forms of iron and aluminium. Incidentally, similar shifts in the pI with increasing valence of ambient cations have also been observed with normal human lymphoid and transformed Burkitt lymphoma cell lines and with frog kidney cells (Kiremidjian & Stotzky 1975, 1976). Reductions in electronegativity and shifts in measured pI to higher pH values also occurred when bacteria were suspended in dilute soil extracts (1:5, soil/water), dilute sea-water (1:100, sea-water/distilled water), and various microbial metabolites or culture media containing organic constituents.

When the electrophoretic mobility of kaolinite and montmorillonite was measured at the same ionic strength used with bacteria, the clays remained negatively charged at all pH values studied (pH 2.0 to 9.0), regardless of the valency of the ambient cations (Santoro & Stotzky 1967c; Stotzky 1972). However, the electronegativity decreased as the valency of the cations, both on the exchange complex of the clays and in the ambient solution, was increased. In the presence of higher concentrations of tri- and tetravalent cations (e.g. when clays homoionic to aluminium or thorium were suspended in solutions containing polyvalent cations), some charge reversal occurred between pH 4.0 and 8.0, probably reflecting the pH-dependent ionisation of the hydrous oxides of these cations that were attached to the clays. When suspended in dilute soil extracts or sea-water, the electronegativity of the clays decreased, but no charge reversal occurred; in microbial metabolites or culture media, the electronegativity either increased or decreased, depending on the source of the metabolites or the type of medium, but there was no charge reversal.

Although these studies confirmed that there must first be a reduction of the EKP (of like charge) of reacting particulates to enable them to approach each other closely enough for the bonds involved in adsorption and binding to be formed, a paradox is apparent. These in-vitro studies indicated that surface interactions between clays and microbes occur only at pH values below the pI of the cells or in the presence of polyvalent inorganic cations and certain organic compounds. In most natural soils, however, such interactions occur (if, indeed, they do occur) at pH values above the apparent pI of cells and usually in the absence of large quantities of soluble tri- and tetravalent cations (e.g. the dominant soluble cations in most soils are K^+, Na^+, H^+ Ca^{2+} and Mg^{2+}) and of soluble organics. The empirical evidence for surface interactions between microbes and soil particulates *in situ* is strong: lack of movement of large numbers of microbes from surface to underlying soil layers during heavy rains or snow melts or even during flooding; failure to wash substantial numbers of microbes from soil columns in perfusion or leaching experiments; removal of microbes from waste waters in percolation beds; and increased release of microbes from soil by sonication and surfactants.

The usual retention of microbes in soil may be the result of their entrapment in narrow channels between soil particulates, inside soil crumbs, and by surface tension within water films tightly bound to the highly reactive surfaces of clay minerals and particulate organic matter. However, it is difficult to assign too large a role to these phenomena, inasmuch as channels are constantly altered, crumbs are slaked, and water films are rearranged during heavy rains, perfusion, and mechanical dispersal of soil in its preparation for enumeration of microbes. Consequently, physico-chemical surface interactions with clay minerals and other soil particulates must still be considered as major mechanisms for the retention of microbes in soil.

Some microbes in a variety of habitats (e.g. soil, rhizosphere, sediment, oral cavity, gastro-intestinal and genito-urinary tracts) appear to elaborate polymeric organics that serve as 'sticky' substances to adhere the microbes to surfaces (Marshall 1976). However, it has not yet been demonstrated that all microbes produce these substances, especially under the poor nutritional conditions that usually exist in soil. Furthermore, unless these substances are vastly different from most microbial metabolites (e.g. do they contain positively charged residues?) the same physico-chemical requisites (e.g. reduction in EKP) apply to these substances as to other microbial metabolites and to cells (Kiremidjian & Stotzky 1973). Unfortunately, not enough is yet known about these substances and their mechanisms of action to evaluate their importance in surface interactions between microbes and clay minerals *in situ*.

Consequently, it is still necessary to reconcile the apparent paradox of sorption between clay minerals and microbial cells *in vitro* essentially only under the extreme conditions that are, presumably, seldom, if ever, encountered in soil *in situ*, where such surface interactions appear to occur routinely, However, some soluble polyvalent cations and macromolecules are undoubtedly present in soil microhabitats, probably at concentrations higher than those measured on a bulk soil basis. Thus, microbes in these microhabitats may be positively charged or, at least, have a lower net negative charge at the prevailing pH than at comparable pH values *in vitro* which would enable their surface interactions with negatively charged clays. The actual pH at the surface of charged particulates (pH_s) may also be lower than that measured in the bulk suspension (pH_b), and the pH_s of clays may actually be near or below the pI of cells, even though the pH_b is above the pI (Stotzky 1972).

Furthermore, microbes may not interact directly with the clay surface but rather with polymeric hydrous oxides of iron, aluminium and manganese which extend from the clay surface and exhibit amphoteric properties. More studies on the occurrence and mechanisms of sorption between soil particulates and microbial cells *in situ* are obviously needed. The major limitation for the conduct of such studies is the availability of adequate experimental techniques.

7.3.3 Adsorption and Binding of Proteins by Clays

In contrast to results obtained with microbes, the amounts of various proteins adsorbed and bound to montmorillonite homoionic to different cations were, generally, inversely proportional to the valence of the cation saturating the clay (Harter & Stotzky 1971). This suggested that a reduction in the EKP was not a prerequisite for surface interactions between clays and water-soluble polymeric adsorbates. Adsorption and binding occurred with every combination of clay (i.e. homoionic to hydrogen, sodium, calcium, lanthanum, aluminium or thorium) and protein (i.e. catalase, casein, chymotrypsin, edestin, lactoglobulin, lysozyme, ovalbumin, ovomucoid and pepsin) studied, even when the pH_b of the clay suspensions (which ranged from pH 3.0 to 7.0) was above the pI of the proteins (which ranged from pH 1.0 to 11.0) and the net charge on both clays and proteins was negative. The binding isotherms of these proteins were either of the L (Langmuir) or H (high affinity) type, with the exception of catalase which was of the C (constant partition) type (Giles *et al.* 1974).

These results (Harter & Stotzky 1971, 1973) suggested the following.

(1) The pH_s of each clay was lower than the pH_b and, therefore, the pH that a protein with a low pI encountered at the clay surface may not have been as unfavourable for surface interactions as indicated by the measured pH. This possibility is questionable, however, as montmorillonite homoionic to sodium (pH_b 6.2) adsorbed and bound almost as much of proteins with pI values between pH 3.8 and 5.7 as did clay homoionic to hydrogen (pH_b 2.7).

(2) As these studies were conducted in unbuffered systems to prevent changes in the cation saturation of the clays, alterations in either the clay surface and the proteins, or in both, by the prevailing ambient pH could have occurred.

(3) Inasmuch as the pI indicates only the pH at which the net charge of an ampholyte is zero (i.e. the pI of a protein is the sum of the pK values of the component amino acids), the proteins may have had positive charges at some loci even at pH values at or above their pI. If these positively charged loci were sufficiently numerous and close, a relatively flexible adsorbate, such as a protein, might be able to approach negative sites on the clay close enough for some ionic interactions to occur.

(4) A predominantly net negatively charged protein may sorb to positive sites on the clays, as appeared to be the situation with pepsin, the pI of which ranges between pH 0 and 2.0 and which, on the basis of X-ray diffraction analysis and transmission electron microscopy, was apparently bound to positively charged edge sites.

(5) If protein sorption to clays is a cation exchange reaction, as suggested by some workers, then the proteins competed better for exchange sites with monovalent cations, which have a lower energy of bonding and can be more easily replaced, than with cations of higher valence. The possible involvement of cation exchange mechanisms in sorption of the proteins was reinforced by the observations that the pH_b of the clays, especially of the acidic homoionic clays, increased as the amount of protein bound increased and that the electrophoretic mobility of the clays changed, presumably as a result of the covering of the predominantly negatively charged clay surfaces by the proteins.

(6) The amount of clay surface available to proteins was lower with clays homoionic to polyvalent cations, as these clays were better flocculated and more resistant to dispersion (Santoro & Stotzky 1967a, b).

The molecular weight, shape and number of binding sites on the proteins also appeared to be important in their surface interactions with clays (Harter & Stotzky 1971). Although the weight of protein bound generally increased as the molecular weight of the protein increased, the number of moles bound was generally inversely related to molecular weight. Regardless of how the data were expressed, the increase in weight or the decrease in moles of protein bound was not always in direct proportion to molecular weight, indicating that some steric hindrance was involved. For example, the molecular weight of casein was about five times greater than that of chymotrypsin and ovomucoid, but the weight of casein bound by montmorillonite homoionic to hydrogen was only about 15% more than that of the other two proteins. The importance of the number of binding sites on the proteins, resulting from different compositions and sequences of amino acids and tertiary structures, was shown, for example, by the twofold greater binding of chymotrypsin than of lysozyme, even though their molecular weights were essentially the same. Lysozyme apparently had more binding sites, and fewer molecules were necessary to saturate binding sites on the clay.

Although the mechanisms involved in the adsorption and binding of these complex amphoteric molecules by clays are not clearly defined, the pI of the proteins and the pH_b of the systems were apparently not as important as the other factors described. Nevertheless, some proteins showed maximum adsorption and binding at a pH_b near their pI, possibly because of the following.

(1) Repulsive forces among protein molecules, both in solution and on the clay, were at a minimum.

(2) The rate of collision of proteins having a net neutral charge with charged clay surfaces was at a maximum (MacRitchie & Alexander 1963).

(3) Proteins with a net neutral charge contributed less to the surface potential than did charged proteins and, therefore, neutral proteins were at a lower energy level and in a more stable state when attached to clay surfaces than when in the bulk solution (Rao 1972).

(4) Other factors, such as solubility (Theng 1979) were involved.

On the basis of X-ray diffraction and electron microscopic analyses, the proteins, with the exception of catalase and pepsin, appeared to enter the clay interlayers (Harter & Stotzky 1973). In general, the amount of expansion of the clays caused by the proteins was proportional to the amount of protein added rather than to the amount actually bound, suggesting that with the higher concentrations added, the protein molecules were initially distributed uniformly across the interlayer surfaces rather than being restricted to the edges of the interlayers as with the more dilute protein solutions. The X-ray diffraction data also indicated that, in general, several layers of protein were intercalated and denaturation of the proteins did not occur.

Even though the amount of catalase bound was greater than that of the other proteins, it caused essentially no expansion of the clays, probably because catalase was too large (molecular weight = 250 000) to be intercalated. Distinct globules of catalase were observed, by electron microscopy, to be located relatively uniformly and densely over the surface of the clay packets, indicating that binding sites for this protein were distributed over the clay surface. Pepsin, in contrast, was concentrated around the edges of the clay particles and adjacent to 'folds' in the surface rather than being distributed randomly—as were some proteins—or uniformly—as was catalase. This suggested that pepsin was bound to positively charged edge sites (Harter & Stotzky 1973).

7.3.4 Adsorption and Binding of Amino Acids and Small Peptides by Clays

Proteins are composed of numerous amino acids with different pK values, and the relative proportion of negative and positive charges varies within a single protein molecule at different pH values. Furthermore, the large size of proteins facilitates the formation of numerous physical bonds with clays. Consequently, the interpretation of adsorption and binding isotherms of proteins on clays and definition of the mechanisms involved are difficult. Therefore, the adsorption and binding of amino acids and small peptides to montmorillonite and kaolinite homoionic to various cations and the availability of these bound organics to microbes were studied (T. Dashman & G. Stotzky, unpublished observations).

Aspartic acid, cysteine, proline and arginine adsorbed to clays homoionic to hydrogen, calcium, zinc or aluminium but not to clays homoionic to sodium or lanthanum. Glycine adsorbed only to clays homoionic to hydrogen and was completely removed by a single wash with water. Aspartic acid was bound only on montmorillonite homoionic to calcium or zinc; cysteine only on montmorillonite or kaolinite homoionic to zinc; proline only on montmorillonite homoionic to hydrogen, calcium or zinc and on kaolinite homoionic to hydrogen; and arginine only on montmorillonite homoionic to hydrogen or aluminium. Both the adsorption and binding isotherms were mostly of the C type, with some being of the S or H type (Giles *et al.* 1974). The mechanisms of binding were related to the structure of the amino acid and the cation saturating the clays and appeared to involve either cation or anion exchange or the formation of *pi* or chelate complexes.

The peptides studied were preferentially adsorbed and bound on montmorillonite, and the amounts increased, in general, as the molecular weight and chain length of the peptides increased. Peptides of glycine, with chain lengths ranging from two to four monomers, were not bound, with the exception of glycyl-glycine which bound to kaolinite homoionic to sodium or calcium. The acidic peptide, L-aspartyl-glycine, and the basic peptide, alanyl-L-lysine, were bound only on montmorillonite homoionic to hydrogen. Of the heterocyclic peptides, L-histidylglycine was bound only on montmorillonite homoionic to aluminium, zinc or sodium, and L-prolyl-L-phenylalanyl-glycyl-L-lysine was bound only on montmorillonite homoionic to aluminium, zinc, hydrogen, lanthanum or sodium and on kaolinite homoionic to aluminium, zinc, sodium, hydrogen or calcium.

Several amino acids and peptides intercalated montmorillonite after equilibrium adsorption, but after removal of weakly adsorbed molecules by washing with water, only arginine produced detectable expansion of montmorillonite homoionic to aluminium. In contrast to results obtained with proteins, the pH_b and the electrophoretic mobility of the clay–amino acid or clay–peptide complexes were not significantly different from those of the clays alone. The amounts of amino acids or peptides bound to the clays were apparently insufficient to cover enough surface sites to affect significantly the electronegativity or pH_b of the clays.

7.3.5 Microbial Utilisation of Clay–Protein Complexes

When various clay–protein complexes were presented to microbes as a sole source of carbon, the relative availability of the complexed proteins appeared to be dependent on the cation saturating the clay, the location of the protein on the clay, the amount of protein bound, and the characteristics of the individual protein (Stotzky 1972; Gerard & Stotzky 1973). Of the five proteins studied in detail, catalase and pepsin were not utilised when complexed, although they were rapidly utilised in the absence of clay. Casein, chymotrypsin, and lactoglobulin were utilised when complexed but to a considerably lesser extent than when not complexed. The latter three proteins were largely intercalated and, in general, in multilayers. Apparently, only those protein molecules attached to the proteins which were bound directly to the clay were utilised, and once the multilayered molecules were removed, no further utilisation of underlying proteins took place.

Even though the location of catalase and pepsin should have rendered them more available to proteolytic enzymes than the intercalated proteins, their lack of susceptibility to proteolysis suggested that:

(1) the terminal amino acid residues necessary for initiation of cleavage of the polypeptide chain by exopeptidases were either involved in binding or were masked by binding; and/or

(2) binding altered the configuration or conformation of the proteins to make them non-susceptible to the action of endopeptidases.

These factors may also have been involved in preventing the utilisation of protein molecules bound to the clay interlayers, even though the multilayer molecules were susceptible.

The catalytic activity of bound catalase on some homoionic montmorillonites was at least four times greater than that of free catalase, further indicating that binding resulted in a structural change in some proteins (Stotzky 1972). This enhanced enzymic activity could have resulted from a concentration of the enzyme on clay surfaces. However, both the inability of proteolytic enzymes to degrade catalase and its greater enzymic activity when bound suggested that binding altered the shape of the molecule so that more active centres were exposed but sites for attachment and/or activation by proteolytic enzymes were masked. The retention of enzymic activity by bound catalase further indicated that the protein was not denatured as a result of binding.

The reduction in the metabolism of complexed proteins by microorganisms was not a result of the binding by the clays of either the microbial exoenzymes or the products of proteolysis. Control experiments, which included the addition of clay to non-complexed protein and of protein to clay–protein complexes just before the start of the experiments, showed that any binding of proteolytic enzymes or of their products did not account for the amount of reduction in the utilisation of the complexed proteins (Stotzky 1972, 1973; Gerard & Stotzky 1973).

7.3.6 Microbial Utilisation of Clay–Amino Acid or Peptide Complexes

When these complexes were presented to microorganisms as a sole source of carbon, nitrogen, or both carbon and nitrogen, only some complexes were utilised, and then only by some microorganisms, even though the non-complexed amino acids and peptides were readily metabolised (T. Dashman & G. Stotzky, unpublished observations). For example, neither aspartic acid nor cysteine complexed with clays was utilised, whereas complexed proline and arginine were. Proline bound to montmorillonite homoionic to hydrogen or calcium was utilised only as a source of nitrogen (when dextrose was added) but was not utilised as a source of both carbon and nitrogen or as a source of only carbon. Arginine bound to montmorillonite homoionic to hydrogen was available as source of carbon, nitrogen and both carbon and nitrogen, but when bound to montmorillonite homoionic to aluminium, arginine was available only as a source of either nitrogen (when dextrose was present) or carbon (when ammonium nitrate was present) but not of both carbon and nitrogen. These differences in utilisation suggested different types and/or energies of binding of the amino acids to the various homoionic clays, an area that requires additional study.

Furthermore, the mechanisms whereby a microorganism can selectively use either the carbon or nitrogen moieties of an extracellular amino acid are not known and also require study, especially as no exoenzymes capable of degrading amino acids have been reported. This apparent selective utilisation of portions of amino acids external to the cell was in contrast to the action of extracellular proteases that cleave proteins into peptides or amino acids, which are then transferred intact into cells and catabolised intracellularly.

7.3.7 Microbial Utilisation of Humic–Substrate Complexes

Soil humic matter displays a spectrum of susceptibility to microbial degradation, ranging from readily degradable, and thus short-lived, non-aromatic components to highly recalcitrant phenolic compounds that may have half-lives measured in centuries (Sørensen 1975). The availability to microorganisms of many potential substrates is reduced as the result of a variety of factors when these organic materials become associated with humic matter:

(1) the bacteriostatic and fungistatic properties of many phenolic substances that are major components of humic matter;

(2) the physical entrapment of substrates within the humic polymers, thus restricting enzyme-substrate and microbe-substrate contact;

(3) the chemical bonding of substrates to the aromatic structures of humic matter during its polymerisation (humification).

The ease with which model polyphenols can be formed, by either chemical auto-oxidation or the action of phenol oxidase enzymes, has prompted the investigation of the effects of humic-like polymers on substrate utilisation and microbial growth. The degradation of microbial cells (bacteria, fungi, algae), their components (e.g. envelopes, cytoplasm) and their chemical constituents (e.g. proteins, amino acids, amino sugars), both free and linked to synthetic polyphenols, has been measured in soil and axenic cultures (Webley & Jones 1971; Martin & Haider 1980). Bondietti *et al.* (1972) demonstrated that the rate of decay of amino sugars (glucosamine, chitosan) was reduced when they were complexed with model humic acids. Verma *et al.* (1975) showed that when ^{14}C-labelled proteins, peptides and amino acids were linked to model phenolic polymers, their rates of mineralisation (i.e. $[^{14}C]CO_2$ evolution) in soil were reduced by 80 to 90%. They suggested that these proteinaceous substrates may be linked to the phenolic structures through amino or sulphydryl groups. Even just the physical mixing of the polymers and the three nitrogenous substrates reduced their decomposition by 30 to 50%, possibly as a result of hydrogen bonding. Verma and Martin (1976) showed that the breakdown of algal cell components in soil was retarded when they were complexed with synthetic humic polymers and that the reduction in the rate of decomposition of the cytoplasmic fraction was greater than that of the cell wall material. Furthermore, naturally occurring dark-coloured polymers (melanins) present in the cell walls of some fungi appear to protect the cell from microbial degradation (Bloomfield & Alexander 1967; Linhares & Martin 1978).

Studies on the incorporation of pesticides or their degradation products into humic constituents have also shown that humic matter can attenuate the degradability of associated organic substrates (Mathur & Morley 1975; Hsu & Bartha 1976; Wolf & Martin 1976; Bollag *et al.* 1980).

In addition to the inhibitory effects of humic matter on the degradation of intimately associated substrates, humic substances may enhance microbial activity by adsorbing potentially inhibitory metabolites (similar to clays) or by serving as a source of organic growth factors and trace elements. These possibilities have been discussed by Martin *et al.* (1976).

7.3.8 Ecological Implications of the Binding of Substrates and Enzymes to Soil Colloids

Inasmuch as the binding to clays of potential substrates decreased their availability as nutrient or energy sources for microbes *in vitro,* the differential effect of clay minerals saturated with different cations on the availability of organic nutrients could have a marked influence on the activity, ecology and population dynamics of microbes in soil. Furthermore, surface interactions between substrates and clays may prevent the migration of organic substances through soil and may explain, in part, the lower concentrations of organic matter in deeper than in surface horizons. Because microbial activity is less in these deeper horizons, primarily because of limitations in available oxygen, the lower content of organic matter in these horizons may be the result of restricted mobility of organic matter rather than of its degradation by microbes.

Although many in-vitro studies have demonstrated that clays bind organic materials, caution must be used in extrapolating the results of such studies to soil *in situ* until another apparent paradox is resolved: if clays bind organics and these are then relatively resistant to microbial degradation, then essentially all clays *in situ* should be saturated with organics and, with the exception of newly formed clay minerals, no free clay surfaces should be available for further binding. Consequently, either mechanisms exist for the removal of bound organics to provide free clay surfaces for the binding of newly formed organics or such binding no longer occurs *in situ.*

Enzymic degradation of some bound organics is apparently not a major factor in their removal, and thus, mechanical methods of removal, which probably have energies greater than those involved in binding, must be considered. These may include abrasion—such as is exerted in the gut of earthworms or filter-feeders in aquatic sediments, in cultivation, or during disturbance of soil by growing roots and micro- and macrofauna—or release during cycles of soil wetting and drying or freezing and thawing. Moreover, organic substances *in situ* may not be bound directly to 'clean' clay surfaces as in these in-vitro studies, but rather to polymeric hydrous oxides of primarily iron, aluminium and manganese that are usually associated with clays in soil.

As the integrity of these inorganic polymers is pH-dependent and their binding of organics probably involves chemical bonds, the release of such bound organics may occur frequently as the pH of natural microhabitats fluctuates, primarily as a result of microbial metabolism. Once these organic materials have been released from the clays, they could be either degraded by the soil microbiota or again bound to clays. Furthermore, newly formed organics may also bind to organic substances directly bound to clays, and these multilayer organics may be more susceptible to enzymic degradation, as suggested by the in-vitro utilisation of some of the proteins of certain clay–protein complexes (Stotzky 1972; Gerard & Stotzky 1973).

Unfortunately, there has been little study of the mechanisms of binding and removal of organic materials from natural (i.e. 'dirty') clays. Until such mechanisms have been defined, the importance of the binding of organics on clay minerals to microbial events or to the migration of organics in soil *in situ* remains speculative.

Microorganisms which depend on the activities of their extracellular enzymes for their nutrition have an additional, but as yet unproven, relationship with their immediate environment. Clearly, such extracellular enzymes should not be rapidly degraded or denatured in the environment if they are to provide an ecological advantage to their parent cell. However, cell-free enzymes do not survive long after externalisation if they remain in the aqueous phase of soil, and the principal mechanism by which they persist is apparently by becoming associated with clay and organic colloids. This 'persistent' or 'accumulated' (Kiss *et al.* 1975) enzyme fraction is probably responsible for a significant amount of the extracellular hydrolytic catalysis of certain substrates, such as urea (Pettit *et al.* 1976), sugar polymers (Lethbridge *et al.* 1978) and organic phosphorus and sulphur (Pettit *et al.* 1977). In addition, soil-immobilised enzymes may be involved in the degradation of xenobiotic compounds (Burns & Edwards 1980).

Some enzymes, such as catalase (see p. 115), retain their catalytic activity after adsorption to the external surfaces of clays, but many resist destruction only by becoming complexed with humic substances. Crude humic-enzyme fractions with enzymic activity are easily extracted from soil, and the enzyme component retains a high proportion of the stability noted in the original soil (i.e. resistance to proteases, survival during storage at room temperature). The nature of such humic-enzyme associations is not clear, although physical entrapment (Burns *et al.* 1972b), ionic, covalent and hydrogen bonding (Ladd & Butler 1975) and copolymerisation with humic molecules (Rowell *et al.* 1973) have been suggested. The last-named mechanism gains support from the method of forming, and the resultant properties of, amino acid–, peptide– and protein–humic complexes described by Verma *et al.* (1975).

An ecological role for accumulated enzymes has been proposed (Burns 1980, 1982), wherein these comparatively stable colloid-bound extracellular catalysts serve as detectors of substrate for microorganisms residing in the same microhabitat. As a result, microbes do not need continually to produce exoenzymes which would, in any case, have a poor chance of surviving long enough to react with a suitable substrate. Instead, the microbe remains enzymically inactive until it receives a signal (i.e. product released by accumulated enzyme activity) which induces the microorganism to produce additional exoenzymes of the same type to degrade the exogenous substrate. In other words, the initial step(s) in the extracellular biological degradation of certain substrates may be independent of microbial growth and be effected by enzymes contained within humic–enzyme complexes, whilst subsequent degradation occurs through direct microbial activity. This may be an efficient strategy in the utilisation of exogenous substrates in the predominantly oligotrophic soil environment.

7.4 Adsorption of viruses

Inasmuch as some organics are protected to various degrees against biodegradation when bound on clays, such interactions may also protect viruses against both biological and abiological inactivation, thereby enabling their persistence in soils, sediments and waters in the absence of their hosts. This protection may be a contributing factor to the apparent increase in viral infections, primarily by water-borne enteric viruses (Berg *et al.* 1976) and may affect the practice of using soils as a repository for sewage sludge. In addition, viruses may be good models with which to study surface interactions between clays and microorganisms or organic substances, for example, viruses share with bacteria a relatively rigid structure, although viruses are approximately 10 to 100 times smaller. The molecular weight of viruses is approximately 100 to 10 000 times greater than that of most proteins, but the outer covering (capsid) of many viruses is composed primarily of protein subunits (capsomers), although some animal viruses have an outer envelope composed primarily of lipid.

Studies have been initiated with two bacteriophages of *Escherichia coli*, coliphages T7 and T1 (which contain double-stranded DNA, are approximately 50 nm in diameter, and have a molecular weight of 20 to 25 × 10^6), with reovirus type 3, (which contains double-stranded RNA, is 75 to 80 nm in diameter and has a molecular weight of 70 × 10^6), and with herpesvirus hominis type 1 (HSV 1) (which contains double-stranded DNA, is approximately 100 nm in diameter, has a molecular weight of 100 × 10^6 and has a lipid envelope). The results of these studies are summarised in Table 3.

All viruses sorbed to kaolinite and montmorillonite: the sorption of the coliphages was greater to kaolinite at high clay concentrations and to montmorillonite at low clay concentrations (Schiffenbauer & Stotzky 1978, 1980); that of reovirus was greater to montmorillonite (Lipson & Stotzky 1979, 1980); and that of HSV 1 was essentially the same to both clays (Yu & Stotzky 1979). Sorption was enhanced by increasing the ionic strength, especially with divalent cations, indicating that, as with bacteria, a reduction in the EKP improves the probability of successful collisions between the clay and viral populations. Only the sorption of reovirus was clearly correlated with the cation exchange capacity of the clays, although a correlation was suggested with adsorption of coliphage T7 but not of coliphage T1. These observations indicated that reovirus sorbed preferentially to negatively charged sites on the clays, that coliphage T7 adsorbed to both negatively and positively charged sites and

Table 3. Summary results of preliminary studies of surface interactions between viruses and montmorillonite (M) or kaolinite (K).

Sorption of reovirus greater to M than to K; correlated with CEC.
Sorption of coliphages T1 and T7 greater to K than to M; probably to positive edge sites and by tails
Sorption of herpes virus hominis type 1 essentially the same to M and K
Sorption enhanced by increasing ionic strength
'Specificity' of sorption sites
differential affinities
competition among viruses and between viruses and organics
Retention of lytic ability by sorbed viruses
Protection of viruses against inactivation by temperature
M more effective than K
Greater sorption of coliphages to M and K than to stationary cultures of yeasts and bacteria, including actinomycetes and host bacteria
Clay minerals reduce inactivation rates of bacteriophages in lake water

that the other viruses sorbed preferentially to positively charged sites (Stotzky *et al.* 1981).

This apparent specificity of sorption sites on the clays was further suggested in preliminary competition studies. Sorption of the coliphages, for example, was reduced in the presence of some soluble organic compounds (Schiffenbauer & Stotzky 1980), whereas the same amounts of HSV 1 were sorbed and bound by clays previously coated with bovine serum albumin and by clays without protein (Yu & Stotzky 1979), which may indicate hydrophobic interactions with the lipid envelope of this virus. In synthetic estuarine water, the prior sorption of reovirus prevented the subsequent sorption of coliphage T1, and the addition after sorption of the coliphage of either reovirus or the organic medium used for the maintenance of the reovirus removed about 90% of the coliphage from the clays. Conversely, the prior sorption of the coliphage or the presence of the maintenance medium did not interfere with the subsequent sorption of reovirus. Although these latter results indicated a higher affinity of the reovirus than of coliphage T1 for the clays, they also suggested that these two viruses sorb to different sites on the clays (Lipson & Stotzky 1979).

Sorption of the coliphages but not of reovirus appeared to be to the edges of the clays (e.g. sorption of the coliphages but not of reovirus was reduced after pretreatment of the clays with sodium metaphosphate—Lipson & Stotzky 1980; Schiffenbauer & Stotzky) although some viral particles also sorbed to the outer surfaces, as determined by electron microscopy. The coliphages sorbed primarily by their tails (Bystricky *et al.* 1975). The viruses, probably because of their large size, did not intercalate montmorillonite, and drying the clays at 180°C had no effect on subsequent sorption of HSV 1. However, interlayer charges appeared to be important, as sorption of reovirus (the only virus studied so far in this respect) to montmorillonite, but not to kaolinite, was markedly reduced in the presence of potassium, presumably because potassium collapsed the montmorillonite and prevented expression of interlayer charges (Lipson & Stotzky 1979, 1980).

The sorbed viruses retained their lytic ability, but the mechanisms by which sorbed viruses lyse their host cells is not known; for example, whether the viruses desorb from clays in the vicinity of host cells (i.e. the host cells may have a greater affinity for the viruses than the clays) or whether, with animal viruses, the intact clay–virus complex undergoes viropexis.

The clays protected the viruses against inactivation at various temperatures (Stotzky *et al.* 1981) and reduced the rates of inactivation of bacteriophages in lake water (Babich & Stotzky 1980b). Studies on the effects of clays on biodegradation of viruses are in the development phase, and no data are available other than the observation that the coliphages sorbed much less to stationary cultures of yeasts and bacteria, including actinomycetes and the host bacteria, than they did to clays (Schiffenbauer & Stotzky 1978).

These preliminary studies indicate that the sorption of viruses to clays exhibits characteristics of the sorption of both microbial cells and soluble monomeric and polymeric amino acids.

Experimental

7.5 Introduction

A brief discussion of the methodologies used in the study of soil colloids, especially clays, and the concepts behind these studies is presented below. This discussion is not intended as a recipe book of techniques, as many are complex, both in concept and in equipment required. Rather, the purpose of this section is to expose the microbial ecologist to these techniques and to alert the ecologist to necessary modifications in what are, for the most part, standard techniques. Consequently, when assistance is sought from clay mineralogists and others more conversant with the physico-chemical methods used in the study of soil, the importance of and reasons for these modifications can be firmly explained to these specialists.

7.6 Clay mineralogical analysis of natural soils and sediments

The state of clay minerals in soils and sediments may be different from that of the same clays in purified form. As it is the natural state of clays that affects associated microbial events, the methods of separating the clay mineral fraction for analysis must be gentle and designed to maintain the clays in as natural a state as possible. This aim is, in general, contrary to that of the clay mineralogist, who goes to great lengths to 'clean up' the clays before analysis (e.g. treatment with hydrogen peroxide to remove clay-associated organic matter and with dithionite to remove clay-associated iron Jackson 1964). Inasmuch as organic matter, iron and other components associated with clays in nature affect the CEC, surface area, expandability, water retention and other characteristics of clays that influence their effects on microbes in natural habitats, these components should not be removed or altered when attempting to correlate the clay mineralogy of soils or sediments with microbial events (Stotzky 1973).

The soil or sediment sample, after soaking for about 18 h with occasional stirring in water (1:5, soil/water), is poured through a 2 mm sieve or a double layer of cheesecloth to remove large organic and inorganic particulates. The filtrate is allowed to settle for a few hours, and the supernatant is decanted and centrifuged at about 4000 *g* for 15 to 20 min. The sediment, containing the clay mineral fraction, is distributed into test tubes and made homoionic to potassium and to magnesium or calcium by washing three times with centrifugation, with the respective 0.5M-chloride salt and then with distilled water. One sediment subsample is not made homoionic and is evaluated with its natural heteroionic cation complement.

The sediments are made into thick slurries by the addition of distilled water, placed onto microscope slides (which have been cut in half) to yield an orientated film containing approximately 2 mg cm^{-2} of sediment, air-dried for at least 24 h, and then placed over glycerol in an evacuated desiccator. Glyceration of a large number of specimens is best accomplished by placing the slides on a wire rack located near the top of the bottom half of the desiccator which contains a layer of glycerol, evacuating the desiccator with a water pump or a low-vacuum mechanical pump, placing the desiccator on a hot-plate maintained at about 50°C, and then placing wet towelling on the desiccator cover to enhance condensation of the volatilised glycerol. The temperature and cooling should be controlled so that large droplets of glycerol do not form on the underside of the desiccator top and fall onto the specimens. The specimens are ready for X-ray analysis when they appear uniformly moist and glistening. Slides containing reference clay minerals (e.g. montmorillonite and kaolinite) made homoionic to the various cations should be glycerated with the natural specimens, but the reference expandable minerals will usually become glycerated more rapidly than the natural specimens.

Slides containing sediments made homoionic to potassium, but not glycerated, are heated to 300 and 550°C for 2 h before X-ray analysis, to distinguish between vermiculite, chlorite and kaolinite (Jackson 1964). To prevent rehydration of the oven-dried samples, the scattering shield over the sample holder of the X-ray diffractometer is covered with a plastic film and a container of silica gel or other desiccant is placed in the shield.

The glycerated or oven-dried samples are analysed for their clay mineral content by X-ray diffraction analysis, usually with X-rays derived from nickel-filtered copper radiation in a diffractometer operated at 40 kV and 20 mA. The operating conditions, scan speeds, and slit arrangements depend on the instrument used. The samples can also be analysed by powder camera techniques, and the homoionic sediments can be glycerated by mixing directly with glycerol before being deposited on the slides. However, the desiccator method of glyceration and the use of a diffractometer are easier, especially when many samples are involved, and more useful data are obtained.

The identification of the clay minerals in the natural samples obtained by X-ray analysis can be supported by differential thermal analysis and petrographic examination (Lambe & Martin 1956; Stotzky & Martin 1963).

When soil or sediment samples that contain, or are suspected to contain, pathogens of human beings are to be analysed, the samples should be autoclaved (121°C, 15 p.s.i.) before preparation. Autoclaving does not affect the X-ray characteristics of clay minerals (Stotzky & DeMumbrum 1965).

7.7 Preparation of homoionic clay minerals

Mined clay minerals, which are usually relatively pure and can be purchased from a variety of sources, are made homoionic to different cations by washing at least three times, by centrifugation, with a 0.5N solution of the appropriate chloride salt, followed by washing, with centrifugation, with three-times distilled water until the supernatants are chloride-negative ($AgNO_3$ test) and their conductivity remains constant (Harter & Stotzky 1971). Clays can also be made homoionic by mixing clay suspensions with, or passing through columns of, cation exchange resins saturated with the desired cation (Jackson 1969). The homoionic clays can be maintained for long periods without decomposition and change in the saturating cation by storage as a thick slurry (approximately 12% (w/v) obtained by centrifugation) at 4°C (J. Colom & G. Stotzky, unpublished observations; Harter & Stotzky 1971).

Before being made homoionic, clays can be fractionated into different size ranges by dispersal in fresh 0.4N Na_2CO_3 (pH 9.5), followed by differential centrifugation (Tanner & Jackson 1947; Jackson 1969).

7.8 Changes in particle size distribution as a measure of surface interactions between clays and microbial cells

The direct demonstration of surface interactions (e.g. equilibrium adsorption, binding) between clay minerals and microorganisms is difficult, especially as the net charge of these microscopic particles, at physiological pH values, is similar (i.e. negative) and usually of approximately the same magnitude (i.e. −30 to −60 mV). Light microscopy is inadequate because of Brownian movement, convection currents, sample drying, common retention of dyes, and sizes approaching the resolution of the microscope. Electron microscopy is plagued by artefacts resulting from the individual particles being brought together by surface tension as water is evaporated from the sample, although special techniques, such as critical point drying, may resolve this problem. Differential sedimentation or centrifugation is not sufficiently critical, as the densities of any clay–microbe complexes are too similar to one or the other of the component populations (e.g. the specific gravity of bacteria is approximately 1.2 and that of clays is approximately 2.5) to result in any significant distinctions. Although density gradient centrifugation may eventually resolve this, the large difference in density between clays and microbes is a considerable obstacle, especially as many of the materials that could be used as density gradients to cover such a range of specific gravity are toxic to microbes.

Consequently, an indirect method, based on comparing the particle size distribution of mixed populations of particulates to those of the individual component populations, was developed (Santoro & Stotzky 1968). If surface interactions between the component populations occur, the particle size distributions of the mixture should differ from the theoretical arithmetic sum of the distributions of the component populations. The particle size distributions are measured with an electrical sensing zone particle analyser (e.g. Coulter Counter), although newer, more sophisticated and more expensive cytometers and cell sorters can be used.

The basic principle of the electrical particle analyser is that a current flows between two electrodes immersed in an electrolyte containing the particles. One of the electrodes is encased in a glass tube and the other is external to the tube, and a small aperture in the tube provides the electrolyte bridge between the electrodes. When a vacuum is applied to the glass tube, electrolyte and particles from the external suspension are pulled through the aperture. When a clay particle, a microbe, or a complex of the two enters the aperture, an amount of electrolyte equivalent to the volume of the particulate is displaced and the resistance between the two electrodes is proportionately increased. This resistance is detected and measured by an electronic circuit, which, after having been calibrated with particles of known size, will indicate the number and size of the particles passing through the aperture. From these data, particle size distribution curves are plotted.

In most applications of the particle counter, a strong electrolyte is used to disperse the particles so that an accurate total count can be obtained. When studying surface interactions between clays and microbes, however, a weak electrolyte must be used to prevent disruption of any clay microbe complexes. Although the lowest electrolyte concentration that can be used will depend, to some extent, on the model of the instrument and the aperture size, a 0.06N concentration of most electrolytes will provide adequate conductivity, maintain a reasonable signal-to-noise ratio and not cause disruption of the complexes (Santoro & Stotzky 1967a, 1968). When homoionic clays are used, the electrolyte should consist of the chloride salt of the cation saturating the clay.

The size of the aperture depends on the size of the clays and microbes being studied. An aperture is usually accurate to about 2 to 10% of its diameter, and therefore, when working with particles in the size range from approximately 2 to 10 μm, a 100 μm aperture can be used. The use of a smaller aperture for particles in this size range will result in excessive plugging and longer measurement times.

Although this technique can be used to determine whether surface interactions occur between particles, it is only a semi-quantitative technique. It is not possible, at present, to determine from the particle size distribution curves how many individual clay particles and microbial cells are involved in an aggregate.

The technique can also be used to measure flocculation of clay minerals or microbes by different cations, metabolites or other compounds (Santoro & Stotzky 1967b), as well as the germination of fungal spores in the presence of other particles, such as clays or bacterial cells (Santoro *et al.* 1967).

In practice, different concentrations of microbial cells and clay minerals are thoroughly mixed and, after various reaction times, aliquots of the mixtures are added to a constant volume of the appropriate electrolyte and placed on the particle analyser. The reactions between the particulates occur externally to the analyser, and the conditions of the reaction mixtures depend on the variables being studied.

7.9 Measurement of the net charge of clays and microorganisms

The net charge of particulates can be determined by placing them in an electrical field and measuring, by microscopic observation, their direction (i.e. either towards the anode or the cathode) and rate of migration. This technique is usually called micro-electrophoresis or particle or cell electrophoresis, and equipment for such measurements is commercially available (e.g. Rank Bros., Bangham *et al.* 1958; Zeta Meter, Riddick 1968; Shaw 1969). The concepts of micro-electrophoresis are beyond the scope of this discussion, and reference is made to appropriate texts (Bier 1959; Riddick 1968; Shaw 1969).

Basically, the particles are placed in a cylindrical or flat linear glass cell that contains an electrode at each end. When a current is applied across the electrodes, the particles will move either towards the negative (cathode) or positive electrode (anode), depending on their net charge, and the rate of migration is a measure of the amount of charge. The direction and rate of migration is observed directly with a microscope. It is imperative that the microscope be focused at the plane where movement of the particles is a result solely of electrophoretic migration (stationary layer) and not of counter-current flow of the suspending liquid (electro-osmosis).

The rate of migration or mobility is dependent on the ionic strength (μ) of the suspending liquid ($\mu = \frac{1}{2}\ \Sigma C_i Z_i^2$, where C_i and Z_i are the molar concentration and valency, respectively, of all the ions in the solution), and when the electrophoretic mobility of different particles or of the same particles in different solutions is compared, μ must be maintained constant. When μ is too high, thermal overturn of the solution can occur when a sufficiently high voltage is applied, and the stationary layer will be disrupted. Higher μ can be used if the electrophoresis cell is immersed in a temperature-controlled water bath, but measurement of the electrophoretic mobility of clays and microbes is usually conducted at low μ, sometimes just in distilled water.

The net charge of microbial cells and of clay minerals that have a pH-dependent charge (e.g. kaolinite, allophane) depends on the pH of the suspending solution. At some pH values, the particles will not move in the electrical field, and this pH is called the isoelectric point (pI) of the particles and represents the pH at which the number of negative and positive charges is equal. At pH values above the pI, the particles will have a net negative charge, and at pH values below the pI, the particles will have a net positive charge. The pI is not a fixed pH value, as it can be shifted by altering the cation or organic composition of the suspending solution (Santoro & Stotzky 1967c; Stotzky 1972; Kiremidjian & Stotzky 1975, 1976;

see p. 111). Furthermore, only the net charge is measured, and a particle with a net negative charge can still have some positively charged sites and vice versa.

When determining the net charge of clays and microbes, the electrophoretic mobility of the particles should be measured over a range of pH values (usually from pH 2.0 to pH 9.0), to determine both the pI and the charge at different pH values. When such mobility-pH measurements are made, it is imperative that μ be maintained constant at each pH. The data obtained from mobility measurements are expressed as $\mu m/s^{-1}/V^{-1}/cm^{-1}$, which are plotted on the ordinate, and the pH values are plotted on the abscissa.

The mobility measurements are often converted to electrokinetic or zeta potentials (expressed in mV) by application of the Smoluchowski equation (Shaw 1969). However, such conversion is based on a number of assumptions (e.g. that the particle is non-conductive and is essentially spherical), which do not always pertain to clay minerals or to all microbial cells. Consequently, even though the conversion of electrophoretic mobility measurements to electrokinetic or zeta potentials provides an estimate of the relative charge on clay minerals and microbial cells, at a given pH, and, therefore, some prediction can be made of their ability to come close enough for surface interactions to occur, the nature of the particles involved must be considered, and it is usually more accurate to express the data as actual mobility measurements. It is also important that sufficient particles in a suspension be tracked to yield statistically valid data (usually 15 to 30 clay particles or microbial cells per measurement are adequate).

7.10 Equilibrium adsorption and binding of organic substances to clay minerals

The most commonly used technique for obtaining equilibrium adsorption isotherms of organic substances on clay minerals is to mix various concentrations of the desired organic compound, in a constant volume, with a constant concentration and volume of clay (Harter & Stotzky 1971). After an optimum mixing period (usually determined in preliminary studies) at a constant temperature, the clay–organic mixtures are centrifuged at a speed sufficient to pellet the clay (40 000 *g* is usually adequate), preferably in a refrigerated centrifuge. The supernatant is carefully decanted, so as not to disturb the pellet, and the concentration of the organic substance is determined on an aliquot of the supernatant. To correct for adsorption of the organic substance to the centrifuge tube and for degradation of the organic substance during the mixing and centrifugation procedures, control tubes containing no clay are run concomitantly with the experimental tubes. The amount of organic substance adsorbed by the clay at equilibrium is calculated as: amount of organic substance recovered in the control tube less the amount of organic substance recovered in the supernatant from the clay-containing tube. The data can be plotted directly, with the amount of organic substance added indicated on the abscissa and the amount of organic substance adsorbed indicated on the ordinate, or the data can be mathematically transformed, as discussed below.

Mixing can be on a rotating wheel, on a shaker or by any method that facilitates maximum contact between the adsorbent and the adsorbate. A constant concentration of the organic substance can be maintained and the amount of clay varied. Although it might be expected that the shape of the isotherms would be the same whether the concentration of the organic is varied and the amount of clay is constant or vice versa, this is not always so as the ratio of the number of molecules of the organic substance to the number of clay particles influences the probability of collision between the two components and apparently other aspects involved in the surface interactions. It is recommended that both types of isotherms be obtained, as information about the mechanisms involved in the adsorption process can often be derived from a comparison of the two methods.

Regardless of whether the concentration of clay or organic substance is maintained constant, the clays must be hydrated before the organic is added, especially when comparisons are made between types of clay. Expandable clays (e.g. as montmorillonite) adsorb greater quantities of water than non-expandable clays (e.g. kaolinite) and erroneous data will be obtained if adsorption is determined with dry clays added to a constant volume of solution; for example, approximately 20% less apparent adsorption may be obtained with dry than with hydrated montmorillonite. This is because the amount of supernatant recovered from dry montmorillonite, after centrifugation, can be about 20% less as a result of hydration of the clay, and therefore, the amount of organic substance per unit volume of supernatant will be greater. Consequently, if dry clays are used in adsorption studies, the desired amount of clay should first be thoroughly mixed with water, centrifuged and the resultant wet pellet mixed with the organic substance before adsorption is determined as above.

It must be stressed that the centrifugation technique yields only 'apparent' equilibrium adsorption, as many of the organic molecules may not be directly associated with the clays but may only be occluded within the fabric (e.g. edge-to-face association) of the clays in the pellet. Furthermore, some organic molecules, especially large ones, may be carried into the pellet during centrifugation by an 'umbrella effect'. Although it has not been demonstrated that centrifugation overcomes electrostatic repulsion between adsorbents and adsorbates of like charge, this possibility should always be considered.

Comparison of the centrifugation technique with techniques involving filtration of clay–organic mixtures through filter membranes that retain the clay and allow the organic substance to pass through, followed by analysis of the filtrate, has indicated that the centrifugation technique is more reproducible and easier to conduct. Dialysis techniques, wherein the clay–organic mixture is placed in a dialysis sac and the non-adsorbed organic is allowed to diffuse out of the sac into the surrounding sulution, are cumbersome and restricted to low molecular weight organic substances (T. Dashman & G. Stotzky, unpublished observations).

Regardless of the techniques used to determine equilibrium adsorption, this parameter is of little value to the microbial ecologist, who is usually interested in the biological availability or activity of the clay-associated organic substance. Inasmuch as some of the organic molecules apparently adsorbed at equilibrium are only loosely associated with the clays and are readily removed from this association by water or by the solutions of the natural habitats, any biological data obtained will be a function of both loosely and tightly bound molecules as well as of molecules in solution. Consequently, the clay–organic mixtures must be washed, with water or other relevant solvent, until no more free adsorbate is detected (i.e. binding isotherms must be constructed) before meaningful biological studies with the clay–organic complexes are conducted.

To obtain the amount of an organic substance bound to the clays, the pellet from the initial equilibrium adsorption centrifugation is resuspended in a constant volume of water or other solvent, usually by vortexing, and again centrifuged. The supernatant is decanted, and the concentration of the organic substance determined. This procedure is repeated until no more organic can be detected in the supernatant. The amount of organic substance bound is calculated as: amount of organic recovered in the control tube less the total amount of organic recovered in the equilibrium solution and in all subsequent washings. Binding isotherms can then be constructed, with the amount of organic added indicated on the abscissa and the amount of organic bound indicated on the ordinate.

It is important that all of the supernatant after each washing is removed from the tube before resuspension of the pellet in a fresh volume of the washing solution. This requires sufficient centrifugal force to form a compact pellet, and a cotton-tipped swab can be used to remove any droplets or film of supernatant adhering to the tube (usually, only an aliquot of the supernatant is analysed and quantitative recovery of supernatant is not necessary). When the level of the organic substance in the supernatant falls below detectable levels, the supernatant should be concentrated before analysis (placing the supernatant in a dialysis sac, which, is placed in front of an operating fan, is a simple and usually non-destructive means of concentration). After no more organic is detected in the supernatant, the clay–organic complex should be washed once or twice more to remove any of the residual, but not detectable, organic substance.

There are numerous ways of plotting both equilibrium adsorption and binding isotherms, but a discussion of the merits and disadvantages of these different methods is beyond the scope of this chapter. The reader is referred to a recent review (Travis & Etnier 1981) for details and discussions of various methods for constructing isotherms that have been used in sorption studies with clay minerals. One expression of sorption data that has not been used extensively in clay–organic studies (Bondy & Harrington 1979) is the Scatchard plot (Scatchard 1949), which is used in cell biology and can provide information on the number of binding sites on the clay and on the association constant of the organic substance with the clay (Krumins & Stotzky 1980). Furthermore, the overall shape of the isotherms can provide considerable information on the mechanisms of adsorption and binding involved (e.g. Giles *et al.* 1974).

The bound clay–organic complex remains relatively stable if stored as a centrifuged pellet at 4°C (T. Dashman & G. Stotzky, unpublished observations; Gerard & Stotzky 1973; Harter & Stotzky 1973).

Similar techniques can be used to determine the equilibrium adsorption and binding of viruses to clay minerals (Stotzky *et al.* 1981).

7.11 Location of organic substances or viruses in clay–organic substances or clay–virus complexes

To evaluate the biological availability or activity of organic substances and viruses complexed with clay minerals, it is often helpful to know the location of the organics and viruses on the clays. A variety of indirect and direct techniques can be used to determine this.

7.11.1 X-ray Diffraction Analysis

X-ray diffraction analysis is most applicable with expandable clay minerals, as it indicates whether a portion of the bound organic substances is located between the lattices of the clays (viruses are usually too large to intercalate these clays). However, this analysis should also be conducted with non-expandable clays, as some organics can intercalate even these clay minerals (Jackson 1969; Theng 1979). Furthermore, X-ray diffraction analysis can also indicate, in conjunction with an analysis of the shape of the binding isotherms (Giles *et al.* 1974), whether mono- or multilayer binding has occurred; whether a linear molecule has been adsorbed horizontally along its long axis or vertically; or whether a molecule with secondary and tertiary bonds, such as a globular protein, becomes denatured and is adsorbed as a linear molecule between the lattices (Talibudeen 1955; McLaren & Peterson 1961; Harter & Stotzky 1973).

A thick slurry of the clay–organic complex is placed onto a pre-cut glass microscope slide to yield an orientated film containing approximately 2 mg cm^{-2} of clay. The film is dried, usually at an elevated temperature (e.g. 110 or 200°C), to remove water of hydration (the drying temperature depends on the heat stability and the hygroscopicity of the organic). To prevent rehydration of the dried complexes, the scattering shield over the sample holder of the X-ray diffractometer is covered with a plastic film, and a container of silica gel or other desiccant is placed in the shield. Despite such precautions, some hydration may occur, and a sample of the clay (especially if homoionic) without the organic substance should be treated and analysed concomitantly with the clay–organic complexes to provide a baseline for the amount of expansion of the clay under the prevailing relative humidity.

The dried samples are analysed by X-ray diffraction, and the amount of expansion of the clay lattices by the organic substance is measured. Depending on the type of diffractometer, it may not be possible to observe directly the amount of basal expansion, as many diffractometers cannot be operated at low enough angles, and this will have to be calculated from the secondary and tertiary basal spacings (Harter & Stotzky 1971, 1973). The optimum operating conditions, scan speeds, and slit arrangements will depend on the diffractometer used.

In some cases, the X-ray diffraction patterns of the clay–organic complexes after equilibrium adsorption should be compared with the patterns obtained with stable bound complexes (i.e. after ultimate washing), as such a comparison can provide information on whether multilayer equilibrium adsorption occurs and on the orientation of the initially adsorbed organic molecules (T. Dashman & G. Stotzky, unpublished observations).

7.11.2 Electron Microscopy

When the organic compound is apparently too large to intercalate the clay (i.e. when no expansion of the clay is indicated by X-ray diffraction analysis) or when it is suspected that, because of charge relations, the organic may adsorb to positively charged edge and surface sites, electron microscopy can be used to determine the location of the organic substance on the clay (Harter & Stotzky 1973). The stable clay–organic complexes are diluted with distilled water (the optimum concentration of the complexes is determined by trial and error), and a drop is placed on an appropriately coated (e.g. polyvinylformaldehyde; Formvar) grid. The sample is air-dried, shadowed (e.g. with chromium or platinum), and examined by standard electron microscopic techniques (Koehler 1978; Wischnitzer 1980).

When other than stable clay–organic complexes are examined (e.g. complexes after equilibrium adsorption), simple air-drying cannot be used, as the removal of water may result in artefacts caused by the physical apposition of the organic on the clay. In such cases, critical point drying or comparable techniques are necessary to prepare the samples, before shadowing with metal, for examination by electron microscopy (Bystricky *et al.* 1975).

When clay–virus complexes are examined by electron microscopy, it is important to distinguish real surface interactions between the clay and virus particles from apparent interactions that occur not in the reaction mixture but on the grid as a result of the particles being brought together by surface tension as water is evaporated from the droplet. Consequently, after depositing the specimen on the grid and adding a drop of 2.5% (w/v) uranyl acetate, the grid is dehydrated in a graded series of ethyl alcohol:water mixtures and the final absolute alcohol is replaced by liquid carbon dioxide which is then heated to 42°C in a closed chamber, which results in a pressure of 1060 p.s.i. and causes the liquid carbon dioxide to pass through its critical point and change into a gas, which is vented from the chamber. If surface interactions between the clay and viruses occurred in the reaction mixture, rather than during drying of the droplet on the grid, then some adsorbed viruses should be located at various angles to the clay platelets rather than just being on the same plane as the clay, as would occur with air-drying of the specimen. This projection into space can be observed on electron micrographs by the size, shape, and direction of the shadows cast by the viruses (especially by tailed bacteriophages) and the clays after the grids that have been treated by critical point drying have been shadowed with platinum or gold-palladium alloy (Bystricky *et al.* 1975).

7.12 Microbial utilisation and biological activity of clay–organic and clay–virus complexes

The availability to microbes of organic substrates bound to clay minerals can be evaluated best by respirometric techniques (Stotzky 1965), the easiest of which is the measurement of oxygen consumption (Stotzky 1972; Gerard & Stotzky 1973). Although standard manometric techniques (e.g. Warburg or Gilson respirometers—Umbreit *et al.* 1964) are familiar to microbial ecologists, it is important that adequate controls be included. In addition to the usual controls (i.e. only the clay–organic complex, the organic, the clay, the microorganism, and the free organic plus the microbe conditioned to utilising the organic as a sole source of nutrients and energy), other internal controls (e.g. the organic substance, clay and conditioned microorganism added individually to the same respiratory flask; the clay–organic complex, 50% of the amount of the organic bound to the clay in the free form, and the conditioned microbe added individually to the same flask) are necessary to determine whether the exoenzymes of the utilising microbes and/or the products of the enzymic degradation of the organics are adsorbed by the clays. By appropriate evaluation of these control and experimental systems (i.e. the clay–organic complex plus the conditioned microbe), the availability of organics complexed with clay minerals can be determined (T. Dashman & G. Stotzky, unpublished observations; Stotzky 1972; Gerard & Stotzky 1973).

The catalytic activity of enzymes bound to clays is more difficult to measure, as the clays will usually interfere with the colorometric techniques which are widely used in enzyme assays. Furthermore, the clays might bind the products of an enzymatic reaction and, therefore, render the products unreactive to chemical analysis. Consequently, manometric methods should be used whenever possible (Stotzky 1972).

The utilisation by microbes of viruses bound to clay minerals as a source of nutrients and energy can be measured by standard dilution plating techniques and by respirometric methods (Stotzky *et al.* 1981). Very high concentrations of viruses are needed, however, as the energy content of a viral particle is low, and many particles (usually in excess of 1×10^8) are necessary to enable even a single microbial cell division. If insufficient viral particles are available, the microbial utilisation of the viruses can be estimated by measuring the decrease in viral titres in the presence of the microbes. It must be recognised, however, that such decreases in titre can result from inactivation, rather than from degradation, of the virus, and preliminary studies are necessary to determine whether and to what extent the viruses adsorb to the microbial cells (Stotzky *et al.* 1981).

The lytic activity of viruses bound to clays can usually be measured by the same techniques used to measure the activity of free viruses; for example, the lytic activity of sorbed bacteriophages can be determined by measuring the decrease in optical density of a host bacterial suspension inoculated with the clay–phage complex, correcting for the light absorbance of the clay. When plaque formation or cytopathic effects (CPE) are measured, it is important to run controls to ensure that the clays alone do not cause plaque formation or CPE (Stotzky *et al.* 1981).

7.13 Measurement of enzyme activities in soil

There are many difficulties in the measurement and interpretation of enzyme activities in soil (Skujins 1976; Burns 1978). One fundamental and recurring problem, and one which is common to many areas of microbial ecology, is the relevance of determinations *in vitro* to those performed *in situ* (i.e. in an undisturbed or minimally disturbed environment) or even in the laboratory on carefully taken and transported soil cores. Nevertheless the predominant current view in soil enzymology is that basic enzymological procedures should be used when at all possible. Consequently, assays are usually performed in a buffer of sufficient strength to poise and maintain the optimum pH for the duration of the assay, at a constant temperature, with excess substrate, and in shaken rather than in stationary mixtures. However, these conditions are obviously different from those existing in natural soil environments, where the ambient pH may be different from the optimum pH for a particular enzymic activity, substrates are discontinuous in both time (e.g. seasonal fluctuations) and space and they are often physically and chemically associated with insoluble organic debris and inorganic particulates. Thus, differences undoubtedly exist between the maximum potential enzymic activity obtained under strictly defined conditions in the laboratory and those occurring in soil *in situ*. It is difficult, if not impossible, to reconcile the results from one approach with those from the other but measurements *in vitro* have a strong appeal because the conditions of the assay can be described precisely and the result reproduced. Furthermore, measurements of activity in conditions similar to those found *in situ* (e.g. no buffer, low substrate concentration, varying temperature) have a number of drawbacks; most obviously, they rarely lend themselves to the derivation of kinetic constants, such as K_m and V_{max} especially as substrate limitation may occur and the pH may fluctuate during the assay.

A second constraint in soil enzymology is the interpretation of the activity data, as it is often difficult to determine which component(s) contributing to the overall activity of a specific enzyme is actually being measured (Fig. 2). Although enzymic activity associated with proliferating microorganisms is a major contribution to the catalysis of a substrate, functional enzymes are also associated with dead cells, cell debris and humic and clay colloids. The differentiation between these enzyme fractions is difficult, even though many soil enzyme assays attempt to eliminate activity due to microbial growth by incorporating an inhibitor (e.g. toluene, chloroform, sodium azide, an antibiotic) in the reaction mixture, by using γ-irradiated (*c.* 5 Mrad—Burns *et al.* 1978) soil or by employing as brief an assay period as possible (e.g. < 2 h) to minimise cell growth. Enzyme activity associated with only the humic fraction may be assessed by extracting a portion of the protein–humic complex from soil using buffers, inorganic salts, and sonication (Getzin & Rosefield 1971; Burns *et al.* 1972a; Pettit *et al.* 1976; Nannipieri *et al.* 1980), followed by passing the crude extract through a bacteriological filter in order to remove cells, cell debris and particulate organic material. Enzyme

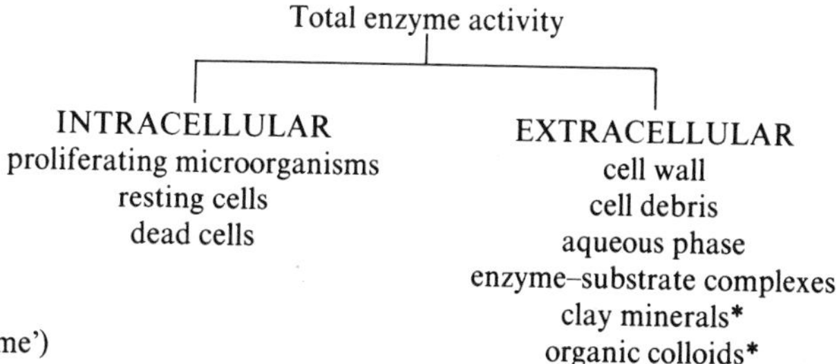

Figure 2. Components of enzyme activity in soil (*'accumulated enzyme' or 'soil enzyme')

activity in the extracted humic fraction (the filtrate) will usually represent < 20% of that measured in the original soil. Difficulty in separating the colloidal humic matter (and thus the humic–enzyme fraction) from the organo-mineral complex, plus inactivation of enzyme during the extraction sequence may account for most of the lost activity. Comparatively little activity is associated with the filtered non-proliferating cells and debris. Current concepts suggest that enzymes bound to the humic matter in soil or in crude extracts (i.e. those that are generally understood by the terms 'soil enzymes', 'accumulated enzymes' and 'persistent enzymes') are co-polymerised with humic polyphenols, an association that provides the extracellular enzymes with a persistence that free enzymes do not have in the hostile degradative, adsorptive and denaturing soil environment. Enzymes adsorbed on or within clays only occasionally retain their activity, even though they may persist as protein (see p 115).

The extraction and purification of a specific enzyme from its humic complex has not been achieved, possibly because it is difficult to separate the enzyme from the humic matter non-destructively, or because even when separation is achieved the enzyme becomes inactive as it depends upon its association with the humic fraction for its stability and even its activity (cf. lipoproteins, glycoproteins). Therefore, it is necessary to characterise fully the enzyme in soil and crude humic–enzyme extracts according to the list in Table 4. In the absence of these characteristics it cannot be assumed that the catalyst is a biological entity. Details of the methodologies associated with the measurement and extraction of soil enzymes are given by Roberge (1981).

Table 4. Characteristics of accumulated enzymes in soil and humic–enzyme extracts.

Maximum rate* (V_{max})
Michaelis constant* (K_m) = substrate concentration giving half maximum rate
pH-activity curves
Temperature-activity curves
Time-activity linearity
Stoichiometry
Temperature coefficient (Q_{10}) = Activity at X°C/ Activity at X − 10°C and is generally < 2
Independence of microbial proliferation: elimination (assay < 2 h, γ-irradiation) or inhibition (sodium azide, antibiotic) of microbes
Lack of correlation with microbial numbers: total or proportion of population
Thermo-labile (destroyed by autoclaving)

*Linearity established by more than one transformation of the Michaelis-Menten equation (for example Lineweaver-Burk, Eadie-Hofstee, Hanes-Woolf, Direct plot).

Any correlation between the activity of individual enzymes and microbial numbers is complex, because the measured activity (even in the absence of a microbial inhibitor) is not entirely due to the extant population (Fig. 2). However, a comparison of changes in microbial numbers (caused by, for example, substrate addition) with concomitant changes in enzyme activity can reveal a direct relationship which will allow the quantification of the various contributions to total activity (Burns 1977).

Nevertheless, traditional methods of assessing the collective activities of a number of enzymes may provide data which are more easily interpreted, and measurements such as carbon dioxide evolution (from cellulose), ammonia production (from protein) or dehydrogenase activity are discussed elsewhere in this volume.

7.14 Conclusions

It is apparent from the foregoing experimental section that methodologies exist which allow the study of soil micro-environments. The results of studies on surface interactions between clay minerals and humic colloids and microbes, viruses and organic solutes, indicate that direct interactions (e.g. sorption of microorganisms) may be as important as indirect interactions (e.g. sorption of substrates, enzymes and inhibitors; buffering capacity) in influencing the activity, population dynamics and survival of microorganisms and viruses in soil. There is no doubt that microbes, their organic and inorganic nutrient and energy sources, their metabolites and their various secretions (e.g. exoenzymes, exopolysaccharides) accumulate at soil surfaces. However, whilst the physical and chemical properties of the potential interactants are known, understanding of the consequences of this accumulation for the ecology of the microbiota is rudimentary. The next decade, however, should see significant advances in knowledge of the microbial ecology of soil micro-environments.

Recommended reading

BABICH H. & STOTZKY G. (1980) Environmental factors that influence the toxicity of heavy metals and gaseous pollutants to microorganisms. *CRC Critical Reviews in Microbiology* **8**, 99–145.

BURNS R. G. (1979) Interaction of microorganisms, their substrates and their products with soil surfaces. In: *Adhesion of Microorganisms to Surfaces* (Eds D. C. Ellwood, J. Melling & P. R. Rutter), pp. 109–38. Academic Press, London.

BURNS R. G. (1980) Microbial adhesion to soil surfaces: consequences for growth and enzyme activities. In: *Microbial Adhesion to Surfaces* (Eds R. C. W. Berkeley, J. M. Lynch, J. Melling, P. R. Rutter & B. Vincent), pp. 249–62. Ellis Horwood, Chichester.

HATTORI T. & HATTORI R. (1976) The physical environment in soil microbiology: an attempt to extend principles of microbiology to soil microorganisms. *CRC Critical Reviews in Microbiology* **4**, 423–61.

MARSHALL K. C. (1976) *Interfaces in Microbial Ecology*. Harvard University Press, Cambridge, Massachusetts.

MARSHALL K. C. (1980) Reactions of microorganisms, ions and macromolecules at interfaces. In: *Contemporary Microbial Ecology* (Eds D. C. Ellwood, J. N. Hedger, M. J. Latham, J. M. Lynch & J. H. Slater), pp. 93–106. Academic Press, London.

MARTIN J. P. & HAIDER K. (1980) Microbial degradation and stabilisation of ^{14}C-labeled lignins, phenols, and phenolic polymers in relation to soil humus formation. In: *Lignin Biodegradation: Microbiology, Chemistry, and Applications* (Eds T. K. Kirk, T. Higuchi & H. M. Cheng), vol. 1, pp. 77–100. CRC Press, Boca Raton, Florida.

STOTZKY G. (1972) Activity, ecology, and population dynamics of microorganisms in soil. *CRC Critical Reviews in Microbiology* **2**, 59–137.

STOTZKY G., SCHIFFENBAUER M., LIPSON S. M. & YU B. H. (1981) Surface interactions between viruses and clay minerals and microbes: mechanisms and implications. In: *Viruses and Wastewater Treatment* (Eds M. Goddard & M. Butler) pp. 199–204. Pergamon Press, Oxford.

References

ARCHIBALD A. R., BADDILEY J. & HEPTINSTALL A. (1973) The alanine ester content and magnesium binding capacity of walls of *Streptococcus aureus* H grown at different pH values. *Biochimica et Biophysica Acta* **291**, 629–34.

BABICH H. & STOTZKY G. (1977a) Reductions in the toxicity of cadmium to microorganisms by clay minerals. *Applied and Environmental Microbiology* **33**, 696–705.

BABICH H. & STOTZKY G. (1977b) Effect of cadmium on fungi and on interactions between fungi and bacteria in soil: influence of clay minerals and pH. *Applied and Environmental Microbiology* **33**, 1059–66.

BABICH H. & STOTZKY G. (1978a) Effects of cadmium on the biota: influence of environmental factors. *Advances in Applied Micriobiology* **23**, 55–117.

BABICH H. & STOTZKY G. (1978b) Effect of cadmium on microbes *in vitro* and *in vivo*: influence of clay minerals. In: *Microbial Ecology* (Eds M. W. Loutit & J. A. R. Miles), pp. 412–16. Springer–Verlag, Berlin.

BABICH H. & STOTZKY G. (1979) Abiotic factors affecting the toxicity of lead to fungi. *Applied and Environmental Microbiology* **38**, 506–13.

BABICH H. & STOTZKY G. (1980a) Environmental factors that influence the toxicity of heavy metals and gaseous pollutants to microorganisms. *CRC Critical Reviews in Microbiology* **8**, 99–145.

BABICH H. & STOTZKY G. (1980b) Reductions in inactivation rates of bacteriophages by clay minerals in lake water. *Water Research* **14**, 185–7.

BANGHAM A. D., HEARD D. H., FLEMANS R. & SEAMAN G. F. (1958) An apparatus for electrophoresis of small particles. *Nature, London* **182**, 642–4.

BERG G., BODILY H. L., LENNETTE E. H., MELNICK J. L. & METCALF T. G. (1976) *Viruses in Water*. American Public Health Association Incorporated, Washington.

BIER M. (1959) *Electrophoresis: Theory, Methods, and Applications*. Academic Press, New York.

BLOOMFIELD B. J. & ALEXANDER M. (1967) Melanins and resistance of fungi to lyses. *Journal of Bacteriology* **93**, 1276–80.

BOLLAG J-M., LIU S-Y. & MINARD R. D. (1980) Cross-coupling of phenolic humus constituents and 2,4-dichlorophenol. *Soil Science Society of America Journal* **44**, 52–6.

BONDIETTI E., MARTIN J. P. & HAIDER K. (1972) Stabilization of amino sugar units in humic-type polymers. *Soil Science Society of America Proceedings* **36**, 597–602.

BONDY S. C. & HARRINGTON M. E. (1979) L-Amino acids and D-glucose bind stereospecifically to a colloidal clay. *Science* **203**, 1243–4.

BOZIAN R. H. & STOTZKY G. (1976) Inhibition of nitrification in soil by SO_2, and the effect of kaolinite and montmorillonite on inhibition. *Agronomy Abstracts* p. 135.

BROWN G., NEWMAN A. C. D., RAYNER J. H. & WEIR A. H. (1978) The structures and chemistry of soil clay minerals. In: *The Chemistry of Soil Constituents* (Eds D. J. Greenland & M. H. B. Hayes), pp. 29–178. John Wiley & Sons, London.

BURNS R. G. (1977) Soil enzymology. *Science Progress (Oxford)* **64**, 275–85.

BURNS R. G. (1978) *Soil Enzymes*. Academic Press, London.

BURNS R. G. (1979) Interaction of microorganisms, their

substrates and their products with soil surfaces. In: *Adhesion of Microorganisms to Surfaces* (Eds D. C. Ellwood, J. Melling & P. R. Rutter), pp. 109–38. Academic Press, London.

BURNS R. G. (1980) Microbial adhesion to soil surfaces: consequences for growth and enzyme activities. In: *Microbial Adhesion to Surfaces* (Eds R. C. W. Berkeley, J. M. Lynch, J. Melling, P. R. Rutter & B. Vincent), pp. 249–62. Ellis Horwood, Chichester.

BURNS R. G. (1982) Carbon mineralisation. In: *Microbial Interactions and Communities* (Eds A. T. Bull & J. H. Slater). Academic Press, London. (In press.)

BURNS R. G. & EDWARDS J. A. (1980) Pesticide breakdown by soil enzymes. *Pesticide Science* **11**, 506–12.

BURNS R. G., EL-SAYED M. H. & MCLAREN A. D. (1972a) Extraction of an urease active organo-complex from soil. *Soil Biology and Biochemistry* **4**, 107–8.

BURNS R. G., GREGORY L. J., LETHBRIDGE G. & PETTIT N. M. (1978) The effect of γ-irradiation on soil enzyme stability. *Experientia* **34**, 301–2.

BURNS R. G., PUKITE A. H. & MCLAREN A. D. (1972b) Concerning the location and persistence of soil urease. *Soil Science Society of America Proceedings* **36**, 308–11.

BYSTRICKY V., STOTZKY G. & SCHIFFENBAUER M. (1975) Electron microscopy of coliphages adsorbed to clay minerals: application of the critical point drying method. *Canadian Journal of Microbiology* **21**, 1278–82.

CLARK A. (1974) *The Chemisorptive Bond*. Academic Press, New York.

FLAIG W., BEUTELSPACHER H. & RIETZ E. (1975) Chemical composition and physical properties of humic substances. In: *Soil Components* (Ed. J. E. Gieseking), pp. 1–211. Springer–Verlag, Berlin.

GERARD J. F. & STOTZKY G. (1973) Smectite-protein complexes vs. non-complexed proteins as energy and carbon sources for bacteria. *Agronomy Abstracts* p. 91.

GETZIN L. W. & ROSEFIELD I. (1971) Partial purification and properties of a soil enzyme that degrades the insecticide malathion. *Biochimica et Biophysica Acta* **235**, 442–53.

GIESEKING J. E. (1975) *Soil Components. Vol. 2. Inorganic Components*. Springer–Verlag, New York.

GILES C. H., SMITH D. & HUITSON A. (1974) A general treatment and classification of the solute adsorption isotherm. *Journal of Colloid and Interface Science* **47**, 755–65.

GRIM R. E. (1968) *Clay Mineralogy*. McGraw–Hill, New York.

HARTER R. D. & STOTZKY G. (1971) Formation of clay–protein complexes. *Soil Science Society of America Proceedings* **35**, 383–9.

HARTER R. D. & STOTZKY G. (1973) X-ray diffraction, electron microscopy, electrophoretic mobility, and pH of some stable smectite-protein complexes. *Soil Science Society of America Proceedings* **37**, 116–23.

HATTORI T. & HATTORI R. (1976) The physical environment in soil microbiology: an attempt to extend principles of microbiology to soil microorganisms. *CRC Critical Reviews in Microbiology* **4**, 423–61.

HAYES M. H. B. & SWIFT R. S. (1978) The chemistry of soil organic colloids. In: *The Chemistry of Soil Constituents* (Eds D. J. Greenland & M. H. B. Hayes), pp. 179–320. John Wiley, New York.

HSU T-S. & BARTHA R. (1976) Hydrolyzable and non-hydrolyzable 3,4-dichloroaniline-humus complexes and their respective rates of degradation. *Journal of Agriculture and Food Chemistry* **24**, 118–22.

JACKSON M. L. (1964) Soil clay mineralogical analysis. In: *Soil Clay Mineralogy* (Eds C. I. Rich & G. W. Kunze), pp. 245–94. University of North Carolina Press, Chapel Hill.

JACKSON M. L. (1969) *Soil Chemical Analysis—Advanced Course*. Published by the author, University of Wisconsin, Madison.

JAMES A. M. (1979) Molecular aspects of biological surfaces. *Chemical Society Reviews* **8**, 389–418.

KIREMIDJIAN L. & STOTZKY G. (1973) Effects of natural microbial preparations on the electrokinetic potential of bacterial cells and clay minerals. *Applied Microbiology* **25**, 964–71.

KIREMIDJIAN L. & STOTZKY G. (1975) Influence of mono- and multivalent ions on the electrokinetic properties of adult *Rana pipiens* kidney cells. *Journal of Cell Physiology* **85**, 125–34.

KIREMIDJIAN L. & STOTZKY G. (1976) Influence of mono- and multivalent cations on the electrokinetic properties of normal human lymphoid and Burkitt lymphoma cells. *Experientia* **32**, 312–14.

KISS S., DRAGAN-BULARDA M. & RADULESCU D. (1975) Biological significance of enzymes in soil. *Advances in Agronomy* **27**, 25–87.

KOEHLER J. K. (1978) *Advanced Techniques in Biological Electron Microscopy II*. Springer–Verlag, New York.

KRUMINS S. & STOTZKY G. (1980) Protein–membrane interactions: equilibrium adsorption and binding of proteins and polyamino acids on erythroblasts transformed by Friend virus. *Cell Biology International Reports* **4**, 1131–41.

KUNC E. & STOTZKY G. (1974) Effect of clay minerals on heterotrophic microbial activity in soil. *Soil Science* **118**, 186–95.

KUNC E. & STOTZKY G. (1977) Acceleration of aldehyde decomposition in soil by montmorillonite. *Soil Science* **124**, 167–72.

KUNC F. & STOTZKY G. (1980) Acceleration by montmorillonite of nitrification in soil. *Folia Microbiologica* **25**, 106–25.

LADD J. N. & BUTLER J. H. A. (1975) Humus-enzyme systems and synthetic, organic polymer analogs. In: *Soil Biochemistry* (Eds E. A. Paul & A. D. McLaren), vol. 4, pp. 143–94. Marcel Dekker, New York.

LAMBE T. W. & MARTIN R. T. (1956) Composition and engineering properties of soils IV. *Highway Research Board Procedings* **35**, 661–77.

LETHBRIDGE G., BULL A. T. & BURNS R. G. (1978) Assay and properties of 1,3-β-glucanase in soil. *Soil Biology and Biochemistry* **10**, 389–91.

LINHARES L. F. & MARTIN J. P. (1978) Decomposition in soil of the humic acid-type polymers (melanins) of *Eurotium echinulatum, Aspergillus Glaucus* sp. and other fungi. *Soil Science Society of America Journal* **42**, 738–43.

LIPSON S. M. & STOTZKY G. (1979) Adsorption of reovirus by clay minerals. *Abstracts of the Annual Meeting of the American Society for Microbiology* p. 188.

LIPSON S. M. & STOTZKY G. (1980) Surface interactions between reovirus and clay minerals. *Abstracts of the Annual Meeting of the American Society for Microbiology* p. 133.

MCLAREN A. D. & PETERSON G. H. (1961) Montmorillonite as a caliper for the size of protein molecules. *Nature, London* **192**, 960–1.

MCLAREN A. D. & SKUJINS J. (1968) The physical environment of microorganisms in soil. In: The *Ecology of Soil Bacteria* (Eds T. R. G. Gray & D. Parkinson), pp. 3–24. Liverpool University Press, Liverpool.

MACRITCHIE F. & ALEXANDER A. E. (1963) Kinetics of adsorption of proteins at interfaces. III. The role of electrical barriers in adsorption. *Journal of Colloid Science* **18**, 464–9.

MACURA J. & STOTZKY G. (1980) Effect of montmorillonite and kaolinite on nitrification in soil. *Folia Microbiologica* **25**, 90–105.

MARSHALL K. C. (1976) *Interfaces in Microbial Ecology*. Harvard University Press, Cambridge, Massachusetts.

MARSHALL K. C. (1980) Reactions of microorganisms, ions and macromolecules at interfaces. In: *Contemporary Microbial Ecology* (Eds D. C. Ellwood, J. N. Hedger, M. J. Latham, J. M. Lynch & J. H. Slater), pp. 93–106. Academic Press, London.

MARTIN J. P., FILIP Z. & HAIDER K. (1976) Effect of montmorillonite and humate on growth and metabolic activity of some actinomycetes. *Soil Biology and Biochemistry* **8**, 409–13.

MARTIN J. P. & HAIDER K. (1980) Microbial degradation and stabilization of ^{14}C-labeled lignins, phenols, and phenolic polymers in relation to soil humus formation. In: *Lignin Biodegradation: Microbiology, Chemistry, and Applications* (Eds T. K. Kirk, T. Higuchi & H-M. Chang), vol. I, pp 77–100. CRC Press, Boca Raton, Florida.

MATHUR S. P. & MORLEY H. V. (1975) A biodegradation approach for investigating pesticide incorporation into soil humus. *Soil Science* **120**, 238–40.

MOORE-LANDECKER E. & STOTZKY G. (1974) Effects of concentration of volatile metabolites from bacteria and germinating seeds on fungi in the presence of selected absorbents. *Canadian Journal of Microbiology* **20**, 97–103.

NANFARA M. & STOTZKY G. (1979) Protection of microorganisms by clay minerals against hypertonic osmotic pressures. *Abstracts of the Annual Meeting of the American Society for Microbiology* p. 189.

NANNIPIERI P., CECCANTI B., CERVELLI S. & MATARESE E. (1980) Extraction of phosphatase, urease proteases, organic carbon and nitrogen from soil. *Soil Science Society of America Journal* **44**, 1011–15.

PETTIT N. M., GREGORY L. J., FREEDMAN R. B. & BURNS R. G. (1977) Differential stabilities of soil enzymes: assay and properties of phosphatase and arylsulphatase. *Biochimica et Biophysica Acta* **485**, 357–66.

PETTIT N. M., SMITH A. R. J., FREEDMAN R. B. & BURNS R. G. (1976) Soil urease: activity, stability and kinetic properties. *Soil Biology and Biochemistry* **8**, 479–84.

RAO S. R. (1972) *Surface Phenomena*. Hutchinson, London.

RIDDICK T. M. (1968) *Control of Colloid Stability through Zeta Potential*. Livingstone Publishing, Wynnewood, Pennsylvania.

ROBERGE M. R. (1978) Methodology of soil enzyme measurement and extraction. In: *Soil Enzymes* (Ed. R. G. Burns), pp. 341–70. Academic Press, London.

ROGERS H. J. (1979) Adhesion of microorganisms to surfaces: some general considerations of the role of the envelope. In: *Adhesion of Microorganisms to Surfaces* (Eds D. C. Ellwood, J. Melling & P. Rutter), pp. 29–56. Academic Press, London.

ROSENZWEIG W. D. & STOTZKY G. (1979) Influence of environmental factors on antagonism of fungi by bacteria in soil: clay minerals and pH. *Applied and Environmental Microbiology* **38**, 1120–6.

ROSENZWEIG W. D. & STOTZKY G. (1980a) Influence of environmental factors on antagonism of fungi by bacteria in soil: nutrient levels. *Applied and Environmental Microbiology* **39**, 354–60.

ROSENZWEIG W. D. & STOTZKY G. (1980b) Prodigiosin and the inhibition of *Aspergillus niger* by *Serratia marcescens* in soil. *Soil Biology and Biochemistry* **12**, 295–7.

ROWELL M. J., LADD J. N. & PAUL E. A. (1973) Enzymically active complexes of proteases and humic acid analogues. *Soil Biology and Biochemistry* **5**, 699–703.

RUTTER P. R. & VINCENT B (1980) The adhesion of micro-organisms to surfaces: physico-chemical aspects. In: *Microbial Adhesion to Surfaces* (Eds R. C. W. Berkeley, J. M. Lynch, J. Melling, P. R. Rutter & B. Vincent), pp. 79–92. Ellis Horwood, Chichester.

SANTORO T. & STOTZKY G. (1967a) Effect of electrolyte composition and pH on the particle size distribution of microorganisms and clay minerals as determined by the electrical sensing zone method. *Archives of Biochemistry and Biophysics* **122**, 664–9.

SANTORO T. & STOTZKY G. (1967b) Influence of cations

on flocculation of clay minerals by microbial metabolites as determined by the electrical sensing zone particle analyzer. *Soil Science Society of America Proceedings* **31**, 761–5.

SANTORO T. & STOTZKY G. (1967c) Effect of cations and pH on the electrophoretic mobility of microbial cells and clay minerals. *Bacteriological Proceedings* A15.

SANTORO T. & STOTZKY G. (1968) Sorption between microorganisms and clay minerals as determined by the electrical sensing zone particle analyzer. *Canadian Journal of Microbiology* **14**, 299–307.

SANTORO T., STOTZKY G. & REM L. T. (1967) The electrical sensing zone particle analyzer for measuring germination of fungal spores in presence of other particles. *Applied Microbiology* **15**, 935–9.

SCATCHARD G. (1949) The attractions of proteins for small molecules and ions. *Annals of the New York Academy of Sciences* **51**, 660–72.

SCHENCK S. & STOTZKY G. (1976) Effect on microorganisms of volatile compounds released from germinating seeds. *Canadian Journal of Microbiology* **21**, 1622–34.

SCHIFFENBAUER M. & STOTZKY G. (1978) Adsorption of coliphages by clay minerals and microbial cells. *Abstracts of the Annual Meeting of the American Society for Microbiology* p. 167.

SCHIFFENBAUER M. & STOTZKY G. (1980) Adsorption and desorption of coliphages T1 and T7 by clay minerals. *Abstracts of the Annual Meeting of the American Society for Microbiology* p. 133.

SCHNITZER M. & KHAN S. U. (1972) *Humic Substances in the Environment*. Marcel Dekker, New York.

SHAW D. J. (1969) *Electrophoresis*. Academic Press, London.

SKUJINS J. (1976) Extracellular enzymes in soil. *CRC Critical Reviews in Microbiology* **4**, 383–421.

SØRENSEN L. H. (1975) The influence of clay on the rate of decay of amino acid metabolites synthesized in soils during the decomposition of cellulose. *Soil Biology and Biochemistry* **7**, 171–7.

STOTZKY G. (1965) Microbial respiration. In: *Methods of Soil Analysis* (Ed. C. A. Black), pp. 1550–70. America Society of Agronomy, Madison, Wisconsin.

STOTZKY G. (1965) Replica plating technique for studying microbial interactions in soil. *Canadian Journal of Microbiology* **11**, 629–36.

STOTZKY G. (1966a) Influence of clay minerals on microorganisms: II. Effect of various clay species, homoionic clays, and other particles on bacteria. *Canadian Journal of Microbiology* **12**, 831–48.

STOTZKY G. (1966b) Influence of clay minerals on microorganisms: III. Effect of particle size, cation exchange capacity, and surface area on bacteria. *Canadian Journal of Microbiology* **12**, 1235–46.

STOTZKY G. (1971) Ecologic eradication of fungi—dream or reality. In: *Histoplasmosis, Proceedings of the Second National Conference* (Eds M. L. Furcolow & E. W. Chick), pp. 477–86. Charles C. Thomas Co., Springfield.

STOTZKY G. (1972) Activity, ecology, and population dynamics of microorganisms in soil. *CRC Critical Reviews in Microbiology* **2**, 59–137.

STOTZKY G. (1973) Techniques to study interactions between microorganisms and clay minerals *in vivo* and *in vitro*. In: *Modern Methods in the Study of Microbial Ecology* (Ed. T. Rosswall), pp. 17–28. Swedish Natural Science Research Council, Stockholm.

STOTZKY G. (1980) Surface interactions between clay minerals and microbes, viruses and soluble organics, and the probable importance of these interactions to the ecology of microbes in soil. In: *Microbial Adhesion to Surfaces* (Eds R. C. W. Berkeley, J. M. Lynch, J. Melling, P. R. Rutter & B. Vincent), pp. 231–47. Ellis Horwood, Chichester.

STOTZKY G., DAWSON J. E., MARTIN R. T. & TER KUILE G. H. H. (1961) Soil mineralogy as a factor in the spread of *Fusarium* wilt of banana. *Science* **133**, 1483–5.

STOTZKY G. & DEMUMBRUM L. E. (1965) Effect of autoclaving on X-ray characteristics of clay minerals. *Soil Science Society of America Proceedings* **29**, 225–7.

STOTZKY G. & MARTIN R. T. (1963) Soil mineralogy in relation to the spread of *Fusarium* wilt of banana in Central America. *Plant and Soil* **18**, 317–38.

STOTZKY G. & POST A. H. (1967) Soil mineralogy as possible factor in geographic distribution of *Histoplasma capsulatum*. *Canadian Journal of Microbiology* **13**, 1–7.

STOTZKY G. & REM L. T. (1966) Influence of clay minerals on microorganisms: I. Montmorillonite and kaolinite on bacteria. *Canadian Journal of Microbiology* **12**, 547–63.

STOTZKY G. & REM L. T. (1967) Influence of clay minerals on microorganisms: IV. Montmorillonite and kaolinite on fungi. *Canadian Journal of Microbiology* **13**, 1535–50.

STOTZKY G. & SCHENCK S. (1976) Volatile organic compounds and microorganisms. *Critical Reviews in Microbiology* **4**, 333–82.

STOTZKY G., SCHIFFENBAUER M., LIPSON S. M. & YU B. H. (1981) Surface interactions between viruses and clay minerals and microbes: mechanisms and implications. In: *Viruses and Wastewater Treatments* (Eds M. Goddard & M. Butler), pp. 199–204. Pergamon Press, Oxford

TALIBUDEEN O. (1955) Complex formation between montmorillonoid clays and amino acids and proteins. *Transactions of the Faraday Society* **51**, 581–90.

TANNER C. B. & JACKSON M. L. (1947) Nomographs of sedimentation times for soil particles under gravity or centrifugal acceleration. *Soil Science Society of America Proceedings* **12**, 60–5.

THENG B. K. G. (1979) *Formation and Properties of Clay-*

Polymer Complexes. Elsevier Scientific Publishing Co., Amsterdam.

TRAVIS C. C. & ETNIER E. L. (1961) A survey of sorption relationships for reactive solutes in soil. *Journal of Environmental Quality* **10**, 8–18.

UMBREIT W. W., BURRIS R. H. & STAUFFER J. F. (1964) *Manometric Techniques*. Burgess Publishing Company, Minneapolis.

VERMA L. & MARTIN J. P. (1976) Decomposition of algal cells and components and their stabilization through complexing with model humic acid-type phenolic polymers. *Soil Biology and Biochemistry* **8**, 85–90.

VERMA L., MARTIN J. P. & HAIDER K. (1975) Decomposition of carbon-14-labeled proteins, peptides and amino acids; free and complexed with humic polymers. *Soil Science Society of America Proceedings* **39**, 279–84.

WARD J. B. & BERKELEY R. C. W. (1980) The microbial cell surface and adhesion. In: *Microbial Adhesion to Surfaces* (Eds R. C. W. Berkeley, J. M. Lynch, J. Melling, P. R. Rutter & B. Vincent), pp. 47–66. Ellis Horwood, Chichester.

WEBLEY D. M. & JONES D. (1971) Biological transformations of microbial residues in soil. In: *Soil Biochemistry* (Eds A. D. McLaren & J. Skujins), vol. 2, pp. 446–84. Marcel Dekker, New York.

WEINBERG S. R. & STOTZKY G. (1972) Conjugation and genetic recombination of *Escherichia coli* in soil. *Soil Biology and Biochemistry* **4**, 171–80.

WISCHNITZER S. (1980) *Introduction to Electron Microscopy*. Pergamon Press, Oxford.

WOLF D. C. & MARTIN J. P. (1976) Decomposition of fungal mycelia and humic-type polymers containing carbon – 14 from ring and side-chain labeled 2,4-D and chloropropham. *Soil Science Society of America Journal* **40**, 700–4.

YU B. & STOTZKY G. (1979) Adsorption and binding of herpesvirus hominis type 1 (HSV 1) by clay minerals. *Abstracts of the Annual Meeting of the American Society for Microbiology*, p. 188.

Chapter 8 • Mineralisation of Carbon

Karl-Erik Eriksson and Susanna Christl Johnsrud

8.1 Degradation of wood

In nature there is a continuous degradation of dead plant material by saprophytic microorganisms. Many different kinds of organisms are involved in the degradation of woody materials but it is mainly a task for fungi as bacteria have only a limited capability. The wood-degrading capability that fungi have depends, in part, upon the organisation of their hyphae which allows the organisms to penetrate their substrates.

The description of wood degradation given in this chapter will be limited to saprophytic organisms, i.e. organisms that degrade dead wood. Pathogenic organisms attacking the living tree will not be discussed. All woody tissues have in common a high lignocellulosic content and a low nitrogen content. Lignin is a complex polymer which is not readily attacked by microorganisms. Similarly the low nitrogen content slows the growth rate and degradation activity of microorganisms on wood. The nitrogen content of wood is only 0.03 to 0.1% of the wood dry weight (Cowling 1970). The lignin content varies from 19.6 ± 0.2% in birch (*Betula verrucosa*) to 28.1 ± 0.2% in pine (*Pinus silvestris*) (Henningsson 1962).

Wood can be divided into sap wood and heart wood. It is generally recognised that sap wood is more susceptible to attack and degradation by microorganisms than heart wood. The resistance of the heart wood to degradation is due to the presence of compounds which inhibit microbial activity. The nature of these compounds varies widely among different tree species, and Erdtman (1939), Rennerfelt (1944) and Rudman and Da Costa (1958) have isolated phenolic and quinonic compounds and tropolones from conifer woods. Käärik (1974b) lists substances which are toxic to some wood-rotting fungi.

The three main components of a wood cell wall, namely, cellulose, hemicellulose and lignin, are degraded by different microorganisms to various extents. Wood is depolymerised and degraded mainly by fungi. Rarely, if ever, is the degradation caused by a monoculture of a fungus. Instead degradation is due to associations and successions of different fungi. In other words the degradation of wood is a complex process involving interactions between different kinds of microorganisms (Käärik 1974b).

8.1.1 Soft-Rot Fungi

Soft-rot fungi principally attack carbohydrates in the wood and lignin is only modified to a limited extent (Seifert 1968). The term soft-rot derives from the fact that there is a softening of the surface layer when wood is attacked by this group of fungi which belong to the Ascomycetes and Fungi Imperfecti. The secondary walls of attacked wood develop cylindrical cavities with conical ends (Nilsson 1974a), whilst some soft-rot fungi also cause an erosion of the cell walls starting from the cell lumen. The latter type of decay is more common in hardwood than in softwood. The formation of secondary cell wall cavities is called type 1 attack, while type 2 attack involves cell wall erosion (Corbett 1965). For further, detailed information concerning the decay of wood by soft-rot fungi the reader is referred to Wilcox (1973), Käärik (1974b) and Nilsson (1974a, b). The lignin-degrading activities of soft-rot fungi have been investigated by Eslyn *et al.* (1975) and discussed by Kirk (1971) and Ander and Eriksson (1978).

In order to ascertain whether soft-rot fungi actually utilise lignin, Haider and Trojanowski (1975) measured the release of [^{14}C]carbon dioxide from ^{14}C-labelled dehydrogenative polymerisates (DHP) with the label in the methoxyl groups or in the side-chains or in the aromatic rings. The results showed that soft-rot fungi had the ability to release [^{14}C]carbon dioxide from the three differently labelled DHP lignins. Thus it is apparent that soft-rot fungi can degrade lignin to a certain extent, especially in hardwoods. They can also cause demethylation of lignin and degrade the side-chains and the aromatic rings to a lesser extent.

8.1.2 Brown-Rot Fungi

Brown-rot fungi principally belong to the Basidiomycetes and mainly decompose the wood polysaccharides usually causing only a slight loss of lignin (Kirk 1973). The brown-rot fungi hyphae are normally localised in the wood cell lumen and penetrate adjacent cells either through existing openings or by producing bore holes in the wood cell walls. In early stages of degradation brown-rot fungi depolymerise cellulose faster than the degradation products are utilised. During the decay process the brown-rot fungi remove cell wall substances first from the S_2-layer of the secondary wall. The S_1-layer of this wall may also be destroyed but the primary wall and the middle lamella are very resistant to degradation by brown-rot fungi due to their high lignin content (Wilcox 1970; Sarkanen & Hergert 1971). The most noted change in the lignin when attacked by brown-rot fungi is a decrease in the methoxyl content (Kirk 1971). In advanced stages of decay, when most of the polysaccharides are consumed, the cell wall collapses causing a decrease in the wood volume.

8.1.3 White-Rot Fungi

White-rot fungi are a heterogeneous group of organisms but have in common the capacity to degrade lignin as well as other wood components. In addition, all have the ability to produce extracellular enzymes which oxidise phenolic compounds related to lignin, a capacity which is used for the identification of white-rot fungi. The relative amounts of lignin and polysaccharide degraded and utilised by these fungi vary, and so does the order of attack. The usual method of wood degradation by white-rot fungi is for the polysaccharide and lignin to be attacked simultaneously. However, there are examples of a specific degradation of the middle lamella (Käärik 1974b) although it has been demonstrated (Ander & Eriksson 1978) that a completely specific, restricted attack on the lignin by white-rot fungi does not occur. This latter observation is not surprising since considerable energy is required to degrade lignin and thus an additional carbon source must be simultaneously metabolised.

A characteristic of white-rot attack is the gradual thinning of cell walls, both in hardwoods and softwoods. Chemical analysis and micromorphological studies (Cowling 1961; Liese & Schmid 1966) indicated that white-rot fungi, in contrast to brown-rot fungi, successively depolymerised cell wall substances only to such an extent that the depolymerisation products can be utilised simultaneously. Also in contrast to the brown-rot fungi whose enzymes diffuse into the inner layer of the cell wall, the white-rot fungi enzymes are restricted to the cell wall layers in the immediate vicinity of the hyphae (Blanchette *et al.* 1978; Eriksson *et al.* 1980a, b; Ruel *et al.* 1981).

8.1.4 Blue-Staining Fungi

Blue-staining fungi are found among the Ascomycetes and Fungi Imperfecti and cause discoloration of wood due to their pigmented hyphae (Käärik 1974b). They have a very limited degradative effect on wood.

8.1.5 Bacteria

Bacteria attack wood slowly in places where the wood has a constantly high moisture content (Boutelje & Kiessling 1964). Compared with fungi, bacteria do not have a penetrating capability and so bacterial invasion usually occurs simultaneously with fungi.

8.2 Degradation of lignin

Lignin is a phenylpropanoid structural polymer of vascular plants which gives the plants rigidity and binds plant cells together (Sarkanen & Ludwig 1971). Lignin also decreases water permeation across cell walls of xylem tissue and protects plants from invasion by pathogenic microorganisms. The white-rot fungi are the most successful group of organisms involved in lignin degradation and remain the only microorganisms known to be capable of totally degrading all the major wood components, including lignin.

A considerable amount of information has now accumulated concerning the mechanisms of lignin degradation by white-rot fungi. To a large extent this is the result of comparative chemical analysis of undegraded lignins and lignins degraded for varying times by specific white-rot fungi. Lignins from spruce, heavily degraded by *Coriolus versicolor* and *Polyporus anceps*, have been characterised in great detail. Analyses have shown that there is a decrease in the number of hydroxyls attached to aliphatic and phenolic compounds, while the number of carboxyl groups increases. There is also an increase in the number of carbonyl groups in the side-chains attached to the aromatic ring (Kirk & Chang 1974). In order to cleave an aromatic ring, microorganisms normally require two hydroxyls adjacent to each other (Ander & Eriksson 1978) or three hydroxyls attached to the ring (Buswell & Eriksson 1979). An increased number of phenolic hydroxyl groups during an early stage of lignin degradation indicates that the aromatic rings are prepared for cleavage. Ring cleavage results in the enrichment of degraded lignin with carboxyl groups concomitant with the loss of phenolic hydroxyls.

Kirk and Chang (1974) distinguished between α,β-unsaturated and aromatic carboxyls within the lignins and observed that a large percentage of the carboxylic groups was not attached to an aromatic ring. This work has also shown that the attacked lignins have been extensively demethylated in spite of the low number of phenolic hydroxyls. The number of aromatic rings were depleted in the degraded lignins, although the lignins had molecular weights in excess of 1700 daltons (Kirk *et al.* 1978b).

From these data, a sequence of degradative reactions for lignins by white-rot fungi was proposed. Kirk *et al.* (1978b) suggested that the fungi first attack exposed surfaces on the lignin polymer. Demethylation reactions, acting upon methoxyl groups within guaiacyl and syringyl residues, are the principal degradative reactions of the initial attack. As a result, the methoxyl content decreases and odiphenolic moieties are formed. These, in turn, are attacked by fungal dioxygenases which open the aromatic rings in the polymer with the formation of aliphatic carboxyl groups.

Recent investigations have demonstrated the critical importance of several culture parameters in lignin metabolism by wood-rotting fungi (Kirk *et al.* 1978a). Keyser *et al.* (1978) showed that metabolism of lignin was maximal at low nitrogen concentrations (2 mM). It has also been demonstrated that lignin was metabolised only in the presence of an easily metabolised additional carbon source, such as glucose, cellobiose or cellulose (Ander & Eriksson 1978). Furthermore, the deg-

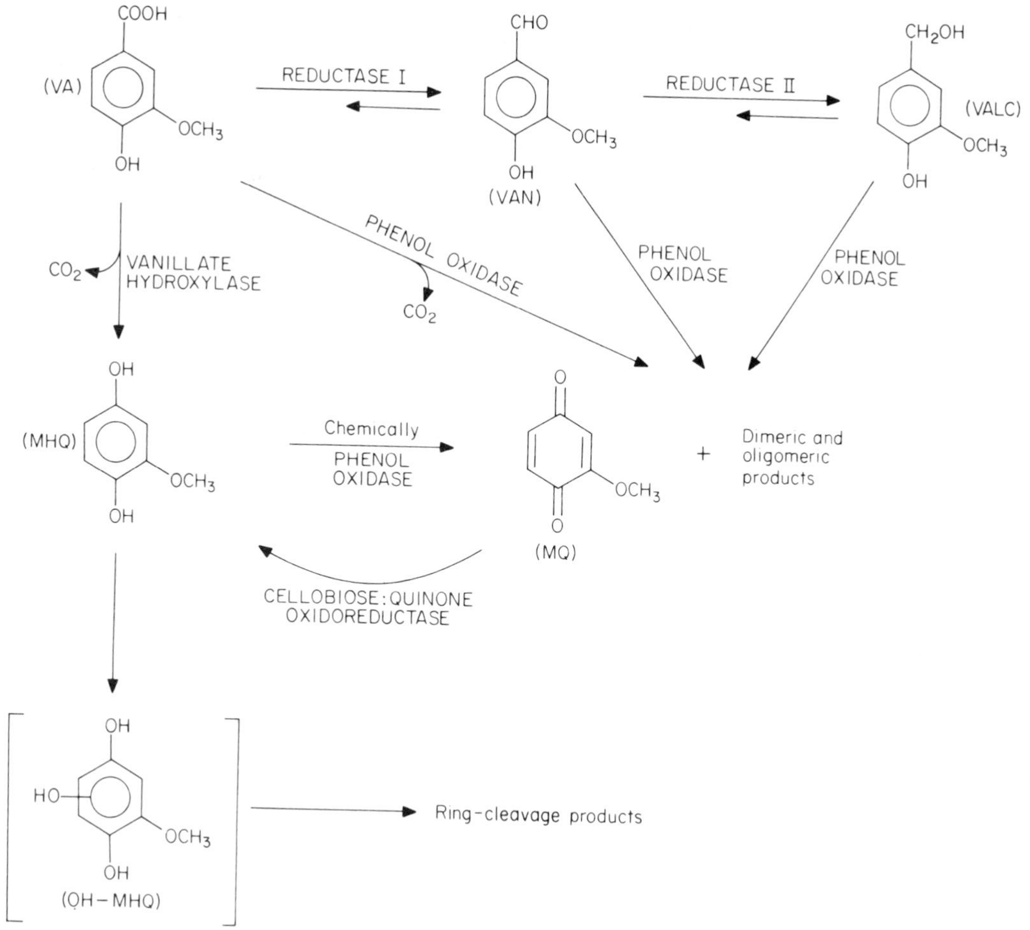

Figure 1. The proposed routes for vanillic acid by *Sporotrichum pulverulentum* (Ander *et al.* 1980). VA, vanillic acid; VAN, vanillin; VALC, vanillyl alcohol; MHQ, methoxydroquinone; MQ, methoxyquinone; OH-MHQ, hydroxylated MHQ—possibly the true substrate for ring cleavage.

radation of the lignin polymer takes place only in standing cultures (Kirk *et al.* 1978a).

It is generally accepted that lignin biosynthesis is catalysed by phenol oxidases (Freudenberg & Neish 1968). Ander and Eriksson (1976) studied the importance of phenol oxidases in lignin degradation by using three different strains of *Sporotrichum pulverulentum namely*:

(1) the wild-type (WT);

(2) a phenol oxidase-less mutant, *Phe* 3 (obtained by UV-irradiation of wild-type spores);

(3) a phenol oxidase-positive revertant, *Rev* 9 (obtained by UV-irradiation of spores from *Phe* 3).

The phenol oxidase-less mutant did not degrade lignin whereas the WT and the revertant did. However, if a highly purified phenol oxidase such as laccase was added to the medium, *Phe* 3 could degrade lignin. These results point to the essential role of phenol oxidases in lignin degradation.

It has been demonstrated in several studies that vanillic acid is always a metabolic product of lignin degradation by white-rot fungi (Kirk *et al.* 1977). Vanillic acid was used as a substrate for the white-rot fungus *Sporotrichum pulverulentum* by Eriksson and co-workers (Ander *et al.* 1980). Their strategy has been to approach the problem of the enzyme mechanisms involved in lignin degradation by initially examining the degradation of small molecules, followed by lignin models and finally examining the complete lignin polymer. It can be seen

(Fig. 1) that vanillic acid is oxidatively decarboxylated to methoxyhydroquinone whilst being simultaneously reduced to vanillin and vanillyl alcohol. The decarboxylation pathway predominates in actively growing cultures whilst reduction occurs in standing cultures. The reduction steps appear to require an energy source in the form of an easily metabolised compound, such as glucose or cellobiose (Ander *et al.* 1980). Vanillic acid is metabolised intracellularly and the carboxylating enzyme, vanillate hydroxylase, has been isolated and characterised (Buswell *et al.* 1979a). Two phenol oxidases, laccase and peroxidase, can also decarboxylate vanillic acid but are not dependent on NAD(P)H for their activity (Krisnangkura & Gold 1979; Ander *et al.* 1980).

It has also been demonstrated that *Sporotrichum pulverulentum* produces an aromatic ring-cleaving enzyme which does not function unless three hydroxyl groups are attached to the aromatic ring (Buswell & Eriksson 1979). By using differently labelled vanillic acids it was possible to show that decarboxylation took place before the ring was cleaved. Furthermore it was shown that the release of [^{14}C]carbon dioxide from vanillate labelled in the methoxy position occurred after the decarboxylation step (Ander *et al.* 1980).

Quinones are readily formed by the action of phenol oxidases induced during the growth of white-rot fungi on both low-molecular weight phenolic compounds and lignin. Although certain quinones are normal components of electron transport systems, they are highly reactive and are known to inhibit a wide range of metabolic processes. Therefore, reduction of quinoid intermediates is an essential step. It was reported by Westermark and Eriksson (1974a, b) that extracellular reduction of quinones and phenoxy radicals takes place via the extracellular enzyme cellobiose: quinone oxidoreductase using cellobiose as a co-substrate. The cellobiose is thereby oxidised to the corresponding lactone. Buswell *et al.* (1979b) have now found a second, intracellular quinone: oxidoreductase from *Sporotrichum pulverulentum* which reduces quinones to hydroquinones using pyridine nucleotides as electron donors.

The use of model lignin substrates in the study of lignin metabolism (Ander & Eriksson 1978) has been adopted by several other laboratories. Gold and co-workers have studied the degradation of both vanillic acid and its dimer by the white-rot fungus *Phanerochaete chrysosporium* (=*Sporotrichum pulverulentum*) (Yajima *et al.* 1979; Enoki *et al.* 1980).They were able to show direct cleavage of the α,β-bond of β-ether type dimers. Other phenolic dimers have been studied (Ohta *et al.* 1979).

8.3 Degradation of cellulose

The enzyme mechanisms involved in cellulose degradation have been particularly well studied for two fungi, namely the white-rot fungus *Sporotrichum pulverulentum* (Eriksson 1978) and the mould *Trichoderma reesei* (Ryu & Mandels 1980). (The fungus *T. viride* QM 6a and strains derived from it are now referred to as *T. reesei.*)

The fungus *Sporotrichum pulverulentum* produces three different types of hydrolytic enzymes:

(1) five different endo-1,4-β-glucanases which attack at random the 1,4-β-linkages along the cellulose chain;

(2) one exo-1,4-β-glucanase which splits off cellobiose or glucose units from the non-reducing end of cellulose;

(3) two 1,4-β-glucosidases which hydrolyse cellobiose and water-soluble cellodextrins to glucose, and cellobionic acid to glucose and gluconolactone (Eriksson 1978).

It is generally accepted that the same enzyme-systems apply for cellulose hydrolysis by *Trichoderma reesei* (Emert *et al.* 1974). However, a few differences have been recognised, such as the number of hydrolytic enzymes present and the degree to which β-glucosidase activity is associated with the fungal cell wall. The action of the exo-glucanase in *Sporotrichum pulverulentum* differs from the action of the corresponding enzymes in *T. reesei* in that the exo-glucanase from *S. pulverulentum* releases both glucose and cellobiose while the exo-glucanases from *T. reesei* only release cellobiose (Eriksson 1978). Recently a much simpler enzyme combination has been found for *T. reesei* QM 9414, compared with that previously reported (Emert *et al.* 1974), by using a different cultivation technique. In this case one endo-1,4-β-glucanase and two exo-1,4-β-glucanases were obtained (Gritzali & Brown 1979).

However, to degrade crystalline cellulose a synergistic action between endo-glucanases and exo-glucanases seems necessary for both fungi since the crystalline form is not attacked by either of these enzymes alone. In contrast, amorphous cellulose is degraded by both types of enzymes independently (Eriksson 1978; Ryu & Mandels 1980).

In *Sporotrichum pulverulentum*, an important oxidative enzyme involved in cellulose degradation has been discovered in addition to the hydrolytic enzymes described above (Ayers *et al.* 1978). The purified enzyme has been found to be a cellobiose oxidase which oxidises cellobiose and higher cellodextrins to their corresponding onic acids using molecular oxygen. The enzyme is a haemoprotein and contains an FAD group. It is not yet known whether this enzyme can oxidise the reducing end-group formed when a 1,4-β-glucosidic bond is split through the action of the endo-glucanases. It was recently reported by Vaheri (1980) that cultures of *Trichoderma reesei* grown on cellulose produce gluconolactone, cellobionolactone and cellobionic acid. These findings indicate that *T. reesei* synthesises an oxidative enzyme involved in cellulose degradation although further confirmation is necessary.

The fungus *Sporotrichum pulverulentum* has two unconventional pathways involved in cellobiose metabolism. The first concerns the enzyme cellobiose oxidase (described above), while the other involves the enzyme cellobiose:quinone oxidoreductase (Westermark & Eriksson 1974a, b). Although this enzyme seems to be involved in both lignin and cellulose degradation, the highest enzyme yields were obtained when cellulose powder was used as a carbon source. In *S. pulverulentum*, development of cellobiose: quinone oxidoreductase activity and cellulolytic activity occurs simultaneously. The enzyme is relatively specific for its disaccharide substrate while the requirements on the quinone structure are less specific and the enzyme is able to reduce both *ortho-* and *para-*quinones.

Regulation of endo-1,4-β-glucanases in *Sporotrichum pulverulentum* has recently been investigated using a new, sensitive assay method (Eriksson & Hamp 1978). The method is based on the observation that endo-1,4-β-glucanase activity lowers the viscosity of solutions of carboxymethyl cellulose (CMC). The effect of inducers and repressors can be determined, as can the localisation of the enzymes either on the cell wall surfaces or in the surrounding growth medium. The results showed that cellobiose induced the enzymes at concentrations as low as 1 mg l^{-1} and that glucose concentrations as low as 50 mg l^{-1} repressed enzyme formation. Mixtures of inducer and repressor gave rise to a delayed enzyme production compared with the inducer only.

Studies of *Trichoderma reesei* QM 6a using the same technique showed that cellobiose was not an efficient inducer. However, sophorose at a concentration of 1 mg l^{-1} induced the endo-1,4-β-glucanases. The regulation of endo-1,4-β-glucanase production in the two fungi demonstrated several other important differences. Alone, CMC induced enzyme formation in *Sporotrichum pulverulentum* but not in *T. reesei*. Under our experimental conditions, no endo-1,4-β-glucanases were actively excreted into the solution by *T. reesei*; the enzymes were bound to the cell wall. However, *S. pulverulentum* released the enzymes into the medium although initially they appeared to be bound to the cell wall. Gritzali and Brown (1979) showed that sophorose gave rise to active excretion of endo-1,4-β-glucanases into the culture solution of *T. reesei* QM 9414. The differences in the results of the two studies could be due to differences either in the fungal strains or in the cultivation conditions.

The production of fungal cellulases can be hampered by factors other than catabolite repression. Váradi (1972) found that a wide variety of phenols repressed the production of both cellulases and xylanases in the fungi *Schizophyllum commune* and *Chaetomium globosum*. At concentrations of less than 1 mM, vanillic acid, vanillyl alcohol and vanillin considerably repressed the production of these enzymes. Ander and Eriksson (1976) showed that in a mutant of *Sporotrichum pulverulentum* (*Phe* 3) which lacked phenoloxidase, the production of endo-1,4-β-glucanases was repressed greatly in the presence of kraft lignin and phenols at concentrations of 1×10^{-3}M. However, both the wild-type and a phenoloxidase-positive revertant produced the enzymes without significant repression by phenols. Futhermore, if a purified laccase preparation was added to the *Phe* 3 growth medium in the presence of phenols, the endo-1,4-β-glucanase production was normal. These results indicated that kraft lignin and phenols decreased endo-1,4-β-glucanase synthesis in *Phe* 3 due to the absence of phenol oxidising enzymes. Thus phenoloxidases may function by regulating the production of cellulases by oxidising lignin-related phenols which act as repressors of enzyme production when *S. pulverulentum* is growing on wood.

The activity of cellulose-hydrolysing enzymes in culture filtrates of *Sporotrichum pulverulentum* is dependent not only on mechanisms regulating their biosynthesis but also on the presence of specific enzyme inhibitors. One such inhibitor is gluconolactone produced in *S. pulverulentum* by glucose oxidation either by the activity of glucose oxidase or by hydrolytic cleavage of cellobionic acid (Eriksson 1978). The importance of gluconolactone for the regulation of 1,4-β-glucosidases from *S. pulverulentum* has been studied by Deshpande *et al.* (1978). The extracellular β-glucosidase activity resolved into two fractions, since both free 1,4-β-glucosidases and cell-bound enzymes appear in *S. pulverulentum*. The K_i values for the two free β-glucosidases were 3.5×10^{-7} and 1.5×10^{-6}M respectively. A K_i value of 3.2×10^{-5}M was obtained for *Trichoderma reesei* QM 9414 β-glucosidase with gluconolactone as the inhibitor. The corresponding K_i value for the same enzyme for glucose is 1×10^{-3}M (Gritzali & Brown 1979).

8.4 Degradation of hemicelluloses

Dekker and Richards (1976) described in great detail what was known concerning the degradation of different hemicelluloses and the reader is referred to this article for further information.

Endo-1,4-β-mannanases are enzymes which hydrolyse the 1,4-β-D-mannopyranosyl linkages of D-mannans and D-galacto D-mannans. Other enzyme preparations which have the capability of degrading mannans with other structures have been described (Tsujisaka *et al.* 1972). The mannanases produced by different microorganisms have been reported to be both induced and constitutive enzymes (Dekker & Richards 1976). All fungal mannanases degrade mannans in a random manner and are of the endo-type.

In addition to the endo-1,4-β-mannanases, some microorganisms also produce 1,4-β-mannosidases (Ahlgren & Eriksson 1967). These enzymes have a similar function in mannan degradation to the 1,4-β-glucosidases in cellulose degradation, that is, catalysing the release of monomeric sugars from water-soluble dextrins.

Since the structure of xylan polysaccharides vary a great deal, the xylanases also show a variety of different activities (Dekker & Richards 1976). Thus, both endo-1,3-β-xylanases and endo-1,4-β-xylanases are produced by both fungi and bacteria. Xylanases are produced both constitutively and by induction. It has repeatedly been demonstrated (Lyr 1959; Bucht & Eriksson 1968) that fungal xylanases are produced when a xylan-free cellulose preparation is used as the sole carbon source for growth. Cellulases and xylanases seem to be different enzymes and, in those cases where one enzyme has been reported as having two activities, it seems likely that an impure enzyme preparation has been used. It was demonstrated by Björndal (cited in Eriksson & Rzedowski 1969) that D-xylan was present in the mycelium of *Stereum sanguinolentum* and it was suggested that the formation of D-xylanase in this organism, when grown on cellulose, may be self-induced. The xylanase may be produced to degrade old cell wall material which then supplies the organism with an endogenous energy supply. Thus, it appears that in fungi there is often no strict differentiation between adaptive and constitutive D-xylanases.

Xylosidases are also produced by fungi (Ahlgren & Eriksson 1967). They function intracellularly to release monomeric sugars from water-soluble xylodextrins. As far as is known, there are no exo-xylanases (Dekker & Richards 1976). As stated above, crystalline cellulose is enzymically degraded by the synergistic action of endo- and exo-1,4-β-glucanases. Since neither mannan nor xylan exists in nature in a crystalline form, there should be no need for the existence of a pair of endo- and exo-enzymes.

8.5 Degradation of polysaccharides other than cellulose and hemicelluloses

Starch is a plant polysaccharide which serves as the major storage form of energy. The structure of starch is similar to cellulose in that both are polymers of glucose. The significant difference, however, is the steric nature of the glycosidic bonds connecting the glucose residues. In cellulose, the bond is a β-linkage, whereas in starch it is an α-linkage. The repeating disaccharide unit in starch is, therefore, maltose, rather than cellobiose as in cellulose. Two forms of starch exist: one is a long, unbranched chain similar to cellulose. The other is a branched chain polysaccharide with branches occurring at every 20 to 30 glucose residues (Bennett & Frieden 1968).

The hydrolysis of starch by α-amylases is unclear and often debated. Two modes of action of α-amylase have been proposed, namely, multiple attack or preferred attack. In the multiple attack theory it is assumed that all the α-bonds are equally susceptible to hydrolysis whilst in the preferred attack theory a single site of hydrolysis is postulated. These two theories have arisen since it is assumed that the glucosidic bonds are not equally susceptible to hydrolysis, with the bonds near the chain ends especially being more resistant to hydrolysis.

Pectic substances (protopectin) occur notably in the middle lamellae of parenchymatous tissues, where they have a structural function in binding and supporting the cells. Pectinolytic enzymes occur widely in plants and microorganisms and their nature and distribution have been reviewed by Rombouts and Pilnik (1972). Albersheim *et al.* (1960) showed the trans-eliminative splitting of glycosidic links by a lyase. Until that report it was thought that the pectin-degrading enzymes were only hydrolytic in nature.

Experimental

8.6 Degradation of wood

8.6.1 Assay Conditions

Under natural conditions wood is decomposed by the interaction of a diverse community of organisms. The role of many organisms is ill-defined but the most biochemically active are the Basidiomycete fungi and some groups of insects such as beetles and termites. Some of the most important ecological investigations of wood decay are discussed by Swift (1977); see also Chapter 10. Reviews of plant cell wall structures and the biochemistry of its degradation have been published by Albersheim *et al.* (1969), Northcote (1972) and Kirk (1973).

Wood decay is due to the combined action of enzymes secreted from fungal cell walls. Wood-destroying fungi feed on the wood cell walls but the fungi do not utilise all the components equally and this inequality results in different types of decay. A detailed survey of the different types of attack on the wood by microorganisms from the chemical and physical point of view is given by Seifert (1968) and the micromorphological aspects are considered by Liese (1970) and Wilcox (1970).

Wood-rotting fungi are divided into soft-rot, brown-rot and white-rot groups. The blue-staining fungi are associated with wood damage but they do not degrade wood. The role of Basidiomycetes during the natural decay process of wood is well known (Boyce 1961) although recent evidence has suggested that Basidiomycetes cannot act alone to decay wood. Studies of the decay of living trees and dead woody materials indicate that a complex group of microorganisms is involved (Shigo & Hillis 1973; Käärik 1974a).

8.6.2 Transmission and Scanning Electron Microscope Techniques

Studies on the morphology of microbial attack on wood have been undertaken at the micromorphological and ultrastructural levels using light microscopy, electron microscopy (EM) and scanning electron microscopy (SEM). The microscopic characteristics of decay have been thoroughly described by Hubert (1924) and Bavendamm (1936). However, the use of light microscopes does not reveal the fine structure changes caused by microbial attack and so more emphasis has been put on developing EM and SEM techniques. Material for electron microscope examination has to be either sectioned or fragmented. No satisfactory methods exist for the examination of living material. The reader is referred to Haine (1961) and Hall (1966) for an account of the electron microscope; to Pease (1964), Kay (1965), Sjöstrand (1967) and Ruthmann (1970) for preparative methods and the operation of the microscope; and Juniper *et al.* (1970) deal specifically with the EM of plant material. The above applies to the conventional transmission electron microscope.

In SEM the specimen is scanned by a narrow electron beam and the image formed from electrons reflected by the surface or secondarily emitted as a result of excitation. SEM gives valuable information about the three-dimensional shape of the specimen. For a brief introduction to SEM techniques the reader is referred to Echlin (1971) and Nixon (1971) and for more detailed techniques to Heywood (1971). The fungal degradation of wood has been examined by SEM techniques (Eriksson *et al.* 1980b) and by EM techniques (Liese 1970; Wilcox 1970; Ruel *et al.* 1981). An SEM study of the influence of yeast and bacteria on the rate of decay by wood-destroying Basidiomycetes is given by Blanchette *et al.* (1978).

8.6.3 Histochemical Techniques

Various useful histochemical methods of analysis for the lignin and cellulose content of wood are given by Pearse (1960, 1968).

8.6.4 Soil-Block Technique

In the soil-block technique wood blocks are placed on the surface of sterile soil or vermiculite within a chamber where the temperature and humidity conditions are controlled. A large number of replicate blocks are inoculated with a microbial species and incubated under suitable conditions for several months. Periodically some blocks are removed for weight loss determinations. After grinding through a 40 μm mesh screen, the wood is analysed for lignin and reducing sugars (Bethge *et al.* 1971; Effland 1977). This technique may also be useful for the evaluation of fungal attack on preservative-treated and untreated wood kept in soil (The International Research Group on Wood Preservation Document 1978).

8.6.5 The Stake Technique for Testing Wood Preservatives

The field stake technique is used for outdoor exposure tests to determine the efficiency of wood preservatives. Stakes (2 × 5 × 50 cm) of either clear sapwoood or clear heartwood of wood species which can be completely penetrated by preservatives are chosen. Untreated stakes are included as controls. The stakes, buried to half their length, are placed randomly in the field and inspected at yearly intervals. A measure of the effect of fungal attack can be obtained according to the Nordic Wood Preservation Council Standard No. 1.4.2.1/71, which involves determining the bending strength of the wood samples.

8.7 Degradation of lignin

8.7.1 Assay Conditions

In spite of numerous studies the microbiology of lignin degradation is not well understood (Ander & Eriksson 1978). One of the main obstacles to the study of lignin biodegradation has been an inability to find suitable growth conditions which give rapid rates of lignin degradation and high levels of lignin-degrading enzymes. A further problem has been the need for a sensitive, quantitative assay to determine lignin catabolism. Problems concerned with inadequate methodologies have been discussed by Kirk (1971), Kirk *et al.* (1975) and Crawford and Crawford (1976, 1978). The cultivation techniques in use today have been developed mainly by Kirk and co-workers. In addition to lignin, the presence of an easily metabolised carbon source is important (Ander & Eriksson 1975; Hiroi & Eriksson 1976), as is a suitable nitrogen source (Kirk *et al.* 1977; Eriksson & Vallander 1980) and correct oxygen levels (Kirk *et al.* 1977). The difference between static and shaken cultures has also been evaluated (Kirk 1976).

8.7.2 Lignin-Degrading Microorganisms

Cellulose in lignocellulosic materials is more resistant to decay than lignin-free cellulose and hemicelluloses, and pure lignin has not been shown to be degraded significantly by any microorganisms. Until recently (Nilsson 1965; Ander & Eriksson 1978; Kirk *et al.* 1975) only one group of microorganisms has been shown to degrade lignin extensively in lignocellulosic materials. These organisms are all mesophilic moulds, known collectively as the white-rot fungi (Kirk 1971). However, some reports suggest that bacteria also play a role in lignin degradation (Hata 1966; Robinson & Crawford 1978).

The low lignin degradation rates have impeded attempts to understand the enzymology of the process, although some progress has been made (Harkin 1967; Ishihara & Miyazaki 1972; Westermark & Eriksson 1974a; Kirk 1975; Ander *et al.* 1980). However, only a few species of white-rot fungi have been examined for their capability to degrade lignin and it is possible better lignin-degrading species may be found. Furthermore, genetic manipulation techniques might be used to improve the rate of selected lignin degradation.

Reviews on lignin degradation have been published by Kirk (1971), Higuchi (1971) and Ander and Eriksson (1978).

8.7.3 Decay Studies

Decay studies have been used to screen fungi for their capacity to degrade lignin in wood (Kirk & Moore 1972; Ander & Eriksson 1977). The studies report on growth rate, weight loss, activities of phenol-oxidising enzymes and delignification of both hardwoods and softwoods. The most useful technique, which does not rely on radiotracers for following lignin biodegradation in natural lignocelluloses, is the soil-block technique (see section

8.6.4, p. 142) (Highley & Scheffer 1970; Highley 1978; Eriksson *et al.* 1980a).

Rotting experiments with wood chips of birch, pine and spruce have been carried out. Selected white-rot fungi and mutants which lack cellulase activity have been used to determine the rate of lignin degradation in the materials. Even in the absence of an absolutely specific attack on the lignin, it has been demonstrated (Ander & Eriksson 1975) that enough lignin can be degraded to cause a decrease in the energy demand for producing thermomechanical pulp, if the wood chips are pretreated with cellulase-less mutants of white-rot fungi.

The lignin content in wood may be determined by the chlorine consumption method (Kyrklund & Strandell 1967) or by determination of the Klason lignin (Bethge *et al.* 1971). In the latter the dry lignin-containing material is hydrolysed with 72% (v/v) sulphuric acid. The dissolved aromatic material (acid-soluble lignin) is measured spectrophotometrically at 205 nm and the remaining solids make up the Klason lignin fraction. The Klason acid hydrolysis procedure only gives a rough estimate of the lignin content since a considerable amount of acid-soluble lignin is present in many lignocelluloses which are not extracted by this method (Migita & Kawamura 1944; Sarkanen & Ludwig 1971). Lignin in the condensation waters from the steaming of wood chips can be determined by acidic hydrolysis (Cellulosindustrius Centralla-botoriums Analyskommitte 5 (CCA 5) 1944; Schöning & Johansson 1965).

A modified lignin known as kraft lignin is a major industrial waste material which is generated during the alkaline pulping of wood. The commercial products Indulin ATR-CI (a softwood kraft lignin), Indulin AT (a kraft pine lignin) and a similar product, Indulin A, have been used frequently as substrates in lignin degradation studies (Marton *et al.* 1969; Sundman & Näse 1972; Nordström 1974; Tansey *et al.* 1977). Their suitability for this purpose is, however, questionable since kraft lignins differ considerably in structure from natural lignins and are thought to be more resistant to microbial attack (Kirk 1971).

The studies of biodegradation of lignin sulphonates have been hampered since no definitive assay has been developed. The chlorine number method used by Hiroi and Eriksson (1976), the nitroso method developed by Selin and Sundman (1972), the method which evaluates the undegraded lignin sulphonate with the aid of UV absorption (Selin *et al.* 1975; Hüttermann *et al.* 1977) and other methods all suffer from the weakness that they are influenced by lignin structural changes which occur during biodegradation. Results obtained with these methods do not necessarily reflect changes in the lignin or lignin sulphonate content, since lignin biodegradation, as opposed to biodegradation of other biopolymers (i.e. protein, carbohydrates and lipids) is not a hydrolytic reversal of the biosynthesis, but implies several changes in the molecular structure (Kirk 1975).

The degradation of lignins by a large number of lignolytic fungi was visualised directly by Sundman and Näse (1972) who used a simple lignin agar plate test with a solution of ferric chloride and potassium ferricyanide for the rapid detection of lignin degradation. This reagent was coloured green by the phenols present in the different lignin preparations used (Sundman & Näse 1971).

8.7.4 Production of ^{14}C-Labelled Lignin

The lack of a definitive and sensitive lignin degradation assay has been largely solved by the development of assays based on ^{14}C-labelled synthetic lignins (Haider & Trojanowski 1975; Kirk *et al.* 1975) and lignin-[^{14}C]lignocelluloses (Crawford & Crawford 1976; Haider & Trojanowski 1975). The decomposition of ^{14}C-labelled lignins is followed by monitoring [^{14}C]carbon dioxide evolution from growing cultures and/or the release of water-soluble ^{14}C-labelled breakdown products into the growth medium. By utilising ^{14}C-labelled lignin specifically labelled in only the ring, side-chain or methoxyl components, it is also possible to determine the specificity of the attack on lignin by different microorganisms. ^{14}C-radioisotopic labelling methods for the study of lignin biodegradation have been summarised by Crawford and Crawford (1978) and Crawford *et al.* (1980).

8.7.5 ^{14}C-Labelled Natural Lignins

In vascular plants lignin is synthesised by step-wise reactions from carbohydrates to the aromatic amino acid by the shikimate-cinnamate pathways (Higuchi 1980). The ability of plants to incorporate radioactive lignin precursors into new lignin has been used as a method to prepare ^{14}C-labelled lignins for study of lignin biodegradation. This was accomplished by feeding plants [^{14}C]-(U)-L-phenylalanine (Crawford & Crawford 1976), [^{14}C]ferulic acid (Haider & Trojanowski 1975), or [^{14}C]coumaryl alcohol (Haider *et al.* 1977) through cut stems or by injection. The feeding of ^{14}C-labelled precursors has allowed preparation of [^{14}C]lignin-lignocelluloses, i.e. lignocelluloses containing [^{14}C]carbon, principally in their lignin components but not in their carbohydrate components (Crawford *et al.* 1980). Microorganisms capable of growing on ^{14}C-labelled lignocelluloses produce [^{14}C]carbon dioxide which is quantified by liquid scintillation techniques.

8.7.6 Enzymes of Lignin Degradation

The involvement of phenol oxidases in lignin degradation has been discussed since Bavendamm (1928) used gallic and tannic acids to differentiate between white-rot and brown-rot fungi. However, no well-defined relationship between lignolytic activity and phenol oxidase activity of wood-rotting fungi has been established. Ander and Eriksson (1976) using genetic manipulation techniques obtained strong evidence that phenol oxidases are required for fungal degradation of lignin.

The enzymes generally referred to as the phenol oxidases are laccase (E.C. 1.14.18.1) and peroxidase (E.C. 1.11.1.7). Laccase has been reviewed by Levine (1966) and plant polyphenol oxidase, including phenol oxidases of fungi and bacteria, has been discussed by Mayer and Harel (1979).

Cellobiose:quinone oxidoreductase is involved in lignin degradation and the assay method is given by Westermark and Eriksson (1974a, b, 1975). Recently an intracellular NAD(P)H:quinone oxidoreductase enzyme has been discovered by Buswell *et al.* (1979b). Iwahara and Higuchi (1979) have presented preliminary evidence to show that an alcohol oxidase attacks a wide range of lignin model compounds containing α,β-unsaturated primary alcohol structures. The enzyme also attacks the macromolecular structure of various lignin preparations. A detailed description of enzymes involved in lignin degradation is given by Ander and Eriksson (1978).

White-rot fungi metabolise vanillate through protocatechuate (Flaig & Haider 1961; Cain *et al.* 1968) or through methoxyhydroquinone (Kirk & Lorenz 1973; Nishida & Fukuzumi 1978). However, the enzymes catalysing these reactions have not been characterised. The studies of vanillic acid metabolism by the white-rot fungus *Sporotrichum pulverulentum* (Ander *et al.* 1980) showed the involvement of vanillate hydroxylase (catalysing a decarboxylation reaction), a hydroxylating enzyme and an aromatic ring-cleaving enzyme (Buswell & Eriksson 1979). The vanillate hydroxylase has been partially purified by Buswell *et al.* (1979a). The microbial degradation of lignin model compounds and its relevance to lignin degradation has been discussed by Crawford and Crawford (1980).

8.8 Degradation of cellulose

8.8.1 Microorganisms producing Cellulases

Enzymes which degrade cellulose are produced by a number of microorganisms but only a few produce extracellular cellulases capable of converting crystalline cellulose to glucose *in vitro* (Mandels & Weber 1969). Fungi of the Ascomycete and Deuteromycete groups, together with some white-rot Basidiomycetes, contain the three necessary enzyme systems (endo-glucanase, exo-glucanase and β-glucosidase) to hydrolyse extensively native cellulose. *Trichoderma reesei* strain QM 6a has been considered to be the most active cellulase producer. Mutant strains of QM 6a designated QM 9123 and QM 9414 (Mandels & Weber 1969), NG-14 and Rut C-30 (Montenecourt & Eveleigh 1977; Montenecourt *et al.* 1978) and VTT-D-78085 (Nevalainen *et al.* 1980) have higher cellulase yields. Other potent cellulase producers are *T. koningii, Fusarium solani* (Wood & Phillips 1969), *Penicillium funiculosum* (Wood & McCrae 1978) and *Sporotrichum pulverulentum* (Eriksson 1978).

For optimal cellulase production, the growth medium should be carefully chosen although in practice the media of Mandels and Reese (1957) and Norkrans (1950) are frequently used.

8.8.2 Assay of Cellulase Activity

A wide range of methods for the determination of the activity of cellulolytic enzymes exist but there is considerable confusion inherent in the multiplicity of substrates, enzyme names, units and activities. This is not surprising, since the cellulase complex is a multi-enzyme system acting on a substrate where neither the real substrate concentrations nor the rates of reaction are easy to determine.

The assay methods for the determination of cellulase activity have been based on weight loss of insoluble substrates, changes in turbidity of cellulose suspensions, increase in reducing end groups, decrease in the viscosity of cellulose derivatives, colorimetric determinations, measurements of clearance zones in cellulase agar and polarographic assays (Almin & Eriksson 1967; Almin *et al.* 1967; Mandels & Weber 1969; Gould 1969; Green *et al.* 1977).

8.8.3 Enzymes of Importance in Cellulose Degradation

In the enzymatic hydrolysis of cellulose, endo-β-1,4-glucanases acting randomly on the cellulose chain open up non-reducing ends for the action of exo-enzymes (Streamer *et al.* 1975). Soluble cellulose derivatives are convenient substrates for the determination of endo-1,4-β-glucanase activity, since they are not hydrolysed by exo-β-glucanases (Pettersson 1975). Carboxymethyl cellulose (CMC), substituted to different extents, has been most frequently used but the viscosity of CMC-solutions is affected by the pH value, the ionic strength of the solvent and the presence of polyvalent cations (Iwasaki *et al.* 1963). In addition, the solutions change phase from sol, at temperatures above 37°C, to gel, at lower temperatures (Li *et al.* 1965). Hydroxyethyl cellulose (HEC) has been used successfully and is considered preferable by some since it is a neutral compound (Child *et al.* 1973). It has, however, the drawback that the substituents may be of different size due to polymerisation. The activity measurements have been based on either a decrease in viscosity (Almin & Eriksson 1967; Almin *et al.* 1967; Demeester *et al.* 1977) or the production of reducing sugars (Mandels & Weber 1979).

Almin and Eriksson (1967) and Almin *et al.* (1967) have developed a method by which viscosity changes can be recalculated to absolute units which makes it a valuable tool in cellulase studies. The method is, however, cumbersome since it requires a thorough knowledge of the properties of the carboxymethyl cellulose used.

Values from measurements of reducing sugars are expressed as glucose equivalents, determined either by the dinitrosalicylic acid method (Miller 1959), the Nelson–Somogyi method (Nelson 1944; Somogyi 1952), the orcinol method (Vasseur 1948) or the modified Hoffman ferricyanide method described by Grady and Lamar (1959). These analyses give the rate of hydrolysis directly, but they are less sensitive than the viscometric technique, requiring longer incubation times.

Direct measurements of exo-glucanase activity can only be made with purified enzymes, since no specific substrates are available. For the determination of exo-β-glucanase activities, amorphous cellulose or phosphoric acid-swollen cellulose (Walseth 1952) may be used. Crystalline cellulose is only hydrolysed by a mixture of endo- and exo-glucanases.

Exo-glucanase activity in culture filtrates, i.e. in mixtures with endo-glucanases, can be determined by using amorphous cellulose, as follows: the CMC-cellulase activity in the mixtures is determined and the amounts of reducing sugars calculated from a standard curve made up with known number of units of purified endo-1,4-β-glucanases (A). The amount of reducing sugars in the sample is determined (B). 100 µl of sample is added to 1 ml of 1% (w/v) Walseth cellulose and incubated at 40°C for 1 h. The supernatant is analysed for reducing sugars (C). Then: $C-A-B=$ sugars released by the exo-glucanase activity.

Frequently the exo-β-glucanase activity is not measured directly, but only expressed in a semi-quantitative manner as the synergistic effect obtained in a mixture with endo-glucanases and β-glucosidases.

1,4-β-glucosidase enzymes hydrolyse both cellobiose and β-1,4-oligosaccharides to glucose. It is the only cellulolytic enzyme for which specific substrates of low molecular weight are available. Standard assay methods used for measuring the β-glucosidase are cellobiose (Selby & Maitland 1967), salicin (Sternberg *et al.* 1977) and β-glucosides with a chromophore, such as *o*-nitrophenyl-β-D-glucopyranoside (Wood 1968) or *p*-nitrophenyl-β-D-glucopyranoside (Berghem & Pettersson 1973). The activity of β-glucosidase can be expressed in absolute units according to the International Union of Biochemistry (1979).

8.8.4 Saccharification Method

The main technical use of cellulases is for the total hydrolysis of cellulose from various cellulosic materials to produce glucose. The substrate may be an insoluble cellulosic material, such as cotton fibres, microcrystalline cellulose, sulphite pulp or filter paper. The standard method is based on saccharification of filter paper (Mandels & Weber 1969; Mandels *et al.* 1976; Montenecourt *et al.* 1978). Filter paper activity was defined by Mandels *et al.* (1976) as the amount of reducing sugars formed in 1 h when using 50 mg Whatman No. 1 filter paper (1 × 6 cm). This has been further developed by Montenecourt *et al.* (1978) who utilised filter paper antibiotic assay discs in place of the 50 mg filter paper strips. The enzymes are diluted in a 0.05M citrate buffer, pH 4.8, with a final assay volume of 1.5 ml. The assays are carried out at 50°C in a stationary water bath for 1 h. Free reducing sugars are quantified by the dinitrosalicylic (DNS) acid method (Miller 1959) and results expressed as glucose equivalents. Filter paper units or disc units are calculated as defined by Mandels *et al.* (1976); i.e. a filter paper unit is expressed as μM glucose min^{-1} based on the release of 2 mg reducing sugar h^{-1} under the defined assay conditions. A reading of 2 mg glucose h^{-1} is equivalent to 0.185 IUB units (μmol min^{-1}).

The release of dye from a dyed insoluble substrate is a convenient way of measuring the cellulose solubilising activity. Dyed filter paper (Mandels & Weber 1969), dyed Solka floc and dyed Avicel (Leisola & Linko 1976) have been used as substrates in this connection. The method using dyed Avicel gives useful information about the cellulolytic capabilities of a cellulase product. It is also suitable for the screening of cellulase production by microorganisms (Poincelot & Day 1972) and can be applied to automatic analysis (Leisola & Kauppinen 1978).

8.9 Degradation of hemicelluloses

8.9.1 Microorganisms Producing Hemicellulases

Hemicelluloses are a heterogeneous group of polysaccharides composed of hexoses, pentoses and uronic acid. Xylan and glucomannan are the dominant hemicelluloses of hardwood and softwood respectively. Enzymic degradation of these polysaccharides is carried out by enzymes specific for xylans and mannans. George (1974) demonstrated that the hemicellulase production of soft-, brown- and white-rot fungi and the occurrence of the fungi on specific species of wood could be related to their enzyme production. Keilich *et al.* (1970) showed that *Polyporus schweinitzii* and *Chaetomium globosum* were able to degrade glucomannans and glucouronoxylans, with the former being more active on the glucomannan and the latter on the glucouronoxylan. Examples of xylan degradation by brown-rot fungi are given by Lyr (1960), Nilsson (1974a, b), Highley (1976) and Nilsson and Ginns (1979).

Mannanases of fungal origin have been studied by Eriksson and Winell (1968), Hashimoto and Fukumoto (1969), Keilich *et al.* (1970), Tsujisaka *et al.* (1972) and Zouchova *et al.* (1977).

8.9.2 Assay of Hemicellulase Activity

The assay methods for determining xylanase activity are based on measurements of clearance zones in larch xylan agar (Nilsson & Ginns 1979), viscometric methods (Keilich *et al.* 1970) and the increase in reducing sugars (Nelson 1944; Somogyi 1952). The reducing sugar method used for measurements of endo-xylanase activity allows the activity to be calculated in absolute terms (Kubackova *et al.* 1978; Schmidt & Liese 1980).

Mannanase activity can be determined viscometrically (Ahlgren *et al.* 1967), or in a similar way to xylanase activity by measuring increase in reducing sugars, or by release of dye from carob, D-galacto D-mannan, dyed with Remazol brilliant Blue (McCleary 1978).

8.10 Degradation of polysaccharides other than cellulose and hemicelluloses

8.10.1 Microorganisms Producing Amylases and Pectic Enzymes

α-Amylase (E.C. 3.2.1.1.) is an endo-enzyme which hydrolyses the internal 1,4-α-glycosidic links of both amylase and amylopectin constituents of starch. The main bacterial source of α-amylase production is *Bacillus subtilis* and fungal sources are *Aspergillus niger* and *Aspergillus oryzae* (Beckhorn 1967). The enzymes of *A. niger* and *A. oryzae* have been the most widely used for industrial starch degradation, although enzymes from species such as *Mucor, Rhizopus, Penicillium* and others have been used.

Amyloglucosidase (E.C. 3.2.1.3.) hydrolyses β-1,4-glucan linked polysaccharides, removing glucose units from the non-reducing ends of the chain. It thus converts starch and amylodextrins to glucose. Amyloglucosidases of *Aspergillus niger* and

Aspergillus oryzae hydrolyse maltose to glucose (Aschengreen 1969). *Rhizopus niveus* is also a source of amyloglucosidase production (Arima 1964). A detailed description of the development of a commercial process using an *Aspergillus fotidus* mutant as an amyloglucosidase source is given by Underkofler (1969).

The pectinolytic enzymes are classified on the basis of their mode of action on the α-1,4-glycosidic bonds. Plants, fungi and bacteria possess different combinations of these enzymes (Edström & Phaff 1964; Mill 1966; Ishii & Yokotsuka 1971; Rombouts & Pilnik 1972). The pectinolytic activity of some ectomycorrhizal and saprophytic fungi have been studied by Lindeberg and Lindeberg (1977).

Commercially pectolytic enzymes are used for the conversions of natural pectic substances of fruit pulp and fruit juice. The fungal enzymes used for this purpose are produced by various species of *Botrytis, Mucor, Penicillium* and *Aspergillus*.

8.10.2 ASSAY OF AMYLASE AND PECTINASE ACTIVITY

Liberation of maltose from starch is measured by its reduction of 3,5-dinitrosalicylic acid. One amylase unit liberates 1 μmol β-maltose min^{-1} at 25°C and pH 4.8 (Bernfeld 1955). Pectinolytic activity is determined viscometrically using pectin as the substrate (Röhm 1970).

Recommended reading

ANDER P. & ERIKSSON K-E. (1978) Lignin degradation and utilization by microorganisms. In: *Progress in Industrial Microbiology* (Ed. M. J. Bull), vol. 14, pp. 1–58. Elsevier, Amsterdam.

BROWN R. D. & JURASEK L. (1979) Hydrolysis of cellulose: mechanisms of enzymatic and acid catalysis. *Advances in Chemistry Series* 181, pp. 1–399. Washington D.C.

CRAWFORD D. L. & CRAWFORD R. L. (1980) Microbial degradation of lignin. *Enzyme and Microbial Technology* **2**, 11–22.

DECKER L. A. (1977) *Worthington Enzyme Manual.* Worthington Biochemical Corporation, Freehold, New Jersey.

DEKKER R. F. H. & RICHARDS G. N. (1976) Hemicelluloses: Their occurrence, purification, properties, and mode of action. *Advances in Carbohydrate Chemistry and Biochemistry* **32**, 277–352.

ERIKSSON K-E. (1981) Fungal degradation of wood components. *Pure and Applied Chemistry* **6**, 33–43.

KÄÄRIK A. A. (1974b) Decomposition of wood. In: *Biology of Plant Litter Decomposition* (Eds C. H. Dickinson & G. J. F. Pugh), vol. 1, pp. 129–74. Academic Press, London.

KIRK T. K., HIGUCHI T. & CHANG H-M. (1980) *Lignin Biodegradation: Microbiology, Chemistry and Potential Applications*, vols I and II. CRC Press, Boca Raton, Florida.

RYU D. D. Y. & MANDELS M. (1980) Cellulases: biosynthesis and applications. *Enzyme and Microbial Technology* **2**, 91–102.

References

AHLGREN E. & ERIKSSON K-E. (1967) Characterization of cellulases and related enzymes by isoelectric focusing, gel filtration and zone electrophoresis. II. Studies on *Stereum sanguinolentum, Fomes annosus* and *Chrysosporium lignorum. Acta Chemica Scandinavica* **21**, 1193–200.

AHLGREN E., ERIKSSON K-E. & VESTERBERG O. (1967) Characterization of cellulases and related enzymes by isoelectric focusing, gel filtration and zone electrophoresis. I. Studies on *Aspergillus* enzymes. *Acta Chemica Scandinavica* **21**, 937–44.

ALBERSHEIM P., JONES T. M. & ENGLISH P. D. (1969) Biochemistry of the cell wall in relation to infective processes. *Annual Revue of Phytopathology* **7**, 171–94.

ALBERSHEIM P., NEWKOM H. & DEUL H. (1960) Über die Bildung von ungesättigten Abbauprodukten durch ein pectinabbauendes Enzym. *Helvetia Chimica Acta* **43**, 1422–6.

ALMIN K-E. & ERIKSSON K-E. (1967) Enzymic degradation of polymers. I. Viscometric method for the determination of enzyme activity. *Biochimica et Biophysica Acta* **139**, 238–47.

ALMIN K-E., ERIKSSON K-E. & JANSSON C. (1967) Enzymic degradation of polymers. II. Viscometric determination of cellulase activity in absolute terms. *Biochimica et Biophysica Acta* **139**, 248–53.

ANDER P. & ERIKSSON K-E. (1975) Influence of carbohydrates on lignin degradation by the white-rot fungus *Sporotrichum pulverulentum. Svensk Papperstidning* **78**, 643–52.

ANDER P. & ERIKSSON K-E. (1976) The importance of phenol oxidase activity in lignin degradation by the white-rot fungus *Sporotrichum pulverulentum. Archives of Microbiology* **109**, 1–8.

ANDER P. & ERIKSSON K-E. (1977) Selective degradation of wood compounds by white-rot fungi. *Physiologia Plantarum* **41**, 239–48.

ANDER P. & ERIKSSON K-E. (1978) Lignin degradation and utilization by microorganisms. In: *Progress in Industrial Microbiology* (Ed. M. J. Bull), vol. 14, pp. 1–58. Elsevier, Amsterdam.

ANDER P., HATAKKA A. & ERIKSSON K-E. (1980) Vanillic acid metabolism by the white-rot fungus *Sporotrichum pulverulentum*. *Archives of Microbiology* **125**, 189–202.

ARIMA K. (1964) Microbial enzyme production. In: *Global Impacts of Applied Microbiology* (Ed. M. P. Starr), pp. 277–94. John Wiley, New York.

ASCHENGREEN N. H. (1969) Microbial enzymes for alcohol production. *Progress in Biochemistry* **4**, 23–5.

AYERS A. R., AYERS S. B. & ERIKSSON K-E. (1978) Cellobiose oxidase, purification and partial characterization of a hemoprotein from *Sporotrichum pulverulentum*. *European Journal of Biochemistry* **90**, 171–81.

BAVENDAMM W. (1928) Über das Vorkommen und den Nachweis von Oxydasen bei Holzzerstörenden Pilzen. *Zeitschrift für Pflanzenkrankheiten und Pflanzenschutz* **38**, 257–76.

BAVENDAMM W. (1936) Erkennen, Nachweis und Kultur der holzfärbenden und holzzersetzenden Pilze. *Handbuch der Biologischen Arbeitsmethoden* **12**, 927–1134.

BECKHORN E. J. (1967) Production of microbial enzymes. In: *Microbial Technology* (Ed. H. J. Peppler), pp. 366–80. Reinhold Publishing Corporation, New York.

BENNETT T. P. & FRIEDEN E. (1968) *Modern Topics in Biochemistry. Structure and Function of Biological Molecules*. Macmillan Co., New York.

BERGHEM L. E. R. & PETTERSSON L. G. (1973) Purification of a cellulolytic enzyme from *Trichoderma viride* active on highly ordered cellulose. *European Journal of Biochemistry* **37**, 4–30.

BERNFELD P. (1955) Amylases, α and β. In: *Methods in Enzymology* (Eds S. P. Colowick & N. O. Kaplan), vol. 1, pp. 149–58. Academic Press, New York.

BETHGE P. O., RÅDESTRÖM R. & THEANDER O. (1971) Kvantitativ kolhydratbestämning- en detaljstudie. *Svenska Träforskningsinstitutet, Meddelande B* **63**, 1–50.

BLANCHETTE R. A., SHAW C. G. & COHEN A. L. (1978) A SEM study of the effects of bacteria and yeasts on wood decay by brown- and white-rot fungi. *Scanning Electron Microscopy* **11**, 61–7.

BOUTELJE J. B. & KIESSLING H. (1964) On water-stored oak timber and its decay by fungi and bacteria. *Archiv für Mikrobiologie* **49**, 305–14.

BOYCE J. S. (1961) *Forest Pathology*. McGraw-Hill, New York.

BROWN R. D. & JURASEK L. (1979) Hydrolysis of cellulose: mechanisms of enzymatic and acid catalysis. *Advances in Chemistry Series* 181, pp. 1–399. Washington D.C.

BUCHT B. & ERIKSSON K-E. (1968) Extracellular enzyme system utilized by the rot fungus *Stereum sanguinolentum* for the breakdown of cellulose. I. Studies on the enzyme production. *Archives of Biochemistry and Biophysics* **124**, 135–41.

BUSWELL J. A. & ERIKSSON K-E. (1979) Aromatic ring cleavage by the white-rot fungus *Sporotrichum pulverulentum*. *FEBS Letters* **104**, 258–61.

BUSWELL J. A., ANDER P., PETTERSSON B.& ERIKSSON K-E. (1979a) Oxidative decarboxylation of vanillic acid by *Sporotrichum pulverulentum*. *FEBS Letters* **103**, 98–101.

BUSWELL J. A., HAMP S. & ERIKSSON K-E. (1979b) Intracellular quinone reduction in *Sporotrichum pulverulentum* by a NAD(P)H: quinone oxidoreductase. Possible role in vanillic acid catabolism. *FEBS Letters* **108**, 229–32.

CAIN R. B., BILTON R. F. & DARRAH J. A. (1968) The metabolism of aromatic acids by micro-organisms. Metabolic pathways in the fungi. *Biochemical Journal* **108**, 797–832.

CCA5 (1944) Determination of lignin in wood and unbleached pulp. *Svensk Papperstidning* **47**, 9.

CHILD J. J., EVELEIGH D. E. & SIEBEN A. S. (1973) Determination of cellulase activity using hydroxyethylcellulose as substrate. *Canadian Journal of Biochemistry* **51**, 39–43.

CORBETT N. H. (1965) Micromorphological studies on the degradation of lignified cell walls by Ascomycetes and Fungi Imperfecti. *Journal of the Institute of Wood Science* **14**, 18–29.

COWLING E. B. (1961) Comparative biochemistry of the decay of sweetgum sapwood by white-rot and brown-rot fungi. *United States Department of Agriculture Technical Bulletin* 1258.

COWLING E. B. (1970) Nitrogen in forest trees and its role in wood deterioration. Acta Universitas Upsaliensis Dissertation Science 1964.

CRAWFORD D. L. & CRAWFORD R. L. (1976) Microbial degradation of lignocellulose: the lignin component. *Applied and Environmental Microbiology* **31**, 714–17.

CRAWFORD D. L. & CRAWFORD R. L. (1980) Microbial degradation of lignin. *Enzyme and Microbial Technology* **2**, 11–22.

CRAWFORD R. L. & CRAWFORD D. L. (1978) Radioisotopic methods for the study of lignin biodegradation. *Development in Industrial Microbiology* **19**, 35–49.

CRAWFORD R. L., ROBINSON L. E. & CHEH A. M. (1980) ^{14}C-labelled lignins as substrates for the study of lignin biodegradation and transformation. In: *Lignin Biodegradation: Microbiology, Chemistry and Potential Applications* (Eds T. K. Kirk, T. Higuchi & H. M. Chang), vol. 1, pp. 61–76. CRC Press, Boca Raton, Florida.

DECKER L. A. (1977) *Worthington Enzyme Manual*. Worthington Biochemical Corporation, Freehold, New Jersey.

DEKKER R. F. H. & RICHARDS G. N. (1976) Hemicelluloses: Their occurrence, purification, properties, and mode of action. *Advances in Carbohydrate Chemistry and Biochemistry* **32**, 277–352.

DEMEESTER J., COOSEMAN W., BRACKE M. & LANWESS A. (1977) Determination of endo-cellulase activity by viscosimetry. *Biochemical Society Transactions* **5**, 1115–17.

DESHPANDE V., ERIKSSON K-E. & PETTERSSON B. (1978)

Production, purification and partial characterization of 1,4-β-glucosidase enzymes from *Sporotrichum pulverulentum*. *European Journal of Biochemistry* **90**, 191–8.

ECHLIN P. (1971) The application of scanning electron microscopy to biological research. *Philosophical Transactions of the Royal Society of London* (B) **261**, 51–9.

EDSTRÖM R. D. & PHAFF H. S. (1964) Purification and certain properties of pectin trans-eliminase from *Aspergillus fonsecaeus*. *Journal of Biological Chemistry* **239**, 2403–8.

EFFLAND M. J. (1977) Modified procedure to determine acid-insoluble lignin in wood and pulp. *Tappi* **60**, 143–4.

EMERT G. H., GUM E. K. JR., LANG S. A, LIN T. H. & BROWN R. D. JR. (1974) Cellulases in food related enzymes. In: *Advances in Chemistry* (Ed. S. Whitaker), vol. 136, 79–100.

ENOKI A., GOLDSBY G. P. & GOLD M. H. (1980) Metabolism of the lignin model compounds veratrylglycerol-β-guaiacyl ether and 4-ethoxy-3-methoxyphenylglycerol-β-guaiacyl ether by *Phanerochaete chrysosporium*. *Archives of Microbiology* **125**, 227–32.

Enzyme Nomenclature (1979) *Recommendations (1980) of the Nomenclature Committee of the International Union of Biochemistry*. Academic Press, London.

ERDTMAN H. (1939) Phenolic constituents of the pine heartwood, their physiological importance and their retarding action upon the normal digestion of pine heartwood according to the sulfite process. *Annalen* **539**, 116–27.

ERIKSSON K-E. (1978) Enzyme mechanisms involved in cellulose hydrolysis by the rot fungus *Sporotrichum pulverulentum*. *Biotechnology and Bioengineering* **20**, 317–32.

ERIKSSON K-E. (1981) Fungal degradation of wood components. *Pure and Applied Chemistry*, **53**, 33–43.

ERIKSSON K-E., GRÜNEWALD A. & VALLANDER L. (1980a) Studies of growth conditions in wood for three white-rot fungi and their cellulaseless mutants. *Biotechnology and Bioengineering* **22**, 363–76.

ERIKSSON K-E., GRÜNEWALD A., NILSSON T. & VALLANDER L. (1980b) A scanning electron microscopy study of the growth and attack on wood by three white-rot fungi and their cellulase-less mutants. *Holzforschung*, **34**, 207–13.

ERIKSSON K-E. & HAMP S. G. (1978) Regulation of endo-1,4-β-glucanase production in *Sporotrichum pulverulentum*. *European Journal of Biochemistry* **90**, 183–90.

ERIKSSON K-E. & RZEDOWSKI W. (1969) Extracellular enzyme system utilized by the fungus *Chrysosporium lignorum* for the breakdown of cellulose. I. Studies on the enzyme production. *Archives of Biochemistry and Biophysics* **129**, 683–8.

ERIKSSON K-E. & VALLANDER L. (1980) Biomechanical pulping. In: *Lignin Biodegradation Microbiology Chemistry and Potential Applications* (Eds T. K. Kirk, T. Higuchi & H-M. Chang), pp. 213–24. CRC Press, Boca Raton, Florida.

ERIKSSON K-E. & WINELL M. (1968) Purification and characterization of a fungal β-mannanase. *Acta Chemica Scandinavica* **22**, 1924–34.

ESLYN W. E., KIRK T. K. & EFFLAND M. J. (1975) Changes in the chemical composition of wood caused by six soft-rot fungi. *Phytopathology* **65**, 473–6.

FLAIG W. & HAIDER K. (1961) Die Verwertung phenolischer Verbindungen durch Weissfäulepilze. *Archiv für Mikrobiologie* **40**, 212–23.

FREUDENBERG K. & NEISH A. C. (1968) *Constitution and Biosynthesis of Lignin*. Springer Verlag, Berlin.

GEORGE C. J. D. (1974) Fungal enzymic degradation of timber with particular reference to the hemicelluloses. Ph.D. thesis, University of London.

GOULD R. F. (1969) *Advances in Chemistry Series No. 95: Cellulases and Their Applications*, pp. 1–499. American Chemical Society, Washington.

GRADY H. J. & LAMAR M. A. (1959) Glucose determination by automatic chemical analysis. *Clinical Chemistry* **5**, 542–50.

GREEN T. R., HAN Y. W. & ANDERSON (1977) A polarographic assay of cellulase activity. *Analytical Biochemistry* **82**, 404–14.

GRITZALI M. & BROWN R. D. JR. (1979) The cellulase system of *Trichodermar*. Relationships between purified extracellular enzymes from induced or cellulose-grown cells. *Advances in Chemistry* **181**, 237–60.

HAIDER K. & TROJANOWSKI J. (1975) Decomposition of specifically ^{14}C-labelled phenols and dehydropolymers of coniferyl alcohol as models for lignin degradation by soft- and white-rot fungi. *Archives of Microbiology* **105**, 33–41.

HAIDER K., MARTIN J. P. & RIETZ E. (1977) Decomposition in soil of ^{14}C-labelled coumaryl alcohol; free and linked into dehydropolymer and plant lignins and model humic acids. *Soil Science Society of America Journal* **41**, 556–62.

HAINE M. E. (1961) *The Electron Microscope*. Spon, London.

HALL C. E. (1966) *Introduction to Electron Miscroscopy*. 2nd edn. McGraw-Hill, New York.

HARKIN J. M. (1967) Lignin—a natural polymeric product of phenol oxidation. In: *Oxidative Coupling of phenols* (Eds W. I. Taylor & A. R. Battersby), pp. 243–321. Marcel Dekker, New York.

HASHIMOTO Y. & FUKUMOTO J. (1969) Enzymic treatment of coffee beans. I. Purification of mannanase of *Rhizopus niveus* and its action on coffee mannan. *Nippon Nogei Kagaku Kaishi* **43**, 317–22.

HATA K. (1966) Investigation on lignins and lignifications. XXXIII. Studies on lignin isolated from spruce wood decayed by *Poria subacida* B11. *Holzforschung* **20**, 142–7.

HENNINGSSON B. (1962) Studies in fungal decomposition

of pine, spruce and birch pulpwood. *Meddelande Statens Skogsforskningsinstitut* **52**, 3.

HEYWOOD V. H. (1971) Scanning Electron Microscopy. Academic Press, New York.

HIGHLEY T. L. (1976) Hemicellulases of white-rot and brown-rot fungi in relation to host preferences. *Material und Organismen* **11**, 33–46.

HIGHLEY T. L. (1978) How moisture and pit aspiration affect decay of wood by soil block test. *Material und Organismen* **13**, 197–206.

HIGHLEY T. L. & SCHEFFER T. C. (1970) A need for modifying the soil-block test for testing natural decay resistance to white-rot. *Material und Organismen* **5**, 281–92.

HIGUCHI T. (1971) Formation and biological degradation of lignins. In: *Advances in Enzymology* (Ed. F. F. Nord), vol. 34, 207–83. Interscience Publications, New York.

HIGUCHI T. (1980) Lignin structure and morphological distribution in plant cell walls. In: *Lignin Biodegradation: Microbiology, Chemistry, and Potential Applications* (Eds T. K. Kirk, T. Higuchi & H. M. Chang), vol. 1, pp. 1–19. CRC Press, Boca Raton, Florida.

HIROI T. & ERIKSSON K-E. (1976) Microbial degradation of lignin. I. Influence of cellulose on the degradation of lignins by the white-rot fungus *Pleurotus ostreatus*. *Svensk Papperstidning* **79**, 157–61.

HUBERT E. E. (1924) The diagnosis of decay in wood. *Journal of Agricultural Research* **29**, 523–67.

HÜTTERMANN A., GEBAUER M., VOLGER C. & RÖSGER C. (1977) Polymerization und Abbau von Natriumlignosulfonate durch *Fomes annonsus* (Fr.) Cooke. *Holzforschung* **31**, 83–9.

The International Research Group on Wood Preservation Document (1978) Screening techniques for potential wood preservative chemicals. *Swedish Wood Preservation Institute, Report* **136**, 1–145.

ISHIHARA T. & MIYAZAKI M. (1972) Oxidation of milled wood lignin by fungal laccase. *Journal of Japanese Wood Research Society* **18**, 415–19.

ISHII S. & YOKOTSUKA T. (1971) Pectin trans-eliminase with fruit juice clarifying activity. *Journal of Agricultural and Food Chemistry* **19**, 958–61.

IWAHARA S. & HIGUCHI T. (1979) Symposium on biosynthesis and biodegradation of cell wall components. *ACS/CSI Chemical Congress*.

IWASAKI T., TOKUYASU K. & FUNATSU M. (1963) Determination of cellulolytic activity using glycocellulose as the substrate. *Journal of Fermentation Technology* **41**, 340–5.

JUNIPER B. E., COX G. C., GILCHRIST A. J. & WILLIAMS P. R. (1970) *Techniques for Plant Electron Microscopy*. Blackwell Scientific Publications, Oxford.

KÄÄRIK A. A. (1974a) Succession of microorganisms during wood decay. In: *Biological Transformation of Wood by Microorganisms* (Ed. W. Liese), pp. 39–51. Springer Verlag, New York.

KÄÄRIK A. A. (1974b) Decomposition of wood. In: *Biology of Plant Litter Decomposition* (Eds C. H. Dickinson & G. J. F. Pugh), vol. 1, pp. 129–74. Academic Press, London.

KAY D. H. (1965) *Techniques for Electron Microscopy*, 2nd edn. Blackwell Scientific Publications, Oxford.

KEILICH G., BAILEY P. & LIESE W. (1970) Enzymatic degradation of cellulose, cellulose derivatives and hemicelluloses in relation to the fungal decay of wood. *Wood Science and Technology* **4**, 273–83.

KEYSER P., KIRK T. K. & ZEIKUS J. G. (1978) Lignolytic enzyme system of *Phanerochaete chrysosporium*: synthesized in the absence of lignin in response to nitrogen starvation. *Journal of Bacteriology* **135**, 790–7.

KIRK T. K. (1971) Effects of microorganisms on lignin. *Annual Review of Phytopathology* **9**, 185–210.

KIRK T. K. (1973) Chemistry and biochemistry of decay. In: *Wood Deterioration and its Prevention by Preservative Treatments* (Ed. D. D. Nicholas), pp. 149–81. Syracuse University Press, New York.

KIRK T. K. (1975) Chemistry of lignin degradation by wood-destroying fungi. In: *Biological Transformation of Wood by Microorganisms* (Ed. W. Liese), pp. 153–64. Springer Verlag, New York.

KIRK T. K. (1976) Biodelignification research at the U.S. Forest Products Laboratory. In: *Biological Delignification. Present Status—Future Directions*, pp. 31–54. Weyerhäuser Symposium, Washington.

KIRK T. K. & CHANG H-M. (1974) Decomposition of lignin by white-rot fungi. 1. Isolation of heavily degraded lignins from decayed spruce. *Holzforschung* **28**, 217–22.

KIRK T. K., CONNORS W. S., BLEAM R. D., HACKETT W. F. & ZEIKUS J. G. (1975) Preparation and microbial decomposition of synthetic [^{14}C]-lignins. *Proceedings of the National Academy of Science U.S.A.* **72**, 2515–19.

KIRK T. K., HIGUCHI T. & CHANG H-M. (1980) *Lignin Biodegradation: Microbiology, Chemistry and Potential Applications, vols I and II*. CRC Press, Boca Raton, Florida.

KIRK T. K. & LORENZ L. F. (1973) Methoxyhydroquinone, an intermediate of vanillate catabolism by *Polyporus dichrous*. *Applied Microbiology* **26**, 173–5.

KIRK T. K. & MOORE W. E. (1972) Removing lignin in wood with white-rot fungi and digestibility of resulting wood. *Wood Fiber* **4**, 72–9.

KIRK T. K., SCHULTZ E., CONNORS W. J., LORENZ L. F. & ZEIKUS J. G. (1978a) Influence of culture parameters on lignin metabolism by *Phanerochaete chrysosporium*. *Archives of Microbiology* **117**, 277–85.

KIRK T. K., YANG H. H. & KEYSER P. (1978b) The chemistry and physiology of fungal degradation of lignin. In: *Developments in Industrial Microbiology* (Ed. L. A. Underkofler), vol. 19, pp. 51–61. American Institute of Biological Sciences, Washington, D.C.

KIRK T. K., ZEIKUS J. G. & CONNORS W. J. (1977) Advances in understanding the microbiological degradation of lignin. In: *Recent Advances in Phytochemistry* (Eds F. A. Loewers & V. C. Runeckle), vol. 11, pp. 369–94.

KRISNANGKURA K. & GOLD M. H. (1979) Peroxidase catalyzed oxidative decarboxylation of vanillic acid to methoxy-p-decarboxylation. *Phytochemistry* **18**, 2019–21.

KUBACKOVA M., KARACSONYI S., BILISICS L. & TOMAN R. (1978) Some properties of an endo-1,4-D-xylanase from the ligniperdous fungus *Trametes hirsuta*. *Folia Microbiologica* **23**, 202–9.

KYRKLUND B. & STRANDELL G. (1967) A modified chlorine number for evaluation of the cooking degree of high yield pulps. *Paper and Timber* **3**, 99–106.

LEISOLA M. & KAUPPINEN V. (1978) Automatic assay of cellulase activity during fermentation. *Biotechnology and Bioengineering* **20**, 837–46.

LEISOLA M. & LINKO M. (1976) Determination of cellulases with dyed substrates. *Technical Research Centre of Finland, Biotechnical Laboratory*, Tiedonanto 13, Helsinki.

LEVINE W. G. (1966) Laccase, a review. In: *The Biochemistry of Copper* (Eds J. Peisach, P. Aisen & W. E. Blumberg), pp. 371–87. Academic Press, New York.

LI E. H., FLORA R. M. & KING K. W. (1965) Individual roles of cellulase components derived from *Trichoderma viride*. *Archives of Biochemistry and Biophysics* **111**, 439–47.

LIESE W. (1970) Ultrastructural aspects of woody tissue disintegration. *Annual Revue of Phytopathology* **8**, 231–58.

LIESE W. & SCHMID R. (1966) Untersuchungen über den Zellwandabbau von Nadelholz durch *Trametes pini*. *Holz als Roh- und Werkstoff* **24**, 454–60.

LINDEBERG G. & LINDEBERG M. (1977) Pectinolytic ability of some mycorrhizal and saprophytic hymenomycetes. *Archives of Microbiology* **115**, 9–12.

LYR H. (1959) Formation of ecto-enzymes by wood-destroying and wood-inhabiting fungi on various culture media. IV. Xylan and glucose as the carbon source. *Archiv für Mikrobiologie* **34**, 418–33.

LYR H. (1960) Die Bildung von Ektoenzymen durch holzzerstörende und holbewohnende Pilze auf verschiedenen Nahrboden. V. Ein komplexes Medium als C-Quelle. *Archiv für Mikrobiologie* **35**, 258–78.

MCCLEARY B. V. (1978) A simple assay procedure for β-D-mannanase. *Carbohydrate Research* **67**, 213–21.

MANDELS M., ANDREOTTI R. & ROCHE C. (1976) Measurement of saccharifying cellulase. *Biotechnology and Bioengineering Symposium* No. 6, 21–33.

MANDELS M. & REESE T. (1957) Induction of cellulase in *Trichoderma viride* as influenced by carbon sources and metals. *Journal of Bacteriology* **73**, 269–78.

MANDELS M. & WEBER J. (1969) The production of cellulases. *Advances in Chemistry Series* **95**, 391–414.

MARTON J., STERN A. M. & MARTON T. (1969) Decolorization of Kraft black liquor with *Polyporus versicolor*, a white-rot fungus. *Tappi* **52**, 1975–81.

MAYER A. M. & HAREL E. (1979) Polyphenol oxidases in plants. *Phytochemistry* **18**, 193–215.

MIGITA W. & KAWAMURA I. (1944) Chemical analysis of wood. *Bulletin of the Agricultural Chemical Society of Japan* **20**, 348–52.

MILL P. S. (1966) The pectic enzymes of *Aspergillus niger*. *Biochemical Journal* **99**, 557–65.

MILLER G. L. (1959) Use of dinitrosalicylic acid reagent for the determination of reducing sugars. *Analytical Chemistry* **31**, 426–8.

MONTENECOURT B. S. & EVELEIGH D. E. (1977) Preparation of mutants of *Trichoderma reesei* with enhanced cellulase production. *Applied and Environmental Microbiology* **34**, 777–82.

MONTENECOURT B. S., EVELEIGH D. E., EDMUND G. K. & PARCELLS J. (1978) Antibiotic disks. An improvement in the filter paper assay for cellulase. *Biotechnology and Bioengineering* **20**, 297–300.

NELSON N. (1944) A photometric adaptation of the Somogyi method for the determination of glucose. *Journal of Biological Chemistry* **153**, 375–80.

NEVALAINEN K. M. H., PALVA E. T. & BAILEY M. J. (1980) A high cellulase-producing mutant strain of *Trichoderma reesei*. *Enzyme and Microbial Technology* **2**, 59–60.

NILSSON T. (1965) Mikroorganismer i flisstacker. *Svensk Papperstidning* **68**, 495–9.

NILSSON T. (1974a) Comparative study on the cellulolytic activity of white-rot and brown-rot fungi. *Material und Organismen* **9**, 173–98.

NILSSON T. (1974b) The degradation of cellulose and the production of cellulase, xylanase, mannanase and amylase by woodattacking microfungi. *Studie Forestalia Suecica* **114**, 61 p.

NILSSON T. & GINNS J. (1979) Cellulolytic activity and the taxonomic position of selected brown-rot fungi. *Mycologia* **71**, 170–7.

NISHIDA A. & FUKUZUMI T. (1978) Formation of coniferyl alcohol from ferulic acid by the white-rot fungus *Trametes*. *Phytochemistry* **17**, 417–19.

NIXON W. C. (1971) The general principles of scanning electron microscope. *Philosophical Transactions of the Royal Society of London* (B) **261**, 45–50.

Nordic Wood Preservation Council Standard for testing of wood preservatives, mycological test, field test—a field test with stakes. *NWPC STANDARD* No. 1.4.2.1/71.

NORDSTRÖM U. M. (1974) Bark degradation by *Aspergillus fumigatus*. Growth studies. *Canadian Journal of Microbiology* **20**, 283–98.

NORKRANS B. (1950) Studies in growth and cellulolytic enzymes of *Tricholoma*. *Symbolae Botanicae Upsaliensis* **11**, 1–126.

NORTHCOTE D. H. (1972) Chemistry of the plant cell wall. *Annual Review of Plant Physiology* **23**, 113–32.

OHTO M., HIGUCHI T. & IWAHARA S. (1979) Microbial degradation of dehydrodiconiferyl alcohol, a lignin substructure model. *Archives of Microbiology* **121**, 23–8.

PEARSE A. G. E. (1960) *Histochemistry Theoretical and Applied*, 2nd edn. Churchill, London.

PEARSE A. G. E. (1968) *Histochemistry Theoretical and Applied*, 2nd edn. vol. 1., Churchill, London.

PEASE D. C. (1964) *Histological Techniques for Electron Microscopy*, 2nd edn. Academic Press, New York.

PETTERSSON L. G. (1975) The mechanism of enzymatic hydrolysis of cellulose by *Trichoderma viride*. In: *Symposium on Enzymatic Hydrolysis of Cellulose* (Eds M. Bailey, T-M. Enari M. Linko), pp. 255–80. Aulanko, Finland.

POINCELOT R. P. & DAY P. R. (1972) Simple dye release assay for determining cellulolytic activity of fungi. *Applied Microbiology* **23**, 875–9.

RENNERFELT E. (1944) Investigations on the toxicity to rot fungi of the phenolic components of pine heartwood. *Meddelande Skongsförsöksanstalten* (*Stockholm*) **33**, 331–64.

ROBINSON L. E. & CRAWFORD R. L. (1978) Degradation of [^{14}C]-labelled lignins by *Bacillus megaterium*. *FEMS Microbiology Letters* **4**, 301–2.

RÖHM (1970) Pectinglycosidase. Technical note Röhm, Darmstadt.

ROMBOUTS F. M. & PILNIK W. (1972) Research on pectin depolymerases in the sixties—A literature review. *Critical Reviews in Food Technology* **3**, 1–26.

RUDMAN P. & DA COSTA E. W. (1958) The causes of natural durability in timber. The role of toxic extractives in the resistance of silvertop ash (*Eucalyptus sieberiana*) to decay. *Australia Commonwealth Scientific and Industrial Research Organization, Division of Forest Products and Technology, Paper 1*, 8.

RUEL K., BARNOUD F. & ERIKSSON K-E. (1981) Micromorphological and ultrastructural aspects of spruce wood degradation by wild-type *Sporotrichum pulverulentum* and its cellulase-less mutant Cel 44. *Holzforschung* **35**, 157–71.

RUTHMANN A. (1970) *Methods in Cell Research*. Bell, London.

RYU D. D. Y. & MANDELS M. (1980) Cellulases: biosynthesis and applications. *Enzyme and Microbial Technology* **2**, 91–102.

SARKANEN K. V. & HERGERT H. L. (1971) Classification and distribution. In: *Lignins: Occurrence, Formation, Structure and Reactions* (Eds K. V. Sarkanen & C. H. Ludwig), pp. 43–94.

SARKANEN K. V. & LUDWIG C. H. (1971) *Lignins: Occurrence, Formation, Structure and Reactions*. Interscience, New York.

SCHMIDT O. & LIESE W. (1980) Variability of wood degrading enzymes of *Schizophyllum commune*. *Holzforschung* **34**, 67–72.

SCHÖNING A. & JOHANSSON G. (1965) Absorptiometric determination of acid-soluble lignin in semichemical bisulfite pulps and in some woods and plants. *Svensk Papperstidning* **68**, 607–13.

SEIFERT K. (1968) Zur Systematik der Holzfäulen, ihre chemischen und physikalischen Kennzeichen. *Holz als Roh- und Werkstoff* **26**, 208–15.

SELBY K. & MAITLAND C. C. (1967) The cellulase of *Trichoderma viride* cellulolytic complex. *Biochemical Journal* **104**, 716–24.

SELIN J-F. & SUNDMAN V. (1972) Analysis of fungal degradation of lignosulphonates in solid media. *Archiv für Mikrobiologie* **81**, 383–5.

SELIN J-F., SUNDMAN V. & RÄIHÄ M. (1975) Utilization and polymerization of lignosulfonates by wood-rotting fungi. *Archives of Microbiology* **103**, 63–70.

SHIGO A. L. & HILLIS W. E. (1973) Heartwood, discolored wood and microorganisms in living trees. *Annual Revue of Phytopathology* **13**, 197–222.

SJÖSTRAND F. S. (1967) Electron microscopy of cells and tissues. *Instrumentation and Techniques* **1**, 1–462.

SOMOGYI M. (1952) Notes on sugar determination. *Journal of Biological Chemistry* **195**, 19–23.

STERNBERG D., VIJAYAKUMAR P. & REESE E. T. (1977) *β*-Glucosidase: microbial production and effect on enzymatic hydrolysis of cellulose. *Canadian Journal of Microbiology* **23**, 139–47.

STREAMER M., ERIKSSON K-E. & PETTERSSON B. (1975) Extracellular enzyme system utilized by the fungus *Sporotrichum pulverulentum* (*Chrysosporium lignorum*) for the breakdown of cellulose. Functional characterization of five endo-1,4-*β*-glucanases and one exo-1,4-*β*-glucanase. *European Journal of Biochemistry* **59**, 607–13.

SUNDMAN V. & NÄSE L. (1971) A simple plate test for direct visualization of biological lignin degradation. *Paper and Timber* **53**, 67–71.

SUNDMAN V. & NÄSE L. (1972). The synergistic ability of some wood degrading fungi to transform lignins and lignosulfonates on various media. *Archiv für Mikrobiologie* **86**, 339–48.

SWIFT M. J. (1977) The ecology of wood decomposition. *Science Progress, Oxford* **64**, 175–99.

TANSEY M. R., MURRMANN D. N., BEHNKE B. K. & BEHNKE E. R. (1977) Enrichment isolation and assay of growth of thermophilic and thermotolerant fungi in lignin-containing media. *Mycologia* **69**, 463–76.

TROJANOWSKI J. & LEONOWICZ A. (1969) The biodeterioration of lignin by fungi. *Microbios* **3**, 247–51.

TSUJISAKA Y., HIYAMA K., TAKENISHI S. & FUKUMOTO J. (1972) Hemicelluloses. III. Purification and some properties of mannase from *Aspergillus niger*. *Nippon Nogei Kagaku Kaishi* **46**, 155–61.

UNDERKOFLER L. A. (1969) Development of a commercial enzyme process: glucoamylase. In: *Advances in Chemistry Series No. 95: Cellulases and Their Applications* (Ed. R. F. Gould), pp. 343–58. American Chemical Society, Washington.

VAHERI M. (1980) Några nya aspekter på nedbrytning av cellulosa med *Trichoderma reesei*. In: *Biotekniska Processer Baserade på Lignocellulosahaltiga Material* Otnäs, Finland. (In press.)

VÁRADI J. (1972) The effect of aromatic compounds on cellulase and xylanase production of fungi *Schizophyllum commune* and *Chaetomium globosum*. In: *Biodeterioration of Materials* (Eds A. H. Walters & E. H. Hueck-van der Plas), vol. 2, 129–35. Applied Science Publishers, London.

VASSEUR E. (1948) Spectrophotometric study on the Orcinol reaction with carbohydrates. *Acta Chemica Scandinavica* **2**, 693–701.

WALSETH C. S. (1952) Occurrence of cellulases in enzyme preparations from microorganisms. *Tappi* **35**, 228–33.

WESTERMARK U. & ERIKSSON K-E. (1974a) Carbohydrate-dependent enzyme quinone reduction during lignin degradation. *Acta Chemica Scandinavica*, Series B **28**, 204–8.

WESTERMARK U. & ERIKSSON K-E. (1947b) Cellobiose: quinone oxidoreductase, a new wood-degrading enzyme from white-rot fungi. *Acta Chemica Scandinavica*, Serie B **28**, 209–14.

WESTERMARK U. & ERIKSSON K-E. (1975) Purification and properties of cellobiose:quinone oxidoreductase from *Sporotrichum pulverulentum*. *Acta Chemica Scandinavica*, Serie B **29**, 419–24.

WILCOX W. W. (1970) Anatomical changes in wood cell walls attacked by fungi and bacteria. *The Botanical Review* **36**, 2–18.

WILCOX W. W. (1973) Degradation in relation to wood structure. In: *Wood Deterioration and its Prevention by Preservative Treatments* (Ed. D. D. Nicholas), vol. 1, pp. 107–48. Syracuse University Press, New York.

WOOD T. M. (1968) Cellulolytic enzyme system of *Trichoderma koningii*. Separation of components attacking native cotton. *Biochemical Journal* **109**, 217–27.

WOOD T. M. & MCCRAE S. I. (1978) The mechanism of cellulase action with particular reference to the C_1 component. In: *Proceedings of Bioconversion Symposium* (Ed. I. I. T. Hauskhas), pp. 111–41. New Delhi.

WOOD T. M. & PHILLIPS D. R. (1969) Another source of cellulase. *Nature, London* **222**, 986–7.

YAJIMA Y., ENOKI A., MAYFIELD M. B. & GOLD M. H. (1979) Vanillate hydroxylase from the white-rot Basidiomycete *Phanerochaete chrysosporium*. *Archives of Microbiology* **123**, 319–321.

ZOUCHOVA Z., KOCOUREK J. & MUSILEK V. (1977) Separation and properties of α-mannosidase and mannanase from the Basidiomycete *Phellinus abietis*. *Folia Microbiologica* **22**, 98–105.

Chapter 9 · Nitrogen Mineralisation in Soils and Sediments

Charles M. Brown

9.1 Introduction

The breakdown of organically combined nitrogen in the soil is a central part of the nitrogen cycle and results in the provision of inorganic nitrogen in a form that is most readily assimilated by plants. A major reason for attempting to measure this process in soil is the need to calculate the nitrogen available for plant growth and thus to be able to devise an economic strategy for fertiliser application and use. In aquatic sediments, with the possible exception of those in the deep sea, the same process takes place resulting in the release of inorganic nitrogen into the overlying water which serves as the nitrogen source for algal and plant growth. In soils the process is known as nitrogen mineralisation, whilst in sediments it is more commonly termed nutrient (nitrogen) regeneration. The former term is used in this chapter since it parallels the closely coupled process of carbon mineralisation (see Chapter 8.

The immediate product of nitrogen mineralisation is usually ammonia (ammonium ion), generated by the process of ammonification, which in turn may be converted to nitrate by the process of nitrification (see Chapter 11) depending on the prevailing redox conditions (and the presence of suitable microorganisms). At one extreme in well-aerated soils which are low in readily utilisable organic carbon, nitrate appears to be the sole product. Alternatively in aquatic sediments with low or zero oxygen tensions and high organic concentrations (and so a high rate of microbial metabolism), ammonia is detected as the major product. Between these two extremes nitrate and ammonia are both produced in amounts which depend upon the local environmental conditions.

The rate of nitrogen mineralisation may be estimated by measuring the change in the ammonia and nitrate concentrations when soil or sediment samples are incubated with or without an organic nitrogen-containing substrate. The changes observed, which may in practice be negative, reflect the balance between the amounts of nitrogen mineralised and/or fixed and the amounts assimilated by microorganisms and plants; lost as gaseous products as a result of the denitrification process; and lost by leaching processes (in soils) or by diffusion into overlying water (in sediments). Thus in order to assess the true rate of nitrogen mineralisation, parallel estimates of these other processes are required. In recent years the use of ^{15}N-labelled substrates has facilitated these estimations. Nevertheless, the relative lack of sensitivity of the [^{15}N]-nitrogen isotope methods (compared with other radioactive isotopes), equipment costs (the methods require the use of a mass spectrometer or an emission spectrometer), and the need for more complex analytical manipulations have lessened the impact of this approach. Thus many experimenters continue to rely on conventional methods of analysis.

Most of the nitrogen present in soils and sediments is associated with rapidly degraded and biochemically recognisable components: amino acids, amino sugars, purines, pyrimidines and other nitrogen-containing organic compounds. The detailed chemistry of soil and sediment nitrogen compounds, however, is not completely understood for in some instances as much as 50% of the total nitrogen is atypical. For example, amino acids in their free form are readily degraded in the soil yet analyses suggest that they can also persist for long periods. This suggests that amino acids adsorb to and complex with clay minerals (Stotzky 1980) or they may combine chemically with organic colloids to form part of the recalcitrant humic fraction. In either case the bound amino acids are much more resistant to microbial degradation (Bondietti *et al.* 1972).

A wide spectrum of bacteria and fungi are involved in nitrogen mineralisation, with the production of extracellular proteases and nucleases being common features of these organisms. It is also probable that nitrogen mineralisation in natural environments is the result of the combined activities of different populations of microorganisms, but the author knows of few attempts to isolate and characterise the process as it is carried out by microbial communities.

The aim of this chapter is to summarise the available methods for the study of nitrogen mineralisation and to illustrate these with selected examples rather than providing an exhaustive review of the literature. For the latter the reader is referred to Alexander (1977).

Experimental

9.2 Nitrogen-containing compounds in soils and sediments

The total nitrogen content of soils and sediments can be determined either by the Kjeldahl digestion procedure or with a commercial carbon-hydrogen-nitrogen analyser. Total nitrogen values in soil range from 0.03% (w/w) nitrogen for a sandy loam (Doner *et al.* 1975) to over 1.0% (w/w) nitrogen for tropical soils with a high organic content (Sowden *et al.* 1976), but these figures vary considerably within a soil according to agricultural practice. Stanley *et al.* (1978) quoted figures of 0.5 to 0.6 (w/w) nitrogen for inshore marine sediments, while deep sea sediments may contain less than 0.01% (w/w) nitrogen. As stated earlier most of the nitrogen is usually present in combination with organic compounds.

Sowden *et al.* (1976) described a useful method for analysing the combined component, which involved shaking air-dried soil with 0.5M-NaOH under nitrogen gas for 24 h followed by acidification to pH 2.0 with 6M-HCl. After a further 24 h, the coagulated humic acids were removed by centrifugation leaving a soluble fulvic acid fraction. After lyophilisation these fractions were hydrolysed by refluxing in 6M-HCl for 24 h. The results of Sowden *et al.* (1976) showed that in 10 different nitrogen-rich soils 98% of the total nitrogen was amenable to hydrolysis and approximately 50% of the nitrogen in the hydrolysate was present in amino acids; a further 8% was in amino sugars, while ammonia accounted for 23%. In a later paper Sowden *et al.* (1977) analysed a larger number of soils with a wider range of total nitrogen contents and concluded that up to 89% of the total nitrogen could be hydrolysed with 6M-HCl. These same authors reported that amino acids constituted between 33 and 41% of the total nitrogen in soils, amino sugars 5 to 7%, while 18 to 32% was ammonia released during hydrolysis. The ammonia probably originated from the decomposition of amino acids, amides and amino sugars, deamination of purine and pyrimidine bases and some release of fixed ammonia from clay minerals. Up to 45% of the total nitrogen was chemically unidentified although at least some of this was available for plant growth and was assumed to occur in the form of lignin-ammonium, quinone-ammonium, quinone-amino acid, phenoxy-amino acid and carbohydrate-amino acid condensation products.

Cheng (1975) and Cheng *et al.* (1975) described a treatment method using hydrofluoric acid which increased the recovery of organic nitrogen to over 90%. This involved shaking soil or sediment samples with 5M-hydrofluoric acid in 0.1M-HCl for 24 h in order to destroy the clay minerals and release the bound nitrogen. After addition of water the mixtures were lyophilised to remove excess hydrofluoric acid and hydrolysed in 6M-HCl.

The small quantities of free amino acids in soils may be estimated by the method described by Wainwright and Pugh (1975). Soil is extracted with 20% (v/v) ethanol for 20 h at 25°C. After lyophilisation the extracted material is treated with 70% (v/v) ethanol and desalted with an ion exchange resin. The amino acids are estimated by two-dimensional paper or thin layer chromatography, or alternatively may be identified by automatic amino acid analyser. A suitable procedure for use with soil extracts or sediment pore waters (see below) has been described by Stanley and Brown (1976).

Purine and pyrimidine bases constitute only a small fraction of the total soil nitrogen. Corliz and Schnitzer (1979) used ion exclusion chromatography and reported between 21 and 138 μg bases (g soil)$^{-1}$ with 211 to 810 μg bases (g humic acid extract)$^{-1}$ and 295 to 1087 μg bases (g fulvic acid extract)$^{-1}$.

The amount of nitrate and ammonia usually constitutes relatively small percentages of the total nitrogen. For example, Doner *et al.* (1975) quoted values of 35 μg NO_3^- – nitrogen (g soil)$^{-1}$ and 0.7 μg NH_4^+ – nitrogen (g soil)$^{-1}$ with an organic nitrogen content of 290 μg nitrogen (g soil)$^{-1}$ since nitrate and ammonia are freely soluble, although ammonia may be adsorbed to clay minerals. A usual method of extraction is to shake a sample with 2M-KCl, centrifuge to remove the debris and assay the supernatant for nitrogen-containing compounds.

For estimations of soluble materials from aquatic sediments a squeezing method is adopted. This involves placing the sediment in a press and squeezing the sediment pore water out through a member filter by applying 5 to 10 p.s.i. pressure of an inert gas. The equipment used in the author's laboratory is detailed in Miller *et al.* (1979). Nitrate concentration in the pore water may be estimated directly using a nitrate specific electrode (Myers & Paul 1968) or colorimetrically after chemical reduction to nitrite. Ammonium may be determined directly by colorimetric methods although the fractional distillation methods of Bremner and Edwards (1965) and Bremner and Keeney (1965) which allow the measurement of both NH_4-nitrogen and NO_3-nitrogen have been widely adopted. In the latter method ammonia is distilled off following the addition of magnesium oxide. The distilled sample is cooled and nitrate is converted to ammonia using Devardas alloy or titanous sulphate as reductant; the ammonia is distilled off. Ammonia in the distillate may be assayed colorimetrically or by direct acid titration. An advantage of this extraction and distillation method is that both ammonia and nitrate (in the form of ammonia) may be retained for subsequent estimation of isotope ratios if [^{15}N]nitrogen is used as a tracer (Bremner & Keeney 1965). For a detailed discussion of the application of [^{15}N]nitrogen to the study of nitrogen cycling in soils the reader is referred to Hauck and Bremner (1976) which includes details of the mass and emission spectrometers used and lists of equipment and suppliers.

9.3 Estimation of the rate of nitrogen mineralisation

As outlined in the introductory paragraphs the practical purpose of estimating nitrogen mineralisation rates is to gain information on nitrogen availability. This may be particularly important, for example, in adopting a fertiliser policy to give maximum crop yield with minimum loss through leaching and the consequent problems of nitrate in ground water. Clearly total nitrogen determinations are of little direct use in this context although a number of methods have been reported to give good correlations with plant growth. Unfortunately different authors have used different methods for measuring total nitrogen with little attempt to find a common approach or procedure; for example, Keeney and Bremner (1976) reported that inorganic nitrogen extracted by boiling water or released on incubation constituted a more reliable index of soil nitrogen availability than ammonia released by alkaline permanganate extraction. Stanford (1978), however, preferred the alkaline permanganate method. Fox and Piekielek (1978a, b), on the other hand, studied a number of extraction procedures and compared their results with those of growth of field grown corn (*Zea mays*) in the same soils. They reported a poor correlation between growth and the levels of either soil nitrate or H_2SO_4/KCl-extractable nitrogen (measured by the Kjeldahl procedure). Better correlation was obtained using values of ammonia and total nitrogen extracted from autoclaved soil (presumably similar to the boiling procedure of Bremner and Keeney (1965)). However, the best correlation was observed with values of UV absorption of soil extracts made with 0.02M sodium bicarbonate. The chemical nature of the extracted material was not studied.

Since plant growth requires inorganic nitrogen most workers have used nitrate and ammonia measurements before and after incubation to estimate mineralisation rates and hence gain information about the availability of nitrogen. For example, Gasser and Kalembasa (1976) in a study of available nitrogen versus growth of rye grass, concluded that nitrogen extracted with hot water or released by aerobic or anaerobic incubation gave the best correlations with plant growth. These results were in general agreement with those of Keeney and Bremner (1967).

Reports of measurements *in situ* of net mineralisation are rare, although the litter bag method (Rixon & Zorin 1978) appears to be applicable. An alternative method for use in soils is described by Westerman and Tucker (1979). In this method aluminium cylinders (25 cm long with an internal diameter of 4.8 cm) were driven into soils and amendments made by mixing them with the top 1.0 cm of the columns, which were then incubated without disturbance for 3 to 12 months when they were sectioned, extracted and analysed in the laboratory. Similar methods may be applied to undisturbed sediment cores obtained with a Craib corer (Craib 1965), a Barnett multiple corer, or a similar device. Incubation of these cores may take place in the laboratory or occur *in vivo* held in position above the sediment surface but immersed in the water for short incubation times (24 to 26 h). In either case measurements of mineralisation involve estimating the release of ammonia into the overlying water.

Most workers have preferred to study samples in the laboratory and the methods used apply both to soils and sediments. In many instances direct chemical estimation of inorganic nitrogen before and after incubation is the method of choice but the use of [^{15}N]nitrogen and [^{14}C]carbon for tracing the breakdown of amino acids and other compounds has been preferred by some workers. Little attempt has been made to standardise incubation conditions and the results reported apply only to those specific samples studied. Many authors, for example, have incubated samples over long time periods and while this may be necessary to obtain data of relevance to nitrogen release for plant growth during a whole growing season, there are obvious problems in interpreting data from this type of experiment; for instance, the mineralisation rate is not a linear function of time. Myers (1975) reported, for tropical soils, ammonification rates of 2.8 μg nitrogen (g soil)$^{-1}$ d^{-1} during the first 7 d of incubation, although this rate decreased to 0.5 μg nitrogen (g soil)$^{-1}$ d^{-1} between 14 and 28 d.

The procedure of Keeney and Bremner (1967) for estimating mineralisation and nitrification rates forms the basis of many methods and is as follows. A mixture of 10 g soil and 30 g quartz sand is incubated for 14 d in a 250 ml capacity wide-necked bottle, thereby allowing incubation under aerobic conditions. 100 ml 2M-KCl is added and, after shaking for 1 h and allowing to settle, 20 ml samples are decanted for analysis of nitrate and ammonia using the fractional distillation procedure described earlier (Bremner 1965; Bremner & Edwards 1965; Bremner & Keeney 1965). The principal variations in this method are the incubation time, temperature, presence of added substrates and inhibitors, and the method of extracting the inorganic nitrogen. A few examples are given to illustrate these variations.

(1) Campbell and Biederbeck (1972) used air-dried soils from 0 to 15 cm depth sieved through a 2 mm mesh and wetted with an atomiser to range from 10 to 22% (w/w) moisture. Some samples were supplemented with 50 μg peptone and the 60 g portions incubated in plastic bags closed with rubber bands. After incubating 5 g subsamples were removed, extracted by shaking for 14 min with 1M potassium sulphate, and then analysed for nitrate and ammonia.

(2) Gasser and Kalembasa (1976) employed both aerobic and anaerobic incubations. 40 g soil samples at 50% water-holding capacity were incubated aerobically at 25°C for 21 d while 5 g samples plus 12.5 ml water (>100% water-holding capacity) were incubated anaerobically at 40°C for 7 d. Nitrogen was extracted by refluxing 5 g subsamples in boiling water for 1 h followed by addition of potassium sulphate and filtering.

(3) Laura (1974) mixed 30 g soil with 2% (w/w) powdered gulmohur leaves as substrate and after wetting to 60% of the water-holding capacity the mixtures were incubated in sealed flasks for 6 months before estimating ammonia-, nitrite- and nitrate-nitrogen released.

(4) Westerman and Tucker (1979) incubated 25 g soil samples in Petri dishes at 28°C for 49 d.

Obviously the choice of detailed method is wide and other organic supplements used include straw (Enwezor 1976), urea (Khandelwal 1977), nitrilotriacetate (Tabatabai & Bremner 1975), slow release nitrogen fertilisers such as 2-oxo-4-methyl-6-ureido-hexa-pyrimidine (CDU) (Kimura & Yamaguchi 1978) and nucleic acids (Greaves & Wilson 1970), while other extraction procedures include the use of 2M-H_2SO_4 (Greaves & Wilson 1970), 10% (w/v) sodium acetate (Enwezor 1976), and 1M-KCl followed by 0.5M-HCl (Broadbent & Nakashima 1971).

The alternative to incubating and extracting samples is to use some form of leaching tube. Stanford and Smith (1972) described a procedure in which 15 g of soil samples plus an equal weight of 20 mesh quartz sand were moistened with a fine spray of water and retained in 50 ml leaching tubes by means of a glass wool pad. The soil with the sand was covered with a 0.6 cm plug of glass wool and inorganic nitrogen removed by leaching with 100 ml 0.01M calcium chloride in 5 to 10 ml increments. The columns were supplemented by the addition of 25 ml nutrient solution and excess water removed by suction. Stanford and Smith (1972) used an incubation period of 2 to 6 weeks after which the columns were leached with 0.01M calcium chloride and the leachates retained for analysis. Guthrie and Duxbury (1978) used similar columns but leached with 1M calcium chloride.

In most instances the Bremner fractional distillation procedure has been used to determine nitrate and ammonia concentrations in extracts and leachates. In combination with ^{15}N-labelled substrates, this method has tremendous potential for measuring nitrogen fluxes in soils and sediments but, so far, few workers have employed ^{15}N-labelled substrates for estimating mineralisation rates (Hauck & Bremner 1976). Usually nitrate and/or ammonium have been used as substrates for mineralisation studies following the initial assimilation (immobilisation) step. Examples of this approach may be found in Broadbent and Nakashima (1971), Westerman and Tucker (1974), McGill *et al.* (1975), Ladd *et al.* (1977a, b), Crasswell (1978) and Wojcik-Wojtkowiak (1978). Jones and Richards (1978) added 95.6% atom excess [^{15}N]-glycine to soils which were sampled at intervals up to 10 weeks in studies which measured both assimilation and mineralisation of the amino acid.

Use has been made of ^{14}C-labelled substrates to trace the breakdown of some nitrogen-containing organic materials. Verma *et al.* (1975) grew cultures of a *Chlorella* sp. and a *Microcoleus* sp. on [^{14}C]glucose and extracted labelled proteins, peptides and amino acids for soil degradation studies. After 4 to 12 weeks incubation some 71 to 95% ^{14}C-labelled substrates added to soil was evolved as [^{14}C]-carbon dioxide. McGill *et al.* (1975) added [^{14}C]-acetate to soil together with [^{15}N]ammonium sulphate. Both ^{14}C- and ^{15}N-labelled compounds were assimilated, initially by fungi and subsequently by bacteria, and served as a method of producing labelled organic material in the soil. These workers extracted the nitrogen compounds from soil with sodium pyrophosphate and 75% (v/v) phenol in water in experiments which indicated a close relationship between nitrogen and carbon transformations (see section 9.4). Mayaudon and Batistic (1978) treated a low nitrogen soil (0.17% (w/w) nitrogen) with ^{14}C-labelled serine, alanine, proline, lysine, leucine, arginine or glutamate and incubated the mixtures at 20°C for 100 d. The ^{14}C-released gave an estimation of amino acids mineralised. These authors also hydrolysed the soil with 6M-HCl and determined the distribution of ^{14}C-labelled compounds by two-dimensional paper chromatography. The distribution of label varied with the source of the [^{14}C]carbon and the product. ^{14}C-labelled amino acids and proteins have also been employed in studies of the degradation of nitrogen-containing compounds complexed with humic materials as outlined below.

9.4 Factors influencing the rate of nitrogen mineralisation

Soils rich in organic nitrogen appear to liberate inorganic nitrogen faster than those which are nitrogen deficient: thus, irrespective of the type of soil, between 1 and 4% of the total nitrogen is likely to be mineralised per annum and made available for plant growth (Alexander 1977). Nitrogen and carbon mineralisation are closely related processes and in unamended soils these occur at relatively similar rates, as judged by the ratio of carbon dioxide-carbon to inorganic nitrogen released (in the range 7 to 15:1). The carbon:nitrogen ratio of supplements has a significant effect on net mineralisation. In general, nitrogen-rich substrates, such as proteins, favour nitrogen release, while nitrogen-poor residues in conjunction with easily degradable carbon substrates, such as glucose, favour carbon dioxide evolution and nitrogen immobilisation due to its assimilation into microbial cells (Jones & Richards 1978). Enwezor (1976) reported that nitrogen mineralisation increased with a decreasing carbon:nitrogen ratio with no mineralisation occurring at a ratio greater than 16:1. A number of workers have reported that the addition of inorganic nitrogen stimulated microbial growth and activity (Parnas 1975). The demonstration of this activity requires the use of either of [^{15}N]ammonium or ^{15}N-enriched organic substrates (Broadbent & Nakashima 1971; Westerman & Kurtz 1973; Westerman & Tucker 1974; Wojcik-Wojtkowiak 1978).

Nitrogen mineralisation is strongly influenced by the soil moisture level and care is usually taken to ensure a known moisture content when incubating samples in the laboratory. Ammonia release is slow at low water levels but is stimulated by increasing the moisture level. In acid environments and in areas having wet/dry cycles, the onset of rainfall is associated with an increased rate or 'flush' of mineralisation (Alexander 1977). Similarly Ladd *et al.* (1977b) reported that intermittent wetting and drying stimulated the breakdown of ^{15}N-enriched supplements and native soil inorganic nitrogen in laboratory incubations. Moisture levels of 50 to 70% of the water-holding capacity appear to be optimum for ammonification (Campbell & Biederbeck 1972).

A number of workers have investigated the effects of added salts on mineralisation in laboratory experiments. Heilman (1975) studied 10 forest soils which were incubated with a range of salts at concentrations from 0.005M to 0.2M. There was marked stimulation of inorganic nitrogen release in the order of $AlCl_3 > CaCl_2 > KCl > K_2SO_4 > K_2HPO_4$. Similarly Westerman and Tucker (1974) reported that dilute solutions of NaCl, $CaCl_2$ and $CuCl_2$ together with ammonium salts, stimulated mineralisation. Agarwal *et al.* (1971) showed stimulation of mineralisation in the order $CaCl_2 > KaCl > NaCl$. Broadbent and Nakashima (1971) and Laura (1974, 1976, 1977) have reported similar results. The temperature of incubation (Chaudhury & Cornfield 1978), pH (Alexander 1977) and the presence of antimicrobial agents (Wainwright & Pugh 1975) have a direct influence on microbial activity and hence on the mineralisation rate.

A further important influence on mineralisation rate is the ability of the microflora to metabolise the nitrogen-containing substrate. With complex substrates, such as CDU (Kimura & Yamaguchi 1978), breakdown occurs only after a lag period, presumably due to the selection of or adaptation of a suitable microbial community. Much of the soil nitrogen, however, is absorbed on clay minerals, or complexed with humic compounds and may be recalcitrant. Sorensen (1975), for example, described experiments in which the amount of ^{14}C-labelled amino acids recovered from the microbial degradation of ^{14}C-labelled barley straw in soil was proportional to the soil clay and silt content. Ladd *et al.* (1977a, b) showed that mineralisation of ^{15}N-enriched substrates and native soil organic nitrogen occurred at a faster rate in sandy soil than in a highly adsorptive clay soil. There are a number of reports of the resistance to degradation of nitrogen-containing compounds when complexed with humic and humic-like materials. In their study of the breakdown of ^{14}C-labelled amino acids and proteins, Verma *et al.* (1975) polymerised some fractions with phenolic compounds including a mixture of hydroxy-phenols and hydroxy-toluenes and a mixture of hydroxy-benzoic acid and hydroxy-cinnamic acids. This linkage to the model polymers lowered the rate of degradation by 80 to 90% compared with the non-complexed control materials. Verma and Martin (1976) also reported that the rates of decomposition of algal cell wall and cytoplasm fractions were decreased by 40 to 70% when complexed with model humic-acid-type phenolic polymers. In a similar type of experiment Nelson *et al.* (1979) grew cultures of bacteria, yeasts and filamentous fungi on [^{14}C]glucose, prepared cell wall and cytoplasmic fractions from the labelled cells, and complexed these with model phenolic polymers. While 50 to 79% of the [^{14}C]carbon of the non-complexed materials was released as [^{14}C]-carbon dioxide within 12 weeks, the rate of decomposition of the complexes was 31 to 56% slower.

9.5 Mineralisation mechanisms

In a natural soil or sediment the breakdown of organic nitrogen-containing compounds is the product of the metabolism of a wide variety of microorganisms and the activity of extracellular particle bound and free enzymes. Almost all heterotrophic bacteria, including actinomycetes, and almost all fungi are able to attack some organic nitrogen-containing compound. These organisms may readily be enumerated by demonstrating alkali production (ammonia release on ammonification) in a peptone or other nitrogen-medium using a MPN tube system (see Chapter 6). Numbers in the range 1×10^5 to 1×10^7 (g soil or sediment)$^{-1}$ are frequently encountered although this type of data has little relevence to the in-situ population since they refer only to the microorganisms' response to substrate present in the medium. An alternative approach is to plate out solutions on general laboratory medium such as nutrient agar, soil extract agar or marine agar, and pick off colonies of organisms at random for experimentation with a range of likely substrates. Roberge and Knowles (1967), for example, plated out dilutions from soil on nutrient agar (with the addition of rose bengal and streptomycin for the isolation of fungi) and tested 40 colonies at random for the ability to hydrolyse urea. Similarly, Paulson and Kurtz (1969) prepared dilution plates on soil extract agar for a study of urea-degrading organisms. Again 40 colonies were picked at random and plated out on soil extract agar supplemented with 2% (w/v) urea and 0.0012% (w/v) phenol red as a pH indicator of urea hydrolysis. In the author's laboratory this method of subculturing from general to specific media has been used extensively for the isolation of bacteria associated with mineralisation as well as other processes involved in nitrogen cycling in aquatic sediments. Greaves and Wilson (1970) incubated glass slides coated with a thin film of a nucleic acid–montmorillonite complex in soils and noted that the slide surfaces became colonised with bacteria. Bacteria isolated at random after serial dilutions and plating were tested for the ability to produce nucleases.

The degradation of proteins involves the excretion of extracellular proteolytic enzymes, a common property of the bacterial genera *Bacillus, Clostridium, Pseudomonas, Serratia* and many actinomycetes, and the fungi *Altenaria, Aspergillus, Mucor, Penicillium* and *Rhizopus*. The products of protein hydrolysis (peptides and amino acids) are readily degraded and assimilated by an even larger range of organisms. Nucleic acids are also degraded first by extracellular enzymes into mononucleotides or short oligonucleotide chains. Soil genera producing extracellular ribonucleases include *Bacillus, Mycobacterium, Pseudomonas, Aspergillus, Cephalosporium, Fusarium, Mucor, Penicillium* and *Rhizopus*, while deoxyribonuclease production is a characteristic of the genera *Arthrobacter, Bacillus, Clostridium, Pseudomonas, Cladosporium* and *Fusarium*. The potential for nucleic acid degradation is very common and 80% or more of the isolates from some soils have been found to excrete nucleases (Alexander 1977). After enrichment of soils with nucleic acids Greaves and Wilson (1970) reported the predominance of pseudomonads able to degrade these substrates rapidly. Nucleic acid degradation products, such as purines and pyrimidine bases, are commonly utilised by the genera *Clostridium, Corynebacterium, Pseudomonas* and *Nocardia*.

Urea is a product of nucleic acid degradation and is also a nitrogen fertiliser and is readily broken down in soils which may become alkaline by virtue of rapid ammonia production. Ureolytic genera include *Bacillus, Clostridium, Corynebacterium, Klebsiella* and *Pseudomonas* together with a wide range of actinomycetes and fungi. Some reports have suggested that up to 30% (Lloyd & Scheaffe 1973) and 69% (Roberge & Knowles 1967) of the soil bacteria are able to hydrolyse urea, while up to 100% fungal isolates from a black spruce humus have this property (Roberge & Knowles 1967). The breakdown of the slow-release nitrogen fertiliser CDU was studied by Kimura and Yamaguchi (1978) over a 5 y period in a soil perfusion apparatus. Two organisms, an *Arthrobacter* sp. (which degraded (−) CDU) and a *Corynebacterium* sp. (which degraded (+) CDU), were enriched in this process and both were required for the breakdown of the substrate to urea and ammonia. Unfortunately this type of community enrichment study is all too rare in this field.

Extracellular proteolytic enzymes, nucleases and urease—all important in nitrogen cycle transformations—may persist in soil in association with the colloidal clays and humic fraction and be in part responsible for mineralisation. For a detailed discussion of soil enzymes the reader is referred to Burns (1978) and just a few examples are given here. Ladd and Butler (1972) described a series of experiments using air-dried, sieved soil which was incubated with a natural protein substrate (100 mg sodium caseinate ml^{-1}) or with synthetic peptides. The use of the latter enabled different proteolytic activities to be estimated: for example, the use of benzoyl-arginine-amide for trypsin-like activity. A typical incubation involved 0.5 g soil suspended in 0.1M to 1.8M Tris borate buffer (pH 8.1) incubated with 2 ml phenylalanyl-leucine as substrate at 40°C in a shaking water bath for 10 to 60 min. The reaction was stopped with 5M-HCl and, after centrifugation, the supernatant assayed for amino nitrogen with ninhydrin. Alternatively, soil and sodium caseinate were incubated for 50°C for 60 min and the reaction stopped with trichloroacetic acid. After centrifugation peptide material in the supernatants was assayed by the Lowry method using BSA as a standard. In both cases, enzyme activity was reported to be proportional to soil concentration and incubation time. Clearly enzyme activity will be modified in heterogeneous environments by complexing with mineral particles (McLaren & Packer 1970). For example, Griffith and Thomas (1979) reported that pronase lost over 70% of its activity in the presence of montmorillonite but significantly the residual enzyme was more stable in this complexed form. Browman and Tabatabai (1978) estimated phosphodiesterase activity in soils by incubation with *p*-nitrophenyl phosphate. Typically 1.0 g soil in 5 ml buffered substrate was incubated for 1 h at 37°C and the *p*-nitrophenol released by enzyme activity extracted in buffered 0.5M calcium chloride and assayed spectrophotometrically. In these experiments steam sterilisation and treatment with formaldehyde decreased enzyme activity, whereas toluene increased enzyme activity—possibly due to cell lysis and release of enzyme activity. Greaves and Wilson (1970) reported the breakdown of nucleic acid–montmorillonite complexes on glass slides incubated in soil using X-ray diffraction analysis to measure basal spacings, but this assay method is clearly beyond the scope of most laboratories.

The most investigated soil enzyme is probably urease. Many methods use the estimation of ammonia release on incubating toluene-treated soils with urea although methods avoiding toluene have been published (Zantua & Bremner 1975; Pettit *et al.* 1976) and its value seriously criticised. Urease activity (as with other soil enzymes) is in part due to enzyme adsorbed on to soil clay colloids (Paulson & Kurtz 1969) and as humus–urease complexes (Burns *et al.* 1972; Ceccanti *et al.* 1978).

Recommended reading

ALEXANDER M. A. (1977) *Introduction to Soil Microbiology*. John Wiley & Sons, New York.

BARTHOLOMEW W. V. (1965) Mineralization and immobilization of nitrogen in the decomposition of plant and animal residues. In: *Soil Nitrogen* (Eds W. V. Bartholomew & F. E. Clark), pp. 285–306. American Society of Agronomy, Madison.

BONDIETTI E., MARTIN J. P. & HAIDER K. (1972) Stabilisation of amino sugar units in humic-type polymers. *Soil Science Society of America Proceedings* **35**, 917–22.

BREMNER J. M. (1965) Nitrogenous Compounds. In: *Methods of Soil Analysis Part 2* (Ed. C. A. Black), pp. 1179–237.

BREMNER J. M. (1967) Nitrogenous Compounds. In: *Soil Biochemistry* (Eds A. D. McLaren & G. H. Peterson), pp. 19–66. Marcel Dekker, New York.

HAUCK R. D. & BREMNER J. M. (1976) Use of tracers for soil and fertilizer nitrogen research. *Advances in Agronomy* **28**, 219–66.

PARSON J. W. & TINSLEY J. (1975) Nitrogenous substances. In: *Soil Components* (Ed. J. E. Gieseking), vol. 6, pp. 263–304. Springer Verlag, New York.

References

AGARWAL A. S., SINGH B. R. & KANCHIRO Y. (1971) Ionic effects of salts on mineral nitrogen release in an Allophanic soil. *Soil Science Society of America Proceedings* **35**, 454–7.

ALEXANDER M. A. (1977) *Introduction to Soil Microbiology*. John Wiley & Sons, New York.

BARTHOLOMEW W. V. (1965) Mineralisation and immobilisation of nitrogen in the decomposition of plant and animal residues. In: *Soil Nitrogen* (Eds W. V. Bartholomew & F. E. Clark), pp. 285–306. American Society of Agronomy, Madison.

BONDIETTI E., MARTIN J. P. & HAIDER K. (1972) Stabilisation of amino sugar units in humic-type polymers. *Soil Science Society of America Proceedings* **35**, 917–22.

BREMNER J. M. (1965) Nitrogenous compounds. In: *Methods of Soil Analysis Part 2* (Ed. C. A. Black), pp. 1179–237.

BREMNER J. M. (1967) Nitrogenous compounds. In: *Soil Biochemistry* (Eds A. D. McLaren & G. H. Peterson), vol. 1, pp. 19–66. Marcel Dekker, New York.

BROWMAN M. G. & TABATABAI M. A. (1978) Phosphodiesterase activity of soils. *Soil Science Society of America Proceedings* **42**, 284–90.

BREMNER J. M. & EDWARDS A. P. (1965) Determination and isotope-ratio analysis of different forms of nitrogen in soils. 1. Apparatus and procedure for distillation and determination of ammonium. *Soil Science Society of America Proceedings* **29**, 504–7.

BREMNER J. M. & KEENEY D. R. (1965) Steam distillation methods for determination of ammonium, nitrate and nitrite. *Analytica Chimica Acta* **32**, 485–95.

BROADBENT F. E. & NAKASHIMA T. (1971) Effect of added salts on nitrogen mineralization in three California soils. *Soil Science of America Proceedings* **35**, 457–60.

BURNS R. G. (1978) *Soil Enzymes*. Academic Press, London.

BURNS R. G., PUKITE A. H. & MCLAREN A. D. (1972) Concerning the location and persistence of soil urease. *Soil Science Society of America Proceedings* **36**, 308–11.

CAMPBELL C. A. & BIEDERBECK V. O. (1972) Influence of fluctuating temperatures and constant soil moistures on nitrogen changes in amended and unamended loam. *Canadian Journal of Soil Science* **52**, 523–36.

CECCANTI B., NANNIPIERI N., CERVELLI S. & SEQUI P. (1978) Fractionation of humus–urease complexes. *Soil Biology and Biochemistry* **10**, 39–45.

CHAUDHURY M. S. & CORNFIELD A. H. (1978) Nitrogen and carbon mineralization of two Bangladesh soils in relation to temperature. *Plant and Soil* **49**, 317–21.

CHENG C-N. (1975) Extracting and desalting amino acids from soils and sediments: evaluation of methods. *Soil Biology and Biochemistry* **7**, 319–22.

CHENG C-N., SHUFELDT R. C. & STEVENSON T. J. (1975)

Amino acid analysis of soils and sediments: extraction and desalting. *Soil Biology and Biochemistry* **7**, 143–51.
CORLIZ J. & SCHNITZER M. (1979) Purines and pyrimidines in soils and humic substances. *Soil Science Society of America Journal* **43**, 958–61.
CRAIB J. S. (1965) A sampler for taking short undisturbed marine cores. *Journal du Conseil du Exploration de le Mer* **30**, 34–9.
CRASSWELL E. T. (1978) Some factors influencing denitrification and nitrogen immobilization in a clay soil. *Soil Biology and Biochemistry* **10**, 241–5.
DONER H. E., VOLZ M. G., BELSER L. W. & LOKEN J. P. (1975) Short term nitrate losses and associated microbial populations in soil columns. *Soil Biology and Biochemistry* **7**, 261–3.
ENWEZOR W. O. (1976) The mineralization of nitrogen and phosphorus in organic materials of varying C:N and C:P ratios. *Plant and Soil* **44**, 237–40.
FOX R. H. & PIEKIELEK W. P. (1978a) Field testing of several nitrogen availability indexes. *Soil Science Society of America Journal* **42**, 747–50.
FOX R. H. & PIEKIELEK W. P. (1978b) A rapid method for estimating the nitrogen supplying capacity of a soil. *Soil Science Society of America Journal* **42**, 751–3.
GASSER J. K. R. & KALEMBASA S. J. (1976) Soil nitrogen. IX. The effects of leys and organic manures on the available-N in clay and sandy soils. *Journal of Soil Science* **27**, 237–49.
GREAVES M. P. & WILSON M. J. (1970) The degradation of nucleic acids and montmorillonite-nucleic acid complexes by soil microorganisms. *Soil Biology and Biochemistry* **2**, 257–68.
GRIFFITH S. M. & THOMAS R. L. (1979) Activity of immobilized pronase in the presence of montmorillonite. *Soil Science Society of America Journal* **43**, 1138–40.
GUTHRIE T. F. & DUXBURY J. M. (1978) Nitrogen mineralization and dentrification in organic soil. *Soil Science Society of America Journal* **42**, 908–12.
HAUCK R. D. & BREMNER J. M. (1976) Uses of tracers for soil and fertilizer nitrogen research. *Advances in Agronomy* **28**, 219–66.
HEILMAN P. (1975) Effect of added salts on nitrogen release and nitrate levels in forest soils of the Washington coastal area. *Soil Science Society of America Proceedings* **39**, 778–82.
JONES J. M. & RICHARDS B. N. (1978) Fungal development and the transformation of ^{15}N-labelled amino- and ammonium-nitrogen in forest soils under several management regimes. *Soil Biology and Biochemistry* **10**, 161–8.
KEENEY D. R. & BREMNER J. M. (1967) Determination and isotope-ratio analysis of different forms of nitrogen in soil. 6. Mineralizable nitrogen. *Soil Science Society of America Proceedings* **31**, 34–9.
KHANDELWAL K. C. (1977) Effect of a formulation of urea, thiourea and humate on nitrogen availability and yield of wheat. *Plant and Soil* **47**, 717–19.
KIMURA R. & YAMAGUCHI M. (1978) Microbial decomposition of 2-oxo, 4-methyl 6-ureido hexa hydropyrimidine in soil. *Soil Biology and Biochemistry* **10**, 503–8.
LADD J. N. & BUTLER J. H. A. (1972) Short term assays of soil proteolytic enzyme activities using proteins and dipeptide derivatives as substrates. *Soil Biology and Biochemistry* **4**, 19–30.
LADD J. N., PARSONS J. W. & AMATO M. (1977a) Studies of nitrogen immobilization and mineralization in calcareous soils. 1. Distribution of immobilized nitrogen amongst solid fractions of different particle size and density. *Soil Biology and Biochemistry* **9**, 309–18.
LADD J. N., PARSONS J. W. & AMATO M. (1977b) Studies of nitrogen immobilization and mineralization in calcareous soils. 2. Mineralization of immobilised nitrogen from solid fractions of different particle size and density. *Soil Biology and Biochemistry* **9**, 319–25.
LAURA R. D. (1974) Effects of neutral salts in carbon and nitrogen mineralization of organic matter in soil. *Plant and Soil* **4**, 113–27.
LAURA R. D. (1976) Effects of alkali salts on carbon and nitrogen mineralization of organic matter in soil. *Plant and Soil* **44**, 587–96.
LAURA R. D. (1977) Salinity and nitrogen mineralization in soil. *Soil Biology and Biochemistry* **9**, 333–6.
LLOYD A. B. & SCHEAFFE M. J. (1973) Urease activity in soils. *Plant and Soil* **39**, 71–80.
MCGILL W. B., SHIELDS J. A. & PAUL E. A. (1975) Relation between carbon and nitrogen turnover in soil organic fraction of microbial origin. *Soil Biology and Biochemistry* **7**, 55–63.
MCLAREN A. D. & PACKER L. (1970) Some aspects of reactions in heterogenous systems. *Advances in Enzymology* **33**, 245–308.
MAYAUDON J. & BATISTIC L. (1978) Decomposition des acides amines ^{14}C dans le sol. *Soil Biology and Biochemistry* **10**, 557–9.
MILLER D., BROWN C. M., PEARSON T. H. & STANLEY S. O. (1979) Some biologically important low molecular weight organic acids in the sediments of Loch Eil. *Marine Biology* **50**, 375–83.
MYERS R. J. K. (1975) Temperature effects on ammonification and nitrification in a tropical soil. *Soil Biology and Biochemistry* **7**, 83–6.
MYERS R. J. K. & PAUL E. A. (1968) Nitrate ion electrode method for soil nitrate nitrogen determination. *Canadian Journal of Microbiology* **48**, 369–71.
NELSON D. W., MARTIN J. P. & ERVIN J. D. (1979) Decomposition of microbial cells and components in soil and their stabilization through complexing with model humic acid-type phenolic polymers. *Soil Science Society of America Journal* **43**, 84–8.
PARNAS H. (1975) Model for decomposition of organic material by microorganisms. *Soil Biology and Biochemistry* **7**, 161–9.
PARSON J. W. & TINSLEY J. (1975) Nitrogenous sub-

stances. In: *Soil Components* (Ed. J. E. Gieseking), vol. 6, pp. 263–304. Springer Verlag, New York.

PAULSON K. N. & KURTZ L. T. (1969) Locus of urease activity in soil. *Soil Science Society of America Proceedings* **33**, 897–901.

PETTIT N. M., SMITH A. R. J., FREEDMAN R. B. & BURNS R. G. (1976) Soil urease: activity, stability and kinetic properties. *Soil Biology and Biochemistry* **8**, 479–84.

RIXON A. J. & ZORIN M. (1978) Transformations of nitrogen and phosphorus in sheep faeces located in salt bush rangeland and on irrigated pasture. *Soil Biology and Biochemistry* **10**, 347–54.

ROBERGE M. R. & KNOWLES R. (1967) The ureolytic microflora in a black spruce (*Picea mariana* Mill) humus. *Soil Science Society of America Proceedings* **31**, 76–9.

SØRENSEN L. H. (1975) The influence of clay on the rate of decay of amino acid metabolites synthesised in soils during decomposition of cellulose. *Soil Biology and Biochemistry* **7**, 171–7.

SOWDEN F. J., GRIFFITH S. M. & SCHNITZER M. (1976) The distribution of nitrogen in some highly organic tropical volcanic soils. *Soil Biology and Biochemistry* **8**, 55–60.

SOWDEN F. T., CHENG Y. & SEHNITZER M. (1977) The nitrogen distribution in soils formed under widely differing climate conditions. *Geochimica et Cosmochimica Acta* **41**, 1524–6.

STANFORD G. (1978) Evaluation of ammonium release by alkaline permanganate extraction as an index of soil nitrogen availability. *Soil Science* **126**, 244–53.

STANFORD G. & SMITH S. J. (1972) Nitrogen mineralisation potentials in soils. *Soil Science Society of America Proceedings* **36**, 465–72.

STANLEY S. O. & BROWN C. M. (1976) Inorganic nitrogen metabolism in marine bacteria. The intracellular free amino acid pools of a marine pseudomonad. *Marine Biology* **38**, 101–9.

STANLEY S. O., PEARSON T. H. & BROWN C. M. (1978) Marine microbial ecosystems and the degradation of organic pollutants. In: *The Oil Industry and Microbial Ecosystems* (Eds K. W. A. Chater & H. J. Somerville), pp. 60–79. Heyden & Son, London.

STOTZKY G. (1980) Surface interactions between clay minerals and microbes, viruses, and soluble organics, and the probable importance of these interactions to the ecology of microbes in soil. In: *Microbial Adhesion to Surfaces* (Eds R. C. W. Berkeley, J. M. Lynch, J. Melling, P. R. Rutter & B. Vincent), pp. 231–47. Ellis Horwood, Chichester.

TABATABAI M. A. & BREMNER J. M. (1975) Decomposition of nitrilotriacetate (NTA) in soils. *Soil Biology and Biochemistry* **7**, 103–6.

VERMA L. & MARTIN J. P. (1976) Decomposition of algal cells and components and their stabilization through complexing with model humic-acid type phenolic polymers. *Soil Biology and Biochemistry* **8**, 85–90.

VERMA L., MARTIN J. P. & HAIDER K. (1975) Decomposition of ^{14}C-labelled proteins, peptides and amino acids; free and complexed with humic polymers. *Soil Science Society of America Proceedings* **39**, 279–84.

WAINWRIGHT M. & PUGH G. T. F. (1975) Changes in the free amino acid content of soil following treatment with fungicides. *Soil Biology and Biochemistry* **7**, 1–4.

WESTERMAN R. L. & KURTZ L. T. (1973) Pruning effect of ^{15}N-labelled fertilizers on soil nitrogen in field experiments. *Soil Science Society of America Proceedings* **37**, 725–73.

WESTERMAN R. L. & TUCKER T. C. (1974) Effects of salts and salts plus an ^{15}N-labelled ammonium chloride on mineralization of solid nitrogen nitrification and immobilization. *Soil Science Society of America Proceedings* **38**, 602–6.

WESTERMAN R. L. & TUCKER T. C. (1979) *In situ* transformations of ^{15}N-labelled materials in Sonoran Desert soils. *Soil Science Society of America Journal* **43**, 95–100.

WOJCIK-WOJTKOWIAK D. (1978) Nitrogen transformation in soil during humification of straw labelled with ^{15}N. *Plant and Soil* **49**, 49–55.

ZANTUA M. I. & BREMNER J. M. (1975) Comparison of methods of assaying urease activity in soils. *Soil Biology and Biochemistry* **7**, 291–5.

Chapter 10 · Microbial Succession During the Decomposition of Organic Matter

Michael J. Swift

10.1 Introduction

One of the main ecological roles of the heterotrophic microflora is the decomposition of organic matter. By this means the essential nutrient elements, such as nitrogen, phosphorus and sulphur, are transformed from organic to inorganic forms and become available once again to primary producers. Another ecologically essential concomitant of microbially mediated decomposition is the formation of recalcitrant organic molecules, such as humic substances, which contribute to the chemical stability of the ecosystem.

The decomposition of any unit of organic matter, whether it is a seaweed frond, a tree stump, a faecal pellet or a paper bag, is almost invariably due to the activity of a diverse community of microorganisms and invertebrate animals. A particular feature of the interrelationship of the member species of these communities is that they commonly exhibit a succession. This phenomenon has attracted the attention of microbial ecologists, probably because it is thought to provide insight into a number of ecological phenomena of fundamental interest. These include: interspecific competition and other microbe–microbe interactions; the physico-chemical environment as a regulator of distribution and activity; the influence of chemical modifiers, such as antibiotics, on microbial activity; the substrate relationships and capacities of microbial species; dispersal and colonisation patterns; and other processes. In practice, the holistic nature of the successional phenomenon, embracing as it does such a wide range of community determinants, commonly eludes all attempts to interpret the changes observed.

Thus there have been a large number of studies of microbial succession on a wide range of organic materials (Hudson 1968; Dickinson & Pugh 1974; Frankland 1981) which have generated nearly as many theories of the fundamental basis for the patterns involved. These are fully discussed in the reviews quoted above and relevant theoretical material can also be found in Margaleff (1968), Odum (1969) and Horn (1976). This is not the place to review the theoretical basis of microbial successions but it is essential to clarify the concepts somewhat. It is axiomatic to this volume that the understanding gleaned from any experiment is, to a considerable extent, determined by the limitations of the methods used. It is also clear that the choice of method and interpretation of results will depend on the conceptual framework within which the experiment is viewed. This makes it imperative that we define some terms and limitations to our concept of succession to avoid possible confusion. Indeed it is possible that the most important preparation for a study of succession lies not so much in pondering over the techniques available, for they are few, but rather in gaining a clear preconception of the nature of succession.

10.1.1 Types of Succession

We may distinguish between what may be termed ecosystem succession and decomposer succession with some of the main characteristics given in Table 1. Ecosystem succession is a familiar concept which is described and discussed in all textbooks of ecology. The successional pattern of the ecosystem is determined by the dynamics of the primary producers. Colonising plants grow and alter environmental conditions in such a way as to promote their replacement by successional species. The trend of succession is towards a climax which optimises primary production, species diversity and other ecosystem characteristics. A succession of heterotrophs also occurs which is primarily determined by the composition, diversity and productivity of the plant community. At the climax, ecosystem production (P) and respiration (R) are balanced ($P/R = 1$). This occurs whether the succession starts in conditions where P is in excess (autotrophic succession) or where R is greater than P (heterotrophic succession). Ecosystem succession of this type is within the purview of the microbiologist—for instance in studies of microalgal succession in lakes, streams, oceans or in experimental microcosms (Margaleff 1968; Gordon *et al.* 1969; Fogg 1977). This chapter, however, is concerned with decomposer succession whose characteristics are quite different (Table 1). There is a single input of energy to the system which therefore shows a continuous decline in energy content and (secondary) productivity, theoretically to a zero level. In practice decomposition always leaves some organic residue within the usual period of study of a microbial succession. Whilst generalisations have been made for ecosystem succession concerning such features of community structure as diversity and interaction these remain largely speculative for decomposer succession (Swift 1976; Frankland 1981).

10.1.2 Structure of Successional Studies

Succession, as most simply stated, can be defined as a change in the composition of the biological community with time. It is necessary to qualify this simple view of a sequence in a number of ways that enable the most appropriate choice of method to be made.

A varying picture of any microbial community is obtained when it is observed at different scales of resolution. This applies both to the spatial and the temporal intervals between sampling points. For

Table 1. Characteristics of different types of succession. Some features of early and late stages of succession in ecosystem and decomposer successions. List modified after Odum (1969) with those for decomposer succession based on the hypotheses of Swift (1976).

	Ecosystem succession		Decomposer succession	
Ecosystem attributes	Early stage	Late stage	Early stage	Late stage
Total organic matter	low	high	high	low
Mineral elements	mainly inorganic and extrabiotic	mainly organic and intrabiotic	mainly organic and immobilised	mainly inorganic and mineralised
Mineral cycles	open	closed	closed	open
Rate of nutrient flux—organism to environment	rapid	slow	slow	rapid
Nutrient conservation	low	high	high	low
Biochemical diversity	low	high	high	low
Species diversity	low	high	high	low
Internal symbiosis and homeostasis	low	high	high	low
Stability	low	high	high	low
Spatial heterogeneity of habitat	low	high	low	high
Niche specialisation	broad	narrow	narrow	broad
Size of organisms	small	large	large	small
Growth form	r-selected	k-selected	k-selected	r-selected
Life cycles	short, simple	long, complex	long, complex	short, simple

instance if the microflora on the leaf litter of a temperate woodland were analysed every summer, it is probable that no major yearly differences would be noted and no concept of succession developed. If the top layer of the litter was sampled every two or three months, however, a sequential change in the microflora would be observed; alternatively, sampling every year, but in different (annual) horizons would reveal a succession which was related but not identical (Fig. 1). Use of an inappropriate scale in space can also produce misleading sequences. Within any temperate wood, for instance, a sequence of fungal fruiting can be observed during the year. This is not strictly a succession, however, for the fungi may not be utilising the same food base nor in any way influencing one another. Thus it may be concluded that our definition of succession should include some reference to the change in composition of the community with time 'in the same site' or preferably 'on the same resource unit'. Resource here means the organic matter being decomposed (Swift 1976); for practical reasons, as argued below, the natural unit of the resource in question (a leaf, a twig, etc.) is usually the best scale for investigation.

Most studies of succession have been confined to single biological groups—fungi, bacteria, protozoa and other microorganisms. The decomposition of most resources is, however, due to the interaction of all these organisms plus invertebrate animals of a variety of phyla. The capacity to interpret successional changes in any one group may be limited by the information available on other groups—a point which has been all too commonly ignored, particularly by mycologists.

The classical view of autotrophic succession pictures one set of species replacing another which eventually becomes extinct. In microbial ecology the concept of extinction is less commonly utilised; what is usually described is the changes in relative activity or abundance of species which are commonly coexistent over a considerable time span. The succession is, therefore, one of change in dominance rather than absolute presence and/or absence. In some microorganisms—fungi, protozoa, spore-forming bacteria, actinomycetes—periods of dominance or suppression may coincide with physiomorphs, e.g. mycelium as opposed to resting spore. This may in fact represent an overlapping sequence of life cycles with organisms going through phases of dormancy (as spore) and vegetative growth (plus reproduction in some cases) followed by dormancy again (Fig. 2). The different types of successional patterns that these produce may be paralleled in ecosystem succession by what Horn (1976) has termed obligatory successions and competitive hierarchies.

Whichever pattern is demonstrated, and in many cases presumably both can be present, the concept that successional species will influence one another is intrinsic to successional studies, i.e. that the colonising species will create conditions conducive to the growth of a succeeding species, the growth of which may inhibit the further activity of its

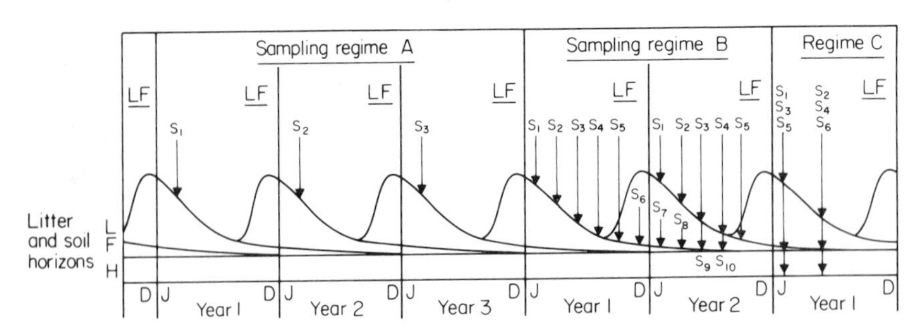

Figure 1. The importance of scale in the design of a sampling programme. The diagram shows three sampling programmes for leaf litter microorganisms which would result in differing conclusions concerning the pattern of community change with time. The programme is imagined as occurring in a forest ecosystem in which the leaf litter input occurs at a single time during the year (LF). The arrows indicate the dates and sites at which sampling for leaf litter microorganisms is carried out. (a) Annual sampling of surface litter three months after litter fall—community composition constant. (b) Sampling of litter layer at two-month intervals following litter fall—reveals a successional change in the leaf litter microflora community. Note that after litter fall in year 2 sampling may be repeated on new litter or may be continued on the same, but buried, layer. (c) Single sampling of different litter and soil horizons three months after litter fall reveals some aspects of the same succession. Further details in text.

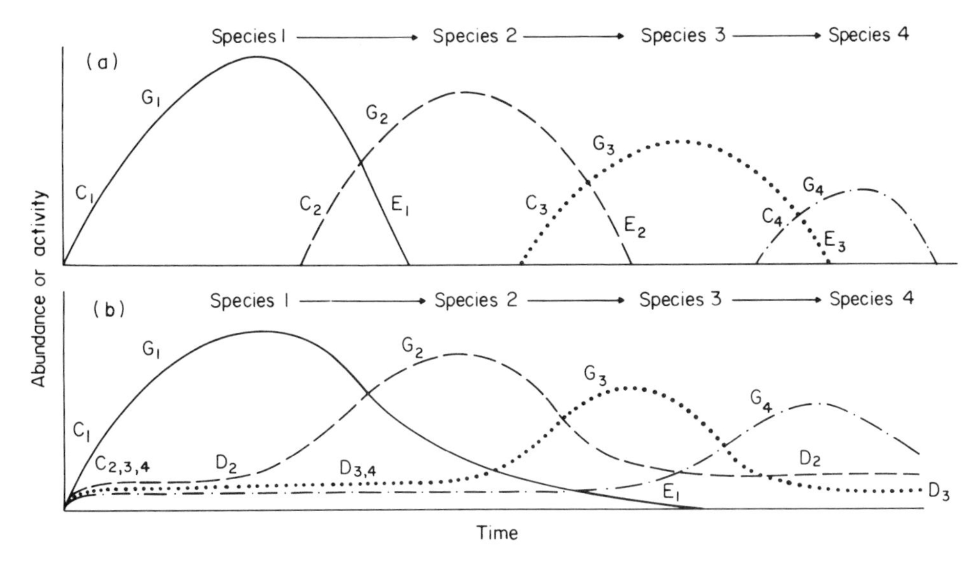

Figure 2. Possible patterns of succession. (a) Succession by species replacement and extinction (obligatory succession—Horn 1976). (b) Succession by change in species dominance (competitive hierarchy—Horn 1976). (C = colonisation phase; G = phase of active growth; E = extinction; D = dormancy; Species 1 = ———; Species 2 = — — —; Species 3 = · · · ·; Species 4 = —·—·—). For further details see text.

predecessor in some way. Thus studies of succession must utilise techniques for determining changes in the chemical and physical state of the resource as well as in the composition and activity of the decomposer community.

Experimental

10.2 Introduction

The study of microbial succession involves little in the way of specialised techniques—the main methods employed are those of observation, isolation and identification described in earlier chapters. What is important, however, is the experimental design of any study. The following account, therefore, concentrates on matters that determine the nature of a successful experimental plan.

We may recognise four distinct problems of method in the study of succession:

(1) the selection and placing of the resource in the environment of study;

(2) the determination of change in the resource after recovery;

(3) the determination of change in the microbial community inhabiting the resource;

(4) the elucidation of the factors regulating the observed changes.

Each of these aspects is dealt with in separate sections below but it should be realised that the choices of method made for the solution of one problem may influence and often constrain the choices relating to secondary problems.

10.3 Choice and placement of resource

The essence of succession studies is to recover and characterise at time T_n a resource unit which was labelled and characterised at time T_0. An immediate problem is to ensure that the initial labelling and characterisation is carried out with minimal disturbance of the resource or the environment in which it is placed. A very wide range of resources has been commonly employed in studies of succession and the choice is usually determined by the overall objectives of the investigation in hand. It is useful, however, to note some aspects of the choice of resource which may affect the results obtained. A distinction may be made, for instance, between the resources employed for study of natural successions and experimental successions. This distinction relates to the aim of the study: the objective in studying natural successions is to follow as accurately as possible a succession which occurs as a commonplace process within the natural environment whereas an experimental succession study is one which either employs a manufactured or treated resource (e.g. purified cellulose or pesticide impregnated leaves) or places a natural resource in an environment that it would not normally occupy (e.g. cut timber placed in contact with the soil).

Among the natural resources that have been studied are a wide range of plant litters (leaves, twigs, branches, roots, reproductive structures—Hudson 1968; Dickinson & Pugh 1974; Swift 1977; Hayes 1979). Much more limited attention has been given to animals, but faeces (Webster 1970; Lodha 1974; Wicklow 1981), feathers (Pugh & Evans 1970) and hair (Griffin 1960) have been studied as well as carcasses (Putman 1978). Semi-natural resources include compost (Chang & Hudson 1967) and sewage (Sykes & Skinner 1971; La Riviere 1977) and various foodstuffs (Skinner & Carr 1976). Experimental resources include cut timber in contact with the ground (Butcher 1968; Käärik 1974) and in marine and freshwater environments (Jones 1974), cellulose (Tribe 1966), chitin (Okafor 1966) and lignin (Jones & Farmer 1967).

10.3.1 Unit Character of Resource

The character of the resource is essentially a sampling problem and any sample taken at a particular time must be replicated. The unit of replication may affect the results obtained. A common method of replication is to take standard weights or sizes of resource (e.g. in studies of leaf litter decomposition, litter bags have standard initial weights of either entire leaves or of standard diameter discs cut from leaves) but vital information may be lost by this approach. All resources have natural units (e.g. individual leaves, twigs, faecal pellets, carcasses). The microbial community of one unit may differ from that of another and this variation may be due to natural features of variation in the units or to opportunistic variation in the microbial colonisation of the units. The former variation embraces such factors as the difference between sun and shade leaves; differences in leaves shed at different times of the year; the effect of the diameter of a twig or branch on its rate of decay; and differences in composition of faecal pellets related to the feed or physiological state of the producing animal. Opportunistic variation defines the effects of differences in time and space of the population structure of colonising propagules which may result in identical resources receiving different colonising microfloras. Where detailed insight is required into the structure of the microbial community and the factors determining it the natural unit should provide the basis of replication. Furthermore, either natural or arbitrary limits may be set to the variation, for instance, by defining diameter classes in studying twig, branch or root decay (Swift *et al.* 1976).

Standardisation of unit size may be experimentally desirable. Where this is done it should be by the selection of approximately equal units (e.g. whole leaves) from a natural population rather than by cutting leaf discs or stem lengths of equal size as has been done in a number of studies. Cutting results in freshly exposed and damaged internal cells which lose nutrients and may have selective effects on the microflora.

For many experimental resources natural units do not exist. In these cases the size and nature of the unit should be chosen with care; for instance in studying the colonisation of wood blocks by freshwater fungi, Sanders and Anderson (1979) showed that the species diversity of the fungal communities was related to the size of the block. A number of purified resources can only be obtained in powdered or particulate form (e.g. chitin or lignin). Presentation to the microflora may in these cases involve suspension in some form of carrier material such as kaolin (Jones & Webley 1968). Care must obviously be taken to ensure that the carrier does not have any selective effect on the microflora. Some resources may be available in alternative forms in terms of their chemical composition or other characteristics (e.g. cellulose may be in the form of textile fibre, paper, cellophane and other forms, all differing in degree of crystallinity and polymerisation). Again the choice must be made with care and always bearing in mind the objectives of the study.

Further information on aspects of choice in relation to specific resources can be obtained from the general references quoted earlier. A general discussion of the way in which the characteristics of a resource (the resource quality) affects the rate of decomposition and the nature of its occupying microflora is given by Swift *et al.* (1979). The unit community concept is discussed by Swift (1976) and the problems associated with particulate organic matter are discussed by Fenchel and Jørgensen (1977).

10.3.2 Initiation and Termination

One of the most difficult aspects to elucidate when studying natural successions is when and by what microbes succession is initiated. In natural successions the actual line of initiation may be a matter of definition by the experimenter; thus all living plant parts have a natural surface microflora (e.g. Parkinson 1967; Dickinson & Preece 1976). Members of the surface microflora are often amongst the first organisms to be detected within the leaf or root during early stages of decomposition. Thus the problem arises of distinguishing between organisms which are incidental and superficial and those which are components of an inhabitant decomposer microflora. The same problem stated another way is that of determining when decomposition starts. Leaf death and decomposition is commonly initiated by pathogens (Hudson 1968) and these may, therefore, be defined as primary colonisers and initiators of succession. In other primary resources, however (twigs and fine roots, for instance), death may be more closely related to physiological variations within the plant itself and thus precede microbial colonisation.

In other resources the problem is somewhat different. Faecal pellets have a characteristic flora immediately after defecation which is quickly succeeded by other microbes. It is also preceded by a succession of microbial communities during passage of the food material through the gut of the animal. The decomposition of plant litter in fresh or marine water is brought about by a microflora which is quite different from that of the terrestrial organisms which may colonise the litter prior to its deposition in water (Jones 1974; Willoughby 1974). In all these cases the investigator must determine what the significant initiation point should be on the basis of the objectives of the investigation concerned. Techniques for the collection of plant litter (litter trapping) are reviewed by Newbould (1969), Chapman (1976) and Swift *et al.* (1979). Techniques for the collection of other resources, such as faecal pellets, can be obtained from the specific references given earlier. In experimental successions these problems are avoided—the experimental resource is generally freed of microbes and succession obviously starts at the time when the resource is placed in contact with a potentially colonising microflora.

Defining the end-point of a succession may be even more difficult. The decomposition of different resources tends to converge in the terminal stages. Thus leaves, twigs, roots, microbial tissues, carcasses and faeces all contribute terminal residues to soil which may be considered together as humus. A problem may arise in distinguishing a specific resource and its inhabitant microflora from its surrounds in the terminal stages of decay. There may be some value in doing so, however, as the microbial and chemical characteristics of a humus particle may be different according to the primary or secondary resource from which it is immediately derived. Very little consideration has been given to this aspect of succession and decomposition.

10.3.3 Environments

Any attempt to simulate or follow a natural succession must ensure that the environment in which the succession takes place is the natural habitat of the process. The environment is important not only because of its physical character—temperature, humidity, light and other factors—but also its biotic content—the population of propagules of the potential colonising microflora. Thus, for instance, the initial stages of succession and decomposition of plant matter take place while the resources are still attached to the plant. Both leaves and branches are colonised before litter-fall; in the case of the latter up to 50% loss in weight may have occurred by this time. Studies in which freshly plucked leaves or branches are placed on the soil surfaces must, therefore, be regarded as experimental rather than natural. The corollary of this is that a dramatic change in environment, such as occurs at litter fall, dispersal of fruits, deposit of faecal pellets, submergence of aquatic detritus, or burial of carcasses, is an important event within any succession and merits particular study. These features, however, have been commonly ignored.

Information on the relationship of environmental factors to the patterns of succession may be obtained by the use of microcosms for successional studies. The best examples of this are those of autotrophic and heterotrophic ecosystem successions in freshwaters (Gordon *et al.* 1969; Cook 1977). Soil microcosms have also been used by a number of workers although this has been less common in recent years (Waksman & Cordon 1939; Dix 1964). Information on methods of manipulating soil physical conditions can be obtained from the review by Griffin (1972) and, for aquatic systems, from that by Golterman *et al.* (1978).

Although experimental successions by their very nature often employ unnatural combinations of resource and environment, care should be taken in the placement of the resource. In experiments on cellulose decomposition in soil, for instance, cellulose strips may be placed either vertically or horizontally in the soil profile depending on the question being investigated.

10.3.4 Labelling

The commonest way of labelling resource units or samples at time T_0 so that they may be recovered at time T_n is by confining them in a mesh bag or cage—a procedure usually termed the litter bag technique. Meshes of varying size may be employed: finer meshes have the advantage of confining litter so that little is lost even when it breaks up into relatively small particles in the terminal stages of decay; on the other hand, fine meshes prevent the entry of larger soil organisms. Swift *et al.* (1979) categorised the decomposer biota in terms of body diameter thus: macrofauna (earthworms, amphipods, millipedes, etc.) >2 mm; mesofauna (dipteran larvae, termites, collembola, acari, etc.) > 100 μm; microfauna and microflora (protozoa, fungi, bacteria) <100 μm. The choice of appropriate mesh size can thus act as a selective agent for the biota.

Despite the very common use of the differential mesh exclusion technique in decomposition studies, it has not been exploited at all in studies of microbial succession. However, this would be a worthwhile approach for it would enable exploration of the effects of different animal groups on the patterns of microbial succession, a factor likely to be important. It is clear from recent studies that grazing by soil animals can alter the composition and activity of the decomposer microflora (Parkinson *et al.* 1978; Hanlon & Anderson 1980; Lussenhop *et al.* 1980). Similar effects may also occur in aquatic systems (Barlocher 1981).

The mesh size of an enclosure also affects the internal environment (e.g. fine mesh bags tend to retain moisture levels beyond that of unconfined litter). A comparison of the decay rate of confined and unconfined leaf litter by Witkamp and Olson (1963) showed that unconfined oak leaves decayed nearly twice as fast as leaves in a mesh bag that purportedly enabled the access of decomposers of all sizes. However, in a similar experiment Woodwell and Marples (1968) found only a small difference of rate in favour of the unconfined litter. The litter bag technique has been discussed by Gilbert and Bocock (1962), Chapman (1976) and Swift *et al.* (1979).

Larger resources, such as branches, twigs or roots, may be labelled by tethering them by a length of nylon thread to a marker post set in the soil. A similar approach has been used for leaf litter by some workers because of the environmental drawbacks of litter bags mentioned above. Variations in the enclosure technique have been used for other resources such as faeces or carcasses and in environments other than soil (see general references cited above).

10.4 Change in the state of the resource

Information relevant to the interpretation of a microbial succession may be obtained from the assessment of changes in the physical condition and chemical composition of the resource with time. In particular the rate of decomposition of the resource is a valuable parameter. Information related to chemical change may be particularly significant if a distinction can be made between the constituents of the resource and of the microbial biomass.

10.4.1 Establishing the Time Scale

The choice of the correct time interval for sampling may be of crucial importance for the type of microbial pattern revealed, as the comments in section 10.1 and Fig. 1 have explained. The discussion of initiation and termination given in the previous section is also relevant.

In some types of successional study it is possible to observe the same set of resource units at successive time intervals (Fig. 3a). An example of this is the observation of the sequence of fungal fruiting on dung (Harper & Webster 1964) or other resources. In most cases, however, the sampling techniques require that the resource unit be destroyed at the time of sampling (see below 10.5, section 4). Thus a replicated programme allowing for destructive sampling must be instituted. In most cases samples can be placed at T_0 and sampled at regular intervals (T_1, T_2, T_3, ... T_n) thereafter (Fig. 3b). In some cases, however, this is not practicable: for instance, the decomposition of wood may take five, ten or even more years from litter fall. An alternative to an experiment lasting

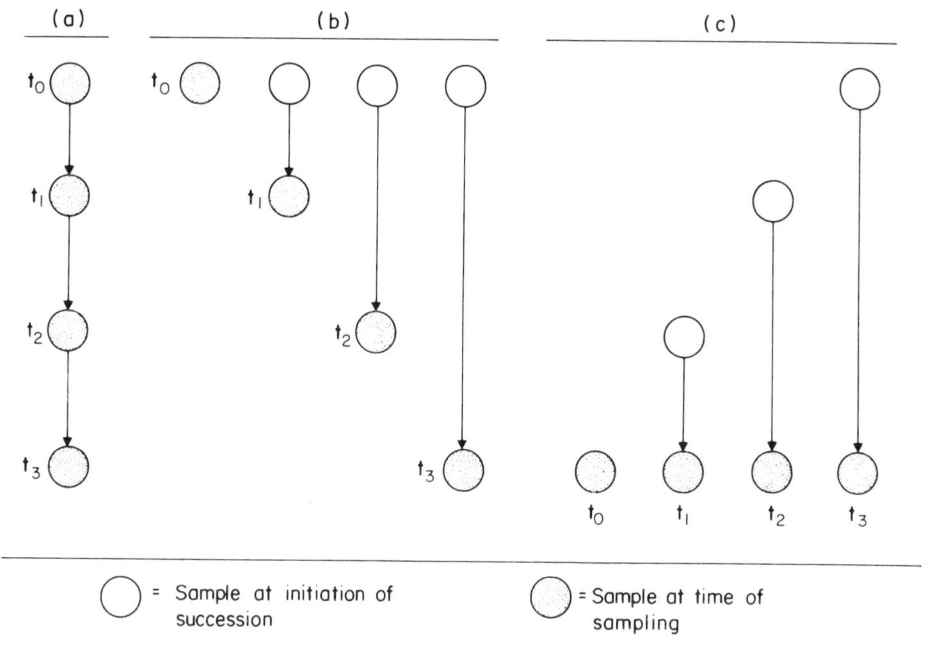

Figure 3. Different designs of sampling programme. (a) Sequential non-destructive sampling of the same sample of resources. (b) Sequential destructive sampling of replicated resource samples. (c) Contemporaneous destructive sampling of replicated resource samples in which succession was initiated at different times. In all cases replication of the sampling units within samples is not shown in this diagram.

this number of years is to obtain resources in different stages of decomposition (i.e. at different times from the start of decay) and analyse them contemporaneously (Fig. 3c). A similar approach may be used with leaf litter by sampling different horizons of litter and soil which represent litter inputs of successive years or seasons (see Fig. 1c—Visser & Parkinson 1975). In these instances it should be noted that an exact time base can rarely be established although an indirect estimate of the extent of decomposition may be possible (see next section).

10.4.2 Change in Physical State

Generally the most useful measurement of the change in physical state is to estimate the change in mass of the resource and thence the rate of decomposition. Two important problems associated with this are that of assessing the way(s) in which weight loss has occurred and that of separating the weight of resource from the biomass of the inhabitant microflora.

Weight may be lost from a decomposing resource by chemical transformation (catabolism) or by loss of chemically intact fragments (export). The latter is generally a physical process which is brought about by animal feeding activity. Animals either directly export the material within their intestines or their feeding produces material of small particle size (comminution) which then 'falls out' of the resource and its enclosure. As the microbiologist is usually more concerned with the direct catabolic effect of organisms on the resource it is useful to distinguish export losses from catabolism.

One approach is to ensure that the latter do not occur by confining the resource in a fine mesh enclosure, but this inevitably introduces both environmental and biological bias (section 10.3.4). If animals are allowed access to the resource, an approximate estimate of their activity can be made by assessing the change in volume of the resource. Thus tunnelling in wood can be measured by estimating the proportional area affected in successive cross-sections of a piece of wood. Then the volume of wood removed (T) can be calculated from:

$$T = \bar{A}.l$$

where $\bar{A}$ is the mean of the total area of tunnels

exposed in a series of cross-sections and l is the length of branch affected by the animals. Estimation of the total volume (V) and weight of the branch (m) gives the weight per unit volume of wood remaining, the relative density, d where:

$$d = m/(V\text{–T}).$$

By estimating d at successive time intervals the change in weight due to microbial decay (catabolism) can be estimated. A similar estimate can be made for planar resources, such as leaves or strips of cellulose, by using weight per unit area to allow for pieces removed by animal feeding. It must be recognised, however, that these estimates are approximate: firstly, not all animal feeding is measurable by these means, e.g. in leaves in addition to the pieces bitten out of the whole leaf and thus measurable as loss in area, there will be surface erosion which is not measurable; secondly, losses in weight per unit volume may occur by leaching as well as by catabolism. This, of course, is particularly important in aquatic environments. More detailed discussion of the relative importance of catabolism, comminution and leaching is given by Swift *et al.* (1979).

The distribution between microbial biomass and resource mass is also a technically difficult problem. The methods for estimating microbial biomass are reviewed in Chapter 6. The choice of method is to some extent determined by the resource type and organisms being studied but it should be noted that for this particular purpose general, non-discriminatory methods, such as ATP or hexosamine determinations, may well be suitable as the main purpose is to distinguish between microbe and substrate biomass. It will only generally be thought necessary to do this if detailed information on decomposition activity, particularly chemical change is required.

Care should be taken for possible weight inaccuracies due to the accumulation in the resource of extraneous material, such as mineral particles from soil. These should be brushed off before weighing. Washing off the material should be avoided as weight losses due to leaching can occur.

The choice of temperature for drying will be governed by the type of resource under study. Whilst high temperatures (100°C) are probably satisfactory for plant material, lower temperatures are necessary for soil and animal samples because of the presence of volatile materials. In some cases freeze-drying may be preferable. Allen *et al.* (1974, 1976) give advice in relation to specific materials.

In cases where non-destructive sampling is to be employed, indirect estimates of weight may be made by sacrificing replicates or sub-samples to estimate moisture contents. In cases where the time scale is not fixed estimates of the state of decay may be made by measuring the relative density of whole resources (Healey & Swift 1971; Swift *et al.* 1976).

10.4.3 Change in Chemical Composition

Chemical changes in the resource have commonly been invoked as important factors in determining the successional change in the microbial community. In particular the major components of the plant cell wall (cellulose, hemicellulose, lignin, etc.) have been suggested as regulators of activity in terms of the correlation of the patterns of chemical change with the apparent enzymic capacities of the successive groups of inhabitant fungi (Garrett 1963). Other authors (Hudson 1968; Swift 1976; Frankland 1981) have pointed out that these nutritional correlations are inadequate to explain the observed patterns of succession. Nonetheless, information on the patterns of chemical change in the resource are essential to the interpretation of the sequence of activity in the microbial community.

Methods for the analysis of both organic and inorganic constituents of plant tissues are given by Allen *et al.* (1974, 1976). Again the value of distinguishing microbial from resource components is emphasised (see Chapter 6, for a starting point for relevant references in this respect). Highly detailed analyses of cell wall constituents may now be made by use of sophisticated separative techniques (Keegstra *et al.* 1973; Montgomery 1982) and use may be made of these under some circumstances.

10.4.4 Calculation of Decomposition Rates

Chapman (1976) and Swift *et al.* (1979) both discuss the question of the appropriate form in which to summarise a decomposition rate. The most convenient and commonly utilised form is as a negative exponential usually calculated from the fraction of weight remaining and is given by:

$$k = \frac{\log_e (W_T/W_o)}{T - T_o}$$

where k = decomposition rate; W_T and W_o are the mass of litter at the given time (T) and the initial time (T_o) respectively. Despite the general usefulness of this form of curve it should always be remembered that more insight into the pattern of resource change and to the biological activity that brings it about, may come from plotting the basic data as a decay curve than from summarising it as a single mathematical expression. The deviation from a standard format may be more instructive than agreement.

10.5 Community change

According to the strictest definition, succession takes place when species B is found to occupy a spatial location previously inhabited by species A but in which the presence of A can no longer be detected. In practice a change in relative abundance and/or activity of the two species in the same general habitat (i.e. on the same resource) is taken as evidence of succession (see Section 10.1). Thus an essential aspect of successional investigation is the identification of microbial species in microhabitats but methods for measuring abundance and activity are also pertinent. The secret of successful study and interpretation lies in the selection of the appropriate methods from the wide choice available. Detailed information on the methods and the criteria on which to base a choice are given elsewhere in this book (see Chapters 1 to 4) and will not be discussed here; a few fundamental points are, however, worth reiterating.

In most cases the only way of identifying the presence of a microbial species in a given habitat is by its isolation and culture under defined conditions. This is usually done by plating fragments of the resource onto an agar medium either with or without the use of pre-treatments, such as serial washing (see Chapter 2). As emphasised elsewhere in this book all such methods are necessarily selective. Choice of any particular isolation technique or medium may well run the risk of a failure to detect significant components of the microbial community. Thus a combination of media is generally desirable. Furthermore, isolation methods will fail to distinguish between active cells or hyphae and those which are dormant. Indeed, in the fungi there is a distinct risk that a species present in the form of dormant propagules, such as chlamydospores, may be preferentially isolated or enumerated on agar plates over a species present as active growing mycelium. Such methods commonly detect the past history of the resource rather than present activity.

Isolation techniques should be supplemented by direct observation. Microbial communities may be viewed under either the light or the electron microscope either directly or after some preparative manipulation, such as the sectioning of the resource (Jones & Griffiths 1964; Anderson 1978) or mounting it in some appropriate form. A range of techniques for the direct observation of soil microorganisms are reviewed by Parkinson *et al.* (1971). Aquatic organisms may be viewed simply in suspension or particle surfaces may be scanned (Meadows 1971) after appropriate sampling. Macroscopic fruiting in terrestrial fungi can be promoted by placing the resource in damp chambers and by subjecting them to a variety of environmental stimuli (Manachère 1980). Jones (1971) has described methods of promoting the sporulation and fructification of aquatic fungi.

Direct observation techniques suffer in general from the difficulties of positive identification: the cells or hyphae of different species may be indistinguishable from one another. On the other hand physiologically different forms of the same organism (e.g. vegetative cells or hyphae compared with dormant spores) may be detected. The use of vital stains in such preparations may give additional information on the viability and activity of microbial cells or hyphae (Babiuk & Paul 1970). Reproductive structures, particularly in the myxomycetes or true fungi, are identifiable and many detailed successions have been described on the basis of sequences in fruiting. As Harper and Webster (1964) pointed out, however, fruiting is only one of many activities in the life of a fungus and may be quite misleading in terms of presence or absence or indeed dominance.

10.6 Mechanisms of community change

Jacques Monod said that 'the study of the growth of bacterial cultures does not constitute a specialised subject or branch of research (but) covers actually our whole discipline' (Monod 1949). The same could be said of the relation between the 'subject' of succession and microbial ecology as a whole. As emphasised earlier, the progress of succession subsumes a very wide range of ecological phenomena. The main value of the study of succession then rests in providing a structure for the investigation of these phenomena—competition, colonisation strategies, niche determination, environmental influences on these processes and many other facets. Conversely, the interpretation of a pattern of succession is dependent on isolating and identifying the individual factors which determine the observed changes in community composition and structure.

The usual approach to resolving these questions is to perform controlled experiments on individual species, or pairs or groups of species, selected from within the successional communities. The experiments are best performed in controlled environments simulating the conditions within which the successional changes take place and utilising resources or substrates which are chemically related to the various stages of resource change that occur within the natural succession. It would be impracticable and futile to list all the different types of experiment that have been carried out to resolve successional phenomena. Thus the references below merely seek to provide a broad cover of some of the more significant factors that have been investigated in the last twenty years or so.

The success of any organism in taking its place in a succession will depend both on its capacity and opportunity. The capacity of an organism is based in its genotype; the opportunity arises when that genetic potential is given phenotypic expression. The opportunity depends on the occurrence of an environment conducive to success and on the past history of the organism in the sense that it must be able to deploy sufficient intrinsic resources (e.g. colonising propagules, food supply and other factors). The spectrum of intrinsic factors which contribute to this success have been analysed by Garrett (1956, 1963) and summed up in two concepts 'competitive saprophytic ability' and 'inoculum potential' (see also Frankland 1981 for a good discussion of these). Garrett's scheme provided a great stimulus for research and many of the experiments mentioned below deal with one or more of the characteristics which he thought important. Garrett summarised a number of the studies carried out by himself and co-workers on the colonisation of wheat straw buried in soil (Garrett 1956, 1970).

Perhaps the most exemplary experimental study is that of Harper and Webster (1964) on the succession of fungi fruiting on rabbit dung. They studied mycelial growth rates, spore germination and time taken to fruit, in relation to the competitive influence both of the fungi and bacteria. They were able to demonstrate the complexity of the interactions involved in successional phenomena as well as discovering a new form of antagonism between hyphae. Growth rates and antibiotic effects were postulated by Garrett as two of the main features of competitive saprophytic ability and several studies have followed this up. Wastie (1961) and Gibbs (1967), for example, showed the possible importance of a toxic environment on succession in pine logs. The enzymic capacity of organisms is also an important feature of their ability to occupy a particular resource at a particular time; many authors have explored this with laboratory tests (Flanagan & Scarborough 1974) or inoculations into litter and subsequent analysis (Hering 1967), or a combination of the two (Frankland 1969). Environmental aspects have less commonly been considered: Hogg (1966) transferred leaves between different soil enviroments to investigate the effect on the pattern of succession and also measured spore viability in the soil.

The 'opportunistic' feature of succession is more difficult to experiment with but the importance of seasonality in relation to the air spora has been demonstrated by both Meredith (1959) and Dowding (1969) for somewhat different woody resources. The ideas of Garrett (1970) on the inoculum potential of propagule populations could be usefully incorporated into successional experiments of a similar type to those involving plant pathogens (Wastie 1962; Chou & Preece 1968). Experiments into inter-specific competition have been carried out by inoculating pairs or groups of species into natural habitats or resources (Harper & Webster 1964; Rayner 1978).

The examples above are concerned largely with fungi. The manipulation of bacteria and protozoa in pure culture is aided by the possibility of utilising continuous as well as batch culture systems (Veldkamp 1977; see also Chapter 20). It should be noted, however, that the choice of continuous culture versus batch culture should be made with care, bearing in mind the type of system being simulated (see Chapter 1).

This is an incomplete and selective list of the types of experiments that have been carried out to assist the interpretation of successional phenomena. Reference to the papers mentioned should, however, help to clarify possible approaches to the problems of method that are encountered. Nonetheless it should be said that there has been a sad lack of ingenuity in such experiments in the last fifteen years and the time is ripe for fresh approaches.

Recommended reading

DICKINSON C. H. & PUGH G. J. F. (1974) *Biology of Plant Litter Decomposition*. Academic Press, London.

FENCHEL T. M. & JØRGENSEN B. B. (1977) Detritus food chains of aquatic ecosystems: the role of bacteria. *Advances in Microbial Ecology* **1**, 1–58.

FRANKLAND J. C. (1981) Mechanisms in fungal succession. In: *The Fungal Community: Its Organisation and Role in the Ecosystem* (Ed. D. T. Wicklow & G. C. Carroll), pp. 403–26. Marcel Dekker, New York.

GARRETT S. D. (1956) *Biology of the Root-infecting Fungi*. Cambridge University Press, Cambridge.

GRIFFIN D. M. (1972) *Ecology of Soil Fungi*. Chapman and Hall, London.

HARPER J. E. & WEBSTER J. (1964) An experimental analysis of the coprophilous fungal succession. *Transactions of the British Mycological Society* **47**, 511–30.

HAYES A. J. (1979) The microbiology of plant litter decomposition. *Science Progress, Oxford* **66**, 25–42.

HORN H. S. (1976) Succession. In: *Theoretical Ecology: Principles and Applications* (Ed. R. M. May), 1st edn, pp. 187–204. Blackwell Scientific Publications, Oxford.

HUDSON H. J. (1968) The ecology of fungi on plant remains above the soil. *New Phytologist* **67**, 837–74.

SWIFT M. J. (1976) Species diversity and the structure of microbial communities. In: *The Role of Aquatic and Terrestrial Organisms in Decomposition Processes* (Eds J. M. Anderson & A. MacFadyen), pp. 185–222. Blackwell Scientific Publications, Oxford.

WICKLOW D. T. (1981) The coprophilous fungal community: a mycological community for examining ecological ideas. In: *The Fungal Community: its Organisation and Role in the Ecosystem* (Eds D. T. Wicklow and G. C. Carroll), pp. 47–76. Marcel Dekker, New York.

References

ALLEN S. E., GRIMSHAW H. M., PARKINSON J. A. & QUARMBY C. (1974) *Chemical Analysis of Ecological Materials*. Blackwell Scientific Publications, Oxford.

ALLEN S. E., GRIMSHAW H. M., PARKINSON J. A., QUARMBY C. & ROBERTS J. D. (1976) Chemical analysis. In: *Methods in Plant Ecology* (Ed. S. B. Chapman), pp. 411–66. Blackwell Scientific Publications, Oxford.

ANDERSON J. M. (1978) The preparation of gelatine—embedded soil and litter sections and their application to some soil ecological studies. *Journal of Biological Education* **12**, 82–8.

BABIUK L. A. & PAUL E. A. (1970) The use of fluorescein isothiocyanate in the determination of the bacterial biomass of grassland soil. *Canadian Journal of Microbiology* **16**, 57–62.

BARLOCHER F. (1981) Fungi on the food and in the faeces of *Gammarus pulex*. *Transactions of the British Mycological Society* **76**, 160–5.

BUTCHER J. A. (1968) The ecology of fungi infecting untreated sapwood of *Pinus radiata*. *Canadian Journal of Botany* **46**, 1577–89.

CHANG Y. & HUDSON H. (1967) The fungi of wheat straw compost I. Ecological studies. *Transactions of the British Mycological Society* **50**, 649–66.

CHAPMAN S. B. (1976) *Methods in Plant Ecology*. Blackwell Scientific Publications, Oxford.

CHOU M. C. & PREECE T. F. (1968) The effect of pollen grains on infections caused by *Botrytis cinerea* Fr. *Annals of Applied Biology* **62**, 11–22.

COOK G. D. (1977) Experimental aquatic laboratory ecosystems and communities. In: *Aquatic Microbial Communities* (Ed. J. Cairns), pp. 59–104. Garland Publishing Inc., London.

DICKINSON C. H. & PREECE T. F. (1976) *Microbiology of Aerial Plant Surfaces*. Academic Press, London.

DICKINSON C. H. & PUGH G. J. F. (1974) *Biology of Plant Litter Decomposition*. Academic Press, London.

DIX N. J. (1964) Colonisation and decay of bean roots. *Transactions of the British Mycological Society* **47**, 285–92.

DOWDING P. (1969) The dispersal and survival of spores of fungi causing bluestain in pine. *Transactions of the British Mycological Society* **52**, 125–37.

FENCHEL T. M. & JØRGENSEN B. B. (1977) Detritus food chains of aquatic ecosystems: the role of bacteria. *Advances in Microbial Ecology* **1**, 1–58.

FLANAGAN P. W. & SCARBOROUGH A. M. (1974) Physiological groups of decomposer fungi on Tundra plant remains. In: *Soil Organisms and Decomposition in Tundra* (Ed. A. J. Holding, O. W. Heal, S. F. MacLean Jnr. & P. W. Flanagan), pp. 159–81. Tundra Biome Steering Committee, Stockholm.

FOGG G. E. (1977) Physiology under the worst possible conditions. *Journal of the Institute of Biology* **24**, 73–9.

FRANKLAND J. C. (1969) Fungal decomposition of bracken petioles. *Journal of Ecology* **57**, 25–36.

FRANKLAND J. C. (1981) Mechanisms in fungal succession. In: *The Fungal Community: its Organisation and Role in the Ecosystem* (Eds. D. T. Wicklow & G. C. Carroll), pp. 403–76. Marcel Dekker, New York.

GARRETT S. D. (1956) *Biology of the Root-infecting Fungi*. Cambridge University Press, Cambridge.

GARRETT S. D. (1963) *Soil Fungi and Soil Fertility*. Pergamon Press, Oxford.

GARRETT S. D. (1970) *Pathogenic Root-infecting Fungi*. Cambridge University Press, Cambridge.

GIBBS J. N. (1967) The role of host vigour in the susceptibility of pines to *Fomes annosus*. *Annals of Botany* **31**, 803–15.

GILBERT D. J. & BOCOCK K. L. (1962) Some methods of studying the disappearance and decomposition of leaf litter. In: *Progress in Soil Zoology* (Ed. P. W. Murphy), pp. 348–52. Butterworths, London.

GOLTERMAN H. L., CLYMO R. S. & OHNSTAD M. A. M. (1978) *Methods for Physical and Chemical Analysis of Freshwaters*. IBP Handbook No. 8 2nd edn. Blackwell Scientific Publications, Oxford.

GORDON R. W., BEYERS R. J., ODUM E. P. & EAGON R. G. (1969) Studies of a simple laboratory microecosystem: bacterial activities in a heterotrophic succession. *Ecology* **50**, 86–100.

GRIFFIN D. M. (1960) Fungal colonisation of sterile hair in contact with soil. *Transactions of the British Mycological Society* **43**, 583–96.

GRIFFIN D. M. (1972) *Ecology of Soil Fungi*. Chapman and Hall, London.

HANLON R. D. G. & ANDERSON J. M. (1980) Influence of macroarthropod feeding activities on microflora in decomposing oak leaves. *Soil Biology and Biochemistry* **12**, 255–61.

HARPER J. E. & WEBSTER J. (1964) An experimental analysis of the coprophilous fungal succession. *Transactions of the British Mycological Society* **47**, 511–30.

HAYES A. J. (1979) The microbiology of plant litter decomposition. *Science Progress, Oxford* **66**, 25–42.

HEALEY I. N. & SWIFT M. J. (1971) Aspects of the accumulation and decomposition of wood in the litter of a coppiced Beech-Oak woodland. In: *Organisms du Sol et Production Primaire*, C. R. IV College Pedobiologique Institute Zoological Society, pp. 417–30. INRA, Paris.

HERING T. F. (1967) Fungal decomposition of oak leaf litter. *Transactions of the British Mycological Society* **50**, 267–73.

HOGG B. M. (1966) Micro-fungi on leaves of *Fagus sylvatica*. 2. Duration of survival spore viability & cellulolytic activity. *Transactions of the British Mycological Society* **49**, 193–205.

HORN H. S. (1976) Succession. In: *Theoretical Ecology*: Principles and Applications (Ed. R. M. May), 1st edn, pp. 187–204. Blackwell Scientific Publications, Oxford.

HUDSON H. J. (1968) The ecology of fungi on plant remains above the soil. *New Phytologist* **67**, 837–74.

JONES D. & FARMER V. C. (1967) The ecology and physiology of soil fungi involved in the degradation of lignin and related compounds. *Journal of Soil Science* **18**, 74–84.

JONES D. & GRIFFITHS E. (1964) The use of thin soil sections for the study of soil micro-organisms. *Plant and Soil* **20**, 232–40.

JONES D. & WEBLEY D. M. (1968) A new enrichment technique for studying lysis of fungal cell walls in soil. *Plant and Soil* **28**, 147–57.

JONES E. B. G. (1971) Aquatic fungi. In: *Methods in Microbiology* (Ed. C. Booth), vol. 4, pp. 335–65. Academic Press, London.

JONES E. B. G. (1974) Aquatic fungi: freshwater and marine. In: *Biology of Plant Litter Decomposition* (Eds C. H. Dickinson & G. J. F. Pugh), pp. 337–84. Academic Press, London.

KÄÄRIK A. (1974) Decomposition of wood. In: *Biology of Plant Litter Decomposition* (Eds C. H. Dickinson & G. J. F. Pugh), pp. 129–74. Academic Press, London.

KEEGSTRA K., TALMADGE K. W., BAUER W. D. & ALBERSHEIM P. (1973) The structure of plant cell walls. III. A model of the walls of suspension-cultured sycamore cells bared on the inter-connections of the macromolecular components. *Plant Physiology* **51**, 188–96.

LA RIVIERE J. W. M. (1977) Microbial ecology of liquid waste treatment. *Advances in Microbial Ecology* **1**, 215–60.

LODHA B. C. (1974) Decomposition of digested litter. In: *Biology of Plant Litter Decomposition* (Eds C. H. Dickinson & G. J. F. Pugh), pp. 213–41. Academic Press, London.

LUSSENHOP J., KUMAR R., WICKLOW D. T. & LLOYD J. E. (1980) Insect effects on bacteria and fungi in cattle dung. *Oikos* **34**, 54–8.

MANACHÈRE G. (1980) Conditions essential for controlled fruiting of macromycetes; a review. *Transactions of the British Mycological Society* **75**, 255–70.

MARGALEFF R. (1968) *Perspectives in Ecological Theory*. University of Chicago Press, London.

MEADOWS P. S. (1971) The attachment of bacteria to solid surfaces. *Archiv für Mikrobiologie* **75**, 374–80.

MEREDITH D. S. (1959) Infection of pine stumps by *Fomes annosus* and other fungi. *Annals of Botany* **23**, 455–76.

MONOD J. (1949) The growth of bacterial cultures. *Annual Review of Microbiology* **3**, 371–94.

MONTGOMERY R. (1982) The role of polysaccharase enzymes. In: *Decomposition by Basidiomycetes* (Eds J. C. Frankland, J. N. Hedger & M. J. Swift). Cambridge University Press, Cambridge. (In press.)

NEWBOULD P. J. (1969) *Methods for Estimating the Primary Production of Forests*: *IBP Handbook No. 2*. Blackwell Scientific Publications, Oxford.

ODUM E. P. (1969) The strategy of ecosystem development. *Science* **164**, 262–70.

OKAFOR N. (1966) Ecology of micro-organisms on chitin buried in soil. *Journal of General Microbiology* **44**, 311–26.

PARKINSON D. (1967) Soil micro-organisms and plant

roots. In: *Soil Biology* (Eds N. A. Burgess & F. Raw), pp. 449–78. Academic Press, London.

PARKINSON D., GRAY T. R. G. & WILLIAMS S. T. (1971) *Methods for Studying the Ecology of Soil Micro-organisms.* Blackwell Scientific Publications, Oxford.

PARKINSON D., VISSER S. & WHITTAKER J. B. (1978) Effects of collembolan grazing on fungal colonisation of leaf litter. *Soil Biology and Biochemistry* **11**, 529–35.

PUGH G. J. F. & EVANS M. D. (1970) Keratinophilic fungi associated with birds. I. Fungi isolated from feathers, nests and soils. *Transactions of the British Mycological Society* **54**, 233–43.

PUTNAM R. J. (1978) Flow of energy and organic matter from a carcase during decomposition. *Oikos* **31**, 58–68.

RAYNER A. D. M. (1978) Interactions between fungi colonising hardwood stumps and their possible role in determining patterns of colonisation and succession. *Annals of Applied Biology* **89**, 131–4.

SANDERS P. F. & ANDERSON J. M. (1979) Colonisation of wood blocks by aquatic hyphomycetes. *Transactions of the British Mycological Society* **73**, 103–7.

SKINNER F. A. & CARR J. G. (1976) *Microbiology in Agriculture, Fisheries and Food.* Academic Press, London.

SWIFT M. J. (1976) Species diversity and the structure of microbial communities. In: *The Role of Aquatic and Terrestrial Organisms in Decomposition Processes* (Eds J. M. Anderson & A. MacFadyen), pp. 185–222. Blackwell Scientific Publications, Oxford.

SWIFT M. J. (1977) The ecology of wood decomposition. *Science Progress, Oxford* **64**, 179–203.

SWIFT M. J., HEAL O. W. & ANDERSON J. M. (1979) *Decomposition in Terrestrial Ecosystems: Studies in Ecology, Vol. 5.* Blackwell Scientific Publications, Oxford.

SWIFT M. J., HEALEY I. N., HIBBERD J. K., SYKES J. M., BAMPOE V. & NESBITT M. E. (1976) The decomposition of branch-wood in the canopy and floor of a mixed deciduous woodland. *Oecologia* **26**, 139–49.

SYKES G. & SKINNER F. A. (1971) *Microbial Aspects of Pollution.* Academic Press, London.

TRIBE H. (1966) Interactions of soil fungi on cellulose film. *Transactions of the British Mycological Society* **49**, 457–66.

VELDKAMP H. (1977) Ecological studies with the chemostat. *Advances in Microbial Ecology* **1**, 59–94.

VISSER S. & PARKINSON D. (1975) Fungal succession on aspen poplar leaf litter. *Canadian Journal of Botany* **53**, 1640–51.

WAKSMAN S. A. & CORDON T. C. (1939) Thermophilic decomposition of plant residues in composts by pure and mixed cultures of micro-organisms. *Soil Science* **47**, 217–25.

WASTIE R. L. (1961) Factors affecting competitive saprophytic colonisation of the agar plate by various root-infecting fungi. *Transactions of the British Mycological Society* **44**, 145–59.

WASTIE R. L. (1962) Mechanism of action of an infective dose of *Botrytis* spores on bean leaves. *Transactions of the British Mycological Society* **45**, 465–73.

WEBSTER J. (1970) Coprophilous fungi. *Transactions of the British Mycological Society* **54**, 161–80.

WICKLOW D. T. (1981) The coprophilous fungal community: a mycological community for examining ecological ideas. In: *The Fungal Community: its Organisation and Role in the Ecosystem* (Eds D. T. Wicklow & G. C. Carroll), pp. 47–76. Marcel Dekker, New York.

WILLOUGHBY L. G. (1974) Decomposition of litter in fresh water. In: *Biology of Plant Litter Decomposition* (Eds C. H. Dickinson and G. J. F. Pugh), pp. 659–82. Academic Press, London.

WITKAMP M. & OLSON J. (1963) Breakdown of confined and nonconfined oak litter. *Oikos* **14**, 138–47.

WOODWELL G. M. & MARPLES T. G. (1968) The influence of chronic gamma irradiation on production and decay of litter and humus in an oak-pine forest. *Ecology* **49**, 456–65.

Chapter 11 • Nitrification

James I. Prosser and Dimity J. Cox

11.1 Introduction

Nitrification is the oxidation of reduced forms of nitrogen, usually ammonium, to nitrate. Schloesing and Muntz (1877) were the first to demonstrate its biological nature by preventing conversion of ammonium to nitrate in soil percolation columns by addition of chloroform. Several years later Winogradsky (1890) isolated the bacteria responsible and classified them into two groups: the first converts ammonium to nitrite; and the second oxidises nitrite to nitrate.

Our present knowledge of nitrification reflects problems of methodology. Analysis of ammonium, nitrite and nitrate is relatively easy and has been facilitated by the development of automated assays. Production of nitrite and nitrate during nitrification can be followed readily and many investigations consider nitrification as a purely chemical process and, indeed, nitrification is still studied by some without consideration of its biological nature. The study of biological aspects of nitrification is limited by difficulties in obtaining and maintaining pure cultures of nitrifying bacteria (much early work was carried out with mixed cultures), their slow growth and poor cell yields.

11.2 Biology of nitrifying bacteria

Classification of nitrifying bacteria is discussed by Walker (1975, 1978) and Watson (1974). They constitute the family Nitrobacteraceae and are classified into genera on the basis of inorganic energy source and cell morphology.

There are three genera of nitrite oxidisers, namely *Nitrobacter*, *Nitrospina* and *Nitrococcus*, of which *Nitrobacter* is by far the most commonly isolated. Cells are Gram-negative, short wedge- or pear-shaped rods which may possess a single flagellum. Many strains of both nitrite-oxidisers and ammonia-oxidisers possess internal membrane systems and the presence of cytochromes results in yellow or red suspensions. Growth in liquid culture often gives rise to a surface pellicle and in shaken cultures some strains grow as clumps of cells.

Ammonia oxidisers are placed by Watson (1974) in four genera, namely, *Nitrosomonas, Nitrosococcus, Nitrosospira* and *Nitrosolobus* and a fifth, *Nitrosovibrio*, had been described by Harms *et al.* (1976). Most work has been carried out on the genus *Nitrosomonas*, which consists of ellipsoidal or short rod-shaped cells, sometimes flagellated and which grow in clumps in shaken culture. However, interest has recently focused on the other genera. MacDonald (1979) found that the genus *Nitrosolobus* was the dominant organism in Rothamsted soil, Bhuiya and Walker (1977) and Walker and Wickramasinghe (1979) isolated *Nitrosospira* from acid soils, whilst Belser and Schmidt (1978a) estimated the relative numbers of different genera in soil. Different genera exhibit different growth kinetics (Belser & Schmidt 1980) and their distribution in different environments may be significant.

11.3 Biochemistry of nitrification

The biochemistry of nitrifying bacteria has been reviewed by Wallace and Nicholas (1969) and Aleem (1970). More recent reviews concerning the bioenergetics of nitrifiers in relation to other chemolithotrophs can be found in Aleem (1977) and Kelly (1978).

Nitrifying bacteria are chemolithotrophs and obtain all their energy from the oxidation of inorganic compounds, while cellular carbon is obtained principally by the fixation of carbon dioxide. The nature of their energy sources leads to low cell yields compared to those of heterotrophs and biochemical studies have been hindered by difficulties in obtaining sufficient cell biomass for analysis.

11.3.1 Ammonia Oxidation

Work on the effect of pH on ammonia oxidation (Suzuki *et al.* 1974; Drozd 1976) indicates that ammonia (NH_3) crosses the cytoplasmic membrane rather than ammonium (NH_4^+). The first step in its oxidation to nitrite is the energy-dependent hydroxylation to hydroxylamine (NH_2OH). This reaction probably involves cytochrome P_{460} which is found along with cytochrome *o* in the soluble fraction of cell-free extracts. Energy is yielded by the oxidation of hydroxylamine to nitrite by the following route involving a membrane-bound electron transport chain containing flavins and cytochromes *a*, *b* and *c* (Aleem 1977):

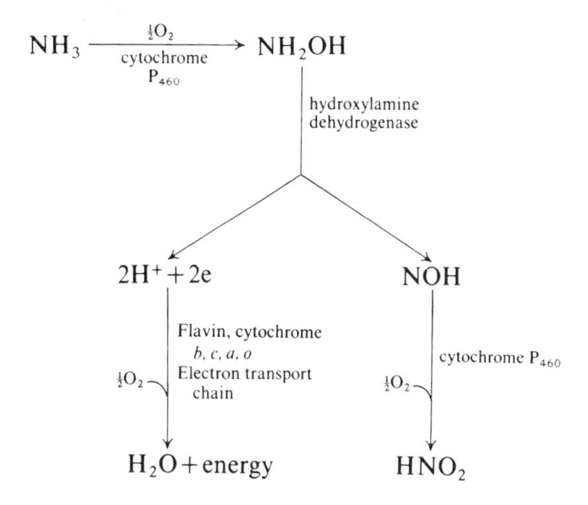

The precise nature of the intermediates in hydroxylamine oxidation is not known (Hooper 1978) but the NH_2OH/NO_2 couple has a redox potential of 66 mV. Theoretically this allows coupling to the electron transport chain at the flavoprotein level and possibly two sites for ATP production. The limited evidence, however, suggests coupling at the cytochrome *c* level with ATP generation at 'site III' in the electron transport chain and a maximum P/O ratio of 1.

Reducing power in ammonia oxidisers, as in other chemolithotrophs, is generated by reverse electron flow coupled to the oxidation of ammonia or, more probably, hydroxylamine. Thus, ATP formed at site III is used to drive reverse electron flow through sites I and II to yield reduced NAD(P).

These features help to explain the poor growth yields of ammonia oxidisers compared with, for example, a heterotroph utilising a hexose sugar. Firstly, 18 mol ATP and 12 mol reduced NAD(P) are required to fix 6 mol carbon dioxide to form 1 mol hexose sugar. Secondly, formation of the reducing equivalents also consumes energy such that in the order of 5 mol ATP are required for the reduction of 1 mol NAD(P).

11.3.2 Nitrite Oxidation

Nitrite oxidation has been studied more extensively than ammonia oxidation, particularly with respect to the coupling of oxidation to the electron transport chain and the involvement of cytochromes. The initial step is the hydration of nitrite followed by dehydrogenation with the reaction sequence (Aleem 1977):

$$NO_2^- + H_2{}^{18}O \rightarrow NO_2^- \cdot H_2{}^{18}O$$
$$NO_2^- \cdot H_2{}^{18}O + 2\ \text{cyt}\ a_1 \cdot Fe^{3+} \rightarrow N^{18}O_3 + 2\ \text{cyt}\ a_1 \cdot Fe^{2+} + 2H^+$$
$$2\ \text{cyt}\ a_1\ Fe^{2+} + 2\ \text{cyt}\ aa_3 \cdot Fe^{3+} \rightarrow 2\ \text{cyt}\ a_1 \cdot Fe^{3+} + 2\ \text{cyt}\ aa_3 \cdot Fe^{2+}$$
$$2\ \text{cyt}\ aa_3 \cdot Fe^{2+} + 2H + \tfrac{1}{2}O_2 \rightarrow 2\ \text{cyt}\ aa_3 \cdot Fe^{3+} + H_2O.$$

Oxidation of nitrite to nitrate, therefore, occurs by the addition of an oxygen atom from water and not from molecular oxygen. Electrons and protons are transferred to molecular oxygen via cytochrome oxidase components with the generation of ATP.

Reducing equivalents are formed by ATP-driven reverse electron flow from cytochrome a_1, through cytochrome *c* and flavoproteins, to NAD. Cellular carbon is obtained from carbon dioxide and ribulose bisphosphate carboxylase, the key enzyme in this process, is believed to reside, in part, in polyhedral inclusion bodies termed carboxysomes (Bock *et al.* 1974; Peters 1974). Some properties of the enzyme in a nitrite oxidiser have been compared with other autotrophs by Harrison *et al.* (1979).

As in ammonia oxidisers, carbon dioxide fixation demands a high energy input and it has been calculated that 15 mol nitrite are required for fixation of 1 mol carbon dioxide to the level of a hexose sugar. When further metabolic processes, such as protein synthesis, are considered this value rises to 21 mol nitrite.

Nitrobacter spp. are also capable of heterotrophic and mixotrophic growth. Smith and Hoare (1968) demonstrated growth with acetate as the sole carbon and energy source and Bock (1976) found growth on pyruvate, formate and acetate in the presence of three different nitrogen sources. In all cases doubling times were much greater than those of nitrite-grown cells. Differences in internal structure and in enzyme levels between heterotrophically and autotrophically grown cells were also observed and the nitrite oxidase enzyme was shown to be inducible. Mixotrophic growth has also been demonstrated by Smith and Hoare (1968) and Steinmüller and Bock (1977) who investigated the effect of culture filtrates of *Pseudomonas fluorescens* on nitrite oxidation. Stimulation of oxidation was found, the extent of which depended on the *Nitrobacter* strain and conditions under which the culture filtrates were obtained.

11.4 Growth of nitrifying bacteria

A major feature of nitrifying bacteria is their relatively slow rate of growth. In liquid culture, maximum specific growth rates for both groups are in the order of 1.0 to 2.0 d^{-1}, the fastest reported being 2.2 d^{-1} (doubling time 8 h) for *Nitrosomonas europaea* (Skinner & Walker 1961). In the soil rates are even lower: for example Morrill and Dawson (1962) reported a range of 0.14 to 0.51 d^{-1} and MacDonald (1979) a range of 0.2 to 1.9 d^{-1}. Growth yields are also low: for example, Skinner and Walker (1961) obtained a yield of 0.71 g dry weight l^{-1} for *N. europaea* growing in continuous culture. Again, growth yields in soil appeared to be lower than in liquid culture (Belser 1979).

Growth is assumed to follow Monod kinetics while ammonium and nitrite oxidations have been shown to obey Michaelis-Menten kinetics (Laudelout & van Tichelen 1960; Loveless & Painter 1968). Saturation constants for the oxidation of inorganic nitrogen compounds lie in the range 1 to 10 $\mu g\ ml^{-1}$. A more comprehensive description of growth functions and their modification to account for different environmental factors, is given by Bazin *et al.* (1976).

11.4.1 Effect of Environmental Factors

The effect of environmental factors on nitrification is reviewed by Painter (1970), Focht and Chang (1975) and Focht and Verstraete (1977). The pH values are particularly important and the favourable range for both *Nitrosomonas* spp. and *Nitrobacter* spp. is pH 7.0 to 9.0 with inhibition below pH 6.0, although slow rates of nitrification have been reported at pH values as low as 4.5. In this respect growth of ammonia oxidisers is autoinhibitory since 2 mol H^+ ions are produced for each mol ammonia oxidised. Inhibition of ammonia oxidation at high and low pH appears to be due to free ammonia and free nitrous acid respectively (Anthonisen *et al.* 1976). Inhibition of *Nitrobacter* spp. at low pH is due to nitrous acid, while high pH results in dissociation of nitrite from an active site of the enzyme complex as a result of competition by hydroxyl ions (Boon & Laudelout 1962).

Both substrate and product inhibition have been reported. High ammonium concentrations inhibit both ammonia and nitrite oxidisers, but the latter are more sensitive. Aleem and Alexander (1958) found inhibition of *Nitrobacter agilis* at concentrations as low as 10 μg NH_4^+-N ml^{-1}. The effect is pH dependent as inhibition results from free ammonia and is reduced in soil, possibly because of chemical adsorption of ammonium ions and the greater buffering capacity. Nitrite inhibition is also pH dependent because of the involvement of nitrous acid. Ammonia oxidisers are inhibited at concentrations of 1.4 mg NO_2^--N ml^{-1} while lower concentrations may increase growth lag periods. Nitrite oxidisers are slightly more sensitive to nitrite and are also inhibited by high levels of nitrate.

Other inhibitors are discussed by Painter (1970). Of particular importance are chlorate, which specifically inhibits nitrite oxidation, and 2-chloro-6-trichloromethyl pyridine (also known as nitrapyrin and N-serve) which inhibits ammonia oxidation.

The optimum temperature for nitrification lies in the range 25 to 35°C with no significant growth below 5°C or above 40°C, although Laudelout and van Tichelen (1960) reported maximum growth of *Nitrobacter winogradskyi* at 42°C and some growth up to 49°C. The Q_{10} values for ammonia and nitrite oxidation fall in the range 1.7 to 2.7. Aeration of cultures of nitrifying organisms is important not only to provide oxygen but also for the supply of carbon dioxide, unless carbonate is used to neutralise acid production. Saturation constants for oxygen uptake are in the range 0.3 to 1.0 mg O_2 l^{-1}, with *Nitrobacter* spp. slightly more sensitive to oxygen limitation than *Nitrosomonas* spp.

11.5 Nitrification in soil

Nitrifying bacteria are present in untreated (no ammonium fertiliser) agricultural soils at levels of 1×10^3 to 1×10^4 g^{-1} but fertilised soils may contain 1×10^5 to 1×10^6 g^{-1}. In other words, the numbers of bacteria appear to be limited by the ammonium concentration, as well as by their inherent low growth rates and low growth yields. Furthermore, numbers are greater in alkaline and neutral soils and, therefore, may be increased in acid soils by the application of lime; as in liquid culture little nitrification occurs in soils below pH 4.5. Lees and Quastel (1946b) showed that nitrification occurred on the surface of soil particles, rather than in the soil solution, and surface growth has been used to explain different pH optima for *Nitrobacter agilis* activity in soil and in liquid culture.

Nitrification proceeds most readily in well-aerated soils and nitrifier populations are greatest near the soil surface (Ardakani *et al.* 1974b). Sabey (1969) and Seifert (1969, 1970, 1972) studied the effect of soil moisture and one half to two thirds of the water holding capacity seems most favourable. The temperature range for nitrification in soil is similar to that in laboratory cultures. There is evidence for adaptation in particular environments to particular temperatures and pH ranges.

In general plants do not appear to significantly affect nitrification, although Moore and Waid (1971) found inhibition by washings of plant roots, particularly those of ryegrass. There is also evidence of inhibition of nitrification in climax ecosystems, although as yet this evidence is inconclusive (Belser 1979).

Ammonia and nitrite oxidisers are always found growing in combination. Nitrite rarely accumulates which suggests that ammonia oxidation is the rate-limiting step. At high pH and high ammonium concentrations, which arise following application of anhydrous ammonia as a fertiliser, nitrite oxidation may be selectively inhibited resulting in the accumulation of nitrite. This is undesirable as nitrite is phytotoxic and also affects microbial activity in the soil (Bancroft *et al.* 1979).

11.5.1 HETEROTROPHIC NITRIFICATION

In addition to autotrophic nitrifying bacteria, many heterotrophic microorganisms can convert ammonium to nitrite and ammonium, and nitrite and organic nitrogen compounds to nitrate. Species capable of such transformations are listed by Focht and Verstraete (1977) and include a wide range of bacteria (including actinomycetes) and fungi. Nitrification by these organisms occurs after growth has ceased and rates of nitrite and nitrate production are several orders of magnitude lower than those of autotrophs. Heterotrophic nitrification can be studied by selectively inhibiting the autotrophic nitrifier population and it may be important at temperatures, pH values and moisture contents which are unfavourable for the growth of autotrophic nitrifying bacteria.

11.6 Significance of nitrification in natural environments

At one time nitrate was thought to be the preferred form of nitrogen for plant growth, and conversion to nitrate of ammonium ions released by decomposition was, therefore, considered desirable. It is now known that plants can also utilise ammonium ions and, on an energetic basis, its more reduced nature makes it a more favourable source of nitrogen. In addition, positively charged ammonium ions are adsorbed to soil surfaces thereby becoming much less susceptible than nitrate to leaching. A high rate of nitrification following the application of ammonium fertiliser results in a significant loss of nitrogen and generally only 30 to 70% of the applied nitrogen is used by plants. For these reasons nitrification is now actively discouraged when applying ammonium-based fertilisers. The most commonly used inhibitor has been N-serve (2-chloro-6-(trichloromethyl)-pyridine) and its use has reduced nitrogen losses in agriculture. More recently carbon disulphide has been used as a nitrification inhibitor (Ashworth *et al.* 1977). Originally this was injected as a gas but is more efficiently applied as potassium ethylxanthate which breaks down in the soil, releasing carbon disulphide.

Nitrification rates in excess of nitrate utilisation by plants, whilst being uneconomical also increase the levels of nitrate in run-off waters. This is particularly important in areas of intensive farming and concentrations greater than 10 μg NO_3^--N ml^{-1} are considered dangerous. High nitrate levels can lead to eutrophication and can cause met-haemoglobinaemia in human infants and animals which imbibe contaminated water.

Experimental

11.7 Isolation of pure cultures of nitrifying bacteria

Pure cultures of nitrifying bacteria are obtained from liquid enrichment cultures or from solid media. However, the standard techniques applicable primarily to heterotrophs must be modified because of the slow growth rates and poor growth yields of nitrifiers.

11.7.1 Plate Method

Nitrifying bacteria were first isolated by Winogradsky (1890) from colonies growing on silica gel solidified medium containing inorganic mineral salts and either ammonium or nitrite as the energy source. Beijerinck (1896) adopted a similar approach, replacing silica gel with agar washed to remove soluble organic components. In both cases heterotrophs failed to grow since the media lacked organic carbon yet media preparation was long and tedious. In addition, nitrifiers form small colonies which are visible only by low power light microscopy and are difficult to remove and distinguish from contaminant colonies.

Soriano and Walker (1968) describe a method for isolating ammonia oxidisers which minimises these difficulties. They solidified inorganic nutrient medium with commercially available purified (Merck) or special Noble (Difco) agar which reduced the number of contaminants without completely preventing their growth. Heterotrophs developed colonies within 1 to 2 d and microscopic ammonia oxidiser colonies appeared after 3 to 4 d. Within 7 to 8 d nitrifier colonies reached a diameter of 20 to 30 μm and could be subcultured. After 1 to 2 weeks the sub-cultured colonies may reach 0.2 mm in diameter. Soriano and Walker (1968) also described a micromanipulator for removing colonies with glass capillary pipettes which enabled the transfer of one or several colonies to fresh liquid medium.

Isolation of pure cultures from soil by this method takes approximately 6 weeks which is half the time required by the dilution method described in section 11.7.2. It also offers the potential for isolating different strains of ammonia oxidisers and is the method of choice if the techniques of culturing nitrifiers on solid media can be mastered.

11.7.2 Enrichment and Dilution Method

The enrichment and dilution method was originally hampered by the belief that growth of nitrifiers, particularly ammonia oxidisers, in liquid media required the presence of particulate matter to which the cells could adsorb. However, such particles reduced suspended growth to such an extent that nitrifiers in the liquid phase were usually outnumbered by heterotrophs.

Lewis and Pramer (1958) were the first to use particle-free media for the isolation of *Nitrosomonas* species. They prepared enrichment cultures either by successive transfers of fully grown cultures to fresh medium or by the addition of more ammonium ions after the oxidation of the initial amount present. Using the latter method they found an initial increase in heterotrophs, a decrease in numbers following the first addition of fresh ammonium and relatively constant numbers following a second addition. Ammonia oxidisers were found to outnumber heterotrophs by a factor of 50. Serial dilutions of enrichment cultures from 1×10^{-1} to 1×10^{-10} were made into fresh medium and tests for purity and nitrite production carried out after 30 d incubation. It is now more usual to inoculate tubes of fresh medium with 1.0 ml of a dilution adjusted to give approximately 1 cell ml^{-1}.

Soriano and Walker (1968) reported that a period of 12 weeks was required for the isolation of nitrifiers from soil by this method but the use of liquid media was more convenient. Belser and Schmidt (1978a) used this technique to obtain different ammonia oxidiser strains but generally the method only results in the isolation of the fastest growing strains. Judging the required dilution factor is also difficult. Viable counts take several weeks and generally underestimate numbers, while total counts, using counting chambers, may overestimate numbers of nitrifiers, particularly if they are difficult to distinguish from heterotrophs.

11.8 Maintenance and preservation of cultures

Nitrifying bacteria are routinely cultured in liquid media. Indeed technical problems and poor growth on solid media make the use of liquid cultures more practical.

Nutritional requirements of nitrifying bacteria have been reviewed by Painter (1970) and Walker (1975). All media contain magnesium, calcium, phosphate, a trace quantity of iron and some incorporate other trace elements. Ammonium is usually provided as ammonium sulphate or ammonium chloride and nitrite as the sodium or potassium salt. The pH is adjusted to the range pH 7.0 to 8.0 and cultures are incubated aerobically at 25 to 30°C.

Growth of ammonia oxidisers is complicated by acid production which necessitates alkali addition, usually as sodium hydroxide or sodium carbonate. The medium of Skinner and Walker (1961) incorporates phenol red which turns yellow below pH 7.0, indicating the need for neutralisation. Early workers used particulate buffers, such as calcium carbonate, and it was thought that such particles were necessary for growth.

However, growth of *Nitrobacter* species in particle-free medium was first demonstrated by Goldberg and Gainey (1955) and growth of *Nitrosomonas* species by Engel and Alexander (1958). As a result clear media are now always used. Ammonium oxidation is photosensitive (Hooper & Terry 1974) and maximum growth requires incubation in the dark. Nitrifying bacteria are also extremely sensitive to carbon disulphide which is released by certain types of rubber (Powlson & Jenkinson 1971) and silicone rubber should be used in culture vessels and in all equipment used for media preparation.

Growth of ammonia oxidisers is assessed by the production of acid, nitrite, nitrate, in the presence of nitrite oxidisers or, more reliably, by the removal of ammonium. Growth of nitrite oxidisers is determined by the depletion of nitrite. Standard techniques (see Chapter 6) are used for determining total numbers of nitrifiers but viable counts give rise to problems which are discussed in section 11.10.1.

A major problem during the growth of nitrifiers is the risk of contamination by heterotrophs, which grow on organic impurities in the medium, microbial cell debris, and on extracellular products released by nitrifiers. The problem is exacerbated by long incubation periods, increasing the risk of contamination, and slow growth of nitrifiers compared to heterotrophic contaminants. Strict aseptic techniques are particularly important and regular purity checks are necessary. These generally involve plating onto nutrient or peptone agar, although some workers use a wider range of organic media. Heterotrophs usually develop on these plates within 3 to 4 d but incubation should be continued for several weeks. Microscopic examination of cultures is also useful in detecting contamination.

Freeze drying nitrifying bacteria is rarely attempted because of difficulties in obtaining sufficient quantities of biomass. Instead cultures are usually stored in liquid medium, in sealed containers which are either refrigerated or frozen. This results in the loss of viability and regular subculturing is necessary every 4 to 6 weeks.

11.9 Chemical analysis of ammonium, nitrite and nitrate

The growth of nitrifying bacteria is usually determined by spot tests for ammonium, nitrite and nitrate. Quantitative experimentation requires accurate estimation of the concentrations present and for this either colorimetric methods or ion selective electrodes are used.

11.9.1 Spot Tests

The presence of ammonium is indicated semi-quantitatively by an orange-brown precipitate after the addition of a drop of Nesslers reagent. Nitrite may be detected by the addition of a drop of Griess Ilosvay's reagent I (sulphanilic acid) followed by a drop of Griess Ilosvay's reagent II (8-amino-naphthylene 2-sulphonic acid). Development of a pink or red colour indicates the presence of nitrite.

Both tests are sensitive to concentrations below 1.0 μg N ml^{-1} but at concentrations above 50 μg NO_2^-–N ml^{-1}, the red colour produced by the Griess Ilosvay's reagents may develop into a straw colour and be mistaken as a negative result. In the absence of nitrite, nitrate may be determined by the Griess Ilosvay's reagents after addition of a pinch of zinc dust which reduces nitrate to nitrite.

11.9.2 Colorimetric Methods

Analysis of soil requires an initial extraction, usually with 2M-potassium chloride. Ammonium may be estimated using Nesslers reagent or by distillation methods (Bremner 1965). Nitrite may be analysed usually with Griess Ilosvay's reagents (Bremner 1965). Nitrate plus nitrite may be estimated using the nitrite method after the reduction of nitrate by copper-cadmium granules. Nitrate may be calculated by subtraction of the previously determined nitrite concentration from this total. Alternatively, nitrate may be determined by the phenoldisulphonic acid method (Snell & Snell 1949).

Analysis of ammonium, nitrite and nitrate has been greatly facilitated by the development of automated colorimetric techniques. These utilise methods similar to those above and generally operate in the ranges 1 to 50 and 0.2 to 1.0 μg N ml^{-1}. In the order of 40 to 60 samples may be analysed per hour and the reproducibility and convenience of these methods has provided greater accuracy in nitrification studies.

11.9.3 Ion Selective Electrodes

Ammonium may be estimated by a probe consisting of a pair of electrodes, one of which is located immediately behind a gas permeable membrane in a very thin layer of electrolyte, while the other acts as a reference electrode. Ammonium ions are converted to ammonia by the addition of alkali to give a solution of pH 12.5. Ammonia is measured by the pH change of the internal filling solution caused by the equilibration of ammonia across the gas-permeable membrane. Equilibration takes 2 to 3 min and the electrode operates in the range 0.1 to 100 μg $NH_4^+ - N$ ml^{-1}.

Nitrite may be measured by a liquid ion-exchange electrode which measures the voltage generated by selective exchange of the ion exchange solution with nitrate ions. Its main disadvantage is interference from nitrite which must be removed by precipitation and therefore has limited use for nitrification studies.

Such electrodes are in an early stage of development and it is hoped that improvements in design will overcome their disadvantages. They would be particularly useful for studying nitrification in liquid culture if they could be operated reliably for long periods of time.

11.10 Enumeration of nitrifying bacteria in soil

The most commonly used method for estimating numbers of nitrifying bacteria is the most-probable-number method (see Chapter 6). The alternatives are a modified dilution-plate method, fluorescent antibody techniques and activity measurements which are discussed later.

11.10.1 Most-Probable-Number Method

The most-probable-number (MPN) method is normally used because of difficulties in culturing nitrifiers on solid media and the relative ease with which growth may be assessed in liquid culture in the presence of other organisms.

The most commonly used procedure is described by Alexander and Clark (1965). Inorganic medium containing ammonium or nitrite is inoculated with samples from a series of dilutions of a soil suspension. The inoculated medium is placed in test tubes or in the wells of haemagglutination trays. The latter, which offers a saving in space and materials may be covered with a plastic film to reduce evaporation over the long incubation periods required. After incubation at 25 to 30°C growth is assessed by spot tests for ammonium, nitrite and nitrate or, for ammonia oxidisers, by acid production although this method is less reliable.

Incubation periods range from 21 to 90 d, with 21 d recommended by Alexander and Clark (1965). Matulewich *et al.* (1975) found that maximum numbers of ammonia oxidisers were obtained after 55 d. Tubes containing nitrite, however, were still giving new positive results after 100 d incubation. It is, therefore, impossible to recommend an incubation period for nitrite oxidisers until more is known about the factors affecting the length of the lag phase.

Lengthy incubation periods also result in changes in the growth medium. Matulewich *et al.* (1975) found a reduction in medium volume from 7.0 ml to 2.0 ml over 100 d and these changes can affect the spot tests for ammonium, nitrite and nitrate. Belser and Schmidt (1978a) found variations in both the final counts of ammonia oxidisers and the incubation time required with three different media. The medium of Soriano and Walker (1968), modified by the addition of trace elements, was found to be most suitable, giving maximum numbers after approximately three weeks. This represented about 94% of those obtained by the fluorescent antibody technique. In addition, Belser and Schmidt (1978a) isolated pure cultures from tubes immediately they gave a positive result and found the different media selected for different genera.

Using inocula from a dilution series based on a $10\times$ factor and taking 5 tubes inoculated at each dilution, the 95% confidence limits may be $\pm 30\%$ of the real value (Cochran 1950). The range of the confidence limit may be reduced by either increasing the number of tubes inoculated at each original dilution or by reducing the dilution factor or both. However, the precision of the MPN method is still not good.

The disadvantages of the method may be summarised as follows:

(1) intrinsic statistical error;

(2) medium selectivity;

(3) lengthy incubation periods;

(4) variation in required incubation time, with numbers sometimes not reaching a maximum;

(5) difficulty in distinguishing between different genera.

These problems are not specific to nitrifying bacteria but, until recently, the absence of suitable alternatives has restricted study of their growth in soil.

11.10.2 Dilution Plate Method

This method is rarely used because of the tedious nature of the technique, the small size of nitrifier colonies and overgrowth by heterotrophs if organic compounds are present. Problems of lengthy incubation periods and media selectivity also apply.

Soriano and Walker (1973) outlined a method based on one described by Winogradsky and Winogradsky (1933). A soil suspension is spread onto a base layer of silica gel and is covered by ammonia oxidiser medium containing powdered calcium carbonate and solidified with silica gel. Ammonia oxidiser colonies produce acid which dissolves the calcium carbonate and may be recognised as clear zones. The method is more tedious then the MPN technique and cannot be used to count nitrite oxidisers since these do not produce acid. It has the advantage that colonies may be picked off and genera identified microscopically, enabling the estimation of the relative numbers of different ammonia oxidisers.

11.10.3 Fluorescent Antibody Technique

A recent, major advance in enumerating nitrifiers in soil has been the application of fluorescent antibody techniques, described for nitrite oxidisers by Fliermans *et al.* (1974) and for ammonia oxidisers by Belser and Schmidt (1978b). The technique does not require long incubation periods, is more precise than other methods and distinguishes between different genera.

The high specificity of fluorescent antibodies and the serological diversity of nitrifying bacteria does present problems. Initial studies indicated the existence of two serotypes for nitrite oxidisers, corresponding to *Nitrobacter winogradskyi* and *Nitrobacter agilis* but Belser (1979) characterised at least seven different serotypes. Josserand and Cleyet-Marel (1979) found that fluorescent antibodies against *N. winogradskyi* and *N. agilis* did not label nitrite oxidisers isolated from their soils. They isolated a further two strains which were immunologically distinct. The situation is even more complicated for ammonia oxidisers with several immunologically distinct strains within different genera.

Ideally the study of a particular environment requires the use of antibodies which are effective against every strain present. The isolation and purification of nitrifiers is such a tedious and time-consuming process that this would outweigh the advantages of the technique. The problem may be reduced if only the important strains are considered but strains isolated from enrichment cultures may not be those which are important in the soil. Belser (1979) suggested that serotypes may not vary greatly between different localities and that once the important fluorescent antibodies have been obtained these may be used generally. If this suggestion is confirmed the fluorescent antibody technique will realise its full potential.

11.11 Incubation studies

Nitrification in soil has been traditionally examined by standard incubation studies. A quantity of soil is adjusted to the required moisture content, ammonium is added and the soil is incubated at 25 to 30°C for several weeks. Water lost by evaporation is replaced and samples are removed regularly and analysed for ammonium, nitrite and nitrate and, less frequently, the ammonia and nitrite oxidisers are counted.

The technique has been used widely to study chemical aspects of nitirification in soil but errors arise due to the heterogeneous nature of the system which makes uniform mixing of water and ammonium difficult. Removal of samples also affects the system and it represents a closed environment, the disadvantages of which are discussed below.

Initial population sizes, growth rates and saturation constants have been determined from similar studies in liquid culture (Knowles *et al.* 1965). Values are estimated by fitting data on rates of ammonium and nitrite conversion to simple growth equations. This has never been attempted in soil incubation studies and errors resulting from heterogeneity in the system may prevent such determinations. In addition, calculation of the initial population size requires the assumption of a negligible lag phase, and this may not be valid in soil. The general topic of activity measurements and their relationship to the efficiency of different counting methods is discussed by Belser (1979).

11.12 Reperfusion studies

Many of the disadvantages of incubation studies were overcome by the reperfusion technique illustrated in Fig. 1 (Lees & Quastel 1946a; Quastel & Scholefield 1957).

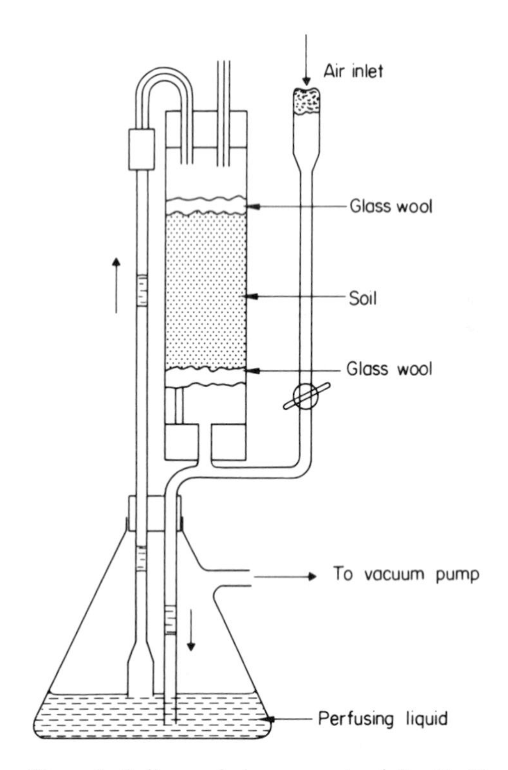

Figure 1. Soil reperfusion apparatus (after Parkinson *et al.* 1971.)

Air-dried soil contained in a vertical glass column is saturated with a solution of ammonium sulphate. A drop of this same solution from a reservoir, is allowed to percolate through the column after which it is returned to the reservoir and a further portion is added to the top of the column. Aeration is effected by air dragged behind the percolating liquid and the system is operated at constant temperature. The solution is, therefore, recirculated during which oxidation of ammonium occurs and the term reperfusion is used to avoid confusion with continuous-flow culture systems.

The system enables a homogeneous distribution of the soil solution and the maintenance of a constant moisture content, which is restricted to a level just below waterlogging. Nitrification is followed by analysis of the perfusate which has the composition of the soil solution. The perfusion solution and the gas phase composition may be changed during an experiment. The technique has been used for fundamental studies on nitrification as well as determining the effect of environmental factors or inhibitors on the process. The latter may be added to the perfusing solution and their effect monitored. It is an inexpensive and simple system which gives reproducible results and provides greater accuracy than incubation studies.

Lees and Quastel (1946a) stated the disadvantages of reperfusion studies as being, first, the necessity of working at a moisture content which is near waterlogging and, second, the reduction in the perfusate volume as samples are removed. In addition, the soil may only be analysed at the end of an experiment, unless similar columns are sacrificed at different stages. Thus the technique is useful for following changes in substrates and metabolic products but does not allow the direct observation of population changes or changes in the soil itself, both of which are possible with incubation studies.

11.13 Continuous-flow methods

A serious drawback of both incubation and reperfusion studies is that they represent closed systems. Metabolic products accumulate during the course of an experiment and the system is always in a transient state because of changes in substrate concentration. As with all closed systems, it is impossible to maintain a constant growth rate other than the maximum rate, and that only for a limited period. Continuous-flow techniques involve the continuous addition of fresh substrates coupled with the removal of effluent rather than its recycling. These systems provide greater experimental control of the growth conditions and metabolic activity and allow the study of submaximal growth rates which are more likely to occur in soil where nutrients are usually present at limiting concentrations (see Chapters 16 and 20).

11.13.1 Soil Columns

Soil columns have been used to study various soil processes, including nitrification (Macura & Malek 1958; Macura & Kunc 1965). This system is similar to that of Lees and Quastel (1946a) except that the ammonium sulphate solution is supplied to the column at a constant rate and the effluent is not recycled. Aeration is effected by application of positive pressure to the top of the column or suction to the base and nitrification is followed by analysis of the effluent. The technique has all the advantages of the reperfusion technique and many of those associated with continuous-flow systems. Steady states may be established and constant rates of nitrification can be maintained for long periods. Transient behaviour may be studied in a controlled manner by changing the substrate concentration or flow rate and other processes, such as denitrification and urea transformations, may be studied in the same experimental system.

Continuous-flow systems are more representative of the soil environment, where ammonium is produced continuously by the decomposition of organic matter and nitrite and nitrate are removed by leaching, although these processes obviously do not occur with the same regularity and constancy as in laboratory systems. The real value of the systems, however, is that nitrification may be studied under controlled experimental conditions, providing fundamental knowledge which may be applied to the more complex and variable natural environment.

Ardakani *et al.* (1973) modified the basic experimental system in two ways. Firstly, they diluted the soil sample with sand to allow an unrestricted flow of liquid and prevent complete conversion of the substrate within too short a length of column. Secondly, the soil was contained in a polypropylene cylinder which had 26 solution-collecting units each with a small sampling port closed by a bung (Fig. 2). When a negative pressure was applied to the collection units the soil solution at different depths was sampled, and soil itself was removed from the sampling ports, and replaced by non-incubated soil. Although the sampling process disturbs the system, this procedure does provide a means of following population changes at different depths and relating the changes to the varying ammonium, nitrite and nitrate concentrations.

Ardakani *et al.* (1974b) also studied nitrification in a large scale field plot with an area of 40 m^3, surrounded by a concrete wall and containing soil to a depth of 150 cm. The soil was first sprinkled with a solution of potassium nitrite and calcium sulphate to give a flow rate of 4.0 cm d^{-1}, and the nitrite was replaced by ammonium. Ammonium, nitrite and nitrate were analysed in samples from 30 collection units placed at different depths and the number of nitrifiers estimated in soil cores. This allowed calculation of rate constants for ammonium and nitrite oxidations. Despite greater variability and heterogeneity in terms of soil particle size, bulk density, temperature and other parameters, the values obtained were similar to those from laboratory systems, although the authors emphasise

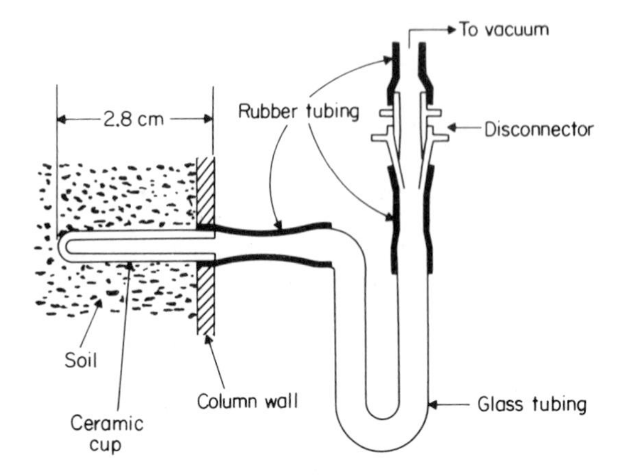

Figure 2. Soil solution collection units used to sample at different depths of a soil column (after Ardakani *et al.* 1973).

that the system did not represent a large scale version of a continuous-flow column.

11.13.2 Glass Bead Columns

In order to obtain even greater control the system of Macura and Kunc (1965) was modified by replacing soil with glass beads together with a small number of marble chips in order to neutralise acid produced during ammonium oxidation (Bazin & Saunders 1973; Prosser & Gray 1977a, b; Cox 1979). The marble chips are not necessary if nitrite oxidation is being studied. Bazin and Saunders (1973) inoculated the columns with pure cultures of nitrifying bacteria and supplied sterile, inorganic ammonia oxidiser medium at a constant rate. This eliminated several experimental variables and permitted the study of the basic interactions between nitrifying bacteria and a much more critical testing of various hypotheses. Growth occurred on the inert, uniform and defined composition glass beads, and use of pure cultures prevented unwanted interference from heterotrophs or other strains of nitrifying bacteria. The growth medium was chemically defined and the temperature and pH held to constant values. At the end of an experiment nitrifier growth on the beads was observed by scanning electron microscopy and steady states were established over a range of flow rates.

11.14 Chemostat studies

Skinner and Walker (1961) and Lovelock and Nicholas (1968) demonstrated the growth of *Nitrosomonas europaea* in continuous-flow culture (see Chapter 20). With the exception of a few studies on nitrification in sewage treatment, involving mixed cultures, this work has not been followed up and almost all of the work in the biochemistry and kinetics of nitrification in liquid culture has been carried out in batch culture. The potential of chemostats in studying the biochemistry, physiology and ecology of nitrification has yet to be realised.

11.15 Mathematical modelling

In recent years mathematical models have been used increasingly to study soil microbial processes, particularly the nitrogen cycle. The use of such models and the interaction of nitrification models with those of other nitrogen transformations has been reviewed in detail (Bazin *et al.* 1976).

Two approaches have been adopted.

(1) The empirical approach which involves fitting arbitrary equations to experimental data and accepting the equations which give the best fit. Such models are useful in describing large quantities of data in terms of a few parameters and generally involve some form of regression analysis.

(2) The mechanistic approach in which the model represents hypotheses on the way a particular system works and leads to experiments designed to test the proposed hypotheses. Such models are explanatory as well as descriptive.

The distinction between empirical and mechanistic models is not always clear, but the latter are discussed more fully since they offer the greater potential for elucidating the basic mechanisms involved in the control of nitrification in the soil.

The first step in model construction is the formulation of hypotheses regarding the process under examination. Some theoretical studies, like many experimental investigations, treat nitrification solely as a chemical process, with ammonia and nitrite oxidation described by first order rate kinetics. Other models consider it to be a single-stage oxidation of ammonium to nitrate. Such simplifying assumptions may be valid under certain conditions but are unlikely to lead to an understanding of the underlying mechanisms of nitrification and most models include hypotheses regarding the growth of populations of ammonia and nitrite oxidisers. Their inclusion increases the complexity of the model and a balance must be achieved between the need to include all the relevant information on growth and the danger of concealing some important factors.

The growth rate is described by either the Monod equation (Monod 1942) or the logistic equation (Pielou 1969). The former is based on the Michaelis-Menten equation for enzyme kinetics and has the form:

$$\frac{dx}{dt} = \frac{\mu_{\max} \cdot s}{(K_s + s)} \cdot x$$

where x represents biomass concentration; μ_{max} is the maximum specific growth rate; s represents growth-limiting substrate concentration; and K_s is the saturation constant. The growth rate is proportional to the growth-limiting substrate concentration at low values compared to the K_s value, and reaches a maximum at high concentrations.

The logistic equation assumes an inverse relationship between growth rate and population size and has the form:

$$\frac{dx}{dt} = \mu x \quad 1 - \frac{x}{k}$$

where μ is the intrinsic rate of increase, i.e. the specific growth rate; and k represents the carrying-capacity of the system. This equation satisfactorily describes situations where growth is limited by the availability of space and is most frequently applied to growth on particulate matter where the growth rate decreases due to overcrowding.

Laudelout *et al.* (1974) used two sets of equations to describe growth and nitrification. The first incorporated the Monod term to describe population growth and the second set of equations used Michaelis-Menten terms to descibe ammonium or nitrite oxidation. The lack of such a distinction in other models creates confusion, although the distinction may be unnecessary where growth and substrate oxidation are tightly coupled. However, in soil columns saturated with nitrifiers, nitrification apparently proceeds without growth as the population size is constant and few cells are found in the effluent from the columns (Ardakani *et al.* 1973, 1974a). To account for this McLaren (1971) introduced a Michaelis-Menten-type function, termed waste metabolism, which described substrate conversion which was not coupled to growth. He also introduced a maintenance energy term and considered competition for space between ammonia and nitrite oxidisers (McLaren & Ardakani 1972).

The basic equations for growth and substrate conversion may be modified to account for other factors, such as ion exchange and hydrodynamic dispersion (Cho 1971), temperature (Laudelout *et al.* 1974), pH and partial pressure of oxygen (Laudelout *et al.* 1976), moisture content (Seifert 1970, 1972) and inhibition (Domsch & Paul 1974). Obviously a model which considered all these factors could not be tested critically and the controlling mechanisms would be concealed by the complexity of the model. When formulating such models it is necessary to decide which are the important factors under study and to test the model in experimental systems in which variation of all other factors is eliminated.

The next stage is to use the model to make predictions which may be compared with experimental data. Most nitrification models consist of a series of differential equations, each describing changes of a variable, e.g. ammonium concentration, *Nitrosomonas* species biomass, pH, with time. Simulation of the model requires the simultaneous solution of these equations. While analytical solution is desirable, and should always be attempted, it is usually not possible without introducing restrictive, simplifying assumptions. The alternative is to use an iterative numerical approximation method with the aid of digital computers. The equations may be solved with programmes written in standard programming languages, or in simpler simulation languages or standard packages, although the last mentioned reduce programming flexibility.

Simulation requires the provision of values for all the constants contained in the model equations, e.g. the maximum specific growth rates, growth yields and others. Ideally these values should be obtained by different experiments to those used to test the model. Often this is not possible and standard values are frequently accepted.

Ordinary differential equations are suitable for homogeneous systems but the heterogeneous nature of soil introduces spatial as well as temporal variation. This requires use of partial differential equations (Cho 1971; Saunders & Bazin 1973) but makes analytical solution more complex, while numerical approximation methods are not as convenient to use as those for solving ordinary differential equations.

For nitrification in a continuous-flow glass bead column, Prosser and Gray (1977b) overcame these problems by using a finite difference approach. The column is considered to be divided into a number of compartments, each homogeneous and described by ordinary differential equations. Effluent from each compartment constitutes the input to the compartment immediately below. The biomass is considered to be fixed to the beads and is not transferred between compartments. This approach allows simulation of models with complex growth functions and permits study of transient behaviour which is not possible by conventional methods.

11.16 Conclusions

Experimental methods have been described as they were developed historically and may be considered as successful attempts to remove variations due to environmental factors. However, each development from field and incubation studies, through soil and glass bead columns, to theoretical models, moves further away from the natural soil habitat and the later techniques have been criticised for their lack of applicability and relevance.

For a thorough scientific investigation of soil nitrification it is better to consider methods in the reverse order to that described. Firstly precise and quantitative hypotheses should be proposed relating biochemical and physiological parameters to cell and population growth and activity; this is likely to require some form of mathematical modelling. Initially, at least, such hypotheses cannot be tested critically in complex natural environments. Any lack of fit or deviations from model predictions may be explained by the heterogeneous nature of soil and variation in uncontrolled or ill-defined factors. Testing may only be carried out in simple, well-defined systems such as chemostats or glass bead columns. These systems may be used to investigate growth kinetics of nitrifying bacteria and the effect of isolated environmental factors, such as pH, temperature and surfaces, in a precise and controlled manner. Knowledge of the fundamental mechanisms obtained may be applied to natural systems, a step which is, indeed, necessary for a complete understanding of the real systems.

The major problems in studying nitrification are the slow growth rates and poor growth yields of nitrifying bacteria, and the consequent difficulties in their culture, isolation and enumeration. This had led to a distinction, in experimental terms, between chemical nitrification and biological nitrification and the separation of nitrification rates from growth rates. While fluorescent antibody techniques may solve some of the problems of enumeration, nitrification studies will still involve tedious techniques and lengthy incubation periods. This should not discourage the study of biological aspects of nitrification in preference to the convenience of chemical analyses since ignorance of the biology of the nitrifying organisms is dangerous in applied as well as pure studies.

Recommended reading

ALEEM M. I. H. (1970) Oxidation of inorganic nitrogen compounds. *Annual Review of Plant Physiology* **21**, 67–90.

ALEEM M. I. H. (1977) Coupling of energy with electron transfer reactions in chemolithotrophic bacteria. In: *Microbial Energetics* (Eds B. A. Haddock & W. A. Hamilton), pp. 351–81. Cambridge University Press, Cambridge.

BAZIN M. J., SAUNDERS P. T. & PROSSER J. I. (1976) Models of microbial interactions in the soil. *CRC Critical Review in Microbiology* **4**, 463–98.

BELSER L. W. (1979) Population ecology of nitrifying bacteria. *Annual Reviews of Microbiology* **33**, 309–33.

FOCHT D. D. & VERSTRAETE W. (1977) Biochemical ecology of nitrification and denitrification. *Advances in Microbial Ecology* **1**, 135–214.

PAINTER H. A. (1970) A review of literature on inorganic nitrogen metabolism in microorganisms. *Water Research* **4**, 393–450.

WALKER N. (1975) Nitrification and nitrifying bacteria. In: *Soil Microbiology* (Ed. N. Walker), pp. 133–46. Butterworths, London.

WALLACE W. & NICHOLAS D. J. D. (1969) The biochemistry of nitrifying organisms. *Biological Reviews* **44**, 359–91.

References

ALEEM M. I. H. (1970) Oxidation of inorganic nitrogen compounds. *Annual Review of Plant Physiology* **21**, 67–90.

ALEEM M. I. H. (1977) Coupling of energy with electron transfer reactions in chemolithotrophic bacteria. In: *Microbial Energetics* (Eds B. A. Haddock & W. A. Hamilton), pp. 351–81. Cambridge University Press, Cambridge.

ALEEM M. I. H. & ALEXANDER M. (1958) Cell free nitrification by Nitrobacter. *Journal of Bacteriology* **76**, 510–14.

ALEXANDER M. & CLARK F. E. (1965) Nitrifying bacteria. In: *Methods of Soil Analysis* (Ed. C. A. Black), pp. 1477–83. American Society of Agronomy, Madison, Wisconsin.

ANTHONISEN A. C., LOEHR R. C., PRAKASAM T. B. S. & SRINATH E. G. (1976) Inhibition of nitrification by ammonia and nitrous acid. *Journal of Water Pollution Control Federation* **48**, 835–52.

ARDAKANI M. S., REHBOCK J. T. & MCLAREN A. D. (1973) Oxidation of nitrite to nitrate in a soils column. *Soil Science Society of America Proceedings* **37**, 53–6.

ARDAKANI M. S., REHBOCK J. T. & MCLAREN A. D. (1974a) Oxidation of ammonium to nitrate in a soil column. *Soil Science Society of America Proceedings* **38**, 96–9.

ARDAKANI M. S., SCHULZ R. K. & MCLAREN A. D. (1974b) A kinetic study of ammonium and nitrite oxidation in a soil field plot. *Soil Science Society of America Proceedings* **38**, 273–7.

ASHWORTH J., BRIGGS G. G., EVANS A. A. & MATULA J. (1977) Inhibition of nitrification by nitrapyrin, carbon disulphide and trithiocarbonate. *Journal of the Science of Food and Agriculture* **28**, 673–83.

BANCROFT K., GRANT I. F. & ALEXANDER M. (1979) Toxicity of NO_2: effect of nitrite on microbial activity in an acid soil. *Applied and Environmental Microbiology* **38**, 940–4.

BAZIN M. J. & SAUNDERS P. T. (1973) Dynamics of nitrification in a continuous flow system. *Soil Biology and Biochemistry* **5**, 531–43.

BAZIN M. J., SAUNDERS P. T. & PROSSER J. I. (1976) Models of microbial interactions in the soil. *CRC Critical Reviews in Microbiology* **4**, 463–98.

BEIJERINCK M. W. (1896) Kulturversuche mit Amöben auf festem Substrate. *Zentrablatt für Bakteriologie, Parasitenkunde, Infektionskrankheiten und Hygiene* **19**, 257–67.

BELSER L. W. (1979) Population ecology of nitrifying bacteria. *Annual Reviews of Microbiology* **33**, 309–33.

BELSER L. W. & SCHMIDT E. L. (1978a) Diversity in the ammonia oxidizing nitrifier population of a soil. *Applied and Environmental Microbiology* **36**, 584–8.

BELSER L. W. & SCHMIDT E. L. (1978b) Serological diversity within a terrestrial ammonia-oxidizing population. *Applied and Environmental Microbiology* **36**, 589–93.

BELSER L. W. & SCHMIDT E. L. (1980) Growth and oxidation kinetics of three genera of ammonia-oxidising nitrifiers. *FEMS Microbiology Letters* **7**, 213–16.

BHUIYA Z. H. & WALKER N. (1977) Autotrophic nitrifying bacteria in acid tea soils from Bangladesh and Sri Lanka. *Journal of Applied Bacteriology* **42**, 253–7.

BOCK E. (1976) Growth of *Nitrobacter* in the presence of organic matter. II. Chemoorganotrophic growth of *Nitrobacter agilis*. *Archives of Microbiology* **108**, 305–12.

BOCK E., DUVEL D. & PETERS K-R. (1974) Charakterisierung eines phagenähnlichen Partikels aus Zellen von *Nitrobacter*. I. Wirts-Partikelbeziehung und Isolierung. *Archiv für Mikrobiologie* **97**, 115–27.

BOON B. & LAUDELOUT H. (1962) Kinetics of nitrite oxidation by *Nitrobacter winogradskyi*. *Biochemical Journal* **85**, 440–7.

BREMNER J. M. (1965) Inorganic forms of nitrogen. In: *Methods of Soil Analysis* (Ed. C. A. Black), pp. 1179–232. American Society of Agronomy, Madison, Wisconsin.

CHO C. M. (1971) Convective transport of ammonium with nitrification in soil. *Canadian Journal of Soil Science* **51**, 339–50.

COCHRAN W. G. (1950) Estimation of bacterial densities by means of the 'most probable number'. *Biometrics* **6**, 105–16.

COX D. J. (1979) *Nitrification in a Continuous Flow Column*. Ph.D. thesis, University of London.

DOMSCH K. H. & PAUL W. (1974) Simulation and experimental analysis of the influence of herbicides in soil nitrification. *Archiv für Mikrobiologie* **97**, 283–301.

DROZD J. W. (1976) Energy-coupling and respiration in *Nitrosomonas europaea*. *Archives of Microbiology* **110**, 257–62.

ENGEL M. S. & ALEXANDER M. (1958) Culture of *Nitrosomonas europaea* in media free of insoluble constituents. *Nature, London* **181**, 136.

FLIERMANS C. B., BOHLOOL B. B. & SCHMIDT E. L. (1974) Autecological study of the chemoautotroph *Nitrobacter* by immunofluorescence. *Applied Microbiology* **27**, 124–29.

FOCHT D. D. & CHANG A. C. (1975) Nitrification and denitrification processes related to waste water treatment. *Advances in Applied Microbiology* **19**, 153–86.

FOCHT D. D. & VERSTRAETE W. (1977) Biochemical ecology of nitrification and denitrification. *Advances in Microbial Ecology* **1**, 135–214.

GOLDBERG S. S. & GAINEY P. L. (1955) Role of surface phenomena in nitrification. *Soil Science* **80**, 43–53.

HARMS H., KOOPS H. P. & WEHRMANN H. (1976) An ammonia-oxidizing bacterium, *Nitrosovibrio tenuis*, nov. gen. nov. sp. *Archives of Microbiology* **108**, 105–11.

HARRISON D., ROGERS L. J. & SMITH A. J. (1979) D-ribulose 1,5-bisphosphate carboxylase of the nitrifying bacterium, *Nitrobacter agilis*. *FEMS Microbiology Letters* **6**, 47–51.

HOOPER A. B. (1978) Nitrogen oxidation and electron transport in ammonia-oxidising bacteria. In: *Microbiology* (Ed. D. Schlessinger), pp. 346–7. American Society of Microbiology, Washington.

HOOPER A. B. & TERRY K. R. (1974) Photoinactivation of ammonia oxidation in *Nitrosomonas*. *Journal of Bacteriology* **119**, 899–906.

JOSSERAND A. & CLEYET-MAREL J. C. (1979) Isolation from soils of *Nitrobacter* and evidence for novel serotypes using immunofluorescence. *Microbial Ecology* **5**, 197–205.

KELLY D. P. (1978) Bioenergetics of chemolithotrophic bacteria. In: *Companion to Microbiology* (Eds A. T. Bull & P. M. Meadow), pp. 363–86. Longman, London.

KNOWLES G., DOWNING A. L. & BARRETT M. J. (1965) Determination of kinetic constants for nitrifying bacteria in mixed cultures, with aid of an electronic computer. *Journal of General Microbiology* **38**, 263–76.

LAUDELOUT H., LAMBERT R., FRIPIAT J. L. & PHAM M. L. (1974) Effect de la température sur la vitesse de l'oxydation de l'ammonium en nitrate par des cultures mixtes de nitrifiants. *Annales de Microbiologie* **125**(B), 75–84.

LAUDELOUT H., LAMBERT R. & PHAM M. L. (1976) Influence du pH et de la pression partielle d'oxygène sur la nitrification. *Annales de Microbiologie* **127**(A), 367–82.

LAUDELOUT H. & VAN TICHELEN L. (1960) Kinetics of nitrite oxidation by *Nitrobacter winogradskyi*. *Journal of Bacteriology* **79**, 39–42.

LEES H. & QUASTEL J. H. (1946a) Biochemistry of nitrification in soil. 1. Kinetics of, and the effects of poison on soil nitrification as studied by a soil perfusion technique. *Biochemical Journal* **40**, 803–15.

LEES H. & QUASTEL J. H. (1946b) Biochemistry of nitrification in soil. 2. The site of soil nitrification. *Biochemical Journal* **40**, 815–23.

LEWIS R. F. & PRAMER D. (1958) Isolation of *Nitrosomonas* in pure culture. *Journal of Bacteriology* **76**, 524–8.

LOVELESS J. E. & PAINTER H. A. (1968) The influence of metal ion concentration and pH value on the growth of a *Nitrosomonas* strain isolated from activated sludge. *Journal of General Microbiology* **52**, 1–14.

LOVELOCK H. R. J. & NICHOLAS D. J. D. (1968) Growth of *Nitrosomonas europaea* in batch and continuous culture. *Archiv für Mikrobiologie* **61**, 302–9.

MACDONALD R. M. (1979) Population dynamics of the nitrifying bacterium *Nitrosolobus* in soil. *Journal of Applied Ecology* **16**, 529–35.

MCLAREN A. D. (1971) Kinetics of nitrification in soil: Growth of the nitrifiers. *Soil Science Society of America Proceedings* **35**, 91–5.

MCLAREN A. D. & ARDAKANI M. S. (1972) Competition between species during nitrification in soil. *Soil Science Society of America Proceedings* **36**, 602–6.

MACURA J. & KUNC F. (1965) Continuous flow method in microbiology. V. Nitrification. *Folia Microbiologia* **10**, 125–35.

MACURA J. & MALEK I. (1958) Continuous flow method for the study of microbiological processes in soil samples. *Nature, London* **182**, 1796–7.

MATULEWICH V. A., STROM P. F. & FINSTEIN M. S. (1975) Length of incubation for enumerating nitrifying bacteria present in various environments. *Applied Microbiology* **29**, 265–8.

MONOD J. (1942) *Recherches sur la Croissance des Cultures Bactériennes*. Hermann et Cie, Paris.

MOORE D. R. E. & WAID J. S. (1971) The influence of washings of living roots on nitrification. *Soil Biology and Biochemistry* **3**, 69–83.

MORRILL L. G. & DAWSON J. E. (1962) Growth rates of nitrifying chemoautotrophs in soil. *Journal of Bacteriology* **83**, 205–6.

PAINTER H. A. (1970) A review of literature on inorganic nitrogen metabolism in microorganisms. *Water Research* **4**, 393–450.

PARKINSON D., GRAY T. R. G. & WILLIAMS S. T. (1971) *Methods for Studying the Ecology of Soil Microorganisms*. IPB Handbook No. 19. Blackwell Scientific Publications, Oxford.

PETERS K-R. (1974) Charakterisierung eines phagenähnlichen Partikels aus Zellen von *Nitrobacter*. II. Struktur und Grösse. *Archiv für Mikrobiologie* **97**, 129–40.

PIELOU E. C. (1969) *An Introduction to Mathematical Ecology*. Wiley Interscience, New York.

POWLSON D. S. & JENKINSON D. S. (1971) Inhibition of nitrification in soil by carbon disulphide from rubber bungs. *Soil Biology and Biochemistry* **3**, 267–9.

PROSSER J. I. & GRAY T. R. G. (1977a) Nitrification studies at non-limiting substrate concentrations. *Journal of General Microbiology* **102**, 111–17.

PROSSER J. I. & GRAY T. R. G. (1977b) Use of finite difference method to study a model system of nitrification at low substrate concentration. *Journal of General Microbiology* **102**, 119–28.

QUASTEL J. H. & SCHOLEFIELD P. G. (1957) Study of soil metabolisms with the perfusion technique. *Methods in Enzymology* **4**, 336–42.

SABEY B. R. (1969) Influence of soil moisture tension on nitrate accumulation in soils. *Soil Science Society of America Proceedings* **33**, 262–6.

SAUNDERS P. T. & BAZIN M. J. (1973) Nonsteady state studies of nitrification in soil: Theoretical considerations. *Soil Biology and Biochemistry* **5**, 545–57.

SCHLOESING T. & MUNTZ A. (1877) Sur la nitrification par les ferments organisés. *Comptes Rendues de l'Academie des Sciences, Paris* **84**, 301–3.

SEIFERT J. (1969) Nitrification in an air-dried and remoistened soil. *Rostlinna vyroba* **15**, 181–4.

SEIFERT J. (1970) The influence of moisture on the degree of nitrification in soil. *Acta Universitatis Carolinae-Biologica* **1969**, 353–60.

SEIFERT J. (1972) The influence of moisture on the degree of nitrification in soil. II. *Acta Universitatis Carolinae-Biologica* **1970**, 467–70.

SKINNER F. A. & WALKER N. (1961) Growth of *Nitrosomonas europaea* in batch and continuous culture. *Archiv für Mikrobiologie* **38**, 339–49.

SMITH A. J. & HOARE D. S. (1968) Acetate assimilation by *Nitrobacter agilis* in relation to its 'obligate autotrophy'. *Journal of Bacteriology* **95**, 844–55.

SNELL F. D. & SNELL C. T. (1949) *Colorimetric Methods of Analysis*, 3rd Edn, vol. II. Van Nostrand, Princeton.

SORIANO S. & WALKER N. (1968) Isolation of ammonia-oxidizing autotrophic bacteria. *Journal of Applied Bacteriology* **31**, 493–7.

SORIANO S. & WALKER N. (1973) The nitrifying bacteria in soils from Rothamsted classical fields and elsewhere. *Journal of Applied Bacteriology* **36**, 523–9.

STEINMÜLLER W. & BOCK E. (1977) Enzymatic studies on autotrophically, mixotrophically and heterotrophically grown *Nitrobacter agilis* with special reference to nitrite oxidase. *Archives of Microbiology* **115**, 51–4.

SUZUKI I., DULAR U. & KWOK S. C. (1974) Ammonia or ammonium as substrate for oxidation by *Nitrosomonas europaea* cells and extracts. *Journal of Bacteriology* **120**, 556–8.

WALKER N. (1975) Nitrification and nitrifying bacteria. In: *Soil Microbiology* (Ed. N. Walker), pp. 133–46. Butterworths, London.

WALKER N. (1978) On the diversity of nitrifiers in nature. In: *Microbiology* (Ed. D. Schlessinger), pp. 346–7. American Society of Microbiology, Washington.

WALKER N. & WICKRAMASINGHE K. N. (1979) Nitrification and autotrophic nitrifying bacteria in acid tea soils. *Soil Biology and Biochemistry* **11**, 231–6.

WALLACE W. & NICHOLAS D. J. D. (1969) The biochemistry of nitrifying organisms. *Biological Reviews* **44**, 359–91.

WATSON S. W. (1974) *Gram-negative chemolithotrophic bacteria.* In: *Bergey's Manual of Determinative Bacteriology* (Eds R. E. Buchanan & N. E. Gibbons), 8th edn, pp. 450–6. Williams & Wilkins, Baltimore.

WINOGRADSKY S. (1890) Recherches sur les organismes de la nitrification. *Annales de l'Institute Pasteur, Paris* **4**, 213–31.

WINOGRADSKY S. & WINOGRADSKY H. (1933) Nouvelles recherches sur les organismes de la nitrification. *Annales de l'Institute Pasteur, Paris* **50**, 350–432.

Chapter 12 · Denitrification

Dennis D. Focht

12.1 Introduction

Denitrification is the reduction of nitrogen oxides, usually nitrite or nitrate, to gaseous products such as nitrous oxide and molecular nitrogen. The reductive pathway is generally accepted to be as follows (Payne 1973):

$$NO_3^- \rightarrow NO_2^- \rightarrow NO \rightarrow N_2O \rightarrow N_2.$$

The process is irreversible and represents a loss of nitrogen from terrestrial and aquatic ecosystems. The process is carried out solely by bacteria, which use nitrogen oxides as terminal electron acceptors in lieu of molecular oxygen. Considerable confusion has occurred in the literature by improperly referring to denitrifiers as facultative anaerobes. Facultative anaerobes in the strict sense are organisms that can grow aerobically by utilising the normal complements of a full respiratory cytochrome system, but which grow anaerobically by shifting over to a completely different electron transport system in which organic compounds serve as electron acceptors (i.e. fermentation). Denitrifiers for the most part are not fermentative and cannot generate ATP by substrate-level phosphorylation. Thus, growth under anaerobic conditions is solely dependent upon the presence of nitrogen oxides. Perhaps a better term for describing denitrifiers would be facultative aerobes.

12.2 Diversity

12.2.1 Denitrifiers

The genera of denitrifying bacteria is diverse. Indeed, since the review articles of Payne (1973) and Focht and Verstraete (1977), a number of additional species have been reported to denitrify and a recent comprehensive list has been compiled by Ingraham (1981). Most interesting are the nitrogen-fixing bacteria *Azospirillum lipoferum* (Eskew *et al.* 1977; Neyra *et al.* 1977) and rhizobia of the cowpea miscellany and *Rhizobium japonicum* (Zablotowicz *et al.* 1978). The photo-organotrophic *Rhodopseudomonas sphaeroides* has also been reported to denitrify (Satoh *et al.* 1976) as has the obligate anaerobe *Propionibacterium acidipropionici* (van Gent-Ruijters *et al.* 1975). Moreover, the genus *Bacillus* contains many species of denitrifiers which are facultative anaerobes. Thus there are some exceptions to the definition proposed in the introduction.

The few ecological studies on the isolation and identification of denitrifiers (Gamble *et al.* 1977; Vives & Parés 1975), indicate that in soil the primary genera are *Pseudomonas* and *Alcaligenes*. These two genera also comprise a significant fraction of total soil and aquatic bacteria, at least as estimated by the viable plate count method. However, their actual quantitative contribution to denitrification in soil has not been tested. Thermophilic *Bacillus* species have been implicated in denitrification studies where maximal rates occur at temperatures of 65°C or higher (Nommik 1956). Chemolithotrophic denitrification by *Thiobacillus denitrificans* in nitrogen-deficient orchard soils amended with sulphur was suggested by Martin and Ervin (1953) and demonstrated in soil columns treated with sulphur (Mann *et al.* 1972). Finally chemolithotrophic (or more properly mixotrophic) denitrification with hydrogen as the energy source is known to occur in *Paracoccus denitrificans* (formerly *Micrococcus denitrificans*—Verhoeven *et al.* 1954) and *Alcaligenes eutrophus* (Davis *et al.* 1969). However, it is becoming increasingly clear that many bacteria are capable of growing autotrophically with hydrogen as the energy source including *Rhizobium japonicum* (Hanus *et al.* 1979), which is also known to be a denitrifier (Zablotowicz *et al.* 1978).

12.2.2 Nitrate-Respiring Organisms

Many bacteria are also capable of reducing nitrate to nitrite but are incapable of further dissimilatory reduction of nitrite. These bacteria are usually referred to as nitrate-respirers and they show a greater generic diversity than the denitrifiers. In fact, genera which contain species of denitrifiers are represented by species that are nitrate-respirers. Many aerobes, facultative anaerobes, and obligate anaerobes are nitrate-respirers. Comprehensive lists have been compiled by Payne (1973), Hall (1978) and Ingraham (1981). Of particular interest are the nitrate-respiring bacteria which are able to reduce nitrite to ammonia since this reductive process, unlike denitrification, conserves nitrogen.

Dissimilatory nitrate reduction is not akin to assimilatory nitrate reduction (apart from the metabolites that are probably formed) because the former process, unlike the latter, occurs only under anaerobic conditions and involves participation of nitrate reductase with a respiratory cytochrome system that generates ATP through oxidative phosphorylation. Further dissimilatory reduction of nitrite does not appear to involve coupling with oxidative phosphorylation (Payne 1973; Prakash & Sadana 1973). It is conceivable that nitrite reduction is a means of detoxification since nitrite is known to be toxic to denitrifiers and nitrate-respirers. However, Caskey and Tiedje (1979) suggest an alternative possibility which is that nitrite serves as an electron sink for anaerobic or facultative anaerobic bacteria in regenerating acetyl phosphate or acetyl CoA used in substrate level phosphorylation. The possible environmental significance of these bacteria and their competition with denitrifiers for nitrogen oxides is discussed in section 12.6.1.

12.2.3 Truncated Denitrifying Organisms

Since the discovery by Yuoatt (1954) and by Vagnai and Klein (1974) of nitrate reductase-deficient denitrifiers, many other denitrifying bacteria which lack other nitrogen oxide reductases have been isolated. According to Ingraham (1981) it appears that nitrate reductase-lacking bacteria are the more prevalent biotypes among truncated denitrifiers, although this is by no means certain. The next most common group appears to be that containing those lacking nitrous oxide reductase. Ingraham (1981) suggested that nitrous oxide reductase may be plasmid coded since many strains upon continuous subculturing eventually lose the ability to reduce nitrous oxide and that the absence of this enzyme has never been reported among strains immediately isolated from soil. Zablotowicz *et al.* (1978) noted that gas formation, presumably dinitrogen, was highly variable among each rhizobium strain and was inversely related to the concentration of nitrous oxide in the culture's gas phase. However, most investigators characteristically isolate denitrifiers from most-probable-number tubes (see section 12.5.1) which show vigorous gas formation. Since nitrous oxide is about 25 times more soluble than dinitrogen and thus not likely to displace as much liquid from inverted Durham tubes, selection of denitrifiers on the basis of gas formation would be biased against any such nitrous oxide reductase-deficient strains occurring in nature. Garcia (1977) reported a considerable diversity among denitrifying *Bacillus* species isolated from the rhizosphere of rice, particularly several strains that were deficient in nitrous oxide reductase. Finally Pichinoty *et al.* (1978) found that *Bacillus licheniformis* lacked both nitrate and nitrous oxide reductase. They also isolated denitrifiers from soil that utilised nitrous oxide, but not nitrate, as a terminal electron acceptor for growth. The environmental implications of truncated denitrifiers is unclear at present, although further study may explain the variability in the proportion of nitrogen gases formed during denitrification among different soils.

12.3 Substrates and products

12.3.1 Reductant

Since denitrification is a respiratory process, it is not surprising that it is increased upon the addition of reductant. The lack of available reductant in many ecosystems, such as secondary waste effluent or soils low in organic matter, is attributable to environmental problems associated with the discharge of nitrate into rivers and lakes or its leaching through soil to ground waters. Such problems can be ameliorated by proper management techniques. Reducing the completion of the secondary treatment stage leaves enough residual organic matter to supply the necessary reductant for complete conversion of nitrate to dinitrogen (Wuhrmann & Mechsner 1973). Similarly, in irrigated agricultural soils or in instances where waste effluent is added to soil, nitrate removal can be regulated by the application frequency to allow for adequate periods of drying, during which oxidation occurs, and wetting, during which reduction occurs (Avnimelech & Raveh 1974; Bouwer *et al.* 1974). The effect described by Birch (1958) that is brought about during wetting and drying cycles in soil also appears to cause greater losses of nitrogen from the system through nitrification and subsequent denitrification (Avnimelech 1971; Paul & Meyers 1971).

12.3.2 Oxidant Concentration

There is considerable confusion as to whether denitrification is increased by the addition of nitrogenous oxides, particularly nitrate. In pure cultures of *Pseudomonas aeruginosa* and *Escherichia coli*, the K_m values for nitrate reductase are 16 μM and 500 μM respectively (Fewson & Nicholas 1961; Nason 1962). The lower value is considerably less than the concentrations normally found in soil or, for that matter detectable by most methods normally used to measure nitrate. Many studies show adherence to apparent first order Michaelis-Menten kinetics in soil, while others do not (Focht & Verstraete 1977). Bowman and Focht (1974) and Kohl *et al.* (1976) showed that a zero order response to nitrate concentration could be the result of limiting availability of the reductant. Thus, when a carbon source was added, the reaction was observed to follow a kinetic pattern that was dependent upon nitrate concentration. Recently Phillips *et al.* (1978) and Reddy *et al.* (1978) advanced the concept that first order kinetics of denitrification could be due to limitations of available nitrate as a result of concentration gradients developed by the slow liquid diffusion of nitrate throughout the complete zone of activity in unshaken vessels. This appears to be a very useful concept inasmuch as the K_m for respiratory oxidations are quite low (μM). Moreover, it would seem logical that during the evolution of respiratory processes, bacteria that were most efficient and successful in extracting oxygen or other oxidised compounds would be those that had lower respiratory Michaelis-Menten constants. Thus the apparent adherence of respiratory processes at high concentrations of oxidant to anything other than a zero order state should be considered in the context of rate limitations brought about by diffusion as opposed to enzyme saturation.

12.3.3 Oxidant Competition

Since oxygen is the preferred electron acceptor to nitrate because of its higher energy potential, it would seem logical that metabolism of all the other nitrogen oxides produced during denitrification would be governed by some basic thermodynamic considerations. In many instances, a sequential reduction of $NO_3^- \rightarrow NO_2^- \rightarrow N_2O \rightarrow N_2$ is observed, while in others nitrite is not found as a transient intermediate. In other cases, none of the transient intermediates between nitrate and dinitrogen are observed (Focht & Verstraete 1977). In fact nitric oxide is rarely observed as a transient intermediate at all in cultures or soil except under acidic conditions and, for the most part, is observed only with cell-free extracts.

Reference to redox potentials in ideal systems is not always helpful in predicting which oxidants would be better electron acceptors. The E_h couple for nitrous oxide: gaseous dinitrogen for example is 1.7 volts (pH 7.0, 25°C—Pourbaix 1966), which places it above the E_h couple for oxygen: water. However, nitrous oxide is a much less efficient oxidant than nitrate or oxygen (Koike & Hattori 1975a,b)—possibly yielding only one ATP per two electron transfer—and is preferentially reduced only after nitrate or nitrite are used up in soil and in culture (Blackmer & Bremner 1978; Firestone *et al.* 1979; Focht *et al.* 1979; Smith *et al.* 1979). Moreover, nitrate is known to repress further reduction of nitric oxide in *Pseudomonas* spp. (Miyata 1971), while nitrite represses both nitrate and nitrous oxide reductases (Payne 1973). Problems in resolving anomalies between cultures or soil incubations are no doubt due to differences in physical or chemical factors which probably alter the kinetics of each step in the pathway differently.

12.3.4 Mixogenic Denitrification

Nitrite, which is generally reduced more rapidly than nitrate, is also much more reactive chemically than nitrate. Nitrous oxide can be generated by the reaction of nitrite and hydroxylamine, both of which are products in the reduction of nitrate to ammonium and in the oxidation of ammonium to nitrite. Thus the production of nitrous oxide from fungi (Bollag & Tung 1972), nitrate-respiring bacteria (Tiedje 1982a), and *Nitrosomonas europaea* (Yoshida & Alexander 1970; Ritchie & Nicholas 1972; Blackmer *et al.* 1980) may occur in this manner. However, the possibility that *Nitrosomonas europaea* may also be a denitrifier as suggested by Ritchie and Nicholas (1972) and Payne (1973) cannot be precluded.

The terms 'chemogenic' and 'biogenic' denitrification are self explanatory. The term 'mixogenic denitrification' is proposed to describe that process whereby nitrite is generated biologically and subsequently reacts chemically to form nitrous oxide or another gaseous nitrogen product.

Only within the last eight years have gas chromatographic methods been improved to the point where it has become feasible and routine to measure low concentrations of nitrous oxide in the atmosphere above culture flasks. These improvements in analytical procedures for measuring gaseous products may reveal that nitrous oxide is produced by more microorganisms than previously considered.

12.4 Environmental factors

12.4.1 PH

Denitrification is most rapid in the slightly alkaline range between pH 7.0 and 8.0 (Wiljer & Delwiche 1954; Nommik 1956) though the limits with respect to pH are quite wide ranging, from pH 3.5 (Cady & Bartholomew 1960) to pH 11.2 (Prakasam & Loehr 1972). Generally, more nitrous oxide is produced under acidic conditions. Acidic soils also release more nitric oxide (Cady & Bartholomew 1960; Bollag *et al.* 1973; Garcia 1973) although a considerable amount is produced by the non-biological reduction of nitrite.

12.4.2 Temperature

Optimal rates of denitrification have been reported to occur at 65°C by Nommik (1956) who attributed this high optimum to the predominance of thermophilic *Bacillus* species. However, Keeney *et al.* (1979) found that gaseous nitrogen evolution exceeded the amount of nitrate added at temperatures of 50°C or higher. Though they were unable to demonstrate significant chemodenitrification to dinitrogen in sterilised soils amended with nitrite and incubated at high temperatures, they did observe considerable quantities of nitric oxide, a gas which was not observed in appreciable quantities in non-sterile soil. They concluded that thermophilic denitrification was a combination of biological and chemical reactions in which nitrate-respiring bacteria (*Clostridium* species or *Bacillus* species) generated nitrite which reacted chemically and biologically to form gaseous products.

It is not clear what effect, if any, temperature has upon the relative rates of nitrogenous oxide reduction though it appears to be similar with respect to both the production and reduction of nitrous oxide (Nommik 1956; Bailey & Beauchamp 1973b; Focht & Verstraete 1977). However, Bailey and Beauchamp (1973b) found that the rates of nitrate reduction were more drastically affected by low temperatures than nitrite reduction. No reduction of nitrate occurred at 5°C during a 22 d incubation period, whereas considerable nitric oxide production was observed from the addition of nitrite. Chemodenitrification was offered as one possible alternative, although the other explanation that nitrite repressed further reduction was preferred by Focht and Verstraete (1977) because of the neutral soil pH and because nitrite is known to repress nitric oxide reduction (Payne 1973). In the light of similar observations by Keeney *et al.* (1979) with sterilised soil at high temperature, chemodenitrification can no longer be ignored as inconsequential under conditions that favour nitrite accumulation (i.e. extreme temperatures) in neutral or alkaline soils.

12.4.3 Oxygen

Aeration, without question, is the greatest and most variable factor that affects the denitrification process in soil. Considerable controversy existed in the early literature regarding 'aerobic denitrification' largely because what was perceived to be an aerobic environment in fact was not. Collins (1955) found that the geometry of the culture flask was an important factor governing the rate of denitrification under 'aerobic conditions'. It was correctly concluded that oxygen diffusion from the atmosphere above to the solution throughout was the major factor involved. Diffusion of gases through liquid is a comparatively slow process by comparison to diffusion in air (about 100 000 times less). An undisturbed soil suspension or bacterial culture will generally be depleted of all the oxygen below a depth of 1 mm or less from the surface (Focht 1979). Consequently, the dissolved oxygen concentration, not the gaseous concentration, determines whether conditions are conducive for denitrification. The oxygen concentration below which denitrification occurs ranges from 0.7 $\mu g\ ml^{-1}$ (Goering & Cline 1970) to 2 $\mu g\ ml^{-1}$ (Wheatland *et al.* 1959), although oxygen becomes limiting to respiration at a lower value of 0.1 $\mu g\ ml^{-1}$ (Chance 1957).

12.4.4 Interactions

Misra *et al.* (1974) noted a synergistic effect of oxygen and temperature upon denitrification. At 19.5°C the rate constant for nitrate reduction was reduced more than 10-fold from a gaseous oxygen concentration of 0 to 20%, whilst a temperature of 34.5°C over the same oxygen concentration range effected a decrease to only 75% of the higher value. Higher temperatures cause a reduction in the soluble oxygen concentration as well as an increase in biological rate processes. This interaction may explain why denitrification in thermic semi-arid soils is characterised as a rapid and transient process confined to the uppermost 10 cm since this zone is subjected to the highest temperatures and most variable change in temperature and aeration upon irrigation (Focht 1978; Focht & Stolzy 1978; Focht *et al.* 1979; Ryden *et al.* 1979b).

Experimental

12.5 Bacterial enumeration

12.5.1 Most Probable Number

The commonest method used for enumeration of denitrifiers is the most-probable-number (MPN) procedure (see Chapter 6, p. 88) in spite of the rather large error ($\pm$ 0.5 log encompasses the 95% confidence interval). The procedure, as it is used for denitrifiers, is outlined in Focht and Joseph (1973) and described in further detail by Alexander (1965a,b) and Tiedje (1982a). It consists of diluting the soil sample in a logarithmic series of sterile dilution bottles up to 1×10^{-9}. Normally five sterile, screw-capped tubes of nitrate broth are inoculated asceptically with 0.1 ml from the dilution bottles, and the caps tightened to exclude oxygen. If the investigator has some idea of the range of bacterial numbers in the sample, it is not necessary to inoculate all the dilutions. At the end of a 14 d incubation, the last three logarithmic tubes near the dilution to extinction are read and removed for analysis. The presence of nitrite is indicative of nitrate-respiring bacteria, while no nitrate or nitrite is presumptive evidence for denitrifiers. This method is not only capable of enumerating denitrifying and nitrate-respiring bacteria, but also all other heterotrophic types that will grow in the medium. Even in instances where the denitrifying populations may constitute a small fraction of the total, they can still be enumerated by removing the tubes from the less diluted series to analyse for nitrate or nitrite. However, it is not possible with this method to enumerate nitrate-respiring bacteria if their numbers are half or less than denitrifiers since any nitrite produced by the former group in the less diluted series will be used by denitrifiers, which would of course be present in all the lower dilutions.

Several modifications to the procedure of Focht and Joseph (1973) have since proved to be more beneficial. Inverted Durham tubes are inserted for trapping gaseous products, particularly dinitrogen, that are evolved during the course of denitrification. The addition of succinic acid (5 g l^{-1}) followed by neutralisation to pH 7.0 is also recommended since this causes more vigorous dinitrogen production in pure culture. Succinate enhances the selection of aerobic bacteria since it cannot be fermented. The use of Durham tubes to note gas formation and a pH indicator dye (e.g. phenol red or bromothymol blue) to note an increase in pH (Valera & Alexander 1961) without assaying nitrate disappearance are equivocal indices for denitrification (Focht & Joseph 1973). However, when the formation of gas is noted with subsequent disappearance of nitrate, the presumption of denitrification is greatly strengthened. Volz (1977) suggested the use of nitrite in place of nitrate, although there does not appear to be any advantage gained in doing so since truncated denitrifiers are excluded and since nitrite is toxic to many bacteria. A lower nitrate concentration, as recommended by Tiedje (1982a), seems warranted in ensuring the complete conversion of intermediates—namely nitrite and nitrous oxide—to dinitrogen.

The disappearance of nitrate and nitrite with concomitant gas formation does not prove that these occurrences are the result of denitrification (Tiedje 1982a). Bubble formation may be the result of carbon dioxide or hydrogen evolved by nitrite-respiring bacteria, such as *Clostridium* spp., which reduce nitrate to ammonia. Consequently, Tiedje (1982a) recommended the use of Hungate tubes with a butyl rubber septum and locking cap. Prior to incubation, 1.0 ml of acetylene is injected into the 10 ml gas space above the medium in order to block further reduction of nitrous oxide to dinitrogen. At the end of 14 d incubation, samples are withdrawn with a syringe and analysed for nitrate and nitrite on a spot plate. Tubes that are negative for both are then sampled for nitrous oxide production by withdrawing 1.0 ml of gas with a syringe and injecting it into a gas chromatograph for analysis. If nitrous oxide is not observed, the culture is considered to be a dissimilatory ammonia producer. This procedure is definitive but is time consuming since each GLC assay requires about 5 min. Gamble *et al.* (1977) found that the gas bubble produced in tubes where nitrate or nitrite was absent was almost always due to denitrification when the medium of Focht and Joseph (1973) was used. Consequently, scientists who do not have access to a gas chromatograph can be somewhat reassured that gas production and disappearance of nitrate and nitrite are good indications of denitrification when nitrate broth is used.

Several other media have been used for enumerating denitrifiers. W. H. Caskey (personal communication) found that a richer medium containing carbohydrate (tryptic soy-nitrate broth) was more selective for nitrate-respirers than for denitrifiers, but noted only a 30% agreement between presumptive and confirmative methods for denitrification with tryptic soy-nitrate broth. Tiedje (1982a) thus concluded that a carbohydrate-free protein medium, such as that used by Focht and Joseph (1973) is best for enumeration of denitrifiers.

Although the choice of media undoubtedly biases the selection of those bacteria which are the best adapted, a medium must be chosen which provides the greatest nutritional diversity or at least one which is a close approximation of the environment from which the sample was obtained. For example, one may justifiably consider peptone media to be more representative of rhizosphere bacteria since amino acid dependency is well established among rhizosphere bacteria (Alexander 1977). On the other hand, the use of tryptic soy media or other vegetable digest high in carbohydrates may be more realistic of conditions which commence when fresh plant residues are incorporated into soil. Defined or nutritionally poor media are discouraged since they always give lower counts of total bacterial numbers as well as denitrifiers by as much as one to three orders of magnitude.

12.5.2 Microtitre Plates

Microtitre dilution plates are a recent development that give greater precision by virtue of the greater numbers of inoculations. The procedure is similar to the MPN method whereby several serial dilutions are made. However, a single plate, rather than several test tubes, is inoculated simultaneously with 25 to 50 small needles from a single reservoir containing the dilution. After a suitable period of anaerobic incubation, nitrate or nitrite disappearance is assessed by adding acid and an appropriate indicator such as diphenylamine. This method (Rowe *et al.* 1977) has considerable promise over the MPN tube method because inoculation is rapid and the precision is greater. The one disadvantage is that denitrification cannot be confirmed by gas production.

12.5.3 Spread Plates

A viable plate count method has also been reported for enumerating denitrifiers (Mycielski & Gucwa 1977). After inoculation of the nitrite agar, the plates are overlaid with another layer of nitrite agar (to impede oxygen diffusion) and incubated. Alternatively, the agar overlay can be omitted if an anaerobic incubator is available. When colonies are visible, the plate is flooded with a reagent, and colourless zones around the colonies are scored as denitrifiers. Nitrite, rather than nitrate, is used on the presumption that the former is more selective for denitrifiers. However, nitrite is toxic even to denitrifiers, so that the method selects for those bacteria which have a high tolerance to nitrite. The greatest problems with viable plate count procedures, however, are the highly variable growth rates and colony sizes among soil isolates. Consequently, many smaller 'satellite' colonies appear around larger ones and cannot be scored with respect to nitrite disappearance since the entire zone surrounding the larger colony encompasses the smaller colonies as well. Longer incubation periods do not appreciably increase the size of the smaller colonies and one cannot be certain that they are denitrifiers.

12.5.4 Importance to Rate Processes

Denitrifying bacteria are ubiquitous in soil and contribute a significant fraction of the normal heterotrophic bacterial flora. Since they are for the most part aerobes, their numbers have no more quantitative meaning than would any other method such as the viable plate count method. A recent study (Focht 1978) illustrated the futility of counting denitrifiers in soil and attempting to correlate these numbers with any known parameter such as organic matter content, moisture content, texture or plant cover. This raises the immediate question, What use is an enumeration procedure which gives no quantitative meaning to the environment from which the sample originated? In terms of simply going out to the field to count denitrifiers, the answer may well be that it represents an exercise in futility. Nevertheless, there are instances in which valuable information can be gained by knowing the relative proportion of denitrifiers to other physiological groups.

The ability to enumerate the growth of denitrifiers as well as the rest of the heterotrophic flora enables the investigator to monitor any changes in population dynamics. Bowman and Focht (1974), for example, found that the optimal rates of denitrification coincided with denitrifying populations of 100% of the total heterotrophic bacterial populations in soil at optimum glucose and nitrate concentrations. At the highest glucose concentration or highest nitrate concentration, the rates were greatly diminished. Not only were total numbers less, but the percentage of denitrifiers was less than 5% in some cases. Thus they were able to explain that high glucose concentrations impeded denitrification because of inhibition to denitrifiers.

The factors that govern the establishment of the two groups of bacteria are important since nitrate respiration is a reductive process which, unlike denitrification, conserves soil nitrogen. Volz *et al.* (1975) found that the numbers of nitrite-oxidising bacteria (presumably *Nitrobacter* spp.) were considerably larger than ammonia-oxidisers, and that nitrate, when added to soil, increased the numbers of *Nitrobacter* spp. Normally *Nitrosomonas* spp. outnumber *Nitrobacter* spp. in soil because an equimolar amount of ammonia supplies three times as much energy for incorporation into biomass as does a nitrite. Volz *et al.* (1975) concluded that the high proportion of nitrate-respirers to denitrifiers and the higher proportion of *Nitrobacter* spp. with respect to *Nitrosomonas* spp. were caused by the cyclic reduction of nitrate to nitrite by the nitrate-respirers in anaerobic microsites, whereupon nitrite diffused to aerobic sites where it was reoxidised by *Nitrobacter* spp.

12.6 Activity *in vitro*

The choice of the size, geometry and the method of sealing the vessel is largely dependent upon what the investigator chooses to measure. For MPN nitrate broth tubes, it is satisfactory to tighten the screw caps without eliminating the initial oxygen concentration since the purpose is strictly to ascertain the presence or absence of nitrate at the end of the incubation period. However, measurements of the denitrification potential of soil, as it most closely resembles field conditions, requires the evacuation of the atmosphere above the flask immediately after adding the soil since three general phases of denitrifying activity occur as follows:

(1) a period in which there is a slow but constant reduction of both nitrate and nitrous oxide;

(2) a period in which nitrate reduction activity increases with a subsequent increase in nitrous oxide;

(3) a period of increased nitrous oxide reduction during which nitrous oxide is rapidly converted to dinitrogen (Firestone *et al.* 1979; Smith *et al.* 1979).

Since the activities (i.e. enzyme concentrations) are not constant during the incubation, it is necessary to make several measurements. The earlier time period is indicative of the immediate denitrifying capacity of the soil as when it is irrigated or rain-fed. The later time periods represent induction or enzyme synthesis *de novo*. The specific regulatory control of the nitrogen oxide reductases is not clearly understood, although it seems that there are small quantities always present, which may explain the low but immediate denitrifying activity upon an immediate shift to anoxic conditions. In the genus *Bacillus*, synthesis of cytochrome a_3, the terminal cytochrome for oxygen reduction, ceases while synthesis of the terminal cytochrome for nitrate reduction increases (Downey & Kiszkiss 1969).

The best method for measuring these activities is not obvious because of conflicting effects relating to enzyme kinetics and gas diffusion in ponded or saturated soils. To be certain that the rate is not dependent upon nitrate concentration, nitrate is frequently added to ensure enzyme saturation. This has the added benefit (or required necessity) in instances where the indigenous soil nitrate is too low to give any measurable response by either analysis of substrate disappearance or product formation. Although the addition of nitrate is to be recommended for measuring nitrate reductase activity, high concentrations of nitrate impede reduction of nitrous oxide (Blackmer & Bremner 1978; Firestone *et al.* 1979; Focht *et al.* 1979). Consequently, if measurements of nitrous oxide reductase activity are required then the substrate nitrous oxide should be added. Garcia (1974) used nitrous oxide reduction as a valid and specific criterion for denitrifying activity and showed an initial stage represented by a slow but constant rate that was attributed to the existing activity and a second more rapid stage which was defined as the maximum denitrifying capacity of the soil.

12.6.1 Nitrate Disappearance

Analytical methods for measuring nitrate have been discussed in other reviews (Bremner 1965a; Focht 1978; Tiedje 1982a).

The loss of nitrate under carefully controlled and well-understood conditions can generally be assumed to be due to denitrification in the absence of exogenous substrate although exceptions do occur. A discussion of these exceptions, therefore, is germaine to resolving when disappearance of nitrate is a valid criterion for measuring denitrification.

The free energy for the reduction of nitrate to ammonium is -143 kcal mol^{-1} NO_3^- as compared with -134 kcal mol^{-1} when nitrate is reduced to dinitrogen. However, denitrification provides more energy (-26.8 kcal) than reduction to ammonium (-17.9 kcal) per electron transfer. Thus Delwiche and Bryan (1976) concluded that denitrification was the more efficient reaction where the reductant was limiting. Without exception, dissimilatory nitrate reduction to ammonium in soils is of consequence only when glucose or other available carbon source is added (Stanford *et al.* 1975; Buresh & Patrick 1978; Caskey & Tiedje 1979). Caskey and Tiedje (1979) conclusively showed that nitrate reduction to ammonium was insignificant by comparison to denitrification when no exogenous substrate was added, but that half of the nitrate was reduced to ammonium upon addition of glucose. Moreover, they showed by sequential time samplings that the pathway did not involve direct assimilation into organic nitrogen. Since ammonium was not formed upon the addition of acetate and occurred only with the liberation of hydrogen with glucose as substrate, they properly concluded that *Clostridium* was the likely bacterial genera involved in dissimilatory nitrate reduction to ammonium. This point once again reinforces previous comments regarding the disappearance of nitrate in MPN tubes of nitrate broth. When organic matter (e.g. nutrient broth) has a low carbon : nitrogen ratio, the system is reductant-limiting and denitrification is strongly favoured; where the carbon : nitrogen ratio is high (e.g. tryptic soy broth), the oxidant is limiting and reduction to ammonium is favoured.

There have been many reports which indicate that nitrate reduction to ammonium in sediments is a very significant mechanism (Keeney *et al.* 1971; Koike & Hattori 1978; Sorenson 1978) which accounts for more than half of the added nitrate even in the absence of an exogenous reductant. In highly anoxic environments, the greatest impediment to the growth and maintenance of obligate anaerobes is in obtaining oxidants for regeneration of acetyl phosphate and acetyl CoA. Thus nitrogen oxides serve as electron sinks for making ATP from substrate level phosphorylation. Therefore, it is not too surprising that growth yields of *Clostridium* spp. are increased upon the addition of nitrate under anaerobic growth (Hasan & Hall 1977; Caskey & Tiedje 1979).

Sediments and rice paddies are more anoxic environments than arable soils and may represent the dissimilatory reductive sequence of nitrate starting from 'the bottom', as opposed to denitrification which starts 'at the top'. For example, when a well-aerated soil shifts to anoxic conditions, the use of inorganic electron acceptors by the bacterial flora follows a characteristic sequence in the order oxygen, nitrate, manganese and ferric ion. This sequence is directly related to the amount of energy that can be coupled with pyridine nucleotide reduction. Since soils are frequently subjected to wetting and drying cycles, the microflora tend to be predominantly aerobes with facultative and obligate anaerobes constituting a smaller proportion. However, if soil is flooded for a long time, as in rice cultivation, the composition of the microflora changes to predominantly anaerobic types except at aerobic microsites near the root and sediment interface. Consequently, when nitrate is added to an anaerobic system where there are plenty of oxidised organic compounds (i.e. acids produced by fermentation), bacteria which have a partial respiratory system can quickly change to an oxidative metabolism since there is a distinct thermodynamic and competitive advantage over those bacteria which are strictly fermentative and unable to use inorganic electron acceptors. This is analogous to the advantages that denitrifying bacteria have over non-denitrifying anaerobes when the environment changes from aerobic to anaerobic.

In practice, nitrate is never added to rice paddies because it is denitrified in the anoxic zone located below the oxidised flood water. Conceivably, if nitrate were added to the deeper, more anoxic zones that were devoid of denitrifiers, it ought to be reduced to ammonium rather than dinitrogen. Although this possibility has not been tested, there is an example of its occurrence in nature namely the ruminant gut. Imbibing of water containing a high nitrate concentration can be fatal to ruminants because of the rapid reduction of nitrate to nitrite by nitrite-respiring anaerobic bacteria (Lee 1970). Although there is no evidence that any of the nitrogen oxides are reduced further to ammonium or dinitrogen, the gut microflora do not contain any of the known denitrifying genera (Bryant 1964).

12.6.2 Gaseous Product Appearance

Unquestionably, the production of gaseous products, nitrous oxide and dinitrogen, is the most definitive evidence for denitrification. During the last 15 years the standard method of analysis of gaseous nitrogen compounds has depended upon the use of gas liquid chromatographic systems. Few, if any, scientists measure gaseous nitrogen compounds by respirometry, a common procedure of three decades ago (Allen & van Niel 1952). The older method is tedious and less definitive and requires the use of two separate series of flasks, one with and one without an alkali well, in order to correct for carbon dioxide evolution. Furthermore the gaseous nitrogenous oxides cannot be distinguished.

Several gas liquid chromatographic methods have been developed for denitrification studies. Two separate columns are generally used for distinguishing between dinitrogen, nitrous oxide and nitric oxide although the last named is only occasionally produced in soil or culture (Bailey & Beauchamp 1973a; Garcia 1973, 1977). A Poropak Q column is used for separation of nitrous oxide from the other gases, while a molecular sieve 0.5 nm is used for the detection of dinitrogen. Neither column is sufficient by itself since nitrous oxide will not pass through a molecular sieve at normal operating temperatures, and dinitrogen, oxygen and argon appear as one peak on a Poropak Q column. Many variations have been suggested to resolve dinitrogen and nitrous oxide using a single sample injection. These methods employ switching valves (Smith 1977; Thompson 1977), or involve splitting the sample flow between the two columns, or running the sample through the detector twice: first on a Poropak Q column and then through a molecular sieve column (Beard & Guenzi 1976). Chromosorb 101 and 102 waxes have also been used for resolution of all atmospheric gases on one sample injection (Thompson 1977), but this usually involves a considerably longer operating time since the starting temperature requires a refrigeration system (−70°C) to separate dinitrogen, oxygen and argon and a temperature programmer to gradually increase the temperature. In terms of cost and simplicity, most investigators who are not concerned with numerous sample sizes generally use two separate injections from the same sample onto separate columns of Poropak Q and molecular sieve 0.5 nm.

The use of very high purity helium may appear warranted when one is interested in measuring dinitrogen evolution by gas chromatography since standard helium mixtures usually contain about 1 to 5 ml l^{-1} dinitrogen. Practically it is very difficult to measure dinitrogen concentrations less than 1 ml l^{-1} because of ambient contamination from dinitrogen that is introduced during the sampling procedure. A locking valve syringe is recommended when dinitrogen is to be measured, although ambient contamination of dinitrogen in the needle cannot be avoided. The insertion of a two-way valve off a branch in the carrier gas line immediately before the gas sample loop is recommended so that the sample loop can be purged prior to each analysis.

The most popular detector used in denitrification studies is the thermal conductivity detector, since it is non-specific and non-destructive. The newer models afford a much higher degree of stability and sensitivity than those used in earlier denitrification studies. The minimum detectable limit for nitrous oxide with this detector (0.5 ml sample size) is about 50 ppm (v/v) from soil or culture gas mixtures. If carbon dioxide is removed from the atmosphere, by either an alkali well placed inside the vessel or by scrubbing the gas sample through an ascarite precolumn, as low as 10 ppm (v/v) can be detected. The electron capture detector is better suited to concentrations below 50 ppm (v/v) with a lower limit of detectability to 0.03 ppm (v/v) for nitrous oxide, which is 0.1 of the ambient concentration. Electron capture does not respond to dinitrogen, but ambient concentrations of oxygen will cause quenching of the detector such that a flat base line is not evident when the nitrous oxide peak emerges. The ultrasonic, or coulometric detector, originally thought to be the best overall for denitrification studies on the basis of its slightly greater stability and response to dinitrogen and nitrous oxide, in reality represents only a very slight improvement over thermal conductivity at considerably greater cost. The most sensitive and versatile detector is a quadrupole mass spectrometer, which is interfaced to a gas chromatograph.

12.6.3 [^{15}N]Nitrogen

[^{15}N]Nitrogen isotopes have been useful in establishing the importance of nitrification and denitrification processes in the losses of nitrogen from flooded soils (Broadbent & Tusneem 1971; Patrick & Reddy 1976). Nevertheless, the inherent insensitivity of dual focusing magnetic mass spectrometers generally precludes direct measurement of [^{15}N]dinitrogen such that these authors calculated denitrification losses by estimating the [^{15}N]nitrogen remaining in the soil. Moreover, the ambient concentration of $^{15}N^{14}N$ is sufficiently high (5.7 ml l^{-1}) to require extreme precision or large amounts of gas formation from [^{15}N]nitrogen compounds to detect a difference. Although small changes would be more easily noticed in $^{15}N^{15}N$ (ambient concentration = 11 μl l^{-1}) from small production rates, the isotope mass spectrometer lacks the sensitivity to detect such a small quantity. The problems related to interpretation of gaseous [^{15}N]nitrogen data have been considered by Hauck *et al.* (1958).

Accurate measurement of low levels of [^{15}N]nitrogen gases (6 pg for $^{15}N^{15}N$) is now possible with a quadrupole mass spectrometer which is interfaced to a gas liquid chromatograph. Although the precision of the quadrupole mass spectrometer is inferior to the double-focusing magnetic mass spectrometer it is about 4 to 5 orders of magnitude more sensitive. Moreover, the gas liquid chromatographic separation enables rapid measurement of low levels of dinitrogen and nitrous oxide, a feature not possible with the dual focusing isotope spectrometer. Many of the problems discussed by Bremner (1965b) and Cheng and Bremner (1965) concerning the quantitative instability of nitrous oxide and interferences from other gases are irrelevant to the quadrupole GLC-MS procedure. Undoubtedly this system is the most powerful, sensitive and rapid analytical tool for identifying and confirming an unknown compound. The sensitivity is exceeded only by [^{13}N]nitrogen tracer studies (see section 12.6.4). A more detailed account of GLC-MS procedures for denitrification studies is given by Focht (1978) and Focht *et al.* (1980).

12.6.4 [^{13}N]NITROGEN

[^{13}N]Nitrogen is a radioactive isotope with a half life of 9.96 min. It decays by positron emission of β and γ rays to [^{13}C]carbon. When produced directly by bombardment of [^{16}O]oxygen and proton (water target) [^{13}N]nitrate is the primary product (Tiedje *et al.* 1979). The short-lived [^{15}N]nitrate compound is immediately purified by high pressure liquid chromatography and added to the reaction vessel. A detailed account of this methodology for denitrification studies is described by Tiedje *et al.*(1979) and Gersberg *et al.* (1976). [^{13}N]nitrogen has limited uses because of the very complex experimental design and expensive equipment required, especially a cyclotron. Nevertheless for those with access to such facilities, [^{13}N]nitrogen isotopes offer a distinct advantage over all GLC methods, including [^{15}N]nitrogen isotopes, namely a far greater sensitivity (detection of 1×10^{-21} mol compared with 1×10^{-12} mol for [^{15}N]nitrogen even when the quadrupole mass spectrometer is considered. This feature is important when measuring the kinetics of denitrification in soil at low nitrate concentrations (less than 1×10^{-6}M). In order to obtain a measurable quantity of gas production for measurement by thermal conductivity, the nitrate concentration must be at least 1 μg N ml^{-1}, and preferably greater in order to observe significant changes during initial time increments. The K_m for the nitrate and nitrous oxide reductases in culture and soil are below this concentration (Yoshinari *et al.* 1977). The [^{13}N]nitrogen method offers a distinct advantage over the [^{15}N]nitrogen method since background contamination from dinitrogen is non-existent, while the lower limits of the quadrupole GLC-MS are generally determined by the amount of ambient dinitrogen gas, specifically ^{15}N^{15}N, that is not removed during the initial flushing.

12.6.5 PURE CULTURE ACTIVITY

Analysis of substrate disappearance and product formation during denitrification in pure cultures is a more difficult task because of problems related to contamination. Although Hungate tubes (see Chapter 1) are suitable for growing denitrifying bacteria anaerobically, the method has its limitations.

Firstly, when the air space above the test tube is flushed using a sterile stainless steel tube connected to a pre-filtered gas line, it is not possible to replace and tighten the cap without including oxygen. Although the oxygen is consumed, kinetic studies are not accurate since the cells use the oxygen before they begin to denitrify. Secondly, the small gas volume of the Hungate tube (10 ml) enables no more than a single 1 ml sample to be withdrawn; however, this can be obviated by inoculating many tubes and removing 2 or 3 for sampling on each occasion. Nephelometry flasks (250 ml) fitted with gas-tight serum caps and gassed free of oxygen are most suitable for combined growth and denitrification studies since several samples can be withdrawn without appreciable errors in partial pressure changes. Nevertheless, this method is not recommended when rich media are used since contamination is a frequent problem during flushing or sampling. The use of a sterilised needle, fitted to 0.22 μm membrane filters on the gassing manifold solves the problem as long as the serum cap is sterile when the manifold needle is introduced into the flask. Tiedje (1982b) recommended pre-sterilised disposable plastic syringes for each injection to insure asepsis during sampling. The easiest way to prevent contamination is to use defined minimal media since most of the contaminants, notably *Bacillus* spp., are Gram-positive bacteria which do not grow or grow very poorly on minimal media. Unfortunately, the use of minimal media limits pure culture studies to the Gram-negative rods, notably *Pseudomonas* and *Alcaligenes* species.

12.7 Field estimates

12.7.1 INDIRECT

In two classical review articles Allison (1955, 1966) concluded that 10 to 30% of the total nitrogen budget was unaccounted for and attributed this discrepancy to denitrification. Similarly a 10-year field study conducted by Pratt *et al.* (1972) resulted in an unaccounted loss of nitrogen upon nitrogen fertiliser applications beyond the yield response; this was also attributed to loss through denitrification.

Although it seems logical to conclude that denitrification accounted for the losses from the system, there are many uncertainties in measuring the various nitrogen components of the soil. Thus estimating denitrification by a difference method is unsatisfactory. The most accurate figure is the input of fertiliser, since inputs from natural sources (rainfall, air pollution and asymbiotic nitrogen fixation) are relatively small (5 to 20 kg nitrogen ha^{-1} yr^{-1}). Thus at a moderate fertilisation rate of 100 kg ha^{-1}, a 10% loss through denitrification would be balanced by aerial inputs, although this degree of precision in field research is optimistic. The most accurate output factor is plant removal by cropping, while leaching losses are subject to tremendous variability (Biggar & Nielsen 1976), despite estimating losses by the use of the nitrate/chloride ratio. Chloride is biologically inert and serves as an internal standard.

[^{15}N]Nitrogen fertiliser provides more definitive information over a short term, but is still subject to the same inherent errors in measuring nitrogen fluxes. Sampling the soil for residual [^{15}N]nitrogen in the large soil organic nitrogen pool presents other sources of error, namely sample variability and bulk density. Even with enclosed lysimeters it is not unusual to collect samples which would yield more [^{15}N]nitrogen than was added by assuming an average bulk density and organic nitrogen content. Nitrogen balances are more reassuring over a several year period because seasonal estimations can be misleading if management or cropping practices are changed. Presumably a constant management practice over several years is more conducive to a quasi-steady state in which the soil nitrogen pool can be assumed to be constant.

It is because of spatial variability and uncertainty that denitrification rates calculated by the difference method have left room for doubt. Consequently there has recently been considerable interest in measuring denitrification directly in order to verify whether or not the assumption regarding nitrogen losses calculated by the difference method are realistic.

12.7.2 Direct

Since the atmosphere contains 78% (v/v) dinitrogen and small changes would be difficult to detect, it is not possible to measure denitrification in the field directly without the use of [^{15}N]nitrate. Many studies have thus focused upon measuring nitrous oxide fluxes by infrared analysis (Arnold 1954; Denmead 1979; Freney *et al.* 1979) or gas chromatography (Rasmussen *et al.* 1976; Focht 1978; Ryden *et al.* 1978). Considerable uncertainty exists about the contribution of fertilisers and nitrous oxide to the destruction of stratospheric ozone (McElroy *et al.* 1976) even though nitrous oxide represents a small fraction of the total gaseous product formation (CAST report 1976).

Rolston *et al.* (1976, 1979) attempted to measure denitrification directly with [^{15}N]nitrate by determining the concentration gradients within the soil profile. Unfortunately, the addition of nitrate to soil generally impeded the reduction of nitrous oxide to dinitrogen, thus giving rise to high nitrous oxide fluxes. Moreover, Rolston *et al.* (1979) noted that there was a high degree of uncertainty in estimating the diffusion coefficient and the concentration gradients.

Fedorova *et al.* (1973), Balderston *et al.* (1976) and Yoshinari and Knowles (1976), observed that acetylene inhibited further reduction of nitrous oxide to dinitrogen. It was shown that an acetylene concentration of 0.01 Pa was sufficient to block completely reduction of nitrous oxide, and that the blockage was reversible upon removal of the acetylene. Acetylene blockage of nitrous oxide reduction has since proved to be a very useful technique not only in measuring denitrification *in vitro*, but *in situ* as well. The value of the method has been confirmed in studies with argon atmosphere (Ryden *et al.* 1979a), with [^{15}N]nitrate (Ryden *et al.* 1979b), and [^{13}N]nitrate (Tiedje *et al.* 1979), all of which show that the recovery of dinitrogen in untreated flasks is identical to the recovery of nitrous oxide in acetylene-treated flasks.

Since it is difficult to exclude dinitrogen from atmospheres containing argon or helium, investigators who do not have access to [^{15}N]- and [^{13}N]nitrogen methodology must use high levels of nitrate to ensure significant generation of dinitrogen. These problems have been discussed previously (see section 12.6). The acetylene inhibition technique, however, makes it feasible to measure gaseous products from relatively low nitrate concentrations, particularly if the gas chromatograph is fitted with an electron capture detector. Two separate series of flasks are used in which one is treated with acetylene and the other is left untreated. The amount of dinitrogen evolved is calculated by the difference between the nitrous oxide concentrations in the treated and untreated flasks. Kinetic studies by gaseous product formation are thus possible with reliable experimental precision (Klemmedtsson *et al.* 1977; Yoshinari *et al.* 1977).

The acetylene inhibition method represents the most direct way of measuring denitrification *in situ*. The first successful demonstration of its use in the field was by Ryden *et al.* (1979b) in which duplicate sets of boxes (Fig. 1) were placed over the soil to a depth of 10 cm. A constant stream of acetylene gas from a depth of 1 m was injected below one pair of boxes to inhibit the reduction of nitrous oxide, while the other pair of boxes was untreated. The boxes and probes were left intact for a period of 3 to 4 h, to enable sufficient collection of nitrous oxide by the apparatus shown in Fig. 2. The quantities of gases evolved by denitrification and totalled over the entire growing season were in good agreement with denitrification as calculated by the difference method showing roughly a third of the added fertiliser was denitrified. An in-situ method for measuring denitrification in sediments by the acetylene inhibition method is described by Chan and Knowles (1979).

The acetylene inhibition method, however, is recommended only for short incubation periods not to exceed 4 h if there is concern about the effects of environmental perturbations. Acetylene is known to inhibit non-selectively many types of bacteria and fungi and inhibition of the nitrifiers appears to be one serious drawback (Hynes & Knowles 1978; Walter *et al.* 1979). However, the methods developed by Ryden *et al.* (1978, 1979a, b) involve a short time period (2 to 4 h) compared with the very slow doubling time of nitrifiers in soil (about 30 h—Belser & Schmidt 1978). Moreover, any inhibitory effect of acetylene would not be noted until nitrate concentrations became limiting. Thus the objection to the acetylene inhibition method is valid only in instances where the soil is constantly treated with acetylene. Consequently it is important when making integrated flux measurements not to subject the same site to acetylene addition more than once.

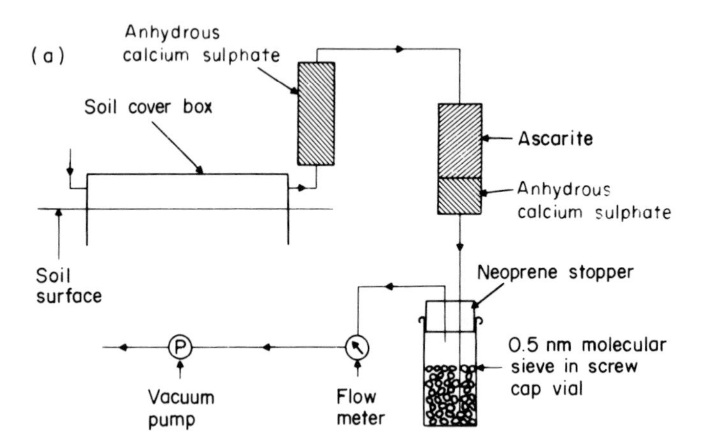

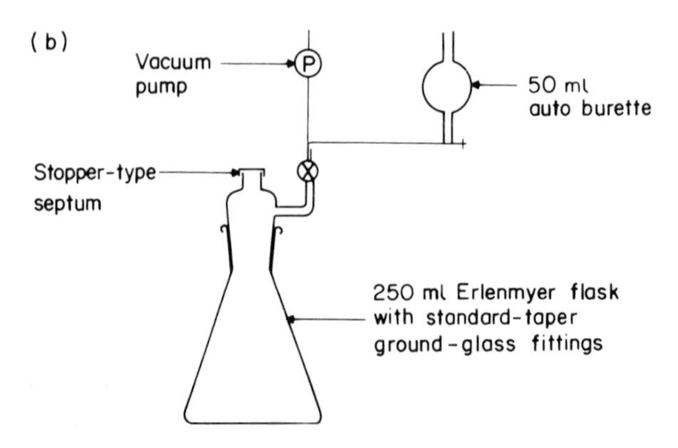

Figure 1. Schematic representation of (a) field collection and adsorption devices, and (b) laboratory removal of nitrous oxide (after Ryden *et al.* 1978).

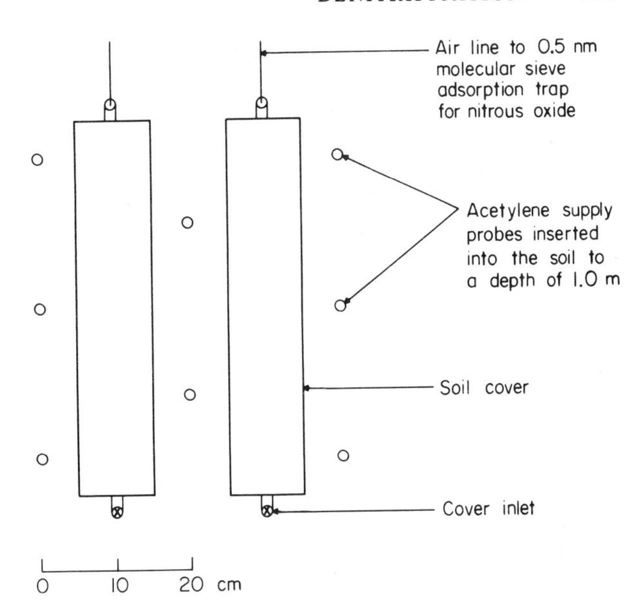

Figure 2. Schematic representation of cover boxes and acetylene probes for direct measurement of denitrification in the field (after Ryden *et al.* 1979b).

Several days treatment with acetylene, even at a 1% (v/v) concentration, does not ensure complete blockage of nitrous oxide reduction (Yoemans & Beauchamp 1978). The resumption of nitrous oxide reduction in the longer term may be due to several factors:

(1) complete reduction of nitrate, and the removal of nitrite which allows nitrous oxide reduction to restart since nitrate is known to inhibit nitrous oxide reduction;

(2) adaptation of a denitrifying population inherently insensitive to acetylene inhibition;

(3) a larger biomass or enzyme concentration which is no longer saturated by acetylene.

The exact mechanism of acetylene inhibition is not clear. Despite the triple bond and structural similarity to nitrous oxide there is no evidence for competitive inhibition.

Recommended reading

Belser L. W. & Schmidt E. L. (1978) Nitrification in soils. In: *Microbiology-1978* (Ed. D. Schlesinger), pp. 348–51. American Society for Microbiology, Washington, D.C.

Delwiche C. C. & Bryan B. A. (1976) Denitrification. *Annual Reviews of Microbiology* **30**, 241–62.

Focht D. D. & Verstraete W. (1977) Biochemical ecology of nitrification and denitrification. In: *Advanced Microbial Ecology* (Ed. M. Alexander), vol. 1, pp. 135–214. Plenum Press, New York.

Focht D. D. (1978) Methods for analysis of denitrification in soil. In: *Nitrogen in the Environment* (Eds D. R. Nielsen & J. G. MacDonald), vol. 2, pp. 433–501. Academic Press, New York.

Hall J. B. (1978) Nitrate-reducing bacteria. In: *Microbiology-1978* (Ed. D. Schlessinger), pp. 296–8. American Society for Microbiology, Washington, D.C.

Ingraham J. L. (1980) Microbiology and genetics of denitrifiers. In: *Denitrification, Nitrification and Atmospheric Nitrous Oxide* (Ed. C. C. Delwiche), pp. 45–65. John Wiley, New York.

References

Alexander M. (1965a) Most-probable-number method for microbial populations. In: *Methods of Soil Analysis, Part 2* (Ed. C. A. Black), pp. 1467–72. American Society of Agronomy, Madison, Wisconsin.

Alexander M. (1965b) Denitrifying bacteria. In: *Methods of Soil Analysis, Part 2* (Ed. C. A. Black), pp. 1484–6. American Society of Agronomy, Madison, Wisconsin.

Alexander M. (1977) *Introduction to Soil Microbiology*. John Wiley, New York.

Allen M. B. & van Niel C. B. (1952) Experiments on bacterial denitrification. *Journal of Bacteriology* **64**, 397–412.

Allison F. E. (1955) The enigma of soil nitrogen balance sheets. *Advances in Agronomy* **7**, 213–50.

Allison F. E. (1966) The fate of nitrogen applied to soils. *Advances in Agronomy* **18**, 219–59.

ARNOLD P. W. (1954) Losses of nitrous oxide from soil. *Journal of Soil Science* **5**, 116–28.

AVNIMELECH Y. (1971) Nitrate transformations in peat. *Soil Science* **111**, 113–18.

AVNIMELECH Y. & RAVEH A. (1974) The control of nitrate accumulation in soils by induced denitrification. *Water Research* **8**, 553–5.

BAILEY L. D. & BEAUCHAMP E. G. (1973a) Gas chromatography of gases emanating from a saturated soil system. *Canadian Journal of Soil Science* **53**, 112–24.

BAILEY L. D. & BEAUCHAMP E. G. (1973b) Effects of temperature on NO_3^- and NO_2^-reduction, nitrogenous gas production, and redox potential in a saturated soil. *Canadian Journal of Soil Science* **53**, 213–18.

BALDERSTON W. L., SHERR B. & PAYNE W. J. (1976) Blockage by acetylene of nitrous oxide reduction in *Pseudomonas perfectomarinus. Applied and Environmental Microbilogy* **31**, 504–8.

BEARD W. E. & GUENZI W. D. (1976) Separation of soil atmospheric gases by gas chromatography with parallel columns. *Soil Science Society of America Journal* **40**, 319–21.

BELSER L. W. & SCHMIDT E. L. (1978) Nitrification in soils. In: *Microbiology-1978* (Ed. D. Schlessinger), pp. 348–51. American Society for Microbiology, Washington, D.C.

BIGGAR J. W. & NIELSEN D. R. (1976) Spatial variability of the leaching characteristics of a field soil. *Water Resources Research* **12**, 78–84.

BIRCH H. F. (1958) The effect of soil drying on humus decomposition and nitrogen availability. *Plant and Soil* **10**, 9–32.

BLACKMER A. M. & BREMNER J. M. (1978) Inhibitory effect of nitrate on reduction of N_2O to N_2 by soil microorganisms. *Soil Biology and Biochemistry* **10**, 187–91.

BLACKMER A. M., BREMNER J. M. & SCHMIDT E. L. (1980) Production of nitrous oxide by ammonia-oxidizing chemoautotrophic microorganisms in soil. *Applied and Environmental Microbiology* **40**, 1060–6.

BOLLAG J. M., DRZYMALA S. & KARDOS L. T. (1973) Biological versus chemical nitrite decomposition in soil. *Soil Science* **116**, 44–50.

BOLLAG J. M. & TUNG G. (1972) Nitrous oxide release by soil fungi. *Soil Biology and Biochemistry* **4**, 271–6.

BOUWER J., LANCE J. C. & RIGGS M. S. (1974) High-rate land treatment: II. Water quality and economic aspects of the Flushing Meadows Project. *Journal of Water Pollution Control Federation* **46**, 844–59.

BOWMAN R. A. & FOCHT D. D. (1974) The influence of glucose and nitrate concentrations upon denitrification rates in sandy soils. *Soil Biology and Biochemistry* **6**, 297–301.

BREMNER J. M. (1965a) Inorganic forms of nitrogen. In: *Methods of Soil Analysis, Part 2* (Ed. C. A. Black), pp. 1179–237. *American Society of Agronomy*, Madison, Wisconsin.

BREMNER J. M. (1965b) Isotope-ratio analysis of nitrogen in nitrogen – 15 tracer investigations. In: *Methods of Soil Analysis, Part 2* (Ed. C. A. Black), pp. 1256–86. American Society of Agronomy, Madison, Wisconsin.

BROADBENT F. E. & TUSNEEM M. E. (1971) Losses of nitrogen from some flooded soils in tracer experiments. *Soil Science Society of America Proceedings* **35**, 922–6.

BRYANT M. P. (1964) Some aspects of the bacteriology of the rumen. In: *Principles and applications in aquatic microbiology* (Eds H. Heukelekian & N. C. Dondero), pp. 366–93, John Wiley, New York.

BURESH R. J. & PATRICK W. H. JR. (1978) Nitrate reduction to ammonium in anaerobic soil. *Soil Science Society of America Journal* **42**, 913–18.

CADY F. B. & BARTHOLOMEW W. V. (1960) Sequential products of anaerobic denitrification in Norfolk soil. *Soil Science Society of America Proceedings* **24**, 477–82.

CASKEY W. H. & TIEDJE J. M. (1979) Evidence for clostridia as agents of dissimilatory reduction of nitrate to ammonium in soils. *Soil Science Society of America Journal* **43**, 931–6.

CAST (Council for Agricultural Science and Technology) (1976) Effect of increased nitrogen fixation on stratospheric ozone. CAST Rep. No. 53, Iowa State University, Ames, Iowa.

CHAN Y. K. & KNOWLES R. (1979) Measurement of denitrification in two freshwater sediments by an *in situ* acetylene inhibition method. *Applied and Environmental Microbiology* **37**, 1067–72.

CHANCE B. (1957) Cellular oxygen requirements. *Federation Proceedings of the Federated American Societies of Experimental Biology* **16**, 671–80.

CHENG H. H. & BREMNER J. M. (1965) Gaseous forms of nitrogen. In: *Methods of Soil Analysis, Part 2* (Ed. C. A. Black), pp. 1287–323. American Society of Agronomy, Madison, Wisconsin.

COLLINS F. M. (1955) Effect of aeration on the formation of nitrate-reducing enzymes by *Ps. aeruginosa. Nature, London* **1975**, 173–4.

DAVIS D. H., DOUDOROFF M., STANIER R. Y. & MANDEL M. (1969) Proposal to reject the genus *Hydrogenomonas*: taxonomic implications. *International Journal Systematic Bacteriology* **19**, 375–90.

DELWICHE C. C. & BRYAN B. A. (1976) Denitrification. *Annual Reviews of Microbiology* **30**, 241–62.

DENMEAD O. T. (1979) Chamber systems for measuring nitrous oxide emission from soils in the field. *Soil Science Society of America Journal* **43**, 89–95.

DOWNEY R. J. & KISZKISS D. F. (1969) Oxygen and nitrate induced modification of the electron transfer system of *Bacillus stearothermophilus. Microbios* **2**, 145–53.

ESKEW D. L., FOCHT D. D. & TING I. P. (1977) Nitrogen fixation, denitrification, and pleomorphic growth in a highly pigmented *Spirillum lipoferum. Applied and Environmental Microbiology* **34**, 582–5.

FEDEROVA R. T., MILEKLINA E. I. & ILYUKHINA N. I.

(1973) Possibility of using the 'gas-exchange' method to detect extraterrestrial life: Identification of nitrogen-fixing organisms. *Izvestiya Akademii Nauk SSSR, Seriya Biologicheskaya* **6**, 797–806.

FEWSON C. A. & NICHOLAS D. J. (1961) Nitrate reductase from *Pseudomonas aeruginosa*. *Biochimica et Biophysica Acta* **49**, 335–49.

FIRESTONE M. K., SMITH M. S., FIRESTONE R. B. & TIEDJE J. M. (1979) The influence of nitrate, nitrite, and oxygen on the composition of the gaseous products of denitrification in soil. *Soil Science Society of America Journal* **43**, 1140–4.

FOCHT D. D. (1978) Methods for analysis of denitrification in soil. In: *Nitrogen in the Environment* (Eds D. R. Nielsen & J. G. MacDonald), vol. 2, pp. 433–501. Academic Press, New York.

FOCHT D. D. (1979) Microbial kinetics of nitrogen losses in flooded soils. In: *Nitrogen and Rice* (Ed. N. C. Brady), pp. 119–34. The International Rice Research Institute, Los Banos, Philippines.

FOCHT D. D. & JOSEPH H. (1973) An improved method for the enumeration of denitrifying bacteria. *Soil Science Society of America Proceedings* **37**, 698–9.

FOCHT D. D. & STOLZY L. H. (1978) Long-term denitrification studies in soils fertilized with $(^{15}NH_4)_2SO_4$. *Soil Science Society of America Journal* **42**, 894–8.

FOCHT D. D., STOLZY L. H. & MEEK B. D. (1979) Sequential reduction of nitrate and nitrous oxide under field conditions as brought about by organic amendments and irrigation management. *Soil Biology and Biochemistry* **11**, 37–46.

FOCHT D. D., VALORAS N. & LETEY J. (1980) Use of interfaced gas chromatography-mass spectrometry for detection of concurrent mineralization and denitrification in soil. *Journal of Environmental Quality* **9**, 218–23.

FOCHT D. D. & VERSTRAETE W. (1977) Biochemical ecology of nitrification and dentrification. In: *Advances in Microbial Ecology* (Ed. M. Alexander), vol. 1, pp. 135–214. Plenum Press, New York.

FRENEY J. R., DENMEAD O. T. & SIMPSON J. R. (1979) Nitrous oxide emission from soils at low moisture contents. *Soil Biology and Biochemistry* **11**, 167–73.

GAMBLE R. N., BETLACH M. R. & TIEDJE J. M. (1977) Numerically dominant denitrifying bacteria from world soils. *Applied and Environmental Microbiology* **33**, 926–39.

GARCIA J. L. (1973) Séquence des produits formés au cours de la dénitrification dans les sols de rizières du Sénégale. *Annales de Microbiologie Institute Pasteur* **124**(B), 351–62.

GARCIA J. L. (1974) Réduction de l'oxyde nitreux dans les sols de rizières du Sénégale: Mesure de l'activité dénitrificante. *Soil Biology and Biochemistry* **6**, 79–84.

GARCIA J. L. (1977) Analyse de differents groupes composant la microflore dénitrificante des sols de rizière du Sénégale. *Annales de Microbiologie Institute Pasteur* **128**(A), 433–46.

GERSBERG R., KROHN K., PEEK N. & GOLDMAN C. R. (1976) Denitrification studies with ^{13}N-labeled nitrate. *Science* **192**, 1229–31.

GOERING J. J. & CLINE J. D. (1970) A note on denitrification in seawater. *Limnology and Oceanography* **15**, 306–8.

HALL J. B. (1978) Nitrate-reducing bacteria. In: *Microbiology-1978* (Ed. D. Schlessinger), pp. 296–8. American Society for Microbiology, Washington, D.C.

HANUS F. J., MAIER R. J. & EVANS H. J. (1979) Autotrophic growth of H_2-uptake positive strains of *Rhizobium japonicum* in an atmosphere supplied with hydrogen gas. *Proceedings of the National Academy of Science, U.S.A.* **76**, 1788–92.

HASAN M. & HALL J. B. (1977) Dissimilatory nitrate reduction in *Clostridium tertium*. *Zhurnal Allegeme Mikrobiologie* **17**, 501–6.

HAUCK R. D., MELSTED S. W. & YANKWICH P. E. (1958) Use of N-isotope distribution in nitrogen gas in the study of denitrification. *Soil Science* **87**, 287–91.

HYNES R. K. & KNOWLES R. (1978) Inhibition by acetylene of ammonia oxidation in *Nitrosomonas europaea*. *FEMS Microbiology Letters* **4**, 319–21.

INGRAHAM J. L. (1981) Microbiology and genetics of denitrifiers. In: *Denitrification, Nitrification and Atmospheric Nitrous Oxide* (Ed. C. C. Delwiche), pp. 45–65. John Wiley, New York.

KEENEY D. R., FILLERY I. R. & MARX G. P. (1979) Effect of temperature on the gaseous nitrogen products of denitrification in a silt loam soil. *Soil Science Society of America Journal* **43**, 1124–8.

KEENEY D. R., HERBERT R. A. & HOLDING A. J. (1971) Microbiological aspects of the pollution of fresh water with inorganic nutrients. In: *Microbial Aspects of Pollution* (Eds G. Sykes & F. A. Skinner), pp. 181–200. Academic Press, London.

KLEMMEDTSSON L., SVENSSON B. H., LINDBERG T. & ROSSWALL T. (1977) The use of acetylene inhibition of nitrous oxide reductase in quantifying denitrification in soils. *Swedish Journal of Agricultural Research* **7**, 179–85.

KOHL D. H., VITAYATHIL F., WHITLOW P., SHEARER G. & CHIEN S. H. (1976) Denitrification kinetics in soil systems: The significance of good fits to mathematical forms. *Soil Science Society of America Journal* **40**, 249–53.

KOIKE I. & HATTORI A. (1975a) Growth yield of a denitrifying bacterium, *Pseudomonas denitrificans*, under aerobic and denitrifying conditions. *Journal of General Microbiology* **88**, 1–10.

KOIKE I. & HATTORI A. (1975b) Energy yield of denitrification: An estimate from growth yield in continuous cultures of *Pseudomonas denitrificans* under nitrate-nitrite, and nitrous oxide-limited conditions. *Journal of General Microbiology* **88**, 11–19.

KOIKE I. & HATTORI A. (1978) Denitrification and ammonia formation in anaerobic coastal sediments. *Applied and Environmental Microbiology* **35**, 278–82.

LEE D. H. K. (1970) Nitrates, nitrites, and methemoglobinemia. *Environmental Research* **3**, 484–511.

MCELROY M. B., ELKINS J. W., WOFSKY S. C. & YUNG Y. L. (1976) Sources and sinks for atmospheric N_2O. *Review of Geophysics and Space Physics* **14**, 143–50.

MANN L. D., FOCHT D. D., JOSEPH H. A. & STOLZY L. H. (1972) Increased denitrification in soils by additions of sulfur as an energy source. *Journal of Environmental Quality* **1**, 329–32.

MARTIN J. P. & ERVIN J. O. (1953) Nitrogen losses during oxidation of sulfur in soils. *California Citrograph* **39**, 54–6.

MISRA C., NIELSEN D. R. & BIGGAR J. W. (1974) Nitrogen transformations in soil during leaching. II. Steady state nitrification and nitrate reduction. *Soil Science Society of America Proceedings* **38**, 294–304.

MIYATA M. (1971) Studies on denitrification. XIV. The electron donating system in the reduction of nitric oxide and nitrate. *Journal of Biochemistry* **70**, 205–13.

MYCIELSKI R. & GUCWA D. (1977) Nitrite agar method for the isolation and enumeration of denitrifying bacteria. *Acta Microbiology Polonica* **26**, 317–18.

NASON A. (1962) Symposium on metabolism of inorganic compounds. II. Enzymatic pathways of nitrate, nitrite, and hydroxylamine metabolism. *Bacteriology Reviews* **26**, 16–41.

NEYRA C. A., DÖBEREINER J., LALANDE R. & KNOWLES R. (1977) Denitrification by N_2-fixing *Spirillum lipoferum*. *Canadian Journal of Microbiology* **23**, 300–5.

NOMMIK H. (1956) Investigations on denitrification in soil. *Acta Agricultura Scandinavia* **6**, 195–228.

PATRICK W. H. & REDDY K. R. (1976) Nitrification-denitrification reactions in flooded soils and sediments: dependence on oxygen supply and ammonium diffusion. *Journal of Environmental Quality* **5**, 469–72.

PAUL E. A. & MYERS R. J. K. (1971) Effect of soil moisture stress on uptake and recovery of tagged nitrogen by wheat. *Canadian Journal of Soil Science* **51**, 37–43.

PAYNE W. J. (1973) Reduction of nitrogenous oxides by microorganisms. *Bacteriology Reviews* **37**, 409–52.

PHILLIPS R. E., REDDY K. R. & PATRICK W. H. JR. (1978) The role of nitrate diffusion in determining the order and rate of denitrification in flooded soils II. Theoretical analysis and interpretation. *Soil Science Society of America Journal* **42**, 272–8.

PICHINOTY F., GARCIA J-L, JOB C. & DURAND M. (1978) La dénitrification chez *Bacillus licheniformis*. *Canadian Journal of Microbiology* **24**, 45–9.

POURBAIX M. (1966) *Atlas of Electrochemical Equilibria in Aqueous Solutions*. Pergamon Press, Oxford.

PRAKASAM T. B. S. & LOEHR R. C. (1972) Microbial nitrification and denitrification in concentrated wastes. *Water Research* **6**, 859–69.

PRAKASH O. & SADANA J. C. (1973) Metabolism of nitrate in *Achromobacter fischeri*. *Canadian Journal of Microbiology* **19**, 15–25.

PRATT P. F., JONES W. W. & HUNSAKER V. E. (1972) Nitrate in deep soil profiles in relation to fertilizer rates and leaching volume. *Journal of Environmental Quality* **1**, 97–102.

RASMUSSEN R. A., KRASNEC J. & PIEROTTI D. (1976) N_2O analysis in the atmosphere via electron capture-gas chromatography. *Geophysical Research Letters* **3**, 615–18.

REDDY K. R., PATRICK W. H. JR & PHILIPS R. E. (1978) The role of nitrate diffusion in determining the order and rate of denitrification in flooded soils. I. Experimental results. *Soil Science Society of America Journal* **42**, 268–72.

RITCHIE G. A. F. & NICHOLAS D. J. D. (1972) Identification of the sources of nitrous oxide produced by oxidative and reductive processes in *Nitrosomonas europaea*. *Biochemical Journal* **126**, 181–9.

ROLSTON D. E., BROADBENT F. E. & GOLDHAMMER D. A. (1979) Field measurement of denitrification: II Mass balance and sampling uncertainty. *Soil Science Society of America Journal* **43**, 703–8.

ROLSTON D. E., FRIED M. & GOLDHAMMER D. A. (1976) Denitrification measured directly from nitrogen and nitrous oxide gas fluxes. *Soil Science Society of America Proceedings* **40**, 259–66.

ROWE R., TODD R. & WAIDE J. (1977) Microtechnique for most-probable number analysis. *Applied and Environmental Microbiology* **33**, 675–80.

RYDEN J. C., LUND L. J. & FOCHT D. D. (1978) Direct in-field measurement of nitrous oxide flux from soils. *Soil Science Society of America Journal* **42**, 731–7.

RYDEN J. C., LUND L. J. & FOCHT D. D. (1979a) Direct measurement of denitrification loss from soils. I. Laboratory evaluation of acetylene inhibition of nitrous oxide reduction. *Soil Science Society of America Journal* **43**, 104–10.

RYDEN J. C., LUND L. J., LETEY J. & FOCHT D. D. (1979b) Direct measurement of denitrification loss from soils. II. Development and application of field methods. *Soil Science Society of America Journal* **43**, 110–18.

SATOH T., HOSHINO Y. & KITAMURA H. (1976) *Rhodopseudomonas sphaeroides* forma sp. *denitrificans*, a denitrifying strain as a subseries of *Rhodopseudomonas sphaeroides*. *Archives of Microbiology* **108**, 265–9.

SMITH K. A. (1977) Gas-chromatographic analysis of the soil atmosphere. *Advances in Chromatography* **15**, 197–231.

SMITH M. S. & FIRESTONE M. K. (1978) The acetylene inhibition method for short term measurement of soil denitrification and its evaluation using nitrogen-13. *Soil Science Society of America Journal* **42**, 611–15.

SMITH M. S., FIRESTONE M. K. & TIEDJE J. M. (1979)

Phases of denitrification following oxygen depletion in soil. *Soil Biology and Biochemistry* **11**, 261–7.

SORENSEN J. (1978) Capacity for denitrification and reduction of nitrate to ammonia in a coastal marine sediment. *Applied and Environmental Microbiology* **35**, 301–5.

STANFORD G., LEGG J. O. DZIENIA S. & SIMPSON E. C. JR. (1975) Denitrification and associated nitrogen transformation in soils. *Soil Science* **120**, 147–52.

THOMPSON B. (1977) *Fundamentals of Gas Analysis by Gas Chromatography*. Varian Associates Inc., Palo Alto, California.

TIEDJE J. M., FIRESTONE R. B., FIRESTONE M. K., BETLACH M. R., SMITH M. S. & Caskey W. H. (1979) Methods for the production and use of nitrogen-13 in studies of denitrification. *Soil Science Society of America Journal* **43**, 709–15.

TIEDJE J. M. (1982a) Denitrification. In: *Methods of Soil Analysis. Part 2, Chemical and Microbiological Properties* (Eds D. R. Keeney & R. A. Miller). American Society of Agronomy, Madison, Wisconsin. (In press.)

TIEDJE J. M. (1981b) Use of nitrogen-13 and nitrogen-15 in studies on the dissimilatory fate of nitrate. In: *Genetic Engineering of Symbiotic N_2 Fixation and Conservation of Fixed Nitrogen* (Ed. J. M. Lyons). Plenum Press, New York. (In press.)

VAGNAI S. & KLEIN D. A. (1974) A study of nitrite-dependent dissimilatory micro-organisms isolated from Oregon soils. *Soil Biology and Biochemistry* **6**, 335–9.

VALERA C. L. & ALEXANDER M. (1961) Nutrition and physiology of denitrifying bacteria. *Plant and Soil* **15**, 268–80.

VAN GENT-RUIJTERS J. L. W., DE VRIES W. & STOUTHAMER A. H. (1975) Influence of nitrate on fermentation pattern, molar growth yields, and synthesis of cytochrome *b* in *Propionobacterium pentosaceum*. *Journal of General Microbiology* **88**, 36–48.

VERHOEVEN W., KOSTER A. L. & VAN NIEVELT M. C. A. (1954) Studies of true dissimilatory nitrate reduction. III. *Micrococcus denitrificans* Biejerinck, a bacterium capable of using molecular hydrogen in denitrification. *Antonie van Leeuwenhoek* **20**, 273–84.

VIVES J. & PARÉS R. (1975) Enumeracion y caracterizacion de la flora desnitricante quimioorganotrofa en una pradera experimental. *Microbiology Espanana* **28**, 43–53.

VOLZ M. G. (1977) Denitrifying bacteria can be enumerated in nitrite broth. *Soil Science Society of America Journal* **41**, 549–51.

VOLZ M. G., BELSER L. W., ARDAKANI M. S. & MCLAREN A. D. (1975) Nitrate reduction and nitrite utilization by nitrifiers in an unsaturated Hanford sandy loam. *Journal of Environmental Quality* **4**, 179–82.

WALTER H. M., KEENEY D. R. & FILLERY I. R. (1979) Inhibition of nitrification by acetylene. *Soil Science Society of America Journal* **43**, 195–6.

WHEATLAND A. B., BARRET M. J. & BRUCE A. M. (1959) Some observations on denitrification in rivers and estuaries. *Journal Proceedings of the Institute of Sewage Purification* 149–59.

WILJER J. & DELWICHE C. C. (1954) Investigations on the denitrifying process in soil. *Plant and Soil* **5**, 155–69.

WUHRMANN K. & MECHSNER K. (1973) Discussion by K. Wuhrmann, p. 682, Fig. 1, on paper by Dawson R. W. & Murphy K. L. (1973). Factors affecting biological denitrification of wastewater. In: *Advances in Water Pollution Research* (Ed. S. H. Jenkins), pp. 671–80. Pergamon Press, New York.

YOEMANS J. C. & BEAUCHAMP E. G. (1978) Limited inhibition of nitrous oxide reduction in soil in the presence of acetylene. *Soil Biology and Biochemistry* **10**, 517–19.

YOSHIDA T. & ALEXANDER M. (1970) Nitrous oxide formation by *Nitrosomonas europaea* and heterotrophic microorganisms. *Soil Science Society of America Proceedings* **34**, 880–2.

YOSHINARI T., HYNES R. & KNOWLES R. (1977) Acetylene inhibition of nitrous oxide reduction and measurement of denitrification and nitrogen fixation in soil. *Soil Biology and Biochemistry* **9**, 177–83.

YOSHINARI T. & KNOWLES R. (1976) Acetylene inhibition of nitrous oxide reduction by denitrifying bacteria. *Biochemistry and Biophysics Research Communications* **69**, 705–10.

YOUATT J. B. (1954) Denitrification of nitrite by a species of *Achromobacter*. *Nature, London* **173**, 826–7.

ZABLOTOWICZ R. M., ESKEW D. L. & FOCHT D. D. (1978) Denitrification in *Rhizobium*. *Canadian Journal of Microbiology* **24**, 757–60.

Chapter 13 • Nitrogen Fixation

David W. Smith

13.1 Biochemistry of nitrogen fixation

Nitrogen fixation may be characterised as the reduction of nitrogen at the 0 (elemental) oxidation state to the −3 (ammonia) oxidation state. The biological process is catalysed by nitrogenase, a large, multiple unit, anaerobic metallo-enzyme. The product ammonia is retained by the nitrogen-fixing cell by the formation of glutamine from glutamate and ammonia (Wolk *et al.* 1976). One consequence of these reactions is that ammonia is only rarely released by healthy nitrogen-fixing cells. Thus newly fixed nitrogen must pass through an organic form before entering the nitrogen cycle. (The question of nitrogen excretion by healthy cells is still somewhat unclear, but has recently received some attention—Wiebe *et al.* 1975; Fogg 1977; Sharp 1977.)

Biochemical investigations of cell extracts and purified enzymes are made difficult by the extreme oxygen sensitivity of nitrogenase. Irreversible inactivation occurs upon even brief exposure to oxygen (Wong & Burris 1972), and so all extractions, purifications and analyses must be done under anaerobic conditions.

Nitrogenase binds nitrogen gas and releases ammonia stoichiometrically as its product (Wolk *et al.* 1976; Brill 1979). This reduction presumably occurs in conventional two electron steps but the intermediates remain bound to the enzyme (Dalton & Mortenson 1972). The enzyme contains two major proteins: component I and component II with molecular weights of 190 000 to 220 000 and 55 000 to 64 000 respectively (Winter & Burris 1976), depending on the organism examined. Component I (sometimes referred to as the MoFe protein) contains 32 atoms of iron and 2 atoms of molybdenum per molecule, while component II (referred to as the Fe protein) has 4 atoms of iron and no molybdenum (Brill 1979). It now appears that component I is actually composed of two proteins (referred to as I_α and I_β—Brill 1979) and a cofactor termed FeMo-co which appears to be a part of the active site (Shah & Brill 1977). The reaction mechanism of the whole complex involves transfer of electrons from ferredoxin to component II which, after binding Mg·ATP, is able to reduce component I to a low potential. The actual reduction of substrates occurs on component I, probably by transfer of electrons through iron at the active site (Winter & Burris 1976). This system has been elegantly fractionated and analysed to deduce the enzymatic mechanism. However, as Winter and Burris (1976) point out, no compound other than component II which has bound Mg·ATP has been found which is capable of fully reducing component I (Zumft *et al.* 1974).

Nitrogenase has a range of substrates, being able to reduce many small molecules which contain triple bonds (N_2, HCN, C_2H_2,HN_3). All these reductions require considerable energy (12 to 15 ATP $(N_2)^{-1}$—Ljones 1974) and reducing power ($6\bar{e}(N_2)^{-1}$). It is, therefore, interesting to note that a large portion of the electrons processed by nitrogenase go to the reduction of protons to hydrogen, even in cells which are actively fixing nitrogen (Brill 1979). Some nitrogen-fixing bacteria have a hydrogenase which can be used to oxidise this hydrogen and thereby save some of the potential electron loss (Dixon 1972). Schubert and

Evans (1976), for example, have suggested that *Rhizobium* strains which have hydrogenase activity fix nitrogen more efficiently and extensively than strains which cannot reoxidise the hydrogen. These basic relations are shown in Fig. 1.

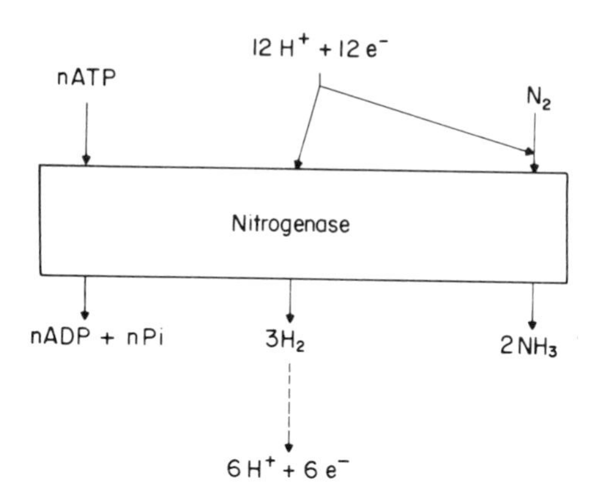

Figure 1. Simultaneous reduction of nitrogen and H^+ by nitrogenase. The dotted line represents the potential saving of electrons obtained by some nitrogen-fixing bacteria by the reoxidation of hydrogen (after Brill 1979).

A final area of biochemical investigation of reactions concerns inhibitors. Hydrogen is a specific inhibitor of nitrogen reduction but appears to have no effect on the reduction of any other substrates except that it stimulates hydrogen cyanide reduction (Rivera-Ortiz & Burris 1975). These workers also pointed out that the reverse is not true: hydrogen reduction still occurs even in an atmosphere of pure nitrogen. Interestingly, nitrogen is a competitive inhibitor of acetylene reduction, but acetylene is a non-competitive inhibitor of nitrogen reduction (Rivera-Ortiz & Burris 1975), perhaps reflecting a need for nitrogenase to accumulate six electrons before reducing nitrogen but only needing two electrons to reduce acetylene (Winter & Burris 1976).

As noted above, nitrogenase is a strictly anaerobic enzyme. It is, therefore, interesting to find that bacteria capable of nitrogen fixation range from obligate anaerobes to cyanobacteria which produce oxygen. Obligate anaerobes, such as species of *Clostridium* (Rosenblum & Wilson 1949), *Desulfovibrio* (Postgate 1970; Reiderer-Henderson & Wilson 1970; Dicker & Smith 1980c) and photosynthetic bacteria (Arnon & Yoch 1974) have no difficulty protecting their nitrogenase since the whole organism is dependent on the absence of oxygen. The aerobic, heterotrophic, nitrogen-fixing bacteria can be divided into two groups: those which are facultative anaerobes and only fix nitrogen when oxygen is absent; and those which are obligate aerobes and fix nitrogen under aerobic conditions. Examples of the facultative group are *Klebsiella* (Pengra & Wilson 1958) and *Bacillus* (Line & Loutit 1971). Cannon *et al.* (1974) inserted *Klebsiella* sp. nitrogen-fixation genes into *Escherichia coli*. The recipient was able to reduce nitrogen, but only when grown anaerobically. The most frequently studied obligately aerobic, nitrogen-fixing bacteria are *Azotobacter* spp. Several species have impressive nitrogen fixation rates under aerobic conditions. It has been proposed (Drozd & Postgate 1970) and generally accepted that *Azotobacter* spp. protect their anaerobic nitrogenase by accelerated respiration which leaves the interior of the cells anaerobic. This increased respiration is reflected in increased synthesis and invagination of the cytoplasmic membrane, the site of respiratory oxygen consumption (Oppenheim & Marcus 1970). Many cyanobacteria fix nitrogen and, since these organisms also perform oxygenic photosynthesis, it would seem very difficult for them to protect the anaerobic enzyme from their own metabolism. For the purpose of this analysis, the cyanobacteria may be divided into two groups, reflecting two different mechanisms of nitrogenase protection. A sophisticated system is found in the filamentous cyanobacteria containing heterocysts, exemplified by the genus *Anabaena* and summarised by Stewart (1974). These organisms apparently protect their nitrogenase by locating it in the heterocysts which are only synthesised under nitrogen-limited conditions when nitrogen fixation is required (see below). The heterocysts have a modified physiology in that they lack photosystem II and therefore do not evolve oxygen. Intimate physical connections exist between the heterocysts and the surrounding vegetative cells so that the heterocysts release nitrogenous products of nitrogen fixation and receive products of carbon dioxide fixation. The unicellular, nitrogen-fixing cyanobacteria, exemplified by the genus *Gloeocapsa* obviously do not have the capability of protection through differentiation. As Stewart (1974) summarised, there must be a careful intracellular localisation so that the oxygenic photosynthetic activity is adequately separated from the anaerobic nitrogenase.

13.2 Regulation of nitrogen fixation

The high energy cost of nitrogen fixation has resulted in the evolution of efficient regulatory mechanisms so that bacteria which have a supply of combined nitrogen will not reduce nitrogen gas. This regulation can be considered in two parts: regulation of nitrogenase synthesis and control of nitrogenase after synthesis. The first aspect, synthetic control, has been studied in considerable detail with the discovery of some 14 genes in the *nif* cluster in *Klebsiella pneumoniae* (MacNeil *et al.* 1978; Roberts *et al.* 1978). There are a variety of gene products and sophisticated interactions in this system, but the major conclusion is that in the presence of excess ammonia, nitrogenase synthesis ceases. Similar findings of repression by ammonia have been made in other bacterial systems (Daesch & Mortenson 1972; Brotonegoro 1974). However, the possibility has been raised that the substrate, nitrogen, is an inducer for the nitrogenase system. But induction does not occur. Indeed, the value of such a system would be questionable since it is virtually impossible for an organism to be in an environment devoid of nitrogen. The nitrogenase system may, therefore, be characterised as derepressible but not inducible. The regulation of nitrogenase after synthesis is less clear. Recently Yoch and Cantu (1980) have described an elegant and sophisticated system in *Rhodospirillum rubrum* involving interconversions among three enzyme configurations, mediated by ammonia and/or glutamate. However, as Yoch and Cantu (1980) point out, this regulatory ability is apparently not found in heterotrophic nitrogen-fixing bacteria. The activity of nitrogenase within a cell can be affected, but only indirectly. The simplest is competition with nitrogen assimilation processes demanding less reducing power, namely the assimilation of ammonium and nitrate (Sorger 1969; Brotonegoro 1974; Nambiar & Shethna 1977; Dicker & Smith 1980b). There is also correlation between nitrogenase activity and intracellular ATP/ADP ratios (Dalton & Mortenson 1972), but the mechanism for such an effect has not been described.

13.3 Ecology

The subject of this chapter has been traditionally referred to as 'non-symbiotic nitrogen fixation'. This description is unfortunate since it obscures the considerable variety of intimate associations which have been observed between nitrogen-fixing prokaryotes and non-nodulated plants. Five specific examples of such relations are: *Paspalum notatum–Azotobacter paspali* (Dobereiner *et al.* 1972); corn–*Spirillum* (Patriquin & Dobereiner 1978); rice–cyanobacteria (Yoshida & Ancajas 1973); dune grass–*Azotobacter* (Ralph 1978); and marsh grass–*Azotobacter* (Patriquin 1978b). Dobereiner (1974) examined many other rhizosphere–bacterial associations, many of which are more than passive relationships due to spatial proximity. Dobereiner and Day (1976, cited in Patriquin 1978b) reported cortical penetration of root cells of the grass *Digitaria decumbens* by a nitrogen-fixing *Spirillum* species. Similarly Patriquin (1978b) found that nitrogen-fixing bacteria (isolated but unidentified) penetrated the cortex of the marsh grass *Spartina alterniflora*, occasionally to the endodermis. Perhaps the most intimate association between bacteria and plant cells is that described by Dobereiner *et al.* (1972) with the grass *Paspalum notatum*. These associations are significant contributors to the nitrogen cycle in their environments with nitrogen fixation potentials comparable to those found in the legume–*Rhizobium* system (Abd-El-Malek 1971; von Bulow & Dobereiner 1975; Nelson *et al.* 1976; Patriquin 1978b; Teal *et al.* 1979). As will be discussed below, rate measurements of nitrogen fixation in nature must be viewed as measurements of the potential capability with the significance of their absolute values minimised.

Experimental

13.4 Use of Isotopes

Both [^{15}N]- and [^{13}N]nitrogen are used for the measurement of nitrogen fixation. The use of the stable, heavy isotope, [^{15}N]nitrogen, can be summarised as follows:

$$^{15}N \underset{i}{\rightarrow} {}^{15}NH_3 \underset{ii}{\rightarrow} {}^{15}NH_2-R \underset{iii}{\rightarrow} {}^{15}NH_3 \underset{iv}{\rightarrow} {}^{15}N_2$$

where $^{15}NH_2-R$ represents organic amines. That is, samples are incubated with [^{15}N]nitrogen gas for an appropriate period of time during which reactions i and ii are catalysed, primarily intracellularly, by the nitrogen-fixing bacteria. After the incubations are terminated, samples are prepared for isotope analysis to determine the amount of heavy nitrogen which has been reduced to the level of ammonia. There are two different analytical principles for performing the isotope analysis: mass spectrometry and optical emission spectroscopy.

The process of mass spectrometry has been used in field nitrogen studies for many years (Delwiche & Wyler 1956; Burris & Wilson 1957) and is well summarised and described in detail by Bremner (1965). Sample preparation requires conversion of [^{15}N]nitrogen compounds to [$^{15}N_2$]nitrogen gas for analysis followed by:

(1) removal of unreacted [$^{15}N_2$]nitrogen gas;

(2) digestion (usually Kjeldahl) to convert ^{15}N-R to $^{15}NH_4^+$ (reaction iii);

(3) oxidation to $^{15}N_2$ (reaction iv) via the Rittenberg hypobromite procedure.

The relatively new technique of optical emission spectroscopy has seen its primary use in plant studies (Muhammad & Kumazawa 1972) and in nitrification and denitrification studies (Schell 1978) but it is applicable to nitrogen fixation as well. The conversion of ^{15}N-R to [$^{15}N_2$] nitrogen gas may be accomplished as for mass spectrometry. From this point the sample handling is very important with precautions summarised by Schell (1978). The key practical differences between emission spectroscopy and mass spectrometry are that:

(1) emission instruments are dedicated to nitrogen analysis;

(2) emission analysis is non-destructive and samples may be reanalysed if desired (Meyer *et al.* 1974).

The short-lived (10 min) radioisotope [^{13}N]-nitrogen has provided valuable insights into the immediate products of nitrogenase in the laboratory (Wolk *et al.* 1976; Thomas *et al.* 1977) but has not yet been applied to field nitrogen fixation studies. Field use of the isotope has been successful for denitrification studies (Gersberg *et al.* 1976; Firestone *et al.* 1979). Presumably as the isotope becomes more widely available it will be applied to field measurements of nitrogen fixation.

All these procedures involve sample incubations with nitrogen gas, the natural substrate. Therefore, there is great specificity and minimum interference from other reactions. However, tracer methods all require sophisticated equipment, such as a mass spectrometer, an emission spectrophotometer or cyclotron (to prepare [^{13}N]nitrogen), and sample preparation may be costly and time-consuming. At the experimental level there is concern about the problems associated with the addition of the tracers. As with all such additions, there are large problems (often unresolved) concerning the mixing of substrate and the disruption of the sample (especially with sediment) during handling.

13.5 Acetylene reduction

The acetylene reduction approach to the measurement of nitrogen fixation takes advantage of the diversity of substrates that are reduced by nitrogenase. Acetylene is readily converted to ethylene by microorganisms contained in natural samples (Whitney *et al.* 1975; Patriquin & Keddy 1978; Teal *et al.* 1979; Dicker & Smith 1980a), pure cultures (Stewart 1969; Drozd & Postgate 1970; Dicker & Smith 1980c) or isolated nitrogenase (Dalton & Mortenson 1972; Brill 1975). Samples are incubated with acetylene and the gas evolved analysed for ethylene with the gas chromatograph.

The instrumentation and reagents for the acetylene reduction procedure are much simpler to use and less expensive than for the tracer methods. As a result, it is possible to analyse large numbers of samples rapidly with excellent sensitivity. As is the case when using tracers, the problem of sample alteration is still unavoidable. However, the acetylene reduction method has additional problems. All of these difficulties are related to the fact that acetylene is not the natural substrate for nitrogenase. Careful consideration must therefore be given to solubility differences: acetylene is much more water-soluble than ethylene, but both are more soluble than nitrogen. It is difficult to be sure that the substrate (acetylene) reaches all the nitrogen-fixing bacteria or that the product (ethylene) is completely released from the sample. These problems are magnified for aqueous or slurried samples, although careful manipulation of the surface area exposed to acetylene can mitigate this problem (Flett *et al.* 1976). There is, naturally, a desire to convert acetylene reduction rates to equivalent nitrogen fixation rates. However, since acetylene is an artificial substrate for nitrogenase, the correct stoichiometric conversion must be attempted with great care. Indeed, values ranging from 3 to 25 mol acetylene reduced (mol reduced nitrogen)$^{-1}$ have appeared in the literature (Hardy *et al.* 1973). Recently, Potts *et al.* (1978) performed careful, simultaneous measurements using [^{15}N]dinitrogen and acetylene reduction. Their results confirmed that direct conversion may not be straightforward. The different reactions of nitrogenase with acetylene and nitrogen contribute to this variation in conversion values. As noted above, acetylene and dinitrogen inhibit the reduction of each other. This confusion is compounded by the observation that when nitrogenase reduces acetylene there is little hydrogen production from hydrogen ions, whereas hydrogen production always accompanies nitrogen reduction (Tjepkema & Winship 1980). This difference in hydrogen production indicates that the efficiencies of electron transfer to acetylene and nitrogen are rather different. It is, therefore, reasonable to preincubate samples for a few hours in an inert atmosphere after collection and before acetylene addition (Patriquin 1978a; Dicker & Smith 1980a). The preincubation removes or at least greatly lowers the concentration of nitrogen so the competition and inhibition are minimised.

At this point the reader may feel overwhelmed by the negative aspects of the acetylene reduction technique. Nevertheless, despite all the difficulties and precautions necessary, the consensus of opinion is that the basic acetylene reduction procedure is the most convenient, sensitive, reproducible and valuable of the available field procedures. A large number of investigators have found acetylene reduction useful in an impressive array of systems. The fact that we have such detailed knowledge of the shortcomings of this technique is a reflection of its widespread application and general acceptance. There is much to beware of, but there is also much to be gained by the careful and considered application of the ability of nitrogenase to reduce acetylene.

13.6 Sampling

With tracer methods or acetylene reduction, there are two general procedures for making measurements: chambers *in situ* and sample removal.

The chamber *in situ* procedure consists of enclosing a portion of soil or sediment within a chamber or inverted funnel (Balandreau & Dommergues 1973). This process clearly causes the minimum disturbance in the sampled area and substrate addition and gas withdrawals are made at intervals through the top. However, when the acetylene reduction method is used in this way, the solubility problems must be considered as well as the many diffusion limitations presented by soil. It is not usually clear how efficient gas exchange has been or how deeply the gas has penetrated. If there is significant subsurface fixation, the chamber method will seriously underestimate the system activity. In marsh sediments the pattern of acetylene reduction with depth can be quite variable, with maximum rates often occurring several cm below the surface (Hanson 1977; Teal *et al.* 1979; Dicker & Smith 1980a; Smith 1980).

The second procedure used for making measurements of nitrogen fixation consists of removing small samples and enclosing them in bottles. This method allows maximum flexibility in sample manipulation and amendment. Also, a large number of samples from different locations within an area can be readily examined. The disadvantages of sample removal are the usual concerns about bottle effects on nitrogen fixation. That is, the incubated sample (and its nitrogen-fixing bacteria) can change dramatically during the course of an enclosed incubation (see below concerning lag). For nitrogen-fixing bacteria these changes can either stimulate or depress apparent nitrogen fixation activity; for example, anaerobic bacteria from a sub-surface micro-environment can be damaged or killed during handling. An example of stimulation is the possible interaction between different bacterial communities within the bottle. Nitrifying bacteria may consume excess ammonia, thereby derepressing nitrogenase synthesis in their nitrogen-fixing neighbours. Evidence that such a relationship may be important has been obtained (Dicker & Smith 1980b).

The solubility problems inherent in acetylene reduction studies are also relevant when considering the size of the sample used in the incubation. Smith (1980) found a dramatic effect of sample size on acetylene reduction rate in marsh sediment samples. Normalised rates decreased exponentially as a function of increasing sample size. This pattern almost certainly results from changes in the ratio of surface area to sample volume in the incubated samples. The smaller samples have a higher ratio and, since diffusion and gas solubilities depend on the available surface area, also have greater gas exchange and higher apparent rates of nitrogen fixation. This observation leads to two conclusions:

(1) quantitative measurements of natural fixation rates are sensitive to many physical and chemical factors;

(2) most field studies of nitrogen fixation performed with enclosed incubations and the acetylene reduction technique (i.e., the vast majority) are probably only measuring the activity of organisms in the surface portions of their samples.

A widespread but not yet well-explained phenomenon in the application of acetylene reduction to natural samples is the variable appearance of a lag between the beginning of an incubation and the commencement of ethylene production (Dobereiner *et al.* 1972; Hanson 1977; Patriquin & Keddy 1978; Teal *et al.* 1979; Dicker & Smith 1980a). There are several possible explanations for this lag, some of which have been discussed in relation to marsh nitrogen fixation (Smith 1980). There are two major possibilities: either the lag is unique to the use of acetylene reduction, or the lag results from the incubation itself and not specifically from the acetylene procedure. Some are of the opinion that the lag is directly related to the use of acetylene (Patriquin 1978b), but most workers seem to feel it is a more general consequence of the incubations (Dobereiner *et al.* 1972; Herbert 1975; Teal *et al.* 1979; Dicker & Smith 1980a). In addition to the varied opinions on the cause of the lag, there is a considerable discrepancy as to what to do about it when analysing the data. The two fundamental positions are:

(1) to consider the lag as an adaptation period to be disregarded (Teal *et al.* 1979; Dicker & Smith 1980a, b);

(2) to consider that the lag reflects physiological changes in the bottle so that the period of no ethylene production is an accurate measure of the fixation ability in the community (David & Fay 1977; Patriquin 1978a).

The choice between these alternatives should be at least partly determined by a consideration of the purpose of the individual experiments. If true natural rates are sought, then it seems the shorter term, minimum change approach is preferable: i.e. the lag should be thought of as a true rate of zero. If, on the other hand, system potentials are sought rather than absolute rates, then the lag should be disregarded and the period of maximum ethylene production measured.

13.7 Conclusions

In conclusion it should be noted that nitrogen fixation is widespread among different bacterial types and in different environments. Precise quantitative estimates are difficult to obtain, but it is clear that the so-called non-symbiotic nitrogen-fixing bacteria are of quantitative importance in many areas. The techniques for studying nitrogen fixation in nature are rather easily applied, especially acetylene reduction. There are, nonetheless, many potential dangers in experimental detail and especially in data interpretation. If an investigator is careful and consistent, then much can be learned about the ecologically vital process of nitrogen fixation.

Acknowledgement

The work reported in the Delaware marsh system was supported by the Sea Grant College Program of the University of Delaware.

Recommended reading

BREMNER J. M. (1965) Total nitrogen. In: *Methods of Soil Analysis* (Ed. C. A. Black), vol. 9, pp. 1149–78. American Society of Agronomy, Madison, Wisconsin.

BRILL W. J. (1975) Regulation and genetics of bacterial nitrogen fixation. *Annual Review of Microbiology* **29**, 109–29.

BRILL W. J. (1979) Nitrogen fixation: basic to applied. *American Scientist* **67**, 458–66.

BRILL W. J. (1980) Biochemical genetics of nitrogen fixation. *Microbiological Reviews* **44**, 449–67.

DALTON H. & MORTENSON L. E. (1972) Dinitrogen (N_2) fixation (with a biochemical emphasis). *Bacteriological Reviews* **36**, 231–60.

DOBEREINER J. (1974) Nitrogen-fixing bacteria in the rhizosphere. In: *The Biology of Nitrogen Fixation* (Ed. A. Quispel), pp. 86–120. North-Holland Publishing Company, Amsterdam.

FLETT R. J., HAMILTON R. D. & CAMPBELL N. E. R. (1976) Aquatic acetylene-reduction techniques: Solutions to several problems. *Canadian Journal of Microbiology* **22**, 43–51.

POTTS M., KRUMBEIN W. E. & METZGER J. (1978) Nitrogen fixation rates in anaerobic sediments determined by acetylene reduction, a new ^{15}N field assay, and simultaneous total N ^{15}N determination. In: *Environmental Biogeochemistry and Geomicrobiology. Vol. 3: Methods, Metals and Assessment* (Ed. W. E. Krumbein), pp. 753–69. Ann Arbor Science Publishers, Ann Arbor, Michigan.

SMITH D. W. (1980) An evaluation of marsh nitrogen fixation. In: *Estuarine Perspectives* (Ed. V. S. Kennedy), pp. 135–42. Academic Press, New York.

VAN BERKUM P. & BOHLOOL B. B. (1980) Evaluation of nitrogen fixation by bacteria in association with roots of tropical grasses. *Microbiological Reviews* **44**, 491–517.

WINTER H. C. & BURRIS R. H. (1976) Nitrogenase. *Annual Review of Biochemistry* **45**, 409–26.

WOLK C. P., THOMAS J. C. & SHAFFER P. W. (1976) Pathway of nitrogen metabolism after fixation of ^{13}N-labelled nitrogen gas by the cyanobacterium *Anabaena cylindrica*. *Journal of Biological Chemistry* **251**, 5027–34.

References

ABD-EL-MALEK Y. (1971) Free-living nitrogen-fixing bacteria in Egyptian soils and their possible contribution to soil fertility. *Plant and Soil*, Special Volume, 423–42.

ARNON D. I. & YOCH D. C. (1974) Photosynthetic bacteria. In: *The Biology of Nitrogen Fixation* (Ed. A. Quispel), pp. 168–201. North-Holland Publishing Company, Amsterdam.

BALANDREAU J. & DOMMERGUES Y. (1973) Assaying nitrogenase (C_2H_2) activity in the field. *Ecological Research Committee Bulletin (Stockholm)* **17**, 247–54.

BREMNER J. M. (1965) Total nitrogen. In: *Methods of Soil Analysis* (Ed. C. A. Black), vol. 9, pp. 1149–78. American Society of Agronomy, Madison, Wisconsin.

BRILL W. J. (1975) Regulation and genetics of bacterial nitrogen fixation. *Annual Review of Microbiology* **29**, 109–29.

BRILL W. J. (1979) Nitrogen fixation: basic to applied. *American Scientist* **67**, 458–66.

BRILL W. J. (1980) Biochemical genetics of nitrogen-fixation. *Microbiological Reviews* **44**, 449–67.

BROTONEGORO S. (1974) Nitrogen fixation and nitrogenase activity of *Azotobacter chroococcum*. Ph.D. thesis, Agricultural University of Wageningen, Netherlands.

BURRIS R. H. & WILSON P. W. (1957) Methods for measurement of nitrogen fixation. In: *Methods in Enzymology* (Eds S. P. Colowick & N. O. Kaplan), vol. 4, pp. 355–66. Academic Press, New York.

CANNON F. C., DIXON R. A., POSTGATE J. R. & PRIMROSE S. B. (1974) Plasmids formed in nitrogen-fixing *Escherichia coli–Klebsiella pneumoniae* hybrids. *Journal of General Microbiology* **80**, 241–51.

DAESCH G. & MORTENSON L. E. (1972) Effect of ammonia on the synthesis and function of the N_2-fixing enzyme system in *Clostridium pasteurianum*. *Journal of Bacteriology* **110**, 103–9.

DALTON H. & MORTENSON L. E. (1972) Dinitrogen (N_2) fixation (with a biochemical emphasis). *Bacteriological Reviews* **36**, 231–60.

DAVID K. A. V. & FAY P. (1977) Effects of long-term treatment with acetylene on nitrogen-fixing microorganisms. *Applied and Environmental Microbiology* **34**, 640–6.

DELWICHE C. C. & WYLER J. (1956) Non-symbiotic nitrogen fixation in soil. *Plant and Soil* **7**, 113–29.

DICKER H. J. & SMITH D. W. (1980a) Acetylene reduction (nitrogen fixation) in a Delaware, USA, salt marsh. *Marine Biology* **57**, 241–50.

DICKER H. J. & SMITH D. W. (1980b) Physiological ecology of acetylene reduction (nitrogen fixation) in a Delaware salt marsh. *Microbial Ecology* **6**, 161–71.

DICKER H. J. & SMITH D. W. (1980c) Enumeration and relative importance of acetylene-reducing (nitrogen-fixing) bacteria in a Delaware salt marsh. *Applied and Environmental Microbiology* **39**, 1019–25.

DIXON R. O. D. (1972) Hydrogenase in legume root nodule bacteroids: occurrence and properties. *Archiv für Mikrobiologie* **85**, 193–201.

DOBEREINER J. (1974) Nitrogen-fixing bacteria in the rhizosphere. In: *The Biology of Nitrogen Fixation* (Ed. A. Quispel), pp. 86–120. North-Holland Publishing Company, Amsterdam.

DOBEREINER J. & DAY J. M. (1976) Associative symbioses in tropical grasses: characterization of microorganisms and dinitrogen-fixing sites. In: *International Symposium on N_2 Fixation-Interdisciplinary Discussion*, 1974, pp. 518–38. Washington State University, Pullman, Washington.

DOBEREINER J., DAY J. M. & DART P. J. (1972) Nitrogenase activity and oxygen sensitivity of the *Paspalum notatum–Azotobacter paspali* association. *Journal of General Microbiology* **71**, 103–16.

DROZD J. & POSTGATE J. R. (1970) Effects of oxygen on acetylene reduction, cytochrome content and respiratory activity of *Azotobacter chroococcum*. *Journal of General Microbiology* **63**, 63–73.

FIRESTONE M. K., SMITH M. S., FIRESTONE R. B. & TIEDJE J. M. (1979) The influence of nitrate, nitrite, and oxygen on the composition of the gaseous products of denitrification in soil. *Soil Science Society of America Journal* **43**, 1140–4.

FLETT R. J., HAMILTON R. D. & CAMPBELL N. E. R. (1976) Aquatic acetylene-reduction techniques: Solutions to several problems. *Canadian Journal of Microbiology* **22**, 43–51.

FOGG G. E. (1977) Excretion of organic matter by phytoplankton. *Limnology and Oceanography* **22**, 576–7.

GERSBERG R., KROHN K., PEEK N. & GOLDMAN C. R. (1976) Denitrification studies with ^{13}N-labeled nitrate. *Science* **192**, 1229–31.

HANSON R. B. (1977) Comparison of nitrogen fixation activity in tall and short *Spartina alterniflora* salt marsh soils. *Applied and Environmental Microbiology* **33**, 596–602.

HARDY R. W. F., BURNS R. C. & HOLSTEN R. D. (1973) Applications of the acetylene-ethylene assay for measurement of nitrogen fixation. *Soil Biology and Biochemistry* **5**, 47–81.

HERBERT R. A. (1975) Heterotrophic nitrogen fixation in shallow estuarine sediments. *Journal of Experimental Marine Biology and Ecology* **18**, 215–25.

LINE M. A. & LOUTIT M. W. (1971) Non-symbiotic nitrogen-fixing organisms from some New Zealand tussock-grasslands soils. *Journal of General Microbiology* **66**, 309–18.

LJONES T. (1974) The enzyme system. In: *The Biology of Nitrogen Fixation* (Ed. A. Quispel), pp. 617–38. North-Holland Publishing Company, Amsterdam.

MACNEIL T., MACNEIL D., ROBERTS G. P., SUPIANO M. A. & BRILL W. J. (1978) Fine-structure mapping and complementation analysis of *nif* (nitrogen fixation) genes in *Klebsiella pneumoniae*. *Journal of Bacteriology* **136**, 253–66.

MEYER G. W., MCCASLIN B. D. & GAST R. G. (1974) Sample preparation and 15-N analysis using a Statron NOI–5 optical analyzer. *Soil Science* **117**, 378–85.

MUHAMMAD S. & KUMAZAWA K. (1972) Use of optical spectrographic ^{15}N-analyses to trace nitrogen applied at the heading stage of rice. *Soil Science and Plant Nutrition* **18**, 143–6.

NAMBIAR P. T. C. & SHETHNA Y. I.(1977) Effect of NH_4^+ on acetylene reduction (nitrogenase) in *Azotobacter vinelandii* and *Bacillus polymyxa*. *Journal of Indian Industry and Science* **59**, 155–68.

NELSON A. D., BARBER L. E., TJEPKEMA J., RUSSELL S. A., POWELSON R., EVANS H. J. & SEIDLER R. J. (1976) Nitrogen fixation associated with grasses in Oregon. *Canadian Journal of Microbiology* **22**, 523–30.

OPPENHEIM J. & MARCUS L. (1970) Correlation of ultrastructure in *Azotobacter vinelandii* with nitrogen source for growth. *Journal of Bacteriology* **101**, 286–91.

PATRIQUIN D. G. (1978a) Factors affecting nitrogenase activity (acetylene reducing activity) associated with excised roots of the emergent halophyte *Spartina alterniflora* Loisel. *Aquatic Botany* **4**, 193–210.

PATRIQUIN D. G. (1978b) Nitrogen fixation (acetylene reduction) associated with cord grass, *Spartina alterniflora* Loisel. *Ecological Research Committee Bulletin (Stockholm)* **26**, 20–7.

PATRIQUIN D. G. & DOBEREINER J. (1978) Light microscopy observations of tetrazolium-reducing bacteria in the endorhizosphere of maize and other grasses in Brazil. *Canadian Journal of Microbiology* **24**, 734–42.

PATRIQUIN D. G. & KEDDY C. (1978) Nitrogenase activity (acetylene reduction) in a Nova Scotian salt marsh: its association with angiosperms and the influence of some edaphic factors. *Aquatic Botany* **4**, 227–44.

PENGRA R. M. & WILSON P. W. (1958) Physiology of nitrogen fixation by *Aerobacter aerogenes*. *Journal of Bacteriology* **75**, 21–5.

POSTGATE J. R. (1970) Nitrogen fixation by sporulating sulphate reducing bacteria including rumen strains. *Journal of General Microbiology* **63**, 137–9.

POTTS M., KRUMBEIN W. E. & METZGER J. (1978) Nitrogen fixation rates in anaerobic sediments determined by acetylene reduction, a new ^{15}N field assay, and simultaneous total N ^{15}N determination. In: *Environmental Biogeochemistry and Geomicrobiology. Vol. 3: Methods, Metals and Assessment* (Ed. W. E. Krumbein), pp. 753–69. Ann Arbor Science Publishers, Ann Arbor, Michigan.

RALPH R. D. (1978) Dinitrogen fixation by *Azotobacter* in the rhizosphere of *Ammophila breviligulata*. Ph.D. thesis, University of Delaware, USA.

REIDERER-HENDERSON M. A. & WILSON P. W. (1970) Nitrogen-fixation by sulphate reducing bacteria. *Journal of General Microbiology* **61**, 27–31.

RIVERA-ORTIZ J. M. & BURRIS R. H. (1975) Interactions among substrates and inhibitors of nitrogenase. *Journal of Bacteriology* **123**, 537–45.

ROBERTS G. P., MACNEIL T., MACNEIL D. & BRILL W. J. (1978) Regulation and characterization of protein products coded by the *nif* (nitrogen fixation) genes of *Klebsiella pneumoniae*. *Journal of Bacteriology* **136**, 267–79.

ROSENBLUM E. G. & WILSON P. W. (1949) Fixation of isotopic nitrogen by *Clostridium*. *Journal of Bacteriology* **57**, 413–14.

SCHELL D. M. (1978) Chemical and isotopic methods in nitrification studies. In: *Microbiology 1978* (Ed. D. Schlessinger), pp. 292–5. American Society for Microbiology, Washington.

SCHUBERT K. R. & EVANS H. J. (1976) Hydrogen evolution: a major factor affecting the efficiency of nitrogen fixation in nodulated symbionts. *Proceedings of the National Academy of Science USA* **73**, 1207–11.

SHAH V. K. & BRILL W. J. (1977) Isolation of an iron-molybdenum cofactor from nitrogenase. *Proceedings of the National Academy of Science USA* **74**, 3249–53.

SHARP J. H. (1977) Excretion of organic matter by marine phytoplankton: do healthy cells do it? *Limnology and Oceanography* **22**, 381–99.

SMITH D. W. (1980) An evaluation of marsh nitrogen fixation. In: *Estuarine Perspectives* (Ed. V. S. Kennedy), pp. 135–42. Academic Press, New York.

SORGER G. J. (1969) Regulation of nitrogen fixation in *Azotobacter vinelandii* OP: the role of nitrate reductase. *Journal of Bacteriology* **98**, 56–61.

STEWART W. D. P. (1969) Biological and ecological aspects of nitrogen fixation by free-living microorganisms. *Proceedings of the Royal Society of London (B)* **172**, 367–88.

STEWART W. D. P. (1974) Blue-green algae. In: *The Biology of Nitrogen Fixation* (Ed. A. Quispel), pp. 202–64. North-Holland Publishing Company, Amsterdam.

TEAL J. M., VALIELA I. & BERLO D. (1979) Nitrogen fixation by rhizosphere and free-living bacteria in salt marsh sediments. *Limnology and Oceanography* **24**, 126–32.

THOMAS J., MEEKS J. C., WOLK C. P., SHAFFER P. W., AUSTIN S. M. & CHIEN W. S. (1977) Formation of glutamine from [^{13}N]ammonia, [^{13}N]dinitrogen, and [^{14}C]glutamate by heterocysts isolated from *Anabaena cylindrica*. *Journal of Bacteriology* **129**, 1545–55.

TJEPKEMA J. D. & WINSHIP L. J. (1980) Energy requirement for nitrogen fixation in actinorhizal and legume root nodules. *Science* **209**, 279–80.

van Berkum P. & Bohlool B. B. (1980) Evaluation of nitrogen fixation by bacteria in association with roots of tropical grasses. *Microbiological Reviews* **44**, 491–517.

Von Bulow J. F. W. & Dobereiner J. (1975) Potential for nitrogen fixation in maize genotypes in Brazil. *Proceedings of the National Academy of Science USA* **72**, 2389–93.

Whitney D. E., Woodwell G. M. & Howarth R. W. (1975) Nitrogen fixation in Flax Pond: a Long Island salt marsh. *Limnology and Oceanography* **20**, 640–3.

Wiebe W., Johannes R. E. & Webb K. L. (1975) Nitrogen fixation in a coral reef community. *Science* **188**, 257–9.

Winter H. C. & Burris R. H. (1976) Nitrogenase. *Annual Review of Biochemistry* **45**, 409–26.

Wolk C. P., Thomas J. C. & Shaffer P. W. (1976) Pathway of nitrogen metabolism after fixation of ^{13}N-labelled nitrogen gas by the cyanobacterium *Anabaena cylindrica*. *Journal of Biological Chemistry* **251**, 5027–34.

Wong P. P. & Burris R. H. (1972) Nature of oxygen inhibition of nitrogenase from *Azotobacter vinelandii*. *Proceedings of the National Academy of Science USA* **69**, 672–5.

Yoch D. C. & Cantu M. (1980) Changes in the regulatory form of *Rhodospirillum rubrum* nitrogenase as influenced by nutritional and environmental factors. *Journal of Bacteriology* **142**, 899–907.

Yoshida T. & Ancajas R. R. (1973) Nitrogen-fixing activity in upland and flooded rice fields. *Soil Science Society of America Proceedings* **37**, 42–6.

Zumft W. G., Mortenson L. E. & Palmer G. (1974) Electron paramagnetic resonance of nitrogenase. Investigation of the redox behavior of azoferredoxin and molybdoferredoxin proteins with potentiometric and rapid-freeze techniques. *European Journal of Biochemistry* **46**, 525–35.

Chapter 14 • Phosphorus Cycle

John W. B. Stewart and Robert B. McKercher

14.1 Introduction

Phosphorus is a constituent of meteorites, igneous rocks, soil, lakes, rivers, the sea and all living organisms. It cycles readily between inorganic and organic systems. Phosphorus is thought (Hutchinson 1970) to be the most ecologically important element because:

(1) it is concentrated by living organisms to levels considerably above that of the source;

(2) it functions as an agent of energy transfer;

(3) a deficiency of available phosphorus is more likely to limit reproduction and hence productivity than any other material, except water.

In life systems, phosphorus and sulphur perform group and energy transformation functions. The reasons generally advanced to explain this (Wald 1962; Halstead & McKercher 1975) are that the atoms of living tissue are those whose nature allows formation of covalent multiple bonds, whereas atoms that are reluctant or unable to form covalent bonds are generally excluded from physiological functions. Atoms associated with life are mostly small and tend to form flexible bonds that show high energy characteristics, form a variety of linkages with other atoms and have an intrinsic instability that facilitates atom exchange. Both phosphorus and sulphur may expand covalent linkages beyond four through their 3*d* orbitals and thus increase number and types of bond linkages. They are unique also among third period elements in that they have a multiple bond. This characteristic is responsible for the large number of resonance structures available to phosphorus compounds and, in turn, for the high free-energy changes associated with phosphorus-group transfer reactions.

Phosphorus occurs naturally on the Earth's surface, principally as orthophosphate (PO_4^{3-}), but under reducing conditions, such as those that occur in lake sediments, phosphine (PH_3) can be present, and phosphonates $(C_nH_{n+1})_4POH$ have been found recently in top soils in a cool moist environment. Phosphorus is never found in its elemental form in nature.

Considerable progress has been made (Cole *et al.* 1977) in the elucidation of specific parts of phosphorus cycling in the soil ecosystem, e.g. ion diffusion, inorganic phosphorus reactions, but progress has been limited by the extreme reactivity of phosphorus and its existence as part of a large number of inorganic and organic compounds. The development of methods to extract and separate these compounds from each other and from soil components has proved difficult (Halstead & McKercher 1975).

The aim of this chapter is to consider various approaches and techniques that have been used to elucidate phosphorus cycling in soils. Much has been gained from consideration of the effects of phosphorus cycling on the composition of specific soil components and the changes in them across environmental gradients. This, in turn, has led to proposed mechanisms of phosphorus cycling, which have been tested by computer simulation modelling and laboratory experimentation. The discussion will be limited to these topics and to methods used in research into the phosphorus cycle. This type of research has developed comparatively recently and has shown potential for the more rapid development of the understanding of soil phosphorus transformations. This is not to say that the application of microbiological and biochemical procedures have not or will not be required. However, soil has proved to be a medium in which such procedures have shown limited success, mainly because of the heterogeneity of its material and the nature of the association between its organic and inorganic constituents (the organo-mineral complexes).

A starting point in the study of phosphorus in soil plant systems could be the conceptual and very simplified view of the phosphorus cycle of a native prairie grassland ecosystem (Fig. 1). This shows that, although most of the soil phosphorus is present as abiotic organic and inorganic phosphorus, a considerable amount is in soil fauna and microorganisms. Since plant roots require that their phosphorus be in a solution phase, the phosphorus-supplying power of the soil will depend in large measure on factors affecting its release from these three fractions.

14.2 Soil organic phosphorus

Soil organic phosphorus has been reviewed by Anderson (1967), Hayman (1975), Halstead and McKercher (1975) and Dalal (1977). To date only about 50 to 70% of the organic phosphorus in soil has been identified, most of it as phosphate esters. However, recent work has pointed to the fact that under reducing conditions phosphines can be a source of phosphorus loss from soil to the atmosphere (Cheng *et al.* 1979), and phosphorus has been detected as a major component of acid rain in an Amazonian rain forest (Herrera 1979; Jordan *et al.* 1980). Phosphonates (C—P) have been found in moist top soils in New Zealand (Newman & Tate 1980), and polyphosphate (P—O—P—Anderson & Russell 1969; Pepper *et al.* 1976) and other non-orthophosphate phosphorus (Beever & Burns 1976) occur in soil and soil microbial tissue. Presumably, other P—C, P—N and P—S linkages exist, but esterphosphates must be considered the predominant form of organic phosphorus in soils. Esterphosphates, such as inositol phosphates (up to 60% of the total organic phosphorus), nucleic acid phosphorus (5 to 10%) and phospholipid phosphorus (<1%), are bonded to soil organic and inorganic components by mechanisms such as adsorption and are generally less a part of the organic complex than organic nitrogen or sulphur (Scott & Anderson 1976).

Soil organic matter content and composition has been related to the availability of phosphorus in soil. Walker and Adams (1958, 1959), Walker *et al.* (1959) and Smeck (1973) suggested that the phosphorus content of the parent material ultimately controlled the organic carbon, phosphorus, nitrogen and sulphur contents of soil. Walker (1965) and Walker and Syers (1976) developed this further, suggesting that phosphorus availability to organisms would control nitrogen fixation and that this, in turn, would ultimately control organic

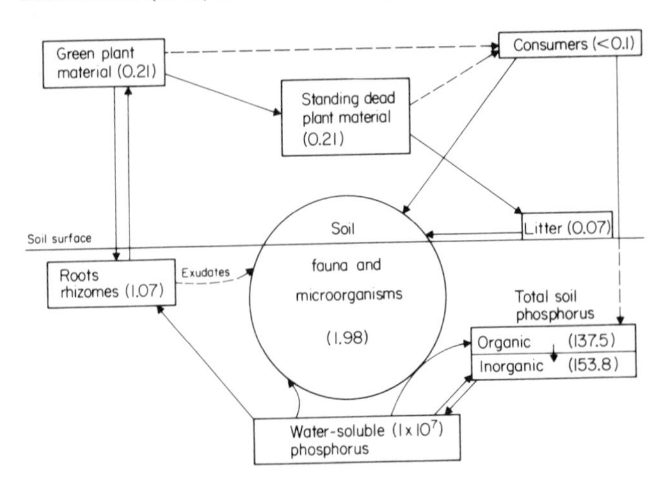

Figure 1. Phosphorus in components of a native grassland ecosystem showing nutrient flow. Numbers in parentheses = phosphorus, expressed in g m^{-2} to a depth of 30 cm (after Halm *et al.* 1972).

matter accumulation. After examining the phosphorus content of several chrono-, topo- and climosequences, they also suggested that during periods of substantial phosphorus loss by weathering and leaching, available soil phosphorus would control organic matter loss (Fig. 2). During the period of increasing availability of phosphorus carbon/organic phosphorus ratios would be constant as the carbon increased, but during decreasing availability of soil phosphorus ratios would increase. In support of this, Floate (1971) found that in phosphorus-deficient hill soils carbon/organic phosphorus ratios were very wide, and Halstead and McKercher (1975) pointed out the range of variability of carbon/organic phosphorus ratios compared with carbon/nitrogen ratios. Cole and Heil (1981) reviewed literature on organic carbon/nitrogen/phosphorus ratios over a narrow range of temperature and moisture conditions of North American soils and found that organic carbon and nitrogen accumulations in mature soils are closely linked to the phosphorus contents of the original parent material. They suggested that microbial growth processes are the principal means for adjusting the supply of nitrogen to the supply of phosphorus.

Other studies on the interrelationship of carbon, nitrogen, sulphur and phosphorus in soil organic matter (McGill 1979) reject the idea that a strict stoichiometric relationship exists between carbon, nitrogen, sulphur and organic phosphorus. Instead, McGill (1979) proposed that the mechanisms stabilising organic carbon, nitrogen, sulphur and organic phosphorus are not necessarily common to all four elements and that mechanisms and pathways of mobilisation are specific to the organic material containing the various elements. This proposal implies that any element may accumulate

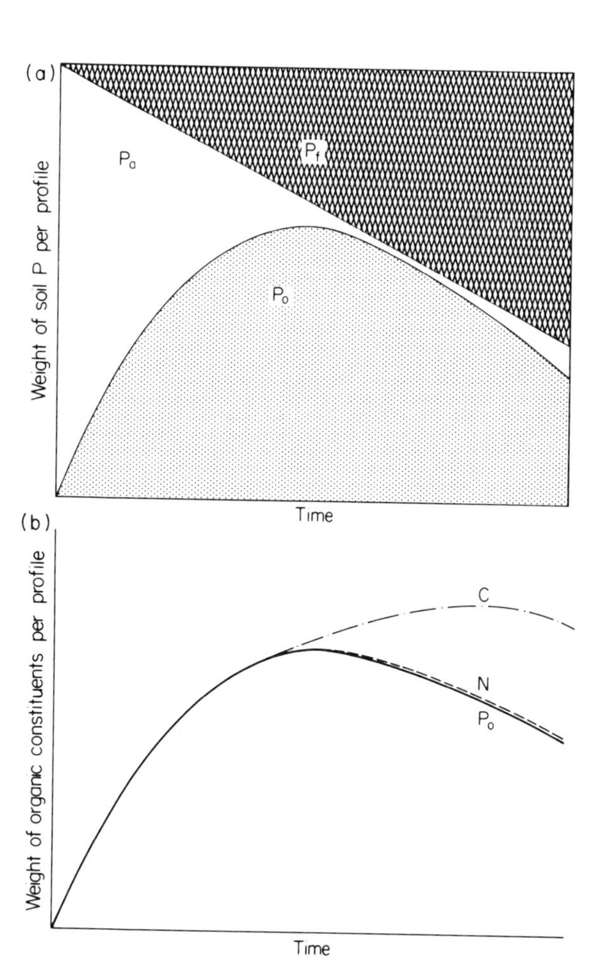

Figure 2. Changes in the forms of (a) soil phosphorus and (b) constituents of soil organic matter during soil development (Floate 1971): P_f, inorganic phosphorus insoluble in 2M-H_2SO_4 (bound by fixed-iron and aluminium); P_a, inorganic phosphorus soluble in 2M-H_2SO_4 (bound by soluble-iron, aluminium, calcium, and apatite); P_o, organic phosphorus.

or decrease at a rate independent of other elements. McGill (1979) proposed a dichotomous system (Fig. 3) in which nitrogen and part of the soil sulphur became stabilised as a result of direct association with carbon (nitrogen- and carbon-bonded sulphur) and mineralised as a result of carbon oxidation (classical biological mineralisation) to provide energy. In contrast, sulphur and phosphorus exist as esters (C—O—S and C—O—P) and are stabilised by the interaction of the ester with soil components. These elements may be mineralised by the organism's need for a specific element. McGill (1979) termed the latter process a biochemical mineralisation, since it operates largely outside the cell and is controlled by the need for the element released (Eivazi & Tabatabai 1977). It must be recognised that certain organic phosphorus and microbial phosphorus components are more rapidly metabolised than others. Szember (1960) and McKercher and Tollefson (1978) found that DNA-phosphorus was about as available as KH_2PO_4 – phosphorus to barley plants but that phospholipid-phosphorus was metabolised less rapidly. The concept proposed by McGill (1979) is useful, as it allows for the variability found in soil organic matter composition and sets the stage for predicting the stability of organic phosphorus in different soils.

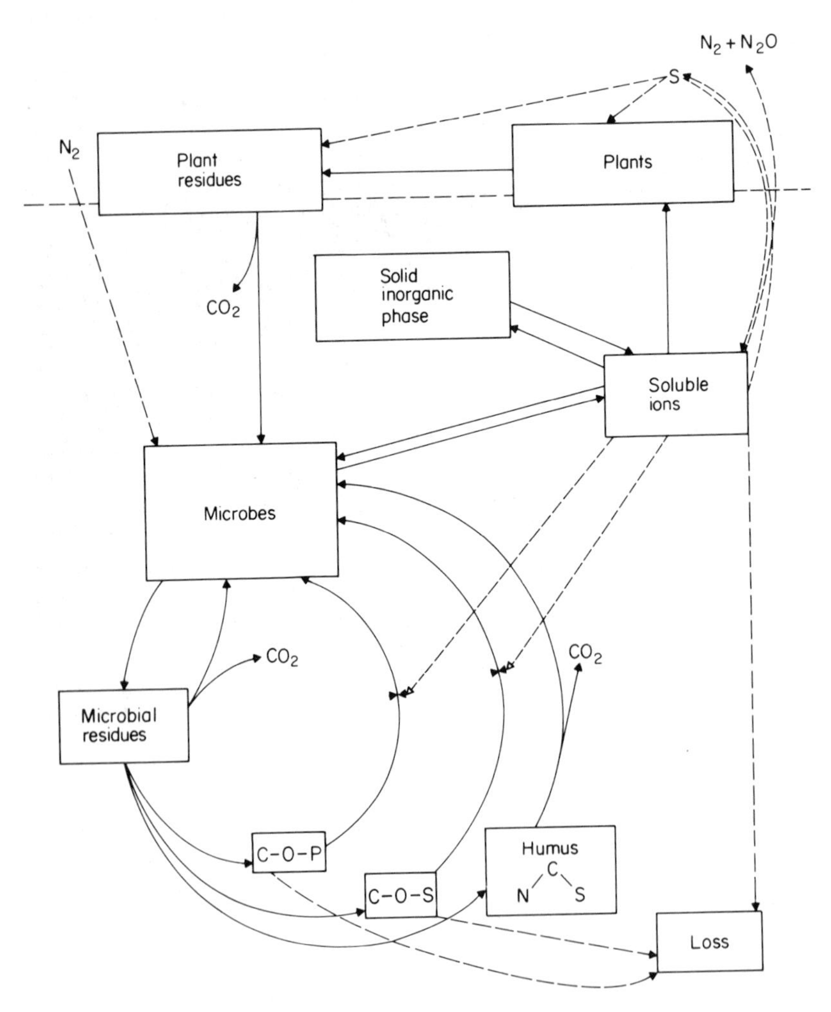

Figure 3. Schematic illustration of interrelations of carbon, nitrogen, sulphur and phosphorus cycling within soil–plant systems (after McGill 1979).

14.3 Inorganic phosphorus

A basic assumption of this discussion is that the phosphorus supply of inorganic phosphorus to the plant can be explained by diffusion of water-soluble forms without contact exchange (Barber 1962). As the phosphorus concentration at the root surface is reduced by plant uptake, a concentration gradient is established in the surrounding water films and phosphorus ions flow towards the root. Replenishing a soluble-phosphorus-depleted zone will depend on the supply of phosphorus from slightly soluble phosphate minerals, phosphate-absorbing surfaces, and organic phosphorus mineralisation. Chemical forms and solubility relationships of phosphorus in soil have been thoroughly reviewed by Larsen (1967) and Olsen and Flowerday (1971) and their implication for phosphorus cycling has been discussed in detail by Cole *et al.* (1977) and Nye (1977).

Solution phosphorus equilibrates rapidly with the labile fraction of absorbed phosphorus. It is essential to define both the quantity of labile inorganic phosphorus and the concentration of solution phosphorus in any soil, as the ratio of labile inorganic phosphorus to that in solution can vary widely, depending on the surface area and adsorption capacity of the soil. Methods for determining the labile inorganic phosphorus include ion-resin equilibration (Amer *et al.* 1955; Sibbesen 1977) and isotope exchange techniques (Olsen & Watanabe 1963; Sadler and Stewart 1977). Measuring the amount of labile inorganic phosphorus and the concentration of solution phosphorus has proved to be more useful in the studies of phosphorus cycling than identifying specific inorganic phosphorus minerals in soil.

14.4 Microbial Phosphorus

Understanding of the role of microorganisms in the phosphorus cycle is far from complete, although progress in elucidating the chemical nature and modes of dephosphorylation of soil organic phosphate has been made in the last decade (Cosgrove 1977). Application of modern methods of extraction, concentration, separation and identification have been demonstrated, but the origin, mode of accumulation and dynamics of soil organic phosphorus remain unknown (Anderson 1975; Halstead & McKercher 1975).

Recent methods of estimating soil microbial biomass from direct microscopic observation (Van Veen & Paul 1979), carbon dioxide respirometry after chloroform treatment (Anderson & Domsch 1978a, b) and extrapolation from ATP content (Paul & Johnson 1977; Nannipieri *et al.* 1978; Jenkinson *et al.* 1979; Oades & Jenkinson 1979) have greatly improved the accuracy and ease of measuring biomass carbon (Jenkinson & Ladd 1981). Estimation of microbial phosphorus also has been improved by measuring the change in the quantity of organic phosphorus and inorganic phosphorus removed by specific extraction agents following chloroform treatment (Chauhan *et al.* 1979, 1981; Stewart *et al.* 1980; Hedley & Stewart 1982). As the type of phosphorus compounds contained in a microbial cell (RNA 30 to 50%; DNA 5 to 10%; acid-soluble inorganic and organic phosphorus 15 to 20% comprising orthophosphate, metaphosphate, sugar and adenosine phosphates, various phosphorylated coenzymes and phospholipids $<10\%$) react with different intensity with soil colloidal material when the cell wall is lysed by chloroform, a variable proportion of the released phosphorus is recovered in a mild extraction reagent and will vary from soil to soil, depending on the reactivity of the soil colloidal surfaces (Chauhan *et al.* 1981). Hedley and Stewart (1982) developed a sequential extraction technique that extracts and separates available inorganic phosphorus, labile inorganic phosphorus and organic phosphorus; moderately labile inorganic and organic phosphorus; aggregate-protected inorganic phosphorus and organic phosphorus, stable inorganic phosphorus, and more resistant organic phosphorus. The microbial phosphorus contribution to the more labile fractions can be estimated by treating the soil with chloroform and extracting the released phosphorus. Use of this fractionation technique has made it possible not only to study more closely the net changes in phosphorus amounts caused by the long-term depletion of soil phosphorus by crops but also to identify accurately smaller changes occurring during short-term incubations, which were designed to promote both the immobilisation and mineralisation of soil phosphorus. During further short-term incubation experiments (Stewart & Hedley 1980) in which [^{33}P]- and [^{32}P]phosphorus were used to label specific soil phosphorus pools, measurements of various phosphorus isotopes and soil extracts and in growing plants were used to explain the movement of phosphorus between different forms in the soil and to predict the long-term effects of such transformations. Stewart and Hedley (1980) claim that the dynamics of phosphorus movement are measurable and that phosphorus mineralisation and immobilisation occur simultaneously and stress the important role of the microbial population in the redistribution of phosphorus in soils.

Although the mechanisms of phosphorus uptake by a growing microbial population have been examined (Beever & Burns 1976; Burns & Beever 1977), the release of phosphorus from the microbial cell is less well understood. Cole *et al.* (1977) used abiotic and environmental factors, such as freezing and thawing in the early spring and heating and drying of the soil, in their simulation model to decrease the size of the microbial population, providing a return of phosphorus to solution. Equally important but less well understood are the interactive effects of bacteria, fungi, and microbial grazers such as amoeba or nematodes (Darbyshire 1975). Bacteria and bacterial-grazer interactions may have special significance in the rhizosphere region of soil in relation to the supply of nutrients for plant food. Fenchel and Harrison (1976) examined earlier work by Barsdate *et al.* (1974) and postulated that bacteria cycled phosphorus in and out of solution at rates much higher than indicated by the observed net mineralisation.

Microbiologists have long recognised that individual soils may have sharply dissimilar microfloras, depending upon climate, soil type, physical and chemical properties, and the type and amount of plant cover (Clark 1949; Clark & Paul 1970; Chapin *et al.* 1978). Although bacteria have most of the enzymes necessary to degrade the structural components of litter, fungi appear to be the primary invaders and, in grassland soils of the temperate zone, constitute the majority (>66% of the microbial biomass) of the soil microbial population, followed by bacteria (Clark & Paul 1970; Gyllenberg & Eklund 1974; Gray 1976). While developing a comprehensive simulation model of carbon and nitrogen dynamics, McGill *et al.* (1981) recognised the different responses fungi and bacteria make to abiotic and environmental factors and that the generally higher carbon/nitrogen ratio in fungi than in bacteria adversely affects the rates of decomposition of dead fungal biomass. Hedley and Stewart (1982) found that fungal phosphorus was more easily extracted and appeared to be more labile than bacterial phosphorus. It appears, therefore, that phosphorus flows at different rates through the microbial population to soil than nitrogen. Unfortunately, methods have not yet been developed to separate fungal and bacterial phosphorus from within an active soil microbial biomass.

Bacteria and fungi are 20 to 50 times more abundant in the rhizosphere than in bulk soil (Coleman 1976; Newman 1978; Rovira 1979). Competition by rhizoplane microflora can influence nutrient flow of phosphorus into roots at low soil solution phosphorus concentrations. Conversely, endotrophic mycorrhizae (vesicular-arbuscular [VA] mycorrhizae) can greatly enhance phosphorus uptake in soils low in solution phosphorus (Gerdemann 1975; Tinker 1975a, b). The mechanism of phosphorus uptake by VA mycorrhizae appears to be a consequence of increased surface area, i.e. an extended depletion zone provided by the hyphae combined with an increased ability to take up phosphorus when it is present in the soil in low concentrations. Thus, phosphorus uptake by plant roots is a balance between the demand of the rhizosphere population for phosphorus and the transfer of some of the immobilised phosphorus directly to the plant by VA mycorrhizae. The concentration of phosphorus in the vicinity of the root depends on the soil inorganic phosphorus chemistry plus diffusion to the root (Nye 1977). Lowering the solution phosphorus concentration can cause an increase in acid phosphatase activity at the root surface (Gould *et al.* 1979).

Several studies have shown that earthworms can enrich surface soils in phosphorus (Barley 1961). Sharpley and Syers (1976) examined the surface casts of earthworms (predominantly *Allolobophora caliginosa*) from a permanent pasture in New Zealand and found that they contained a higher pool of readily exchangeable inorganic phosphorus than the surrounding soil and that the rate of mineralisation of easily extractable organic phosphorus was higher than in the surrounding soil. The observed seasonal variations in cast production were attributed to fluctuations in soil moisture and temperature. The total phosphorus content of the casts remained constant over the year, but the inorganic phosphorus content decreased during the colder months. This decline in phosphorus was attributed to lower soil temperatures which reduced microbial activity and phosphatase enzyme activity. Sharpley and Syers (1977) estimated that 9 and

13 kg ha^{-1} yr^{-1} of inorganic phosphorus and organic phosphorus, respectively, accumulated in surface casts in this soil representing a significant amount of phosphorus cycling. McKercher *et al.* (1979) examined the biomass numbers and phosphorus content of three soil invertebrate taxa, excluding earthworms which were not present in this ecosystem, in a native grassland ecosystem and found that the soil invertebrates accounted for a small fraction of the total soil biomass but postulated that their role in phosphorus transformation was relatively more important.

Experimental

14.5 Recent developments

Application of modern analytical techniques, including isotope labelling, together with the introduction of new techniques, have enabled a variety of approaches to the study of phosphorus cycling. Most important has been the development of team research, which allows groups of investigators to look simultaneously at chemical, physical and biological interactions in the redistribution of phosphorus in soils.

Studies of the complete phosphorus cycle depend on the rigorous evaluation of all the information required for each stage in the cycle. In this evaluation, simulation models for quantifying and assimilating information have proved invaluable for pinpointing areas where research is needed. Similarly the study of simplified systems (microcosms) has been important in investigating possible phosphorus flows between soil fauna, microorganisms and labile soil phosphorus pools. Isotopic labelling of specific soil pools has provided a means of examining the actual phosphorus flows rather than the net flows obtained by conventional means. Estimates of microbial and faunal biomass, activity and phosphorus content have shown the mode of redistribution of phosphorus in soils. Improved chemical characterisation of organic phosphorus plus continued progress in the investigation of the factors controlling phosphatase activity have clarified mineralisation processes. Application of methods of identifying inorganic phosphorus compounds in soils taken across environmental gradients, from catenary sequences (a sequence of soils of similar age and parent material, but with different characteristics due to variations in relief and in drainage), or from controlled growth chamber incubations have given insight into the forms and availability of inorganic phosphorus in soils. Integration of these effects can be achieved by statistical or simulation models.

14.6 Microcosms in phosphorus cycling studies

The dynamic nature of ecosystems is the product of variations in environmental factors and interactions among species. Given the complicated nature of interactions among producers, herbivores and bacteria and the lack of understanding about how limiting nutrients affect species interactions, it is not surprising that describing system dynamics has been more successful than predicting them. Difficulties of interpreting observations in natural ecosystems have led to the study of simplified systems (microcosms) where the addition of selected bacterial, fungal, actinomycete, and nematode populations to sterilised soils has allowed quantitative sampling of the components and control of environmental variables (Witkamp 1976; Anderson *et al.* 1979a; Elliott *et al.* 1979). The microcosms, incorporating a few species of producers, consumers and decomposers, provide systems with some of the complex dynamic phenomena of natural habitats. Cole *et al.* (1978) studied phosphorus immobilisation and mineralisation in soil incubations, simulating rhizospheres with combinations of bacterial, amoebal and nematode populations. They found that bacteria quickly assimilated and retained much of the labile inorganic phosphorus as carbon substrates were metabolised. Most of this bacterial phosphorus was mineralised and returned to the phosphorus pool through the action of amoebal grazing. Nematode effects on phosphorus mineralisation were small, except for indirect effects on amoebal activity. This agreed with data of Coleman *et al.* (1978) which suggested that nematodes can feed on bacteria but eat only amoeba when given a choice between bacteria and amoeba in soil. However, the authors warned that this was true only for the few species of nematodes, amoeba, and bacteria examined and that the work needed to be extended to cover a greater range of soil microflora and fauna.

14.7 Simulation models as research tools

A simulation model is a complex series of mathematical equations with which numerical solutions are predicted by the hypothesis (Kowal 1971). If the predicted states disagree with measured states at a chosen level of statistical significance, the hypothesis is rejected. Viewed in this light, simulation modelling is nothing more or less than the application of the scientific method to problems too complex for the predictions of hypotheses to be worked out without the aid of a computer. However, comparing model output and data will remain subjective until manageable statistical tests for validation are developed.

Constructing simulation models stimulates the precise formulation of hypotheses, aids the interpretation of complex sets of data, facilitates the organisation of information, and can contribute to the design of critical experiments (Woodmansee 1978).

Structured models of microbial populations are becoming more important in ecological research. Unstructured models represent populations by a single variable, such as numbers or biomass. Structured models take simultaneous account of several population variables, such as number, mean cell size and total biomass (Hunt 1977).

Hunt *et al.* (1977) modelled the growth of bacteria limited by either carbon, nitrogen or phosphorus. Their model distinguished between storage compartments for the three elements and a pool of small precursor molecules, as well as the structural and synthetic components. The model was adjusted to fit published data on the chemical composition of bacteria growing in a chemostat limited by either carbon, nitrogen or phosphorus. The model successfully predicted the kinds of waste products produced under various conditions, the transient response to a change in dilution rate and the response to a change in the make-up of the medium, even though these observations were not used to formulate the model or establish parameter values.

Phosphorus-cycling simulation models (Cole *et al.* 1977; Chapin *et al.* 1978; Harrison 1978; Mishra *et al.* 1979) have allowed the integration of the effects of moisture, soil properties, plant phenology, and microbial decomposition of organic matter on phosphorus flows. These models revealed gaps in knowledge of key processes, such as organic phosphorus mineralisation, the influence of phosphatases on phosphorus mineralisation and influence of the microbial biomass on the redistribution of inorganic and organic phosphorus in soils, thereby giving direction to future research. Use of simulation models is an important technique already being used to improve both the understanding of natural processes (Smith 1976, 1979) and the management of agro-ecosystems (Kafkafi *et al.* 1978) and will remain an important tool in ecological research.

14.8 Use of isotopes to measure mineralisation

Although isotopes of phosphorus have proved useful for tracing the fate of applied fertiliser phosphorus and measuring the size of the soil labile inorganic pool (Amer *et al.* 1955; Fried & Broeshart 1967; Larsen 1967), their use in evaluating organic phosphorus mineralisation is a comparatively recent development. Soil receives a large amount of phosphorus annually from the mineralisation of roots, litter, and animal wastes (Halm *et al.* 1972). Dalal (1979) and Blair and Boland (1978) added [^{32}P]phosphorus to soil and measured its utilisation by growing crops. Similarly, Bettany *et al.* (1974), Till and Blair (1978), Sweet *et al.* (1979) and Stewart and Hedley (1980) estimated the mineralisation of soil organic sulphur and phosphorus by labelling the labile inorganic sulphur and phosphorus pools with [^{35}S]sulphur and [^{32}P]phosphorus, respectively, and examining the rate of dilution of this material with [^{32}S]sulphur and [^{31}P]phosphorus with time. In general, the use of isotopes indicates that the actual rate of mineralisation of soil organic matter has been underestimated by more conventional means. The use of isotopes allows the actual flow of the element between various pools to be documented, whereas other methods measure net flow rates. By labelling the soil labile inorganic phosphorus with [^{33}P]phosphorus and the microbial pool with [^{33}P]labelled bacteria, Stewart and Hedley (1980) were able to measure the dynamics of phosphorus movement between labile and stable phosphorus sources under various soil perturbations, including such treatments as freezing and thawing of the soil.

14.9 Chemical characterisation of organic phosphorus

Anderson (1975), Halstead and McKercher (1975) and Dalal (1977) reviewed conventional methods of assessing organic phosphorus in soils, and noted the difficulties in separating inorganic and organic phosphorus compounds from soil and that values obtained from chemical separations may vary with the method employed. Some investigators have used the ease of extraction in various reagents to identify labile inorganic and organic phosphorus pools rather than attempting separate chemical identification of specific forms.

It is almost certain that virtually all organic phosphorus can be removed from soil. The major problem is that most methods of removal require solubilisation of the phosphorus from the original soil matrix and the changes in portions of the phosphorus caused by these processes are largely unknown. It seems that in soil there is a variable reserve of organic phosphorus which is associated with humic material. Thus, the solubilisation of organic phosphorus must be related to some extent to the constitution of the humic materials, whilst humic material itself is known to change during soil genesis (Anderson 1979). The problem presented by solubilisation procedures for organic phosphorus recovery, may be avoided by isolating some of the phosphorus by precipitation as suggested by Tinsley and Ozsavasci (1974) who used titanium for this purpose.

Steady progress has been made in applying modern methods of detection to the chemical identification of organic phosphorus compounds in soil. Gel filtration and high-pressure liquid chromatography have been used to improve the recovery of inositol hexaphosphate and ATP (Gerritse 1978). Mass spectroscopy has been used to identify separate soil inositol phosphorus compounds (L'Annunziata & Fuller 1976; L'Annunziata *et el.* 1977). Nuclear magnetic resonance techniques have been used to identify orthophosphate monoesters and phosphonates in grassland soils (Newman & Tate 1980). Methods of assessing phosphatase activity have been improved (Gerritse & Van Dijk 1978), as has our understanding of the factors affecting phosphatase inhibition and distribution in soils (Juma & Tabatabai 1977, 1978; Spier & Ross 1978; Makboul & Ottow 1979; Speirs & McGill 1979). Methods have been developed for isolating and determining bacterial DNA in soils (Torsvik & Goksøyr 1978; Torsvik 1980). Redistribution of phosphorus in soil infiltration of organic phosphorus compounds (Rolston *et al.* 1975) and via microbial activity (Stewart *et al.* 1980) has been studied.

Bowman and Cole (1978b) characterised organic phosphorus by fractionating soil organic matter according to the chemical extractant used and categorised it into four groups:

(1) a labile pool, extracted by 0.5M-$NaHCO_3$;

(2) a moderately labile pool consisting of acid soluble organic phosphorus and alkali insoluble inorganic phosphorus;

(3) a moderately resistant fraction soluble in alkali and acid (fulvic-acid phosphorus);

(4) a highly resistant fraction soluble in alkali, but insoluble in acid (humic-acid phosphorus).

Chauhan *et al* (1979) found that, although this method can be used to follow changes in the amount and composition of organic phosphorus following carbon additions to incubating soil, it did not clearly show the dynamics of soil organic phosphorus. Hedley and Stewart (1982) followed a similar procedure in developing a sequential extraction method (Fig. 4). The extractions were considered to remove, in sequence:

(1) the most biologically available inorganic phosphorus by resin extraction (Amer *et al.* 1955; Bowman *et al.* 1978);

(2) labile inorganic phosphorus and organic phosphorus adsorbed on the soil surface, plus a small amount derived from the biomass (0.5M–$NaHCO_3$) (Bowman & Cole 1978a);

(3) inorganic and organic phosphorus released from the biomass (0.5M-$NaHCO_3$/$CHCl_3$);

(4) inorganic phosphorus held more strongly by chemisorption to iron and aluminium components of soil surfaces, and organic phosphorus possibly held in a similar manner (0.1M-NaOH) (Ryden *et al.* 1977; McLaughlin *et al.* 1977);

(5) protected organic phosphorus and inorganic phosphorus held at the internal surfaces of soil aggregates (0.1M-NaOH/ sonication);

(6) more stable inorganic phosphorus removed by 1M-HCl which is thought to be mainly apatites (Williams *et al.* 1971) but could include occluded phosphate in more weathered soils (Williams *et al.* 1980);

(7) more chemically stable organic phosphorus (perchlorate oxidation).

The advantages of this method are that it distinguishes between the inorganic phosphorus into fractions (labile, secondary, occluded and primary minerals) that have been commonly described in the identification of phosphorus compounds in soil (Peterson & Corey 1966; Sadler & Stewart 1974, 1975), while simultaneously providing information on labile and stable organic phosphorus forms and indicating the amount of microbial phosphorus. The amount of microbial phosphorus recovered in the sodium bicarbonate

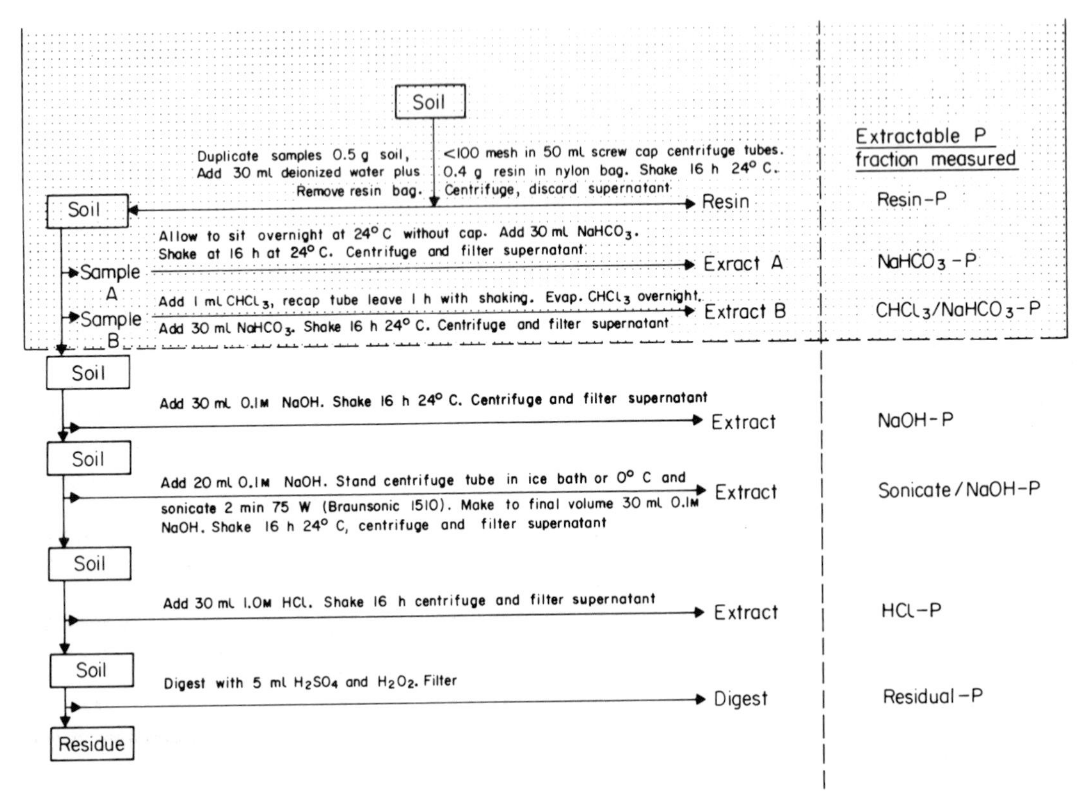

Figure 4. Flow chart of the fractionation of soil phosphorus into various labile and stable inorganic phosphorus and organic phosphorus forms, including an estimate of microbial phosphorus (after Hedley & Stewart 1982). To estimate microbial phosphorus only the shaded part of the fractionation scheme need be followed.

extract following chloroform treatment varies from soil to soil. Hedley and Stewart (1982) recovered approximately 37% of the microbial phosphorus in this extraction from a chernozemic soil of neutral pH. Because the percentage recovery of microbial phosphorus from more weathered soil is less, the method has to be calibrated for recovery rate of each soil type.

The transformations of the more labile organic compounds in soils can be evaluated by adapting the sodium bicarbonate method of Olsen *et al.* (1954). This method was primarily designed to evaluate the inorganic phosphorus available to plants by simulating root extraction in calcareous soil but did not measure the organic phosphorus extracted at the same time. Halm *et al.* (1972) showed that soluble sodium bicarbonate–organic phosphorus was positively correlated with soil phosphatase activity and microbial biomass during the growing season. Bowman and Cole (1978a) evaluated the sodium bicarbonate extraction method and found that labile compounds such as RNA, nucleotides and glycerophosphates were recoverable from the extract, whereas sodium phytate, a relatively resistant compound, was not.

Other approaches to the assessment of labile inorganic and organic phosphorus availability in soils have included a bioassay method based on the relationships between the internal phosphorus status of the plant and the uptake of [^{32}P]phosphorus from dilute phosphate solutions (Bowen 1971; Harrison & Helliwell 1979).

14.10 Recommended methods for the study of phosphorus cycling

Knowledge of the degradation of plant tissues, microbial bodies, soil organic matter, and the biogeochemical cycles of carbon, nitrogen, sulphur and phosphorus can be obtained only after one has an understanding of the rate of microbial decay

(Jenkinson & Rayner 1977; Paul & McGill 1977; Paul & Van Veen 1978). In many earlier studies of phosphorus cycling in soils, only parts of the cycle were studied. Although this can provide useful information, there is always the risk when dealing with a complex and dynamic system that important aspects are missed. Likewise, unless unlimited resources are available, decisions have to be made on which parts of the phosphorus cycle can be assumed to be stable under the specific environmental conditions being studied and therefore need not be closely monitored.

The phosphorus cycle, as shown in the expanded flow diagram (Fig. 5), can be studied by the methods listed in Table 1. Although further refinement of these techniques is necessary, application of these methods allows measurement of the dynamics of phosphorus movement with the same ease and accuracy as measuring nitrogen and carbon flow in their cycles. The methods shown here were developed in the study of phosphorus movement in calcareous neutral soils and in many cases need modification for more acidic environments. Such modifications for inorganic phosphorus fractionations have already been described (Williams & Walker 1969; Williams *et al.* 1980).

Some controversy exists about the measurement of biological activity within the soil system. For this reason the main method referenced is that of trapping carbon dioxide in NaOH solution in order to measure respiration. Partitioning of the carbon dioxide produced between the various soil microbial and faunal organisms remains a problem, as does the partitioning within any one soil population between maintenance energy requirements and growth. Controversy also exists about estimating biomass by the ATP method, since various microbial carbon/ATP ratios have been proposed. However, as it has been shown (C. V. Cole, personal communication) that the carbon/ATP ratio will change with the metabolic state of the organisms, the reliance on carbon/ATP ratios as a basis for predicting microbial biomass is questionable. The limitations of carbon/ATP ratios as a biomass indicator have been thoroughly discussed in a recent review (Karl 1980).

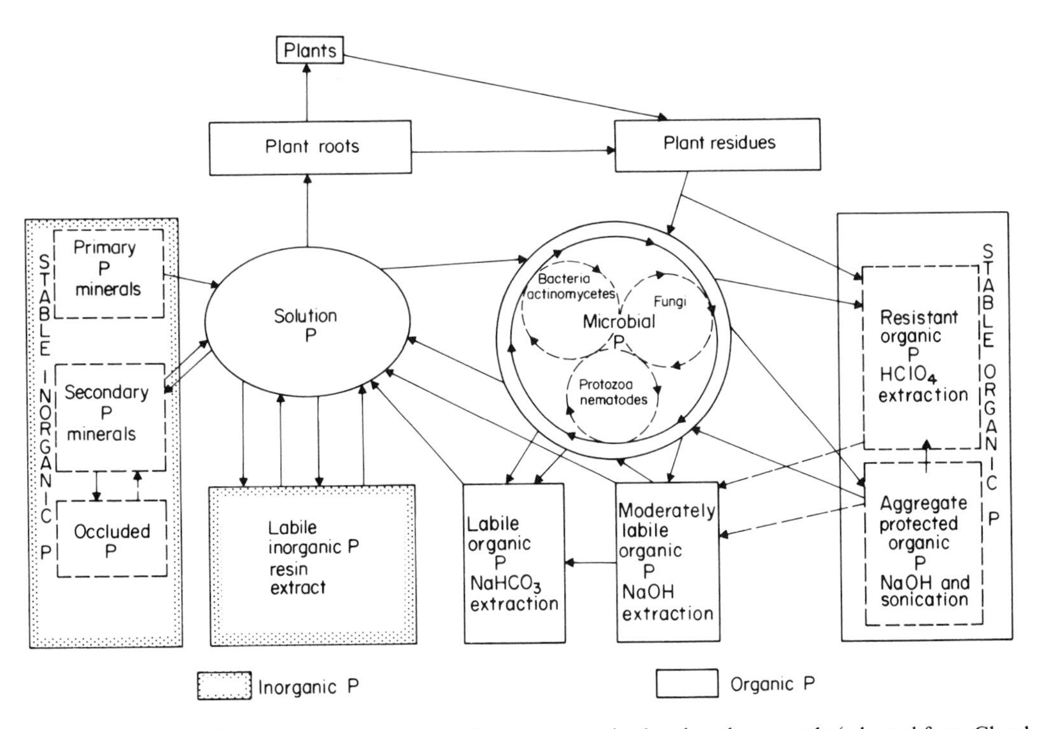

Figure 5. Schematic illustration of the measurable components in the phosphorus cycle (adapted from Chauhan *et al.* 1981 and Hedley & Stewart 1982). [^{32}P]phosphorus isotopes were used to label the solution phosphorus and labile inorganic phosphorus, and [^{33}P]phosphorus was used to trace microbial phosphorus, thus allowing measurement of phosphorus dynamics under various soil perturbations.

Table 1. Recommended analytical methods.

Method	Reference
Total mineral and organic phosphorus	Saunders & Williams (1955) method modified to use 1M-H_2SO_4; Anderson (1960); Halstead & McKercher (1975)
Total phosphorus plus division into stable and labile inorganic phosphorus and organic phosphorus fractions	Hedley & Stewart (1982)
Inorganic phosphorus fractions	Peterson & Corey (1966) modified for calcareous soils by Sadler & Stewart (1975); Hedley & Stewart (1981)
Resin extractable phosphorus	Amer *et al.* (1955); Sibbesen (1977)
Organic phosphorus fractions	Hedley & Stewart (1981)
Inositol phosphorus, lipid phosphorus and nucleo-phosphates	Halstead & McKercher (1975)
$NaHCO_3$-extractable phosphorus	
(i) inorganic phosphorus	Olsen *et al.* (1954)
(ii) organic phosphorus	Bowman & Cole (1978a)
Microbial phosphorus ($CHCl_3$ treatment + $NaHCO_3$ extraction following resin distribution)	Hedley & Stewart (1982)
Microbial activity—CO_2 production trapped in NaOH in closed containers	—
Microbial biomass—bacteria	Babiuk & Paul (1970) modified by Van Veen & Paul (1979)
—fungi	Paul & Johnson (1977) modified by Van Veen & Paul (1979)
Microbial biomass—total	Anderson & Domsch (1978b)
Protozoa and nematodes	Anderson *et al.* (1979b); Oostenbrink (1970); Darbyshire *et al.* (1974)
Phosphatases	Gerritse & van Dijk (1978)
ATP	Paul & Johnson (1977); Jenkinson & Oades (1979)

14.11 Discussion

The state of our knowledge of various stages of the phosphorus cycle is still uneven and far from complete. Some progress has been made in the past decade both in chemically identifying and quantifying phosphorus compounds and their flow, but a large percentage of the organic phosphorus in soils still cannot be chemically identified. Many workers have taken a different approach by attempting to identify flows of labile organic phosphorus to stable organic phosphorus, or to inorganic phosphorus forms, rather than identifying specific compounds. If this approach is successful, it will be necessary to identify more correctly the compounds that constitute labile and stable organic phosphorus.

Some progress has been made in quantifying microbial phosphorus in calcareous neutral soils and this technique will have to be applied to more weathered soils with higher phosphorus adsorption capacities. Differences have been noted in the ability of bacteria and fungi both to accumulate phosphorus and to redistribute phosphorus to inorganic and organic forms on death of the organisms. The action of bacterial grazers and nematodes in the cycling of phosphorus has been studied in microcosms and now needs to be studied in complete soil systems. In this review no consideration has been given to the redistribution of nutrients that occurs through grazing of plant materials by large herbivores and subsequent distribution in soil, as this has been the subject of other reviews (Batzli 1978; Floate 1981).

Although some progress has been made in the understanding of the chemical nature and modes of dephosphorylation of soil organic phosphorus forms (Cosgrove 1977), much remains to be discovered. Lack of solution phosphorus has been shown to increase the rate of labile organic phosphorus mineralisation, but the rate of this reaction in different soil systems needs to be clarified. In this regard, the increased use of isotopes to label soil phosphorus pools has been important, as has been the use of microcosms to provide details of possible pathways involved.

Computer simulation models have provided a means of integrating rate constants with environmental and abiotic effects. As more accurate rate constants become available from experimental studies, they can be used in models to give direction to further research. An overall objective of such models would be to gain the ability not only to explain observed immobilisation or mineralisation of phosphorus in soils under different ecological and management systems but also to predict future changes in present and proposed systems. The ability to predict change in soil organic matter quality will become an important tool in the management practices of agricultural land.

Much more accurate information is needed on root behaviour in soils and on the flows of phosphorus in the root rhizosphere. Significant progress has been made in understanding the action of VA mycorrhizal fungi in phosphorus uptake, but the interaction of bacteria, fungi, actinomycetes, microbial grazers and nematodes near to the root and their combined effects on phosphorus uptake need clarification.

Finally, perhaps the most important development in research into phosphorus cycling has been the integration of methods developed to study microbial, inorganic and organic phosphorus separately. Although these methods have been partly successful when applied separately to fertilised crop soils, they often applied to only part of the picture and at times gave erroneous results.

Acknowledgements

This manuscript was completed at Colorado State University while John Stewart was on sabbatical leave. He wishes to acknowledge with thanks the assistance provided by the editorial, drafting and typing staff of the Natural Resource Ecology Laboratory at CSU.

Recommended reading

Anderson G. (1975) Organic phosphorus compounds. In: *Soil Components* (Ed. J. E. Gieseking), vol. 1, pp. 305–32. Springer-Verlag, New York.

Anderson J. M. & MacFayden A. (1976) *The Role of Terrestrial and Aquatic Organisms in Decomposition Processes.* Blackwell Scientific Publications, Oxford.

Cole C. V., Innis G. I. & Stewart J. W. B. (1977) Simulation of phosphorus cycling in semi-arid grasslands. *Ecology* **58**, 1–15.

Cosgrove D. J. (1977) Microbial transformation in the phosphorus cycle. In: *Advances in Microbial Ecology* (Ed. M. Alexander), vol. 1, pp. 95–134. Plenum Press, New York.

Dalal R. C. (1977) Soil organic phosphorus. *Advances in Agronomy* **29**, 83–113.

Dawes I. W. & Sutherkland I. W. (1976) *Microbial Physiology*. Blackwell Scientific Publications, London.

Fenchel T. &. Blackburn T. H. (1979) *Bacteria and Mineral Cycling*. Academic Press, London.

Halstead R. L. & McKercher R. B. (1975) Biochemistry and cycling of phosphorus. In: *Soil Biochemistry* (Eds E. A. Paul & A. D. McLaren), vol. 4, pp. 31–63. Marcel Dekker, New York.

Harley J. L. & Russell R. S. (1979) *The Soil-Root Interface*. Academic Press, London.

Hayman D. S. (1975) Phosphorus cycling in soil microorganisms and plant roots. In: *Soil Microbiology, A Critical Review* (Ed. N. Walker), pp. 67–91. Butterworths, London.

Roswall T. (1973) *Modern Methods in the Study of Microbial Ecology*. Bulletin **17**, Ecological Research Committee, Swedish Natural Science Research Council, Stockholm.

References

Amer F., Bouldin D. R., Black C. A. & Duke F. R. (1955) Characterization of soil phosphorus by an ion exchange resin adsorption and P^{32} equilibration. *Plant and Soil* **6**, 391–408.

Anderson D. W. (1979) Processes of human formation and transformation in soils of the Canadian Great Plains. *Journal of Soil Science* **30**, 77–89.

Anderson G. (1960) Factors affecting the estimation of phosphate-esters in soils. *Journal of the Science of Food and Agriculture* **11**, 497–503.

Anderson G. (1967) Nucleic acid derivatives and organic phosphates. In: *Soil Biochemistry* (Eds A. D. McLaren & G. H. Peterson), pp. 67–91. Marcel Dekker, New York.

Anderson G. (1975) Organic phosphorus compounds. In: *Soil Components* (Ed. J. E. Gieseking), vol. 1, pp. 305–32. Springer-Verlag, New York.

ANDERSON G. & RUSSELL J. D. (1969) Identification of inorganic polyphosphates in alkaline extracts of soil. *Journal of the Science of Food and Agriculture* **20**, 78–81.

ANDERSON J. M. & MACFAYDEN A. (1976) *The Role of Terrestrial and Aquatic Organisms in Decomposition Processes*. Blackwell Scientific Publications, Oxford.

ANDERSON J. P. E. & DOMSCH K. H. (1978a) Mineralization of bacteria and fungi in chloroform-fumigated soils. *Soil Biology and Biochemistry* **10**, 207–13.

ANDERSON J. P. E. & DOMSCH K. H. (1978b) A physiological method for the quantitative measurement of microbial biomass in soils. *Soil Biology and Biochemistry* **10**, 215–21.

ANDERSON R. V., COLEMAN D. C., COLE C. V., ELLIOTT E. T. & MCCLELLAN J. F. (1979a) The use of soil microcosms in evaluating bacteriophagic nematode responses to other organisms and effects. *Journal of Environmental Studies* **13**, 175–82.

ANDERSON R. V., GOULD W. D., INGHAM R. E. & COLEMAN D. C. (1979b) A staining method for nematodes: Determination of nematode resistant stages and direct counts from soil. *Transactions of the American Microscopical Society* **98**, 213–18.

BABIUK L. A. & PAUL E. A. (1970) The use of fluroescein isothiocyanate in the determination of the bacterial biomass of grassland soils. *Canadian Journal of Microbiology* **16**, 57–62.

BARBER S. A. (1962) A diffusion and mass-flow concept of soil nutrient availability. *Soil Science* **93**, 39–49.

BARLEY K. P. (1961) Abundance of earthworms in agricultural land and their possible significance in agriculture. *Advances in Agronomy* **13**, 249–68.

BARSDATE R. J., PRENTKI R. T. & FENCHEL T. (1974) Phosphorus cycles of model ecosystems: Significance for decomposer food chains and the effect of bacterial grazers. *Oikos* **25**, 239–51.

BATZLI G. O. (1978) The role of herbivores in mineral cycling. In: *Environmental Chemistry and Cycling Processes* (Eds D. C. Adriano & I. L. Brisbin), pp. 95–112. Department of Energy Series 45, United States Department of Energy, Washington, D.C.

BEEVER R. E. & BURNS D. J. W. (1976) Microorganisms and the phosphorus cycle: Some physiological considerations. In: *Prospects for Improving Efficiency of Phosphorus Utilizations* (Ed. G. J. Blair), pp. 113–18. University of New England, Armidale, N.S.W.

BETTANY J. R., STEWART J. W. B. & HALSTEAD E. H. (1974) Assessment of available soil sulphur in a ^{35}S growth chamber experiment. *Canadian Journal of Soil Science* **54**, 309–15.

BLAIR G. J. & BOLAND O. W. (1978) The release of phosphorus from plant material added to soil. *Australian Journal of Soil Research* **16**, 101–11.

BOWEN G. D. (1971) Early detection of phosphate deficiency in plants. *Communications in Soil Science and Plant Analysis* **1**, 293–8.

BOWMAN R. A. & COLE C. V. (1978a) Transformations of organic phosphorus substrates in soils evaluated by $NaHCO_3$ extraction. *Soil Science* **125**, 49–54.

BOWMAN R. A. & COLE C. V. (1978b) An exploratory method for fractionation of organic phosphorus from grassland soils. *Soil Science* **125**, 95–101.

BOWMAN R. A., OLSEN S. R. & WATANABE F. S. (1978) Greenhouse evaluation of residual phosphate by four phosphorus methods in neutral and calcareous soils. *Soil Science Society of America Journal* **42**, 451–4.

BURNS D. J. W. & BEEVER R. E. (1977) Kinetic characterization of the two phosphate uptake systems in the fungus *Neurospora crassa*. *Journal of Bacteriology* **132**, 511–19.

CHAPIN F. S., BARSDATE R. J. & BAREL D. (1978) Phosphorus cycling in Alaska coastal tundra: A hypothesis for the regulation of nutrient cycling. *Oikos* **31**, 188–99.

CHAUHAN B. S., STEWART J. W. B. & PAUL E. A. (1979) Effect of carbon additions on soil labile inorganic, organic and microbially held phosphate. *Canadian Journal of Soil Science* **59**, 387–96.

CHAUHAN B. S., STEWART J. W. B. & PAUL E. A. (1981) Effect of labile inorganic phosphate status and organic carbon additions on the microbial uptake of phosphorus in soils. *Canadian Journal of Soil Science* **61**, 373–85.

CHEN MIN (1979) Kinetics of phosphorus absorption by *Corynebacterium bovis*. *Microbial Ecology* **1**, 164–75.

CHENG C. N., DUNN S. J. & FOCHT D. D. (1979) Volatilization of phosphate P in soils. 71st Annual Meeting American Society of Agronomy, Fort Collins, Colorado. *Agronomy Abstracts* p. 155.

CHRISTIE P., NEWMAN E. I. & CAMPBELL R. (1978) The influence of neighboring plants on each others endomycorrhizas on root surface microorganisms. *Soil Biology and Biochemistry* **10**, 521–7.

CLARK F. E. (1949) Soil microorganisms and plant roots. *Advances in Agronomy* **1**, 241–88.

CLARK F. E. & PAUL E. A. (1970) The microflora of grassland. *Advances in Agronomy* **22**, 375–435.

COLE C. V., ELLIOTT E. T., HUNT H. W. & COLEMAN D. C. (1978) Trophic interactions in soils as they affect energy and nutrient dynamics. V. Phosphorus transformations. *Microbial Ecology* **4**, 381–7.

COLE C. V. & HEIL R. D. (1981) Phosphorus effects on terrestrial nitrogen cycling. In: *Terrestrial Nitrogen Cycles, Processes, Ecosystem Strategies and Management Impact* (Eds F. E. Clark & T. Roswall), *Ecological Bulletin (Stockholm)* **33**, 363–74.

COLE C. V., INNIS G. I. & STEWART J. W. B. (1977) Simulation of phosphorus cycling in semi-arid grasslands. *Ecology* **58**, 1–15.

COLEMAN D. C. (1976) A review of root production systems and their influence in terrestrial ecosystems. In: *The Role of Terrestrial and Aquatic Organisms in*

Decomposition Processes (Eds J. M. Anderson & A. MacFayden), pp. 417–34, Blackwell Scientific Publications, Oxford.

COLEMAN D. C., ANDERSON R. V., COLE C. V., ELLIOTT E. T., WOODS L. & CAMPION M. K. (1978) Trophic interactions in soils as they affect energy and nutrient dynamics. IV. Flows of metabolic and biomass carbon. *Microbial Ecology* **4**, 373–80.

COSGROVE D. J. (1977) Microbial transformation in the phosphorus cycle. In: *Advances in Microbial Ecology* (Ed. M. Alexander), vol. 1, pp. 95–134. Plenum Press, New York.

DALAL R. C. (1977) Soil organic phosphorus. *Advances in Agronomy* **29**, 83–113.

DALAL R. C. (1979) Mineralization of carbon and phosphorus from carbon-14 and phosphorus-32 labelled plant material added to soil. *Soil Science Society of America Journal* **43**, 913–17.

DARBYSHIRE J. (1975) Soil protozoa—animalcules in the subterranean micro-environment. In: *Soil Microbiology A Critical Review* (Ed. N. Walker), pp. 147–63. Butterworths, London.

DARBYSHIRE J. F. WHEATLEY R. E., GREAVES M. P. & INKSON R. H. E. (1974) A rapid method for estimating bacterial and protozoan population in soil. *Revue d'Écologie et de Biologie du Sol* **11**, 465–75.

DAWES I. W. & SUTHERLAND I. W. (1976) *Microbial Physiology*. Blackwell Scientific Publications, Oxford.

EIVAZI F. & TABATABAI M. A. (1977) Phosphatases in soils. *Soil Biology and Biochemistry* **8**, 167–72.

ELLIOTT E. T., COLE C. V., COLEMAN D. C., HUNT H. W. & MCCLELLAND J. F. (1979) Amoebal growth in soil microcosms: A model system of C, N and P trophic dynamics. *International Journal of Environmental Studies* **13**, 169–74.

FENCHEL T. & BLACKBURN T. H. (1979) *Bacteria and Mineral Cycling*. Academic Press, London.

FENCHEL T. & HARRISON P. (1976) The significance of bacterial grazing and mineral cycling for the decomposition of particular detritus. In: *The Role of Terrestrial and Aquatic Organisms in Decomposition Processes* (Eds J. M. Anderson & A. MacFayden), pp. 285–9. Blackwell Scientific Publications, Oxford.

FLOATE M. G. S. (1971) Plant nutrient at hill land. *Hill Farm Research Organisation Annual Report* 1968–71, pp. 15–35.

FLOATE M. G. S. (1981) Effects of grazing by large herbivores on nitrogen cycling in agricultural ecosystems. In: *Terrestrial Nitrogen Cycles. Processes Ecosystem Strategies and Management Impact* (Eds F. E. Clark & T. Rosswall) *Ecological Bulletin (Stockholm)* **33**, 585–601.

FRIED M. & BROESHART H. (1967) *The Soil-Plant System in Relation to Inorganic Nutrition*. Academic Press, New York.

GERDEMANN J. W. (1975) Vesicular-arbuscular mycorrhizae. In: *The Development and Function of Roots* (Eds J. G. Torrey & D. T. Clarkson), pp. 575–91. Academic Press, New York.

GERRITSE R. G. (1978) Assessment of a procedure for fractionating organic phosphates in soil and organic materials using gel filtration and H.P.L.C. *Journal of the Science of Food and Agriculture* **29**, 577–86.

GERRITSE R. G. & VAN DIJK H. (1978) Determination of phosphatase activities of soils and animal wastes. *Soil Biology and Biochemistry* **10**, 545–51.

GOULD W. D., COLEMAN D. C. & RUBINK A. J. (1979) Effect of bacteria and amoeba on rhizosphere phosphatase activity. *Applied and Environmental Microbiology* **37**, 943–6.

GRAY T. R. G. (1976) Survival of vegetative microbes in soil. In: *The Survival of Vegetative Microbes* (Eds T. R. G. Gray & J. R. Postgate), pp. 327–64. Cambridge University Press, London.

GYLLENBERG H. G. & EKLUND E. (1974) Bacteria. In: *Biology of Plant Litter Decomposition* (Eds C. H. Dickinson & G. J. F. Pugh), pp. 245–68. Academic Press, New York.

HALM B. J., STEWART J. W. B. & HALSTEAD R. L. (1972) The phosphorus cycle in a native grassland ecosystem. In: *Isotopes and Radiation in Soil Plant Relationships Including Forestry*, pp. 571–86. International Atomic Energy Agency, Vienna.

HALSTEAD R. L. & MCKERCHER R. B. (1975) Biochemistry and cycling of phosphorus. In: *Soil Biochemistry* (Eds E. A. Paul & A. D. McLaren), vol. 4, pp. 31–63. Marcel Dekker, New York.

HARLEY J. L. & RUSSELL R. S. (1979) *The Soil-Root Interface*. Academic Press, London.

HARRISON A. F. (1978) Phosphorus cycles of forest and upland grassland systems and some effects of land management practices. In: *Phosphorus in the Environment: Its Chemistry and Biochemistry*, CIBA Foundation Symposium 57, pp. 175–95. Elsevier/North-Holland, Amsterdam.

HARRISON A. F. & HELLIWELL D. R. (1979) A bioassay for comparing phosphorus availability in soils. *Journal of Applied Ecology* **16**, 497–505.

HAYMAN D. S. (1975) Phosphorus cycling in soil microorganisms and plant roots. In: *Soil Microbiology, A Critical Review* (Ed. N. Walker), pp. 67–91. Butterworths, London.

HEDLEY M. J. & STEWART J. W. B. (1982) Method to measure microbial phosphorus in soils. *Soil Biology and Biochemistry*. (In press.)

HERRERA R. A. (1979) Nutrient distribution and cycling in an Amazon Caatinga forest on podsols, in southern Venezuela. Ph.D. thesis, University of Reading, England.

HUNT H. W. (1977) A simulation model for decomposition in grasslands. *Ecology* **58**, 469–84.

HUNT H. W., COLE C. V., KLEIN D. A. & COLEMAN D. C. (1977) Simulating the effect of predators on the growth and composition of bacteria. *Ecological Bulletin (Stockholm)* **25**, 409–19.

HUTCHINSON G. E. (1970) The biosphere. *Scientific American* **223**, 45–53.

JENKINSON D. S., DAVIDSON S. A. & POWLSON D. S. (1979) Adenosine triphosphates and microbial biomass in soil. *Soil Biology and Biochemistry* **11**, 521–8.

JENKINSON D. S. & LADD J. N. (1981) Microbial biomass in soil—measurement and turnover. In: *Soil Biochemistry* (Eds E. A. Paul & J. N. Ladd), vol. 5, pp. 415–71. Marcel Dekker, New York.

JENKINSON D. S. & OADES J. M. (1979) A method for measuring adenosine triphosphate in soil. *Soil Biology and Biochemistry* **11**, 193–9.

JENKINSON D. S. & POWLSON D. S. (1976) The effects of biocidal treatments on metabolism in soil. V. A method for measuring soil biomass. *Soil Biology and Biochemistry* **8**, 209–13.

JENKINSON D. S. & RAYNER J. H. (1977) The turnover of soil organic matter in some of the Rothamsted classical experiments. *Soil Science* **123**, 298–305.

JORDAN C. F., GOLLEY F., HALL J. & HALL J. (1980) Nutrient scavengings of rainfall by the canopy of an Amazonian rain forest. *Biotropica* **12**, 61–6.

JUMA N. G. & TABATABAI M. A. (1977) Effects of trace elements on phosphatase activity in soils. *Soil Science Society of America Journal* **41**, 343–6.

JUMA N. G. & TABATABAI M. A. (1978) Distribution of phosphomonoesterases in soils. *Soil Science* **126**, 101–8.

KAFKAFI V., BAR-YOSEF B. & HADAS A. (1978) Fertilization decision model—a synthesis of soil and plant parameters in a computerized program. *Soil Science* **125**, 261–8.

KARL D. M. (1980) Cellular nucleotide measurements and applications in microbial ecology. *Microbiological Reviews* **44**, 739–97.

KOWAL N. E. (1971) A rationale for modeling dynamic ecological systems. In: *Systems Analysis and Simulation in Ecology* (Ed. B. C. Patten), vol. 1, pp. 123–94. Academic Press, New York.

L'ANNUNZIATA M. F. & FULLER W. H. (1976) Evaluation of mass spectral analysis of soil inositol, inositol phosphates and related compounds. *Soil Science Society of America Journal* **40**, 672–8.

L'ANNUNZIATA M. F., GONZALEZ I. J. & OLIVARES O. L. A. (1977) Microbial epimerization of myo-inositol to chiro-inositol in soil. *Soil Science Society of America Journal* **41**, 733–6.

LARSEN S. (1967) Soil phosphorus. *Advances in Agronomy* **19**, 151–210.

MCGILL W. B. (1979) A concept regarding comparative C, N, S and P cycling through soil organic matter. 71st Annual Meeting American Society of Agronomy, Fort Collins, Colorado. *Agronomy Abstracts*, p. 165.

MCGILL W. B., HUNT H. W., WOODMANSEE R. G. & REUSS J. O. (1981) Dynamics of C and N in grassland soils. In: *Terrestrial Nitrogen Cycles. Processes, Ecosystem Strategies and Management Impact* (Eds F. E. Clark & T. Roswall). *Ecological Bulletin (Stockholm)*, **33**, 49–115.

MCKERCHER R. B. & TOLLEFSON T. S. (1978) Barley response to phosphorus from phospholipids and nucleic acids. *Canadian Journal of Soil Science* **58**, 103-5.

MCKERCHER R. B., TOLLEFSON T. S. & WILLARD J. R. (1979) Biomass and phosphorus contents of some soil invertebrates. *Soil Biology and Biochemistry* **11**, 387–91.

MCLAUGHLIN J. R., REYDEN J. C. & SYERS G. K. (1977) Development and evaluation of a kinetic model to describe phosphate sorption by hydrous ferric oxide gels. *Geoderma* **18**, 295–307.

MAKBOUL H. E. & OTTOW J. C. G. (1979) Alkaline phosphatase activity and Michaelis constant in the pressure of different clay minerals. *Soil Science* **128**, 129–35.

MISHRA B., KHANNA P. K. & ULRICH B. (1979) A simulation model for organic phosphorus transformation in a forest soil ecosystem. *Ecological Modelling* **6**, 31–46.

NANNIPIERI P., JOHNSON R. L. & PAUL E. A. (1978) Criteria for measurement of microbial growth and activity in soil. *Soil Biology and Biochemistry* **10**, 223–9.

NEWMAN E. J. (1978) Root microorganisms: Their significance in the ecosystem. *Biological Reviews* **53**, 511–54.

NEWMAN R. H. & TATE K. R. (1980) Soil phosphorus characterisation by ^{31}P nuclear magnetic resonance. *Communications in Soil Science and Plant Analysis* **11**, 835–42.

NYE P. H. (1977) The rate-limiting step in plant nutrient absorption from soil. *Soil Science* **123**, 292–7.

OADES J. M. & JENKINSON D. S. (1979) Adenosine triphosphate content of the soil microbial biomass. *Soil Biology and Biochemistry* **11**, 201–4.

OLSEN S. R., COLE C. V., WATANABE F. S. & DEAN L. A. (1954) *Estimation of Available Phosphorus in Soils by Extraction with Sodium Bicarbonate*. United States Department of Agriculture Circular No. 939. United States Government Printing Office, Washington, D.C.

OLSEN S. R. & FLOWERDAY A. D. (1971) Fertilizer phosphorus interactions in alkaline soils. In: *Fertilizer Technology and Use* (Eds R. A. Olsen, T. J. Army, J. J. Hanway & V. J. Kilmer), pp. 153–85. Soil Science Society of America Incorporated, Madison, Wisconsin.

OLSEN S. R. & WATANABE F. S. (1963) Diffusion of phosphorus as related to soil texture and plant uptake. *Soil Science Society of America Proceedings* **27**, 648–53.

OLSEN S. R. & WATANABE F. S. (1970) Diffusive supply of phosphorus in relation to soil textural variations. *Soil Science* **110**, 318–27.

OOSTENBRINK M. (1970) Comparison of techniques for population estimation of soil and plant nematodes. In: *Methods of Study in Soil Ecology* (Ed. J. Phillipson), pp. 249–55. UNESCO, Paris.

PAUL E. A. & JOHNSON R. L. (1977) Microscopic counting and adenosine 5′-triphosphate measurement in determining microbial growth in soils. *Applied and Environmental Microbiology* **34**, 263–9.

PAUL E. A. & MCGILL W. B. (1977) Turnover of microbial biomass, plant residues and soil humic constituents under field conditions. In: *Soil Organic Matter Studies*, vol. 1, pp. 149–57. International Atomic Energy Agency, Vienna.

PAUL E. A. & VAN VEEN J. (1978) The use of tracers to determine the dynamic nature of organic matter. *Transactions of the 11th International Congress of Soil Science* **3**, 61–102.

PEPPER I. L., MILLER R. H. & GHONISKAR C. P. (1976) Microbial inorganic polyphosphates: Factors influencing their accumulation in soil. *Soil Science Society of America Journal* **40**, 872–5.

PETERSON G. W. & GOREY R. B. (1966) A modified Chang and Jackson procedure for routine fractionation of inorganic soil phosphate. *Soil Science Society of America Proceedings* **30**, 563–5.

ROLSTON D. E., RAUSCHKOLB R. S. & HOFFMAN D. L. (1975) Infiltration of organic phosphate compounds in soil. *Soil Science Society of America Proceedings* **39**, 1089–94.

ROSWALL T. (1973) *Modern Methods in the Study of Microbial Ecology*. Bulletin **17**, Ecological Research Committee, Swedish Natural Science Research Council, Stockholm.

ROVIRA A. D. (1979) Biology of the soil-root interface. In: *The Soil–Root Interface* (Eds J. L. Harley & R. S. Russell), pp. 145–60. Academic Press, New York.

RYDEN J. C., MCLAUGHLIN J. R. & SYERS J. K. (1977) Mechanisms of phosphate sorption by soils anhydrous ferric oxide gel. *Journal of Soil Science* **28**, 72–92.

RYDEN J. C. & SYERS J. K. (1977) Origin of the labile phosphate pool in soils. *Soils Science* **123** 353–61.

SADLER J. M. & STEWART J. W. B. (1974) *Residual Fertilizer Phosphorus in Western Canadian Soils; A review*. Saskatchewan Institute of Pedology Publication No. R136, Saskatoon, Saskatchewan, Canada.

SADLER J. M. & STEWART J. W. B. (1975) Changes with time in form and availability of residual fertilizer phosphorus in a catenary sequence of chernozemic soils. *Canadian Journal of Soil Science* **55**, 149–59.

SADLER J. M. & STEWART J. W. B. (1977) Labile residual fertilizer phosphorus in chernozemic soils. I. Solubility and quantity/intensity studies. *Canadian Journal of Soil Science* **57**, 65–73.

SAUNDERS W. M. H. & WILLIAMS E. G. (1955) Observations on the determination of organic phosphorus in soils. *Journal of Soil Science* **6**, 254–67.

SCOTT N. M. & ANDERSON G. (1976) Organic sulphur fractions in Scottish soils. *Journal of Science and Food Agriculture* **27**, 358–66.

SHARPLEY A. N. & SYERS J. K. (1976) Potential role of earthworm casts for the phosphorus enrichment of run-off waters. *Soil Biology and Biochemistry* **8**, 341–6.

SHARPLEY A. N. & SYERS J. K. (1977) Seasonal variation in casting activity and in the amounts and release to solution of phosphorus forms in earthworm casts. *Soil Biology and Biochemistry* **9**, 227–31.

SIBBESEN E. (1977) A simple ion-exchange procedure for extracting plant-available elements from soil. *Plant and Soil* **46**, 665–9.

SMECK N. E. (1973) Phosphorus: an indicator of pedogenic weathering processes. *Soil Science* **115**, 199–206.

SMITH O. L. (1976) Nitrogen, phosphorus and potassium utilization in the plant-soil system: an analytical model. *Soil Science Society of America Journal* **40**, 704–15.

SMITH O. L. (1979) Application of a model of the decomposition of soil organic matter. *Soil Biology and Biochemistry* **11**, 607–18.

SPEIR T. W. & ROSS D. J. (1978) Soil phosphatase and sulphatase. In: *Soil Enzymes* (Ed. R. G. Burns), pp. 197–250. Academic Press, London.

SPEIRS G. A. & MCGILL W. B. (1979) Effects of phosphorus additions and energy supply on acid phosphatase production and activity in soils. *Soil Biology and Biochemistry* **11**, 3–8.

STEWART J. W. B. & HEDLEY M. J. (1980) Phosphorus immobilization, mineralization and redistribution in soils. 72nd Annual Meeting American Society of Agronomy, Detroit. *Agronomy Abstracts*, p. 176.

STEWART J. W. B., HEDLEY M. J. & CHAUHAN B. S. (1980) The immobilization, mineralization and redistribution of phosphorus in soils. In: *Proceedings of the Western Canada Phosphate Symposium* (Ed. J. T. Harapiak), pp. 276–306. Alberta Soil Science Society, Edmonton, Canada.

SWEET J. J., COLE C. V. & COLEMAN D. C. (1979) Bacterial effects on phosphorus mineralization as measured by isotopic dilution. 71st Annual Meeting American Society of Agronomy, *Agronomy Abstracts*, p. 165.

SZEMBER A. (1960) Influence on plant growth of the breakdown of organic phosphate compounds by micro-organisms. *Plant and Soil* **13**, 147–58.

TANSEY M. R. (1977) Microbial facilitation of plant mineral nutrition. In: *Microorganisms and Minerals* (Ed. E. D. Weinberg), pp. 343–85. Marcel Dekker, New York.

TILL A. R. & BLAIR G. J. (1978) The utilization by grass of sulphur and phosphorus by clover litter. *Australian Journal of Agriculture Research* **29**, 235–42.

TINKER P. B. (1975a) Effects of vesicular-arbuscular mychorrizas on higher plants. In: *Symbiosis* (Eds D. H. Jennings & D. L. Lee), pp. 325–49. Cambridge University Press, London.

TINKER P. B. (1975b) Soil chemistry of phosphorus and mycorrhizal effects on plant growth. In: *Endomychorrhizas* (Eds F. E. Sanders, B. Mosse & P. B. Tinker), pp. 353–71. Academic Press, New York.

TINSLEY J. & OZSAVASCI C. (1974) Studies of soil organic phosphorus using titanic chloride. *Transactions of the 10th International Congress of Soil Science* **2**, 332–40.

TORSVIK V. L. (1980) Isolation of bacterial DNA from soil. *Soil Biology and Biochemistry* **12**, 15–21.

TORSVIK V. L. & GOKSØYR J. (1978) Determination of bacterial DNA in soil. *Soil Biology and Biochemistry* **10**, 7–12.

VAN VEEN J. A. & PAUL E. A. (1979) Conversion of bivolume measurement of soil organisms grown under various moisture tensions to biomass and their nutrient content. *Applied and Environmental Microbiology* **37**, 686–92.

WALD G. (1962) Life in the second and third periods; or why phosphorus and sulfur form high-energy bonds? In: *Horizons in Biochemistry* (Eds M. Kasha & B. Pullman), pp. 156–67. Academic Press, New York.

WALKER T. W. (1965) The significance of phosphorus in pedogenesis. In: *Experimental Pedology* (Eds S. G. Hallsworth & D. V. Crawford), pp. 295–315. Butterworths, London.

WALKER T. W. & ADAMS A. F. R. (1958) Studies on soil organic matter. 1. Influence of phosphorus content of parent materials on accumulation of carbon, nitrogen, sulfur and organic phosphorus in grassland soils. *Soil Science* **85**, 307–18.

WALKER T. W. & ADAMS A. F. R. (1959) Studies on soil organic matter. 2. Influence of increased leaching on various stages of weathering on levels of carbon, nitrogen, sulfur and organic and total phosphorus. *Soil Science* **87**, 135–40.

WALKER T. W. & SYERS J. K. (1976) The fate of phosphorus during pedogenesis. *Geoderma* **15**, 1–19.

WALKER T. W., THAPA B. K. & ADAMS A. F. R. (1959) Studies on soil organic matter. 3. Accumulation of carbon, nitrogen, sulfur, organic and total P in improved grassland soils. *Soil Science* **87**, 135–40.

WILLIAMS J. D. H., MAYERS T. & NRIAGU J. O. (1980) Extractability of phosphorus from phosphate minerals common in soils and sediment. *Soil Science Society of America Journal* **44**, 462–5.

WITKAMP M. (1976) Microcosm experiments on elemental transfer. *International Journal of Environmental Studies* **10**, 59–63.

WOODMANSEE R. G. (1978) Critique and analysis of the grassland ecosystem model ELM. In: *Grassland Simulation Model* (Ed. G. S. Innis), pp. 257–81. Springer-Verlag, New York.

Chapter 15 • Primary Production of the Oceans

Ian Morris

15.1 Introduction

In 1961 Steele wrote that 'there is no subject that can properly be called primary production'. In essence this view reflects an awareness of the vast complex of physical, chemical and biological processes which affect the relationship between primary producers and their environment. A proper understanding of primary production depends on such an awareness of the peripheral processes, including grazing, sinking, nutrient cycling and movements of water masses. However, it is possible to take a more restricted and simplistic view of primary production. It is this approach which is emphasised in this chapter.

In a restricted sense, primary production can be defined as the process by which the inorganic materials of the environment are converted into the organic matter of cell material. However, primary production is generally considered in an even more restricted sense, since it is the conversion of inorganic carbon to organic carbon which is usually emphasised. Thus, at its simplest, a definition of primary production is the process by which organisms convert inorganic carbon (CO_2 or HCO_3^-) to the organic carbon of living cells.

In natural waters the vast majority of such primary production depends on photosynthesis. The activity of chemosynthetic microorganisms is confined to a few specialised environments such as selected transition zones in sediments and specialised oxic–anoxic interfaces in lakes or marine basins. Primary production of natural waters depends on the activity of two groups of organisms: the macroscopic plants which grow at the edges of lakes and oceans and the microscopic algae which constitute the photoplankton population. This chapter deals only with phytoplankton productivity.

15.2 Some historical comments

The history of primary production studies is recent. Only in the past fifty to sixty years have there been attempts to measure quantitatively the distribution of phytoplankton in natural waters and the rates at which they convert inorganic nutrients to cellular material. The earlier history was one of microscopy and taxonomy involving, for the most part, descriptions of various forms and the assigning of Latin or Greek names to them. The name of C. G. Ehrenberg is particularly important in this early work. However, it is the work of Hooker (1847) describing diatoms on the ice in Antarctica which is generally viewed as one of the more important early accounts of marine phytoplankton. The first general review of marine phytoplankton in English was that of Gran (1912). Taylor (1980) has recently written that 'any serious marine phytoplanktologist should consider (that review) as required reading if he wishes to avoid the embarrassment of proudly claiming discovery of some feature known a century before'—a lesson too few of us are willing to learn.

Recognition of the algal flora which make up the phytoplankton populations of natural waters, therefore, preceded any attempts to measure the activities of such organisms. In 1927 Gaarder and Gran published their account of 'investigations of production of plankton in the Oslo fjord' and initiated the modern studies of phytoplankton productivity. Such studies of phytoplankton in natural waters have involved measurements of both the biomass of the phytoplankton and of the activities of that biomass. Correlations between such measurements and those of the accompanying physical and chemical properties of the water have formed the basis (and continue to do so) of our supposed understanding of primary production in natural waters. In historical terms, the use by Harvey (1934) of plant pigment concentrations as indices of phytoplankton biomass was a significant contribution. The later use of the radioisotope of [^{14}C]carbon by Steemann Nielsen (1952) made a comparable contribution to measurements of phytoplankton activity.

Most of the work of the past fifty years has been concerned with methodology. To a large extent the history of primary production studies is one of a constant search for improvements of techniques leading to greater convenience and sensitivity. An analysis of these developments and of the current status of methods is presented later in the chapter (see sections 15.5 to 15.9, pp. 243–7). For the moment, I want to present briefly some of the more essential characteristics of primary production in aquatic systems.

Interest in primary production has focused on two main aspects:

(1) a survey of geographical variation in values of primary production with particular emphasis on the oceans;

(2) a search for some understanding of the environmental factors controlling the growth and distribution of the primary producers.

15.3 Geographical variation in primary production

During the past twenty years there has developed considerable interest in determining values for primary production on a global scale. The aim of such efforts has been twofold: to enable a calculation of the annual value for primary productivity in all natural waters and to provide a broad understanding of the differences between the major regions. As might be expected, greatest interest has focused on the oceans and the reviews of Ryther (1959, 1963, 1969), Steemann Nielsen (1963), Koblentz-Mishke *et al.* (1970), Russel-Hunter (1970), Menzel (1974), Morris (1974) and DeVooys (1980) are examples of this approach.

Table 1 gives examples of the data frequently presented in such reviews. There is a growing body of opinion, with which this author agrees, that little significance can be attached to the precise numbers produced by such calculations. The bulk of this chapter is a commentary on some of the major doubts and problems surrounding measurements of primary production; doubts which apply at a fundamental level and not simply involving trivia of technique. Thus, it is difficult to assess the validity of the precise productivity values presented in Table 1. However, some general awareness of the different broad regions of primary productivity can emerge from such data. The areas of open ocean are, in general, regions of extremely low primary productivity. Their significant contribution to the total production arises from their vast areas and not from any enhanced productivity per unit area. Higher productivity values have been measured in coastal waters and the highest in specialised regions, such as upwelling areas.

For the purposes of comparison, Table 2 summarises some values for primary production from freshwater and terrestrial systems. Again, the significance of the precise number is to be doubted. However, it is valid to emphasise the fact that primary production on land is considerably higher than that in aquatic systems. This difference does not arise from any greater photosynthetic efficiency of land-plants. Rather, it follows from the greater photosynthetic biomass which can be maintained above a unit area of land than that which can be maintained beneath a unit area of water (Morris 1974).

Table 1. Some values for primary production from various regions of the oceans of the world.

Ocean	Percentage area of ocean	Mean productivity (g C m^{-2} yr^{-1})	Total productivity (1×10^9 tons C yr^{-1})	Reference
Open ocean, sub-tropical	40.3	25.6	3.79	Koblentz-Mishke *et al.* (1970)
Transition between sub-tropical and polar	22.7	51.1	4.22	,,
Equatorial divergence and polar oceanic waters	23.4	73.0	6.31	,,
Inshore waters	10.6	124.1	4.80	,,
Neritic waters*	3.0	365.0	3.90	,,
TOTAL			23.00	
Opean ocean	90.0	50.0	16.30	Ryther (1969)
Coastal zone	9.9	100.0	3.60	,,
Upwelling areas	0.1	300.0	0.10	,,
TOTAL			20.00	

*Neritic being the shallow waters of the coast including estuaries.

Table 2. A comparison of some primary production values from various types of communities. The mean productivity values are expressed as g C m^{-2} d^{-1}.

Community	Mean productivity	Reference
Oligotrophic lakes	0.05–0.3	De Vooys (1980)
Mesotrophic lakes	0.25–1.0	,,
Eutrophic lakes	0.6–8.0	,,
Littoral seaweed (maximum)	2.1–2.8	Westlake (1963)
Tropical rain forest (maximum)	7.6	,,
Temperate gram	8.1	,,
Open ocean	1.0	cited by Phillipson (1966)
Coastal zone	0.5–3.0	,,
Intensive agriculture	10.0–25.0	,,
Deserts	0.5	,,

15.4 Factors regulating primary production in aquatic systems

Our understanding of the environmental factors influencing phytoplankton ecology and aquatic primary productivity comes from two main approaches.

(1) The first originates with experimental studies in laboratory cultures of algae derived from natural phytoplankton populations. Such an approach allows one to specify the way in which changes in particular environmental factors modify the growth of algae under controlled laboratory conditions.

(2) The second makes correlations between the observed distribution and activity of phytoplankton and the accompanying physical and chemical conditions of the surrounding waters.

From these two approaches, three environmental factors are generally assumed to interact in a direct way with the algal cells of the phytoplankton population: light intensity, nutrient concentration and temperature. However, confining any discussion to these three characteristics of the aquatic environment is simplistic. The problems of sinking, grazing, movements of water masses and other processes are all superimposed on the other factors. It would not be appropriate to attempt any kind of summary of the main observations in the limited space available here. The reviews of Yentsch (1964, 1974), Strickland (1965), Eppley (1972) and Morris (1974) are examples of such summaries. Also, the books of Hutchinson (1967), Parsons *et al.* (1977), Morris (1980) and Raymont (1980) contain analyses of the phytoplankton ecology relating to the controlling environmental factors.

Although any comprehensive summary might be misplaced (and impossible), it is appropriate to present an example of the types of problems which accompany attempts to understand the environmental regulation of primary production.

Inevitably, these comments reflect the personal interest and bias of this particular author. One of the major problems is the fact that casual statements frequently depend on correlations. That is, an aim of the phytoplanktologist might be the identification of those mechanisms which cause any observed changes in productivity from one region to another or from one time to another. However, such mechanisms are inferred from correlations between different sets of observations. The sophisticated nature of the measurements and of the correlations do not alter that main point. An example can be seen in the attempts to understand the relationship between primary production and nutrient concentration. Thus, the apparently obvious statement that the low productivity of open oceans (and oligotrophic lakes) results from the low levels of certain essential nutrients in the surface waters, depends on correlations between the distribution of such nutrients and the density of phytoplankton. Yet, the precise link between nutrient concentration and phytoplankton growth remains unclear. The main effect appears to be on the size of the phytoplankton population which can be supported in waters with varying nutrient concentrations. There are no apparent indications of physiological nutrient deficiency in the phytoplankton from nutrient deficient waters (Morris *et al.* 1971b; Goldman *et al.* 1979).

The species composition of phytoplankton populations also differs in nutrient-poor regions compared with that in nutrient-rich waters. Based on superficial correlations, it would be tempting to suggest that species selection depends on the ability of certain species to utilise low nutrient concentrations. Dugdale (1967) introduced a possible means of investigating such a proposal by relating the rate of nutrient assimilation (or phytoplankton growth) to nutrient concentration by a Michaelis-Menten curve (see Chapter 20, p. 344). The kinetic parameters which can be derived from such a relationship can be used to describe the relationship between cell growth and nutrient concentration of any particular species. This exciting concept stimulated much work during the last decade (Eppley & Coatsworth 1968; Fuhs *et al.* 1972; Azam *et al.* 1974; Droop 1974; Eppley & Renger 1974; Conway *et al.* 1976; Dugdale 1977; McCarthy & Goldman 1978). In general, this work has not permitted the development of a mechanistic explanation of the relationship between nutrient concentration and phytoplankton growth and selection.

The search for more precise causal mechanisms in phytoplankton ecology has generally involved an increasing emphasis being placed on the physiology and biochemistry of the photosynthetic cells floating in natural waters. Such an emphasis has dominated nuch of the work on primary production in the past twenty years (Yentsch 1964, 1974). To a large extent, it was a natural development since field studies of phytoplankton ecology could be based solidly on laboratory studies of algal biochemistry. Also, convenient and sensitive techniques appeared to exist for measuring the biomass of phytoplankton and their fixation of inorganic carbon (see section 15.9, p. 245) as well as ambient physical (light and temperature) and chemical (nutrient concentrations) properties of the water. Such conditions might be expected to yield an understanding of the factors controlling primary production in aquatic systems greater than that for other levels in the aquatic ecosystem. However, the success of the physiological approach might be doubted (Yentsch 1974).

In recent years, one of the most important developments has been the increasing realisation that studies of phytoplankton physiology need to be accompanied by a proper understanding of the physical properties of the water masses in which phytoplankton float. Of particular interest is the question of scale—both in time and in space. There is growing evidence that conventional scales of discrete measurements tend to produce average values for properties which, in fact, show variability over scales not normally detected. The introduction of continuous monitoring of properties such as temperature, salinity, chlorophyll and nutrients (Lorenzen 1966; Platt 1972a) has allowed the detection of such variability. Such studies have led to the important concepts of patchiness (Platt *et al.* 1970; Platt 1972b; Platt & Denman 1975, 1980; Wroblewski *et al.* 1975) and frontal systems (Pingree *et al.* 1975; Holligan 1979).

For the remainder of the chapter, discussion will focus on the specific problems of measuring primary production. At this point, it is worth re-emphasising the fact that the search for those mechanisms controlling primary production must not be confined to the level of cell physiology. Mechanisms based on interactions between the physical forces operating in water masses and the floating cells of the phytoplankton populations are also important. Any proper understanding will depend on bringing the two approaches together.

Experimental

15.5 Introduction

There are two aspects to primary production measurements:

(1) determination of phytoplankton biomass;

(2) measurement of the rates at which that biomass converts inorganic carbon to cell material.

Statements about primary production generally emphasise the second aspect. The term primary production is frequently equated with primary productivity, thus implying that it is a measurement of a rate of a process. Thus, the units generally take the form of weight of carbon fixed per volume of water per time (e.g. g C m^{-3} h^{-1}). Frequently, the values are integrated over the euphotic zone (generally assumed to extend down to regions with approximately 1% of the surface irradiance) and the time lengthened to a day or a year (e.g. g C m^{-2} d^{-1}).

The emphasis on the rate of carbon assimilation is also adopted in this chapter. However, it will become clear that the problems of making reliable measurements of the biomass are important aspects of primary production studies.

In terrestrial ecosystems (as well as for macrophytes at the edges of bodies of water) it is possible to measure the net production of plant material in a direct way. Selected areas can be cropped before and after varying periods of time so that the actual increase in weight of plant material per unit area can be determined. Such a direct approach cannot be adopted with phytoplankton. Sinking, movement of water and other processes will ensure that the phytoplankton populations at any one sampling site change. Thus, the rate of assimilation has to be measured under somewhat artificial conditions and over periods which might be relatively short compared with the doubling times of the phytoplankton biomass.

Although the vast majority of primary production studies focus on the assimilation of carbon, other elements have received some attention. In marine systems, the assimilation of nitrogen has been emphasised (Dugdale & Goering 1967; Eppley & Coatsworth 1968; Eppley *et al.* 1969; McCarthy *et al.* 1975; Conway 1977). In freshwater systems greater attention has been focused on phosphorus (Fuhs *et al.* 1972; Lean & Nalewajko 1976; Titman & Kilham 1976). These areas of interest reflect, largely, the general opinion that nitrogen is most frequently the nutrient limiting phytoplankton growth in the oceans, and phosphorus the limiting nutrient in lakes. Some studies have also measured the assimilation of silicon (reflecting the importance of this element in the growth of diatoms—Goering *et al.* 1973; Azam & Chisholm 1976; Chisholm *et al.* 1978). Most of these studies on elements other than carbon have been designed to comment on some specialised aspect of phytoplankton physiology. Little emphasis has been placed on using such assimilation rates as estimates of primary production.

Thus, measurements of primary productivity are measurements of rates of photosynthesis: that is, the light-dependent fixation of carbon dioxide and the evolution of oxygen.

15.6 Water sampling and incubations

Measurements of phytosynthetic rates involve incubations of water samples in vessels of some kind. Water can be collected from any required depth with a number of water sampling bottles such as Niskin or Van Dorn bottles. The need to avoid metal surfaces for physiological studies resulted in the change from the older Nansen bottles (so important in hydrographic studies) to the later polyvinyl chloride plastic types.

Three types of incubation can be made.

(1) *In situ*.

(2) Simulated *in situ*. Whereby incubations under simulated conditions *in situ* are made on deck where the containers are covered with neutral density light filters designed to provide a range of light intensities. The aim of such incubations is to mimic the change in light intensity down through the water column. A popular design of such experiments is to collect water samples from light depths (depths determined by the attenuation of light in the water) and incubate under neutral density filters which correspond to the relative light intensity (% of surface irradiance) from which the sample was taken. The samples are incubated under natural illumination in incubators flushed with running water (generally from the surface of the ocean or lake).

(3) Under artificial illumination. Incubations under artificial illumination also involve the use of neutral density filters for modifying the light intensity. However, the incubations are made in light-proof incubators equipped with artificial

lighting—generally fluorescent tubes—and also flushed with running surface water.

The use of artificial illumination is best confined to comparative studies of some aspect of phytoplankton physiology requiring controlled environmental conditions. It is of little use for actual determinations of values of primary production since the light intensity of fluorescent tubes is generally 10 to 30% of that of a bright sunny day.

The choice between in-situ and simulated in-situ incubations is generally made on the basis of feasibility. In-situ experiments require the boat to remain in one place for the period of the experiment. Simulated in-situ experiments can be performed while the boat is under way which is an important consideration. The main doubt surrounding simulated in-situ incubations revolves around the problems of the spectral quality of the light. Neutral density filters are designed to have no effect on the spectral quality of the incident irradiance. However, the spectrum is drastically altered when light passes through the water column. It is difficult to assess the significance of such a difference between in-situ and simulated in-situ conditions. This author has not seen any consistent differences when measuring rate and pattern of [^{14}C]carbon assimilation by marine phytoplankton under these two experimental situations.

Incubations are generally made between 100% and 1% of the incident light intensity. Such a range is designed to permit integration of any productivity measurements over the water column so that the value can be expressed per unit area. However, the rigid use of light-depths and their corresponding neutral density filters is to be discouraged. It ignores other interesting features of the water column such as position of the thermocline and location of any sub-surface chlorophyll maximum.

One of the major questions concerning incubations of water samples in the ways described above (whether *in situ* or under simulated in-situ conditions) is that of the length of incubation. Conflicting factors interact to determine such a decision. In one sense, it might be argued that the rate of a process such as photosynthesis should be measured over a period as short as possible yielding a value approximating to an instantaneous rate. Two factors influence the ability and advisability of making such short-term measurements. First, the 'rates of photosynthesis and respiration in natural waters are often measured from small changes in a large content of dissolved oxygen or carbon dioxide' (Talling 1973). Prolonged incubation times (often up to 24 h) are generally used to overcome this problem. Second, in the one technique with sufficient sensitivity, the [^{14}C]carbon technique, short-term rates are difficult to relate to net production (see p. 246).

Nevertheless, prolonged incubations have their problems, generally associated with the changes which can occur in bottles when natural water samples are incubated for periods up to 24 h. Growth of bacteria on the walls of the container, selective growth or death of some phytoplankton species and grazing are all examples of the changes which might occur. Possibly, the most disappointing feature of most measurements of primary production is the willingness to use prolonged incubation periods without performing time-courses. Indeed, the apparent willingness to measure rates of a particular process without an emphasis on initial rates determined from a time-series is an unfortunate characteristic of most reported values of primary production.

More detailed comments on the problems associated with measurements of primary production will be made in the sections dealing with the particular techniques generally employed.

15.7 Changes in oxygen concentration

The original measurement of photosynthesis in phytoplankton populations involved measuring changes in oxygen concentration using the Winkler technique (Gaarder & Gran 1927; Jenkin 1937). The essentials of the technique are described in Carritt and Carpenter (1966). Strickland and Parsons (1968) and Vollenweider (1969). Minor modifications to the technique can be found in Bradbury and Hambly (1952), Drew and Robertson (1974) and Bryan *et al.* (1976).

The major problem with this technique is one of sensitivity. The Winkler technique can measure 0.02 to 0.15 mg 0_2 l^{-1} (Strickland 1965; Golterman 1975; Hall & Moll 1975) and, frequently, determinations of the rates depend on measuring small changes in a large initial value.

The use of oxygen electrodes is an alternative to the Winkler technique. Some of the better examples of the use of these electrodes with natural phytoplankton populations can be found in Harris (1973), Harris and Lott (1973) and Harris and Piccinin (1977). The use of such a technique depends on having cell densities which are much higher than normal in natural waters and such an approach can yield important information on the physiology of the algae under study. The main doubts, however, centre around the validity of concentrating cell suspensions and the review of Harris (1978) contains a critical discussion of this approach.

15.8 Changes in dissolved inorganic carbon

Few measurements of primary production depend on measuring changes in the dissolved inorganic carbon. As with the techniques involving changes in oxygen concentration, the measurements often involve detecting small changes in a large initial value. Also, the interactions of the species of carbon dioxide in the aqueous phase introduce complications (discussed in detail by Talling (1973)).

15.9 The [^{14}C]carbon technique

Since its introduction by Steemann Nielsen (1952), the use of the radioisotope [^{14}C]carbon has become the most widely used method of measuring primary production. It has a convenience and sensitivity which permits routine measurements to be made over relatively short time periods and in nutrient-poor waters. However, the method is not without its problems. Indeed, it might be suggested that the convenience of the technique has resulted in the production of data without adequate investigation of their meaning and of the underlying processes. Peterson (1980) has written a history of the [^{14}C]carbon method and its use in measurements of aquatic primary productivity. Carpenter and Lively (1980) summarised some of the more important uncertainties and problems with the technique and its relationship to phytoplankton growth.

Standard procedures adopted in the [^{14}C]carbon technique are described by Strickland and Parsons (1968) and Vollenweider (1969). However, the appearance of a recipe-like description of the procedure should not discourage the beginner from undertaking proper investigations of the process.

Basically the procedure is simple. A known amount of sodium [^{14}C]bicarbonate is added to a bottle containing a water sample and the bottle incubated in ways described earlier (section 15.6, p. 243). At the end of the incubation period, the contents of the bottle are filtered through membrane or glass-fibre filters and the radioactivity on the filters determined. It is a genuine tracer technique, since the amount of bicarbonate present in the ^{14}C-labelled solution to be added is generally very small relative to the amount present in the water. Behind this simplicity, however, lies a vast potential for errors and misunderstanding. The main aim of this section is to identify the potential problems, both technical and conceptual, which can accompany the use of the [^{14}C]carbon technique for measuring primary production.

15.9.1 Quality of the [^{14}C]Bicarbonate Solution

The solution of sodium [^{14}C]bicarbonate can vary considerably in quality and there are three potential problems:

(1) the solution may contain less radioactivity than is indicated;

(2) there may be contaminating particulate matter;

(3) there may be contaminating dissolved organic compounds.

It is now becoming common practice to count standards of each [^{14}C]carbon solution used in the experimental system. A small volume (e.g. 0.01 ml) of a solution containing 1 μCi ml^{-1} is added to a scintillation vial containing a little phenethanolamine (a substance which can absorb the carbon dioxide is needed since most scintillation fluids are acidic), scintillant added and the radioactivity determined in a scintillation counter.

Contaminating particulate matter can be removed by preliminary filtration—a precaution which is recommended for each new solution of [^{14}C]bicarbonate. Also, a zero-time determination (the [^{14}C]carbon solution added to a water sample and filtered immediately) can be a useful indication of the presence of any contaminating material.

Contamination of the [^{14}C]bicarbonate with dissolved organic compounds is particularly serious in studies of excretion in phytoplankton and has probably contributed significantly to some of the high percentage excretion values which have been reported (Williams *et al.* 1972; Williams & Yentsch 1976; Sharp 1977).

15.9.2 Periods of Incubation

Some remarks about the problems of prolonged incubations and the absence of time-courses have already been made (section 15.6, p. 243). There is, however, a special problem which might result. Harris and Piccinin (1977) and Marra (1978) have reported the way in which short-term measurements of photosynthesis (over periods of approximately 10 min) yield higher rates at saturating irradiances than do more prolonged incubations. This appears to be the result of the phenomenon of photo-inhibition due to prolonged incubations in glass bottles. This concept has profound implications for photosynthesis in environments where turbulence can cause short-term fluctuations in light intensities. That is, values for photosynthesis at the surface might be much higher than is indicated from an experiment involving prolonged incubations.

15.9.3 Filtration

Most investigators use Millipore membrane filters (0.45 μm pore size) for the final filtration but there is some tendency to use glass-fibre filters. For the latter filtration times are much shorter and they are frequently used for chlorophyll determination. However, it appears likely that some of the smaller algae pass through the glass-fibre filters—particularly of the Gelman GF A/E type—and this can be a serious problem when significant numbers of algae of less than 1 μm are present (Johnson & Sieburth 1979; Waterbury *et al.* 1979). In our experience the GF/F type of glass-fibre filters give data comparable to those obtained with membrane filters. Filtration of large volumes (Arthur & Rigler 1967) and use of high filtration pressures (Herbland 1974) might cause cell disruption.

15.9.4 Acidification

In the original description of the [^{14}C]carbon technique, Steemann Nielsen (1952) used fuming hydrochloric acid to remove surplus [^{14}C]-bicarbonate from the filters. Alternative techniques involve rinsing the filter with filtered water and/or dilute hydrochloric acid. The experience of the author of this chapter suggests that adequate rinsing combined with drying of the filters before adding the scintillation fluid is adequate for removal of the remaining bicarbonate. Again, a zero-time blank and a time-course are the best methods of ensuring proper technique.

15.9.5 Counting of Radioactivity

Self-absorption problems associated with planchet counting of membrane filters have received some attention (Jitts 1957; Thomas 1964; Paasche 1969; Wood 1971) but the introduction of liquid scintillation counting avoids many of these problems (Lind & Campbell 1969; Pugh 1970, 1973).

In work with natural phytoplankton populations, there appears to be a willingness to accept levels of incorporated [^{14}C]carbon which give low counts per minute (not apparent when data are reported as $g\,C\,m^{-2}\,d^{-1}$). In part, this results from an apparent reluctance to vary the amount of [^{14}C]bicarbonate added and hence the final specific activity. Such a variation is recommended so as to ensure significant counts when samples containing few algae are measured or when shorter periods of incubations are used.

15.9.6 Net and Gross Photosynthesis

It is surprising that, after almost thirty years of use, there is still controversy over the question of whether the [^{14}C]carbon technique measures net or gross photosynthesis (gross photosynthesis equals net photosynthesis plus respiration). There are a number of different problems:

(1) determining respiration rates in dilute phytoplankton populations is not simple and, in general, the available techniques lack the required sensitivity;

(2) the question of whether a rate of respiration measured in the dark remains unaltered in the light continues to be uncertain (Brown 1953; Glidewell & Raven 1975; Bidwell 1977);

(3) similarly, there is uncertainty about the refixation of any respired carbon dioxide (Raven 1972).

In the absence of reliable measurements of phytoplankton respiration in the field, it has been tempting to extrapolate from studies with algal cultures (Ryther 1954, 1959) demonstrating the rate of respiration to be approximately 5 to 10% of the photosynthetic rate. In essence, the doubts about using a correction factor of 5 to 10% centres around the variability of such a factor with the physiological state of the cells (Ryther 1954).

Even with an estimate of respiration, the uncertainty about the [^{14}C]carbon technique measuring net or gross photosynthesis remains. Ryther (1956) concluded that the technique measures net photosynthesis, whereas Steemann Nielsen and Hansen (1959) suggested that this was so only over prolonged periods. Most attempts to resolve the problem have involved a comparison of the [^{14}C]carbon and oxygen methods of measuring photosynthesis. Harris (1980) cites the comparisons of Ryther (1954, 1956), Ryther and Vaccaro (1954), McAllister (1961), Antia *et al.* (1963), Thomas (1964) and Harris and Piccinin (1977) as evidence for the view that over a period of 4 h, the [^{14}C]carbon technique measures something between net and gross photosynthesis.

15.9.7 The Dark Bottle

Convention demands that measurements of primary production include a dark bottle. The [^{14}C]carbon fixed in the dark is subtracted from that in the light. The reason centres around the idea that [^{14}C]carbon dioxide fixation in the dark does not represent net fixation (Steeman Nielsen 1952). Morris *et al.* (1971a) have questioned this conventional wisdom particularly in oligotrophic waters where the dark fixation can be a significant proportion of fixation in the light and does contribute to the production of biomass.

15.10 Photosynthesis, primary production and phytoplankton growth

One of the major uncertainties surrounding our interpretation of any rate of phytoplankton photosynthesis, determined, for example, by the [^{14}C]carbon technique concerns the use of measurements in a small volume of water over a period of a few hours in the light. The assumptions required to extrapolate from such a number to any value integrated over an area and extended to periods of days or years are many and various.

One other problem remains. The usual view of primary production in aquatic systems resembles that for terrestrial environments in emphasising conversion of energy and matter. An alternative approach views a population of phytoplankton as a suspension of microbes with an emphasis on establishing the growth rate of such a population. Adopting such an approach would mean that primary production and phytoplankton growth should be measured as rates at which the cell number in the population is increasing. Calculating growth rates from rates of photosynthesis, determined by the methods described in this chapter, is not easy. There are three major problems:

(1) determining the loss functions, such as respiration, so as to ensure that a net rate has been measured;

(2) normalising a net rate of assimilation to a biomass parameter to permit calculation of population doubling times;

(3) the problems of extrapolating from a photosynthetic rate measured over a few hours in the light to a growth rate of a microbial population growing and dividing in the light and the dark.

In the past, the first two of these problems have been emphasised. However, this author supports an approach which views a phytoplankton population as a suspension of microorganisms and emphasises the importance of determining growth rates as doubling times. With such an approach, the third problem mentioned above is fundamental, even under conditions when the first two problems are minimal. Recently, two reviews have emphasised this aspect of the problem. Eppley (1981) pointed out that the existence of unbalanced growth in phytoplankton populations makes it impossible to extrapolate from rates of assimilation of an element to division rates of the phytoplankton population. That is, for a certain time during the light period the rate of and nature of assimilation of carbon and other elements would not reflect the net chemical composition of the cells after several generations of growth and division. Morris (1981) illustrated a special nature of this fundamental problem by emphasising rates of synthesis of storage products relative to those polymers more closely linked to the division process. When conditions create enhanced synthesis of storage products during the light period, growth rates calculated from rates of carbon assimilation in the light can seriously overestimate actual doubling times. Also, the extent of the discrepancy between growth rates estimated from carbon assimilation and the actual rates can vary with different patterns of carbon metabolism.

It might be suggested that incubations of 24 h (covering light and dark periods) can solve the problem. However, problems arising from prolonged incubations have been mentioned already. Also, it is unclear whether incubations over 24 h improve estimates of growth rates of populations with doubling times of longer than one day. Rather, it might be hoped that existing approaches to primary productivity be improved so that precise instantaneous measurements can be made of those cellular processes which can indicate growth rates of the phytoplankton populations.

15.11 Conclusions

(1) Primary productivity is the rate at which phytoplankton populations convert the inorganic materials dissolved in the water to new cell material. Emphasis is usually placed on the assimilation of carbon.

(2) Such a definition constitutes a restricted view of the subject. Many other factors regulate the growth and distribution of the algae in a phytoplankton population. Problems such as grazing, sinking, scales of measurement and distribution, regulation of species composition and interaction with physical and chemical properties of the water are important considerations in our attempts to understand aquatic primary productivity in the broader sense.

(3) Work on primary productivity has considered two aspects of the problem: geographical variation in the productivities of different natural waters and an assessment of the environmental factors which control that variation.

(4) A consideration of primary productivity includes both the amount of phytoplankton material (e.g. cell number or chlorophyll *a* concentration) and the activity of the population (rates of photosynthesis).

(5) Calculations of global aquatic primary productivity and identification of major regions involve assumptions which cast doubt on the precise values generally reported. However, broad differences between the major types of water masses can be identified.

(6) Our understanding of the environmental factors controlling primary productivity in natural waters derives from studies of laboratory cultures and from correlations between the distribution of phytoplankton and the prevailing physical and chemical properties of the water. Light intensity and nutrient concentration interact directly with the algal cells. Temperature has a profound indirect effect through its role in altering the stability of the water column (and thus the supply of nutrients to the euphotic zone). However, precise mechanistic explanations of the way in which environmental factors control the distribution, activity and species composition of phytoplankton populations are lacking.

(7) Rates of photosynthesis in natural phytoplankton populations are measured either as rates of oxygen evolution or rates of inorganic carbon assimilation. The existence of the radioisotope [^{14}C]carbon affords a convenient and sensitive tracer technique for measuring inorganic carbon assimilation. This is the most common method used for measuring primary productivity in natural waters.

(8) Despite its apparent convenience and sensitivity, the [^{14}C]carbon technique is not without its problems. Some of the problems are technical, whilst others are conceptual. Most of the technical problems can be overcome with appropriate care over methodology. Many result from an apparent willingness to make single-point determinations in a routine manner and an apparent reluctance to undertake time-courses with appropriate experimental and control treatments.

(9) More fundamental problems surround the relationship between rates of [^{14}C]carbon assimilation and those of phytoplankton growth. Such problems include uncertainty about whether the [^{14}C]carbon technique measures net or gross photosynthesis; difficulties of measuring loss functions and the problems associated with biomass measurements which would permit the calculation of specific assimilation rates. Also, there is a fundamental difficulty in extrapolating from a rate of carbon assimilation measured over a few hours in the light to a growth rate of a population of microalgae growing and dividing in alternating light/dark periods.

Acknowledgements

This work was completed while the author was at the Bigelow Laboratory for Ocean Sciences, West Boothbay Harbour, Maine, USA, and was supported in part by NSF Grant OCE 77–18722.

Recommended reading

BOLIN B., DEGENS E. T., KEMPE S. & KLEPNER P. (1980) *The Global Carbon Cycle*. John Wiley & Sons, New York.

GOLDBERG E. D., MCCAVE I. N., O'BRIEN J. J. & STEELE J. H. (1974) *The Sea*. John Wiley & Sons, New York.

GOLDBERG E. D., MCCAVE I. N., O'BRIEN J. J. & STEELE J. H. (1977) *The Sea, Ideas and Observations on Progress in the Study of the Seas*. John Wiley & Sons, New York.

HUTCHINSON G. E. (1967) *A Treatise on Limnology*, vol. 2. John Wiley & Sons, New York.

LEITH M. & WHITTAKER R. M. (1977) *Primary Production in the Biosphere*. Springer-Verlag, Berlin.

MORRIS I. (1980) *The Physiological Ecology of Phytoplankton.* Blackwell Scientific Publications, Oxford.

MURRAY J. & JJART J. (1912) *The Depths of the Ocean.* MacMillan, London.

RAYMONT J. E. G. (1980) *Plankton and Productivity in the Oceans,* 2e. Pergamon Press, Oxford.

RUSSEL-HUNTER W. D. (1970) *Aquatic Productivity.* Collier-MacMillan, London.

References

ANTIA N. J., MCALLISTER C. D., PARSON T. R., STEPHENS K. & STRICKLAND J. D. M. (1963) Further measurements of primary production using a large-volume plastic sphere. *Limnology and Oceanography* **15**, 402–7.

ARTHUR C. R. & RIGLER F. H. (1967) A possible source of error in the carbon method of measuring primary productivity. *Limnology and Oceanography* **12**, 121–4.

AZAM F. & CHISHOLM S. W. (1976) Silicic acid uptake and incorporation by natural marine phytoplankton populations. *Limnology and Oceanography* **21**, 427–35.

AZAM F., HEMMINGSON B. B. & VOLCANI B. E. (1974) Role of silicon in diatom metabolism. V. Silicic acid transport and metabolism in the heterotrophic diatom *Nitzschia alba. Archiv für Mikrobiologie* **97**, 103–14.

BIDWELL R. G. S. (1977) Photosynthesis and light and dark respiration in freshwater algae. *Canadian Journal of Botany* **55**, 809–18.

BOLIN B., DEGENS E. T., KEMPE S. & KLEPNER P. (1980) *The Global Carbon Cycle.* John Wiley & Sons, New York.

BRADBURY J. H. & HAMBLY A. N. (1952) An investigation of errors in the amperometric and starch indicator methods for the titration of millinomal solutions of iodine and thiosulphate. *Australian Journal of Science Research, Series A* **5**, 541–4.

BROWN A. H. (1953) The effects of light on respiration using isotopically enriched oxygen. *American Journal of Botany* **40**, 719–29.

BRYAN J. R., RILEY J. P. & WILLIAMS P. J. LEB. (1976) A Winkler procedure for making precise measurements of oxygen concentration for productivity and related studies. *Journal of Experimental Marine Biology and Ecology* **21**, 191–8.

CARPENTER E. J. & LIVELY J. S. (1980) Review of estimates of algal growth using ^{14}C tracer techniques. In: *Primary Productivity in the Sea* (Ed. P. G. Falkowski), pp. 161–78. Plenum Press, New York.

CARRITT D. E. & CARPENTER J. H. (1966) Comparison and evaluation of currently employed modifications of the Winkler method for determining dissolved oxygen in seawater. A NASCO Report. *Journal of Marine Research* **24**, 286–318.

CHISHOLM S. W. AZAM F. & EPPLEY R. W. (1978) Silicic acid incorporation in marine diatoms on light-dark cycles: use as an assay for phased cell division. *Limnology and Oceanography* **23**, 518–29.

CONWAY H. L. (1977) Interactions of inorganic nitrogen in the uptake and assimilation by marine phytoplankton. *Marine Biology* **39**, 221–32.

CONWAY H. L., HARRISON P. J. & DAVIS C. O. (1976) Marine diatoms grown in chemostats under silicate or ammonium limitation. II. Transient response of *Skeletonema costatum* to a single addition of the limiting nutrient. *Marine Biology* **35**, 187–99.

DEVOOYS C. G. N. (1980) Primary production in aquatic environments. In: *The Global Carbon Cycle* (Eds B. Bolin, E. T. Degens, S. Kempe & P. Klepner), pp. 259–92. John Wiley & Sons, New York.

DREW E. A. & ROBERTSEN W. A. A. (1974) A simple field version of the Winkler determination of dissolved oxygen. *New Phytology* **73**, 793–6.

DROOP M. R. (1974) The nutrient status of algal cells in continuous culture. *Marine Biological Association of the United Kingdom Journal* **54**, 825–55.

DUGDALE R. C. (1967) Nutrient limitation in the sea: dynamics indentification and significance. *Limnology and Oceanography* **12**, 685–95.

DUGDALE R. C. (1977) Modelling. In: *The Sea, Ideas and Observations on Progress in the Study of the Seas* (Eds E. D. Goldberg, I. N. McCave, J. J. O'Brien & J. H. Steele), pp. 789–806. John Wiley & Sons, New York.

DUGDALE R. C. & GOERING J. J. (1967) Uptake of new and regenerated forms of nitrogen in primary productivity. *Limnology and Oceanography* **12**, 196–206.

EPPLEY R. W. (1972) Temperature and phytoplankton growth in the sea. *Fishery Bulletin (US)* **70**, 1063–85.

EPPLEY R. W. (1981) Relationship between assimilation and phytoplankton growth. In: *Algal Physiology and Phytoplankton Ecology.* (In press.)

EPPLEY R. W. & COATSWORTH J. L. (1968) Nitrate and nitrite uptake by *Ditylum brightwellii.* Kinetics and mechanisms. *Journal of Phycology* **4**, 151–6.

EPPLEY R. W., COATSWORTH J. L. & SOLORZANO L. (1969) Studies of nitrate reductase in marine phytoplankton. *Limnology and Oceanography* **14**, 194–205.

EPPLEY R. W. & RENGER E. M. (1974) Nitrogen assimilation of an oceanic diatom in nitrogen-limited continuous culture. *Journal of Phycology* **10**, 15–23.

FUHS G. S., DEMMERLE S. D., CANELLI E. & MIU CHIU (1972) Characterization of phosphorus-limited plankton algae. *American Society of Limnology and Oceanography Symposium* **1**, 113–33.

GAARDER T. & GRAN H. H. (1927) Investigation of the production of plankton in the Oslo fjord. *Rapport des Procédés Verbal Conseil International pour l'Exploration de la Mer* **42**, 3–48.

GLIDEWELL S. M. & RAVEN J. A. (1975) Measurements of simultaneous O_2 evolution and uptake in *Hydrodictyon africanum. Journal of Experimental Botany* **26**, 479–88.

GOERING J. J., NELSON D. M. & CARTER J. A. (1973) Silicic acid uptake by natural populations of marine phytoplankton. *Deep-Sea Research* **20**, 777–89.

GOLDBERG E. D., MCCAVE I. N., O'BRIEN J. J. & STEELE J. H. (1974) *The Sea*. John Wiley & Sons, New York.

GOLDBERG E. D., MCCAVE I. N., O'BRIEN J. J. & STEELE J. H. (1977) *The Sea, Ideas and Observations on Progress in the Study of the Seas*. John Wiley & Sons, New York.

GOLDMAN J. C., MCCARTHY J. J. & PEAVEY D. G. (1979) Growth rate influence on the chemical composition of phytoplankton in oceanic waters. *Nature, London* **279**, 210–15.

GOLTERMAN H. L. (1975) *Physiological Limnology. An approach to the Physiology of Lake Ecosystems*. Elsevier, Amsterdam.

GRAN H. H. (1912) Pelagic plant life. In: *The Depths of the Ocean* (Eds J. Murray & J. Hjort), pp. 307–86. Macmillan, London.

HALL A. A. S. & MOLL R. (1975) Methods of assessing aquatic primary production. In: *Primary Production in the Biosphere* (Eds M. Leith & R. M. Whittaker), pp. 19–55. Springer-Verlag, Berlin.

HARRIS G. P. (1973) Diel and anjual cycles of net plankton photosynthesis in Lake Ontario. *Journal of the Fisheries Research Board of Canada* **30**, 1779–87.

HARRIS G. P. (1978) Photosynthesis, productivity and growth. The physiological ecology of phytoplankton. *Ergebnisse der Limnologie* **10**, 1–171.

HARRIS G. P. (1980) The measurement of photosynthesis in natural populations of phytoplankton. In: *The Physiological Ecology of Phytoplankton* (Ed. I. Morris), pp. 129–87. Blackwell Scientific Publications, Oxford.

HARRIS G. P. & LOTT J. N. A. (1973) Light intensity and photosynthetic rates in phytoplankton. *Journal of the Fisheries Research Board of Canada* **30**, 1771–8.

HARRIS G. P. & PICCININ B. B. (1977) Photosynthesis by natural phytoplankton populations. *Archiv für Hydrobiologie* **80**, 405–57.

HARVEY H. W. (1934) Measurements of phytoplankton populations. *Marine Biological Association of the United Kingdom Journal* **19**, 761–800.

HERBLAND A. (1974) Influence de la depression de filtration sur la mesure simultanée de l'assimilation et de l'excretion organique de phytoplancton. *Cahiers ORSTOM Série Océanographie* **12**, 173–7.

HOLLIGAN P. M. (1979) Dinoflagellate blooms associated with tidal fronts around the British Isles. In: *Second International Symposium on Toxic Dinoflagellates*, pp. 249–56. Elsevier, Amsterdam.

HOOKER J. D. (1847) *The Botany of the Antarctic Voyage of H. M. Discovery Ships 'Erebus' and 'Terror' in the years 1839–1843 under the Command of Captain Sir James Clark Ross*. Reeve Bros, London. (Reprinted J. Cramer, Weinheim, 1963.)

HUTCHINSON G. E. (1967) *A Treatise on Limnology*, vol. 2. John Wiley & Sons, New York.

JENKIN P. M. (1937) Oxygen production by the diatom *Coscinodiscus exctricus* Ehr. in relation to submarine illumination in the English channel. *Marine Biological Association of the United Kingdom Journal* **22**, 301–43.

JITTS H. R. (1957) The ^{14}C method for measuring CO_2 uptake in marine productivity studies. *CSIRO Australian Division of Fishery and Oceanographic Reports* **8**, 1–12.

JOHNSON P. W. & SIEBIRTH J. MCN. (1979) Chroococcord cyanobacteria in the sea: a ubiquitous and diverse phototrophic biomass. *Limnology and Oceanography* **24**, 928–35.

KOBLENTZ-MISHKE O. J., VALKOVINSKY V. V. & KABANOVA J. G. (1970) Plankton primary production of the world ocean. In: *Scientific Exploration of the South Pacific* (Ed. W. S. Wooster) National Academy of Sciences, Washington.

LEAN D. R. S. & NALEWAJKO C. (1976) Phosphate exchange and organic phosphorus excretion by freshwater algae. *Journal of the Fisheries Research Board of Canada* **33**, 1312–23.

LEITH M. & WHITTAKER R. M. (1977) *Primary Production in the Biosphere*. Springer-Verlag, Berlin.

LIND O. T. & CAMPBELL R. S. (1969) Comments on the use of liquid scintillation for routine determination of ^{14}C activity in production studies. *Limnology and Oceanography* **14**, 787–9.

LORENZEN C. J. (1966) A method for the continuous measurement of *in vivo* chlorophyll concentration. *Deep-Sea Research* **13**, 223–37.

MARRA J. (1978) Effect of short-term variations in light intensity on photosynthesis of a marine phytoplankter: A laboratory simulation study. *Marine Biology* **46**, 191–202.

MCALLISTER C. D. (1961) Decontamination of filters in the ^{14}C method of measuring marine photosynthesis. *Limnology and Oceanography* **6**, 447–50.

MCCARTHY J. J. & GOLDMAN J. R. (1978) Steady state growth and ammonium uptake of a fast-growing marine diatom. *Limnology and Oceanography* **23**, 695–703.

MCCARTHY J. J., TAYLOR W. R. & TAFT J. L. (1975) The dynamics of nitrogen and phosphorus cycling in the open waters of the Chesapeake Bay. In: *Marine Chemistry in the Coastal Environment* (Ed. T. M. Church), pp. 664–81. ACS Symposium.

MENZEL D. W. (1974) Primary productivity, dissolved and particulate organic matter and the sites of oxidation of organic matter. In: *The Sea* (Eds E. D. Goldberg, I. N. McCave, J. J. O'Brien & J. H. S. Steele), pp. 659–78. John Wiley & Sons, New York.

MORRIS I. (1974) The limits to the productivity of the sea. *Science Progress, Oxford* **61**, 99–115.

MORRIS I. (1980) *The Physiological Ecology of Phytoplankton*. Blackwell Scientific Publications, Oxford.

MORRIS I. (1981) Photosynthetic products, physiological state and phytoplankton growth. In: *Algal Physiology and Phytoplankton Ecology*. (In press.)

MORRIS I., YENTSCH C. M. & YENTSCH C. S. (1971a) The relationship between light carbon dioxide fixation and dark carbon dioxide fixation by marine algae. *Limnology and Oceanography* **16**, 854–8.

MORRIS I., YENTSCH C. M. & YENTSCH C. S. (1971b) The physiological state of phytoplankton from low-nutrient sub-tropical water with respect to nitrogen, as measured by the effect of ammonium on dark carbon dioxide fixation. *Limnology and Oceanography* **16**, 859–68.

MURRAY J. & JJART J. (1912) *The Depths of the Ocean.* MacMillan, London.

PAASCHE E. (1969) Light-dependent cocolith formation in the two forms of *Coccolithus pelagicus* with remarks on the ^{14}C zero-thickness counting efficiency of cocolithophorids. *Archiv für Microbiologie* **67**, 199–208.

PARSONS T. R., TAKAHASHI M. & HARGRAVES B. (1977) *Biological Oceanographic Processes*, 2e. Pergamon Press, Oxford.

PETERSON B. J. (1980) Aquatic primary productivity and the ^{14}C-CO_2 method: A history of the productivity problem. *Annual Review of Ecology and Systematics* **11**, 359–85.

PHILLIPSON J. (1966) *Ecological Energetics* Institute of Biology Studies in Biology No. 1. Edward Arnold, London.

PINGREE R. D., PUGH P. R., HOLLIGAN P. M. & FOSTER G. R. (1975) Summer phytoplankton blooms and red tides along tidal fronts in the approaches to the English Channel. *Nature, London* **258**, 672–7.

PLATT T. (1972a) The feasibility of mapping the chlorophyll distribution in the Gulf of St. Lawrence. *Fisheries Research Board of Canada Technical Report* 332.

PLATT T. (1972b) Local phytoplankton abundance and turbulence. *Deep Sea Research* **19**, 183–7.

PLATT T. & DENMAN K. (1975) Spectral analysis in ecology. *Annual Review of Ecology and Systematics* **6**, 189–210.

PLATT T. & DENMAN K (1980) Patchiness in phytoplankton distribution. In: *The Physiological Ecology of Phytoplankton* (Ed. I. Morris), pp. 413–31. Blackwell Scientific Publications, Oxford.

PLATT T., DICKIE L. M. & TRITES R. W. (1970) Spatial heterogeneity of phytoplankton in a near-shore environment. *Journal of the Fisheries Research Board of Canada* **27**, 1453–73.

PUGH P. R. (1970) Liquid scintillation of ^{14}C-diatom material on filter papers for use in productivity studies. *Limnology and Oceanography* **15**, 652–5.

PUGH P. R. (1973) An evalution of liquid scintillation counting techniques for use in aquatic primary production studies. *Limnology and Oceanography* **18**, 310–19.

RAVEN J. A. (1972) Endogenous inorganic carbon sources in plant photosynthesis. II. Comparison of total CO_2 production in the light with measured CO_2 evolution in the light. *New Phytology* **71**, 995–1014.

RAYMONT J. E G. (1980) *Plankton and Productivity in the Oceans,* 2e. Pergamon Press, Oxford.

RUSSEL-HUNTER W. D. (1970) *Aquatic Productivity.* Collier-MacMillan, London.

RYTHER J. H. (1954) The ratio of phytosynthesis to respiration in marine plankton algae and its effect upon the measurement of productivity. *Deep-Sea Research* **2**, 134–9.

RYTHER J. H. (1956) Photosynthesis in the ocean as a function of light intensity. *Limnology and Oceanography* **1**, 61–70.

RYTHER J. H. (1959) Potential productivity of the sea. *Science* **130**, 602–8.

RYTHER J. H. (1963) Geographic variations in productivity. In: *The Sea* (Ed. M. N. Hill), vol. 2, pp. 347–80. John Wiley & Sons, New York.

RYTHER J. H. (1969) Photosynthesis and fish production in the sea. *Science* **166**, 72–6.

RYTHER J. H. & VACCARO R. F. (1954) A comparison of the oxygen and methods of measuring marine photosynthesis. *Journal du Conseil Permanent International pour L'exploration de la Mer.* **20**, 25–34.

SHARP J. H. (1977) Excretion of organic matter by marine phytoplankton. Do healthy cells do it? *Limnology and Oceanography* **22**, 38–99.

SMITH A. E. & MORRIS I. (1980) Pathways of carbon assimilation in phytoplankton from the Antarctic Ocean. *Limnology and Oceanography* **25**, 865–72.

STEELE J. H. (1961) Primary production. In: *Oceanography*, pp. 519–38. American Association for the Advancement of Science.

STEEMAN NIELSEN E. (1952) The use of radioactive carbon-[^{14}C] for measuring organic production in the sea. *Journal du Conseil Permanent International pour L'exploration de la Mer* **18**, 117–40.

STEEMAN NIELSEN E. (1963) Productivity, definition and measurement. 2. Fertility of the oceans. In: *The Sea* (Ed. M. N. Hill), vol 2, pp. 129–64. John Wiley & Sons, New York.

STEEMAN NIELSEN E. & HANSEN V. K. (1959) Light adaptation in marine phytoplankton populations and its interrelation with temperature. *Physiological Plant Pathology* **12**, 353–70.

STRICKLAND J. D. H. (1965) Production of organic matter in the primary stages of the marine food chain. In: *Chemical Oceanography* (Eds J. P. Riley & G. Skirrow), pp. 477–610. Academic Press, New York.

STRICKLAND J. D. H. & PARSONS T. R. (1968) A practical handbook of seawater analysis. *Fisheries Research Board of Canada Bulletin* **167**, 1–311.

TALLING J. F. (1973) The application of some electrochemical methods to the measurement of photosynthesis and respiration in fresh waters. *Freshwater Biology* **3**, 335–62.

TAYLOR J. P. R. (1980) Basic biological features of phytoplankton cells. In: *Physiological Ecology of Phytoplankton* (Ed. I. Morris), pp. 3–55. Blackwell Scientific Publications, Oxford.

THOMAS W. H. (1964) An experimental evaluation of the ^{14}C method for measuring phytoplankton production, using cultures of *Dunaliella primolecta* Butcher. *Fisheries Bulletin* **63**, 273–92.

TITMAN D. & KILHAM S. S. (1976) Phosphate and silicate growth and uptake kinetics of the diatoms *Asterionella formosa* and *Cyclotella meneghiniana* in batch and semi-continuous culture. *Journal of Phycology* **12**, 375–83.

VOLLENWEIDER R. A. (1969) Primary production in aquatic environments. IBP Handbook **12**. Blackwell Scientific Publications, Oxford.

WATERBURY T. B., WATSON S. W., GUILLARD R. R. L. & BRAND L. E. (1979) Widespread occurrence of a unicellular marine plankton cyanobacterium. *Nature, London* **277**, 293–4.

WESTLAKE G. F. (1963), Comparison of plant productivity. *Biological Review* **38**, 385–425.

WILLIAMS P. J. LEB., BERMAN T. & HOLM-HANSEN O. (1972) Potential sources of error in the measurement of low rates of planktonic photosynthesis and excretion. *Nature, London* **236**, 91–92.

WILLIAMS P. J. LeB. & YENTSCH C. S. (1976) An examination of photosynthetic production, excretion of photosynthetic products, and heterotrophic utilization of dissolved organic compounds with reference to results from a coastal sub-tropical sea. *Marine Biology* **35**, 31–40.

WOOD K. G. (1971) Self absorption corrections for the ^{14}C method with $BaCO_3$ for measurement of primary productivity. *Ecology* **52**, 491–8.

WROBLEWSKI J. S., O'BRIEN J. J. & PLATT T. (1975) On the physical and biological scales of phytoplankton patchiness in the ocean. *Mémoires de la Société Royale Scientifique Liège, 6e ser* **7**, 43–58.

YENTSCH C. S. (1964) Primary production. *Oceanography and Marine Biology Annual Review* **1**, 157–75.

YENTSCH C. S. (1974) Some aspects of the environmental physiology in marine phytoplankton: A second look. *Oceanography and Marine Biology Annual Review* **12**, 41–75.

Part 3
Microbe–Microbe Interactions

Chapter 16 · Microbial Ecology in the Laboratory: Experimental Systems

J. Howard Slater and David J. Hardman

16.1 Introduction

Experimental microbial ecology is a subject area which is difficult to define precisely and, indeed, it frequently generates conflicting attitudes and strategies for research amongst microbial ecologists. In one sense most aspects of the laboratory study of microbial physiology, biochemistry and genetics come within the purview of microbial ecology since, in however limited a way, these branches of microbiology have to account for the microorganisms' environment. Most microbiologists are aware of the effects of environmental conditions on community, population and cellular processes, even though the control or importance of these parameters is rarely considered with sufficient rigour. On the other hand, others see microbial ecology in terms of either the natural history of a range of natural habitats or the activity of microbial populations *in situ*. This spectrum of understanding of the discipline of microbial ecology reflects 'the broad nature of the subject and the diverse, individual disciplines which need to be drawn together' (Ellwood *et al.* 1980). A fundamental base line, however, is that microbial ecology is concerned with the behaviour of microorganisms with respect to the environment in which those organisms either live and grow or happen to be at the time of study. In terms of experimental microbial ecology this definition can be interpreted in two extreme ways.

First there are those microbial ecologists who consider that measurements and data collection can only be made satisfactorily through field studies: that is measurements *in situ*. Satisfactory measurements and information here means accurate in terms of the quantitative behaviour of microorganisms in their natural surroundings. Certainly this is a valid approach, particularly for some processes such as the overall cycling of elements (Jørgensen 1980) or at the complete community level (Fry 1982). However, at the population, cellular and sub-cellular levels, measurements *in situ* are likely to be technically difficult, if not completely impossible with current methods and instrument technology. Furthermore, relating particular microbial changes to particular environmental parameters may be exceedingly difficult because of the heterogeneity of the factors involved and the inability to control environmental variables.

Second other experimentalists attempt to isolate and define precisely, specific microbial systems and activities under laboratory-based conditions.

The advantage of a laboratory approach is that it is possible to control the appropriate parameters and examine the effect of a single parameter on the population under study. However, this is necessarily a reductionist approach and the experimental techniques are likely to eliminate significant factors relevant to the microbial activity under study. It is important to know what parameters are likely to be significant and this requires a thorough understanding of the properties of the original habitat. Even if this preliminary assessment is made accurately and has been thoroughly examined in the laboratory, laboratory-based studies still suffer from the problem of whether or not it is valid to extrapolate the results back to the natural habitat.

Ideally there ought not to be a conflict between these two approaches and probably the only satisfactory way in most instances involves a synthesis of field work and laboratory studies. However, it has to be said that many field studies contribute little to a detailed understanding of the behaviour of microorganisms in their natural habitats in terms of mechanisms, causes and effects. Similarly many laboratory-based ecological studies are worthless because insufficient attention has been paid to the properties of the natural habitats. For example, Jannasch (1967) wrote: 'Although detailed information on physiological and biochemical properties of various marine microorganisms under *laboratory* conditions is available, there are no quantitative data describing growth in nutrient conditions characteristic of their *natural* environment. Natural sea-water is characterised by extremely low nutrient concentrations, compared with media commonly used for studies of bacterial growth. Most results obtained by the classical techniques for assessing microbial activities in such impoverished environments are not amenable to a meaningful interpretation.' It could be added that then, as now, wild extrapolations are unfortunately made despite the insights to these problems by workers such as Jannasch.

With these problems in mind, this chapter is concerned with the problems of and approaches to laboratory-based experimentation relevant to ecological needs, especially with respect to the growth of microbial populations.

16.2 The problems of simulating natural habitats in the laboratory

The primary requirement of laboratory systems is to incorporate features which bear a close resemblance, if they are not a complete replication, to the properties of the natural habitat. Jannasch (1967), as previously noted, drew attention to the type of problem which has to be dealt with: namely a judicious choice of nutrients and their appropriate concentrations. There simply is no point in attempting to examine the physiological properties of an oligotrophic microorganism if it is grown in a nutrient-rich environment.

The traditional approach of the microbial physiologist or biochemist has been to grow 'easy' microorganisms in simple systems in a homogeneous fashion. Normally pure cultures are needed and grown under fully aerobic (or completely anaerobic) conditions to give homogeneous cell suspensions at high densities. Generally these are perfectly acceptable procedures for obtaining adequate quantities of organism for further study in this context. Translating these procedures and methods to ecologically oriented studies is the subject of this chapter since account has to be made of the heterogeneity of almost all natural habitats, a non-uniformity which affects many important properties of natural environments and hence the behaviour of microbial populations.

16.2.1 Mixtures of Microorganisms

As has been discussed in more detail elsewhere (Slater & Bull 1978; Slater & Lovatt 1982), most environments contain many different types of microorganisms, and this includes the so-called extreme environments, such as high-temperature hot springs, salty solar lakes and environments with a low water activity (Kushner 1978, 1980). For example, many different species of the genera *Halobacterium* and *Halococcus* grow together in a solar lake near the Red Sea where the salinity is as high as 15% (w/v) and furthermore the temperature can rise to 56°C (Kushner 1980). In certain alkaline (pH value = 11.0), salt (salinity = 30% w/v) lakes heterotrophic bacteria, photosynthetic bacteria, cyanobacteria, eukaryotic algae and zooflagellates coexist (Imhoff *et al.* 1979). In the 'normal' environments, such as soil or freshwater habitats, there may be many hundreds of different microbial species. It is important to recognise, therefore, that laboratory systems ought to examine the significance of growing mixtures of microorganisms in contrast to the pure cultures traditionally adopted (Bull 1980; Slater 1981). The relationships between different microbial populations may be important in several ways.

Metabolic interactions between different microorganisms

Nutritional interactions showing a mutualistic or commensal relationship occur frequently. In model systems recognition of their importance is crucial in retaining key populations, i.e. organisms with a required capability, within the experimental system which otherwise might be lost or might be unable to function. The various categories of nutritional interaction have recently been summarised in detail (Slater 1981; Bull & Slater 1982) but the following examples serve to illustrate the major categories.

(1) Mixed assemblages of microorganisms may be important in either providing essential growth metabolites, such as amino acids (Nurmikko 1956; Megee *et al.* 1972) or vitamins (Jensen 1957), or removing toxic end-products of metabolism, such as methanol (Wilkinson *et al.* 1974), hydrogen (Bryant *et al.* 1967) or hydrogen sulphide (Gray *et al.* 1973).

(2) The growth kinetic properties of mixed populations compared with axenic cultures may be significant and explain the advantages which appear to be gained by the mixture (Osman *et al.* 1976; Bull & Brown 1979). This is an area which needs to be explored much more fully, particularly with respect to determining the magnitude of the kinetic changes which might be possible and the detailed mechanisms behind these changes.

(3) It is becoming apparent that the transformation of many compounds occurs as a result of the combined activities of more than one microbial population. This can include processes which depend on a cometabolic step, such as the degradation of cycloalkanes (Beam & Perry 1973, 1974), parathion (Daughton & Hsieh 1977) or some halogenated alkanoic acids (Slater & Lovatt 1982). Synergistic metabolism is also important where part of a metabolic sequence is accomplished by one species and other steps by one or more other populations, such as in the cases of the catabolism of diazinon (Gunner & Zuckerman 1968), linear alkylbenzenesulphonates (Johanides & Hršak 1976), chlorinated benzoic acids (Hartmann *et al.* 1979; Reineke & Knackmuss 1979), benzoic acid (Cossar *et al.* 1981) and chlorinated propionamide (J. H. Slater & A. H. Filipiuk, unpublished observations). The last example illustrates the dilemma of devising experimental systems solely on the basis of pure cultures since, from the soil samples examined, it proved impossible to isolate a pure culture directly able to use this substrate as the sole carbon and energy source. (This is not to say that such an organism does not exist in nature in other soil samples or habitats.) A mixture of organisms was isolated and the tentative conclusion was that one organism produced an amidase which acted to yield a chloroalkanoic acid which it could not further metabolise. Other microorganisms, possibly three species, were able to dechlorinate the acid but not the amide. Similar mechanisms are likely to be extremely important in the biodegradation of more complex natural products such as lignin and cellulose. This is a relatively unexplored area although there is some indication that these processes do occur.

(4) The overall rate of biodegradation of compounds may be significantly greater by mixtures of several organisms compared with the sum of the rates of transformation by the individual organisms. For example, this has been demonstrated in the case of the herbicide dalapon where the maximum rate of degradation achieved by a seven-membered community (Senior *et al.* 1976) was 20% greater than the sum of the individual rates by the dalapon-degrading members of the community (Senior 1977).

In all these instances and many others (Slater 1981; Slater & Lovatt 1982), the interactions between the populations which, by and large, have been elucidated from laboratory-based enrichments and analyses, are likely to occur in nature with the same organisms participating. However much more experimental evidence in the natural habitat is required.

Genetic interactions between different microorganisms

In part, of course, the metabolic interactions described reflect differences between the genetic composition of different organisms living in the same habitat and so a form of genetic interaction. Furthermore, this situation could provide the basis for subsequent interactions involving the transfer of genetic information between different organisms. A theoretical example of the evolution of a single organism with a complete metabolic pathway from a two-membered mixed culture is given in Table 1, events which might involve gene transfer mediated by transduction, transformation or plasmid-mediated gene transfer.

Table 1. The selection of a novel metabolic sequence in a single organism from an interacting microbial community.

Pathway

$$A \xrightarrow{a} B \xrightarrow{b} C \xrightarrow{c} D \xrightarrow{d} E$$

Organism	Capability
X	Requires E for growth No growth on A Contains enzymes a and b Intermediate C accumulates; no growth on C
Y	Requires E for growth No growth on A Contains enzymes c and d Growth on C since it can produce E
X + Y	Growth on A Complete pathways; enzymes a, b, c and d present
Z	Growth on A Complete pathway; enzymes a and b transferred from X to Y
Z	Growth on A Complete pathway; enzymes c and d transferred from Y to X

Plasmid-mediated gene transfer has been observed in mixed cultures of two pseudomonads, *Pseudomonas* sp. strain B13 and *Pseudomonas putida* mt-2, in evolving, in chemostat culture, a single organism capable of utilising 3-chloro-, 4-chloro- and 3,5-dichlorobenzoate as the sole carbon and energy source. Neither parent organism alone was able to grow on any of the three compounds (Hartmann *et al.* 1979; Reineke & Knackmuss 1979; Knackmuss 1981). Similarly there is evidence in the 2-chloropropionamide (2CPA) microbial community that after a prolonged period of chemostat growth a simple organism able to use 2CPA as the carbon and energy source evolved, probably as a result of the plasmid-mediated transfer of the amidase gene to the pseudomonad with the dehalogenase function (J. H. Slater & A. H. Filipiuk, unpublished observations). Morrison *et al.* (1978) demonstrated the transfer of streptomycin resistance by transduction using the generalised transducing phage F116 between populations of *Pseudomonas aeruginosa*. Other examples of gene transfer in natural and laboratory environments are given by Slater and Godwin (1980).

Genetic interactions of this sort are clearly of considerable importance in nature and a great deal is known about the basis of these mechanisms. In natural environments, however, much less is known about the significance or frequency of these events; for example, how does the frequency of plasmid transfer under ideal laboratory conditions compare with the frequency of transfer of the same plasmid in a soil environment? Even less is known about the behaviour of these mechanisms in laboratory growth systems, and the effect of factors, such as the organisms' growth rates, on gene transfer rates.

16.2.2 Mixtures of Substrates

Most defined laboratory studies tend to concentrate on growth on a single carbon source with little attention devoted to the effects of the presence of multiple carbon and energy sources. A similar disregard for the presence and availability of different forms of all the basic elements required for growth is also apparent. In natural environments growth on a single carbon and energy source may occur, but microbial populations usually encounter a diversity of potential carbon and energy sources. It has long been known that the availability of two or more carbon sources frequently leads to sequential use of the compounds and characteristic lag phases in the population's growth between the exhaustion of one substrate and the beginning of the use of another—a phenomenon known as diauxie (Monod 1942). Alternatively microbes may use different carbon sources simultaneously (Standing *et al.* 1972) and do not show a diauxic growth response. These differences and the sequence of substrate utilisation reflect differences in individual organism's mechanisms for controlling the operation of catabolic sequences. Therefore, the presence of mixed substrates influences overall microbial physiology and subsequently the behaviour of an organism within its particular niche. A realistic appreciation of a microorganism's activity and function necessitates detailed knowledge of how a microbial population responds to mixtures of substrates. Technically in either closed or open laboratory growth systems these mixed substrate studies simply require appropriate growth media design and will not be discussed in detail in the experimental section. A number of pertinent examples may be found in Harder and Dijkhuizen (1976; see Chapter 20).

The availability of mixed substrates influences processes such as succession (see Chapter 10), and can have major effects on the outcome of interactions between different microbial populations. For example, the availability of different energy sources has a great influence on the outcome of competition between different organisms with slightly different nutritional characteristics, such as the obligate and facultative chemolithotrophic thiobacilli (Kuenen & Gottschal 1982). Furthermore, it is now clear that cometabolic processes play a major role in the transformation of many organic compounds, particularly xenobiotic compounds (Horvath 1972; Alexander 1979; Chou & Bohonos 1979; Slater & Lovatt 1982), processes which lead to important environmental implications, such as novel product formation and the establishment of microbial community structure. In this context the lattice of interactions and their metabolic basis can only be appreciated by studying mixed substrate utilisation by microorganisms, and preferably using mixed culture systems as well.

16.2.3 Growth at Interfaces and in Solute Gradients

Microorganisms in their natural habitats encounter the physico-chemical effects, as well as biologically generated effects, of regions of the interfaces between solids, liquids and gases, in varying combinations depending on the nature of the habitat. At one extreme these effects may lead to discrete compartmentalisation, as, for example, between a substrate which is immiscible with water and the aqueous phase. Alternatively, broad or narrow regions, depending on the individual characteristics of the particular components in question, with concentration gradients may be established providing a continuous spectrum of unique habitats. Interface regions are exceptionally important in ecological studies because of their ubiquity and because the scale of interface regions are comparable to the size of individual microorganisms or colonies of microorganisms. The characteristics of interfaces have been described in detail elsewhere and will not be considered here (Zviaginsev 1973; Burns 1980; Fletcher *et al.* 1980; Marshall 1980; Rutter & Vincent 1980). Water body surfaces have microlayers, usually representing a minute component of the total water volume, which are extremely significant to the microflora and its activity (Norkrans 1979). At these regions high concentrations of bacteria accumulate with a population composition which may be significantly different from that of the remainder of the water body. It represents the region of gaseous exchange between the liquid and the atmosphere with correspondingly sharp gradients across the thin surface microlayer. Furthermore certain compounds also appear to concentrate within the microlayer. Thus it is clear that these regions are substantially different from a submerged region of the water body and similar considerations apply to other interfaces.

A solid surrounded by an aqueous environment generates a surface charge which depends on the nature of the surface and the type of aqueous environment. A charged surface attracts oppositely charged ions from the surrounding environment and repels similarly charged ions leading to the formation of gradients of charged solutes and particles (Fletcher *et al.* 1980). These regions may be significant in creating different microbial microhabitats because of the comparable microbial sizes with the dimensions of these regions. In addition the charges influence the attachment of the bacteria attempting to grow in these heterogeneous gradients, and since most surfaces and bacteria carry a net negative charge, there is a repulsion effect to be overcome. Nevertheless there is now considerable evidence to suggest that in normal, low-nutrient aquatic environments, the metabolic activity of microbes freely suspended in the aqueous phase is so low that they are, effectively, metabolically dormant (Marshall 1980). Thus most metabolically active organisms occur only when growing attached to or close to surfaces where nutrient accumulation permits adequate metabolism. Thus the role of solid surfaces, whether these are organic or inorganic, dead or living plant or animal material, mineral particles, sand grains or man-made structures may be of crucial importance, particularly for oligotrophic microbes. Notwithstanding this, few attempts have been made to examine the physiology or biochemistry of microorganisms growing attached to surfaces, although it is not technically too difficult (see section 16.6.5, p. 266). Furthermore, surface growth may be important in promoting the transfer of genetic information between individuals which are relatively firmly juxtaposed as a consequence of growing on a surface.

Experimental

16.3 Introduction

The introduction to this chapter on laboratory-based systems for growing microbial populations has emphasised some of the fundamental problems which have to be considered in developing suitable systems. The remainder of the chapter will consider these systems, many of which have been applied to ecological-type studies and some which have not but which could have some utility. These descriptions will not generally include detailed information on the source of components or the methods of manufacture—details which may be found in the primary references. In many instances, however, these details are not crucial and subject only to the preference of the experimenter. Thus the aims are to indicate the variety of experimental systems which could be explored and to discuss the advantages or disadvantages of each particular system. Furthermore, this discussion is not exhaustive since it is apparent that the range of systems described with their idiosyncracies is extensive and the result of specific experimental requirements. However, it is hoped that the major types are covered.

Finally, the experimental section is concerned with laboratory-based growth systems and the discussion predominantly excludes such important factors as growth media design (see Chapter 1), methods of population analysis (see Chapter 6), use of suitable recovery media and detailed analysis of mixed populations. Obviously the successful application of these growth systems demands an intelligent analytical programme.

16.4 The choice between batch (closed) and continuous (open) culture systems

Much has been written about the advantages and disadvantages of the two major laboratory growth systems—batch and continuous culture. In many cases there is apparently a vested interest displayed by the competing protagonists which sometimes extends beyond valid argument leading to exaggerated claims. This is particularly true of those who advocate the use of continuous-flow culture systems in ecological studies. In the first place the difficulties of simulating the growth conditions in the laboratory are often underestimated whichever culture system is used. Probably the best that can be justifiably claimed is that the various culture systems have features which may reflect some property of the natural environment's growth conditions. Thus the competing claims between open and closed growth systems are almost irrelevant by comparison with the much greater complexity of natural growth situations.

Batch culture systems are the usual, traditional methods with the advantage of technical simplicity, cheapness and the potential for establishing replicates. The systems are characterised by continuously changing growth conditions (Bull 1974; Pirt 1975) and the need to start with high concentrations of all the growth substrates required for the populations under consideration. These features are often, in comparison with the advantages of continuous-flow culture systems, presented as disadvantages. Certainly batch culture systems favour the selection of microorganisms with r-selection (or zymogenous) properties at the expense of K-selection (or autochthonous) properties (Jannasch 1974; Carlile 1980; Slater & Lovatt 1982). Nevertheless these organisms represent part of the natural microflora and must have some function within their natural environment. One likely difficulty, however, is that the organisms selected during batch culture enrichment are less likely to be the functionally significant organisms, certainly in those environments where nutrient concentrations and substrate availability are low: this common condition of most natural environments means that K-selection organisms probably are more significant but the less likely to be enriched in closed culture systems. Moreover if oligotrophic organisms are selected, their study, in terms of physiological and biochemical properties, in closed culture is likely to produce information which is quantitatively irrelevant compared with the organism growing under natural conditions.

Continuous-flow systems do, however, possess certain properties which reflect ecologically important characteristics. Continuous-flow culture systems of the type known as chemostats were introduced by Monod (1950) and Novick and Szilard (1950) with the fundamental characteristic that microbial growth is restricted by the availability of one known substrate, the growth-limiting substrate. The basic principles of chemostat growth will not be considered in detail in this chapter and may be found elsewhere (Herbert *et al.* 1956; Tempest 1970; Bull 1974; Pirt 1975; Slater 1979; see Chapters 1 and 20). For a number of years it has been argued that the following properties of continuous-flow culture systems enable useful ecological parameters to be modelled (Veldkamp & Jannasch 1972; Veldkamp & Kuenen 1973; Jannasch & Mateles 1974).

(1) For chemostats the medium used in the system must be designed to contain one substrate at a concentration which limits the growth of the microbial population. That is, in batch culture using this medium the growth-limiting substrate would be the first to be exhausted at the end of the closed culture growth cycle. The growth-limiting concentration may be any value below a maximum concentration above which some other nutrient may become the growth-limiting substrate. An important feature of the chemostat system is that this substrate always remains at a growth-limiting concentration during the whole growth time in the apparatus. Most usually the growth-limiting substrate concentration chosen enables high biomass concentrations to be obtained in order to facilitate physiological and biochemical analyses. However, it is possible to use very low substrate concentrations, including those not present at limiting values to obtain very low population densities (Fig. 1).

(2) One consequence of having a growth-limiting substrate and the capacity to alter the rate of supply of fresh medium is that it is possible to grow populations at submaximal values. Normally in batch systems the organisms grow at their maximum specific growth rate for those conditions since for the bulk of the growth period all the substrates are present in considerable excess. In most natural environments it is highly likely that for much of the time populations are growing at rates considerably below their maximum potential. Thus this property of chemostat systems enables a systematic

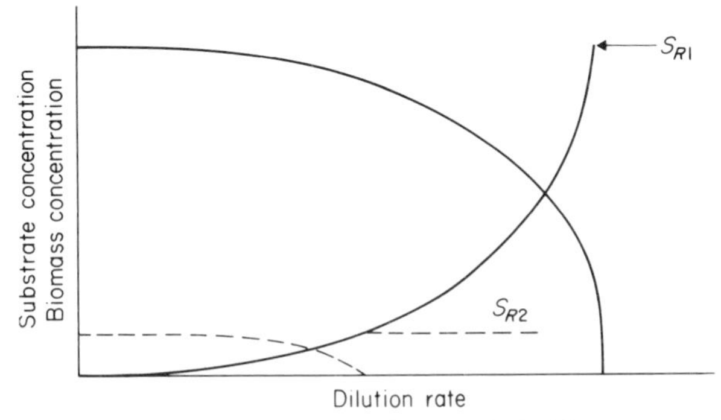

Figure 1. The influence of initial growth-limiting substrate concentration (S_R) on biomass concentration at different dilution rates.

examination on the influences of an organism's growth rate on various physiological parameters. Growth rate has major effects on all microbial activities including its basic macromolecular composition (Herbert 1961); enzyme specific activities (Clarke & Lilly 1969; Dean 1972); enzyme complement (Almengor-Hecht & Bull 1978; Bushell & Bull 1981; Hardman & Slater 1981); and intracellular pool constituents and sizes (Brown & Stanley 1972). In many cases the ecological significance of these physiological characteristics in response to growth rate have not been examined or even appreciated.

(3) The continuous supply of fresh medium at a washout rate coupled with a constant culture volume enables a constant dilution rate (≡ the specific growth rate) to be maintained which, after an appropate equilibrium time, results in the establishment of a steady state culture. That is, it is possible to maintain constant conditions over an indefinite time period which results in a divorce between the long-term effects of environmental changes which occurred due to the past growth conditions of the organisms. This feature is extremely important in biochemical and physiological studies and in certain types of ecological study such as competition (Powell 1958; Slater 1979; see Chapter 20) and prey–predator relationships (Curds & Bazin 1977). It is probably a disadvantage in many other cases because few natural habitats maintain constant physico-chemical or indeed biological conditions for any length of time. Most environments are subject to many transient regimes of various parameters such as light/dark; temperature; nutrient concentrations; and many others. Much remains to be done to devise ways of using continuous-flow culture systems to deal with both regular and irregular transient conditions. Certainly there is scope for such developments and it could lead to an important extension of the range of uses for continuous culture systems. Some attempts have been made to examine light/dark regimes and their associated transient conditions (van Gemerden 1974). It is a technically simple matter to pulse compounds into a chemostat system at regular intervals and examine the transient behaviour of a population immediately after the stepwise introduction of the compound and its subsequent dilution from the system. Continuous-flow culture systems may be used with alternating periods of supply of different nutrients (Lovatt *et al.* 1978; Slater & Lovatt 1982). Analysis of these

experimental procedures and many similar variations to generate transient conditions will prove to be rewarding in the future as important departures from the pasts' over-emphasis of the steady-state culture.

(4) Continuous culture systems provide the best (but not exclusive) method of isolating and examining stable mixed cultures or microbial communities (Daughton & Hsieh 1977; Slater 1978; Slater & Lovatt 1982). Furthermore the system is a good experimental system for looking at many different microbial interactions, such as competition (Godwin & Slater 1979; see Chapter 20).

16.5 The basic continuous-flow culture system

There are many different designs of continuous culture system ranging from simple glass systems, usually home-made, to highly complex manufactured systems with extensive control and monitoring systems. The detailed construction will not be reviewed here and may be found elsewhere (Evans *et al.* 1970). One system which has been widely copied is that of Baker (1968) although other simple systems have been described (Marcus & Halpern 1966; Connor & Wilson 1972). For most ecological studies a primary requirement is normally for a simple system which operates reliably for long periods of time, perhaps up to several months. Clearly the more complex the system exploited, the greater is the probability that system failures will occur: experimental microbial ecologists should not be seduced into using unnecessary monitoring and control systems simply because it is fashionable to use them. In many cases their use is redundant and adds little to the value of the experimental design.

The basic chemostat design requires a culture vessel, usually somewhere between 1 and 3 l working culture volume. Smaller culture vessels are used but often these are beset with the unfavourable effects of wall growth which is more significant with smaller vessels, and evaporation difficulties due to the aeration of a smaller culture volume. There must be a reservoir of fresh medium, usually a 20 l vessel, and a receptacle to collect the waste culture from the culture vessel. A pump, usually a peristaltic flow inducer, is required to transfer fresh medium to the growth vessel and some means (e.g. effluent culture weir) to remove culture and maintain a constant culture volume. Finally, for most cultures, including anaerobic systems, some means of storing the culture to maintain a homogeneous culture is required. Figure 2 illustrates a simple chemostat system.

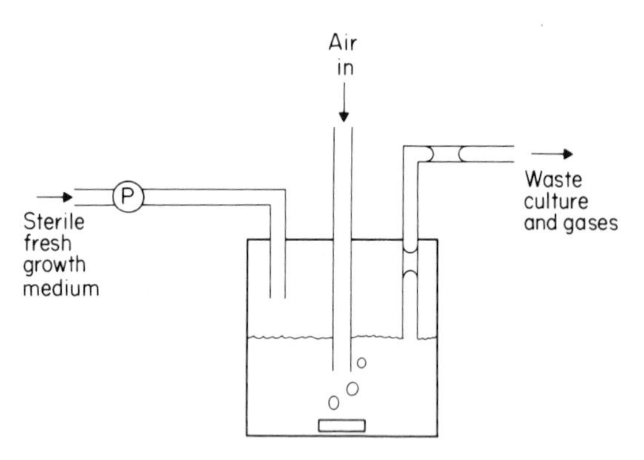

Figure 2. Schematic diagram of a simple continuous-flow culture system.

Continuous-flow culture systems need not be complex, highly sophisticated pieces of equipment, to attain the required characteristics of open growth systems. Indeed, there is usually good reason with the longer term (i.e. of the order of months) operations of many ecologically based experiments to use simple, and consequently more reliable, system designs. It is, however, necessary in many instances to include additional components particularly with respect to monitoring and controlling important parameters such as temperature, dissolved oxygen, redox potential, etc.

The kinetics of growth in chemostat systems may be found elsewhere (Herbert *et al.* 1956; Jannasch 1967; Tempest 1970; Slater 1979).

16.6 Examples of the application of continuous-flow culture systems

16.6.1 The Simple Single-Stage Chemostat

The simple single-stage chemostat system has been widely applied to the enrichment of microorganisms from different habitats under different regimes of enrichment (Jannasch 1967; Veldkamp & Jannasch 1972; Bull 1974; Bull & Brown 1979). Their use will not be discussed at length but it must be recognised that these methods may be particularly valuable for enriching for certain organisms or communities under defined growth conditions (see Chapter 1). For example, the advantage of using continuous culture enrichment is well illustrated with the isolation of *Thiomicrospira denitrificans*, an organism which is very similar, in physiological terms, to *Thiobacillus denitrificans*, but which is not enriched in batch culture systems (Timmer-ten Hoor 1975, 1981; see Chapter 20).

Many variations of the simple, single-stage chemostat have been designed, including tower and column fermenter systems. A modification of potential use in ecological studies may be the air-lift fermenter (Fig. 3) which is widely used in various large-scale industrial fermentations, normally in the batch culture mode, but to date have been rarely used as smaller-scale laboratory systems. This despite the fact that air-lift fermenters have the advantage of a simple construction, may be reliably used and are relatively cheap to construct (Kiese *et al.* 1980). Furthermore they can be easily adapted to continuous culture. Their basic feature is the separation of the culture in the main part of the column with an internal draft tube enabling the circulation of growth medium, culture and gases. In some respects the air-lift fermenters are similar to the successful soil perfusion system (see Chapter 1) and warrant further detailed analysis, particularly with respect to the inclusion of solid substrates or inert surfaces (see section 16.6.5, p. 266). Kiese *et al.* (1980) noted that the method of bottom aeration ensured that solid metal sulphides were easily kept in suspension during the growth of a *Thiobacillus* species.

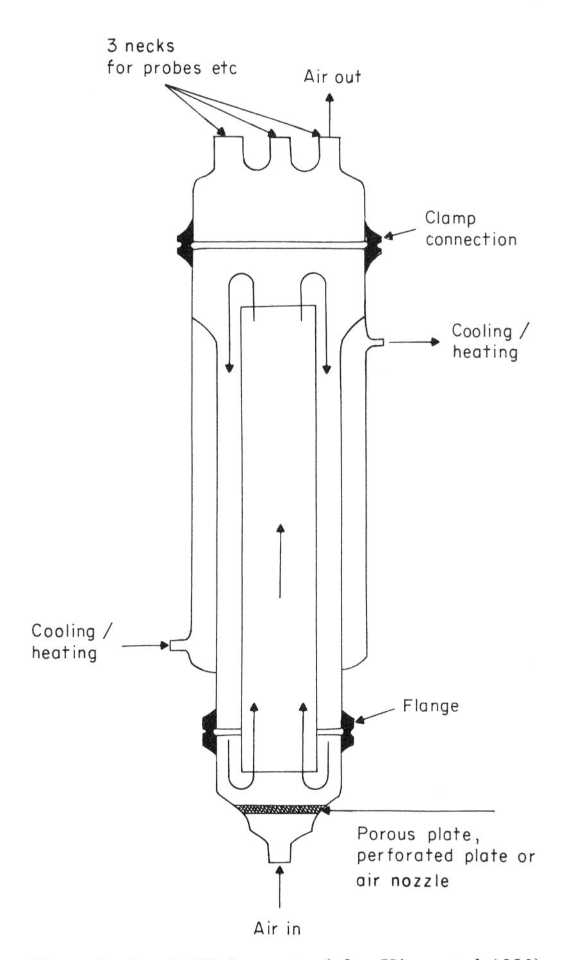

Figure 3. An air-lift fermenter (after Kiese *et al.* 1980).

16.6.2 Single-Stage Chemostat with Organism Feedback

An open growth system with some means of retaining organism within the growth vessel or, alternatively, a mechanism for returning part or all of the biomass leaving the growth vessel, is termed a chemostat with feedback (Pirt & Kurowski 1971). Retention of organisms within the culture vessel may be achieved by some kind of filter or baffle system to separate organisms from the waste liquid (nutrient) component of the growth system (Fig. 4). Alternatively there may be some system, such as a sedimentation chamber (Fig. 4), separating the organisms and a pump system to return them to the growth vessel. The detailed growth kinetics of this system have been developed (Pirt 1969) and tested (Pirt & Kurowski 1971). The system has principally been used for processes which require elevated microbial process rates (compared with single-stage systems), such as in effluent disposal plants and systems dealing with toxic waste materials. There is little evidence that the system has been applied to ecological studies: however, feedback growth systems are analogous to many natural habitats where growing organisms are retained within a defined locality while simple nutrients perfuse through that environment. In most natural situations, retention of organisms is due to attachment on surfaces (see section 16.2.3, p. 259) and this may confer important physiological characteristics on the attached population. However, chemostats enable a homogeneous population to be grown in suspension subjected to the influence of retaining organism biomass. Thus, it should be possible to exploit these systems to analyse the behaviour of organisms growing under these con-

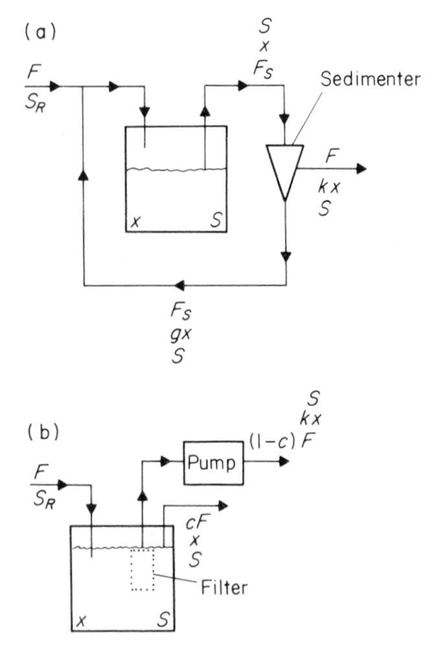

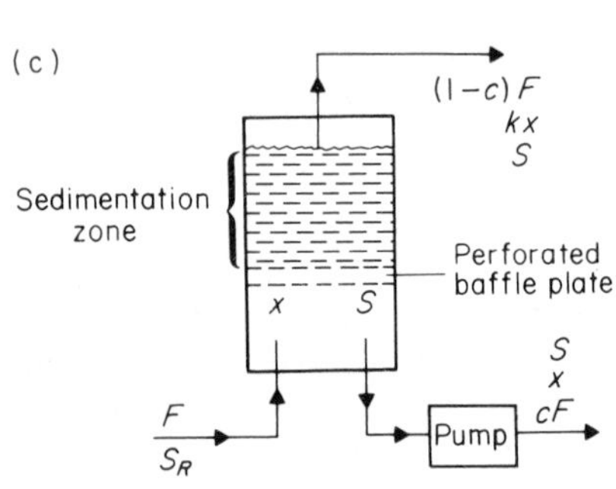

Figure 4. Single-stage chemostats with various methods of organism feedback. F, medium flow rate; F_S, flow rate of culture from culture vessel; c, g, k are constants; S, growth-limiting substrate concentration; S_R, initial growth-limiting substrate concentration; x, biomass concentration. (a) System with an external sedimentation system to concentrate biomass before returning it to the fermenter. (b) System with an internal filter which removes biomass before spent culture liquid is removed from the culture vessel. (c) System with an internal sedimentation mechanism (after Pirt & Kurowski 1971).

ditions, just as homogeneous, single-stage chemostat systems have provided useful, ecologically relevant information. Further application of feedback systems to ecological studies is warranted. The production of cellulases by *Trichoderma reesei* has been examined in continuous-flow culture with cell recycle (Ghose & Sahai 1979) and it was shown that the procedure resulted in as much as a 33% increase in the rate of cellulolysis. Although this study was designed from a biotechnological standpoint, it is most probable that organism recycle mimics the behaviour of cellulolytic fungal populations growing on a solid substrate in nature. Thus, the growth kinetics, physiology and biochemical properties of the fungus may be better simulated in open growth systems with feedback. One advantage of this particular system is that it enables the production of a homogeneous cell suspension which may be used for biochemical studies, a procedure which may be precluded from many heterogeneous systems with organisms growing on surfaces.

16.6.3 Continuous-Flow Culture Systems in Series-Staged Systems

It is possible, with the judicious choice of growth conditions and growth medium composition, to use single-stage continuous-flow culture systems to investigate the influence of several parameters simultaneously on population and community growth. For example, various environmental conditions with respect to a chemolithotrophic energy supply (thiosulphate) and an organic energy supply (acetate) have been established by using different, varying concentrations of the two energy sources in the medium supplied to a growth vessel (Kuenen & Gottschal 1982). This single-stage system has then been used to examine competition and survival between an obligate and a facultative chemolithotrophic thiobacillus. However, an alternative approach is to use a series of separate continuous-flow culture systems which may be linked in a variety of ways. Multistage systems result in the transfer of materials—organisms as well as substrates—from one vessel to the next, with the pattern of transfer, and hence the effect the different conditions have on organism growth, depending on the method of linkage of the various units. Column or tower fermenter systems may also be staged; for example Falch and Gaden (1969) have described a stirrer tower fermenter separated into four compartments by horizontal perforated plates which allow some transfer of culture between the compartments whilst retaining a degree of physical separation. Similar systems have been described by Kitai *et al.* (1969) and Páca and Grégr (1976).

The simplest system involves n stages (all homogeneous stirred vessels) supplied with nutrients from one end only, systems which have been termed unidirectional by Wimpenny (1981). A number of modifications to this basic system can be envisaged, such as the introduction of additional media supplies to any of the stages or the challenging of any of the populations with a toxic compound. Presumably the behaviour of the populations will depend on the physiological state of the organisms in the different stages. A further modification is to make the system bidirectional by the simultaneous supply and counterflow of different nutrients from each end. In practice, such a system has been achieved by Lovitt and Wimpenny (1979, 1981) and has been termed a gradostat. The gradostat has been used to investigate the physiology of mixtures of microorganisms, as well as monocultures, growing in opposing gradients of selected solutes. Clearly the capacity to establish steady state gradients of different nutrients down the various stages of the gradostat is relevant to many natural habitats, although the full potential of the system has yet to be explored. One difficulty with the system is its relative complexity and it may well be that the full utility of systems such as this needs the development of reliable and accurate small-scale systems: such systems, which will also be important in establishing the ability to set up replicate cultures and enrichments, are becoming available.

Lovitt and Wimpenny (1981) have examined the growth of *Paracoccus denitrificans* in opposing gradients of nitrate and succinate as well as *Escherichia coli* growing in gradients of oxygen (aerobic respiration) and glucose with nitrate (anaerobic respiration). A similar system had previously been described by Cooper and Copeland (1973). In common with Lovitt and Wimpenny's system, the continuous-series estuarine micro-ecosystem contained five stages with a freshwater medium entering at one end and salt water medium at the other. Six of these systems were used to provide nutrient exchange and overall retention characterisation similar to those found in Trinity Bay, Texas. Parameters such as previous product and overall community respiration were analysed as a function of changes in the freshwater input to the system. For example, it was demonstrated that these parameters were dependent on the freshwater contact of the vessels closest to the freshwater input end of the assay, whereas they were independent of the freshwater complement at the high salinity end of the multistage system. Under conditions of normal freshwater flow the majority of the microbial activity was due to heterotrophs, whereas under simulated drought conditions, autotrophs became the dominant member of the community. Addition of industrial effluent affected the freshwater populations more than the salt water ones and resulted in significant decreases in species diversity, primary production and respiration. It is, however, important to note that Cooper and Copeland (1973) cautioned that although the composition of the micro-ecosystem's microflora was qualitatively similar, the quantitative distribution was very dissimilar to the Trinity Bay distribution. It is important not to assume that an increase in complexity of laboratory-based experiments systems increases the similarity with the natural environment.

Systems based on these principles have also been used to examine the relationship between methanogenic and sulphate-reducing populations in anoxic environments (Senior *et al.* 1980), but to date there has been little detailed examination of these interesting systems.

Staged systems may also be used to set up simple food chains and examine trophic structure; for example, Lampert (1976) described a system in which algae (*Scenedesmus acutus*) were continuously grown in a chemostat whose effluent was passed through a diluting system before entering an assay of experimental vessels containing a population of *Daphnia pulex*. Experiments were undertaken either with a range of replicate experimental feeding vessels with the same rate of supply and concentration of the alga or with different algal feeding rates.

Staged systems may be used to examine the production of specific cellular components and the influence of environmental conditions on the process. Jensen (1972), for example, examined protease production by *Bacillus subtilis* in a two-stage fermenter system, and gave detailed information on the construction of such systems. In this study the second stage was used to increase the overall retention time of the complete system (compared with a single-stage system) with a reasonably fast throughput of growth medium. Thus the second stage served as a region of little active growth, conditions under which protease production was maximised. Clearly, staged systems can be adapted to examine the behaviour of populations growing at extremely low growth rates which may be characteristic of many natural habitats.

16.6.4 Cage or Dialysis Cultures

As has been previously remarked, one of the unsatisfactory features of laboratory-based open growth systems is that none appear to reflect quantitatively the distribution and behaviour of natural microbial communities. Some attempts have been made to grow defined or unknown populations of organisms within confined and biologically isolated environments within the natural habitat. These are known as cage or dialysis cultures and are most effective for aquatic systems. The microorganism under study is retained in a compartment which is open to the dissolved nutrients of the surrounding environment as well as the metabolic products of the natural flora and fauna (Jensen 1980; Skipnes *et al.* 1980). These systems have been used to determine the growth potential of marine phytoplankton and examine the effect of heavy metal pollution on the population *in situ* (Eide & Jensen 1979). The important principle is that the experimental population must have access to the nutrients available to the normal microflora. In the simplest version of this culture system, the population is simply enclosed in dialysis tubing and suspended in the aquatic system under study (Jensen 1980; Gowland & Slater 1982) although this technique can be considerably modified (Morrison *et al.* 1978). A more complex system is to mount the dialysis bags on a rotor which may be suspended in the water and, together with glass beads enclosed in the bags, the rotation ensures good mixing and elimination of solute diffusion gradients across the dialysis bags (Fig. 5).

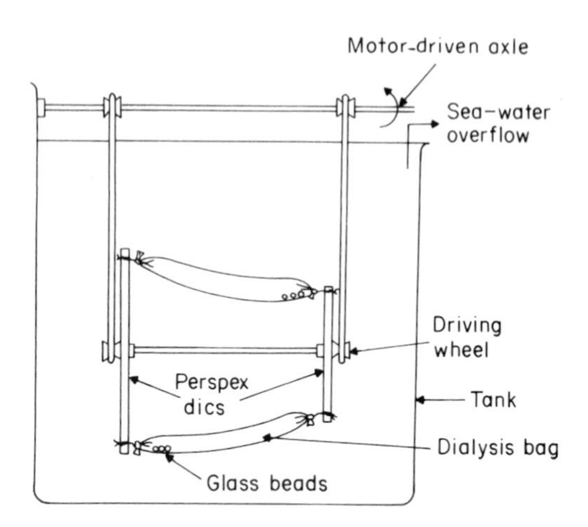

Figure 5. A dialysis cage culture system for use in running water (after Jensen 1980).

More conventional dialysis culture systems have been used to study the growth of populations which are sensitive to their own metabolic products (Pan & Umbreit 1972) and to examine the nature of interactions between microbial populations (Harrison *et al.* 1976). A typical system is shown in Fig. 6.

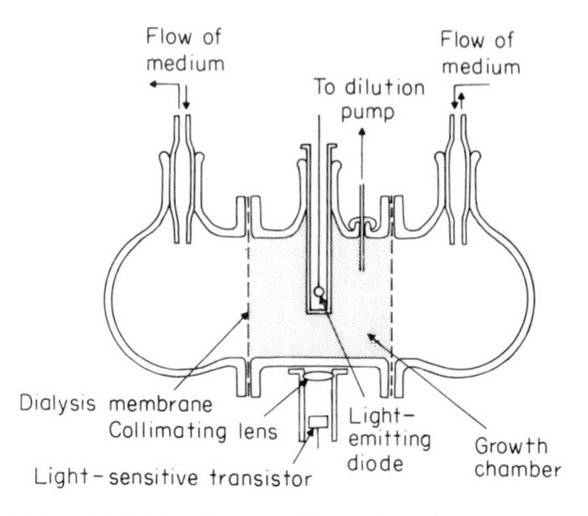

Figure 6. A dialysis cage culture adapted for use as a turbidostat (after Jensen 1980).

A simple growth kinetic analysis is given by Jensen (1980), specifically accounting for the development of gradients across the dialysis membranes. However, more detailed theoretical and experimental analysis of these systems is required in order to fulfil their obvious usefulness as laboratory-based or field growth systems. In other studies, attempts have been made to modify chemostat systems in order to use conventional apparatus in the field (de Noyelles & O'Brien 1974).

16.6.5 Growth on Surfaces

Growth on surfaces, whether these are inert non-growth substrates or solid growth substrates, is a major feature of all natural environments, including aquatic systems (Marshall 1980; see section 16.2.3). The reasons for and mechanisms of microbial attachment to surfaces have been discussed in detail elsewhere (Marshall 1968, 1980; Marshall *et al.* 1971; Berkeley *et al.* 1980). Growth at surfaces occurs either as films of varying thickness or as flocs centred on a particle which may be suspended in the aqueous phase. The growth properties and characteristics have been discussed in detail elsewhere (Atkinson 1974; Atkinson & Fowler 1974). With few exceptions, such as the nitrification processes (Bazin & Saunders 1973), little attention has been paid to modelling growth on surfaces even though suitable (or adaptable) systems have been developed in other contexts.

Soil perfusion flasks which contain a column of particles, usually soil particles, represent a commonly used, closed-culture system enabling surface growth effects to be considered (Lees & Quastel 1946; Lynch & Poole 1979; see Chapter 1). Macura (1961) developed a continuous-flow variation of the perfusion flask system which is widely used and extensively modified for other systems than for studying the microbiology of soil (Fig. 7). The column, which may contain soil, glass beads or other inert material such as plastic strips, is continuously perfused with fresh growth medium from the top downwards and is aerated from the base of the column. In the original system a constant flow rate was achieved by flow through an appropriate bore capillary tube from a Mariotte

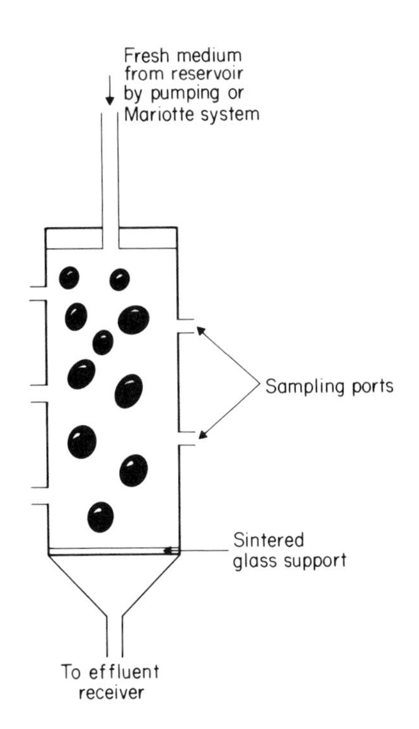

Figure 7. A Macura column (after Macura 1961).

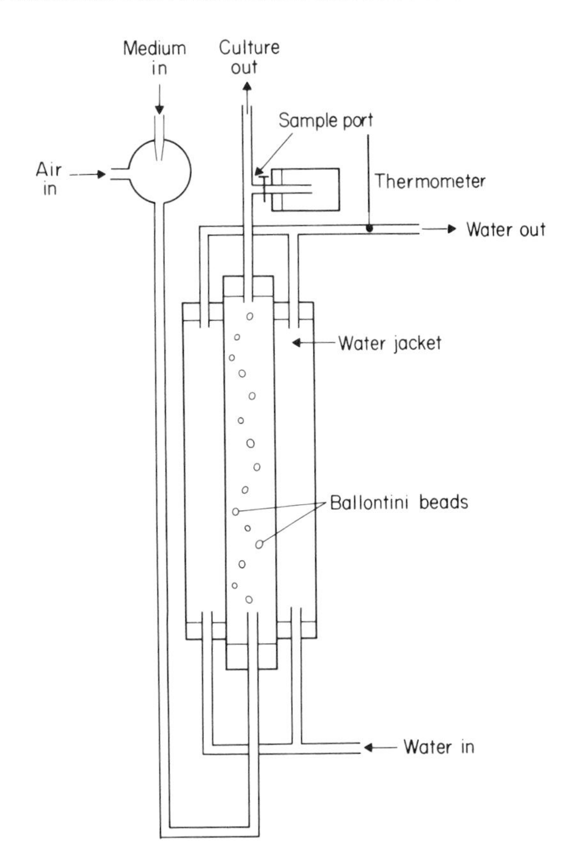

Figure 8. A continuous-flow culture system for growing organisms attached to surfaces (after Gowland & Slater 1982).

bottle fitted with a constant head device. For simple systems this is entirely adequate but it is now more usual to maintain the flow of fresh growth medium by a flow inducer pump. A convenient modification of the system has been developed by Gowland and Slater (1982), shown in Fig. 8. These systems are characterised by the maintenance of the surface within the culture system: there is no continuous addition of fresh surfaces although some or all of the surfaces may be replaced at chosen intervals. It is, however, possible to devise a system to add continuously a particular substrate or inert particulate material to the culture vessel. Owens and Evans (1972), for example, described the construction of an apparatus for dosing a fermenter, semi-continuously, with slurries (Fig. 9). The important feature is the pinch valve system, usually operated by a solenoid valve, which periodically opens for a prescribed period of time to allow the addition of an appropriate volume of slurry into the growth vessel. Simple microprocessor control units may be readily constructed to provide a range of times for valve opening and length of valve-open state: this is a facility which enables effectively the continuous addition of particulate materials at a range of flow rates to be achieved. Similar results may be obtained, for example, for the continuous addition of cellulose-containing growth medium or the addition of particulate matter when growing rumen organisms in continuous culture (Slyter 1975).

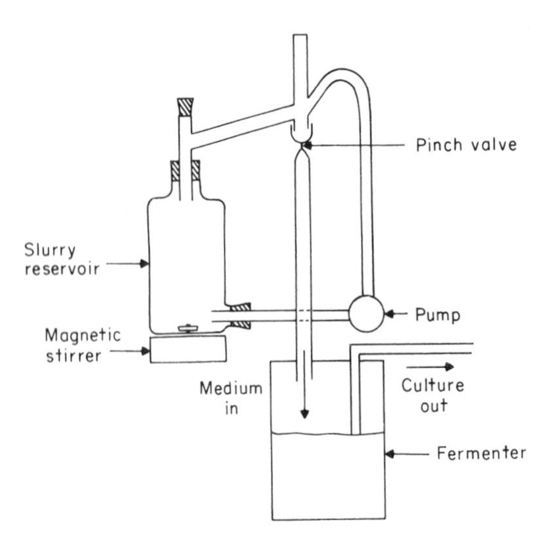

Figure 9. A simple system for continuously dosing a culture with a particulate substrate medium or a medium containing inert particles.

Other workers have, however, found that it is feasible to use normal flow inducers to transfer particle-containing medium from a reservoir to the growth vessel. Jannasch and Pritchard (1972) examined the effect of kaolinite particles, at a concentration of 9×10^5 particles ml^{-1}, on the outcome of enrichment for marine microorganisms growing on valeric acid as the sole carbon and energy source. In the presence of kaolinite, a different organism was enriched compared with that which predominated under identical enrichment conditions in the absence of kaolinite. Similarly Maigetter and Pfister (1975) examined the effect of the presence or absence of kaolinite on the interaction between *Chromobacterium lividium* and a *Pseudomonas* species in a single-stage chemostat system involving the continuous addition of kaolinite. Generally, however, this approach is fraught with problems: the particles tend to sediment in medium supply lines leading to blockages and, more seriously, the liquid and solid phases move at different rates leading to non-steady-state conditions in the growth vessel.

In the case of materials such as glass beads or strips of plastic, retention is not difficult to achieve. Indeed, conventional chemostats may be easily adapted to include suitable materials; for example, Brown *et al.* (1977) and Wardell *et al.* (1980) have examined the influence of the nature of the growth-limiting substrates (glucose, glycerol, acetate or nitrogen) and the growth rate on the selection of surface growth on glass microscope slides or aluminium foil strips contained within a conventional homogeneous chemostat system.

In general, the retention of particles within a culture system does not present problems and is achieved by the use of appropriate retaining filters, such as scintered glass discs. It is generally considered that a more serious problem lies in maintaining the particles in suspension and many microbial film fermenters used in industry operate on the fluidised bed principle (Atkinson & Knights 1975). The key operating principle is to have a second pump operating at an appropriate flow rate to recirculate the growing culture and provide the necessary force to support the particles within the main column. Normally the recirculation rate is such that the contents are completely mixed providing the same conditions throughout the fermenter, in terms of particle density. Their utility in terms of ecological studies has not been evaluated but they can clearly be adapted to provide heterogeneous particle densities, different packing values and other differences which may well be suitable models of soil and sediment systems.

16.6.6 TURBIDOSTATS

As we have observed elsewhere in this chapter, there are many variations on the basic open growth system. With the various chemostat systems described in the previous sections the growth rates of the populations are either directly or indirectly controlled by an external factor, namely the availability of a growth-limiting substrate. An alternative, major group of open growth systems depend on an internal control where the microorganisms' growth rates are determined by their own inherent properties: that is, the rate of growth is dictated by the organisms' own maximum capacity to grow. Open growth systems which allow this form of growth control are known as turbidostats and steady-state conditions are achieved by controlling the population density rather than growth rate (Myers & Clark 1944; Munson 1970; Watson 1972). Normally, turbidostats operate via some means of continuously monitoring the population density and using the signal generated to control the rate of supply of fresh growth medium to the culture vessel. It could, however, be by any parameters under internal control, such as a metabolic product, carbon dioxide production, acid (or alkaline) production, etc. This would obviate the need for the proliferation of terms such as pH-stat or carbon dioxide-stat (Watson 1972). Whatever parameter is chosen for control, the principle is the same: an organism (population) generated parameter is maintained at a constant value through appropriate monitoring. Ferrara *et al.* (1975) have described in detail a turbidostat with improved methods of removing samples, preventing wall growth problems and maintaining its automatic operation. This latter point includes a detailed description of the electronic system for the turbidometric control. The system was tested by growing *Euglena gracilis* as a test organism. A similar system has been described by Wardley-Smith *et al.* (1975) for the continuous growth of a luminous bacterium *Photobacterium phosphoreum* in a system which was termed a luminostat and which was controlled by measuring the culture absorbance and the luminescence.

16.7 Microcosms

The last ten years has seen a developing interest in much more complex laboratory systems, generally known as microcosms, particularly with respect to their use in the study of the environmental fate of potentially hazardous materials, such as pesticides. Microcosms are perhaps best defined as experimental systems which attempt to mimic natural environments in the laboratory as closely as possible—and certainly more closely than the systems described previously in this chapter. These systems may then be used to examine quantitatively the fate of compounds in the controllable system in the laboratory. It must be noted that there is much controversy over microcosm definition, some even including simple batch shake flasks within the definition (Matsumura 1979). Much more important than the semantic argument is the probability that, however carefully microcosms are established, they do not reflect natural environments. This needs a systematic investigation but it is almost certain that the claims of their reproducibility and exactness to natural environments are false.

Many systems have been described (Metcalf *et al.* 1971; Isensee *et al.* 1973; Cole *et al.* 1976; Bourquin *et al.* 1977; Giddings & Eddlemon 1977; Giddings *et al.* 1979; Pritchard *et al.* 1979) and only an example will be illustrated here. Pritchard *et al.* (1979) have described a simple system, termed an eco-core, to mimic the sediment–water interface (Fig. 10). These are established by obtaining intact and, it is hoped undisturbed, sediment cores which

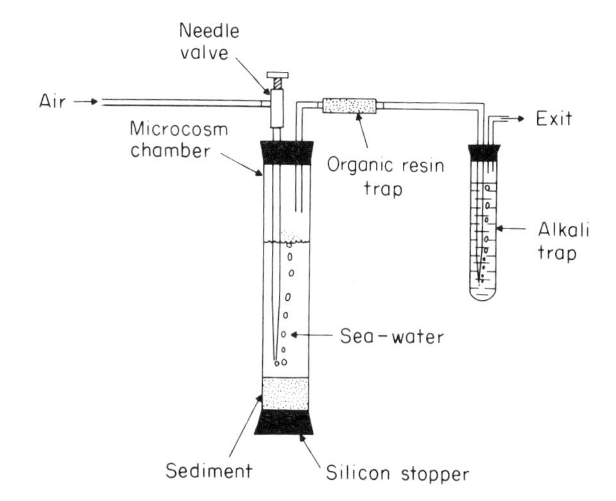

Figure 10. A simple eco-core microcosm (after Pritchard *et al.* 1979).

are transferred to the culture vessel and carefully overlayed with natural water or growth medium as appropriate. The system may be used as a closed system (Fig. 10) or simply adapted for open culture use by continuously perfusing fresh water over the core's surface. Pritchard *et al.* (1979) examined the rate of methylparathion degradation in these systems and observed that the rate of degradation in the eco-core was higher than the rate of methylparathion degradation in cores taken directly from the environment. This probably indicates that disturbing the sediments, however carefully they are removed, does substantially affect the microbial community structure. Nevertheless, microcosm studies may be a valuable adjunct to simpler laboratory studies and field studies.

Recommended reading

BULL A. T. (1974) Microbial growth. In: *Companion to Biochemistry* (Eds A. T. Bull, J. R. Lagnado, J. O. Thomas & K. F. Tipton), pp. 415–42. Longman, London.

BULL A. T. (1980). Biodegradation: some attitudes and strategies of microorganisms and microbiologists. In: *Contemporary Microbial Ecology* (Eds D. C. Ellwood, J. N. Hedger, M. J. Latham, J. M. Lynch & J. H. Slater), pp. 107–36. Academic Press, London.

JANNASCH H. W. & MATELES R. I. (1974). Experimental bacterial ecology studied in continuous culture. *Advances in Microbial Physiology* **11**, 165–212.

KNACKMUSS H-J. (1981) Degradation of halogenated and sulfonated hydrocarbons. In: *Microbial Degradation of Xenobiotics and Recalcitrant Compounds* (Eds R. Hutter & T. Leisinger). Academic Press, London. (In press.)

MONOD J. (1950) The technique of continuous culture. Theory and applications. *Annals Institute Pasteur, Paris* **79**, 390–410.

PIRT S. J. (1975) *The Principles of Microbe and Cell Cultivation.* Blackwell Scientific Publications, Oxford.

SLATER J. H. (1979) Population and community dynamics. In: *Microbial Ecology—A Conceptual Approach* (Eds J. M. Lynch & N. J. Poole), pp. 45–63. Blackwell Scientific Publications, Oxford.

VELDKAMP H. & JANNASCH H. W. (1972). Mixed culture studies with the chemostat. *Journal of Applied Chemistry and Biotechnology* **22**, 105–23.

References

ALEXANDER M. (1979) Role of cometabolism. In: *Microbial Degradation of Pollutants in Marine Environments* (Eds A. W. Bourquin & P. H. Pritchard), pp. 67–75. United States Environmental Protection Agency, Gulf Breeze.

ALMENGOR-HECHT L. & BULL A. T. (1978) Continuous-flow enrichment of strains of *Erwinia caratovora* having high specificity for highly methylated pectin. *Archives of Microbiology* **119**, 163–6.

ASHBY R. E. & BULL A. T. (1977) A column fermenter for the continuous cultivation of microorganisms attached to surfaces. *Laboratory Practice* **26**, 327–9.

ATKINSON B. (1974) *Biochemical Reactors.* Pioneer Press, London.

ATKINSON B. & FOWLER H. W. (1974) The significance of microbial film in fermenters. In: *Advances in Biochemical Engineering Vol. 3* (Eds T. K. Ghose, A. Fiechter & N. Blakebrough), pp. 221–77. Springer-Verlag, Berlin.

ATKINSON B. & KNIGHTS A. J. (1975). Microbial film fermenters: their present state and future applications. *Biotechnology and Bioengineering* **17**, 1245–67.

BAKER K. (1968). Low cost continuous culture apparatus. *Laboratory Practice* **17**, 817–24.

BAZIN M. J. & SAUNDERS P. T. (1973) Dynamics of nitrification in a continuous flow system. *Soil Biology and Biochemistry* **5**, 531–43.

BEAM H. W. & PERRY J. J. (1973) Co-metabolism as a factor in microbial degradation of cycloparaffinic hydrocarbons. *Archiv für Mikrobiologie* **91**, 87–90.

BEAM H. W. & PERRY J. J. (1974) Microbial degradation of cycloparaffinic hydrocarbons via cometabolism and commensalism. *Journal of General Microbiology* **82**, 163–9.

BERKELEY R. C. W., LYNCH J. M., MELLING J., RUTTER P. R. & VINCENT B. (1980) *Microbial Adhesion to Surfaces.* Ellis Horwood, Chichester.

BOURQUIN A. W., HOOD M. A. & GARNAS R. L. (1977) An artificial microbial ecosystem for determining effects and fate of toxicants via salt-marsh environment. *Developments in Industrial Microbiology* **18**, 185–91.

BROWN C. M., ELLWOOD D. C. & HUNTER J. R. (1977) Growth of bacteria at surfaces: influence of nutrient limitation. *FEMS Microbiology Letters* **1**, 163–6.

BROWN C. M. & STANLEY S. O. (1972) Environment-mediated changes in the cellular content of the 'pool' constituents and their associated changes in cell physiology. *Journal of Applied Chemistry and Biotechnology* **22**, 363–89.

BRYANT M. P., WOLIN E. A., WOLIN M. J. & WOLFE R. S. (1967) *Methanobacillus omelianskii*, a symbiotic association of two species of bacteria. *Archiv für Mikrobiologie* **59**, 20–31.

BULL A. T. (1974) Microbial growth. In: *Companion to Biochemistry* (Eds A. T. Bull, J. R. Lagnado, J. O. Thomas & K. F. Tipton), pp. 415–42. Longman, London.

BULL A. T. (1980) Biodegradation: some attitudes and strategies of microorganisms and microbiologists. In: *Contemporary Microbial Ecology* (Eds D. C. Ellwood, J. N. Hedger, M. J. Latham, J. M. Lynch & J. H. Slater), pp. 107–36. Academic Press, London.

BULL A. T. & BROWN C. M. (1979) Continuous culture applications to microbial biochemistry. In: *Microbial Biochemistry, International Review of Biochemistry* (Ed. J. R. Quayle), vol. 21, pp. 177–226. University Park Press, Baltimore.

BULL A. T. & SLATER J. H. (1982) Microbial interactions and community structure. In: *Microbial Interactions and Communities* (Eds A. T. Bull & J. H. Slater). Academic Press, London. (In press.)

BURNS R. G. (1980) In: *Microbial Adhesion to Surfaces* (Eds R. C. W. Berkeley, J. M. Lynch, J. Melling, P. R. Rutter & B. Vincent). Ellis Horwood, Chichester.

BUSHELL M. E. & BULL A. T. (1981) Anaplerotic metabolism of *Aspergillus nidulans* and its effect on biomass synthesis in carbon limited chemostat. *Archives of Microbiology* **128**, 282–7.

CARLILE M. J. (1980) From prokaryote to eukaryote: gains and losses. In: *The Eukaryotic Microbial Cell* (Eds G. W. Gooday, P. Lloyd & A. P. J. Trinci), pp. 1–40. Cambridge University Press, Cambridge.

CHOU T-W. & BOHONOS N. (1979) Diauxic and cometabolic phenomena in biodegradation evaluations. In: *Microbial Degradation of Pollutants in Marine Environments* (Eds A. W. Bourquin & P. H. Pritchard), pp. 76–88. United States Environmental Protection Agency, Gulf Breeze, Florida.

CLARKE P. H. & LILLY M. D. (1969) The regulation of enzyme synthesis during growth. In: *Microbial Growth* (Eds P. M. Meadow & S. J. Pirt), pp. 113–59. Cambridge University Press, Cambridge.

COLE L. J., METCALF R. L. & SANBORN J. R. (1976) Environmental fate of insecticides in terrestrial model ecosystems. *International Journal of Environmental Studies* **10**, 7–14.

CONNOR P. M. & WILSON K. W. (1972) A continuous-flow apparatus for assessing the toxicity of substances to marine animals. *Journal of Experimental Marine Biology and Ecology* **9**, 204–15.

COOPER D. C. & COPELAND B. J. (1973) Responses of continuous-series estuarine microecosystems to point-source input variations. *Ecological Monographs* **43**, 213–36.

COSSAR D., BROWN C. M. & WATKINSON R. J. (1981) Some properties of a marine microbial community using benzoate. *Society for General Microbiology Quarterly* **8**, 47.

CURDS C. R. & BAZIN M. J. (1977) Protozoan predation in batch and continuous culture. *Advances in Aquatic Microbiology* **1**, 115–76.

DAUGHTON C. G. & HSIEH D. P. H. (1977) Parathion utilization by bacterial symbionts in a chemostat. *Applied and Environmental Microbiology* **34**, 175–84.

DEAN A. C. R. (1972) Influence of environment on the control of enzyme synthesis. *Journal of Applied Chemistry and Biotechnology* **22**, 245–59.

DE NOYELLES F. & O'BRIEN W. J. (1974) The *in situ* chemostat—a self contained continuous culturing and water sampling system. *Limnology and Oceanography* **19**, 326–30.

EIDE I. & JENSEN A. (1970) Applications of *in situ* cage cultures of phytoplankton for monitoring heavy metal pollution in two Norwegian fjords. *Journal of Experimental Marine Biology and Ecology* **37**, 271–86.

ELLWOOD D. C., HEDGER J. N., LATHAM M. J., LYNCH J. M. & SLATER J. H. (1980) *Contemporary Microbial Ecology*. Academic Press, London.

EVANS C. G. T., HERBERT D. & TEMPEST D. W. (1970) The continuous cultivation of microorganisms. II. Construction of a chemostat. In: *Methods in Microbiology* (Eds J. R. Norris & D. W. Ribbons), vol. 2, pp. 277–327. Academic Press, London.

FALCH E. A. & GADEN E. L. (1969) A continuous, multistage tower fermenter. I. Design and performance tests. *Biotechnology and Bioengineering* **11**, 927–43.

FERRARA R., GRASSI S. & DEL CARRATORE G. (1975) An automatic homocontinuous culture apparatus. *Biotechnology and Bioengineering* **17**, 985–95.

FLETCHER M., LATHAM M. J., LYNCH J. M. & RUTTER P. R. (1980) The characteristics of interfaces and their role in microbial attachment. In: *Microbial Adhesion to Surfaces* (Eds R. C. W. Berkeley, J. M. Lynch, J. Melling, P. R. Rutter & B. Vincent), pp. 67–78. Ellis Horwood, Chichester.

FRY J. C. (1982) The analysis of microbial interactions and communities *in situ*. In: *Microbial Interactions and Communities* (Eds A. T. Bull & J. H. Slater). Academic Press, London. (In press.)

GHOSE T. K. & SAHAI V. (1979) Production of cellulases by *Trichoderma reesei* QM 9414 in fed-batch and continuous-flow culture with cell recycle. *Biotechnology and Bioengineering* **21**, 283–96.

GIDDINGS J. M. & EDDLEMON G. K. (1977) The effects of microcosm size and substrate type on aquatic microcosm behaviour and transport. *Archives of Environmental Contamination and Toxicology* **6**, 491–505.

GIDDINGS J. M., WALTON B. T., EDDLEMON G. K. & OLSON K. G. (1979) Transport and fate of autovacene in aquatic microcosms. In: *Microbial Degradation of Pollutants in Marine Environments* (Eds A. W. Bourquin & P. H. Pritchard), pp. 312–20. United States Environmental Protection Agency, Gulf Breeze, Florida.

GODWIN D. & SLATER J. H. (1979) The influence of the growth environment on the stability of a drug resistance plasmid in *Escherichia coli* K12. *Journal of General Microbiology* **111**, 201–10.

GOWLAND P. C. & SLATER J. H. (1982) Transfer and stability of plasmids in *Escherichia coli* K12. *Microbial Ecology*. (In press.)

GRAY B. H., FOWLER C. F., NUGENT N. A., RIGOPOULOS N. & FULLER R. C. (1973) Reevaluation of *Chloropseudomonas ethylica* 2K. *International Journal of Systematic Bacteriology* **23**, 256–64.

GUNNER H. B. & ZUCKERMAN B. M. (1968) Degradation of 'Diazinon' by synergistic microbial action. *Nature, London* **217**, 1183–4.

HARDER W. & DIJKHUIZEN L. (1976) Mixed substrate utilization In: *Continuous Culture 6: Applications and New Fields* (Eds A. C. R. Dean, D. C. Ellwood, C. G. T. Evans & J. Melling), pp. 297–314. Ellis Horwood, Chichester.

HARDMAN D. J. & SLATER J. H. (1981) The dehalogenase complement of a soil pseudomonad grown in closed and open cultures on haloalkanoic acids. *Journal of General Microbiology* **127**, 399–405.

HARRISON D. E. F., WILKINSON T. G., WREN S. J. & HARWOOD J. H. (1976) Mixed bacterial cultures as a basis for continuous production of SCP from C1 compounds. In: *Continuous Culture 6: Applications and New Fields* (Eds A. C. R. Dean, D. C. Ellwood,

C. G. T. Evans & J. Melling), pp. 122–34. Ellis Horwood, Chichester.

HARTMANN J., REINEKE W. & KNACKMUSS H-J. (1979) Metabolism of 3-chloro, 4-chloro- and 3,5-dichlorobenzoate by a pseudomonad. *Applied and Environmental Microbiology* **37**, 421–8.

HERBERT D. (1961) The chemical composition of microorganisms as a function of their environment. In: *Microbial Reaction to Environment* (Eds G. G. Meynell & H. Gooder), pp. 391–416. Cambridge University Press, Cambridge.

HERBERT D., ELSWORTH R. & TELLING R. C. (1956) The continuous culture of bacteria: a theoretical and experimental study. *Journal of General Microbiology* **14**, 601–22.

HORVATH R. S. (1972) Microbial cometabolism and the degradation of organic compounds in nature. *Bacteriological Reviews* **36**, 146–55.

IMHOFF J. F., SAHL H. G., SOLIMAN G. S. H. & TRÜPER H. G. (1979) The Wadi Natrun: chemical composition and microbial biomass developments in alkaline brines of eutrophic desert lakes. *Geomicrobiological Journal* **1**, 219–34.

ISENSEE A. R., KEARNEY P. C., WOOLSON E. A., JONES G. E. & WILLIAMS V. P. (1973) Distribution of alkyl arsenicals in model ecosystem. *Environmental Science Technology* **7**, 841–5.

JANNASCH H. W. (1967) Growth of marine bacteria at limiting concentrations of organic carbon in seawater. *Limnology and Oceanography* **12**, 264–71.

JANNASCH H. W. (1974) Steady state and the chemostat in ecology. *Limnology and Oceanography* **19**, 716–20.

JANNASCH H. W. & MATELES R. I. (1974) Experimental bacterial ecology studied in continuous culture. *Advances in Microbial Physiology* **11**, 165–212.

JANNASCH H. W. & PRITCHARD H. P. (1972) The role of inert particulate matter in the activity of aquatic microorganisms. *Memorie dell' Istituto Italiano di Idrobiologia* **29**, 289–308.

JENSEN A. (1980) The use of phytoplankton cage cultures for *in situ* monitoring of marine pollution. *Rapport et Procès-verbaux des Réunions. Conseil Permanent International pour L'exploration de la Mer* **179**, 322–32.

JENSEN D. E. (1972) Continuous production of extracellular protease by *Bacillus subtilis* in a two stage fermenter. *Biotechnology and Bioengineering* **14**, 647–62.

JENSEN H. L. (1957) Decomposition of chlorosubstituted aliphatic acids by soil bacteria. *Canadian Journal of Microbiology* **3**, 151–64.

JOHANIDES V. & HRŠAK (1976) Changes in mixed bacterial cultures during linear alkyl-benzenesulphonate (LAS) biodegradation. In: *Abstracts of Papers, Fifth International Symposium on Fermentation* (Ed. H. Dellweg), p. 426. Verlag Versuchs- und Lehranstalt für Spiritusfabrikation und Fermentationstechnologie, Berlin.

JØRGENSEN B. B. (1980) Mineralization and the bacterial cycling of carbon, nitrogen and sulphur in marine sediments. In: *Contemporary Microbial Ecology* (Eds D. C. Ellwood, J. N. Hedger, M. J. Latham, J. M. Lynch & J. H. Slater), pp. 239–51. Academic Press, London.

KIESE S., EBER H. G. & ONKEN U. (1980) A simple laboratory airlift fermentor. *Biotechnology Letters* **2**, 345–50.

KITAI A., TONE H. & OZAKI A. (1969) Performance of a perforated plate column as a multistage continuous fermentor. *Biotechnology and Bioengineering* **11**, 911–26.

KNACKMUSS H-J. (1981) Degradation of halogenated and sulfonated hydrocarbons. In: *Microbial Degradation of Xenobiotics and Recalcitrant Compounds* (Eds R. Hutter & T. Leisinger). Academic Press, London. (In press.)

KUENEN J. G. & GOTTSCHAL J. (1982) Competition among chemolithotrophs and their interaction with heterotrophic bacteria. In: *Microbial Interactions and Communities* (Eds A. T. Bull & J. H. Slater). Academic Press, London. (In press.)

KUSHNER D. J. (1978) *Microbial Life in Extreme Environments*. Academic Press, London.

KUSHNER D. J. (1980) Extreme environments. In: *Contemporary Microbial Ecology* (Eds D. C. Ellwood, J. N. Hedger, M. J. Latham, J. M. Lynch & J. H. Slater), pp. 29–54. Academic Press, London.

LAMPERT W. (1976) A directly coupled, artificial two-step food chain for long-term experiments with filter-feeders at consistent food concentrations. *Marine Biology* **37**, 349–55.

LEES H. & QUASTEL J. H. (1946) Biochemistry of nitrification in soil. *Biochemistry Journal* **40**, 803–15.

LOVATT D., SLATER J. H. & BULL A. T. (1978) The growth of a stable mixed culture on picolinic acid in continuous-flow culture. *Society for General Microbiology Quarterly* **6**, 27–9.

LOVITT R. W. & WIMPENNY J. W. T. (1979) The gradostat: a tool for investigating microbial growth and interactions in solute gradients. *Society for General Microbiology Quarterly* **6**, 80.

LOVITT R. W. & WIMPENNY J. W. T. (1981) The gradostat: a bidirectional compound chemostat and its application in microbiological research. *Journal of General Microbiology*. (In press.)

LYNCH J. M. & POOLE N. J. (1979) *Microbial Ecology: a Conceptual Approach*. Blackwell Scientific Publications, Oxford.

MACURA J. (1961) Continuous flow method in soil microbiology. I. Apparatus. *Folia Microbiologica* **6**, 328–35.

MAIGETTER R. Z. & PFISTER R. M. (1975) A mixed bacterial population in a continuous culture with and without kaolinite. *Canadian Journal of Microbiology* **21**, 173–80.

MARCUS M. & HALPERN Y. S. (1966) A simple and

reliable chemostat for enzyme studies. *Israel Journal of Medical Science* **2**, 81–3.

MARSHALL K. C. (1968) Interaction between colloidal montmorillonite and cells of *Rhizobium* species with different ionogenic surfaces. *Biochimica et Biophysica Acta* **156**, 179–86.

MARSHALL K. C. (1980) Bacterial adhesion in natural environments. In: *Microbial Adhesion to Surfaces* (Eds R. C. W. Berkeley, J. M. Lynch, J. Melling, P. R. Rutter & B. Vincent), pp. 187–96. Ellis Horwood, Chichester.

MARSHALL K. C., STOUT R. & MITCHELL R. (1971) Mechanism of the initial events in the sorption of marine bacteria to surfaces. *Journal of General Microbiology* **68**, 337–48.

MATSUMURA F. (1979) Microcosms. In: *Microbial Degradation of Pollutants in Marine Environments* (Eds A. W. Bourquin & P. H. Pritchard), pp. 520–4. United States Environmental Protection Agency, Gulf Breeze, Florida.

MEGEE R. D., DRAKE J. F., FREDRICKSON A. G. & TSUCHIYA H. M. (1972) Studies in intermicrobial symbiosis. *Saccharomyces cerevisiae* and *Lactobacillus casei. Canadian Journal of Microbiology* **18**, 1733–42.

METCALF R. L., SANGHA G. K. & KAPOOR I. P. (1971) Model ecosystem for the evaluation of pesticide biodegradability and ecological magnification. *Environmental Science Technology* **5**, 709–13.

MONOD J. (1942) *Recherches sur la Croissance des Cultures Bactériennes*. Hermann et Cie, Paris.

MONOD J. (1950) The technique of continuous culture. Theory and applications. *Annals Institute Pasteur, Paris* **79**, 390–410.

MORRISON W. D., MILLAR R. V. & SAYLER G. S. (1978) Frequency of F116-mediated transduction of *Pseudomonas aeruginosa* in a freshwater environment. *Applied and Environmental Microbiology* **36**, 724–30.

MUNSON R. J. (1970) Turbidostats. In: *Methods in Microbiology* (Eds J. R. Norris & D. W. Ribbons), vol. 2, pp. 349–76. Academic Press, London.

MYERS J. & CLARK L. B. (1944) Culture conditions and the development of the photosynthetic mechanism. II. An apparatus for the continuous culture of *Chlorella*. *Journal of General Physiology* **28**, 103–12.

NORKRANS B. (1979) Role of surface microlayers. In: *Microbial Degradation in Marine Environments* (Eds A. W. Bourquin & P. H. Pritchard), pp. 201–13. United States Environmental Protection Agency, Gulf Breeze, Florida.

NOVICK A. & SZILARD L. (1950) Experiments with the chemostat on spontaneous mutation of bacteria. *Proceedings of the National Academy of Sciences, U.S.A.* **36**, 708–18.

NURMIKKO V. (1956) Biochemical factors affecting symbiosis among bacteria. *Experientia* **12**, 245–9.

OSMAN A., BULL A. T. & SLATER J. H. (1976) Growth of mixed microbial populations on orcinol in continuous culture. In: *Abstracts of Papers, Fifth International Fermentation Symposium* (Ed. H. Dellweg), p. 124. Verlag Versuchs- und Lehranstalt für Spiritusfabrikation und Fermentationstechnologie, Berlin.

OWENS J. D. & EVANS M. R. (1972) An apparatus for dosing laboratory fermentors with suspensions. *Journal of Applied Bacteriology* **35**, 91–7.

PÁCA J. & GRÉGR V. (1976) Design and performance characteristics of a continuous multistage tower fermenter. *Biotechnology and Bioengineering* **18**, 1075–90.

PAN P. & UMBREIT W. W. (1972) Growth of obligate autotrophic bacteria on glucose in a continuous flow-through apparatus. *Journal of Bacteriology* **109**, 1149–55.

PIRT S. J. (1969) Microbial growth and product formation. In: *Microbial Growth* (Eds P. M. Meadow & S. J. Pirt), pp. 199–221. Cambridge University Press, Cambridge.

PIRT S. J. (1975) *The Principles of Microbe and Cell Cultivation*. Blackwell Scientific Publications, Oxford.

PIRT S. J. & KUROWSKI W. M. (1971) An extension of the theory of the chemostat with feedback of organisms. Its experimental realization with a yeast culture. *Journal of General Microbiology* **63**, 357–66.

POWELL E. O. (1958) Criteria for growth of contaminants and mutants in continuous culture. *Journal of General Microbiology* **18**, 249–68.

PRITCHARD P. H., BOURQUIN A. W., FREDERICKSON H. L. & MAZIARZ T. (1979) System design factors affecting environmental fate studies in microcosms. In: *Microbial Degradation of Pollutants in Marine Environments* (Eds A. W. Bourquin & P. H. Pritchard), pp. 251–72. United States Environmental Protection Agency, Gulf Breeze, Florida.

REINEKE W. & KNACKMUSS H-J. (1979) Construction of haloaromatic utilizing bacteria. *Nature, London* **277**, 385–6.

RUTTER P. R. & VINCENT B. (1980) The adhesion of microorganisms to surfaces: physico-chemical aspects. In: *Microbial Adhesion to Surfaces* (Eds R. C. W. Berkeley, J. M. Lynch, J. Melling, P. R. Rutter & B. Vincent), pp. 79–92. Ellis Horwood, Chichester.

SENIOR E. (1977) Characterisation of a microbial association growing on the herbicide Dalapon. Ph.D. thesis, University of Kent.

SENIOR E., BALBA M. T. M., LINDSTROM E. B. & NEDWELL D. B. (1980) Analysis of interacting anaerobic microbial associations by use of multistage open culture systems. *Society for General Microbiology Quarterly* **8**, 40–1.

SENIOR E., BULL A. T. & SLATER J. H. (1976) Enzyme evolution in a microbial community growing on the herbicide Dalapon. *Nature, London* **263**, 476–9.

SKIPNES O., EIDE I. & JENSEN A. (1980) Cage culture turbidostat: a device for rapid determination of algal growth rate. *Applied and Environmental Microbiology* **40**, 318–25.

SLATER J. H. (1978) The role of microbial communities in the natural environment. In: *The Oil Industry and Microbial Ecosystems* (Eds K. W. A. Chater & H. J. Somerville), pp. 137–54. Heyden & Sons, London.

SLATER J. H. (1979) Population and community dynamics. In: *Microbial Ecology: A Conceptual Approach* (Eds J. M. Lynch & N. J. Poole), pp. 45–63. Blackwell Scientific Publications, Oxford.

SLATER J. H. (1981) Microbial interactions and microbial communities. In: *Mixed Culture Fermentations* (Eds M. E. Bushell & J. H. Slater), pp. 1–24. Academic Press, London.

SLATER J. H. & BULL A. T. (1978) Interactions between microbial populations. In: *Companion to Microbiology* (Eds A. T. Bull & P. M. Meadow), pp. 181–206. Longman, London.

SLATER J. H. & GODWIN D. (1980) Microbial adaptation and selection. In: *Contemporary Microbial Ecology* (Eds D. C. Ellwood, J. N. Hedger, M. J. Latham, J. M. Lynch & J. H. Slater), pp. 137–60. Academic Press, London.

SLATER J. H. & LOVATT D. (1982) The significance of microbial communities in biodegradation. In: *Biochemistry of Microbial Degradation* (Ed. D. T. Gibson). Marcel-Dekker Inc, New York. (In press.)

SLYTER L. L. (1975) Automatic pH control and soluble and insoluble substrate input for continuous culture of rumen organisms. *Applied Microbiology* **30**, 330–2.

STANDING C. N., FREDRICKSON A. G. & TSUCHIYA H. M. (1972) Batch and continuous culture transients for two substrate systems. *Applied Microbiology* **23**, 354–9.

TEMPEST D. W. (1970) The continuous cultivation of microorganisms. I. Theory of the chemostat. In: *Methods in Microbiology* (Eds J. R. Norris & D. W. Ribbons), vol. 2, pp. 259–76. Academic Press, London.

TIMMER-TEN HOOR A. (1975) A new type of thiosulphate oxidizing, nitrate reducing microorganism: *Thiomicrospira denitrificans* sp. nov. *Netherlands Journal of Sea Research* **9**, 344–50.

TIMMER-TEN HOOR A. (1981) Cell yield and bioenergetics of *Thiomicrospira denitrificans* compared with *Thiobacillus denitrificans*. *Antonie van Leeuwenhoek* **47**, 231–43.

VAN GEMERDEN H. (1974) Coexistence of organisms competing for the same substrate: an example among the purple sulphur bacteria. *Microbial Ecology* **1**, 104–19.

VELDKAMP H. & JANNASCH H. W. (1972) Mixed culture studies with the chemostat. *Journal of Applied Chemistry and Biotechnology* **22**, 105–23.

VELDKAMP H. & KUENEN J. G. (1973) The chemostat as a model system for ecological studies. *Bulletin of the Ecological Research Commission Stockholm* **17**, 347–55.

WARDELL J. N., BROWN C. M. & ELLWOOD D. C. (1980) A continuous culture study of the attachment of bacteria to surfaces. In: *Microbial Adhesion to Surfaces* (Eds R. C. W. Berkeley, J. M. Lynch, J. Melling, P. R. Rutter & B. Vincent), pp. 221–30. Ellis Horwood, Chichester.

WARDLEY-SMITH B., WHITE D. C. & LOWE A. E. (1975) The continuous culture of luminous bacteria: a luminostat. *Journal of Applied Bacteriology* **39**, 337–43.

WATSON T. G. (1972) The present state and future prospects of the turbidostat. *Journal of Applied Chemistry and Biotechnology* **22**, 229–43.

WILKINSON T. G., TOPIWALA H. H & HAMER G. (1974) Interactions in a mixed bacterial population growing on methane in continuous culture. *Biotechnology and Bioengineering* **16**, 41–9.

WIMPENNY J. W. T. (1981) Spatial order in microbial ecosystems. *Biological Reviews* (Cambridge Philosophical Society) **56**. (In press.)

ZVIAGINSEV D. Z. (1973) *The Interaction of Microorganisms with Solid Surfaces*. University Press, Moscow.

Chapter 17 • Microbe–Microbe Interactions at Surfaces

J. William Costerton and Kou-Joan Cheng

17.1 Introduction

The direct examination of bacteria growing in natural environments has shown that many of them grow adhering to surfaces in soil (Bae *et al.* 1972) and water (Fletcher & Floodgate 1973; Geesey *et al.* 1977) and adhering to dead or living plants (Akin & Amos 1975; Dazzo & Brill 1977) and animals (Bauchop *et al.* 1975; Cheng *et al.* 1979a). This preference for the adherent mode of growth has been explained in terms of high nutrient concentrations at surfaces (Zobell 1943), the maintenance of favourable nutrient conditions (Cheng *et al.* 1977) or protection from antibacterial agents, such as surfactants (Govan 1975) and phagocytic eukaryotic cells (Schwartzmann & Boring 1971). More recently, quantitative determinations have shown that adhering, sessile bacteria outnumbered planktonic organisms in most aquatic systems (Geesey *et al.* 1978). Once adherent to a surface by means of an exopolysaccharide glycocalyx (Costerton *et al.* 1978), a bacterium in a favourable nutrient situation divides and the daughter cells remain enmeshed in a common exopolysaccharide matrix. Further cell division of these adherent bacterial cells produces a microcolony: these are seen very frequently in direct examinations of adherent bacterial populations where they can be distinguished by the morphological similarity of their component cells (Fig. 1).

Bacterial microcolonies may form on suspended particles (Paerl 1975) or on surface components that detach to become suspended particles (e.g. distal cells of stratified epithelia—Marrie *et al.* 1980). Hydraulic forces operating on thick adherent populations often detach microcolonies that are then described as planktonic (Fig. 2). Traditional microbiological methods are unsuitable for the study of bacteria in this adherent mode of growth, which may be predominant in nature, because they have largely been developed for examination of the single planktonic or motile bacterial cell. Methods for the removal of bacterial cells from surfaces to which they are attached are in their infancy; enumeration by plate counts or MPN methods counts aggregates of bacteria as single colony-forming units, and physiological methods ignore

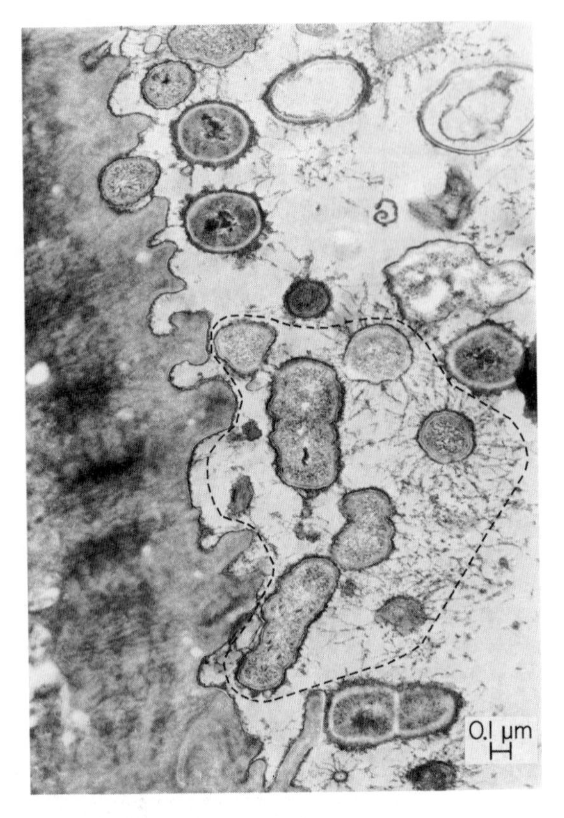

Figure 1. Transmission electron micrograph (TEM) of a section of a ruthenium red-stained preparation of bovine rumen epithelium. Note the formation of glycocalyx-enclosed microcolonies (one delineated by a dotted line) by morphologically similar bacterial cells within this autochthonous adherent population.

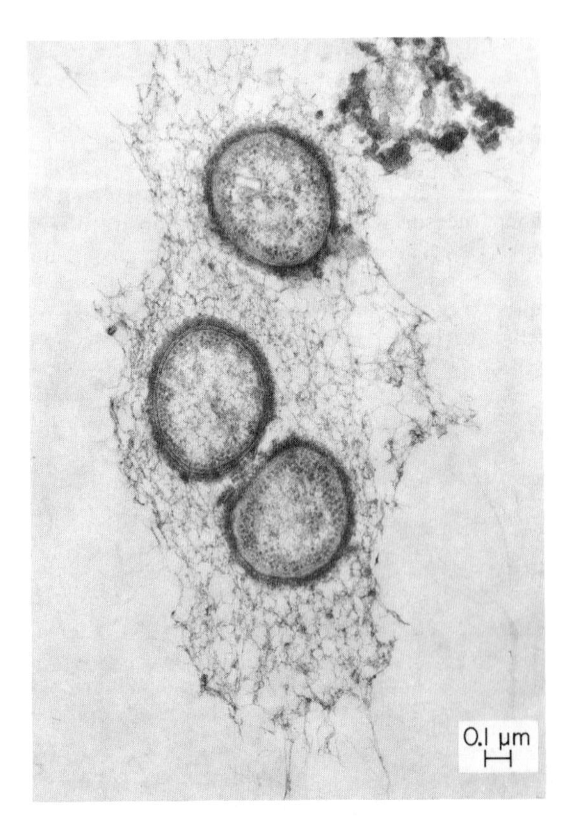

Figure 2. TEM of a section of a ruthenium red-stained preparation of material recovered by filtration from an alpine stream. Note that the morphologically similar bacterial cells in this distinct microcolony are joined to each other, and to some electron-dense detritus, by a fine matrix of ruthenium red-positive glycocalyx fibres.

the ion-exchange characteristics of the copious exopolysaccharide that envelopes surface-attached microcolonies. As isolates from nature are transferred *in vitro*, the classical methods become more applicable because mutants that do not produce extensive exopolysaccharide grow more rapidly and come to predominate (Doggett *et al.* 1964). However, it is often a serious error to extrapolate from studies of these naked strains to the extensive, enclosed microcolonies which are found in nature.

17.1.1 Microbial Microcolonies on Inert Surfaces

The concentration of solutes near surfaces in aquatic environments (Hendrici 1933; Zobell 1943) makes this location seem ideal for bacterial growth, and the sorption of many bacteria to these surfaces, initially by reversible and subsequently by irreversible mechanisms (Marshall *et al.* 1971), has been noted (Fletcher & Floodgate 1973). The aqueous phase of these environments may vary from a clear mountain stream to the oral cavity of a mammal, but the rock surface exposed to the former (Geesey *et al.* 1978) and the enamel surface exposed to the latter (Gibbons & van Houte 1975) are both extensively colonised by adhering bacteria. The quantification of both sessile and planktonic populations by direct methods, such as epifluorescence microscopy, has shown that the adherent bacteria on 1 cm^2 of submerged surface outnumber the floating bacteria in 500 ml of flowing water in 84 different streams examined (Costerton & Geesey 1979a). Adherent populations become more specific, and usually much more numerous, when the colonised surface provides a nutritional advantage: for example, when the cuticular surface of a water plant exudes organic nutrients (Allen 1969) or when the surface is an insoluble substrate (e.g. cellulose). The specific adhesion to insoluble nutrient surfaces has been used to recover certain nutritive types of bacteria: for example, when cellulose (Patterson *et al.* 1975) or starch (Minato & Suto 1976) were added to rumen fluid, or when heavy oil was suspended in a stream (Wyndham & Costerton 1981a). In all cases the bacteria capable of degrading the particular substrate adhered to it and were specifically recovered.

17.1.2 Microbial Microcolonies on Tissue Surfaces

The surfaces of animal cells appear to be composed of a felt-like matrix of polysaccharides and glycoproteins (Roseman 1974) which is called the glycocalyx. This surface can easily be colonised by bacteria whose outermost component is a similar glycocalyx (Costerton *et al.* 1978). Plant surfaces may also be specifically colonised when the plant releases a lectin (e.g. trifolin) that interacts with both the plant surface polysaccharides and the exopolysaccharide glycocalyx of a particular *Rhizobium* species (Dazzo & Brill 1977). This process binds these bacteria to the root hair surface and initiates the formation of nitrogen-fixing nodules. Other bacteria bind specifically to plant surfaces (Nissen 1973) and facilitate the uptake of a particular substrate (e.g. choline sulphate). In animal systems many tissues provide an excellent nutritional niche for bacteria and are colonised heavily and often specifically (Foglesong *et al.* 1975; Savage 1977; McCowan *et al.* 1978) by populations of autochthonous bacteria which are quite distinct from those in the adjoining fluids and materials. As an example, the bacterial population growing on the bovine rumen wall is quite distinct taxonomically from the rumen fluid population. The wall population takes advantage of its partly aerobic, urea-rich, proteinaceous habitat to develop facultative respiration, and ureolytic and proteolytic activities which are important, and even essential, to the normal physiological functioning of this organ (Cheng & Costerton 1980). Similarly, the distal human urethra bears an autochthonous adherent bacterial population that occupies this strategic site and excludes pathogens that would otherwise enter the system to infect the bladder (Marrie *et al.* 1980). These autochthonous bacterial populations of the gut (Fig. 1) and of the urethra (Fig. 3) adhere to the tissue surface and, as they replicate in this favourable environment, form confluent masses of microbial growth. Within this structure discrete microcolonies are easily discerned and identified on the basis of their morphological similarity and exopolysaccharide production.

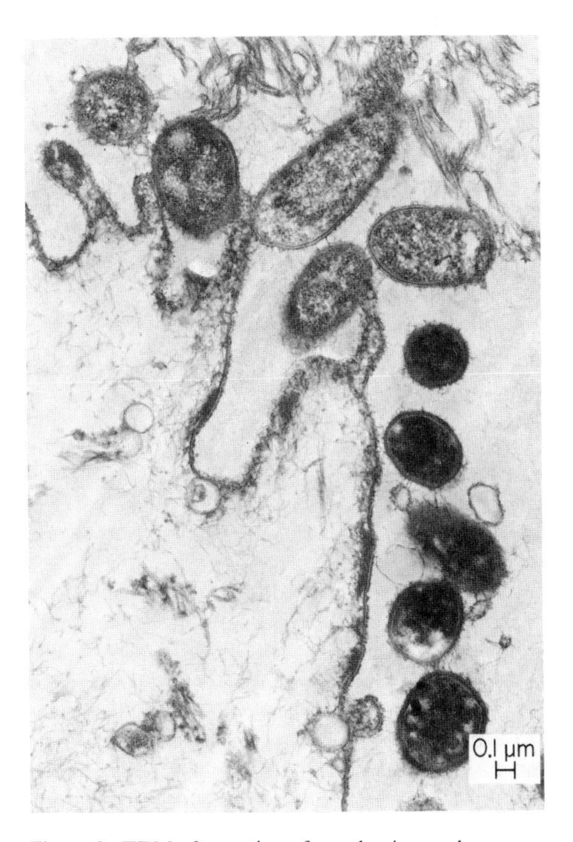

Figure 3. TEM of a section of a ruthenium red-stained preparation of epithelial cells sedimented from the urethral urine of a healthy female volunteer. Note the presence of only two bacterial morphotypes in this area of the adherent population and that the Gram-negative cells (at the top) form a glycocalyx-enclosed microcolony quite distinct from that formed by the Gram-positive cells (on the right).

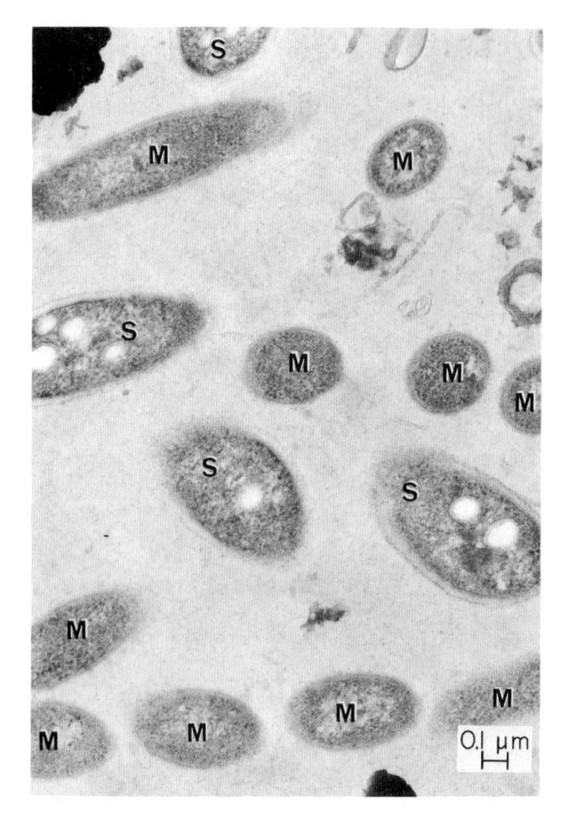

Figure 4. TEM of a section of a ruthenium red-stained preparation from a mixed culture of *Syntrophomonas wolfei* (S) and *Methanospirillum hungatii* (M) showing the formation of a mixed microcolony by these organisms that cooperate to produce methane from butyrate.

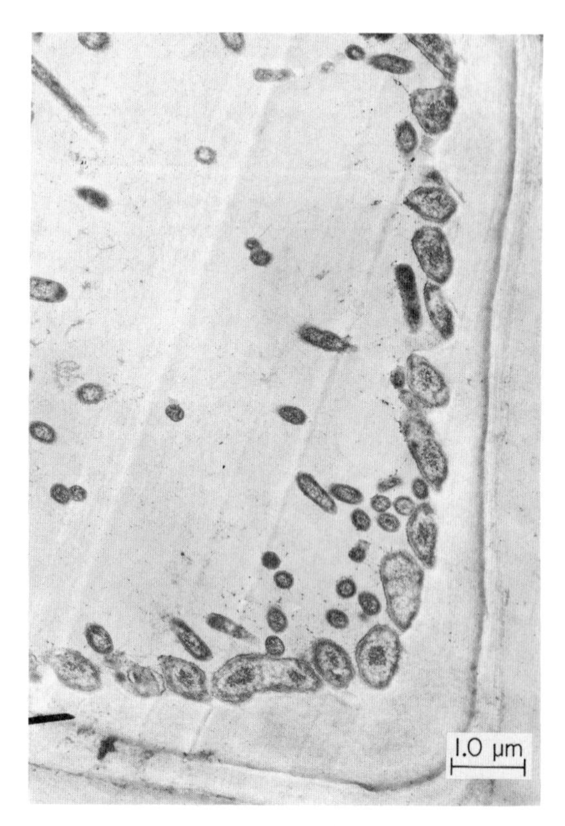

Figure 5. TEM of a section of a ruthenium red-stained preparation of rumen contents showing the colonisation of a fragment of plant cell wall by a distinct bacterial morphotype that has produced or occupied 'pits' on one side of the fragment to form a matrix-enclosed planar microcolony. A second very distinct morphotype has then been attracted, from the hundreds of species in this rumen fluid, to colonise the surface of the initial coloniser so that a stratified bacterial consortium is formed.

17.1.3 The Organisation of Microbial Microcolonies into Communities

Many degradative processes are carried out by aggregates composed of different bacterial species. These metabolic processes proceed by passing an intermediate molecular product from the cells of one species to those of another. Clearly, physical juxtaposition of the cooperative cells would be *de rigeur* and communities of different species were predicted by Slater (1978), Whittenbury (1978) and Slater and Lovatt (1982). Methane can be formed from butyrate when the butyric acid oxidising *Syntrophomonas wolfei* passes hydrogen directly to *Methanospirillum hungatii* to allow methane production (M. J. MacInerney, M. P. Bryant, R. B. Hespell & J. W. Costerton, unpublished observations). It is important to note that these two species in methane-producing butyrate cultures form mixed microcolonies in which the cells of both species are closely associated (Fig. 4). Direct observations of natural systems often show morphological evidence of consortium formation. For example, plant cell wall fragments are heavily colonised by one bacterial type (Fig. 5), drawn by specific adhesion from a very mixed rumen fluid population, while another bacterial species then adheres to the exopolysaccharide glycocalyx of the initial colonisers. This results in the development of a layered consortium.

17.1.4 The Role of Microbial Microcolonies in Corrosion Fouling and Disease

Microcolony growth not only enables a bacterium to persist in a nutritionally suitable niche but the enveloping exopolysaccharide protects the colony from chemical agents, such as acids (Duggan & Lundgren 1965) and surfactants (Govan 1975), and from biological agents, such as amoebae and bacteriophages. This mode of growth enables anaerobes, such as the sulphate-reducing bacteria, to sustain anaerobiosis in a microniche, even in a pipe carrying oxygenated water. If these microcolonies develop next to the metal, they can act as foci for corrosion (Costerton & Geesey 1979b). Similarly, dental caries are produced when microcolonies of adhering anaerobic bacteria which comprise dental plaque, produce acid resulting in the etching of the enamel (Gibbons & van Houte 1975). Diseases of compromised patients caused by opportunistic pathogens, such as *Pseudomonas aeruginosa,* are similar in that these organisms withstand host defence mechanisms, such as phagocytic cells (Schwartzmann & Boring 1971) and antibodies (Baltimore & Mitchell 1980), by the production of large exopolysaccharide-enclosed microcolonies on the tissue surfaces in the infected organs (Marrie *et al.* 1979; Lam *et al.* 1980). These diseases are characteristically slow-developing and non-disseminating, and are very resistant to conventional antibiotic therapy (Fang *et al.* 1978; Rabin *et al.* 1980). These diseases constitute an increasingly significant health hazard as the number of compromised human beings increases sharply. Similarly, bacteria within adhering microcolonies in industrial aquatic systems are resistant to biocides (Costerton & Geesey 1979b): for example, adhering microcolonies of *Serratia marcescens* have been shown to grow and thrive on the plastic walls of vessels containing concentrated Hibitane (T. J. Marrie & J. W. Costerton, unpublished observations).

17.1.5 The Chemical Nature of Bacterial Exopolysaccharides

Bacterial exopolysaccharides are extensively condensed during dehydration for electron microscopy. The thickness and organisation of this surface structure was not appreciated until Birdsell *et al.* (1975) used lectins to stabilise the surface teichoic acids of a Gram-positive bacterium and Bayer and Thurow (1977) used specific antibodies to prevent the collapse of the exopolysaccharide of a Gram-negative organism during dehydration. This technique (Mackie *et al.* 1979) shows that the exopolysaccharide is a very thick, radially structured mass of fine fibres at the cell surface (Fig. 6). The highly organised structure of the hydrated exopolysaccharide has also been established by X-ray diffraction (Moorhouse *et al.* 1977) showing that the fibres are regularly spaced with a centre-to-centre distance of 0.4 nm. Specific antibodies have been used to stabilise the exopolysaccharide of pathogenic bacteria growing on the surface of an infected tissue and this shows extensive numbers

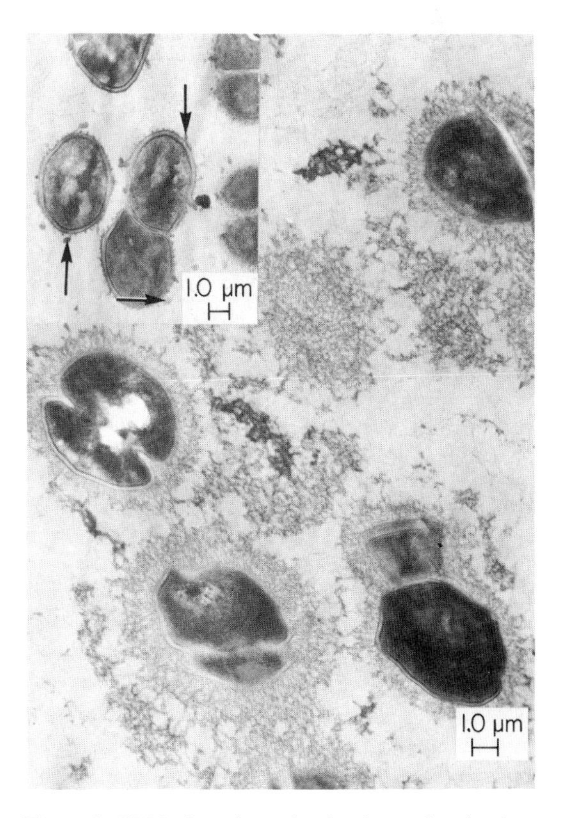

Figure 6. TEM of sections of ruthenium red-stained preparations of a pure culture of a group B *Streptococcus* sp. The exopolysaccharide of cells simply fixed and stained by normal methods (inset) is radically condensed during dehydration to form electron-dense masses at the cell surface (arrows) while that of cells treated with specific antibodies is stabilised and protected from condensation to preserve the extensive radial arrangement of the native hydrated glycocalyx.

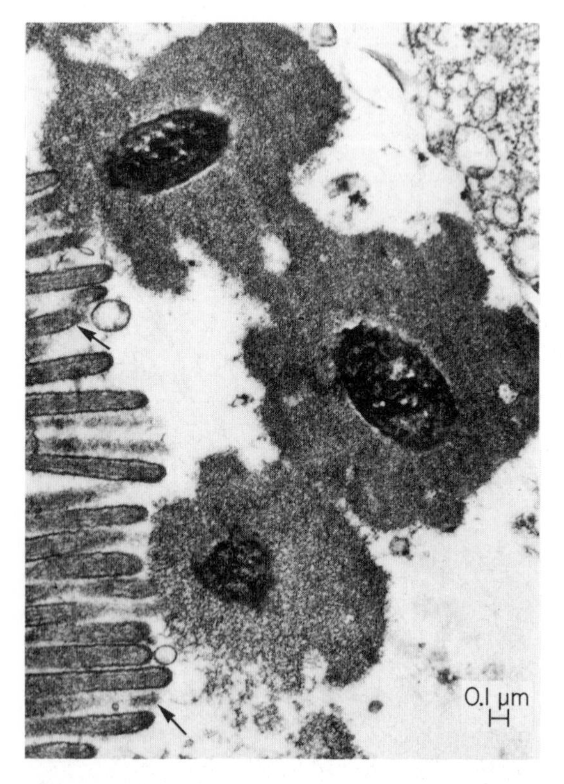

Figure 7. TEM of a section of a ruthenium red-stained preparation of the ileal epithelium of a newborn calf experimentally infected with a K30 + K99 − non-piliated strain of enteropathogenic *Escherichia coli*. The glycocalyx of the infecting bacteria was stabilised by treatment with specific antibodies and its extensive dimensions are preserved so that we can perceive its role in adhesion to the infected tissue, in microcolony formation, and in protection of the bacteria from surfactants and other host defence factors. The glycocalyx (arrows) of the ileal tissue is not stabilised by these specific antibodies.

of fine fibres surrounding the organisms (Fig. 7). The chemical nature of the bacterial glycocalyx has been controversial because early workers attempted to recover the glycocalyx polymers from disrupted cells or from colonies which contained many dead and lysed cells (Costerton *et al.* 1979). More recent studies have used preparations carefully recovered from the surface of living cells and a wide variety of different polymers have been described (Sutherland 1977), many of which are anionic carbohydrate polymers, including many uronic acids. Most of the bacterial exopolysaccharides stain strongly with ruthenium red and they can be stabilised by the addition of lectins or specific antibodies. It is now possible to visualise them in their true configurations and it has been found that they surround almost all of the bacteria examined directly in natural or pathogenic situations (Costerton *et al.* 1981).

Experimental

17.2 Methods for the examination of microbial microcolonies

The slow growth of bacterial microcolonies and the effect of surfaces on growth, have made these populations more difficult to study than bacteria growing dispersed in a liquid culture medium. For example, it is difficult to recognise a growth curve for adhering organisms. However, the frequency of attached microbial colony growth in most natural habitats, especially those associated with disease, compels the development of new methods for the study of the distribution, morphology and physiological activity *in situ* of these bacteria if we are to understand the basic growth characteristics of these colonies.

17.2.1 Morphological Examination of Adhering Microcolonies on Inert Surfaces

The nature of the substratum to which microbial microcolonies adhere can severely limit their microscopic examination. The opacity of most natural substrata poses a serious problem for light microscopy whilst the hardness of most of these materials makes embedding and section cutting for Transmission electron microscopy (TEM) almost impossible. Early workers compromised by using unnatural substrata, such as glass, so that the colonisation processes could be examined by light microscopy, or embedding plastic (Costerton *et al.* 1978) if the colonisation sequences were to be examined by TEM. However, these substances differ from natural substrata in topography, chemical and physical properties and it is these factors which are important in bacterial colonisation. More recently epi-illumination methods have been developed for fluorescence light microscopy and Scanning electron microscopy (SEM) is also proving invaluable. These methods require that the colonised surface is stained and/or dehydrated, manipulations which may lead to some loss of bacteria. If the nature of the preparation artefacts is borne in mind, useful topographical information concerning the distribution of the colonising bacteria may be obtained.

Light microscopy

The opacity of natural substrata can be counteracted by staining attached bacteria with either fluorescent compounds that have an affinity for nucleic acids or fluorescein-labelled antibodies that have an affinity for bacterial surface antigens (Diem *et al.* 1977) and examining the preparations by epifluorescence light microscopy. This in-situ method is not entirely quantitative because loosely adhering bacteria are lost in the treatment process and it does not distinguish living from dead bacteria. Furthermore it is a difficult method because the stained bacteria lie at different levels on natural rough surfaces. Accordingly, sessile, attached bacterial populations are routinely enumerated by scraping the biofilm from a measured area, homogenising and recovering the bacteria on the smooth surface of a nucleopore filter. The organisms are stained with fluorescent dyes and enumerated by epifluorescence light microscopy (Zimmerman & Meyer-Reil 1974; Geesey *et al.* 1978; Costerton 1980). An aquatic bacterial population can be conveniently divided into sessile, particle-associated, and planktonic subpopulations by scraping the first from surfaces, by filtering particles on a large pore size filter, and by trapping the plankton on a fine filter (Costerton & Geesey 1979a). In the rare instances in which it can be used (e.g. with transparent materials), light microscopy offers the simplest and most accurate method of visualising adhering bacteria in relation to topographic features, especially when used with phase or differential interference (Nomarski) optical systems (Fig. 8). The staining of bacterial exopolysaccharide by specific antibodies (Bayer & Thurow 1977) renders the glycocalyx visible by phase or Nomarski light microscopy and shows the role of the polysaccharide in microcolony formation and adhesion.

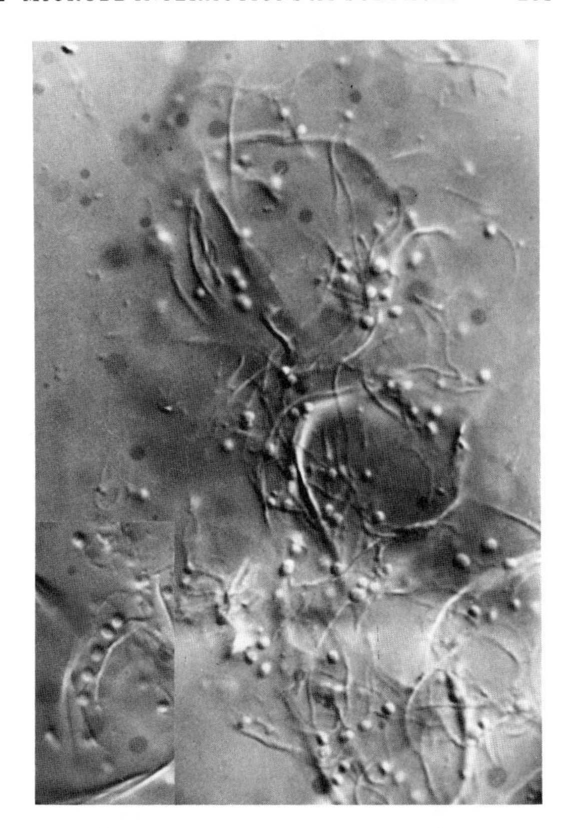

Figure 8. Light micrograph of the adhesion of live cells of a pure culture of *Ruminococcus albus* to cellulose fibres *in vitro*. Differential interference (Nomarski) optics were used to make this micrograph and the alignment of bacterial cells on the cellulose fibres is clearly shown (inset).

Transmission electron microscopy

The simplest solution to the problem of embedding and cutting sections of a colonised substrate is to use embedding plastic as a test substrate. The adhering bacterial microcolonies may be fixed, stained and dehydrated on this plastic and then embedded in the same plastic for sectioning and examination (Fig. 9). Wyndham and Costerton (1981a) used hydrocarbon-impregnated nucleopore filters which were exposed in an oil-degrading environment, colonised, embedded and sectioned to reveal the presence of well-developed, adhering microcolonies of hydrocarbon-oxidising bacteria (Fig. 10). If the inert substratum is amenable to embedding and sectioning, the colonised surface can be visualised in relation to the microcolonies and much useful data obtained from sections cut at a right angle to the surface (Figs 5 and 10). The micrographs reveal the position of the adhering bacteria with respect to the surface (Akin & Amos 1975) and the arrangement of the exopolysaccharide fibres between the bacterium and the substra-

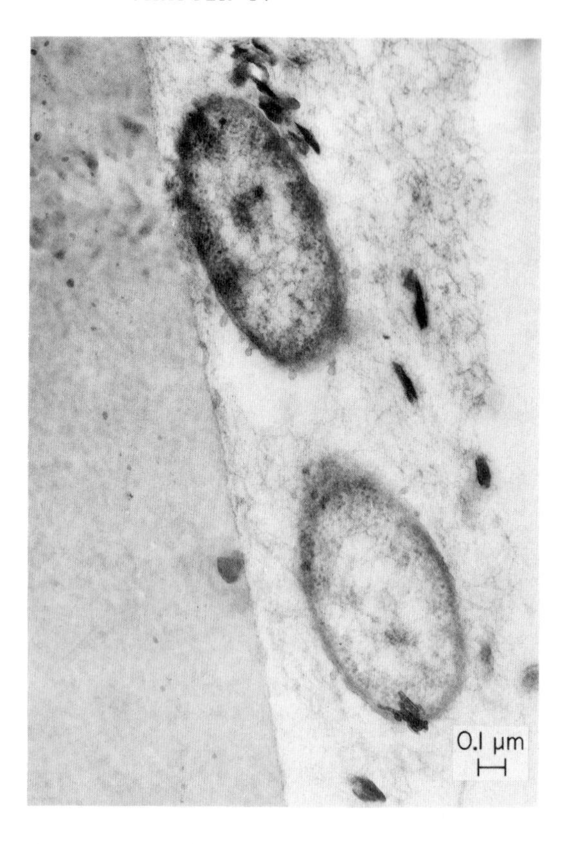

Figure 9. TEM of a section of a ruthenium red-stained preparation of the biofilm that developed when an epoxy disc was exposed to the flowing water of a subalpine stream. Note the fibrous glycocalyx produced by the adherent bacteria that anchors them to the surface, and to each other, and serves also to trap electron-dense clay platelets.

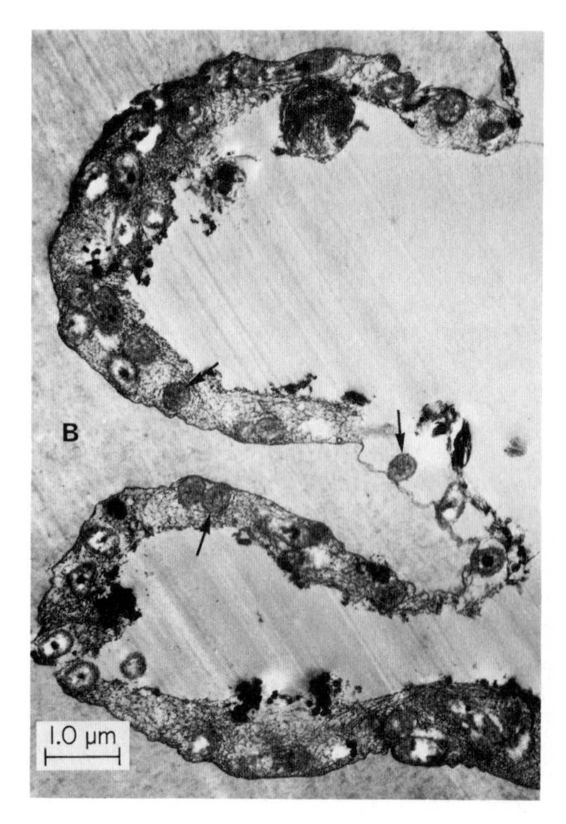

Figure 10. TEM of a section of a ruthenium red-stained preparation of a bitumen surface (B) exposed to the flowing water of a river with a very large population of bitumen-decomposing bacteria. Note the colonisation of the bitumen surface by bacteria (arrows) enclosed in a confluent fibrous glycocalyx.

tum which are preserved from the dehydration-condensation artefact (Cheng *et al.* 1980). If a surface cannot be embedded, sectioned and viewed *in situ* by TEM methods, the microbial film can often be removed by sterile scraping and the morphological state of the cells and structure of the microcolonies can be determined (Fig. 11).

Scanning electron microscopy

Scanning electron microscopy is ideally suited for the examination of bacterial microcolonies on inert surfaces because the electron beam impinges on the surface to be examined and the image is formed by scattered and secondary emitted electrons. However, the exopolysaccharides of adherent bacterial cells are often badly preserved and poorly resolved, even if the dehydration artefacts are partially prevented by critical point drying (Cohen *et al.* 1968) and if the specimens are chemically metallised by the use of osmium tetroxide and thiocarbohydrazide (Malick & Wilson 1975). For extensive exopolysaccharide envelopes, this material condenses to form an amorphous mass (Lam *et al.* 1980) from which bacterial cells protrude (Fig. 12). This material is less well preserved during preparation for SEM than by preparation for TEM, even when the exopolysaccharide is stabilised by specific antibodies to prevent dehydration (Mackie *et al.* 1979).

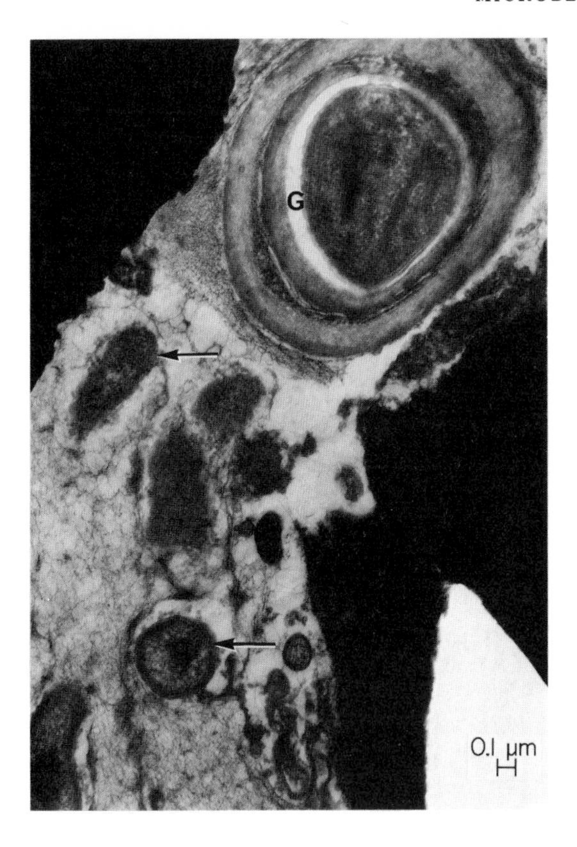

Figure 11. TEM of a section of a ruthenium red-stained preparation of material scraped from the natural cobble substratum of a boreal stream. Orientation is lost in these crude preparations, and detritus is common, but it can be deduced that intact bacteria (arrows) are present in a confluent glycocalyx and that blue-green bacteria (G) are present in well-developed slime tubes.

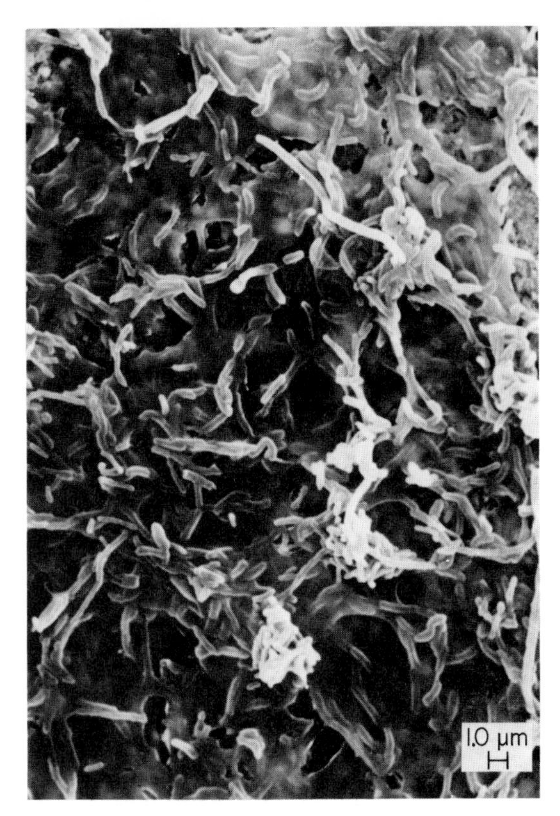

Figure 12. Scanning electron micrograph (SEM) of a critical point-dried thiocarbohydrazide-treated preparation of the biofilm developed on a metal surface after 64 h exposure to a high nutrient medium. Even though the enveloping glycocalyx of these adherent cells is radically condensed by dehydration it is sufficiently extensive to bury all but the uppermost bacterial cells.

17.2.2 Recovery of Bacteria from Adherent Microcolonies on Inert Surfaces

The bacterial film can be recovered by sterile scraping from natural substrata in most ecosystems, a process which is most efficient with thick films (Geesey *et al.* 1978). The film can then be disrupted by mild sonication and mechanical agitation until it yields a dispersed population of single cells and discrete microcolonies. The total bacterial cells present in this dispersed suspension can be enumerated by epifluorescence and related to the area from which the film was removed. The various physiological groups of living bacteria can be enumerated using plating and most-probable-number (MPN) techniques (Colwell 1979; see Chapter 6).

17.2.3 Measurement of the Physiological Activity of Microcolonies on Inert Surfaces

Since adherent bacteria predominate in most aquatic systems (Costerton & Geesey 1979a) and are the only users of dissolved organic nutrients at the concentrations normally found in the environment (Sepers 1977), it is important to know if adherent bacteria remove dissolved organic nutrients at a rate which is proportional to their numbers. Ladd *et al.* (1979) studied colonised polished rock squares and smooth metal studs of known surface areas in natural streams, and measured the uptake of [^{14}C]glutamate by:

(1) the sessile bacteria on the colonised surface;

(2) the sessile bacteria removed from the surface and mechanically dispersed;

(3) the unattached, planktonic population.

The rate of glutamate uptake per unit cell for the dispersed sessile bacterial population was slightly more active than the undisturbed adherent population, suggesting that the exopolysaccharide surrounding microcolonies marginally impeded diffusion of the substrate. However, the attached bacteria were always more active in substrate uptake than free-living bacteria. Higher rates of substrate uptake by attached bacteria seems to be a general phenomenon (Ladd *et al.* 1979; Wyndham & Costerton 1981b).

17.2.4 Morphological Examination of Adherent Microcolonies on Tissue Surfaces

Examinations of healthy animals have shown that many tissues are colonised by specific autochthonous bacterial populations (Savage *et al.* 1968; Fuller & Brooker 1974; Cheng & Costerton 1980) which are taxonomically distinct from the populations in the organ's contents. Moreover, the bacterial population can be integrated into the normal physiological function of the organ (Cheng & Costerton 1980). The attached bacterial populations of healthy tissues change with environmental stress (Johanson *et al.* 1980) and age (Marrie *et al.* 1978) and give rise to specific adherent populations of pathogens in disease states (Marrie *et al.* 1979; Lam *et al.* 1980). The autochthonous and pathogenic bacterial populations form exopolysaccharide-enclosed microcolonies and, as before, their examination is complicated by the opacity of tissues to light and electrons. In addition, the inaccessibility of many colonised tissue surfaces in human beings and the presence of mucus layers at the surfaces of many tissues presents difficulties in examining the attached microcolonies.

Light microscopy

Light microscopy is the ideal way to examine adherent microcolonies on tissues, at least for those whose opacity allows direct observation. Differential interference (Nomarski) microscopy is helpful because of the much reduced contrast of out-of-focus levels of the specimen. The 'Quellung' reaction produced by specific antibodies is especially useful in demonstrating the exopolysaccharide capsule (Bayer & Thurow 1977). Bacterial genera have been identified in plant tissues by the use of fluorescent antibodies (Lalonde & Knowles 1975). In human body fluids pathogenic bacteria are routinely identified and enumerated by this method (Marrie *et al.* 1979). Light microscopy, especially epifluorescence microscopy of fluorescein-stained cells, is often used to monitor bacterial adhesion to tissue culture cells (Hartley *et al.* 1978).

Transmission electron microscopy

Most plant and animal tissues are amenable to embedding and sectioning and adherent bacterial microcolonies are retained in the colonised tissues of plants (Cheng *et al.* 1980) and animals (McCowan *et al.* 1978). The bacterial exopolysaccharide is condensed by dehydration during the preparation of the specimens for TEM so that, when stained with ruthenium red (Figs 1 and 5), the capsule appears as electron-dense fibrils which are partially stabilised when anchored to adjacent surfaces. The exopolysaccharide can be stabilised by exposure to specific antibodies so that it does not collapse during dehydration (Mackie *et al.* 1979; Figs 6 and 7). It is important to note that pili do not stain with ruthenium red and are not usually resolved in sectioned material because they are thinner than the resolution (2.5 nm) of the TEM sections. One serious problem in the TEM of colonised tissues involves the mucus which overlies many animal tissues. Mucus hardens during fixation and dehydration and forms a coherent layer in which many bacterial microcolonies are embedded (Fig. 13). The simple hydraulic forces involved in processing tissues for TEM often remove the mucus from the tissues surface together with its encased microcolonies. Thus its loss yields deceptive data on the extent of bacterial colonisation of tissues. It is often instructive to sediment the supernatants of fixation and dehydration solutions and to embed and examine the particles recovered from these solutions. The numbers obtained in this way may be used to correct for losses of bacterial microcolonies.

When a colonised tissue is not accessible for direct sampling, for example, with the human

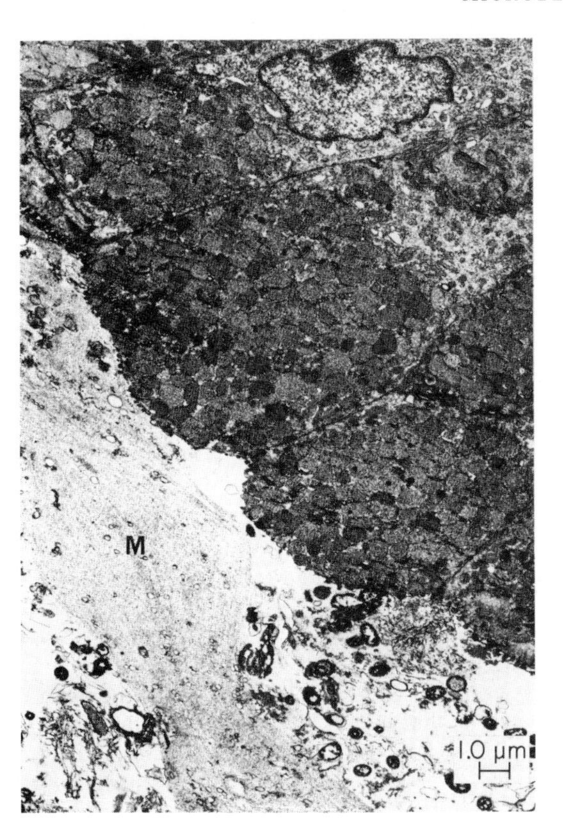

Figure 13. TEM of a section of a ruthenium red-stained preparation of the bovine ileum showing the fortuitous retention of the mucus (M) overlying this tissue and the relationship of autochthonous bacteria to this mucus layer. The removal of this mucus layer, during processing for TEM and SEM, removes many of these tissue-adherent bacteria.

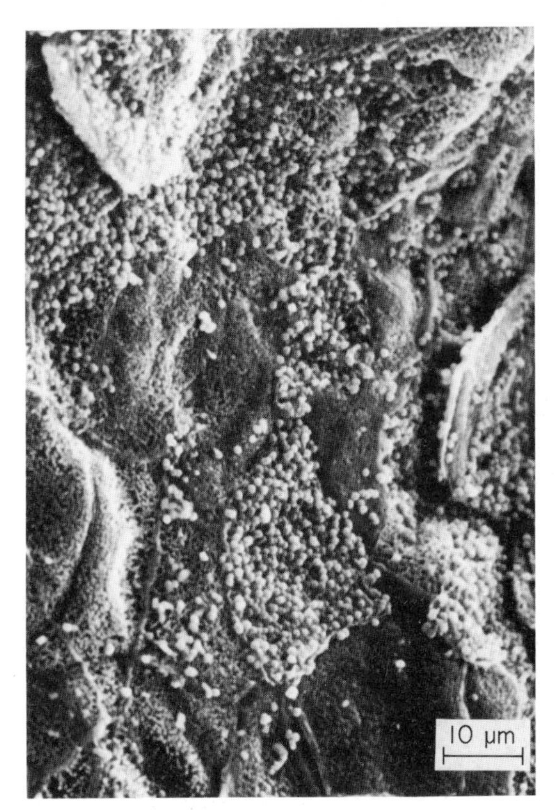

Figure 14. SEM of a critical point-dried thiocarbohydrazide-treated preparation of the epithelium of the bovine rumen. In this sparsely colonised region of the rumen epithelium this technique shows the relationship of the colonising bacteria to many topographic features such as epithelial cell boundaries and sloughing epithelial cells.

urethra, sloughed cells of the urethral epithelial tissue can be recovered from urethral urine and examined: approximately 30% are colonised on one side (Marrie *et al.* 1980; Fig. 3). Similarly, attached bacteria can be recovered from sloughed epithelial cells of the human buccal cavity (Johanson *et al.* 1980) and from the bovine tongue (McCowan *et al.* 1979).

Scanning electron microscopy

The usefulness of SEM depends on the type of tissue to be examined, in that the adherent bacteria on a sparsely colonised tissue are clearly seen in relation to topographical features (Fig. 14) even though the fibres of the bacterial exopolysaccharide are condensed and unresolved. If the layer of adherent bacteria is thick, its relationship to surface features of the tissue is masked and, if the tissue is covered with mucus, only the outer surface of the mucus layer is seen. Bacteria within the microcolonies enmeshed in the mucus are only clearly seen when the mucus layer is broken during preparation (Fig. 15), and the mucus layer and its bacteria may be lost entirely during extensive washing before or after fixation and dehydration (Fig. 16). A useful method for the preparation of colonised tissues for SEM (Costerton 1980) has been described but it must be stressed that the adherent microcolonies attached to tissue surfaces by hydrated matrices of exopolysaccharide fibres are profoundly altered by dehydration. Thus colon-

Figure 15. SEM of a critical point-dried thiocarbohydrazide-treated preparation of the ileum of the mouse showing the retention of much of the mucus layer. The surface (S) of this layer contains few bacteria but, where the mucus layer is fractured many embedded bacteria are seen (arrows), and their relationship to the tissue surface (T) can be deduced.

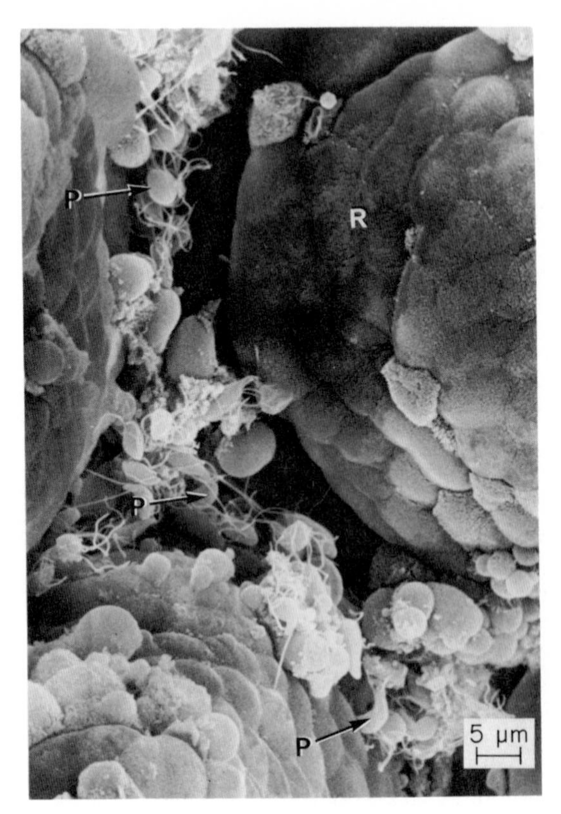

Figure 16. SEM of a critical point-dried thiocarbohydrazide-treated preparation of the ileum of the mouse identical to that shown in Fig. 15 except that the tissue was washed in phosphate buffered saline before processing for SEM. Note the almost complete removal of the bacteria-containing mucus layer from the surface of this tissue and the exposure of deeper tissue surfaces to which protozoa (P) are attached.

ised mucus-covered surfaces must be seen with the mucus layer intact, broken and missing in order to piece together the complex relations between adherent microcolonies and the tissue surface.

17.2.5 Recovery of Bacteria from Adherent Microcolonies on Tissue Surfaces

Bacterial microcolonies often become structurally integrated into colonised animal tissues by growing into the intercellular spaces of the tissue where they adhere tenaciously (McCowan *et al.* 1979); as a result, simple swabbing of the tissue surface fails to remove these organisms. The colonised tissue must be washed to remove adventitious bacteria, excised under sterile conditions, homogenised without damage to the bacteria and plated out under both aerobic and anaerobic conditions in order to recover the bacterial components (Cheng & Costerton 1980). Using these techniques, up to 1.0×10^8 cells cm^{-2} of adherent bacteria were recovered from the bovine rumen epithelium. It has been shown that these organisms differ taxonomically from the rumen fluid bacteria and comprise a partly facultative, ureolytic, proteolytic population developed in response to the presence of oxygen, urea and a keratin substratum at the tissue surface. These bacterial recovery techniques can easily be monitored by light microscopy and TEM to assess the effects of washing and of homogenisation methods.

17.2.6 Measurement of the Physiological Activity of Bacterial Microcolonies on Tissue Surfaces

Some bacterial enzyme activities can be detected, but not quantified, by the appearance of a product (e.g. inorganic phosphate) which can be deposited as an insoluble, electron-dense salt (e.g. lead phosphate) and detected by TEM. This method has been used by Fay *et al.* (1979) to detect the activity of alkaline phosphatase of adherent bacteria both at the surface and within the epithelial cells of the colonised rumen epithelium. If the tissue can be shown to lack a certain enzyme, by studies, for example, of the same tissue of gnotobiotic animals or by sterilisation of the tissue by starvation and antibiotics (Cheng *et al.* 1979b), the enzyme activity of colonised blocks of tissue, due to bacterial activity, can be determined (McCowan *et al.* 1980). This method has been used to study the distribution of ureolytic bacteria in the bovine rumen (McCowan *et al.* 1980) and the recorded activities correlate well with physiological data and with the enzyme activities of bacteria recovered from the sites examined.

17.3 Conclusions

Recent examinations of many ecosystems have shown that adherent bacterial populations are important by virtue either of their numerical predominance or their special physiological activities. The growth of these bacteria in exopolysaccharide-enclosed microcolonies which adhere to solid surfaces poses many difficult problems in their examination by morphological and physiological methods. However, to ignore them and to continue to extrapolate from standard laboratory cultures to ecosystems where adherent bacteria predominate is unwise. Each ecosystem ought to be examined, at the outset, to determine whether or not adherent bacteria are numerically or physiologically important, using the criteria previously outlined (Costerton & Geesey 1979a). If adherent bacteria play a minor role, as appears to be the case in the deep oceans, then classic microbiological methods may be used. However, in the absence of this assurance, microbial ecologists must join in the difficult and frustrating task of adapting basic microbiological methods to examine adherent populations on the surfaces that they colonise or we risk serious error and irrelevance.

Recommended reading

Cheng K-J. & Costerton J. W. (1980) Adherent rumen bacteria—their role in the digestion of plant material, urea and epithelial cells. In: *Digestive Physiology and Metabolism in Ruminants* (Eds Y. Ruckebush & P. Thivend), pp. 227–50. MTP Press, Lancaster.

Costerton J. W. & Geesey G. G. (1979) Microbial contamination of surfaces. In: *Surface Contamination* (Ed. K. L. Mittal), pp. 211–21. Plenum Publishing, New York.

Costerton J. W., Geesey G. G. & Cheng K-J. (1978) How bacteria stick. *Scientific American* **238,** 86–95.

Costerton J. W., Irvin R. T. & Cheng K-J. (1981) The role of bacterial surface structures in pathogenesis. *CRC Critical Reviews in Microbiology,* pp. 303–38. CRC Press, Boca Raton, Florida.

References

Akin D. E. & Amos H. E. (1975) Rumen bacterial degradation of forage cell walls investigated by electron microscopy. *Applied Microbiology* **29,** 692–9.

Allen H. L. (1969) Chemoorganatrophy in epiphytic bacteria with reference to macrophytic release of dissolved organic matter. In: *The Structure and Function of Freshwater Microbial Communities* (Ed. J. Cains), pp. 277–80. American Microscopical Society, Washington, D.C.

Bae H. C., Cota-Robles E. H. & Casida L. E. (1972) Microflora of soil as viewed by transmission electron microscopy. *Applied Microbiology* **23,** 637–48.

Baltimore R. S. & Mitchell M. (1980) Immunological investigation of mucoid strains of *Pseudomonas aeruginosa*: comparison of susceptibility by opsonic antibody in mucoid and nonmucoid strains. *Journal of Infectious Diseases* **141,** 238–47.

Bauchop T., Clarke R. T. J. & Newhook J. C. (1975) Scanning electron microscope study of bacteria associated with the rumen epithelium of sheep. *Applied Microbiology* **30,** 668–75.

Bayer M. E. & Thurow H. (1977) Polysaccharide capsule of *Escherichia coli*: microscopic study of its size, structure, and sites of synthesis. *Journal of Bacteriology* **130,** 911–36.

Birdsell D. C., Doyle R. J. & Morgenstern M. (1975) Organization of teichoic acid in the cell wall of *Bacillus subtilis*. *Journal of Bacteriology* **121,** 726–34.

Cheng K-J., Akin D. E. & Costerton J. W. (1977) Rumen bacteria: interaction with dietary components and response to dietary variation. *Federation Proceedings* **36,** 193–7.

CHENG K-J., BAILEY C. B. M., HIRONAKA R. & COSTERTON J. W. (1979b) A technique for depletion of bacteria adherent to the epithelium of the bovine rumen. *Canadian Journal of Animal Science* **59,** 207–9.

CHENG K-J. & COSTERTON J. W. (1980) Adherent rumen bacteria—their role in the digestion of plant material, urea and epithelial cells. In: *Digestive Physiology and Metabolism in Ruminants* (Eds Y. Ruckebush & P. Thivend), pp. 227–50. MTP Press, Lancaster.

CHENG K-J., FAY J. P., HOWARTH R. E. & COSTERTON J. W. (1980) Sequence of events in the digestion of fresh legume leaves by rumen bacteria. *Applied and Environmental Microbiology* **40,** 613–25.

CHENG K-J., MCCOWAN R. P. & COSTERTON J. W. (1979a) Adherent epithelial bacteria in ruminants and their roles in digestive tract function. *American Journal of Clinical Nutrition* **32,** 139–48.

COHEN A. L., MARLOW D. P. & GARNER G. E. (1968) A rapid critical point using fluorocarbon ('freons') as intermediate and transitional fluids. *Journal of Microscopy* **7,** 331–42.

COLWELL R. R. (1979) Enumeration of specific populations by the most-probable-number (MPN) method. In: *Native Aquatic Bacteria: Enumeration, Activity, and Ecology* (Eds J. W. Costerton & R. R. Colwell), pp. 56–64. ASTM Press, Philadelphia.

COSTERTON J. W. (1980) Some techniques involved in the study of adsorption of microorganisms to surfaces. In: *Adsorption of Microorganisms to Surfaces* (Eds G. Bitton & K. C. Marshall), pp. 403–32. John Wiley & Sons, New York.

COSTERTON J. W., BROWN M. R. W. & STURGESS J. M. (1979) The cell envelope: its role in infection. In: *Pseudomonas aeruginosa: Clinical Manifestations and Current Therapy* (Ed. R. G. Doggett), pp. 41–62. Academic Press, New York.

COSTERTON J. W. & GEESEY G. G. (1979a) Which population of aquatic bacteria should we enumerate? In: *Native Aquatic Bacteria: Enumeration, Activity, and Ecology* (Eds J. W. Costerton & R. R. Colwell), pp. 7–18. ASTM Press, Philadelphia.

COSTERTON J. W. & GEESEY G. G. (1979b) Microbial contamination of surfaces. In: *Surface Contamination* (Ed. K. L. Mittal), pp. 211–21. Plenum Publishing, New York.

COSTERTON J. W., GEESEY G. G. & CHENG K-J. (1978) How bacteria stick. *Scientific American* **238,** 86–95.

COSTERTON J. W., IRVIN R. T. & CHENG K-J. (1981) The role of bacterial surface structures in pathogenesis. *CRC Critical Reviews in Microbiology,* pp. 303–38. CRC Press, Boca Raton, Florida.

DAZZO F. B. & BRILL W. J. (1977) Receptor site on clover and alfalfa roots for *Rhizobium*. *Applied and Environmental Microbiology* **33,** 132–6.

DIEM H. G., GODBILLON G. & SCHMIDT E. L. (1977) Application of the fluorescent antibody technique to the study of an isolate of Beijerinckia in soil. *Canadian Journal of Microbiology* **23,** 161–5.

DOGGETT R. W., HARRISON G. M. & WALLIS E. S. (1964) Comparison of some properties of *Pseudomonas aeruginosa* isolated from infections of persons with and without cystic fibrosis. *Journal of Bacteriology* **87,** 427–31.

DUGGAN P. R. & LUNDGREN D. G. (1965) Energy supply for the chemoautotroph *Ferrobacillus ferooxidans*. *Journal of Bacteriology* **89,** 825–34.

FANG L. S. T., TOLKOFF-RUBIN N. E. & RUBIN R. H. (1978) Efficacy of single-dose and conventional amoxicillin therapy in urinary-tract infection localized by the antibody-coated bacteria technique. *New England Journal of Medicine* **298,** 413–16.

FAY J. P., CHENG K-J. & COSTERTON J. W. (1979) Production of alkaline phosphatase by epithelial cells and adherent bacteria of the bovine rumen and abomasum. *Canadian Journal of Microbiology* **25,** 932–6.

FLETCHER M. M. & FLOODGATE G. D. (1973) Electron microscopic demonstration of an acidic polysaccharide involved in the adhesion of a marine bacterium to solid surfaces. *Journal of General Microbiology* **72,** 325–34.

FOGLESONG M. A., WALKER D. H., PUFFER J. S. & MARKOVETZ A. J. (1975) Ultrastructural morphology of some prokaryotic microorganisms associated with the hindgut of cockroaches. *Journal of Bacteriology* **123,** 336–45.

FULLER R. & BROOKER B. E. (1974) Lactobacilli which attach to the crop epithelium of the fowl. *American Journal of Clinical Nutrition* **27,** 1305–12.

GEESEY G. G., MUTCH R., COSTERTON J. W. & GREEN R. B. (1978) Sessile bacteria: an important component of the microbial population in small mountain streams. *Limnology and Oceanography* **23,** 1214–23.

GEESEY G. G., RICHARDSON W. T., YEOMANS H. G., IRVIN R. T. & COSTERTON J. W. (1977) Microscopic examination of natural sessile bacterial populations from an alpine stream. *Canadian Journal of Microbiology* **23,** 1733–6.

GIBBONS R. J. & VAN HOUTE J. (1975) Bacterial adherence in oral microbial ecology. *Annual Reviews of Microbiology* **29,** 19–44.

GOVAN J. R. W. (1975) Mucoid strains of *Pseudomonas aeruginosa*: the influence of culture medium on the stability of mucus production. *Journal of Medical Microbiology* **8,** 513–22.

HARTLEY C. L., ROBBINS C. M. & RICHMOND M. H. (1978) Quantitative assessment of bacterial adhesion to eukaryotic cells of human origin. *Applied Bacteriology* **24,** 91–7.

HENDRICI A. T. (1933) Studies of freshwater bacteria. I. A direct microscopic technique. *Journal of Bacteriology* **25,** 277–86.

JOHANSON W. G., HIGUCHI J. H., CHANDHURI T. R. & WOODS D. E. (1980) Bacterial adherence to epithelial cells in bacillary colonization of the respiratory tract. *American Reviews of Respiratory Disease* **121,** 55–63.

LADD T. I., COSTERTON J. W. & GEESEY G. G. (1979) Determination of the heterotrophic activity of epilithic microbial populations. In: *Native Aquatic Bacteria: Enumeration, Activity and Ecology* (Eds J. W. Costerton & R. R. Colwell), pp. 180–95. ASTM Press, Philadelphia.

LALONDE M. & KNOWLES R. (1975) Demonstration of the isolation of non-infective *Alnus crispa* var. *mollis* Fern. nodule endophyte by morphological immunolabelling and whole cell composition studies. *Canadian Journal of Microbiology* **21,** 1901–20.

LAM J., CHAN R., LAM K & COSTERTON J. W. (1980) Production of mucoid microcolonies by *Pseudomonas aeruginosa* within infected lungs in cystic fibrosis. *Infection and Immunity* **28,** 546–56.

MCCOWAN R. P., CHENG K-J., BAILEY C. B. M. & COSTERTON J. W. (1978) Adhesion of bacteria to epithelial cell surfaces within the reticulorumen of cattle. *Applied and Environmental Microbiology* **35,** 149–55.

MCCOWAN R. P., CHENG K-J. & COSTERTON J. W. (1979) Colonization of a portion of the bovine tongue by unusual filamentous bacteria. *Applied and Environmental Microbiology* **37,** 1224–9.

MCCOWAN R. P., CHENG K-J. & COSTERTON J. W. (1980) Adherent bacterial populations on the bovine rumen wall: distribution patterns of the adherent bacteria. *Applied and Environmental Microbiology* **39,** 233–41.

MACKIE E. B., BROWN K. N., LAM J. & COSTERTON J. W. (1979) Morphological stabilization of capsules of group B Streptococci, types Ia, Ib, II, and III, with specific antibody. *Journal of Bacteriology* **138,** 609–17.

MALICK L. E. & WILSON B. W. (1975) Modified thiocarbohydrazide procedure for scanning electron microscopy: routine use for normal, pathological, or experimental tissues. *Stain Technology* **50,** 265–9.

MARRIE T. J., HARDING G. K. M. & RONALD A. R. (1978) Anaerobic and aerobic urethral flora in healthy females. *Journal of Clinical Microbiology* **8,** 67–72.

MARRIE T. J., HARDING G. K. M., RONALD A. R., DIKKEMA J., LAM J., HOBAN S. & COSTERTON J. W. (1979) Influence of antibody coating of *Pseudomonas aeruginosa*. *Journal of Infectious Diseases* **139,** 357–61.

MARRIE T. J., LAM J. & COSTERTON J. W. (1980) Bacterial adhesion to uroepithelial cells—a morphological study. *Journal of Infectious Diseases*, **142**, 239–46.

MARSHALL K. C., STOUT R. & MITCHELL R. (1971) Mechanism of the initial sorbtion of marine bacteria to surfaces. *Journal of General Microbiology* **68,** 337–48.

MINATO H. & SUTO T. (1976) Technique for fractionation of bacteria in rumen microbial ecosystem. I. Attachment of rumen bacteria to starch granules and elution of bacteria attached to them. *Journal of General and Applied Microbiology* **22,** 259–76.

MOORHOUSE R., WINTER W. T., ARNOTT S. & BAYER M. E. (1977) Conformation and molecular organization in fibers of capsular polysaccharide from *Escherichia coli* M41 mutant. *Journal of Molecular Biology* **109,** 373–91.

NISSEN P. (1973) Kinetics of ion uptake in higher plants. *Physiologia Plantarum* **28,** 113–20.

PAERL H. W. (1975) Microbial attachment to particles in marine and freshwater ecosystem. *Microbial Ecology* **2,** 73–83.

PATTERSON H., IRVIN R., COSTERTON J. W. & CHENG K-J. (1975) Ultrastructure and adhesion properties of *Ruminococcus albus*. *Journal of Bacteriology* **122,** 278–87.

RABIN H. R., HARLEY F. L., BRYAN L. E. & ELFRING G. L. (1980) Evaluation of a high dose tobramycin and ticarcillin treatment protocol in cystic fibrosis based on improved susceptibility criteria and antibiotic pharmacokinetics. In: *Perspectives in Cystic Fibrosis* (Ed. J. M. Sturgess), pp. 370–5. Canadian Cystic Fibrosis Foundation, Toronto.

ROSEMAN S. (1974) Complex carbohydrates and intercellular adhesion. In: *Biology and Chemistry of Eukaryotic Cell Surfaces* (Eds E. Y. C. Lee & E. E. Smith), pp. 317–37. Academic Press, New York.

SAVAGE D. C. (1977) Microbial ecology of the gastrointestinal tract. *Annual Reviews of Microbiology* **31,** 107–33.

SAVAGE D. C., DUBOS R. & SCHAEDLER R. W. (1968) The gastrointestinal epithelium and its autochthonous bacterial flora. *Journal of Experimental Medicine* **127,** 67–76.

SCHWARTZMANN S. & BORING J. R. (1971) Antiphagocytic effect of slime from a mucoid strain of *Pseudomonas aeruginosa*. *Infection and Immunity* **3,** 762–7.

SEPERS A. B. J. (1977) The utilization of dissolved organic compounds in aquatic environments. *Hydrobiologia* **52,** 39–54.

SLATER J. H. (1978) The role of microbial communities in the natural environment. In: *The Oil Industry and Microbial Ecosystems* (Eds K. W. A. Chater & H. J. Sommerville), pp. 137–54, Heyden and Son, London.

SLATER J. H. & LOVATT D. (1982) Biodegradation and the significance of microbial communities. In: *Biochemistry of Microbial Degradation* (Ed. D. T. Gibson). Marcel Dekker, New York. (In press.)

SUTHERLAND I. W. (1977) Bacterial polysaccharides—their nature and production. In: *Surface Carbohydrates of the Prokaryotic Cell* (Ed. I. W. Sutherland), pp. 27–96. Academic Press, New York.

WHITTENBURY R. (1978) Bacterial nutrition. In: *Essays in Microbiology* (Eds J. R. Norris & M. H. Richmond), pp. 16/1–16/32. John Wiley & Sons, Chichester.

WYNDHAM R. C. & COSTERTON J. W. (1981a) Microbial degradation of bituminous hydrocarbons and *in situ* colonization of bitumen surfaces within the Athabasca oil sands deposit. *Applied and Environmental Microbiology* **41**, 783–90.

WYNDHAM R. C. & COSTERTON J. W. (1981b) Heterotrophic potentials and hydrocarbon degradation potentials of sediment microorganisms within the Athabasca oil sands deposit. *Applied and Environmental Microbiology* **41**, 791–800.

ZIMMERMAN R. & MEYER-REIL L. A. (1974) A new method for fluorescence staining of bacterial populations on membrane filters. *Kiel Meeresforschung* **30**, 24–7.

ZOBELL C. E. (1943) The effect of solid surfaces upon bacterial activity. *Journal of Bacteriology* **46**, 39–56.

Chapter 18 • Lichens

Brian W. Ferry

18.1 Introduction

Lichen ecology has developed rapidly in the last three decades, both as a result of the general resurgence of interest in lichenology as a whole and the increased availability of sophisticated techniques. Advances in 'pure' (field) ecology tend to relate to the former stimulus, and those in physiological (laboratory) ecology to the latter. This review is concerned with both aspects because there is overlap in methodology and because both approaches are essential to an understanding of the subject. In contrast to other chapters in this volume, methods in lichen ecology have much more in common with those for higher plants than for microorganisms. This reflects the extent to which the composite lichen plant differs from the simple sum of its microbial components. Consequently studies on isolated, cultured microorganisms are of little relevance in lichen ecology.

18.2 The lichen and its environment

18.2.1 The Lichen Thallus

Both field and laboratory studies demand that particular attention is paid to identification and accessibility of reference material. This is especially important with lichens because of the difficulties of growing them in culture and the consequent reliance placed on fresh material, either used in, or collected from, the field. Hawksworth (1974) provides some helpful guidance on these matters and also a comprehensive list of floras for identification (Hawksworth 1977).

Some knowledge of the basic structure of lichens is vital, not only for identification purposes but also to help in understanding how lichens interact with their environment. Readers are directed to general texts (Jahns 1973; Hale 1974; Henssen & Jahns 1974). At least as important is the need for an awareness of the extent of intraspecific morphological and anatomical variation, whether this be genotypically or phenotypically determined (Hawksworth 1973a; Weber 1977; Kunkel 1980). Environmentally induced, phenotypic variation will always be of interest to lichen ecologists in terms of the causal factors involved. However, all variants will react in a unique manner to a given set of environmental conditions and therefore the source and pre-treatment of any material taken from the field requires very careful monitoring.

Two points will suffice to illustrate the extent of this problem. It has been clearly demonstrated that lichens exhibit both spatial and temporal variations in the proportion of algal tissue (number of cells) they contain. This will affect photosynthetic capacity which is difficult to measure (Harris 1971; Millbank 1976; Collins & Farrar 1978), let alone control in experimental material. Secondly, it has been shown that lichens can compensate for environmental change to a degree, perhaps through changes in algal cell numbers. Larson and Kershaw (1975c), Kershaw and MacFarlane (1980) and Tegler and Kershaw (1980) have investigated this phenomenon of acclimation in relation to seasonal changes of temperature and light but it is not known how general this phenomenon is or how rapidly it occurs.

Ultrastructural studies on lichens (Peveling 1973) have become increasingly important, both in taxonomy and to some extent in ecology. The same is true of lichen chemistry, the study of lichen substances, particularly in relation to taxonomy (Santesson 1973). There is also evidence of environmentally associated variation of lichen substances within species (Culberson & Culberson 1967; Sheard 1978). It is not always clear whether these ultrastructural and chemical variations are the consequences of, or adaptations to, environmental pressures and whether they influence performance.

One other problem requires consideration. Almost any field-collected material will exhibit physical damage, especially when separated from its substrate or reduced to convenient-sized fragments for experimental purposes. Physiological processes such as respiration are certain to be affected by such treatments and some attempt should therefore be made to minimise or at least standardise the damage.

18.2.2 Lichens and Microclimate

The growth and development of lichens in the field is influenced by a complex set of environmental factors, of which the two most important are those of microclimate and substrate (Ahmadjian & Hale 1973; Farrar 1973; Smith 1975; Seaward 1977). In certain habitats or regions, environmental pollution (Ferry *et al.* 1973) and biotic factors, such as grazing (Richardson & Young 1977), may be of overriding importance, whilst presumably interspecific competition must always be of significance (Topham 1977).

The relationship with microclimate is particularly strong. On the one hand lichens are to be observed growing in habitats subject to extremes of moisture availability, temperature and insolation, whilst on the other hand they are small, gametophytic plants, lacking sophisticated structural features for stabilising their water contents, i.e. they are poikilohydric (Blum 1973; Kappen 1973). A lichen's moisture content will mirror that of its immediate environment and will tend to fluctuate widely. The situation is not simple because both air and substrate are potential sources of moisture, the relative importance of each depending on thallus morphology and degree of contact with the substrate. A further complication results from interactions of microclimatic factors. Moisture levels and temperature tend to be related such that lichens are usually either moist and cool or dry and warm. Likewise, under conditions of high insolation, lichens are likely to be dry rather than moist. However, in the final analysis what matters to the lichen is the thallus moisture content and temperature, and the quantity and quality of light reaching the algal layer. These thallus factors and their fluctuations caused by diurnal, seasonal and other less predictable circumstances, determine the metabolic activity of the lichen, particularly in relation to its carbon balance and therefore its growth (Blum 1973; Farrar 1973).

Remarkably, not only do most lichens manage to grow in these fluctuating environments but they also apparently need them (Smith 1975). Direct evidence for this derives from the very few successful attempts to culture lichens in the laboratory. Aquatic lichens which require more-or-less constant immersion are exceptional (Ried 1960a, b,c). Although the emphasis is usually placed on fluctuating moisture regime (Farrar 1976b), it is not clear that this is always the critical factor (Harris & Kershaw 1971). Why lichens should require microclimatic fluctuations for their continued integrity is not clear. How they survive them is, in part, understood and appears to involve the ability to survive temperature extremes and high levels of insolation, to suspend most metabolic activity at low water contents and, perhaps most important, to regain rapidly full metabolic competence on being returned to favourable conditions. Polyol metabolism seems to be central to the mechanism(s) involved (Farrar 1973; Smith 1975).

Within the general range of reactions to microclimate there is clear evidence of physiological and, to a degree, morphological adaptation amongst lichens of different habitats and geographical regions (Kappen 1973). Lichens of hot deserts (Rogers 1977) and polar regions (Lindsay 1977) are good examples, and adaptations have also been noted within a species occupying closely adjacent but distinct habitats (Harris 1971). Usually the degree of physiological adaptation has been assessed in terms of net carbon dioxide fixation (or some component of it such as carbon transfer between symbionts), but nitrogenase activity (MacFarlane & Kershaw 1977) and phosphate uptake (Farrar 1973) have also been monitored.

18.2.3 Lichen Substrates

Despite the emphasis that has always been placed on substrate ecology, surprisingly little is known of how substrate factors actually influence lichens. Barkman (1958), in his extensive analysis of the ecology of corticolous lichens in Europe, identified various factors as being important, ranging from the physical and chemical characteristics of bark, including pH, to the microclimate associated with bark. Wirth (1972) provides a similar analysis for saxicolous lichens. Brodo (1973) refers to two particular sets of problems associated with studies on substrate ecology. One concerns the multiplicity of variables inevitably present in any environment and the certainty of interactions occurring between these, both of which make identification of key factors very difficult. Other problems are associated with the choice of techniques and expression of data. It is obviously important to be able to measure environmental factors in the field, and lichen ecologists must take advantage of the ever-increasing range of sophisticated microtechniques now available. But what, for example, constitutes a relevant measure of bark pH and what components of substrate mineral status and water regime influence lichens?

One approach to dealing with these complex situations—experiments with laboratory-cultured lichens—is only slowly becoming feasible. Ordination methods (Whittacker 1973) and simulation models (Kershaw & Harris 1971; Harris 1972) are useful alternatives and can provide valuable information. The transplant technique which has been well exploited in air pollution studies on lichens (Ferry *et al.* 1973; *The Lichenologist* 1974 to 1981) has been little used in basic ecological studies (Richardson 1967; Schumm 1975; Armstrong 1976, 1977a). The main problem with this technique would seem to be the setting up of adequate controls (Ferry & Coppins 1979). It has also been suggested that observations on the colonisation of artificial substrates by lichens might constitute a useful approach (Brightman & Seaward 1977), perhaps even involving manufacture of substrates to particular chemical and physical specifications. Nevertheless, none of these latter techniques is a satisfactory substitute for the fully controlled laboratory experiment using cultured lichens.

Certain other aspects of lichen-substrate interrelationships are deserving of increased attention. Relatively little work has been done on colonisation and establishment (Pyatt 1973; Bailey 1976; Topham 1977), yet it is probable that conditions for these phases of development are more exacting and critical in the lichen life-cycle than others. Likewise conditions for the formation of reproductive propagules might be quite specific. Substrate factors may well be directly involved in all of these processes.

Perhaps one problem in particular stands out as being largely unresolved, and that concerns substrates as sources of nutrients for lichens. It has been argued (Farrar 1973) that lichens need little in the way of inorganic nutrients, because of their necessarily slow rates of accumulation of carbon and consequent slow growth rates, and that the meagre supply of nutrients available in substrates, particularly rock and bark, or in washout from the air, should suffice. Therein lies the problem. Which of these two potential sources operates? There is evidence for a nutritional role for both (Tuominen & Jaakola 1973) although the balance between them may vary with species and substrate. It is difficult to reconcile general observations on lichen-substrate ecology with the notion that no nutrients are derived from the substrate. Many lichens have distinct substrate affinities and, even if allowance is made that some of these might be responses to microclimate or cases of chemical tolerance (Lange & Ziegler 1963), others are likely to be nutritional requirements. The facts that soluble nutrients are detectable in substrates (Barkman 1958; Brodo 1973) and substrate run-off (Carlisle *et al.* 1967), and that lichens are directly involved in solubilising minerals (Syers & Iskander 1973; Ascaso *et al.* 1976; Jones *et al.* 1980), reinforce the case for a nutritional role for substrates. It might be expected that crustose species, especially those immersed in their substrates, would derive a high proportion of nutrients from this source, and that fruticose species might be more reliant on aerial sources.

18.2.4 Lichen Phytosociology and Phytogeography

The sensitivity of lichens to microclimatic and substrate factors has been emphasised. Where gradients and discontinuities of habitat factors occur one observes species zonation and the presence of distinct communities. The value of ordination methods and simulation models in unravelling these relationships has been mentioned. Another approach to their study is the phytosociological one adopted by Barkman (1958) and Wirth (1972) in Europe and, more recently, in the United Kingdom by James *et al.* (1977). The value of this approach is discussed in the latter paper, and in the more general text of Hawksworth (1974).

Whilst it is appropriate to concentrate on the lichens immediate environment, this, in turn, must be influenced by the more regional climate. Lichen ecology then becomes, at this level, essentially lichen geography, with perhaps some emphasis being placed on the distribution of species rather than communities. Studies on this scale have progressed hardly at all (Seaward 1977), although mapping data for the British Isles is now sufficiently detailed to allow the impending publication of a lichen atlas. Such studies demand the accurate accumulation and coordination of vast amounts of field data, a task at present clearly beyond the resources of lichenologists in global terms. The ideal aim would seem to be a blend of the phytogeographical and phytosociological approaches—perhaps to map communities rather than species.

18.2.5 Man's Impact on Lichens

Phytosociology and large-scale mapping techniques have proved useful in assessing the extent of man's impact on lichens. Richardson (1975) and Gilbert (1977) provide a useful summary of the variety of man's influences in the British Isles. Generally they fall into two categories, direct habitat removal, to a degree counterbalanced by the provision of new habitats, and environmental pollution. Of the first type, woodland clearance has undoubtedly been the most important (Hawksworth *et al.* 1974; Rose 1974, 1976). The provision of rock habitats, in the form of buildings, walls, tombstones, etc., in lowland Britain, can be viewed as a positive contribution (Brightman & Seaward 1977). Ahti (1977) summarises evidence for similar provision in the boreal coniferous regions of north-west Europe. Air pollution, in relation to lichens, has received much attention in the last two decades (Ferry *et al.* 1973). Sulphur dioxide in the air emerges as a factor of major geographical importance in industrial countries. Ozone may be of comparable importance in global terms (James 1973; Heath 1975). In more local situations fluorides (Gilbert 1973) and heavy metals (Nieboer *et al.*, 1977) can have a marked deleterious effect on lichen communities. Hypertrophication of habitats, by road dust and agricultural fertilisers, is essentially a form of pollution, but it can result in the increase of distinctive communities, e.g. the *Xanthorion*, where these were previously rare in, or absent from, particular areas (James 1973; Coppins 1976). For further information on air pollution and lichens readers are referred to *The Lichenologist* in which lists of relevant publications have appeared since 1974.

Experimental

18.3 Mapping techniques

Studies of lichens which lead to the production of distribution maps provide vital, basic information for lichen ecologists. Since 1964 the British Lichen Society has organised the mapping of lichens in the British Isles, based on the Ordnance Survey's National Grid 10 km squares, and species distribution maps have appeared regularly in issues of *The Lichenologist*. Such maps have also been included in papers concerned, either wholly or partly, with air pollution studies (Hawksworth *et al.* 1973, 1974) which have provided much of the stimulus for the mapping scheme. The scale of mapping varies according to the objectives involved. Hawksworth and Rose (1970) produced a zone map for England and Wales based on 10 km square records, while Rose (1973), working in south-east England, used 1 km squares. Morgan-Hughes and Haynes (1973) recorded on individual trees in their study around the Fawley Oil Refinery in Southern England. Many other workers involved in air pollution studies have generated their own maps, generally of large scale, and usually centred on individual towns or isolated pollution sources.

Readers should consult the comprehensive bibliography of Hawksworth (1973b) and *The Lichenologist* (1974 to 1981) for details of these papers. Hawksworth (1973b) discusses in some detail the application of mapping techniques to pollution studies and emphasises the value of recording negative sites/areas, and the enhancement of data quality by the inclusion of site details and an indication of species performance.

The British Lichen Society maps have also proved valuable in relating lichen distributions to variations in regional climate (Coppins 1976) and, to some extent, to changes in woodland and heathland management, increased land utilisation and the spread of man-made substrates (Hawksworth *et al.* 1974). In these latter studies it has sometimes proved possible to map past distributions of species and to compare them with present day ones, a testimony to the meticulous recording of past lichenologists. The current disappearence of the elm from the British countryside, important to species like *Bacidia incompta* and *Caloplaca luteoalba*, could be usefully studied in this manner.

18.4 Phytosociological studies

Papers concerned with the phytosociological approach to the classification of lichen communities are few in number. Nevertheless, amongst these are some works of major importance which provide a very useful base for further studies (Barkman 1958; Wirth 1972; James *et al.* 1977). The methods involved in phytosociology in general are well reviewed by Shimwell (1971) and Whittacker (1973) and, in relation to lichens in particular, by Hawksworth (1974) and James *et al.* (1977).

Three areas of controversy or confusion seem to pervade the lichen phytosociological literature. The first concerns the different systems adopted, there being two main ones in recent and current usage, the Uppsala system of Du Rietz and the Zürich-Montpellier system of Braun-Blanquet. Shimwell (1971) offers a detailed, critical analysis of both. In some early studies the former system has been used by British (Laundon 1967; Hawksworth 1969, 1972) and North American (Brodo 1968) workers, and the latter system by some continental workers (Barkman 1958; Klement 1955, 1960; Wirth 1972). In more recent studies in the British Isles (Rose & James 1974) and North America (Neal & Kershaw 1973) the Zürich-Montpellier system has been favoured, and Hawksworth (1974) suggests that it might become generally adopted.

The second area of confusion concerns nomenclatural and taxonomic problems which arise in naming and classifying lichen communities as opposed to individual species. Barkman (1973) and Hawksworth (1974) provide basic guide lines, whilst Barkman *et al.* (1976) have produced a 'Code of Phytosociological Nomenclature' which should become internationally accepted. James *et al.* (1977) indicate the nature and extent of the problems which currently arise.

The third point of controversy is concerned with the recording of data. Most of the workers referred to above have used subjective methods for choosing sampling sites and then recorded tables of relevé data using either the Domin or Braun-Blanquet scales or modifications of them. These scales are explained by Shimwell (1971) and Hawksworth (1974). James *et al.* (1977) make a case for using subjective, intuitive methods for choosing sampling sites, especially for preliminary reports. Neal and Kershaw (1973), on the other hand, used a procedure which generated randomly distributed sites. They favoured this objective approach and quoted the evaluation of Moore *et al.* (1970) in support of their view.

One further important point is worth making. The aim of the phytosociologist is to delimit and classify communities as if they were individual, discrete species, whilst at the same time needing to appreciate that they are likely to be noda selected from a continuum of vegetation types (James *et al.* 1977). Providing this is realised then phytosociology remains a valid and useful approach in community ecology.

18.5 Ordination techniques

Lichen ecologists are interested in all variations in lichen communities, both continuous and discrete, with respect to the environmental factors responsible for their existence. Much early work of this nature for epiphytic communities is summarised in Barkman (1958). More recently, it has proved valuable to use ordination techniques to analyse complex situations where there are many interacting variables and particularly where both lichen vegetation and environmental factors tend to exist as gradients. Ordination techniques are critically analysed by Whittacker (1973), under the general headings of direct and indirect gradient analysis.

The former analysis is suited to studies where environmental gradients are discernible and can be measured and the latter to studies where such gradients are not obvious or may be absent. Sheard and Jonescu (1974) provide a brief, simplified summary of ordination techniques, and also of classification techniques, and comments on their value in lichen ecology.

Ordination techniques, generally using principal components analysis, but often in conjunction with other forms of analysis, have been used successfully to identify gradients in a variety of lichen communities, and often to identify tentatively the relevant environmental gradients. In north-west Canada Kershaw and his colleagues have worked extensively on the terricolous lichen communities of a raised beach system (Kershaw & Rouse 1973; Neal & Kershaw 1973; Larson & Kershaw 1974; Pierce & Kershaw 1976) and spruce woodland (Maikawa & Kershaw 1976). The communities of objectively delimited sites were assessed by various means, a Domin scale in less critical studies and percentage cover and/or biomass where more detailed information was required. By means of principal components analysis of data, community gradients were identified which could variously be related to age of raised beach, microtopography, soil moisture and water table levels, pH, depth of peat and phases in post-fire colonisation. Rogers (1972a,b) and Rogers and Lange (1972), using similar techniques, related the distribution of terricolous lichens of Australian arid lands to rainfall and seasonal temperature. Webber (1978), in his study of terricolous tundra lichens in Alaska, used the polar ordination method developed by Bray and Curtis (1957), together with cluster analysis (Sokal & Sneath 1973), to relate vegetational noda to soil moisture, hydrogen sulphide and soluble phosphate levels. Lechowicz and Adams (1974a), in their studies on *Cladonia* spp. in Canada, used a combination of principal components analysis and a modified Bray and Curtis (1957) technique to relate lichen distribution to moisture, light and temperature gradients. Similarly Puckett and Finegan (1980) used principal component analysis to establish patterns of metal content in lichen thalli and See and Bliss (1980) used Bray and Curtis, combined with reciprocal, averaging ordinations to relate lichen distributions to soil pH and moisture content.

Yarranton (1967), Fletcher (1973a,b), Bates (1975) and Larson (1980b) used ordination techniques in their studies of saxicolous lichen communities. Yarranton (1967) used plotless samples in which he recorded contacts of lichen thalli at points objectively fixed by a grid, whilst Bates (1975) estimated local frequency at subjectively fixed sites. Both workers analysed their data by conventional principal components analysis (Kendall 1957). Fletcher (1973a,b) measured percentage cover of species and analysed his data by an ordination technique of reciprocal averaging (Hill 1973) which he considered superior to other methods. Finally Larson (1980b) used a Domin scale to assess his species and an analysis involving reciprocal averaging after Gauch *et al.* (1977). Yarranton (1967) considered soil depth, associated with field cover and boulder size, to be important and Fletcher (1973a,b) identified wetness/dryness, microtopography and bird perch areas as being important. Likewise Bates (1975) related species distribution to salt spray and sun/shade and Larson (1980b) related the distribution of *Umbilicaria* species to slope and exposure.

Studies of corticolous lichens, utilising ordination techniques, have been largely restricted to North America. Beals (1965), Jonescu (1970) and Adams and Risser (1971) all used the polar ordination technique of Bray and Curtis (1957). Beals (1973) and Sheard and Jonescu (1974) discuss the relative merits of this technique versus principal components analysis. Beals (1965) identified light, bark pH and water-holding capacity as being of probable significance, and Adams and Risser (1971) identified regional variations of climate as being important. More recently Jesberger and Sheard (1973) and Sheard and Jonescu (1974) compared the value of ordination techniques (principal components analysis) and classification methods (agglomerative polythetic classification—Orloci 1967). They found excellent agreement in terms of the results obtained and concluded that they could be mutually corroborative. Besides some specific factor(s) related to tree species, they identified general moisture, light intensity, bark pH, nutrient content and age as being important in their studies.

18.6 Measurement of lichen abundance

The quantitative description of vegetation has essentially two aims:

(1) to provide information, in the form of either absolute or relative amount of plant material

present, which can be related to environmental gradients or discontinuities;

(2) to provide an absolute measure of vegetation bulk as biomass which can be used to calculate productivity, essential to studies on ecosystem energetics.

Methods for obtaining these measures are thoroughly discussed in Greig-Smith (1964), Kershaw (1973) and Goldsmith and Harrison (1976).

Biomass, as g m^{-2} or kg ha^{-1}, is clearly the ideal measure for both of the above purposes, the main drawback being that the direct and most accurate method of measurement is destructive. For this reason it has been relatively little used in lichen ecology. Larson and Kershaw (1974) measured biomass of terricolous lichens growing on raised beaches. Williams *et al.* (1978) and Webber (1978) included biomass data for terrestrial lichens in detailed studies of the Alaskan tundra, and Kjelvik and Karenlampi (1975) and Wiegolaski (1975) summarised similar detailed work for Fennoscandia. Coppins and Shimwell (1971) and Crittenden (1975) have recorded biomass of terrestrial lichens, on English moorlands and Arctic tundra respectively. Some workers have measured the biomass of epiphytic fruticose and foliose lichens in relation to either growth rate (Platt & Amsler 1955), food chain (Edwards *et al.* 1960) or mineral cycling (Pike 1978) studies. Ahti (1977) refers to estimates of lichen biomass, made by several workers in relation to reindeer and caribou management. In all of these studies biomass was measured by drying samples to constant weight at 70 to 100°C, following detachment of the lichens from their substrates.

Clearly crustose lichens present special difficulties in this regard. Fletcher (1972a) developed a method for estimating the biomass of saxicolous crustose lichens by dissolving them from their substrates with acid dichromate, making the assumption that the carbon content of lichens is constant. This assumption is probably not justified and the method has not been adopted by other workers.

A few attempts have been made to estimate oven dry weight biomass indirectly, from either air dry weight measurements (Kärenlampi 1971a; Kärenlampi *et al.* 1975), or measurements of podetia diameters in *Cladonia* spp. (Yarranton 1975). Although such methods have the merit of being non-destructive, inevitably they lack accuracy.

Percentage cover has been widely used in ecological studies on lichens and is particularly suited to the two-dimensional form of crustose and foliose species. In many phytosociological and related studies it has proved sufficient to estimate cover subjectively by means of Domin or Braun-Blanquet scales (Bliss 1980), or even a simple DAFOR scale as used by Watson (1918a,b, 1919, 1925, 1932, 1936) in his pioneer studies of lichen communities in the British Isles. The limitations of such methods are discussed in Goodall (1952), Greig-Smith (1964), Kershaw (1973) and Goldsmith and Harrison (1976), the single clear advantage being speed. These same workers discuss the value of, and the problems associated with, the pin-hit technique used by many workers where more accurate, objectively based data was required (Sheard & Ferry 1967; Sheard 1968; Ferry & Sheard 1969; Fletcher 1973a,b). Larson and Kershaw (1974), Lechowicz and Adams (1974a), Maikawa and Kershaw (1976) and Webber (1978) have all used this method in studies on terricolous lichens, mainly fruticose forms. Here the possibility of measuring cover repetition (Goldsmith & Harrison 1976), which might be convertible to biomass, exists. Workers are generally agreed that straight percentage cover cannot be converted to biomass because of thickness variations in thalli.

Frequency of occurrence has also been much used by lichen ecologists, particularly in relation to mapping (Hawksworth 1973b) and pollution (Ferry *et al.* 1973) studies and occasionally elsewhere (Alvin 1960; Jesberger & Sheard 1973; Webber 1978). It has the merit of being rapid but is not without problems, particularly those associated with sample and quadrat size.

Both cover and frequency measurements are non-destructive and can be used for repeated measurements in time. The same is true of density but this particular measure has found little favour with lichen ecologists. Individual thalli can vary greatly in size and, in the case of fruticose species, may not be discernible. Yarranton (1975) was, however, able to record numbers of individuals of *Cladonia stellaris* and to show changes with time.

Finally, brief mention must be made of sampling methods. Quadrats have been employed almost universally to define individual samples, except in cases where natural units, e.g. whole trees, have been used. The distribution of samples may be subjectively determined (Kjelvik & Kärenlampi 1975; Wiegolaski 1975; James et al. 1977; Webber 1978) or objectively determined. In the latter case line or belt transects have proved especially appropriate (Alvin 1960; Sheard 1968; Ferry & Sheard 1969; Rogers 1972a; Fletcher 1973a; Neal & Kershaw 1973; Kershaw 1974b; Larson & Kershaw 1974), whilst some workers have divided investigation areas into strata (Sheard & Ferry 1967; Kershaw & Rouse 1973; Lechowicz & Adams 1974a) or sites (See & Bliss 1980) which can be subjectively or objectively delimited. Most workers have subsequently distributed quadrats in a random fashion within belt transects or strata, or systematically along line transects. These and other methods of sampling, together with problems of sample number and quadrat size, are discussed in Goldsmith and Harrison (1976).

18.7 Measurement of lichen growth rates

Growth is the most fundamental and comprehensive response any plant can make to its environment. Both higher plants and fungi in general are amenable to growth rate measurements because:

(1) they grow reasonably quickly;

(2) they can often be grown in controlled environments.

Generally lichens fulfil neither of these requirements and consequently measurements of growth rates have been few and, until recently, of somewhat limited ecological value. Most of the early work was aimed at obtaining measures of long-term, averaged growth rates. Over the last decade some improvement in the level of resolution obtainable with some techniques has been made (Hale 1973; Topham 1977). Various methods, both direct and indirect, have been used. Hooker and Brown (1977) discuss the relative merits of the two most useful techniques for short-term studies, direct measurement on the thallus itself and photography.

Hakulinen (1966) and Hale (1970) measured lobe tip growth directly, the former using base marks on the substrate, the latter using base marks on the non-growing parts of thallus lobes. Armstrong (1973, 1974, 1975, 1976, 1977a,b) has developed this method in his studies of saxicolous lichens. Using a lens fitted with a micrometre scale, he measured growth in relation to base marks scratched on the substrate. In two of his studies transplanted thalli were used (Armstrong 1976, 1977a).

Frey (1959) was one of the first workers to use the photographic technique to record changes in lichen thalli with time, but in his case to follow succession and establishment. Hakulinen (1966), Phillips (1969), Hale (1970), Seaward (1976), Hooker and Brown (1977), Fisher and Proctor (1978) and Hooker (1980) have all adopted this method specifically to measure growth rates. Hooker and Brown (1977) concluded that the direct and photographic methods are of comparable resolution, 0.5 to 1 mm, but that the latter is quicker and provides a permanent record which can be assessed at leisure in the laboratory. The photographic technique is, however, more expensive and perhaps demanding more expertise. Both methods require that the lichen be in a dry state when measurements are made and that the substrate be reasonably flat.

Other methods, employed by various workers in the past, are of less value in critical short-term studies. Tracing thalli at intervals onto suitable transparent materials (Hale 1954, 1959; Rydzak 1956, 1961; Brodo 1964, 1965; Seaward 1976) is inherently inaccurate, especially where expanding bark substrates are involved. Estimating the age of thalli from dated substrates such as gravestones (Beschel 1958) and twigs (Platt & Amsler 1955; Degelius 1964; Hale 1974) takes no account of initial colonisation time, nor of yearly, seasonal and other variations in growth rate. Farrar (1974) offers a variation of this method which relies on measuring and plotting frequency of thallus size of a species, where comparison can be made between two populations growing on identical substrates, differing only in their known ages. Provided such populations can be found the method has the merit of being quick. Finally mention should be made of certain methods which have been used for assessing the age of, or measuring the growth rates of, podetia of *Cladonia* spp., based either on the number of branch whorls produced (Andreev 1954) or the length of the longest living internode (Ahti 1961; Scotter 1963; Prince 1974). Again one must conclude that they are too imprecise for short-term critical studies.

Opinion varies as to the most suitable units for expressing lichen growth rates. Linear growth, as mm yr^{-1} extension of thallus lobe or margin, has been most widely used. The measure is directly comparable with measuring radial growth rates of cultured fungal mycelia and suffers from the same shortcoming, that linear growth cannot easily be related to biomass increase. Woolhouse (1968) proposed the use of relative growth rate, expressed as an increase in area per unit pre-existing area per unit time ($cm^2\ cm^{-2}\ yr^{-1}$). Several workers have commented on the value of this measure (Hale 1973; Farrar 1974; Topham 1977) and Woolhouse himself agrees that one problem is measurement of the pre-existing area where the central part of the thallus is missing or moribund. It is even doubtful whether all of the living thallus contributes, by translocation of photosynthate, to the growing margin.

A few workers have attempted to measure growth rates of lichens in terms of biomass increases in the field. This is only possible with thalli which are free of substrates and even then thalli have to be weighed air dry, a measure which is bound to be inconsistent (Miller 1966; Kärenlampi 1971a; Williams *et al.* 1978).

Using the more accurate direct or photographic methods discussed earlier, various workers have been able to measure short-term growth responses to season (Phillips 1963; Hale 1970; Armstrong 1973, 1975; Showman 1976), rainfall (Fisher & Proctor 1978), aspect (Hakulinen 1966; Armstrong 1975), nutrient supply (Hakulinen 1966; Jones & Platt 1969), height above sea level (Hakulinen 1966), level of air pollution (Seaward 1976) and thallus age (Armstrong 1974).

Finally brief mention must be made of the technique of lichenometry, the ageing of substrates from measurements of lichen thalli growing on them (Beschel 1961; Webber & Andrews 1973; Richardson 1975). The technique has found particular application in dating glaciers but is likely to be of value only where such long time-scales are involved.

18.8 Lichen colonisation

The ability to colonise new substrates is obviously important. Colonisation may well be the most critical phase in terms of microclimatic, nutritional and substrate requirements, and susceptibility to air pollution and yet the process remains largely a mystery as recent reviews testify (Pyatt 1973; Bailey 1976). Colonisation can be divided for convenience into dispersal (liberation and transport) of propagules from old substrates, and impaction and establishment (germination) on new substrates. Our knowledge of all of these events in lichens is very fragmentary. Equally uncertain is the colonisation role, if any, of some of the several types of propagule produced by lichens.

In laboratory experiments using wind tunnels, Brodie and Gregory (1953) and Bailey (1964, 1966a) subjected dry or wetted thalli of several common sorediate species to varying wind velocities to test for soredial liberation. In addition Bailey (1966a, 1968) looked at the effectiveness of mist drops, water drop impaction and water flow. In general soredia were released, particularly following wetting. Other evidence, comprising records of soredia in snow fall (Du Rietz, 1931) and in spore traps (Pettersson 1940; Bailey 1966a; Rudolph 1970) indicated the efficacy of wind as a transport agent. From a number of simple observations it appears that many invertebrates, including Collembola, Acari, Coleoptera, Lepidoptera (larvae), Neuroptera (lacewing larvae) and Hymenoptera (ants), disperse soredia on their bodies (Gerson & Seaward 1977). Bailey and James (1979) were able to recognise soredia in washings and brushings from the feet of birds. Clearly animals are likely to be effective dispersal agents over short to moderate distances, and possibly long distances in the case of birds. Evidence that sorediate species are the most efficient long distance colonisers derives from global distributional studies of sorediate and non-sorediate species (Poelt 1970, 1972; Hale 1974). Man cannot be ignored as a transport agent (Ahti 1977), although specific transport by man has not been demonstrated. In an interesting study of short-distance colonisation, Tapper (1976) derived a mathematical formula to define the spread of two sorediate species along an avenue of trees, making the assumptions that soredia were the propagules involved and that wind was the transport agent. In a sense this is the only complete colonisation study to date.

Isidia have generally been regarded as being effective colonisation propagules (Barkman 1958), yet evidence for this is lacking. Kershaw and Millbank (1970), following their observations on the development of squamulose isidia in *Peltigera aphthosa* var. *variolosa*, successfully grew detached specimens of these structures in a growth chamber. There is no clear evidence that the isidia of this or any other species are dispersed from parent thalli, although Nienburg (1919) and Armstrong (1978) provide indirect evidence for this in two other isidiate species.

Thallus fragments are likely to be important reproductive propagules in many lichens, but again evidence is scant. There are no reports of detachment of such fragments from parent thalli, although quite clearly this does occur. Thus, thallus fragments have been identified in snow fall (Du Rietz 1931), spore traps (Pettersson 1940; Bailey 1966a; Rudolph 1970), on various invertebrates (Gerson & Seaward 1977) and on the feet of birds (Bailey & James 1979). In relation to establishment, Dibben (1971) observed that macerated thallus material of several *Cladonia* spp. grew on sterilised soil in a phytotron. Vagant or erratic forms (Rogers 1977; Weber 1977) seem to be a rather special case of whole thallus dispersal.

The role, if any, of mycobiont ascospores and conidia in colonisation is unclear. Several workers have observed ascospore liberation (Scott 1959, 1964; Bailey & Garrett 1968; Pyatt 1968b, 1969, 1973; Garrett 1971) from a number of species, and ascospores, identifiable as lichen structures, have been recorded in the air spora (Ingold 1965; Rudolph 1970). More effort has been directed towards investigating germination of ascospores (Ahmadjian 1961, 1967; Bailey 1966b; Fox 1966; Garrett 1968; Pyatt 1973) and conidia (Vobis 1977). The techniques employed in all these studies were those standard for non-lichenised fungi (Booth 1971; Ingold 1971). An overriding consideration is whether ascospores and conidia ever initiate new lichen plants for which purpose an algal partner would be required. The observations that certain lichen genera liberate ascospores with algal cells attached (Poelt 1973), and that the slug *Limax flavus* voids viable ascospores and algal cells of *Lecania erysibe* together (McCarthy & Healey 1978) constitute the only good evidence for a colonisation role for ascospores. Alternative functions for ascospores and conidia are discussed in Bailey (1976) together with the possible problem of continued vegetative propagation leading to a loss of genetic plasticity.

18.9 Lichen competition

Competition between living organisms for environmental resources is widespread and lichens must presumably be involved in such interactions. However, evidence for this is rather sparse, both with regard to the existence of competition and the mechanisms involved (Topham 1977). Lichens must presumably compete with each other, both inter- and intraspecifically, for space, and with other plants, perhaps mainly for light. In this latter instance lichens seem likely to succumb to shading effects, although vigorous types like *Peltigera* spp. may overtop or otherwise compete successfully with grasses and bryophytes. Allelopathic inhibition by lichens of grass seed germination (Rondon 1966; Pyatt 1967, 1968a), root growth of *Cucumis* sp. (Miller *et al.* 1965) and moss protonemal establishment (Keever 1957) have been reported.

Competition amongst saxicolous lichens has been investigated by Frey (1959), Beschel (1961), Beschel and Weideck (1973) and Hawksworth and Chater (1979) using a photographic technique, and Pentecost (1980) by direct observation. All of these workers concentrated on describing marginal contacts between lichen thalli, mostly in crustose species. Pentecost (1980) produced quantitative data, in which large numbers of interspecific interactions of crustose species growing on a calcareous wall and on rhyolyte tuft were assessed and analysed statistically. He recorded interactions as did Stebbing (1973) for encrusting invertebrates on seaweeds, in which he recognised overgrowth, truce and epiphytic situations. Other workers mentioned the occurrence of inhibition zones (Beschel & Weideck 1973), and intraspecific fusion (Frey 1959) and repulsion of thalli (Beschel 1961). The possibility that some of these interactions might involve biochemical mechanisms has not been explored.

The competitive ability of lichens, compared with most other groups of plants, is bound to be low because of their slow growth rates and small size. To a considerable degree lichens avoid direct competition with other plants by exploiting their own distinctive habitats. Where avoidance is not complete, e.g. corticolous lichens being subjected to canopy shade and muscicolous and terricolous species growing amongst mosses and other low vegetation, the general effect is for tolerant lichen communities to be selected (Degelius 1940; Frey 1959). Clearly the ability of lichens to compete with each other will vary with changing environmental conditions, including those associated with cyclical vegetational changes (Maikawa & Kershaw 1976), higher plant succession (Alvin 1960) and those, like air pollution, imposed by man (Gilbert 1974). Given a particular set of environmental conditions, the ability of a lichen species to compete with other lichens will be determined by speed and density of colonisation, growth rate, thallus form, outcome of marginal contact interactions and rate of senescence (Topham 1977; Pentecost 1980). Data at present available on these attributes is too fragmentary to be of value in calculating competitive indices.

18.10 Successional phenomena involving lichens

Directional and cyclical changes of higher vegetation with time are well documented in the literature (Miles 1978). A number of workers have considered the involvement of lichens in such phenomena, including Topham (1977). She emphasises that lichens are rarely, if ever, contributory to primary successions in the classical sense of modifying their environments to the advantage of future different plant generations (Odum 1969); rather that they are 'carried along' with the more important higher vegetation.

This seems to be distinctly true of epiphytic lichen communities, a view held by Yarranton (1972). With the aid of simple, quantitative (Kershaw 1964) ordination (Yarranton 1972) and phytosociological methods (Hale 1955; Rose & James 1974; James *et al.* 1977), many epiphytic communities have been described which relate to stages in successional and cyclical phenomena in higher vegetation. The contribution of terricolous lichen communities to vegetational change is probably more important, especially at high latitudes where lichens often predominate (Ahti 1977). In such regions, and using ordination techniques, lichen communities have been related to stages in successional and cyclical phenomena associated with woodlands (Robinson 1959; Lambert & Maycock 1968; Lechowicz & Adams 1974a; Maikawa & Kershaw 1976) and raised beaches (Kershaw & Rouse 1973). Elsewhere similar demonstrations, based on the use of simple descriptive or quantitative methods, have been made for heathlands (Gimingham 1951, 1964; Coppins & Shimwell 1971), shingle (Scott 1965) and sand dunes (Alvin 1960). As stated by Syers and Iskander (1973), Richardson (1975) and Topham (1977) the involvement of saxicolous lichen communities in the lithosphere, particularly in relation to pedogenesis, continues to be a subject of much debate. What is clear is that freshly exposed rock surfaces do become colonised by distinct lichen communities, different from those on rocks of great age (Schauer 1969; Orwin 1970, 1972) and such differences can be used to date rock exposures, notably glacial moraines (Gilbert *et al.* 1969; Miller & Birkeland 1974). These studies have tended to involve the use of simple quantitative methods.

The extent to which lichens do alter their environments is in need of further study. There is evidence to suggest that epiphytic lichens can change the water relations (Barkman 1958) and pH (Barkman 1958; Kershaw 1964) of bark, although it is doubtful whether such effects can influence successional change. Terricolous lichen communities must contribute organic matter to developing soils, yet there is a surprising lack of data to support this. Shields (1957) and Shields *et al.* (1957) have demonstrated increased carbon and nitrogen content in lichen-encrusted soils of hot arid regions. Protection from erosion is another probable role of lichens in such areas (Weber 1962, 1967; Cameron & Black 1966). Finally, saxicolous lichen communities are known to be able to accelerate biogeophysical and biogeochemical weathering of rock to produce a more substantial weathered rind than would otherwise exist (Syers & Iskander 1973). Recently Ascaso *et al.* (1976), using X-ray diffraction and spectrophotometric methods, and Hallbauer and Jahns (1977), using SEM and X-ray microanalysis, have provided unequivocal evidence of chemical and morphological alterations to rock minerals by lichens. The significance of these activities in the very long-term build up of soil is uncertain (Topham 1977).

The occurrence of distinctive lichen communities associated with particular stages of vegetational development is established. One characteristic of the lichen communities associated with later successional stages is their greater species diversity (Robinson 1959; Degelius 1964; Gilbert *et al.* 1969; Coppins & Shimwell 1971; Rose 1974, 1976; Rose & James 1974). Topham (1977) suggests that it might be of value to classify lichen species as being either r- or k-selected, as has been done for higher plants and animals (MacArthur & Wilson 1967; Pianka 1970), and she lists certain biological characteristics which could be important in such a classification.

18.11 Ecosystem studies

Productivity, defined as the rate of production of new material, and usually expressed as biomass (g or kg) or energy (J, c or C) per unit area (m^2 or ha) per unit time (yr), is a fundamental measure in ecosystem studies. Like other green plants lichens are primary producers, and it is their net primary production which is potentially available to consumer organisms, both herbivores and decomposers. For a given interval of time, net primary production comprises change in biomass (standing crop) plus losses due to death or shedding of parts plus losses to herbivores (Phillipson 1966; Chapman 1976). All three components need to be accounted for in basic ecosystem studies, and ideally gross primary production as well, to provide some measure of the lichens net production efficiency. Data on all aspects of lichen productivity are meagre, the best available being biomass changes with time, particularly in relation to the lichen–reindeer or caribou food-chains.

Williams *et al.* (1978) in their studies of terrestrial lichens of the Alaskan Arctic tundra, used two methods for estimating biomass increase. For *Dactylina arctica* thalli were divided into new, old and decomposing portions and the contribution of each of these portions to total dry weight was measured on two successive occasions. This provided information on biomass changes for individual thalli and populations. For *Cetraria richardsonii* it was possible to weigh individual, tagged thalli on successive occasions, measurements being made after 24 h in a dessicator, the thalli then being returned to the field.

Most lichens are not susceptible to the morphological analysis described above for *Dactylina arctica*, except perhaps for certain *Cladonia* spp. Prince (1974) estimated biomass increase in two *Cladonia* spp. from total dry weight measurements of samples of podetia divided by their mean age; the latter estimate relying on the observation of Andreev (1954) that one new internode is produced each year. Yarranton (1975) describes a method for estimating biomass increase in *Cladonia stellaris* from measurements of podetia diameters recorded on photographs on successive occasions. Kärenlampi (1970, 1971a,b, 1973) and Kärenlampi *et al.* (1975) used a method of estimating changes in *Cladonia* spp. by periodically weighing thalli in an air-dry state, the thalli being kept in the field in perforated, plastic boxes.

All of the above methods suffer from some lack of precision and also the fact that no account is taken of losses to herbivores and decomposers. Kjelvik and Kärenlampi (1975) and Wiegolaski (1975) in their studies of the Fennoscandian tundra estimated, on successive occasions, the biomass of various components of the terricolous vegetation, both in terms of species, plant groups and plant parts. Lichens were included with other cryptogams. Flow diagrams constructed for each sampling time enabled them to estimate production as biomass increase plus decomposition losses. In the case of Wiegolaski (1975), where enclosures prevented losses to herbivores, the estimates were essentially of net primary production. Rosswall *et al.* (1975) working in the same region estimated yearly decomposition of lichens, among other plants, with litter bags. Finally, Harris (1972) has explored the possibility of using a simulation model to predict productivity in the corticolous species *Parmelia caperata*.

There is abundant evidence that lichens serve as food sources for animals, but very little quantitative data to indicate how important this activity is. Again studies on the lichen–reindeer or caribou food chain predominate. Richardson (1975) briefly reviews the literature on a variety of lichen–animal interrelationships.

Gerson and Seaward (1977) reviewed the literature on invertebrate–lichen associations and list Collembola (springtails), certain Acari (mites), some species of Psocoptera (bark lice) and gastropod Mollusca as being the most important invertebrate herbivores on lichens. Quantitative data is limited to a few estimates of numbers of Collembola and Acari associated with certain lichens or lichen communities and even fewer estimates of fresh weight biomass of some groups of invertebrates. One of the most detailed studies (Colman 1939) gives numbers for several invertebrate groups associated with the marine lichen *Lichina pygmaea*.

Amongst the vertebrates, only certain mammals seem to be significant herbivores on lichens, most notably reindeer and caribou which rely on lichens for their winter food supply. Quantitative data on the lichen–reindeer or caribou food-chain, including lichen biomass and productivity, lichen intake by reindeer and reindeer productivity is reviewed in Gaare and Skogland (1975), Wiegolaski (1975), Ahti (1977) and Richardson and Young (1977). Much emphasis has been placed on the need to control grazing intensity, particularly of domesticated stock, and in this respect the computer simulation studies of Bunnell *et al.* (1973, 1975) are of interest.

It has been suggested that lichen-dominated ecosystems would be ideal for studying such phenomena as energy flow, nutrient cycling and animal population dynamics, because of their relative simplicity (Barrett & Kimmel 1972). The lichen–reindeer or caribou food-chain is clearly one such system for which much data is already available. Other systems involving invertebrate herbivores have yet to be studied in depth. Readers are referred to Petrusewicz and MacFayden (1970) and Grodzinski *et al.* (1975) for details of basic methods, and to Golley (1961) and Wiegolaski (1975) for an indication of what can be achieved.

Finally, it should be noted that lichens provide protection from predators for invertebrates, amphibians and reptiles, and a moderated climate for invertebrates, as well as nesting material for species of birds (Gerson & Seaward 1977; Richardson & Young 1977). The only notable quantitative investigation on these aspects is a protectional one, that of industrial melanism in the moth *Biston betularia* (Kettlewell 1961).

18.12 Measurement of microclimatic factors in the field

Information on microclimate, especially water, temperature and light regimes, is of vital importance in both field and laboratory studies on lichens. The general aim in the field is to monitor microclimatic gradients and fluctuations, whereas in the laboratory both control and measurement of these factors are desirable. A continuing aim must be to simulate, as closely as possible, field microclimate in laboratory studies. Readers are directed to Platt and Griffiths (1965), Wadsworth (1968), Kubĭn (1971), Szeicz (1975) and Painter (1976) for detailed methods in (micro)climatology, and to Monteith (1972) for information on available instruments. The value of making measurements directly on, or within, the lichen thallus is worth emphasising.

The moisture component of microclimate comprises precipitation, dew and mist droplets and water vapour, all of which are important to lichens (Barkman 1958; Wirth 1972; Blum 1973). Precipitation was recorded or noted by Rogers (1972b) in distributional studies, by Harris (1972) in productivity simulation studies, by Kershaw and Larson (1974) in modelling evapo-transpiration, by Kärenlampi *et al.* (1975) to predict thallus moisture levels and by Armstrong (1975) and Fisher and Proctor (1978) in growth studies. Lange *et al.* (1970a,b), in a field study on gas exchange in various lichens, measured dew with dew bars. Reference to early studies on air humidity measurements in the field is made by Barkman (1958). A filter paper resistance probe, first used by Harris (1969), and subsequently by Kershaw and Rouse (1971a,b), Harris (1972) and Kershaw and Larson (1974), served to measure air humidity in very close contact with lichen thalli and substrates. In a similar study Kershaw and Field (1975) used wet bulb sensors for the same purpose.

Measurement of air temperature, near or at the thallus surface, has usually been accomplished with copper-constantin thermocouples (Field *et al.* 1974; Kershaw & Larson 1974; Kershaw 1975a; Kershaw & Field 1975; Kunkel 1980; MacFarlane & Kershaw 1980a; Tegler & Kershaw 1980). In some of these studies measurements were also made with the device inserted in the thallus, a technique used by Lange *et al.* (1970a,b). The latter authors also used a platinum resistance thermometer to measure air temperatures, while other workers have employed thermistors for this purpose (Bliss & Hadley 1964; Lechowicz & Adams 1974b; Lechowicz *et al.* 1974). Radiation problems can arise with all of these instruments and need to be dealt with. Earlier field measurements of temperature with mercury in glass thermometers are mentioned in Barkman (1958).

Light levels in the field are particularly difficult to measure because of variations in both intensity and quality and perhaps some uncertainty about what needs to be measured. Solarimeters (pyranometers) which measure total radiation on a surface were used by Lange *et al.* (1970a,b), while pyrheliographs (pyrheliometers or actinometers) measuring direct solar radiation were used by Lechowicz and Adams (1973, 1974b) and Lechowicz *et al.* (1974). Kershaw and Rouse (1971a) and Kershaw and Field (1975) used a net radiometer which measures net (incident less reflected) radiation at the earth's surface. Normally these instruments cover the band 300 to 3000 nm. Many workers have used light meters which, although measuring visible radiation between 300 and 700 nm, have the main drawback of being non-uniform in their spectral response (Bliss & Hadley 1964; Harris 1972; Lechowicz & Adams 1974b; Lechowicz *et al.* 1974; Kjelvik *et al.* 1975). Barkman (1958) refers to earlier field work in which light meters and chemical methods were used. More recently some workers (Kunkel 1980; Tegler & Kershaw 1980) have used quantum (photon) sensors which measure photosynthetically useful radiation in the field.

18.13 Control and measurement of microclimatic factors in the laboratory

Relevant laboratory studies include those concerned with purely physical responses to water and temperature regimes and those concerned with physiological responses, e.g. gas exchange and nitrogenase activity, to various microclimate factors. Both will be considered here.

Many workers have studied water uptake or loss by lichens, involving either liquid water or moist air, and commencing usually with either dry (dessicated) or saturated thalli. Water gain or loss is followed by weighing thalli at intervals. Blum (1973) summarises much of this work. Specific studies with humid air streams include those of Heatwole (1966) and Showman and Rudolph (1971), who controlled humidity with saturated salt solutions and monitored thallus weight continuously, and those of Hoffman and Gates (1970), Larson and Kershaw (1976) and Larson (1979), who used ambient air of measured relative humidity and weighed thalli at intervals. Smyth (1934) compared rates of water loss from the upper and lower surfaces of *Peltigera polydactyla* by vaselining appropriate surfaces. Other workers have studied water distribution in thalli using either dissection (Smith 1960) or staining techniques (Showman & Rudolph 1971).

In the majority of physiological studies in which an infra-red gas analyser (IRGA) has been used to measure gas exchange, thalli were initially saturated and their final, usually reduced, moisture contents checked by weighing at the end of an experimental run (Ried 1960c; Lange 1966a,b, 1969a,b,; Schulze & Lange 1968; Lange & Kappen 1972; Lechowicz & Adams 1973, 1974b; Lechowicz *et al.* 1974; Larson & Kershaw 1975a,b,c; Kershaw 1977a,b; Kershaw *et al.* 1977; Kershaw & MacFarlane 1980; Larson 1980a,b; MacFarlane & Kershaw 1980a). A similar level of course monitoring was used in studies on nitrogenase activity by Kershaw (1974a), MacFarlane and Kershaw (1977) and Crittenden and Kershaw (1978). Rather more precise and continuous measurement of thallus moisture content in IRGA experiments was possible with a sensitive hydrometer, first used by Harris (1971) and subsequently by Kershaw and Rouse (1971a) and Kershaw (1972). A similar continuous weighing device was used by Eickmeyer and Adams (1973). In other IRGA studies air streams of known relative humidity have been used in which monitoring varied from the use of thermocouple psychrometers and lithium chloride sensors placed in the main air stream (Bertsch 1966; Lange *et al.* 1970a,b) to the emplacement of a hair hygrometer next to the lichen thallus (Bertsch 1966; Lange 1969b). A particularly sophisticated control system, regulating both relative humidity and temperature, was used by Lange *et al.* (1977) in field studies on desert lichens. Method details are given in Lange *et al.* (1969), Koch *et al.* (1971) and Schulze *et al.* (1972). Control of thallus moisture content is obviously limited with gas exchange and related methods where a closed system is used, e.g. manometric methods, and where thalli are necessarily saturated, e.g. in oxygen electrode and [^{14}C]-bicarbonate fixation methods.

Temperature control in laboratory experiments with lichens has invariably involved the use of water baths or constant temperature cabinets or rooms. The need to monitor thallus temperature closely is evident from several studies and for this purpose copper-constantin thermocouples are employed (Lange 1969b; Hoffman & Gates 1970; Eickmeier & Adams 1973; Lechowicz & Adams 1974b; Lechowicz *et al.* 1974; Kershaw 1975a,b, 1977a,b; Larson & Kershaw 1975c,d, 1976; Lange *et al.* 1977; MacFarlane & Kershaw 1977, 1980a; Kershaw & MacFarlane 1980; Tegler & Kershaw 1980). Thermistors have also been widely used (Lange 1965; Kershaw & Rouse 1971a; Lange & Kappen 1972; Maikawa & Kershaw 1975).

Control and measurement of light in laboratory studies on lichens has often been inadequate. The aim should be to obtain an accurate measure of light intensity and spectral composition at the thallus surface and to simulate natural daylight as closely as possible. Many workers give some indication of the light source and a measure of luminous flux density, as lumen ft^{-2} (foot candles) or lumen m^{-2} (lux), obtained with a light meter. Radiant flux density, as watts m^{-2}, was measured with a solarimeter (pyranometer) by Kershaw (1975b), Larson and Kershaw (1975c), Lange *et al.* (1977) and Larson (1979) and with a net radiometer by Eickmeier and Adams (1973), Lechowicz and Adams (1973, 1974b) and Lechowicz *et al.* (1974). This measure can usefully be combined with a spectroradiometric analysis of the light source (Harris 1971; Lechowicz & Adams 1973; Kershaw 1975b; Larson & Kershaw 1975c). Finally, in a number of studies a quantum sensor has been used to measure photosynthetically useful radiation, as microeinsteins m^{-2} s^{-1}, arriving at the thallus surface (Lechowicz & Adams 1974b; Lechowicz *et al.* 1974; Kershaw 1975b; Larson & Kershaw 1975c; Larson 1979; MacFarlane & Kershaw 1980a,b; Tegler & Kershaw 1980).

18.14 Measurement of substrate factors

Although considerable attention has been paid by lichen ecologists to the physical and chemical properties of substrates, surprisingly few measurements of some of these have been made. Moisture regime and pH have been considered most frequently, temperature and mineral status less often.

Barkman (1958) and Brodo (1973) summarised much of the literature on bark moisture regime, and emphasised some problems associated with its measurement. Moisture content has been estimated by air-drying (Des Abbeyes 1932; Culberson 1955; Hale 1955), drying in a dessicator (Hilitzer 1925; Knebel 1936) and drying at about 100°C (Billings & Drew 1938; Young 1938; Le Blanc 1962; Margot 1965; Kalgutkar & Bird 1969). Data has been variously expressed on a unit (dry) weight, volume and surface area basis. Some workers have estimated water uptake and loss in bark samples, although the value of such results is much reduced if freshly exposed bark surfaces are not sealed to prevent water exchange. Because of the uncertainty about which of these measures, if any, is relevant in terms of moisture accessibility to lichen thalli, a technique which allows all possible forms of data expression would seem advisable. There have been very few helpful studies of the moisture regime of rock substrates, an exception being that of Wirth (1972). The problems mentioned above for bark are likely to apply to rock also. Soil moisture has always been of interest to plant ecologists and pedologists and hence methods developed in these disciplines are available (Black 1965; Wadsworth 1968; Shepley 1973; Ball 1976). In studies on terricolous lichens Larson and Kershaw (1974), Webber (1978) and Williams *et al.* (1978) estimated soil moisture by drying samples at 100°C, whilst, in similar studies, Kershaw and Rouse (1971a,b) and Rouse and Kershaw (1973) used neutron probes. Kershaw (1974b) estimated water table level in augered holes.

Barkman (1958) and Brodo (1973) also summarised much information on bark pH values, and discussed the problems of acquiring and using such data. Colorimetric methods, used by some earlier workers, are too inaccurate, and electrometric methods should be used. A widely used technique is that of steeping flaked or ground pieces of bark for some hours in distilled water, prior to measuring pH with a meter (Billings & Drew 1938; Hale 1955; Kershaw 1964; Kalgutkar & Bird 1969; Jesberger & Sheard 1973). Alternatively, Carlisle *et al.* (1967) measured the pH of stem flow on tree boles. The only substantial study of rock substrate pH, in relation to lichens, is that of Mattick (1932) who compared results from both colorimetric and electrometric methods. Werner (1956) also determined the pH values of certain granites and schists in relation to lichens. Soils have inevitably received more attention because of readily available standard techniques. Mattick (1932), Brodo (1968), Rogers (1972b), Kershaw (1974b), Pierce and Kershaw (1976) and Webber (1978) all measured the pH of soil slurries with a pH meter.

Substrate temperatures have rarely been measured in field studies on lichens, although it is clear that lichen and substrate temperatures often differ (Kershaw, 1975a) and that consequently substrates may act as significant heat sources or sinks, affecting not only thallus temperature but also moisture content (Hoffman & Gates 1970; Wirth 1972). Brodo (1959) studied bark temperature in relation to lichen distribution and Rudolph (1963) recorded rock surface temperatures in Antarctica. Kershaw (1974b) measured soil temperature with the aid of thermistor probes.

The literature on substrate–lichen mineral status relationships is voluminous, much of it being summarised in Barkman (1958), Brodo (1973), James (1973), Syers and Iskander (1973), Tuominen and Jaakola (1973) and Richardson and Nieboer (1980). A wide range of, often sophisticated, methods are available for the elemental (mineral) analysis of biological materials and soils (Black 1965; Allen *et al.*, 1976) and rock (Zussman 1977) but most remain to be exploited by lichen ecologists. Atomic absorption spectroscopy was used by Nieboer *et al.* (1972) in studies on heavy metal uptake by lichens, and by Pierce and Kershaw (1976), Bates (1975) and See and Bliss (1980) to estimate levels of metals in soil and rock. A number of workers have used X-ray fluorescence spectroscopy to estimate metals and sulphur in a variety of lichens and substrates (Olkkonen & Patomaki 1973; Olkkonen 1974; Olkkonen & Takala 1975, 1976; Takala & Olkkonen 1976; Tomassini *et al.* 1976; Laaksovirta *et al.* 1976; Puckett & Finegan

1980). Hallbauer and Jahns (1977) and Lawrey (1977) used energy dispersive X-ray analysis to estimate elemental concentrations in lichens and rock particles. Finally, several simpler chemical techniques have been used by Rogers (1972b) and See and Bliss (1980) to estimate levels of sodium, potassium, nitrate, phosphate and sulphate in soils.

18.15 Measurement of physiological responses

Distribution and growth of lichens are essentially long-term responses to environmental factors. Physiological responses are generally more immediately measurable and, of the various techniques available for accomplishing this end, those directly concerned with net assimilation rate (NAR) and nitrogen fixation will be considered here.

A wide range of techniques for measuring photosynthetic (and net assimilation) rates are thoroughly reviewed in Sesták *et al.* (1971). In a series of studies on desert lichens Lange (1965, 1966a,b, 1969a,b,) and Lange *et al.* (1970a,b, 1977) used an IRGA to measure carbon dioxide exchange and, as in other similar studies on a variety of lichens (Ried 1960a,c,d; Bliss & Hadley 1964: Bertsch 1966; Schulze & Lange 1968; Lange & Kappen 1972; Eickmeier & Adams 1973; Lechowicz & Adams 1973, 1974b; Lechowicz *et al.* 1974; Collins & Farrar 1978), they employed an open air flow system. Some of this work was carried out in the field (Bliss & Hadley 1964; Schulze & Lange 1968; Lange *et al.* 1970a,b, 1977). In other IRGA studies on carbon dioxide exchange in lichens Harris (1971), Kershaw and Rouse (1971a), Kershaw (1972) and Kershaw and Larson (1974) used a closed air flow system. Although simpler in terms of plumbing, closed systems suffer the potential disadvantage of a gradually changing carbon dioxide concentration, unless the lichen is operating at carbon dioxide compensation point as in part of the study of Collins and Farrar (1978). Later, Larson and Kershaw (1975b) described a new, discreet sampling method which had the distinct advantage of enabling a much larger number of readings to be obtained in a given time. Samples of air taken from lichen incubation chambers were introduced into a very simple air flow system so as to produce a pulse reading on passing through the IRGA sample cell. Under conditions of positive NAR these workers determined that a fall in carbon dioxide concentration to 170 ppm was not limiting to their lichens. However, such a threshold value should not be universally assumed for all lichens and conditions (Green & Snelgar 1981). The method has since been used in a number of studies (Kershaw 1975b, 1977a,b; Larson & Kershaw 1975c,d; Kershaw & MacFarlane 1980; Larson 1980a; MacFarlane & Kershaw 1980a; Tegler & Kershaw 1980).

Another method for estimating carbon dioxide concentrations is with a katharometer detector in conjunction with gas liquid chromatography, but sensitivity comparable with the IRGA is only obtainable if hydrogen is used as the carrier gas. The method has not been used in lichen studies. However, in a combined study on carbon dioxide exchange and nitrogenase activity, Crittenden and Kershaw (1978) reduced carbon dioxide to methane and estimated the latter, together with ethylene derived from acetylene reduction, by gas liquid chromatography. A single sample served for both assays.

In some early carbon dioxide exchange studies several workers (Stocker 1927; Smyth 1934; Stålfelt 1939; Ried 1953, 1960d) absorbed carbon dioxide in barium hydroxide solution and titrated with acid but this method is insensitive compared with the physical methods mentioned above. Other workers have used a colorimetric method in which air is bubbled into a bicarbonate buffer containing a suitable pH indicator (Butin 1954; Lange 1953, 1956; Fletcher 1972b). Used in the form of an open air flow system sensitivity is comparable with the barium hydroxide method mentioned above, and carbon dioxide concentration can be monitored continuously. Used as closed system, after Ålvik (1939), light compensation points can be estimated. Kjelvik *et al.* (1975) described a colorimetric method for estimating carbon dioxide by absorption in a hydrazine compound but admit to the relative inaccuracy of the technique.

In other studies oxygen exchange has been monitored, most often manometrically (Scholander *et al.* 1952; Ensgraber 1954; Pearson 1970; Rogers 1971; Smith & Molesworth 1973; Farrar 1976a,b,c; Fletcher 1976; Williams *et al.* 1978) or, more rarely, polarographically with an oxygen electrode (Baddeley *et al.* 1971; Rogers 1971). Both are sensitive, the polarographic technique particularly so, but this latter unfortunately requires that the lichen thallus be fully infiltrated with water. Winkler's method of oxygen analysis used by Ried (1953) also requires full immersion of the thallus in solution.

As an alternative to measuring gas exchange,

fixation of [^{14}C]bicarbonate has been monitored in a number of physiological studies on lichens, most of no direct ecological significance. The basic technique is used by Smith and Drew (1965) and is well described in Bednar and Smith (1966) and Richardson and Smith (1968). The papers of Smith and Molesworth (1973) and Farrar (1976a,b,c), concerned with water relations and carbon fixation, are of obvious ecological significance.

The acetylene reduction technique, evaluated in Hardy *et al.* (1968) and described in detail in Postgate (1972) has considerably stimulated work on nitrogen fixation, particularly field studies. Much of the early data is summarised by Millbank Kershaw (1973) and Millbank (1976). In more recent laboratory studies nitrogenase activity has been measured in relation to thallus moisture (Henrikson & Simu 1971; Kershaw 1974a; MacFarlane & Kershaw 1977), temperature (Kallio 1973; Kelly & Becker 1975; Maikawa & Kershaw 1975; Kershaw *et al.* 1977; MacFarlane & Kershaw 1977, 1980a;) and light (Kallio *et al.* 1972; Kelly & Becker 1975; MacFarlane *et al.* 1976; MacFarlane & Kershaw 1977, 1980b). Englund and Myerson (1974), Kallio and Kallio (1975), Huss-Danell (1977) and Alexander *et al.* (1978), all working in tundra regions, used the acetylene reduction technique under field conditions.

18.16 Symbiont isolation, lichen resynthesis and culture

Ahmadjian (1967) remarked that 'When we consider a lichen, we seem to be dealing with three plants; a fungus, an alga and a composite form'. This is true in all respects and it can, therefore, be argued that experiments with isolated symbionts are likely to be irrelevant in lichen ecology. At best they serve to emphasise the extent of the morphological and physiological differences that exist between isolated symbionts and those in the intact thallus. The real value of work with isolated symbionts is in lichen taxonomy. Methods for their isolation and culture are given in Ahmadjian (1967, 1973b).

Resynthesis experiments with lichens have met with very little success to date, completion of the life cycle from ascospore to ascospore being accomplished by one species only (Ahmadjian & Heikkila 1970; Ahmadjian 1973a). Attempts to maintain and grow lichens in phytotrons have been a little more successful (Kershaw & Millbank 1969; Pearson 1970; Dibben 1971; Harris & Kershaw 1971; Galun *et al.* 1972) and such procedures are becoming routine in some laboratories (Kershaw *et al.* 1977; MacFarlane & Kershaw 1977; Kershaw & MacFarlane 1980; Tegler & Kershaw 1980). The value of having a supply of cultured lichen material is that it should be more constant than material brought in at different times from the field. The intrinsic slow growth rates of lichens would seem to preclude any possibility of producing 'single cell' resynthesised lichen material for use in physiological experiments.

Acknowledgement

The author wishes to thank Mr Peter James for his helpful and constructive reading of this chapter.

Recommended reading

AHMADJIAN V. & HALE M. E. (1973) *The Lichens*. Academic Press, New York.

BARKMAN J. J. (1958) *Phytosociology and Ecology of Cryptogamic Epiphytes*. Van Gorcum, Assen.

BROWN D. H., HAWKSWORTH D. L. & BAILEY R. H. (1976) *Lichenology: Progress and Problems*. Academic Press, London.

CHAPMAN S. B. (1976) *Methods in Plant Ecology*. Blackwell Scientific Publications, Oxford.

FERRY B. W., BADDELEY M. S. & HAWKSWORTH D. L. (1973) *Air Pollution and Lichens*. University of London, Athlone Press, London.

HAWKSWORTH D. L. (1973a) Ecological factors and species delimitation in the lichens. In: *Taxonomy and Ecology* (Ed. V. H. Heywood), pp. 31–69. Academic Press, London.

KERSHAW K. A. & MACFARLANE J. D. (1980) Physiological-environmental interactions in lichens. X. Light as an ecological factor. *New Phytologist* **84**, 687–702.

LARSON D. W. (1980a) Seasonal change in the pattern of net CO_2 exchange in *Umbilicaria lichens. New Phytologist* **84**, 349–69.

LARSON D. W. (1980b) Patterns of species distribution in an *Umbilicaria* dominated community. *Canadian Journal of Botany* **58**, 1269–79.

MACFARLANE J. D. & KERSHAW K. A. (1980a) Physiological-environmental interactions in lichens. IX. Thermal stress and lichen ecology. *New Phytologist* **84**, 669–86.

SEAWARD M. R. D. (1977) *Lichen Ecology*. Academic Press, London.

SEE M. G. & BLISS L. C. (1980) Alpine lichen-dominated communities in Alberta and the Yukon. *Canadian Journal of Botany* **58**, 2148–70.

TEGLER B. & KERSHAW K. A. (1980) Studies on lichen-dominated systems. XXIII. The control of seasonal rates of net photosynthesis by moisture, light and temperature in *Cladonia rangiferina*. *Canadian Journal of Botany* **58**, 1851–8.

WIEGOLASKI F. E. (1975) Primary productivity of alpine meadow communities. In: *Fennoscandia Tundra Ecosystems. Part 1. Plants and Microorganisms* (Ed. F. E. Wiegolaski), pp 121–8. Springer-Verlag, New York.

References

ADAMS B. D. & RISSER P. G. (1971) Some factors influencing the frequency of bark lichens in North Central Oklahoma. *American Journal of Botany* **58**, 752–7.

AHMADJIAN V. (1961) Studies on lichenized fungi. *Bryologist* **64**, 168–79.

AHMADJIAN V. (1967) *The Lichen Symbiosis*. Blaisdell, Waltham, Massachusetts.

AHMADJIAN V. (1973a) Resynthesis of lichens. In: *The Lichens* (Eds V. Ahmadjian & M. E. Hale), pp. 565–79. Academic Press, New York.

AHMADJIAN V. (1973b) Methods of Isolating and Culturing Lichen Symbionts and Thalli. In: *The Lichens* (Eds V. Ahmadjian & M. E. Hale), pp. 653–9. Academic Press, New York.

AHMADJIAN V. & HALE M. E. (1973) *The Lichens*. Academic Press, New York.

AHMADJIAN V. & HEIKKILA H. (1970) The culture and synthesis of *Endocarpon pusillum* and *Staurothele clopima*. *Lichenologist* **4**, 259–67.

AHTI T. (1961) Taxonomic studies on reindeer lichens (*Cladonia*, subgenus *Cladina*). *Suomalaisen eläin-ja kasvitieteellisen seuran vanamon tiedonannot* **32**, 1–160.

AHTI T. (1977) Lichens of the boreal coniferous zone. In: *Lichen Ecology* (Ed. M. R. D. Seaward), pp. 145–82. Academic Press, London.

ALEXANDER V., BILLINGTON M. & SCHELL D. M. (1978) Nitrogen fixation in arctic and alpine tundra. In: *Vegetation and Production Ecology of an Alaskan Arctic Tundra* (Ed. L. L. Tieszen), pp. 539–58. Springer-Verlag, New York.

ALLEN S. E., GRIMSHAW H. M., PARKINSON J. A., QUARMBY C. & ROBERTS J. D. (1976) Chemical analysis. In: *Methods in Plant Ecology* (Ed. S. B. Chapman), pp. 411–66. Blackwell Scientific Publications Oxford.

ÅLVIK G. (1939) Über Assimilation und Atmung einiger Holzgewachse im Westnorwegischen Winter. *Meddel. No 22 Vestlandets forstlige Forsoksstation (Bergen)* **6**, 1–266.

ALVIN K. L. (1960) Observations of the lichen ecology of South Haven Peninsula, Studland Heath, Dorset. *Journal of Ecology* **48**, 331–9.

ANDREEV V. N. (1954) Prirost kormovykh lishainikov i priemy ego regulirovaniia. *Trudȳ Botanicheskogo instituta. Akademiy nauk SSSR, series 3, Geobotanika* **9**, 11–74.

ARMSTRONG R. A. (1973) Seasonal growth and growth rate colony size relationships in six species of saxicolous lichens. *New Phytologist* **72**, 1023–30.

ARMSTRONG R. A. (1974) Growth phases in the life of a lichen thallus. *New Phytologist* **73**, 913–18.

ARMSTRONG R. A. (1975) The influence of aspect on the pattern of seasonal growth in the lichen *Parmelia glabratula* spp. *fuliginosa* (Fr. ex Duby) Lond. *New Phytologist* **75**, 245–51.

ARMSTRONG R. A. (1976) The influence of the frequency of wetting and drying on the radical growth of three saxicolous lichens in the field. *New Phytologist* **77**, 719–24.

ARMSTRONG R. A. (1977a) The response of lichen growth to transplantation to rock surfaces of different aspect. *New Phytologist* **78**, 473–8.

ARMSTRONG R. A. (1977b) Studies on the growth rates of lichens. In: *Lichenology: Progress and Problems* (Eds D. H. Brown, D. L. Hawksworth & R. H. Bailey), pp. 309–22. Academic Press, London.

ARMSTRONG R. A. (1978) The colonization of slate rock surface by a lichen. *New Phytologist* **81**, 85–8.

ASCASO C., GALVAN J. & ORTEGA C. (1976) The pedogenic action of *Parmelia conspersa, Rhizocarpon geographicum* and *Umbilicaria pustulata*. *Lichenologist* **8**, 151–72.

BADDELEY M. S., FERRY B. W. & FINEGAN E. J. (1971) A new method of measuring lichen respiration: Response of selected species to temperature, pH and sulphur dioxide. *Lichenologist* **5**, 18–25.

BAILEY R. H. (1964) Studies on the dispersal of lichen propagules. M.Sc. thesis, University of London.

BAILEY R. H. (1966a) Studies on the dispersal of lichen soredia. *Journal of the Linnaean Society of London, Botany* **59**, 479–90.

BAILEY R. H. (1966b) Notes on the germination of lichen ascospores. *Revue bryologique et lichénologique* **34**, 852–3.

BAILEY R. H. (1968) Dispersal of lichen soredia in water trickles. *Revue bryologique et lichénologique* **36**, 314–15.

BAILEY R. H. (1976) Ecological aspects of dispersal and establishment in lichens. In: *Lichenology: Progress and Problems* (Eds D. H. Brown, D. L. Hawksworth & R. H. Bailey), pp. 215–48. Academic Press, London.

BAILEY R. H. & GARRETT R. M. (1968) Studies on the discharge of ascospores from lichen apothecia. *Lichenologist* **4**, 57–65.

BAILEY R. H. & JAMES P. W. (1979) Birds and the dispersal of lichen propagules. *Lichenologist* **11**, 105–6.

BALL D. F. (1976) Site and soils. In: *Methods in Plant Ecology* (Ed. S. E. Chapman), pp. 297–367. Blackwell Scientific Publications, Oxford.

BARKMAN J. J. (1958) *Phytosociology and Ecology of Cryptogamic Epiphytes*. Van Gorcum, Assen.

BARKMAN J. J. (1973) Synusial approaches to classification. In: *Ordination and Classification of Communities* (Ed. R. H. Whittacker), pp. 435–91. W. Junk, The Hague.

BARKMAN J. J., MORAVEC J. & RAUSCHERT S. (1976) Code of phytosociological nomenclature. *Vegatatio* **33**, 131–85

BARRETT G. W. & KIMMEL R. G. (1972) Effects of DDT on the density and diversity of tardigrades. *Proceedings of the Iowa Academy of Science* **78**, 41–2.

BATES J. W. (1975) A quantitative investigation of saxicolous bryophyte and lichen vegetation of Cape Clear Island, County Cork. *Journal of Ecology* **63**, 143–62.

BEALS E. W. (1965) Ordination of some corticolous cryptogamic communities in South-central Wisconsin. *Oikos* **16**, 1–8.

BEALS E. W. (1973) Ordination: mathematical elegance and ecological naivete. *Journal of Ecology* **61**, 22–36.

BEDNAR T. W. & SMITH D. C. (1966) Studies in the physiology of lichens. VI. Preliminary studies of photosynthesis and carbohydrate metabolism of the lichen *Xanthoria aureola*. *New Phytologist* **65**, 211–20.

BERTSCH A. (1966) Über den CO_2-gaswechsel einiger Flechten nach Wasserdampfaufnahme. *Planta* **68**, 157–66.

BESCHEL R. E. (1958) Flechtenvereine der Städte, Stadtflechten und ihr Wachstum. *Bericht des Naturwissenschaftlich-medizinischen Vereins in Innsbruck* **52**, 1–158.

BESCHEL R. E. (1961) Dating rock surfaces by lichen growth and its application to glaciology and physiography (lichenometry). In: *Geology of the Arctic* (Ed. G. O. Raasch), pp. 1044–62. University of Toronto Press, Toronto.

BESCHEL R. E. & WEIDECK A. (1973) Geobotanical and geomorphological reconnaissance in West Greenland, 1961. *Arctic and Alpine Research* **5**, 311–19.

BILLINGS W. D. & DREW W. B. (1938) Bark factors affecting the distribution of corticolous bryophytic communities. *American Midland Naturalist* **20**, 302–30.

BLACK C. A. (1965) *Methods of Soil Analysis*. American Society of Agronomy, Madison, Wisconsin.

BLISS L. C. & HADLEY E. B. (1964) Photosynthesis and respiration of alpine lichens. *American Journal of Botany* **51**, 870–4.

BLUM O. (1973) Water relations. In: *The Lichens* (Eds V. Ahmadjian & M. E. Hale), pp. 371–400. Academic Press, New York.

BOOTH C. (1971) *Methods in Microbiology*, Vol. 4. Academic Press, London.

BRAY J. R. & CURTIS J. T. (1957) An ordination of the upland forest communities of southern Wisconsin. *Ecological Monographs* **27**, 325–49.

BRIGHTMAN F. H. & SEAWARD M. R. D. (1977) Lichens of man-made substrates. In: *Lichen Ecology* (Ed. M. R. D. Seaward), pp. 253–94. Academic Press, London.

BRODIE H. J. & GREGORY P. H. (1953) The action of wind in the dispersal of spores from cup-shaped plant structures. *Canadian Journal of Botany* **31**, 402–10.

BRODO I. M. (1959) A study of lichen ecology in central New York. M.S. thesis, Cornell University, Ithaca.

BRODO I. M. (1964) Field studies of the effects of ionising radiations on lichens. *Bryologist* **67**, 76–87.

BRODO I. M. (1965) Studies on growth rates of corticolous lichens on Long Island, New York. *Bryologist* **68**, 451–6.

BRODO I. M. (1968) The lichens of Long Island, New York: A vegetational and floristic analysis. *Bulletin of New York State Museum Scientific Service* **410**, 1–330.

BRODO I. M. (1973) Substrate ecology. In: *The Lichens* (Eds V. Ahmadjian & M. E. Hale), pp. 401–41. Academic Press, New York.

BROWN D. H., HAWKSWORTH D. L. & BAILEY R. H. (1976) *Lichenology: Progress and Problems*. Academic Press, London.

BUNNELL F. L., DAULPHINE D. C., HOLBORN R., MILLER D. R., MILLER F. L., MCEWAN E. H., PARKER G. R., PETERSEN R., SCOTTER G. W. & WALTERS C. T. (1975) Preliminary report on computer simulation of barren-ground caribou management. In: *Proceedings 1st International Reindeer/Caribou Management Symposium*, Fairbanks, Alaska.

BUNNELL F. L., KÄRENLAMPI L. & RUSSELL D. E. (1973) A simulation model of lichen-*Rangifer* interactions in northern Finland. *Report of the Kevo Subarctic Research Station* **10**, 1–8.

BUTIN H. (1954) Physiologisch-ökologische Untersuchungen über den Wasserhaushalt und die Photosynthese bei Flechten. *Biologisches Zentralblatt* **73**, 459–502.

CAMERON R. E. & BLACK G. B. (1966) Desert algae: soil crusts and diaphanous substrata as algal habitats. *NASA Technical Report* 32–291, 1–41.

CARLISLE A., BROWN A. H. F. & WHITE E. J. (1967) The nutrient content of tree stem flow and ground flora litter and leachates in a sessile oak (*Quercus petraea*) woodland. *Journal of Ecology* **55**, 615–27.

CHAPMAN S. B. (1976) *Methods in Plant Ecology*. Blackwell Scientific Publications, Oxford.

COLLINS C. R. & FARRAR J. F. (1978) Structural resistances to mass transfer in the lichen *Xanthoria parietina*. *New Phytologist* **81**, 71–83.

COLMAN J. (1939) On the faunas inhabiting intertidal seaweeds. *Journal of the Marine Biological Association of the United Kingdom* **24**, 129–83.

COPPINS B. J. (1976) Distribution patterns shown by epiphytic lichens in the British Isles. In: *Lichenology: Progress and Problems* (Eds D. H. Brown, D. L. Hawksworth & R. H. Bailey), pp. 249–78. Academic Press, London.

COPPINS B. J. & SHIMWELL D. W. (1971) Cryptogam complement and biomass in dry *Calluna* heath of different ages. *Oikos* **22**, 204–9.

CRITTENDEN P. D. (1975) Nitrogen fixation by lichens on glacial drift in Iceland. *New Phytologist* **74**, 41–9.

CRITTENDEN P. D. & KERSHAW K. A. (1978) A procedure for the simultaneous measurement of net CO_2-exchange and nitrogenase activity in lichens. *New Phytologist* **80**, 393–401.

CULBERSON W. L. (1955) The corticolous communities of lichens and bryophytes in the upland forests of Northern Wisconsin. *Ecological Monographs* **25**, 215–31.

CULBERSON W. L. & CULBERSON C. F. (1967) Habitat selected by chemically differentiated races of lichens. *Science* **158**, 1195–7.

DEGELIUS G. (1940) Studien über die Konkurrenzverhältnisse der Laubflechten auf nakten Fels. *Acta Horti Gothoburgensis* **14**, 195–219.

DEGELIUS G. (1964) Biological studies of epiphytic vegetation on twigs of *Fraxinus excelsior*. *Acta Horti Gothoburgensis* **27**, 11–55.

DES ABBEYES H. (1932) Contribution à l'étude des qualités écologiques du substratum des lichens: Hygrometrie des écorces. *Seances de la Société française de biologique. Paris* **109**, 1096–9.

DIBBEN M. J. (1971) Whole lichen culture in a phytotron. *Lichenologist* **5**, 1–10.

DU RIETZ G. E. (1931) Studier over vinddriften på snöfalt i de skandivaviska fjallen. *Botaniska Notiser* 1931, 31–46.

EDWARDS R. Y., SOOS J. & RITSEY R. W. (1960) Quantitative observations on epidendric lichens used as food by caribou. *Ecology* **41**, 425–31.

EICKMEIER W. G. & ADAMS M. S. (1973) Net photosynthesis and respiration of *Cladonia ecmocyna* (Ach.)Nyl. from the Rocky Mountains and comparison with three eastern alpine species. *American Midland Naturalist* **89**, 58–69.

ENGLUND B. & MYERSON H. (1974) In site measurement of nitrogen fixation at low temperatures. *Oikos* **25**, 283–7.

ENSGRABER A. (1954) Über der Einfluss der Antrocknung auf die Assimilation und Atmung von Moosen und Flechten. *Flora (Jena)* **141**, 433–75.

FARRAR J. F. (1973) Lichen physiology: progress and pitfalls. In: *Air Pollution and Lichens* (Eds B. W. Ferry, M.S. Baddeley & D. L. Hawksworth), pp. 238–82. University of London, Athlone Press, London.

FARRAR J. F. (1974) A method for investigating lichen growth rates and succession. *Lichenologist* **6**, 151–5.

FARRAR J. F. (1976a) Ecological physiology of the lichen *Hypogymnia physodes*. I. Some effects of constant water saturation. *New Phytologist* **77**, 93–103.

FARRAR J. F. (1976b) Ecological physiology of the lichen *Hypogymnia physodes*. II. Effects of wetting and drying cycles and the concept of physiological buffering. *New Phytologist* **77**, 105–13.

FARRAR J. F. (1976c) Ecological physiology of the lichen *Hypogymnia physodes*. III. The importance of the rewetting phase. *New Phytologist* **77**, 115–25.

FERRY B. W., BADDELEY M. S. & HAWKSWORTH D. L. (1973) *Air Pollution and Lichens*. University of London, Athlone Press, London.

FERRY B. W. & COPPINS B. J. (1979) Lichen transplant experiments and air pollution studies. *Lichenologist* **11**, 63–73.

FERRY B. W. & SHEARD J. W. (1969) Zonation of supralittoral lichens on rocky shores around the Dale Peninsula, Pembrokeshire. *Field Studies* **3**, 41–67.

FIELD G., LARSON D. W. & KERSHAW K. A. (1974) Studies on lichen-dominated systems. VIII. The instrumentation of a raised beach ridge for temperature and wind speed measurements. *Canadian Journal of Botany* **52**, 1927–34.

FISHER P. J. & PROCTOR M. C. F. (1978) Observations on seasons growth in *Parmelia caperata* and *P. sulcata* in South Devon. *Lichenologist* **10**, 81–90.

FLETCHER A. (1972a) A method for estimating the dry weights of crustaceous saxicolous lichens. *Lichenologist* **5**, 314–16.

FLETCHER A. (1972b) *The Ecology of Marine and Maritime Lichens of Anglesey*. Ph.D. thesis, University of Wales.

FLETCHER A. (1973a) The ecology of marine (littoral) lichens on some rocky shores of Anglesey. *Lichenologist* **5**, 368–400.

FLETCHER A. (1973b) The ecology of maritime (supralittoral) lichens on some rocky shores of Anglesey. *Lichenologist* **5**, 401–22.

FLETCHER A. (1976) Nutritional Aspects of Marine and Maritime Lichen Ecology. In: *Lichenology: Progress and Problems* (Eds D. H. Brown, D. L. Hawksworth & R. H. Bailey), pp. 359–84. Academic Press, London.

FOX C. H. (1966) Experimental studies on the physiology of the phycobiont and mycobiont *Ramalina ecklonii*. *Physiologia Plantarum* **19**, 830–9.

FREY E. (1959) Die Flechtenflora und -Vegetation des National parks im Unterengadin. II. Die Entwicklung der Flechtenvegetation auf photogrammetrisch kontrollierten Dauerflachen. *Ergebnisse der wissenschaftlichen Untersuchung des Schweizerischen Nationalparks* **6**, 237–319.

GAARE E. & SKOGLAND T. (1975) Wild reindeer food habits and range use at Hardangervidda. In: *Fennoscandian Tundra Ecosystems. Part 2. Animals and Systems Analysis* (Ed. F. E. Wiegolaski), pp. 195–205. Springer-Verlag, Berlin.

GALUN M., MARTON K. & BEHR L. (1972) A method for the culture of lichen thalli under controlled conditions. *Archiv für Mikrobiologie* **83**, 189–92.

GARRETT R. M. (1968) Observations on the germination of lichen ascospores. *Revue Bryologique et Lichénologique* **36**, 330–2.

GARRETT R. M. (1971) Studies on some aspects of ascospore liberation and dispersal in lichens. *Lichenologist* **5**, 33–44.

GAUCH H. G., WHITTACKER R. H. & WENTWORTH T. R. (1977) A comparative study of reciprocal averaging and other ordination techniques. *Journal of Ecology* **65**, 157–74.

GERSON U. & SEAWARD M. R. D. (1977) Lichen-invertebrate associations. In: *Lichen Ecology* (Ed. M. R. D. Seaward), pp. 69–120. *Academic* Press, London.

GILBERT O. L. (1973) The effects of airborne fluorides. In: *Air Pollution and Lichens* (Eds B. W. Ferry, M. S. Baddeley & D. L. Hawksworth), pp. 176–91. University of London, Athlone Press, London.

GILBERT O. L. (1974) Lichens and air pollution. In: *The Lichens* (Eds V. Ahmadjian & M. E. Hale), pp. 443–72. Academic Press, London.

GILBERT O. L. (1977) Lichen conservation in Britain. In: *Lichen Ecology* (Ed. M. R. D. Seaward), pp. 415–36. Academic Press, London.

GILBERT O. L., JAMIESON D., LISTER H. & PEDLINGTON A. (1969) Regime of an Afghan glacier. *Journal of Glaciology* **8**, 51–65.

GIMINGHAM C. H. (1951) Contribution to the maritime ecology of St. Cyrus, Kincardineshire. II. The sand dunes. *Transactions and Proceedings of the Botanical Society of Edinburgh* **35**, 387–411.

GIMINGHAM C. H. (1964) Dwarf shrub heaths. In: *The Vegetation of Scotland* (Ed. J. H. Burnett), pp. 232–87. Oliver & Boyd, Edinburgh.

GOLDSMITH F. B. & HARRISON C. M. (1976) Description and analysis of vegetation. In: *Methods in Plant Ecology* (Ed. S. B. Chapman), pp. 85–156. Blackwell Scientific Publications, Oxford.

GOLLEY F. B. (1961) Energy values of ecological materials. *Ecology* **42**, 581–4.

GOODALL D. W. (1952) Some considerations in the use of point quadrats for the analysis of vegetation. *Australian Journal of Scientific Research* **B5**, 1–41.

GREEN T. G. A. & SNELGAR W. P. (1981) Carbon dioxide exchange in lichens. Relationship between net photosynthetic rate and CO_2 concentration. *Plant Physiology* **68**, 199–210.

GREIG-SMITH P. (1964) *Quantitative Plant Ecology*, 2e. Butterworths, London.

GRODZINSKI W., KLEKOWSKI R. Z. & DUNCAN A. (1975) *Methods for Ecological Bioenergetics*. Blackwell Scientific Publications, Oxford.

HAKULINEN R. (1966) Über die Wachstumgeschwindigkeit einiger Lambflechten. *Annales Botanici Fennici* **3**, 167–79.

HALE M. E. (1954) First report on lichen growth rate and succession at Aton Forest, Connecticut. *Bryologist* **57**, 244–7.

HALE M. E. (1955) Phytosociology of corticolous cryptogams in the upland forests of southern Wisconsin. *Ecology* **36**, 45–63.

HALE M. E. (1959) Studies on lichen growth rate and succession. *Bulletin of the Torrey Botanical Club* **66**, 126–9.

HALE M. E. (1970) Single lobe growth rate patterns in the lichen *Parmelia caperata*. *Bryologist* **73**, 72–81.

HALE M. E. (1973) Growth. In: *The Lichens* (Eds V. Ahmadjian & M. E. Hale), pp. 473–94. Academic Press, New York.

HALE M. E. (1974) *The Biology of Lichens* 2e. Edward Arnold, London.

HALLBAUER D. K. & JAHNS H. M. (1977) Attack of lichens on quartzitic rock surfaces. *Lichenologist* **9**, 119–22.

HARDY R. W. F., HOLSTEN R. D., JACKSON E. K. & BURNS R. C. (1968) The acetylene-ethylene assay for N_2 fixation: laboratory and field evaluation. *Plant Physiology* **43**, 1185–207.

HARRIS G. P. (1969) A Study of the Ecology of Corticolous Lichens. Ph.D. thesis, University of London.

HARRIS G. P. (1971) The ecology of corticolous lichens. II. The relationship between physiology and the environment. *Journal of Ecology* **59**, 441–52.

HARRIS G. P. (1972) The ecology of corticolous lichens. III. A simulation model of productivity as a function of light intensity and water availibility. *Journal of Ecology* **60**, 19–40.

HARRIS G. P. & KERSHAW K. A. (1971) Thallus growth and the distribution of stored metabolites in the phycobionts of the lichens *Parmelia sulcata* and *P. physodes*. *Canadian Journal of Botany* **49**, 1367–72.

HAWKSWORTH D. L. (1969) The lichen flora of Derbyshire. *Lichenologist* **4**, 105–93.

HAWKSWORTH D. L. (1972) The natural history of Slapton Ley Nature Reserve. IV. Lichens. *Field Studies* **3**, 535–78.

HAWKSWORTH D. L. (1973a) Ecological factors and species delimitation in the lichens. In: *Taxonomy and Ecology* (Ed. V. H. Heywood), pp. 31–69. Academic Press, London.

HAWKSWORTH D. L. (1973b) Mapping studies. In: *Air Pollution and Lichens* (Eds B. W. Ferry, M. S. Baddeley & D. L. Hawksworth), pp. 38–76. University of London, Athlone Press, London.

HAWKSWORTH D. L. (1974) *Mycologist's Handbook*. Commonwealth Mycological Institute, London.

HAWKSWORTH D. L. (1977) A bibliographic guide to the lichen floras of the world. In: *Lichen Ecology* (Ed. M. R. D. Seaward), pp. 437–502. Academic Press, London.

HAWKSWORTH D. L. & CHATER A. O. (1979) Dynamism and equilibrium in a saxicolous lichen mosaic. *Lichenologist* **11**, 75–80.

HAWKSWORTH D. L., COPPINS B. J. & ROSE F. (1974) Changes in the British lichen flora. In: *The Changing Fauna and Flora of Britain* (Ed. D. L. Hawksworth), pp. 47–78. Academic Press, London.

HAWKSWORTH D. L. & ROSE F. (1970) Qualitative scale for estimating sulphur dioxide air pollution in England and Wales using epiphytic lichens. *Nature, London* **227**, 145–8.

HAWKSWORTH D. L., ROSE F. & COPPINS B. J. (1973)

Changes in the lichen flora of England and Wales attributable to pollution of the air by sulphur dioxide. In: *Air Pollution and Lichens* (Eds B. W. Ferry, M. S. Baddeley & D. L. Hawksworth), pp. 330–67. University of London, Athlone Press, London.

HEATH R. L. (1975) Ozone. In: *Responses of Plants to Air Pollution* (Eds J. B. Mudd & T. T. Kozlowski), pp. 23–56. Academic Press, New York.

HEATWOLE H. (1966) Moisture exchange between the atmosphere and some lichens of the genus *Cladonia*. *Mycologia* **58**, 148–56.

HENRIKSSON E. & SIMU B. (1971) Nitrogen fixation by lichens. *Oikos* **22**, 119–21.

HENSSEN A. & JAHNS H. M. (1974) *Lichenes*. Georg Thieme Verlag, Stuttgart.

HILITZER A. (1925) Etude sur la végétation épiphyte de la Boheme. *Spisy vydávané Přírodorédeckou fakultón Karlovy University* **41**, 1–202.

HILL M. O. (1973) Reciprocal averaging; an eigenvector method of ordination. *Journal of Ecology* **61**, 237–49.

HOFFMAN G. R. & GATES D. M. (1970) An energy budget approach to the study of water loss in cryptogams. *Bulletin of the Torrey Botanical Club* **97**, 361–6.

HOOKER T. N. (1980) Lobe growth and marginal zonation in crustose lichens. *Lichenologist* **12**, 313–24.

HOOKER T. N. & BROWN D. H. (1977) A photographic method for accurately measuring the growth of crustose and foliose lichens. *Lichenologist* **9**, 65–76.

HUSS-DANELL K. (1977) Nitrogen fixation by *Stereocaulon paschale* under field conditions. *Canadian Journal of Botany* **55**, 585–92.

INGOLD C. T. (1965) *Spore Liberation*. Clarendon Press, Oxford.

INGOLD C. T. (1971) *Fungal Spores. Their Liberation and Dispersal*. Clarendon Press, Oxford.

JAHNS H. M. (1973) Anatomy, morphology and development. In: *The Lichens* (Eds V. Ahmadjian & M. E. Hale), pp. 3–58. Academic Press, New York.

JAMES P. W. (1973) The effect of air pollutants other than hydrogen fluoride and sulphur dioxide on lichens. In: *Air Pollution and Lichens* (Eds B. W. Ferry, M.S. Baddeley & D. L. Hawksworth), pp. 143–75. University of London, Athlone Press, London.

JAMES P. W., HAWKSWORTH D. L. & ROSE F. (1977) Lichen communities in the British Isles: A preliminary conspectus. In: *Lichen Ecology* (Ed. M. R. D. Seaward), pp. 295–414. Academic Press, London.

JESBERGER J. A. & SHEARD J. W. (1973) A quantitative survey and multivariate analysis of corticolous lichen communities in the southern boreal forest of Saskatchewan. *Canadian Journal of Botany* **51**, 185–201.

JONES D., WILSON M. J. & TAIT J. M. (1980) Weathering of a basalt by *Pertusaria corallina*. *Lichenologist* **12**, 277–90.

JONESCU M. E. (1970) Lichens on *Populus tremuloides* in west-central Canada. *Bryologist* **73**, 557–78.

KALGUTKAR R. M. & BIRD C. D. (1969) Lichens found on *Larix lyellii* and *Pinus albicaulis* in southwestern Alberta, Canada. *Canadian Journal of Botany* **47**, 627–48.

KALLIO S. (1973) The ecology of nitrogen fixation in *Stereocaulon paschale*. *Report of the Kevo Subarctic Research Station* **10**, 34.

KALLIO S. & KALLIO P. (1975) Nitrogen fixation in lichens at Kevo, North Finland. In: *Fennoscandian Tundra Ecosystems. Part 1. Plants and Microorganisms* (Ed. F. E. Wiegolaski), pp. 292–304. Springer-Verlag, Berlin.

KALLIO P., SUOHEN S. & KALLIO H. (1972) The ecology of nitrogen fixation in *Nephroma arcticum* and *Solorina crocea*. *Report of the Kevo Subarctic Research Station* **9**, 7–14.

KAPPEN L. (1973) Response to extreme environments. In: *The Lichens* (Eds V. Ahmadjian & M. E. Hale), pp. 311–80. Academic Press, New York.

KÄRENLAMPI L. (1970) Morphological analysis of the growth and productivity of the lichen, *Cladonia alpestris*. *Report of the Kevo Subarctic Research Station* **7**, 9–15.

KÄRENLAMPI L. (1971a) Studies on the relative growth rate of some fruticose lichens. *Report of the Kevo Subarctic Research Station* **7**, 33–9.

KÄRENLAMPI L. (1971b) On methods for measuring and calculating energy flow through lichens. *Report of the Kevo Subarctic Research Station* **7**, 40–6.

KÄRENLAMPI L. (1973) Biomass and estimated yearly net production of the ground vegetation at Kevo. In: *Primary Production and Production Processes* (Eds L. C. Bliss & F. E. Wiegolaski), pp. 111–14. Edmonton–Oslo: IBP Tundra Biome Steering Committee 1973.

KÄRENLAMPI L., TAMMISOLA J. & HURME H. (1975) Weight increase of some lichens as related to carbon dioxide exchange and thallus moisture. In: *Fennoscandian Tundra Ecosystems. Part I. Plants and Microorganisms* (Ed. F. E. Wiegolaski), pp. 135–7. Springer-Verlag, Berlin.

KEEVER C. (1957) Establishment of *Grimmia laevigata* on bare granite. *Ecology* **38**, 422–9.

KELLY B. B. & BECKER V. E. (1975) Effects of light intensity and temperature on nitrogen fixation by *Lobaria pulmonaria, Sticta wiegelli, Leptogium cyanescens* and *Collema subfurvum*. *Bryologist* **78**, 350–8.

KENDALL M. G. (1957) *A Course in Multivariate Analysis*. Griffin, London.

KERSHAW K. A. (1964) Preliminary observations on the distribution and ecology of epiphytic lichens in Wales. *Lichenologist* **2**, 263–76.

KERSHAW K. A. (1972) The relationship between moisture content and net assimilation rate of lichen thalli and its ecological significance. *Canadian Journal of Botany* **50**, 543–55.

KERSHAW K. A. (1973) *Quantitative and Dynamic Plant Ecology*, 2e. Edward Arnold, London.

KERSHAW K. A. (1974a) Dependence of level of nitrogenase activity on the water content of the thallus in *Peltigera canina, P. evansiana, P. polydactyla* and *P. praetextata. Canadian Journal of Botany* **52**, 1423–7.

KERSHAW K. A. (1974b) Studies on lichen-dominated systems. X. The sedge meadows of the coastal raised beaches. *Canadian Journal of Botany* **52**, 1947–72.

KERSHAW K. A. (1975a) Studies on lichen-dominated systems. XII. The ecological significance of thallus colour. *Canadian Journal of Botany* **53**, 660–7.

KERSHAW K. A. (1975b) Studies on lichen-dominated systems. XIV. The comparative ecology of *Alectoria nitidula* and *Cladina alpestris. Canadian Journal of Botany* **53**, 2608–13.

KERSHAW K. A. (1977a) Physiological-environmental interactions in lichens. II. The pattern of net photosynthetic acclimation in *Peltigera canina* (L.) Willd. var. *praetextata* (Floerke in Somm.) Hue. and *P. polydactyla* (Neck) Hoffm. *New Phytologist* **79**, 377–90.

KERSHAW K. A. (1977b) Physiological-environmental interactions in lichens. III. The rate of net photosynthetic acclimation in *Peltigera canina* (L.) Willd. var. *praetextata* (Floerke in Somm.) hue, and *P. polydactyla* (Neck.) Hoffm. *New Phytologist* **79**, 391–402.

KERSHAW K. A. & FIELD G. F. (1975) Studies on lichen-dominated systems. XV. The temperature and humidity profiles in a *Cladonia alpestris* mat. *Canadian Journal of Botany* **53**, 2614–20.

KERSHAW K. A. & HARRIS G. P. (1971) Simulation studies and ecology: use of the model. In: *Statistical Ecology* (Eds G. P. Patil, E. C. Pielou & W. E. Waters), vol. 3, pp. 23–42. Pennsylvania State University Press, University Park, Pennsylvania.

KERSHAW K. A. & LARSON D. W. (1974) Studies on lichen-dominated systems. IX. Topographic influences on microclimate and species distribution. *Canadian Journal of Botany* **52**, 1935–45.

KERSHAW K. A. & MACFARLANE J. D. (1980) Physiological-environmental interactions in lichens. X. Light as an ecological factor. *New Phytologist* **84**, 687–702.

KERSHAW K. H., MACFARLANE J. D. & IGSIANCZAY M. J. (1977) Physiological-environmental interactions in lichens. V. The interaction of temperature with nitrogenase activity in the dark. *New Phytologist* **79**, 409–16.

KERSHAW K. & MILLBANK J. W. (1969) A controlled environment lichen growth chamber. *Lichenologist* **4**, 83–7.

KERSHAW K. A. & MILLBANK J. W. (1970) Isidia as vegetative propagules in *Peltigera aphthosa* var. *variolosa* (Massal.) Thoms. *Lichenologist* **4**, 214–17.

KERSHAW K. A. & ROUSE W. R. (1971a) Studies on lichen-dominated ecosystems. I. The water relations of *Cladonia alpestris* in spruce–lichen woodland and in northern Ontario. *Canadian Journal of Botany* **49**, 1389–99.

KERSHAW K. A. & ROUSE W. R. (1971b) Studies on lichen-dominated ecosystems. II. The growth pattern of *Cladonia alpestris* and *Cladonia rangiferina. Canadian Journal of Botany* **49**, 1401–10.

KERSHAW K. A. & ROUSE W. R. (1973) Studies on lichen-dominated systems. V. A primary survey of a raised beach system in northwestern Ontario. *Canadian Journal of Botany* **51**, 1285–307.

KETTLEWELL H. B. D. (1961) The phenomenon of industrial melanism in Lepidoptera. *Annual Review of Entomology* **6**, 245–62.

KJELVIK S. & KÄRENLAMPI L. (1975) Plant biomass and primary production of Fennoscandian Subarctic and subalpine forests and of alpine willow and heath ecosystems. In: *Fennoscandian Tundra Ecosystems. Part 1. Plants and Microorganisms* (Ed. F. E. Wiegolaski), pp. 111–20. Springer-Verlag, New York.

KJELVIK S., WIEGOLASKI F. E. & JAHREN A. (1975) Photosynthesis and respiration studied by field technique at Hardangervidda, Norway. In: *Fennoscandian Tundra Ecosystems. Part 1. Plants and Microorganisms* (Ed. F. E. Wiegolaski). pp. 184–93. Springer-Verlag, New York.

KLEMENT O. (1955) Prodromus der mitteleuropäischen Flechtengesellschaften. *Reprium nov Spec Regni veg Beih* **135**, 5–194.

KLEMENT O. (1960) Zur Flechtenvegetation der Oberpfalz. *Bericht der Bayerischen botanischen Gesellschaft zur Erforschung der heimischen Flora* **28**, 250–75.

KNEBEL G. (1936) Monographie der Algenreihe der Prasiolales, insbesondere von *Prasiola crispa. Hedwigia* **75**, 1–120.

KOCH W., LANGE O. L. & SCHULZE E-D. (1971) Ecophysiological investigations on wild and cultivated plants in the Negev Desert. *Oecologia* **8**, 296–309.

KUBÍN S. (1971) Measurement of Radiant Energy. In: *Plant Photosynthetic Production. Manual of Methods* (Eds Z. Šesták, J. Čatský & R. G. Jarvis), pp. 702–65. W. Junk, The Hague.

KUNKEL G. (1980) Microhabitat and structural variation in the *Aspicilia desertorum* group (lichenized Ascomycetes). *American Journal of Botany* **67**, 1137–45.

LAAKSOVIRTA K., OLKKONEN H. & ALAKUIJALA P. (1976) Observations of the lead content of lichen and bark adjacent to a highway in southern Finland. *Environmental Pollution* **11**, 247–55.

LAMBERT M. J. & MAYCOCK P. F. (1968) The ecology of terricolous lichens of the Northern Conifer-Hardwood Forests of central Eastern Canada. *Canadian Journal of Botany* **46**, 1043–78.

LANGE O. L. (1953) Hitze- und Trockenresistenz der Flechten in Beziehung zu ihrer Verbreitung. *Flora (Jena)* **140**, 39–97.

LANGE O. L. (1956) Zur methodik der kolorimetrischen CO_2-Bestimmung nach Alvik. *Berichten der Deutschen Botanischen Gesellschaft* **69**, 49–60.

LANGE O. L. (1965) Der CO_2-Gaswechsel von Flechten bei tiefen Temperaturen. *Planta* **64**, 1–19.

LANGE O. L. (1966a) Der CO_2-Gaswechsel von Flechten

nach Erwärmung im feuchten Zustand. *Berichten der Deutschen Botanischen Gesellschaft* **78**, 441–54.

LANGE O. L. (1966b) CO_2-Gaswechsel der Flechte *Cladonia alcicornis* nach langfristigem Aufenthalt bei tiefen Temperaturen. *Flora (Jena)* **156**, 500–2.

LANGE O. L. (1969a) Die funktionellen Anpassungen der Flechten an die ökologischen Bedingungen arider Gebiete. *Berichten der Deutschen Botanischen Gesellschaft* **82**, 3–22.

LANGE O. L. (1969b) Experimentell-ökologische Untersuchungen an Flechten der Negev-Wüste. I. CO_2-Gaswechsel von *Ramalina maciformis* (Del.) Borg unter kontrollierten Bedingungen in laboratorium. *Flora (Jena)* **158**, 324–59.

LANGE O. L., GEIGER I. L. & SCHULZE E-D. (1977) Ecophysiological investigations on lichens in the Negev Desert. V. A model to simulate net photosynthesis and respiration of *Ramalina maciformis*. *Oecologia* **28**, 247–59.

LANGE O. L. & KAPPEN L. (1972) Photosynthesis of lichens from Antarctica. *Antarctic Research Series* **20**, 83–95.

LANGE O. L., KOCH W. & SCHULZE E-D. (1969) CO_2-Gaswechsel und Wasserhaushalt von Pflanzen in der Negev-Wüste em Ende der Trockenzeit. *Berichten der Deutschen Botanischen Gesellschaft* **82**, 39–61.

LANGE O. L., SCHULZE E-D. & KOCH W. (1970a) Experimentellökologische Untersuchungen an Flechten der Negev-Wüste. II. CO_2-Gaswechsel und Wasserhaushalt von *Ramalina maciformis* (Del.) Bory am natürlichen Standort während der sommerlichen Trockenperiode. *Flora (Jena)* **159**, 38–62.

LANGE O. L., SCHULZE E-D. & KOCH W. (1970b) Experimentellökologische Untersuchungen an Flechten der Negev-Wüste. III. CO_2-Gaswechsel und Wasserhaushalt von Krusten- und Blatten-flechten am natürlichen Standort während der sommerlichen Trockenperiode. *Flora (Jena)* **159**, 525–38.

LANGE O. L. & ZIEGLER H. (1963) Der Schwermetallgehalt von Flechten aus dem *Acrosporetum sinopicae* auf Erzschlackenhalden des Hurzes. 1. Eisen und Kupfer. *Mitteilungen der Floristsoziologischen Arbeitsgemeinschaft* **10**, 156–83.

LARSON D. W. (1979) Lichen water relations under drying conditions. *New Phytologist* **82**, 713–31.

LARSON D. W. (1980a) Seasonal change in the pattern of net CO_2 exchange in *Umbilicaria* lichens. *New Phytologist* **84**, 349–69.

LARSON D. W. (1980b) Patterns of species distribution in an *Umbilicaria* dominated community. *Canadian Journal of Botany* **58**, 1269–79.

LARSON D. W. & KERSHAW K. A. (1974) Studies on lichen-dominated systems. VII. Interaction of the general lichen heath with edaphic factors. *Canadian Journal of Botany* **52**, 1163–76.

LARSON D. W. & KERSHAW K. A. (1975a) Acclimation in arctic lichens. *Nature, London* **254**, 421–3.

LARSON D. W. & KERSHAW K. A. (1975b) Measurement of CO_2 exchange in lichens: a new method. *Canadian Journal of Botany* **53**, 1535–41.

LARSON D. W. & KERSHAW K. A. (1975c) Studies on lichen-dominated systems. XIII. Seasonal and geographical variation of net CO_2 exchange of *Alectoria ochroleuca*. *Canadian Journal of Botany* **53**, 2598–607.

LARSON D. W. & KERSHAW K. A. (1975d) Studies on lichen-dominated systems. XVI. Comparative patterns of net CO_2 exchange in *Cetraria nivalis* and *Alectoria ochroleuca*. *Canadian Journal of Botany* **53**, 2884–92.

LARSON D. W. & KERSHAW K. A. (1976) Studies on lichen-dominated systems. XVIII. Morphological control of evaporation in lichens. *Canadian Journal of Botany* **54**, 2061–73.

LAUNDON J. R. (1967) A study of the lichen flora of London. *Lichenologist* **3**, 277–327.

LAWREY J. D. (1977) X-ray emission microanalysis of *Cladonia cristatella* from a coal strip-mining area in Ohio. *Mycologia* **69**, 855–60.

LE BLANC F. (1962) Hydrométrie des écorces et épiphytisme. *Revue Canadienne de Biologie* **21**, 41–5.

LECHOWICZ M. J. & ADAMS M. S. (1973) Net photosynthesis of *Cladonia mitis* (Sand.) from sun and shade sites on the Wisconsin Pine Barrens. *Ecology* **54**, 413–19.

LECHOWICZ M. J. & ADAMS M. S. (1974a) Ecology of *Cladonia* lichens. I. Preliminary assessment of the ecology of terricolous lichen–moss communities in Ontario and Wisconsin. *Canadian Journal of Botany* **52**, 55–64.

LECHOWICZ M. J. & ADAMS M. S. (1974b) Ecology of *Cladonia* lichens. II. Comparative physiology and ecology of *C. mitis*, *C. rangiferina* and *C. uncialis*. *Canadian Journal of Botany* **52**, 411–22.

LECHOWICZ M. J., JORDAN W. P. & ADAMS M. S. (1974) Ecology of *Cladonia* lichens. III. Comparison of *C. caroliniana*, endemic to southeastern North America, with three northern *Cladonia* species. *Canadian Journal of Botany* **52**, 565–73.

LINDSAY D. C. (1977) Lichens of cold deserts. In: *Lichen Ecology* (Ed. M. R. D. Seaward), pp. 183–210. Academic Press, London.

MACARTHUR R. H. & WILSON F. O. (1967) *The Theory of Island Biogeography*. University Press, Princeton.

MCCARTHY P. M. & HEALEY J. A. (1978) Dispersal of lichen propagules by slugs. *Lichenologist* **10**, 131.

MACFARLANE J. D. & KERSHAW K. A. (1977) Physiological-environmental interactions in lichens. IV. Seasonal changes in nitrogenase activity of *Peltigera canina* (1.) Willd. var. *praetextata* (Floerke in Somm.) Hue., and *P. canina* (L.) Willd. var. *rufescens* (Weiss) Mudd. *New Phytologist* **79**, 403–8.

MACFARLANE J. D. & KERSHAW K. A. (1980a) Physiological-environmental interactions in lichens. IX. Thermal stress and lichen ecology. *New Phytologist* **84**, 669–86.

MacFarlane J. D. & Kershaw K. A. (1980b) Physiological-environmental interactions in lichens. XI. Snowcover and nitrogenase activity. *New Phytologist* **84**, 703–10.

MacFarlane J. D., Maikawa E., Kershaw K. A. & Oaks A. (1976) Environmental-physiological interactions in lichens. I. The interaction of light/dark periods and nitrogenase activity in *Peltigera polydactyla*. *New Phytologist* **77**, 705–11.

Maikawa E. & Kershaw K. A. (1975) The temperature dependance of thallus nitrogenase activity in *Peltigera canina*. *Canadian Journal of Botany* **53**, 527–9.

Maikawa E. & Kershaw K. A. (1976) Studies on lichen-dominated systems. XIX. The postfire recovery sequence of black spruce–lichen woodland in the Abitou Lake Region, N.W.T. *Canadian Journal of Botany* **54**, 2679–87.

Margot J. (1965) Evolution de la végétation epiphytique du peuplier en relation avec l'âge et les modification de l'écorce. *Annales Universitatis Mariae Curie-Sklodowska* **C22**, 159–63.

Mattick F. (1932) Bodenreaktion und Flechtenverbreitung. *Beihefte Botanischen Zentralblatt* **49**, 241–71.

Miles J. (1978) *Vegetation Dynamics*. Chapman & Hall, London.

Millbank J. W. (1976) Aspects of nitrogen metabolism in lichens. In: *Lichenology: Progress and Problems* (Eds D. H. Brown, D. L. Hawksworth & R. H. Bailey), pp. 441–56. Academic Press, London.

Millbank J. W. & Kershaw K. A. (1973) Nitrogen metabolism. In: *The Lichens* (Eds V. Ahmadjian & M. E. Hale), pp. 289–309. Academic Press, New York.

Miller A. G. (1966) Lichen growth and species composition in relation to duration of thallus wetness. B.Sc. Thesis, Queen's University, Kingston.

Miller C. D. & Birkeland P. W. (1974) Probable preneoglacial age of the type Temple Lake Moraine, Wyoming: discussion and additional relative age data. *Arctic and Alpine Research* **6**, 301–6.

Miller E. V., Griffin C. E., Schaeffers T. & Gordon M. (1965) Two types of growth inhibitors in extracts from *Umbilicaria papulosa*. *Botanical Gazette* **126**, 100–7.

Monteith J. L. (1972) *Survey of Instruments for Micrometeorology*. Blackwell Scientific Publications, Oxford.

Moore J. J., Fitzsimmons P., Lamb E. & White J. (1970) A comparison and evaluation of some phytosociological techniques. *Vegetatio* **20**, 1–20.

Morgan-Hughes D. I. & Haynes F. N. (1973) Distribution of some epiphytic lichens around an Oil refinery at Fawley, Hampshire. In: *Air Pollution and Lichens* (Eds B. W. Ferry, M.S. Baddeley & D. L. Hawksworth), pp. 89–108. University of London, Athlone Press, London.

Neal M. W. & Kershaw K. A. (1973) Studies on lichen-dominated systems. III. Phytosociology of a raised beach system near Cape Henrietta Maria, northern Ontario. *Canadian Journal of Botany* **51**, 1115–25.

Nieboer E., Ahmed H. M., Puckett K. J. & Richardson D. H. S. (1972) Heavy metal content of lichens in relation to distance from a nickel smelter in Sudbury, Ontario. *Lichenologist* **5**, 292–304.

Nieboer E., Puckett K. J., Richardson D. H. S., Tomassini F. D. & Grace B. (1977) Ecological and physicochemical aspects of the accumulation of heavy metals and sulphur in lichens. In: *International Conference on Heavy Metals in the Environment*, Symposium Proceedings **2**, (Ed. T. C. Hutchinson), 331–52, Toronto.

Nienburg W. (1919) Studien zur Biologie der Flechten, I,II,III. *Zeitschrift für Botanik* **11**, 1–38.

Odum E. P. (1969) The strategy of ecosystem development. *Science* **164**, 262–70.

Olkkonen H. (1974) Isotooppiherätteisen rontgenflouresenssimenetelmän käytto jäkälien saastukepitoisuuksien analysoinnissa. *Helsingin Yliopiston Kasvitieteen Perusop Tuksen Laitoksen Monistaita* **2**, 31–9.

Olkkonen H. & Patomaki L. (1973) Isotooppiherätteisen röntgenfluoresenssimenetelmän sopivuudesta kasvinäytteiden hivenaipe- ja saastukeanalyysiin. *Savon Luonto* **5**, 49–51.

Olkkonen H. & Takala K. (1975) Total sulphur content of an epiphytic lichen as an index of air pollution and the usefulness of the X-ray fluorescence method in sulphur determinations. *Annales Botanici Fennici* **12**, 131–4.

Olkkonen H. & Takala K. (1976) Biological monitoring of air pollution in the urban area of Kuopio. In: *Proceedings of the Second National Meeting on Biophysics and Biotechnology in Finland* (Eds A. L. Kairento, E. Riihimaki & P. Tarkka), pp. 215–17. Helsinki.

Orloci L. (1967) An agglomerative method for classification of plant communities. *Journal of Ecology* **55**, 193–206.

Orwin J. (1970) Lichen succession on recently deposited rock surfaces. *New Zealand Journal of Botany* **8**, 452–77.

Orwin J. (1972) The effect of environment on assemblages of lichens growing on rock surfaces. *New Zealand Journal of Botany* **10**, 37–47.

Painter R. (1976) Climatology and Environmental Measurement. In: *Methods in Plant Ecology* (Ed. S. B. Chapman), pp. 369–410. Blackwell Scientific Publications, Oxford.

Pearson L. C. (1970) Varying environmental factors in order to grow intact lichens under laboratory conditions. *American Journal of Botany* **57**, 659–64.

Pentecost A. (1980) Aspects of competition in saxicolous lichen communities. *Lichenologist* **12**, 135–44.

Petrusewicz K. & MacFayden A. (1970) *Productivity of Terrestrial Animals: Principles and Methods*. Blackwell Scientific Publications, Oxford.

Pettersson B. (1940) Experimentelle Untersuchungen

über die euanenomochore Verbreitung der Sporenpflanzen. *Acta Botanica Fennica* **25**, 1–103.

PEVELING E. (1973) Fine Structure. In: *The Lichens* (Eds V. Ahmadjian & M. E. Hale), pp. 147–82. Academic Press, New York.

PHILLIPS H. C. (1963) Growth rate of *Parmelia isidiosa* (Müll. Arg.) Hale. *Journal of the Tennessee Academy of Science* **38**, 95–6.

PHILLIPS H. C. (1969) Annual growth rates of three species of foliose lichens determined photographically. *Bulletin of the Torrey Botanical Club* **96**, 202–6.

PHILLIPSON J. (1966) *Ecological Energetics*. Edward Arnold, London.

PIANKA E. R. (1970) On r- and k-selection. *American Naturalist* **104**, 592–7.

PIERCE W. & KERSHAW K. A. (1976) Studies on lichen-dominated systems. XVII. The colonization of young raised beaches in NW Ontario. *Canadian Journal of Botany* **54**, 1672–83.

PIKE L. H. (1978) The importance of epiphytic lichens in mineral cycling. *Bryologist* **81**, 247–57.

PLATT R. B. & AMSLER F. P. (1955) A basic method for the immediate study of lichen growth rates and succession. *Journal of the Tennessee Academy of Science* **30**, 177–83.

PLATT R. B. & GRIFFITHS J. F. (1965) *Environmental Measurement and Interpretation*. Reinhold, London.

POELT J. (1970) Das Konzept der Artenpaare bei den Flechten. *Vorträge aus dem Gesamtgebiet der Botanik (Deutsche botanische gesellschaft)* **4**, 187–98.

POELT J. (1972) Die taxonomische Behandlung von Artenpaaren bei den Flechten. *Botaniska notiser* **125**, 77–81.

POELT J. (1973) Systematic evaluation of morphological characters. In: *The Lichens* (Eds V. Ahmadjian & M. E. Hale), pp. 91–115. Academic Press, New York.

POSTGATE J. R. (1972) The acetylene reduction test for nitrogen fixation. In: *Methods in Microbiology* (Eds J. R. Norris & D. W. Ribbons), vol. 6B, pp. 343-56. Academic Press, London.

PRINCE C. R. (1974) Growth rates and productivity of *Cladonia arbuscula* and *C. impexa* on the Sands of Forvie, Scotland. *Canadian Journal of Botany* **52**, 431–3.

PUCKETT K. J. & FINEGAN E. J. (1980) An analysis of the element content of lichens from the Northwest Territories, Canada. *Canadian Journal of Botany* **58**, 2073–89.

PYATT F. B. (1967) The inhibitory influence of *Peltigera canina* on the germination of graminaceous seeds and the subsequent growth of the seedlings. *Bryologist* **70**, 326–9.

PYATT F. B. (1968a) The effect of sulphur dioxide on the inhibitory influence of *Peltigera canina* on the germination and growth of grasses. *Bryologist* **71**, 97–101.

PYATT F. B. (1968b) An investigation into conditions influencing ascospore discharge and germination in lichens. *Revue bryologique et lichénologique* **36**, 323–9.

PYATT F. B. (1969) Studies on the periodicity of spore discharge and germination in lichens. *Bryologist* **72**, 48–53.

PYATT F. B. (1973) Lichen propagules. In: *The Lichens* (Eds V. Ahmadjian & M. E. Hale), pp. 117–45. Academic Press, New York.

RICHARDSON D. H. S. (1967) The transplantation of lichen thalli to solve some taxonomic problems in *Xanthoria parietina* (L.) Th.Fr. *Lichenologist* **3**, 386–91.

RICHARDSON D. H. S. (1975) *The Vanishing Lichens*. David and Charles, Newton Abbott.

RICHARDSON D. H. S. & NIEBOER E. (1980) Surface binding and accumulation of metals in lichens. In: *Cellular Interactions in Symbiosis and Parasitism* (Eds C. B. Cook, P. W. Pappas & E. D. Rudolph), pp. 75–94. Ohio State University Press, Columbus, Ohio.

RICHARDSON D. H. S. & SMITH D. C.(1968) Lichen physiology. IX. Carbohydrate movement from the *Trebouxia* symbiont of *Xanthoria aureola* to the fungus. *New Phytologist* **67**, 61–8.

RICHARDSON D. H. S. & YOUNG C. M. (1977) Lichens and Vertebrates. In: *Lichen Ecology* (Ed. M. R. D. Seaward), pp. 121–44. Academic Press, London.

RIED A. (1953) Photosynthese und Atmung bei xerostabilen und xerolabilen Krustenflechten in der Wachwirkung Vorausgegangener Entquellungen. *Planta* **41**, 436–8.

RIED A. (1960a) Thallusbau und Assimilationshaushalt von Laub- und Krustenflechten. *Biologisches Zentralblatt* **79**, 129–51.

RIED A. (1960b) Stoffwechsel und Verbreitungsgrenzen von Flechten. I. Flechtenzonierungen an Bachufern und ihre Beziehungen zur järlichen über Flutungsdauer und zum Mikrolima. *Flora (Jena)* **148**, 612–38.

RIED A. (1960c) Stoffwechsel und Verbreitungsgrenzen von Flechten. II. Wasser- und Assimilationshaushalt, Entquellungs- und Submersion-resistenz von Krustenflechten benechbarter Standorte. *Flora (Jena)* **149**, 345–85.

RIED A. (1960d) Nachwirkungen der Entquellung auf den Gaswechsel von Krustenflechten. *Biologisches Zentralblatt* **79**, 657–78.

ROBINSON H. (1959) Lichen succession in abandoned fields in the Piedmont of N. Carolina. *Bryologist* **62**, 254–9.

ROGERS R. W. (1971) Distribution of the lichen *Chondropsis semiviridis* in relation to its heat and drought resistance. *New Phytologist* **70**, 1069–77.

ROGERS R. W. (1972a) Soil surface lichens in arid and semi-arid south-eastern Australia. II. Phytosociology and geographic zonation. *Australian Journal of Botany* **20**, 215–27.

ROGERS R. W. (1972b) Soil surface lichens in arid and semi-arid south-eastern Australia. III. The relationship between distribution and environment. *Australian Journal of Botany* **20**, 301–16.

ROGERS R. W. (1977) Lichens of hot arid and semi-arid lands. In: *Lichen Ecology* (Ed. M. R. D. Seaward), pp. 211–52. Academic Press, London.

ROGERS R. W. & LANGE O. L. (1972) Soil surface lichens in arid and sub-arid south-eastern Australia. I. Introduction and Floristics. *Australian Journal of Botany* **20**, 197–213.

RONDON Y. (1966) Action inhibitrice de l'extrait du lichen *Roccella fucoides* (Dicks.) Vain sur la germination. *Bulletin Société botanique de France* **113**, 1–2.

ROSE F. (1973) Detailed mapping in south-east England. In: *Air Pollution and Lichens* (Eds B. W. Ferry, M.S. Baddeley & D. L. Hawksworth), pp. 77–88. University of London, Athlone Press, London.

ROSE F.(1974) The epiphytes of oak. In: *The British Oak; its History and Natural History* (Eds M. G. Morris & F. H. Perring), pp. 250–73. Classey, Faringdon.

ROSE F. (1976) Lichenological indicators of age and environmental continuity in woodlands. In: *Lichenology: Progress and Problems* (Eds D. H. Brown, D. L. Hawksworth & R. H. Bailey), pp. 279–308. Academic Press, London.

ROSE F. & JAMES P. W. (1974) Regional studies on the British lichen flora. I. The corticolous and lignicolous species of the New Forest, Hampshire. *Lichenologist* **6**, 1–72.

ROSSWALL T., VEUM A. K. & KÄRENLAMPI L. (1975) Plant litter decomposition at fennoscandian tundra sites. In: *Fennoscandian Tundra Ecosystems. Part 1. Plants and Microorganisms* (Ed. F. E. Wiegolaski), pp. 268–78. Springer-Verlag, Berlin.

ROUSE W. R. & KERSHAW K. A. (1973) Studies on lichen-dominated systems. VI. Interrelations of vegetation and soil moisture in the Hudson Bay Lowlands. *Canadian Journal of Botany* **51**, 1309–16.

RUDOLPH E. D. (1963) Vegetation of Hallett Station area, Victoria Land, Antarctica. *Ecology* **44**, 585–6.

RUDOLPH E. D. (1970) Local dissemination of plant propagules in Antarctica. In: *Antarctic Ecology* (Ed. M. W. Holdgate), vol. 2, pp. 812–17. Academic Press, New York.

RYDZAK J. (1956) A method of studying growth in lichens. *Annales Universitatis Mariae Curie-Sklodowska* **10**, 87–91.

RYDZAK J. (1961) Investigations on the growth rates of lichens. *Annales Universitatis Mariae Curie-Sklodowska* **16**, 1–15.

SANTESSON J. (1973) Identification and isolation of lichen substances. In: *The Lichens* (Eds V. Ahmadjian & M. E. Hale), pp. 633–52. Academic Press, New York.

SCHAUER T. (1969) Die Flechtenvegetation der Kiesfläche auf der Garchinger Heide nördlich von München. *Herzogia* **1**, 181–6.

SCHOLANDER P. F., FLAGG W., WALTER V. & IRVING L. (1952) Respiration in some arctic and tropical lichens in relation to temperature. *American Journal of Botany* **39**, 707–13.

SCHULZE E-D. & LANGE O. L. (1968) CO_2-Gaswechsel der Flechte *Hypogymnia physodes* bei tiefen temperaturen im Freiland. *Flora (Jena)* **158**, 180–4.

SCHULZE E-D., LANGE O. L. & LEMBKE G. (1972) A digital registration system for net photosynthesis and transpiration measurements in the field and an associated analysis of errors. *Oecologia* **10**, 151–66.

SCHUMM F. (1975) Beiträge zur photosyntheseleistung der Flechten und ihre Eignung als Maß zur Inikation der Immissionbelastung der Luft. Doctoral Dissertation, University of Hochenheim.

SCOTT G. D. (1959) Observations on spore discharge and germination in *Peltigera praetextata* (Flk.)Vain. *Lichenologist* **1**, 109–11.

SCOTT G. D. (1964) Studies of the lichen symbiosis. 2. Ascospore germination in the genus *Peltigera*. *Zeitschrift für allgemeine Mikrobiologie* **4**, 326–36.

SCOTT G. R. M. (1965) The shingle succession at Dungeness. *Journal of Ecology* **53**, 21–31.

SCOTTER G. W. (1963) Growth rates of *Cladonia alpestris, C. mitis* and *C. rangiferina* in the Taltson River Region, N.W.T. *Canadian Journal of Botany* **41**, 1199–202.

SEAWARD M. R. D. (1976) Performance of *Lecanora muralis* in an urban environment. In: *Lichenology: Progress and Problems* (Eds D. H. Brown, D. L. Hawksworth & R. H. Bailey), pp.323–58. Academic Press, London.

SEAWARD M. R. D. (1977) *Lichen Ecology*. Academic Press, London.

SEE M. T. & BLISS L. C. (1980) Alpine lichen-dominated communities in Alberta and the Yukon. *Canadian Journal of Botany* **58**, 2148–70.

SESTÁK Z., ĆATSKÝ J. & JARVIS P. G. (1971) *Plant Photosynthetic Production. Manual of Methods*. W. Junk, The Hague.

SHEARD J. W. (1968) The zonation of lichens on three rocky shores of Inishowen, Co. Donegal. *Proceedings of the Royal Irish Academy* **66B**, 101–12.

SHEARD J. W. (1978) The comparative ecology and distribution and within-species variation of the lichenized Ascomycetes *Ramalina cuspidata* and *R. siliquosa* in the British Isles. *Canadian Journal of Botany* **56**, 939–52.

SHEARD J. W. & FERRY B. W. (1967) The lichen flora of the Isle of May. *Transactions and Proceedings of the Botanical Society of Edinburgh* **40**, 268–87.

SHEARD J. W. & JONESCU M. E. (1974) A multivariate analysis of the distribution of lichens on *Populus tremeloides* in West-Central Canada. *Bryologist* **77**, 514–30.

SHEPLEY A. V. (1973) *Soil Studies*. Pergamon Press, Oxford.

SHIELDS L. M. (1957) Algal and lichen floras in relation to nitrogen content of certain volcanic and acid range soils. *Ecology* **38**, 661–3.

SHIELDS L. M., MITCHELL C. & DROUET F. (1957) Alga- and lichen-stabilized surface crusts as nitrogen sources. *American Journal of Botany* **44**, 489–98.

SHIMWELL D. W. (1971) *The Description and Classification of Vegetation*. Sidgwick and Jackson, London.

SHWMAN R. E. (1976) Seasonal growth of *Parmelia caperata*. *Bryologist* **79**, 360–3.

SHOWMAN R. E. & RUDOLPH E. D. (1971) Water relations in living, dead and cellulose models of the lichen *Umbilicaria papulosa*. *Bryologist* **74**, 444–50.

SMITH D. C. (1960) Studies in the physiology of lichens. 3. Experiments with dissected discs of *Peltigera polydactyla*. *Annals of Botany* **24**, 186–99.

SMITH D. C. (1975) Symbiosis and the biology of lichenized fungi. In: *Symbiosis* (Eds D. H. Jennings & D. L. Lee), pp. 373–405. Cambridge University Press, Cambridge.

SMITH D. C. & DREW E. A. (1965) Studies in the physiology of lichens. V. Translocation from the algal layer to the medulla in *Peltigera polydactyla*. *New Phytologist* **64**, 195–200.

SMITH D. C. & MOLESWORTH S. (1973) Lichen physiology. XIII. Effects of rewetting dry lichens. *New Phytologist* **72**, 525–33.

SMYTH E. S. (1934) A contribution to the physiology and ecology of *Peltigera canina* and *Peltigera polydactyla*. *Annals of Botany* **48**, 781-818.

SOKAL R. R. & SNEATH P. H. A. (1973) *Numerical Taxonomy*. W. H. Freeman, San Francisco.

STÅLFELT M. G. (1939) Der Gasaustausch der Flechten. *Planta* **29**, 11–31.

STEBBING A. R. D. (1973) Competition for space between the epiphytes of *Fucus serratus* L. *Journal of the Marine Biological Association of the United Kingdom* **53**, 247–61.

STOCKER O. (1927) Physiologische und ökologische Untersuchungen an Laub- und Strauchflechten. *Flora (Jena)* **121**, 334–415.

SYERS J. K. & ISKANDER I. K. (1973) Pedogenetic significance of lichens. In: *The Lichens* (Eds V. Ahmadjian M. E. Hale), pp. 225–48. Academic Press, New York.

SZEICZ G. (1975) Instruments and their exposure. In: *Vegetation and the Atmosphere* (Ed. J. L. Monteith), pp. 229–74. Academic Press, London.

TAKALA K. & OLKKONEN H. (1976) Lead content of the lichen *Pseudevernia furfuracea* in the urban of Kuopio, central Finland. In: *Proceedings of Kuopio meeting on Plant Damages caused by Air Pollution* (Ed. L. Kärenlampi), pp. 64–7. University of Kuopio, Finland.

TAPPER R. (1976) Dispersal and changes in the local distributions of *Evernia prunastri* and *Ramalina farinacea*. *New Phytologist* **77**, 725–34.

TEGLER B. & KERSHAW K. A. (1980) Studies on lichen-dominated systems. XXIII. The control of seasonal rates of net photosynthesis by moisture, light and temperature in *Cladonia rougiferina*. *Canadian Journal of Botany* **58**, 1851–8.

TOMASSINI F. D., PUCKETT K. J., NIEBOER E., RICHARDSON D. H. S. & GRACE B. (1976) Determination of copper, iron, nickel and sulphur by X-ray fluorescence in lichens from the Mackenzie Valley, Northwest Territories, and the Sudbury District, Ontario. *Canadian Journal of Botany* **54**, 1591–603.

TOPHAM P. B. (1977) Colonization, growth, succession and competition. In: *Lichen Ecology* (Ed. M. R. D. Seaward), pp. 31–68. Academic Press, London.

TUOMINEN Y. & JAAKOLA T. (1973) Absorption and accumulation of mineral elements and radioactive nuclides. In: *The Lichens* (Eds V. Ahmadjian & M. E. Hale), pp. 185–223. Academic Press, New York.

VOBIS G. (1977) Studies on the germination of lichen conidia. *Lichenologist* **9**, 131–6.

WADSWORTH R. M. (1968) *The Measurement of Environmental Factors in Terrestrial Ecology*. Blackwell Scientific Publications, Oxford.

WATSON W. (1918a) Cryptogamic vegetation of the sand-dunes of the west coast of England. *Journal of Ecology* **6**, 126–43.

WATSON W. (1918b) The bryophytes and lichens of calcareous soils. *Journal of Ecology* **6**, 189–98.

WATSON W. (1919) The bryophytes and lichens of fresh water. *Journal of Ecology* **7**, 71–83.

WATSON W. (1925) The bryophytes and lichens of arctic-alpine vegetation. *Journal of Ecology* **13**, 1–23.

WATSON W. (1932) The bryophytes and lichens of moorland. *Journal of Ecology* **20**, 284–313.

WATSON W. (1936) The bryophytes and lichens of British woods. *Journal of Ecology* **24**, 139–61, 446–78.

WEBBER P. J. (1978) Spatial and temporal variation of the vegetation and its productivity. In: *Vegetation and Production Ecology of an Alaskan Arctic Tundra* (Ed. L. L. Tieszen), pp. 37–112. Springer-Verlag, New York.

WEBBER P. J. & ANDREWS J. T. (1973) Lichenometry: a commentary. *Arctic and Alpine Research* **5**, 295–302.

WEBER W. A. (1962) Environmental modification and the taxonomy of the crustose lichens. *Svensk botanisk tidskrift* **56**, 293–333.

WEBER W. A. (1967) Environmental modification in crustose lichens. II. Fruticose growth forms in *Aspicilia*. *Aquilo Ser. Botanica* **6**, 43–51.

WEBER W. A. (1977) Environmental modification and lichen taxonomy. In: *Lichen Ecology* (Ed. M. R. D. Seaward), pp. 9–30. Academic Press, London.

WERNER R. G. (1956) Etudes ecologiques sur les lichens des terrains schisteux maritimes. *Bulletin de Science, Nancy (NS)* **15**, 137–52.

WHITTACKER R. A. (1973) *Handbook of Vegetation Science. V. Ordination and Classification of Communities*. W. Junk, The Hague.

WIEGOLASKI F. E. (1975) Primary productivity of alpine meadow communities. In: *Fennoscandian Tundra Eco-*

systems. Part 1. Plants and Microorganisms (Ed. F. E. Wiegolaski), pp. 121–8. Springer-Verlag, New York.

Williams M. E., Rudolph E. D., Schofield E. A. & Prasher D. C. (1978) The role of lichens in the structure, productivity, and mineral cycling of the wet coastal Alaskan tundra. In: *Vegetation and Production Ecology of an Alaskan Arctic Tundra* (Ed. L. L. Tieszen), pp. 185–206. Springer-Verlag, New York.

Wirth V. (1972) Die Silikatflechten-Gemeinschaften im anseralpinen Lentraleuropa. *Dissert Bot* **17**, 1–306.

Woolhouse H. W. (1968) The measurement of growth rates in lichens. *Lichenologist* **4**, 32–3.

Yarranton G. A. (1967) A quantitative study of the bryophyte and macrolichen vegetation of the Dartmoor granite. *Lichenologist* **3**, 392–408.

Yarranton G. A. (1972) Distribution and succession of epiphytic lichens on black spruce near Cochrane, Ontario. *Bryologist* **75**, 462–80.

Yarranton G. A. (1975) Population growth in *Cladonia stellaris* (Opiz) Pouz. and Vezda. *New Phytologist* **75**, 99–110.

Young C. (1938) Acidity and moisture in tree bark. *Proceedings of the Indiana Academy of Science* **47**, 106–15.

Zussman J. (1977) *Physical Methods in Determinative Mineralogy*. Academic Press, London.

Chapter 19 • Protozoal Symbionts

Dale S. Weis

19.1 Introduction

The last comprehensive treatment of protozoal symbionts was that of Ball (1969) although reviews with more limited objectives have appeared in many places during the last decade or so. The evolutionary implications of symbiosis and the serial endosymbiosis theory have been reviewed by Karakashian and Karakashian (1965), Margulis (1970, 1976), Taylor (1974, 1979), Smith (1979), Whatley *et al.* (1979) and Muscatine (1981). The physiology and biochemistry of intracellular parasitism have been discussed by Trager (1974), Van den Bossche (1976), Curds (1977), Gutteridge and Coombs (1977) and Bannister (1979). Host–symbiont interactions have received the attention of Smith (1974) and Muscatine *et al.* (1975). McLaughlin and Zahl (1966), Karakashian (1970b), Taylor (1973) and Trench (1979) have reviewed algal–invertebrate associations. Preer (1975) discussed the symbionts of *Paramecium aurelia*. The genetics of intracellular symbiosis have been reviewed by Karakashian and Siegel (1965) and Preer and Preer (1977). Symbiosis in *Paramecium bursaria* has received the attention of Karakashian (1975). Kessler (1976) reviewed the systematics of the genus *Chlorella*.

Symbiosis is defined as a stable, persistent and intimate association between several partners. The following discussion is limited primarily to endocellular, hereditary symbioses: i.e. stable, self-perpetuating associations in which the symbiont is located inside the cell of the host. Infection is defined as that process whose end point is symbiosis. Nothing short of an established, permanent, stable symbiosis is considered a successful infection.

This chapter will focus principally on methods for studying cell–cell interaction in a model symbiotic system, *Paramecium bursaria–Chlorella* sp. A secondary aim of this account is to present sufficient examples of prokaryotic and eukaryotic symbionts of protozoa to indicate the variety and evolutionary impact of such associations.

19.2 Natural history of prokaryotic symbionts

Unlike eukaryotic symbionts which have been investigated for over a century, prokaryotic symbionts (bacteria-like or cyanobacteria-like) have received the serious attention of biologists only since Sonneborn studied the kappa plasmagenes of *Paramecium aurelia* (Sonneborn 1943). Since Sonneborn's pioneer work, enlarged into an extensive complex of symbionts by his co-workers and others, new potential complexes of prokaryotic symbionts have been discovered in *Euplotes* sp. and *Paramecium caudatum*. At present prokaryotic symbionts have been found in representatives of all the major groups of protozoa and are even being added to by the synthetic genius of man. It is the objective of this section to present sufficient examples to demonstrate the variety and wide distribution of these symbionts and to discuss a few examples which are better understood and/or seem to occupy a key position or exemplify prokaryotic symbiont evolution.

19.2.1 Flagellates

The consortium between a cyanobacterium and the colourless biflagellate protist *Cyanophora paradoxa* has been a puzzle to physiologists and microbiologists. Controversy surrounding the identity of the host has been terminated by its firm assignment to the cryptomonads (Pickett-Heaps 1972). The identity of the symbiont, however, remains controversial. Trench *et al.* (1978) and Trench and Siebens (1978) have presented strong evidence that the blue-green bodies in *C. paradoxa* are cyanobacteria and not chloroplasts from red algae. However, the genome size of the symbiotic cyanobacteria is only about 10% that of free-living cyanobacteria (Stainer & Cohen-Bazire 1977). *C. paradoxa* was found to be an obligate phototroph, unable to grow in continuous darkness and apparently auxotrophic for cobalamin (vitamin B_{12}). The cyanobacterial symbionts, which dwell in individual, tightly appressed host vacuoles, produce a rudimentary prokaryotic cell wall and apparently supply all the host's needs, chiefly in the form of glucose, and perhaps sucrose, by their photosynthetic activity (Trench *et al.* 1978).

The bacterioid symbionts of flagellates which dwell in haematophagous insects ensure that their flagellate hosts, *Blastocrithidia culicis* and *Crithidia oncopelti*, require no exogenous haemin (Chang 1975). Studies on the isolated DNA of the symbiont show little homology with host DNA but considerable homology with kinetoplast DNA (Twan & Chang 1975). The genome size of this DNA is considerably less than that of *Escherichia coli* with a copy number of 10, similar to the lambda bacteroids in *Paramecium aurelia* (Soldo & Godoy 1973). Another unrelated observation is that the presence of endosymbionts (which are not obligatory in this system) alters the quantity of saccharide ligands in the *C. oncopelti* surface membrane as evidenced by a shift in concanavalin A (con A) agglutinability (Dwyer & Chang 1976).

19.2.2 Rhizopods

The thecate amoeba, *Paulinella chromatophora*, harbours cyanobacteria which presumably contribute photosynthate to the host. Although this has never been demonstrated experimentally, it has been established that the cyanobacteria lie within host vesicles and that there are hexagonal particles (which might be virions) in the nucleus and cytoplasm (Kies 1974). Division synchrony between the symbiont and the host has also been established (Kies 1974).

In a series of reports Jeon and co-workers have announced the progress of an extraordinary event: the evolution of a specific and obligatory symbiosis between *Amoeba proteus* and an unidentified bacterium (Jeon & Jeon 1976; Jeon & Ahn 1978; Ahn & Jeon 1979). The brief facts available are that the endosymbiotic bacteria were present in large host vesicles and were not digested by the host below 26°C. If the temperature was raised above 26°C, bacteria became enclosed and digested in vacuoles and the host died. In further experiments, Ahn and Jeon (1979) infected a fresh strain of *A. proteus* with bacteria of the symbiotic strain (XD). These bacteria avoided digestion and reached the carrying capacity of the amoeba (40 000 bacteria per amoeba). Initially, the bacteria caused a high rate of mortality among the hosts. This harmful effect declined with time as did the bacterial rate of multiplication and a stable symbiosis was established with the host within about six months. Does this represent an actual evolution of symbiosis or only a laboratory curiosity? Given the apparent pre-adaptation of *A. proteus* for symbiosis, as indicated by reports that they are already inhabited by other bacteroid symbionts (Ball 1969), and the rapid selection possible in bacterial populations, such an evolutionary sequence is certainly possible. We await the further analysis of the evolution of these monsters.

Whatley (1976) working with natural symbiosis in *Pelomyxa palustris*, showed that the relative concentration of the three bacterial types normally present varied according to the stage in the life cycle. The large bacteria, which have a bizarre morphology unknown among free-living bacteria, divided under anaerobic conditions. The small bacteria (two types) divided under aerobic conditions. Physical contact between the host nuclei and the large bacteria was provided by an elaborate arrangement of tubules linking bacterial vacuoles with each other and with the nuclear membrane (Fig. 1). John and Whatley (1976) suggest that a major role of the large bacteria is to carry out respiration since the amoebae are devoid of mitochondria.

19.2.3 SPOROZOANS

Sporozoans, because of complexities in their life cycles and their own host dependencies, do not seem likely hosts for symbionts. Indeed, reports have been limited to the Gregarines where the

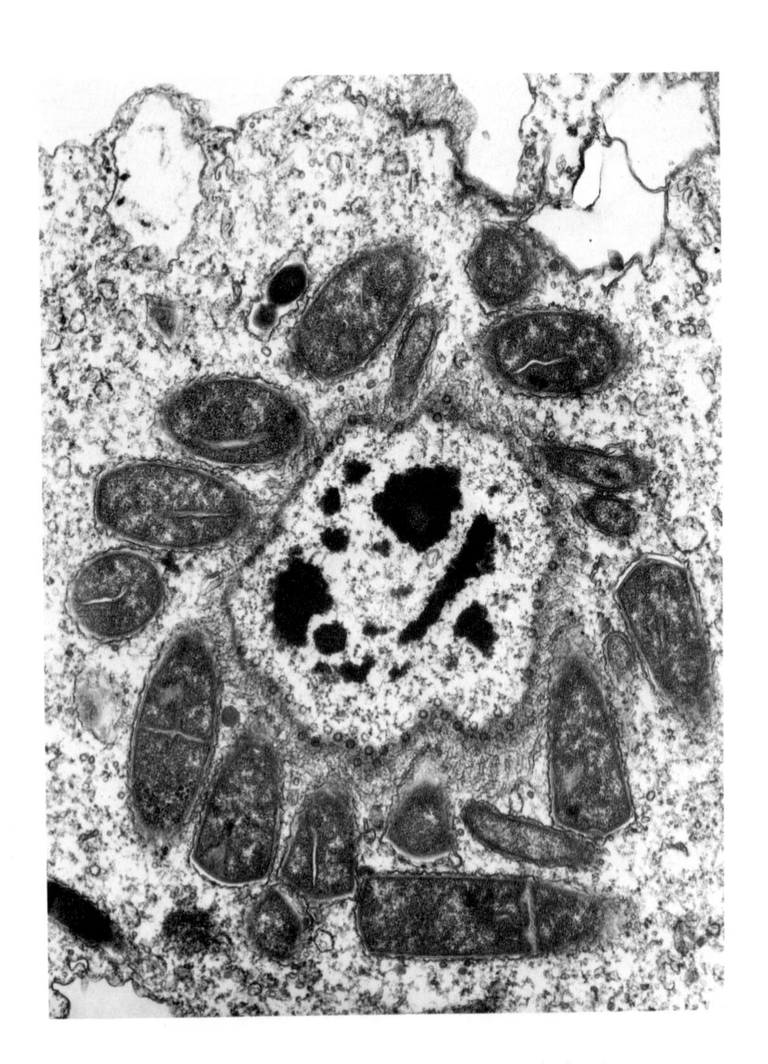

Figure 1. The large bacteroid symbionts of *Pelomyxa palustris*, one of three types of bacteroids present in this host. Note the unique morphology and the vacuoles surrounding each bacteroid. These vacuoles are linked by tubules with each other and with the nuclear membrane. (Electron micrograph courtesy of J. M. Whatley.)

bacterial symbionts are parasitic and not confined in vacuoles (Devauchelle & Vinckier 1976).

19.2.4 Ciliates

The discovery of genes in the cytoplasm of *Paramecium aurelia* by Sonneborn (1943) has led to the dominance of this area of protozoan genetics. It was Preer (1950) who established this territory for symbiosis by making visible the cytoplasmic genes which turned out rather to be bacteria-like symbionts. These early discoveries were added to from time to time so that there is now available a rather large repertoire of prokaryotic symbionts in *P. aurelia*. Similar prokaryotic symbionts have been found, some with killer capability, in *Euplotes minuta* (Heckmann *et al.* 1967), *E. crassus* (Rosati & Verni 1975; Rosati *et al.* 1976) and *E. aediculatis* (Heckmann 1975). Considerable controversy surrounds the nature of the bacteroids in these various ciliate hosts. Soldo and Godoy (1973) have determined the molecular complexity of *P. aurelia* symbiont lambda DNA. Their results indicate that there are 10 to 20 copies of the genome per lambda, a situation unknown among free-living bacteria but common in organelles. The low molecular complexity and multicopy genome suggest that lambda symbionts have lost a great deal of their autonomy and are approaching organelle status in terms of the serial endosymbiosis theory. Others maintain that the endosymbionts are bacteria adapted to endozoic life but sufficiently autonomous to be so identified (Dilts 1977; Preer & Preer 1977; Quackenbush 1977). Only careful genome analysis would seem to offer the possibility of resolution of this controversy.

Among the bacteroid symbionts of *Paramecium aurelia*, only kappa and lambda are infective and only under special conditions. None of the *P. aurelia* symbionts has yielded to constant culture *in vitro*. Free and vesicular location of these cytoplasmic symbionts are both represented. Kappa is distinguished by the presence of defective phages (R bodies) which apparently came into the ciliate along with the symbiont. Treatment of kappa with ultraviolet light causes the induction of R bodies which are associated with killing. The symbionts of *P. aurelia* are all Gram-negative rods and most have a killer effect or produce a toxin. Mu releases no toxin but kills its mate at conjugation (Siegel 1954). In this *P. aurelia* complex ordinarily only one kind of endosymbiont is present in a ciliate. The possible selective advantage or any contribution of these infectious elements to the host is unknown. Two prominent symbionts of *P. aurelia*, kappa and pi, show little homology in DNA–DNA hybridisation studies (Dilts 1977), suggesting that the various bacteroid symbionts are unrelated.

The hypotrichs (two species of *Euplotes*) and *Paramecium caudatum* have yielded multiple prokaryotic symbionts. At least four morphologically distinct types of symbionts have been found in the cytoplasm of several stocks of *Euplotes crassus* with one or two types in each strain (Rosati *et al.* 1976). *E. crassus* also has bacteroid symbionts in the macronucleus (Rosati & Verni 1975). The bacteroid cytoplasmic endosymbiont in *E. aediculatus* was eliminated by penicillin treatment, after which the host failed to divide. Reinfection of the host allowed resumption of normal growth (Heckmann 1975). Another series of bacteroid symbionts has been found in the macro- and micronuclei of *P. caudatum*. Osipov reported two different symbionts, both Gram-negative, one inhabiting the micronucleus (Osipov 1973) and the other limited to the macronucleus (Osipov *et al.* 1974). Ciliates infected with either symbiont were incapable of conjugation, the missing step being the failure to synthesise mating substances (Osipov *et al.* 1974). Ciliates infected in the macronucleus underwent hypertrophy in the perinuclear space. Infection of the micronucleus caused the hypertrophy of that body. Double infection (macro- and micronucleus) was rare (Osipov *et al.* 1974). Osipov *et al.* (1975) have suggested that the analysis of the interrelation of endonucleosymbiont and the nuclei may be an effective method for studying the structure and functional activity of ciliate nuclei. That is, the bacteroids must recognise differences between macro- and micronuclei, differences that are related to the nature of the differentiation of the nuclei (Osipov *et al.* 1975). On the basis of this recognition function, they should be valuable probes of nuclear dimorphism at an ultrastructural level. Golikova (1978) has found intranuclear parasitic bacteroids in micro- and macronuclei of the ciliate *P. bursaria*. The infected micronucleus undergoes hypertrophy at high levels of infection and chromosomes are no longer visible. Infection of macronuclei also leads to moderate hypertrophy. These results are quite similar to those in the closely related *P. caudatum* previously discussed. Finally, Berezina (1976) has reported rod-shaped bacterial endosymbionts of the cytoplasm of the ciliate *Blepharisma japonicum*.

19.3 Natural history of eukaryotic symbionts

Prokaryotic symbionts of protozoa tend to be identifiable elements in the association. These elements, although often highly integrated into the host metabolism, as in *Crithidia oncopelti* or *Cyanophora paradoxa*, still retain their identity. There is little physical integration of symbiont and host. However, in certain eukaryotic associations with protozoa integration is the key characteristic; for example, the 'red water' ciliate *Mesodinium rubrum* and its cryptomonad symbiont are in such intimate physical contact that it has been possible only recently to define the limits of each cell. There are similar examples in the chrysophyte–heterotrophic dinoflagellate associations where the symbiotic chrysophyte has lost most of its cellular apparatus in becoming a symbiont. As might be expected these associations are apparently obligatory for both partners. On the other hand, associations with *Chlorella* spp. are facultative for both partners and the symbiont is always discretely sequestered in a special host vacuole.

19.3.1 FLAGELLATES

Several species of heterotrophic dinoflagellates have been discovered as hosts for algae (Tomas & Cox 1973; Jeffrey 1976; Jeffrey & Vesk 1976; Withers *et al.* 1977). Little is known about the interrelation of host and symbiont and controversy has surrounded the identity of the latter. It is only recently, with the aid of detailed electron microscopy, that the intermingling of the elements of two different cells within the plasmalemma of the host cell has been confirmed (Jeffrey & Vesk 1976). In *Peridinium balticum* and *Kryptoperidinium foliaceum* the symbiont—most likely a chrysophyte (Withers *et al.* 1977), although this placement of the symbiont is disputed by Fine and Loeblich (1976)—has lost its cell wall. The eukaryotic nucleus, chloroplasts and associated ribosome-dense cytoplasm, all belonging to the symbiont, are separated by a single membrane from the rest of the host cell (Jeffrey & Vesk 1976—Fig. 2). Starch grains occurring in the dinoflagellate (host) portions of the cytoplasm are obviously the end-product of symbiont photosyn-

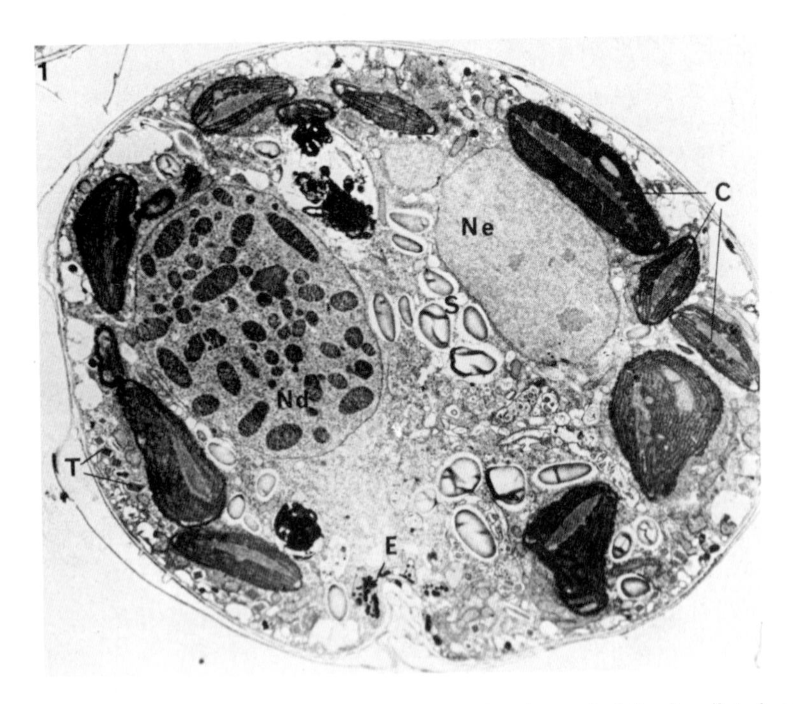

Figure 2. Section through *Peridinium foliaceum* showing typical dinoflagellate (mesocaryotic) nucleus (Nd) and a second lobed endosymbiont (chrysophyte) nucleus (Ne): note chloroplasts (C), which tend to be arranged around the cell periphery, and typical dinoflagellate organelles and inclusions, such as eyespot (E), trichocysts (T), starch grains (S) present (× 4800) (from Jeffrey & Vesk 1976, courtesy of the editor, *Journal of Phycology*).

thesis which has been transported to the heterotrophic host (Jeffrey 1976).

19.3.2 RHIZOPODS

Small amoebae (species of *Vahlkampfia* and *Endamoeba*) have been reported in the giant amoeba *Pelomyxa palustris* and in *Amoeba proteus* (Hollande & Guilcher 1945). Other examples of symbionts in rhizopods include radiolaria and foraminifera as hosts. The studies on radiolaria (Hallande & Carré 1974) and foraminifera (Lee *et al.* 1979a, b) are particularly noteworthy. Schmaljohann and Röttger (1978) first identified symbionts in certain foraminifera (*Hoterostegina depressa*) as diatoms without cell walls. Lee *et al.* (1979b) successfully isolated and cultivated *in vitro* the diatom symbionts showing that they produced quite normal frustules when free of the host. Lee and co-workers also successfully isolated and cultivated two *Chlamydomonas* species (*C. hedleyi* and *C. provasolii*) and one dinoflagellate (*Symbiodinium microadriaticum*) from foraminiferan hosts (Lee *et al.* 1979a). All three algal symbionts are auxotrophic for biotin, thiamine and vitamin B_{12}, suggesting that in their natural habitat they receive these vitamins through the feeding activities of the host which must obtain them from the bacteria on which it feeds (Lee & Bock 1976). Hollande and Carré (1974) have described the dinoflagellate symbionts in radiolarians and, as in foraminifera, there is a tendency towards the loss of the cell wall. Anderson (1978) has presented isotopic evidence for assimilation of organic substances by the host from its zooxanthellae. Anderson (1978) has also presented cytological evidence suggesting that each radiolarian cell is surrounded by a silicon shell which segregates the host cell body from the zooxanthellae in the peripheral ectoplasm.

19.3.3 SPOROZOANS

Sporozoans seem unlikely hosts for eukaryotic symbionts because of their specialised life cycle and indeed reports of such associations have been scanty. In the Gregarines, where hosts are found for prokaryotic symbionts, Devauchelle and Vinckier (1976) have also reported a microsporidian, *Nosema vivieri*, in the cytoplasm.

19.3.4 CILIATES

Freshwater and marine ciliates have diverged in the source from which they have drawn their eukaryotic symbionts. Freshwater ciliates are for the most part limited to the genus *Chlorella*, whereas the dinoflagellates are the dominant group from which eukaryotic symbionts of marine ciliates are drawn. A potential divergence exists between freshwater and marine ciliate associations with respect to the relative autonomy of the two partners: freshwater associations tend to be facultative for both partners, whereas, within the limits of the small number of marine associations examined, mutually obligatory associations predominate. One of the most interesting and thoroughly studied marine associations is between the ciliate *Mesodinium rubrum* and its cryptomonad symbiont. In research on this association by White *et al.* (1977) and Taylor (1979) it has been discovered that the symbiont is considerably modified compared to free-living cryptomonads. This severe modification is expressed in:

(1) the reduction of its periplast to a single membrane giving the symbiont an irregular shape;

(2) the apparent proliferation of chloroplasts while retaining a single nucleus;

(3) the complete loss of its flagellar apparatus;

(4) the apparent temporary loss of the Golgi system.

In comparison with other algal invertebrate symbioses, this degree of structural modification is unique and most closely resembles the situation in certain chrysophyte symbionts of heterotrophic dinoflagellates, such as *Peridinium balticum* and *Kryptoperidinium foliacenum* (Fig. 3).

There is a report of a species of *Leptomonas* parasitising the macronucleus of *Paramecium trichium* (Gillies & Hanson 1962). This is an apparently fatal parasitism which can be transferred from *P. trichium* to a number of other *Paramecium* species. Bannister (1979) reports several ciliates (Suctorida) which are intracellular parasites of other ciliates, e.g. *Sphaerophyra sol* in *Stylonychia* sp. and *Paramecium* sp. and *Pottsia infusorium* in folliculinid ciliates. Finally, *Chlorella* spp. have been reported as symbionts in the ciliates, *P. bursaria*, *Stentor polymorphus*, *S. igneus*, *Climacostomum virens*, *Frontonia* sp., *Vorticella* sp., *Euplotes* sp., *Holosticha* sp., *Ophrydium* sp. and *Prorodon* sp. (Diller & Kohnaris 1966; Sud 1968; Graham & Graham 1978; D. Fischer-Defoy, personal communication). One universal feature in this group of symbioses is the relegation of the algal symbionts to special host vacuoles (Sud 1968). There are several questions which must be answered to see if

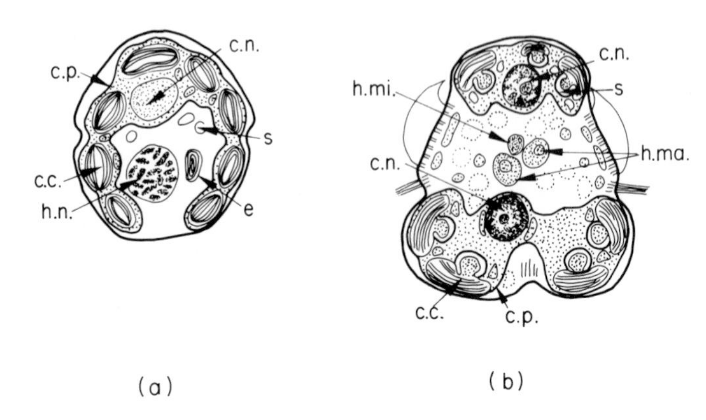

Figure 3. Cryptic eukaryotic cytobionts (symbionts) diagrammatic. Key: cc., cytobiont chloroplast; c.n., cytobiont nucleus; c.p., cytobiont plasmalemma; e., eyespot; h.n., host nucleus; h.ma., host macronucleus; h.mi., host micronucleus; s., starch.
(a) *Kryptoperidinium foliaceum*; arrangement of the symbiont within the host.
(b) *Mesodinium rubrum*: a cell with two symbionts (after Taylor 1979, with permission).

there are other common features. What is the phylogenetic relation between the various symbioses? Are the symbiotic chlorellae interchangeable from host to host? Are the chlorellae identical? These questions are answerable using available techniques of symbiont isolation, cultivation and infection and should provide valuable information.

19.4 Evolutionary implications of prokaryotic and eukaryotic symbionts of protozoa

There appears to be a fundamental divergence in the evolutionary direction taken by prokaryotic symbionts of protozoa compared with most eukaryotic symbionts. This impression is based on several more or less well established facts which can be generalised to apply to prokaryotic symbionts.

(1) The prokaryotic symbiont retains a sharply separate identity from the host. Indeed, this statement applies equally well to organelles (peroxisomes, hydrogenosomes, mitochondria and chloroplasts) which, according to the serial endosymbiosis theory (Margulis 1970), are the end-products in this line of evolution.

(2) The host and the symbiont are not equal partners in the relationship, one partner, the host, being clearly dominant.

(3) There is a progressive loss of symbiont autonomy, e.g. none of the *Paramecium aurelia* complex has been successfully cultivated *in vitro*.

(4) There is a shrinkage (perhaps progressive) of the symbiont genome compared to similar free-living prokaryotic species.

(5) In the earlier phases of coevolution (e.g. *Paramecium aurelia* complex) the role, if any, of the symbiont in the host economy is obscure.

(6) In advanced phases of coevolution, assuming serial endosymbiosis, the symbiont (organelle) retains only one or a few biological functions of importance to the host and the role of the organelle in serving the host becomes more sharply defined.

(7) Symbiosis is often obligatory for the symbiont but facultative for the host.

The above speculative generalisations may be compared with similar generalisations concerning eukaryotic symbionts of protozoa.

(1) It is sometimes difficult to untangle physically the symbiont from the host and it is sometimes unclear which is the dominant partner (e.g. *Peridinium balticum*–chrysophyte or *Mesodinium rubrum*–dinoflagellate).

(2) The relationships between the symbiont and the host tend to be reciprocal. They may be reciprocally obligate, as is apparent in *Mesodinium rubrum*, or they may be reciprocally facultative, as in *Paramecium bursaria*.

(3) The symbiont tends to remain capable of independent existence—for example, chlorellae in *Paramecium bursaria* are capable of cultivation *in vitro* as is the host.

(4) There is no evidence for continuing progressive evolutionary change in the biological role of the symbiont in the host economy.

(5) There is evidence for an elaborate recognition mechanism, complete with signals and receptors, for ensuring reinfection of host by symbiont and maintenance of symbiont clonal purity in *Paramecium bursaria* and probably in the other mutually facultative eukaryotic symbioses (other ciliate–chlorellae associations).

If the suggestion can be entertained that certain prokaryotic symbionts are evolving in the direction of becoming host organelles, or represent stages in such an evolution, what is the putative direction of evolution of eukaryotic symbionts? Of course, in certain individual cases organelle status may also be the end-point of evolution of eukaryotic symbionts (e.g. chromoplasts from eukaryotic algae—Whatley *et al.* 1979). However, it is possible to speculate that an evolutionary tendency of eukaryotic–eukaryotic symbioses is towards some sort of mosaicism and the various eukaryotic–protozoan symbioses represent more or less successful stages in such a sequence. The desirability for the ciliate host of the continued, optimal functioning and genetic purity of its symbiont population, has perhaps provided the selection pressure for evolving a surveillance system rather sophisticated for a unicellular organism. Primitive recognition systems, particularly mating systems, are well established phenomena in microorganisms, such as *Chlamydomonas* spp. and ciliates. However, little attention has been paid to recognition associated with particle feeding. Nevertheless, evidence is accumulating indicating that discrimination in particle feeders is a general occurrence (Barna & Weis 1973; Lee & Bock 1976; Berger 1977). An organism such as *Paramecium bursaria*, a particle feeder with a mating response system, would seem to be ideally preadapted to the development of a complex recognition system in relation to its symbionts. The lack of the development of such systems in some other particle-feeding ciliates with mating systems (e.g. *Paramecium aurelia*, *P. caudatum*, *Tetrahymena* sp.) however, remains unexplained.

Experimental

19.5 Methods for the study of cell–cell interaction in protozoal symbioses: *Paramecium bursaria* as a model system

19.5.1 Introduction

The green ciliate, *Paramecium bursaria*, has figured prominently in symbiosis research since the green bodies associated with certain freshwater protozoa and invertebrate metazoa were first described as algae by Von Siebold in 1849 (Loefer 1936). Brandt in 1881 introduced the term *Zoochlorella* to designate these algae but later withdrew the generic name when Entz in 1881 reported that the algal cells from crushed *Stentor* sp. were capable of growth outside the host (Loefer 1936). On the basis of this observation and his recognition of the morphological similarity of the symbionts of species of *Hydra*, *Stentor* and *Paramecium* to free-living *Chlorella vulgaris*, Beijerinck in 1890 renamed the genus *Chlorella*, intending it to include both free-living and symbiotic forms (Loefer 1936). The subsequent discovery that laboratory-grown *Paramecium bursaria* survives loss of its symbionts (Pringsheim 1928) and that isolated algae can be cultivated axenically (Loefer 1934) suggests that algae of symbiotic origin may exist in nature as free-living forms readapted to the extracellular state. The development of a method of separation of host and symbiont and their independent cultivation, has allowed study of the morphology and physiology of each symbiont and their phylogenetic relationships to free-living counterparts. This method has also allowed the study of the conditions most favourable to reinfection of host by symbiont and a study of the structural, physiological, biochemical and behavioural coadaptations of symbiont and host. This method has also allowed the study of the specificity, recognition and cell–cell interactions involved during infection and in steady state symbiosis and in the future may allow the study of the ecological implications of this symbiosis. This major method and the associated ancillary techniques will be the subject of this section.

19.5.2 Techniques for Separating Host and Symbiont

Isolation of Chlorella-free (aposymbiotic) ciliates

The methods reported as successful in producing aposymbiotic ciliates include prolonged dark cultivation (Siegel 1960; Karakashian 1963), cultivation in darkness at elevated temperatures (Siegel 1960) and treatment with X-rays at dosages of 100 to 300 Mrad (Wichterman 1948). These methods all suffer, more or less, from the drawback of unpredictability. In all these methods, after treatment, it is necessary to isolate single cells to test for the presence of intracellular algae. In order to do this, each isolated cell is allowed to produce a clone in the light on bacterised lettuce medium (Sonneborn 1950). If algae are present, the clones become green.

Another method which perhaps suffers less from this drawback and has been claimed to produce large numbers of aposymbiotic hydra (Pardy 1976) and paramecia (Reisser 1976) employs the photosystem II inhibitor 3-(3,4-dichlorophenyl)-1,1-dimethylurea (DCMU) (Pardy 1976; Reisser 1976) and higher light intensity (Pardy 1976).

Isolation of symbiotic algae

Loefer (1934) reported the first successful axenic culture of algae from *Paramecium bursaria*. Loefer's method took advantage of the persistence of live symbiotic algae in dead *P. bursaria* cultures. The green sediment in such culture tubes was streaked on 1% (w/v) starch or nitrate agar plates and the resulting green colonies picked off and streaked on mineral agar medium (Loefer 1936). This technique has also been successful in the author's laboratory. However, it should be noted that the cell used to establish the ciliate clone used as a source of symbiotic algae must have been washed free of external algae. In the author's laboratory penicillin (800 μg ml^{-1}) and streptomycin (2 μg ml^{-1}) are added to the algal medium agar plates used to isolate colonies. A successful variation of this method uses axenic ciliate cultures as the source of algal material (Weis 1978). Another method employing intact *P. bursaria* cells for isolating symbiotic chlorellae depends on washing individual cells free of contaminating bacteria and algae (Siegel 1960; Weis 1978). Washed cells are individually plated on algal medium where they rupture, liberating the algae. The resulting colonies are picked off and transferred to slants of the same medium.

Still another method depends on axenic ciliate cultures (Weis 1978). In this method, performed aseptically throughout, large numbers of algae can be made available for experimentation immediately after isolation. The ciliates are harvested by centrifugation at 100 ***g*** into a small volume of 1% (w/v) methyl cellulose. The packed ciliates are lysed, either by use of a tissue homogeniser or by rapid expulsion from a syringe fitted with a No. 26 Becton-Dickenson needle having an internal bore diameter of 0.2 mm. The lysate is transferred to a discontinuous sucrose gradient (10%, 25% and 50% (w/v)) and centrifuged at 1500 ***g*** for 15 min. The pellet which contains uncontaminated algae is washed three times in a phosphate-buffered, dilute salts medium (DBS—Weis 1978) and either used directly for experiments or plated out to obtain pure cultures.

19.5.3 Techniques for the Separate Cultivation *In Vitro* of the Host and the Symbiont

The standard medium for cultivating aposymbiotic ciliates is lettuce infusion medium inoculated with *Enterobacter cloacae* (Sonneborn 1950). Following isolation, the algal-free ciliates are inoculated into this medium and incubated at 15 to 25°C, illuminated by 500 foot-candle fluorescent light. Light incubation is used to help expose possible algal contaminants in the ciliates since a single intracellular algal contaminant is thought to be sufficient to re-establish symbiosis (Weis & Ayala 1979).

Frequent failures of early investigators (and contemporary investigators as well) to establish symbiotic algae from *Paramecium bursaria* in pure culture *in vitro* suggests that these algae may be nutritionally fastidious. The experience from several laboratories with algae isolated from *Hydra* sp. and *P. bursaria* is that it is not always possible to grow them successfully following an ostensibly successful, live isolation. Furthermore, even a successful isolation and cultivation of symbiotic algae on a certain medium, perhaps a mineral medium, may end in failure to survive after 5 to 10 successive transfers at approximately monthly intervals, again suggesting depletion of some growth factor required in minute amounts. Systematic study of the nutritional requirements of symbiotic algae from *P. bursaria* is being carried out by Reisser and Pado and independent reports indicate that the symbiotic algae of *Paramecium* sp. are auxotrophic for thiamine (Pado 1969) or thiamine and cobalamin (Reisser 1975). The possibility of amino acid requirements for certain strains is indicated by positive responses to vitamin-free caseine hydrolysate (Bacto casitone) (D. S. Weis, unpublished observations). Further analysis of growth factor requirements of isolated symbiotic algae is an important theoretical and practical problem of cultivation. A thorough knowledge of such requirements would be of immense value in attempting to unravel the line of evolutionary divergence of symbiotic chlorellae from free-living forms. It would also make possible the evaluation of contemporary ecological interaction between symbiotic and free-living chlorellae, a subject to be discussed later (p. 336). Fortunately, a solid framework of *Chlorella* systematics exists which could be used in comparisons due especially to the monumental researches of Kessler and his colleagues (Hellmann & Kessler 1974; Kessler 1976; Kerfin & Kessler 1978) with further contributions by Shihira and Krauss (1963) and Fott and Nováková (1969).

19.5.4 TECHNIQUES FOR STUDYING HOST AND SYMBIONT ADAPTATIONS

Since the observations of Beijerinck (1890), the morphological similarity of free-living and symbiotic chlorellae is generally acknowledged. Any structural adaptations of symbiotic algae to symbiosis would most likely be at the ultrastructural level. Karakashian (1970a) has noted that the ultrastructure of free-living *Chlorella* spp. is quite variable. Subtle changes, perhaps in cell surfaces, superimposed on a general morphological plasticity in the symbiotic forms would be difficult to detect by morphological observation. This is not to minimise the importance of ultrastructural studies on the host–symbiont association which have been of immense value and have established the perialgal vacuole with its unique features as perhaps the prime host adaptation to symbiosis and the home of the symbiont *in situ* (Vivier *et al.* 1967; Karakashian *et al.* 1968).

Three changes in physiology and ultrastructure have been discovered which are potential adaptations to symbiotic life, they are:

(1) sugar release by the alga;
(2) auxotrophy of the alga;
(3) host elaboration of non-digestive perialgal vacuoles.

In hydra and other invertebrates, lichens and *Paramecium bursaria*, the algae *in situ* maintain a high level of sugar or polyhydric alcohol release to the tissue or cell of the host. Freshly isolated symbiotic algae (and zooanthellae) release photosynthate to the exterior but this ability *in vitro* of lichens and marine invertebrates is rapidly and completely lost (Smith 1979). However, in algae from *P. bursaria* and in a freshwater sponge, sugar release continues *in vitro* and apparently does not decline (Weis 1979). Indeed, a symbiotic strain isolated by Loefer (1934) recently, after 45 years of cultivation *in vitro*, released sugar at the rate of approximately 5 to 10 μg (mg dry weight)$^{-1}$ h^{-1}. The detection of the sugar was by the reduction of ferricyanide method (Park & Johnson 1949), with thin layer chromatography providing the sugar identification. Some implications of this sugar release by algae *in vitro* will be considered in a later section (p. 332). The auxotrophy of symbiotic algae *in vitro* has already been discussed (p. 328). There is at present no experimental evidence to suggest how the growth factor requirements of the symbiont are met or how they might function as a regulatory system. Since the nutritionally fastidious host almost certainly matches any growth factor requirement of the symbiont with the possible exception of vitamin B_{12}, it might be expected that host and symbiont are supplied through the feeding activities of the host. That is, *P. bursaria* remains a particle feeder in order to supply the host and the symbiont with their growth factor requirements.

The perialgal vacuole revealed by electron microscopy (Karakashian *et al.* 1968) is remarkable in its capacity to harbour the resident symbiont. The alga survives within the vacuole, apparently unaffected by the digestive capacity of the host. In order to explain the resistance of the algae to digestion, it has been suggested that the cell wall of symbiotic algae in *Hydra* spp. is particularly resistant to digestion (Smith 1979). However, the fact is that exogenous algae ingested by *Paramecium bursaria* are digested and assimilated *en masse* in *Paramecium bursaria* (Weis 1976). Furthermore the algae *in situ* remain susceptible to digestion since either in continuous darkness or in the presence of 2×10^{-6}M DCMU in the light, it is possible to observe the apparent digestion of symbionts by *Paramecium* spp. (Weis 1976) and by *Hydra* spp. (Oschman 1966). Reisser (1976) has presented excellent electron micrographs showing that in the dark or in the presence of DCMU alterations occur in the perialgal vacuole membrane. Canals project from the membrane which enlarges and becomes loose fitting around the enclosed alga, suggesting that new membrane is being added to the vacuole membrane much as in the initial formation of food vacuoles (Fig. 4). Karakashian and Karakashian (1973) demonstrated the absence of digestive enzymes from perialgal vacuoles. S. Karakashian (personal communication) showed that stain from thorotrast-stained lysomes did not penetrate the perialgal vacuoles although it appeared in the developing food vacuoles. The evidence, although it is far from conclusive, points to a fundamental functional dichotomy between food vacuoles and perialgal vacuoles in *P. bursaria*, with food vacuoles perhaps being reversibly produced from perialgal vacuoles presumably by changes in the chemical composition of the vacuole membrane induced by algae. Another recently discovered property of symbiotic algae *in situ* is their susceptibility to virus attack (Kawakami & Kawakami 1978). According to these authors, only algae in food vacuoles, but not those in perialgal vacuoles or external to the host, are susceptible to viral attack.

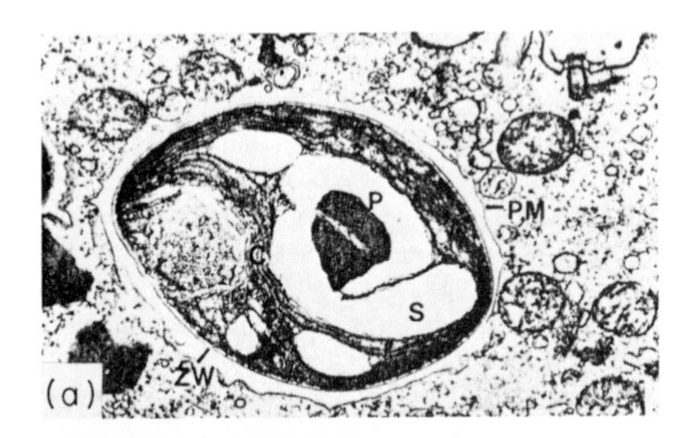

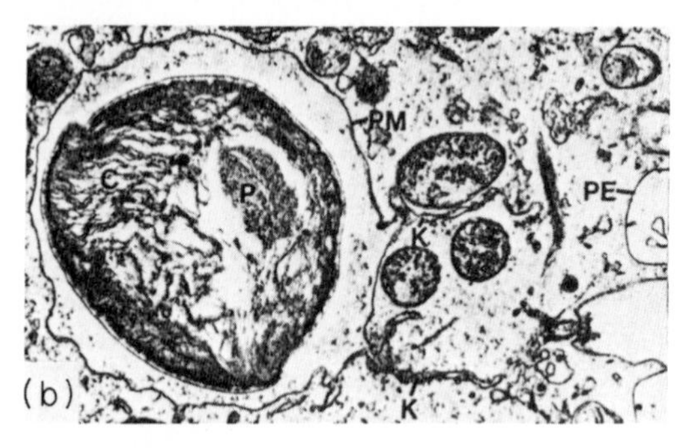

Figure 4. The effect of dark incubation of the *Paramecium bursaria–Chlorella* association on the development of the perialgal vacuole membrane. (a) *Chlorella* symbiont in *Paramecium bursaria* maintained in alternating 14 h light: 10 h dark. (b) A similar symbiont in *P. bursaria* maintained in continuous darkness. Key: ZW, cell wall; P, pyrenoid; S, starch; all structures of the algal symbiont. PM, perialgal vacuole membrane; PE, pellicle of the ciliate host. Note the channel formation and the loosening of the perialgal vacuolar membrane (×9900) (from Reisser 1976, courtesy of the editor, *Archives of Microbiology*).

19.5.5 Methods for Studying Host–Symbiont Interactions

Partner specificity

The most direct method for studying host–symbiont specificity in *Paramecium* spp. is to observe the success of infection by various combinations of algae and ciliates (Oehler 1922; Hämmerling 1946). Hämmerling (1946) observed that infection of the aposymbiotic *Stentor polymorphus* with chlorellae from *Paramecium* spp. was less successful than the homologous infection of *S. polymorphus* with *Stentor* algae, suggesting a degree of partner specificity. Schultze (1951) enlarged on this by showing that whereas homologous infections in *Stentor* sp. and *Paramecium* sp. always succeeded, cross-infection between *S. polymorphus* algae and aposymbiotic *P. bursaria* failed or was unstable. However, algae from *S. igneus* stably infected *P. bursaria* aposymbionts. On the other hand, this author was not successful in producing symbiosis in *Paramecium* sp. with free-living chlorellae. A further indication of host specificity was shown by Tartar (1953) working with *Stentor coeruleus–Stentor polymorphus* chimaeras. The algae were rejected by these chimaeras unless the graft was predominantly *S. polymorphus* (the normal host for chlorellae). More recent studies of partner specificity include that of Siegel (1960) who found differences in infectivity among different symbiont strains. Karakashian (1963) also found specificities in both host and symbiont but showed that algae from *Hydra* sp. formed a stable, growth-promoting association in *P. bursaria* equal to that produced by algae from *Paramecium* spp. Bomford (1965) tested host specificity with 21 aposymbiotic stocks, belonging to four syngens, challenged with several strains of free-living and symbiotic chlorellae. No specificity differences were demonstrated among the host ciliates with respect to the strains of algae they were competent to maintain. Karakashian and Karakashian (1965) came to the conclusion that some free-living algae were capable of forming temporary associations with *P. bursaria* but were not capable of establishing long-term, mutualistic symbioses. Hirshon (1969) came to a similar conclusion, although it was also found that a free-living strain of *Chlorella kessleri* was capable of establishing itself as a symbiont in *P. bursaria*.

Perhaps at this point a few words of caution about carrying out these experiments are appropriate. First, the obvious must be stated: the ciliates must be completely free of algae, since even one alga is sufficient to repopulate a cell and result in the development of a green clone. The procedures for achieving aposymbiotic cells have already been discussed (p. 327) and they all have one requirement in common—patience. To the author's knowledge, a rapid procedure for producing and confirming aposymbiotic ciliates does not exist.

Second, a single criterion must be established in

order to judge a successful infection. This criterion must be so sharp and rigorous that there is no room for an equivocal result. In the author's studies the production of a green clone was chosen as the only admissible evidence, although this criterion has a drawback in that it takes *P. bursaria* 10 d to go from a single aposymbiotic cell to a green clone. However, the use of this criterion has revealed a pattern of infectivity among algae which suggests that certain algal properties are related to the ability of the algae to infect susceptible host cells.

Finally, a standard procedure for the achievement of symbiosis must be established. The author suggests the following procedure: an aliquot (1.0 ml) phosphate buffered dilute saline (DBS—p. 328) is carefully layered on to 10 ml late exponential phase culture of aposymbiotic ciliates contained in a 10.0 ml volumetric flask. The ciliates swim through the medium concentrate in the upper DBS layer and are largely freed of exogenous bacteria. The concentrated ciliates are counted and allowed to disgorge food vacuoles by suspension in the same medium for 2 h. In the meantime algal cells from late exponential or early stationary phase cultures are harvested by centrifugation, washed and resuspended in the same mineral medium. The algae and ciliates are mixed at a specific ratio and for a definite time period. After exposure to the algae, the ciliates are washed free of any residual, exogenous algae by micropipetting from well to well of a nine-well spot plate containing 0.3 ml of DBS per well and transferred to fresh growth medium to give 1 ciliate per well of a 96-well microtitre tray. The single ciliate cultures are incubated and after 10 d the trays are scored for the number of green and white clones using a dissection microscope. The ratio of green clones to total clones is called the frequency of infection.

The procedure can be simplified if it is desired only to know if a particular alga can infect. In this case a saturating number of algae (10 000 algae per ciliate) is added to the aposymbionts and exposure is maintained for 4 to 5 d. The ciliates are washed free of exogenous algae and the procedure given above followed.

Using this procedure a large number of free-living and symbiotic algae (from *Paramecium* and *Hydra* species and freshwater sponges) were tested for infectivity for *Paramecium bursaria* Syn. 2 aposymbiotic ciliates. The results shown in Table 1 indicate that *P. bursaria* aposymbionts respond to algae from *Paramecium* spp. only. The data on Con A agglutination and sugar release suggest a biochemical basis for this. Since only algal strains capable of sugar release and relatively resistant to agglutination with Con A were infective it is suggested that it is not sufficient for an alga to have the right lineage; it must also be able to excrete maltose and perhaps display the right cell surface residues (Weis 1979).

Table 1. A comparison of symbiotic and free-living algae in regard to Con A agglutinability, sugar release and identity, and infectivity for aposymbiotic syn. 2 *Paramecium bursaria*.

Algal strain	Source	*Chlorella* species	Resistance to Con A agglutination	Sugar release	Sugar identity	Infectivity for *P. bursaria*
1	*P. bursaria*	U*	–	trace	U*	–
2	*P. bursaria*	U	+	+	maltose	+
8/13-1	*P. bursaria*	U	+	+	maltose	+
8/13-2	*P. bursaria*	U	+	–	—	–
Utex 130	*P. bursaria*	U	+	+	maltose	+
NC 64A	*P. bursaria*	U	+	+	maltose	+
Utex 838	freshwater sponge	U	+	+	glucose	–
77C	hydra	U	+	–	–	–
77W2	hydra	U	+	–	–	–
211-13	free-living	*C. protothecoides*	–	–	–	–
211-8c	free-living	*C. fusca var. vacuolata*	–	–	–	–
211-30	free-living	*C. lobophora*	+	–	–	–
211-81	free-living	*C. vulgaris* Beijerinck	–	–	–	–
211-11c	free-living	*C. vulgaris* Beijerinck	+	–	–	–
C 1.6.7	free-living	*C. sorokiniana*	–	–	–	–

*U, unidentified. See text for details

Isolation of non-infective algal variants

Non-infective algae from strains of a symbiotic origin were isolated from the green sediment at the bottom of dead or ageing *Paramecium bursaria* cultures (Weis 1978). This material is probably rich in non-infective variants because these are egested by the ciliate into the medium where they may multiply more rapidly than the wild type. The method for isolating pure cultures of symbiotic algae was followed as described on p. 327. Isolated colonies were tested for their ability to infect the homologous aposymbiotic ciliates and shown to be non-infective strains. These variants were later tested for Con A-mediated agglutinability and their capacity to release maltose. The variants isolated by this procedure were classified into two groups on the basis of the Con A-mediated agglutination titre:

(1) those requiring Con A at a concentration of 40 μg ml^{-1} or more for strong agglutination (Con A resistant);

(2) those agglutinating appreciably at 2 μg Con A ml^{-1} (Con A sensitive).

Although some of these variant strains released trace amounts of maltose, none was capable of sugar release at even 10% of the rate observed in infective algae. Thus there is a strong correlation between inability, or poor ability, to release maltose and non-infectivity for *P. bursaria*. This suggests that maltose release (5 to 10 μg (mg dry wt)$^{-1}$ h^{-1}) may act as a signal to the host, a possibility to be discussed in the next section.

Conceptual approaches for studying host–symbiont recognition during infection

Experiments on the Con A agglutinability of algae of a symbiotic origin have shown that non-infectivity was not correlated with sensitivity to Con A-mediated agglutination. However, infectivity was associated with resistance to Con A-mediated agglutination (Weis 1978). This suggests that the density of cell surface glucose residues, as well as maltose released by the alga, might constitute algal recognition signals to the host (Weis 1979). Indirect evidence to support this hypothesis was obtained from experiments which had another objective: to determine those conditions most favourable for the maximum rate of infection. These experiments were carried out as previously indicated (p. 331) and showed that the mean number of algae required to produce the maximum infection frequency was approximately 1000 algae per ciliate (Fig. 5). The same experiments showed that when algae were not limiting, a mean exposure of ciliates to algae of approximately 6 h was required to produce maximum infection frequency (Fig. 5). Subjecting the same data to a Poisson analysis revealed, however, that a single alga was theoretically sufficient to initiate infection of a ciliate (Fig. 6).

On the basis of this data and some additional results from split-exposures of ciliates to algae (Fig. 7), a hypothesis was constructed which stated that an aposymbiotic ciliate in long-term culture lost its ability to recognise symbiotic algae (Karakashian 1975; Weis & Ayala 1979). To regain this ability,

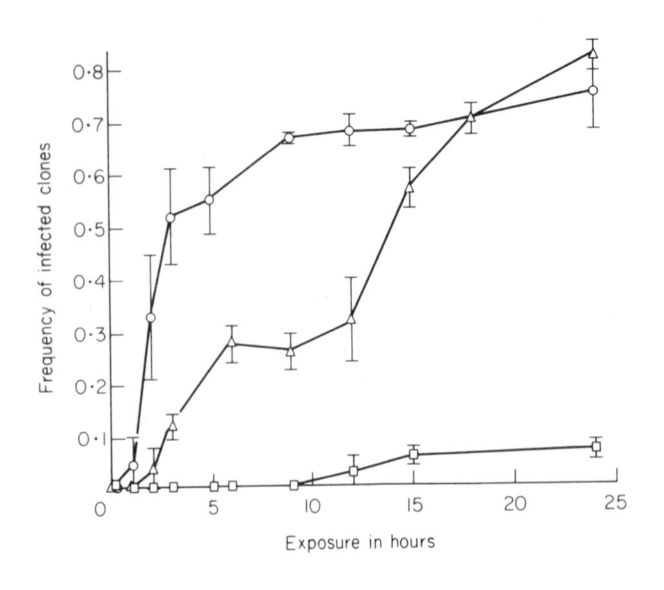

Figure 5. The effect of the algal/ciliate ratio on the frequency of infected clones. Key: (○) 3×10^4 algae per ciliate; (△) 1×10^3 algae per ciliate; (□) 1×10^2 algae per ciliate. Notice that about 1000 algae per ciliate seem to be required for maximal infection frequency (after Weis & Ayala 1979, with permission).

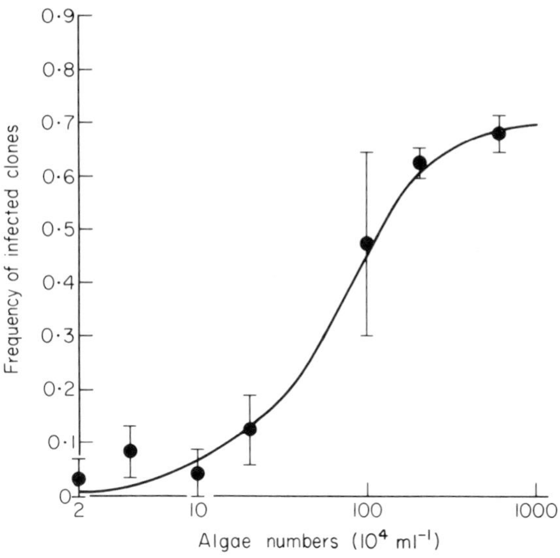

Figure 6. The relation of $\log_{10}$ concentration of symbiotic algae to the frequency of infected clones fitted to the theoretical Poisson distribution curve for $2 > N > 0$. The exposure period of aposymbiotic ciliates to algae was 12 h (after Weis & Ayala 1979, with permission).

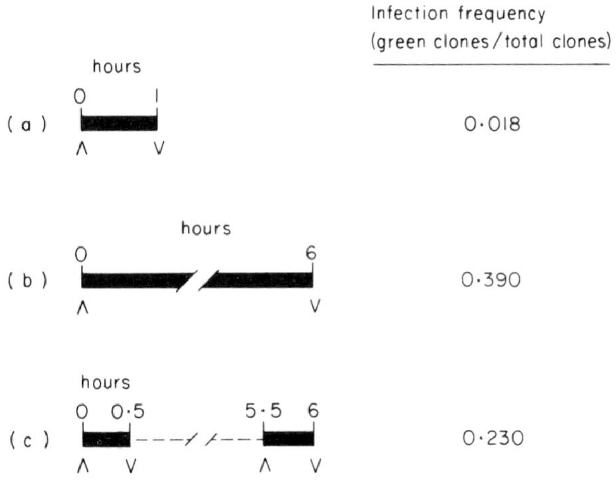

Figure 7. The frequency of infected clones after (a) 60 min, and (b) 360 min, uninterrupted exposures to algae and (c) after two-30 min exposures separated by 300 min in the absence of external algae. (∧) represents algae added; (∨) represents algae removed (after Weis 1980, with permission).

it must be subjected to an induction triggered by approximately 1000 symbiotic algae or a product of the algae, perhaps maltose, and requires exposure to the algae for approximately 6 h. This proposed first induction phase of infection, which resembles the resorting phase in *Hydra* sp. infection (Jolley & Smith 1980), would be required, theoretically, only by ciliates whose recognition mechanism had completely decayed. Again theoretically, freshly isolated aposymbiotic ciliates should not require induction or require less. A method by which fresh aposymbionts could be made available in large numbers is very much desired and at present the method of Pardy (1976) (pp. 327–8) is most promising. Once the recognition capability has been re-established in the aposymbiotic ciliate population then, according to the proposal, the second phase of infection, recognition, can proceed. According to hypothesis, this phase requires as few as one alga per ciliate and is essentially instantaneous. It is not known whether infection proceeds in this way or if it can be completed by a single alga following induction.

What are the algal requirements for the successful completion of infection? Attempted infections with variant symbiotic chlorellae have indicated that successful completion of infection requires an alga with a low density of cell surface glucose residues (or perhaps a high density of some other cell wall constituent such as protein) and high rate of maltose release. However, a mutant alga capable of enough maltose release but with a high density of surface glucose residues, and so non-infective for the host, is not available to test the hypothesis. Therefore, the requirement for a specific density of cell surface glucose (or protein) residues has not been rigorously tested. Con A-coated algae were markedly less infective than untreated algae for aposymbiotic ciliates (Fig. 8). If the exposure of aposymbiotic ciliates to Con A-coated algae was limited to 30 min, adding a second, small dose of untreated algae after 15 h eliminated the inhibition of infection (Fig. 8). However, if Con A-treated algae were used in the second addition, there was only a partial restoration of the inhibition (Fig. 8). Since Con A at the concentration employed has no effect on maltose excretion, this result is compatible with the hypothesis that cell surface ligands, presumably obscured by binding Con A, may be involved as signals to the host during the later (recognition?) phase of infection.

The possible role of maltose release as a signal during infection was investigated by use of the photosystem II inhibitor, 3-(3,4-dichlorophenyl)-1,-1-dimethylurea (DCMU) which at a concentration of 1×10^{-6}M inhibited maltose excretion by algae grown *in vitro* by 90%. At this concentration DCMU also inhibited infection by 87% when present throughout the 6 h exposure of ciliates to algae (Fig. 9). Restricting DCMU to the first 30 min of exposure of the ciliates to the algae and introducing a second small dose of untreated algae resulted in a nearly equal inhibition of infection (Fig. 9). When DCMU was introduced 6 h after exposure of ciliates to algae, considerable inhibition of infection was evident (Fig. 9), suggesting that maltose release was involved as a recognition signal throughout the infection period with its primary role being played during the initial contact of algae and ciliates (the induction phase). The results of these preliminary experiments are compatible with the hypothesis that cell surface ligands and maltose excretion may be important signals during infection, and suggest that maltose excretion, but not cell surface ligands, is involved during the period of initial contact of algae and ciliates (induction).

Techniques for studying nutrient exchange between symbiont and host

Muscatine (1965), working with green *Hydra* sp., and Smith *et al.* (1969), working with lichens, and their associate Mews (1980) have developed the

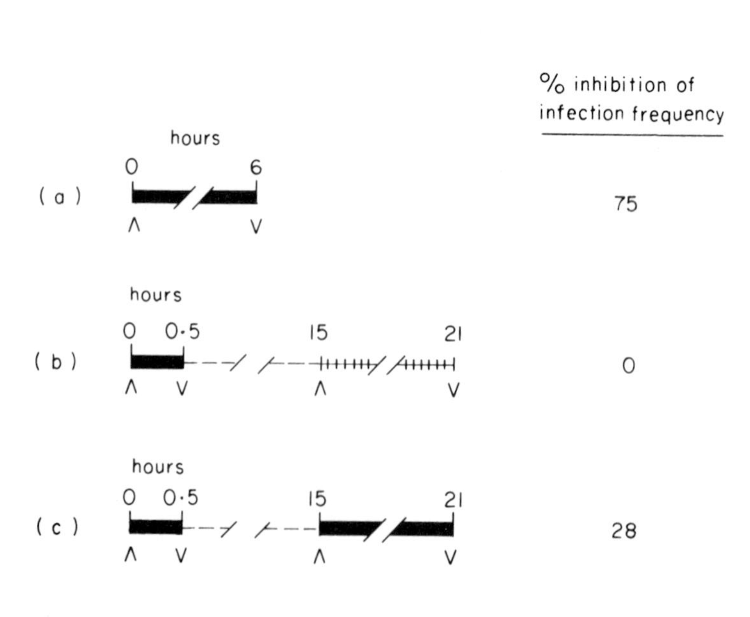

Figure 8. A comparison of the infectivity of Con A-treated and untreated algae for aposymbiotic ciliates. (a) Ciliates exposed to Con A-treated algae for 6 h. (b) Ciliates exposed to Con A-treated algae for 0.5 h, washed free of exogenous algae, incubated in the absence of exogenous algae for 14.5 h, then ciliates were exposed to 100 untreated algae per ciliate (insufficient for induction) for an additional 6 h. (c) Ciliates treated as in (b) except that at 15 h ciliates were exposed to Con A-treated algae rather than untreated algae. (■), ciliates exposed to Con A-treated algae; (┼┼┼), ciliates exposed to untreated algae; (---), ciliates incubated in the absence of algae; (∧), algae added; (∨), algae removed (after Weis 1980, with permission).

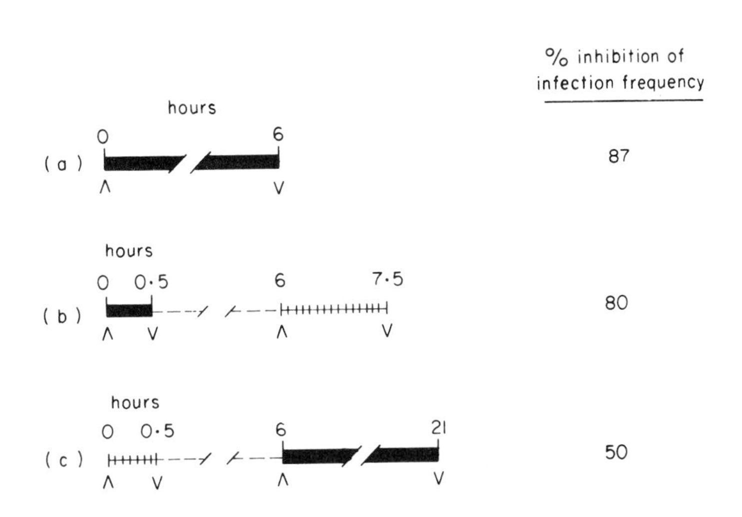

Figure 9. A comparison of the infectivity of DCMU-treated and untreated algae for aposymbiotic ciliates. (a) Ciliates exposed to DCMU-treated algae for 6 h. (b) Ciliates exposed to DCMU-treated algae during the initial 30 min only, washed free of exogenous algae, incubated for an additional 5.5 h in the absence of exogenous algae, then 100 untreated algae per ciliate were added to the ciliate suspension and incubation resumed for an additional 1.5 h. (c) Ciliates exposed to untreated algae during the initial 30 min only, washed free of exogenous algae, incubated in the absence of exogenous algae for an additional 5.5 h, then the ciliates were exposed to 100 DCMU-treated algae per ciliate for 15 h. (■), ciliates exposed to DCMU-treated algae; (HHH) = ciliates exposed to untreated algae at 100 algae per ciliate; (---), ciliates incubated in the absence of algae; (∧), algae added; (∨), algae removed (after Weis 1980, with permission).

use of ^{14}C-labelled algal photosynthate and followed its subsequent release from the algae either to the external medium *in vitro* or to the host tissues *in situ*. There have been a few studies of nutrient exchange in *Paramecium bursaria* using this technique (Muscatine *et al.* 1967; Brown & Nielsen 1974) and it is hoped that the full exploitation of this technique in *Paramecium bursaria* will soon be undertaken. The use of this technique and subsequent paper chromatography and autoradiography, led to the discovery (Muscatine 1965) that in *Hydra* sp. symbiosis, a single major algal excretory product, maltose, was involved. Muscatine *et al.* (1967) were also able to demonstrate the same sugar in *P. bursaria* symbiosis and similar methods have demonstrated that other sugars and polyols were specific products in lichen symbioses (Smith *et al.* 1969). A feature of lichen and marine invertebrate symbioses is the rapid loss of the capacity of the algal symbionts to release sugar *in vitro* (Smith 1979). This implies that during infection of aposymbionts in these systems there must be an induction mechanism to promote the full rate of sugar release. Efforts to augment the decreased rate of sugar release in marine invertebrates with host extracts have met with some success (Trench 1971), suggesting that host factors were involved in the induction of the full rate of sugar release *in vivo*.

The growth factor requirements of the alga which are met by host predation are a cheap price for the host to pay for the massive influx of photosynthate. However, supplying the algal requirements is an absolute contractual obligation which the ciliate cannot avoid and, since the ciliate itself probably requires these same substances, it must continue to depend on an external supply, probably particulate, for its growth factors. In other words, feeding must continue in *Paramecium bursaria* if for no other reason than to supply itself and its symbiont with growth factors as was suggested for foraminifera symbiosis (Lee *et al.* 1979a).

It is not known how nutrient exchange in *Paramecium bursaria* is regulated. If the algal growth factor requirements were known precisely, it might be possible to test the effect of externally supplied growth factors on algal sugar release using a ^{14}C-labelling technique. The mechanism for host regulation of algal cell number is one of the major unsolved problems in *Paramecium bursaria* symbiosis. This experimental approach may result in an understanding of host–symbiont balancing (see below).

A proposal for studying host–symbiont regulation (balancing)

The previous discussion should have made clear the bias of the author that partner specificity, mutual recognition, signalling, nutrient exchange and balancing are integrated at a fundamental biochemical level. Perhaps the most difficult questions are: How do these processes become packaged into this miniature system in such a functional manner? And is the balance maintained? To the author's knowledge these questions have not yet been tested by suitable experiments.

19.5.6 Suggestions for Studying the Systematics, Evolution and Ecology of Host and Symbiont in Relation to the External Environment

What is perhaps most striking about some symbioses is the amount of exchange at all levels between the partners. An attempt has been made here to treat these exchanges comprehensively in one symbiotic association. But *Paramecium bursaria, Hydra* spp., and lichens are not completely isolated from the wider environment and one would like to indulge in a few speculations about the *P. bursaria–Chlorella* sp. symbiosis—particularly its origin, its evolutionary destiny and how host and symbiont might interact with their free-living counterparts. Of course, much more basic information is needed before such speculations are substantiated. For instance, a complete description of the systematics of symbiotic chlorellae would be extremely valuable. The availability of an extensive systematic study of free-living chlorellae, established particularly by Kessler (1976) has paved the way for such a description. At present a study is in progress by W. Reisser who has taken advantage of the morphological-physiological criteria described by Kessler (1976) to try to relate symbiotic algae to free-living forms. The work has already shown that an algal isolate from *P. bursaria* appears most closely related to two taxa, *Chlorella vulgaris* Beijerinck and *Chlorella sorokiniana* Shihira and Krauss (Reiser 1975). These two species are particularly interesting since numerous variants from the mean GC/AT ratio occur, suggesting that this is an evolutionarily active group (Hellmann & Kessler 1974). Perhaps the most likely speculation at the moment is that symbiotic algae of *P. bursaria* evolved from chlorellae closely related to the present free-living species *C. sorokiniana*. The occurrence of growth factor requirements and sugar release in a free-living strain indicates that either it had a symbiotic origin or that it was a suitable candidate for symbiosis. It is certainly possible that some strains of free-living algae had a symbiotic origin and have subsequently re-adapted to the free-living mode of life. It is indeed possible that there may have been several endozoic/free-living cycles in the evolution of some chlorellae.

The freshwater ciliates which have developed associations with a *Chlorella* sp. include a number of distantly related genera, including *Paramecium, Stentor, Frontonia, Climacostomum, Vorticella, Euplotes* and *Prorodon*. It is unlikely that these *Chlorella* sp. symbioses developed independently in each ciliate. It seems more probable that symbiosis developed first in one ciliate (perhaps *Paramecium* spp.) or, at most, in a few and the symbiont was subsequently exploited by the other ciliates. This suggestion implies that certain features of symbiotic algae from *Paramecium* spp., such as maltose excretion and certain cell surface properties, should also be found in algae from the secondarily exploiting ciliates. It also implies that specific host adaptations to symbiosis found in *Paramecium* spp., such as the unique perialgal vacuole membrane with its possible recognition capability, might be missing or highly modified in these other symbioses. Other associations which parallel symbiosis in *P. bursaria* are the *Hydra* sp. symbioses. First, in these systems, as in *P. bursaria*, the major algal excretory product is maltose. Second, there has been success in several of the attempts at cross-infection. Further studies, specifically on sugar release *in vitro* and surface properties of hydra algae, are essential. However, the author would like to offer the suggestion that in *Hydra* sp. as in ciliates other than *P. bursaria*, the symbiont has been secondarily acquired from *P. bursaria* and therefore what has been said about variation in the ciliate host adaptations to symbiosis would also apply to *Hydra* sp. That is, *Hydra*, with all the potential inherent in multicellularity, has probably evolved new and unique ways to adapt to its adopted symbiont and one might expect a quite different and perhaps more sophisticated host recognition-surveillance system as suggested by Pool (1979).

19.6 Protozoal symbionts as models for studying cell–cell interactions

Cell–cell interaction involving recognition of specific cell surface ligands coexists with multicellularity. The multicellular organism must be capable of distinguishing self cells from not-self cells and, depending on the sophistication of its developmental process, must be capable of detecting subtle differences between surface ligands on self cells. Although unicells in general demonstrate primitive recognition-response systems, such as mating systems in yeasts, *Chlamydomonas* sp. and ciliates, associations of eukaryotic symbionts with protozoa and particularly algal–ciliate associations may be unique in the extent to which they have resorted to recognition systems to maintain the steady state and ensure reinfection. Such elaborate surveillance might be necessary in *Paramecium bursaria*, e.g. in which the algal population approaches 1000 algae per ciliate with the inevitable occurrence of variants. In addition, being a particle feeder, discrimination between exogenous and resident chlorellae is also necessary to prevent the intrusion of non-productive chlorellae.

The unique experimental value of the *Paramecium–Chlorella* system is as an advanced recognition-response system in a compact, manipulable form. In *Paramecium bursaria* it should be possible to isolate the ligands on the algal cell surface and the corresponding receptors for the bound ligands and for diffusible substances released by the symbionts, in the ciliate membrane system.

Acknowledgement

The author would like to express his appreciation to Professor D. C. Smith, Professor L. Muscatine and Dr D. J. Patterson for their critical reviews of the manuscript.

Recommended reading

BALL G. H. (1969) Organisms living on and in protozoa. In: *Research in Protozoology* (Ed. T. Chen), vol. 3, pp. 567–718. Pergamon Press, Oxford.

GUTTERIDGE W. E. & COOMBS G. H. (1977) *Biochemistry of Parasitic Protozoa*. University Park Press, Baltimore.

MARGULIS L. (1970) *Origin of Eukaryotic Cells*. Yale University Press, New Haven.

MUSCATINE L., POOL R. R. & TRENCH R. K. (1975) Symbiosis of algae and invertebrates: aspects of the symbiont surface and the host–symbiont interface. *Transactions of the American Microscopical Society* **94**, 450–69.

SMITH D. C. (1979) From extracellular to intracellular: the establishment of a symbiosis. *Proceedings of the Royal Society of London B* **204**, 115–30.

TAYLOR F. J. R. (1979) Symbionticism revisited: a discussion of the evolutionary impact of intracellular symbioses. *Proceedings of the Royal Society of London B* **204**, 267–86.

TRAGER W. (1970) *Symbiosis-Selected Topics in Modern Biology*. Van Nostrand-Reinhold, New York.

References

AHN T. I. & JEON K. W. (1979) Growth and electron microscopic studies on an experimentally established bacterial endosymbiosis in amoeba. *Journal of Cell Physiology* **98**, 49–58.

ANDERSON O. R. (1978) Fine structure of a symbiont-bearing colonial radiolarian, *Collosphaera globularis* and ^{14}C isotopic evidence for assimilation of organic substances from its zooxanthellae. *Journal of Ultrastructure Research* **62**, 181–9.

BALL G. H. (1969) Organisms living on and in protozoa. In: *Research in Protozoology* (Ed. T. Chen), vol. 3, pp. 567–718. Pergamon Press, Oxford.

BANNISTER L. H. (1979) The interrelations of intracellular protista and their host cells with special reference to heterotropic organisms. *Proceedings of the Royal Society of London B* **204**, 141–63.

BARNA I. & WEIS D. S. (1973) The utilization of bacteria as food for *Paramecium bursaria*. *Transactions of the American Microscopical Society* **92**, 434–40.

BEIJERINCK M. W. (1980) Kulturversuche mit Zoochlorellen, Leichengonidien, und anderen Algen. *Botanisches Zeitung* **48**, 725–85.

BEREZINA I. G. (1975) An electron microscope study of endosymbionts in *Blepharisma japonicum*. *Acta Protozoologica* **13**, 365–70.

BERGER J. (1977) 'Imprinting' in *Didinium*: The effect of prior diet on its feeding behaviour. *Proceedings of the Vth International Congress of Protozoology, New York*, Abstract 32.

BOMFORD R. (1965) Infection of alga-free *Paramecium bursaria* with strains of *Chlorella*, *Scenedesmus* and a yeast. *Journal of Protozoology* **12**, 221–4.

BROWN J. A. & NIELSEN P. J. (1974) Transfer of photosynthetically produced carbohydrate from endosymbiotic chlorellae to *Paramecium bursaria. Journal of Protozoology* **21**, 569–70.

Cernichiari E., Muscatine L. & Smith D. C. (1969) Maltose excretion by the symbiotic algae of *Hydra viridis. Proceedings of the Royal Society B* **173**, 557–76.

Chang K-P. (1975) Haematophagous insect and haemoflagellate as hosts for prokaryotic endosymbionts. In: *Symbiosis* (Eds D. Jennings & D. Lee), pp. 407–28. Cambridge University Press, Cambridge.

Curds C. R. (1977) Microbial interactions involving protozoa. In: *Aquatic Microbiology* (Eds F. Skinner & J. Shewan), pp. 69–105. Academic Press, London.

Devauchelle G. & Vinckier D. (1976) Occurrence of microorganisms in parasite gregarines. *Zeitschrift fuer Parasitenkunde* **48**, 297–8.

Diller W. F. & Kohnaris D. (1966) Description of a zoochlorella-bearing form of *Euplotes, E. daidaleus* n. sp. (Ciliophora, Hypotrichida). *Biological Bulletin* **131**, 437–45.

Dilts J. A. (1977) Chromosomal and extrachromosomal DNA from four bacterial endosymbionts derived from stock 51 of *Paramecium tetraurelia. Journal of Bacteriology* **129**, 885–94.

Dwyer D. M. & Chang K-P. (1976) Surface membrane carbohydrate alterations of a flagellated protozoan mediated by bacterial endosymbiotes. *Proceedings of the National Academy of Science U.S.A.* **73**, 852–6.

Fine K. E. & Loeblich A. R. (1976) Endosymbiosis in the marine dinoflagellate *Kryptoperidinium foliaceum. Journal of Protozoology* **23**, 88.

Fott B. & Nováková M. (1969) A monograph of the genus *Chlorella—the freshwater species.* In: *Studies in Phycology* (Ed. B. Fott), pp. 10–74. Academia, Prague.

Gillies C. & Hanson E. D. (1962) A new species of *Leptomonas* parasitizing the macronucleus of *Paramecium trichium. Journal of Protozoology* **10**, 467–73.

Golikova M. N. (1978) Intranuclear parasitic bacteria in micro and macro nuclei of the ciliate *Paramecium bursaria. Tsitologiya* **20**, 576–80.

Graham L. E. & Graham J. M. (1978) Ultrastructure of endosymbiotic *Chlorella* in a *Vorticella. Journal of Protozoology* **25**, 207–10.

Gutteridge W. E. & Coombs G. H. (1977) *Biochemistry of Parasitic Protozoa.* University Park Press, Baltimore.

Hämmerling J. (1946) Über die Symbiose von *Stentor polymorphus. Biologie Zentralblatt* **65**, 52–61.

Heckmann K. (1975) Omikron, ein essentieller Endosymbiont von *Euplotes aediculatis. Journal of Protozoology* **22**, 97–104.

Heckmann K., Preer J. R. & Straehling W. H. (1967) Cytoplasmic particles in the killers of *Euplotes minuta* and their relationship to the killer substance. *Journal of Protozoology* **14**, 360–3.

Hellmann V. & Kessler E. (1974) Physiologische und biochemische Beiträge zur Taxonomie der Gattung *Chlorella.* VIII. Die Basenzusammensetzung der DNS. *Archiv für Mikrobiologie* **95**, 311–18.

Hirshon J. B. (1969) The response of *Paramecium bursaria* to potential endocellular symbionts. *Biological Bulletin* **136**, 33–42.

Hollande A. & Carré D. (1974) Les xanthelles des radiolaires sphaerocollides, des acanthaires et de *Velella velella*: infrastructure–cytochemie–taxonomie. *Protistologica Tome X, fascicile* **4**, 373–601.

Hollande A. & Guilcher Y. (1945) Les amibes du genre Pelomyxa: ethologie. *Bulletin Société Zoologique France* **70**, 53–6.

Jeffrey S. W. (1976) Dinoflagellates which contain chrysophyte endosymbionts. *CSIRO Annual Report 1975/6*, 33–5.

Jeffrey S. W. & Vesk M. (1976) Further evidence for a membrane-bound endosymbiont within the dinoflagellate *Peridinium foliaceum. Journal of Phycology* **12**, 450–5.

Jeon K. W. & Ahn T. I. (1978) Temperature sensitivity: a cell characteristic determined by obligate endosymbionts in amoebas. *Science* **202**, 625–7.

Jeon K. W. & Jeon M. S. (1976) Endosymbiosis in amoeba: recently established endosymbionts have become required cytoplasmic components. *Journal of Cell Physiology* **89**, 337–44.

John P. & Whatley F. R. (1976) *Paracoccus denitrificans* Davis (*Micrococcus denitrificans* Beijerinck) as a mitochondrion. *Advances in Botanical Research* **4**, 51–115.

Jolley E. & Smith D. C. (1978) The green hydra symbiosis. I. Isolation, culture and characteristics of the *Chlorella* symbiont of 'European' *Hydra viridis. New Phycology* **81**, 637–45.

Jolley E. & Smith D. C. (1980) The green hydra symbiosis. II. The biology of the establishment of the association. *Proceedings of the Royal Society B* **207**, 311–33.

Karakashian M. W. (1975) Symbiosis in *Paramecium bursaria.* In: *Symbiosis* (Eds D. Jennings & D. Lee), pp. 145–73. Cambridge University Press, Cambridge.

Karakashian M. W. & Karakashian S. J. (1973) Intracellular digestion and symbiosis in *Paramecium bursaria. Experimental Cell Research* **81**, 111–19.

Karakashian S. J. (1963) Growth of *Paramecium bursaria* as influenced by the presence of algae symbionts. *Physiological Zoology* **36**, 52–68.

Karakashian S. J. (1970a) Morphological plasticity and the evolution of algal symbionts. *Annals of New York Academy of Science* **175**, 474–87.

Karakashian S. J. (1970b) Invertebrate symbioses with *Chlorella.* In: *Biochemical Coevolution*, Proceedings of 29th Annual Biology Colloquium (Ed. K. Chambers), pp. 33–52. Oregon State University Press, Corvallis.

Karakashian S. J. & Karakashian M. W. (1965) Evolution and symbiosis in the genus *Chlorella* and related algae. *Evolution* **19**, 368–72.

Karakashian S. J., Karakashian M. W. & Rudzinska

M. A. (1968) Electron microscopic observations on the symbiosis of *Paramecium bursaria* and its intracellular algae. *Journal of Protozoology* **15**, 113–28.

KARAKASHIAN S. J. & SIEGEL R. W. (1965) A genetic approach to endocellular symbiosis. *Experimental Parasitology* **17**, 103–22.

KAWAKAMI H. & KAWAKAMI N. (1978) Behaviour of a virus in a symbiotic system, *Paramecium bursaria*—zoochlorella. *Journal of Protozoology* **25**, 217–25.

KERFIN W. & KESSLER E. (1978) Physiological and biochemical contributions to the taxonomy of the genus *Chlorella* XI. DNA hybridization. *Archives of Microbiology* **116**, 97–103.

KESSLER E. (1976) Comparative physiology, biochemistry and the taxonomy of *Chlorella* (Chlorophyceae). *Plant Systematics and Evolution* **125**, 129–38.

KIES L. (1974) Electronen mikroskopische Untersuchungen an *Paulinella chromatophora* Lauterborn, einer Thekamöbe mit blau-grünen Endosymbionten (cyanellen). *Protoplasma* **80**, 69–89.

LEE J. J. & BOCK W. D. (1976) The importance of feeding in the species of Soritid foraminifera with algal symbionts. *Bulletin of Marine Science* **26**, 530–7.

LEE J. J, MCENERY M. E., KAHN E. C. & SCHUSTER F. L. (1979a) Symbiosis and the evolution of larger foraminifera. *Micropaleontology* **25**, 118–40.

LEE J. J., MCENERY M. E., SHILO M. & REISS Z. (1979b) Isolation and cultivation of diatom symbionts from larger foraminifera (protozoa). *Nature, London* **280**, 57–8.

LOEFER J. B. (1934) Bacteria-free culture of *Paramecium*. *Science* **80**, 206–7.

LOEFER J. B. (1936) Isolation and growth characteristics of the *'Zoochlorella'* of *Paramecium bursaria*. *American Midlands Naturalist* **70**, 184–8.

MCLAUGHLIN J. J. A. & ZAHL P. A. (1966) Endozoic algae. In: *Symbiosis* (Ed. S. Henry), vol. 1, pp. 257–97. Academic Press, New York.

MARGULIS L. (1970) *Origin of Eukaryotic Cells*. Yale University Press, New Haven.

MARGULIS L. (1976) Genetic and evolutionary consequences of symbiosis. *Experimental Parasitology* **39**, 277–349.

MEWS L. K. (1980) The green hydra symbiosis. III. The biotrophic transport of carbohydrate from alga to animal. *Proceedings of the Royal Society B* **209**, 377–401.

MUSCATINE L. (1965) Symbiosis of hydra and algae. III. Extracellular products of the algae. *Comparative Biochemistry and Physiology* **16**, 77–92.

MUSCATINE L. (1981) The establishment of photosynthetic eukaryotes as endosymbionts in animal cells. In: *On the origin of chloroplasts* (Ed. J. Schiff), Elsevier, North-Holland. (In press.)

MUSCATINE L., COOK C. B., PARDY R. L. & POOL R. R. (1975) Uptake, recognition and maintenance of symbiotic *Chlorella* by *Hydra viridis*. In: *Symbiosis* (Eds D. Jennings & D. Lee), pp. 175–203. Cambridge University Press, Cambrige.

MUSCATINE L., KARAKASHIAN S. J. & KARAKASHIAN M. W. (1967) Soluble extracellular product of algae symbiotic with a ciliate, a sponge and a mutant hydra. *Comparative Biochemistry and Physiology* **20**, 1–12.

MUSCATINE L., POOL R. R. & TRENCH R. K. (1975) Symbiosis of algae and invertebrates: aspects of the symbiont surface and the host–symbiont interface. *Transactions of the American Microscopical Society* **94**, 450–69.

OEHLER R. (1922) Die Zellverbindung von *Paramecium bursaria* mit *Chlorella vulgaris* und anderen Algen. *Arbeits aus den Staatsinstitut für Experimentelle Therapie (Frankfurt)* **15**, 3–19.

OSCHMAN J. L. (1966) Apparent digestion of algae and nematocysts in the gastrodermal phagocytes of *Chlorohydra viridissima*. *American Zoologist* **6**, 320.

OSIPOV D. V. (1973) Species infective specificity of omega particles—symbiotic bacteria of the micronucleus of *Paramecium caudatum*. *Tsitologiya* **15**, 211–17.

OSIPOV D. V., RAUTIAN M. S. & SKOBLO I. I. (1974) Loss of the capacity for sexual reproduction in *Paramecium caudatum* cells infected with intranuclear symbiotic bacteria. *Soviet Genetics* **10**, 859–65.

OSIPOV D. V., SKOBLO I. I. & RAUTIAN M. S. (1975) Iota particles, macronuclear symbiotic bacteria of ciliate *Paramecium caudatum* clone M-115. *Acta Protozoologica* **14**, 263–80.

PADO R. (1969) Mutual relations of protozoans and symbiotic algae in *Paramecium bursaria*. III. Nutritive requirements of the *Chlorella* clone occurring as an endosymbiont of *Paramecium bursaria*. *Acta Societatis Botanicorum Poloniae* **38**, 39–55.

PARDY R. (1976) The production of aposymbiotic green hydra by the photodestruction of their symbiotic algae. *Biological Bulletin* **151**, 225–35.

PARK J. T. & JOHNSON M. J. (1949) A submicro determination of glucose. *Journal of Biological Chemistry* **181**, 149–51.

PICKETT-HEAPS J. (1972) Cell division in *Cyanophora paradoxa*. *New Phytologist* **71**, 561–7.

POOL R. R. (1979) The role of algal antigenic determinants in the recognition of potential algal symbionts by cells of *Chlorohydra*. *Journal of Cell Science* **35**, 367–79.

PREER J. R. (1950) Microscopically visible bodies of the cytoplasm of the 'killer' strains of *Paramecium aurelia*. *Genetics* **35**, 344–62.

PREER J. R. (1975) The hereditary symbionts of *Paramecium aurelia*. In: *Symbiosis* (Eds D. Jennings & D. Lee), pp. 125–44. Cambridge University Press, Cambridge.

PREER L. B. & PREER J. R. (1977) Inheritance of infectious elements. *Cell Biology* **1**, 319–71. Academic Press, New York.

PRINGSHEIM E. G. (1928) Physiologische Untersuchungen an *Paramecium bursaria*. Ein Beitrag zur Symbioseforschung. *Archiv für Protistenkunde* **64**, 289–418.

QUACKENBUSH R. L. (1977) Phylogenetic relationship of bacterial endosymbionts of *Paramecium aurelia*. Polynucleotide sequence relationships of S1 kappa and its mutants. *Journal of Bacteriology* **129**, 895–900.

REISSER W. (1975) Zur Taxonomie einer auxotrophen *Chlorella* aus *Paramecium bursaria* Ehrenberg. *Archives of Microbiology* **104**, 293–5.

REISSER W. (1976) Die stoffwechselphysiologischen Beziehungen zwischen *Paramecium bursaria* Ehrbg. und *Chlorella* spec. in der *Paramecium bursaria*-Symbiose II. Symbiose-Spezifische Merkmale der Stoffwechselphysiologie und der Cytologie des Symbioses Verbandes und ihre Regulation. *Archives of Microbiology* **111**, 161–70.

ROSATI G. & VERNI F. (1975) Macronuclear symbionts in *Euplotes crassus* (Ciliata, Hypotrichida). *Bollettino di Zoologia* **42**, 231–2.

ROSATI G., VERNI F. & LUPORINI P. (1976) Cytoplasmic bacteria-like endosymbionts in *Euplotes crassus* (Dujardin) (Ciliata Hypotrichida). *Monitore Zoologica Italiano* **10**, 449–60.

SCHMALJOHANN R. & RÖTTGER R. (1978) The ultrastructure and taxonomic identification of the symbiotic algae of *Heterostegina depressa* (Foraminifera: Nummulitidae). *Journal of the Marine Biology Association of the U.K.* **58**, 227–37.

SCHULTZE K. L. (1951) Experimentelle Untersuchungen über die Chlorellen-Symbiose bei Ciliaten. *Biologie Generalis, Archive für die Allgemeinen Fragen der Lebensforschung* **19**, 281–98.

SHIHIRA I. & KRAUSS R. W. (1963) *Chlorella: Physiology and Taxonomy of 41 Isolates*. University of Maryland Press, College Park, Maryland.

SIEGEL R. W. (1954) Mate-killing in *Paramecium aurelia*, variety 8. *Physiological Zoology* **27**, 89–100.

SIEGEL R. W. (1960) Hereditary endosymbiosis in *Paramecium bursaria*. *Experimental Cell Research* **19**, 239–52.

SMITH D. C. (1974) Transport from symbiotic algae and symbiotic chloroplasts to host cells. In: *Transport at the Cellular Level* (Ed. G. Jennings), pp. 485–520. Cambridge University Press, Cambridge.

SMITH D. C. (1979) From extracellular to intracellular: the establishment of a symbiosis. *Proceedings of the Royal Society B* **204**, 115–30.

SMITH D. C., MUSCATINE L. & LEWIS D. H. (1969) Carbohydrate movement from autotrophs to heterotrophs in parasitic and mutualistic symbiosis. *Biological Review* **44**, 17–90.

SOLDO A. T. & GODOY G. A. (1973) The molecular complexity of *Paramecium* symbiont lambda DNA: evidence for the presence of a multicopy genome. *Journal of Molecular Biology* **73**, 93–108.

SONNEBORN T. M. (1943) Genes and cytoplasm. The determination and inheritance of the killer character in variety 4 of *Paramecium aurelia*. II. The bearing of determination and inheritance of characters in *Paramecium aurelia* on problems of cytoplasm in inheritance, *Pneumococcus* transformations, mutations and development. *Proceedings of the National Academy of Science U.S.A.* **29**, 329–43.

SONNEBORN T. M. (1950) Methods in the general biology and genetics of *Paramecium aurelia*. *Journal of Experimental Zoology* **113**, 87–148.

STANIER R. Y. & COHEN-BAZIRE G. (1977) Phototrophic prokaryotes: the cyanobacteria. *Annual Review of Microbiology* **31**, 225–74.

SUD G. C. (1968) Volumetric relationships of symbiotic zoochlorellae to their hosts. *Journal of Protozoology* **15**, 605–7.

TARTAR V. (1953) Chimeras and nuclear transplantations in ciliates, *S. coeruleus* × *S. polymorphus*. *Journal of Experimental Zoology* **124**, 63–103.

TAYLOR D. L. (1973) Algal symbionts of invertebrates. *Annual Review of Microbiology* **27**, 171–87.

TAYLOR F. J. R. (1974) Implications and extensions of the serial endosymbiosis theory of the origin of eukaryotes. *Taxonomy* **23**, 229–58.

TAYLOR F. J. R. (1979) Symbionticism revisited: a discussion of the evolutionary impact of intracellular symbioses. *Proceedings of the Royal Society B* **204**, 267–86.

TOMAS R. N. & COX E. R. (1973) Observations on the symbiosis of *Peridinium balticum* and its intracellular alga. I. Ultrastructure. *Journal of Phycology* **9**, 304–23.

TRAGER W. (1970) *Symbiosis-Selected Topics in Modern Biology*. Van Nostrand-Reinhold, New York.

TRAGER W. (1974) Some aspects of intracellular parasitism. *Science* **183**, 269–74.

TRENCH R. K. (1971) The physiology and biochemistry of zooxanthellae symbiotic with marine coelenterates. VII. The effect of homogenates of host tissues on the excretion of photosynthetic products *in vitro* by zooxanthellae from two marine coelenterates. *Proceedings of the Royal Society B* **177**, 251–64.

TRENCH R. K. (1979) The cell biology of plant–animal symbioses. *Annual Review of Plant Physiology* **30**, 485–531.

TRENCH R. K., POOL R. R., LOGAN M. & ENGELLAND A. (1978) Aspects of the relation between *Cyanophora paradoxa* (Korschikoff) and its endosymbiotic cyanelles *Cyanocyta korschikoffiana* (Hall & Claus). I. Growth, ultrastructure, photosynthesis and the obligate nature of the association. *Proceedings of the Royal Society B* **202**, 423–43.

TRENCH R. K. & SIEBENS H. C. (1978) Aspects of the relation between *Cyanophora paradoxa* (Korschikoff) and its endosymbiotic cyanelles *Cyanocyta korschikoffiana* (Hall & Claus). IV. The effects of rifampicin, chloramphenicol and cycloheximide on the synthesis with ribosomal nucleic acids and chlorophyll. *Proceedings of the Royal Society B* **202**, 473–82.

TWAN R. S. & CHANG K. P. (1975) Isolation of intracellular symbiotes by immune lysis of flagellate protozoa and characterization of their DNA. *Journal of Cell Biology* **65**, 309–23.

VAN DEN BOSSCHE H. (1976) *Biochemistry of Parasites and Host-Parasite Relationships*. Elsevier-North Holland Biomedical Press, Amsterdam.

VIVIER E., PETITPREZ A. & CHIVE A. F. (1967) Observations ultrastructurales sur les chlorelles symbiotes de *Paramecium bursaria*. *Protistologica* **3**, 325–34.

WEIS D. S. (1976) Digestion of added homologous algae by *Chlorella*-bearing *Paramecium bursaria*. *Journal of Protozoology* **23**, 527–9.

WEIS D. S. (1978) Correlation of infectivity and concanavalin A agglutinability of algae exsymbiotic from *Paramecium bursaria*. *Journal of Protozoology* **25**, 366–70.

WEIS D. S. (1979) Correlation of sugar release and concanavalin A agglutinability with infectivity of symbiotic algae from *Paramecium bursaria* from aposymbiotic *P. bursaria*. *Journal of Protozoology* **26**, 117–19.

WEIS D. S. (1980) Hypothesis: free maltose and cell surface sugars are signals in the infection of *Paramecium bursaria* by algae. In: *Endocytobiology* (Eds W. Schwemmler & H. Schenk), pp. 105–12. Walter de Gruyter, Berlin.

WEIS D. S. & AYALA A. (1979) Effect of exposure period and algal concentration on the frequency of infection of aposymbiotic ciliates by symbiotic algae from *Paramecium bursaria*. *Journal of Protozoology* **26**, 245–8.

WHATLEY J. M. (1976) Bacteria and nuclei in *Pelomyxa palustris*: comments on the theory of serial endosymbiosis. *New Phytologist* **76**, 111–20.

WHATLEY J. M., JOHN P. & WHATLEY F. R. (1979) From extracellular to intracellular: the establishment of mitochondria and chloroplast. *Proceedings of the Royal Society B* **204**, 165–87.

WHITE A. W., SNEATH A. G. & HELLEBUST J. A. (1977) A red tide caused by the marine ciliate *Mesodinium rubrum* in Passamoquoddy Bay, including pigment and ultrastructure studies of the endosymbiont. *Journal of the Fisheries Research Board of Canada* **34**, 413–16.

WICHTERMAN R. (1948) The biological effects of X-rays in mating types and conjugation of *Paramecium bursaria*. *Biological Bulletin* **93**, 113–27.

WITHERS N. W., COX E. R., TOMAS R. & HAXO F. T. (1977) Pigments of the dinoflagellate *Peridinium balticum* and its photosynthetic endosymbiont. *Journal of Phycology* **13**, 354–8.

Chapter 20 • Microbial Competition in Continuous Culture

J. Gijs Kuenen and Wim Harder

20.1 Introduction

Natural environments generally are heterogeneous open systems in which populations of microorganisms rarely exist in complete isolation from each other. In fact it is now axiomatic that different kinds of organisms can coexist in nature in the same habitat because each type of organism has its own functional status in the ecosystem. Gause (1934) recognised many years ago that, not only are the constraints imposed upon a particular population of one or more species by the physical and chemical environment important, but so too are the activities of neighbouring populations. Various types of interactions between pairs of microorganisms have been encountered, both negative and beneficial (Slater & Bull 1978). Of these interactions competition is probably the most important as it is the basis, as Gause (1934) observed, of the struggle for existence and occurs when populations of different organisms are limited in terms of growth rate or final population size by the common dependence on an external factor required for growth.

It is now generally accepted that in nature competition most often occurs at submaximal growth rates because substrates for microbial growth become available in low concentrations (Veldkamp & Jannasch 1972). However, the maximum specific growth rate of a microorganism which is expressed when all nutrients are in excess may also have a selective advantage in those natural environments in which fluctuations in nutrient concentrations occur. The faster growing organism will be able to build up a larger population during periods of excess substrate and thereby gradually increase their proportion of the total microbial biomass. The importance of this type of competition in quantitative terms in nature is difficult to assess. The experimental approach to this problem is the classical batch type enrichment culture, which, in the hands of Beijerinck and Winogradski, led to the discovery of the various metabolic types among microorganisms and their significance in nature. It is still an important method and the reader is referred to the reviews by

Veldkamp (1970) and Whittenbury (1978) for further information. Reproducible enrichments of this type have been largely confined to organisms of pronounced metabolic specificities and it can therefore be regarded as a method to obtain and study competition of specialist organisms.

The wide application of batch enrichment techniques has for a long time overshadowed the fact that enrichment techniques for organisms lacking pronounced metabolic specificity were practically non-existent. These became available with the introduction of the chemostat. In this chapter microbial competition and selection in homogeneous continuous-flow system is discussed. Heterogeneous continuous-flow systems have also been used successfully in studies on microbial competition. Examples are soil columns (Macura 1961; Bazin *et al.* 1976), the method described by Mulder and van Veen (1965) for the enrichment and study of iron bacteria and the studies of Wirsen and Jannasch (1978) on a *Thiovolum* sp. Continuous-flow systems in which suspended particles were present in the culture have also been used (Jannasch & Pritchard 1972). Competition in continuous culture has been the subject of many recent reviews (Veldkamp & Jannasch 1972; Meers 1973; Jannasch & Mateles 1974; Harder *et al.* 1977; Slater & Bull 1978).

20.2 Theoretical considerations of competition in continuous culture

20.2.1 Introduction

A short summary of the theory of continuous culture is given below. For an extensive discussion of the subject the reader is referred to the review of Tempest (1970) (see also Chapter 1).

Continuous cultivation of microorganisms in the chemostat permits the study of microorganisms growing at submaximal growth rates. In principle, a microorganism multiplying in a continuous culture grows at a constant submaximal specific growth rate (μ) when the rate of dilution (D) of the culture with fresh medium is constant. This medium must contain all ingredients required for growth in excess except one, the growth-limiting nutrient. Under the above conditions a steady state is established where $\mu = D$. The concentration of the growth-limiting substrate (s) in the steady state ($\bar{s}$) is given by the empirical formula due to Monod (1942):

$$\mu = \mu_{max}\left[\frac{s}{K_s + s}\right] \tag{1}$$

where μ_{max} is the maximum specific growth rate and K_s a saturation constant numerically equal to s at $\frac{1}{2}\mu_{max}$. For the steady state it follows that:

$$\bar{s} = K_s\left[\frac{D}{\mu_{max} - D}\right] \tag{2}$$

It is important to note that equation (2) specifies a growth-limiting concentration which is determined by the dilution rate (D) and by the values of μ_{max} and K_s for a particular organism: in other words by its typical substrate-saturation curve (Fig. 1). The steady state substrate concentration is neither affected by the concentration of microorganisms in the culture nor by the concentration of the growth-limiting substrate (S_R) in the inflowing medium. Assuming a growth rate-independent yield (Y= weight of organisms formed per unit amount of substrate consumed), the steady state concentration of microorganisms ($\bar{x}$) in the chemostat will be given by:

$$\bar{x} = Y(S_R - \bar{s}) = Y\left[S_R - K_s \cdot \frac{D}{(\mu_{max} - D)}\right] \tag{3}$$

20.2.2 Competition for Growth-Limiting Substrates

Consider two organisms A and B possessing growth characteristics for a common growth-limiting substrate as described in Fig. 1. In a steady state culture of organism A, the growth-limiting substrate concentration ($\bar{s}$) is maintained at a level characteristic for organism A at that particular dilution rate. If, as in the case of Fig. 1(a), organism B is introduced into a culture of A, the specific growth rate of organism B must be lower than that of organism A at that substrate concentration. This holds for all dilution rates. As a result, organism B is washed out of the culture since it cannot grow at the required rate. In the case of Fig. 1(b) μ_B would be higher than μ_A at high dilution rates and organism A would be washed out. The reverse would be true at dilution rates below the crossing point of the two curves. Theoretically, coexistence can occur at the crossing point, but it can be shown mathematically that this is essentially an unstable condition (Frederickson 1977). It is important to recall that $\bar{s}$ is not dependent on the yield of organisms (eqn 2) and therefore Y has no influence

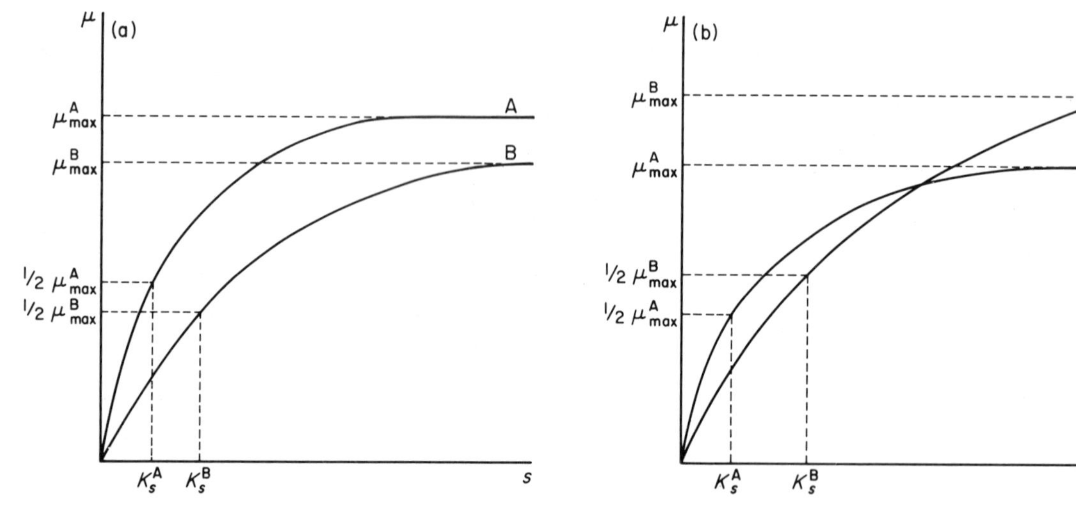

Figure 1. The μ/s relationship for organisms A and B. (a) $K_s^A < K_s^B$ and $\mu_{max}^A > \mu_{max}^B$; (b) $K_s^A < K_s^B$ and $\mu_{max}^A < \mu_{max}^B$ (after Veldkamp 1970).

on the outcome of the competition in the chemostat. One of the crucial requirements for the validity of the arguments as discussed here is that no other interactions occur between organisms A and B (Powell 1958).

The theory can easily be extended to competition between more than two organisms for one growth-limiting substrate and of course also holds for parent strains and their mutants. Recently, mathematical models have also been presented for organisms competing for two or more substrates (Taylor & Williams 1975; Yoon *et al.* 1977). These mathematical treatments are based on Monod kinetics with simple modifications to account for the influence of the metabolism of one substrate on that of the other. Examples of such modifications of the organisms' growth rates are given by:

$$\mu = \frac{\mu_{max1}s_1}{K_1 + s_1 + a_2 s_2} + \frac{\mu_{max2}s_2}{K_2 + s_2 + a_1 s_1} \quad (4)$$

(Yoon *et al.* 1977)

or

$$\mu = \mu_{max}\left[\frac{s_1}{s_1 + K_1}\right]\left[\frac{s_2}{s_2 + K_2}\right] \quad (5)$$

(Taylor & Williams 1975)

where μ_{max1} and μ_{max2} are the maximum specific growth rates of the organism on substrate 1 and 2, respectively; s_1 and s_2 are the concentrations of the two substrates, K_1 and K_2 are the affinity constants of the organism for the two substrates; and a_1 and a_2 are constants.

The most important conclusion from these mathematical exercises is that competition between different organisms for mixed substrates is, as expected, based on the organism's growth characteristics, K_s and μ_{max}, and it also depends on the yield (Y) which, in the case of one growth-limiting nutrient, does not affect selection. (We will come back to this point in an example of competing *Thiobacillus* species during mixed substrate supply.) Another important conclusion is that in the absence of other interactions, coexistence is possible for a number of organisms up to the number of growth-limiting substrates provided, resulting in a stable mixed culture in contrast to that with a single growth-limiting substrate.

The theoretical models of growth and competition in the chemostat, as based on Monod kinetics, can often account for the behaviour of pure and mixed cultures in spite of the fact that a number of simplifications are made. However, many deviations have been described.

First, not all organisms show a typical Monod growth response (Dijkhuizen & Harder 1975). Furthermore, the growth yield is often not constant (Stouthamer 1979) and may vary with the growth rate due to changes in the cell composition, changes in the efficiency of substrate utilisation or maintenance energy requirement. If in a culture not all the cells are viable, the outcome of the competition may be different for a variety of reasons, the most important being that organisms must grow at a rate higher than the dilution rate in order to maintain themselves in the culture (Pirt 1975). Another deviation from the idealised behaviour may be due to population effects, such as when growth-stimulating or growth-inhibitory substances are excreted, changing the growth characteristics of the organism(s). A special case is the growth of organisms on toxic compounds which become growth inhibitory at relatively low levels. The substrate saturation curve then shows a clear maximum at a low concentration of the substrate and above that declines to zero (Fig. 2). Empirical formulae have been proposed to describe the kinetics of growth under these conditions. The Haldane equation is the one most commonly used (Edwards 1970; Pawlowski & Howell 1973):

$$\mu = \mu_{max}\left[\frac{s}{(K_s+s)(1+s/K_i)}\right]$$

where K_i = the inhibition constant and μ_{max} = the theoretical maximum specific growth rate.

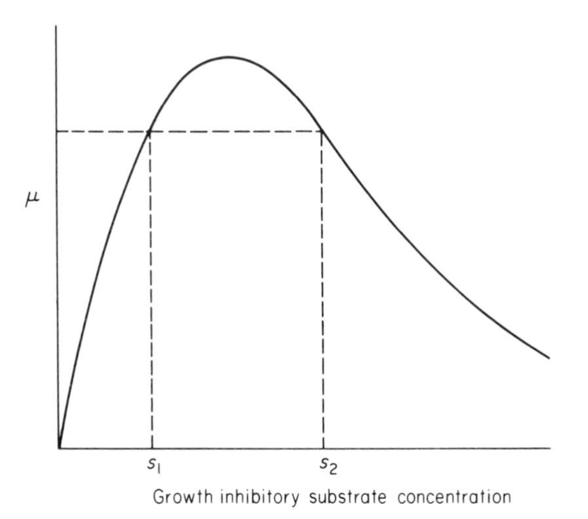

Figure 2. Relation between the specific growth rate and the concentration of a toxic substrate.

As pointed out by Veldkamp and Jannasch (1972) and by Harder *et al.* (1977), steady states in the chemostat at submaximal growth rates are theoretically possible at two discrete substrate concentrations, s_1 and s_2 (Fig. 2). Steady states are only possible at substrate concentrations, such as s_1, which is below the concentration allowing maximal growth rate. At concentration s_2 a minor change in the flow rate will lead either to an increase in the concentration of the toxic substrate or to a decrease. In this case the result will be growth inhibition, a subsequent increase of $\bar{s}$, further growth inhibition until eventually the culture washes out. In the former case the substrate concentrations will go down, the growth rate will subsequently increase, lowering even further the substrate concentration. This will eventually lead to the establishment of a new steady state at s_1. Growth at concentration s_2 is possible only by controlling the substrate concentration directly, for example by automatic monitoring with a substrate-specific electrode.

20.3 Experimental competition in continuous culture

20.3.1 SELECTION AT DIFFERENT SUBSTRATE CONCENTRATIONS

Novick and Szilard (1950) first used the chemostat for the selective cultivation of microorganisms while Jannasch (1967) pioneered the use of the chemostat for the selection of organisms from nature. Jannasch (1967) performed enrichment experiments with four different concentrations of the limiting carbon source in the medium reservoir (S_R) and four different dilution rates. The results (Table 1) indicated that organism selection was not only dependent upon the concentration of the growth-limiting carbon source in the culture as governed by the dilution rate but, in contrast to the theory (see section 20.2.2 p. 343), also upon the population density of the culture which is controlled by the concentration of the carbon source in the

Table 1. Genera of marine bacteria enriched from offshore sea-water at four dilution rates and four different concentrations of the carbon source. A = no dominant population developed; B = heavy wall growth; C = visible turbidity (after Jannasch 1967).

Dilution rate (h^{-1})	Concentration of lactate in the reservoir ($mg\ l^{-1}$) 0.1	1.0	10	100
0.05	B	*Vibrio* sp.	A,B,C,	A,B,C
0.10	*Achromobacter* sp.	*Micrococcus* sp.	*Spirillum* sp.	
0.25	A	*Pseudomonas* sp.	*Pseudomonas* sp.	*Aerobacter* sp.
0.50	A	A	*Aerobacter* sp.	*Aerobacter* sp.

reservoir. The basis for selection at the different dilution rates was investigated as follows. A *Spirillum* sp. and a *Pseudomonas* sp. which became the dominant populations in a lactate-limited chemostat at low and high dilution rates, respectively, were isolated in pure culture. The relationship between growth rate and lactate concentration was determined for both organisms and the results (Fig. 3) indicated that the *Spirillum* sp. had a higher affinity for lactate (lower K_s value) than the *Pseudomonas* sp., whereas the latter organism had the higher μ_{max}. Thus the outcome of the enrichment experiments could be explained on the basis of competition of organisms for a growth-limiting nutrient (see section 20.2.2, p. 345): if the chemostat was operated at dilution rates below the point of intersection of the curves ($D<0.55\ h^{-1}$) the *Spirillum* sp. ultimately displaced the *Pseudomonas* sp. from a mixture of the two and became the dominant population. The reverse applied if the chemostat was operated at dilution rates above $D=0.55\ h^{-1}$).

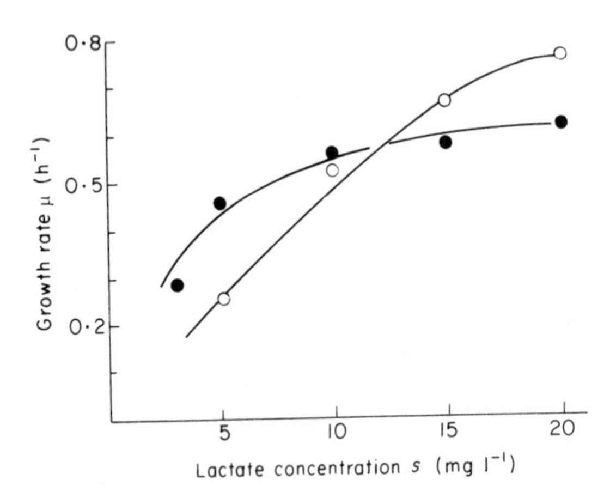

Figure 3. Specific growth rate of a marine *Spirillum* sp. and *Pseudomonas* sp. as a function of lactate concentration. *Pseudomonas* sp., $K_s=9\times10^{-5}$M; *Spirillum* sp., $K_s=3\times10^{-5}$M (after Jannasch 1967).

Selections in the chemostat from fresh water have been carried out by Veldkamp and his coworkers (Kuenen *et al.* 1977). These experiments were performed using carbon-limiting conditions (lactate, glutamate or methanol), energy limitation (thiosulphate) or phosphate or iron limitation. With any of these substrates, small, thin and often spiral-shaped bacteria were selected at low dilution rates. At high dilution rates longer, thicker rod-shaped bacteria were selected. In a number of instances the relationship between growth rate and growth-limiting substrate concentration for organisms isolated at low or high dilution rates on a particular substrate, was determined and in all cases it was found that the curves for the two organisms crossed as shown in Fig. 3. Thus it appears that the outcome of selection of organisms in a nutrient-limited continuous culture can often be explained on the basis of competition for the growth-limiting nutrient. Competition studies between a *Spirillum* sp. isolated at low phosphate concentrations and of a pseudomonad which was isolated at higher phosphate concentrations were also carried out using different nutrient limitations. It was shown that the *Spirillum* sp. outcompeted the pseudomonad at low concentrations of any of the following substrates: lactate, succinate, alanine, asparagine, magnesium, potassium or ammonium. Experiments designed to test the metabolic versatility of the two organisms revealed no differences with respect to the utilisation of 20 different carbon sources as growth substrates (Kuenen *et al.* 1977).

It is highly probable that the principles deduced from chemostat studies also apply to selection in nature. For example in Veluwemeer, a large freshwater lake in Central Holland, the population of green algae, which was dominant for many years, suddenly decreased and the cyanobacterium *Oscillatoria agardhii* became the dominant planktonic organism. Studies by van Liere (1979) and Mur *et al.* (1977) clearly showed that competition for the limiting factor, light, was the basis for this takeover by the cyanobacterium. These authors performed competition experiments between *Scenedesmus protuberans* and *Oscillatoria agardhii* in light-limited continuous cultures at different values of incident light irradiation and different dilution rates. At low light irradiation ($1\ W\ m^{-2}$) and a dilution rate of $0.01\ h^{-1}$ (Fig. 4) *Oscillatoria agardhii* expressed a higher growth rate than the green alga which was progressively washed out as the *Oscillatoria* sp. population increased. A second experiment was performed at a higher incident irradiance ($24\ W\ m^{-2}$) and a dilution rate of $0.03\ h^{-1}$. This irradiance provided optimum conditions for both organisms when growing separately. Initially (Fig. 5) the population of *Scenedesmus* sp. increased more rapidly than that of *Oscillatoria* sp. because the μ_{max} of the former

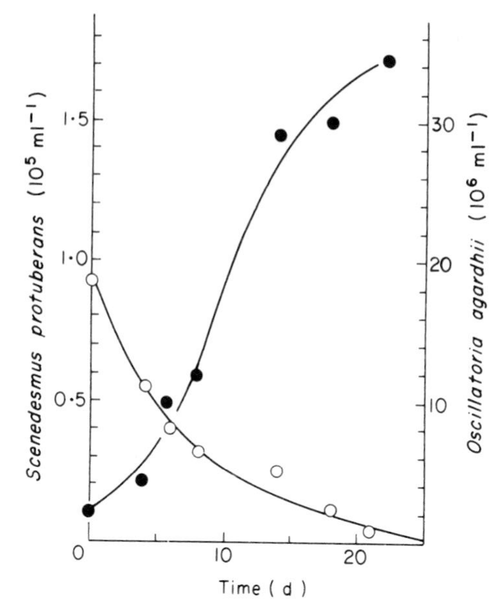

Figure 4. Competition experiment between *Oscillatoria agardhii* Gomont (●) and *Scenedesmus protuberans* Fritsch (○). The incident irradiance was 1 W m^{-2} and the dilution rate was 0.01 h^{-1} (after van Liere 1979).

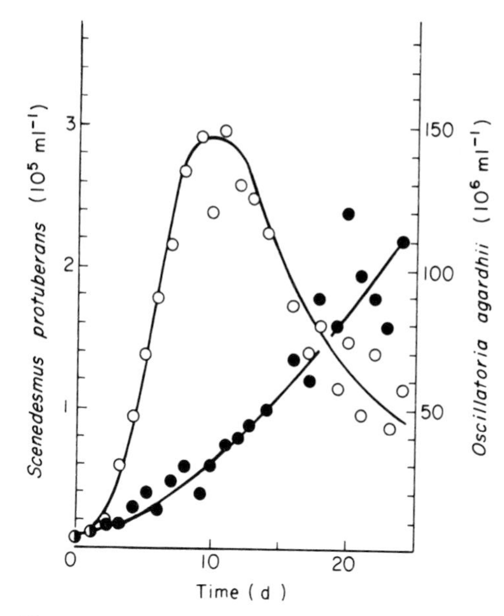

Figure 5. As Fig. 4 except that the incident irradiance was 24 W m^{-2} and the dilution rate was 0.03h^{-1} (after van Liere 1979).

organisms was higher. As the total biomass concentration in the culture increased, the culture became light-limited. The specific light-energy uptake rate of the cells decreased which caused *Scenedesmus* sp. to grow at a rate lower than the dilution rate and it was progressively eliminated from the mixture. A third and very convincing competition experiment was performed at an incident irradiance of 39 W m^{-2} and a dilution rate of 0.03 h^{-1} (Fig. 6). In this case the initial average irradiance was growth inhibitory to the cyanobacterium but not to the green algae. Thus *Scenedesmus* sp. was able to form the dominant population and *Oscillatoria* sp. was completely eliminated from the culture. After 38 d, when *Scenedesmus* sp. had approached steady state conditions under light-limitation, the culture was reinoculated with *Oscillatoria* sp. Under these light-limiting conditions *Oscillatoria* sp. was able to grow faster than *Scenedesmus* and the green alga was subsequently progressively eliminated from the mixed culture. Thus, as a consequence of mutual shading caused by the dense *Scenedesmus* sp. population, light conditions more favourable to

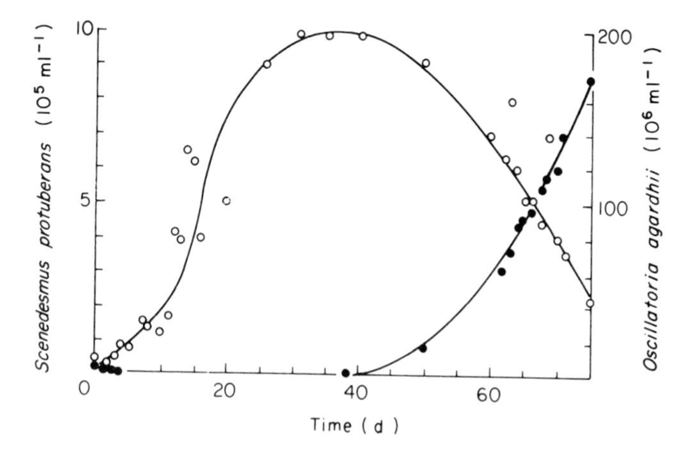

Figure 6. As Fig. 4 except that the incident irradiance was 39 W m^{-2} and the dilution rate was 0.03 h^{-1}. At $t = 38$ d *Oscillatoria agardhii* was reinoculated into the culture vessel (after van Liere 1979).

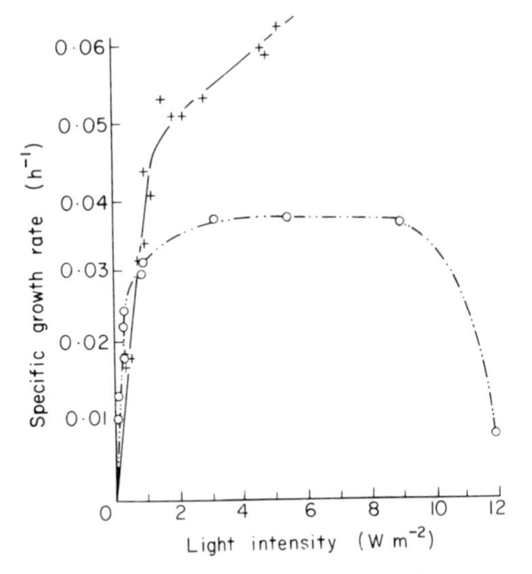

Figure 7. The μ/I relationship for *Scenedesmus protuberans* (+) and *Oscillatoria agardhii* (○) (after Mur *et al.* 1977).

Oscillatoria sp. prevailed and this organism became dominant. It is not surprising that the curves relating specific growth rate and incident light intensity for the two organisms (Fig. 7) cross in a way similar to the general picture (Fig. 1b).

The rate at which competition for one growth-limiting factor leads to one type of organism becoming dominant depends very much on the nature of the growth-limitation and on the dilution rate applied. It can range from days to months, although in many cases a dominant population will develop within the time required for ten volume changes (see Powell (1958) for a detailed mathematical treatment).

The low μ_{max}-low K_s type of organism was considered by Jannasch (1967) to represent the autochthonous (Winogradski 1949) component of the microbial population of natural waters since as a rule these environments are poor in nutrients. But very little information is available concerning their physiology and ecology because selection from these and other environments by batch culture procedures selects against such organisms.

20.3.2 Selection in Response to more than one Environmental Factor

Unfortunately there are few reports in the literature of microbial competition in response to more than one environmental factor. In the following paragraphs the effects of temperature and of a light/dark regime on the outcome of the competition will be considered.

Harder and Veldkamp (1971) studied competition between a facultatively psychrophilic *Pseudomonas* sp. (optimum growth temperature = 30°C) and an obligately psychrophilic *Spirillum* sp. (optimum growth temperature = 14°C) with respect to two environmental parameters namely substrate (lactate) concentration and temperature. At temperature extremes of 16°C and −2°C, one of the organisms grew faster at all lactate concentrations (equivalent to given dilution rates), the higher temperature leading to the dominance of the facultative psychrophile whilst the lower temperature favoured selection of the obligate psychrophile (Fig. 8). At intermediate temperatures selection was dependent upon the concentration of the growth-limiting substrate at which competition occurred, because the curves describing the relationship between growth rate and lactate concentration for the two organisms crossed. This indicated that K_s and μ_{max} changed independently with temperature, a situation which was also found for a *Spirosoma* sp. (Wirsen & Jannasch 1970).

Another example of selection on the basis of more than one parameter can be seen in the competition of two phototrophic bacteria, *Chromatium vinosum* and *Chromatium weissei* grown under sulphide limitation and continuous light conditions in the chemostat (van Gemerden 1974). As can be seen from Fig. 9, *C. vinosum* became the dominant organism at all dilution rates. This does not provide an adequate explanation for the occasional occurrence of *C. weissei* in high numbers in nature. However, in nature the sulphide concentration, in contrast to that in chemostat cultures, may vary significantly due to the day–night cycle. When a light and dark rhythm was introduced as an additional variable in similar competition experiments in the chemostat, *C. weissei* was able to maintain itself in the culture at high levels. It should be emphasised that in this case the continuous culture, by virtue of the dark and light regime, was essentially maintained in a non-steady state condition. The explanation was that *C. weissei* oxidised the sulphide that accumulated during the dark period more rapidly than did *C. vinosum*. This allowed *C. weissei* to utilise a significant portion of the available sulphide in the early light stages after the dark period. During this early oxidation of

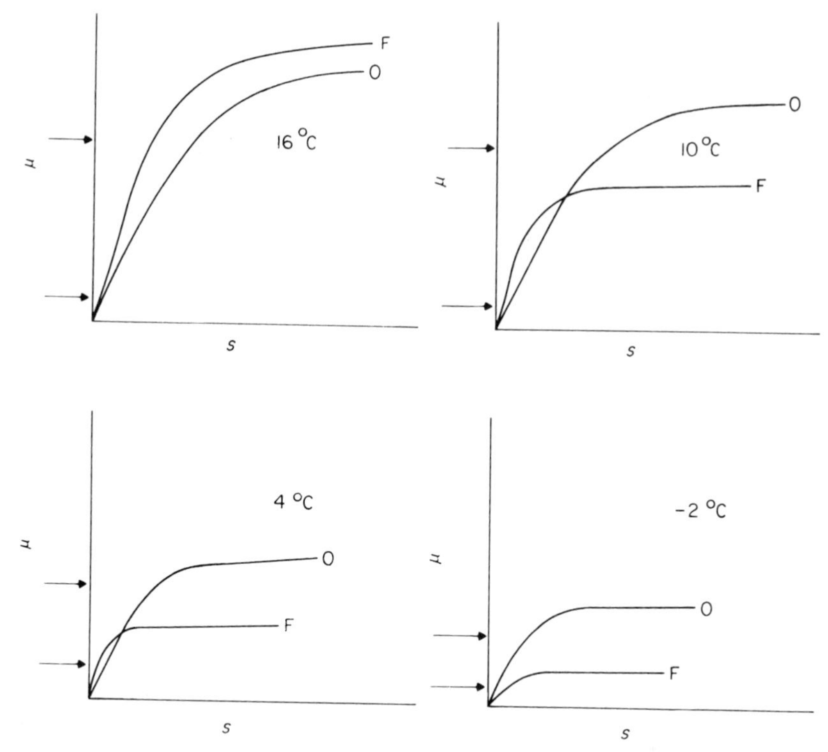

Figure 8. Specific growth rate of an obligately psychrophilic *Pseudomonas* sp. (O) and a facultatively psychrophilic *Spirillum* sp. (F) as a function of lactate concentration at different temperatures. The curves are schematic based on two measurements each at the growth rates indicated by the arrows (after Harder & Veldkamp 1971).

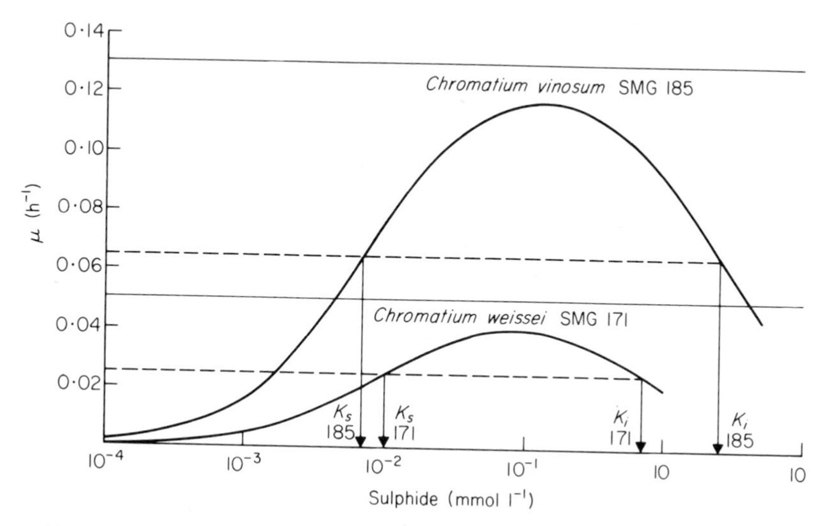

Figure 9. Relation between the specific growth rate and the sulphide concentration for *Chromatium vinosum* and *Chromatium weissei*. Solid horizontal lines represent the theoretical μ_{max} (after van Gemerden 1974).

sulphide, its product sulphur was stored intracellularly as were polysaccharide reserve materials. In the later stage, when sulphide concentrations declined, *C. weissei* continued to grow at the expense of the intracellular sulphur and the reserve polymers at a rate which was higher than that predicted from the μ/sulphide concentration curve. The principle illustrated in these experiments may explain the coexistence of similar organisms in natural habitats.

20.3.3 Selection of Mixed Populations

Some instances have been recorded in the literature in which selection in continuous culture in the presence of one growth-limiting substrate led to stable mixed microbial populations. This is probably a reflection of the occurrence of communities of interrelated organisms in nature; a relationship which maximises their corporate chance of survival. These studies have either been carried out in connection with the production of single cell protein (Harder *et al.* 1977) or the breakdown of herbicides (Slater & Bull 1978).

Wilkinson *et al.* (1974) studied the nature of the interaction of a mixed microbial population enriched on methane as the sole source of carbon and energy. The population consisted of a methane-utilising *Pseudomonas* sp., a methanol-utilising *Hyphomicrobium* sp. and two other organisms. It appeared that the *Hyphomicrobium* sp. had a beneficial effect on the growth of the methane oxidiser because it metabolised the methanol excreted into the growth medium by the methane utiliser. In pure culture the methane oxidiser was unable to grow because accumulation of methanol quickly inhibited its growth. The function of the two other organisms was believed to be the removal of products of growth or lysis.

A more complex microbial community growing on the herbicide Dalapon (2,2-dichloropropionic acid) was studied by Senior *et al.* (1976). This community consisted of three primary Dalapon utilisers, two bacteria and a fungus, and a group of three different bacteria and a yeast which were initially unable to grow on Dalapon and were therefore termed secondary utilisers. The yeast was not very tightly integrated into the community and was quickly eliminated from the culture when the dilution rate was increased. The remaining six-membered culture was extremely stable. After approximately 120 d one of the secondary utilisers acquired the capacity to grow on Dalapon and thus became a primary utiliser. This is an example of enzyme evolution *in vitro* (see section 20.5.3, p. 355).

20.4 Microbe–microbe interactions and competition

20.4.1 Introduction

In the previous sections it was stressed that in the study of microbial competition, the essential assumption was that, apart from the competition for the growth-limiting substrate, no additional interactions occurred. Indeed, in a large number of competition experiments it has been shown that such interactions, if they occur, do not play an essential role. However, in a limited number of experiments results have been encountered which could not be explained solely on the basis of competition, but were a consequence of other interactions which often resulted in the coexistence of different species growing on a single growth-limiting substrate.

A clear definition of biological interaction is difficult to give. It generally implies an almost infinite number of relations between organisms and their biotic environment. Table 2 shows the possibilities listed by Bungay and Bungay (1968).

Table 2. Common terms used to describe microbial interactions (after Bungay & Bungay 1968).

Interaction	Definition
Neutralism	Absence of any interaction between members of a mixed population
Commensalism	One member of a mixed population benefits from another member which is unaffected itself
Mutualism	Both members of a mixed population benefit from the presence of each other
Competition	A conflict for nutrients, space or some other factor which results in all members of the mixed population growing less well compared with their growth characteristics alone
Amensalism	One population adversely affects the growth of another population whilst itself being unaffected by the other population
Parasitism	One organism consumes another organism, often in a subtle, non-debilitating relationship
Predation	One organism ingests another organism and consumes it, often in a violent, destructive relationship

Another term often encountered in the literature, but not listed here, is syntrophism which is a special case of mutualism. This term is used when an organism growing on a substrate can only do so if it is in close association with a second organism which grows on the metabolic products excreted by the first organism. The terms in Table 2 should be considered as simplifications of microbial interactions and not as rigid definitions, since it is likely that in nature many interactions are mixtures of two or more of these extreme cases. A few reports have appeared in the literature which may be considered as classical examples of competition combined with commensalism or with mutualism.

20.4.2 Commensalism

One well-studied example of competition and commensalism concerns a mixed culture of the yeast *Saccharomyces cerevisiae* and a riboflavin-requiring strain of *Lactobacillus casei* (Megee *et al.* 1972). The yeast excreted riboflavin which allowed *L. casei* to grow on glucose when this was the growth-limiting substrate. Figure 10 shows that coexistence of the two species was possible in the chemostat over a wide range of dilution rates. At high dilution rates *L. casei* was washed out. When riboflavin was added to the inflowing medium at a concentration of approximately 16 μg l^{-1}, *L. casei* competed successfully against the yeast at a dilution rate of 0.06 h^{-1}. Apparently by adding the riboflavin this system was reduced to a simple case of competition for the growth-limiting glucose. In Fig. 10 the experimental data are shown as symbols and the continuous lines represent results from a mathematical model developed to describe this competition with comensalism system. The model fits the data adequately. The virtue of such a mathematical exercise is that it shows that the assumptions of the model, namely that this is a case of competition and commensalism, is indeed consistent with the experimental data. When a medium without riboflavin and with a relatively high pH was used the same system could be transformed into a case of competition and mutualism. In these media the yeast and the lactic acid bacterium competed for glucose, the yeast produced riboflavin and *L. casei* lowered the pH which was favourable for growth of *S. cerevisiae*.

The same group of workers has described a case of commensalism without competition (Lee *et al.* 1976). *Lactobacillus plantarum* and *Propionibacterium shermanii* were able to coexist at dilution rates below 0.07 h^{-1} in a glucose-limited chemostat. The first organism metabolised glucose to lactic acid which was used by the propionic acid bacterium. Although *P. shermanii* was able to use glucose it fermented lactic acid in preference to glucose, thus reducing a potential commensalism plus competition system to a system where commensalism was the dominant interaction. This interpretation was supported by a mathematical analysis.

In practice, when performing enrichments in the chemostat (see section 20.7, p. 359), examples of commensalism are frequently encountered. It is a common observation that selection experiments in the chemostat rarely lead to pure cultures, but normally show persistent, small numbers of satellite populations which presumably thrive on metabolic products or products from the lysis of the dominant species. A pertinent example of commensalism (see section 20.3.3, p. 350) is the selection of a mixed culture growing on 2,2-dichloropropionic acid which consisted of primary and secondary utilisers (Senior *et al.* 1976) and was described earlier. Another example pertains to continuous selection of autotrophic bacteria. These organisms

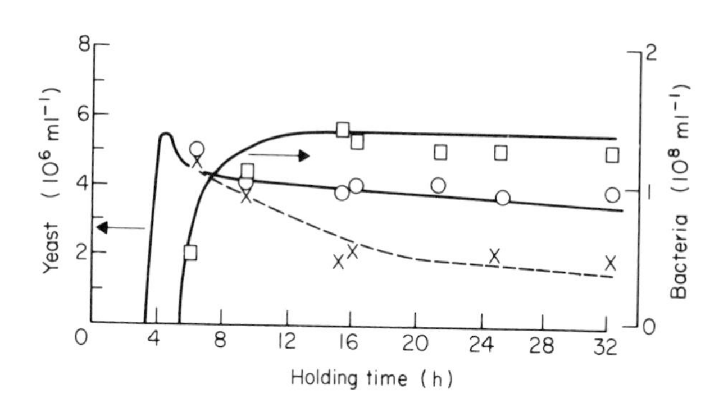

Figure 10. Commensalism plus competition symbiosis for *Saccharomyces cerevisiae* and *Lactobacillus casei* (after Megee *et al.* 1972). Steady state organism concentrations in the chemostat as functions of the holding time (i.e. the reciprocal of the dilution rate). At a small holding time the organisms grow at a high specific rate and vice versa. (S_R (glucose) = 4.9 g l^{-1}; pH = 5.5) (□, *L. casei*; ○, *S. cerevisiae*, total count; x, *S. cerevisiae*, viable count; solid lines, model.)

very often excrete organic compounds (often glycollate): for instance, enriching for *Thiobacillus* sp. in a thiosulphate-limited chemostat, results in mixed cultures consisting of 80 to 90% of the obligately chemolithotrophic thiobacilli with 10 to 20% of the heterotrophic bacteria growing on excretion products.

20.4.3 MUTUALISM

Although the microbiological literature provides many examples of mutualism (Bryant *et al.* 1967; Pfennig & Biebl 1976) only a few cases have been studied in any detail in continuous culture. A theoretical analysis has been given by Meyer *et al.* (1975). They based their model on relationships between two organisms, B_1 and B_2, which compete for substrate S_3 and excrete compounds S_2 and S_1, respectively, each one of which is an essential nutrient for the growth of the other organism. According to this model stable coexistence of organisms B_1 and B_2 in continuous culture in the presence of the common substrate S_3 is possible under conditions in which not only the products S_1 and S_2 but also the common substrate S_3 limit the growth of both organisms. Furthermore, it has been shown by de Freitas and Frederickson (1978) that stable coexistence of two species competing for one growth-limiting substrate is also possible when the two organisms excrete auto-inhibitory substances.

Another type of interaction is sometimes referred to as protocooperation (Frederickson 1977) or syntrophism (Pfennig & Biebl 1976). One example is the interspecies hydrogen transfer encountered in the two-membered mixed culture commonly known as *Methanobacillus omelanskii* (Bryant *et al.* 1967) growing on ethanol. In this mixture one organism ferments ethanol to acetate and molecular hydrogen, but can continue to grow only at extremely low partial pressures of hydrogen. The second organism converts the hydrogen to methane, thereby keeping the partial pressure of hydrogen sufficiently low to allow the first organism to ferment ethanol.

Another example has been mentioned in section 20.3.3 (p. 350), where the selection of a mixed culture growing on methane was discussed. The population consisted of a methane-utilising pseudomonad and three other organisms one of which was a *Hyphomicrobium* species. The latter organism played an essential role in the mixed culture by removing small amounts of methanol which were produced during methane oxidation by the pseudomonad and which otherwise would have inhibited its growth.

20.4.4 PREDATION

Another example of the coexistence of two species competing for a single growth-limiting substrate has been described by Jost *et al.* (1973). It was shown that *Azotobacter vinelandii* and *Escherichia coli* coexisted in a glucose-limited chemostat in the presence of a common predator, *Tetrahymena pyriformis* (Fig. 11). When the predator was not present in the mixture *E. coli* rapidly became dominant and *A. vinelandii* was eliminated. One explanation was that *T. pyriformis* was feeding preferentially on *E. coli*. Another possibility was that the predator, while feeding on both organisms, was excreting metabolic products which provided a mixture of organic compounds allowing the coexistence of the two bacterial species. In our laboratory we have often observed that protozoa easily develop in chemostat enrichment cultures that have been inoculated with natural samples. This often causes dramatic oscillations in the density of the culture and changes in the composition of the bacterial population. It should be realised that in the presence of a predator the actual growth rate of the bacteria may be much higher than the dilution rate, since the density of the bacteria will never reach the level found in the absence of the predator. As a consequence the substrate concentration in the culture will always be higher than that in the absence of the predator.

20.5 Competition and physiology

20.5.1 INTRODUCTION

The numerous investigations of the last decade that have been carried out on selection of microorganisms in the chemostat have established unequivocally that they are based on differences in specific growth rate at a certain substrate concentration (Fig. 1). There is little doubt that the phenomena observed in these model systems must also be the governing principles in the natural environment. The question arises as to what the physiological basis is for the competitiveness of an organism; i.e. which particular characteristic determines whether

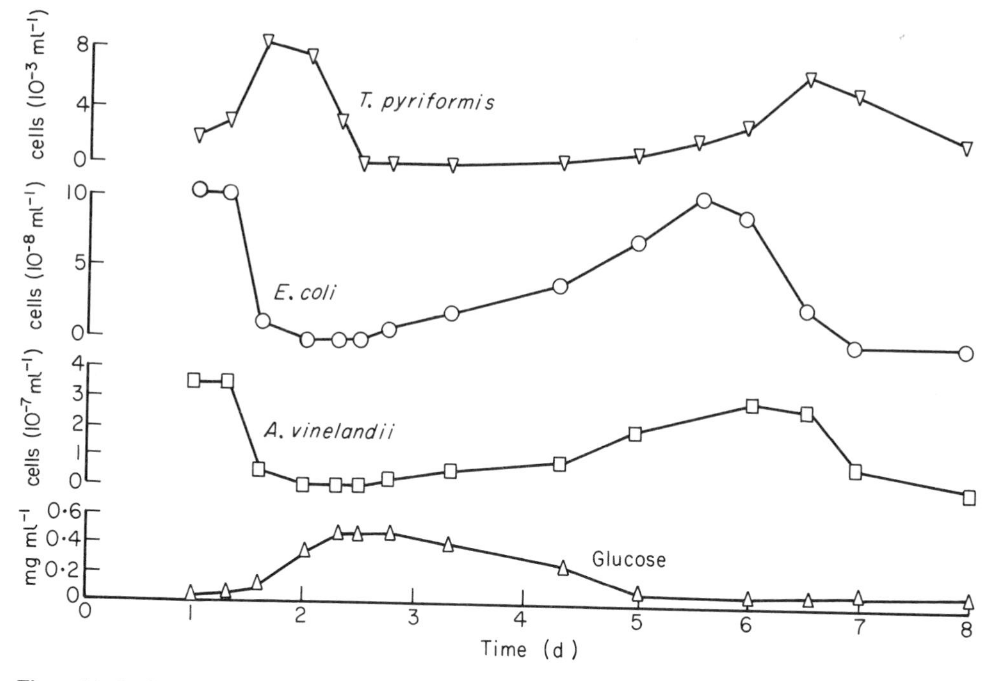

Figure 11. Oscillations in the *Azotobacter vinelandii–Escherichia coli–Tetrahymena pyriformis*–glucose food web (after Jost *et al.* 1973). (Holding time, 7.4 h; S_R (glucose), 49 mg ml^{-1}.)

an organism will grow rapidly or slowly at a given substrate concentration. An answer to this important question may lead towards an understanding of the functional status, or ecological niche, of the great diversity of organisms coexisting in nature.

There is no unifying concept or principle which is able to explain the differences in growth rate. Instead it appears that a spectrum of properties leads to the phenomenological expression of low or high affinity for a substrate, or a high or low growth rate. Some phenomena can now be explained on the basis of properties of enzymes, others can be traced back to more complex properties of the cell. We will briefly summarise some aspects.

20.5.2 Transport

During growth at low nutrient concentrations the rate of transport of the growth-limiting substrate across the membrane may become the limiting factor. Competing species may differ in their ability to transport the limiting nutrient into the cell. Matin and Veldkamp (1978) studied a *Pseudomonas* sp. and a *Spirillum* sp. with crossing substrate saturation curves (Fig. 1b). The *Spirillum* sp. which had been selected at low dilution rates (low K_s), appeared to possess a transport system for the growth-limiting substrate, lactate, which had a lower apparent K_m than that of the *Pseudomonas* sp. The surface to volume ratio of an organism is directly related to transport capacity. In the same study Matin and Veldkamp confirmed earlier observations of Kuenen *et al.* (1977) indicating that an organism with a selective advantage at low dilution rates had a higher surface to volume ratio than an organism with a relatively high μ_{max} (Table 3). This difference would allow the organism with the high surface to volume ratio to scavenge low concentrations of the growth-limiting nutrient. In fact both organisms increased the surface to volume ratio at decreasing growth rates but the increase in the *Spirillum* sp. was much more pronounced than that in the *Pseudomonas* sp. Other properties which may have contributed to the competitiveness of the *Spirillum* sp. at low growth rate were its larger overcapacity to respire lactate and its lower maintenance energy requirement.

Table 3. Surface to volume ratios of four couples of chemostat-grown spiral and rod-shaped bacteria competing for the growth-limiting substrate. These ratios were calculated from the dimensions of the bacterial cells, treating them as cylinders. Dimensions of the bacterial cells were estimated from photomicrographs and electron micrographs (after Kuenen *et al.* 1977).

Organism	Limiting substrate	Specific growth rate (h^{-1})	Average length (μm)	Average diameter (μm)	Surface to volume (μm^{-1})
1. *Pseudomonas* sp.	Lactate	0.10	2.9	1.10	4.3
Spirillum sp.	Lactate	0.10	3.5	0.55	8.0
2. *Pseudomonas* sp.	Lactate	0.15	2.3	1.00	4.9
Spirillum sp.	Lactate	0.15	2.6	0.66	6.8
3. Unidentified rod	Phosphate	0.20	2.8	1.10	4.3
Spirillum sp.	Phosphate	0.20	3.8	0.60	7.2
4. *Thiobacillus thioparus*	Thiosulphate	0.10	2.0	0.40	11.0
Thiomicrospira pelophila	Thiosulphate	0.10	2.5	0.20	21.0

20.5.3 Substrate-Capturing Enzymes

When an organism is grown under conditions of nutrient limitation, its response often is to increase the concentration (activity) of the first enzyme in the pathway for the metabolism of that substrate. If the level of the enzyme in question is the rate-limiting step for growth, the result is that the organism grows faster at the same substrate concentration, or at least maintains the same growth rate at a lower substrate concentration. This was clearly demonstrated by Novick (1961) who showed that in a lactose-limited chemostat culture of *Escherichia coli*, mutants were selected that were constitutive or even super-constitutive for β-galactosidase (Fig. 12). In these mutants the level of this enzyme increased to 5 to 6 times the level of that usually found in constitutive strains and the enzyme comprised approximately 20% of the total protein content. Figure 13 shows the effect of the increased enzyme levels on the μ/s curve of *E. coli* as postulated by Novick (1961). As pointed out above, this result can be rationalised by assuming that β-galactosidase constituted the rate-

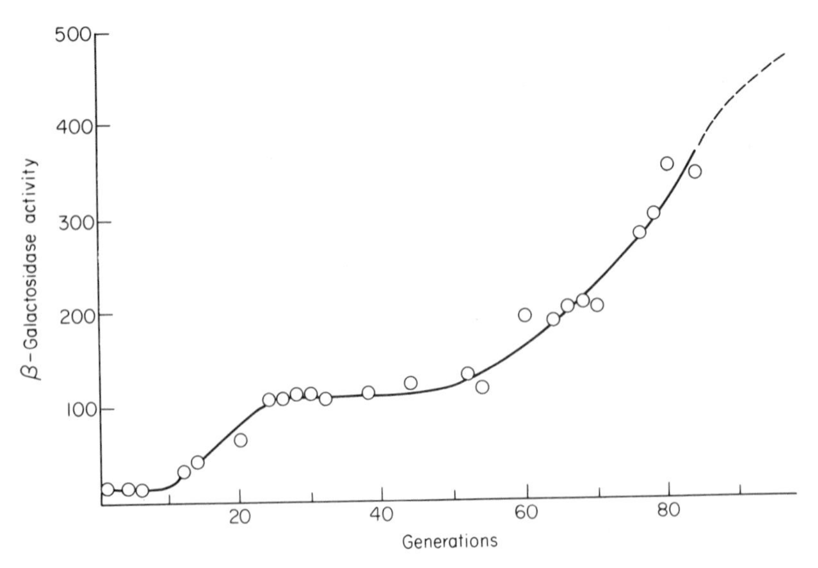

Figure 12. Rise in β-galactosidase activity of *Escherichia coli* strain E102 grown in a lactose-limited chemostat (after Novick 1961).

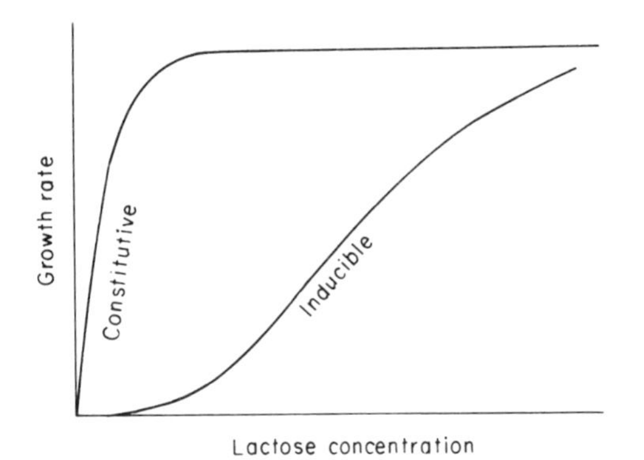

Figure 13. Hypothetical relationship between growth rate and lactose concentration when the growth-limiting enzyme is constitutive and when it is inducible (after Novick 1961).

limiting step in the rate of lactose utilisation. This is not unlikely since the enzyme has a relatively high K_m for lactose (3×10^{-3}M). As a result of the increase of β-galactosidase activity the K_s for growth on lactose for the organism decreased.

An increase in the level of the enzyme is one way in which lower extracellular concentrations of lactose can be created by the organism. Another possibility would be an improvement in the affinity of the enzyme for the substrate. This has been demonstrated elegantly by Hartley *et al.* (1972) who used a parent and mutant strain of *Klebsiella aerogenes* possessing ribitol dehydrogenases with different K_m values for xylitol, a compound that is also oxidised by this enzyme. The mutant having a K_m for xylitol of 0.5M rapidly outcompeted the parent strain (K_m = 1M) under xylitol-limitation in the chemostat. Obviously both enzymes have very low affinities for xylitol but this does not detract from the argument. In section 20.7 (p. 361) another example is discussed concerning the chemostat of a mutant with an improved (lower K_m) enzyme compared with the parent strain.

An entirely different response to growth limitation by a certain nutrient may be the induction of an isoenzyme having a lower K_m or the synthesis of a new enzyme (or enzyme system) possessing a lower K_m. In both cases the general effect is that the organism's affinity for the growth-limiting substrate is increased. An example of such a response is the induction of the glutamine synthetase with glutamate synthase system under ammonium-limitation. Under conditions of ammonium excess this enzyme system does not operate and ammonia uptake proceeds via glutamate dehydrogenase which has a much higher K_m for ammonium ions than glutamine synthetase/glutamate synthase system (Tempest *et al.* 1970).

20.5.4 Loss of Metabolic Pathways or Redundant Genetic Information

In the search for a physiological explanation for the competitiveness of an organism, Zamenhof and Eichhorn (1967) provided some very interesting clues, demonstrating that sometimes very subtle differences between organisms can account for their success in competition. It was shown that the loss of a certain biosynthetic function led to the dominance of a mutant over the parent strain. A histidine-requiring mutant (*his*$^-$) of *Bacillus subtilis* outcompeted the wild type (*his*$^+$) in a glucose-limited chemostat in the presence of excess histidine. Apparently carbon and energy economy of the auxotrophic mutant gave rise to the growth advantage enjoyed by the mutant (Fig. 14). This result becomes even more striking when it is realised that also in the wild-type strain histidine synthesis was repressed in the presence of excess histidine. In these experiments the possibility that the mutant produced an inhibitor against the *his*$^+$ parent strain was excluded. Similar results were obtained in competition experiments using an indole-requiring mutant and its parent. Recently the general validity of this phenomenon has been confirmed by Mason and Slater (1979) who showed that a tyrosine-requiring mutant of *Escherichia coli* rapidly outcompeted the prototrophic strain in a glucose-limited chemostat containing excess tyrosine. In contrast, in tyrosine-limited chemostat cultures of the mutant strain in the presence of excess glucose, prototrophic revertants always developed and subsequently outcompeted the mutant.

The energetic economy in auxotrophs is expected to be greater when the block occurs early in the biosynthetic pathway. In that case a relatively large number of reactions have been abolished so that in such mutants not even traces of intermediates will accumulate. Zamenhof and Eichhorn (1967) showed that a mutant possessing such an early block did have an advantage over an organism having a late block in the synthesis of the same metabolite. Thus an anthranilate-requiring mutant (*ant*$^-$) was able to outgrow a tryptophan-requiring mutant (*try*$^-$), lacking tryptophan synthetase (Fig. 15), under glucose-limitation in a medium supplemented with tryptophan. In spite of the convincing evidence it remains hard to imagine that these seemingly minor differences have such a dramatic effect on the outcome of the competition.

Another phenomenon related to the energetic economy of the cell has been studied during selection of mutants from microorganisms possessing drug resistance plasmids (Godwin and Slater 1979). *Escherichia coli* K12 containing the plasmid TP120 coding for resistance to ampicillin, streptomycin, sulphonamide and tetracycline, was grown in a chemostat under either carbon and energy limitation or phosphate limitation. After some time mutants arose which had lost the resistance to one or more drugs, and which were more competitive than the parent population. The mutant populations did not only show higher growth rates at growth-limiting concentrations of glucose or phosphate, but also displayed significantly higher maximum specific growth rates. In all cases, except one, the mutation did not lead to a complete loss of the plasmid. This was somewhat unexpected since in theory full elimination of the plasmid should have the highest energy-sparing effect. In fact, Slater and Bull (1978) had described earlier observations showing that an F′*lac* plasmid-containing, β-galactosidase-negative mutant of *E. coli*, when grown under lactose-limitation in the chemostat, rapidly produced a dominant mutant population which lacked the plasmid, but had incorporated the functional β-galactosidase gene into the bacterial genome.

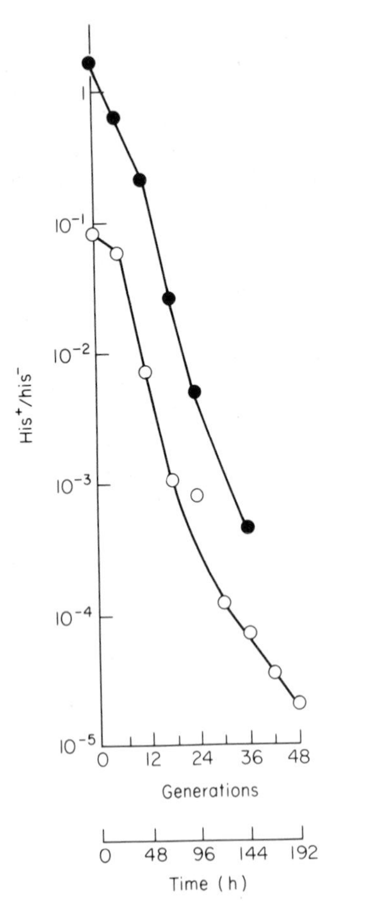

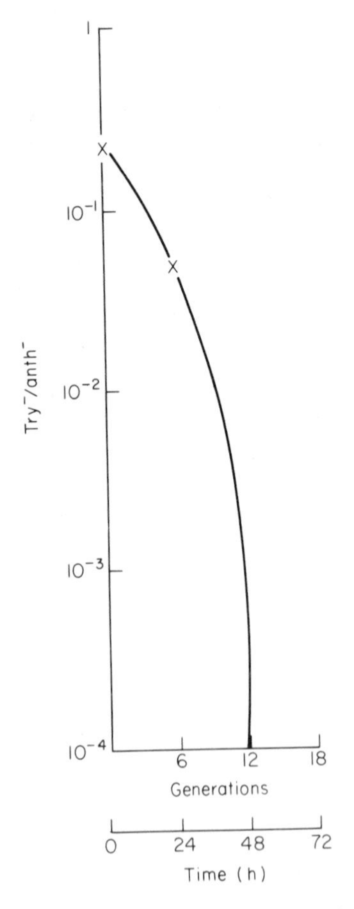

Figures 14 (left) and 15 (right). Change in the ratio of cell numbers of *Bacillus subtilis* strains when grown together in continuous culture. The two curves in Fig. 14 refer to two different starting population ratios. See text for strain designation. Mean viability was over 90% (after Zamenhof & Eichorn 1967).

20.5.5 Changes in Substrate Affinity as a Result of Mixed Substrate Utilisation

Some studies on the utilisation of mixed substrates by a variety of different types of bacteria have shown that under growth-limitation by mixed substrates, these compounds are consumed simultaneously, rather than in a diauxic pattern (Silver & Mateles 1969; Dijkhuizen & Harder 1979a, b; Gottschal & Kuenen 1980b).

Law and Button (1977) measured steady state concentrations of glucose and arginine in cultures of a marine coryneform bacterium during growth in the chemostat under limitation by the two compounds. Figure 16 shows that the glucose concentration was different in glucose-limited and in glucose plus arginine-limited cultures. This implies that the apparent K_s for glucose was clearly different under the two growth conditions.

Similar conclusions may be drawn from the work of Gottschal *et al.* (1979) and Gottschal and Kuenen (1980b). They studied a facultative chemolithotroph, *Thiobacillus* strain A2, grown in the chemostat under dual limitation by acetate and thiosulphate. Both from their pure culture studies and from competition experiments it can be inferred that the μ/s relationship for acetate in *Thiobacillus* strain A2 was dependent on the ratios of thiosulphate and acetate provided to the cultures.

The above observations illustrate the ecological importance of mixotrophic metabolism since the presence of a second substrate will have a bearing on the competiveness of a bacterium. That this is indeed the case has been shown recently by several investigators (Gottschal *et al.* 1979; Laanbroek *et al.* 1979; Smith & Kelly 1979). An example will be discussed in the following section.

Experimental

20.6 Equipment

In general, selection and competition experiments do not require technically complicated or highly sophisticated equipment. Various types and sizes of simple all-glass chemostat vessels have been described in the literature and can be used for this purpose. We have used either an all-glass culture vessel, which is described in detail by Veldkamp and Kuenen (1973), or a glass jar fitted with a butyl or silicone rubber stopper (Fig. 17). Temperature control is most conveniently achieved by placing the fermenter in an incubator room but if such a facility is not available the jar may be fitted with a waterjacket. Depending on the conditions, pH control may or may not be incorporated in the system. At the low population densities generally employed, pH control is not essential for most studies. A number of workers have recommended the use of fermenter vessels with built-in baffles for aerobic studies: in the experience of the present authors such an additional facility is not necessary in population dynamic studies where the culture density is always relatively low since various tubes extending into the culture function as baffles.

There are a few points requiring special attention. Since the outcome of selection and competition studies is critically dependent on the dilution rate, a constant flow rate of fresh medium and a constant culture volume must be ensured. The first requirement can be met by using a reliable peristaltic pump, such as the LKB Vario Perpex pump or a Watson-Marlow pump, and the second by preventing evaporation of culture liquid by moistening all the air supplied to the vessel. In

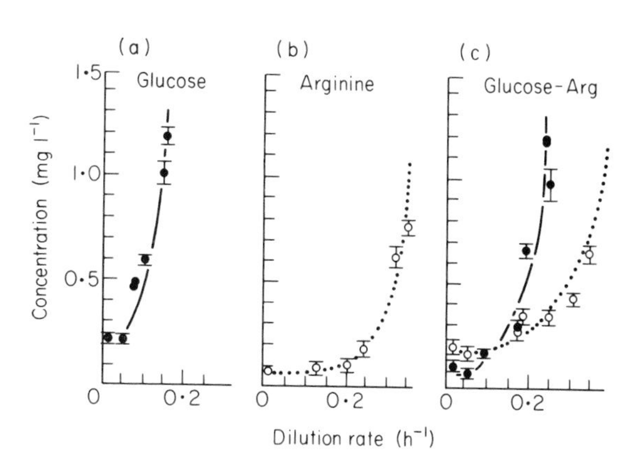

Figure 16. μ/s relationship of a marine coryneform showing limiting substrate concentrations at steady state during (a) glucose-; (b) arginine- and (c) glucose plus arginine-limited growth. Error bars show standard deviation among triplicate determinations from each of several samples collected over a 2 to 3 week interval. •, glucose; ○, arginine (after Law & Button 1977).

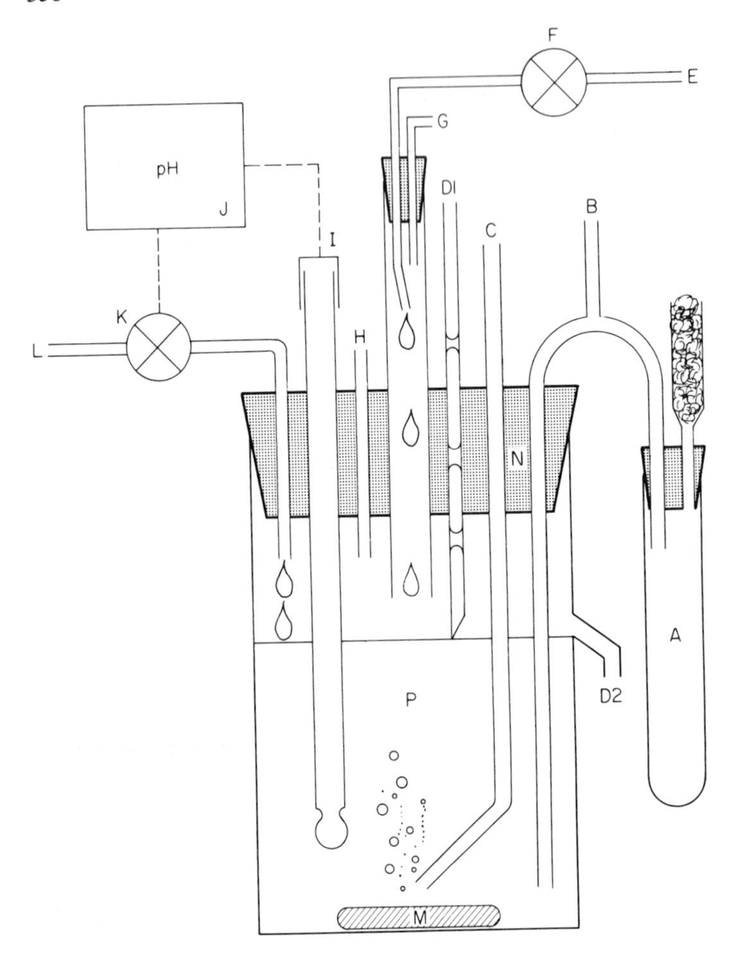

Figure 17. Schematic drawing of a low-cost, 500 ml glass fermenter (300 ml working volume) for small-scale continuous cultivation of microorganisms. The fermenter consists of a glass jar with a rubber bung. Stirring is accomplished with a magnetic stirrer bar. A, sample tube with rubber bung with cotton plug. Sampling is done by creating a small overpressure by blocking the air outlet (D1, 2). B, sterile air to empty tubing before sampling. During sampling B is blocked with tubing clamp. C, sterile air for aeration. D1, overflow tube to keep the culture volume constant. Note that the tube is cut at an angle of 60°. D1 serves also as air outlet. D2, overflow tube used as alternative to D1. E, supply of sterile medium via tubing pump F. F, tubing pump. G, sterile air supply to protect the inflowing medium from contamination with the microorganisms from the culture. H, entry port connected to suitable flask for inoculation or addition of sterile liquids. I, autoclavable pH electrode connected to pH meter (J) (optional). J, pH meter connected to electrode and to pump (or valve) (K) to add sterile acid or alkali from (L) (optional). M, teflon-coated magnetic bar for stirring. N, rubber bung. P, culture.

studies in which pH control is used, care should be taken to include the volume of titrant used for controlling the pH in the total flow of liquid through the culture, when calculating the dilution rate.

Several small chemostats suitable for competition studies are now commercially available; for example, the New Brunswick Scientific Co., New Brunswick, USA, markets two small bench-top fermenters: the multigen and bioflow. Both fermenters use a culture vessel with a working volume of 350 ml. The bioflow is available in a form ready for use, while the multigen requires additional equipment for continuous operation. Gallenkamp, UK, offers a 200 to 800 ml modular fermenter suitable for continuous operation, while LH Engineering, UK, markets equipment such as the type 4-IL multistation fermentation unit which can be easily adapted for use as a four-station continuous culture system. Finally, the all-glass chemostat described by Veldkamp and Kuenen (1973) is available from Elgebee, Leek, The Netherlands. Additional information and a detailed description of more sophisticated equipment is given by Evans *et al.* (1970).

20.7 Selection and competition studies in practice

In this section three examples are given of selection and competition experiments in order to illustrate the methods used. The first deals with selection of organisms from nature, the second considers selection of a mutant organism from a population of the yeast *Saccharomyces cerevisiae* and the third describes competition between the facultatively chemolithotrophic *Thiobacillus* strain A2, an obligately chemolithotrophic *Thiobacillus* sp. and a heterotrophic *Spirillum* sp. for inorganic and organic substrates.

Before embarking on a detailed description of how these selection experiments have been conducted, a few general remarks are in order. The first and possibly most important is that in competition studies carried out under limitation by one or more substrates it is essential to establish experimentally that the supposedly limiting factor in the growth medium is in fact growth-limiting. In the case of carbon-limitation this can conveniently be done by measuring total organic carbon in the culture supernatant. When other factors are growth-limiting, specific chemical determination must be performed. A convenient check on these chemical assays is to inject a small volume of a concentrated solution of the growth-limiting component into the culture through the inoculation port. If the component is growth-limiting, the culture density should increase (section 20.2, eqn 3, p. 343). A second general point is that the practice of performing competition studies requires that the different organisms can be enumerated and counted separately. It is therefore mandatory to develop techniques which enable discrimination between the competing organisms in the culture. Several techniques may be used, the most convenient one is plating on specific agar media on which the various organisms show different colony types or a different growth response. A more rapid technique is the microcolony counting or slide culture technique developed by Postgate *et al.* (1961), which, in addition to the virtues of plating, allows discrimination between organisms of different morphology. When the size of the different organisms is sufficiently different, they can conveniently be counted directly using a Coulter counter equipped with a chanelliser system. In a number of instances these methods are inadequate for following the progress of competition. In these cases either measurement of metabolic processes in the mixed culture or activities of one or more enzymes in cell-free extracts of the culture may be used as an indicator of the progress of competition. Further practical details can be found in the description of the various selection experiments and in section 20.3.

Selections in the chemostat from fresh water have been carried out by Kuenen *et al.* (1977) as follows. Two small continuous culture vessels (Fig. 18) are connected to the same medium reservoir containing an appropriate medium with a selected growth-limiting factor. Prior to sterilisation the two chemostats are filled to approximately half the working volume with the medium minus the growth-limiting factor so that when the pumps are started its concentration starts to increase slowly. After adding a small amount of complete medium to the vessels (approximately 10% of the working volume), both chemostats are inoculated with a natural sample which has been previously filtered through membrane filters of 1.2 μm pore size in order to remove predators such as protozoa or amoeba. Organisms are allowed to grow in the vessels, while care is taken to ensure that the environmental conditions, such as temperature, pH, dissolved oxygen, are the same in the two cultures. After approximately 24 h the two pumps are started and their flow rates slowly increased over a period of 48 h to give different dilution rates in the two chemostats: e.g. one low dilution rate ($0.03\ h^{-1}$) and the other high ($0.3\ h^{-1}$). During this increase in dilution rates of the culture, it may be necessary to reinoculate the fermenters. The course of the competition is monitored by making serial dilutions of the culture and plating on appropriate media at convenient time intervals. As a rule, different bacterial populations become dominant in the two vessels after approximately five to ten volume changes. In our experience this procedure has never led to pure cultures: even after 6 months operation, the different organisms accompanying the dominant population generally account for approximately 5% of the total counts. This is most probably due to the fact that the dominant population either excretes small amounts of metabolic products (particularly under non-carbon-limiting conditions) or contains non-viable cells or products of lysis which can support the growth of other populations. Therefore the initial enrichment procedure is usually followed by purification on plates of the dominant population occurring at the two dilution rates.

The results of the enrichment can be checked by carrying out competition experiments with the pure cultures. In the experiments described above spiral-shaped bacteria were almost invariably isolated at low dilution rates whereas at high dilution rates rod-shaped organisms constituted the dominant population (section 20.3.1, p. 345). The dominant organisms were isolated in pure culture, mixed and again introduced in two chemostats run at $D=0.03$ and $0.3\ h^{-1}$, respectively. As expected,

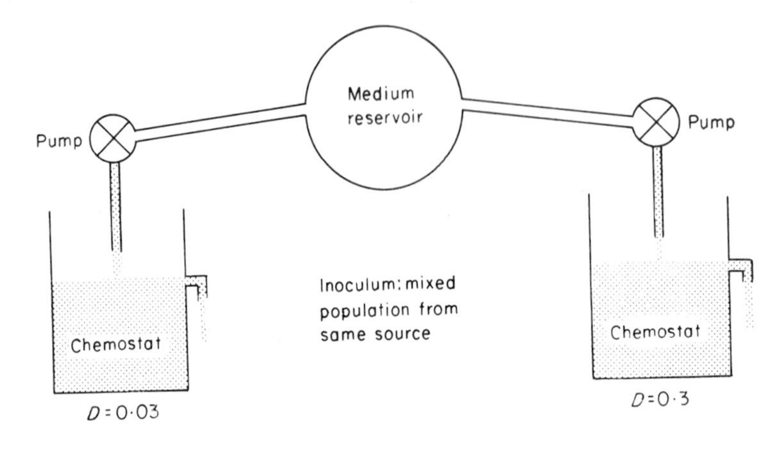

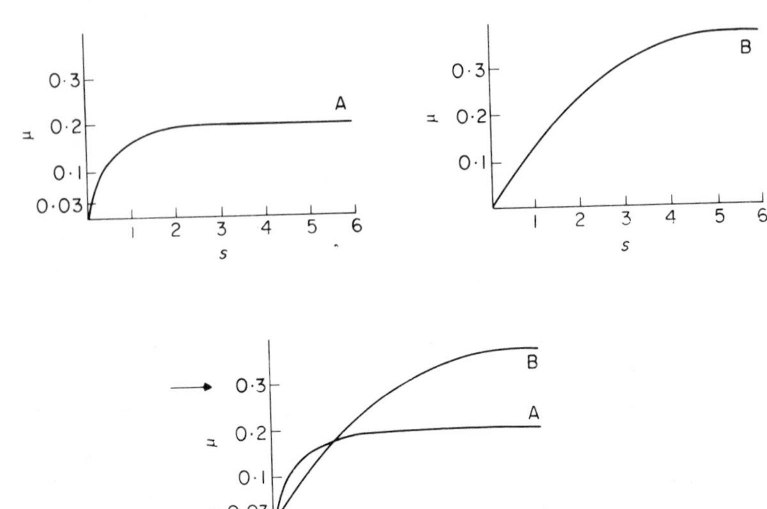

Figure 18. Diagrammatic representation of the method used in selecting organisms on the basis of substrate affinity. Two chemostats are fed the same medium from the reservoir. The dilution rates in the left and right fermenters are $D=0.03\ h^{-1}$ and $D=0.3\ h^{-1}$, respectively. Both fermenters are inoculated with a mixed population from the same (natural) source. In the left fermenter A becomes dominant (low μ_{max}, low K_s), in the right fermenter B (high μ_{max}, high K_s). If the experiment is repeated with mixed, pure cultures of A and B (lower diagram) A becomes dominant at low and B becomes dominant at high dilution rates (after Veldkamp & Kuenen 1973).

the *Spirillum* sp. outgrew the rod in one vessel and the reverse occurred in the other.

The relationship of growth rate and phosphate concentration of a *Spirillum* sp. (S) and a rod-shaped bacterium (R) is shown in Fig. 19. The curves for the two organisms cross and it is evident that if competition between them is carried out under phosphate limitation at a phosphate concentration to the left of the crossing point, the *Spirillum* sp. will outgrow the rod, while the reverse would be true at concentrations to the right of the crossing point. The physiological basis for selective advantage of the low K_s, low μ_{max}-type organism at the lower substrate concentrations has been discussed in section 20.5 (p. 352). For a discussion of methods available for the determination of K_s values, the reader is referred to Pirt (1975).

An elegant demonstration of the use of continuous-flow techniques in the selection of a mutant organism from a population of the yeast *Saccharomyces cerevisiae* has been given by Francis and Hansche (1972, 1973). These experiments also provided an example of competition as a mechanism in enzyme evolution because they involved an attempt to enhance genetically certain functional properties of the enzyme acid phosphatase. Wild-type *S. cerevisiae* was grown under phosphate-limitation at a constant dilution rate ($0.12\ h^{-1}$) using β-glycerophosphate as the only source of phosphorus. Under these conditions acid phospha-

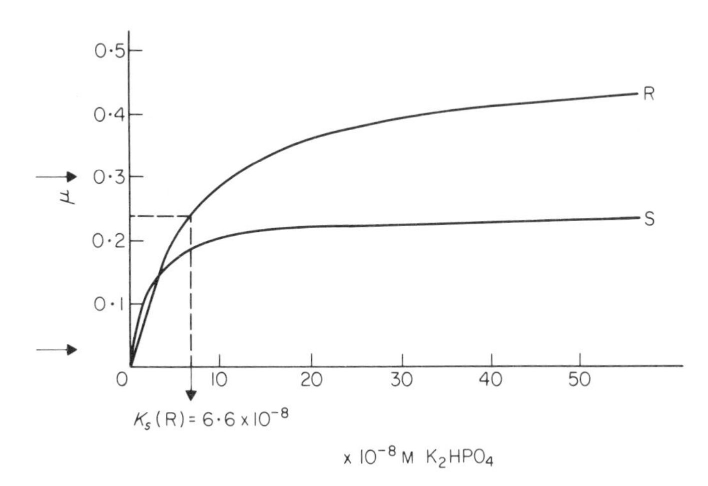

Figure 19. Specific growth rate of a freshwater rod-shaped bacterium (R) and a *Spirillum* sp. (S) as a function of phosphate concentration. The curves are schematic and based on two measurements each at the growth rates indicated by the arrows. Rod-shaped bacterium, $K_s = 6.6 \times 10^{-8}$M; *Spirillum* sp., $K_s = 2.7 \times 10^{-8}$M (after Veldkamp & Kuenen 1973, and Kuenen *et al.* 1977).

tase, which is present in the yeast cell wall and has a pH optimum of 4.2, plays a key role in furnishing the cells with inorganic phosphate by its release from the growth-limiting phosphorus-containing compound. A selective pressure was applied on the yeast population by growing the organisms in a chemostat at pH 6.0, where the phosphatase activity was only 25% of its maximum at pH 4.2. The conditions therefore favoured the selection of mutant organisms with an improved substrate affinity for orthophosphate and with a modified activity of the phosphatase in response to unfavourable pH conditions.

In one experiment the parent strain S288C was maintained in steady state under the selective conditions for about 180 generations before it was displaced by mutant M1. This mutant had a significantly lower K_s for β-glycerophosphate (Table 4), but there were no changes in μ_{max} of the mutant organism, the V_{max} or the pH optimum of the mutant's enzyme. There was, however, a decrease in the K_m of the phosphatase and it seems likely that the change in the K_s value of the mutant was due to this alteration in the K_m of the enzyme for β-glycerophosphate. The mutant M1 stayed in the culture for a further 220 generations when it was, in turn, replaced by mutant M2. This second mutational event was more extensive than the first one since M2 was altered in both its affinity for the substrate and its maximum specific growth rate (Table 4). The new mutant had a substantial increase in acid phosphatase activity at pH 6.0 and the pH optimum was shifted from 4.2 to 4.8. It is important to stress that the lower K_s of the organism for β-glycerophosphate is due to the increase of V_{max} of the phosphatase. The precise molecular details underlying these mutations have not yet been elucidated. However, these experiments and those of other workers (Clarke 1974; Hartley 1974; Senior *et al.* 1976) convincingly demonstrate the role of competition in selecting for a mutant organism better suited for growth in a particular environment.

Table 4. Changes in growth and enzyme constants of acid phosphatase-producing *Saccharomyces cerevisiae* (from Slater & Bull 1978, after Francis & Hanasche 1972).

	Strain		
	S288C	M1	M2
Organism characteristics			
μ_{max} (h^{-1})	0.23	0.23	0.26
K_s(mg β-glycerophosphate l^{-1})	4.6	1.9	0.3
Enzyme characteristics			
V_{max} (μmol β-glycerophosphate min^{-1} $cell^{-1} \times 10^{-10}$)	6.7	5.5	10.7
K_m(M $\times 10^{-5}$)	1.5	1.3	1.3
pH optimum	4.2	4.2	4.8

Competition between a specialised obligate chemolithotroph, a heterotrophic *Spirillum* sp. and a versatile facultative chemolithotroph (mixotroph) under mixed substrate limitation was recently studied by Gottschal *et al.* (1979). These workers investigated the potential of facultatively chemolithotrophic thiobacilli to compete successfully with specialised chemolithotrophs (for reduced sulphur compounds) in order to understand their occurrence and role in nature. Previously it had been suggested (Rittenberg 1972) that these organisms should be able to compete successfully with a specialist thiobacillus under mixed substrate conditions. The validity of this suggestion was studied as follows. When the facultatively chemolithotrophic *Thiobacillus* strain A2 and the obligately chemolithotrophic *Thiobacillus neapolitanus* grown separately in thiosulphate-limited chemostats at a dilution rate of 0.05 h^{-1}, were mixed on a 1:1 ratio and cultivated in thiosulphate-limited mixed culture at the same dilution rate, the numbers of *T. neapolitanus* cells increased with time. In further experiments it was found that *T. neapolitanus* outcompeted *Thiobacillus* strain A2 at both higher and lower dilution rates under thiosulphate-limited conditions. However, when the competition experiment was carried out under conditions where both an inorganic energy source (thiosulphate) and acetate were present in the reservoir, the outcome of the selection was dependent on the ratio of the concentrations of thiosulphate and acetate. At a thiosulphate concentration (S_R) of 40 mM, *Thiobacillus* strain A2 was able to outcompete *T. neapolitanus* when acetate was present in the mixture in excess of 10mM. In further experiments it was demonstrated that *Thiobacillus* strain A2 could also successfully outcompete a heterotrophic *Spirillum* sp. strain G7 under conditions where, in addition to acetate, thiosulphate was present in the reservoir. From these experiments it was predicted that *Thiobacillus* strain A2 should be able to compete successfully with both *T. neapolitanus* and the *Spirillum* sp. when grown in continuous culture under mixed (thiosulphate + acetate) substrate conditions. The predictions were borne out in practice when competition between the three organisms for acetate and thiosulphate as the growth-limiting substrates was studied in the chemostat (Fig. 20). With all test mixtures in the inflowing medium, *Thiobacillus* strain A2 was able to maintain itself in the culture. The left-hand part of the figure shows that as the acetate/thiosulphate ratio increased from 0:40 to 5:30, *Thiobacillus* strain A2 increased in relative numbers, whereas *T. neapolitanus* decreased and *Spirillum* sp. strain G7 could not maintain itself in the culture. The right-hand part of the figure shows that *Thiobacillus*

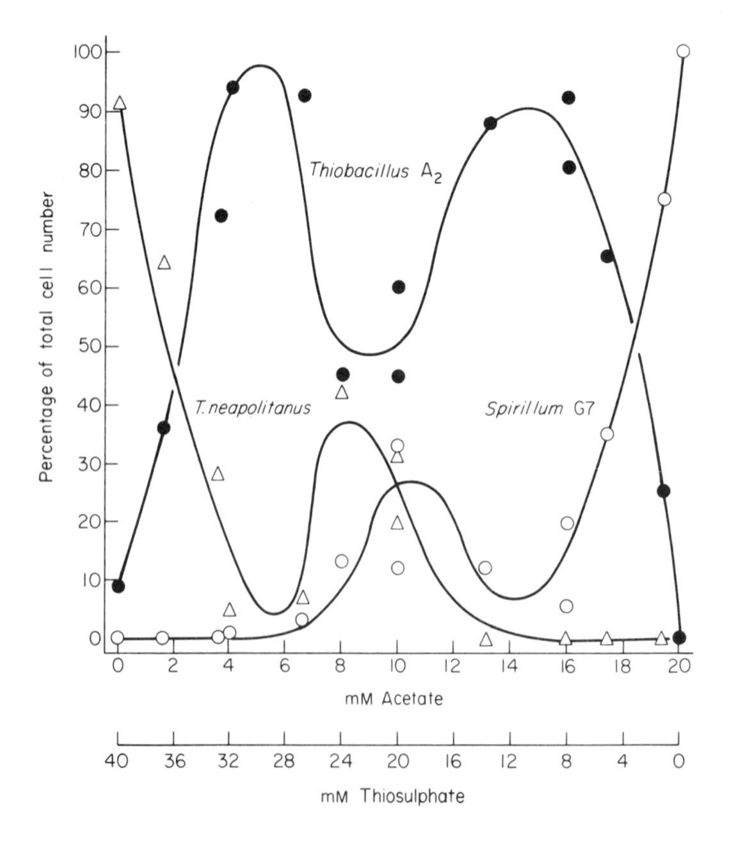

Figure 20. Competition between *Thiobacillus* strain A2, *Thiobacillus neapolitanus* and *Spirillum* sp. strain G7 for thiosulphate and acetate as growth-limiting substrates in the chemostat at a dilution rate of 0.075 h^{-1}. Concentrations of these substrates in the inflowing medium ranged from 0 to 20 mM for acetate and 40 to 0 mM for thiosulphate. After steady states with different ratios between acetate and thiosulphate concentrations in the feed had been established, relative cell numbers of *Thiobacillus* strain A2 (●), *T. neapolitanus* (△) and *Spirillum* sp. strain G7 (○) were determined (after Gottschal *et al.* 1979).

A2 increased in numbers with decreasing acetate/thiosulphate ratios from 20:0 to 15:10; the spirillum showed a concomitant decrease, whereas *T. neapolitanus* was unable to grow. When the medium composition was as shown in the middle part of the figure, coexistence of all three organisms was observed, even during prolonged cultivation (20 volume changes). The importance of yield (Y) during competition for mixed substrates can best be rationalised by considering one example of these experiments. If in a mixed chemostat culture of *T. neapolitanus* and *Thiobacillus* A2, the latter would have a very high yield on acetate, its density will be relatively high. As a consequence it will be able to consume, per unit of biomass, more of the thiosulphate than would have been possible if *Thiobacillus* A2 had possessed a lower yield coefficient. Thus, *T. neapolitanus* would be washed out from the culture at a faster rate when *Thiobacillus* A2 has a higher yield.

These experiments clearly indicated that an ecological niche for facultatively chemolithotrophic thiobacilli might occur in nature since it has been argued that in many environments growth-limiting concentrations of inorganic and organic energy sources may be available simultaneously. If this is true, it should be possible to enrich for such organisms in continuous culture under mixed substrate conditions. Very recently Gottschal and Kuenen (1980a) demonstrated that this is indeed possible (Table 5), thereby justifying the earlier surmises of the selective advantage of the mixotrophic way of life. At present it is not known why enrichments from a marine environment so far have only yielded mixtures of obligate chemolithotrophs and heterotrophs, whereas in contrast enrichments from freshwater environments always lead to dominant populations of mixotrophic bacteria, provided the thiosulphate/acetate ratio in the inflowing medium is sufficiently high.

20.8 Complications

In theory the chemostat described in section 20.7 (p. 359) is simple to operate. However, in practice it is generally not so easy and the experimenter quickly finds that apparently trivial difficulties interfere with its proper functioning. This has led many authors to proffer words of caution such as 'all chemostats are continuous fermenters, but some are more continuous than others' (Solomons 1972). Very often culture disturbances are caused by technical problems (Veldkamp 1976), some of which are discussed below.

(1) In most chemostat designs the fresh medium enters the growth vessel by an inlet which has some kind of device to discourage growth of organisms in the medium supply line. Most investigators have opted for a design in which the direct introduction of medium into the culture liquid is avoided. Even with a liquid break earlier in the supply line in order to prevent organisms from growing back into the reservoir, accumulation of microorganisms

Table 5. Results of enrichment cultures after 15 to 20 volume changes in the chemostat and thiosulphate plus acetate limitation at a dilution rate of 0.05 h^{-1}. The mixotrophs were facultative chemolitho(auto)trophs in experiments 1 to 4 and a chemolithotrophic heterotroph in experiment 5 (after Gottschal & Kuenen 1980a).

Enrichment	Inoculum	Growth-limiting substrate concentration(mM)	Dominant population	Secondary population
1	canal water	thiosulphate(30) and acetate (5)	mixotroph	heterotroph
2	canal water	thiosulphate(10) and acetate (15)	mixotroph	heterotroph
3	ditch water	thiosulphate(30) and acetate (15)	mixotroph	autotroph and heterotroph
4	ditch water	thiosulphate(20) and acetate (10)	mixotroph	heterotroph
5	ditch water	thiosulphate(10) and acetate (15)	mixotroph	heterotroph
6	sea-water	thiosulphate(30) and acetate (5)	autotroph and heterotroph	heterotroph

inside the medium inlet tube immersed in the culture occurs, especially at low medium supply rates. The method most frequently used is the dropwise introduction of medium through a shielded medium inlet tube not in direct contact with the culture liquid (Fig. 17), although this also has drawbacks. The most significant of these is that it leads to a discontinuous addition of medium, a factor which becomes important at very low dilution rates when, for example, one drop of fresh medium may be added to the culture every few minutes. The effect of this discontinuous addition of medium on the physiology of microorganisms in the chemostat may be important (Brooks & Meers 1973), but unfortunately an adequate solution is not readily available. Another disadvantage of dropwise addition of medium occurs when volatile substrates (e.g. methanol) are used. An unknown amount of substrate evaporates during the time required for the drop to build up and fall from the inlet into the culture and is carried away with the exhaust gases from the vessel. In some cases direct introduction of medium into the culture is used: growth in the inlet is prevented by using media of very low ($pH < 3.5$) or very high ($pH > 10.0$) pH values. The pH of the culture can be adjusted subsequently. Alternatively, the fresh medium can be separated into two parts having extreme and opposite pH values and be administered to the culture with two pumps. However, this procedure also has disadvantages, such as the need for more equipment, uncontrolled changes in S_R as a result of small variations in the relative flow rates of the two pumps, and possible formation of precipitates at the point where the acid or alkaline flow of liquid enters the culture.

(2) The correct operation of any chemostat requires homogeneity of the culture. This means that good mixing is required (Solomons 1972), that wall growth is absent (Topiwala & Hamer 1971) and that clumping of microorganisms must be avoided. Often clumping of organisms is due to a medium imbalance, although there are also instances in which aggregation of cells is an intrinsic property of the organism. In the latter case the sad but best advice is to use another organism if that option is possible.

In recent years it has become fashionable to try and measure many culture parameters. Very often this is achieved by using electrodes which are inserted in the culture (e.g. for measuring the concentration of various ions, P_{O_2}, P_{CO_2}, pH or devices for measuring incident light or temperature). The general effect of all these devices is that there is a significant increase in surface area in the culture vessel. A word of caution may be in order here since the effect of wall growth on the culture dynamics is related to the total available surface area.

(3) Any liquid stream entering the culture in addition to fresh medium (i.e. titrant for pH control) must be taken into account when the dilution rate is calculated. As pointed out by Solomons (1972) the dilution rate is based on the volume of liquid in the culture and not on the fermenter contents (the volume of rapidly stirred aerobic cultures may be 25% air and this should be taken into account). At low dilution rates culture evaporation must be avoided by equilibrating all gases passed through the culture with water at the same temperature as the culture. From this discussion it may also be inferred that in order to keep the liquid volume constant it is necessary to maintain a constant stirring rate and a constant flow of gas through the culture.

(4) It is obvious that the inoculum to be used in population dynamic studies must contain the organism(s) able to grow under the conditions employed in the fermenter. In studies with bacteria it is important that predators present in the inoculum be removed: this is conveniently done by passing the inoculum through a 1.2 μm pore size membrane filter. Theoretically the size of the inoculum is of no importance when starting a chemostat culture. In practice, however, it may be crucial, particularly when organisms are sensitive to excessive aeration or in the case of strict anaerobes when redox conditions in the vessel are initially not optimal for growth. In particular in population dynamic studies it may be important to reinoculate the culture after several days running. The substrate concentration(s) will then have dropped to low levels and organisms present in the inoculum which may have been inhibited by the initial substrate concentration present will then meet more favourable conditions. This procedure is also recommended when toxic substrates are used

20.9 Conclusions

In recent years microbial ecology has increasingly developed from an era of qualitative descriptions towards a much deeper understanding of the dynamics of complex microbial populations in nature. The study of simple mixed cultures and microbial communities is fundamental to this development and it is particularly in this respect that the application of continuous-flow techniques has brought experimental microbial ecology into a higher gear.

Such techniques not only allow an assessment of environmental factors which play an essential role in microbial competition and selection in nature, but have also indicated the importance of whole integrated microbial communities in the turnover of organic matter in nature or indeed in the cycle of the various important elements. Application of continuous cultures has been particularly rewarding in the study of the behaviour of microbial communities in response to a variety of environmental stresses. These may be entirely natural (i.e. substrate-limitation, temperature, pH, Po_2 and others) or due to man's interference with the biosphere, such as the introduction of xenobiotic materials. We are only beginning to understand the ecophysiological basis of survival value and it is our belief that a wider application of the methods discussed in this chapter will throw more light on the behaviour of microbial populations in nature.

Recommended reading

BUNGAY H. R & BUNGAY M. L. (1968) Microbial interactions in continuous culture. *Advances in Applied Microbiology* **1**, 269–90.

FREDERICKSON A. G. (1977) Behavior of mixed cultures of microorganisms. *Annual Review of Microbiology* **31**, 63–89.

HARDER W., KUENEN J. G. & MATIN A. (1977) Microbial selection in continuous culture. *Journal of Applied Bacteriology* **43**, 1–24.

JANNASCH H. W. & MATELES R. I. (1974) Experimental bacterial ecology studied in continuous culture. *Advances in Microbial Physiology* **11**, 165–212.

MEERS J. L. (1973) Growth of bacteria in mixed cultures. *CRC Critical Reviews in Microbiology* **2**, 139–84.

PIRT S. J. (1975) *Principles of Microbe and Cell Cultivation.* Blackwell Scientific Publications, Oxford.

SLATER J. H. & BULL A. T. (1978) Interactions between microbial populations. In: *Companion to Microbiology* (Eds A. T. Bull & P. M. Meadow), pp. 181–206. Longman, London.

TEMPEST D. W. (1970). The continuous cultivation of micro-organisms. I. Theory of the chemostat. In: *Methods in Microbiology* (Eds J. R. Norris & D. W. Ribbons), vol. 2, pp. 259–76. Academic Press, London.

VELDKAMP H. (1976) Continuous culture in microbial physiology and ecology. In: *Patterns of Progress* (Ed. J. Gordon Cook). Meadowfield, Durham.

VELDKAMP H. & JANNASCH H. W. (1972) Mixed culture studies with the chemostat. *Journal of Applied Chemistry and Biotechnology* **22**, 105–23.

References

BAZIN M. J., SAUNDERS P. Y & PROSSER J. M. (1976) Models of microbial interaction in the soil. *CRC Critical Reviews in Microbiology* **4**, 463–98.

BROOKS J. D. & MEERS J. L. (1973) The effect of discontinuous methanol addition on the growth of a carbon-limited culture of *Pseudomonas*. *Journal of General Microbiology* **77**, 513–19.

BRYANT M. P., WOLIN E. A. & WOLFE R. S. (1967) *Methanobacillus omelanskii*, a symbiontic association of two species of bacteria. *Archiv für Mikrobiologie* **59**, 20–31.

BUNGAY H. R. & BUNGAY M. L. (1968) Microbial interactions in continuous culture. *Advances in Applied Microbiology* **10**, 269–90.

CLARKE P. H. (1974) The evolution of enzymes for the utilization of novel substrates. In: *Evolution in the Microbial World* (Eds M. J. Carlile & J. J. Skehel), pp. 183–217. Cambridge University Press, London.

DE FREITAS M. J. & FREDERICKSON A. G. (1978) Inhibition as a factor in maintenance of diversity of microbial systems. *Journal of General Microbiology* **106**, 307–21.

DIJKHUIZEN L. & HARDER W. (1975) Substrate inhibition in *Pseudomonas oxalaticus* OX1: a kinetic study of inhibition by oxalate and formate using extended cultures. *Antonie van Leeuwenhoek* **41**, 135–46.

DIJKHUIZEN L. & HARDER W. (1979a) Regulation of autotrophic and heterotrophic metabolism in *Pseudomonas exaliticus* OX1. Growth on mixtures of acetate and formate in continuous culture. *Archives of Microbiology* **123**, 47–53.

DIJKHUIZEN L. & HARDER W. (1979b) Regulation of autotrophic and heterotrophic metabolism in *Pseudomonas oxalaticus* OX1. Growth on mixtures of oxalate and formate in continuous culture. *Archives of Microbiology* **123**, 55–63.

EDWARDS V. H. (1970) The influence of high substrate concentrations on microbial kinetics. *Biotechnology and Bioengineering* **12**, 679–712.

EVANS C. G. T., HERBERT D. & TEMPEST D. W. (1970) The continuous cultivation of microorganisms: 2. Construction of a chemostat. In: *Methods in Microbiology* (Eds J. R. Norris & D. W. Ribbons), vol. 2, pp. 277–327. Academic Press, London.

FRANCIS J. C. & HANSCHE P. E. (1972) Direct evolution of metabolic pathways in microbial populations. I. Modification of the acid phosphatase pH optimum in *Saccharomyces cerevisiae*. *Genetics* **70**, 59–73.

FRANCIS J. C. & HANSCHE P. E. (1973) Direct evolution of metabolic pathways in microbial populations. II. A repeatable adaption in *Saccharomyces cerevisiae*. *Genetics* **74**, 259–65.

FREDERICKSON A. G. (1977) Behavior of mixed cultures of microorganisms. *Annual Review of Microbiology* **31**, 63–89.

GAUSE G. F. (1934) *The Struggle for Existence*. Dover Publications, New York.

GODWIN D. & SLATER J. H. (1979) The influence of the growth environment on the stability of a drug resistance plasmid in *Escherichia coli* K12. *Journal of General Microbiology* **111**, 201–10.

GOTTSCHAL J. C., DE VRIES S. & KUENEN J. G. (1979) Competition between the facultatively chemolithotrophic *Thiobacillus* A2, an obligately chemolithotrophic *Thiobacillus* and a heterotrophic Spirillum for inorganic and organic substrates. *Archives of Microbiology* **121**, 241–9.

GOTTSCHAL J. C. & KUENEN J. G. (1980a) Selective enrichment of facultatively chemolithotrophic Thiobacilli and related organisms in continuous culture, *FEMS Microbiology Letters* **7**, 241–7.

GOTTSCHAL J. C. & KUENEN J. G. (1980b) Mixotrophic growth of *Thiobacillus* A2 on acetate and thiosulfate as growth limiting substrate in the chemostat. *Archives of Microbiology* **126**, 33–42.

HARDER W., KUENEN J. G. & MATIN A. (1977) Microbial selection in continuous culture. *Journal of Applied Bacteriology* **43**, 1–24.

HARDER W. & VELDKAMP H. (1971) Competition of marine psychrophile bacteria at low temperatures. *Antonie van Leeuwenhoek* **37**, 51–63.

HARTLEY B. S. (1974) Enzyme families. In: *Evolution in the Microbial World* (Eds M. J. Carlile & J. J. Skehel), pp. 151–82. Cambridge University Press, London.

HARTLEY B. S., BURLEIGH B. D., MIDWINTER G. G., MOORE C. H., MORRIS H. R., RIGBY P. W. J., SMITH M. J. & TAYLOR S. S. (1972) Where do enzymes come from? In: *Enzymes: Structure and Function* (Eds J. Drenth, R. A. Oosterbaan & C. Veeger), pp. 151–76. Elsevier-North Holland. Amsterdam.

JANNASCH H. W. (1967) Enrichment of aquatic bacteria in continuous culture. *Archiv für Mikrobiologie* **59**, 165–73.

JANNASCH H. W. & MATELES R. I. (1974) Experimental bacterial ecology studied in continuous culture. *Advances in Microbial Physiology* **11**, 165–212.

JANNASCH H. W. & PRITCHARD H. (1972) The role of inert particulate matter in the activity of aquatic microorganisms. *Memorie dell Istituto Italiano di Idrobiologia* **29**, 289–308.

JOST J. L., DRAKE J. F., FREDERICKSON A. G. & TSUCHIYA H. M. (1973) Interactions of *Tetrahymena pyriformis*, *E. coli*, *Azobacter vinelandii* and glucose in a minimal medium. *Journal of Bacteriology* **113**, 834–40.

KUENEN J. G., BOONSTRA J., SCHRÖDER H. G. J. & VELDKAMP H. (1977) Competition for inorganic substrates among chemoorganotrophic and chemolithotrophic bacteria. *Microbial Ecology* **3**, 119–30.

LAANBROEK H. J., SMIT A. J., KLEIN NULEND G. & VELDKAMP H. (1979) Competition for L-glutamate between specialized and versatile *Clostridium* species. *Archives of Microbiology* **120**, 61–6.

LAW A. T. & BUTTON D. K. (1977) Multiple-carbon-source-limited growth kinetics of a marine coryneform bacterium. *Journal of Bacteriology* **129**, 115–23.

LEE I. H., FREDERICKSON A. G. & TSUCHIYA H. M. (1976) Dynamics of mixed cultures of *Lactobacillus plantarum* and *Propionibacterium shermanii*. *Biotechnology and Bioengineering* **18**, 513–26.

MACURA J. (1961) Continuous flow method in soil microbiology. I. Apparatus. *Folia Microbiologica* **6**, 328–34.

MASON T. G. & SLATER J. H. (1979) Competition between an *Escherichia coli* tyrosine auxotroph and a prototrophic revertant in glucose- and tyrosine-limited chemostat. *Antonie van Leeuwenhoek* **45**, 253–63.

MATIN A. & VELDKAMP H. (1978) Physiological basis of the selective advantage of a *Spirillum* sp. in a carbon-limited environment. *Journal of General Microbiology* **105**, 187–97.

MEERS J. L. (1973) Growth of bacteria in mixed cultures. *CRC Critical Reviews in Microbiology* **2**, 139–84.

MEGEE R. D., DRAKE J. F., FREDERICKSON A. G. & TSUCHIYA H. M. (1972) Studies in intermicrobial symbiosis. *Saccharomyces cerevisiae* and *Lactobacillus casei*. *Canadian Journal of Microbiology* **18**, 1733–42.

MEYER J. S., TSUCHIYA H. M. & FREDERICKSON A. G. (1975) Dynamics of mixed populations having complementary metabolism. *Biotechnology and Bioengineering* **17**, 1065–81.

MONOD J. (1942) *Recherches sur la Croissance des Cultures Bactériennes*. Hermann, Paris.

MULDER E. G. & VAN VEEN W. L. (1965) Anreicherung von Organismen der *Spearotilus leptothrix*-Gruppe. In: *Anreicherungskultur und Mutantenauslese* (Ed. H. G. Schlegel), pp. 28–46. Guster Fischer Verlag, Stuttgart.

MUR L. R., GONS H. J. & VAN LIERE L. (1977) Some experiments on the competition between algae and blue-green bacteria in light-limited environments. *FEMS Microbiology Letters* **1**, 335–8.

NOVICK A. (1961) Bacteria with high levels of specific enzymes. In: *Growth in Living Systems* (Ed. M. X. Zarrow), pp. 93–106. Basic Books Inc., New York.

NOVICK A. & SZILARD L. (1950) Experiments with the chemostat on spontaneous mutation of bacteria., *Proceedings of the National Academy of Sciences U.S.A.* **36**, 708–19.

PAWLOWSKI H. & HOWELL J. A. (1973) Mixed culture biooxidation of phenol. I. Determination of kinetic parameters. *Biotechnology and Bioengineering* **15**, 889–96.

PFENNIG N. & BIEBL H. (1976) *Desulfuromonas acetoxidans*, gen. nov. and sp. nov., a new anaerobic, sulfur-reducing, acetate-oxidizing bacterium. *Archives of Microbiology* **110**, 3–12.

PIRT S. J. (1975) *Principles of Microbe and Cell Cultivation*. Blackwell Scientific Publications, Oxford.

POSTGATE J. R., CRUMPTON J. E. & HUNTER J. R. (1961) The measurement of bacterial viabilities by slide culture. *Journal of General Microbiology* **24**, 15–24.

POWELL E. O. (1958) Criteria for growth of contaminants and mutants in continuous culture. *Journal of General Microbiology* **18**, 259–68.

RITTENBERG S. C. (1972) The obligate autotroph—the demise of a concept. *Antonie van Leeuwenhoek* **38**, 457–78.

SENIOR E., BULL A. T. & SLATER J. H. (1976) Enzyme evolution in a microbial community growing on the herbicide Dalapon. *Nature, London* **263**, 476–9.

SILVER R. S. & MATELES R. I. (1969) Control of mixed-substrate utilization in continuous cultures of *Escherichia coli*. *Journal of Bacteriology* **97**, 535–43.

SLATER J. H. & BULL A. T. (1978) Interactions between microbial populations. In: *Companion to Microbiology* (Eds A. T. Bull & P. M. Meadow), pp. 181–206. Longman, London.

SMITH A. L. & KELLY D. P. (1979) Competition in the chemostat between an obligately and a facultatively chemolithotrophic *Thiobacillus*. *Journal of General Microbiology* **115**, 377–84.

SOLOMONS G. L. (1972) Improvements in the design and operation of the chemostat. *Journal of Applied Chemistry and Biotechnology* **22**, 217–28.

STOUTHAMER A. H. (1979) The search for correlation between theoretical and experimental growth yields. In: *Microbial Biochemistry* (Ed. J. R. Quayle), pp. 1–47. University Park Press, Baltimore.

TAYLOR P. A. & WILLIAMS P. J. LE B. (1975) Theoretical studies on the coexistence of competing species under continuous flow conditions. *Canadian Journal of Microbiology* **21**, 90–8.

TEMPEST D. W. (1970) The continuous cultivation of micro-organisms. I. Theory of the chemostat. In: *Methods in Microbiology* (Eds J. R. Norris & D. W. Ribbons, vol. 2, pp. 259–76. Academic Press, London.

TEMPEST, D. W., MEERS J. L. & BROWN C. M. (1970) Synthesis of glutamate in *Aerobacter aerogenes* by a hitherto unknown route. *Biochemical Journal* **117**, 405–7.

TOPIWALA H. H. & HAMER G. (1971) Effect of wall growth in steady state continuous cultures. *Biotechnology and Bioengineering* **13**, 919–22.

VAN GEMERDEN H. (1974) Coexistence of organisms competing for the same substrate: an example among the purple sulfur bacteria. *Microbial Ecology* **1**, 104–19.

VAN LIERE L. (1979) On *Oscillatoria agardhii gomont*, Experimental Ecology and Physiology of a Nuisance Bloom-forming Cyanobacterium. Ph.D. thesis, University of Amsterdam.

VELDKAMP H. (1970) Enrichment cultures of prokaryotic organisms. *Methods in Microbiology* (Eds J. R. Norris & D. W. Ribbons), vol. 3a, pp. 305–61. Academic Press, London.

VELDKAMP H. (1976) Continuous culture in microbial physiology and ecology. In: *Patterns of Progress* (Ed. J. Gordon Cook). Meadowfield, Durham.

VELDKAMP H. & JANNASCH H. W. (1972) Mixed culture studies with the chemostat. *Journal of Applied Chemistry and Biotechnology* **22**, 105–23.

VELDKAMP H. & KUENEN J. G. (1973) The chemostat as a model system for ecological studies. In: *Modern Methods in the Study of Microbial Ecology* (Ed. T. Rosswall). *Bulletins from the Ecological Research Committee (Stockholm)* **17**, 347–355.

WHITTENBURY R. (1978 Bacterial nutrition. In: *Essays in Microbiology* (Eds J. R. Norris & M. H. Richmond), pp. 16/1–16/32. John Wiley & Sons, Chichester.

WILKINSON T. G., TOPIWALA H. H. & HAMER G. (1974) Interactions in a mixed bacterial population growing on methane in continuous culture. *Biotechnology and Bioengineering* **16**, 41–59.

WINOGRADSKI S. (1949) Microbiologie du sol. *Oeuvres Complètes*. Masson, Paris.

WIRSEN C. O. & JANNASCH H. W. (1970) Growth response of *Spirosoma* sp. to temperature shifts in continuous culture. *Bacteriological Proceedings G* **118**, 2.

WIRSEN C. O. & JANNASCH H. W. (1978) Physiological and morphological observations on *Thiovulum* sp. *Journal of Bacteriology* **136**, 765–74.

YOON H., KLINZING G. & BLANCH H. W. (1977) Competition for mixed substrates by microbial populations. *Biotechnology and Bioengineering* **19**, 1193–210.

ZAMENHOF S. & EICHHORN H. H. (1967) Study of microbial evolution through loss of biosynthesis functions: Establishment of 'defective' mutants. *Nature, London* **216**, 456–8.

Chapter 21 • Bacteriocins

Kimber G. Hardy

21.1 Classification and nomenclature

Bacteriocins are antibacterial proteins produced by bacteria. The most well-studied bacteriocins are the colicins, which are produced by *Escherichia coli* and closely related genera belonging to the *Enterobacteriaceae*, such as *Shigella* and *Salmonella*. About 40% of *E. coli* strains are colicinogenic (*Col*$^{+}$). These strains are immune to the colicin they produce. The genes coding for colicins and for colicin immunity are on *Col* plasmids, many of which are transmissible by conjugation.

Colicins have a narrow spectrum of action, killing only strains of enterobacteria. They are classified into about 20 types according to their effects on a set of colicin-insensitive strains of enterobacteria. Colicins of a particular type are defined as all those colicins which are ineffective against a particular colicin-insensitive strain (indicator strain). Colicins were originally classified in this way by Fredericq (1957), and Davies and Reeves (1975a,b) have recently used genetically characterised indicator strains to re-examine a variety of colicin-producing strains isolated by Fredericq (1957).

Classifying the colicins produced by natural isolates of enterobacteria is a complicated procedure involving a large number of indicator strains. The difficulties arise because many colicin-insensitive strains are resistant to more than one type of colicin. In addition, a strain can produce several different types of colicin.

Colicins and *Col* plasmids are given names such as K-K235 which indicates the colicin type, K, and the strain which was first shown to produce the particular colicin, *Escherichia coli* strain K235. Colicin immunity is used to subdivide colicin types. Colicins E2 and E3 are both members of the E group of colicins because they are ineffective against the colicin E-resistant indicator strain. However, a strain harbouring a *Col*E2 plasmid is immune to colicin E2 but not to colicin E3. At least seven colicin E immunity groups can be distinguished. Similarly, *Col*Ib^{+} bacteria are immune to colicin Ib but not to colicin Ia.

Bacteriocins have been found in many different bacterial genera. They are given names according to the genus or species of the strain which produces them. Examples are staphylococcins and megacins (produced by *Bacillus megaterium*). It is becoming clear that many bacteriocins, in addition to colicins, are coded for by genes on plasmids. Plasmids have recently been shown to code for staphylococcins, streptococcins and clostridicins (Tagg *et al.* 1976).

Bacteria produce a wide range of proteinaceous inhibitors so that it is difficult to decide how many of them should be called bacteriocins. Colicins are a relatively homogeneous group of substances; they are all plasmid-determined proteins which have a narrow spectrum of action. Other groups of bacteria produce a variety of autolytic and bacteriolytic enzymes which are more difficult to classify. Many bacteriocins produced by Gram-positive bacteria do not have a narrow spectrum of action. In addition, defective phages are often called bacteriocins.

21.2 Structure of bacteriocins

Bacteriocins may be either simple proteins or proteins linked to lipid and carbohydrate. Several colicins can exist in two forms; either as simple proteins or proteins covalently linked to lipopolysaccharide components of the cell wall. Inhibitors of DNA synthesis, such as ultraviolet irradiation or mitomycin C, greatly increase the colicin titres produced by cultures of Col^{+} bacteria. Colicins purified from induced cultures are simple proteins. Colicins linked to lipopolysaccharide which forms part of the cell wall O antigen have been isolated from uninduced cultures. Other types of bacteriocin isolated from induced cultures of Gram-negative bacteria are proteins similar to colicins. The molecular weights of various bacteriocins are listed in Table 1.

Several colicins are composed of two proteins; for example colicin E3 consists of a protein with a molecular weight of 50 000 which is associated with a peptide which has a molecular weight of 10 000. The bacteriocidal activity of the molecule is associated with the larger subunit although it is less active against sensitive cells than when it is linked to the smaller subunit. An important property of the small peptide is that it confers immunity on the cells which have a particular *Col* plasmid. The combination of the two subunits is sometimes called the colicin complex. Only a few colicins or other bacteriocins have been examined specifically in order to find out whether they are composed of two subunits so this type of structure may be a more general feature of colicins than is indicated by the data listed in Table 1. Several colicins have similar amino-acid sequences indicating that they have evolved from a common ancestor.

Defective bacteriophages (Lotz 1976) can give bacteriocin-like reactions. They are sometimes called bacteriocins although they are clearly distinct from the low molecular types of colicin. Defective bacteriophages which have antibacterial properties are often found in strains of *Pseudomonas aeruginosa*. Some of these resemble the contractile tails of *Pseudomonas* phages and in fact have immunological cross-reactions with functional temperate phages found in *Pseudomonas* species. Other types of aeruginocin or pyocin (*P. aeruginosa* was formerly called *P. pyocyanea*) resemble non-contractile phage tails or are low molecular weight bacteriocins like colicin.

Several bacteriocins have been purified from Gram-positive bacteria (Tagg *et al.* 1976; Konisky 1978). The composition of some of these is listed in Table 1. Unlike the colicins and many other bacteriocins produced by Gram-negative bacteria, bacteriocins from Gram-positive bacteria are often uninducible by mitomycin C or other agents and are consequently more difficult to purify. Several of the bacteriocins produced by staphylococci, streptococci and lactobacilli are composed of protein, lipid and carbohydrate and tend to aggregate. In each case, bacteriocidal activity is destroyed by proteases.

Table 1. Examples of bacteriocins (after Konisky 1978).

	Molecular weight	Composition
Colicins		
D-CA23	90 000	protein
B-K260	92 000	protein
E1-K30	56 000	protein
E2-P9	50 000 + 10 000[a]	protein
E3-CA38	50 000 + 10 000[a]	protein
Ia-CA53	80 000	protein
Ib-P9	80 000	protein
K-K235	45 000	protein
M-K260	27 000	protein, 5 to 20% phosphatidyl ethanolamine
O-111	69 000	protein
Cloacin DF13	56 000 + 8000[a]	protein
Other Bacteriocins		
Butyricin 7423	32 500	protein
Marcescin JF246	64 000	protein
Megacin A-216	51 000	protein
Perfringocin 11105	76 000	protein
Pesticin α-A122	65 000	protein, 1% hexose
Pyocin S2	72 000	protein
Staphylococcin 414	*c.* 12 000	protein, 46% lipid, 24% carbohydrate, 1.2%

[a]the lower molecular weight subunit is the immunity protein.

21.3 Bacteriocinogenic plasmids

Phage-tail bacteriocins are often specified by chromosomal genes, but whenever the location of the gene coding for a low molecular weight bacteriocin has been investigated it has always proved to be on a plasmid. Plasmids coding for colicins have been studied in greatest detail. There are two groups of *Col* plasmids, conjugative and non-conjugative (Hardy 1975). Conjugative *Col* plasmids have molecular weights of 60 to 90×10^6 and are able to transfer copies of themselves by conjugation. Plasmids coding for colicins B and V specify F-like sex pili (pili resembling those specified by the F plasmid found in *Escherichia coli* K-12). Conjugative plasmids coding for colicin I usually specify type I sex pili. Many conjugative *Col* plasmids code for more than one type of colicin and some also code for resistance to antibiotics. Non-conjugative *Col* plasmids have molecular weights of about 5×10^6 and code for only one type of colicin.

Plasmids have also been found to code for low molecular weight aeruginocins (pyocins), a marcescin (produced by *Serratia marcescens*), a pesticin (produced by *Yersinia pestis*), a straphylococcins, streptococcins, a perfringocin (produced by *Clostridium perfringens*) and an agrocin, a nucleotide derivative produced by *Agrobacterium radiobacter*. Many other bacteriocins are probably plasmid-determined because the ability to produce them is frequently lost when cells are treated with plasmid-curing agents such as acridine orange, ethidium bromide or acriflavine (Tagg *et al.* 1976). The ability to produce lactocins and megacins is frequently lost from cultures treated with curing agents.

21.4 Bacteriocin synthesis

Bacteriocins are usually detected on solid growth media. Bacteria growing on the surface of a medium release bacteriocins which diffuse into the surrounding agar. When the bacteriocinogenic colonies are killed and overlaid with agar containing a bacteriocin-sensitive strain, a zone of inhibition is seen around the colonies.

The mechanisms controlling the synthesis of colicins seem to be similar to those which control temperate phages, such as λ, in lysogenic bacteria. Most cells in a culture of lysogenic bacteria are not producing phage particles; synthesis of most phage proteins is repressed by a phage-specified repressor. A few lysogenic cells are spontaneously induced; in these cells the phage repressor ceases to operate effectively (for reasons that are still unclear) and infective phage particles are produced. These are released upon lysis of the bacterial cell. Colicin synthesis is also repressed in most *Col*$^+$ cells. Usually only about 0.01% of cells in broth cultures are producing colicin, although this proportion can increase to about 10% as broth cultures enter the stationary phase. There is as yet no biochemical evidence for a colicin-repressor analogous to the repressor coded for by the *c*I gene of bacteriophage λ which keeps prophages in the repressed state.

Colicin synthesis is induced by inhibitors of DNA synthesis such as ultraviolet irradiation and mitomycin C. Several prophages, including the λ prophage, are also inducible by these agents. Under appropriate conditions induction results in at least 95% of *Col*$^+$ cells producing colicin. Induction of certain types of colicin, such as E2, invariably kills the producing cell but the synthesis of other colicins, for example colicin Ib, can be induced without killing the producing cells. Neither the mechanism of induction nor the reason for cell death which accompanies the synthesis of certain types of colicin is understood. The number of individual cells producing a bacteriocin can sometimes be detected using the lacuna technique (see section 21.9, p. 375). Almost all bacteriocins isolated from Gram-negative bacteria are inducible, but bacteriocins produced by Gram-positive bacteria are not inducible or are only slightly inducible.

21.5 Effects of bacteriocins on sensitive cells

Bacteriocins are bacteriocidal rather than bacteriostatic under most conditions. They can kill cells in a variety of ways but most bacteriocins are bacteriocidal because of their effects on the cytoplasmic membrane. Other bacteriocins damage either RNA or DNA in sensitive cells or damage the bacterial cell wall.

The mechanism of action of several colicins has been studied in detail (Hardy 1975; Holland 1975; Konisky 1978). The lethal effect of these bacteriocins is confined to members of the *Enterobacteriaceae* related to *Escherichia coli*, such as the genera *Salmonella, Shigella, Enterobacter, Citrobacter* and *Klebsiella*. A strain harbouring a particular *Col* plasmid is immune to the colicins it codes for but the bacterium may also be inherently insensitive to many other types of colicin. Smooth strains which have long lipopolysaccharide side-chains extending from the cell surface are sensitive to few colicins. Rough strains, such as *E. coli* K-12, which lack these side chains, are usually sensitive to almost all colicins.

Colicins bind to proteins (receptors) in the outer membranes of sensitive bacteria and most colicins bind to different receptors. Many colicin receptors also function as receptors for metabolites which are transported into the cell. The protein receptors for colicins B, E, K and M are involved in transporting ferri-enterochelin, vitamin B_{12}, nucleosides and ferrichrome, respectively. Several colicin receptors are also receptors for bacteriophages.

Because many colicin receptors are necessary for the uptake of metabolites, colicin-resistant mutants may be at a disadvantage because they are unable to accumulate metabolites from their environment.

Colicin receptors appear to be involved in transporting colicins through the cell wall; receptors are not essential for colicin action if the cell wall is disrupted to allow access of colicins to the cytoplasmic membrane.

Most colicins are bacteriocidal because they increase the permeability of the cytoplasmic membrane to ions, especially cations. The release of important cations, such as K^+, from colicin-treated cells dissipates the potential difference across the membrane so that all the reactions (including the active transport of many amino acids and sugars) which depend on the energised state of the membrane are inhibited. A further consequence of the loss of membrane potential is that the ATP in the cells is used up in attempts to maintain the membrane potential. The decrease in the intracellular concentration of ATP inhibits many energy-dependent reactions in the cell.

Two colicins, E2 and E3, have a different mode of action. Colicin E2 degrades the cell's DNA and colicin E3 degrades the 16S rRNA molecule of bacterial ribosomes. Because these colicins are such large molecules it was originally thought they probably activated a membrane-bound nuclease while remaining outside the cells themselves. However, it has recently become clear that the colicins themselves are nucleases (Konisky 1978). The colicin proteins are probably split at the cell surface and the C-terminal parts, which have the nuclease activities, are then transferred through the cytoplasmic membrane. In addition, the colicin-immunity protein which is firmly bound to the C-terminal end of the colicins must be removed before the colicins can act as nucleases. Presumably, this protein is removed during the passage of the colicins through the cell wall. Although the immunity protein must be removed from colicins E2 and E3 before they can act as nucleases *in vitro* or, presumably, inside sensitive cells, the immunity protein does not prevent colicins from binding to sensitive cells.

Bacillus stearothermophilus is resistant to colicin E3, presumably because it lacks the appropriate receptors and other wall components which allow entry of the colicin. But ribosomes isolated from *B. stearothermophilus* are as sensitive to colicin E3 *in vitro* as are those from *Escherichia coli*.

Several other sorts of bacteriocin seem to have a mechanism of action similar to those colicins which inhibit many reactions dependent on the energised state of the cytoplasmic membrane (Konisky 1978). A staphylococcin and two streptococcins act in this way. A bacteriocin produced by *Clostridium butyricum* was found to inhibit the membrane-bound ATPase of sensitive cells. Colicin M and a pesticin were found to weaken the cell wall so that the cells lysed if they were in a hypotonic medium. Megacin A produced by *Bacillus megaterium* was found to be a phospholipase.

21.6 Ecology

The high frequency of genes for bacteriocins in natural populations of bacteria implies that they are highly successful genes. What is the basis of this success? How effective are bacteriocins as antibiotics in natural environments? The ecological significance of bacteriocins has, in the past, received much less attention than the biochemical aspects. However, experiments have recently been made on the ecology of strains producing colicins, agrocins, streptococcins and staphylococcins and bacteriocins from *Bacteroides* spp.

Experiments have been made to determine whether colicins influence the succession of strains in the alimentary tracts of man or animals but the results are ambiguous. The synthesis of several types of colicin is lethal but other members of the clone, which are colicin-immune, could benefit from the decreased competition which results from the colicin released by a small proportion of the cells which die in the process.

Strains of *Escherichia coli* inhabiting the alimentary tracts of man or animals can be either resident strains, which can be repeatedly isolated for several weeks or months, or transient strains which survive for only a few days. Most investigations have shown that resident strains in the human alimentary tract are more likely to be colicinogenic than transient strains (Hardy 1975). It has also been found that resident strains are more likely to be colicin-resistant than are transient strains.

An alternative approach has been to ingest cultures of bacteria to determine whether Col^+ or Col^- strains are more likely to become established and whether Col^+ strains can eliminate colicin-sensitive strains which were previously resident in the intestine. The results of a series of experiments along these lines by Cooke *et al.* (1972) were generally in agreement with the hypothesis that colicinogeny increased the ability of *Escherichia coli* to become established in the alimentary tract. However, colicin-sensitive strains were not always displaced when an ingested Col^+ strain became established, and the former occasionally became established despite the presence of colicinogenic strains. The *E. coli* strains used in these experiments had a variety of different serotypes. In addition to colicinogeny, there are of course many other features, determined by chromosomal or plasmid genes, which could influence the ability to colonise the alimentary tract.

In an attempt to control these variables, Smith and Huggins (1976, 1978) compared the ability of Col^+ and Col^- forms of the same strain to survive in the human alimentary tract. Col^- forms of a strain were made by treating them with curing agents to remove *Col* plasmids. Or Col^+ strains were made by transferring a plasmid into a strain by conjugation. By comparing the ability of these strains to survive, differences in chromosomal genes which could affect persistence were controlled.

When mixtures of *Col*V$^+$ and *Col*V$^-$ strains were ingested by two healthy people, it was found that the *Col*V$^+$ form persisted for much longer than the *Col*V$^-$ form. The *Col*V plasmid used in these experiments was not a conjugative plasmid so the enhanced survival was not due to transfer from the *Col*V$^+$ to the *Col*V$^-$ form. Both forms were made resistant to nalidixic acid so that they could be readily recovered from faeces. Although certain types of *Col*V plasmid markedly increased survival, a *Col*E plasmid did not enhance the ability of the strain to survive, and only certain types of *Col*V plasmids enhanced the residence time of strains in the intestine.

Although these experiments showed that certain *Col*V plasmids increase the residence time of *Escherichia coli*, the authors point out that the effect cannot necessarily be ascribed to the colicin produced by the strain because there may be other plasmid genes which have this effect. *Col*V plasmids usually have molecular weights of at least 60×10^6 and could, therefore, comprise about 100 genes, some of which could affect survival.

Similarly, the greater frequency of colicin production amongst resident strains could be because plasmid genes other than those for colicins enhance survival, or simply because the longer a strain persists in the alimentary tract, the more likely it is that plasmids will be transferred into it, irrespective of whether they significantly increase its survival. Many *Col* plasmids code for drug-resistance determinants. About 40% of conjugative I-like R plasmids from *Salmonella typhimurium* strains produce colicin I. The survival of genes coding for colicins will be closely correlated with that of any closely linked determinants.

Germ-free animals have also been used to determine whether colicins confer a selective advantage. In one such experiment (Kelstrup & Gibbons 1969), groups of mice which had been colonised by a *Col*E2$^+$ strain of *Escherichia coli* K-12 were housed with other mice which had been colonised with either a Col^-, colicin-sensitive strain of *E. coli* K-12 or with a Col^-, colicin-resistant strain of *E. coli* K-12. After about 40 d, both groups of mice which were originally colonised by Col^- bacteria were found to have predominantly (more than 90%) Col^+ bacteria in their alimentary tracts. Because the colicinogenic bacteria became predominant in competition with colicin E2-resistant bacteria, the selective advantage conferred by the non-conjugative *Col*E2 plasmid could not be ascribed to colicin production. Bacteriocinogenic strains of *E. coli* or of *Bacteroides* spp. (Booth *et al.* 1977) are often found to coexist with sensitive strains in the alimentary tract.

The apparent ineffectiveness of bacteriocins could be because the proteins are inactivated by proteases, but analyses of protease activities in various parts of the alimentary tract of the pig show that the activities in the colon are not high enough to inactivate colicins.

Bacteriocin-producers are common on skin.

Examples of antagonism between strains of *Staphylococcus aureus* have been well documented but the relative importance of bacteriocins in this interference is unclear. About 20% of staphylococci and micrococci found on skin produce diffusible inhibitors. There are several instances of a resident strain preventing the colonisation of the skin by other strains, but the best-documented example of an antagonistic strain of *S. aureus* which prevented colonisation (strain 502A) did not produce a bacteriocin (Shinefield *et al.* 1974). The mechanism of the highly effective antagonism of strain 502A is not understood. On the other hand, Selwyn (1975) found that dermatological patients who had various types of skin lesion were less likely to acquire a secondary infection if they were already colonised by bacteriocin-producers (chiefly members of the *Micrococcaceae*). Bacteriocins produced by Gram-positive bacteria have a wide spectrum of action.

Recent experiments have shown that bacteriocins produced by *Streptococcus mutans* are effective in dental plaque (Rogers *et al.* 1979; van der Hoeven & Rogers 1979). *S. mutans* is believed to be an important species in initiating dental caries. In comparison with non-bacteriocinogenic variants, bacteriocinogenic strains of *S. mutans* became established more quickly in plaque and eventually formed a greater proportion of the total population of microorganisms in plaque.

A strain of *Agrobacterium radiobacter* producing a nucleotide derivative called agrocin 84 is used extensively in Australia and California to control *Agrobacterium tumefaciens* which causes crown gall disease in fruit trees and other dicotyledonous plants (Kerr 1980).

Experimental

21.7 Testing for bacteriocin production

Bacteriocins are usually detected by growing bacteria on solid growth media, either as spots or streaks, and then overlaying them with bacteria (indicator strains) of the same or related species. The bacteriocins diffuse into the agar so that an inhibition zone surrounds the bacteriocinogenic bacteria after incubation of plates overlaid with a sensitive strain. Chloroform is often used to kill the organisms before overlaying or cross streaking with the indicator strain, although a few bacteriocins are inactivated by chloroform.

An alternative method is to dilute the culture to be tested with molten agar which is poured as a layer over the surface of a suitable nutrient agar plate. This layer is then overlaid with more agar so that the individual colonies are formed underneath the surface of the agar. When the colonies are formed, a layer of agar seeded with an indicator strain, is added. This method can be used to detect bacteriocin-producers in a mixed bacterial population, for example, from faeces or sewage. It is also useful for detecting bacteriocinogenic recipients in genetic experiments in which a bacteriocinogenic plasmid has been transferred by conjugation or by some other means. This procedure is only suitable if the strains being tested are not obligate aerobes.

Bacteriocins have also been detected by growing the test strains on an agar plate and then inverting the agar so that the indicator strain is added to the other side of the agar.

Very different amounts of bacteriocins are produced on different types of growth media so it is usually necessary to screen colonies growing on several types of nutrient agar to find out which is most suitable. For colicins, the author has found that Diagnostic Sensitivity Test Agar (Oxoid CM 261) is the most suitable for screening colonies of natural isolates for colicins. Some strains producing colicin V form large inhibition zones on this medium but no zones at all on either Nutrient Agar (Oxoid CM 3) or Blood Agar Base (Oxoid CM 271). Others have noted a similar effect of growth medium on the production of other types of bacteriocin. The reasons for the effects of growth media on colicin production are not understood.

Growth media also affect the sensitivity of the indicator strain; for example, colicins V, I and B are much more effective against indicator strains grown in media which have low concentrations of iron. The numbers of receptors for these colicins is regulated by iron concentrations; in iron-deficient media the number of receptors is greatly increased (Konisky 1978). Strains of *Streptococcus mutans* and *S. salivarius* were more resistant to bacteriocins when sucrose was included in their growth media because they formed an extracellular layer of polysaccharide which protected them (Rogers 1974).

The host strain of a bacteriocinogenic plasmid can also affect the size of the inhibition zone. Smaller inhibition zones are produced by colicinogenic bacteria which have colicin receptors than by bacteria which are colicin-resistant and cannot adsorb colicin.

When analysing inhibition zones produced on agar plates it is necessary to distinguish bacteriocins from other types of inhibitor which may be produced. Catalase can be included in growth media to prevent inhibition zones caused by hydrogen peroxide. The spectrum of action of the inhibitor against a range of related and unrelated bacteria is usually determined. Bacteriocins produced by Gram-negative bacteria have a narrow activity spectrum but bacteriocins produced by Gram-positive bacteria may be active against a wide range of Gram-positive genera. The sensitivity of the bacteriocin to proteases can usually be simply determined by adding enzymes to plates or to bacteriocins in a suitable buffer.

If the inhibitor is sensitive to proteases, it may be a bacteriophage, an autolytic or bacteriolytic enzyme or a bacteriocin. The distinction between certain lytic enzymes and bacteriocins is unclear. A few substances which have a bacteriolytic mode of action, such as colicin M and pesticin, are termed bacteriocins. Colicin M and probably pesticin are plasmid-determined, but a wide variety of bacteriolytic and autolytic enzymes are produced by bacteria and most of these are not called bacteriocins.

Bacteriocins can usually be distinguished from temperate bacteriophages by dropping dilutions of a cell-free supernatant onto a plate inoculated with a sensitive strain which is then incubated. Successive dilutions of bacteriocins produce gradually less inhibition of growth of the indicator, but bacteriophages form discrete plaques at high dilutions. The inhibition zones produced by colonies releasing phages or defective phages are much smaller than those produced by most bacteriocins. Phages and defective phages can be distinguished from low molecular weight bacteriocins by ultracentrifugation.

21.8 Bacteriocin typing

Bacteriocin typing schemes have been devised for several species of human pathogenic bacteria so that epidemiological investigations can be made to trace sources of infection. Two typing methods have been used:

(1) the determination of the sensitivity of the test strains to a range of bacteriocins;

(2) the analysis of the types of bacteriocin produced by the test strains.

A priori it would seem that the latter approach would be more useful. Bacteriocins are plasmid-determined so that a strain could readily gain or lose its bacteriocinogenic properties. In addition, many strains do not produce bacteriocins and could not, therefore, be typed. In practice, however, this approach has been used successfully to make typing schemes for *Shigella sonnei* and *Pseudomonas aeruginosa*. Bacteriocin typing schemes which use standard colicin-producers to test the sensitivity of strains are not without their difficulties; for example, the transfer of a plasmid into a strain can also change the latter's sensitivity pattern. The general problems associated with bacteriocin typing schemes have been discussed by Mayr-Harting *et al.* (1972) and Meitert and Meitert (1978).

Two widely used typing schemes which depend on bacteriocin production by the test strains are those for *Shigella sonnei* (Gillies 1978) and *Pseudomonas aeruginosa* (Govan 1978). In the typing scheme for *S. sonnei*, the strain to be typed is streaked down the centre of a suitable agar plate which is then incubated for 24 h. The bacterial growth is removed with a glass microscope slide and residual bacteria killed by chloroform vapour. A total of 15 indicator strains are streaked at right angles to the test organism and the plate is re-incubated for 18 to 24 h. Both the frequency and range of colicin types produced by *S. sonnei* are large enough to make colicin typing a useful epidemiological tool, although some potential drawbacks have been noted (Meitert & Meitert 1978). Among these is the finding that the frequency of colicin production by predominant strains in the population can be rather low at times (about 40%) so the most important strains cannot be typed. In addition, changes in the types of strains during an epidemic can occur if they receive additional *Col* plasmids. The standardised conditions necessary to ensure reproducible results have been described (Gillies 1978).

Pyocin (or aeruginocin) typing of *Pseudomonas aeruginosa* also depends on determining the activity spectrum of pyocins, which are often defective phages, on a set of indicator strains (Govan 1978). The procedure has been used in many parts of the world and only about 8% of strains cannot be typed because they do not produce bacteriocins.

Bacteriocin typing schemes devised for *Clostridium perfringens* (Mahoney 1979), *Corynebacterium diphtheriae* (Saragea *et al.* 1979) and *Serratia marcescens* (Traub 1978) depend on the sensitivity of test strains to a range of standard bacteriocin-producers. These schemes have not so far been used extensively.

21.9 Titration of bacteriocins

Three types of biological assay have been used to titrate bacteriocins:

(1) the critical dilution method;
(2) the survivor count method;
(3) the diffusion zone method.

Details of these methods have been described by Mayr-Harting *et al.* (1972).

21.9.1 Critical Dilution Method

A twofold dilution series of the bacteriocin preparation is made and a small volume of each is dropped onto the surface of an agar plate which has been inoculated with a sensitive strain. Pipettes which release 0.02 ml volumes, as used in the Miles and Misra technique, can be used. The inoculum is usually added in the form of an overlay on the surface of a nutrient agar growth medium. The plates are incubated for a standardised time. The bacteriocin titre is expressed either as the highest dilution which completely inhibits the indicator or as the highest dilution which produces detectable inhibition of the indicator.

21.9.2 Survivor Count Method

This is used to determine the titre of a bacteriocin preparation in terms of lethal or killing units per ml of preparation. A lethal unit is the minimum amount of colicin required to kill a sensitive cell. Bacteriocins are mixed with a known number of viable cells for a sufficient time to allow adsorption. The number of viable cells remaining is determined by making a viable count. At the end of the adsorption period bacteriocins are assumed to be distributed amongst the cells according to the Poisson distribution. Bacteria which have adsorbed a lethal unit will subsequently be unable to form a colony. The number of lethal units can, therefore, be determined from the equation:

$$\frac{n}{n_0} = e^{-m}$$

where n_0 is the number of viable cells before adsorption of colicin; n is the number which have survived; m is the multiplicity of lethal units; and e is the exponential function. The number of colicin molecules which constitute a lethal unit ranges from one to several hundred for different colicins.

21.9.3 Diffusion Zone Method

Colicin is added to wells cut in nutrient agar plates which have been inoculated with indicator bacteria. The plates are kept at 4°C for a standardised period of time (12 to 48 h) to allow the bacteriocin to diffuse into the agar. Some colicins, such as V, diffuse more rapidly than others, such as I. The plates are incubated for an appropriate time (12 to 48 h) to allow the indicator to grow. The diameter of the inhibition zone is measured and is usually found to be proportional to the logarithm of the bacteriocin concentration (Mayr-Harting *et al.* 1972; Hardy *et al.* 1974). This method has several advantages over the others if many samples are to be assayed.

21.9.4 Lacuna Counts

Lacunae are clearings produced in a lawn of confluent growth of indicator bacteria by the bacteriocin released from individual cells (Ozeki *et al.* 1959; Hardy *et al.* 1974). To determine the number of lacuna-forming cells in a culture, it must first be killed, usually with chloroform. The cells are then mixed with sensitive indicator bacteria and soft agar which is poured over the surface of a suitable solid growth medium. Sufficient indicator bacteria are added so that they form a lawn of confluent growth after incubation. Cells releasing bacteriocin form small clearings, up to 3 mm diameter, in this lawn.

The method has been used mainly to determine the number of colicin-producing cells which exist in a culture at any one time. In the case of cultures of *Escherichia coli* containing plasmid *Col*E2-P9, the percentage of lacuna-forming cells was found by Hardy *et al.* (1974) to range between about 0.01% (in exponential phase) and about 10% on approaching the stationary phase. After induction by mitomycin C or ultraviolet irradiation more than 90% of the cells form lacunae. Not all bacteriocinogenic strains form lacunae, either because individual cells do not form sufficient bacteriocin to produce a visible clearing or because the bacteriocin remains largely cell-bound and does not diffuse away from individual cells.

21.10 Isolation of bacteriocins

The bacteriocins produced by Gram-negative bacteria have generally proved easier to isolate and characterise than those produced by Gram-positive strains. The titres of colicins and other bacteriocins produced by Gram-negative bacteria can usually be greatly increased by mitomycin C or other inducing agents. Colicins produced by induced cultures are usually released into the growth medium but some colicins remain largely cell-bound when they have been induced. Colicins are released without cell lysis. Herschman and Helinski (1967) found that more than 90% of the induced colicins E2-P9 and E3-CA38 were cell-bound but could be removed from cell pellets by washing with 1.0M NaCl.

If *Escherichia coli* strains are grown in a minimal salts medium, the nature and concentration of the carbon source can greatly influence the yield of colicin, presumably because of the effects of catabolite repression. Yields of extracellular colicin can also be increased by using a colicin-resistant strain which does not bind colicins. The bacteriocin-producing strain must be chosen with care because many strains produce several types of bacteriocin and/or bacteriophages which can greatly complicate subsequent purification. For example, the strain originally found to produce colicin K-K235, *E. coli* K-235, also produced two other types of colicin, I and X, and was lysogenic for two kinds of bacteriophage, P2 and P4 (Mayr-Harting *et al.* 1972). The use of lysogenic strains as sources of bacteriocins is avoided because they may lyse after being treated with inducing agents.

The bacteriocins produced by Gram-positive bacteria are generally of low titre which cannot be greatly increased by inducing agents. In addition, they are often firmly attached to the walls of the producing cells and some are produced only on solid growth media (Tagg *et al.* 1976). The various methods used to purify bacteriocins are those which are used to purify many other sorts of protein and can be found in standard texts on protein purification.

21.11 The ecology of bacteriocinogenic bacteria

Several experiments described in section 21.6 (p. 371) failed to find a selective advantage conferred by bacteriocin production. When bacteriocinogenic strains were found to survive better than other strains, it could not be concluded that the bacteriocins themselves were responsible.

The methods described by Smith and Huggins (1978) provide a good basis for further experiments designed to identify any selective advantage conferred by genes coding for colicins or other genes on *Col* plasmids. The effect of plasmids on the ability of an ingested strain to survive in the human alimentary tract was determined. A host strain was chosen which was a good recipient for many kinds of plasmid and which generally survived in the human alimentary tract for about 6 to 10 d after ingestion. The strain was made resistant to nalidixic acid so that its presence in faeces could be easily determined by plating samples on MacConkey agar containing this compound. Two strains were given simultaneously, a Col^+ and Col^- form of the same bacterium. Colicin-resistant mutants of both forms were usually used. Certain *Col*V plasmids were found to enhance greatly the ability of *Escherichia coli* to survive in the alimentary tract. A *Col*E plasmid and several other types of plasmid either did not significantly affect survival or reduced the ability of bacteria to persist. In addition, one type of *Col*V plasmid, *Col*V I-94, did not increase the persistence of bacteria in the alimentary tract.

Experiments of this type provide the basis for further work which could determine whether colicins are responsible for the selective advantage of certain Col^+ strains. In particular, the effects of mutant *Col* plasmids defective in colicin production could be compared with those of the wild-type plasmid. Several results suggest that *Col* plasmid genes, other than those for colicins, can influence the survival of strains; for example, many *Col*V plasmids code for an iron-uptake system based on the synthesis of a hydroxamate (Stuart *et al.* 1980; Williams 1980) and also have genes coding for resistance to bacteriocidal effects of complement (Binns *et al.* 1979).

Plasmids which have derepressed conjugal transfer systems appear to reduce the survival of *Escherichia coli* in the alimentary tract (Smith & Huggins 1978), so *Col* plasmids having transfer genes of this type should perhaps be avoided in experiments designed to test the ecological significance of colicins.

21.12 Genetic aspects of bacteriocins

Bacteriocins are usually specified by bacterial

plasmids and it is often useful to be able to transfer these plasmids from one strain to another or to remove them from strains.

Most genetic work has been carried out with *Col* plasmids from enterobacteria. Conjugative *Col* plasmids can be readily transferred by mating in broth cultures (Clowes & Hayes 1968). A recipient strain is chosen which has a marker, such as resistance to an antimicrobial drug, so that recipients can be selected from the mixture by adding appropriate dilutions to plates containing the drug.

Two methods can be used to select Col^+ recipients. Samples of the mixture can be included in a layer of agar on the surface of a plate containing the appropriate drug to prevent growth of the donor (Fredericq 1957), as described in section 21.7 (p. 373). Col^+ recipients can also be selected by the colicin-immunity conferred by *Col* plasmids. Plates containing colicin can be made by spreading a Col^+ strain over the surface of a plate and incubating it so that it forms a confluent lawn of growth. This is killed by treatment with chloroform vapour and an overlay of approximately 10 ml nutrient agar containing an antibacterial drug to select the recipient strain is added. A few hours is allowed for colicin diffusion and dilutions of the mating mixture are spread over the surface. This technique depends on using a colicin-sensitive recipient strain which may, of course, be killed by the colicin released by the donor strain in the mating mixture. Some Col^+ donors do not release sufficient colicin to kill most cells of the recipient strains, but with other Col^+ donors, trypsin should be included in the mating mixture to prevent the recipient being killed. The frequency of mutation to colicin-resistance is usually high (*c.* 1×10^6 cell^{-1} generation^{-1}), so the selection of colicin-resistant mutants of the recipient strain on the colicin-containing plates can sometimes be a problem.

Non-conjugative *Col* plasmids can be transferred at high frequency by mobilisation: i.e. a conjugative plasmid is used to promote transfer of the non-conjugative plasmid. Transfer by mobilisation is highly efficient if a suitable conjugative plasmid is used; for example, if an F plasmid is used to promote the transfer of the non-conjugative *Col*E1-K30, more than 90% of F^+ recipients will also be *Col*E1$^+$. If the mating is interrupted after 3 min, about 25% of *Col*E1$^+$ recipients will be F^-. Plasmids can also be transferred by transduction using the generalised transducing phages P1 or P22 or by transformation of $CaCl_2$-treated *Escherichia coli.*

Curing agents are used to remove plasmids from bacteria. Acridine orange is often used to remove plasmids from enterobacteria (Hirota 1956). Mitomycin C is particularly effective for removing many plasmids found in *Pseudomonas*, and ethidium bromide, acridine orange and acriflavine have been used to remove many types of plasmid found in Gram-positive bacteria (Tagg *et al.* 1976). Sodium dodecyl sulphate can also be used as a plasmid-curing agent (Tomoeda *et al.* 1968).

Recommended reading

HARDY K. G. (1975) Colicinogeny and related phenomena. *Bacteriological Reviews* **39**, 464–515.

HOLLAND I. B. (1975) Physiology of colicin action. *Advances in Microbial Physiology* **12**, 56–139.

KONISKY J. (1978) The bacteriocins. In: *The Bacteria* (Eds L. N. Ornston & J. R. Sokatch), vol. 6, pp. 71–136. Academic Press, New York.

LOTZ W. (1976) Defective bacteriophages: the phage tail-like particles. *Progress in Molecular and Subcellular Biology* **4**, 53–102.

MAYR-HARTING A., HEDGES A. J. & BERKELEY R. C. W. (1972) Methods for studying bacteriocins. In: *Methods in Microbiology* (Eds J. R. Norris & D. W. Ribbons), vol. 7A, pp. 315–422. Academic Press, London.

REEVES P. (1972) *The Bacteriocins.* Chapman & Hall, London.

TAGG J. R., DAJANI A. S. & WANNAMAKER L. W. (1976) Bacteriocins of Gram-positive bacteria. *Bacteriological Reviews* **40**, 722–56.

References

BINNS M. M., DAVIES D. L. & HARDY K. G. (1979) Cloned fragments of plasmid Col V, I-K94 specifying virulence and serum resistance. *Nature, London* **279**, 778–81.

BOOTH S. J., JOHNSON J. L. & WILKINS T. D. (1977) Bacteriocin production by strains of Bacteroides isolated from human feces and the role of these strains in the bacterial ecology of the colon. *Antimicrobial Agents and Chemotherapy* **11**, 718–24.

CLOWES R. C. & HAYES W. (1968) *Experiments in Microbial Genetics.* Blackwell Scientific Publications, Oxford.

COOKE E. M., HETTIARATCHY I. G. I. & BUCK A. C. (1972) Fate of ingested *Escherichia coli* in normal persons. *Journal of Medical Microbiology* **5**, 361–9.

DAVIES J. K. & REEVES P. (1975a) Genetics of resistance to colicins in *Escherichia coli* K-12: cross-resistance among colicins of group A. *Journal of Bacteriology* **123**, 102–17.

DAVIES J. K. & REEVES P. (1975b) Genetics of resistance to colicins in *Escherichia coli* K-12: cross-resistance among colicins of group B. *Journal of Bacteriology* **123**, 96–101.

FREDERICQ R. (1957) Colicins. *Annual Review of Microbiology* **11**, 7–22.

GILLIES R. R. (1978) Bacteriocin typing of *Enterobacteriaceae*. In: *Methods in Microbiology* (Eds T. Bergan & J. R. Norris), vol. 2, pp. 80–6. Academic Press, London.

GOVAN J. R. W. (1978) Pyocin typing of *Pseudomonas aeruginosa*. In: *Methods in Microbiology* (Eds T. Bergan & J. R. Norris), vol. 10, pp. 61–91. Academic Press, London.

HARDY K. G. (1975) Colicinogeny and related phenomena. *Bacteriological Reviews* **39**, 464–515.

HARDY K. G., HARWOOD C. R. & MEYNELL G. G. (1974) Expression of colicin factor E2-P9. *Molecular and General Genetics* **131**, 313–31.

HERSCHMAN H. R. & HELINSKI D. R. (1967) Purification and characterization of colicin E_2 and E_3. *Journal of Biological Chemistry* **242**, 5360–8.

HIROTA Y. (1956) Artificial elimination of the F factor in *Bacterium coli* K-12. *Nature, London* **178**, 92.

HOLLAND I. B. (1975) Physiology of colicin action. *Advances in Microbial Physiology* **12**, 56–139.

KELSTRUP J. & GIBBONS R. J. (1969) Inactivation of bacteriocins in the intestinal canal and oral cavity. *Journal of Bacteriology* **99**, 888–90.

KERR A. (1980) Biological control of crown gall through production of agrocin 84. *Plant Disease* **64**, 25–30.

KONISKY J. (1978) The bacteriocins. In: *The Bacteria* (Eds L. N. Ornston & J. R. Sokatch), vol. 6, pp. 71–136. Academic Press, New York.

LOTZ W. (1976) Defective bacteriophages: the phage tail-like particles. *Progress in Molecular and Subcellular Biology* **4**, 53–102.

MAHONEY D. E. (1979) Bacteriocin, bacteriophage and other epidemiological typing methods for the genus *Clostridium*. In: *Methods in Microbiology* (Eds T. Bergan & J. R. Norris), vol. 13, pp. 1–30. Academic Press, London.

MAYR-HARTING A., HEDGES A. J. & BERKELEY R. C. W. (1972) Methods for studying bacteriocins. In: *Methods in Microbiology* (Eds J. R. Norris & D. W. Ribbons), vol. 7A, pp. 315–422. Academic Press, London.

MEITERT T. &. MEITERT E. (1978) Usefulness, applications and limitations of epidemiological typing methods to elucidate nosocomial infections and the spread of communicable diseases. In: *Methods in Microbiology* (Eds T. Bergan and J. R. Norris), vol. 10, pp. 1–37. Academic Press, London.

OZEKI H., STOCKER B. A. D. & DE MARGERIE H. (1959) Production of colicins by single bacteria. *Nature, London* **184**, 337–9.

REEVES P. (1972) *The Bacteriocins*. Chapman & Hall, London.

ROGERS A. H. (1974) Bacteriocin production and susceptibility among strains of *Streptococcus mutans* grown in the presence of sucrose. *Antimicrobial Agents and Chemotherapy* **6**, 547–50.

ROGERS A. H., VAN DER HOEVEN J. S. & MIKX F. H. M. (1979) Effect of bacteriocin production by *Streptococcus mutans* on the plaque of gnotobiotic rats. *Infection and Immunity* **23**, 571–6.

SARAGEA A., MAXIMESCU P. & MEITERT E. (1979) *Corynebacterium diphtheriae*: microbiological methods used in clinical and epidemiological investigations. In: *Methods in Microbiology* (Eds T. Bergan & J. R. Norris), vol. 13, pp. 61–176. Academic Press, London.

SELWYN S. (1975) Natural antibiosis among skin bacteria as a primary defence against infection. *Journal of Dermatology* **93**, 487–93.

SHINEFIELD H. R., RIBBLE J. C., BORIS M., EICHENWALD H. F., ALY R. & MAIBACH H. (1974) Interference between strains of *S. aureus*. *Annals of the New York Academy of Sciences U.S.A.* **236**, 444–55.

SMITH H. W. & HUGGINS M. B. (1976) Further observations on the association of the colicine V plasmid of *Escherichia coli* with pathogenicity and with survival in the alimentary tract. *Journal of General Microbiology* **92**, 335–50.

SMITH H. W. & HUGGINS M. B. (1978) The effect of plasmid-determined and other characteristics on the survival of *Escherichia coli* in the alimentary tract of two human beings. *Journal of General Microbiology* **109**, 375–9.

STUART S. J., GREENWOOD K. T. & LUKE R. K. J. (1980) Hydroxamate-mediated transport of iron controlled by ColV plasmids. *Journal of Bacteriology* **143**, 35–42.

TAGG J. R., DAJANI A. S. & WANNAMAKER L. W. (1976) Bacteriocins of Gram-positive bacteria. *Bacteriological Reviews* **40**, 722–56.

TOMOEDA M., INUZUKA M., KUBO N. & NAKAMA S. (1968) Effective elimination of drug resistance and sex factors in *Escherichia coli* by sodium dodecylsulphate. *Journal of Bacteriology* **95**, 1078–89.

TRAUB W. H. (1978) Bacteriocin typing of clinical isolates of *Serratia marcescens*. In: *Methods in Microbiology* (Eds T. Bergan & J. R. Norris), vol. 11, pp. 223–42. Academic Press, London.

VAN DER HOEVEN J. S. & ROGERS A. H. (1979) Stability of the resident microflora and the bacteriocinogeny of *Streptococcus mutans* as factors affecting its establishment in specific pathogen-free rats. *Infection and Immunity* **23**, 206–12.

WILLIAMS P. H. (1980) A novel iron uptake system specified by colicin V plasmids: an important component in the virulence of invasive strains of *Escherichia coli*. *Infection and Immunity* **26**, 925–32.

Chapter 22 • Bdellovibrios—Intraperiplasmic Growth

Sydney C. Rittenberg

22.1 Introduction

The bdellovibrios' normal life cycle consists of two phases: a motile, free-living, non-growing phase, and a growth phase that takes place in the periplasmic space of other bacteria. The study of bdellovibrio growth is thus the study of its interactions with the bacterium serving as its substrate, and the methodology employed for such studies must be focused on two-membered cultures. The exploitation of such two-membered cultures for laboratory investigations of the bdellovibrios' intraperiplasmic development is the major concern of this chapter. Recent reviews dealing specifically with bdellovibrio methodology (Starr & Stolp 1976), ecology (Varon & Shilo 1979) and intraperiplasmic growth (Rittenberg & Thomashow 1979; Thomashow & Rittenberg 1979) are available. They can be consulted for additional background information, for literature references to undocumented generalised statements in the text, and for detailed procedures for the isolation and enumeration of bdellovibrios.

Electron microscope observations have been important for obtaining the initial qualitative overview of the intraperiplasmic growth cycle, and phase contrast microscopy offers the most rapid and convenient manner of monitoring the progress of many types of experiments. However, the microscopic procedures as well as the biochemical methods and instrumentation applied in bdellovibrio research have been standard and will not be considered. Some success, but little understanding, has been achieved in growing wild-type bdellovibrios in complex media containing bacterial extracts (Horowitz *et al.* 1974) and axenic growth of appropriate bdellovibrio mutants is easily and reproducibly obtainable (Seidler & Starr 1969b). However, a discussion of axenic growth of bdellovibrios will not be included because its study requires no fundamentally different procedures or approaches than those applied to the study of *Escherichia coli* or other common heterotrophic bacteria. This is not to imply that the methodologies available for the latter organisms are necessarily available for investigations of the bdellovibrio. For example, the most pressing methodological problem for bdellovibrio research is the development of a genetic system that can be exploited experimentally.

The study of two-membered systems is not

completely foreign to bacteriology. In particular, the methodology employed for the study of phage development most closely resembles that of value in investigating intraperiplasmic growth. It is important to emphasise that despite similarities there are very fundamental differences in the two systems. Phage have an absolute dependency on the metabolic activities of their hosts, and the latter must be in an environment that permits their growth for phage development to occur. In contrast, bdellovibrios have no dependence on the metabolic processes of the bacteria they attack, neither for energy generation nor for biosynthetic or degradative functions. They will grow with the pre-existing components of the attacked cell as their sole source of all nutrients and their growth rates or growth yields will not be greatly increased by the presence of additional nutrients in the environment (Varon & Shilo 1969b; Rittenberg & Hespell 1975). A second important difference is that the bdellovibrio, in contrast to the phage, is a complete bacterium very similar in chemical composition and in many metabolic processes to the cell it attacks. This creates methodological problems of distinguishing between components and activities of the two bacteria which do not usually exist in phage research.

22.2 The intraperiplasmic growth cycle

22.2.1 Attack

The free-living bdellovibrio is a small (*c.* 0.3 × 1.5 μm), aerobic, mesophilic bacterium, highly motile by virtue of a single sheathed polar flagellum. Swimming in a liquid menstruum, it hits a potential substrate cell with considerable force. The collision is apparently random, and all attempts to demonstrate a chemotactic attraction between the bdellovibrio and a potential substrate cell have been negative (Straley & Conti 1977). In an effective attack it remains attached after collision. Little is known about the mechanism of attachment although it has been suggested that the lipopolysaccharide (LPS) of the substrate cell is involved (Varon & Shilo 1969a).

Only Gram-negative bacteria can serve as substrates for bdellovibrio and the two reported exceptions have been shown to be erroneous (Tudor & Conti 1977). A particular bdellovibrio strain may attack a narrow or broad range of substrate organisms (Stolp & Starr 1963; Taylor *et al.* 1974) although the basis of this limited specificity is unknown.

22.2.2 Penetration and Bdelloplast Formation

Some 10 min after effective attachment, the bdellovibrio passes through the outer membrane and peptidoglycan layers of the attacked cell and lodges in its periplasmic space. As judged by the cinematographic record (Stolp 1967) and observations by phase microscopy, the actual passage through the outer cell boundaries is very rapid. Usually, but not always, entry is associated with a rounding up of the attacked cell. The rounded cell containing the bdellovibrio is referred to as the bdelloplast.

Penetration follows a series of coordinated enzymic processes that are initiated upon attachment. These processes result in the production of an entry pore for the bdellovibrio, and in the subsequent stabilisation of the bdelloplast. Identified activities (Thomashow & Rittenberg 1978a, b, c) include:

(1) a solubilisation of about 25% of the amino sugars of the substrate cell's LPS;

(2) excision of about 15% of the peptidoglycan sugars;

(3) solubilisation of an additional 15% of the diaminopimelic acid residues of the peptidoglycan without concomitant solubilisation of the sugars;

(4) deacetylation of the sugars in the residual substrate cell peptidoglycan;

(5) removal of the Braun lipoprotein;

(6) covalent addition of long-chain fatty acids to the bdelloplast peptidoglycan structure.

The first two activities which are complete in about 20 min of a 3 to 4 h cycle are believed to be directly involved in penetration, i.e. breaching the outer membrane and producing a pore in the peptidoglycan layer. The remaining processes probably function in stabilising the bdelloplast wall. Although many of the details remain to be worked out, it is quite clear that the bdellovibrio expends a substantial amount of biochemical work to produce a stable growth chamber (Rittenberg & Thomashow 1979).

22.2.3 Early Damage to the Substrate Cell

Biochemical processes initiated by the bdellovibrio upon attachment or during the entry process largely or completely destroy the attacked substrate cell as a functioning metabolic unit. Within a few minutes after the bdellovibrio has become attached, the substrate cell has lost its potential to synthesise RNA, protein and DNA (Varon *et al.* 1969); its cytoplasmic membrane becomes permeable to small molecules (Rittenberg & Shilo 1970) that normally require energy-driven mechanisms for transport, and its respiratory mechanisms are completely inactivated (Rittenberg & Shilo 1970). Neither the order of the above-mentioned events nor the biochemical mechanisms responsible for them have been determined. It is clear, nevertheless, that shortly after bdellovibrio attack the substrate cell is a non-viable unit incapable of energy generation or biosynthetic activity.

A regulated degradation of nucleic acids of the substrate cell is also initiated early in bdellovibrio development (Matin & Rittenberg 1972; Hespell *et al.* 1975). DNA breakdown is a consequence of the sequential activity of a series of DNAases synthesised by the bdellovibrio (Rosson & Rittenberg 1979). By about 60 min into a normal development cycle the substrate cell DNA has been almost completely converted to intermediate-size fragments (mol. wt. $=5\times10^5$ dalton fragments for *Escherichia coli* DNA) that are retained within the bdelloplast with less than 5% of the DNA escaping from the bdelloplast chamber as low molecular weight fragments. The initial DNA degradation pattern is very similar to that occurring during phage T4 attack on *E. coli* (Kutter & Wiberg 1968; Warren & Bose 1968). The pattern of RNA breakdown, although somewhat slower, is similar to the DNA pattern. By 90 min into a typical growth cycle, substrate cell ribosomes have disappeared and 23S and 16S ribosomal RNA have been replaced by fragments of mol. wt. 1×10^4 to 1×10^5 daltons (Hespell *et al.* 1975). The enzymology of these degradative processes has not yet been investigated.

The degradation of substrate cell macromolecules is catalysed largely or completely by enzymes synthesised by the bdellovibrio without significant involvement of those pre-existing in the substrate cell. This conclusion follows from the finding that bdellovibrio growth on heat-killed *Esherichia coli* proceeds with normal kinetics (Hespell 1978; Rosson 1978). Such cells are almost completely lacking in DNAase activity. Further, when normal *E. coli* cells are used as substrate organisms, pre-existing *E. coli* deoxynucleases are either inhibited or degraded very early during bdellovibrio attack (Rosson & Rittenberg 1979).

22.2.4 Growth

After about 60 min in a 3 to 4 h cycle, growth *per se* of the bdellovibrio is first detectable in a variety of ways. The respiration rate and activities of tricarboxylic acid cycle enzymes of cultures (Hespell 1976) start to increase at about this time and materials derived from substrate cell macromolecules are incorporated into bdellovibrio DNA (Matin & Rittenberg 1972) and RNA (Hespell *et al.* 1975). The energy required for biosynthesis is obtained by the bdellovibrio from its electron transport system (Simpson & Robinson 1968) fuelled by the oxidation of amino acids (Hespell *et al.* 1973), fatty acids (Kuenen & Rittenberg 1975) and the ribose moiety of ribonucleotide monophosphates (Hespell & Odelson 1978), all produced from the macromolecules of the substrate cell.

The macromolecules of the substrate cell also provide in large part the monomeric units for the biosynthesis of the corresponding polymers of the bdellovibrio. Thus, bdellovibrio DNA (Matin & Rittenberg 1972), RNA (Hespell *et al.* 1975), lipids (Kuenen & Rittenberg 1975) and proteins (Rittenberg & Langley, cited in Rittenberg & Thomashow 1979) are synthesised almost completely from unaltered or slightly altered monomeric units pre-existing in the corresponding pools of the substrate cell. Two exceptions to this statement are known: the peptidoglycan components of the substrate cell are not used for synthesis of the corresponding bdellovibrio polymer (Thomashow & Rittenberg 1978a); and only the lipid A portion of the substrate cell's LPS appears to be incorporated into the bdellovibrio's LPS (Nelson 1979).

Although used only to a limited extent during intraperiplasmic growth, the bdellovibrios have the genetic potential for alteration or biosynthesis *de novo* of a range of monomeric units. For example, during growth on substrate cells of normal composition there is some synthesis of DNA from materials in the substrate cell's RNA pool (Hespell *et al.* 1975; Rosson 1978). This biosynthetic potential is amplified when the bdellovibrios are grown on substrate cells having abnormally high protein or RNA to DNA ratios (Pritchard *et al.* 1975). Likewise, although the fatty acid composition of the bdellovibrio closely mimics that of the substrate cell on which it grew, there is evidence that chain shortening and hydrogenation of pre-existing fatty acids as well as synthesis *de novo* from acetate occurs (Kuenen & Rittenberg 1975).

Growth leads to the formation of a bdellovibrio filament whose ultimate length is directly proportional to substrate cell size (Kessel & Shilo 1976). At the end of growth the filament fragments into unit-size cells and flagella are formed. A lytic enzyme(s) is then produced which solubilises the bdelloplast wall releasing the motile progeny, thus completing the growth cycle.

The growth process takes place with a very high overall efficiency. Based on the relation of substrate cell carbon assimilated to that respired to carbon dioxide, a Y_{ATP} (g dry wt cell material formed per mole of ATP produced) of 20 to 25 has been calculated (Rittenberg & Hespell 1975). This exceptionally high efficiency of ATP utilisation has been attributed to a combination of factors unique to the intraperiplasmic growth process: the availability of a complete and balanced nutrient supply; the conservation of pre-existing monomeric units and high energy phosphate bonds; the closely regulated breakdown of substrate cell polymers to provide these monomers at a rate equivalent to the biosynthetic demand; and the close coupling of energy generation and utilisation.

Experimental

22.3 Isolation, enumeration and maintenance of cultures

22.3.1 Isolation

Plating on lawns of suitable substrate cells for plaque development is the basic procedure for isolation of bdellovibrios from nature. The plating methods mimic phage procedures. Water, sewage or soil suspensions appropriately diluted are mixed in soft agar with a suspension of substrate cells and the mixture is overlaid on a layer of non-nutrient agar in a Petri dish and incubated. The choice of substrate organism for the plating procedure will vary with the specific attributes desired in the isolated bdellovibrio. The overlay medium may be one that supports growth of the substrate organism used or may be a non-nutrient gel if an adequate number of pregrown substrate cells are added. The latter procedure is preferable since it minimises the possibility of phage development and growth of unwanted bacteria added with the inoculum. In either case, plates should be examined after overnight incubation and then daily for 4 to 7 d. Bdellovibrio plaques, in contrast to phage plaques, develop late and increase in size with continued incubation. To insure isolation of pure clones, material from single plaques should be transferred serially through several platings.

Since the numbers of bdellovibrios detected in most environments examined to date have been small, isolation by plaque formation may be preceded by concentration or enrichment procedures. To concentrate bdellovibrios from suspensions of natural materials, slow speed centrifugation to remove rapidly sedimentable particles from water, sewage or soil suspensions, followed by collection of the finer particles, including bdellovibrios, by sedimentation or filtration has been used (Starr & Stolp 1976; Varon & Shilo 1979). To obtain enrichments of a bdellovibrio population, a suspension of substrate cells in buffer is inoculated with the material under investigation. The culture is incubated until clearing is evident and then examined microscopically. If bdellovibrios are present, the culture is plated at optimum dilution for plaque development.

Depending on the nature of the inocula and to a lesser extent on the organism used as the substrate cell, the presence of bacteriophage, protozoa or myxobacterial-like organisms may complicate enrichment attempts. Phage and protozoa are most likely to appear when polluted waters are used as inocula and frequently overgrow the bdellovibrios. The unavoidable addition of nutrients with the inoculum as well as utilisation of nutrients released from lysing cells may permit sufficient growth of the substrate organism to support a massive phage development. Phage development can be avoided by using suspensions of non-viable cells as the substrate for bdellovibrio growth (see below). In theory it should be possible to control the development of protozoa by the addition of selective inhibitors such as cycloheximide to the enrichment culture, but in practice this seldom works (Staples & Fry 1973). We have observed development of myxobacteria in enrichment cultures only when using polluted sea-water samples as the inocula. They usually are in the minority if bdellovibrios are present and do not complicate the isolation of the latter organism unduly.

22.3.2 Enumeration

The basic procedure for counting isolated bdellovibrios in two-membered cultures or in purified suspensions is the plaque assay, although direct microscopic counting (Snellen *et al.* 1978) and Coulter counting (Patinkin 1975) can also be used. The plaque assay procedure is generally less tedious and more accurate than other methods. It presents no special problems if the obvious variables are controlled, e.g. temperature, agar concentration in the two layers and choice and concentration of substrate organisms.

Although the counting of bdellovibrios in pure culture is a routine matter, their enumeration in natural materials presents all the difficulties encountered in dealing with any group of bacteria unevenly distributed in a heterogeneous milieu containing mixed populations as well as difficulties peculiar to the bdellovibrios (Staples & Fry 1973; Starr & Stolp 1976; Varon & Shilo 1979). Enumeration experiments have usually involved detecting viable bdellovibrios by plaque assay, sometimes preceded by a variety of pretreatments of the samples including differential centrifugation or filtration to remove interfering organisms and homogenisation to disperse the bdellovibrios. Less frequently, enumeration has been by variations of the most-probable-number procedure. The use of either type of method carries with it all the problems of overgrowth by other bacteria and competition by phages or protozoa already mentioned in discussing isolation procedures. Pretreatments of samples by any means introduces the problem of loss of bdellovibrios. Over and above these types of difficulties, either type of viable counting method is flawed as a truly quantitative procedure because of strain differences in the bacterial species attacked by the bdellovibrios. For example, in the examination of the bdellovibrio populations in sewage samples, average counts using *Serratia marcescens*, *Escherichia coli* or *Achromobacter* spp. as substrate cells were in the ratio of 3:144:900 bdellovibrio (Staples & Fry 1973). With the procedures available there is no way of deciding what the relation is between the highest value obtained with any substrate cell (or the sum of the values with all substrate cells) and the actual populations *in situ*.

A possible alternative approach to enumeration not yet investigated would be direct counting by the fluorescent antibody technique since many bdellovibrio strains share common antigens (Kramer & Westergaard 1977). Whether one could concentrate low density samples (e.g. unpolluted sea-water—Taylor *et al.* 1974) sufficiently to make direct counting possible cannot be decided *a priori*. Additionally, this type of approach would have unique difficulties since it would be expected that in any environment in which bdellovibrios are growing, a certain proportion of them would be within the bdelloplasts and thus inaccessible to antibodies.

22.3.3 Routine Growth Procedures

To grow stock cultures or to obtain cells for experimental purposes, a bdellovibrio culture is usually grown on a developing substrate cell culture. A medium that will support growth of the latter is inoculated with a relatively large number of substrate cells and a much smaller number of bdellovibrios. Upon incubation the substrate cells initially outgrow the bdellovibrios and are in excess until late in culture development. To obtain maximum bdellovibrio yields from this type of culture, the input ratio of bdellovibrios to substrate cells should be small enough for the latter to exhaust available nutrients before all are attacked by the bdellovibrios. The slower-growing the substrate organism, the smaller the input ratio should be. By way of example, DNB medium (Shilo & Bruff 1965), a dilute nutrient broth inoculated at an input ratio of 1×10^{-3} to 1×10^{-4} with bdellovibrios and *Escherichia coli* (1 to 2×10^{8} ml^{-1} final concentration) and incubated overnight, yields a fully developed bdellovibrio culture containing in the vicinity of 2×10^{9} bdellovibrio and fewer than 100 viable *E. coli* ml^{-1}. Since a few substrate cells apparently always escape destruction, even in a concentrated bdellovibrio population, very small inocula should be used in shifting a bdellovibrio culture from one type of substrate cell to another to dilute these survivors to extinction or else plaques should be developed with the new substrate cell and picked.

Two general precautions must be observed in relation to the culturing method described above.

(1) The growth medium must not contain components that inhibit the motility of the bdellovibrio strain employed. For certain strains of bdellovibrio these may be as innocuous as NaCl (0.15M) or phosphate (50 mM) (Varon & Shilo 1968).

(2) Media of usual nutrient concentrations

should not be employed. Undiluted nutrient broth inoculated with bdellovibrios and *Escherichia coli* in the ratio mentioned above rarely yields good bdellovibrio cultures. The failure probably results from the accumulation of toxic wastes released by the rapidly growing substrate cells rather than from the high numbers of substrate cells obtainable since one can use over 1×10^{10} washed substrate cells ml^{-1} in buffer or in DNB to generate large bdellovibrio populations (Thomashow 1979).

Pregrown substrate cells may also be used for routine growth of bdellovibrio cultures. This approach is particularly convenient if one has large-scale culture facilities available. Cells harvested from such cultures can be stored frozen for long periods without any decrease in their suitability as substrates for bdellovibrio growth (M. Varon personal communication).

22.3.4 Preparation and Standardisation of Bdellovibrio Suspensions

Bdellovibrio cultures, in contrast to the typical bacterial culture, contain not only the desired cells and soluble compounds but also particulate debris and unlysed substrate cells. For many purposes, a slow-speed centrifugation (1500 ***g***, 3 to 5 min) of the culture, which removes much of the larger debris and most intact substrate cells, followed by a higher-speed centrifugation (10 000 ***g*** for 20 min) and one or two washings of the sediment gives a suspension of adequate clarity. Suspensions can be passed through filters of appropriate pore size or banded in Ficoll gradients (Varon & Shilo 1970) to insure a more complete removal of intact substrate cells. For some experimental purposes, special procedures (see p. 388) may be needed to demonstrate that contaminating materials carried over, along with harvested bdellovibrios, are reduced to an acceptable level.

Depending on the purposes at hand, the final suspensions may be prepared in complete medium, in dilute buffer or in distilled water. In the last two situations, Ca^{2+} (1 mM) and Mg^{2+} (0.1 mM) should be added to the solution. These ions appear necessary for effective attachment of the bdellovibrio to its substrate cell (Seidler & Starr 1969a) and their carryover in the washed bdellovibrio and substrate cell suspensions may not provide an adequate concentration.

Standardisation of final suspensions is most conveniently achieved by turbidity measurements that are compared to standard curves relating turbidity to viable counts by plaque assay or to direct counts. This procedure may give grossly inaccurate results if bdellovibrios are harvested from old cultures or if suspensions are held at physiological temperatures for any extended period of time. Bdellovibrios have a very high endogenous respiration rate and upon starvation rapidly lose viability and cell carbon, and change in optical properties (Hespell *et al.* 1974). To obtain the most accurate indirect estimate of cell numbers (viable and non-viable) DNA contents of suspensions are determined and related to standard curves. The DNA content of bdellovibrios grown on a particular type of substrate cell is fairly constant and DNA is conserved under starvation conditions.

22.3.5 Maintenance and Preservation of Stock Cultures

In general, bdellovibrio cultures held at either incubation or room temperature after growth is complete lose viability very rapidly (Burger *et al.* 1968; Varon & Shilo 1968). Washed suspensions in solutions containing an energy source survive much longer under these conditions (Hespell *et al.* 1974). Refrigeration of either type of preparation prolongs culture viability and one can use cultures stored for several weeks at 4°C as working stocks. The longer the storage period, the more sluggish the remaining viable bdellovibrios become and the more variable the experimental results when such cells are used as inocula. For experiments requiring synchronous cultures it is essential that stored cultures are revitalised by several serial transfers and that highly motile bdellovibrios freshly released from the bdelloplasts (nascent cells) are employed.

Long-term preservation of stocks is achieved by standard procedures (Starr & Stolp 1976). Our laboratory maintains its stocks either as lyophiles containing skimmed milk or in liquid suspensions containing 15% (w/v) glycerol at −70°C.

22.4 Study of intraperiplasmic growth

22.4.1 Kinetics of Attack—Attachment and Penetration

The kinetics of attachment under laboratory conditions have been investigated by using pregrown radioactively labelled bdellovibrios (see p. 385) as the experimental organisms (Varon & Shilo 1968). The labelled bdellovibrios were mixed with unlabelled substrate cells to start an experiment. Samples of the mixture were removed at intervals, passed through a filter of appropriate size (1.2 μm for *Escherichia coli*) and the retained radioactivity determined. Unattached bdellovibrios passed through the filter while attached bdellovibrios were retained along with the substrate cells.

The kinetics of penetration were measured in the same manner as the kinetics of attachment, with one additional step (Varon & Shilo 1968). That is, the time samples removed from the cultures were given a brief treatment in a high speed blender before filtration. Attached bdellovibrios that had not penetrated the substrate cell were sheared off and these, along with the unattached bdellovibrios, passed through the filters while penetrated bdellovibrios and the substrate cells were retained.

An alternative approach for studying the kinetics of bdellovibrio attack was developed by Varon and Ziegler (1978) using a luminous *Photobacterium* sp. as the substrate organism for growth of a marine bdellovibrio. They showed that the light-emitting system of the *Photobacterium* sp. was damaged concomitantly with the penetration of the bdellovibrio. The resulting decrease in light intensity of the culture, which can be measured sensitively and accurately, is thus a measure of the attack rate. The method has the virtue of permitting continuous measurements without disturbing or processing the culture in any way. More important, the sensitivity of the method permits the measurement of substrate cell densities as low as 1×10^3 organisms ml^{-1} or less. Laboratory experiments can, therefore, be conducted at population levels characteristic of the unpolluted marine environment. Since a variety of non-marine luminous bacteria exist (Nealson & Hastings 1979), the technique is probably applicable for studies of freshwater and soil bdellovibrios.

22.4.2 Alternate Approaches in Choice of Media

Two growth environments and hence two different culture systems have been used to study intraperiplasmic development. The first, used almost exclusively in the initial studies, employed complete media in which the bdellovibrios multiplied on substrate cells also growing during the course of the experiments. This general approach followed from the original concept of the bdellovibrio–substrate cell relation as that of an obligate parasite and host (Stopl & Starr 1963). It was assumed intuitively that some vital function(s) of the latter was required for the development of the bdellovibrio, a view strengthened by the failure of all bdellovibrio isolates to grow axenically in complex media. It is now known that mutations conferring the ability to grow axenically occur at a frequency that suggests a single gene is involved (Seidler & Starr 1969b), and it has been suggested that the gene controls a regulatory signal rather than replacing a required metabolic function or nutrient (Horowitz *et al.* 1974). Regardless of the validity of this interpretation, the use of a complete medium introduces the problem of distinguishing which of the two growing organisms (or to what degree) is responsible for the changes being measured. This experimental approach, although adequate for descriptive studies, has contributed little to a precise understanding of the physiology and biochemistry of intraperiplasmic growth *per se*.

The second approach (Guelin *et al.* 1967) has been to grow the bdellovibrios on pregrown substrate cells suspended in a non-nutrient menstruum. It has been shown that the typical bdellovibrio growing intraperiplasmically neither requires continuing metabolic functions of the attacked cell nor profits significantly from an additional exogenous supply of nutrients. Strains may exist for which these two generalisations do not hold, but for all the others the use of pregrown substrate cells suspended in buffer as the bdellovibrio growth medium permits a much clearer interpretation of experimental data and is the approach of choice.

A medium consisting exclusively of pregrown bacterial cells can be varied widely to suit the needs of the experiment. Radioactive or non-radioactive probes can be introduced by culturing wild-type or mutant substrate cells in an appropriate fashion; for example U-[^{14}C]-labelled wild-type *Escherichia coli* cells were used as the substrate to study the efficiency of bdellovibrio growth (Rittenberg & Hespell 1975), and uniformly labelled bdellovibrios obtained by growth on such *E. coli* proved valuable in the study of bdellovibrio respiration (Hespell *et al.* 1974). Further examples include: the use of [^{3}H] or [^{14}C]thymidine-labelled *E. coli* to study DNA metabolism (Matin & Rittenberg 1972); the evaluation of the level of contamination in harvested bdellovibrios (see p. 388); and the use of a double mutant auxotrophic for diaminopimelic acid and incapable of catabolising glucosamine in order to label both the sugar and amino acid portions of the substrate cell peptidoglycan in studies on penetration (Thomashow & Rittenberg 1978a, b, c). As an example of the use of a non-radioactive probe, abnormal fatty acids were introduced into an *E. coli* mutant auxotrophic for unsaturated fatty acids in investigations of the lipid metabolism of the bdellovibrio (Kuenen & Rittenberg 1975).

An obvious advantage of using *Escherichia coli* as the substrate organism is the ease of its manipulation and the large number of mutants available. However, it may be necessary or desirable to use other substrate organisms to obtain a specific change in media composition; for example *Pseudomonas putida* was used when it was desirable for the substrate organism to have a different DNA buoyant density (Matin & Rittenberg 1972) or a different LPS composition (Nelson 1979) than the bdellovibrio strain being investigated.

The quantitative composition of the growth medium can also be altered by manipulating the growth of the substrate cell. Cells with abnormally high DNA to protein ratio obtained by unbalanced growth of *Escherichia coli* in the presence of methotrexate proved valuable in one investigation (Pritchard *et al.* 1975). One can conceive of experiments in which RNA-depleted cells or cells containing carbohydrate, lipid or inorganic inclusions might be useful for studying the intraperiplasmic growth process, although these possibilities have not yet been exploited.

Compounds can also be added to the buffer to serve as probes rather than for nutritional purposes. Thus, the addition of lactose or orthonitrophenylgalactoside to a bdellovibrio culture, growing on a cryptic *Escherichia coli* strain constitutive for β-galactosidase and lacking lactose permease, made it possible to show early membrane damage during bdellovibrio attack (Rittenberg & Shilo 1970). As other examples of this approach, exogenously added radioactive uracil and leucine were used as probes to study decay of the RNA and protein synthesising potentials of attacked substrate cells (Varon *et al.* 1969), and radioactive glutamate was similarly employed in investigating the energy substrates for intraperiplasmic growth (Hespell *et al.* 1974). Antibiotics may also be added to inhibit specific aspects of bdellovibrio metabolism (Varon & Shilo 1968). If the compound added to the buffer can be metabolised by the substrate cells, its addition can be delayed until that metabolic potential has been destroyed by the bdellovibrio attack—typically after about 1 h in a 3 h growth cycle. Also, since intraperiplasmic development can be interrupted by chilling a culture and reinitiated by warming, bdelloplasts can be removed from one environment by centrifugation or filtration and placed into another.

Another important type of variation in the medium is the substitution of non-viable for viable substrate cells. To clearly evaluate the bdellovibrios' role in some specific metabolic process, ultraviolet-irradiated or heat-treated substrate cells may be preferable to viable ones. Bdellovibrio attachment to (Varon & Shilo 1968) and growth on both types of treated cells (Hespell 1978; Rosson 1978) may be very similar to that observed with untreated viable cells if the killing process is properly controlled.

22.4.3 The Single-Cycle Culture

The main application of the single-cycle culture is the study of the kinetics of processes taking place during intraperiplasmic growth. It has been used, for example, to study the breakdown of substrate cell DNA and RNA and for bdellovibrio synthesis of these macromolecules (Matin & Rittenberg 1972; Hespell *et al.* 1975). This type of culture involves only one growth cycle of the input bdellovibrios and is synchronous, providing nascent bdellovibrios and an appropriate substrate organism are used. Samples of such a culture taken at intervals during its development give populations of substrate cells and bdellovibrios at successive stages of breakdown and growth, respectively.

A synchronous, single-cycle culture is achieved by the simple expedient of mixing buffer suspensions of bdellovibrios and substrate cells at an input ratio of 1.5 to 2.0 bdellovibrios per substrate cell. With active bdellovibrio populations and appropriate concentrations of substrate cells (1 to 5×10^9 for *Escherichia coli*), essentially all of the latter will be converted to the bdelloplast stage (a bdellovibrio-penetrated cell) within 20 min. Since the duration of the intraperiplasmic growth cycle of a typical bdellovibrio growing on an average-sized bacterium, e.g. *E. coli*, is about 3 to 3.5 h, the growing bdellovibrios in such a culture are in synchrony with each other to approximately $\pm$ 10% of the intraperiplasmic growth cycle.

The most important variable needing control to

achieve good synchronous cultures is the physiological state of the bdellovibrio population. Typically, a nascent bdellovibrio undergoes morphological and physiological changes after release from the bdelloplast. The most easily observable changes in at least some strains are elongation and straightening of the cell and a decrease in motility. More subtle changes occur, probably induced by the wastes released by the lysing substrate cells and by the rapid endogenous metabolism of the bdellovibrio as its energy substrates are exhausted. The effects of the ageing process, whatever their nature, are clearly seen in synchronised, multicycle cultures, both in complete medium and in buffer suspensions. In both situations the first cycle of bdellovibrio growth is always significantly longer (50 to 100% greater) than the second and subsequent cycles (Varon & Shilo 1969a). Therefore, in order to get good synchronous cultures young freshly released and freshly harvested cells must be used.

Although the effect of the physiological state of the substrate cell has not been systematically investigated, this variable also has an effect (Varon & Shilo 1969a) and should be controlled in any experimental programme. With *Bdellovibrio bacteriovorus* 109J consistent results have been obtained using late log or very early stationary phase cells as substrate. The other basic variables are the type and concentration of buffer, pH, concentration of added salts, if any, and temperature of incubation. These variables require the same degree of control that should be exercised in any growth studies; the precise choices made of these parameters will vary with the strain of bdellovibrio used and with the experiences of each laboratory.

22.4.4 The Multicycle Culture

In a multicycle culture the input ratio is significantly less than one and the bdellovibrio must go through several cycles of growth before all substrate cells are attacked and consumed. The initial cycles of growth may be fairly synchronous if nascent bdellovibrios are used as inocula. Consequently, this type of culture has been used for the determination of burst size, i.e. the average yield of progeny bdellovibrios per substrate cell (Seidler & Starr 1969a; Varon & Shilo 1969b). For this purpose it makes little difference whether the experiments are done in a complete growth medium, permitting growth of the substrate organism, or in buffer; in either case, to insure rapid attachment of the bdellovibrios the substrate cells must be at a high concentration (Varon & Shilo 1968).

The multicycle culture is the method of choice for all studies in which one is interested in initial and final states, for example for determination of overall growth yields (Rittenberg & Hespell 1975), or for mass balance studies to determine the fate of any specific or general substrate cell component during intraperiplasmic growth (Pritchard *et al.* 1975). Since the input ratio may be as small as desired in a multicycle culture, experiments can be done without having to correct for materials introduced in the bdellovibrio inoculum. This is not true for the single-cycle culture in which the input bdellovibrio (growing on *Escherichia coli*) have a mass of about 15 to 20% of that of the starting substrate cells (input ratio 2:1) and 25% of that of the progeny bdellovibrios. On the other hand, the rapid loss of synchrony in multicycle cultures and the large background of unattached cells renders this technique essentially useless for following the kinetics of intraperiplasmic growth processes.

For the types of investigations mentioned in the preceding paragraph it is essential that the experiments be done in a non-nutrient suspending solution incapable of supporting growth or metabolic activities of the substrate cell. Otherwise, one is confronted with the difficult task of determining how much multiplication of the substrate organism occurred and to what extent extracellular compounds were directly utilised by the bdellovibrio or were first incorporated by the substrate organism. Even using a non-nutrient menstruum, metabolic activities of the unattacked substrate cells have to be considered in interpreting results. In multicycle cultures with very low input ratios the vast majority of substrate cells remain unattacked until the last one or two cycles of bdellovibrio growth (Rittenberg & Hespell 1975). Not only will these cells have an endogenous metabolism and a changing composition, they may also be metabolising materials released by lysis of attacked cells. In experiments in which these phenomena complicate interpretation of results, it might be desirable to use heat or ultraviolet-killed cells as substrates.

The one additional variable to be considered in multicycle cultures not already discussed for single-cycle cultures is the input ratio. If it is desirable to have a specific number of cycles, the input ratio required can be calculated knowing the average burst size for the substrate organism used. If the number of cycles above some minimum is not important, it is easy to calculate an input ratio that permits inoculation of cultures at the end of a normal working day and have freshly completed cultures available at any desired time the next morning.

22.4.5 Estimation and Avoidance of Substrate Cell Debris

In some investigations (e.g. determination of cell yields or incorporation of a particular compound, or class of compounds, from the substrate cell) it may be necessary to reduce contaminating material from substrate cell debris in the harvested bdellovibrios. As already noted, intact substrate cells and unlysed bdelloplasts can be removed by slow speed centrifugation or filtration, and soluble debris by washing. For the smaller particulate debris (largely membraneous) banding of the bdellovibrios on Ficoll or sucrose gradients or their sedimentation through 65% sucrose may achieve the desired cleanliness.

All purification steps including slow speed centrifugation or filtration result in the loss of bdellovibrios which must be taken into account in any balance study. One can determine the amount of loss if the cultures being harvested are fully developed and contain negligible numbers of intact substrate cells or unlysed bdelloplasts. The only sedimentable DNA (10 000 g min^{-1}) in such cultures is that packaged inside the intact bdellovibrios (Matin & Rittenberg 1972; Rosson & Rittenberg 1979). Thus the ratio of the amount of sedimentable DNA (determined either by chemical assays or radioactivity measurements of incorporated U-[^{14}C] or [^{3}H]thymidine) in the purified suspension to that in the culture at the time of harvesting is a measure of the fraction of the total population recovered. One can also measure the effectiveness of a series of purification steps by following the ratio of the suspected impurity to that of DNA in the successive stages of the process (Kuenen & Rittenberg 1975). A decreasing ratio that approaches a constant value indicates that either the contaminating material has been effectively removed or that the terminal purification steps have not improved the cleanliness of the bdellovibrio preparation. For maximum sensitivity in applying this methodology it is helpful if both the impurity of interest and the bdellovibrio DNA are radioactive (e.g. U-[^{14}C]-labelled substrate cells and [^{3}H]thymidine-labelled bdellovibrio DNA).

Some types of contaminating materials, for example lipids, could conceivably be so strongly absorbed to the bdellovibrio cell surface that they behave in a purification process as if they were an integral component of the bdellovibrio. A test has been devised for the presence of such materials and applied to fatty acids derived from the phospholipids (Kuenen & Rittenberg 1975) or lipid A (Nelson 1979) of the substrate cells. The test assumes that an integrated bdellovibrio component, in contrast to an absorbed one, will not be removed by passage of the bdellovibrio through the outer membrane and peptidoglycan layers of the substrate cell during several successive cycles of growth.

22.5 Unsolved problems of methodology

There are at least three major unsolved problems of methodology whose resolution would greatly facilitate and expand our potential for investigating the bdellovibrio group. One of these has already been discussed: the development of additional techniques for enumerating bdellovibrios in natural environments. It is difficult to anticipate much progress in the understanding of bdellovibrio ecology and assessing their role in various environments unless some knowledge is acquired concerning the relation between the numbers obtained by available counting procedures and the actual populations *in situ*.

A second important experimental problem is the development of procedures for freeing the bdellovibrio undamaged from the bdelloplast chamber during the course of its growth. Were such a procedure available, it might be possible to investigate a variety of questions that, at present, can be approached only indirectly, if at all. One could, for example, determine whether bdellovibrio growth initiated intraperiplasmically could be completed in artificial media. This is a possible prediction that follows from the suggestion (Horowitz *et al.* 1974) that the dependence of the bdellovibrio on the substrate cell has a regulatory rather than a nutritional basis.

The author and co-workers have attempted to free the bdellovibrio from the bdelloplast by several procedures, including mechanical breakage of the bdelloplast wall with a French press following the method used in releasing the forespore of *Bacillus* sp. (Andreoli *et al.* 1973), and by digestion of the bdelloplast wall with a purified lytic enzyme produced by the bdellovibrio at the end of its growth cycle (Thomashow & Rittenberg 1978a). Using the French press, apparently intact bdellovibrios have been recovered in small numbers from 30 to 60 min bdelloplasts but not from later in the growth cycle (S. Suehiro, unpublished observations). Our efforts with the enzymic approach have been completely unsuccessful to date. Apparently the lytic enzyme acts only from within the bdelloplast chamber and alteration of the outermost layer of the bdelloplast wall may be necessary before exogenously added lytic enzyme can reach its peptidoglycan target.

The third and most pressing methodological problem for bdellovibrio research is undoubtedly the development of a genetic system that can be exploited experimentally. A most interesting and important facet of bdellovibrio biology is the precise control the bdellovibrio exercises over its growth process. A good understanding of these control mechanisms requires the potential for genetic manipulation of the organism. There has been only one published abstract reporting efforts in this direction (Lania *et al.* 1976) although several laboratories have worked towards this goal without progress. The nature of the intraperiplasmic development cycle makes it unlikely that either conjugation or transformation mechanisms will be functional for wild-type bdellovibrios whose growth occurs only within the periplasm. Development of the former mechanism seems excluded because of the unlikelihood of two bdellovibrios invading the same substrate cell (Rittenberg & Thomashow 1979) and the latter because of the barrier the bdelloplast wall and the degradative enzymes in the periplasmic space would present to exogenous DNA before it reached the growing bdellovibrio. Neither restriction, of course, would exist for those mutant bdellovibrios capable of axenic growth. However, the existence of several types of virulent bdellophages (Varon 1974) suggests the possibility that lysogenic bdellophages also exist and there is some evidence (Sagi & Levisohn 1976; L. Thomashow, unpublished observations) that this is the case. It would appear, therefore, that the development of a transducing system is a possibility.*

Recommended reading

Rittenberg S. C. & Thomashow M. F. (1979) Intraperiplasmic growth—life in a cozy environment. In: *Microbiology 1979* (Ed. D. Schlessinger), pp. 80–6. American Society for Microbiology, Washington, D.C.

Shilo M. (1969) Morphological and physiological aspects of the interaction of *Bdellovibrios* with host bacteria. *Current Topics in Microbiology and Immunology* **50**, 174–204.

Starr M. P. & Seidler R. J. (1971) The bdellovibrios. *Annual Review of Microbiology* **25**, 649–78.

Starr M. P. & Stolp H. (1976) Bdellovibrio methodology. *Methods in Microbiology* **9**, 217–44.

Stolp H. (1973) The bdellovibrios: bacterial parasites of bacteria. *Annual Reviews of Phytopathology* **11**, 53–76.

Thomashow M. F. & Rittenberg S. C. (1979) The intraperiplasmic growth cycle—the life style of the bdellovibrios. In: *Developmental Biology of Prokaryotes* (Ed. J. H. Parish), pp. 115–38. Blackwell Scientific Publications, Oxford.

Varon M. & Shilo M. (1979) Ecology of aquatic bdellovibrios. *Recent Advances in Aquatic Microbiology* **2**, 1–48.

References

Andreoli A. J., Suehiro S., Sakiyama D., Takemoto J., Vivanco E., Lara J. C. & Klute M. C. (1973) Release and recovery of forespores from *Bacillus subtilis*. *Journal of Bacteriology* **115**, 1159–66.

Burger A., Drews G. & Ladwig R. (1968) Wirkskreis und Infektionscyclus eines neuisolierten *Bdellovibrio bacteriovorus* Stammes. *Archiv für Mikrobiologie* **61**, 261–79.

Guelin A., Lepine P. & Lamblin D. (1967) Pouvoir bactéricide des eaux polluées et rôle de *Bdellovibrio bacteriovorus*. *Annales Institut Pasteur* **113**, 660–5.

Hespell R. B. (1976) Glycolytic and tricarboxylic acid cycle enzyme activities during intraperiplasmic growth of *Bdellovibrio bacteriovorus* on *Escherichia coli*. *Journal of Bacteriology* **128**, 677–80.

Hespell R. B. (1978) Intraperiplasmic growth of *Bdellovibrio bacteriovorus* 109J on heat-treated *Escherichia coli*. *Journal of Bacteriology* **133**, 1156–62.

Hespell R. B., Miozzari G. F. & Rittenberg S. C. (1975) Ribonucleic acid destruction and synthesis

*We have recently found (W. Cover, S. Suehiro & E. Martin, unpublished observations) that a brief EDTA treatment of the bdelloplast makes the altered peptidoglycan accessible to the lytic enzyme. Successive treatment with these two reagents results in the premature release of viable bdellovibrios at any desired stage in their development (E. Ruby, unpublished observations).

during intraperiplasmic growth of *Bdellovibrio bacteriovorus. Journal of Bacteriology* **123**, 481–91.

HESPELL R. B. & ODELSON D. A. (1978) Metabolism of RNA-ribose by *Bdellovibrio bacteriovorus* during intraperiplasmic growth on *Escherichia coli. Journal of Bacteriology* **136**, 936–46.

HESPELL R. B., ROSSON R. A., THOMASHOW M. F. & RITTENBERG S. C. (1973) Respiration of *Bdellovibrio bacteriovorus* strain 109J and its energy substrates for intraperiplasmic growth. *Journal of Bacteriology* **113**, 1280–8.

HESPELL R. B., THOMASHOW M. F. & RITTENBERG S. C. (1974) Changes in cell composition and viability of *Bdellovibrio bacteriovorus* during starvation. *Archives of Microbiology* **97**, 313–27.

HOROWITZ A. T., KESSEL M. & SHILO M. (1974) Growth cycle of predacious bdellovibrios in a host-free extract system and some properties of the host extract. *Journal of Bacteriology* **117**, 270–82.

KESSEL M. & SHILO M. (1976) Relationship of bdellovibrio elongation and fission to host cell size. *Journal of Bacteriology* **128**, 477–80.

KRAMER T. T. & WESTERGAARD J. B. (1977) Antigenicity of bdellovibrios. *Applied Environmental Microbiology* **33**, 967–70.

KUENEN J. G. & RITTENBERG S. C. (1975) Incorporation of long-chain fatty acids of the substrate organism by *Bdellovibrio bacteriovorus* during intraperiplasmic growth. *Journal of Bacteriology* **121**, 1145–57.

KUTTER E. M. & WIBERG J. S. (1968) Degradation of cytosine-containing bacterial and bacteriophage DNA after infection of *Escherichia coli* B with bacteriophage T4D wild type and with mutants defective in genes 46, 47, and 56. *Journal of Molecular Biology* **38**, 395–411.

LANIA L., COCCHIARASO V., SANTONASTASO V. & GUERRINI E. F. (1976) Trasformazione genetica in *Bdellovibrio bacteriovorus. Atti Associazione Genetica Italiana* **21**, 74–6.

MATIN A. & RITTENBERG S. C. (1972) Kinetics of deoxyribonucleic acid destruction and synthesis during growth of *Bdellovibrio bacteriovorus* strain 109D on *Pseudomonas putida* and *Escherichia coli. Journal of Bacteriology* **111**, 664–73.

NEALSON K. H. & HASTINGS J. W. (1979) Bacterial bioluminescence: its control and ecological significance. *Microbiological Reviews* **43**, 496–518.

NELSON D. R. (1979) Incorporation of substrate cell derived lipid A components into the lipopolysaccharide of intraperiplasmically grown *Bdellovibrio bacteriovorus* 109J. Ph.D. thesis, University of California, Los Angeles, California.

PATINKIN D. (1975) Sizing of bdellovibrio during growth. *Journal of Bacteriology* **124**, 564–6.

PRITCHARD M. A., LANGLEY D. & RITTENBERG S. C. (1975) Effects of methotrexate on intraperiplasmic and axenic growth of *Bdellovibrio bacteriovorus. Journal of Bacteriology* **121**, 1131–6.

RITTENBERG S. C. & HESPELL R. B. (1975) Energy efficiency of intraperiplasmic growth of *Bdellovibrio bacteriovorus. Journal of Bacteriology* **121**, 1158–65.

RITTENBERG S. C. & SHILO M. (1970) Early host damage in the infection cycle of *Bdellovibrio bacteriovorus. Journal of Bacteriology* **102**, 149–60.

RITTENBERG S. C. & THOMASHOW M. F. (1979) Intraperiplasmic growth—life in a cozy environment. In: *Microbiology 1979* (Ed. D. Schlessinger), pp. 80–6. American Society for Microbiology, Washington, D.C.

ROSSON R. A. (1978) Deoxyribonucleic acid metabolism during intraperiplasmic growth of *Bdellovibrio bacteriovorus* strain 109J on *Escherichia coli*: Kinetics and enzymes of deoxyribonucleic acid degradation and synthesis. Ph.D. thesis, University of California, Los Angeles, California.

ROSSON R. A. & RITTENBERG S. C. (1979) Regulated breakdown of *Escherichia coli* deoxyribonucleic acid during intraperiplasmic growth of *Bdellovibrio bacteriovorus. Journal of Bacteriology* **140**, 620–33.

SAGI B. & LEVISOHN R. (1976) Isolation of temperate bacteriophage for *Bdellovibrio bacteriovorus. Israel Journal of Medical Sciences* **12**, 709–10.

SEIDLER R. J. & STARR M. P. (1969a) Factors affecting the intracellular growth of *Bdellovibrio bacteriovorus* developing within *E. coli. Journal of Bacteriology* **97**, 912–23.

SEIDLER R. J. & STARR M. P. (1969b) Isolation and characterization of host-independent bdellovibrios. *Journal of Bacteriology* **100**, 769–85.

SHILO M. (1969) Morphological and physiological aspects of the interaction of bdellovibrios with host bacteria. *Current Topics in Microbiology and Immunology* **50**, 174–204.

SHILO M. & BRUFF B. (1965) Lysis of Gram-negative bacteria by host-dependent ectoparasitic *Bdellovibrio bacteriovorus* isolates. *Journal of General Microbiology* **40**, 317–28.

SIMPSON F. J. & ROBINSON J. (1968) Some energy-producing systems in *Bdellovibrio bacteriovorus* strain 6–5 S. *Canadian Journal of Biochemistry* **46**, 865–73.

SNELLEN J. E., MARR A. G. & STARR M. P. (1978) A membrane filter direct count technique for enumerating bdellovibrio. *Current Microbiology* **1**, 117–22.

STAPLES D. G. & FRY J. C. (1973) Factors which influence the enumeration of *Bdellovibrio bacteriovorus* in sewage and river water. *Journal of Applied Bacteriology* **36**, 1–11.

STARR M. P. & SEIDLER R. J. (1971) The bdellovibrios. *Annual Review of Microbiology* **25**, 649–78.

STARR M. P. & STOLP H. (1976) Bdellovibrio methodology. *Methods in Microbiology* **9**, 217–44.

STOLP H. (1967) *Bdellovibrio bacteriovorus* (Pseudomonadaceae). Parasitischer Befall und Lysis von *Spirillum serpens.* Encyclopaedia Cinematographica Film E. 1314, Institut für den wissenschaftlichen Film, Göttingen.

STOLP H. (1973) The bdellovibrios: bacterial parasites of bacteria. *Annual Reviews of Phytopathology* **11**, 53–76.

STOLP H. & STARR M. P. (1963) *Bdellovibrio bacteriovorus* gen. et sp. n., a predatory ectoparasitic and bacteriolytic microorganism. *Antonie van Leeuwenhoek* **29**, 217–48.

STRALEY S. C. & CONTI S. F. (1977) Chemotaxis by *Bdellovibrio bacteriovorus* toward prey. *Journal of Bacteriology* **132**, 628–40.

TAYLOR V. I., BAUMANN P., REICHELT J. L. & ALLEN R. D. (1974) Isolation enumeration, and host range of marine bdellovibrios. *Archives of Microbiology* **98**, 101–14.

THOMASHOW L. S. (1979) Isolation, ultrastructure and composition of sheathed flagella from *Bdellovibrio bacteriovorus*. Ph.D. thesis, University of California, Los Angeles, California.

THOMASHOW M. F. & RITTENBERG S. C. (1978a) Intraperiplasmic growth of *Bdellovibrio bacteriovorus* 109J: solubilization of *Escherichia coli* peptidoglycan. *Journal of Bacteriology* **135**, 998–1007.

THOMASHOW M. F. & RITTENBERG S. C. (1978b) Intraperiplasmic growth of *Bdellovibrio bacteriovorus* 109J: N-deacetylation of *Escherichia coli* peptidoglycan amino sugars. *Journal of Bacteriology* **135**, 1008–14.

THOMASHOW M. F. & RITTENBERG S. C. (1978c) Intraperiplasmic growth of *Bdellovibrio bacteriovorus* 109J: attachment of long-chain fatty acids to *Escherichia coli* peptidoglycan. *Journal of Bacteriology* **135**, 1015–23.

THOMASHOW M. F. & RITTENBERG S. C. (1979) The intraperiplasmic growth cycle—the life style of the bdellovibrios. In: *Developmental Biology of Prokaryotes* (Ed. J. H. Parish), pp. 115–38. Blackwell Scientific Publications, Oxford.

TUDOR J. J. & CONTI S. F. (1977) Characterization of bdellocysts of *Bdellovibrio* species. *Journal of Bacteriology* **131**, 314–22.

VARON M. (1974) The bdellophage three-membered parasite system. *CRC Critical Reviews in Microbiology* **4**, 221–41.

VARON M., DRUCKER I. & SHILO M. (1969) Early effects of bdellovibrio infection on the synthesis of protein and RNA of host bacteria. *Biochemical and Biophysical Research Communications* **37**, 518–24.

VARON M. & SHILO M. (1968) Interaction of *Bdellovibrio bacteriovorus* and host bacteria. I. Kinetic studies of attachment and invasion of *E. coli* B by *Bdellovibrio bacteriovorus* 109. *Journal of Bacteriology* **95**, 744–53.

VARON M. & SHILO M. (1969a) Attachment of *Bdellovibrio bacteriovorus* to cell wall mutants of *Salmonella* and *Escherichia coli*. *Journal of Bacteriology* **97**, 977–9.

VARON M. & SHILO M. (1969b) Interaction of *Bdellovibrio bacteriovorus* and host bacteria. II. Intracellular growth and development of *Bdellovibrio bacteriovorus* strain 109 in liquid cultures. *Journal of Bacteriology* **99**, 136–41.

VARON M. & SHILO M. (1970) Methods for separation of bdellovibrio from mixed bacterial population by filtration through millipore filters or by gradient differential centrifugation. *Revue Internationale d'Oceanographie Medicale* **8–9**, 145–52.

VARON M. & SHILO M. (1979) Ecology of aquatic bdellovibrios. *Recent Advances in Aquatic Microbiology* **2**, 1–48.

VARON M. & ZIEGLER B. P. (1978) Bacterial predator–prey interaction at low prey density. *Applied and Environmental Microbiology* **36**, 11–17.

WARREN R. J. & BOSE S. K. (1968) Bacteriophage-induced inhibition of host functions. I. Degradation of *Escherichia coli* deoxyribonucleic acid after T4 infection. *Journal of Virology* **2**, 327–34.

Part 4
Microbe–Plant Interactions

Chapter 23 · The Rhizosphere

James M. Lynch

23.1 Introduction

23.1.1 Anatomy of the Rhizosphere

Hiltner (1904) originally defined the rhizosphere as the zone of bacterial growth which was stimulated around the roots of legumes by the release of nitrogen compounds by the nodules. We now extend the definition to the zone of soil in which microbial activity is affected by the roots of any plant species which may release organic substrates. The rhizoplane refers to the zone of attachment of microorganisms to the root surface itself. However, microbial colonisation across the regions of the soil–root interface (rhizosphere to rhizoplane to epidermis/cortex) is often difficult to distinguish and it seems more reasonable to consider it as a continuum (Darbyshire & Greaves 1973; Old & Nicolson 1978). Balandreau and Knowles (1978) have recently termed the epidermis/cortex region the endorhizosphere. Fig. 1 illustrates the various regions of the root that influence microbes. Colonisation is minimal at the region around the root cap, probably because microbes need time to colonise this youngest part of the root, but it is greatest in the zone of elongation behind the root tip (Schippers & van Vuurde 1978).

The microbiology of the rhizosphere cannot be fully appreciated without an understanding of root anatomy and physiology in relation to soil conditions; some excellent texts on this subject are available (Nye & Tinker 1977; Russell 1977; Bowen 1980).

23.1.2 Substrates for Microbes

A wide range of organic materials are released by roots, including carbohydrates, amino acids, organic acids, enzymes and vitamins (Table 1), but it is probable that all constituents of cells occur in the rhizosphere.

Several regions of the root produce potential substrates for microorganisms and collectively these have been referred to in the past as root exudates. However it is now possible to be more specific about the terminology (Rovira *et al.* 1979). Exudates are compounds of low molecular weight which leak from root cells. Secretions vary in molecular weight and are actively pumped from the roots by metabolic processes. Lysates are

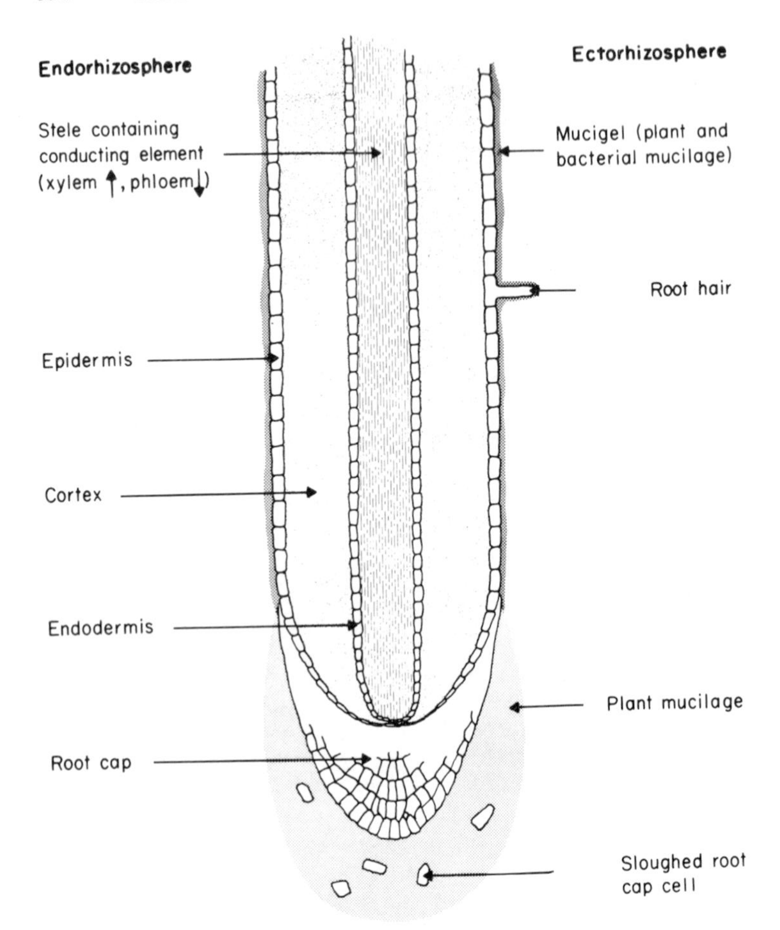

Figure 1. Sites for microbial colonisation of plant roots.

released from roots during their autolysis. Plant mucilages originate from the root cap (secreted by Golgi bodies), the primary cell wall between epidermal and sloughed root cap cells, and epidermal cells (including root hairs). Microbial mucilages are released by rhizosphere microorganisms. Collectively plant and microbial mucilages, microbial cells and their products together with associated organic and mineral matter are referred to as mucigel.

The microbial population in and around the roots includes bacteria, fungi, yeasts and protozoa. Some are free-living, others are symbiotic (notably the root nodule bacteria and mycorrhizal fungi) and some are parasitic. This review is concerned mainly with the free-living organisms and the others are discussed elsewhere in this book (see Chapters 25 to 27). Many investigators have shown that fungi are the dominant contributors to soil biomass, but Vancura and Kunc (1977) have demonstrated that bacteria can dominate the rhizosphere by selectively inhibiting bacteria or fungi with antibiotics (Anderson & Domsch 1978). There have been claims that particular genera, such as *Pseudomonas*, can predominate in the rhizosphere of some soils (Rouatt & Katznelson 1961).

The rhizosphere population could be regarded as a stable community around a particular plant species in a specific soil or, alternatively, as an unstable succession of populations. However, there is little experimental evidence to support either option although Newman *et al.* (1979) have demonstrated that one plant species can influence the abundance of root surface bacteria and mycorrhizae of another.

Table 1. Organic compounds released by plant roots (after Rovira 1965).

Sugars	arabinose, fructose, glucose, maltose, raffinose, rhamnose, sucrose, xylose
Amino acids	α-alanine, β-alanine, γ-aminobutyric acid, arginine, asparagine, aspartic acid, cysteine/cystine, glutamic acid, glutamine, glycine, homoserine, leucine/*iso*-leucine, lysine, methionine, phenylalanine, proline, serine, threonine, tryptophan, tyrosine, valine
Vitamins	*p*-aminobenzoic acid, biotin, choline, inositol, *n*-methylnicotinic acid, niacin, pantothenate, pyridoxine, thiamin
Organic acids	acetic, butyric, citric, fumaric, glycolic, malic, propionic, oxalic, succinic, tartaric, valeric
Nucleotides	adenine, guanine, uridine/cytidine
Enzymes	amylase, invertase, phosphatases, polygalacturonase, proteases
Miscellaneous	auxins, flavonone, glycosides, saponin, scopolotin (6-methoxy 7-hydroxycoumarin)

23.1.3 Significance to Plants

It is normal in plant physiological studies to work with plants exposed to casual laboratory microorganisms, which are highly variable between different laboratories, and to ignore any effect of the microbes on the plant. There are many aspects of the plant's physiology which could be affected by microbes and effects on absorption of ions have been studied in greatest detail (Barber 1978). Under different conditions uptake can be promoted or inhibited by microbes which can also modify the associated metabolic reactions, such as the incorporation of inorganic phosphate into nucleic acids. Interpretations of studies of the sensitivity of plants to metabolic inhibitors may also be difficult if microbes are also sensitive leading to changes in their population around the root. Microorganisms may also produce plant growth regulators (Brown 1974; Lynch 1976) although it is not known whether they can produce these in sufficient concentration to affect the plants.

The rhizosphere microflora may be particularly critical in effecting the breakdown of organic matter and basic minerals and in the cycling of nutrients. It may reduce or promote the uptake of nutrients but there have been very few studies of this type with plants growing in soil, with the exception of mycorrhizal studies showing that phosphate uptake is usually promoted by the presence of the fungi (see Chapter 26).

The normal rhizosphere community structure does not include pathogens and may resist even their invasion if it remains stable. Many soil factors such as water potential and plant treatments such as application of foliar sprays modify the community. Until the normal rhizosphere is better understood the significance of such effects remains in doubt.

The manipulation of the rhizosphere microflora by inoculation of seeds with beneficial microorganisms has been attempted for many years (Brown 1974). Early studies in the Soviet Union concerned the use of azotobacterin to stimulate nitrogen fixation and phosphobacterin to promote phosphorus mineralisation (Mishustin & Naumova 1962). More recently inoculation has been concerned with promoting a rhizosphere that is resistant to invasion by pathogens (Brown 1974). While these approaches remain feasible, none has been used on a commercial scale.

Experimental

23.2 Microscopy, autoradiography and root mapping

The visualisation of the rhizosphere is important for an understanding of its function. Microscopic (both light and electron), autoradiographic and agar plate techniques have been employed successfully. Generally most of the studies in soil have been concerned with the rhizoplane microflora and no really satisfactory assessments of the relative contribution of this to the total rhizosphere microflora have been made. In solution culture studies it is possible to assess the total effective rhizosphere provided the microbes on the roots, in solution and on the walls of the culture vessel are counted (Barber & Lynch 1977). If total counts are required, shaking the roots (and culture vessel) with glass beads and a detergent (such as 0.5% (w/v) Manoxol OT) is a useful means of detaching microbes from surfaces. However, this method can reduce viability and for viable counts either ultrasonic removal or, more drastically, treating with a blender are possible methods.

Rovira *et al.* (1974) considered different methods of quantitatively assessing the rhizosphere microflora. They found that an insignificant number (3 to 4%) of bacteria was removed by washing the roots but that the remainder were removed by shaking in distilled water with glass beads for 15 min. When compared with plate counts using tryptic soy agar, the direct counts were greater by a factor of 10. This is a particular problem with plants grown in soil where there is a mixed population of bacteria and hence different parts of the community are selected by different media. With gnotobiotic cultures, where the bacteria are known to grow satisfactorily on a particular medium, this is of less importance. Plate counts are more time consuming and less precise but direct counts need more concentration by the investigator, more subjective judgement and the conversion from numbers to biomass needs assumptions on microbial cell size and the distribution pattern on the root surface. To assess fungal mycelium the line intercept method can be used and by measuring the hyphal diameter, the biovolume and biomass can be found.

23.2.1 Light Microscopy

Although phase contrast and dark field microscopy of unstained preparations are sometimes useful, classical microscopy using phenol-aniline blue stain (Jones & Mollison 1948) provides one of the best means of quantifying bacteria and fungi on roots (Fig. 2). Fluorescence microscopy using ultraviolet light has become available more recently and sometimes distinguishes living and dead cells, although this can be difficult to interpret, especially with acridine orange as the stain. The lack of transmitted light usually means that relatively long time exposures are needed for photography.

The procedures for staining are outlined in Table 2 but the precise methods vary depending on the microscope used and the investigators involved, who often disagree on the value of different stains (Anderson & Slinger 1975; Jenkinson *et al.* 1976). Recently van Vuurde and Elenbaas (1978) have obtained good contrast between fluorescence of microorganisms and fluorescence of root and soil structures after successively staining roots with the 0.05%(w/v) fluorochromes coriphosphine for 15 min, 0.01%(w/v) Congo red for 10 min and 0.01%(w/v) acridine orange for 2 min.

Table 2. Stains used for soil and rhizosphere microorganisms. nd, not determined.

Stain	Mean ratio of counts in rhizosphere and soil: Anderson & Slinger (1975)	Jenkinson *et al.* (1976)
Phenol-aniline blue: 14 ml 5% (v/v) phenol; 1 ml 6% (w/v) aniline blue; 4 ml glacial acetic acid. Stain, rinse in water. The stain must be filtered before use and the sample washed in water to remove excess stain	0.5	2.3
Acridine orange: 0.2% (w/v) acridine orange. Rinse with 1% (w/v) sodium pyrophosphate. Irradiate with u.v.	nd	1.2
Fluorescein isothiocynate: 1.3 ml 0.5M-carbonate/bicarbonate buffer, pH 7.2; 5.7 ml 0.85% (w/v) physiological saline; 5.3 mg crystalline fluorescein isothyocyanate. Mix, stain at 37°C, rinse in buffer, pH 9.6. Irradiate with u.v.	0.8	2.2
Differential fluorescent stain: Europium chelate (Europium (III) thenoyltrifluoroacetonate), 2 mM and fluorescent brightener [disodium salt of 4,4′-bis(4-anilino 6-bis(2-hydroxy-ethyl)amino S- triazin-2-ylamino) 2,2′-stilbene disulphonic acid], 25 μM, in 50% ethanol. Stain. Irradiate with u.v.	1.0	1.0

The preparation of fluorescent antisera is particularly useful in tracing the activity of specific organisms (Schmidt 1973). The microbial cells are injected into rabbits and γ-globulin is separated from the supernatant of blood samples. This is conjugated with fluorescein isothiocyanate and used to stain the microorganism. Further details are given in Parkinson *et al.* (1971). One potentially useful approach to light microscopy which has been little used is to incubate root sections in malate-phosphate buffer containing 2,3,5-triphenyltetrazolium chloride medium overnight. This shows metabolically active bacteria in and between the cortical cells and in the stele as red

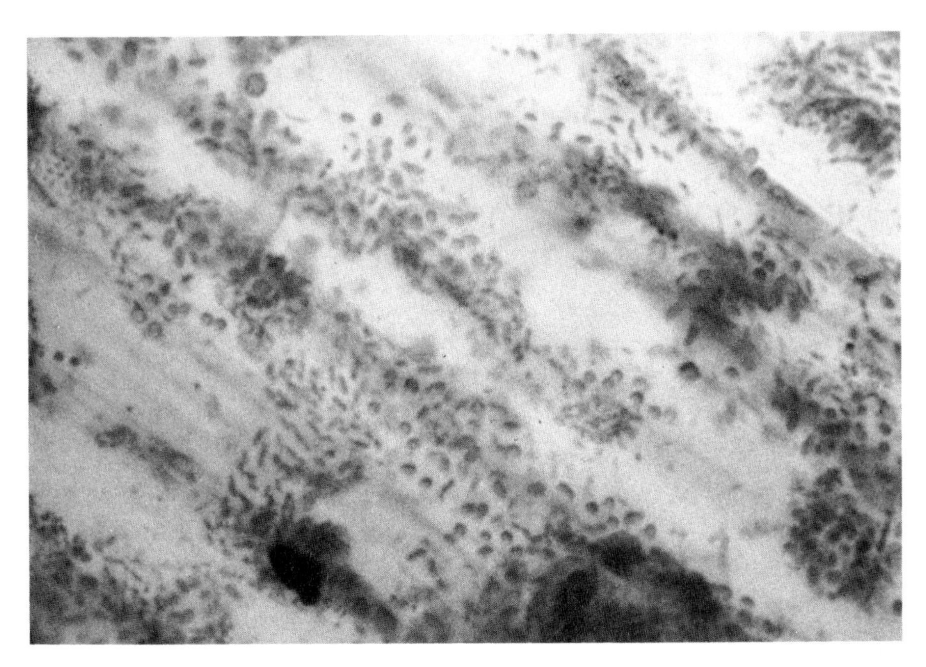

Figure 2. Microbial colonisation of plant roots stained with phenol-aniline blue (× 1800).

pigments, sometimes associated with crystals (Patriquin & Dobereiner 1978). Wada *et al.* (1978) used nitrotetrazolium blue to show microorganisms on plant surfaces which yields blue crystals at the site of metabolically active cells. Such techniques might be used profitably in conjunction with fluorescence methods.

23.2.2 Electron Microscopy

Many electron micrographs of the rhizosphere have been published (Fig. 3). Transmission electron microscopy (TEM) has shown the continuum nature of the rhizosphere referred to above (Foster & Rovira 1978). It has also been valuable in demonstrating the thickness (1 to 10 μm) of the mucigel layer (Bowen & Rovira 1976) and the position of mineral particles in the rhizosphere. A particular problem with TEM is that thin sections have to be prepared showing only a small area of root. By contrast, scanning electron microscopy (SEM) provides less detail but greater areas of the root can be seen (Fig. 4). Besides being of limited value for quantitative studies, a potential drawback of both TEM and SEM is that sample preparation may modify the arrangement of microorganisms in the rhizosphere; for example, by using a simple dehydration procedure (freeze-drying) it has been demonstrated by this SEM method that between the root cap and root hair zone there is a spongy surface network consisting of cavities bounded by aggregates of rhizosphere fibrils which are cross-linked by individual rhizoplane fibrils (Leppard 1975). The holes may provide attachment sites for bacteria as they are of similar size. SEM has also been useful in illustrating the non-uniform distribution of bacteria on the root surface. This supports the quantitative conclusions (total cover *c.* 5 to 10%) of Newman and Bowen (1974) using light microscopy and a statistical technique for the detection of the pattern of colonisation of rhizoplane bacteria, showing the size of areas with different densities of colonisation. Newman and Bowen suggested that the causes of these discrete patterns arose from differences in initial colonisation, ability to spread, the amount of exudate release, variation in soil contact and grooves in the roots providing attachment sites. Neither age nor root diameter appear to be related to the pattern of colonisation.

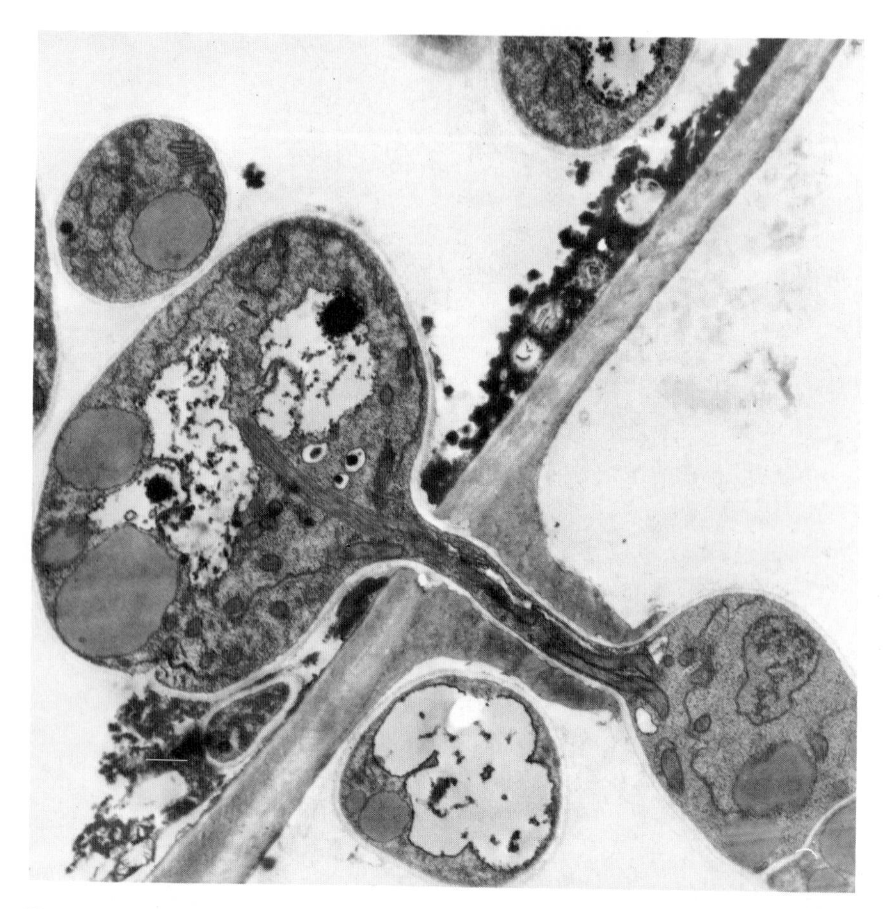

Figure 3. Transmission electron micrograph of the penetration of a cortical cell wall of wheat (*Triticum aestivum*) by the take-all fungus (*Gaeumannomyces graminis*). The cytoplasm is necrotic and a lignotuber has been formed at the point of entry. The hole in the cell wall has been enlarged beyond the size needed for passage of the hypha, suggesting enzymic action (× 12 600—from Faull & Campbell 1979, with permission of J. L. Nicklin and R. Campbell and the National Research Council of Canada.)

23.2.3 Autoradiography

Autoradiography of roots following the uptake of labelled phosphate (Barber *et al.* 1968) has shown that in the presence of microorganisms [^{32}P]phosphate accumulates around the epidermis (Fig. 5b) whereas the distribution of [^{32}P]phosphate is more uniform across the root in sterile conditions (Fig. 5a). The distribution of [^{32}P]phosphate could be readily quantified if the silver grains were counted. Other isotopes, including [^{14}C]-labelled microbes, might be usefully employed in the study of the rhizosphere. This idea has received little exploitation but Warembourg and Billies (1979) pulse-labelled plants with [^{14}C]carbon dioxide and by inoculating axenically the roots with a bacterium producing a polysaccharide of molecular weight 51 000, it was possible to trace bacterial growth by measuring the radioactivity appearing in the polysaccharide following gel filtration.

23.2.4 Root Mapping

A method of following the colonisation of viable cells in the rhizosphere, e.g. after seed inoculation, is to place the roots on a suitable agar medium and either incubate directly or trace the root pattern on the underside of the plate with a felt-tip pen before removing the roots and then incubate the plate

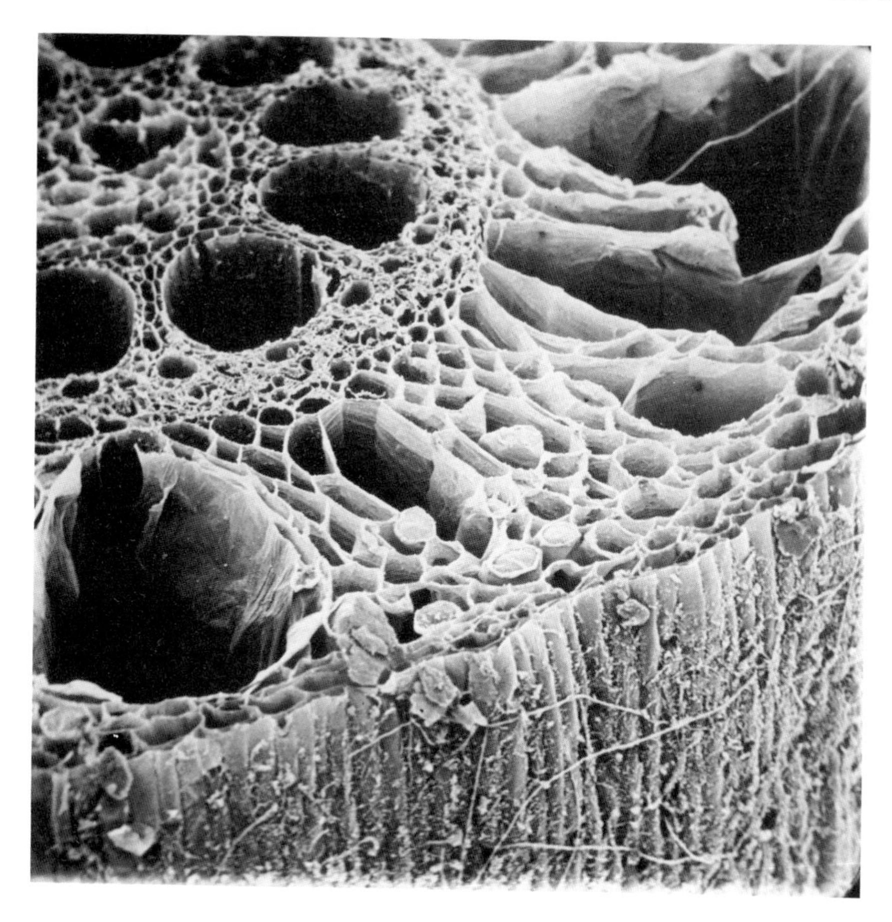

Figure 4. Scanning electron micrograph of maize (*Zea mays*) adventitious root 5 cm from the tip. Note the prominent xylem vessels in the stele at the centre of the root and the presence of air spaces (arenchyma) in the cortex (× 192). (With permission of M. C. Drew and R. Campbell.)

(Fig. 6). This method was employed successfully by Jackson and Brown (1966) to follow the colonisation of wheat roots by *Azotobacter chroococcum* following seed inoculation.

23.3 Gnotobiotic culture of plants

When sterile plant cultures are inoculated with known microorganisms, they are said to be gnotobiotic, contrasting with the axenic state which merely implies that the roots are not contaminated (Lynch & White 1977). A variety of gnotobiotic methods have been employed to study the rhizosphere in solution culture and in soil, using the following stages.

(1) Sterilisation of the culture vessels.

(2) Seed sterilisation. Various chemicals, e.g. bromine, ethanol, calcium and sodium hypochlorites, hydrogen peroxide, hydroxylamine hydrochloride, mercuric chloride, peracetic acid and sulphuric acid have been used on different seeds. The efficiency of sterilisation can vary between seed stocks and the year of harvest, probably depending on the aerial microflora at the time the grain was formed.

(3) Culture of sterile seedlings. Sloppy agar (Barber 1967) is sometimes used and this can also provide a check on the sterility of the seedlings, although malt and nutrient agar are preferable for checking contamination.

(a)

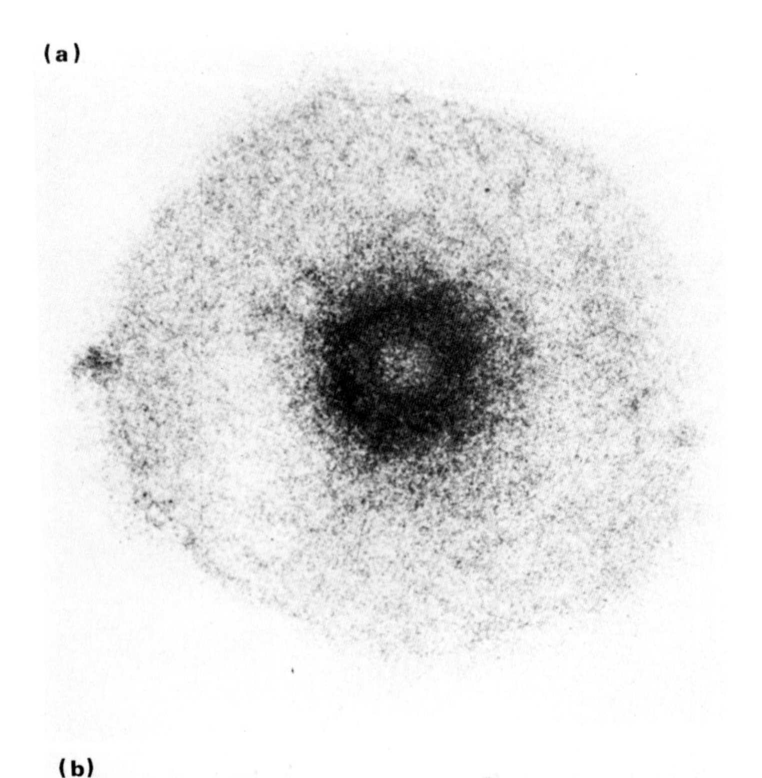

(b)

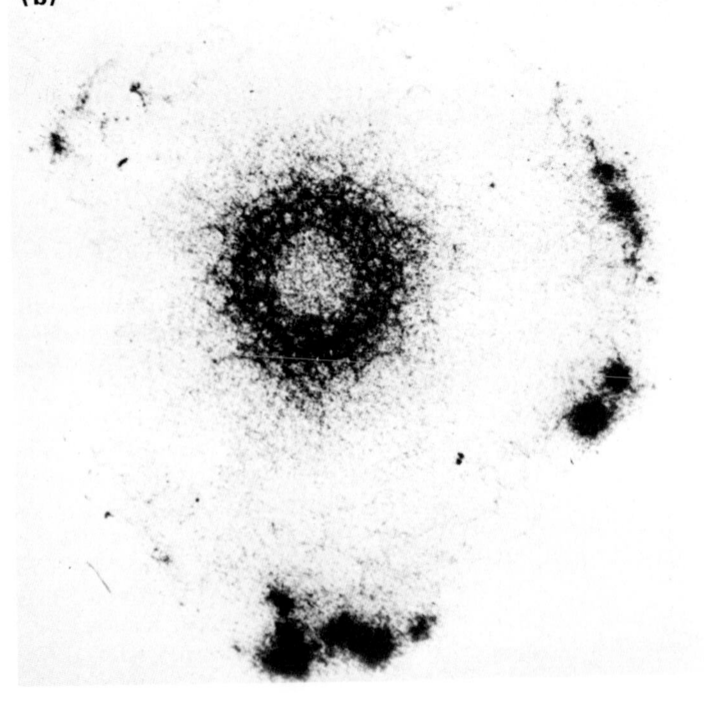

Figure 5. Autoradiographs showing the distribution of phosphate in transverse sections of roots from barley (*Hordeum vulgare*) plants which had been supplied for 6 h with 3×10^{-6}M potassium phosphate containing 1 μg [^{32}P]phosphorus l^{-1}. (a) plant grown in the absence of microorganisms; (b) plant infected at the ambient laboratory level (from Barber *et al.* 1968, with permission of D. A. Barber and Macmillan Journals Ltd).

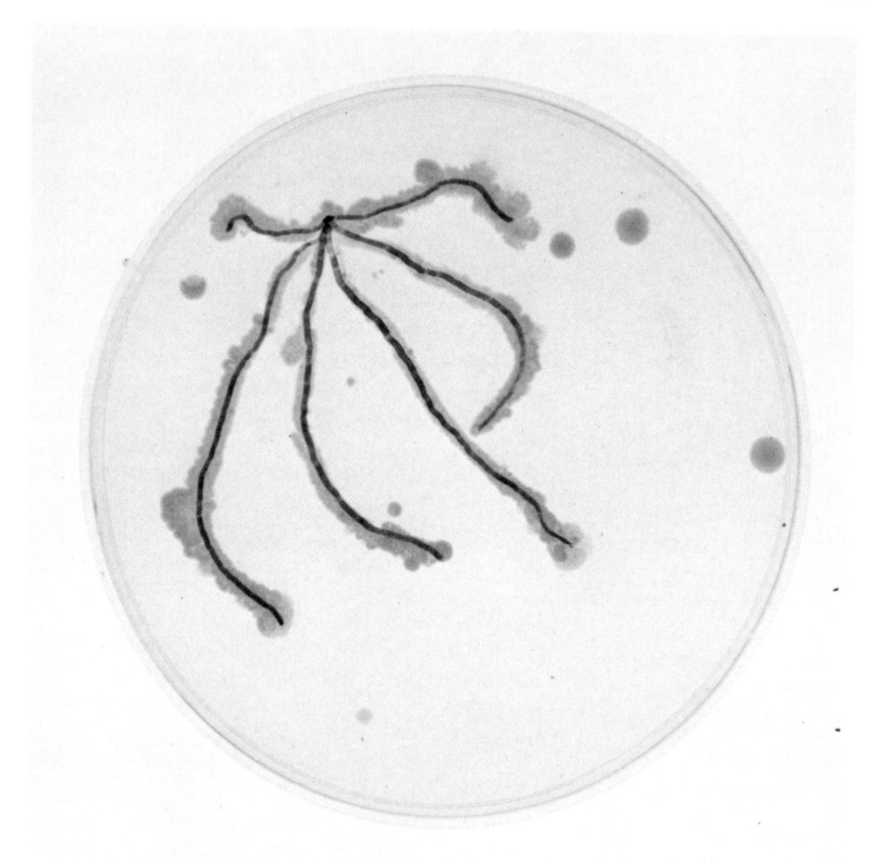

Figure 6. Root map of microbial growth on nutrient agar around barley roots.

(4) Growth of plants in sterilised nutrient solution (which may contain a solid support of sand or glass beads) or sterilised soil (usually γ-irradiated at 2.5 to 5 Mrad). This is the stage usually most sensitive to contamination and suitable air filters are necessary. Plants may be grown in tubes stopped with cotton wool (Trolldenier & Marckwordt 1962; Bowen & Rovira 1966; Lynch & White 1977), fruit preserving (Kilner) jars (Barber 1967; Darbyshire & Greaves 1973) or plastic film isolators adapted from gnotobiotic animal experiments (Hale 1969). It is not essential to enclose the shoots in a sterile container, but this can reduce the risk of contamination. It is quite common to grow the roots in a Kilner jar and the shoots in a similar inverted container, the two being joined with sterile adhesive tape. However, many arrangements are possible, depending on the type of microbe–root interaction being studied.

23.4 Measurement of the release of organic compounds by roots

Before microbial growth in the rhizosphere can be analysed, the nature and quantity of substrates available to the microorganisms must be known. The simplest approach is to grow plants in sterile conditions in solution culture so that no extraneous carbon (e.g. from rubber bungs) can enter the solution. The total carbon in solution can be analysed by a total organic carbon analyser and, following concentration by rotary evaporation or freeze drying, the components can be determined. Carbohydrates are usually the major constituent of the material released by roots and hence the carbon/nitrogen ratio (*c.* 30:1) is much greater than in microbial cells (Table 3). However, a complete carbon and nitrogen budget for the rhizosphere has not been attempted and the contribution of the various components given in Table 1 is unknown. When microorganisms are added to the culture solution, the biomass due to growth can be measured by microscopy and determination of the mean cell dimensions (Barber & Lynch 1977). The biomass that would be expected from the amount of substrate released by the roots can be calculated assuming a growth yield of about 0.35 g biomass (g substrate)$^{-1}$. Typically, however, the observed biomass is greater than the calculated value by an order of magnitude, possibly suggesting that microorganisms enhance the release of substrates by roots.

Table 3. Release of organic materials by roots of barley (*Hordeum vulgare*) grown for 21 d in solution culture under axenic conditions in the presence and absence of glass Ballotini beads of diameter 1 mm (after Barber & Gunn 1974).

	Root environment	
Organic compound released (units per plant)	Ballotini beads absent	Ballotini beads present
Amino acids (μmol)	0.14	0.23
Carbohydrates (mg)	1.45	3.03
Total material (mg)	1.59	3.26

23.4.1 THE [^{14}C]CARBON DIOXIDE GROWTH CHAMBER

Substrate release can be more reliably estimated by growing plants in solution or soil contained in a sealed plant growth chamber where [^{14}C]carbon dioxide is the sole source of carbon dioxide for the plant (Barber & Martin 1976). It is important to realise that such cabinets are difficult and expensive to construct. The labelled photosynthate passes through the plant/soil system and can be fractionated within the sealed root container and a carbon budget obtained. A significant proportion of the material released from roots is utilised by microorganisms and respired as [^{14}C]carbon dioxide which is trapped. These studies have demonstrated that about 20% of the fixed carbon dioxide is released by the roots and accounts quantitatively for the microbial biomass determined in the root region (Barber & Lynch 1977).

23.4.2 PULSE LABELLING

Another, but more equivocal, measure of fixed carbon release is to use a [^{14}C]carbon dioxide pulse labelling technique which needs less sophisticated equipment. In its simplest form a bell jar could be employed but in practice more elaborate means have been used (Warembourg & Billies 1979). The major drawbacks are that the lack of control of [^{14}C]carbon content in the [^{14}C]carbon dioxide can lead to radiation damage and that the plants become non-uniformly labelled so that total carbon budgets are difficult to balance because [^{14}C]carbon release varies in different regions of the root. However the technique has been employed, for example, in comparing the effects of light on carbon release (Warembourg & Billies 1979).

23.4.3 CHEMOTAXIS

Allen and Newhook (1973) have demonstrated a chemotactic response of the pathogen *Phytophthora cinnamoni* to ethanol which can be produced by plant roots during brief periods of fermentative metabolism, such as may be induced by waterlogging. When *P. cinnamoni* zoospores were placed in gradients of ethanol by linking solutions of different concentrations with capillary tubes, they moved up the gradient and accumulated in numbers proportional to the concentration of ethanol at the source. This did not happen in isotropic solutions of ethanol or in its absence. This elegant approach could be applied to investigate chemotaxis of other microorganisms to different exudates.

23.4.4 GENETIC MANIPULATION

The modification of the rhizosphere by genetic manipulation of the plant has been demonstrated by Neal *et al.* (1973). Substitution of the 5B chromosome with another 5B chromosome in one wheat cultivar changed the rhizosphere microbial characteristics (including total numbers, root rot, cellulolytic, pectinolytic, amylolytic and ammonifying bacteria), which presumably resulted from differences in the type and form of carbon released by the genetically manipulated plant such that the rhizosphere was more similar to that donor plant which provided the substitute 5B chromosome. This approach should be considered in plant breeding programmes, in order to obtain, for example, more efficient rhizosphere population and characteristics.

23.5 Microbe and plant competition for nutrients

Most competition studies have used plants grown in solution culture under gnotobiotic conditions with a uniform root environment and a restricted microflora (Barber 1978). This approach has led to the development of important concepts on the microbial regulation of ion uptake by plants. The most extensively studied ion has been phosphate, mainly because of its importance to plants coupled with its low concentration in the soil solution but also because it is a major nutrient for microorganisms. Results obtained by Barber in the United Kingdom and Bowen and Rovira in Australia appeared to be in conflict. However a collaborative study showed that the uptake period and age of plant were critical in determining the significance of microbe-induced ion uptake. With 8-day old barley seedlings, bacteria stimulated phosphate uptake over a 30 min period, but reduced it over 24 h. With 21-day old plants, bacteria reduced phosphate uptake over both periods (Barber *et al.* 1976). This demonstrated the importance of standardising experimental details if different observations were to be compared. Other ions which have been studied in solution culture are nitrate and ammonium (Barber 1971), rubidium, calcium and potassium (Trolldenier & Marckwordt 1962; Barber & Frankenburg 1971) and manganese (Barber & Lee 1974).

The effects of microorganisms on nutrient uptake in soil have been demonstrated by using radiation-sterilised soil. Microorganisms appear to compete with plants for phosphorus (Benians & Barber 1974) and manganese (Loutit & Brooks 1970). However, caution must be exercised in the generalisations and conclusions drawn since soil bacteria appear to solubilise phosphorus (Gerretsen 1948; Duff *et al.* 1963) and this might be critical in the provision of phosphorus to plants in some soils.

23.6 Microbial production of plant growth regulators

Early claims that seed inoculation stimulated plant growth by the provision of nutrients from the bacteria (Mishustin & Naumova 1962) have since been refuted by Brown (1974). She has claimed that bacteria provide growth regulators based on the observation that microorganisms growing in pure culture on glucose as an energy source can produce indoleacetic acid (from tryptophan), gibberellin-like compounds and cytokinins (Lynch 1976). However, most of the evidence is based on the detection of these compounds in microbial culture supernatants by various bioassays. The bioassays are only a guide and chemical methods, such as gas liquid chromatography coupled with mass spectrometry, should be used to confirm the identity of the substances. Few investigators have tested the effect of plant growth regulators applied to seedlings or roots under sterile conditions and some studies have found no positive advantage to the plant of such treatments (Lynch & White 1977; Harper & Lynch 1979).

23.7 The chemostat as a rhizosphere

The root provides substrates to the rhizosphere continuously and can be likened to a continuous-flow culture system. However the population is not diluted and it is probably more appropriate to consider it as a fed batch culture (Pirt 1975). Thus:

Overall rate of energy consumption = Consumption for growth + Consumption for maintenance

$$\frac{\mu x}{Y} = \frac{\mu x}{Y_G} + mx$$

where μ is the specific growth rate; x is the biomass; Y is the observed growth yield; Y_G is the true growth yield when no energy is used for maintenance purposes; and m is the maintenance coefficient.

It would be dangerous to carry this analogy too far until we know more about the rate and nature of substrate flow (see section 23.4) because the analysis would depend on a steady flow of substrate of constant chemical composition.

Assuming that the chemostat can provide a model for the rhizosphere, some interactions of the associated community have been described (Anderson *et al.* 1978; Cole *et al.* 1978; Coleman *et al.* 1978a,b; Herzberg *et al.* 1978). These studies have attempted to examine:

(1) the effects of starvation on the utilisation of root exudates;

(2) trophic relationships in microcosms; and

(3) biotic activities in relation to carbon, nitrogen and phosphorus transformations.

Glucose with amino acid and vitamin supplements was considered as a substrate to represent exudates. A synthetic community was set up consisting of *Pseudomonas* sp. (zymogenous or fermentative, true rhizosphere inhabitants), *Arthrobacter* sp. (autochthonous or indigenous, a spore-forming soil bacterium), *Acanthamoeba* sp. (a protozoan and a predator of bacteria) and *Mesodiplogaster* sp. (a nematode and predator of protozoa and bacteria). It was concluded that:

(1) the *Pseudomonas* sp. grew better than the *Arthrobacter* sp;

(2) the nematode and amoebal grazing of bacteria and the nematode grazing of amoebae set up a food web (Fig. 7);

(3) predators accelerated the rate of decomposition by releasing nutrients located in microbial biomass;

(4) the amoeba (but not the nematode) recycled bacterial phosphorus to the soil pool of phosphorus.

This approach can be criticised since glucose was used as the substrate whereas polysaccharides are probably the major natural substrates in the real rhizosphere, and the culture was diluted making it a continuous rather than a fed batch system. Furthermore, the community was constructed rather than isolated (Slater & Bull 1978) and was not truly representative of a natural population. However the approach is interesting and it is to be hoped that work will be undertaken with natural substrates and communities in the future.

23.8 Mathematical models

The simplified approach to the growth kinetics of the rhizosphere/rhizoplane region has been considered in a mathematical model. Newman and Watson (1977) analysed the change in substrate concentration with time, considering the root as an idealised cylinder, as follows:

The rate of change of substrate concentration	= The rate of supply of substrate due to diffusion from the root	+ The rate of substrate supply from indigenous sources	− The rate of used substrate due to microbial growth

$$\frac{ds}{dt} = F_D + F_I - F_{G'} \quad (1)$$

Now

$$F_D = \frac{D}{r}\frac{d}{dr}\left(r\frac{ds}{dt}\right) \quad (2)$$

where D is the coefficient of diffusion; r is the radial distance from the root axis; and s is the substrate concentration.

$$F_I = mx_0 \quad (3)$$

where m is the maintenance coefficient; and x_0 is

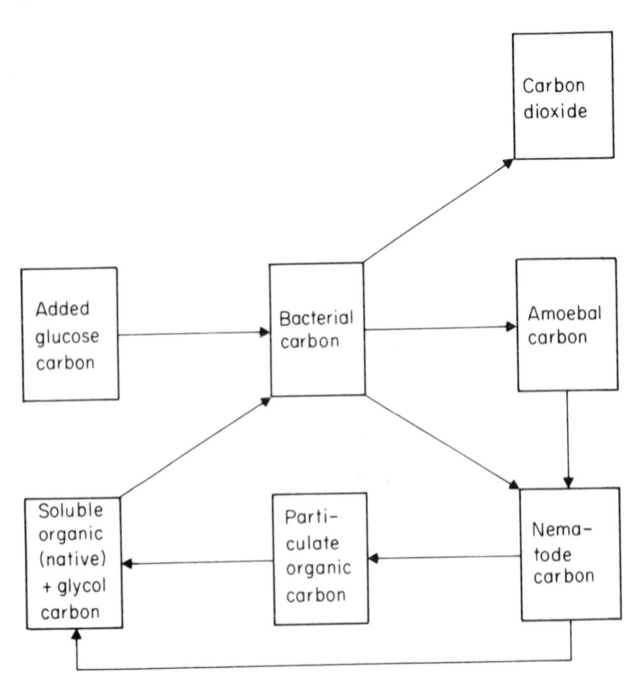

Figure 7. Carbon food web in rhizosphere microcosm. Asterisk indicates artefact from propylene oxide sterilisation (after Coleman *et al.* 1979, with permission).

the initial biomass concentration. It is important to note that this term is only one way of expressing the initial (or indigenous) concentration of available material and may be a considerable oversimplification.

The rate of substrate used for growth, $F_{G'}$, is a function of two components, namely, the rate of substrate utilisation for biomass production, F_G, and the rate of substrate utilisation for maintenance energy purposes, F_M.

That is,
$$F_{G'} = F_G + F_M \quad (4)$$

Thus
$$F_G = \frac{1}{Y_G} \cdot \frac{dx}{dt} \quad (5)$$

where Y_G is the true growth yield (i.e. substrate used only for new biomass production per unit of existing biomass); and dx/dt is the rate of growth of the population.

Now
$$\frac{dx}{dt} = \mu x$$

and by substituting the Monod function, we have:

$$\frac{dx}{dt} = \frac{\mu_{max} \cdot s}{(K_s\theta + s)} \cdot x \quad (6)$$

where μ_{max} is the maximum specific growth rate of the population; s is the growth-limiting substrate concentration; K_s is the saturation constant; x is the biomass concentration; and θ is the soil water content. This latter term is included since normally, in the present context, x and s are measured in terms of amounts per unit volume of soil and K_s is normally determined as an amount per unit volume of soil solution.

And
$$F_M = mx \quad (7)$$

where m is the maintenance coefficient; and x is the biomass concentration.

Thus equation (4) may be written as:

$$F_{G'} = \frac{x}{Y_G} \cdot \frac{\mu_{max} \cdot s}{(K_s\theta + s)} + mx \quad (8)$$

And substituting equations (2), (3) and (8) in equation (1), we have:

$$\frac{ds}{dt} = \frac{D}{r}\frac{d}{dr}\left(r\frac{ds}{dt}\right) + mx_0 - \frac{x}{Y_G} \cdot \frac{\mu_{max} \cdot s}{(K_s\theta + s)} - mx \quad (9)$$

Note that the term F_G may be simplified to incorporate both individual components by measuring Y, the observed growth yield. In this case no distinction is made between substrate used for new biomass production or maintenance requirements. Thus:

$$F_{G'} = \frac{1}{Y} \cdot \frac{dx}{dt} \quad (10)$$

and so by substitution in equation (1) with equations (2), (3), (6) and (10), we have:

$$\frac{ds}{dt} = \frac{D}{r}\frac{d}{dr}\left(r\frac{ds}{dt}\right) + mx_0 - \frac{x}{Y} \cdot \frac{\mu_{max} \cdot s}{(K_s\theta + s)} \quad (11)$$

Similarly it is possible to derive equations (e.g. equation (6)) for the biomass concentrate rate of change.

These equations which are coupled and nonlinear cannot be solved analytically in the general case; numerical methods of solution must be employed.

Newman and Watson (1977) assumed a wide range of input parameters for the model and then held all factors constant except the variables under consideration. They predicted, for example, the variation in microbial concentration with time at various distances from the root and the variation in substrate concentration with distance from the root surface after various times.

The major weakness is that the model is not supported by measurements and the authors also pointed at the many limitations of their assumptions. They assumed that the maintenance coefficient decreases with growth rate, although this has never been substantiated. However, they obtained reasonably good fits to the model using a very limited amount of data, which should stimulate further experimental work.

23.9 The rhizosphere in the field

Most studies of the rhizosphere have been made under laboratory conditions. Some of the early investigators (Timonin 1946) demonstrated that viable counts of specific organisms were greater near roots than in the bulk soil. However, this gives no idea of the biomass (the total amount of living organisms) around roots.

To assess biomass microscopically is tedious and recently techniques have become available for indirect measurements, by assaying ATP in soils (Paul & Johnson 1977; Jenkinson & Oades 1979; Oades & Jenkinson 1979) or determining the flush of carbon dioxide produced from fumigated soil on reinoculation (Jenkinson & Powlson 1976; see Chapter 26). We have modified the latter technique to measure the microbial and faunal biomass associated with roots in the field (Lynch & Panting 1980a,b). Using intact soil cores with a diameter of 7.6 cm and a depth of 5 cm placed in closed bottles (Fig. 8), half the replicates were fumigated and the remainder untreated. The carbon dioxide produced after 10 d from the cores was measured by gas chromatography and the biomass, B, calculated from:

$$B = \frac{X - x}{k}$$

where X is the carbon in carbon dioxide produced in 10 d from the fumigated cores; x is the carbon in carbon dioxide produced in the same time from untreated cores; k is the proportion of cells mineralised to carbon dioxide produced: a value of 0.41 was assumed (Anderson & Domsch 1978). It was shown that the microbial and faunal biomass increased with increasing root biomass in the cores and this was particularly apparent when the cores of grassland were compared with an adjacent fallow area (Table 4).

Figure 8. The fumigation and respiration method of biomass determination (after Jenkinson & Powlson 1976, modified by Lynch & Panting 1980a).

Nitrogenase (Balandreau & Knowles 1978), denitrification (Woldendorp 1963; Day *et al.* 1978) and sulphate reduction (Dommergues & Jacq 1972) have also been shown to be enhanced around roots in soil.

Table 4. Comparison of the biomass (mg C (100 g dry soil)$^{-1}$) in the surface 5 cm of a clay soil (Denchworth series) under grass and arable cultivation (oil seed rape direct-drilled) as measured by the fumigation and respiration technique (from Lynch & Panting 1980b).

System	Root biomass	Microbial and faunal biomass
Grass	227	206
Arable	8	72

23.10 Conclusion

The rhizosphere is an interesting habitat for the study of microbial activity because of its relevance to crop production. The use of and development of quantitative methods will facilitate this. Present work is moving away from describing the natural history of the rhizosphere and towards making assessments of its significance. It is known that plants grow well in the absence of microorganisms providing they are supplied with adequate inorganic nutrients, but a complete evaluation of the significance or otherwise of the presence of microorganisms around roots has yet to be made. Furthermore, the direct control of the microorganisms within the rhizosphere in order to improve their beneficial effects on plant growth and productivity has yet to be satisfactorily attempted.

Recommended reading

Bowen G. D. (1980) Misconceptions, concepts and approaches in rhizosphere biology. In: *Contemporary Microbial Ecology* (Eds D. C. Ellwood, J. N. Hedger, M. J. Latham, J. M. Lynch & J. H. Slater), pp. 283–304. Academic Press, London.

Bowen G. D. & Rovira A. D. (1976) Microbial colonization of plant roots. *Annual Review of Phytopathology* **14**, 121–44.

Brown M. E. (1974) Seed and root bacterization. *Annual Review of Phytopathology* **12**, 181–97.

Dommergues Y. R. & Krupa S. V. (1978) *Interactions Between Non-pathogenic Soil Micro-organisms and Plants*. Elsevier, Amsterdam.

Drew M. C. & Lynch J. M. (1980) Soil anaerobiosis, micro-organisms and root function. *Annual Review of Phytopathology* **18**, 37–67.

Elliott L. F., Campbell C. M., Lynch J. M. & Tittemore D. (1982) Bacterial colonization of plant roots. In: *Microbial-Plant Interactions* (Ed. R. L. Todd). Soil Science Society of America, Madison, Wisconsin. (In press.)

Harley J. L. & Russell R. S. (1979) *The Soil–Root Interface*. Academic Press, London.

Newman E. I. (1978) Root micro-organisms: their significance in the ecosystem. *Biological Reviews* **53**, 511–54.

Nye P. H. & Tinker P. B. (1977) *Solute Movement in the Soil–Root System*. Blackwell Scientific Publications, Oxford.

Russell R. S. (1977) *Plant Root Systems: Their Function and Interaction with the Soil*. McGraw-Hill, London.

References

Allen R. N. & Newhook F. J. (1973) Chemotaxis of zoospores of *Phytophthora cinnanomi* to ethanol in capillaries of soil pore dimensions. *Transactions of the British Mycological Society* **61**, 287–302.

Anderson J. P. E. & Domsch K. (1978) A physiological method for the quantitative measurement of microbial biomass in soils. *Soil Biology and Biochemistry* **10**, 215–21.

Anderson J. R. & Slinger J. M. (1975) Europium chelate and fluorescent brightener staining of soil propagules and their photomicrographic counting. II. Efficiency. *Soil Biology and Biochemistry* **7**, 205–9.

Anderson R. V., Elliott E. T., McClellan J. F., Coleman D. C., Cole C. V. & Hunt H. W. (1978) Trophic interactions in soils as they affect energy and nutrient dynamics. III. Biotic interactions of bacteria, amoebae and nematodes. *Microbial Ecology* **4**, 361–71.

Balandreau J. & Knowles R. (1978) The rhizosphere. In: *Interactions Between Non-Pathogenic Soil Micro-organisms and Plants* (Eds Y. R. Dommergues & S. V. Krupa), pp. 243–68. Elsevier, Amsterdam.

Barber D. A. (1967) The effect of micro-organisms on the absorption of inorganic nutrients of intact plants. I. Apparatus and culture technique. *Journal of Experimental Botany* **18**, 163–9.

Barber D. A. (1971) The influence of micro-organisms on the assimilation of nitrogen by plants from soil and fertilizer sources. In: *Nitrogen-15 in Soil–Plant Studies*, pp. 91–101. International Atomic Energy Agency, Vienna.

Barber D. A. (1978) Nutrient uptake. In: *Interactions Between Non-pathogenic Soil Micro-organisms and Plants* (Eds Y. R. Dommergues & S. V. Krupa), pp. 131–62. Elsevier, Amsterdam.

Barber D. A., Bowen G. D. & Rovira A. D. (1976) Effects of microorganisms on the absorption and distribution of phosphate in barley. *Australian Journal of Plant Physiology* **3**, 801–8.

Barber D. A. & Frankenburg U. C. (1971) The contribution of microorganisms to the apparent absorption of ions by roots grown under non-sterile conditions. *New Phytologist* **70**, 1027–34.

Barber D. A. & Gunn K. B. (1974) The effect of mechanical forces on the exudation of organic substances by the roots of cereal plants grown under sterile conditions. *New Phytologist* **73**, 39–45.

Barber D. A. & Lee R. B. (1974) The effect of micro-organisms on the absorption of manganese by plants. *New Phytologist* **73**, 97–106.

Barber D. A. & Lynch J. M. (1977) Microbial growth in the rhizosphere. *Soil Biology and Biochemistry* **9**, 305–8.

Barber D. A. & Martin J. K. (1976) The release of organic substances by cereal roots in soil. *New Phytologist* **76**, 69–80.

Barber D. A., Sanderson J. & Russell R. S. (1968) Influence of micro-organisms on the distribution in roots of phosphate labelled with phosphorus-32. *Nature, London* **217**, 644.

Benians G. J. &. Barber D. A. (1974) The uptake of phosphate by barley plants from soil under aseptic and non-sterile conditions. *Soil Biology and Biochemistry* **6**, 195–200.

Bowen G. D. (1980) Misconceptions, concepts and approaches in rhizosphere biology. In: *Contemporary Microbial Ecology* (Eds D. C. Ellwood, J. N. Hedger, M. J. Latham, J. M. Lynch & J. H. Slater), pp. 283–304. Academic Press, London.

Bowen G. D. & Rovira A. D. (1966) Microbial factor in short-term phosphate uptake studies with plant roots. *Nature, London* **211**, 665–6.

Bowen G. D. & Rovira A. D. (1976) Microbial colonization of plant roots. *Annual Review of Phytopathology* **14**, 121–44.

Brown M. E. (1974) Seed and root bacterization. *Annual Review of Phytopathology* **12**, 181–97.

Cole C. V., Elliott E. T., Hunt H. W. & Coleman D. C. (1978) Trophic interactions in soils as they affect energy and nutrient dynamics. V. Phosphorus transformations. *Microbial Ecology* **4**, 381–7.

Coleman D. C., Anderson R. V., Cole C. V., Elliott E. T., Woods L & Campion M. K. (1978a) Trophic interactions in soils as they affect energy and nutrient dynamics. IV. Flows of metabolic and biomass carbon. *Microbial Ecology* **4**, 373–80.

COLEMAN D. C., COLE C. V., HUNT H. W. & KLEIN D. A. (1978b) Trophic interactions in soils as they affect energy and nutrient dynamics. I. Introduction. *Microbial Ecology* **4**, 345–9.

DARBYSHIRE J. F. & GREAVES M. P. (1973) An improved method for the study of the interrelationships of soil microorganisms and plant roots. *Soil Biology and Biochemistry* **2**, 65–71.

DAY P. R., DONER H. E. & MCLAREN A. D. (1978) Relationships among microbial populations and rates of nitrification and denitrification in a Hanford soil. In: *Nitrogen in the Environment* (Eds D. R. Nielsen & J. G. MacDonald), vol. 2, pp. 305–63. Academic Press, New York.

DOMMERGUES Y. & JACQ V. (1972) Microbiological transformations of sulphur in the rhizosphere and spermosphere. *Annales Agronomique* **23**, 201–15.

DOMMERGUES Y. R. & KRUPA S. V. (1978) *Interactions Between Non-pathogenic Soil Micro-organisms and Plants*. Elsevier, Amsterdam.

DREW M. C. & LYNCH J. M. (1980) Soil anaerobiosis, micro-organisms and root function. *Annual Review of Phytopathology* **18**, 37–67.

DUFF R. B., WEBLEY D. M. & SCOTT R. O. (1963) Solubilization of minerals and related materials by 2-ketogluconic acid-producing bacteria. *Soil Science* **95**, 105–14.

ELLIOTT L. F., CAMPBELL C. M., LYNCH J. M. & TITTEMORE D. (1982) Bacterial colonization of plant roots. In: *Microbial–Plant Interactions* (Ed. R. L. Todd). Soil Science Society of America, Madison, Wisconsin. (In press.)

FAULL J. L. & CAMPBELL R. (1979) Ultrastructure of the interaction between the take-all fungus and antagonistic bacteria. *Canadian Journal of Botany* **57**, 1800–1808.

FOSTER R. C. & ROVIRA A. D. (1978) The ultrastructure of the rhizosphere of *Trifolium subterraneum* L. In: *Microbial Ecology* (Eds M. W. Loutit & J. A. R. Miles), pp. 278–90. Springer-Verlag, Berlin.

GERRETSEN R. C. (1948) The influence of micro-organisms on the phosphate intake by the plant. *Plant and Soil* **1**, 51–85.

HALE M. G. (1969) Loss of organic compounds from roots. I. Cultural conditions for axenic growth of peanut (*Arachis hypogaea* L.) *Plant and Soil* **31**, 463–72.

HARLEY J. L. & RUSSELL R. (1979) *The Soil–Root Interface*. Academic Press, London.

HARPER S. H. T. & LYNCH J. M. (1979) Effects of *Azotobacter chroococcum* on barley seed germination and seedling development. *Journal of General Microbiology* **112**, 45–51.

HERZBERG M. A., KLEIN D. A. & COLEMAN D. C. (1978) Trophic interactions in soils as they affect energy and nutrient dynamics. II. Physiological responses of selected rhizosphere bacteria. *Microbial Ecology* **4**, 351–9.

HILTNER L. (1904) Über neuere Erfahrungen und Probleme auf dem Gebiet der Bodenbakteriologie und unter besonderer Berücksichtigung der Gründungen und Brache. *Arbeiten der Deutschen Landwirtschaftsgesellschaft Berlin* **98**, 59–78.

JACKSON R. M. & BROWN M. E. (1966) Behaviour of *Azotobacter chroococcum* induced into the plant rhizosphere. *Supplement to Annales de l'Institut Pasteur* **111**, 103–12.

JENKINSON D. S. & OADES J. M. (1979) A method for measuring adenosine triphosphate in soil. *Soil Biology and Biochemistry* **11**, 193–9.

JENKINSON D. S. & POWLSON D. S. (1976) The effects of biocidal treatments on metabolism in soil. V. A method for measuring soil biomass. *Soil Biology and Biochemistry* **8**, 209–13.

JENKINSON D. S., POWLSON D. S. & WEDDERBURN R. W. M. (1976) The effects of biocidal treatments on metabolism in soil. III. The relationship between soil biovolume, measured by optical microscopy, and the flush of decomposition caused by fumigation. *Soil Biology and Biochemistry* **8**, 189–202.

JONES P. C. T. & MOLLISON J. E. (1948) A technique for the quantitative estimation of soil micro-organisms. *Journal of General Microbiology* **2**, 54–69.

LEPPARD G. G. (1975) A physical relationship between soil microbes and the surface of the wheat root. *Naturwissenschaften* **62**, 41.

LOUTIT M. W. & BROOKS R. R. (1970) Rhizosphere organisms and molybdenum concentration in plants. *Soil Biology and Biochemistry* **2**, 131–5.

LYNCH J. M. (1976) Products of soil micro-organisms in relation to plant growth. *CRC Critical Reviews in Microbiology* **5**, 67–107.

LYNCH J. M. & PANTING L. M. (1980a) Cultivation and the soil biomass. *Soil Biology and Biochemistry* **12**, 29–33.

LYNCH J. M. & PANTING L. M. (1980b) Variations in the size of the soil biomass. *Soil Biology and Biochemistry*. (In press.)

LYNCH J. M. & WHITE N. (1977) Effects of some non-pathogenic micro-organisms on the growth of gnotobiotic barley plants. *Plant and Soil* **47**, 161–70.

MISHUSTIN E. N. & NAUMOVA A. N. (1962) Bacterial fertilizers, their effectiveness and mode of action. *Microbiology* (English translation) **31**, 442–52.

NEAL J. L., LARSON R. I. & ATKINSON T. G. (1973) Changes in rhizosphere populations of selected physiological groups of bacteria related to substitution of specific pairs of chromosomes in spring wheat. *Plant and Soil* **39**, 209–12.

NEWMAN E. I. (1978) Root micro-organisms: their significance in the ecosystem. *Biological Reviews* **53**, 511–54.

NEWMAN E. I. & BOWEN H. J. (1974) Patterns of distribution of bacteria on root surfaces. *Soil Biology and Biochemistry* **6**, 205–9.

NEWMAN E. I., CAMPBELL R., CHRISTIE P., HEAP A. J. & LAWLEY R. (1979) Root micro-organisms in mixtures and monocultures of grassland plants. In: *The Soil–Root Interface* (Eds J. L. Harley & R. S. Russell), pp. 161–73. Academic Press, London.

NEWMAN E. I. & WATSON A. (1977) Microbial abundance in the rhizosphere: a computer model. *Plant and Soil* **48**, 17–56.

NYE P. H. & TINKER P. B. (1977) *Solute Movement in the Soil–Root System.* Blackwell Scientific Publications, Oxford.

OADES J. M. & JENKINSON D. S. (1979) Adenosine triphosphate content of the soil microbial biomass. *Soil Biology and Biochemistry* **11**, 201–4.

OLD K. M. & NICOLSON T. H. (1978) The root cortex as part of a microbial continuum. In: *Microbial Ecology* (Eds M. W. Loutit & J. A. R. Miles), pp. 291–4. Springer-Verlag, Berlin.

PARKINSON D., GRAY T. R. G. & WILLIAMS S. T. (1971) *Methods for Studying the Ecology of Soil Micro-organisms.* Blackwell Scientific Publications, Oxford.

PATRIQUIN D. G. & DOBEREINER J. (1978) Light microscopy observations of tetrazolium-reducing bacteria in the endorhizosphere of maize and other grasses in Brazil. *Canadian Journal of Microbiology* **24**, 734–42.

PAUL E. A. & JOHNSON R. L. (1977) Microscopic counting and adenosine 5′-triphosphate measurement in determining microbial growth in soils. *Applied and Environmental Microbiology* **34**, 263–9.

PIRT S. J. (1975) *Principles of Microbe and Cell Cultivation.* Blackwell Scientific Publications, Oxford.

ROUATT J. W. & KATZNELSON H. (1961) A study of the bacteria on the root surface and in the rhizosphere soil of crop plants. *Journal of Applied Bacteriology* **24**, 164–71.

ROVIRA A. D. (1965) Plant root exudates and their influence upon soil micro-organisms. In: *Ecology of Soil-borne Plant Pathogens—Prelude to Biological Control* (Eds K. F. Baker & W. C. Snyder), pp. 170–86. John Murray, London.

ROVIRA A. D., FOSTER R. C. & MARTIN J. K. (1979) Note on terminology: Origin, nature and nomenclature of the organic materials in the rhizosphere. In: *The Soil–Root Interface* (Eds J. L. Harley & R. S. Russell), pp. 1–4. Academic Press, London.

ROVIRA A. D., NEWMAN E. I., BOWEN H. J. & CAMPBELL R. (1974) Quantitative assessment of the rhizoplane microflora by direct microscopy. *Soil Biology and Biochemistry* **6**, 211–16.

RUSSELL R. S. (1977) *Plant Root Systems: Their Function and Interaction with the Soil.* McGraw-Hill, London.

SCHIPPERS B. & VAN VUURDE J. W. L. (1978) Studies of microbial colonization of wheat roots and the manipulation of the rhizosphere microflora. In: *Microbial Ecology* (Eds M. W. Loutit & J. A. R. Miles), pp. 295–8. Springer-Verlag, Berlin.

SCHMIDT E. L. (1973) Fluorescent antibody techniques for the study of microbial ecology. In: *Modern Methods for the Study of Microbial Ecology* (Ed. T. Rosswall), vol. 17, pp. 67–76. Bulletins of the Ecological Research Committee, Stockholm.

SLATER J. H. & BULL A. T. (1978) Interactions between microbial populations. In: *Companion to Microbiology* (Eds A. T. Bull & P. M. Meadow), pp. 181–206. Longmans, London.

TIMONIN M. I. (1946) Microflora of the rhizosphere in relation to the manganese deficiency disease of oats. *Soil Science Society of America Proceedings* **11**, 284–92.

TROLLDENIER G. & MARCKWORDT U. (1962) Untersuchungen über den Einfluss der Bodenmikroorganismen auf die Rubidium- und Calcium-Aufnahme im Nährlösung wachsender Pflanzen. *Archiv für Mikrobiologie* **43**, 148–51.

VANCURA V. & KUNC F. (1977) The effect of streptomycin and actidione on respiration in the rhizosphere and non-rhizosphere soil. *Zentralblatt für Bakteriologie Parasitenkunde Infektionskrankheiten und Hygiene Abteilung 2.* **132**, 472–8.

VUURDE J. W. L. VAN & ELENBAAS P. F. M. (1978) Use of fluorochromes for direct observation of micro-organisms associated with wheat roots. *Canadian Journal of Microbiology* **24**, 1272–5.

WADA H., SAITO M. & TAKAI Y. (1978) Effectiveness of tetrazolium salts in microbial ecological studies in submerged soil. *Soil Science and Plant Nutrition (Tokyo)* **24**, 349–56.

WAREMBOURG F. R. & BILLIES G. (1979) Estimating carbon transfers in the plant rhizosphere. In: *The Soil–Root Interface* (Eds J. L. Harley & R. S. Russell), pp. 183–96. Academic Press, London.

WOLDENDORP J. W. (1963) The influence of living plants on denitrification. *Mededeelingen van de Landbouwhogeschool, Wageningen* **63**, 1–100.

Chapter 24 • The Phylloplane and Other Aerial Plant Surfaces

Colin H. Dickinson

24.1 The phylloplane

It has long been recognised that the aerial surfaces of green plants provide habitats for a large number of microorganisms. In the nineteenth century bacteriologists and mycologists isolated numerous species from these habitats, although they were never studied with the same enthusiasm as were the roots and soil. One of the first systematic studies of the phylloplane was carried out by Potter (1910) who simply pressed leaves against a solidified culture medium and then incubated the Petri dishes. A number of organisms grew on the medium and, with some inspiration, he wondered if they might play a role in controlling the growth of foliar pathogens. Potter's lead was not followed up by others and interest in the phylloplane was minimal in the first half of the twentieth century. By contrast, soil microbiology flourished during this period and this led to the development of several methods for the biological control of soil-borne plant pathogens (Baker & Cook 1974). These successes in soil microbiology stimulated interest in the ecology of microorganisms on aerial plant surfaces as the possibilities for similar methods of disease control were explored.

The systematic ecological studies of the microflora on cereal leaves by Last (1955) and on tropical tree leaves by Ruinen (1961) led to a spate of investigations of the phylloplane and it is now possible to obtain an overall view of its microbial populations. Most of the investigations of leaf surface organisms have concentrated on the fungi which live in this habitat. A wide range of fungal species have been found on leaves, including yeasts and yeast-like organisms and filamentous fungi with saprobic and pathogenic tendencies.

Yeasts have been regularly found to be the most abundant fungi on the surfaces of leaves. The actual populations of yeasts on leaves can, however, fluctuate considerably over quite short time intervals. Heavy rain, desiccation and extremes of temperatures can all have drastic affects on the viable counts of yeasts, but the residual populations are able to multiply rapidly when the environmental conditions become more favourable. The yeast populations on leaves include representatives of the Ascomycotina, the Basidiomycotina and the Fungi Imperfecti. Genera which have been regularly recorded include *Candida, Cryptococcus, Rhodotorula, Sporobolomyces, Tilletiopsis* and *Torulopsis*. *Aureobasidium* which can grow either as a yeast or as a filamentous fungus, has also been found to be a major component of many phylloplane populations.

Amongst the filamentous fungi recorded on aerial plant surfaces, the sooty moulds stand out as producing particularly spectacular and colourful colonies. These fungi are found in many parts of the world but especially luxuriant colonies are found in damp, cool-temperature regions, such as New Zealand. They have complex life cycles and their nutritional relationships with the leaves and stems on which they grow have not been fully investigated (Hughes 1976). They seem, however, to have evolved vegetative and reproductive structures which enable them to take advantage of the stable surfaces provided by long-lived aerial plant organs. Most sooty moulds seem to be dependent on the honeydew produced by various plant-sucking insects but others, including many species of the Chaetothyriaceae, live on plants which are not infested with such pests. It is not known whether these sooty moulds live on leaf leachates or on the exogenous nutrients deposited on leaves from elsewhere.

Several other fungal genera which grow extensively on leaves and other surfaces develop physiological connections with the underlying host tissues. The genus *Erysiphe* is perhaps the best known example of this group of pathogenic epiphytes. Whilst these fungi are clearly significant pathogens their biology is of interest to microbial ecologists as they can be a major influence on aerial plant surface ecosystems.

Most of the cultural methods which have been used for the study of the leaf surface microflora result in the isolation of large numbers of filamentous fungi. It would, however, be rash to assume that all these diverse species are able to grow on healthy green leaves. Many records merely indicate the presence of propagules of the species and careful study is necessary to determine if and when these commence active growth. Some only germinate once the leaf has been damaged or has begun to senesce and others can only grow if substantial supplies of nutrients, such as are provided by pollen and honeydew, are available. A number of species which are regularly encountered are parasitic organisms capable of extensive epiphytic growth leading to penetration via wounds or senescent tissue. If these categories of fungi are eliminated from the species lists produced by isolation methods, few filamentous fungi are left as possible phylloplane inhabitants (Dickinson 1976).

Two species which have been considered as such are *Alternaria alternata* and *Cladosporium cladosporioides*. These fungi are amongst the most common species isolated from nearly every plant whose phylloplane flora has been studied to date. They are both well adapted to life on leaf surfaces (Dickinson & O'Donnell 1977) and a *Cladosporium* sp. has been shown to be capable of budding in a yeast-like manner (Dickinson & Bottomley 1980). In favourable environmental conditions both species can grow extensively on leaves and other photosynthetic tissues producing sooty mould-like colonies. Both species have, however, also been shown to have parasitic tendencies (O'Donnell & Dickinson 1980), although these seem to be less important than their ability to grow saprobically on dead leaves as primary saprophytes. Further investigations will probably show that other fungi now considered to be epiphytes are in fact weak parasites which become primary saprophytes on newly dead tissues. Hence there is some doubt as to which of the filamentous fungi commonly recorded on aerial plant surfaces are long-term epiphytes in these habitats.

Despite the obvious potential offered by the extensive aerial plant surfaces for the development of bacteria relatively little is known about either the species or the numbers which inhabit these habitats. Burri (1903) found several million bacteria per gram of leaf tissue and noted that these were actively growing and distinct from the soil and air flora. Until recently, however, this pioneer work was not developed further. As with many other aspects of microbial ecology, this neglect has been mainly due to problems in bacterial taxonomy. The difficulties of classifying leaf surface bacteria by traditional methods have been discussed by Billing (1976) and these have led to the recent emphasis on numerical taxonomic methods for qualitative studies (Austin *et al.* 1978; Ercolani 1978). Bacteria recorded to date on leaf surfaces include Gram-negative taxa, *Pseudomonas, Xanthomonas, Flexibacter, Erwinia* and *Acinetobacter*; Gram-positive taxa, *Staphylococcus, Bacillus* and *Micrococcus*; and Gram-variable taxa, *Corynebac-*

terium and *Listeria*. In a study of the leaves of *Lolium perenne* different bacterial taxa were found to predominate at successive sampling dates, which suggests that this population may fluctuate rather like that of the yeasts (Dickinson *et al.* 1975). Little is known about the origin of the inocula which initiate the colonisation of newly exposed leaves or about the factors which affect the development of the populations.

Several other groups of microorganisms have been found on green leaves but few have yet been the subject of a systematic investigation. Unicellular algae have been seen on leaf surfaces (Allen 1973; Bernstein & Carroll 1977) and it is likely that these organisms flourish on long-lived leaves in very humid environments. Similarly moss and fern spores can germinate and grow on leaves, although whether many of these can complete their life cycle in the phylloplane is as yet unknown. Lichens are usually thought of as especially slow-growing organisms but even these symbionts can become established on some long-lived leaves in tropical climates (Ruinen 1961).

The development of a complex phylloplane microflora is, in some instances, a direct consequence of the activities of the leaf surface microfauna, especially the aphids which produce copious quantities of honeydew (van der Burg 1974). In contrast, other leaf surface animals, such as the protozoa, probably depend on the phylloplane microbes for their nutrition and perhaps even for the microhabitats in which they live (Ruinen 1961).

24.2 Stems, flowers and fruits

The above account is based on studies of the phylloplane, which has been the most intensively investigated habitat of any of the aerial plant surfaces. Knowledge of the microorganisms inhabiting the surfaces of herbaceous and woody stems, flowers and fruits is relatively poor, except in a few instances where there has been an exceptional stimulus for their study.

Petals and other floral parts are almost always short-lived, they are infrequently attacked by pathogens and knowledge of their epiphytic saprobic microfloras is poor. Exceptions to this are the vine and the apple where there has been economic and academic interest in the yeast flora of the flowers and the fruits, in relation to the subsequent fermentation processes (Davenport 1976). More is known about fruit surface microfloras and there has recently been emphasis on the microflora of soft fruits whose shelf life is naturally limited to a few days by their rapid maturation, senescence and deterioration (Dennis 1976).

Little is known about the microflora of bark despite the fact that this long-lived, if somewhat inert and inhospitable substrate obviously becomes colonised by algae, such as species of *Pleurococcus*, and numerous lichens (Dickinson 1976). Carroll *et al.* (1980) have recently carried out a systematic study of the epiphytes on conifer bark and they have found it to be inhabited by a small number of fungal species, some of which appeared to be able to establish themselves within the superficial bark cells. One microhabitat on woody stems which has received special attention is the leaf scar which is the infection court for a number of pathogens, such as *Nectria galligena* on apple. Swinburne (1973) found that apple leaf scars were colonised by a number of saprophytic fungi and bacteria. These latter included *Bacillus subtilis*, which is antagonistic to *Nectria galligena*, and this discovery has enabled a system of biological control to be devised.

24.3 Flower and leaf buds

Not all aerial plant surfaces are exposed, inhospitable habitats. Buds provide a more protected habitat and they have been shown, in several instances, to harbour substantial populations of bacteria and fungi (Leben 1971). The microflora of the bud, or the 'gemmisphere' as it has been termed, includes pathogens, such as *Pseudomonas glycinae* which subsequently attacks the expanded leaves, yeasts which become dominant on the leaves, and filamentous fungi. The numbers of organisms increased as the buds swelled prior to bud break but colonisation was mainly restricted to the peripheral scales and the floral and leaf primordia were relatively free from organisms (Andrews & Kenerley 1980).

24.4 Factors affecting epiphytic populations

It is frequently suggested that the physical environment has more pronounced effects on the epiphytes associated with aerial plant surfaces than it has on the corresponding populations around the roots. This supposition is based in part on the extreme weather fluctuations which can occur over short time intervals and also on the fact that aerial surfaces are exposed to sunlight and wind, which do not directly affect the root surface microflora. As mentioned above, a number of investigators have demonstrated violent changes in the phylloplane populations and it would appear that some of these result from fluctuations in the environment. The plant itself can also influence the local microclimate of the individual leaf, the whole plant and the community.

Other factors affecting microbial epiphytes include the chemistry of the host plant's surfaces, which depends partly on their composition and partly on the debris which collects on the surfaces (Jeffree *et al.* 1976). Epiphytic microbes themselves can influence the development of their associates by either acting as a source of nutrients or by producing inhibitors which limit the growth of other organisms (Blakeman & Brodie 1976). Foliar pathogens can also influence the quality of the surface environment as a consequence of their effects on the physiology of the host. In this respect necrotrophic pathogens are likely to produce more dramatic effects than their biotrophic counterparts.

Experimental

24.5 Selection of material, transport and storage

24.5.1 Selection of Material

Some of the most intractable problems which face microbial ecologists studying higher plants or animals concern the selection of material for examination. Two complementary difficulties which are often encountered are those of ensuring that any sample is representative of the whole tissue, the plant or the community, and that comparable samples are collected over a period of time. Such problems have received scant attention as regards roots and other subterranean organs, perhaps because of the considerable difficulties involved in discovering the age and physiological state of individual rootlets and their spatial distribution patterns in the soil. These excuses are less acceptable when the tissues being studied are displayed in the aerial environment. Nevertheless, there are still considerable problems inherent in any study of epiphytes on the aerial surfaces of higher plants. The patterns of plant development must be determined and methods may need to be found to measure the metabolic state of the tissues being studied. The effects of regional weather patterns can be significant, as are local variations in microclimate caused by the host plants themselves.

Many studies of phylloplane microbes have been conducted with little regard for such considerations. Leaves have been selected without considering their height above the ground, their lateral proximity to the canopy periphery or their compass bearing with regard to the main axis of the plant. Andrews *et al.* (1980) have shown that the first two factors affected the distribution of filamentous fungi and yeasts and all three were of some importance in determining the distribution of the bacteria. The actual age of the leaves in days is rarely recorded and their physiological state has only been noted in terms of whether or not senescence has begun. The number of leaves comprising a sample often appears to be decided in an arbitrary manner. Discs or segments have been cut out without any checks as to whether these represent a reliable subsample from the whole leaves. The mid-rib and leaf margins have often been deliberately avoided when cutting out discs without any obvious justification. Sampling is often conducted according to arbitrary calendar dates rather then being related to host plant development. Weather patterns can also affect the results obtained as samples taken after heavy rain may yield lower populations than other samples taken just prior to the rain. Such unscientific practices should be discouraged, although it is recognised that improvements in technique will require a great deal more preliminary work than is customary at present.

These problems are most acute in those general investigations designed to elucidate the microflora colonising aerial tissues over an extended period of time. By contrast, other types of approach are more satisfactory in that they yield reliable data concerning the magnitude and distribution of one or two components of the epiphytic microflora or they are specifically designed to minimise the complications discussed above.

Studies of individual saprobic microorganisms are rare but such autecological investigations are commonplace in plant pathology. More microbial ecologists could usefully adopt this approach which would lead to a detailed understanding of at least some saprobic organisms. An indication of the benefits which might be obtained can be seen in studies of *Pseudomonas mors-prunorum*, which is both a virulent stem and branch pathogen and a phylloplane inhabitant of cherry trees. Crosse (1959) attempted to obtain a reliable estimate of the inoculum potential, i.e. the population of the bacterium, on a stand of cherry trees and found that it was necessary to collect 8 leaves from each of 24 branches. These were washed whole since discs excluded the leaf apices where bacteria accumulated and they did not allow for differences in the distribution of the bacteria on the leaves of various cultivars.

An alternative, synecological approach which can yield reliable data involves the study of comparable batches of plants which differ only in that a treatment has been applied to some of the batches. Assuming that the treatment does not have a major affect on the growth of the plants then sample selection is simplified. This type of approach has been used in studies of the effects of pesticides on phylloplane populations (Dickinson & Wallace 1976; Andrews & Kenerley 1978). The main problem with such experiments is not the collection of samples but the subsequent processing in which pesticides can play a significant role. A similar system was used by Fokkema *et al.* (1979) to study the response of the phylloplane populations to added nutrients in the form of a sugar solution.

24.5.2 Transport and Storage

Apart from some general microbiological considerations, such as the collection of material into sterile or clean containers, transporting it to the laboratory without undue delay, and storing it at low temperatures if it is not processed immediately, few investigators have laid any emphasis on these stages in their procedure.

In some studies where the plants are near the laboratory there seems to be little wrong with the accepted methodology but difficulties can arise if the study involves long distance transport or a delay between sample collection and processing. Perhaps the most significant factor involved is humidity and especially the creation of a continuous high humidity in polythene bags. Millar and Richards (1974) have drawn attention to the dangers involved and they suggested that some processing might usefully be undertaken at the collection site. A simple alternative was suggested by Dickinson and Wallace (1976) who collected cereal tillers, instead of just the flag leaves which were being examined. These tillers were brought back in standing water at the ambient humidity and temperature and the leaves were only picked off just prior to processing.

If the plant material has to be kept for a period before processing, it can be kept at a low temperature, *c.* 5°C, but if the above suggestion is adopted then this may cause an unnecessary check to microbial activity.

24.6 Microscopic studies of epiphytes *in situ*

Roots may develop in an environment which encourages extensive microbial development but aerial plant surfaces are considered by many to be aesthetically more attractive and practically more convenient as regards the study of their epiphytic microfloras. The rhizoplane microflora is often difficult to distinguish from the rhizosphere and soil microfloras. This is especially so for most annual feeding roots, which do not have a clearly defined epidermis. By contrast the epiphytic microflora on aerial plant tissues is relatively easily distinguished from the air spora and it inhabits a variety of well-defined surfaces. These differences between roots and aerial tissues show up most clearly when one considers the number of microscopic examination techniques which have been devised for the study of their respective epiphytic floras.

A wide range of microscopic methods have been used to look directly or indirectly at the populations inhabiting aerial surfaces (Table 1). Amongst this profusion of methods several points stand out. The methods used for studying epiphytic yeasts and bacteria should not involve processes which might disturb loosely attached, single-celled organisms. Most microscopic methods suffer from the disadvantage that the organisms cannot be identified beyond the group or generic level. Methods which rely on the use of the microscope are likely to be relatively tedious. In a study of cereal leaves Dickinson and Wallace (1976) spent over six months recording one experiment and even then they were only able to measure the microflora on about 7 mm^2 of leaf surface per sample from each 20 m^2 plot! In this respect the modification described by Bottomley (1980) permits a very much larger sample to be processed (Table 1) but even then there is a large discrepancy between the scale of the microbial ecology and that of the higher

Table 1. Microscopic studies of aerial plant surface epiphytes.

DIRECT MICROSCOPY			INDIRECT MICROSCOPY	
Scanning microscopy	Light microscopy		Light microscopy	
	Transmission illumination	Incident illumination	Impression examined directly	Impression processed before examination
Methods (Royle 1976; Andrews & Kenerley 1978) Systematic records (Mishra & Dickinson 1981)	Tissue bleached and cleared (Daft & Leben 1966; Pugh & Buckley 1971a; Ruscoe 1971; McBride & Hayes 1977; Andrews & Kenerley 1978) Tissue sectioned (Carroll 1979) Tissue surface scraped and scrapings mounted for examination (Carroll *et al.* 1980) Trichomes detached and examined (Lindsey & Pugh 1976b)	Epifluorescence microscopy (Bernstein & Carroll 1977; MacNamara *et al.* 1977)	Agar and adhesive tape impressions (Diem 1974) Sellotape impressions (Beech & Davenport 1971; Fokkema & Lorbeer 1974) Cellulose nitrate or acetate (collodion, celloidin) impressions (Norse 1972b; van der Burg 1974; Dickinson *et al.* 1974; Wildman & Parkinson 1979) Plastic material (Allen 1973)	Cellulose nitrate impression, redissolved, sample aliquots filtered through Millipore filter which is then cleared, stained and examined (Bottomley 1980)

plant ecology or crop treatments. Such considerations make it imperative that the leaf sampling process is efficiently organised, that comparable areas of the leaf are studied at each sampling date and that appropriate statistical techniques are applied to the data obtained.

24.6.1 Scanning Electron Microscopy

In recent years improvements in scanning electron microscopy (SEM) techniques have resulted in the publication of many impressive pictures of epiphytic microorganisms (Fig. 1). These improvements in technique, notably involving better fixation and critical point drying of tissues, have not, however, made it any easier to obtain quantitative data from this instrument. Numerous studies of leaves and other aerial tissues have thus been graced with useful illustrations, but most population data has still been obtained using light microscopy. Such illustrations have, of course, value in themselves in that they enable us to gain a vivid impression of the way in which epiphytic microorganisms grow and colonise aerial plant surfaces. Scanning microscopy is also useful in that the great range of magnification enables bacteria to be examined more effectively than is possible using the light microscope. The stereoscopic images obtained also improve our understanding of the filamentous organisms on surfaces; e.g. Mishra and Dickinson (1981) found it easier to score the amounts of hyphae on leaves using impression films but the presence of fungal conidiophores was more easily detected using the SEM.

To date little quantitative data has been obtained from the use of SEM. This is probably due to limitations on access to suitable instruments, as the collection of such data is very time consuming. In addition, there are problems in creating suitable scoring systems which can be operated efficiently within the limitations imposed by the instrument. There is a further problem in that it is difficult or impossible to know if the organisms being scored are alive or dead and less information can be obtained regarding their likely identity than would be gained using light microscopy on impression films.

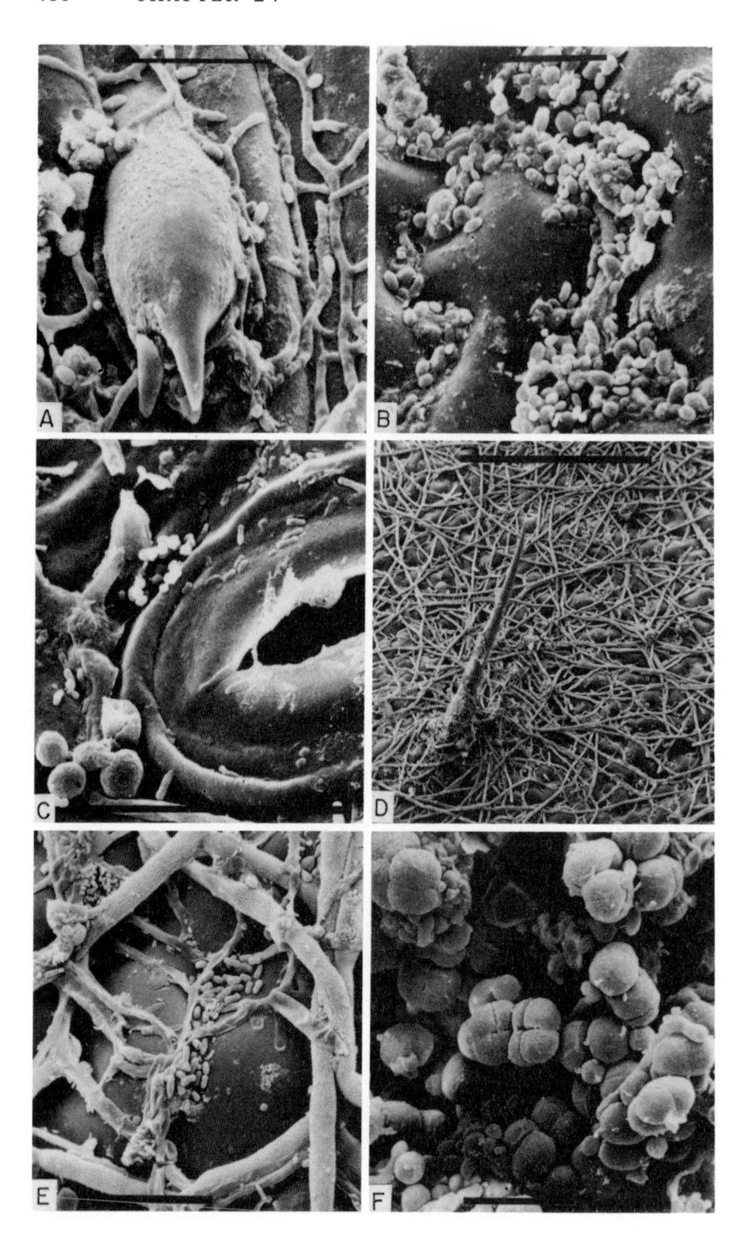

Figure 1. Scanning electron micrographs of aerial plant surfaces. All material was fixed in 2.5% (v/v) gluteraldehyde buffered with cacodylate at pH 7.0, dried in a critical point apparatus, mounted on stubs and coated with gold for examination in a Jeol SM1 scanning microscope. Scale bars indicate magnification, A, B, C and E = 20 μm and D and F = 400 μm.
A. Adaxial surface of a *Dactylis glomerata* leaf with a short, pointed trichome prominent. Showing extensive hyphal development with scattered yeast cells and fungal spores.
B. Adaxial frond surface of *Dryopteris filix-mas*. Numerous yeast cells occupy the depressions between epidermal cells.
C. Adaxial leaf surface of *Hedera helix* showing part of a stoma. At higher magnifications numerous bacteria can be seen, usually grouped in small clusters or colonies.
D. Adaxial surface of *Mercurialis perennis* leaf. Community dominated by filamentous fungi which have colonized the whole area. The base of the trichome is particularly densely clothed in microbial structures.
E. As D but at higher magnification which shows that amongst the fungi there are a lot of bacteria which are associated with the hyphae and with the epidermal cell junctions.
F. *Acer pseudoplatanus* bark. The bark surface is dominated by an extensive colony of *Pleurococcus* which is itself colonised by yeasts and bacteria.

24.6.2 Light Microscopy—Transmission Illumination

Numerous methods have been devised for examining epiphytic microorganisms directly on aerial plant tissues using light microscopy. Where the tissues are relatively thin and the main pigment present is chlorophyll then the techniques involved are usually simple (Preece 1971). The chlorophyll can be bleached out with a gaseous or liquid reagent or can be dissolved out with alcohol, chloroform or pyridine and the tissue is then cleared with chloral hydrate, lactophenol or another reagent with a similar refractive index. These two processes can also be carried out simultaneously with a reagent mixture such as lactophenol-ethanol (Dickinson & Crute 1974). The tissues are then stained in cotton blue, periodic-Schiff, phenol-acetic-aniline blue, Ziehl's carbol-fuchsin or any other stain which differentiates between the host and the microorganism.

Such methods have been mainly developed for the study of plant pathogens, which are usually firmly attached to or established within their host's tissues. They are less useful for examining loosely attached bacteria, yeasts, fungal spores or algae on aerial surfaces. These methods are also difficult or impossible to apply where the host tissues are extremely thick or where their surfaces are convoluted or covered with numerous trichomes. Tissues containing substantial amounts of lignin, suberin or sclerenchyma or with a thick waxy cuticle may not be readily cleared or made sufficiently translucent to allow adequate light penetration. In addition, some photosynthetic tissues turn brown when attempts are made to remove their chlorophyll.

Such difficulties have usually resulted in microbial ecologists avoiding the particular tissues, but in a few instances ingenious methods have been devised to circumvent these problems. Thin surface layers can be removed from tissues by either mechanical sectioning or the use of a pectinase preparation (Preece 1962). Epiphytes on twigs have been examined by scraping off the superficial bark, mounting the scrapings under a coverslip and recording the microbial structures in a number of microscope fields (Carroll *et al.* 1980). The surface microflora of conifer needles has been studied by examining numerous transverse sections through the needles and recording the distribution and size of the microbial colonies around the perimeter of each section (Carroll 1979). A novel approach was used by Lindsey and Pugh (1976b) who found that they could detach the peltate trichomes of a *Hippophaë* sp. by shaking leaves in water. These trichomes were collected by centrifugation and then stained and mounted for microscopic examination.

24.6.3 Light Microscopy—Incident Illumination

High magnification incident illumination has been mainly used in conjunction with fluorescence microscopy which appears to offer exciting possibilities for the study of microorganisms in natural habitats. Bernstein and Carroll (1977) used this method to study the epiphytes on Douglas fir needles but most of the microbial structures they saw could only be identified to rather broad categories, such as darkly pigmented fungal hyphae or hyaline microbial colonies. They noted this problem and attributed it to the low resolution inherent in the optical system and the poor depth of focus available. MacNamara *et al.* (1977) found that this system of microscopy worked well if the host tissues were intact but damaged areas fluoresced and this interfered with their observations of microbial activity.

24.6.4 Indirect Microscopy—Tissue Impressions

Leaf impressions can, in many instances, provide a simple, elegant and effective way of separating epiphytes from their hosts, whilst at the same time maintaining their distribution patterns as they occur *in situ* (Fig. 2). This method works well with planar, glabrous tissues and some of our most detailed knowledge of epiphytic populations has been obtained by painstaking studies of such films (Dickinson & Wallace 1976).

Materials used to obtain impressions include agar, transparent adhesive tape and a variety of cellulose-based compounds. These materials have been applied in many different ways but there seems good reason to regard spray applications, preferably in two stages, as the optimum system (Dickinson *et al.* 1974). The first spray, which is directed above rather than at the tissues, is a light application designed to hold the epiphytes in place and the second, heavier, application builds up the thickness of the film so as to enable it to be stripped off the tissues and handled without tearing. All the materials used permit the microorganisms within the impression to be stained without any difficulty.

This method has been widely used (Table 1) but there have, to date, been few systematic investigations into its efficiency in removing all epiphytic microbes. This aspect could be checked by examining stripped leaves using SEM. Another methodological problem is that it is difficult to obtain uniformly thin impression films of the leaf margins and petioles. Thus these microhabitats can become neglected in favour of the lamina which is more conveniently handled. This problem assumes significance because many of the microbial propagules in the air spora are more readily impacted on to the leaf margins and petioles than on to the lamina. A similar problem occurs where the study involves senescing tissues, which usually support considerable microbial activity. Impression films made

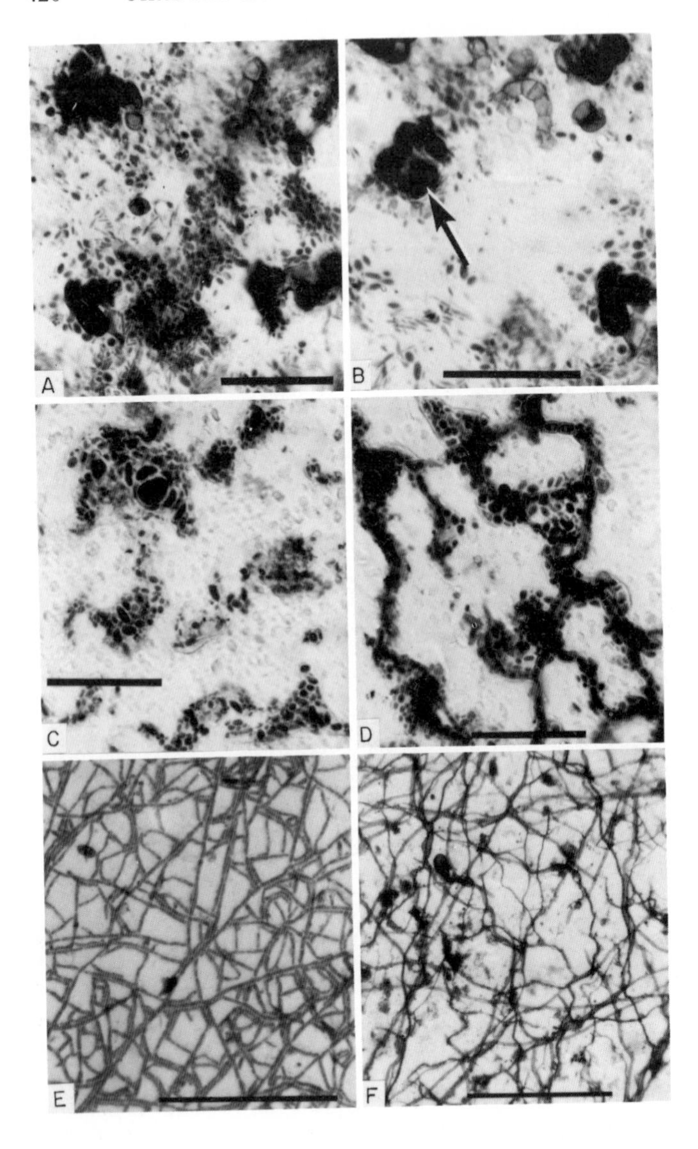

Figure 2. Light micrographs of microorganisms inhabiting aerial plant surfaces. Preparations were made by spraying the surfaces with cellulose nitrate, stripping off the film and staining it in 0.1% (v/v) trypan blue in lactophenol. Scale bars indicate magnification, A to D = 20 μm; E and F = 100 μm.
A. Adaxial surface of *Hedera helix* leaf. A heterogeneous mixture of fungal spores, yeast cells and bacteria, with no obvious spatial distribution pattern.
B. As A. Yeast cells and bacteria plus a short length of thick-walled hypha and a small algal colony (arrowed).
C. Adaxial surface of *Dryopteris felix-mas* frond. Extensive yeast colonies with considerable variation in cell size in each colony, and discrete distribution related to pattern of epidermal cells.
D. As C. Hyphae growing along depressions between epidermal cells budding off spores in a similar manner to that seen in *Aureobasidium* species. Yeast cells also common in the grooves.
E. Adaxial surface of *Rhododendron ponticum* leaf showing part of a very extensive hyphal network which did not appear to be producing any spores. Few other organisms can be seen at this magnification.
F. As E. Hyphal network colonising leaf though not the same fungus as in E. More yeasts and other organisms present than in E.

with cellulose-based compounds are difficult to remove from such dying tissues and hence recording may have to cease just when most activity occurs.

Impression films are a valuable guide to the form and distribution of microbes on aerial plant surfaces but quantitative data can only be obtained by scoring the number of cells and lengths of hyphae per unit area of film. This is a severe limitation when the object of the study is to examine crops or large stands of natural vegetation (Dickinson & Wallace 1976). Hence efforts have been made to assess the microbial content of the films using a more rapid process. Bottomley (1980) has attempted to compare several crop treatments by assessing the chitin content of impression films from the flag leaves of wheat. This method was unsuccessful in that the data obtained did not correlate with a direct microscopic assessment of the films. A second method was, however, more successful, if somewhat less convenient. In this method impression films are taken from a large number of leaves, their area is measured and the bulk sample is dissolved in amyl acetate. The microbes are transferred to water by centrifugation. Finally an aliquot of the resulting suspension is filtered through a Millipore filter which is cleared and stained. Counts and measurements are made on known areas of the filter and these can be related to the area of film originally processed. Using this method which is similar in principle to that of Hanssen *et al.* (1974) a larger and more representative sample of leaves can be examined than was possible when the emphasis was on direct microscopy of the original film.

Whatever method is used to assess the amounts of microbial material contained within the impression films there is the further problem of determining the physiological state of the various pieces of hyphae and yeast cells. Frankland (1975) has discussed the problems and possibilities involved and it is imperative that her suggestions, involving the use of phase contrast microscopy and assessments of the presence of cell contents, be adopted for studies of the phylloplane.

Another promising approach involves the use of double-sided, sticky, cellulose tape which is pressed against the leaf to obtain a print of its surface mycoflora (Langvad 1980). This tape can then be stuck to a microscope slide, covered with a thin film of an agar medium and then incubated in a moist environment. Many fungi will grow into the medium and by comparing results from tapes which have been stained and examined microscopically with those on which fungi were cultured information can be obtained about the viability of particular components of the mycoflora.

24.7 Culture methods for the enumeration of microorganisms

24.7.1 Dilution Plate Methods

The majority of phylloplane studies have been based on cultural techniques involving a variation of the dilution plate method. In these studies the general assumption has been made that the microbial populations on leaves can be dislodged by a washing process which involves shaking samples in a liquid medium. In some instances, where the leaf being examined is more or less planar and glabrous, this assumption may be justified, at least as far as certain groups of microorganisms are concerned. More complex aerial plant tissues, such as the corrugated leaves of the genus *Ammophila* or the very hairy leaves of *Stachys* species, have been generally neglected by microbiologists, due no doubt to the technical problems which would be encountered if attempts were made to wash all epiphytic organisms off their surfaces. Washing methods are generally satisfactory when dealing with single-celled organisms or spores but little is known about their efficiency for the removal of hyphal networks.

The washing technique itself is an important element in the success of these methods. Crosse (1959) has shown that the numbers of bacteria dislodged from leaves increases with the time of shaking up to about 4 h, but Dickinson *et al.* (1975) found that maximum numbers of bacteria were removed from *Lolium* leaves after 20 min shaking. Dickinson (1971) has discussed some of the factors which can affect the efficiency of the washing process and there is obviously a need for careful checks to be made whenever this technique is used.

The methods used to cultivate organisms in the suspensions obtained have mainly involved either overpour plates, where the suspension is incorporated into a molten agar culture medium, or spread plates, where the suspension is spread over the surface of a prepoured, and usually dried, agar medium. Overpour plates are normally used for bacteria and filamentous fungi but the spread plate has become widely accepted as the best method for culturing yeasts following Flannigan's (1973) report that these organisms are remarkably heat sensitive. The media used in fungal studies have been mainly simple concoctions with a bias towards semi-natural formulations, such as potato dextrose agar, malt extract agar and yeast extract agar and yeast extract–apple juice agar (Beech & Davenport 1971). Most of these mixtures probably provide richer supplies of nutrients than are available in the field and this may make it difficult to compare the populations as recorded by cultural techniques and those seen using direct examination methods. For example, Dickinson and Bottomley (1980) demonstrated that *Cladosporium cladosporioides* can multiply in a yeast-like manner in nutrient-poor conditions whereas it forms hyphal colonies on agar media. Dickinson *et al.* (1975) compared a range of media for the isolation of bacteria and they found that the largest numbers of colonies grew on a glucose–yeast extract medium.

Incubation conditions can have as pronounced a selective effect on the populations recorded as can the medium employed. In many studies of aerial plant surfaces incubation has been at 25°C in the dark for one or two weeks. A simple consideration of the conditions which prevail in these habitats would, however, suggest that the organisms experience fluctuating temperatures with alternating light and dark over the long periods during which they can develop. In some recent studies note has been taken of these points and more appropriate incubation conditions have been devised (Dennis 1976; Dickinson & Wallace 1976; Pennycook & Newhook 1978; Wildman & Parkinson 1979).

Populations have usually been calculated by counting colonies on appropriate dilutions and then relating these to the area or weight of tissue washed by use of the dilution factor. A modification which can be used to estimate numbers of individuals of various groups of microorganism involves a plate-dilution frequency and most-probable-number technique (Andrews & Kennerley 1978).

24.7.2 Isolation from Washed Tissues

In addition to the populations of bacteria, yeasts and spores which can be readily washed off leaves there are other organisms which appear to be firmly attached to aerial plant surfaces, due perhaps to their extensive vegetative growth. A number of species of filamentous fungi behave in this way and they can be grown from plant tissues which have been exhaustively washed. Other fungi have a more complex relationship with their host plants in that they can both grow on its surface and also invade the living tissues. These have been considered to be endophytes by Pugh and Buckley (1971b) but it is probably more accurate to consider them as parasites, even if their development inside the plant is very limited (O'Donnell & Dickinson 1980). Most of the fungi concerned are not likely to be considered significant pathogens as far as their effects on the host are concerned but their limited parasitism is probably of greater significance in relation to their own biology as they compete for the plant's tissues during senescence.

A number of cultural methods have been used to study these two groups of organisms. One method which facilitates the isolation of both epiphytic and internal colonisers involves washing leaves in several aliquots of sterile liquid until all detachable propagules have been removed. The leaf tissue is then plated on an agar medium and after incubation the organisms growing from it are recorded. This technique is analogous to the root washing method devised by Harley and Waid (1955) to study the colonisation of the rhizoplane. When applied to leaves there is no doubt that a different spectrum of fungi are recorded than is found if the initial washing suspensions are cultured (Dickinson & Wallace 1976). What is less easy to determine is whether the individual colonies growing around the washed pieces have developed from epiphytic mycelium or from an internal, parasitic infection. It is also difficult to determine if the organisms are growing from the upper or the lower surface and this prompted Lindsey and Pugh (1976b) to explore the use of the resin Bedacryl 122X to seal one or the other surface prior to plating out the washed discs.

24.7.3 Use of Surface Sterilants

Another method used to study the organisms which have invaded aerial tissues involves the use of biocides which kill off all epiphytic organisms. Internal colonists can thus grow out on to the dead leaf tissue or on to an agar medium without experiencing undue competition. The chemicals which have been employed as surface sterilising agents include mercuric chloride (Pugh & Buckley 1971a; Norse 1972a), silver nitrate (Wildman & Parkinson 1979) and sodium hypochlorite (O'Donnell & Dickinson 1980). This technique is frequently used by plant pathologists as it is one of the simplest ways of obtaining cultures of many pathogens. In open-ended investigations into microbial ecology, however, it is less acceptable since it is difficult to determine the optimum treatment conditions in terms of the strength of the sterilant and the length of time the tissue should be exposed to its action. Careful checks must be made whenever this technique is employed and the aim in most instances must be to achieve the minimum concentration and exposure period which will allow the broadest spectrum of organisms to develop. In autecological studies its use is more acceptable as the effects of a sterilant on a specific fungus can be determined (O'Donnell & Dickinson 1980).

Comparative data for the phylloplane have been obtained by washing leaves and by using surface sterilants. Studies of the genera *Acer* (Pugh & Buckley 1971a), *Eucalyptus* (MacCauley & Thrower 1966), *Nicotiana* (Norse 1972a), *Nothofagus* (Ruscoe 1971) and *Populus* (Wildman & Parkinson 1979) have yielded data which suggest that a number of the filamentous fungi which were initially thought to be saprobic epiphytes can colonise their host's internal tissues. The extent of this internal colonisation generally increased as the host tissues approached senescence. Pugh and Buckley (1971a) attempted to produce concrete proof that species of *Aureobasidium* and *Cladosporium* had invaded *Acer* leaves by using pectinase to macerate the leaves. The maceration was allowed to proceed until only the veins remained intact. These were then washed and plated on to an agar medium. Both fungi grew from a significant number of veins.

These cultural techniques are undoubtedly useful in that they not only provide some information as to site of activity of the organisms but they also permit the species composition of the population to be determined. It is, however, necessary in most instances for such cultural data to be supported by direct observations which will provide unequivocal proof as to the exact location of the microbial colonists.

24.7.4 Spore Fall Method

Amongst the cultural techniques used for studying epiphytic microorganisms one method is unique in that it has only been applied extensively to the aerial surfaces of green plants. This method is used to search for members of the Sporobolomycetaceae, a group of basidiomycete yeasts which produce ballistospores. The ballistospores are forcibly discharged from their parent cells and then they become dispersed by the action of gravity and wind. Kluyver and van Niel (1924), Derx (1930) and Buller (1933) capitalised on this feature of their reproduction and developed the spore fall method which has since been extensively employed to study the ecology of these yeasts.

Pennycook and Newhook (1978) examined various aspects of this process before adopting a method involving exposure of a leaf over an agar medium for a period of 12 h light: 12 h dark at 23°C, after which the leaf was removed before a further 72 h incubation at 16°C. They used Difco cornmeal agar and counted the individual colonies which developed beneath each leaf. In our experience this method works well except that dense populations of the genus *Sporobolomyces* can rapidly overgrow the rather slow-growing colonies of *Tilletiopsis* which, because of their lack of pigmentation, tend to be more difficult to see even if growing alone.

Using this method Last (1955, 1970), van der Burg (1974), Dickinson and Wallace (1976) and others have shown that *Sporobolomyces* spp., and to a lesser extent *Tilletiopsis* spp. and *Bullera* spp., occur on a wide range of hosts. On dicotyledons large numbers of cells have been recorded on adaxial leaf surfaces whereas on monocotyledons the populations are similar on both surfaces but greatest at the distal end of the lamina. The spore fall method has shown the presence of distribution patterns on individual leaves, with high concentrations of the yeasts occurring along the veins and at the margins. It is not sufficiently sensitive to show more detailed aspects of their distribution, such as the fact that they congregate in the grooves above the anticlinal epidermal cells walls (van der Burg 1974).

The elegance and simplicity of the spore fall method has great attractions but Dickinson and Wallace (1976) and others have drawn attention to its obvious weaknesses. These were explored in detail by Pennycook and Newhook (1978) and they concluded that it should not be used for estimating population densities. It is useful for rapid checks on whether these ballistosporic yeasts are actively growing and sporulating. It can also be helpful in showing whether they are growing especially well, when, for example, the leaf surface is being bathed in aphid honeydew (van der Burg 1974), or particularly badly, when, for instance, a broad-spectrum fungicide has decimated the populations (Dickinson & Wallace 1976).

24.7.5 Homogenisation and Maceration

Relatively few investigations of the microflora of aerial plant surfaces have been based on techniques involving homogenisation or maceration of the whole tissue (Lindsey 1976). Exceptions to this have involved studies where the prime aim has been to examine the bacterial and yeast components of the microflora (Di Menna 1959; Hislop & Cox 1969; Warren 1976) which were assumed to be mainly present on the tissue's surfaces, or where the tissues being examined had a complex morphology (Dennis 1976; McBride & Hayes 1977) which did not readily permit samples to be taken with respect to a known surface area.

Studies of a number of soft fruits have been described by Dennis (1976) who reviewed the problems which result from the use of maceration methods. Apart from a number of technical problems which will eventually be solved there are two fundamental difficulties in any such procedure. The destruction of living host tissues results in the release into the growth medium of a number of metabolites, and their oxidised derivatives, which would normally be well separated from the habitat in which the phylloplane fungi grow. It is almost impossible to identify what the consequences of this might be in terms of the development of the microflora. In addition the mechanical destruction of the host tissues is likely to be accompanied by the break up of some microbial structures, resulting in loss of viability, or fragmentation, leading to an artificial increase in the numbers of propagules. Despite these potential problems it is of interest that Hislop and Cox (1969) found that there were no consistent differences between the numbers of bacteria, moulds and yeasts recorded on apple leaves using a maceration technique as against those obtained using a simple shaking method.

In some studies maceration has been carried out following a washing process, which is supposed to have removed all the readily detachable organisms from the tissue's surface. Such a scheme has been advocated for determining the total yeast microflora of a tissue (Beech & Davenport 1971). Maceration can also be used after surface sterilisation. In both instances, the foregoing criticisms are still valid and data from such combined techniques should be treated with caution.

24.8 Culture methods for qualitative studies of epiphytes

24.8.1 Imprint Techniques

One of the first techniques used to culture microbes from plant surfaces involves pressing the tissue momentarily against an appropriate solidified medium which is then incubated. Dickinson *et al.* (1975) increased the spectrum of organisms isolated from *Lolium* leaves by employing this technique, which yielded numerous actinomycetes whose presence had not been discovered using a washing and dilution plating method. Lindsey and Pugh (1976a) have provided some useful comparative data regarding the value of the impression technique, which was shown to be an effective method for recording the presence of *Cladosporium* and *Sporobolomyces* species. Comment was made before on an indirect imprint technique, developed by Langvad (1980), which has some potential as a rapid method for assessing microbial distribution whilst the leaves are still attached in the field.

Andrews and Kenerley (1978) found that the leaf imprint technique underestimated leaf populations as compared with a leaf washing method. They attributed this to the occurrence of many of the organisms in colonies or mixed groups. They also found imprinting was particularly inefficient for assessing bacteria, including actinomycetes.

This technique has, however, only been used occasionally in recent investigations as its obvious limitations preclude its widespread adoption (Beech & Davenport 1971). It is useful as a field technique when it is necessary to show beyond doubt that an organism is present on the tissues in the natural environment. It is also of limited use for studying the ecological distribution of organisms on plant surfaces. An ingenious modification which increases its potential in this respect is the balloon print method described by Rusch and Leben (1968). With this technique a leaf is pressed against the sterile surface of a partially inflated balloon, the balloon is then fully inflated and the leaf imprint is transferred to an agar medium. Hence the leaf organisms are spaced more widely than would be the case with a direct imprint and this will lessen the competition between adjacent colonies on the agar medium.

24.8.2 Moist Chamber Incubation

Many students of plant litter decomposition have employed a moist chamber technique in which debris is kept under conditions of high humidity and at a constant and generally warm temperature for an extended period. These conditions favour the sporulation of many fungi and their presence can then be detected using a dissecting microscope. Using this technique Webster (1959), Kendrick and Burges (1962), Frankland (1966) and others have recorded associations of fungi which grow on various types of litter as it becomes progressively decomposed. Quantitive data can be obtained by examining a large number of pieces of litter and scoring the presence or absence of particular species on each. Such data are, however, only a guide to possible trends in the fungal populations as the appearance of conidiophores may indicate that a species has colonised a large part of the tissue or that it has merely germinated and then immediately sporulated, owing to the conditions in the debris being unfavourable for extensive growth.

There are, in addition, two fundamental weaknesses in the moist chamber method itself. The first of these is that many fungi colonising plant debris sporulate either sporadically or not at all under the conditions provided by this technique. For example, the majority of the Basidiomycotina which produce toadstool-like sporophores are unlikely to fruit and the sexual fruit bodies of many Ascomycotina only develop when the debris reaches a certain state of decay. In addition, those fungi that can be induced to sporulate are unlikely to be all equally encouraged by whatever incubation conditions are emplyed. The successful colonisation of agar plates by soil fungi is affected by temperature (Dickinson & Kent 1972), and light, humidity and the competitive effects of other organisms can all differentially influence the sporulation of invidual species.

A second problem with the moist chamber technique is that it involves transferring debris from one set of environmental conditions in the field, over which it itself exerts a substantial influence, to another set of conditions in the laboratory. This can mean, for instance, that litter from the L layer, which is both regularly wetted and dried and subject to widely fluctuating temperatures, and litter from the H layer, which is permantly wet and at a relatively steady temperature, are both held under the same constant conditions of humidity and temperature. Thus incubation conditions are not likely to be appropriate for both the L and H layer populations. In addition, the laboratory incubation conditions are unlikely to take account of the biotic element which influences the conditions prevailing in the field.

On a more constructive note it may be argued that this method enables potentially important members of the saprobic mycoflora to be identified. This may be acceptable when positive records are obtained but the absence of a name from a species list cannot be taken as proof that that fungus is not growing in that particular substrata. It should be noted that this method need not involve the introduction of further competition and that those fungi which sporulate do so at the expense of nutrients obtained from the debris itself. This is the case if the moist environment is created from using inert material, such as sand or asbestos, but many workers have employed filter paper and there is no doubt that this can introduce an important selective advantage for cellulolytic species.

All these considerations apply with equal or greater force when the technique is used for studies of living, aerial plant tissues. When such tissues are being studied thare is an additional problem in that discs, segments or whole organs are cut from the parent plant and this act leads to their premature senescence. Hence the tissues will begin to change and mycoflora which develops will be more relevant to the early stages of senescence than to the phylloplane.

Sharma *et al.* (1974) compared the use of moist chambers with the dilution plate and leaf impression methods and they found that several fungi were exclusively recorded using the former method. The fungi involved, notably species of *Chaetomium*, *Colletotrichum* and several myxomycetes, were unlikely to have been significant phylloplane inhabitants but it is of interest that they were present on the leaf surfaces prior to senescence. This technique has also been used by Dickinson (1967), who found that more species were recorded on *Pisum* leaf discs incubated in moist chambers than were found on washed discs plated out on an agar medium. Ruscoe (1971) employed the damp chamber method to examine whole untreated leaves and leaf discs which had been serially washed or surface sterilized. A smaller range of fungi were recorded on all these samples than were found on washed leaves which had been incubated on dextrose peptone agar. There were no obvious differences between the floras on the washed and the surface-sterilised discs.

In conclusion, it may be argued that this method is inappropriate for studies of epiphytes on living tissues and efforts would be better directed towards improving the efficiency of those techniques which are more soundly based.

24.9 Strategies for studying epiphytic microorganisms

It is now widely accepted that studies of microbes in natural habitats should involve several complementary methods which will explore different facets of the populations. As can be seen by the foregoing discussion there is a wide variety of methods available for the study of epiphytic microorganisms on aerial plant surfaces. In such studies combinations of four or five methods are now commonplace and this can be seen to be the culmination of a long-term trend towards a more comprehensive approach (Lindsey 1976).

The application of such combinations of methods, however, is time consuming; for example, in an experiment to determine the effects of five fungicides on wheat flag leaf microorganisms it was necessary to employ six people on each sampling date in order to finish the processing which had to be completed on that day. Some short cuts are possible but until more information is available concerning the response of the microbial populations to various treatments then such studies will continue to be positively laborious. There are more prospects for improvements in the recording processes. Methods such as that described by Bottomley (1980) (section 24.6.4) enable large areas to be sampled efficiently. Other improvements may involve the use of electronic counting and sizing instruments, which might be used to record organisms washed off tissues into suspension, and devices which would enable areas to be selected out at random on an SEM screen.

24.10 Processing the data

24.10.1 Bases for Expression of Data

Counts or measurements of microorganisms can be related to one of several parameters which define the amount of tissue sampled. In many studies discs or segments of a known size are cut from the tissues and hence the counts can be related to comparable area measurements. If whole leaves or petals are processed, an area measurement can be obtained by making photocopy silhouettes which are cut out and weighed, or by using area measuring instruments.

Such measurements can seem to provide a sound basis for comparing treatments or different hosts but results based on area should be viewed with extreme caution. Even planar, glabrous, gland-free leaves have microscopic surface corrugations due to the convex bulges of the individual epidermal cells and the irregular nature of the stomatal apertures. More convoluted leaves, needles or twigs or those organs bearing one or more type of trichome or gland can be so complex that area measurements of the type discussed above are almost meaningless.

This difficulty has led a number of investigators to use fresh or dry weight measurements as the basis for their sampling procedure. Weight may be useful if the tissues being sampled are very similar in morphology and anatomy but Dickinson *et al.* (1975) found that there can be very poor correlation between area and weight measurements, even when the same leaf off the same host is sampled throughout the growing season.

There is no obvious and simple solution to this problem and each host or tissue must be considered separately so as to determine the most appropriate basis for the expression of the results, bearing in mind the underlying purpose of the investigation and the nature of the surfaces being studied.

24.10.2 Data Processing

In many studies of the ecology of epiphytic microorganisms published to date the data have been presented in a descriptive fashion rather than being treated in a statistical manner. Such an approach is not satisfactory and it is especially unfortunate when large amounts of data, which have been collected by rather laborious methods, are available for analysis. More attention must be paid to the design of investigations and experiments so as to allow such data to be rigorously analysed.

Wildman and Parkinson (1979) have discussed some of the problems involved in handling the type of data obtained from cultural and direct examination techniques. They found that records from washed or surface-sterilised leaf discs needed to be transformed before analysis. Data from microscopic studies of leaf impressions were analysed using non-parametric tests, such as the Kruskall-Wallis one-way analysis of variance and the Mann-Whitney U test—Wilcoxon Rank Sum W test. Analysis of variance models were employed by Andrews *et al.* (1980) to test various hypotheses regarding the factors which influence the distribution of microorganisms in a complex tree canopy. More detailed analyses of the distribution of organisms in the various microhabitats provided by leaves can be based on the methods of pattern analysis used by higher plant ecologists. Using these methods Allen (1973) showed that colonies of the alga *Phycopeltis* were significantly associated with the margins of individual epidermal cells and that, on a larger scale, they were abundant on the abaxial leaf margins. Statistical methods which can be applied to microscopic data indicating distribution patterns of organisms on individual tissues and within a canopy have been discussed by Bernstein and Carroll (1977) and Carroll (1979). Microbial cells were found to be significantly associated with the midrib groove and stomatal zones of Douglas fir needles and the microbial cell volumes were found to vary with the age of the needles and their height in the canopy (Carroll 1979).

24.11 Future studies of aerial surface epiphytes

In future studies of epiphytic populations more attention should be paid to assessing biomass or microbial cell volumes (Carroll 1979) rather than simply quoting lengths of hyphae and numbers of yeast cells. Estimates of the standing crop of microbial epiphytes and of relative production should also be made to enable their role in the whole ecosystem to be determined.

As regards the individual epiphytic organisms, urgent questions requiring attention concern the biology of the filamentous fungi which can become established on green tissues. It seems likely that the yeast-like fungi are saprobic epiphytes but there is increasing evidence that many of the filamentous fungi have parasitic tendencies or are able to become established as endophytes within living host tissues.

The nutrition of the phylloplane inhabitants also needs further study. It has not yet been established whether the phylloplane bacteria or yeasts can cause a significant deterioration of any component of the cuticle. In addition there is still too little information concerning the quantities of leachates or exudates which leak out on to the leaf surface.

The effects of environmental factors on the phylloplane populations are still largely inferred from investigations in which climatic changes occurred by chance. Experimental proof is needed of the effects of heavy rain, long periods of high humidity and prolonged high temperatures. The influence of pollutants, including noxious gases and agricultural chemicals, is the subject of continuing study.

The climax of all these studies will be sufficient understanding of the nature and behaviour of the saprobic populations on plant surfaces such that antagonistic interactions with pathogenic organisms can be predicted, tested and employed in the fight against disease.

Recommended reading

BAKER K. F. & COOK R. J. (1974) *Biological Control of Plant Pathogens*. W. H. Freeman, San Francisco.

BLAKEMAN J. P. (1981) *Microbial Ecology of the Phylloplane*. Academic Press, London.

DICKINSON C. H. & PREECE T. F. (1976) *Microbiology of Aerial Plant Surfaces*. Academic Press, London.

LAST F. T. & DEIGHTON F. C. (1965) The non-parasitic microflora on the surfaces of living leaves. *Transactions of the British Mycological Society* **48**, 83–99.

PREECE T. F. & DICKINSON C. H. (1971) *Ecology of Leaf Surface Micro-organisms*. Academic Press, London.

RUINEN J. (1961) The phyllosphere. I. An ecologically neglected milieu. *Plant and Soil* **15**, 81–109.

References

ALLEN T. F. (1973) A microscopic pattern analysis of an epiphyllous tropical alga. *Journal of Ecology* **61**, 887–99.

ANDREWS J. H. & KENERLEY C. M. (1978) The effects of a pesticide program on non-target epiphytic microbial populations of apple leaves. *Canadian Journal of Microbiology* **24**, 1058–72.

ANDREWS J. H. & KENERLEY C. M. (1980) Microbial populations associated with buds and young leaves of apple. *Canadian Journal of Botany* **58**, 847–55.

ANDREWS J. H., KENERLEY C. M. & NORDHEIM E. V. (1980) Positional variation in phylloplane microbial populations within an apple tree canopy. *Microbial Ecology* **6**, 71–84.

AUSTIN B., GOODFELLOW M. & DICKINSON C. H. (1978) Numerical taxonomy of phylloplane bacteria isolated from *Lolium perenne*. *Journal of General Microbiology* **104**, 139–55.

BAKER K. F. & COOK R. J. (1974) *Biological Control of Plant Pathogens*. W. H. Freeman, San Francisco.

BEECH F. W. & DAVENPORT R. R. (1971) A survey of methods for the quantitative examination of the yeast flora of apple and grape leaves. In: *Ecology of Leaf Surface Micro-organisms* (Eds T. F. Preece & C. H. Dickinson), pp. 139–57. Academic Press, London.

BERNSTEIN M. E. & CARROLL G. C. (1977) Microbial populations on Douglas fir needle surfaces. *Microbial Ecology* **4**, 41–52.

BILLING E. (1976) The taxonomy of bacteria on the aerial parts of plants. In: *Microbiology of Aerial Plant Surfaces* (Eds C. H. Dickinson & T. F. Preece), pp. 223–73. Academic Press, London.

BLAKEMAN J. P. (1981) *Microbial Ecology of Phylloplane*. Academic Press, London.

BLAKEMAN J. P. & BRODIE I. D. S. (1976) Inhibition of pathogens by epiphytic bacteria on aerial plant surfaces. In: *Microbiology of Aerial Plant Surfaces* (Eds C. H. Dickinson & T. F. Preece), pp. 529–59. Academic Press, London.

BOTTOMLEY D. (1980) Studies of *Alternaria* and *Cladosporium* as pathogens of *Triticum aestivum*. Ph.D. thesis, University of Newcastle-upon-Tyne, England.

BULLER A. R. H. (1933) *Sporobolomyces*, a basidiomycetous yeast-genus. In: *Researches on Fungi*, vol. 5, pp. 171–206. Longmans Green, London.

BURRI R. (1903) Die Bakterienvegetation auf der Oberfläche normal entwickelter Pflanzen. *Zentralblatt für Bakteriologie, Parasitenkunde, Infektionskrankheiten und Hygiene* (*Abteilung II*) **10**, 756–63.

CARROLL G. C. (1979) Needle microepiphytes in a Douglas fir canopy: biomass and distribution patterns. *Canadian Journal of Botany* **57**, 1000–7.

CARROLL G. C., PIKE L. H., PERKINS J. R. & SHERWOOD M. (1980) Biomass and distribution patterns of conifer twig microepiphytes in a Douglas fir forest. *Canadian Journal of Botany* **58**, 624–30.

CROSSE J. E. (1959) Bacterial canker of stone-fruits. IV. Investigation of a method for measuring the inoculum potential of cherry trees. *Annals of Applied Biology* **47**, 306–17.

DAFT G. C. & LEBEN C. (1966) A method for bleaching leaves for microscope investigation of microflora on the leaf surface. *Plant Disease Reporter* **50**, 493.

DAVENPORT R. R. (1976) Distribution of yeasts and yeast-like organisms on aerial surfaces of developing apples and grapes. In: *Microbiology of Aerial Plant Surfaces* (Eds C. H. Dickinson & T. F. Preece), pp. 325–60. Academic Press, London.

DENNIS C. (1976) The microflora on the surface of soft fruits. In: *Microbiology of Aerial Plant Surfaces* (Eds C. H. Dickinson & T. F. Preece), pp. 419–32. Academic Press, London.

DERX H. G. (1930) Etude sur les Sporobolomycètes. *Annales mycologici* **28**, 1–23.

DICKINSON C. H. (1967) Fungal colonization of *Pisum* leaves. *Canadian Journal of Botany* **45**, 915–27.

DICKINSON C. H. (1971) Cultural studies of leaf saprophytes. In: *Ecology of Leaf Surface Micro-organisms* (Eds T. F. Preece & C. H. Dickinson), pp. 129–37. Academic Press, London.

DICKINSON C. H. (1976) Fungi on the aerial surfaces of higher plants. In: *Microbiology of Aerial Plant Surfaces* (Eds C. H. Dickinson & T. F. Preece), pp. 293–324. Academic Press, London.

DICKINSON C. H., AUSTIN B. & GOODFELLOW M. (1975) Quantitative and qualitative studies of phylloplane bacteria from *Lolium perenne*. *Journal of General Microbiology* **91**, 157–66.

DICKINSON C. H. & BOTTOMLEY D. (1980) Germination and growth of *Alternaria* and *Cladosporium* in relation to their activity in the phylloplane. *Transactions of the British Mycological Society* **74**, 309–19.

DICKINSON C. H. & CRUTE I. R. (1974) Influence of seedling age and development on the infection of lettuce by *Bremia lactucae*. *Annals of Applied Biology* **76**, 49–61.

DICKINSON C. H. & KENT J. W. (1972) Critical analysis of fungi in two sand-dune soils. *Transactions of the British Mycological Society* **58**, 269–80.

DICKINSON C. H. & O'DONNELL J. (1977) Behaviour of phylloplane fungi on *Phaseolus* leaves. *Transactions of the British Mycological Society* **68**, 193–9.

DICKINSON C. H. & PREECE T. F. (1976) *Microbiology of Aerial Plant Surfaces*. Academic Press, London.

DICKINSON C. H. & WALLACE B. (1976) Effects of late applications of foliar fungicides on activity of micro-organisms on winter wheat flag leaves. *Transactions of the British Mycological Society* **66**, 103–12.

DICKINSON C. H., WATSON J. & WALLACE B. (1974) An impression method for examining epiphytic micro-organisms and its application to phylloplane studies. *Transactions of the British Mycological Society* **63**, 616–19.

DIEM H. G. (1974). Micro-organisms of the leaf surface: estimation of the mycoflora of the barley phyllosphere. *Journal of General Microbiology* **80**, 77–83.

DI MENNA M. E. (1959) Yeasts from the leaves of pasture plants. *New Zealand Journal of Agricultural Research* **2**, 394–405.

ERCOLANI G. L. (1978) *Pseudomonas savastanoi* and other bacteria colonizing the surface of olive leaves in the field. *Journal of General Microbiology* **109**, 245–57.

FLANNIGAN B. (1973) An evaluation of dilution plate methods for enumerating fungi. *Laboratory Practice* August 1973, 530–1.

FOKKEMA N. J., DEN HOUTER J. G., KOSTERMAN Y. J. C. & NELLIS A. L. (1979) Manipulation of yeasts on field-grown wheat leaves and their antagonistic effect on *Cochliobolus sativus* and *Septoria nodorum*. *Transactions of the British Mycological Society* **72**, 19–29.

FOKKEMA N. J. & LORBEER J. W. (1974) Interactions between *Alternaria porri* and the saprophytic mycoflora of onion leaves. *Phytopathology* **64**, 1128–33.

FRANKLAND J. C. (1966) Succession of fungi on decaying petioles of *Pteridium aquilinum*. *Journal of Ecology* **54**, 41–63.

FRANKLAND J. C. (1975) Estimation of live fungal biomass. *Soil Biology and Biochemistry* **7**, 339–40.

HANSSEN J. F., THINGSTAD T. F. & GOKSØYR J. (1974) Evaluation of hyphal lengths and fungal biomass in soil by a membrane filter technique. *Oikos* **25**, 102–67.

HARLEY J. L. & WAID J. S. (1955) A method of studying active mycelia on living roots and other surfaces in soil. *Transactions of the British Mycological Society* **38**, 104–18.

HISLOP E. C. & COX T. W. (1969) Effects of captan on the non-parasitic microflora of apple leaves. *Transactions of the British Mycological Society* **52**, 223–35.

HUGHES S. J. (1976) Sooty moulds. *Mycologia* **68**, 693–820.

JEFFREE C. E., BAKER E. A. & HOLLOWAY P. J. (1976) Origins of the fine structure of plant epicuticular waxes.

In: *Microbiology of Aerial Plant Surfaces* (Eds C. H. Dickinson & T. F. Preece), pp. 119–58. Academic Press, London.

KENDRICK W. B. & BURGES A. (1962) Biological aspects of the decay of *Pinus sylvestris* leaf litter. *Nova Hedwigia* **4**, 313–58.

KLUYVER A. J. & VAN NIEL C. B. (1924) Über Spiegelbilder erzeugende Hefenarten und die neue Hefengattung *Sporobolomyces. Zentralblatt für Bakteriologie, Parasitenkunde und Infektionskrankheiten und Hygiene (Abteilung II)* **63**, 1–20.

LAMB R. J. & BROWN J. F. (1970) Non-parasitic mycoflora on leaf surfaces of *Paspalum dilatatum, Salix babylonica* and *Eucalyptus stellulata. Transactions of the British Mycological Society* **55**, 383–90.

LANGVAD F. (1980) A simple and rapid method for qualitative and quantitative study of the fungal flora of leaves. *Canadian Journal of Microbiology* **26**, 666–70.

LAST F. T. (1955) Seasonal incidence of *Sporobolomyces* on cereal leaves. *Transactions of the British Mycological Society* **38**, 221–39.

LAST F. T. (1970) Factors associated with the distribution of some phylloplane microbes. *Netherlands Journal of Plant Pathology* **76**, 140–3.

LAST F. T & DEIGHTON F. C. (1965) The non-parasitic microflora on the surfaces of living leaves. *Transactions of the British Mycological Society* **48**, 83–99.

LEBEN C. (1971) The bud in relation to the epiphytic microflora. In: *Ecology of Leaf Surface Micro-organisms* (Eds T. F. Preece & C. H. Dickinson), pp. 117–27. Academic Press, London.

LINDSEY B. I. (1976) A survey of the methods used in the study of microfungal succession on leaf surfaces. In: *Microbiology of Aerial Plant Surfaces* (Eds C. H. Dickinson & T. F. Preece), pp. 217–22. Academic Press, London.

LINDSEY B. I. & PUGH G. J. F. (1976a) Succession of microfungi on attached leaves of *Hippophaë rhamnoides. Transactions of the British Mycological Society* **67**, 61–7.

LINDSEY B. I. & PUGH G. J. F. (1976b) Distribution of fungi over the surfaces of attached leaves of *Hippophaë rhamnoides. Transactions of the British Mycological Society* **67**, 427–33.

MCBRIDE R. P. & HAYES A. J. (1977) Phylloplane of european larch. *Transactions of the British Mycological Society* **69**, 39–46.

MACCAULEY B. J. & THROWER L. B. (1966) Succession of fungi in leaf litter of *Eucalyptus regnans. Transactions of the British Mycological Society* **49**, 509–20.

MACNAMARA O., PRING R. J. & RICHMOND D. V. (1977) Estimation of rust infestions. *Annual Report Long Ashton Research Station for 1976*, pp. 114–15.

MILLAR C. S. & RICHARDS G. M. (1974) A cautionary note on the collection of plant specimens for mycological examination. *Transactions of the British Mycological Society* **63**, 607–10.

MISHRA R. R. & DICKINSON C. H. (1981) Phylloplane and litter fungi of *Ilex aquifolium* L. *Transactions of the British Mycological Society*, **77**, 329–37.

NORSE D. (1972a) Fungi isolated from surface-sterilized tobacco leaves. *Transactions of the British Mycological Society* **58**, 515–18.

NORSE D. (1972b). Fungal populations of tobacco leaves and their effect on the growth of *Alternaria longipes. Transactions of the British Mycological Society* **59**, 261–71.

O'DONNELL J. & DICKINSON C. H. (1980) Pathogenicity of *Alternaria* and *Cladosporium* isolates on *Phaseolus. Transactions of the British Mycological Society* **74**, 335–42.

PENNYCOOK S. R. & NEWHOOK F. J. (1978) Spore fall as a quantitative method in phylloplane studies. *Transactions of the British Mycological Society* **71**, 453–6.

POTTER M. C. (1910) Bacteria in their relation to plant pathology. *Transactions of the British Mycological Society* **3**, 150–68.

PREECE T. F. (1962) Removal of apple leaf cuticle by pectinase to reveal the mycelium of *Venturia inaequalis* (Cooke) Wint. *Nature, London* **193**, 902–3.

PREECE T. F. (1971) Some environmental and microscopical procedures useful in leaf surface studies. In: *Ecology of Leaf Surface Micro-organisms* (Eds T. F. Preece & C. H. Dickinson), pp. 245–53. Academic Press, London.

PREECE T. F. & DICKINSON C. H. (1971) *Ecology of Leaf Surface Micro-organisms.* Academic Press, London.

PUGH G. J. F. & BUCKLEY N. G. (1971a) The leaf surface as a substrate for colonization by fungi. In: *Ecology of Leaf Surface Micro-organisms* (Eds T. F. Preece & C. H. Dickinson), pp. 431–45. Academic Press, London.

PUGH G. J. F. & BUCKLEY N. G. (1971b) *Aureobasidium pullulans*: an endophyte in sycamore and other trees. *Transactions of the British Mycological Society* **57**, 227–31.

ROYLE D. J. (1976) Scanning electron microscopy of plant surface micro-organisms. In: *Microbiology of Aerial Plant Surfaces* (Eds C. H. Dickinson & T. F. Preece), pp. 569–606. Academic Press, London.

RUINEN J. (1961) The phyllosphere. I. An ecologically neglected milieu. *Plant and Soil* **15**, 81–109.

RUSCH V. & LEBEN C. (1968) Epiphytic microflora: the balloon print isolation technique. *Canadian Journal of Microbiology* **14**, 486–7.

RUSCOE Q. W. (1971) Mycoflora of living and dead leaves of *Nothofagus truncata. Transactions of the British Mycological Society* **56**, 463–74.

SHARMA K. R., BEHERA N. & MUKERJI K. G. (1974) A comparison of three techniques of the assessment of phylloplane microbes. *Transactions of the Mycological Society of Japan* **15**, 223–33.

SWINBURNE T. R. (1973) Microflora of apple leaf scars in relation to infection by *Nectria galligena. Transactions of the British Mycological Society* **60**, 389–403.

VAN DER BURG A. C. (1974) *The Occurrence of Sporobolomyces roseus, Red Yeast, on Leaves of Phragmites australis.* Graduate Press, Amsterdam.

WARREN R. C. (1976) Microbes associated with buds and leaves: Some recent investigations on deciduous trees. In: *Microbiology of Aerial Plant Surfaces* (Eds C. H. Dickinson & T. F. Preece), pp. 361–74. Academic Press, London.

WEBSTER J. (1959) Succession of fungi on decaying cocksfoot culms. Part I. *Journal of Ecology* **44**, 517–44.

WILDMAN H. G. & PARKINSON D. (1979) Microfungal succession on living leaves of *Populus tremuloides. Canadian Journal of Botany* **57**, 2800–11.

Chapter 25 • Leguminous Root Nodules

Frank B. Dazzo

25.1 Introduction

The genus *Rhizobium* is characterised as Gram-negative non-sporulating motile rods which live in the soil and are capable of infecting, nodulating and establishing a nitrogen-fixing symbiosis with roots of various leguminous plants. Six species of *Rhizobium* are defined according to the legume host(s) which they nodulate. Legumes nodulated by the same rhizobial species are placed within a cross-inoculation group. Speciation of rhizobia, based on cross-inoculation groups, is in accord with the high degree of host specificity displayed in this plant–microorganism symbiosis, especially with the temperate legumes. The basis for this host specificity is unknown; current theories propose that recognition involves the unique interaction between saccharide sequences on the bacterial surface and carbohydrate-binding lectins on the host root (Bohlool & Schmidt 1974; Dazzo & Hubbell 1975). However, some rhizobia which typically nodulate the tropical legumes display little host specificity and are, therefore, considered promiscuous. These latter legumes are part of the 'cowpea miscellany' cross-inoculation group. Table 1 is a list of rhizobial species, a few representative strains which are commonly used in research and their corresponding legume cross-inoculation groups.

Table 1. The genus *Rhizobium*.

Species	Strains in common use	Legume genus cross-inoculation groups
R. leguminosarum	PRE, 300, 128C53	*Pisum, Lathyrus, Vicia, Lens*
R. trifolii	TA1, 0403	*Trifolium*
R. phaseoli	127K17	*Phaseolus* (most species)
R. meliloti	2011, 41, 102F51	*Melilotus, Medicago, Trigonella*
R. japonicum	3I1b110, 3I1b123, 61A76, 3I1b138	*Glycine*
R. lupini	NZP2257, 250	*Lupinus, Ornithopus*
R. species 'cowpea miscellany'	32H1	*Vigna, Macroptilium*

Because of the large requirement of fossil fuel energy to manufacture nitrogen fertilisers industrially, scientists have focused greater attention on biological processes of nitrogen fixation. Next to hydrogen and oxygen (as water), nitrogen is the plant nutrient which most commonly limits crop production and legumes can offset this major limitation by forming root-nodule symbioses with nitrogen-fixing rhizobia. Hardy and Holsten (1972) have estimated that rhizobia in symbiosis with legumes fix 9×10^7 tonnes of nitrogen annually.

The *Rhizobium*–legume symbiosis offers the greatest promise of all systems for providing the additional protein needed for the future.

Rhizobia are routinely cultivated at neutral pH in either a yeast extract–mannitol medium or in a simple, chemically defined medium containing mannitol as a carbon source, glutamate as a nitrogen source, a few vitamins and mineral salts. Rhizobia are classified either as fast-growers (*Rhizobium trifolii, R. leguminosarum, R. phaseoli, R. meliloti*) with typical generation times of 2 to 4 h, or as slow-growers (*R. japonicum, R. lupini* and most of the *Rhizobium* 'cowpea miscellany') with generation times of 6 to 8 h or longer.

25.2 Process of nodulation

The infection of small-seeded legume roots by rhizobia can be observed by using a simple slide-culture chamber devised by Fahraeus (1957). Rhizobia accumulate on the root hairs which project laterally from the root epidermis. Within 2 to 3 d the root hairs undergo a marked deformation, or curling, at their growing tip, forming what is known as the shepherd's crook (Fig. 1). The rhizobia within the overlap of the shepherd's crook penetrate the root hair wall and occupy the lumen of an open refractile tube called the infection thread. The tip of the infection thread grows to the base of the root hair where it enters the outer cortical cells of the root. As the infection thread continues to grow through the root tissue, inner cortex cells are stimulated to grow and divide, eventually forming an organised mass of infected plant tissue which protrudes from the root surface as a visible nodule (Fig. 2). Rhizobia are released from the infection thread within some of the nodule cortex cells and may become enlarged and pleomorphic to form bacteroids. It is the bacteroids that reduce nitrogen (Fig. 3). Each *Rhizobium* species produces bacteroids of a characteristic shape within its host. They range from pleomorphic rods (*Rhizobium japonicum* in soybean), club or Y forms (*R. leguminosarum, R. trifolii* and *R. meliloti* in peas, clover or lucerne, respectively) to spherical

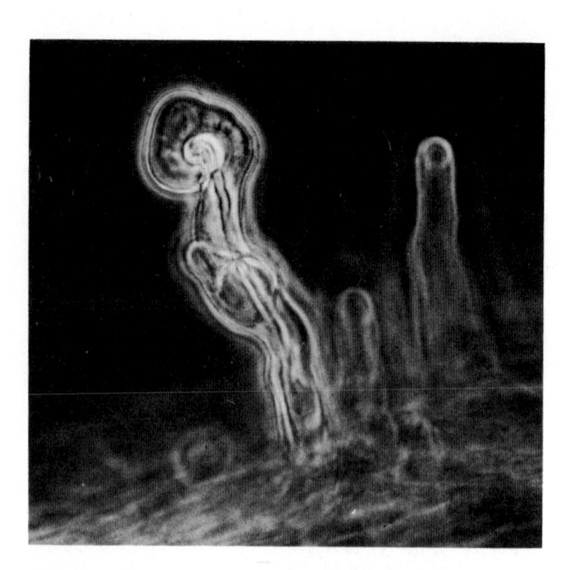

Figure 1. White clover (*Trifolium repens*) root hair infected with *Rhizobium trifolii* 0403 in a Fahraeus slide assembly. Note the tightly curled root hair tip (Shepherd's crook) and the internal, refractile infection thread (× 1732).

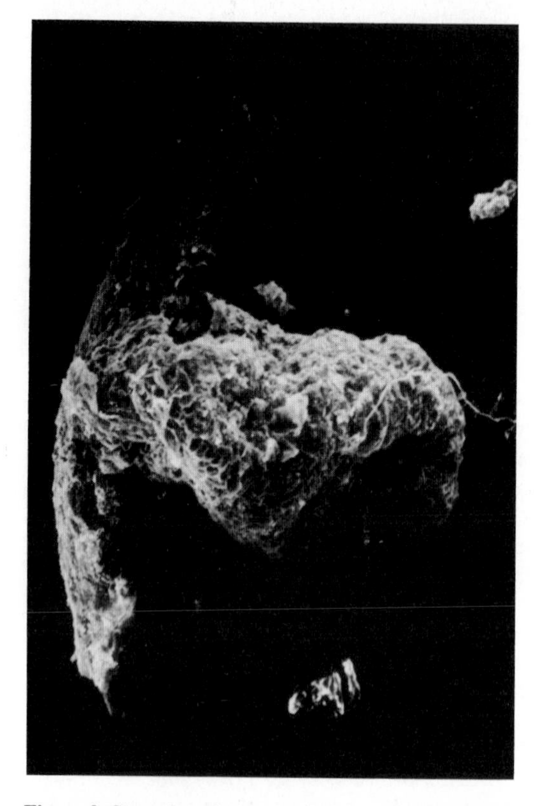

Figure 2. Scanning electron micrograph of a nitrogen-fixing nodule protruding from the root of black medic (*Medicago lupulina*) which was infected by the microsymbiont *Rhizobium meliloti* (natural specimen) (× 37).

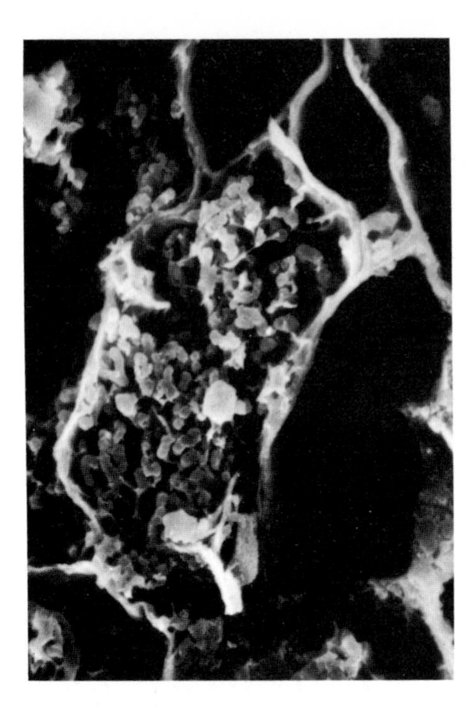

Figure 3. Scanning electron micrograph of a root nodule cell (centre) heavily infected with the microsymbiont *Rhizobium meliloti* (× 1600).

shapes (*Rhizobium* spp. in peanut). The infection process may abort at any stage, resulting in the failure of the plant to derive fixed nitrogen from the symbiosis. Thus, rhizobial strains which fail to infect the root hairs, incite nodule formation or fix nitrogen within nodules are considered non-infective, non-nodulating or ineffective, respectively.

The competitiveness of a rhizobial strain is measured by its relative ability to infect and nodulate a legume host in the presence of other nodulating strains within the soil. A *Rhizobium* strain is considered competitive when it produces a large proportion of the nodules on a plant growing in a soil or substrate heavily infested with other infective strains (Burton 1979). Occasionally, two strains may occupy the same nodule (Johnston & Beringer 1975). Many agricultural soils either lack rhizobia or contain rhizobial strains which are competitive but ineffective. Therefore, it is advisable to inoculate a legume at planting with the host-specific rhizobial inoculant to insure that an efficient nitrogen-fixing symbiosis results. Some, but not all, of the commercially available inoculants consist of competitive and effective rhizobial cultures which have been mixed with a neutralised peat base.

Experimental

25.3 Cultivation of rhizobia

There are a variety of solid culture media which are available for the isolation and cultivation of rhizobia. Rhizobia are generally cultivated aerobically at 25 to 30°C. Fast-growing rhizobia develop colonies in 3 to 6 d, and slow-growing rhizobia in 7 to 10 d. The medium most commonly used for rhizobia is yeast extract–mannitol agar (YEM), which has the following composition (g in 1.0 l distilled water): mannitol, 10; yeast extract, 0.5; NaCl, 0.1; K_2HPO_4, 1; KH_2PO_4, 1; $MgSO_4 \cdot 7H_20$, 0.18; and agar, 15. Also, 0.1 g $CaCl_2$ and 4 mg $FeCl_3 \cdot 6H_2O$ can be added. The medium is adjusted to pH 6.8 and autoclaved. Various additives which have been found useful are given below (Burton 1978).

(1) Congo red: add 2.5 ml sterile 1% (w/v) congo red solution per litre of medium prior to pouring plates. In most cases, rhizobial colonies kept in the dark remain white while colonies of *Agrobacterium* spp. and other soil bacteria absorb the red dye.

(2) Rose bengal-oligomycin: add 1 ml 2% (w/v) oligomycin solution in ethanol and 4 ml 0.3% (w/v) rose bengal solution in water to 400 ml of the medium to suppress the growth of fungi and actinomycetes when enumerating rhizobia in non-sterile peat inoculant or inoculated seed.

(3) Brilliant green, congo red, sodium azide and pentachloronitrobenzine (PCNB): add 1 ml brilliant green (1 mg 10 ml^{-1}); 1 ml of 0.25% (w/v) congo red solution; 1 ml of filtered sodium azide (1 mg 10 ml^{-1} water); and 1 ml of PCNB (0.5 g 100 ml^{-1} acetone plus 1 drop of Tween 85) to 400 ml of the YEM. Plates should be incubated in the dark and a positive control of the *Rhizobium* culture should be plated for comparison. Variations of this selective and differential medium are discussed by Barber (1979).

In many cases, a chemically defined medium may be preferred. Such a medium is BIII (Bishop *et al.* 1976) which consists of (g in 1.0 1 distilled water): mannitol, 10; sodium glutamate, 1.1; K_2HPO_4, 0.23; $MgSO_4 \cdot 7H_2O$, 0.1; and 1 ml 1000 × stock trace element solution solidified with 13 g Difco purified agar. Mannitol may be replaced with 10 g glycerol for some slow-growing species or to suppress extracellular polymer production. The medium is adjusted to pH 7.0 and autoclaved. Filter-sterilised vitamin stock solution (1 ml) is added before pouring the plates. The 1000 × trace element stock contains (g in 1.0 l distilled water): $CaCl_2$, 5.0; H_3BO_3, 0.145; $FeSO_4 \cdot 7H_2O$, 0.125; $CoSO_4 \cdot 6H_2O$, 0.059; $CuSO_4 \cdot 5H_2O$, 0.005; $MnCl_2 \cdot 4H_2O$, 0.0043; $ZnSO_4 \cdot 7H_2O$, 0.108; Na_2MoO_4, 0.125; nitrilotriacetic acid, 7. The pH of the 1000 × trace element stock solution should be adjusted to 5.0 before adding nitrilotriacetic acid in order to prevent precipitation. The 1000 × vitamin stock contains (mg in 1.0 litre 50 mM Na_2HPO_4–NaH_2PO_4 buffer, pH 7.0): riboflavin, *p*-aminobenzoic acid, nicotinic acid, biotin, thiamine-HCl, pyridoxine-HCl and calcium pantothenate, all at 20, and inositol, 120.

25.4 Isolation of rhizobia from root nodules

In order to isolate rhizobia from root nodules the soil is carefully loosened from around the legume root and the root removed. Using a razor blade, the root is cut so that a 1 to 2 mm segment remains attached to a root nodule. The nodule is washed by rubbing gently between the fingers in order to remove any adhering soil. The nodule is placed in a tube containing 2 ml sterile distilled water and washed thoroughly. The nodule is momentarily exposed to 95% (v/v) ethanol and immersed in 0.1% (w/v) mercuric chloride acidified with HCl (5 ml concentrated HCl l^{-1}) for 4 to 10 min (according to size). The nodule is thoroughly washed with sterile water and aseptically crushed with a sterile glass rod to make a milky suspension of the rhizobia. The suspension is streaked on to BIII and/or YEM agar plates and incubated at 30° C for approximately 7 d. As a modification of the above, various chambers have been devised to streamline surface sterilisation and washing of root nodules for field-oriented ecological work (Gault *et al.* 1973). During isolation a loopful of the suspension may be transferred to a microscope slide, air-dried, heat-fixed, stained for 1 min with crystal violet solution (as in the Gram stain), rinsed with running water, dried and then examined at × 1000 magnification in order to observe the characteristic bacteroid forms of the rhizobia (Fig. 4). The pink colour of the nodules is due to leghaemoglobin which can be quantified by optical absorption of pyridine haemochromogen (Bergersen 1961) or by a more sensitive fluorometric assay of this haeme protein (LaRue & Child 1979).

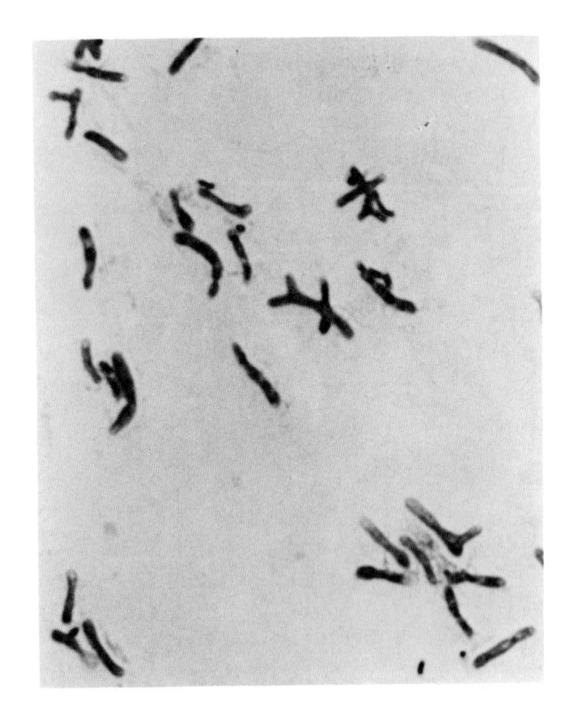

Figure 4. Bacteroids of *Rhizobium meliloti* from a nodule squash of alfalfa, stained with crystal violet. Note the enlarged and Y cell forms (× 4420).

Many species produce large (3 to 10 mm) mucoid colonies which range from clear to whitish opaque (fast-growers), or small (1 to 3 mm) translucent colonies (slow-growers). A typical result from clover nodule squashes streaked on BIII agar and incubated for 3 to 5 d is an almost pure culture of large, mucoid, whitish colonies characteristic of *Rhizobium trifolii* (Fig. 5). The colony can be restreaked to confirm culture purity and inoculated on a suitable host to confirm that the isolate is the required *Rhizobium* species.

25.5 Enumeration of rhizobia from soil

Rhizobial populations in soil are most frequently enumerated by the indirect plant infection technique which depends on the ability of the rhizobial cell to incite nodule formation on the roots of the legume host. In general, the plant infection count compares well with a regular plate count if the seedlings are grown on agar (e.g. Jensen tube method—Vincent 1970). Many legumes can be

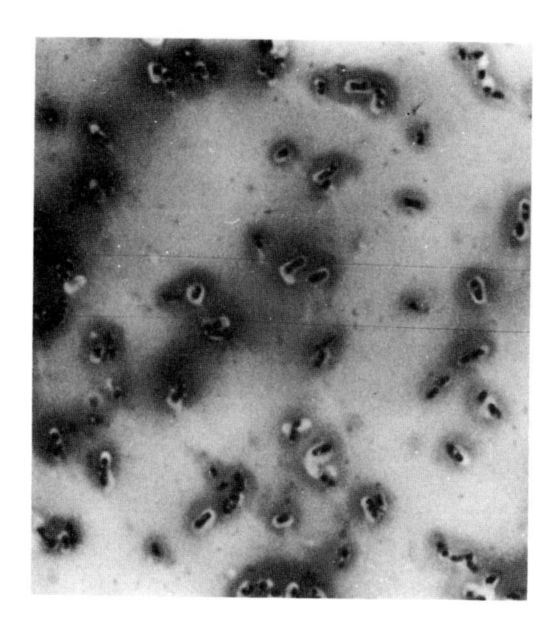

Figure 5. Pure culture of *Rhizobium trifolii* isolated on BIII agar from a clover nodule squash and then stained with crystal violet. Note the distinct capsular halo surrounding the cell and the basophilic, amorphous slime layer which extends outward from the capsular periphery.

Figure 6. Jensen tube of white clover nodulated by *Rhizobium trifolii* after 1 month incubation. The seedling on the right is 3 d old.

tested on agar in enclosed tubes or with the roots enclosed and the shoots outside the container. Enclosed tubes work well with *Trifolium* species (clovers) for *Rhizobium trifolii, Medicago sativa* (alfalfa lucerne) for *R. meliloti, Vicia* species (vetch) for *R. leguminosarum, Ornithopus sativa* (serradella) for *R. lupini, Glycine soja* (wild soybean) for *R. japonicum, Macroptilium atropurpureum* for a wide variety of the tropical rhizobia ('cowpea miscellany') and *Lotus corniculatus* (birdsfoot trefoil) for lotus rhizobia.

Small seeds are surface-sterilised and spread on the surface of sterile water agar (1.5% (w/v) agar). For clover, overnight pretreatment of the sterilised seeds on plates in the refrigerator often enhances root hair infection by *Rhizobium* spp. (Nutman *et al.* 1970). These plates are removed from the refrigerator and incubated at room temperature in the dark for seed germination. Germination of some hard-coated seeds (e.g. *Glycine soja*) may require scarification using fine emery cloth or exposure (20 to 30 min) to concentrated H_2SO_4. The primary root radicles of clover seedlings will grow approximately 1 cm in the humid air-space within the plate in 24 to 36 h. Larger seeds may be surface-sterilised and germinated on sterile moistened filter paper in Petri dishes. Once the radicles emerge, the seedlings can be carefully and aseptically transferred by spatula to the surface of agar slopes (0.7 to 1% (w/v) purified agar) containing a nitrogen-free plant nutrient medium. Five replicates of serial dilutions of soil (0.1 ml each) are used to inoculate the roots of the seedlings. These Jensen tubes are closed, incubated under adequate plant growth conditions and scored for root nodulation after 1 month (Fig. 6) (clover root nodules may appear as early as 1 week). Quantification requires reference to a most-probable-number table. The ingredients of several neutral, phosphate-buffered nitrogen-free plant nutrient media are given in Table 2.

Table 2. Nitrogen-free plant nutrient media. (Components in g unless specified.)

Jensen (1942)		Fahraeus (1957)		Burton (1978)	
$CaHPO_4$	1.0	$CaCl_2$	0.10	KCl	0.75
K_2HPO_4	0.2	$MgSO_4 \cdot 7H_2O$	0.12	$CaSO_4 \cdot 2H_2O$	0.19
$MgSO_4 \cdot 7H_2O$	0.2	KH_2PO_4	0.10	$MgSO_4 \cdot 7H_2O$	0.19
NaCl	0.2	$Na_2HPO_4 \cdot 2H_2O$	0.15	$Ca_3(PO_4)_2$	0.19
$FeCl_3$	0.1	Ferrous citrate formula	0.005	$FePO_4$	0.19
Water	1.0 l	Trace element solution*	0.1 ml	Trace element solution†	5.0 ml
		Water	1.0 l	Water	995.0 ml

* Trace element stock solution contains: $MnCl_2$, $CuSO_4$, $ZnSO_4$, H_3BO_3 and Na_2MoO_4 all at 1.0 g l^{-1} distilled water.

† Trace element stock solution contains (g in 1.0 l distilled water): H_3BO_3, 0.57; $MnSO_4 \cdot H_2O$, 0.31; $ZnSO_4 \cdot 6H_2O$, 0.09; $CuSO_4 \cdot 5H_2O$, 0.08; molybdic acid (85% MoO_3), 0.016; and $CoCl_2 \cdot 6H_2O$, 0.0008.

Figure 7. Soybeans grown in cellophane pouches and nodulated by *Rhizobium japonicum* (by permission of R. Maier & W. Brill).

Many large-seeded legumes, e.g. *Phaseolus vulgaris*, require a system where the shoot lies outside of the chamber in order to enumerate their rhizobia. In this case, the cellophane pouch method is very efficient and is becoming increasingly popular as a chamber to obtain nodulated roots for a variety of studies (Weaver & Frederick 1972). The pouch is supplied with a blotting paper wick which can be autoclaved after wetting and placed vertically on a standard gramophone record rack with stiff cardboard supports (Fig. 7). Seeds are surface-sterilised, planted in the trough formed by the paper wick and nitrogen-free nutrient solution added alternately with water when needed. The roots are inoculated after they penetrate the perforated trough of the paper wick. Three replicate chambers may be formed by a heat-sealing process which provides a space-saving option of the pouch container.

25.6 Identification of rhizobial strains

Specific strains of rhizobia are routinely identified by serology, phage lysis patterns and/or specific tolerance or resistance to antibiotics. Agglutination, immunodiffusion and immunofluorescence are serological methods commonly employed to identify rhizobial strains in nodules and soil. Reviews which discuss the preparation of specific antisera for rhizobia and their use in diagnostic serology are available (Vincent 1970; Dudman 1977). Immunofluorescence is becoming increasingly important as a tool to type strains within nodules and to study the autecology of rhizobia in soil (Fig. 8a) and on their hosts (Fig. 8b—Bohlool & Schmidt 1968; Schmidt *et al.* 1968).

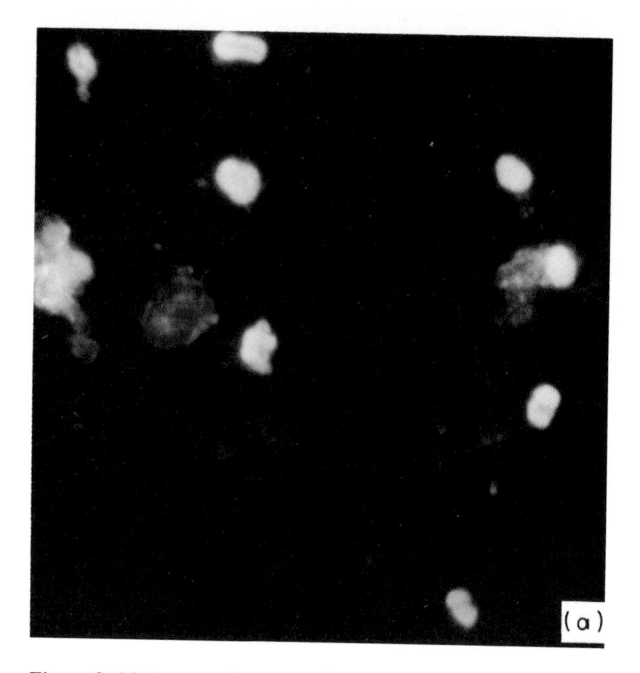

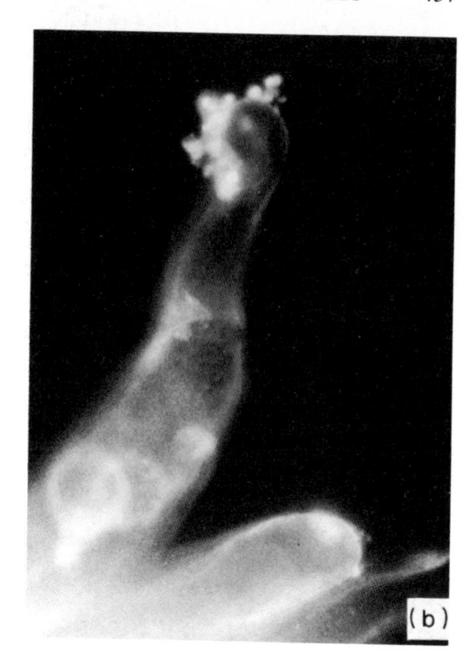

Figure 8. (a) Immunofluorescence photomicrograph of *Rhizobium trifolii* 0403 in soil (× 4849). (b) Immunofluorescence photomicrograph of *Rhizobium trifolii* 0403 on the tip of a white clover root hair growing in soil (× 2088).

Lytic bacteriophages for rhizobia (Fig. 9) are readily isolated from the rhizosphere of the corresponding legume host, especially if the soil has been heavily inoculated with the rhizobial strain of interest (Barnet 1972). Lytic bacteriophages vary in plaque morphology and can be detected in BIII or YEM soft agar overlays seeded with an exponentially growing culture of the rhizobia. Preadsorption of the phage suspension with the rhizobia for 10 min prior to seeding the soft agar inproves the plating efficiency. Phage cultures can be purified by repeated picking of single plaques. Stocks of rhizobiophage can be maintained in a sterile broth (Vincent 1970) containing (g in 1.0 l distilled water): sucrose, 2.5; K_2HPO_4, 0.5; $MgSO_4 \cdot 7H_20$, 0.2; $CaSO_4$, 0.16; NaC1, 0.1; $FeC1_3 \cdot 6H_20$, 0.02; and yeast extract 0.5. High titre stocks (*c.* 1×10^{10} p.f.u. ml^{-1}) can be obtained by adding 2 ml of phage broth to a plate showing confluent lysis, scraping the soft agar overlay into a centrifuge tube, removing the debris by centrifugation at 12 000 *g* for 30 min and treating the supernatant with chloroform. For phage typing, several phage suspensions can be spotted on plates containing semi-solid agar seeded with single strains of rhizobia (Vincent 1970).

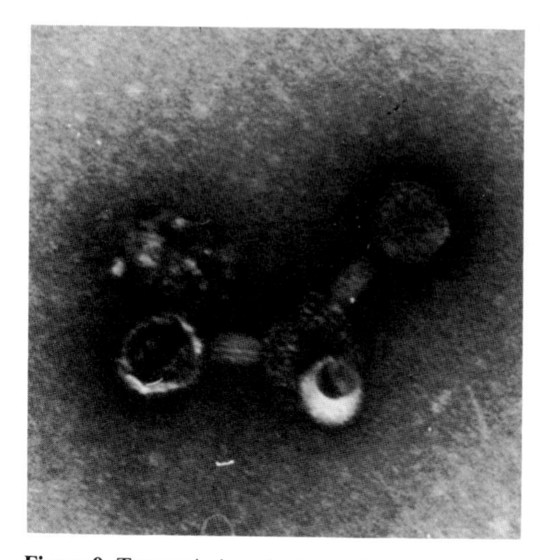

Figure 9. Transmission electron micrograph of bacteriophage BY 06 which infects *Rhizobium trifolii* and is negatively stained with uranyl acetate.

Resistance to antibiotics and other inhibitors has been used for strain identification (Kuykendall & Weber 1979). Whenever these genetic markers are employed, it is imperative to ensure that pleiotropic complications affecting the performance of the rhizobial symbiont do not arise as a result of mutation to antibiotic resistance. Another complication is the low grade of intrinsic resistance of wild type strains to many antibiotics which may be diagnostic under certain conditions (Josey *et al.* 1979). Furthermore, such antibiotic resistances can be transferred between rhizobial strains in soil as well as in a nodule (Pariiskaya & Gorelova 1976; Johnston & Beringer 1977).

Although tedious, a more definitive approach to the identification of rhizobial strains is to match the total profile of cell protein with that of known strains using two-dimensional gel electrophoresis (Roberts *et al.* 1980). In this procedure, the proteins from exponentially growing cells are separated in polyacrylamide gels by isoelectric focusing in the first dimension and then according to their molecular weight, in the presence of sodium dodecylsulphate in the second dimension. The important advantage of this technique is that definitive strain identification is based on the overall genetic background of the cell rather than a few selected properties.

25.7 Methods to study the infection process

The infection process of small-seeded legumes can be studied by the Fahraeus technique (Fahraeus 1957). This procedure involves aseptic germination and growth of a seedling in a thin layer of nitrogen-free medium containing a *Rhizobium* sp. on a glass slide assembly supported by a cover slip and inserted in a sterile tube containing 20 ml nutrient solution. The slide assembly can be removed at any time, blotted and the roots examined microscopically. The Fahraeus slide assembly is prepared as follows (Fig. 10). A clean microscope slide is placed into a glass Petri dish and a cover slip rested on the upper left-hand corner of the slide. The unit is autoclaved. Fahraeus nutrient medium (20 ml) is added to a glass tube (10 cm long by 38 mm diameter, fire polished edge) which is covered with a 50 ml beaker and autoclaved. In order to assemble the slide, a seedling is held near the cotyledon with a finely tapered pair of forceps and the 1 cm seedling root placed on the centre of the slide. Alternatively, two seedlings can be aligned 1 cm apart on the slide. The rhizobial suspension (0.2 ml with or without 0.3% (w/v) agar) is pipetted on to the slide near its bottom edge. The cover slip (22 × 40 mm, No. 1 thickness) is picked up by the side (see Fig. 10) with the sterile forceps and carefully

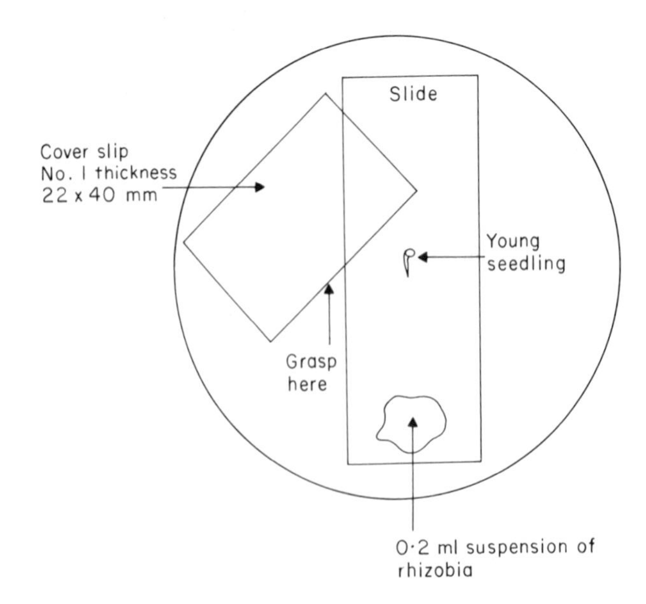

Figure 10. Schematic diagram of the steps to assemble a Fahraeus slide culture.

lowered on to the seedling root. The space should fill with suspended inoculum without trapping air bubbles. Any bubbles which may form eventually rise to the top of the cover slip when the slide is picked up at the top right side and transferred vertically to the large test tube containing the nutrient medium. The 50 ml beaker closure is quickly replaced (Fig. 11(a) and (b)). The slide should rest slanted in the tube such that the seedling side is directed towards the top. The assembly is incubated under suitable conditions for plant growth.

The Fahraeus slide technique is an artificial method since roots are in a hydroponic environment and may behave differently in soil. For instance, some plants produce secondary water roots which are anatomically different from roots grown in natural soil. However, studies with this technique are usually conducted on the seedling's primary root. Two other artificial conditions which require consideration are illumination of roots and crowding of seedlings. Nutman (1970) found that infection of clover root hairs by *Rhizobium trifolii* was enhanced by some light and that the number of infections per plant dropped significantly only if four or more seedlings were present per vessel.

The success of this technique can be attributed to the fact that only the host and the appropriate *Rhizobium* sp. are necessary for expression of recognition, host specificity and root hair infection. An elegant time-lapse movie of the infection process using Fahraeus slides is available (Nutman *et al.* 1973). The technique can be adapted to study the interaction of any microbe with the roots of small-seeded plants (Umali-Garcia *et al.* 1980). Root hair infection of large-seeded legumes (e.g. soybean) can be observed after inoculated roots are grown in Leonard jar assemblies (Fig. 13—Rao & Keister 1978).

Dazzo *et al.* (1976) modified the Fahraeus slide technique to quantify the adsorption of rhizobial cells to target clover root hairs (Fig. 12). Seedlings (24 h old) are inoculated with 1×10^7 cells plant $^{-1}$, incubated for 12 h in a growth chamber, rinsed with Fahraeus solution and examined by phase contrast microscopy. Root hairs of these seedlings are straight; adsorbed bacterial cells are refractile and easily distinguishable. The number of bacteria adsorbed to root hairs approximately 200 μm in length (diameter of the microscope field) is determined and the data are compared statistically. Root hairs are examined only along the optical median plane. This direct assay of adsorption scores only those bacterial cells in firm physical contact with cell walls of root hairs in a uniform state of physiological development. The assay eliminates several sources of error in measuring rhizobial adhesion, such as unadsorbed bacterial

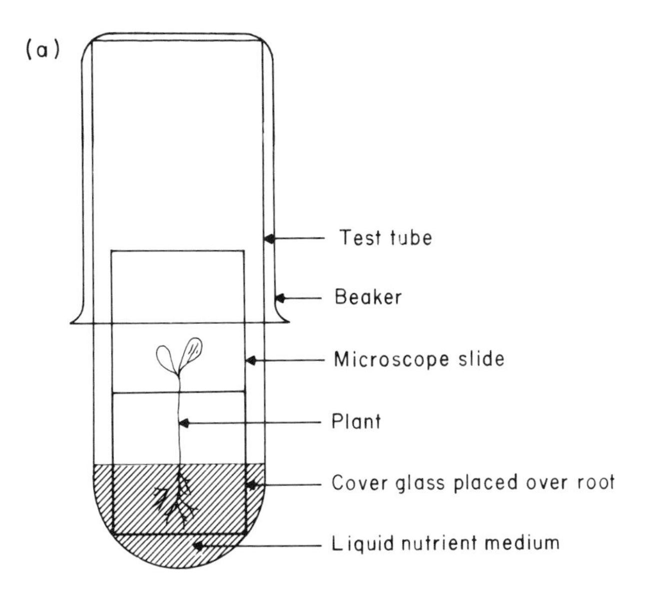

Figure 11. (a) Complete assembly of a plant slide culture (by permission of D. Hubbell). (b) Assembled slide culture using two clover seedlings.

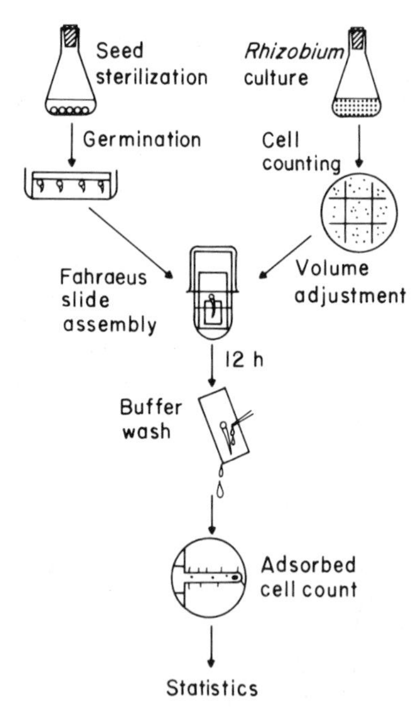

Figure 12. Schematic diagram of the protocol of the *Rhizobium*-root adsorption assay (after Dazzo *et al.* 1976).

cells, cells adsorbed to undifferentiated epidermal cells, non-dispersable flocs and the variability of root surface areas among individual plants.

Simple hydroponic chambers which provide easy access to sterile root exudate or roots (Dazzo & Brill 1978, 1979) can be prepared aseptically. This is achieved by placing blocks of surface-sterilising seeds embedded in 1% (w/v) purified agar on the top of stainless steel wire mesh supports fitted within enclosed sterile glass dishes or tubes containing a plant nutrient medium filled to within 2 to 3 mm below the wire support. The root radicles grow through the wire mesh into the rooting medium while the agar blocks provide support for the seedlings and prevent seed coats and seed exudates from contaminating the solution below. Root exudate can be sampled by hypodermic syringe and needle.

25.8 Methods to study root nodulation

Nodulation of legume roots will occur in Fahraeus slides, Jensen tubes and cellophane pouch assemblies described above. Another commonly used assembly for studying nodulation is the Leonard jar (Burton 1978). These assemblies are protected from surface contamination by covering the tops with gravel and are watered from below with nitrogen-free nutrient solution by capillary action of an inert wick (Fig. 13). The assembly consists of a 1 l glass bottle with the bottom removed, a 500 ml wide-mouthed jar and a rubber stopper equipped with a heat-resistant glass tube traversing the stopper to support a nylon or polypropylene wick approximately 20 cm in length.

After fitting the wetted stopper into the neck of the bottle, it is inverted and placed on top of the widemouthed jar. The inverted bottle is filled with 1000 g washed, moderately coarse sand. The wide-mouthed jar is filled with 400 ml nutrient solution to check the operation of the wick. The dry sand should be completely moistened in approximately 1 h. The bottles are covered with a glass Petri dish top or aluminium foil and autoclaved for 2 h. Before planting, the sandy surface is moistened with distilled water and holes are made where seeds are to be planted. The Leonard jar accomodates 12 alfalfa or clover seedlings or 6 soybean seedlings. Inoculated seeds should be planted aseptically. Non-inoculated surface-sterilised seeds should be planted as negative controls. Some of the same seeds inoculated with a culture of known effective rhizobia should be planted as positive controls. Extreme care should be taken not to spread nodule bacteria from one jar to another. Fine sterile gravel spread over the sand will serve to keep the surface dry and reduce cross-contamination.

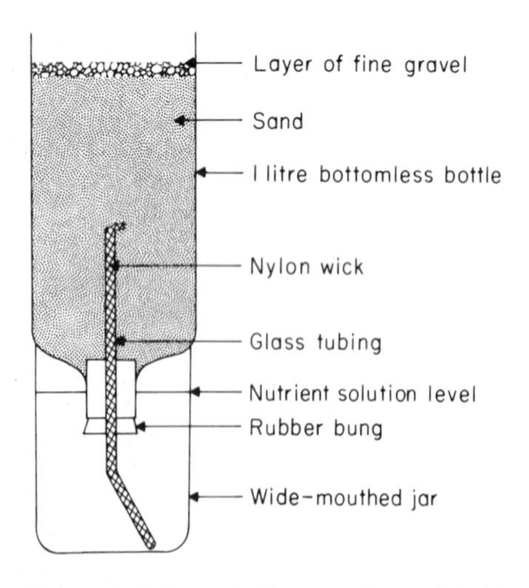

Figure 13. Schematic diagram of a modified Leonard jar (by permission of J. Burton).

Rolfe *et al.* (1980) have developed a plate technique which is suitable for screening symbiotically defective mutants of *Rhizobium*. Aseptically germinated seedlings are laid flat on the surface of N-free plant nutrient agar in sterile Petri plates and the roots are inoculated with a suspension of rhizobia. The plates are stacked vertically in a plant growth room so the plant grows normally and becomes nodulated (Fig. 14).

25.9 Methods to study nitrogen fixation

Initial tests to select rhizobial strains which form nitrogen-fixing root nodules are generally carried out in a greenhouse or growth chamber (Vincent 1970; Wacek & Brill 1976; Burton 1978). In general, the growth response and nitrogen fixing activities are compared between inoculated and uninoculated legumes grown on nitrogen-free medium. These studies show that *Rhizobium* strains vary widely in their ability to fix nitrogen and stimulate plant growth.

The only way of knowing with certainty that a strain of *Rhizobium* will fix nitrogen with a particular variety of the legume host is to allow them to develop together in a favourable environment and then to measure the nitrogen fixed. If the particular symbiotic strain is effective, root nodules will form and provide the plant with fixed nitrogen resulting in luxuriant plant growth. If the strain of *Rhizobium* is ineffective, the plants will be nitrogen-limited and remain small and yellow. Analysis of the total Kjeldahl nitrogen of the plants is useful, although there is usually a good correlation between dry weights of the plants and total nitrogen under these controlled conditions (Burton *et al.* 1972). Effective nodulation has been measured by determining nodule size, leghaemoglobin content of nodules and crop yields (Fred *et al.* 1932). The enzyme nitrogenase, besides reducing nitrogen to ammonia, is capable of reducing acetylene to ethylene (Schöllhorn & Burris 1966) and many applications of this acetylene reduction assay have been used to detect and quantify nitrogen fixing ability in legumes (Hardy *et al.* 1968).

In its simplest form, the acetylene reduction assay can be performed by placing freshly obtained nodulated roots in a serum bottle and sealing with a serum stopper. Acetylene is added to give a pressure of 0.1 atmospheres and the bottle incubated for a minimum of 1 to 2 h. A sample of gas is withdrawn and the ethylene quantified by gas chromatography using a Porapak N or equivalent column and a flame-ionization detector. Enzyme specific activity is expressed as nmol ethylene produced (mg nodule weight)$^{-1}$ h^{-1}. When possible, nodulated roots should be shaken free from soil and assayed without washing. Detached nodules can be assayed but they are less active. Rhizobia can be readily isolated from the nodules after this assay is completed. The controls of non-nodulated roots usually show small amounts of

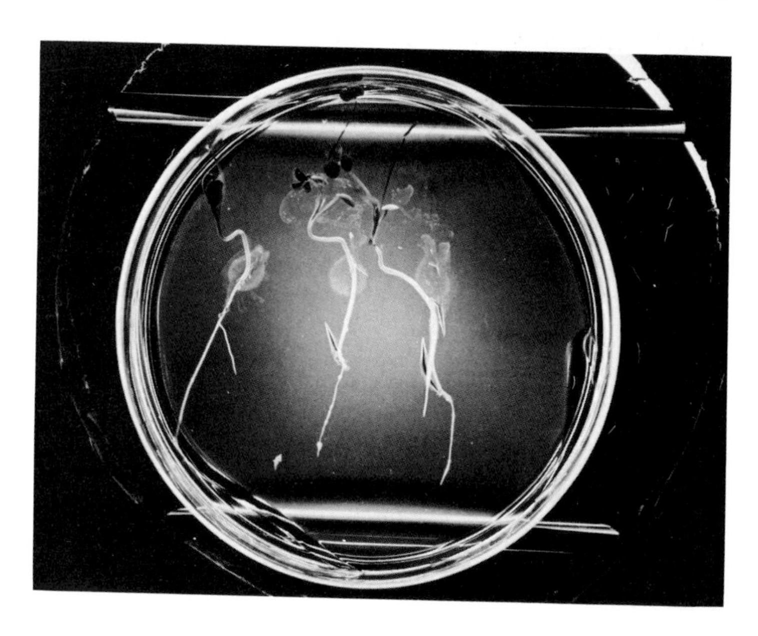

Figure 14. White clover plants nodulated by *Rhizobium trifolii* 0403 on a Petri dish of nitrogen-free plant medium.

ethylene derived from background ethylene in the added acetylene and ethylene produced by the plant and free-living microbes on the root or in adhering soil. The theoretical conversion ratio of 3:1 for ethylene produced to nitrogen fixed is rarely obtained, mainly because nitrogenase also reduces protons in a competing reaction, but other factors contribute to invalidate the theoretical ratio. Thus, calculations of nitrogen fixation rates based on acetylene reduction assays require some nitrogen-based measurement (e.g. [^{15}N]nitrogen and mass spectrometry) to determine experimentally the correct conversion factor. In the field, acetylene reduction assays should be conducted throughout the legume growing season if extrapolation of the total amount of nitrogen fixed is desired since fixation rates will change with the ontogeny of the plant. A more detailed discussion of the acetylene reduction technique appears in Chapter 13.

The Wacek effectiveness assay (Wacek & Brill 1976) is perhaps most appropriate for examining the nitrogen-fixing activity of very large numbers of rhizobial isolates under strict, bacteriologically controlled conditions (as is necessary when screening several thousand survivors of bacterial mutagenesis for ineffective or non-nodulating mutant strains—Maier & Brill 1976). This maintenance-free procedure can be used for either small- or large-seeded legumes. Sterile seedlings are placed in sterile serum bottles filled with vermiculite and a buffered, nitrogen-free plant nutrient medium. The seedlings are inoculated by adding 1 ml rhizobia suspension containing 1×10^7 to 1×10^9 cells) to each sterilised bottle containing the vermiculite. Alternatively, peat-based rhizobial inoculant can be mixed (1% w/v) with the vermiculite prior to filling the serum bottle and planting (F. B. Dazzo & J. C. Burton, unpublished observations). A sterile, clear plastic bag is placed halfway over the bottle and tied with the wire provided (Fig. 15); the bag should be sufficiently loose to allow for gas exchange. These bags minimise bacterial contamination of the plant and reduce moisture evaporation from the bottle so that no watering is required during the 14 d incubation period. After 14 d, the plastic bag is removed, the plant is cut at the base of the stem and a serum stopper is placed on the bottle (Fig. 14). Alternatively, the plastic bag can be pulled over the entire bottle and the bottom securely sealed. Acetylene is injected on to the bottles to give 0.3 or 0.4 atmospheres of pressure and the bottle incubated for 1 to 2 h at 25°C after which 0.4 to 0.5 ml gas samples are removed and injected into a gas chromatograph to determine the amount of ethylene produced by each plant. Nodules can be harvested and rhizobia can be isolated from them after the gas has been sampled.

Figure 15. Soybean plant growing in the vermiculite-serum bottle and prepared for the effectiveness assay by the acetylene-reduction technique (by permission of T. Wacek and W. Brill).

25.10 Electron microscopic examination

There are several recent publications which detail special techniques to process infected and nodulated roots for transmission electron microscopy. Only a few representative publications are cited here (Kijne 1975; Napoli & Hubbell 1975; Newcomb 1976; Verma *et al.* 1978; Callaham 1979).

In order to study root hair infection, seedlings can be carefully removed from Fahraeus slide assemblies or Jensen tubes and processed. They are washed in buffer, fixed in glutaraldehyde, post-fixed in osmium tetroxide and dehydrated in graded ethanol followed by acetone or propylene oxide. The materials are then infiltrated with resin under vacuum, transferred to flat, rectangular disposable embedding moulds and polymerised. The block of embedded seedling is removed from the mould, viewed under phase contrast microscopy and areas containing infection threads can be identified. These areas are cut out of the block, sectioned, post-stained with lead citrate and/or uranyl acetate and examined (Napoli & Hubbell 1975). This technique was used to produce the ultrathin section of *Rhizobium trifolii* near the curvature of a deformed clover root hair shown in Fig. 16.

Fahraeus slide assemblies, Jensen tubes and cellophane pouches are ideal for obtaining root nodules for electron microscopy since their development can be readily observed. Another technique is to use aeroponic cultures, where seedling roots grow into a relatively large dark tank and are continuously bathed in a mist of dilute plant nutrient medium inoculated with the rhizobia (Zobel *et al.* 1976). Better results for electron microscopy are obtained if nodule slices are fixed instead of whole nodules, since the latter are often infiltrated poorly.

For scanning electron microscopy, roots or nodules are rinsed in buffer, fixed in 5% (w/v) glutaraldehyde in 0.1M sodium cacodylate buffer (pH 7.2), post-fixed in 2% (w/v) osmium tetraoxide and dehydrated progressively in an ethanol series. Fixed specimens are critical-point dried in pressurised liquid carbon dioxide and sputter coated on stubs with a 20 to 30 nm layer of gold (Dazzo & Brill 1979). This technique was used to produce Figs 2 and 3 of black medic nodules taken from the field and exposed by cutting the nodule in half to reveal a host cell infected with *Rhizobium meliloti*.

25.11 Preparation and use of legume inoculants

Good carriers for inoculants have the following properties (Burton 1979):

(1) they are non-toxic to rhizobia;
(2) they have good adsorption qualities;
(3) they are easily pulverised and sterilised;
(4) they have good adhesion to seed;
(5) they are readily available at moderate cost.

Most legume inoculant cultures are prepared by adding a broth culture of *Rhizobium* sp. to a finely ground carrier base such as peat, mixtures of peat and soil or compost, various coals or other organic materials. Peat has proved to be the best. A peat-based inoculant is prepared in the following manner. To a 500 g capacity container is added 60 ml broth culture containing at least 1×10^9 viable rhizobia ml^{-1}, 125 g finely pulverised, heat-treated sedge peat (7% moisture content) and sufficient calcium carbonate to give a final pH of 6.8 to 7.0. The container is closed and shaken vigorously to mix the broth culture with the peat and the contents spread in a thin layer on paper for 2 to 3 d at 22 to 24°C. The inoculant is milled to break up aggregates, packaged, sealed in flexible polyethylene bags to provide gas exchange and a barrier against moisture loss and cured or matured at 24°C for 2 to 4 weeks. During curing, the viable population may increase by as much as 10 times (Burton 1967). The final moisture content should be 35 to 40%. Rhizobial survival is maximal when the sealed inoculant is stored at 4°C or lower. A minimum count of 1×10^7 viable rhizobia g^{-1} is

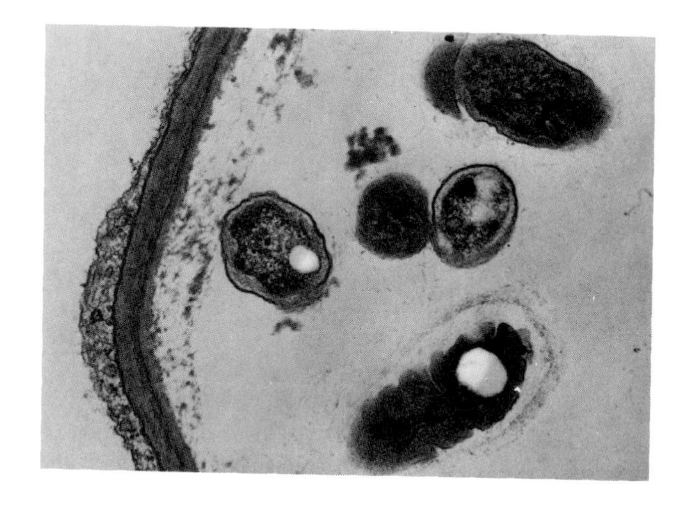

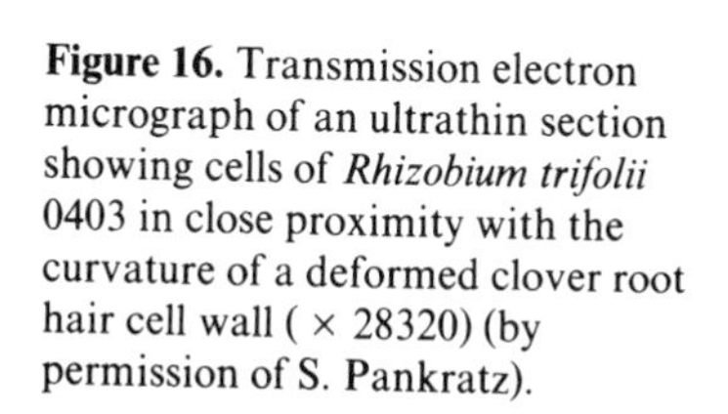

Figure 16. Transmission electron micrograph of an ultrathin section showing cells of *Rhizobium trifolii* 0403 in close proximity with the curvature of a deformed clover root hair cell wall (× 28320) (by permission of S. Pankratz).

considered satisfactory although a count of 5 to 10×10^8 g^{-1} is more desirable.

If microbial antagonists are abundant in the peat, 20 g peat may be sterilised prior to mixing with the broth culture by autoclaving for 4 h at 121°C, or for 1 h on 3 consecutive days, or by irradiating with 2.5 Mrad of γ rays.

It is best to inoculate seeds just before planting. This is done by mixing the seeds with a slurry containing 25 g peat inoculant plus 100 ml water or a 40% (w/v) gum arabic adhesive solution, followed by mixing the seeds with a dry inoculant to form a uniform, adhering seed coating. There is no authoritative, unequivocal answer to how many rhizobia are required for each seed to bring about effective nodulation under field conditions (Burton 1967). The minimum standards in Australia (Vincent 1970) and in Canada (Burton 1978) are 300 to 1000 viable rhizobia to seed, respectively. Clearly larger numbers of rhizobia increase the chance of a successful infection.

Assessment of the need for inoculation must be determined as a prerequisite for microbial ecology programmes designed for *Rhizobium* strain selection. There are many intermediate situations in the field where the need for inoculation will depend on the legume, the competitiveness of indigenous rhizobia and the effectiveness of the strains with the introduced host, and on the previous history of husbandry in the area. Date (1977) has described a simple three-treatment inoculation experiment to determine the need for inoculation. These treatments are:

(1) an uninoculated control to check for the presence or absence of native rhizobia and their effectiveness with the host;

(2) an inoculated treatment using a marked strain known to be effective with the host in question;

(3) an inoculated treatment as in (2) above plus nitrogen fertiliser to ensure that the host will grow when nitrogen is supplied.

The results of this experiment will determine how necessary and how suitable the inoculant strain is in providing fixed nitrogen for legume production under field conditions.

25.12 Culture collections of rhizobia

A few of the major collections are listed in Table 3. Subcultures are available for research purposes without charge and are identified by a coded strain number. A few culture catalogues have been published (e.g. Rothamsted Culture Collection, NiFTAL Tropical Rizobia Culture Collection).

Table 3. *Rhizobium* culture collections.

Curator	Location
J. C. Burton & R. S. Smith	The Nitragin Company, Milwaukee, Wisconsin, 53209, USA
D. Weber	United States Department of Agriculture, Beltsville, Maryland, 20705, USA
J. Halliday	NifTAL Project, University of Hawaii, Paia, Maui, Hawaii, 96779, USA
A. Pattison	Soil Microbiology Department, Rothamsted Experimental Station, Harpenden, Hertfordshire, England
F. Bergersen	Microbiology Section, CSIRO, Division of Plant Industry, Canberra, ACT 2600, Australia

Acknowledgements

This chapter is dedicated to Dr Joseph C. Burton, the Nitragin Company, Milwaukee, Wisconsin, in recognition of his outstanding contributions to the science of rhizobiology. I would like to thank Estelle Hrabak, Karel Schubert and David Hubbell for their comments on this manuscript, Stuart Pankratz for providing Fig. 12 and Winston Brill for a preprint of the two-dimensional gel electrophoresis technique. Portions of the work described here were supported by Grant No. PCM 80–21906 from the National Science Foundation, Grant No. 79–00099 from the Competitive Grant Office of the Science and Education Administration of the United States Department of Agriculture, by the Agricultural Experiment Station (Journal Article No. 9386) and the College of Natural Science, Michigan State University, East Lansing, Michigan.

Recommended reading

BERGERSEN F. J. (1981) *Methods for Evaluating Biological Nitrogen Fixation*. John Wiley & Sons, New York.

BURTON J. C. (1979) Rhizobium species. In: *Microbial Technology*, 2e, vol. 1, pp 29–58. Academic Press, New York.

DUDMAN W. F. (1977) Serological methods and their application to dinitrogen-fixing organisms. In: *A Treatise on Dinitrogen Fixation, Section IV: Agronomy and Ecology* (Eds R. Hardy & A. Gibson), pp. 487–508. John Wiley & Sons, New York.

NUTMAN P. S. (1970) The relation between nodule bacteria and the legume host in the rhizosphere and the process of infection. In: *Ecology of Soil Borne Plant Pathogens* (Eds K. F. Baker & W. C. Snyder), pp. 231–47. University of California Press, Berkeley.

SKINNER K. J. (1976) Nitrogen fixation. *Chemistry and Engineering News*, October, 22–35.

VINCENT J. B. (1970) *A Manual for the Practical Study of Root-nodule Bacteria. IBP Handbook No. 15*. Blackwell Scientific Publications, Oxford.

References

BARBER L. E. (1979) Use of selective agents for recovery of *Rhizobium meliloti* from soil. *Soil Science Society of America Journal* **43**, 1145–8.

BARNET Y. (1972) Bacteriophages of *Rhizobium trifolii*. I. Host range and morphology. *Journal of General Virology* **13**, 1–15.

BERGERSEN F. J. (1961) Haemoglobin content of legume root nodules. *Biochimica et Biophysica Acta* **50**, 576–8.

BERGERSEN F. J. (1981) *Methods for Evaluating Biological Nitrogen Fixation*. John Wiley & Sons, New York.

BISHOP P. E., GUEVARA J. G., ENGELKE J. A. & EVANS H. J. (1976) Relationship between glutamine synthetase and nitrogenase activity in the symbiotic association between *Rhizobium japonicum* and *Glycine max*. *Plant Physiology* **57**, 542–6.

BOHLOOL B. B. & SCHMIDT E. L. (1968) Nonspecific staining: its control in immunofluorescence examination of soil. *Science* **162**, 1012–14.

BOHLOOL B. & SCHMIDT E. L. (1974) Lectins: possible basis of host specificity in the *Rhizobium*–legume association. *Science* **185**, 269–71.

BURTON J. C. (1967) *Rhizobium* culture and use. In: *Microbial Technology* (Ed. H. J. Peppler), pp. 1–33. Van Nostrand-Reinhold, Princeton, New Jersey.

BURTON J. C. (1978) Monitoring quality in legume inoculants and preinoculated seed. *Proceedings IX Reunion Latinoamericana sobre Rhizobium* (Ed. M. Valdes) pp. 308–25. Instituto Polytecnico Nacional, Mexico.

BURTON J. C. (1979) *Rhizobium* species. In: *Microbial Technology*, 2e. vol. 1, pp. 29–58. Academic Press, New York.

BURTON J. C., MARTINEZ C. J. & CURLEY R. L. (1972) *Methods of Testing and Suggested Standards for Legume Inoculants and Preinoculated Seed*. The Nitragin Company, Milwaukee, Wisconsin.

CALLAHAM D. A. (1979) A structural basis for infection of root hairs of *Trifolium repens* by *Rhizobium trifolii*. M.S. thesis, University of Massachusetts, Amherst.

DATE R. A. (1977) The development and use of legume inoculants. In: *Biological Nitrogen Fixation in Farming Systems of the Tropics* (Eds A. Ayanaba & P. J. Dart), pp. 170–80. John Wiley & Sons, New York.

DAZZO F. B. & BRILL W. J. (1978) Regulation by fixed nitrogen of host–symbiont recognition in the *Rhizobium*–clover symbiosis. *Plant Physiology* **62**, 18–21.

DAZZO F. B. & BRILL W. J. (1979) Bacterial polysaccharide which binds *Rhizobium trifolii* to clover root hairs. *Journal of Bacteriology* **137**, 1362–73.

DAZZO F. B. & HUBBELL D. H. (1975) Cross-reacting antigens and lectin as determinants of host specificity in the *Rhizobium*-clover association. *Applied Microbiology* **30**, 1017–33.

DAZZO F. B. NAPOLI C. A. & HUBBELL D. H. (1976) Adsorption of bacteria to roots as related to host specificity in the *Rhizobium*-clover symbiosis. *Applied and Environmental Microbiology* **32**, 166–71.

DUDMAN W. F. (1977) Serological methods and their application to dinitrogen-fixing organisms. In: *A Treatise on Dinitrogen Fixation, Section IV: Agronomy and Ecology* (Eds R. Hardy & A. Gibson), pp. 487–508. John Wiley & Sons, New York.

FAHRAEUS G. (1957) The infection of clover root hairs by nodule bacteria studied by a single glass slide technique. *Journal of General Microbiology* **16**, 374–81.

FRED E. B., BALDWIN I. L. & MCCOY E. (1932) *Root Nodule Bacteria*. University of Wisconsin Press, Madison, Wisconsin.

GAULT R., BYRNE P. & BROCKWELL J. (1973) Apparatus for surface sterilization of individual legume root nodules. *Laboratory Practice* **22**, 292–3.

HARDY R. W. & HOLSTEN R. D. (1972) Global nitrogen cycling: pools, evolution, transformations, transfers, quantitation, and research needs. In: *The Aquatic Environment: Microbial Transformation and Water Management Implications*, (Eds L. J. Guarraia & R. K. Ballentine) pp. 87–132. U.S. Government Printing Office, Washington D.C.

HARDY R. W., HOLSTEN R. D., JACKSON E. & BURNS R. (1968) The acetylene–ethylene assay for N_2 fixation: laboratory and field evaluation. *Plant Physiology* **43**, 1185–207.

JENSEN H. L. (1942) Nitrogen fixation in leguminous plants. I. General characters of root nodule bacteria isolated from species of *Medicago* and *Trifolium* in Australia. *Proceedings of the Linnean Society of New South Wales* **66**, 98–108.

JOHNSTON A. W. B. & BERINGER J. E. (1975) Identification of *Rhizobium* strains in plant root nodules using genetic markers. *Journal of General Microbiology* **87**, 343–50.

JOHNSTON A. W. & BERINGER J. E. (1977) Chromosomal recombination between *Rhizobium* species. *Nature, London* **267**, 611–13.

JOSEY D. P., BEYON J. L., JOHNSTON A. W. & BERINGER J. E. (1979) Strain identification in *Rhizobium* using intrinsic antibiotic resistance. *Journal of Applied Bacteriology* **46**, 343–50.

KIJNE J. W. (1975) The fine structure of pea root nodules. I. Vacuolar changes after endocytotic host cell infection by *Rhizobium leguminosarum*. *Physiological Plant Pathology* **5**, 75–9.

KUYKENDALL L. D. & WEBER D. F. (1979) Genetically marked *Rhizobium* identifiable as inoculum strain in nodules of soybean plants grown in fields populated with *Rhizobium japonicum*. *Applied and Environmental Microbiology* **35**, 915–19.

LARUE T. & CHILD J. J. (1979) Sensitive fluorimetric assay for leghemoglobin. *Analytical Biochemistry* **92**, 11–15.

MAIER R. J. & BRILL W. J. (1976) Ineffective and non-nodulating mutant strains of *Rhizobium japonicum*. *Journal of Bacteriology* **127**, 763–9.

NAPOLI C. A. & HUBBELL D. H. (1975) Ultrastructure of *Rhizobium*-induced infection threads in clover root hairs. *Applied Microbiology* **30**, 1003–9.

NEWCOMB N. (1976) A correlated light and electron microscopic study of symbiotic growth and differentiation in *Pisum sativum* root nodules. *Canadian Journal of Botany* **54**, 2163–86.

NUTMAN P. S. (1970) The relation between nodule bacteria and the legume host in the rhizosphere and the process of infection. In: *Ecology of Soil Borne Plant Pathogens* (Eds K. F. Baker & W. C. Snyder), pp. 231–47. University of California Press, Berkeley.

NUTMAN P. S., DONCASTER C. C. & DART P. J. (1973) *Infection of Clover by Root-Nodule Bacteria*. Film available from the British Film Institute, London.

NUTMAN P. S., ROUGHLEY R. J., DART P. J. & SUBBA-RAO N. S. (1970) Effect of low-temperature pretreatment on infection of clover root hairs by *Rhizobium*. *Plant and Soil* **33**, 257–9.

PARIISKAYA A. N. & GORELOVA O. P. (1976) Factors of multiple resistance to antibiotics in rhizobia. *Microbiologiya* **45**, 747–50.

RAO V. & KEISTER D. L. (1978) Infection threads in the root hairs of soybean (*Glycine max*) plants inoculated with *Rhizobium japonicum*. *Protoplasma* **97**, 311–16.

ROBERTS G. P., LEPS W. T., SILVER L. E. & BRILL W. J. (1980) Use of two-dimensional polyacrylamide gel electrophoresis to identify and classify *Rhizobium* strains. *Applied and Environmental Microbiology* **39**, 414–22.

ROLFE B. G., GRESSHOFF P. M. & SHINE J. (1980) Rapid screening for symbiotic mutants of *Rhizobium* and white clover. *Plant Science Letters* **19**, 277–84.

SCHMIDT E. L., BANKHOLE R. O. & BOHLOOL B. B. (1968) Fluorescent antibody approach to the study of rhizobia in soil. *Journal of Bacteriology* **95**, 1987–92.

SCHÖLLHORN R. & BURRIS R. H. (1966) Study of intermediates in nitrogen fixation. *Federation of American Societies for Experimental Biology* **25**, 710.

UMALI-GARCIA M., HUBBELL D. M., GASKINS M. H. & DAZZO F. B. (1980) Association of *Azospirillum* with grass roots. *Applied and Environmental Microbiology* **39**, 219–26.

VERMA D. P., KAZAZIAN V., ZOGBI V. & BAL A. K. (1978) Isolation and characterization of the membrane envelope enclosing the bacteroids on soybean root nodules. *Journal of Cell Biology* **78**, 919–36.

VINCENT J. B. (1970) *A Manual for the Practical Study of Root-nodule Bacteria. IBP Handbook No. 15*. Blackwell Scientific Publications, Oxford.

WACEK T. & BRILL W. J. (1976) A simple rapid assay for screening nitrogen-fixing ability in soybeans. *Crop Science* **16**, 519–23.

WEAVER R. & FREDERICK L. (1972) A new technique for most-probable-number counts of *Rhizobium*. *Plant and Soil* **36**, 219–22.

ZOBEL R. W., DEL TREDICI P. & TORREY J. G. (1976) A method for growing plants aeroponically. *Plant Physiology* **57**, 344–6.

Chapter 26 • Mycorrhizae

Conway Ll. Powell

26.1 Introduction

Mycorrhizal fungi are amongst the most common soil organisms in natural ecosystems and are the subject of numerous studies. Nevertheless, the abundance of mycorrhizae in soils and their influence on plant growth have received little attention in standard microbiology and plant nutrition texts (Russell 1973; Alexander 1977). Mycorrhizae can be classified as endomycorrhizal (vesicular-arbuscular, orchidaceous or ericoid) in which the fungus mainly grows inside the root, or as ectomycorrhizal in which the fungus mainly grows around the root surface. In this chapter, arbutoid mycorrhizae (a group of minor importance) have been considered to be ectomycorrhizal.

26.2 Vesicular-arbuscular mycorrhizae

Vesicular-arbuscular (VA) mycorrhizae probably evolved with the Devonian land flora (Nicolson 1975) and are now formed by most angiosperms, gymnosperms and ferns (Baylis 1975; Trappe 1977): few angiosperm families are consistently non-mycorrhizal (Gerdeman 1968). Mycorrhizae develop when a hypha from a spore or an already infected root (Powell 1976) contacts a suitable root, forms an appressorium and then a network of inter- and intracellular hyphae in the cortex. Large vesicles are often borne terminally on intercellular hyphae. Arbuscles develop within a few days of root infection (Rich & Bird 1974) as hyphae penetrate mechanically and enzymically into cortical cells (Kinden & Brown 1975c). As the arbuscle hyphae bifurcate repeatedly (Cox & Sanders 1974) they are enveloped by a host-derived encasement layer (Scannerini & Bonfante-Fasolo 1979) and the continuously invaginating host plasmalemma (Cox & Sanders 1974; Kinden & Brown 1975a, b; Dexheimer *et al.* 1979).

The main effect of mycorrhizal infection on plant growth is the stimulation of phosphorus uptake (Mosse 1973b; Gerdemann 1976) due to exploration by the external hyphae of the soil beyond the root hair and phosphorus depletion zones (Gray & Gerdemann 1969; Rhodes & Gerdemann 1975). Absorbed phosphorus is probably converted into polyphosphate granules in the external hyphae (Callow *et al.* 1978) and passed to the arbuscle for transfer to the host (Schoknecht & Hattingh 1976; White & Brown 1979). This flow of phosphorus occurs in the presence of acid phosphatases (Gianinazzi-Pearson *et al.* 1978; Gianinazzi *et al.* 1979) during the arbuscle lifespan (Cox & Tinker 1976) or senescence (Kinden & Brown 1975c, 1976).

VA mycorrhizal fungi are obligate symbionts and have been grown on excised roots (Mosse & Hepper 1975) or whole plants (Mosse 1962; Mosse & Phillips 1971; Pearson & Tinker 1975) in artificial media but never in pure culture. The endophytes, identified by vesicle and arbuscle formation in roots (Abbott & Robson 1979), dimorphic branching of extramatrical hyphae (Mosse 1959; Nicolson 1959), and production of large chlamydospores and azygospores in soil, are classified as phycomycetes in the family Endogonaceae (Gerdemann & Trappe 1977; Hall & Fish 1979).

Spores are produced singly or in sporocarps at densities ranging from 0 to >100 (g soil)$^{-1}$ (Sutton & Barron 1972) and may be borne epigeally (Daniels & Trappe 1979) but are never abscissed to become airborne although they are occasionally ingested and transported by animals (McIlveen & Cole 1976). In agricultural soils, spore densities are highest in late summer and autumn (Hayman 1970; Saif & Khan 1975) at moderate phosphorus fertiliser levels (Hayman *et al.* 1975; Porter *et al.* 1978) and are depressed by previous non-mycorrhizal crops (Iqbal & Quereshi 1976; Black & Tinker 1979). Spore densities are lower under undisturbed native vegetation (Mosse & Bowen 1968; Hayman & Stovold 1979) in which non-sporing races frequently occur (Baylis 1969; Sparling & Tinker 1978).

There is a lag phase before young roots become infected by mycorrhizal propagules in soil, followed by rapid mycorrhiza colonisation and a final plateau (Sutton 1973; Saif 1977) at an infection level primarily determined by the internal plant phosphorus concentration (Azcon *et al.* 1978; Menge *et al.* 1978c; Jasper *et al.* 1979). Infection levels are usually highest in summer before root senescence begins (Nicolson & Johnston 1979; Rabatin 1979) and generally decrease with increasing phosphorus (Mosse 1973a; Menge *et al.* 1978b) and nitrogen fertilisers (Chambers *et al.* 1980) and at high soil moisture levels (Sondergaard & Laegaard 1977). Infection levels are also depressed by low temperatures (Furlan & Fortin 1973; Schenck & Schroder 1974) and low light intensity or short day length (Daft & El-Giahmi 1978; Hayman 1974).

There is one-way carbon transfer from host to fungus (Ho & Trappe 1973) in the form of sucrose and glucose with no evidence for fungal-specific polyol storage in mycorrhizal roots (Hayman 1974; Bevege *et al.* 1975). There is limited carbon storage in glycogen (Bevege *et al.* 1975) and lipids (Cooper & Losel 1978) and, together with extensive fungal growth and respiration, this must constitute enough of a carbon sink to maintain supplies from the host (Bevege *et al.* 1975) and account for the faster incorporation of ^{14}C-labelled compounds into mycorrhizal than into non-mycorrhizal roots (Losel & Cooper 1979).

In some experiments, mycorrhizal fungi have seemed to tap organic and inorganic phosphorus sources in soil which are normally unavailable to non-mycorrhizal plants (Murdoch *et al.* 1967; Mosse 1973a; Ross & Gilliam 1973; Powell 1979a). Cress *et al.* (1979) showed that mycorrhizal roots had greater affinity than non-mycorrhizal roots for orthophosphate ions, but Sanders and Tinker (1971) and Hayman and Mosse (1972) using ^{32}P-labelled soil and Barrow *et al.* (1977) using heat-treated soil suggested that mycorrhizal and non-mycorrhizal plants tap the same phosphorus sources.

The growth benefit from inoculation generally decreases with increasing phosphorus fertiliser levels (Ross 1971; Mosse 1973a; Menge *et al.* 1978b), although Ross and Gilliam (1973) and Abbott and Robson (1977) found that mycorrhizal responses were also small at very low soil phosphorus levels. Mycorrhizal growth responses differ among host species (Powell & Sithamparanathan 1977) and inoculation stimulates clovers at the expense of associated grasses (Crush 1974; Hall 1978). Plant growth responses to inoculation with a single isolate have frequently been shown in sterilised soil in pots (Mosse 1973b; Gerdemann 1976) but there are large inter- and intra-specific differences in efficiency of phosphorus absorption from soil (Mosse 1972; Powell 1977). Efficient mycorrhizal fungi have stimulated plant growth in unsterilised soil in pots (Mosse *et al.* 1969; Abbott & Robson 1977, 1978; Bagyaraj & Menge 1978; Powell & Daniel 1978) and in the field (Khan 1972, 1975; Black & Tinker 1977; Saif & Khan 1977; Azcon *et al.* 1979; Hayman & Mosse 1979; Owusu-Bennoah & Mosse 1979; Powell 1979b; Powell *et al.* 1979). In some of these experiments, however, control plants have not been inoculated with the indigenous fungi before transplant into the unsterilised test soils, soil inoculum rates have been very high and mycorrhizal growth responses have been short lived. Conclusive long-term benefits from field inoculation have still to be proved.

Increased uptake of phosphorus is not the only effect of VA mycorrhizal fungi on plant growth. Mycorrhizal inoculation stimulates rooting (Barrows & Roncadori 1977) and growth and transplant survival (Bryan & Kormanik 1977; Kormanik *et al.* 1978) of cuttings and seedlings raised in sterilised nursery media. Re-establishment of plant cover on mine spoil and eroded soils is dependant on reinfestation with mycorrhizal fungi (Daft & Hacskaylo 1976; Khan 1978; Hall & Armstrong 1979; Hall 1980). VA mycorrhizae also stimulate plant uptake of zinc, copper, sulphur, potassium and calcium (Gilmore 1971; Gerdemann 1973; Bowen *et al.* 1974; La Rue *et al.* 1975; Powell 1975; Cooper & Tinker 1978; Rhodes & Gerdemann 1978a; Lambert *et al.* 1979), although not as markedly as phosphorus (Cooper & Tinker 1978).

Mycorrhizal infection enhances nodulation in legumes (Crush 1974; Smith & Daft 1977; Carling *et al.* 1978) and generally decreases growth depression and root rots caused by fungal pathogens (Schenck & Kellam 1978; Schonbeck & Dehne 1979). It also depresses root penetration and larval development of nematodes (Sikora 1978; Bagyaraj *et al.* 1979) and alleviates nematode-induced growth depressions (Roncadori & Hussey 1977; Bagyaraj *et al.* 1979). Non-systemic fungicides have little effect on mycorrhizal infection or growth responses (Burpee & Cole 1978; Smith 1978) although soil drenches of benomyl and methyl-bromide reduce mycorrhiza formation and efficiency (Bailey & Safir 1978; Boatman *et al.* 1978; Menge *et al.* 1978a). There is no conclusive evidence for increased water uptake by mycorrhizal compared to non-mycorrhizal plants (B. Ward, personal communication) as originally suggested by Safir *et al.* (1971).

26.3 Orchidaceous mycorrhizae

In nature, orchids only germinate when infected by endomycorrhizal fungi which subsequently colonise the roots of the young plants (Harley 1969; Hadley 1975). The endophytes are mostly of the genus *Rhizoctonia* with perfect stages occurring in the basidiomycetes (mainly Tulasnellales) and ascomycetes (Warcup & Talbot 1966; Warcup 1975), although Hall (1976a) reported mycorrhiza formation in *Corybas macranthus* by VA mycorrhizal fungi. Thin-walled hyphae enter the protocorm (germinating seed) through epidermal hairs and anastomose repeatedly within cortical cells to form peleton coils (Hadley 1975), but they are separated from the host cytoplasm by the invaginating host plasmalemma and encasement layer (Hadley 1975; Strullu & Gourret 1975) as in VA arbuscles. Mycorrhizal fungi do not stimulate germination of the orchid seed before infection occurs (Hadley & Ong 1978), despite the ability of some fungi in pure culture to produce vitamins capable of inducing germination (Harvais & Pekkala 1975).

Mycorrhizal infection stimulates the growth of protocorms (Smith 1966; Hadley 1970; Hadley & Williamson 1972) and saprophytic adult plants (Purves & Hadley 1975) by transfer of carbon (as trehalose—Smith 1967) from undigested fungal peletons (Purves & Hadley 1975) to the orchid. The mycorrhizal fungi can tap a wide range of carbon compounds (including cellulose) in pure culture (Hadley & Ong 1978) or when mycorrhizal with orchid seedlings in soil (Smith 1966, 1967) and carbon transfer from orchid to fungus has never been observed (Smith *et al.* 1969; Hadley & Purves 1974). Orchid endophytes grow well in pure culture with asparagine, arginine and urea, but not ammonium, as the nitrogen source (Stephen & Fung 1971a, b; Hadley & Ong 1978).

Symbiotic germination tests (Hadley 1970; Jonsson & Nylund 1979) suggest a low level of orchid–fungus specificity although there are few data on specificity amongst mature orchid plants in the field (Warcup 1975). *Rhizoctonia* spp. isolates vary from those which can synthesise vitamins, have simple nitrogen requirements and are often pathogenic (Hadley & Ong 1978), to the slower growing vitamin-dependant strains (Hijner & Arditti 1973) which have evolved some dependance on the orchid host (Hadley & Ong 1978) and are likely to be beneficial endophytes. There are no reports of mycorrhiza-enhanced ion uptake in autotrophic adult orchids, although Smith (1966) demonstrated hyphal translocation of [^{32}P]phosphorus by *Rhizoctonia repens* to *Dactylorchis purpurella* protocorms in axenic culture.

26.4 Ericoid mycorrhizae

The Ericaceae and Epacridaceae are normally infected by smooth-walled mycorrhizal fungi which form very dense intracellular coils in the outer cortical cells (Read & Stribley 1975a; Bonfante-fasolo & Gianinazzi-Pearson 1979). The endophytes of *Calluna vulgaris* are slow-growing in culture (Pearson & Read 1973a), normally sterile and restricted to the hair roots: The contention (Rayner 1913, 1922) that the *Calluna vulgaris* endophyte was a *Phoma* sp. systemic in the plant was discounted by Harley (1969) and again by Read (1974) who cultured *Pezizella ericae* (ascomycete) as the perfect stage. *Clavaria* spp. (basidiomycetes) have been suggested as the endophytes of *Rhododendron* spp. and *Azalea* spp. (Seviour *et al.* 1973; Englander & Hull 1980).

The endophyte may account for up to 80% of the mass of the hair roots of *Calluna vulgaris* (Read & Stribley 1975a) providing a minimum of 2000 hyphal connections per cm between soil and host root. As with VA and Orchid mycorrhizae, invaginated host plasmalemma and an encasement material envelope the hyphae in the intracellular complexes (Bonfante-fasolo & Gianinazzi-Pearson 1979).

Mycorrhizal infection by *Pezizella ericae* greatly stimulates nitrogen uptake and plant growth in infertile peat soils (Read & Stribley 1973; Stribley & Read 1974b; Stribley *et al.* 1975) through mineralisation and high specific absorption of organic nitrogen by the mycorrhizal fungus (Stribley & Read 1974b; Pearson & Read 1975; Stribley *et al.* 1975). Growth responses did not occur in very fertile soils (Brook 1952; Morrison 1957; Bannister & Norton 1974). *Pezizella ericae* also translocated [^{32}P]phosphorus in pure culture and accumulated [^{32}P]phosphorus in *Calluna vulgaris* roots (Pearson & Read 1973b).

The endophyte is probably wholly host dependant for carbon supply. Stribley and Read (1974a) found accumulation of host-fed [^{14}C]carbon dioxide in the mycorrhizal fungus as mannitol and trehalose. There was no evidence of fungus-to-host carbon flow (Pearson & Read 1973b) or of host retrieval of fungal carbon through hyphal lysis (Bonfante-fasolo & Gianinazzi-Pearson 1979).

Mycorrhizal infection levels decreased with increasing altitude (Haselwandter 1979) and increasing nitrogen fertiliser level in sand culture (Stribley & Read 1976). There was no difference between Northern and Southern hemisphere endophytes in their ability to stimulate growth in a pot trial (Read 1978) and there have been no reported field inoculation trials. C. Powell (unpublished observation) found that mycorrhizal inoculation of blueberries (*Vaccinium corymbosum*) in a peat soil with a low ericoid mycorrhiza population stimulated shoot and fruit yields by 35% and 32% respectively.

26.5 Ectomycorrhizae

Ectomycorrhizae are formed by only 5% of vascular plants, but predominate in several economically important families (Pinaceae & Fagaceae-Betulaceae—Harley 1969; Trappe 1977) and occur sporadically in other temperate and tropical families (Meyer 1973). Ectomycorrhizal fungi are usually basidiomycetes and ascomycetes (Trappe 1962) and occasionally phycomycetes (Fassi *et al.* 1969). Strict host specificity is rare (Lange 1923) and one plant may form mycorrhizae with several fungi simultaneously (Dominik 1959). A fungus may be specific to one host genus (Malajczuk *et al.* 1975; Molina 1979) although a wider host and geographic range is usual (Marx 1977) and Trappe (1977) estimated that 2000 fungi can probably form ectomycorrhizae with Douglas fir.

The fungus forms a mantle or sheath (20 to 40 μm diameter) around the short roots (Hatch & Doak 1933), with an internal network of hyphae (the Hartig net) penetrating by physical pressure between the elongated host cortical cells (Hatch 1937; Foster & Marks, 1966, 1967), but not penetrating the endodermis (Meyer 1974). Ectomycorrhizal fungi produce auxins, gibberillic acids and cytokinins in varying amounts (Slankis 1958, 1973; Turner 1962; Miller 1963; Bowen 1973) which stimulate repeated branching of the short roots. Mycorrhizal infection greatly increases root–soil contact (Bowen 1973) producing up to 4 m external hyphae (g soil)$^{-1}$ (Burges & Nicholas 1961) capable of solubilising nutrients from soil (Wilde 1954) and translocating them to the host root (Skinner & Bowen 1974).

Ectomycorrhizal fungi are unable to degrade lignin but occasionally degrade cellulose and humic material (Lundeberg 1970; Lamb 1974; Linkins & Antibus 1979; Todd 1979) and grow best on glucose (Melin 1925; Keller 1950; Harley 1969; Hacskaylo 1973). They are usually inhibited but occasionally stimulated in soil or agar by saprophytic bacteria and fungi (Jones 1924; Levisohn 1957; Harley 1969; Bowen & Theodorou 1979) and probably only persist in soil as basidiocarps or in mycorrhizal association (Harley 1969; Lamb & Richards 1974). Ectomycorrhizal fungi have an obligate but varying requirement for thiamin, pantothenic acid, nicotinic acid and biotin (Melin & Nyman 1941; Melin 1953; Harley 1969).

Carbon is translocated from the host to the fungus (Melin & Nilsson 1957; Bjorkman 1960) as glucose, fructose or sucrose and is converted in the sheath into mannitol, trehalose and glycogen (Lewis & Harley 1965a, b, c). These polyols are unavailable to the host root and so maintain a strong source-to-sink carbon flow to the fungus which is irreversible (Lewis 1970), although this is questioned by Hacskaylo (1973).

Ectomycorrhizal infection greatly stimulates plant growth and nutrient uptake (Melin 1917), especially in soils of low to moderate fertility (Hatch 1937; Hacskaylo & Snow 1959; Harley 1969; Bowen 1973). Recent work has shown that ectomycorrhizal inoculation increases plant survival and growth rate in natural soils outside the previous host range (Kessell 1927; Mikola 1970), in sterilised nursery soils (Hacskaylo & Palmer 1957; Henderson & Stone 1970; Marx & Bryan 1975; Marx *et al.* 1976; Wojahn & Iyer 1976; Marx 1979b), in disturbed soils with low mycorrhiza populations (Berry & Marx 1978; Ruehle 1980; Berry *et al.* 1979; Marx & Artman 1979), and in soils containing inefficient mycorrhizal fungi (Benecke & Gobl 1974; Marx *et al.* 1977a; Mexal *et al.* 1979). Many fungal isolates need to be field tested in any selection trial (Trappe 1977) because of the large inter- and intra-specific differences in fungal efficiency (Bjorkman 1970; Laiho 1970; Lamb & Richards 1971; Vozzo & Hacskaylo 1971; Benecke & Gobl 1974; Bega 1979) and Bowen (1973) warned that fungi must also be selected for disease resistance, ease of introduction and persistence.

Mycorrhiza development of short roots is greatest under moderate nitrogen and phosphorus deficiency in soil, adequate soluble carbohydrates in host roots and optimum (20%) soil oxygen concentration (Hatch 1937; Bjorkman 1942; Harley 1969; Marx *et al.* 1977b; Marais & Kotze 1978). Mycorrhiza development was greater under foliar than root feeding due to higher root carbohydrate levels (Dixon *et al.* 1979), and with ammonium than nitrate fertiliser (Bigg & Alexander 1979; Bledsoe *et al.* 1979).

Nitrogen and phosphorus concentrations are often higher in mycorrhizal than non-mycorrhizal plants (Hatch 1937; Mitchell *et al.* 1937) and the uptake and translocation of nitrogen and phosphorus by excised mycorrhiza was temperature and oxygen dependant (Harley & McCready 1950, 1952; Harley *et al.* 1956; Jennings 1964; Carrodus 1966). Pine and beech mycorrhizae with well-developed sheaths had 2 to 9 times the phosphorus-absorbing ability of uninfected short roots (Harley & McCready 1950; Bowen 1973). Phosphate accumulated in the fungal sheath but was translocated to the host as plant concentrations decreased. Ammonium may (Carrodus 1966, 1967) or may not (Lundeburg 1970) be preferred to nitrate as the nitrogen source. There is no evidence of mycorrhizal mineralisation of organic nitrogen (Bowen 1973) or nitrogen fixation (Harley 1969), although Giles and Whitehead (1976, 1977) transferred acetylene reduction capability from *Azotobacter vinelandii* to a *Rhizopogon* sp. ectomycorrhizal with *Pinus radiata*.

Ectomycorrhizae stimulate phosphorus uptake from soluble and insoluble sources in soil (Henderson & Stone 1970; Bowen 1973; Mejstrik & Krause 1973; Malajczuk *et al.* 1975) in which surface and sheath-bound acid phosphatases (Bartlett & Lewis 1973; Williamson & Alexander 1975; Ho & Zak 1979) may be important. Ectomycorrhizal infection also increases the uptake and translocation of calcium, sodium, zinc and some alkali metals (Melin & Nilsson 1955; Wilson 1957; Melin *et al.* 1958; Bowen *et al.* 1974) but not of sulphur (Morrison 1962).

Many ectomycorrhizal fungi produce antimicrobial compounds (Marx 1969, 1970, 1973; Marais & Kotze 1976; Kais & Snow 1979) which offer the host plant a varying amount of resistance against soil-borne pathogens or competitive saprophytes (Gadgil & Gadgil 1975). Even non-mycorrhizal roots may be protected by anti-microbial secretion from adjacent mycorrhizae (Marx 1973) and the mantle is an effective physical barrier to many root pathogens.

Soil temperatures greatly influence fungal growth and ectomycorrhizal colonisation of roots (Moser 1958a; Laiho 1970; Trappe 1977) and survival of mycorrhizal fungi (Marx & Bryan 1971). Most ectomycorrhizal fungi in pure culture have temperature optima of 18 to 27°C (Harley 1969), although influenced by considerable inter- and intra-specific variation, and affected by moisture stress and field conditions (Cline *et al.* 1979). Mycelial cultures store well and are viable after storage at −10°C (France *et al.* 1979) and at 5°C (Marx & Daniel 1976; Marx 1979a). Water stress has little effect *per se* on ectomycorrhizal fungi (Sands & Theodorou 1978; Reid & Bowen 1979) although phosphorus uptake was depressed probably due to restricted growth of mycelial strands in dry soil (Gadgil 1972). Plant protection chemicals and air pollutants may or may not inhibit (Iloba 1978, 1979) the growth of ectomycorrhizal fungi in pure culture (Kelley & South 1978; Smith & Ferry 1979) and the field (Iyer & Wojahn 1976).

Experimental

26.6 Morphology and classification

26.6.1 Vesicular-Arbuscular Mycorrhizae

Roots (fresh, dried, or fixed) can be cleared in potassium hydroxide and stained with trypan or cotton blue for routine determination of VA mycorrhizal infection under the light microscope (Phillips & Hayman 1970). Glucosamine assay and colorimetric methods (Becker & Gerdemann 1977; Hepper 1977) are faster but unsuitable for morphological study. All mycorrhizal associations can be examined by scanning electron microscopy (SEM) and transmission electron microscopy (TEM) using standard fixing, embedding and sectioning procedures (Foster & Marks 1966, 1967; Hadley & Williamson 1971; Kinden & Brown 1975a; Bonfante-fasolo & Gianinazzi-Pearson 1979). VA mycorrhizal fungi are usually classified (Gerdemann & Trappe 1977; Hall & Fish 1979) on spore characters (determined under the light microscope), although Abbott and Robson (1979) suggest that hyphal morphology may also be useful, especially in identifying *Glomus tenuis* (the fine endophyte—Greenall 1963).

26.6.2 Orchidaceous Mycorrhizae

In order to examine orchid mycorrhizae the mycorrhizal protocorms are teased out into 0.02 ml 20% (w/v) glycerine and mounted on a slide for examination at ×150 (Hadley & Williamson 1971) and roots of mature orchids are hand sectioned for microscopical examination (Hadley & Williamson 1972). Perfect states of orchid *Rhizoctonia* isolates are listed by Warcup (1975).

26.6.3 Ericoid Mycorrhizae

Ericoid mycorrhizal roots (densely pigmented) are cleared in potassium hydroxide and 20 volume hydrogen peroxide (2 to 10 min) and stained with cotton or trypan blue (Phillips & Hayman 1970) for microscopical examination. Only one endophyte (from *Calluna vulgaris*) has been identified at the perfect stage (*Pezizella ericae*—Read 1974).

26.6.4 Ectomycorrhizae

Ectomycorrhizal short roots are classified on branching pattern, size, form, colour, surface texture and presence or absence of attached mycelial strands or rhizomorphs (Dominik 1969; Zak 1973). For a discussion of embedding and sectioning techniques for light microscopy see Clowes (1951).

26.7 Fungal biomass

26.7.1 Vesicular-Arbuscular Mycorrhizae

It is generally believed that spore density in soil reflects the soil's mycorrhizal infectivity (Black & Tinker 1979). Soil samples (25 to 100 g) are washed through a series of sieves (700 to 43 μm—Gerdemann & Nicolson 1964). Spores in the retained sievings can be concentrated by sedimentation in gelatinised columns (Mosse & Jones 1968), successive settling and decanting (Sutton & Barron 1972) or by compressed air flotation in 50% (w/v) glycerol (Furlan & Fortin 1975). Large spores (>150 μm diameter) are most easily retrieved by flotation and aspiration from a sucrose gradient (Mertz *et al.*

1979). Spore counting is tedious and is best done in a nematode counting dish under a dissecting microscope (Hayman & Stovold 1979).

Mycorrhizal infectivity of a soil may also be assessed by stepwise dilution (up to 1:10 000) with sterilised soil and examination of mycorrhizal development in six-week-old white clover 'bait plants' at each dilution level. Infectivity is measured as the number of entry points (ml soil)$^{-1}$ (Smith & Bowen 1979) or as a most-probable-number (MPN) value based on the number of replicates mycorrhizal at each dilution level (Powell 1980). These are more time consuming methods than spore counting, but do measure total mycorrhizal infectivity from spores, hyphae and infected root fragments.

Nicolson and Johnston (1979) estimated hyphal biomass in sand by wet sieving and decanting 500 ml field samples and collecting and weighing hyphae from 420 to 75 μm sieves. Sanders *et al.* (1977) stripped and weighed external hyphae from roots grown in soil, although hyphae are easily broken and left behind. Rhodes and Gerdemann (1975) measured hyphal growth into soil by translocation of [^{32}P] spots (40 μCi) applied at increasing distances from onion roots in polystyrene soil chambers.

Mycorrhizal infection is usually assessed microscopically (Nicolson 1959) as the percentage of 1 to 10 mm root segments mycorrhizal in cleared and stained roots lightly squashed on a microscope slide (Black & Tinker 1979; Nicolson & Johnston 1979; Rabatin 1979). Hayman (1978) also assessed density of mycorrhizal infection in each segment and Sanders *et al.* (1977) estimated total mycorrhizal root length. Porter *et al.* (1978) spread the entire root sample in a Petri dish and assessed infection level by eye in ten fields under the dissecting microscope. This method is probably faster but less accurate where roots are lightly stained.

26.7.2 Orchidaceous Mycorrhizae

Infection intensity of orchid mycorrhizae in young protocorms and mature roots is assessed as the number of infected cortical cells as a percentage of cells available for infection (Hadley & Williamson 1972).

26.7.3 Ericoid Mycorrhizae

The distribution and density of ericoid mycorrhizal infection in the root system, and the cortical cell dimensions should be measured for an estimate of fungal contribution to total root volume (up to 80% for *Calluna vulgaris*—Read & Stribley 1975a). Mycorrhizal infection levels are usually assessed as the proportion of infected cortical cells in the terminal 3 mm of 50 randomly selected root tips (Stribley & Read 1976).

26.7.4 Ectomycorrhizae

Initial colonisation of roots by ectomycorrhizae was estimated microscopically by staining *Pinus radiata* roots at 4 to 8 weeks after inoculation with 1% (w/v) cotton blue in lactophenol (Bowen 1973). Infection in older roots is scored by eye as the percentage of short roots (Hatch 1937; Gobl 1965; Lamb & Richards 1974) or root tips which are mycorrhizal (Bjorkmann 1942; Lyr 1963; Meyer & Gottsche 1971). The latter method gives proper emphasis to the increased surface area of branched mycorrhizal short roots. Harley (1969) estimated fungal biomass in individual mycorrhizae by dissection and weighing of the fungal mantle. Skinner and Bowen (1974) assessed fungal growth and mycelial strand penetration in different soil types by inserting soil cores into perspex root boxes for 4 to 6 weeks.

Fungal biomass has been estimated in the field from the number and weight of mycorrhizal roots in a known sample volume at different soil horizons (Gobl 1965; Marks *et al.* 1968; Meyer & Gottsche 1971) and Gobl (1965) assessed sclerotia biomass in sieved 2 ml soil samples. Fruiting body production is greatly influenced by weather and season (Trappe & Fogel 1977).

26.8 Fungal isolation and culture

26.8.1 Vesicular-Arbuscular Mycorrhizae

No VA mycorrhizal endophytes have been grown in pure culture on artificial media, although spores germinate readily and make limited hyphal growth on water agar (Mosse 1959) but fail to germinate on nutrient agars (Daniels & Graham 1976). Spores are sterilised for 2 min in 2% (w/v) chloramine T (with 200 μg ml^{-1} streptomycin—Mosse & Phillips 1971) or 0.5% (w/v) sodium hypochlorite (Daniels & Duff 1978), or for 1 week in filter sterilised streptomycin (200 μg ml^{-1}) and gentamicin (100 μg ml^{-1}) at 4°C in the dark (Mertz *et al.* 1979), and are then rinsed with several changes of sterile distilled water.

Germination was stimulated by: dark storage at 6 to 10°C (Daniels & Graham 1976; Hepper & Smith 1976; Sward *et al.* 1978); optimal pH (Green *et al.* 1976); addition of dialysed soil extract (Daniels & Graham 1976); contact with soil amended agar (Mosse 1959); or placement in nylon mesh bags in soil (Daniels & Duff 1978). Germination was inhibited by addition of manganese at 13.6 $\mu g\ l^{-1}$ or zinc at 700 $\mu g\ l^{-1}$ (Hepper & Smith 1976) or a range of metabolic inhibitors (Hepper 1979). Charcoal may (Watrud *et al.* 1978) or may not (Mosse & Phillips 1971) stimulate germination and hyphal growth through adsorption of toxins produced in agar culture.

VA mycorrhizae have been synthesised in clover roots on test tube slopes of modified Jensen's medium (Mosse 1962) using germinated spores with or without pectinase or *Pseudomonas* sp. contamination for infection establishment (Mosse 1962; Mosse & Phillips 1971). Mycorrhizal infections can also be synthesised *in vitro* on clover plants using sterilised germinated spores in split Petri dishes with soil amended agar (2 g per dish—Pearson & Tinker 1975; Cooper & Tinker 1978). Mosse and Hepper (1975) infected excised root organ cultures of tomato and red clover (growing on White's medium at pH 7.0) with sterilised pregerminated *Glomus mosseae* spores, but were unable to transfer mycorrhizal infection by subculturing.

26.8.2 Orchidaceous Mycorrhizae

Endophytes of orchid mycorrhizae are isolated from mature orchid roots by surface sterilizing 5 mm root pieces in mercuric chloride in 20% (v/v) ethanol with several rinses in sterile water and plating onto peptone dextrose agar (PDA) at pH 4.5 to 5.0 (Smith 1967). Fungi appearing from roots are subcultured, identified and stored indefinitely on PDA (Hadley 1970; Stephen & Fung 1971a). Fungi are transferred to Pfeffer mineral salt medium for determination of nitrogen requirements (Stephen & Fung 1971a) and vitamin heterotrophy (Stephen & Fung 1971b; Hadley & Ong 1978). For symbiosis testing of putative endophytes, orchid seed is surface sterilised in 5% bleaching solution long enough to bleach the seed coat (20 to 60 min—Harvais & Hadley 1967) and sown onto Pfeffer agar slopes (Hadley 1970). A 4 mm diameter mycelial plug on PDA (storage medium) or water agar (preparation medium for symbiosis testing—Hadley & Ong 1978) is placed next to the seed on the agar slant a few days after seed sowing (Hadley 1970). Symbiosis test cultures are maintained at 20°C in darkness. Warcup (1975) details other media low in nitrogen and micronutrients suitable for symbiotic germination of orchid seed.

26.8.3 Ericoid Mycorrhizae

In the maceration method of Pearson and Read (1973a), ericoid hair roots are transferred through 20 to 25 sterile water washes (each lasting 5 min) and macerated in 5 ml distilled water to produce detached cortical cells which are pipetted sparingly over 0.5% (w/v) water agar plates. The endophyte of *Calluna vulgaris* identified as *Pezizella ericae* (Read 1974) was slow growing, normally sterile, characteristically dark and easily masked by rapidly growing saprophytes. Maintenance on malt peptone and Czapek Dox yeast agars produced mycelial cultures suitable for inoculation. For mycorrhiza synthesis *in vitro*, aseptic seedlings are inoculated with a mycelial plug on 0.5% (w/v) distilled water agar amended with organic soil particles (Pearson & Read 1973a).

26.8.4 Ectomycorrhizae

Many ectomycorrhizal fungi are difficult or impossible to culture (Trappe 1977). Pure cultures are attempted by placing pieces of internal pileus tissue from insect-free sporocarps aseptically into test-tube agar slants in the field (Trappe 1977) or in the laboratory as soon as possible after sporocarp collection (Marx 1979b). Fungal symbionts can also be isolated from mycorrhizal roots by vigorous washing (10 distilled water rinses) followed by surface sterilising in 0.7% (w/v) calcium hypochlorite for 10 min or 20 further rinses in sterile distilled water (Chu-Chou 1979). Modified Melin-Norkrans media are widely used (Marx & Bryan 1975) for initial isolations (e.g. M_{19} potato-glucose and M_{74} ammonium phosphate glucose citrate—Trappe 1977) and successful effective ectomycorrhizal isolates may be very slow growing. Isolates are incubated at 20 to 25°C (or 10 to 15°C for high altitude isolates—Trappe 1977) and cleared of bacterial contaminants by subculturing and antibiotics (Palmer 1971). Well-established cultures are maintained in the dark on agar at 5°C for 4 to 6 months between transfers, or indefinitely in water. Fortin and Piche (1979) described hypocotyl explant culture of *Pinus strobus* for routine ectomycorrhizal synthesis *in vitro*.

26.9 Inoculum production and use

26.9.1 VESICULAR-ARBUSCULAR MYCORRHIZAE

Until VA fungi can be obtained in pure culture, the inoculum must be produced by growing suitable host plants (onion, clover, sorghum, maize) inoculated with sterilised or unsterilised spores (1 to 30 per plant), in open pots or large bins or sterilised soil or sand (Khan 1972; Johnson 1977; Abbott & Robson 1978; Menge *et al.* 1978b; Owusu-Bennoah & Mosse 1979; Powell *et al.* 1979). For inoculating plants in pot trials, spores (30 to 50 per plant—Hall 1976b; Johnson 1977), hyphal wefts (collected from soil sievings and pipetted over seedling roots—Johnson 1977) and lyophilised mycorrhizal roots (Crush & Pattison 1975) have been used occasionally. Wet sievings or a slurry of inoculum soil containing infected root segments, spores and hyphae are most commonly used (Mosse *et al.* 1969; Powell 1977; Abbott & Robson 1978; Menge *et al.* 1978a), and chopped mycorrhizal roots have also been frequently employed (Mosse 1972; Hall 1976b; Pichot & Binh 1976; Schultz *et al.* 1979).

For pot trials, the inoculum is usually layered below seed (Jackson *et al.* 1972; Pichot & Binh 1976; Barrow *et al.* 1977; Hall 1978) at rates of 0.5 to 10 g inoculum per plant. Plants have been preinoculated for field or pot trials in dishes of soil or sand infested with mycorrhizal fungi before transplanting to sterilised or unsterilised soils (Khan 1972, 1975; Mosse 1972, 1973a; Powell & Daniel 1978; Azcon *et al.* 1979) but control plants must be inoculated with the indigenous fungi when planting out into unsterilised soils.

Seeds can be inoculated with spores (using 1% (w/v) methyl cellulose sticker—Gaunt 1978; Swaminathan & Verma 1979) or mycorrhizal soil (Hattingh & Gerdemann 1975; Hall 1979; Powell 1979b) bound to the seed coat with methyl cellulose or 10% (w/v) clay mix with the inoculum. Clover seed inoculated 1:30 (w/w) with pulverised, dried mycorrhizal soil by a commercial process has retained mycorrhizal infectivity for up to 4 months storage at 5°C (C. Powell, unpublished observations). Field plots have also been inoculated with mycorrhizal soil on the soil surface (C. Powell, unpublished observations) or in the seed-bed (Black & Tinker 1977; Owusu-Bennoah & Mosse 1979) at 0.5 to 2.0 kg m^{-2}.

Control plants should always receive sieved and/or filtered washings from the mycorrhizal inocula to ensure contamination by the same bacterial flora.

26.9.2 ORCHIDACEOUS MYCORRHIZAE

The fungal inoculum and inoculation of orchid mycorrhizae are discussed in section 26.8.2.

26.9.3 ERICOID MYCORRHIZAE

Ericoid mycorrhizal fungal mycelium grown on peptone or Czapek Dox yeast agar (Pearson & Read 1973a) was centrifuged, washed and macerated and 0.1 to 5.0 mg fresh weight applied to roots of seedlings in open pots (Stribley *et al.* 1975; Stribley & Read 1976). In New Zealand, container-grown blueberry plants (*Vaccinium corymbosum*) are usually non-mycorrhizal and 2-year-old plants have been inoculated with 100 g mycorrhizal soil (from under old mycorrhizal blueberries) at the planting out stage (C. Powell, unpublished observations).

26.9.4 ECTOMYCORRHIZAE

Spontaneous inoculation of planted out pine seedlings by airborne spores of ectomycorrhizal fungi is haphazard and not recommended (Trappe 1977). Nurseries have been routinely inoculated by incorporation of soil and humus (Henderson & Stone 1970; Trappe 1977) or freshly chopped mycorrhizal roots (Henderson & Stone 1970; Sinclair 1974) but that is no control over efficacy of mycorrhizal fungi or incorporation of soil pathogens. Mikola (1970) and Donald (1975) mixed chopped sporocarps with nursery soil and Theodorou (1971) squeezed spores from surface-sterilised sporocarps and found that a suspension of 1×10^6 basidiospores was an effective inoculum for *Pinus radiata*. Basidiospores are best collected from dry ripe sporocarps (up to 1 kg basidiospores can be collected in 3 h) and stored in the dark at 5°C (Marx & Bryan 1975). Spores rapidly initiate mycorrhiza formation when applied loose to the soil surface (at 1×10^6 to 1×10^9 spores m^{-2}—Marx *et al.* 1976) or incorporated with a hydromulch (Marx *et al.* 1979).

However, mycelial cultures of fast-growing fungi such as *Boletus plorans* or *Pisolithus tinctorius* are the best inocula and rapidly synthesise mycorrhizae on young seedlings (Moser 1958b; Marx *et al.* 1976). Mycelial inoculum is produced in 2 l containers of vermiculite and moss moistened with Melin-Norkrans medium (Marx & Bryan 1975) inoculated with mycelial plugs of suitable fungi. After 15 weeks growth at room temperature the vermiculite was completely permeated by *Pisolithus tinctorius* (Marx & Bryan 1975) and the inoculum medium, rinsed with water to leach remaining soluble nutrients, was ready for bulking-up with sterilised soil (1:8) for inoculation of plants in nursery or pot trials.

26.10 Measurement of mycorrhizal growth responses

26.10.1 Vesicular-Arbuscular Mycorrhizae

Shoot dry weight is the most common measure of growth response to mycorrhizal inoculation although shoot fresh weight is occasionally used (Crush 1974; Mosse 1977; Abbott & Robson 1978). Leaf length and number have been measured (Mosse & Hayman 1971; Daft & Hacskaylo 1976) as well as bulb diameter (Mosse & Hayman 1971) and fresh weight (Ames & Linderman 1978) and tiller numbers in cereals (Saif & Khan 1977). Root dry weight, fresh weight or length are often measured (Abbott & Robson 1978; Chambers *et al.* 1980) and root/shoot ratios calculated (Abbott & Robson 1978). Fruit and tuber yields are more important parameters for grain and root crops than vegetative yield (Khan 1972; Ross & Gilliam 1973; Saif & Khan 1977; Black & Tinker 1977; Swaminathan & Verma 1979). Serial harvesting of individual plants in pot (Barrow 1977; Smith & Bowen 1979) and field trials (Khan 1972; Saif & Khan 1977) is useful for correlation of mycorrhizal infection build-up with growth response, and repeated shoot harvest of pasture plants in pot and field trials is the only valid measure of longevity of mycorrhizal response (Hall 1978; Powell & Daniel 1978).

The effect of mycorrhizal inoculation on nodulation and nitrogen fixation of legumes has been assessed visually (Abbott & Robson 1977; Owusu-Bennoah & Mosse 1979), by the removal and weighing of all nodules (Hayman & Mosse 1979; Chambers *et al.* 1980) and by acetylene reduction (Mosse 1977). Phosphorus concentration in roots and/or shoots is usually measured in inoculation trials along with other elements (zinc, copper, nitrogen) where necessary. Uptake of radioactive metabolites and elements for all mycorrhizal associations is discussed in section 26.12.

26.10.2 Orchidaceous Mycorrhizae

The length and breadth of protocorms inoculated with orchidaceous mycorrhiza was assessed microscopically (Smith 1966) for five protocorms in each of three replicate tubes per treatment (Harvais & Hadley 1967). Hadley and Williamson (1971) calculated protocorm volume from length and breadth data and there are no recorded growth responses of autotrophic adult plants to inoculation.

26.10.3 Ericoid Mycorrhizae

Ericoid mycorrhizal responses were initially assessed by measurement of shoot and root dry matter and nitrogen concentration (Read & Stribley 1973). Stribley *et al.* (1975) derived other functions (relative growth rate and specific absorption rate for nitrogen) which were useful in pinpointing levels of nitrogen fertiliser at which mycorrhizal responses were optimised.

26.10.4 Ectomycorrhizae

Shoot dry weight of ectomycorrhizal associations is usually measured in laboratory or glasshouse experiments (Hatch 1937; Henderson & Stone 1970; Bowen 1973; Malajczuk *et al.* 1975), with shoot fresh weight and stem diameter assessed in nursery and field studies (Marx & Bryan 1971; Marx *et al.* 1976; Berry & Marx 1978). Root collar diameter and shoot height are useful indicators of mycorrhiza response in the field (Theodorou & Bowen 1970; Berry & Marx 1978; Marx & Artman 1979) and plant survival rate should be measured where mycorrhizal inoculation of eroded or polluted soils is attempted (Marx & Bryan 1971; Carney *et al.* 1978). A plot volume index derived from plant survival and growth was used for mycorrhizal field assessment by Marx and Artman (1979), and the percentage of cull plants from nursery and glasshouse propagation is an early indicator of mycorrhizal development (Trappe 1977; Marx *et al.* 1979). Element concentrations (especially phosphorus) are usually measured in leaves and/or shoots.

The effect of mycorrhizal inoculation on host disease resistance has been assessed on the occurrence of pathogenic fungi in the root cortex (Corte 1969, cited in Marx 1973) tannin deposition in cortical cells (Sylvia & Sinclair 1979) and antibiotic production in foliage and roots (Marx 1973).

26.11 Pot and field trial design

26.11.1 Vesicular-Arbuscular Mycorrhizae

Plants are normally grown at 1 to 10 per pot in 0.3 to 3 l pots (Abbott & Robson 1977; Barrow *et al.* 1977; Powell & Sithamparanathan 1977) of sieved soil or soil/peat/sand mixtures (Daft & Okusanya 1973; Mosse 1973a; Hall 1978; Schultz *et al.* 1979) on glasshouse benches, growth cabinets (Hayman 1974) or root cooling tanks in hot climates (Abbott & Robson 1978). Cores of undisturbed soil (75 to 100 mm deep) have been retrieved from the field for mycorrhizal inoculation (Crush 1976; Powell & Daniel 1978). Soil containers have been designed to determine spread of mycorrhizal infection (Powell 1979c), autoradiography of [^{32}P]phosphorus uptake zones (Owusu-Bennoah & Wild 1979) and uptake of radioisotopes from agar and soil (Pearson & Tinker 1975; Rhodes & Gerdemann 1975). Soils have been sterilised with formalin (4.8 l of 2% (w/v) solution m^{-2}), methylbromide (1 kg (50 kg)$^{-1}$ soil or 450 g (12 m^2)$^{-1}$ soil surface), gamma irradiation (0.8 to 2.5 Mrad), autoclaving at 121°C for 1 h or by steaming between electrodes at 80°C (Hayman & Mosse 1972; Crush 1976; Barrow *et al.* 1977; Johnson 1977; Hall 1978; Menge *et al.* 1978b; Owusu-Bennoah & Mosse 1979; Schultz *et al.* 1979).

Phosphorus fertilisers are usually mixed through the soil before planting (Ross & Gilliam 1973; Powell 1977) or applied to the soil in solution (Mosse 1972; Barrow *et al.* 1977). Banding fertiliser in the soil (Mosse 1977) or broadcasting on the pot or microplot surface (Crush 1976; Schultz *et al.* 1979), has more field relevance. Modified Long Ashton (Hewitt 1952) and other nutrient solutions are often regularly applied to supply elements other than phosphorus.

Many field inoculation trials have been unsatisfactorily designed. There has often been no treatment replication (Khan 1972, 1975; Saif & Khan 1977; Azcon *et al.* 1979). Owusu-Bennoah and Mosse (1979) used long, narrow blocks (*c.* 2 × 7 m) confounded by soil fertility fluctuation. Powell *et al.* (1979) used very small plots and Black and Tinker (1977) did not state replicate numbers. Plot sizes for field trials have varied from 0.16 m^2 to 500 m^2 (Black & Tinker 1977; Saif & Khan 1977; Powell *et al.* 1979) with 6 × 2 m plots acceptable for most grain and root crops. There should be unplanted guard strips of 1 to 3 m between plots to prevent mycorrhizal spread and experiments should be laid out in randomised (square) block design.

26.11.2 Orchidaceous Mycorrhizae

No pot or field trial methods have been reported for orchid mycorrhizae.

26.11.3 Ericoid Mycorrhizae

Mycorrhizal and non-mycorrhizal heath plants have been grown axenically in stoppered 250 ml crystallising dishes on soil amended agar for several months (Pearson & Read 1973a; Read & Stribley 1973) or in steam sterilised sand or soil in open pots in the glasshouse (Stribley *et al.* 1975; Stribley & Read 1976).

26.11.4 Ectomycorrhizae

In glasshouse experiments on ectomycorrhizae, plants are usually grown in large pots (up to 6 l—Henderson & Stone 1970; Bowen 1973; Malajczuk *et al.* 1975) of forest soil or sand/soil/vermiculite/perlite mixtures (Stack & Sinclair 1975; Giles & Whitehead 1977; Marx 1979b). Lamb and Richards (1974) inoculated *Pinus radiata* in undisturbed soils in 130 mm deep field cores. Styrofoam blocks containing 20 to 80 closely spaced planting holes (100 to 120 ml in volume) are useful for inoculation experiments with young plants (Stack & Sinclair 1975; Marx *et al.* 1977b). Fortin *et al.* (1980) grew young seedlings in flat transparent polyester pouches for visual assessment of rapid mycorrhizal synthesis.

Soils for pot trials have been sterilised by gamma irradiation (Malajczuk *et al.* 1975), and multiple steaming to 85 to 100°C (Skinner & Bowen 1974; Marx *et al.* 1977b; Marx 1979b) and nursery and potting soils are often sterilised *in situ* with methyl bromide at 350 kg ha^{-1} (Henderson & Stone 1970; Marx *et al.* 1979). Fertilisers are usually surface applied (Lamb & Richards 1974) and watered in or mixed with the top 10 cm of soil (Malajczuk *et al.* 1975; Marx *et al.* 1977b; Marx 1979c).

Inoculation trials have been carried out in the nursery in wooden microplots ($1.5 \times 1.0 \times 0.6$ m—Marx 1979c) or in seedling bed plots up to 30×1.5 m (Marx *et al.* 1979). Pines are spaced at 1.0 to 1.25 m intervals in 5×5 m plots for field trials (Berry & Marx 1978) with non-planted guard strips of up to 10 m between plots to minimise mycorrhizal contamination (Marx *et al.* 1977a).

26.12 Physiology

26.12.1 Carbon Translocation

Carbon pathways are best studied using [^{14}C]carbon techniques common for all mycorrhizal associations. Large field plants were pulse labelled with [^{14}C]carbon dioxide (50 μCi liberated from [^{14}C]barium carbonate—Englander & Hull 1980) for periods of between 30 min and 6 weeks in gas-tight transparent containers (Ho & Trappe 1973; Englander & Hull 1980) and putative fungal symbionts confirmed by appearance of [^{14}C]labelling in sporing bodies (Bjorkman 1960). Potted or freshly washed plants were pulse labelled (1–2 h) with [^{14}C]carbon dioxide (10 to 50 μCi (plant)$^{-1}$ by release from [^{14}C]barium carbonate, or [^{14}C]sodium bicarbonate via lactic acid injection into gas-tight containers (Pearson & Read 1973b; Bevege *et al.* 1975; Hirrel & Gerdemann 1979; Losel & Cooper 1979). Excess [^{14}C]carbon dioxide is absorbed by injected potassium hydroxide with a chase period of 18 h or longer. Intact freshly washed roots have been bathed in 2-[^{14}C]-acetate. [^{14}C]glycerol and [^{14}C]sucrose at 2 μCi (g fresh roots)$^{-1}$ in the light for 1 h (Losel & Cooper 1979). Hirrel and Gerdemann (1979) injected [^{14}C]glucose (6.25 μCi in 20 μl water) into soil chambers at 4 cm from the nearest roots to determine [^{14}C]glucose uptake by hyphae in soil.

Lewis and Harley (1965b,c) fed partially dissected, excised beech mycorrhizae (10 mm long) with [^{14}C]sucrose in solution or in 3 mm agar blocks for up to 23 h, and Losel and Cooper (1979) dark-incubated excised onion mycorrhizae with 2 to 4 μCi of 2-[^{14}C]acetate, [^{14}C]glycerol or [^{14}C]sucrose. Pearson and Read (1973b) measured hyphal translocation of [^{14}C]carbon in pure culture by applying 1 μCi in 1% (w/v) glucose solution to the growing front or inoculation point. Smith (1966, 1967) measured [^{14}C]carbon activity in mycorrhizal orchid plants in split agar plates when their hyphae have crossed the diffusion barrier and absorbed [^{14}C]glucose (5 μCi) from an agar well. The diffusion barrier (5 mm high glass or plastic partition) must be coated with anhydrous lanolin to prevent capillary flow of [^{14}C]carbon (or other radiotracers) back along the hyphae to the host root (Read & Stribley 1975b).

Where no identification of ^{14}C-labelled compounds is required, dried roots or shoots are weighed on to planchets and assayed for total radioactivity using gasflow Geiger-Mueller detection (Hirrel & Gerdemann 1979; Englander & Hull 1980). For identification and radioactive counting methods of ^{14}C-labelled hexoses, polyols and lipids, the reader is referred to the detailed methods of Lewis and Harley (1965a), Smith (1967), Pearson and Read (1973b), Bevege *et al.* (1975) and Losel and Cooper (1979).

26.12.2 Phosphorus Uptake

Movement of phosphorus from fungus to host can be demonstrated for all mycorrhizal associations using [^{32}P]phosphorus. Englander and Hull (1980) applied [^{32}P]phosphorus (8 μCi in 3 to 4 μl) to *Clavaria* sp. fruiting bodies in the field and found the [^{32}P]-label in the roots and leaves of associated heath plants 11 to 14 d later. [^{32}P]phosphorus can be mixed with potting soil (at 100 μCi kg^{-1} soil) and mycorrhizal and non-mycorrhizal plants grown for 4 to 10 weeks before sampling for [^{32}P]- and [^{31}P]phosphorus and determination of specific activity (Sanders & Tinker 1971; Hayman & Mosse 1972).

[^{32}P]phosphorus (5 to 13 μCi) can be added to a well or the agar surface on split agar plates for absorption and translocation by mycorrhizal hyphae in pure culture (Pearson & Read 1973b) or to a mycorrhizal host plant on the other side of the diffusion barrier (Smith 1967; Pearson & Read 1973b; Cooper & Tinker 1978). Phosphorus fluxes in hyphae can be calculated once hyphal numbers crossing the barrier, hyphal diameter and total [^{32}P]phosphorus uptake in the host plant are known (Pearson & Tinker 1975). Mejstrik and Krause (1973) estimated [^{32}P]phosphorus uptake in intact mycorrhizal *Pinus radiata* plants by placing the washed roots of axenically grown two-month-old seedlings in aerated ^{32}P-labelled nutrient solution or in sterilized [^{32}P]phosphorus amended humus (38 μCi) for 12 d.

Excised beech mycorrhizae with or without

intact mantles have been exposed to [^{32}P]-phosphorus (20 to 60 μCi l^{-1}) in inorganic culture medium for 0.5 to 7 h (Harley & McCready 1950, 1952; Harley *et al.* 1956) in which effects of oxygen concentration, carbohydrate amendments and metabolic inhibitors were easily assessed. Radioactivity of root or shoot material has been measured by Geiger-Mueller counting after samples were washed and blotted (Harley & McCready 1950), dried and ground (Hayman & Mosse 1972; Englander & Hull 1980), ashed at 400°C (Pearson & Read 1973b), crushed in ethanol on the planchet (Smith 1967) or extracted for various phosphorus fractions (Jennings 1964).

Surface and sheath bound acid phosphatase activity of excised field grown ectomycorrhizae has been assayed by hydrolysis of *p*-nitrophenyl phosphate (Bartlett & Lewis 1973; Williamson & Alexander 1975). Gianinazzi-Pearson and Gianinazzi (1978) assessed soluble acid and alkaline phosphatases in VA mycorrhiza by extraction of roots in 0.1M borate buffer with subsequent hydrolysis of *p*-nitrophenol. MacDonald and Lewis (1978) and Gianinazzi *et al.* (1979) used cytochemical staining of lead-precipitated, enzyme-released phosphate under light and electron microscopy to determine the occurrence of acid and alkaline phosphatases in VA mycorrhizae

Phosphate presence and translocation in arbuscles has been measured by X-ray analysis (SEM and TEM) of freeze-fractured and glutaraldehyde-fixed mycorrhizae (Schoknecht & Hattingh 1976; Walker & Powell 1979; White & Brown 1979).

26.12.3 Uptake of Other Ions

The mycorrhiza-enhanced uptake of several elements other than phosphorus has been studied (Harley 1969) and methods for some elements are presented here. Uptake of [^{35}S]sulphur by intact mycorrhizal and/or non-mycorrhizal plants has been assessed in sand (1.5 to 15 μCi for 3 to 60 h exposure—Gray & Gerdemann 1973), in soil chambers (20 μCi—Rhodes & Gerdemann 1978b) and on split agar plates (10 μCi supplied—Cooper & Tinker 1978). Morrison (1963) measured [^{35}S]sulphur uptake by excised beech mycorrhiza (after the methods of Harley & McCready 1950, 1952) and by intact *Pinus radiata* plants in culture solution after 7 d exposure. Samples for radioactivity detection by Geiger-Mueller counting were washed and acid-digested (Gray & Gerdemann 1973; Cooper & Tinker 1978) and adjusted for self absorption of β rays (Morrison 1963).

Mycorrhizal enhancement of zinc uptake has been measured in intact clover plants growing on split plates (soil and agar) fed 5 μCi [^{65}Zn]zinc with a 10 d exposure (Cooper & Tinker 1978) or from excised roots of *Pinus radiata* and *Araucaria cunninghamii* fed [^{65}Zn]zinc (400 μCi l^{-1}) at 20°C (and 2°C to assess non-metabolic entry or absorption—Bowen *et al.* 1974). Radioactivity was measured by spectrometer or X-ray counter.

Mycorrhizal enhancement of calcium uptake has been measured by injection of [^{45}Ca]calcium (80 μCi) into soil chambers with a 5 d exposure (Rhodes & Gerdemann 1978a) and by injection of [^{45}Ca]calcium (1.2 μCi) into the culture dish of *Boletus variegatus* for translocation into *Pinus sylvestris* in axenic culture over 2 d (Melin & Nilsson 1955). Plant samples were dried and ashed before radioactivity counting by Geiger-Mueller tube. Nitrogen uptake has been assessed in excised mycorrhizae (Carrodus 1966) using the technique developed by Harley and McCready (1950). Stribley and Read (1974b) assessed mycorrhiza-enhanced uptake of organic nitrogen by amending 1200 g peat soil with 1 g [^{15}N]ammonium sulphate and 5 g glucose and allowing 55 to 65% of the [^{15}N]nitrogen to be incorporated into the soil organic fraction by unrestricted microbial activity for 197 d. Mycorrhizal and non-mycorrhizal *Vaccinium macrocarpon* plants were then grown for 6 months in 14 g [^{15}N]-amended peat above agar in axenic culture. [^{15}N]nitrogen enrichment in shoots was determined by mass spectrometer after Kjeldahl digestion.

26.12.4 Hormones

Addition of 2 to 8 ml cell-free fungal culture or synthetic auxins induced dichotomy and morphological change in excised pine root cultures (Slankis 1948, 1958; Turner 1962). Ectomycorrhizal fungi in pure culture synthesised indole acetic acid readily from tryptophan supplied at 1 to 30 mg l^{-1} (Ulrich 1960), or occasionally from alanine and asparagine supplied at 280 mg N l^{-1} (Moser 1959).

Kinetin production by ectomycorrhizal and VA mycorrhizal fungi was assayed by the stimulation of soybean callus compared to that of pure Kinetin at 1 mg l^{-1} (Miller 1963; Slankis 1973; Allen *et al.* 1980). Vitamin production by mycorrhizal fungi in pure culture has been assessed by stimulation of yeast growth (Shemakhanova 1962, cited in Slankis 1973) or by a sensitive bacterial turbidity bioassay (for niocin, pantothenic acid and thiamine concentrations as low as 1×10^{-9} g ml^{-1}—Harvais & Pekkala 1975). Fungal (dry weight) growth responses to vitamins were assessed by their addition to basal culture medium at 5 to 1400 μg l^{-1} (Stephen & Fung 1971b).

Acknowledgements

I record my thanks to the trustees of the Miss E. L. Hellaby Indigenous Grasslands Research Trust for a travel grant to attend the 4th North American Conference on Mycorrhizae at Fort Collins, Colorado, June 1979. I am grateful to the Ruakura typists for preparing the manuscript and to my wife Kathryn for supporting me in writing this review.

Recommended reading

GERDEMANN J. W. (1976) Vêsicular-arbuscular mycorrhizae. In: *The Development and Function of Roots* (Eds J. G. Torrey & D. T. Clarkson), pp. 576–91. Academic Press, London.

GIANINAZZI-PEARSON V. (1976) Les mycorhizes endotrophes: Etat actuel des connaissances et possibilités d'application dans la pratique cultural. *Annual Review of Phytopathology* **8**, 249–56.

MARKS G. C. & KOZLOWSKI T. T. (1973) *Ectomycorrhizae—Their Ecology and Physiology*. Academic Press, New York.

MOSSE B. (1973) Advances in the study of vesicular-arbuscular mycorrhiza. *Annual Review of Phytopathology* **11**, 171–96.

SANDERS F. E., MOSSE B. & TINKER P. B. (1975) *Endomycorrhizas*. Academic Press, London.

TRAPPE J. M. (1977) Selection of fungi for ectomycorrhizal inoculation in nurseries. *Annual Review of Phytopathology* **15**, 203–22.

References

ABBOTT L. K. & ROBSON A. D. (1977) Growth stimulation of subterranean clover with vesicular arbuscular mycorrhiza. *Australian Journal of Agricultural Research* **28**, 639–49.

ABBOTT L. K. & ROBSON A. D. (1978) Growth of subterranean clover in relation to the formation of endomycorrhizas by introduced and indigenous fungi in a field soil. *New Phytologist* **81**, 575–85.

ABBOTT L. K. & ROBSON A. D. (1979) A quantitative study of the spores and anatomy of mycorrhizas formed by a species of *Glomus* with reference to its taxonomy. *Australian Journal of Botany* **27**, 363–75.

ALEXANDER M. (1977) *Introduction to Soil Microbiology*, 2e. John Wiley & Sons, New York.

ALLEN M. F., MOORE T. S. & CHRISTENSEN M. (1980) Phytohormone changes in *Bouteloua gracilis* infected by vesicular-arbuscular mycorrhizae. I. Cytokinin increases in the host plant. *Canadian Journal of Botany* **58**, 371–4.

AMES R. N. & LINDERMAN R. G. (1978) The growth of Easter lily (*Lilium longiflorum*) as influenced by vesicular-arbuscular mycorrhizal fungi, *Fusarium oxysporum* and fertility level. *Canadian Journal of Botany* **56**, 2773–80.

AZCON C., AZCON R. & BAREA J. M. (1979) Endomycorrhizal fungi and *Rhizobium* as biological fertilizers for *Medicago sativa* in normal cultivation. *Nature, London* **279**, 325–7.

AZCON R., MARIN A. D. & BAREA J. M. (1978) Comparative role of phosphate in soil or inside the host on the formation and effects of endomycorrhiza. *Plant and Soil* **49**, 561–7.

BAGYARAJ D. J., MANJUNATH A. & REDDY D. D. R. (1979) Interaction of vesicular-arbuscular mycorrhiza with root knot nematode. *Plant and Soil* **51**, 397–403.

BAGYARAJ D. J. & MENGE J. A. (1978) Interactions between a VA mycorrhiza and *Azotobacter* and their effects on rhizosphere microflora and plant growth. *New Phytologist* **80**, 567–73.

BAILEY J. E. & SAFIR G. R. (1978) Effect of benomyl on soybean endomycorrhizae. *Phytopathology* **68**, 1810–12.

BANNISTER P. & NORTON W. M. (1974) The response of mycorrhizal rooted cuttings of heather (*Calluna vulgaris* (L.)Hull) to variations in nutrient and water regimes. *New Phytologist* **73**, 81–9.

BARROW N. J. (1977) Phosphorus uptake and utilization by tree seedlings. *Australian Journal of Botany* **25**, 571–84.

BARROW N. J., MALAJCZUK N. & SHAW T. C. (1977) A direct test of the ability of vesicular-arbuscular mycorrhiza to help plants take up fixed soil phosphate. *New Phytologist* **78**, 269–76.

BARROWS J. E. & RONCADORI R. W. (1977) Endomycorrhizal synthesis by *Gigaspora margarita* in *Poinsettia*. *Mycologia* **69**, 1173–84.

BARTLETT E. M. & LEWIS D. H. (1973) Surface phosphatase activity of mycorrhizal roots of beech. *Soil Biology and Biochemistry* **5**, 249–59.

BAYLIS G. T. S. (1969) Host treatment and spore production by *Endogone*. *New Zealand Journal of Botany* **7**, 173–4.

BAYLIS G. T. S. (1975) The magnolioid mycorrhiza and mycotrophy in root systems derived from it. In: *Endomycorrhizas* (Eds F. E. Sanders, B. Mosse & P. B. Tinker), pp. 373–89. Academic Press, London.

BECKER W. N. & GERDEMANN J. W. (1977) Colorimetric quantification of vesicular-arbuscular mycorrhizal infection in onion. *New Phytologist* **78**, 289-95.

BEGA R. V. (1979) The effect of six mycorrhizal fungi on nursery growth of five species of conifers. In: *Fourth North American Conference on Mycorrhizae* (Ed. C. P. P. Reid), pp. 1–100. Colorado State University, Fort Collins.

BENECKE U. & GOBL F. (1974) The influence of different mycorrhizae on growth, nutrition and gas exchange of *Pinus mugo* seedlings. *Plant and Soil* **40**, 21–32.

BERRY C. R. & MARX D. H. (1978) Effects of *Pisolithus tinctorius* mycorrhizae on growth of loblolly and virginia pines in the Tennessee Copper Basin. *USDA Forest Service Research Note SE-264-6*.

BERRY C. R., MARX D. H. & RUEHLE J. L. (1979) Survival and growth of pitch, loblolly and pitch × loblolly pine seedlings with *Pisolithus* ectomycorrhizae after one year on coal spoils in Alabama and Tennessee. In: *Fourth North American Conference on Mycorrhizae* (Ed. C. P. P. Reid). Colorado State University, Fort Collins.

BEVEGE D. I., BOWEN G. D. & SKINNER M. F. (1975) Comparative carbohydrate physiology of ecto and endomycorrhizas. In: *Endomycorrhizas* (Eds F. E. Sanders, B. Mosse & P. B. Tinker), pp. 149–74. Academic Press, London.

BIGG W. L. & ALEXANDER I. J. (1979) Effect of ammonium and nitrate nitrogen on formation of coniferous ectomycorrhizas. In: *Fourth North American Conference on Mycorrhizae* (Ed. C. P. P. Reid). Colorado State University, Fort Collins.

BJORKMAN E. (1942) Über die Bedingungen der Mykorrhizabildung bei Kiefer und Fichte. *Symbolae botanicae upsalienses* **6**, 1–191.

BJORKMAN E. (1960) *Monotropa hypopitys* L. an epiparasite on tree roots. *Physiologica Plantarum* **13**, 308–29.

BJORKMAN E. (1970) Mycorrhiza and tree nutrition in poor forest soils. *Studia Forestalia Suecica* **83**, 1–24.

BLACK R. L. B. & TINKER P. B. (1977) Interaction between effects of vesicular-arbuscular mycorrhiza and fertilizer phosphorus on yields of potatoes in the field. *Nature, London* **267**, 510–11.

BLACK R. L. B. & TINKER P. B. (1979) The development of endomycorrhizal root systems. II. Effect of agronomic factors and soil conditions on the development of vesicular-arbuscular mycorrhizal infection in barley and on the endophyte spore density. *New Phytologist* **83**, 401–13.

BLEDSOE C. S., O'SHEA M. & NADKARNI N. (1979) Effects of ammonium or nitrate fertilisation on the nutrition of mycorrhizal Douglas fir seedlings. In: *Fourth North American Conference on Mycorrhizae* (Ed. C. P. P. Reid). Colorado State University, Fort Collins.

BOATMAN N., PAGET D., HAYMAN D. S. & MOSSE B. (1978) Effects of systemic fungicides on vesicular-arbuscular mycorrhizal infection and plant phosphate uptake. *Transactions of the British Mycological Society* **70**, 443–450.

BONFANTE-FASOLO P. & GIANINAZZI-PEARSON V. (1979) Ultrastructural aspects of endomycorrhiza in the Ericaceae. I. Naturally infected hair roots of *Calluna vulgaris* (L.) Hull. *New Phytologist* **83**, 739–44.

BOWEN G. D. (1973) Mineral nutrition of ectomycorrhizae. In: *Ectomycorrhizae—Their Ecology and Physiology* (Eds G. C. Marks & T. T. Kozlowski), pp. 151–206. Academic Press, New York.

BOWEN G. D., SKINNER M. F. & BEVEGE D. I. (1974) Zinc uptake by mycorrhizal and uninfected roots of *Pinus radiata* and *Araucaria cunninghamii*. *Soil Biology and Biochemistry* **6**, 141–4.

BOWEN G. D. & THEODOROU C. (1979) Interactions between bacteria and ectomycorrhizal fungi. *Soil Biology and Biochemistry* **11**, 119–26.

BROOK P. J. (1952) Mycorrhiza of *Pernettya macrostigma*. *New Phytologist* **51**, 388–97.

BRYAN W. C. & KORMANIK P. P. (1977) Mycorrhizae benefit survival and growth of sweet-gum seedlings in the nusery. *Southern Journal of Applied Forestry* **1**, 21–3.

BURGES A. & NICHOLAS D. F. (1961) The use of soil sections in studying the amount of fungal hyphae in soil. *Soil Science* **92**, 25–9.

BURPEE L. L. & COLE H. (1978) The influence of alachlor, trifluralin and diazinon on the development of endogenous mycorrhizae in soybeans. *Bulletin of Environmental Contamination and Toxicology* **19**, 191–7.

CALLOW J. A., CAPACCIO L. C. M., PARISH G. & TINKER P. B. (1978) Detection and estimation of polyphosphate in vesicular-arbuscular mycorrhiza. *New Phytologist* **80**, 125–34.

CARLING D. E., RIEHLE W. G., BROWN M. F. & JOHNSON D. R. (1978) Effects of a vesicular-arbuscular mycorrhizal fungus on nitrate reductase and nitrogenase activites in nodulating and non-nodulating soybeans. *Phytopathology* **68**, 1590–6.

CARNEY J. L., GARRETT H. E. & HEDRICK H. G. (1978) Influence of air pollutant gases on oxygen uptake of pine roots with selected ectomycorrhizae. *Phytopathology* **68**, 1160–3.

CARRODUS B. B. (1966) Absorption of nitrogen by mycorrhizal roots of beech. I. Factors affecting the assimilation of nitrogen. *New Phytologist* **65**, 358–71.

CARRODUS B. B. (1967) Absorption of nitrogen by mycorrhizal roots of beech. II. Ammonia and nitrate as sources of nitrogen. *New Phytologist* **66**, 1–4.

CHAMBERS C. A., SMITH S. E. & SMITH F. A. (1980) Effects of ammonium and nitrate ions on mycorrhizal infection, nodulation and growth of *Trifolium subterraneum*. *New Phytologist* **85**, 47–62.

CHU-CHOU M. (1979) Mycorrhizal fungi of *Pinus radiata* in New Zealand. *Soil Biology and Biochemistry* **11**, 557–62.

CLINE M. L., FRANCE R. C. & REID C. P. P. (1979)

Comparison of inter and intraspecific differences of selected ectomycorrhizal fungi under combined temperature and water stress. In: *Fourth North American Conference on Mycorrhizae* (Ed. C. P. P. Reid). Colorado State University, Fort Collins.

Clowes F. A. L. (1951) The structure of mycorrhizal roots of *Fagus sylvatica*. *New Phytologist* **50**, 1–16.

Cooper K. M. & Losel D. M. (1978) Lipid physiology of vesicular-arbuscular mycorrhiza. I. Composition of lipids in roots of onion, clover and ryegrass infected with *Glomus mosseae*. *New Phytologist* **80**, 143–51.

Cooper K. M. & Tinker P. B. (1978) Translocation and transfer of nutrients in vesicular-arbuscular mycorrhizas. II. Uptake and translocation of phosphorus, zinc and sulphur. *New Phytologist* **81**, 43–52.

Cox G. & Sanders F. E. (1974) Ultrastructure of the host–fungus interface in a vesicular-arbuscular mycorrhiza. *New Phytologist* **73**, 901–12.

Cox G. & Tinker P. B. (1976) Translocation and transfer of nutrients in vesicular-arbuscular mycorrhizas. I. The arbuscle and phosphorus transfer: a quantitative ultrastructural study. *New Phytologist* **77**, 371–8.

Cress W. A., Throneberry G. O. & Lindsey D. L. (1979) Kinetics of phosphorus absorption by mycorrhizal and non-mycorrhizal tomato roots. *Plant Physiology* **64**, 484–7.

Crush J.. R. (1974) Plant growth responses to vesicular-arbuscular mycorrhiza. VII. Growth and nodulation of some herbage legumes. *New Phytologist* **73**, 743–9.

Crush J. R. (1976) Endomycorrhizas and legume growth in some soils of the MacKenzie Basin Canterbury, New Zealand. *New Zealand Journal of Agricultural Research*, **19**, 473–6.

Crush J. R. & Pattison A. C. (1975) Preliminary results on the production of vesicular-arbuscular mycorrhizal inoculum by freeze drying. In: *Endomycorrhizas* (Eds F. E. Sanders, B. Mosse & P. B. Tinker), pp. 485–94. Academic Press, London.

Daft M. J. & El-Giahmi A. A. (1978) Effect of arbuscular mycorrhiza on plant growth. VIII. Effects of defoliation and light on selected hosts. *New Phytologist* **80**, 365–72.

Daft M. J. & Hacskaylo E. (1976) Arbuscular mycorrhizas in the anthracite and bituminous coal wastes of Pennsylvania. *Journal of Applied Ecology* **13**, 523–31.

Daft M. J. & Okusanya B. O. (1973) Effect of *Endogone* mycorrhiza on plant growth. V. Influence of infection on the multiplication of viruses in tomato, petunia and strawberry. *New Phytologist* **72**, 975–83.

Daniels B. A. & Duff D. M. (1978) Variation in germination and spore morphology among four isolates of *Glomus mosseae*. *Mycologia* **70**, 1261–7.

Daniels B. A. & Graham S. O. (1976) Effects of nutrition and soil extracts on germination of *Glomus mosseae* spores. *Mycologia* **68**, 108-16.

Daniels B. A. & Trappe J. M. (1979) *Glomus epigaeus*, a useful fungus for vesicular-arbuscular mycorrhizal research. *Canadian Journal of Botany* **57**, 539–42.

Dexheimer J., Gianinazzi S. & Gianinazzi-Pearson V. (1979) Ultrastructural cytochemistry of the host–fungus interfaces in the endomycorrhizal association. *Glomus mosseae–Alium cepa*. *Zeitschrift für Pflanzenphysiologie* **92**, 191–206.

Dixon R. K., Garrett H. E. & Cox G. S. (1979) Foliage fertilization improves growth and ectomycorrhizal development of containerized black oak seedlings. *Transactions of the Missouri Academy of Science* **13**, 174–87.

Dominik T. (1959) Development dynamics of mycorrhizae formed by *Pinus silvestris* and *Boletus luteus* in arable soils. *Prace Szczecinskiego Towarzystwa Naukowe* **1**, 1–30.

Dominik T. (1969) Key to ectotrophic mycorrhizae. *Folia Forestalia Polonica* Seria A **15**, 309.

Donald D. G. M. (1975) Mycorrhizal inoculation for pines. *South African Forestry Journal* **92**, 27–9.

Englander L. & Hull R. J. (1980) Reciprocal transfer of nutrients between ericaceous plants and a *Clavaria* sp. *New Phytologist* **84**, 661–7.

Fassi B., Fontana A. & Trappe J. M. (1969) Ectomycorrhizae formed by *Endogone lactiflua* with species of *Pinus* and *Pseudotsuga*. *Mycologia* **61**, 412–14.

Fortin J. A. & Piche Y. (1979) Cultivation of *Pinus strobus* root hypocotyl explants for synthesis of ectomycorrhizae. *New Phytologist* **83**, 109–19.

Fortin J. A., Piche Y. & Lalonde M. (1980) Techniques for the observation of early morphological changes during ectomycorrhiza formation. *Canadian Journal of Botany* **58**, 361–5.

Foster R. C. & Marks G. C. (1966) The fine structure of the mycorrhizas of *Pinus radiata* D. Don. *Australian Journal of Biological Science* **19**, 1027–38.

Foster R. C. & Marks G. C. (1967) Observations on the mycorrhizas of forest trees. II. The rhizosphere of *Pinus radiata* D. Don. *Australian Journal of Biological Science* **20**, 915–26.

France R. C., Cline M. L. & Reid C. P. P. (1979) Recovery of ectomycorrhizal fungi after exposure to subfreezing temperatures. *Canadian Journal of Botany* **57**, 1845–8.

France R. C. & Reid C. P. P. (1979) Dark fixation of bicarbonate ion by components of ectomycorrhizae. In: *Fourth North American Conference on Mycorrhizae* (Ed. C. P. P. Reid). Colorado State University, Fort Collins.

Furlan V. & Fortin J. A. (1973) Formation of endomycorrhizae by *Endogone calospora* on *Allium cepa* under three temperature regimes. *Naturaliste Canadien* **100**, 467–77.

Furlan V. & Fortin J. A. (1975) A flotation-bubbling system for collecting Endogonaceae spores from sieved soil. *Naturaliste Canadien* **102**, 663–7.

Gadgil P. D. (1972) Effect of waterlogging on mycorrhizas of radiata pine and Douglas fir. *New Zealand Journal of Forestry Science* **2**, 222–6.

GADGIL R. L. & GADGIL P. D. (1975) Suppression of litter decomposition by mycorrhizal roots of *Pinus radiata*. *New Zealand Journal of Forestry Science* **5**, 33–41.
GAUNT R. E. (1978) Inoculation of vesicular-arbuscular mycorrhizal fungi on onion and tomato seeds. *New Zealand Journal of Botany* **16**, 69–71.
GERDEMANN J. W. (1968) Vesicular-arbuscular mycorrhiza and plant growth. *Annual Review of Phytopathology* **6**, 397–418.
GERDEMANN J. W. (1976) Vesicular-arbuscular mycorrhizae. In: *The Development and Function of Roots* (Eds J. G. Torrey & D. T. Clarkson), pp. 576–91. Academic Press, London.
GERDEMANN J. W. & NICOLSON T. H. (1964) Spores of mycorrhizal *Endogone* species extracted from soil by wet-sieving and decanting. *Transactions of the British Mycological Society* **46**, 235–44.
GERDEMANN J. W. & TRAPPE J. M. (1977) Endogonaceae in the Pacific North West. *Mycological Memoir* (New York Botanical Garden) **5**, 1–76.
GIANINAZZI S., GIANINAZZI-PEARSON V. & DEXHEIMER J. (1979) Enzymatic studies on the metabolism of VA mycorrhiza. III. Ultrastructural localisation of acid and alkaline phosphatase in onion roots infected by *Glomus mosseae* (Nicol. & Gerd.). *New Phytologist* **82**, 127–32.
GIANINAZZI-PEARSON V. (1976) Les mycorhizes endotrophes: Etat actuel des connaissances et possibilités d'application dans la pratique cultural. *Annual Review of Phytopathology* **8**, 249–56.
GIANINAZZI-PEARSON V. & GIANINAZZI S. (1978) Enzymatic studies on the metabolism of vesicular-arbuscular mycorrhiza. II. Soluble alkaline phosphatase specific to mycorrhizal infection in onion roots. *Physiological Plant Pathology* **12**, 45–53.
GIANINAZZI-PEARSON V., GIANINAZZI S., DEXHEIMER J., BERTHEAU Y. & ASIMI S. (1978) Les phosphatases alcalines solubles dans l'association endomycorrhizienne à vesicles et arbuscles *Physiologie Végétale* **16**, 671–8.
GILES K. L. & WHITEHEAD H. C. M. (1976) The uptake and continued metabolic activity of *Azotobacter* with fungal protoplasts. *Science* **193**, 1125–6.
GILES K. L. & WHITEHEAD H. C. M. (1977) Reassociation of a modified mycorrhiza with the host plant roots (*Pinus radiata*) and the transfer of acetylene reduction activity. *Plant and Soil* **48**, 143–52.
GILMORE A. E. (1971) The influence of endotrophic mycorrhizae on the growth of peach seedlings. *Journal of the American Horticultural Society* **96**, 35–7.
GOBL F. (1965) Mykorrhizauntersuchungen in einem subalpinen Fichtenwald. *Mitteilungen aus der Forstlichen Bundesversuchsanstalt Mariabrunn* **66**, 173–95.
GRAY L. E. & GERDEMANN J. W. (1969) Uptake of ^{32}P by vesicular-arbuscular mycorrhizae. *Plant and Soil* **30**, 415–20.
GRAY L. E. & GERDEMANN J. W. (1973) Uptake of ^{35}sulphur by vesicular-arbuscular mycorrhiza. *Plant and Soil* **39**, 687–9.
GREEN N. E., GRAHAM S. O. & SCHENCK N. C. (1976) The influence of pH on the germination of vesicular-arbuscular mycorrhizal spores. *Mycologia* **68**, 929–33.
GREENALL J. M. (1963) The mycorrhizal endophytes of *Griselinia littoralis* (Cornaceae). *New Zealand Journal of Botany* **1**, 389–400.
HACSKAYLO E. (1973) Carbohydrate physiology of ectomycorrhizae. In: *Ectomycorrhizae—Their Ecology and Physiology* (Eds G. C. Marks & T. T. Kozlowski), pp. 207–230. Academic Press, New York.
HACSKAYLO E. & PALMER J. G. (1957) Effects of several biocides on growth of seedling pines and incidence of mycorrhizae in field plots. *Plant Disease Reporter* **41**, 354–8.
HACSKAYLO E. & SNOW A. G. (1959) Relation of soil nutrients and light to prevalence of mycorrhizae. *USDA Station Paper No. 125, North Eastern Forest Service*, 1–13.
HADLEY G. (1970) Non-specificity of symbiotic infection in orchid mycorrhiza. *New Phytologist* **69**, 1015–23.
HADLEY G. (1975) Organisation and fine structure of orchid mycorrhiza. In: *Endomycorrhizas* (Eds F. E. Sanders, B. Mosse & P. B. Tinker), pp. 334–51. Academic Press, London.
HADLEY G. & ONG S. H. (1978) Nutritional requirements of orchid endophytes. *New Phytologist* **81**, 561–9.
HADLEY G. & PURVES S. (1974) Movement of ^{14}carbon from host to fungus in orchid mycorrhiza. *New Phytologist* **73**, 475–82.
HADLEY G. & WILLIAMSON B. (1971) Analysis of the post-infection growth stimulus in orchid mycorrhiza. *New Phytologist* **70**, 445–55.
HADLEY G. & WILLIAMSON B. (1972) Features of mycorrhizal infection in some Malayan orchids. *New Phytologist* **71**, 1111–18.
HALL I. R. (1976a) Vesicular mycorrhizas in the orchid *Corybas macranthus*. *Transactions of the British Mycological Society* **66**, 160.
HALL I. R. (1976b) Response of *Coprosma robusta* to different forms of endomycorrhizal inoculum. *Transactions of the British Mycological Society* **67**, 409–11.
HALL I. R. (1978) Effects of endomycorrhizas on the competitive ability of white clover. *New Zealand Journal of Agricultural Research* **21**, 509–15.
HALL I. R. (1979) Soil pellets to introduce vesicular-arbuscular mycorrhizal fungi into soil. *Soil Biology and Biochemistry* **11**, 85–6.
HALL I. R. (1980) Growth of *Lotus pedunculatus* Cav. in an eroded soil containing soil pellets infested with endomycorrhizal fungi. *New Zealand Journal of Agricultural Research* **23**, 103–5.
HALL I. R. & ARMSTRONG P. (1979) Effect of vesicular-arbuscular mycorrhizas on growth of white clover, lotus and ryegrass in some eroded soils. *New Zealand Journal of Agricultural Research* **22**, 479–84.

HALL I. R. & FISH B. (1979) A key to the Endogonaceae. *Transactions of the British Mycological Society* **73**, 261–70.

HARLEY J. L. (1969) *The Biology of Mycorrhiza*. Leonard Hill, London.

HARLEY J. L. & MCCREADY C. C. (1950) Uptake of phosphate by excised mycorrhizal roots of the beech. I. *New Phytologist* **49**, 388–97.

HARLEY J. L. & MCCREADY C. C. (1952) Uptake of phosphate by excised mycorrhizal roots of the beech. II. *New Phytologist* **51**, 56–64.

HARLEY J. L., MCCREADY C. C., BRIERLY J. K. & JENNINGS D. H. (1956) The salt respiration of excised beech mycorrhizas. II. *New Phytologist* **55**, 1–28.

HARVAIS G. & HADLEY G. (1967) The development of *Orchis purpurella* in asymbiotic and inoculated cultures. *New Phytologist* **66**, 217–30.

HARVAIS G. & PEKKALA D. (1975) Vitamin production by a fungus symbiotic with orchids. *Canadian Journal of Botany* **53**, 156–63.

HASELWANDTER K. (1979) Mycorrhizal status of ericaceous plants in alpine and subalpine areas. *New Phytologist* **83**, 427–31.

HATCH A. B. (1937) The physical basis for mycotrophy in the genus *Pinus*. *Black Rock Forest Bulletin* **6**, 1–168.

HATCH A. B. & DOAK K. D. (1933) Mycorrhizal and other features of the root system of *Pinus*. *Journal of the Arnold Arboreum* **14**, 85–99.

HATTINGH M. J. & GERDEMANN J. W. (1975) Inoculation of Brazilian sour orange seed with an endomycorrhizal fungus. *Phytopathology* **65**, 1013–16.

HAYMAN D. S. (1970) *Endogone* spore numbers and vesicular-arbuscular mycorrhiza in wheat as influenced by season and soil treatment. *Transactions of the British Mycological Society* **54**, 53–63.

HAYMAN D. S. (1974) Plant growth responses to vesicular-arbuscular mycorrhiza. VI. Effect of light and temperature. *New Phytologist* **73**, 71–80.

HAYMAN D. S. (1978) Mycorrhizal populations of sown pastures and native vegetation in Otago, New Zealand. *New Zealand Journal of Agricultural Research* **21**, 271–6.

HAYMAN D. S., JOHNSON A. M. & RUDDLESDIN I. (1975) The influence of phosphate and crop species on *Endogone* spores and vesicular-arbuscular mycorrhiza under field conditions. *Plant and Soil* **43**, 489–95.

HAYMAN D. S. & MOSSE B. (1972) Plant growth responses to vesicular-arbuscular mycorrhiza. III. Increased uptake of labile P from soil. *New Phytologist* **71**, 41–7.

HAYMAN D. S. & MOSSE B. (1979) Improved growth of white clover in hill grasslands by mycorrhizal inoculation. *Annals of Applied Biology* **93**, 141–8.

HAYMAN D. S. & STOVOLD G. E. (1979) Spore populations and infectivity of vesicular-arbuscular mycorrhizal fungi in New South Wales. *Australian Journal of Botany* **27**, 227–33.

HEPPER C. M. (1977) A colorimetric method for estimating vesicular-arbuscular mycorrhizal infection in roots. *Soil Biology and Biochemistry* **9**, 15–18.

HEPPER C. M. (1979) Germination and growth of *Glomus caledonius* spores: the effect of inhibitors and nutrients. *Soil Biology and Biochemistry* **11**, 269–79.

HEPPER C. M. & SMITH G. (1976) Observations on the germination of *Endogone* spores. *Transactions of the British Mycological Society* **66**, 189–94.

HENDERSON G. S. & STONE E. L. (1970) Interactions of phosphorus availability, mycorrhizae, and soil fumigation on coniferous seedlings. *Soil Science Society of America Proceedings* **34**, 314–18.

HEWITT E. J. (1952) *Sand and Water Culture Methods Used in the Study of Plant Nutrition*. Commonwealth Agricultural Bureau, Farnham, England.

HIJNER J. A. & ARDITTI J. (1973) Orchid mycorrhiza: vitamin production and requirements by the symbionts. *American Journal of Botany* **60**, 829–35.

HIRREL M. C. & GERDMANN J. W. (1979) Enhanced carbon transfer between onions infected with a vesicular-arbuscular mycorrhizal fungus. *New Phytologist* **83**, 731–8.

HO I. & TRAPPE J. M. (1973) Translocation of ^{14}C from *Festuca* plants to their endomycorrhizal fungi. *Nature, London* **244**, 30–1.

HO I. & ZAK B. (1979) Acid phosphatase activity of six ectomycorrhizal fungi. *Canadian Journal of Botany* **57**, 1203–5.

ILOBA C. (1978) Side effect of pesticides on non-target micro-organisms—a case study with ectomycorrhizal fungi. *Flora* **167**, 480–4.

ILOBA C. (1979) The effect of fungicide application on the development of ectomycorrihizae in seedlings of *Pinus sylvestris*. *Flora* **168**, 352–7.

IQBAL S. H. & QUERESHI K. S. (1976) The influence of mixed sowing (cereals and crucifers) and crop rotation on the development of mycorrhiza and subsequent growth of crops under field conditions. *Biologia (Pakistan)* **22**, 287–98.

IYER J. G. & WOJAHN K. E. (1976) Effect of the fumigant dazomet on the development of mycorrhizae and growth of nursery stock. *Plant and Soil* **45**, 263–6.

JACKSON N. E., FRANKLIN R. E. & MILLER R. H. (1972) Effects of vesicular-arbuscular mycorrhizae on growth and phosphorus content of three agronomic crops. *Soil Science Society of America Proceedings* **36**, 64–7.

JASPER D. A., ROBSON A. D. & ABBOTT L. K. (1979) Phosphorus and the formation of vesicular-arbuscular mychorrhizas. *Soil Biology and Biochemistry* **11**, 501–5.

JENNINGS D. H. (1964) Changes in the size of orthophosphate pools in mycorrhizal roots of beech with reference to absorption of the ion from the external mycelium. *New Phytologist* **63**, 181–93.

JOHNSON P. N. (1977) Mycorrhizal endogonaceae in a New Zealand forest. *New Phytologist* **78**, 161–70.

JONES F. R. (1924) A mycorrhizal fungus in the roots of

legumes and some other plants. *Journal of Agricultural Research* **29**, 459–70.

JONSSON L. & NYLUND J. E. (1979) *Favolashia dybowskyana* (Singer) Singer (Aphyllophorales). A new orchid mycorrhizal fungus from tropical Africa. *New Phytologist* **83**, 121–8.

KAIS A. G. & SNOW G. A. (1979) Interaction of longleaf pine ectomycorrhizae, benlate, and brown-spot needle blight. *Fourth North American Conference on Mycorrhizae* (Ed. C. P. P. Reid). Colorado State University, Fort Collins.

KELLER H. G. (1950) Die Verwertbarkeit verschiedener Kohlehydrate und Dicarbonsäureu durch *Cenococcum graniforme*. *Schweizerische Zeitschrift fuer Allgemeine Pathologie* **13**, 565–9.

KELLEY W. D. & SOUTH D. B. (1978) In vitro effects of selected herbicides on growth of mycorrhizal fungi. *Weed Science Society of American Meeting, Texas* (Abstract).

KESSELL S. L. (1927) Soil organisms: the dependence of certain pine species on a biological soil factor. *Empire Forestry Journal* **6**, 70–4.

KHAN A. G. (1972) The effect of vesicular-arbuscular mycorrhizal associations on growth of cereals. I. Effects on maize growth. *New Phytologist* **71**, 613–19.

KHAN A. G. (1975) The effect of vesicular-arbuscular mycorrhizal associations on growth of cereals. II. Effects on wheat growth. *Annals of Applied Biology* **80**, 27–36.

KHAN A. G. (1978) Vesicular-arbuscular mycorrhizas in plants colonizing black wastes from bituminous coal mining in the Illawarra region of New South Wales. *New Phytologist* **81**, 53–63.

KINDEN D. A. & BROWN M. F. (1975a) Electron microscopy of vesciular arbuscular mycorrhizae of yellow poplar. I. Characterisation of endophytic structure by scanning electron stereoscopy. *Canadian Journal of Microbiology* **21**, 989–93.

KINDEN D. A. & BROWN M. F. (1975b) Electron microscopy of vesicular-arbuscular mycorrhizae of yellow poplar. II. Intracellular hyphae and vesicles. *Canadian Journal of Microbiology* **21**, 1768–80.

KINDEN D. A. & BROWN M. F. (1975c) Electron microscopy of vesicular-arbuscular mycorrhizae of yellow poplar. III. Host-endophyte interactions during arbuscle development. *Canadian Journal of Microbiology* **21**, 1930–9.

KINDEN D. A. & BROWN M. F. (1976) Electron microscopy of vesicular-arbuscular mycorrhizae of yellow poplar. IV. Host–endophyte interactions during arbuscular deterioration. *Canadian Journal of Microbiology* **22**, 64–75.

KORMANIK P. P., BRYAN W. C. & SCHULTZ R. C. (1978) Endomycorrhizal inoculation during transplanting improves growth of vegetatively propagated yellow-poplar. *The Plant Propagator* **23**, 4–5.

LAIHO O. (1970) *Taxillus involutus* as a mycorrhizal symbiont of forest trees. *Acta Forestalia Fennica* **106**, 1–72.

LAMB R. J. (1974) Effect of D-glucose on utilisation of single carbon sources by ectomycorrhizal fungi. *Transactions of the British Mycological Society* **63**, 295–306.

LAMB R. J. & RICHARDS B. N. (1971) Effect of mycorrhizal fungi on the growth and nutrient status of slash and radiata pine seedlings. *Australian Forestry* **35**, 1–7.

LAMB R. J. & RICHARDS B. N. (1974) Inoculation of pines with mycorrhizal fungi in natural soils. I. Effects of density and time of application of inoculum and phosphorus amendment on mycorrhizal infection. *Soil Biology and Biochemistry* **6**, 167–71.

LAMBERT D. H., BAKER D. E. & COLE H. (1979) The role of mycorrhizae in the interactions of phosphorus with zinc, copper and other elements. *Soil Science Society of America Journal* **43**, 976–80.

LANGE J. E. (1923) Studies in the agarics of Denmark, Part I. *Dansk Botanisk Arkiv* **4**, 1–52.

LA RUE J. H., MCCLELLAN W. D. & PEACOCK W. L. (1975) Mycorrhizal fungi and peach nursery nutrition. *Californian Agriculture* **29**, 6–7.

LEVISOHN I. (1957) Antagonistic effects of *Alternaria tenuis* on certain root-fungi of forest trees. *Nature, London* **179**, 1143–4.

LEWIS D. H. (1970) Physiological aspects of symbiosis between green plants and fungi. *Lichenologist* **4**, 326–36.

LEWIS D. H. & HARLEY J. L. (1965a) Carbohydrate physiology of mycorrhizal roots of beech. I. Identity of endogenous sugars and utilization of exogenous sugars. *New Phytologist* **64**, 224–37.

LEWIS D. H. & HARLEY J. L. (1965b) Carbohydrate physiology of mycorrhizal roots of beech. II. Utilization of exogenous sugars by uninfected and mycorrhizal roots. *New Phytologist* **64**, 238–55.

LEWIS D. H. & HARLEY J. L. (1965c) Carbohydrate physiology of mycorrhizal roots of beech. III. Movement of sugars between host and fungus. *New Phytologist* **64**, 256–69.

LINKINS A. E. & ANTIBUS R. K. (1979) Growth on and metabolism of cellulose and crude oil by selected mycorrhizal fungi which have extracellular cellulases and aryl hydrocarbon hydroxylase. In: *Fourth North American Conference on Mycorrhizae* (Ed. C. P. P. Reid). Colorado State University, Fort Collins.

LOSEL D. M. & COOPER K. M. (1979) Incorporation of ^{14}C-labelled substrates by uninfected and VA mycorrhizal roots of onion. *New Phytologist* **83**, 415–26.

LUNDEBERG G. (1970) Utilization of various nitrogen sources, in particular bound soil nitrogen by mycorrhizal fungi. *Studia Forestalia Suecica* **79**, 1–95.

LYR H. (1963) Über die Abnahme der Mykorrhiza und Knöllchenfrequenz mit zunehmender Bodentiefe. *International Mykorrhiza Symposium* 1960, 303–13.

MACDONALD R. M. & LEWIS M. (1978) The occurrence of some acid phosphatases and dehydrogenases in the vesicular-arbuscular mycorrhizal fungus *Glomus mosseae*. *New Phytologist* **80**, 135–41.

McIlveen W. D. & Cole H. (1976) Spore dispersal of Endogonaceae by worms, ants, wasps and birds. *Canadian Journal of Microbiology* **54**, 1486–9.

Malajczuk N., McComb A. J. & Longeragan J. F. (1975) Phosphorus uptake and growth of mycorrhizal and uninfected seedlings of *Eucalyptus calophylla* R. Br. *Australian Journal of Botany* **23**, 231–8.

Marais L. J. & Kotze J. M. (1976) Ectomycorrhizae of *Pinus patula* as biological deterrents to *Phytophthora cinnamomi*. *South African Forestry Journal* **99**, 35–9.

Marais L. J. & Kotze J. M. (1978) Growth of mycorrhizal and non-mycorrhizal *Pinus patula* Schlecht and Cham. seedlings in relation to soil N and P. *South African Forestry Journal* **107**, 37–42.

Marks G. C., Ditchburne N. & Foster R. C. (1968) Quantitative estimates of mycorrhiza populations in radiata pine forests. *Australian Forestry* **32**, 26–38.

Marks G. C. & Foster R. C. (1973) Structure, morphogenesis and ultrastructure of ectomycorrhizae. In: *Ectomycorrhizae—Their Ecology and Physiology* (Eds G. C. Marks & T. T. Kozlowski), pp. 2–42. Academic Press, New York.

Marks G. C. & Kozlowski T. T. (1973) *Ectomycorrhizae—Their Ecology and Physiology*. Academic Press, New York.

Marx D. H. (1969) The influence of ectotrophic mycorrhizal fungi on the resistance of pine roots to pathogenic infections. II. Production, identification and biological activity of antibiotics produced by *Leucopaxillus cerealis var piceina*. *Phytopathology* **59**, 411–17.

Marx D. H. (1970) The influence of ectotrophic mycorrhizal fungi on the resistance of pine roots to pathogenic infections. V. Resistance of mycorrhizae to infection by vegetative mycelium of *Phytophthora cinnamomi*. *Phytopathology* **60**, 1472–3.

Marx D. H. (1973) Mycorrhizae and feeder root diseases. In: *Ectomycorrhizae—Their Ecology and Physiology* (Eds G. C. Marks & T. T. Kozlowski), pp. 351–82. Academic Press, New York.

Marx D. H. (1977) Tree host range and world distribution of the ectomycorrhizal fungus, *Pisolithus tinctorius*. *Canadian Journal of Microbiology* **23**, 217–23.

Marx D. H. (1979a) *Pisolithus* ectomycorrhizae survive cold storage on short leaf pine seedlings. *USDA Forest Service Research Note SE-281*, 1–4.

Marx D. H. (1979b) Synthesis of ectomycorrhizae by different fungi on northern red oak seedlings. *USDA Forest Service Research Note SE-282*, 1–7.

Marx D. H. (1979c) Synthesis of *Pisolithus* ectomycorrhizae on pecan seedlings in fumigated soil. *USDA Forest Service Research Note SE-283*, 1–4.

Marx D. H. & Artman J. D. (1979) *Pisolithus tinctorius* ectomycorrhizae improve survival and growth of pine seedlings on acid coal spoils in Kentucky and Virginia. *Reclamation Review* **2**, 23–31.

Marx D. H. & Bryan W. C. (1971) Influence of ectomycorrhizae on survival and growth of asceptic seedlings of loblolly pine at high temperature. *Forest Science* **17**, 37–41.

Marx D. H. & Bryan W. C. (1975) Growth and ectomycorrhizal development of loblolly pine seedlings in fumigated soil infested with the fungal symbiont *Pisolithus tinctorius*. *Forest Science* **21**, 245–54.

Marx D. H., Bryan W. C. & Cordell C. E. (1976) Growth and ectomycorrhizal development of pine seedlings in nursery soils infested with the fungal symbiont *Pisolithus tinctorius*. *Forest Science* **22**, 91–100.

Marx D. H. & Daniel W. J. (1976) Maintaining cultures of ectomycorrhizal and plant pathogenic fungi in sterile cold water storage. *Canadian Journal of Microbiology* **22**, 338–41.

Marx D. H., Bryan W. C. & Cordell C. E. (1977a) Survival and growth of pine seedlings with *Pisolithus* ectomycorrhizae after two years on reafforestation sites in North Carolina and Florida. *Forest Service* **23**, 363–73.

Marx D. H., Hatch A. B. & Mendicino J. F. (1977b) High soil fertility decreases sucrose content and susceptibility of loblolly pine roots to ectomycorrhizal infection by *Pisolithus tinctorius*. *Canadian Journal of Botany* **55**, 1569–74.

Marx D. H., Mexal J. G. & Morris W. G. (1979) Inoculation of nursery seedbeds with *Pisolithus tinctorius* spores mixed with hydromulch increases ectomycorrhizae and growth of loblolly pines. *Southern Journal of Applied Forestry* **3**, 175–8.

Mejstrik V. K. & Krause H. H. (1973) Uptake of ^{32}P by *Pinus radiata* roots inoculated with *Suillus luteus* and *Cenococcum graniforme* from different sources of available phosphate. *New Phytologist* **72**, 137–40.

Melin E. (1917) Studier over de norrlandska myrmarkernas vegetation med sarskildhansyn till deras skogsvegetation efter torrlaggning. *Akadamisk Afhandling, Uppsala* 11–426.

Melin E. (1925) *Untersuchungen über die Bedeutung der Baummykorrhiza. Eine Ökologishe physiologische Studie* G. Fischer, Jenner.

Melin E. (1936) Methoden der experimentellen Untersuchung myktotropher Pflanzen. *Hanbuch der biologischen Arbeitsmethoden* **4**, 1015–18.

Melin E. (1953) Physiology of mycorrhizal relations in plants. *Annual Review of Plant Physiology* **4**, 325–46.

Melin E. & Nilsson H. (1955) ^{45}Ca used as an indicator of transport of cations to pine seedlings by means of mycorrhizal mycelia. *Svensk Botanisk Tidskrift* **49**, 119–22.

Melin E. & Nilsson H. (1957) Transport of ^{14}C labelled photosynthate to the fungal associate of pine mycorrhizae. *Svensk Botanisk Tidskrift* **51**, 166–86.

Melin E., Nilsson H. & Hacskaylo E. (1958) Translocation of cations to seedlings of *Pinus virginiana* through mycorrhizal mycelia. *Botanical Gazette* **119**, 241–6.

MELIN E. & NYMAN B. (1941) Über das Wuchstoffbedürfnis von *Boletus granulatus* (L). *Archiv für Mikrobiologie* **12**, 254–9.

MENGE J. A., LABANAUSKAS C. K., JOHNSON E. L. V. & PLATT R. G. (1978b) Partial substitution of mycorrhizal fungi for phosphorus fertilization in the greenhouse culture of citrus. *Soil Science Society of America Proceedings* **42**, 926–30.

MENGE J. A., LEMBRIGHT H. & JOHNSON E. L. V. (1977) Utilization of mycorrhizal fungi in citrus nurseries. *Proceedings of the International Society of Citriculture* **1**, 129–32.

MENGE J. A., MUNNECKE D. E., JOHNSON E. L. V. & CARNES D. W. (1978a) Dosage response of the vesicular-arbuscular fungi *Glomus fasciculatus* and *G. constrictus* to methyl bromide. *Phytopathology* **68**, 1368–72.

MENGE J. A., STEIRLE D., BAGYARAJ D. J., JOHNSON E. L. V. & LEONARD R. T. (1978c) Phosphorus concentrations in plants responsible for inhibition of mycorrhizal infection. *New Phytologist* **80**, 575–8.

MERTZ S. M., HEITHAUS J. J. & BUSH R. L. (1979) Mass production of axenic spores of the endomycorrhizal fungus *Gigaspora margarita*. *Transactions of the British Mycological Society* **72**, 167–9.

MEXAL J. G., MARX D. H. & MORRIS W. G. (1979) Plantation performance of loblolly pine following inoculation with mycorrhizal fungi. In: *Fourth North American Conference on Mycorrhizae* (Ed. C. P. P. Reid). Colorado State University, Fort Collins.

MEYER F. H. (1973) Distribution of ectomycorrhizae in native and man-made forests. In: *Ectomycorrhizae—Their Ecology and Physiology* (Eds G. C. Marks & T. T. Kozlowski), pp. 79–105. Academic Press, New York.

MEYER F. H. (1974) Physiology of mycorrhiza. *Annual Review of Plant Physiology* **25**, 567–86.

MEYER F. H. & GOTTSCHE D. (1971) Distribution of root tips and tender roots of beech. In: *Ecological studies, Analysis and Synthesis* (Ed. H. Ellenberg), vol. 2, pp. 48–52. Springer-Verlag, Berlin.

MIKOLA P. (1970) Mycorrhizal inoculation in afforestation. *International Review of Forest Research* **3**, 123–96.

MILLER C. O. (1963) Kinetin and Kinetin-like compounds. In: *Moderne Methoden der Pflanzenanalyse* (Eds H. F. Linskins & M. V. Tracey), vol. 6, pp. 194–202. Springer-Verlag, Berlin.

MITCHELL H. L., FINN R. F. & ROSENDAHL R. O. (1937) The relation between mycorrhizae and the growth and nutrient absorption of coniferous seedlings in nursery beds. *Black Rock Forest Paper* **1**, 58–73.

MOLINA R. (1979) Pure culture synthesis and host specificity of red alder mycorrhizae. *Canadian Journal of Botany* **57**, 1223–8.

MORRISON T. M. (1957) Host–endophyte relationships in mycorrhizas of *Pernettya macrostigma*. *New Phytologist* **56**, 247–57.

MORRISON T. M. (1962) Uptake of sulphur by mycorrhizal plants. *New Phytologist* **61**, 21–7.

MORRISON T. M. (1963) Uptake of sulphur by excised beech mycorrhizas. *New Phytologist* **62**, 44–9.

MOSER M. (1958a) Der Einfluss tiefer Temperaturen auf das Wachstum und die Lebenstätigkeit höherer Pilze mit spezieller Berucksichtigung von Mykorrhizapilzen. *Sydowia* **12**, 386–99.

MOSER M. (1958b) Die Kunstliche Mykorrhizaimpfung von Forstpflanzen. II. Die Torfstreukultur von Mykorrhizapilzen. *Forstwissenschaftliches Zentralblatt* **77**, 257–64.

MOSER M. (1959) Beitrage zur Kenntnis der Winchsstoffbeziehungen im Bereich ectotropher Mycorrhizen. *Archiv für Mikrobiologie* **34**, 251–60.

MOSSE B. (1959) Observations of the extra-matrical mycelium of a vesicular-arbuscular endophyte. *Transactions of the British Mycological Society* **42**, 439–48.

MOSSE B. (1962) The establishment of vesicular-arbuscular mycorrhiza under aseptic conditions. *Journal of General Microbiology* **27**, 509–20.

MOSSE B. (1970) Honey coloured, sessile *Endogone* spores. II. Changes in fine structure during spore development. *Archiv für Mikrobiologie* **74**, 129–45.

MOSSE B. (1972) Effects of different *Endogone* strains on the growth of *Paspalum notatum*. *Nature, London* **239**, 221–3.

MOSSE B. (1973a) Plant growth responses to vesicular-arbuscular mycorrhiza. IV. In soil given additional phosphate. *New Phytologist* **72**, 127–36.

MOSSE B. (1973b) Advances in the study of vesicular-arbuscular mycorrhiza. *Annual Review of Phytopathology* **11**, 171–96.

MOSSE B. (1977) Plant growth responses to vesicular-arbuscular mycorrhiza. X. Responses of *Stylosanthes* and maize to inoculation in unsterile soils. *New Phytologist* **78**, 277–88.

MOSSE B. & BOWEN G. D. (1968) The distribution of *Endogone* spores in Australian and New Zealand soils, and in an experimental field soil at Rothamsted. *Transactions of the British Mycological Society* **51**, 485–92.

MOSSE B. & HAYMAN D. S. (1971) Plant growth responses to vesicular arbuscular mycorrhiza. II. In unsterilized field soils. *New Phytologist* **70**, 29–34.

MOSSE B., HAYMAN D. S. & ARNOLD D. J. (1973) Plant growth responses to vesicular arbuscular mycorrhiza. V. Phosphate uptake by three plant species from P-deficient soils labelled with ^{32}P. *New Phytologist* **72**, 809–15.

MOSSE B., HAYMAN D. S. & IDE G. J. (1969) Growth responses of plants in unsterilised soil to inoculation with vesicular-arbuscular mycorrhiza. *Nature, London* **224**, 1031–2.

MOSSE B. & HEPPER C. M. (1975) Vesicular-arbuscular mycorrhizal infections in root organ cultures. *Physiological Plant Pathology* **5**, 215–23.

MOSSE B. & JONES G. (1968) Separation of *Endogone* spores from organic soil debris by differential sedimentation on gelatin columns. *Transactions of the British Mycological Society* **51**, 604–8.

MOSSE B. & PHILLIPS J. M. (1971) The influence of phosphate and other nutrients on the development of vesicular-arbuscular mycorrhiza in culture. *Journal of General Microbiology* **69**, 157–66.

MURDOCH C. L., JACKOBS J. A. & GERDEMANN J. W. (1967) Utilization of phosphorus sources of different availability by mycorrhizal and non-mycorrhizal maize. *Plant and Soil* **27**, 329–334.

NICOLSON T. H. (1959) Mycorrhiza in the Graminae. I. Vesicular-arbuscular endophytes with special reference to the external phase. *Transactions of the British Mycological Society* **42**, 421–38.

NICOLSON T. H. (1975) Evolution of vesicular-arbuscular mycorrhizas. In: *Endomycorrhizas* (Eds F. E. Sanders, B. Mosse & P. B. Tinker), pp. 25–34. Academic Press, London.

NICOLSON T. H. & JOHNSTON C. (1979) Mycorrhiza in the Graminae III *Glomus fasciculatus* as the endophyte of pioneer grasses in a maritime sand dune. *Transactions of the British Mycological Society* **72**, 261–8.

OWUSU-BENNOAH E. & MOSSE B. (1979) Plant growth responses to vesicular-arbuscular mycorrhiza. XI. Field inoculation responses in barley, lucerne and onion. *New Phytologist* **83**, 671–9.

OWUSU-BENNOAH E. & WILD A. (1979) Autoradiography of the depletion zone of phosphate around onion roots in the presence of vesicular-arbuscular mycorrhiza. *New Phytologist* **82**, 133–40.

PALMER J. G. (1971) Techniques and procedures for culturing ectomycorrhizal fungi. In: *Mycorrhizae* (Ed. E. Hacskaylo), pp. 132–44. USDA Forest Service Miscellaneous Publication 1189.

PEARSON V. & READ D. J. (1973a) The biology of mycorrhiza in the Ericaceae. I. The isolation of the endophyte and synthesis of mycorrhizas in aseptic culture. *New Phytologist* **72**, 371–9.

PEARSON V. & READ D. J. (1973b) The biology of mycorrhiza in the Ericaceae. II. The transport of carbon and phosphorus by the endophyte and the mycorrhiza. *New Phytologist* **72**, 1325–31.

PEARSON V. & READ D. J. (1975) The physiology of the mycorrhizal endophyte of *Calluna vulgaris*. *Transactions of the British Mycological Society* **64**, 1–7.

PEARSON V. & TINKER P. B. (1975) Measurement of phosphorus fluxes in the external hyphae of endomycorrhizas. In: *Endomycorrhizas* (Eds F. E. Sanders, B. Mosse & P. B. Tinker), pp. 277–88. Academic Press, London.

PHILLIPS J. M. & HAYMAN D. S. (1970) Improved procedures for clearing roots and staining parasitic and vesicular-arbuscular mycorrhizal fungi for rapid assessment of infection. *Transactions of the British Mycological Society* **55**, 158–61.

PICHOT J. & BINH T. (1976) Action des endomycorrhizes sur la croissance et la nutrition phosphaté de l'*Agrostis* en vases de vegetation et sur le phosphore isotopiquement diluable du sol. *Aqronomie Tropicale* **31**, 375–8.

PORTER W. M., ABBOTT L. K. & ROBSON A. D. (1978) Effect of rate of application of superphosphate on populations of vesicular-arbuscular endophytes. *Australian Journal of Experimental Agriculture and Animal Husbandry* **18**, 573–8.

POWELL C. LL. (1975) Potassium uptake by endotrophic mycorrhizas. In: *Endomycorrhizas* (Eds F. E. Sanders, B. Mosse & P. B. Tinker), pp. 461–8. Academic Press, London.

POWELL C. LL. (1976) Development of mycorrhizal infections from *Endogone* spores and infected root segments. *Transactions of the British Mycological Society* **66**, 439–45.

POWELL C. LL. (1977) Mycorrhizas in hill country soils. II. Effect of several mycorrhizal fungi on clover growth in sterilized soils. *New Zealand Journal of Agricultural Research* **20**, 59–62.

POWELL C. LL. (1979a) Effect of mycorrhizal fungi on recovery of phosphate fertiliser from soil by ryegrass plants. *New Phytologist* **83**, 681–94.

POWELL C. LL. (1979b) Inoculation of white clover and ryegrass seed with mycorrhizal fungi. *New Phytologist* **83**, 81–5.

POWELL C. LL. (1979c) Spread of mycorrhizal fungi through soil. *New Zealand Journal of Agricultural Research* **22**, 335–9.

POWELL C. LL. (1980) Mycorrhizal infectivity of eroded soils. *Soil Biology and Biochemistry* **12**, 247–51.

POWELL C. LL. & DANIEL J. (1978) Growth of white clover in undisturbed soils after inoculation with efficient mycorrhizal fungi. *New Zealand Journal of Agricultural Research* **21**, 675–81.

POWELL C. LL., GROTERS M. & METCALFE D. M. (1980) Mycorrhizal inoculation of a barley crop in the field. *New Zealand Journal of Agricultural Research* **23**, 107–9.

POWELL C. LL. & SITHAMPARANATHAN J. (1977) Mycorrhizas in hill country soils. IV. Infection rate in grass and legume species by indigenous mycorrhizal fungi under field conditions. *New Zealand Journal of Agricultural Research* **20**, 489–94.

PURVES S. & HADLEY G. (1975) Movement of carbon compounds between the partners in orchid mycorrhiza. In: *Endomycorrhizas* (Eds F. E. Sanders, B. Mosse & P. B. Tinker), pp. 175–94. Academic Press, London.

RABATIN S. C. (1979) Seasonal and edaphic variation in vesicular-arbuscular mycorrhizal infection of grasses by *Glomus tenuis*. *New Phytologist* **83**, 95–102.

RAYNER M. C. (1913) The ecology of *Calluna vulgaris*. *New Phytologist* **12**, 59.

RAYNER M. C. (1922) Mycorrhiza in the Ericaceae. *Transactions of the British Mycological Society* **8**, 61.

READ D. J. (1974) *Pezizella ericae* sp. nov., The perfect

state of a typical mycorrhizal endophyte of Ericaceae. *Transactions of the British Mycological Society* **63**, 381–2.

READ D. J. (1978) The biology of mycorrhiza in heathland ecosystems with special reference to the nitrogen nutrition of the Ericaceae. In: *Microbial Ecology* (Eds M. W. Loutit & J. A. R. Miles), pp. 324–8. Springer-Verlag, Berlin.

READ D. J. & STRIBLEY D. P. (1973) Effect of mycorrhizal infection on nitrogen and phosphorus nutrition of Ericaceous plants. *Nature, London* **244**, 81–2.

READ D. J. & STRIBLEY D. P. (1975a) Some mycological aspects of the biology of mycorrhiza in the Ericaceae. In: *Endomycorrhizas* (Eds F. E. Sanders, B. Mosse & P. B. Tinker), pp. 105–17. Academic Press, London.

READ D. J. & STRIBLEY D. P. (1975b) Diffusion and translocation in some fungal culture systems. *Transactions of the British Mycological Society* **64**, 381–8.

REID C. P. P. & BOWEN G. D. (1979) Effect of water stress on phosphorus uptake by mycorrhizas of *Pinus radiata*. *New Phytologist* **83**, 103–7.

RHODES L. H. & GERDEMANN J. W. (1975) Phosphate uptake zones of mycorrhizal and non-mycorrhizal onions. *New Phytologist* **75**, 555–61.

RHODES L. H. & GERDEMANN J. W. (1978a) Translocation of calcium and phosphate by external hyphae of vesicular-arbuscular mycorrhizae. *Soil Science* **126**, 125–6.

RHODES L. H. & GERDEMANN J. W. (1978b) Influence of phosphorus nutrition on sulphur uptake by vesicular-arbuscular mycorrhizae of onion. *Soil Biology and Biochemistry* **10**, 361–4.

RICH J. R. & BIRD G. W. (1974) Association of early season vesicular-arbuscular mycorrhizae with increased growth and development of cotton. *Phytopathology* **64**, 1421–5.

RONCADORI R. W. & HUSSEY R. S. (1977) Interaction of the endomycorrhizal fungus *Gigaspora margarita* and root knot nematode on cotton. *Phytopathology* **67**, 1507–11.

ROSS J. P. (1971) Effect of phosphate fertilization on yield of mycorrhizal and non-mycorrhizal soybeans. *Phytopathology* **61**, 1400–3.

ROSS J. P. & GILLIAM J. W. (1973) Effect of *Endogone* mycorrhiza on phosphorus uptake by soybeans from inorganic phosphates. *Soil Science Society of America Proceedings* **37**, 237–9.

RUEHLE J. L. (1980) Growth of containerized loblolly pine with specific ectomycorrhizae after two years on an amended borrow pit. *Reclamation Review* **3**, 95–101.

RUSSELL E. W. (1973) *Soil Conditions and Plant Growth*, 10e. Longman, London.

SAFIR G. R., BOYER J. S. & GERDEMANN J. W. (1971) Mycorrhizal enhancement of water transport in soybean. *Science* **172**, 581–3.

SAIF S. R. (1977) The influence of stage of host development on vesicular-arbuscular mycorrhizae and Endogenaceous spore population in field-grown vegetable crops. I. Summergrown crops. *New Phytologist* **79**, 341–8.

SAIF S. R. & KHAN A. G. (1975) The influence of season and stage of development of plant on *Endogone* mycorrhiza of field grown wheat. *Canadian Journal of Microbiology* **21**, 1020–4.

SAIF S. R. & KHAN A. G. (1977) The effect of vesicular-arbuscular mycorrhizal association on growth of cereals. III. Effects on barley growth. *Plant and Soil* **47**, 17–26.

SANDERS F. E., MOSSE B. & TINKER P. B. (1975) *Endomycorrhizas*. Academic Press, London.

SANDERS F. E. & TINKER P. B. (1971) Mechanism of absorption of phosphate from soil by *Endogone* mycorrhizas. *Nature, London* **233**, 278.

SANDERS F. E., TINKER P. B., BLACK R. L. B. & PALMERLEY S. M. (1977) The development of endomycorrhizal root systems. I. Spread of infection and growth-promoting effects with four species of vesicular-arbuscular endophyte. *New Phytologist* **78**, 257–68.

SANDS R. & THEODOROU C. (1978) Water uptake by mycorrhizal roots of radiata pine seedlings. *Australian Journal of Plant Physiology* **5**, 301–9.

SCANNERINI S. & BONFANTE-FASOLO P. (1979) Ultrastructural cytochemical demonstration of polysaccharides and proteins within the host-arbuscle interfacial matrix in an endomycorrhiza. *New Phytologist* **83**, 87–94.

SCHENCK N. C. & KELLAM M. K. (1978) *The Influence of Vesicular-Arbuscular Mycorrhizae on Disease Development*. Bulletin 798 (technical). Institute of Food and Agricultural Sciences, University of Florida, Gainesville, USA.

SCHENCK N. C. & SCHRODER V. N. (1974) Temperature response of *Endogone* mycorrhiza on soybean roots. *Mycologia* **66**, 600–5.

SCHOKNECHT J. D. & HATTINGH M. J. (1976) X-Ray Micro-analysis of elements in cells of VA mycorrhizal and non-mycorrhizal onions. *Mycologia* **68**, 296–303.

SCHONBECK F. & DEHNE H. W. (1979) Investigations on the influence of endotrophic mycorrhiza on plant diseases. 4. Fungal parasites on shoots, *Olpidium brassicae*, TMV. *Zeitschrift für Pflanzenkrankheiten und Pflanzenschutz* **86**, 103–12.

SCHULTZ R. C., KORMANIK P. P., BRYAN W. C. & BRISTER G. H. (1979) Vesicular-arbuscular mycorrhiza influence growth but not mineral concentrations in seedlings of eight sweetgum families. *Canadian Journal of Forestry Research* **9**, 218–23.

SEVIOUR R. J., WILLING R. R. & CHILVERS G. A. (1973) Basidiocarps associated with ericoid mycorrhizas. *New Phytologist* **72**, 381–5.

SIKORA R. A. (1978) Influence of the endotrophic mycorrhiza (*Glomus mosseae*) on the host–parasite relationship of *Meloidogyne incognita* on tomato. *Zeitschrift für Pflanzenkrankheiten und Pflanzenschutz* **85**, 197–202.

SINCLAIR W. A. (1974) Development of ectomycorrhizae in a Douglas fir nursery. II. Influence of soil fumigation, fertilization and cropping history. *Forestry Science* **20**, 57–63.

SKINNER M. F. & BOWEN G. D. (1974) The penetration of soil by mycelial strands of ectomycorrhizal fungi. *Soil Biology and Biochemistry* **6**, 57–61.

SLANKIS V. (1948) Einfluss von Exudaten von *Boletus variegatus* auf die dichotomische Verzweigung isolierter Kiefernwurzeln. *Physiologia Plantarum* **1**, 390–7.

SLANKIS V. (1958) The role of auxin and other exudates in mycorrhizal symbiosis of forest trees. In: *Physiology of Forest Trees* (Ed. K. V. Thimann), pp. 427–43. Ronald Press, New York.

SLANKIS V. (1973) Hormonal relationships in mycorrhizal development. In: *Ectomycorrhizae—Their Ecology and Physiology* (Eds G. C. Marks & T. T. Kozlowski), pp. 231–98. Academic Press, New York.

SMITH D., MUSCATINE L. & LEWIS D. H. (1969) Carbohydrate movement from autotrophs to heterotrophs in parasitic and mutualistic symbiosis. *Biological Reviews* **44**, 17–90.

SMITH J. R. & FERRY B. W. (1979) The effects of simazine, applied for weed control, on the mycorrhizal development of *Pinus* seedlings. *Annals of Botany* **43**, 93–9.

SMITH S. E. (1966) Physiology and ecology of orchid mycorrhizal fungi with reference to seedling nutrition. *New Phytologist* **65**, 488–99.

SMITH S. E. (1967) Carbohydrate translocation in orchid mycorrhizas. *New Phytologist* **66**, 371–8.

SMITH S. E. & BOWEN G. D. (1979) Soil temperature, mycorrhizal infection and nodulation of *Medicago trunculata* and *Trifolium subterraneum*. *Soil Biology and Biochemistry* **11**, 469–73.

SMITH S. E. & DAFT M. J. (1977) Interactions between growth, phosphate content and nitrogen fixation in mycorrhizal and non-mycorrhizal *Medicago sativa*. *Australian Journal of Plant Physiology* **4**, 403–13.

SMITH T. F. (1978) Some effects of crop protection chemicals on the distribution and abundance of vesicular-arbuscular mycorrhizas. *Journal of the Australian Institute of Agricultural Science* **44**, 82–7.

SONDERGAARD M. & LAEGAARD S. (1977) Vesicular-arbuscular mycorrhiza in some aquatic vascular plants. *Nature, London* **268**, 232–3.

SPARLING G. P. & TINKER P. B. (1978) Mycorrhizal infection in Pennine grassland. I. levels of infection in the field. *Journal of Applied Ecology* **15**, 943–50.

STACK R. W. & SINCLAIR W. A. (1975) Protection of Douglas fir seedlings against *Fusarium* root rot by a mycorrhizal fungus in the absence of mycorrhiza formation. *Phytopathology* **65**, 468–72.

STEPHEN R. C. & FUNG K. K. (1971a) Nitrogen requirements of the fungal endophytes of *Arundina chinensis*. *Canadian Journal of Botany* **49**, 407–10.

STEPHEN R. C. & FUNG K. K. (1971b) Vitamin requirements of the fungal endophytes of *Arundina chinensis*. *Canadian Journal of Botany* **49**, 411–15.

STRIBLEY D. P. & READ D. J. (1974a) The biology of mycorrhiza in the Ericaceae. III. Movement of Carbon-14 from host to fungus. *New Phytologist* **73**, 731–41.

STRIBLEY D. P. & READ D. J. (1974b) The biology of mycorrhiza in the Ericaceae. IV. The effect of mycorrhizal infection on uptake of ^{15}N from labelled soil by *Vaccinium macrocarpon* Ait. *New Phytologist* **73**, 1149–55.

STRIBLEY D. P. & READ D. J. (1976) The biology of mycorrhiza in the Ericaceae. VI. The effects of mycorrhizal infection and concentration of ammonium nitrogen on growth of cranberry (*Vaccinium macrocarpon* Ait) in sand culture. *New Phytologist* **77**, 63–72.

STRIBLEY D. P., READ D. J. & HUNT R. (1975) The biology of mycorrhiza in the Ericaceae. V. The effects of mycorrhizal infection, soil type and partial soil-sterilization (by gamma irradiation) on growth of cranberry (*Vaccinium macrocarpon* Ait). *New Phytologist* **75**, 119–30.

STRULLU D. G. & GOURRET J. P. (1975) Ultrastructure et evolution du champignon symbiotique des racines de *Dactylorchis maculata* (L). Verm. *Journale de Microscopie* **20**, 285–94.

SUTTON J. C. (1973) Development of vesicular-arbuscular mycorrhizae in crop plants. *Canadian Journal of Botany* **51**, 2487–93.

SUTTON J. C. & BARRON G. L. (1972) Population dynamics of *Endogone* spores in soil. *Canadian Journal of Botany* **50**, 1909–14.

SWAMINATHAN K. & VERMA B. C. (1979) Responses of three crop species to vesicular-arbuscular mycorrhizal infection on zinc-deficient Indian soils. *New Phytologist* **82**, 481–7.

SWARD R. J., HALLAM N. D. & HOLLAND A. A. (1978) *Endogone* spores in a heathland area of South-eastern Australia. *Australian Journal of Botany* **26**, 29–43.

SYLVIA D. M. & SINCLAIR W. A. (1979) Primary root elongation of Douglas fir seedlings suppressed and tannin deposition stimulated by *Laccaria laccata*. In: *Fourth North American Conference on Mycorrhizae*, (Ed. C. P. P. Reid). Colorado State University, Fort Collins.

THEODOROU C. (1971) Introduction of mycorrhizal fungi into soil by spore inoculation of seed. *Australian Forestry* **35**, 23–6.

THEODOROU C. & BOWEN G. D. (1970) Mycorrhizal responses of radiata pine in experiments with different fungi. *Australian Forestry* **34**, 183–8.

TODD A. W. (1979) Decomposition of selected soil organic matter components by Douglas fir ectomycorrhizal associations. In: *Fourth North American Conference on Mycorrhizae* (Ed. C. P. P. Reid). Colorado State University, Fort Collins.

TRAPPE J. M. (1962) Fungus associates of ectotrophic mycorrhiza. *Botanical Review* **28**, 538–606.

TRAPPE J. M. (1977) Selection of fungi for ectomycorrhizal inoculation in nurseries. *Annual Review of Phytopathology* **15**, 203–22.

TRAPPE J. M. & FOGEL R. D. (1977) Ecosystematic functions of mycorrhizae. In: *The Below-Ground Ecosystem: A Synthesis of Plant Associated Processes*, pp. 205–14. Range Science Department, Science Series No. 26, Colorado State University, Fort Collins.

TURNER P. D. (1962) Morphological influence of exudates of mycorrhizal and non-mycorrhizal fungi on excised root cultures of *Pinus sylvestris* L. *Nature, London* **194**, 551–2.

ULRICH J. M. (1960) Auxin production by mycorrhizal fungi. *Physiologia Plantarum* **13**, 429–43.

VOZZO J. A. & HACSKAYLO E. (1971) Inoculation of *Pinus caribaea* with ectomycorrhizal fungi in Puerto Rico. *Forest Science* **17**, 239–45.

WALKER G. D. & POWELL C. LL. (1979) Vesicular-arbuscular mycorrhizas in white clover: An SEM and microanalytic study. *New Zealand Journal of Botany* **17**, 55–9.

WARCUP J. H. (1975) Factors affecting symbiotic germination of orchid seed. In: *Endomycorrhizas* (Eds F. E. Sanders, B. Mosse & P. B. Tinker), pp. 87–104. Academic Press, London.

WARCUP J. H. & TALBOT P. H. B. (1966) Perfect states of some Rhizoctonias. *Transactions of the British Mycological Society* **49**, 427–35.

WATRUD L. S., HEITHAUS J. J. & JAWORSKI E. G. (1978) Evidence of production of inhibitor by the vesicular-arbuscular mycorrhizal fungus *Gigaspora margarita*. *Mycologia* **70**, 821–8.

WHITE J. A. & BROWN M. F. (1979) Ultrastructure and X-ray analysis of phosphorus granules in a vesicular-arbuscular mycorrhizal fungus. *Canadian Journal of Botany* **57**, 2812–18.

WILDE S. A. (1954) Mycorrhizal fungi: their distribution and effect on tree growth. *Soil Science* **78**, 23–31.

WILLIAMSON B. & ALEXANDER I. J. (1975) Acid phosphatase localized in the sheath of beech. *Soil Biology and Biochemistry* **7**, 195–8.

WILSON J. M. (1957) A study of the factors affecting the uptake of potassium by the mycorrhiza of beech. D.Phil thesis, Oxford University.

WOJAHN K. E. & IYER J. G. (1976) Eradicants and mycorrhizae. *Tree Plantation Notes* **27**, 12–13.

ZAK B. (1973) Classification of ectomycorrhizae. In: *Ectomycorrhizae—Their Ecology and Physiology* (Eds G. C. Marks & T. T. Kozlowski), pp. 43–78. Academic Press, New York.

Chapter 27 • Pathogenic Interactions of Microbes with Plants

Richard N. Strange

27.1 Plants as an ecological niche for parasitic microorganisms

Green plants are found on almost every land surface of the earth, the only exceptions being environments which are extremely cold or free of water or do not provide a suitable rooting medium. Thus plants provide a substantial ecological niche for those microorganisms which are able to parasitise them. However, the abundance of this potential niche with respect to any individual microbe is more apparent than real since few are able to grow in a wide range of plant species. The reasons for this specificity are a major concern in plant pathology, not only because of the grave social and economic consequences which ensue when major crops are devastated (Large 1940; Klinkowski 1970; Ullstrup 1972; Padmanabhan 1973), but also because advances in our understanding of the biochemistry of resistance and susceptibility to pathogens have begun to give us fascinating glimpses into the subtleties of the phenomenon (Strange 1972; Strobel 1975; Deverall 1977; Hedin 1977). A second important topic deals with the changes in host morphology and physiology caused by the continued presence of the living parasite in susceptible interactions or, in other words, symptom expression (Murai *et al.* 1980). Frequently such alterations ensure the survival of the parasite by allowing it to produce propagules on a massive scale (Dekhuijzen 1976).

27.2 Specificity

A closer look at the phenomenon of specificity reveals that not only do microorganisms generally parasitise a narrow host range but also that they seldom colonise all parts of a host. Furthermore, they may be restricted to plants which have reached a particular stage of development. It is convenient, therefore, to qualify the term specificity with the adjectives host, organ, tissue and age. A good example of host specificity is the eye-spot disease of sugar-cane caused by the fungus *Helminthosporium sacchari*. On susceptible clones this organism characteristically causes an eye-shaped lesion from which a long red runner extends up the leaf but on resistant clones no symptoms occur (Steiner &

Byther 1971). Ergot (caused by various species of *Claviceps*) is a disease of graminaceous plants but the fungus is confined to the ovary and may be said to be organ specific. In powdery mildew infections of cereals caused by *Erysiphe graminis* the association is even more limited in that only one tissue, the epidermis, is penetrated. An extreme example of age specificity is found in ergot of pearl millet where infection is only successful between the emergence of stigmas and shedding of pollen, a period of 2 to 5 d, since pollination causes withering of the stigmas through which invasion normally occurs (Thakur & Williams 1980). A further opportunity for specificity occurs with plant viruses as, apart from those which are normally transmitted exclusively within the propagules of their hosts, they require a vector to which they may be specific. Different strains of tobacco necrosis virus, for example, are transmitted differentially by the chytrid fungus *Olpidium brassicae* (Temmink *et al.* 1970). Furthermore, some viruses can replicate in insects as well as plants and are therefore both animal and plant viruses.

27.3 Gene-for-gene relationships

Our understanding of the evolution of parasite specificity in natural plant populations is speculative but there is little doubt that it is mirrored, albeit on a more dramatic scale and in a shorter time, in cultivated crops. Essentially genetic change in the parasite allows it to invade a plant to which it was formerly avirulent. If, as a consequence of the infection, the plant fails to survive to its reproductive age, the newly evolved parasite will die out along with the plant (Day 1974). In natural populations, therefore, only the more resistant plants and the correspondingly less virulent parasites survive. With crop plants, genetic change on the part of a parasite to virulence can be catastrophic; for example, a new and virulent variant of the fungus *Helminthosporium maydis* swept through the corn belt of the United States in 1970 causing losses estimated at $\$1 \times 10^9$ (Tatum 1971). Intriguingly, virulence was expressed only on those plants which contained a cytoplasmic gene for male sterility (Ullstrup 1972). This gene was present, however, in about 80% of the crop of the United States since it facilitated the production of high-yielding hybrids from this normally self-compatible plant (Hooker 1974). Not surprisingly a disaster on this scale led to fundamental questions concerning the genetic vulnerability of crop plants to infectious disease (National Academy of Science 1972). On a more modest scale many new cultivars often have a short commercial life or even fail to reach the stage of distribution to farmers because of the evolution of variants of parasites with new virulence characteristics (ten Houten 1974). Frustratingly for the plant breeder, the new variants frequently occur in the very species of parasite which the plant was bred to resist! Thus, there is a spiral, new cultivars of crop plants being replaced as soon as new variants of the parasite with the appropriate virulence characteristics occur. Such variants are usually termed physiologic races and the understanding of their genetic relationships with their hosts was pioneered by Flor who formulated the gene-for-gene hypothesis. One definition of the hypothesis is as follows.

'A gene-for-gene relationship exists when the presence of a gene in one population is contingent on the continued presence of a gene in another population and where the interaction between the two genes leads to a single phenotypic expression by which the presence or absence of the relevant gene in either organism may be recognized' (Person *et al.* 1962).

This concept, which was reviewed by Flor (1971) has been of considerable value in the study of the genetic basis of infectious plant disease but the embarrassing fact remains that, despite much genetic information, we still know very little about the products of the genes which determine specificity (i.e. resistance or susceptibility in the host and virulence or avirulence in the parasite). Work is urgently needed to fill this gap in our knowledge so that the breeding of important crop plants for resistance may proceed on a less empirical basis.

27.4 Passive defence

27.4.1 Physical Barriers

For a parasite to become established in its host it must penetrate the plant. The act of penetration may be brought about by the parasite itself or alternatively may occur through some other agency. The multicellular nature of fungi, for example, allows them in many instances to penetrate directly but viruses require wounds which may be made by vectors. Although entry into the plant through natural openings removes the immediate hazard of desiccation, the cell wall surrounding the nutritious protoplast has still to be overcome. Frequently, this is also cuticularised although the cuticle may be thinner than that found on the epidermis. The question, therefore, arises as to the effectiveness of the cuticle and the cell wall below it as barriers to infection. Van den Ende and Linskens (1974) in their review of cutinolytic enzymes in relation to pathogenesis concluded that parasites degraded host cutin more efficiently than a standard cutin preparation from a *Gasteria* sp.; that greater cutinolytic activity occurred in congenial rather than non-congenial host–parasite relationships; and that there was a difference in the activity of the parasite enzyme on the cutin of resistant and susceptible hosts, presumably the activity being greater on the latter. In contrast, Martin (1964) considered that the role of the cuticle in resistance was not very significant.

It has been hypothesized that the variation of polysaccharide composition of plant cell walls might account for the specificity of plant parasites, only those organisms capable of secreting the appropriate combination of enzymes being able to penetrate the plant (Albersheim *et al.* 1969). This notion provided a spur to carbohydrate analysis of plant cell walls which, however, did not support the hypothesis as the cell walls of the major subclasses of plants, although complex, were found to be very similar (Albersheim & Anderson-Prouty 1975). Nevertheless, inhibition of either cutinolytic or cell wall-lytic enzymes inhibits penetration (Anderson & Albersheim 1972; Maiti & Kolattukudy 1979). Thus, the situation now seems to be that although the proper functioning of these enzymes is a prerequisite for virulence, the explanation of specificity normally resides in other determinants.

27.4.2 Biochemical Barriers

Most plants possess compounds which are either antimicrobial or precursors of antimicrobial compounds from which the active component may be released by simple enzyme cleavage. These include lactones, cyanogenic glucosides, sulphur compounds, phenols and their glycosides and saponins (Schönbeck & Schlösser 1976). It seems probable that these compounds are effective in preventing the establishment of many plant parasites with the exception of those that are virulent for the particular plant. These parasites have evolved mechanisms for circumventing or degrading the antimicrobial compounds in their hosts; for example, *Gloeocercospora sorghi*, a parasite of sorghum which may contain as much as 30% of its dry weight as dhurrin, a cyanogenic glucoside, detoxifies the cyanide released on infection (Myers & Fry 1978). Similarly, a variety of the take-all fungus (*Gaeumannomyces graminis* var. *avenae* = *Ophiobolus graminis* var. *avenae*) which was able to attack oats, produced an enzyme, avenacinase, which destroyed the antimicrobial activity of the oat saponin, avenacin (Turner 1961). Conversely, a variety of the fungus which did not attack oats (*G. graminis*) did not produce the enzyme (Turner 1961).

On the basis of the known evidence, therefore, it seems that antimicrobial compounds may play a more important role in the determination of specificity than the cuticle or cell wall since the molecular structure of these compounds is more diverse. This necessitates the development of a strategy on the part of the virulent parasite to deal with the particular compound possessed by its host.

27.5 Active defence

27.5.1 Physical Barriers

Although there is a wealth of evidence that fungal and bacterial parasites of plants produce enzymes capable of degrading cell walls, it is now clear that the cell wall may be modified in response to microbial challenge (Aist 1976; Ride 1978). In some cases cell division is induced, thereby increasing the number of cell walls to be penetrated, while in others an array of modifications occur which include the deposition of papillae below sites of attempted penetration, the accumulation of calcium and silica, impregnation with phenolic acids, suberisation and lignification. As Ride (1978) points out, it is difficult to obtain direct evidence that any of these mechanisms actually represents a barrier to infection and, in most instances, their implication in defence has rested upon the appropriateness of the timing of the response in relation to attempted infection. Such evidence is circumstantial and does not rule out, for example, the possibility that other defence reactions may be occurring simultaneously. One study, that goes some way to meet these difficulties, is that of Manocha (1975). He showed, using the temperature-sensitive allele *Sr6*, which confers resistance of wheat to *Puccinia graminis*, that sheath production around the fungal haustorium was more rapid at the non-permissive than the permissive temperature and also that the sheath inhibited the flow of radioactive leucine to the fungus, implying that its function may be to deprive the fungus of nutrients. Other studies (Ride 1975; Sherwood & Vance 1976; Vance & Sherwood 1976, 1977; Ride & Pearce 1979; Pearce & Ride 1980; Ride 1980a) have strongly implicated lignification as a defence mechanism. Ride (1978) has suggested that the newly formed lignin may provide a barrier to mechanical and enzymic penetration as well as presenting an obstacle to diffusion of water and metabolites between host and parasite. Moreover, the synthesis of lignin may result in the production of precursors which are fungitoxic and may even lignify the parasite (Ride 1978).

27.5.2 Biochemical Barriers

Since the classic experiments of Müller and Börger (1940) over 40 years ago, a tremendous amount of information concerning the accumulation *de novo* of antimicrobial compounds by plants in response to challenge by avirulent microorganisms has been published. Such compounds have been called phytoalexins but the definition of this term has run into difficulties. Very few compounds conform to all the criteria discussed by Cruickshank (1963) but there are now at least 100 compounds which satisfy some of them. Possibly a not too stringent definition, such as 'an antimicrobial compound of plant origin which accumulates in response to challenge with an avirulent microorganism', might be helpful until more is known about these compounds and their occurrence. The compounds are of diverse chemical structure and include polyacetylenes, terpenes, pterocarpans, flavones, isoflavones and stilbenes. There are many reports of their accumulation in response to challenge by avirulent strains of parasites and a lesser number of the lack of such accumulation in challenge by virulent parasites. Originally, phytoalexins were thought to lie at the end of a biosynthetic sequence but it is now becoming clear that they may also be metabolised by the plant (Stoessel *et al.* 1977). Accumulation on challenge by an avirulent organism may, therefore, result from increased synthesis, decreased metabolism or both (Yoshikawa 1978). Conversely, failure of accumulation, e.g. on challenge with a virulent organism, may result from a lack of increased synthesis or from increased metabolism. A central question here is the respective roles of plant and microbe in these regulatory processes.

A number of fungal products, mostly of a polysaccharide nature and termed elicitors, have been purified which cause phytoalexin accumulation when introduced into plants but there are also two examples in which it appears that the plant contains its own endogenous low molecular weight elicitor(s) (Hargreaves & Bailey 1978; G. E. Aguamah & R. N. Strange, unpublished observations). These are presumably sequestered in some way and released only on receipt of an appropriate signal or on injury. Possibly endogenous elicitors mediate the phytoalexin accumulation which occurs in response to a bewildering array of abiotic substances; these include such diverse compounds as heavy metal ions, UV light, cyclic AMP, various peptides, proteins and antibiotics (Keen & Bruegger 1977). Yoshikawa (1978) has reported that abiotic phytoalexin elicitors decreased phytoalexin breakdown in soybeans while a biotic elicitor (a cell wall glucan from *Phytophthora megasperma* var. *sojae*) increased phytoalexin synthesis.

Phytoalexin accumulation often occurs in conjunction with a limited area of host cell necrosis termed hypersensitivity. This reaction of a plant to an incompatible fungus, bacterium or virus was early recognized as being concomitant with resistance. It was also claimed to be the cause of resistance, particularly in infections by obligate biotrophs since, it was reasoned, if such parasites kill the cells they attempt to parasitize, then they themselves by definition could not survive. More recently some workers have considered that hypersensitivity is a consequence of resistance (Kiraly *et al.* 1972) rather than its cause (MaClean *et al.* 1974), but Heath (1976) has given a timely warning on the hazards of generalising about this subject. She suggests that there are many ways in which a cell can die and that each host–parasite interaction should be studied individually, a point taken up by Ingram (1978).

An interesting sequel to the development of the hypersensitive response in virus infections is that the affected plant becomes both locally and systematically resistant to subsequent challenge by virulent virus strains (Ross 1961a,b). Acquired resistance to fungal infections has also been reported and has been studied recently by Kuć's group (Kuć & Caruso 1977). They have found that infection of primary leaves of cucurbits with *Colletotrichum lagenarium* gave pronounced systemic protection against subsequent challenge by the same pathogen for 4 to 5 weeks and a second inoculation extended the time of protection into the fruiting period. The mechanisms underlying this phenomenon are still a mystery although it may be significant that increased titres of a substance that agglutinates hyphae of *C. lagenarium* have been obtained from protected plants (Kuć & Caruso 1977).

27.6 Parasite nutrition

It is axiomatic that for a parasite to be virulent it must be able to use the nutrients provided by the host for growth. In attempts to ascertain the role of nutrition in host–parasite relationships, a number of scientists have prepared auxotrophs of virulent parasites. Often the virulence of such auxotrophs is attenuated but this is not always the case. Unfortunately, the concentration of the required metabolite in the host has seldom been measured. Coplin *et al.* (1974) have met this point in their study of amino acid auxotrophs of *Pseudomonas solanacearum* in which it was found that, where the required amino acid was limiting, as determined by studies *in vitro*, virulence was attenuated. In detailed studies of a tryptophan auxotroph, virulence was restored by supplementing the host plant (tobacco) with tryptophan or genetic transformation of the parasite to prototrophy. The question arises, therefore, as to whether natural auxotrophs occur among plant parasites. Little work has been directed at answering this question although literature describing the stimulatory effect of plant extracts and exudates on parasite growth is abundant. In two instances specific compounds which promote growth of parasites *in vitro* have also been found to increase virulence *in vivo*. These are canker of plum caused by *Rhodosticta quercina* which has a requirement for myo-inositol (Lukezic & DeVay 1964) and headblight of wheat caused by *Fusarium graminearum* which has a requirement for choline and betaine (Strange *et al.* 1978).

27.7 Virulence attributes

A parasite may be said to be virulent if it kills its host's tissues and destroys them. The former may be accomplished by toxins which poison the host and the latter by degradative enzymes. A more subtle form of virulence is the alteration of host physiology and biochemistry to suit the parasite but, since this normally results in far-reaching symptoms, it will be dealt with in the next section.

27.7.1 Toxins

Over a quarter of a century ago Gäumann (1954) claimed that microorganisms are pathogenic only if they are toxigenic. Considerable evidence has been accumulated to support this statement over the intervening period but it is too early to say whether it is universally true. Phytotoxins produced by plant parasites are usually divided into two groups (Rudolph 1976; Scheffer 1976):

(1) a large group in which the toxins affect plants both within and outside the host range of the microorganism;

(2) a much smaller group in which the toxins only affect plants which are susceptible to the toxigenic microorganism.

These two groups have been termed non-specific and host specific, respectively, despite the fact that a toxin cannot have a host but rather may exert its effect selectively on a limited number of species or cultivars.

An example of a non-specific toxin is the compound, tabtoxin, produced by *Pseudomonas tabaci*, the causal agent of wildfire disease of tobacco. In addition to tobacco, in which it causes chlorosis, the toxin also affects many other species of plants including *Chlorella vulgaris*. The compound has a β-lactam structure which on hydrolysis yields the amino acid threonine and tabtoxinine (5-amino 2-aminoethyl 2-hydroxyadipic acid). Two other species of *Pseudomonas* which are pathogenic to oats and timothy also produce tabtoxin and all three species are now known to synthesise a variant of the toxin in which serine is substituted for threonine. It is not yet clear how these compounds exert their toxic effects (Strobel 1977).

There are at least 10 host-specific toxins known but, unfortunately, difficulties have been experienced in the chemical identification of the compounds. One example of extreme interest is the toxin, helminthosporoside, produced by *Helminthosporium sacchari*, a parasite of sugar-cane (Strobel 1975). The compound is an α-galactoside and the spectroscopic data originally obtained were consistent with a cyclopropyl aglycone moiety. However, further work has shown that the structure is probably a digalactoside with a large aliphatic aglycone and a molecular formula of $C_{27}H_{42}O_{12}$ (Beier 1980). The toxin reproduces the characteristic red lesions associated with the disease, an effect which may be eliminated by treatment of the cane leaves with the galactosides, melibiose and raffinose as well as heat. Susceptibility to the disease and sensitivity to the toxin is correlated with the presence of a toxin-binding protein in the plasma membrane. Resistant clones of sugar-cane possess a similar protein which differs from that of susceptible clones by four amino acid residues and does not bind the toxin (Strobel 1975).

One of the most exciting possibilities that awaits the discovery of toxin involvement in plant disease is the use of the toxin to select resistant plants. Sugar-cane clones, for example, are now selected for resistance to *Helminthosporium sacchari* with partially purified preparations of helminthosporoside (Steiner & Byther 1971). Another approach is to use the toxin to select cells from tissue cultures which are insensitive and therefore likely to be resistant to the disease (Gengenbach *et al.* 1977). Conversely, such toxins may prove useful in the future as selective herbicides for weed control.

27.7.2 Enzymes

The cell wall as a physical barrier to penetration by pathogenic microorganisms was discussed earlier (see section 27.4.1) in which it was considered that the proper functioning of the appropriate degradative enzymes was essential for the expansion of pathogenicity. Many parasites possess enzymes which breach the cuticle and the underlying cell wall and in some cases cause lysis of host membranes. These structures are chemically complex and a correspondingly large number of enzymes from plant parasites have been obtained which can cleave the many types of chemical bonds that they contain (Bateman & Basham 1976). Clearly the functioning of such enzymes can and do have a very destructive effect on host tissues.

27.8 Symptom expression

The cell death and destruction described in the previous section are the symptoms associated with relatively unspecialised parasites. More specialised parasites cause little cell death but may fundamentally alter the physiology of their hosts. Often this is achieved by variation of the hormone concentrations of the host. Plant hormones consist of auxins, gibberellins, cytokinins and ethylene. Auxins are involved in cell enlargement, initiation of root formation, inhibition of root growth, bud inhibition and apical dominance, abscission of leaves and fruits, differentiation, fruit growth and parthenocarpy. Gibberellins promote internode growth, induce flowering and sub-apical cell division, reverse dwarfism, induce parthenocarpy and stimulate α-amylase synthesis. Cytokinins stimulate cell division and cell enlargement, retard senescence and induce the movement of metabolites while ethylene can cause epinasty, abscission, chlorosis and root development, among other effects. Many of these phenomena have been construed as symptoms of infectious disease and indeed the spindly growth associated with the bakanae disease of rice caused by *Gibberella fujikuroi* led to the discovery of the gibberellins (Stowe & Yamaki 1957). In this instance there is little doubt that the fungus itself produces the hormone but in other diseases it is less clear whether the parasite is directly responsible for enhanced hormone levels or, in some way, induces the plant to synthesise excess. Whichever mechanism prevails, the end-results are often startling. These may take the form of witches brooms where lateral buds are released from apical dominance (Klämbt *et al.* 1966), gross hypertrophy and hyperplasia and mobilisation of nutrients towards the site of infection to the detriment of the uninfected portions of the plant (Dekhuijzen 1976). Rust infections are a good example of this last effect and no doubt the nutrients accruing around infection sites are largely responsible for enabling the fungus to sporulate prolifically.

Finally, plants contain a hormonal growth inhibitor, abscisic acid, and attempts have been made to implicate this compound in a few of the many diseases in which stunting is an obvious symptom (Pegg 1976).

Experimental

27.9 Estimation of fungi, bacteria and viruses in the host

The quantitative determination of the amount of a parasite in a plant is an important parameter of virulence. Viruses and bacteria are discrete entities and therefore mild homogenisation of an infected plant under appropriate conditions will give a suspension of the infectious agent which may be assayed. Virus suspensions and their dilutions are assayed on local lesion hosts and bacteria by plate counts on a suitable medium. Fungi are filamentous and would be fragmented by homogenisation. A plate count of such a homogenate would, therefore, be considerably influenced by the efficiency of comminution and the ability of the organism to withstand this treatment.

27.9.1 Estimation of Fungi

In the last decade, specific chemical techniques for the estimation of fungi in plant tissues have been published based on chitin (Ride & Drysdale 1972) and ergosterol (Seitz *et al.* 1979). Chitin is a constituent of the cell walls of most fungi with the exception of the Oomycetes. Similarly, ergosterol has been widely reported as a major sterol in fungi. Neither compound is thought to occur in significant concentrations in plants.

Chitin may be hydrolysed in acid to yield glucosamine which may be estimated by chromatography and reaction with ninhydrin on a conventional amino acid analyser (Mayama *et al.* 1975; Wu & Stahmann 1975). Alternatively, treatment with concentrated alkali at high temperatures causes extensive deacetylation and partial depolymerisation to yield a group of compounds known as chitosan. Treatment of chitosan with nitrous acid causes deamination and further depolymerisation to yield a soluble aldehydic product. This is mixed with 3-methyl-2-benzothiazolone hydrazone and reacts to give a blue dye which may be determined spectrophotometerically (Ride & Drysdale 1972).

In the ergosterol assay the sample is extracted with methanol, saponified and partitioned against petroleum ether. The petroleum ether fraction is dried, dissolved in 99%(v/v) methylene chloride/1%(v/v) isopropanol and filtered. The sample is injected into a high pressure liquid chromatography system (HPLC) using either normal or reversed phase chromatography. Ergosterol is detected by its retention time and UV absorption at 282 nm.

A comparison of these assays shows that the chitin assay of Ride and Drysdale (1972) is suitable for the determination of many samples simultaneously since they may be processed as a batch. In contrast, the use of an amino acid analyser for the determination of glucosamine obtained from chitin hydrolysis restricts the number of samples that can be handled at the same time to one, although new (and very expensive) amino acid analysers based on HPLC are available which would speed up the chromatography step. Assay by the ergosterol technique also suffers from the same limitation.

The sensitivity of the Ride and Drysdale (1972) assay is 0.1 μg glucosamine at best which, for a fungus containing 20 μg chitin (mg fungus)$^{-1}$ is equivalent to 5 μg fungus. Assays involving an amino acid analyser may be somewhat more sensitive. The sensitivity of the ergosterol assay is 0.04 to 0.09 μg ergosterol (g sample of inoculated milled rice)$^{-1}$ which is equivalent to 20 to 45 μg fungus with an ergosterol concentration of 2 μg (mg fungus)$^{-1}$—a value which is slightly lower than those of the fungi tested, 2.3 to 5.9 μg ergosterol (mg fungus)$^{-1}$. However, Burnett (1976) has reported that some fungi contain as much as 2%(w/w) ergosterol. The Ride and Drysdale (1972) assay involves a considerable amount of pipetting but the only sophisticated equipment needed is a spectrophotometer. In contrast the other assays require either an amino acid analyser or HPLC, neither of which are yet routine pieces of equipment in plant pathology laboratories. As regards specificity, background levels for some tissues can be high in the Ride and Drysdale assay (J. P. Ride, personal communication) while in the amino acid version of chitin analysis, glucosamine from other compounds, such as glycoproteins, may interfere. None of the assays distinguish between living and dead hyphae and there is evidence that the chitin content of fungi grown *in vitro* varies according to growth conditions and age (Sharma *et al.* 1977). Presumably, this would also be true for the fungus growing *in vivo*. Whether this latter consideration also applies to the ergosterol assay remains to be seen.

Despite the limitations of these assays, they represent a considerable advance over the more subjective methods they supersede and, providing they are used with care and adequate controls, should do much to further our understanding of the growth of pathogenic fungi in plants.

27.9.2 Estimation of Bacteria

The bacterial population of a plant may be estimated by grinding the infected tissue in a mortar in buffer and making serial dilutions which are plated onto a nutrient medium. It is usually important that this should be selective since otherwise phyllosphere organisms and secondary invaders will obscure the results. Kado and Heskett (1970) have published five selective media for plant pathogenic bacteria in the genera *Agrobacterium*, *Corynebacterium*, *Erwinia*, *Pseudomonas* and *Xanthomonas* which are of considerable value both in diagnostic and quantitative work. More recent studies have improved the selectivity and planting efficiency of these media (Gross & Vidaver 1979).

27.9.3 Estimation of Viruses

Virus estimation has been treated extensively by Matthews (1970) and more recently by Gibbs and Harrison (1976). Three basic principles are generally used: infectivity, electron microscopy and, since the coat proteins of viruses are antigenic, serology. Of the infectivity tests, the local lesion assay is the one most widely used and with a suitable indicator plant less than 1×10^6 virus particles (ml virus suspension)$^{-1}$ can be detected. An interesting development is the culture of viruses which can replicate in their insect vectors in monolayers of vector cells (Black 1969). The virus is detected by reacting the cell layers with antiserum to the virus conjugated with fluorescein and viewing with a fluorescence microscope. Dilution end points may be obtained by inoculating the vector monolayers with serial dilutions of virus suspension.

With the electron microscope, numbers of virus particles in a preparation are generally counted by mixing the preparation with a suspension of polystyrene latex particles of known concentration, spraying droplets of the mixture onto an electron microscope grid and photographing. The numbers of virus particles and latex particles can be compared (Williams & Backus 1949; Nixon & Fisher 1958). Sensitivity is about 1×10^8 virus particles ml^{-1}.

A large number of serological techniques have been published and the reader is referred to Gibbs and Harrison (1976) for a critical review of them. Recently, the enzyme-linked immunosorbent assay (ELISA) has been used with considerable success (Clark & Adams 1977). In this technique the wells of a titre plate are coated with antibody to the virus and virus preparation added. After washing, antibody to the virus linked to alkaline phosphatase is added and the plate washed once more before adding a substrate of the enzyme, *p*-nitrophenylphosphate. The coloured reaction product, which may be measured spectrophotometrically at 504 nm, accumulates in those wells that contain the virus. Another recent technique is serologically specific electron microscopy (SSEM) (Derrick & Brlansky 1976; Brlansky & Derrick 1979). Here an electron microscope grid is coated with Parlodion and inverted in antiserum. The grid is placed in virus preparation and those viruses specific to the antibodies in the antiserum are adsorbed and may be viewed under the electron microscope (see Hamilton & Nichols 1978 for an evaluation of the relative merits of ELISA and SSEM).

Very recently Randles *et al.* (1980), working with the viroid-like causal agent of Cadang-Cadang disease, have published a diagnostic technique which may be used where serology is inappropriate. The technique relies upon the synthesis of a tritiated DNA probe complementary to the viroid-like RNA and its hybridisation with the RNA fraction from a diseased plant containing the putative viroid-like RNA.

27.10 Cell wall modifications

27.10.1 Degradative Phenomena

Since the enzymic degradation of plant cell walls is one of the most frequent and obvious signs of infection, it is not surprising that a vast literature dealing with the subject has developed. This has been reviewed by Bateman and Millar (1966) and Bateman and Basham (1976).

Plant cell walls are predominantly composed of carbohydrates consisting of pectic substances, hemicelluloses, cellulose and lignin. Purified cell wall fractions or synthetic carbohydrates corresponding to these natural products have frequently been used as substrates for putative enzymes usually obtained from cultural filtrates of the parasite grown *in vitro*. The question remains, however, as to whether these enzymes are found *in vivo* particularly as they are often both inducible and subject to catabolite repression.

A general technique for testing for the production of hydrolytic enzymes is the cup-plate assay (Dingle *et al.* 1953). The substrate (1%w/v) is incorporated in buffered agar together with salicylanilide (0.01%w/v) to prevent growth of moulds.

The enzyme preparation is placed in wells cut in the agar and, after incubation, enzyme activity is visualised by treatment with a reagent appropriate to the substrate. The diameter of the zone of enzyme activity is proportional to the logarithm of the concentration of the enzyme in the wells. Pectic enzymes may be subdivided into esterases, i.e. those that hydrolyse methanol from methylated polygalacturonic acid chains, and chain-splitting enzymes. The latter may be hydrolases or transeliminases. Pectin methyl esterase activity is usually measured by continuous titration of the polygalacturonic acid released in reaction mixtures (Kertesz 1951) and hydrolases by the determination of the increase in reducing groups (Nelson 1944). Pectin forms a viscous solution in water and a sensitive assay for the measurement of hydrolases which attack this substrate is viscometry (Bateman 1963). Transeliminases may be measured by following the increase of unsaturated reaction products by the increase of absorbance of UV light at 230 to 235 nm (the reaction products of pectic acids absorb maximally at the lower figure and those of pectins at the higher). A useful technique for the detection of a given carbohydrase employs an artificial substrate, such as the *p* or *o*-nitrophenyl derivatives of a given sugar, which gives coloured products on hydrolysis (Jones *et al.* 1972; Mount *et al.* 1979).

Some cellulases cause the loss of crystallinity of native cellulose and these may be detected by loss of birefringence (Bateman 1964) or the absorption of NaOH solution (Kelman & Cowling 1965). Others degrade the polysaccharide chains to cellobiose and glucose. These may be detected either by loss of weight of the substrate (Halliwell 1963) or by a glucose oxidase assay (Halliwell 1965).

A rather different approach has been adopted by Albersheim and his co-workers (Albersheim *et al.* 1967; Bateman *et al.* 1969). They subjected cell wall preparations to various enzymic treatments and determined the remaining polysaccharides by hydrolysis with trifluoracetic acid, conversion of the resulting sugars to their corresponding alditols and acetylation to give products which could be separated and quantified by gas chromatography.

The same principle of depolymerisation and analysis of products has been used for the analysis of cutin (Kolattukudy 1977). Depolymerisation may be brought about by transesterification with methanol containing sodium methoxide or boron trifluoride and analysis of the products by TLC followed by GC-MS (Kolattukudy 1977). Studies of cutinases using this technique to monitor changes in substrate will be of considerable interest.

27.10.2 Synthetic Phenomena

Of the many changes which occur in plant cell walls challenged by pathogenic microorganisms (Ride 1978) lignification is the best documented and on current evidence appears to be the most significant in resistance. Ride and Pearce in a series of papers have described excellent techniques for studying its formation and relation to resistance (Ride 1975; Ride & Pearce 1979; Pearce & Ride 1980; Ride 1980). Lignin, after extraction, was detected by ionisation difference spectra at pH 12.0 and pH 7.0 and characterised by nitrobenzene oxidation followed by TLC, silyation and GLC of the products. Subsequent tests have demonstrated the extreme resistance of the lignin of wheat leaves induced by microbial challenge to degradation *in vitro* by 14 fungal species.

27.11 Stimulants of virulence and microbial growth

Methods for measuring parasite growth *in vivo* have already been discussed (see section 27.9) but as pointed out in section 27.6 it is often of interest to compare growth *in vivo* with growth *in vitro*. The growth kinetics of bacteria in relation to nutrition have been amply reviewed elsewhere (Pirt 1975) but difficulties occur with fungi owing to their filamentous nature. Also, spore germination in fungi is an entirely different phenomenon from hyphal extension and therefore conditions which promote the one process may not necessarily have the same effect on the other. One assay which gives a combined measurement of both speed of spore germination and hyphal extension but does not differentiate between the two has been reported (Strange & Smith 1971; Strange *et al.* 1974). Wells are cut in an agar plate which may contain a nutrient medium, the stimulant preparation placed in a central well and spores of the fungus in peripheral wells equally spaced from the centre well. After incubation hyphal extension towards the centre well is measured under a microscope fitted with a $\times 4$ objective and a graticule eyepiece. This parameter for *Fusarium graminearum* was found to be linearly related to the logarithm of the concentration of the growth stimulant over a concentration range of about two log units. The technique is rapid (subsequent work has shown that 1 to 2 d incubation is sufficient for a number of fungi) and reasonably quantitative. It is ideal for monitoring the purification of active compounds, e.g. the effluents from chromatography columns and may be adapted to direct assay of compounds on paper chromatograms (Strange & Smith 1971).

27.12 Microbial inhibitors

Almost any plant extract will be found to contain some antimicrobial activity so the problem that often faces the plant pathologist is whether this activity is significant in relation to disease resistance. Some inhibitory compounds are preformed but others may be released on enzymic cleavage from a precursor. It is important in the latter case to allow enzyme and substrate time to react in a suitable buffered homogenate before extraction of the inhibitor. Doubt has been thrown on the capacity of some plants or parts of plants to produce phytoalexins but this may mean that the appropriate method of elicitation has yet to be found. Cucurbits and peanut leaves were both reported to be unable to accumulate phytoalexins (Deverall 1976; Ingham 1976) but both have been found to produce them in response to appropriate microbial challenge (A. Alavi & R. N. Strange; G. E. Aguamah & R. N. Strange, both unpublished observations).

Once conditions for production of the inhibitor have been established, it is important to isolate and identify it before going on to test its role in defence of the plant against disease. Usually the plant material is extracted by homogenising in a solvent, which may be water, containing buffered ascorbate to prevent the oxidation of phenols. The homogenate is squeezed through cheesecloth and centrifuged at about 10 000 ***g*** to remove particulate material. If an extract of green material has been made in aqueous buffer, the supernatant will normally be a clear yellow colour since chloroplasts will have sedimented; but if organic solvents were used, chlorophyll is liable to be present in solution.

It is important at this stage to attempt to remove the bulk of the impurities and a useful technique is partitioning between immiscible solvents; for example, many phytoalexins may be extracted in aqueous alcohol, the alcohol removed by film evaporation and the phytoalexin partitioned into ethyl acetate. The pH of the aqueous phase is critical since at low values acid and phenolic compounds will not be ionized and will, therefore, pass easily into the organic phase. At higher pH values the partition coefficient of these compounds is more likely to favour the aqueous phase. This behaviour may be exploited by extracting the ethyl acetate phase containing the inhibitor with sodium carbonate or sodium bicarbonate, acidifying the aqueous phase and re-extracting with ethyl acetate. This procedure normally results in a considerable purification but care must be taken to ensure that the compound is not degraded by the treatment. After drying the organic phase, usually by the addition of anhydrous sodium sulphate, it is reduced to a small volume by film evaporation and spotted onto a silica-gel TLC plate. The plate is developed in a suitable solvent, dried and sprayed with a thick spore suspension of *Cladosporium cucumerinum* or *C. herbarum* in double strength Czapek Dox medium (Homans & Fuchs 1970). After incubation for 2 to 3 d at high humidity, antifungal areas appear as white spots against the dark background provided by the growing fungus. The result may be permanently recorded by photocopying the plate. Many phytoalexins have been isolated by removing the silica gel from an unsprayed portion of the plate corresponding to the inhibitor and rechromatographing until the compound is pure. A more convenient technique is to use HPLC. In the author's laboratory the silica from the active area of the plate is removed by suction: the narrow end of a Pasteur pipette is plugged with glass wool, attached to a water pump and the wide end scraped over the appropriate area. The active material is conveniently eluted from the adsorbent by clamping the pipette vertically (narrow end downwards) and adding a suitable solvent, such as methanol. The UV absorption spectrum of the eluate is recorded and a suitable wavelength (usually λ_{max}) chosen for monitoring the effluent of the HPLC column. A reverse phase column is generally used such as octodecylsilylated silica and an aqueous solvent such as 50%(v/v) acetonitrile. The column, which is usually of a semi-preparative size (i.e. 25.10×1.0 cm), is protected by a precolumn packed with similar or the same material as that in the main column. The advantage of this is that the main column is less likely to be fouled with tightly adsorbing solutes as these will be retained by the precolumn which may be repacked periodically with new adsorbent at a fraction of the cost of a new column. Fractions from the HPLC are collected and those which are active may be detected by the TLC bioassay (see above). Usually at this stage a UV absorbing peak on the recorder can be identified with the retention time of the active compound. It is normally a simple matter to scale up the purification by chromatographing the organic phase obtained in the partitioning stage of the fractionation without the preliminary TLC procedure. Multiple runs usually result in the accumulation of sufficient material for analysis.

Identification of the active compound is generally achieved by mass spectrometry, UV, IR and proton nuclear magnetic resonance spectroscopy. Another useful spectroscopic technique, if the size of the sample permits, is C^{13} nuclear magnetic resonance spectroscopy. For further information on these techniques and the interpretation of the data obtained, the reader is referred to Silverstein *et al.* (1974); McLafferty (1966) and Abraham and Loftus (1978).

The experimenter is now in a position to analyse the compound quantitatively in relatively crude extracts of plant material. This may be done on an analytical HPLC column (i.e. 25.0 × 0.5 cm). A known weight of the pure compound is first chromatographed with a known weight of a standard and the correction factor for the inhibitor, k, obtained which is given by the expression:

$$^{k}\text{Compound} = \frac{\text{Peak Area of Compound} \times \text{Weight of Internal Standard}}{\text{Peak Area of Internal Standard} \times \text{Weight of Compound}}$$

This can be rearranged to give the following expression which can be used to calculate the weights of analytical samples:

$$\text{Weight of Compound} = \frac{\text{Peak Area of Compound} \times \text{Weight of Internal Standard}}{\text{Peak Area of Internal Standard} \times {}^{k}\text{Compound}}$$

Care must be taken to ensure that the heights of neither the compound assayed nor the internal standard are spuriously enhanced by interfering compounds when crude products are assayed.

Recently, this technique has been used to assay an endogenous elicitor of phytoalexin accumulation from ground-nuts (G. E. Aguamah & R. N. Strange, unpublished observations). Leaflets were surface sterilised and the epidermis ruptured along a 0.5 cm line parallel to the veins by a hypodermic needle. Droplets (10 μl) of endogenous elicitor preparations were placed on the wounds, or water for controls, and incubated at high humidity for 48 h. The droplets were collected by means of a Pasteur pipette, mixed with an equal volume of acetonitrile containing the internal standard (butyrophenone) and injected into the HPLC. It seems likely that a similar technique could be adopted for elicitors from microbial sources although it should be borne in mind that cellular concentrations of phytoalexins are liable to be higher than those in diffusion droplets.

Once the concentration of an inhibitor at a given part of a plant is established, it is necessary to ascertain whether there is sufficient material to inhibit the growth of an invading parasite. This problem presents considerable difficulties since most inhibitors are not water soluble. They cannot, therefore, be introduced into the plant without the use of solvents which undoubtedly complicate the issue by damaging the plant. A compromise is to test the activity of the compound *in vitro*. Assay techniques usually involve germination of fungal spores and hyphal extension on solid or in liquid media. Although easily performed, results obtained from them are difficult to interpret (Bailey *et al.* 1976; Skipp & Bailey 1977; Bailey & Skipp 1978; Skipp & Carter 1978). Part of the problem is that inhibition of hyphal extension as a proportion of the controls without phytoalexin often varies markedly with incubation time. Other factors which complicate interpretation are adaptation of the fungus to the phytoalexin and degradation. One point that has not received much attention by experimenters is the composition of the nutrient medium used in the tests although Hargreaves *et al.* (1976) have used synthetic pod nutrients in their studies of wyerone acid metabolism by *Botrytis cinerea* and *Botrytis fabae*.

27.13 Toxins

Toxigenic microorganisms frequently give rise to symptoms at a distance from the location of the parasite itself. These symptoms include wilting, chlorosis, necrosis and water-soaking and may form the basis of bioassays used to monitor the purification of the active compound. Toxins are normally obtained from liquid cultures of the parasite and the selection of the medium can be critical. This, ideally, should allow high titres of toxin to accumulate and should not be toxic to test plants. Care in the maintenance of the parasite is also important since attenuation can occur on prolonged subculture away from the host. In one instance this has been overcome by culturing the fungus on extracts of host tissue and the active factor maintaining toxin production identified (Pinkerton & Strobel 1976). As a last resort, it may be necessary to grow the parasite on the host and obtain the toxin from the diseased material (Goodman *et al.* 1974).

With all assays it is important to test a dilution series of the preparation so that a dose-response curve may be obtained. From this curve the median response may be deduced which can form a useful basis of comparison between different preparations and between fractions during purification. In the latter procedure dry weights of successive fractions should be recorded so that the purification factor may be determined.

Wilting is usually an unequivocal symptom but it is essentially a quantal response. It is necessary, therefore, to test each dilution of a toxin preparation with a number of individual plants or parts of plants in order to determine the dilution factor at which the median response occurs. As this is expensive in terms of consumption of possibly valuable toxin preparation, it is not surprising that various efforts to quantify wilting on a continuum basis have been made; e.g. Johnson and Strobel's wilt-o-meter (Johnson & Strobel 1970) but here variation in the tensile strength of different plants could be a difficulty.

Chlorosis is an easier symptom to quantify. Droplets of toxin may be placed on leaves over a needle wound and the area of chlorosis measured (Barash *et al.* 1975) or, alternatively, the chlorophyll may be extracted and measured (Halloin *et al.* 1969). A necrotic response may be determined in a similar manner to chlorosis.

Water soaking or water clearing results from a malfunctioning of the plasma membrane, and possibly also the tonoplast, allowing water and solutes to move into the intercellular spaces. The leakage of electrolytes may be quantified by measurement of the increased conductance of ambient solutions. Tissue samples are usually exposed to toxin briefly and placed in conductivity water. The rate of leakage is normally linear with time and constant for a given concentration of toxin for at least 3 h. It is also proportional to the concentration of toxin but care must be taken to standardise procedures and to include a known standard in the assay for comparative purposes (Scheffer 1976).

Inhibition of root growth is a classical method for assaying toxins but results are usually obtained only after 3 d which is too long for work with unstable molecules (Luke & Wheeler 1955).

Since malfunctioning of membranes is often a symptom of toxin action it would seem that protoplasts might be ideal material for assay purposes. Protoplasts suspensions could be exposed to a dilution series of toxin solution and the proportion remaining viable after a given period ascertained by the fluorescein diacetate technique (Widholm 1972). Viable protoplasts take up the compound and cleave it to give free fluorescein which is easily visualised under a fluorescence microscope. Alternatively dead protoplasts may be stained with phenosafranine and counted (Widholm 1972).

An important problem with all toxins is to ascertain how they interact with susceptible and resistant tissue. One approach is to label the toxin by growing the toxigenic organism on radioactive substrate (Steiner & Strobel 1971). For example, using radioactive helminthosporoside, Strobel (1973) was able to show that susceptible clones of sugar-cane bound the toxin and that the label was associated with a plasma membrane protein.

Gengenbach *et al.* (1977) have shown that it is possible to select cell cultures of a plant for increasing toxin tolerance and to regenerate plants which maintained this tolerance. Their material was maize containing the *cms-T* cytoplasmic factor which confers susceptibility to *Helminthosporium maydis* race T (section 27.3) and a partially purified toxin preparation from cultural filtrates of the fungus.

27.14 Hormones

Since the roles of auxins, gibberellins, cytokinins, ethylene and abscisic acid in plants are diverse, there are potentially many phenomena which may be used to form the basis of a bioassay. However, this raises the problem of the specificity of the response. For this reason chemical determination is a more satisfactory procedure but the difficulty with this approach is that hormones are present in very low concentrations. Thus, with the exception of ethylene which is gaseous, lengthy procedures are required to remove the impurities before reliable measurements can be made. During purification a considerable proportion of the hormone is liable to be lost but this may be estimated by incorporating a known amount of radio-labelled or deuterated hormone at the beginning of the extraction procedure and determining the loss at the end of the purification by scintillation counting or selected ion monitoring. Absolute purity is not necessary if the means of detecting the compound is sufficiently selective. Indole-3-acetic acid, for example, may be measured by GC-MS after partial purification and derivatisation with *bis*-trimethylsilylacetamide or *bis*-(trimethylsilyl)-trifluoroacetamide to give the *bis*-trimethylsilyl derivative (McDougall & Hillman 1978). Alternatively, the compound may be assayed more rapidly by reacting the partially purified preparation with acetic anhydride to form the fluorescent compound 2-methylindole 2-pyrone and determining it fluorimetrically (Mousdale *et al.* 1978). This procedure is sufficiently sensitive to measure as little as 1 ng (g dry weight)$^{-1}$ in favourable material.

Both HPLC (Reeve & Crozier 1978) and GC-MS (Gaskins & MacMillan 1978) have been used to determine gibberellins after extraction and partial purification. For HPLC, benzyl esters of the acids are prepared since these can be conveniently monitored with a UV detector at 256 nm. Confirmation of the identity of individual gibberellins may be obtained by mass spectrometry. For GC-MS the hormones are converted to volatile compounds by methylation and silylation before injection. Quantification is made by measuring the total ion current profile and identification made by recording retention times on the GC and fragmentation patterns in the MS (Binks *et al.* 1969). The sensitivity of the assay may be further increased by selective ion current monitoring which allows the measurement of amounts as small as 5×10^{-11} to 5×10^{-14} g.

Cytokinins are extracted and purified by a variety of techniques, an ion exchange procedure usually being one of the steps. Cellulose ion exchangers (Parker & Letham 1973) are preferred to the stronger Dowex types since the latter may lead to breakdown of the compounds (Dyson & Hall 1972). After ion exchange chromatography, separation on Sephadex LH-20 has proved a useful technique followed by paper chromatography (Scarbrough *et al.* 1973). More recently, Hashizume *et al.* (1979) have published an interesting and rapid technique for GC-MS determination of cytokinins based on stable isotope dilution with a sensitivity of about 1 ng.

Ethylene is readily quantified by gas chromatography on Alumina F_1 (Ward *et al.* 1978) but obtaining the samples from plant material is more problematical. One method is to condense the gas by passing the vapour from the sample through a gelatin column to dry it before introducing it into a stainless steel trap immersed in liquid oxygen (Swoboda & Lea 1965). The trap is heated and a current passed through it to release the volatile components which are fed directly into the GC. Another method is to trap the gas in mercuric perchlorate from which it is liberated for analysis by the addition of HCl or lithium chloride (Young *et al.* 1952).

Partial purification of abscisic acid is conveniently achieved by partitioning extracts between aqueous solution and an ether phase, use being made of the fact that the partition coefficient of the compound favours the aqueous phase at high pH and the ether phase at low pH (Saunders 1978). After partial purification the compound can be converted to the methyl ester with diazomethane and quantified on GC. A useful check on the identity of a putative methyl abscisic acid peak is to rerun the sample after irradiation with UV. The irradiation treatment causes the formation of the *trans* isomer and it is separated from the natural isomer by GC. If the peak has been correctly identified, it will be smaller in the irradiated sample and a peak corresponding to the *trans* isomer will appear. Electron capture is a very sensitive and selective technique for monitoring the effluent of GC columns for methyl abscisic acid and levels as low as 0.1 ng per injection can be measured (Saunders 1978).

Recently, radioimmunoassays have been developed for indole 3-acetic acid (Pengelly & Meins 1977) and abscisic acid (Walton *et al* 1979). The specificity of this technique obviates much of the labour of purification but caution needs to be exercised in interpreting the results, as a number of similar compounds may interfere.

27.15 Conclusions

It will be clear to the reader that in this review I have been able to do little more than touch on certain aspects of the subject. These aspects are related to my conviction that the dependable control of plant disease will only come about when we understand the nature of the interaction of the host and parasite at the molecular level. This research requires precise and sensitive quantitative techniques which are routinely in use in physics, chemistry, biochemistry and molecular biology laboratories. The plant pathologist should have no inhibitions about borrowing them and making them his own.

Recommended reading

DALY J. M. & URITANI I. (1979) *Recognition and Specificity in Plant Host–Parasite Interactions*. Japan Scientific Societies Press, Tokyo. University Park Press, Baltimore, USA.

DEVERALL B. J. (1977) *Defence Mechanisms of Plants*. Cambridge University Press, Cambridge.

FRIEND J. &. THRELFALL D. R. (1976) *Biochemical Aspects of Plant–Parasite Relationships*. Academic Press, London.

GIBBS A. & HARRISON B. (1976) *Plant Virology: The Principles*. Edward Arnold, London.

HEDIN P. A. (1977) *Host Plant Resistance to Pests*. American Chemical Society, Washington, D.C.

HEITEFUSS R. & WILLIAMS P. H. (1976) *Physiological Plant Pathology*. Springer-Verlag, Berlin.

References

ABRAHAM R. J. & LOFTUS P. (1978) *Proton and Carbon-13 NMR Spectroscopy an Integrated Approach*. Heyden & Sons, London.

AIST J. R. (1976) Cytology of penetration and infection—fungi. In: *Physiological Plant Pathology* (Eds R. Heitefuss & P. H. Williams), pp. 197–221. Springer-Verlag, Berlin.

ALBERSHEIM P. & ANDERSON-PROUTY A. J. (1975) Carbohydrates, proteins, cell surfaces and the biochemistry of pathogenesis. *Annual Review of Plant Physiology* **26**, 31–52.

ALBERSHEIM P., JONES T. M. & ENGLISH P. D. (1969) Biochemistry of the cell wall in relation to infective processes. *Annual Review of Phytopathology* **7**, 171–94.

ALBERSHEIM P., NEVINS D. J., ENGLISH P. D. & KARR A. (1967) A method for the analysis of sugars in plant cell walls by gas–liquid chromatography. *Carbohydrate Research* **5**, 340–5.

ANDERSON A. L. & ALBERSHEIM P. J. (1972) Host-pathogen interactions. V. Comparison of the abilities of proteins isolated from three varieties of *Phaseolus vulgaris* to inhibit the endopolygalacturonases secreted by three races of *Colletotrichum lindemuthianum*. *Physiological Plant Pathology* **2**, 339–46.

BAILEY J. A., CARTER G. A. & SKIPP R. A. (1976) The use and interpretation of bioassays for fungitoxicity of phytoalexins in agar media. *Physiological Plant Pathology* **8**, 189–94.

BAILEY J. A. & SKIPP R. A. (1978) Toxicity of Phytoalexins. *Annals of Applied Biology* **89**, 354–8.

BARASH I., KARR A. L. & STROBEL G. A. (1975) Isolation and characterization of stemphylin, a chromone glucoside from *Stemphylium botryosum*. *Plant Physiology* **55**, 645–51.

BATEMAN D. F. (1963) Pectolytic activities of culture filtrates of *Rhizoctonia solani* and extracts of *Rhizoctonia*-infected tissues of bean. *Phytopathology* **53**, 197–204.

BATEMAN D. F. (1964) Cellulase and the *Rhizoctonia* disease of bean. *Phytopathology* **54**, 1372–7.

BATEMAN D. F. & BASHAM H. G. (1976) Degradation of plant cell walls and membranes by microbial enzymes. In: *Physiological Plant Pathology* (Eds R. Heitefuss & P. H. Williams), pp. 316–55. Springer-Verlag, Berlin.

BATEMAN D. F. & MILLAR R. L. (1966) Pectic enzymes in tissue degradation. *Annual Review of Phytopathology* **4**, 119–46.

BATEMAN D. F., VAN ETTEN H. D., ENGLISH P. D., NEVINS D. J. & ALBERSHEIM P. (1969) Susceptibility to enzymatic degradation of cell walls from bean plants resistant and susceptible to *Rhizoctonia solani* Kuhn. *Plant Physiology* **44**, 641–8.

BEIER R. C. (1980) Carbohydrate chemistry. Synthetic and structural investigation of the phytotoxins found in *Helminthosporium sacchari* and *Rhynchosporium secalis*. Ph.D thesis, Montana State University.

BINKS R., MACMILLAN J. & PRYCE R. J. (1969) Combined gas chromatography–mass spectrometry of the methyl esters of gibberellins A_1 to A_{24} and their trimethylisilylethers. *Phytochemistry* **8**, 271–84.

BLACK L. M. (1969) Insect tissue cultures as tools in plant virus research. *Annual Review of Phytopathology* **7**, 73–100.

BRLANSKY R. J. & DERRICK K. S. (1979) Detection of seedborne plant viruses using seriologically specific electron microscopy. *Phytopathology* **69**, 96–100.

BURNETT J. H. (1976) *Fundamentals of the Fungi*. Edward Arnold, London.

CLARK M. F. & ADAMS A. N. (1977) Characteristics of the microplate method of enzyme-linked immunosorbent assay for the detection of plant viruses. *Journal of General Virology* **34**, 475–83.

COPLIN D. L., SEQUEIRA L. & HANSON R. S. (1974) *Pseudomonas solanacearum*: virulence of biochemical mutants. *Canadian Journal of Microbiology* **20**, 519–29.

CRUICKSHANK I. A. M. (1963) Phytoalexins. *Annual Review of Phytopathology* **1**, 351–74.

DALY J. M. & URITANI I. (1979) *Recognition and Specificity in Plant Host-Parasite Interactions*. Japan Scientific Societies, Tokyo. University Park Press, Baltimore, USA.

DAY P. R. (1974) *Genetics of Host–Parasite Interaction.* W. H. Freeman, San Francisco.

DEKHUIJZEN H. M. (1976) Endogenous cytokinins in healthy and diseased plants. In: *Physiological Plant Pathology* (Eds R. Heitefuss & P. H. Williams), pp. 526–59. Springer-Verlag, Berlin.

DERRICK K. S. & BRLANSKY R. H. (1976) Assay for viruses and mycoplasmas using serologically specific electron microscopy. *Phytopathology* **66**, 815–20.

DEVERALL B. J. (1976) Current perspectives in research on phytoalexins. In: *Biochemical Aspects of Plant–Parasite Relationships* (Eds J. Friend & D. R. Threlfall), pp. 207–37. Academic Press, London.

DEVERALL B. J. (1977) *Defence Mechanisms of Plants.* Cambridge University Press, Cambridge.

DINGLE J., REID W. W. & SOLOMONS G. L. (1953) The enzymatic degradation of pectin and other polysaccharides. II. Application of the 'cup-plate' assay to the estimation of enzymes. *Journal of the Science of Food and Agriculture* **4**, 149–55.

DYSON W. J. & HALL R. H. (1972) N^6-(Δ^2-isopentenyl) adenosine; its occurrence as a free nucleoside in an autonomous strain of tobacco tissue. *Plant Physiology* **50**, 616–21.

FLOR H. H. (1971) Current state of the gene-for-gene concept. *Annual Review of Phytopathology* **9**, 275–96.

FRIEND J. & THRELFALL D. R. (1976) *Biochemical Aspects of Plant–Parasite Relationships.* Academic Press, London.

GASKINS P. & MACMILLAN J. (1978) GC and GC-MS techniques for gibberellins. In: *Isolation of Plant Growth Substances* (Ed. J. R. Hillman), pp. 79–95. Cambridge University Press, Cambridge.

GÜMANN E. (1954) Toxins and plant disease. *Endeavour* **13**, 198–204.

GENGENBACH B. G., GREEN C. E. & DONOVAN C. M. (1977) Inheritance of selected pathotoxin resistance in maize plants regenerated from cell cultures. *Proceedings of the National Academy of Sciences U.S.A.* **74**, 5113–17.

GIBBS A. & HARRISON B. (1976) *Plant Virology: The Principles.* Edward Arnold, London.

GOODMAN R. N., HUANG J. S. & HUANG P. Y. (1974) Host-specific phytotoxic polysaccharide from apple tissue infected by *Erwinia amylovora. Science* **183**, 1081–2.

GROSS D. C. & VIDAVER A. K. (1979) A selective medium for isolation of *Corynebacterium nebraskense* from soil and plant parts. *Phytopathology* **69**, 82–7.

HALLIWELL G. (1963) Cellulose. In: *Methods of Enzymatic Analysis* (Ed. H. U. Bergmeyer), pp. 1132–42. Verlag Chemie, Weinheim.

HALLIWELL G. (1965) Hydrolysis of fibrous cotton and reprecipitated cellulose by cellulolytic enzymes from soil microorganisms. *Biochemical Journal* **95**, 270–81.

HALLOIN J. M., DEZOETEN G. A., GAARD G. R. & WALKER J. C. (1969) The effects of tentoxin on chlorophyll synthesis and plastid structure in cucumber and cabbage. *Plant Physiology* **45**, 310–14.

HAMILTON R. I. & NICHOLS C. (1978) Serological methods for detection of pea seed-borne mosaic virus in leaves and seeds of *Pisum sativum. Phytopathology* **68**, 539–43.

HARGREAVES J. A. & BAILEY J. A. (1978) Phytoalexin production by hypocotyls of *Phaseolus vulgaris* in response to constitutive metabolites released by damaged bean cells. *Physiological Plant Pathology* **13**, 89–100.

HARGREAVES J. A., MANSFIELD J. W. & COXON D. T. (1976) Conversion of wyerone to wyerol by *Botrytis cinerea* and *B. fabae in vitro. Phytochemistry* **15**, 651–3.

HASHIZUME T., SUGIYAMA T., IMURA M., CORY H. T., SCOTT M. F. & MCCLOSKEY J. A. (1979) Determination of cytokinins by mass spectrometry based on stable isotope dilution. *Analytical Biochemistry* **92**, 111–22.

HEATH M. C. (1976) Letter to the editor. Hypersensitivity, the cause or the consequence of rust resistance. *Phytopathology* **66**, 935–6.

HEDIN P. A. (1977) *Host Plant Resistance to Pests.* American Chemical Society, Washington, D.C.

HEITEFUSS R. & WILLIAMS P. H. (1976) *Physiological Plant Pathology.* Springer-Verlag, Berlin.

HOMANS A. L. & FUCHS A. (1970) Direct bioautography on thin layer chromatograms as a method for detecting fungitoxic substances. *Journal of Chromatography* **51**, 327–9.

HOOKER A. L. (1974) Cytoplasmic susceptibility in plant disease. *Annual Review of Phytopathology* **12**, 167–79.

INGHAM J. L. (1976) 3,5,4′ Trihydroxystilbene as a phytoalexin from groundnuts (*Arachis hypogaea*). *Phytochemistry* **15**, 1791–3.

INGRAM D. S. (1978) Cell death and resistance to biotrophs. *Annals of Applied Biology* **89**, 291–5.

JOHNSON T. B. & STROBEL G. A. (1970) The active site on the phytotoxin of *Corynebacterium sependonicum. Plant Physiology* **45**, 761–4.

JONES T. M., ANDERSON A. J. & ALBERSHEIM P. (1972) Studies on the polysaccharide-degrading enzymes secreted by *Fusarium oxysporum* f. sp. *lycopersici. Physiological Plant Pathology* **2**, 153–66.

KADO C. I. & HESKETT M. G. (1970) Selective media for isolation of *Agrobacterium, Corynebacterium, Erwinia, Pseudomonas* and *Xanthomonas. Phytopathology* **60**, 969–76.

KEEN N. T. & BRUEGGER B. (1977) Phytoalexins and chemicals that elicit their production in plants. In: *Host Plant Resistance to Pests* (Ed. P. A. Hedin), pp. 1–26. American Chemical Society, Washington, D.C.

KELMAN A. & COWLING E. B. (1965) Cellulose of *Pseudomonas solanacearum* in relation to pathogenesis. *Phytopathology* **55**, 148–55.

KERTESZ Z. I. (1951) *The Pectic Substances.* Interscience Publishers, New York.

KIRALY Z., BARNA B. & ERSEK T. (1972) Hypersensitivity

as a consequence, not a cause, of plant resistance to infection. *Nature, London* **239**, 456–8.

KLÄMBT D., THIES G. & SKOOG F. (1966) Isolation of cytokinins from *Corynebacterium fascians*. *Proceedings of the National Academy of Sciences U.S.A.* **56**, 52–9.

KLINKOWSKI M. (1970) Catastrophic plant diseases. *Annual Review of Phytopathology* **8**, 37–60.

KOLATTUKUDY P. E. (1977) Lipid polymers and associated phenols, their chemistry, biosynthesis and role in pathogenesis. *Recent Advances in Phytochemistry* **11**, 185–246.

KUĆ J. & CARUSO F. L. (1977) Active co-ordinated chemical defence against disease in plants. In: *Host Plant Resistance to Pests* (Ed. P. A. Hedin), pp. 78–89. American Chemical Society, Washington, D.C.

LARGE E. C. (1940) *The Advance of the Fungi*. Jonathan Cape, London.

LUKE H. H. & WHEELER H. E. (1955) Toxin production by *Helminthosporium victoriae*. *Phytopathology* **45**, 453–8.

LUKEZIC F. L. & DEVAY J. F. (1964) Effect of myo-inositol in host tissues on the parasitism of *Prunus domestica* var. *President* by *Rhodosticta quercina*. *Phytopathology* **54**, 697–700.

MCCLEAN D. J., SARGENT J. S., TOMMERUP I. C. & INGRAM D. S. (1974) Hypersensitivity as the primary event in resistance to fungal parasites. *Nature, London* **294**, 186–7.

MCDOUGALL J. & HILLMAN J. R. (1978) Analysis of indole-3-acetic acid using GC-MS techniques. In: *Isolation of Plant Growth Substances* (Ed. J. R. Hillman), pp. 1–25. Cambridge University Press, Cambridge.

MCLAFFERTY F. W. (1966) *Interpretation of Mass Spectra: An Introduction*. W. A. Benjamin, New York.

MAITI I. B. & KOLATTUKUDY P. E. (1979) Prevention of fungal infection of plants by specific inhibition of cutinase. *Science* **205**, 507–8.

MANOCHA M. S. (1975) Autoradiography and fine structure of host–parasite interface in temperature sensitive combinations of wheat stem rust. *Phytopathologische Zeitschrift* **82**, 207–15.

MARTIN J. T. (1964) Role of cuticle in the defence against plant disease. *Annual Review of Phytopathology* **2**, 81–100.

MATTHEWS R. E. F. (1970) *Plant Virology*. Academic Press, London.

MAYAMA S., REHFELD D. W. & DALY J. M. (1975) A comparison of the development of *Puccinia graminis tritici* in resistant and susceptible wheat based on glucosamine content. *Physiological Plant Pathology* **7**, 243–57.

MOUNT M. S., BERMAN P. M., MORTLOCK R. P. & HUBBARD J. P. (1979) Regulation of endopolygalacturonate transeliminase in an adenosine-3′,5′-cyclic monophosphate deficient mutant of *Erwinia carotovora*. *Phytopathology* **69**, 117–20.

MOUSDALE D. M. A., BUTCHER D. N. & POWELL R. G. (1978) Spectrophotofluorimetric methods of determining indole-3-acetic acid. In: *Isolation of Plant Growth Substances* (Ed. J. R. Hillman), pp. 27–39. Cambridge University Press, Cambridge.

MÜLLER K. & BÖRGER H. (1940) Experimentelle Untersuchungen über die Phytophthora-Resistenz der Kartoffel. *Arbeiten aus der Biologischen Abteilung (Anstalt-Reichsanstalt für Land- und Forstwirtschaft am Kaiserlichen Gesundheitsamte* **23**, 189–231.

MURAI N., SKOOG F., DOYLE M. E. & HANSON R. S. (1980) Relationship between cytokinin production, presence of plasmids and fasciation caused by strains of *Corynebacterium fascians*. *Proceedings of the National Academy of Sciences, U.S.A.* **77**, 619–23.

MYERS D. F. & FRY W. E. (1978) Enzymatic release and metabolism of hydrogen cyanide by *Gloeocercospora sorghi*. *Phytopathology* **68**, 1717–22.

NATIONAL ACADEMY OF SCIENCE (1972) *Genetic Vulnerability of Major Crops*. Washington D.C.

NELSON N. (1944) A photometric adaptation of the Somogyi method for the determination of glucose. *Journal of Biological Chemistry* **153**, 375–80.

NIXON H. L. & FISHER H. L. (1958) An improved spray droplet technique for quantitative electron microscopy. *British Journal of Applied Physics* **9**, 68–70.

PADMANABHAN S. Y. (1973) The great Bengal famine. *Annual Review of Phytopathology* **11**, 11–26.

PARKER C. W. & LETHAM D. S. (1973) Regulation of cell division in plant tissues. XVI. Metabolism of zeatin by radish cotyledons and hypocotyls. *Planta* **114**, 199–218.

PEARCE R. B. & RIDE J. P. (1980) Specificity of induction of the lignification response in wounded wheat leaves. *Physiological Plant Pathology* **16**, 197–204.

PEGG G. F. (1976) Endogenous inhibitors in healthy and diseased plants. In: *Physiological Plant Pathology* (Eds R. Heitefuss & P. M. Williams), pp. 607–16. Springer-Verlag, Berlin.

PENGELLY W. & MEINS F. (1977) A specific radio immunoassay for nanogram quantities of the auxin indole-3-acetic acid. *Planta* **136**, 173–80.

PERSON C., SAMBORSKI D. J. & ROHRINGER R. (1962) The gene-for-gene concept. *Nature, London* **194**, 561–2.

PINKERTON F. & STROBEL G. A. (1976) Serinol as an activator of toxin production in attenuated cultures of *Helminthosporium sacchari*. *Proceedings of the National Academy of Sciences, U.S.A.* **73**, 4007–11.

PIRT S. J. (1975) *The Principles of Microbe and Cell Cultivation*. Blackwell Scientific Publications, Oxford.

RANDLES J. W., BOCCARDO G. & IMPERIAL J. S. (1980) Detection of the Cadang-Cadang associated RNA in African oil palm and Buri palm. *Phytopathology* **70**, 185–9.

REEVE D. R. & CROZIER A. (1978) The analysis of gibberellins by high performance liquid chromatography. In: *Isolation of Plant Growth Substances* (Ed.

J. R. Hillman), pp. 42–77. Cambridge University Press, Cambridge.

RIDE J. P. (1975) Lignification in wounded wheat leaves in response to fungi and its possible role in resistance. *Physiological Plant Pathology* **5**, 125–34.

RIDE J. P. (1978) The role of cell wall alterations in resistance to fungi. *Annals of Applied Biology* **89**, 302–6.

RIDE J. P. (1980) The effect of induced lignification on the resistance of wheat cell walls to fungal degradation. *Physiological Plant Pathology* **16**, 187–96.

RIDE J. P. & DRYSDALE R. B. (1972) A rapid method for the chemical estimation of filamentous fungi in plants. *Physiological Plant Pathology* **2**, 7–15.

RIDE J. P. & PEARCE R. B. (1979) Lignification and papilla formation at sites of attempted penetration of wheat leaves by non-pathogenic fungi. *Physiological Plant Pathology* **15**, 79–92.

ROSS A. F. (1961a) Localized acquired resistance to plant virus infection in hypersensitive hosts. *Virology* **14**, 329–39.

ROSS A. F. (1961b) Systemic acquired resistance induced by localized virus infections in plants. *Virology* **14**, 340–58.

RUDOLPH K. (1976) Non-specific toxins. In: *Physiological Plant Pathology* (Eds R. Heitefuss & P. H. Williams), pp. 270–315. Springer-Verlag, Berlin.

SAUNDERS P. F. (1978) The identification and quantitative analysis of abscisic acid in plant extracts. In: *Isolation of Plant Growth Substances* (Ed. J. R. Hillman), pp. 115–34. Cambridge University Press, Cambridge.

SCARBROUGH E., ARMSTRONG D. J., SKOOG F., FRIHART C. R. & LEONARD N. J. (1973) Isolation of *cis*- zeatin from *Corynebacterium fascians* cultures. *Proceedings of the National Academy of Sciences, U.S.A.* **70**, 3825–9.

SCHEFFER R. P. (1976) Host specific toxins in relation to pathogenesis and disease resistance. In: *Physiological Plant Pathology* (Eds R. Heitefuss & P. H. Williams), pp. 247–69. Springer-Verlag, Berlin.

SCHÖNBECK F. & SCHLÖSSER E. (1976) Preformed substances as potential protectants. In: *Physiological Plant Pathology* (Eds R. Heitefuss & P. H. Williams), pp. 653–78. Springer-Verlag, Berlin.

SEITZ L. M., SAUER D. B., BURROUGHS R., MOHR M. E. & HUBBARD J. D. (1979) Ergosterol as a measure of fungal growth. *Phytopathology* **69**, 1201–3.

SHARMA P. D., FISHER P. J. & WEBSTER J. (1977) Critique of the chitin assay technique for estimation of fungal biomass. *Transactions of the British Mycological Society* **69**, 479–83.

SHERWOOD R. T. & VANCE C. P. (1976) Histochemistry of papillae formed in reed canarygrass leaves in response to non-infecting pathogenic fungi. *Phytopathology* **66**, 503–10.

SILVERSTEIN R. M., BASSLEE G. C. & MORRILL T. C. (1974) *Spectrometric Identification of Organic Compounds*. John Wiley & Sons, New York.

SKIPP R. A. & BAILEY J. A. (1977) The fungitoxicity of isoflavanoid phytoalexins measured using different types of bioassay. *Physiological Plant Pathology* **11**, 101–12.

SKIPP R. A. & CARTER G. A. (1978) Adaptation of fungi to isoflavanoid phytoalexins. *Annals of Applied Biology* **89**, 366–9.

STEINER G. W. & BYTHER R. S. (1971) Partial characterization and use of a host–specific toxin from *Helminthosporium* on sugarcane. *Phytopathology* **61**, 691–5.

STEINER G. W. & STROBEL G. A. (1971) Helminthosporoside, a host specific toxin from *Helminthosporium sacchari*. *Journal of Biological Chemistry* **246**, 4350–7.

STOESSEL A., ROBINSON J. R., ROCK G. L. & WARD E. W. B. (1977) Metabolism of capsidiol by sweet pepper tissue; some possible implications for phytoalexin studies. *Phytopathology* **67**, 64–6.

STOWE B. B. & YAMAKI T. (1957) The history and physiology of the gibberellins. *Annual Review of Plant Physiology* **8**, 181–216.

STRANGE R. N. (1972) Plants under attack. *Science Progress (Oxford)* **60**, 365–85.

STRANGE R. N., DERAMO A. & SMITH H. (1978) Virulence enhancement of *Fusarium graminearum* by choline and betaine and of *Botrytis cinerea* by other constituents of wheat germ. *Transactions of the British Mycological Society* **70**, 201–7.

STRANGE R. N., MAJER J. R. & SMITH H. (1974) The isolation and identification of choline and betaine as the two major components in anthers and wheat germ that stimulate *Fusarium graminearum in vitro*. *Physiological Plant Pathology* **4**, 277–90.

STRANGE R. N. & SMITH H. (1971) A rapid assay of fungal growth stimulants. *Transactions of the British Mycological Society* **56**, 485–8.

STROBEL G. A. (1973) The helminthosporoside-binding protein of sugarcane, its properties and relationship to susceptibility to the eyespot disease. *Journal of Biological Chemistry* **248**, 1321–8.

STROBEL G. A. (1975) A mechanism of disease resistance in plants. *Scientific American* **232**, 80–9.

STROBEL G. A. (1977) Bacterial phytotoxins. *Annual Review of Microbiology* **31**, 205–24.

SWOBODA P. A. T. & LEA C. H. (1965) The flavour volatiles of fats and fat containing foods. II. A gas chromatographic investigation of volatile autoxidation products from sunflower oil. *Journal of the Science of Food and Agriculture* **16**, 680–9.

TATUM L. A. (1971) The southern corn leaf blight epidemic. *Science* **171**, 1113–16.

TEMMINK J. H. M., CAMPBELL R. N. & SMITH P. R. (1970) Specificity and site of *in vitro* acquisition of tobacco necrosis virus by zoospores of *Olpidium brassicae*. *Journal of General Virology* **9**, 201–13.

TEN HOUTEN J. G. (1974) Plant pathology; changing agricultural methods and human society. *Annual Review of Phytopathology* **12**, 1–11.

THAKUR R. P. & WILLIAMS R. J. (1980) Pollination effects on pearl millet ergot. *Phytopathology* **70**, 80–4.

TURNER E. M. C. (1961) An enzymic basis for pathogenic specificity in *Ophiobolus graminis*. *Journal of Experimental Botany* **12**, 169–75.

ULLSTRUP A. J. (1972) The impacts of the southern corn leaf blight epidemic of 1970–1971. *Annual Review of Phytopathology* **10**, 37–50.

VAN DEN ENDE G. & LINSKENS H. F. (1974) Cutinolytic enzymes in relation to pathogenesis. *Annual Review of Phytopathology* **12**, 247–58.

VANCE C. P. & SHERWOOD R. T. (1976) Regulation of lignin formation in reed canary grass in relation to disease resistance. *Plant Physiology* **57**, 915–19.

VANCE C. P. & SHERWOOD R. T. (1977) Lignified papilla formation as a mechanism for protection in reed canarygrass. *Physiological Plant Pathology* **10**, 247–56.

WALTON D., DASHEK W. & GALSON E. (1979) Radioimmunoassay for abscisic acid. *Planta* **146**, 139–45.

WARD T. M., WRIGHT M., ROBERTS J. A., SELF R. & OSBORNE D. J. (1978) Analytical procedures for the assay and identification of ethylene. In: *Isolation of Plant Growth Substances* (Ed. J. R. Hillman), pp. 135–51. Cambridge University Press, Cambridge.

WIDHOLM J. M. (1972) The use of fluorescent diacetate and phenosafranine for determining viability of cultured plant cells. *Stain Technology* **47**, 189–94.

WILLIAMS R. C. & BACKUS R. C. (1949) Macromolecular weights determined by direct particle counting. I. The weight of the bushy stunt virus particles. *Journal of the American Chemical Society* **71**, 4052–7.

WU L. & STAHMANN M. A. (1975) Chromatographic estimation of fungal mass in plant materials. *Phytopathology* **65**, 1032–4.

YOSHIKAWA M. (1978) Diverse modes of action of biotic and abiotic phytoalexin elicitors. *Nature, London* **275**, 546–7.

YOUNG R. E., PRATT H. K. & BIALE J. B. (1952) Manometric determination of low concentrations of ethylene. *Analytical Chemistry* **24**, 551–5.

Part 5
Microbe–Animal Interactions

Chapter 28 · Sampling the Skin Surface

William C. Noble

28.1 Introduction

Although the majority of studies of the skin surface flora have been carried out on man, there is sufficient information available to make this review generally applicable at least to the mammals. Before attempting to review the methods available for studying the skin flora it is necessary to consider the mode and site of growth of this microflora.

Mammalian skin is composed, at its surface, of a mosaic of dead epithelial cells, called squames or corneocytes, which are comparatively flat cells filled with keratin, a protein whose composition differs not only from species to species but between hair, nail and epithelial cells on the individual animals. The skin surface is not flat but has folds, protrusions and crevices which are large in relation to the microflora. This surface also bears appendages such as hairs which protrude through it in varying density. Hairs have associated sebaceous glands which supply a complex lipid mixture to the skin surface via the hair shaft. In some areas, such as the face of man, the hair becomes rudimentary and the sebaceous gland is the predominant organ. In man, eccrine sweat glands, which are thermoregulatory, secreting a very dilute salt solution, have separate openings to the skin surface. In cattle the eccrine ducts empty into the common sebaceous duct and hair follicle. This may result in water and lipid-soluble nutrients becoming available to microbes at the human skin surface separately but in cattle the secretion may appear as an emulsion which percolates throughout the stratum corneum (Lloyd 1978; Noble 1981).

The skin surface is subject to constant erosion. Squames are lost continually from the skin surface as the epidermis is renewed from below; grooming and washing occur in all species and in man the surface may be subject to the application of potentially toxic substances. However, in man neither continual washing nor total lack of washing has much effect on the skin flora which is regulated by the degree of humidity and rate of desquamation (Hartman 1978, 1979).

The skin surface in many, perhaps most, mammals has an acid pH; in man this about pH 5.5 but ranges from about pH 4.0 to 7.0 on different individuals with a normal distribution (Behrendt & Green 1971; Noble 1981). Similar values have been obtained in cattle (Jenkinson & Mabon 1973).

Some microorganisms, such as the aerobes, appear to reproduce at the skin surface, whilst the anaerobes are found in sebaceous ducts (Holland *et al.* 1974; Puhvel *et al.* 1975). Wolff and Plewig (1976) suggested that in man yeasts, such as *Pityrosporum* spp., are found in the mouths of sebaceous ducts, Micrococcaceae are found in the upper part of the duct and the anaerobic *Propionibacterium* spp. deep in the duct. There is general agreement that *Propionibacterium* spp. reproduce in the depths of follicles but many become deposited on the skin surface perhaps as a result of sebum secretion. If the skin is sampled by the sticky tape method, taking repeated samples from the same area, the majority of all organisms appear to be at, or very close to, the skin surface (Lloyd 1978; Noble 1981). Figure 1 shows that about 60% of the available colony-forming units occur on the first four samples from a site. This might be an artefact of sampling however (see below).

In cattle, Lloyd *et al.* (1979) found bacteria mostly associated with hair follicle infundibula or

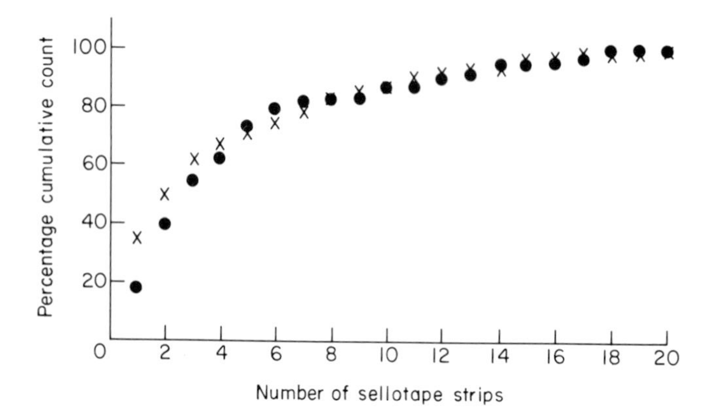

Figure 1. Cumulative percentage of organisms recovered on successive sellotape strips from two areas of the body. It can be seen that about 60% of the available colony forming units are recovered on the first four samples. ×, skin over tibia; ●, skin over scapula.

in areas of disorganised stratum corneum. They also found mixed aggregates of organisms attached to squames. About half the aggregates seen were of mixed cocci and pleomorphic rods while of the 55 studied 3 included yeasts. More than 75% of all the aggregates contained less than 100 cells. In skin sections bacteria are seen in aggregates or micro-colonies in both humans and cattle (Montes & Wilborn 1970; Lloyd 1978).

It is evident that microorganisms forming the skin flora of man and animals are found principally at the surface but also in the depths of the skin in loose squamous layers or in sebaceous follicles. The concept that organisms live in microcolonies has been referred to above. Microcolonies also occur on the skin of man (Holt 1971; Malcolm & Hughes 1980; Noble 1981) but there is no good agreement on their size. In a study of normal humans, Noble and Somerville (1974) reported that in males microcolonies of aerobic organisms gave about 2000 c.f.u. per colony on the forehead, 300 on the periumbilicus and 160 on the forearm. In females values of 600, 260 and 120, respectively, were found. In a later study of operating room personnel lower values were obtained which were of the order of 100 cells (Fig. 2). This may reflect the stricter hygienic precautions observed by operating room staff (Noble 1981) or may throw doubt on previous numbers.

The normal flora of human skin is distributed in an approximately $\log_{10}$ distribution. Figure 3 shows the distribution of 432 counts from skin areas with an approximately similar flora (chest, back, arm, thigh, shin and abdomen). Figure 4 shows the effect of considering only one taxon, however, when there are many sites at which no recoveries

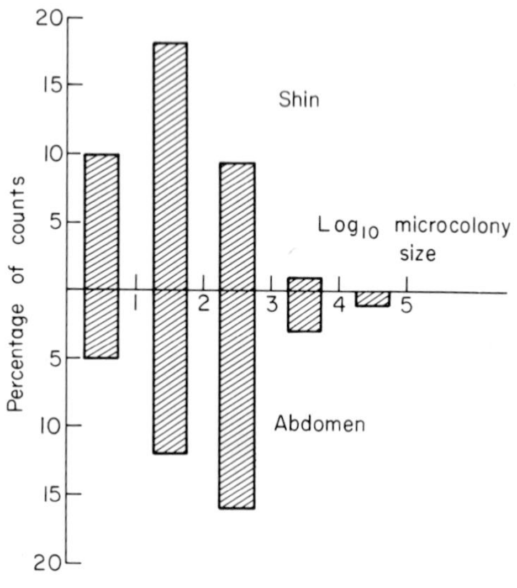

Figure 2. Microcolony size at two sites in 38 male surgical staff.

are made, although an approximate $\log_{10}$ normal distribution is obtained for the positive sites.

The variation in count can be further illustrated by a study of data compiled during investigations into the flora of the hands. A panel of 29 housewives was seen on seven occasions over a period of three months. On each occasion 4 cm^2 of the skin of the dorsum, the palm and the wrist of each woman was sampled by the scrub technique. The results were pooled to give a composite count for each occasion. The results varied from week to week, the standard deviation averaging 13% of the mean count of 12×10^3 viable units cm^{-2} with a range of 7 to 28%. Nevertheless the differences between panellists were significantly greater than those within panel-

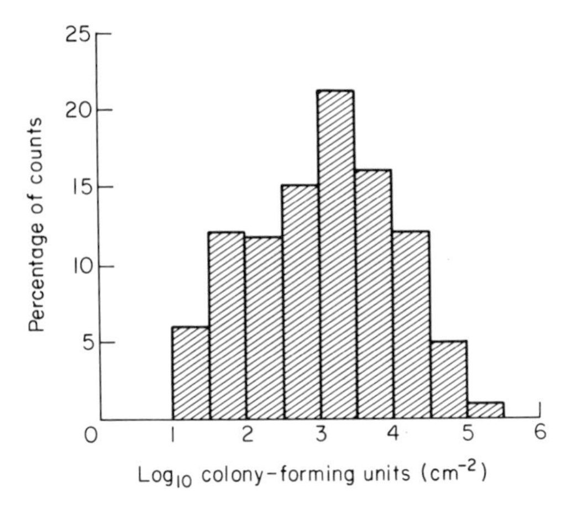

Figure 3. Distribution of total viable cells per square centimetre assessed by the scrub count technique. A total of 432 counts from chest, back, arm, thigh, shin and abdomen contributed to these counts.

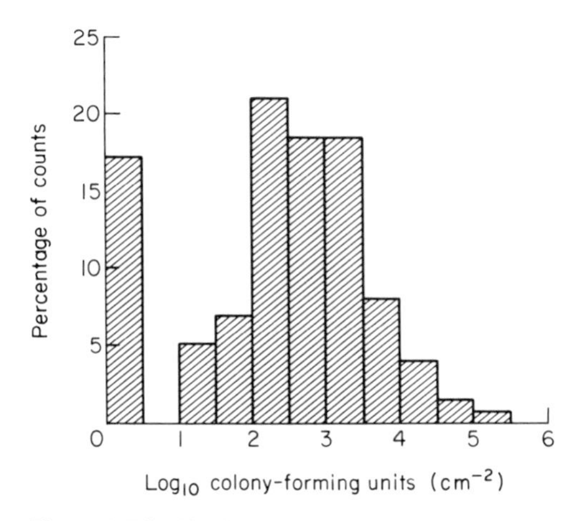

Figure 4. Distribution of total viable cells of Micrococcaceae in the same individuals as Fig. 3.

lists ($F=2.27$, $n_1=28$, $n_2=174$, $1\%>P>0.1\%$). This interhost variation has been a feature of most studies of human and animal skin.

There are consistent and statistically significant differences in the density of colonisation of human males and females. Figure 5 shows the difference in cumulative distribution between the skin flora of Micrococcaceae in adult males and females.

Humans are frequently classifiable into those with coccoid- or coryneform-dominated floras. This is probably a constant feature related in some unknown way to the nutritional status of individual skins. Humans also tend to be either high-count or low-count individuals. That is, persons with a low

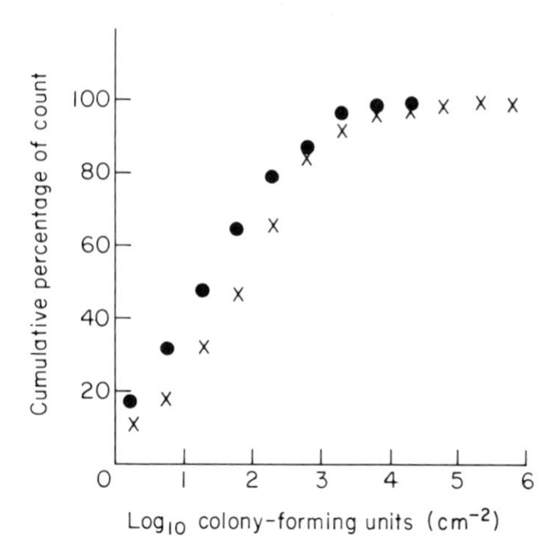

Figure 5. Cumulative distributions of counts of Micrococcaceae on the skin of 38 males (×) and 34 females (●). 50% of counts on females are above 20 c.f.u. cm^{-2} whilst in males 50% are above 80 c.f.u. cm^{-2}.

count on one skin area will tend to have low counts at other areas and the carrier status is maintained high or low over a period of years (Evans 1975). The reasons for this are unknown. There are insufficient studies of other mammals to be able to judge if this is true for other species.

Experimental

28.2 Contact methods

The simplest method of sampling the skin surface is to bring it into direct contact with a growth medium. This is commonly done by overfilling a container, such as a small Petri dish lid, with agar so that the meniscus is above the level of the rim. The surface of solid agar is pressed against the surface to be sampled. This method has the advantage of revealing the local geographical distribution of organisms but it is not known whether all colonies are sampled equally efficiently. Recent applications of this method include that by Brown *et al.* (1975). Modifications of this technique include incorporating a sterile bandage into the medium to strengthen it and to enable it to be everted from the Petri dish.

Sticky tape can be used to sample the surface and this has been particularly recommended for the study of surface infection with fungi (Keddie *et al.* 1961; Milne & Barnetson 1974; Lachappelle *et al.* 1977) since sticky tape removes a partial layer of stratum corneum. The presence of hair interferes with the method however.

The skin surface biopsy technique described by Marks and Dawber (1972) in which skin is attached to a microscope slide by exposing it to cyanoacrylate glue has been applied to studies of superficial fungal infection (Whiting & Bisset 1974). This is a convenient technique since a sample of the skin surface is needed to demonstrate the presence of mycelium.

All surface contact methods suffer the disadvantage that not only is it impossible to tell whether all aggregates have been sampled, but also the method does not distinguish between aggregates of 10×10^7 cells or those of 1×10^7 cells. This may account for the apparently conflicting concept that sebaceous gland ducts are packed with organisms but that most organisms appear to be situated at the skin surface. A small number of cells spread over the skin surface would give rise to several colonies on sticky tape or contact plate whilst a very large number of cells in a sebaceous duct would result in only one colony. Nevertheless the use of contact methods has been crucial to assessments of microcolony size.

28.3 Indirect sampling

For some purposes it will suffice to sample merely for the presence or absence of certain organisms; e.g. in looking for some types of infection where the pathogen is present in large numbers. A sterile massage brush rubbed through the hair and implanted on suitable agar is useful to detect dermatophyte or yeast infection in animals and man since large quantities of infected scales or even spores are usually present in fungal lesions (Goldberg 1963; MacKenzie 1963; Rosenthal & Wapnick 1963; Midgley & Clayton 1972; Noble & Midgley 1978). A roughly quantitative result can be achieved by noting how many colonies appear on the growth medium as this corresponds to the number of contaminated teeth on the massage brush. A variant of this technique is the multipoint sampler described by Wooley *et al.* (1970) which consists of 100 pins set in a 50 × 50 mm block. This has been used in a study of the flora of dogs. The brush technique has been used in studies of infection *versus* contamination with dermatophyte fungi in children. Samples yielding only small numbers of dermatophyte colonies came from children without clinical signs of infection.

Velvet pads may also be used (Holt 1966; Noble 1966). Raahave (1975) studied the efficiency of sampling by this method using organisms seeded onto an agar plate as a source. When the pad was used as a simple replica plate method the efficiency of transfer was about 60% from the original plate to the pad but only 4% from pad to secondary plate. If the pad was rinsed in saline and this used as the inoculum, efficiency rose to 66%. The efficiency of sampling from a natural habitat, such as skin, may be much less than this.

In clinical bacteriology a cotton swab is used to sample the flora; this is a very convenient but inefficient method of sampling. A cotton swab, even if moistened in broth before being rubbed over the skin surface, transfers only about 1% of the available flora to the agar plate (Noble & Somerville 1974). This does not matter when large quantities of a single pathogen are available but may be valueless in situations where pathogens are scarce as occurs with some lesions due to mycobacteria. The efficiency of the method can be increased by using soluble alginate swabs which are dissolved in buffer and can result in about 10% of the available flora being made accessible for study (Higgins 1950; Cain & Steele 1953). Soluble swabs are useful in giving a semi-quantitative count. Noble and Somerville (1974) reported that soluble swab gave about eight times more organisms per square centimetre than did contact plates. Comparisons between individuals are still less than optimal by this method but the relative frequency of different species in the same sample can be studied. The merits of cotton *versus* alginate or rayon swabs probably depend on the nature of the surface being sampled. Skin poses a challenge different from that of, for example, a stainless steel work surface (Angelotti *et al.* 1964; Notermans *et al.* 1976).

28.4 Quantitative methods

All quantitative methods are derived from the cup-scrub method of Williamson (1965) in which a cylinder of glass, stainless steel or teflon is pressed against the skin surface and partially filled with a fluid containing a detergent. The skin area circumscribed by the cylinder is rubbed with a rod releasing bacteria and squames into the sample fluid. After a standard interval, usually 1 min, the fluid is withdrawn from the cup and processed. It is frequently necessary to dilute the fluid before it is inoculated onto agar so that single colonies may be obtained. Counts obtained may range from 1×10^2 c.f.u. cm^{-2} on the leg of a man to 1×10^7 c.f.u. cm^{-2} on the forehead. Counts of 1×10^4 c.f.u. cm^{-2} are common on cattle. This is generally accepted as the best method of sampling for ecological study and a more detailed appraisal will be given of this method rather than other, non-quantitative methods.

From the original studies of Williamson and Kligman (1965) it was found that about 85% of the available aerobic flora was removed in a single 1 min scrub. However, since anaerobes live deep in follicles, Aly *et al.* (1978) found that only about 10% of the anaerobes were collected by this procedure. The detergents most frequently used in the cup-scrub method are Tween 80 (polyoxyethylene sorbitan monooleate) and Triton X100 (*p*-tert-octyl phenoxypolyethoxyethanol) although both these compounds are slightly toxic to some bacteria. Streptococci are particularly susceptible to the detergents. Noble and Somerville (1974) referred to this problem and Bloom *et al.* (1979) recommended that no more than 30 min should elapse between sampling and inoculation on agar. Detergents are used in an attempt to separate the individual cells in the microcolony and hence obtain an accurate viable cell count. However, Lloyd (1978) showed that even after shaking samples with glass beads some squames retained large numbers of bacteria. Prolonged shaking with beads reduced the viability of the organisms.

A number of workers have introduced mechanical modifications of this method without greatly increasing its accuracy and the reasons for this may lie in the ecology rather than methods (see below). Stringer and Marples (1976) described the use of an ultrasonic probe in place of the rod in studies of the human skin flora. Fewer corneocytes were removed by the ultrasonic probe than by conventional methods although the harvest of bacteria was the same. Only aerobic bacteria were studied, however, since it proved impossible to use the ultrasonic probe on the forehead which is the major source of anaerobic bacteria on man. An alternative approach was investigated by Staal and Noordzij (1978) who used a pulsed water spray intended originally for cleaning teeth. Using a 2 mm spray head recovery efficiencies of about 89% were obtained using 100 ml sample fluid, although this large volume would be a disadvantage in practice. Using a 4 mm spray head, however, the recovery efficiency was only 25% even using 100 ml sample fluid. Bibel and Lovell (1976) attempted to standardise the scrubbing technique by using a weighted scraper in a square container.

A particular modification that seeks to sample only the sebaceous follicles was described by Holland *et al.* (1974). A disc is stuck to the skin using cyanoacrylate glue which, when removed, takes some of the skin surface adhering to the disc. This is known as the skin surface biopsy method. The sebaceous follicles project downwards from the skin sample and their bacteria can be studied by agitating the disc in sampling fluid. This method has the merit of immobilising the surface flora in the glue thus enabling the follicular flora alone to be examined.

In man it is generally only practicable to sample about 4 cm^2 of skin surface at a time, since the sample cup must fit tightly to the skin to prevent leakage. This comparatively small sample area leads to a large variation in microbial count even between adjacent sites on a single individual. This is, of course, a reflection of the presence or absence of the microcolonies described earlier. Table 1 shows the results, expressed as $\log_{10}$ colony-forming-units cm^{-2}, obtained by taking two adjacent samples from each forearm of twelve individuals. The correlation coefficient between the paired samples is 0.76 with the mean standard deviation equal to 57% of the mean count. By pooling samples and comparing the left and right arms, however,

Table 1. Variation in microbial flora on the human forearm. A comparison of paired samples from each arm in twelve individuals. Counts expressed as $\log_{10}$ c.f.u. cm^{-2} (W. C. Noble, unpublished observations).

	Right arm		Left arm	
Volunteer	Sample A	Sample B	Sample A	Sample B
1	1.62	1.93	1.56	1.82
2	1.77	0.70	1.30	1.26
3	3.36	3.19	2.74	3.70
4	1.41	1.38	1.81	1.68
5	0.95	0.30	0.01	1.67
6	1.77	2.02	1.98	2.20
7	2.00	2.06	2.02	2.04
8	2.45	2.15	1.04	1.15
9	0.70	0.01	1.97	0.01
10	1.85	2.10	2.06	1.98
11	1.15	1.08	2.10	1.68
12	3.80	3.98	3.26	2.19

the correlation coefficient rises to 0.86 and the mean standard deviation drops to 42% of the mean count. The uneven distribution of organisms on the skin surface drew Keith *et al.* (1979) to remark that determining the bacterial flora at one site does not lead to an accurate prediction of the numbers at an adjacent site, a factor of importance in skin degerming studies.

Wilson (1970) has shown that the method used to sample the skin flora in a study of antibacterial soaps can determine the result obtained. When contact methods were used, there was no apparent difference in the levels of skin flora when respondants used either a germicidal soap or a control soap. However, if the respondants were sampled on the same occasion by the cup-scrub method, the germicidal soap could be shown to reduce the flora by more than 95%, i.e. a significant reduction had been achieved. The explanation for this lies in the microcolony formation discussed earlier. The effect of antibacterial substances is to reduce the size, but not the number of microcolonies, i.e. the total viable cells are reduced in number, but not the aggregates in which they live. There is an impression that the skin has permanent sites of colonisation but this has not been substantiated.

It might seem that biopsy, the surgical removal of a piece of skin, would be the most satisfactory way of obtaining an estimate of the skin flora. Biopsy has, however, several disadvantages:

(1) it cannot easily be used in man;

(2) large biopsies cannot be easily obtained from the living animal (and the smaller the piece of tissue, the greater the inherent variation);

(3) repeat samples cannot be obtained;

(4) the skin is a tough tissue which is not easily ground to a fine enough suspension to release all organisms.

Nevertheless, several workers have used skin biopsy in humans and in experimental burn infections of guinea-pigs. Selwyn and Ellis (1972) obtained very high counts in biopsy specimens from cadavers yet Loebl *et al.* (1974) regarded more than 1×10^4 c.f.u. g^{-1} as evidence of infection in tissue from burned humans. Mustakallio *et al.* (1967) used full thickness epidermal suction blister samples which have the advantage that they do not cause scarring but this method can only be applied to flat, relatively hairless skin areas.

One problem which none of the sampling techniques can truly overcome is that of distinguishing between resident and transient floras. Price (1938) introduced the concept of resident and transient flora in which the resident flora is understood to be occupying a permanent or semi-permanent niche in the habitat and to be multiplying there, whilst the transient flora is a contaminant flora acquired by contact with the environment. Attempts to assess resident status have been made by quantitative methods: the presence of large numbers of cells implying that the organism is resident. Also by sampling repeatedly over a time period the continuing presence of an organism indicates that it is a resident rather than a transient. Both methods have disadvantages, the most obvious being that small numbers of cells may only indicate a low rate of cell division and that the continued presence of an organism may merely reflect continued contamination from the separate source. Evans and Stevens (1976) suggested that it was possible to distinguish between surface and subsurface bacteria by using both swab and cup-scrub methods. Certainly this method appears to work for anaerobic bacteria.

28.5 Conclusions

It is evident then that our concept of the skin flora is intimately bound up with the results of our sampling methods. The skin of man and animals is inhabited by a variety of microbial species which may be distributed at different sites on the skin surface or subsurface in follicles. This flora exists in microcolonies which are the reproductive units but single cells may be spread out over the skin surface. Mixed microcolonies may occur but these can only be seen by microscopic examination. The size of the microcolonies can only be determined by comparing counts which are believed to represent measures of either all the individual viable cells or all the viable aggregates of cells. Since the same area of skin cannot be sampled by two methods, the between-sample variation illustrated in Table 1 would occur even if the sampling methods were perfect, which is not the case. We must accept a vary large measure of error in estimates of microcolony size. The number of organisms inhabiting the skin varies between sites, between hosts and as a result of local climatic conditions.

Non-invasive sampling methods which reveal the distribution and nature of resident and of transient organisms are sorely needed although it is difficult to envisage how they might be devised.

Recommended reading

Aly R. & Maibach H. I. (1981) *Skin Microbiology: its Relevance to Infection.* Springer-Verlag, Heidelberg.

Jenkinson D. McE. (1980) Surface ecosystems and interactions with them which overcome skin defence mechanisms. *Royal Society of Edinburgh* **79B**, parts 1 to 3.

Marples M. J. (1965) *Ecology of Human Skin.* Charles C. Thomas, Springfield, Illinois.

Noble W. C. (1981) *Microbiology of Human Skin.* Lloyd-Luke, London.

References

Aly R. & Maibach H. I. (1981) *Skin Microbiology: its Relevance to Infection.* Springer-Verlag, Heidelberg.

Aly R., Maibach H. I. & Bloom E. (1978) Quantification of anaerobic diphtheroids on the skin. *Acta Dermatovenerologica (Stockholm)* **58**, 501–4.

Angelotti R., Wilson J. L., Litsky W. & Walter W. G. (1964) Comparative evaluation of the cotton swab and Rodac methods for the recovery of *Bacillus subtilis* spore contamination from stainless steel surfaces. *Health Laboratory Science* **1**, 289–96.

Behrendt H. & Green M. (1971) *Patterns of Skin pH from Birth through Adolescence.* Charles C. Thomas, Springfield, Illinois.

Bibel D. J. & Lovell D. J. (1976) Skin flora maps: a tool in the study of cutaneous ecology. *Journal of Investigative Dermatology* **67**, 265–9.

Bloom E., Aly R. & Maibach H. I. (1979) Quantitation of skin bacteria: lethality of the wash solution used to remove bacteria. *Acta Dermatovenereologica (Stockholm)* **59**, 460–3.

Brown J., Wannamaker L. W. & Ferrieri P. (1975) Enumeration of β haemolytic streptococci on normal skin by direct agar contact. *Journal of Medical Microbiology* **8**, 503–12.

Cain R. M. & Steele H. (1953). The use of calcium alginate soluble wool for the examination of cleaned eating utensils. *Canadian Journal of Public Health* **44**, 464–7.

Evans C. A. (1975) Persistent individual differences in the bacterial flora of the skin of the forehead: numbers of Propionibacteria. *Journal of Investigative Dermatology* **64**, 42–6.

Evans C. A. & Stevens R. J. (1976) Differential quantitation of surface and subsurface bacteria of normal skin by the combined use of the cotton swab and the scrub methods. *Journal of Clinical Microbiology* **3**, 576–81.

Goldberg H. C. (1963) 'Brush' technique in animals. *Archives of Dermatology* **92**, 103.

Hartmann A. A. (1978) Restriction of washing and its effect on the normal skin flora. *Archives of Dermatological Research* **263**, 105–14.

Hartmann A. A. (1979) Daily bath and its effect on the normal human skin flora. *Archives of Dermatological Research* **265**, 153–64.

Higgins M. (1950) A comparison of the recovery rate of organisms from cotton wool and calcium alginate wool swabs. *Monthly Bulletin of the Ministry of Health and Public Health Laboratory Service* **9**, 50–1.

Holland K. T., Roberts C. D., Cunliffe W. J. & Williams M. (1974) A technique for sampling micro-organisms from the pilosebaceous ducts. *Journal of Applied Bacteriology* **37**, 289–96.

Holt R. J. (1966) Pad culture studies on skin surfaces. *Journal of Applied Bacteriology* **29**, 625–30.

Holt R. J. (1971) Aerobic bacterial counts on human skin after bathing. *Journal of Medical Microbiology* **4**, 319–27.

Jenkinson D. McE. (1980) Surface ecosystems and interactions with them which overcome skin defence mechanisms. *Royal Society of Edinburgh* **79B**, parts 1 to 3.

Jenkinson D. McE. & Mabon R. M. (1973) The effect of temperature and humidity on skin surface pH and the ionic composition of skin secretions in Ayrshire cattle. *British Veterinary Journal* **129**, 282–95.

Keddie F., Orr A. & Liebes D. (1961) Direct plating on vinyl plastic tape, demonstration of the cutaneous flora of the epidermis by the strip method. *Sabouraudia* **1**, 108–11.

Keith W. A., Smiljanic R. J., Akers W. A. & Keith L. W. (1979) Uneven distribution of aerobic mesophilic bacteria on human skin. *Applied and Environmental Microbiology* **37**, 345–7.

Lachappelle J. M., Gouverneur J. C., Boulet M. & Tennstedt D. (1977) A modified technique (using polyester tape) of skin surface biopsy. Its interest for the investigation of athlete's foot. *British Journal of Dermatology* **97**, 49–52.

Lloyd D. H. (1978) The effect of climate on the microbial ecology of the skin of cattle and sheep. Ph.D thesis, University of Glasgow.

Lloyd D. H., Dick W. D. B. & Jenkinson D. McE. (1979) Location of the microflora in the skin of cattle. *British Veterinary Journal* **135**, 519–26.

Loebl E. C., Marvin J. A., Heck E. L., Curreri P. W. & Baxter C. R. (1974) The use of quantitative biopsy cultures in bacteriologic monitoring of burn patients. *Journal of Surgical Research* **16**, 1–5.

MacKenzie D. W. R. (1963) 'Hairbrush diagnosis' in detection and evaluation of non-fluorescent scalp ringworm. *British Medical Journal* **ii**, 363–5.

Malcolm S. A. & Hughes T. C. (1980) The demonstration of bacteria on and within the stratum corneum using scanning electron microscopy. *British Journal of Dermatology* **102**, 267–75.

Marks R. & Dawber R. P. R. (1972) In situ microbiology

of the stratum corneum. An application of skin surface biopsy. *Archives of Dermatology* **105**, 216–221.

MARPLES M. J. (1965) *Ecology of Human Skin*. Charles C. Thomas, Springfield, Illinois.

MIDGLEY G. & CLAYTON Y. M. (1972) Distribution of dermatophytes and Candida spores in the environment. *British Journal of Dermatology* (Supplement 8) **86**, 69–77.

MILNE J. J. R. & BARNETSON R. ST. C. (1974) Diagnosis of dermatophytoses using vinyl adhesive tape. *Sabouraudia* **12**, 162–5.

MONTES L. F. & WILBORN W. H. (1970) Anatomical location of normal skin flora. *Archives of Dermatology* **101**, 145–59.

MUSTAKALLIO K. K., SALO O. P., KIISTALA R. & KIISTALA U. (1967) Counting the numbers of aerobic bacteria in full thickness human epidermis separated by suction. *Acta Pathologica et Microbiologica Scandinavica* **69**, 477–8.

NOBLE W. C. (1966) *Staphylococcus aureus* on the hair. *Journal of Clinical Pathology* **19**, 570–2.

NOBLE W. C. (1981) *Microbiology of Human Skin*. Lloyd-Luke, London.

NOBLE W. C. & MIDGLEY G. (1978) Scalp carriage of *Pityrosporum* species: the effect of physiological maturity, sex and race. *Sabouraudia* **16**, 229–32.

NOBLE W. C. & SOMERVILLE D. A. (1974) *Microbiology of Human Skin*. W. B. Saunders, London.

NOTERMANS S., HINDLE V. & KAMPELMACHER E. H. (1976) Comparison of cotton swabs versus alginate swabs sampling method in the bacteriological examination of broiler chickens. *Journal of Hygiene (Cambridge)* **77**, 205–10.

PRICE P. B. (1938) The bacteriology of normal skin. A new quantitative test applied to a study of the bacterial flora and the disinfectant action of mechanical cleansing. *Journal of Infectious Diseases* **63**, 301–18.

PUHVEL S. M., REISNER R. M. & AMIRIAN D. A. (1975) Quantification of bacteria in isolated pilosebaceous follicles in normal skin. *Journal of Investigative Dermatology* **65**, 525–31.

RAAHAVE D. (1975) Experimental evaluation of the velvet pad rinse technique as a microbiological sampling method. *Acta Pathologica et Microbiologica Scandinavica B* **83**, 416–24.

ROSENTHAL S. A. & WAPNICK H. (1963) The value of Mackenzie's 'Hair Brush' technique in the isolation of *T. mentagrophytes* from clinically normal guinea-pigs. *Journal of Investigative Dermatology* **41**, 5–6.

SELWYN S. & ELLIS H. (1972) Skin bacteria and skin disinfection reconsidered. *British Medical Journal* **i**, 136–40.

STAAL E. M. & NOORDZIJ A. C. (1978) A new method for the quantitative determination of microorganisms from the skin. *Journal of Society of Cosmetic Chemists* **29**, 607–15.

STRINGER M. F. & MARPLES R. R. (1976) Ultrasonic methods for sampling human skin microorganisms. *British Journal of Dermatology* **94**, 551–5.

WHITING D. A. & BISSET E. A. (1974) The investigation of superficial fungal infections by skin surface biopsy. *British Journal of Dermatology* **91**, 57–65.

WILLIAMSON P. (1965) Quantitative estimation of cutaneous bacteria. In: *Skin Bacteria and their Role in Infection* (Eds H. I. Maibach & G. Hildick-Smith), pp. 3–11. McGraw-Hill, New York.

WILLIAMSON P. & KLIGMAN A. M. (1965) A new method for the quantitative investigation of cutaneous bacteria. *Journal of Investigative Dermatology* **45**, 498–503.

WILSON P. E. (1970) A comparison of methods for assessing the value of antibacterial soaps. *Journal of Applied Bacteriology* **33**, 574–81.

WOLFF H. H. & PLEWIG G. (1976) Ultrastruktur der mikroflora in follikeln und komedonen. *Der Hautarzt* **27**, 432–40.

WOOLEY R. E., HINN J. A. & GRATZEK J. B. (1970) Efficacy of two topical ointments in reducing canine cutaneous flora. *Veterinary Medicine* **65**, 128–30.

Chapter 29 • Microbe–Animal Interactions in the Rumen

Colin G. Orpin

29.1 Digestion in the ruminant

The majority of the organic material in land plants is present as insoluble carbohydrates which are the principal components of the cell walls (Rogers & Perkins 1968). These carbohydrates are mainly cellulose and hemicelluloses and in growing grasses, which constitute the bulk of the diet of many ruminants, cellulose constitutes up to 42% and hemicelluloses up to 51% of the dry weight (Rogers & Perkins 1968). No herbivorous mammals have the ability to digest these polysaccharides, but many prokaryotic and eukaryotic microorganisms synthesise enzyme complexes capable of degrading cellulose and hemicelluloses. These enzymes cleave the polysaccharides into smaller oligosaccharides and eventually disaccharides and monosaccharides which are metabolised by the microorganisms. All mammals that are primarily herbivorous have evolved a system whereby they can digest cellulose and hemicelluloses by means of a symbotic association with microorganisms capable of digesting these polymers. The microorganisms are housed in a modified portion of the alimentary canal, either anterior to the abomasum (foregut fermentation) as in ruminants, colobid monkeys, sloths, hippopotami, macropod marsupials and camels, or posterior (hindgut fermentation) as in the Equidae, elephants, some rodents and lagomorphs. Bauchop (1977) and McBee (1977) have published reviews of foregut fermentation and hindgut fermentation, respectively.

In the ruminant, the microbial fermentation occurs in a series of sacs derived from the embryonic cardiac region of the stomach, called the rumen, reticulum and omasum. In the adult ruminant, the rumen—the largest of the three sacs—may comprise 80% of the total stomach volume and usually has a volume of 6 to 10 l in sheep and 100 to 130 l in cattle (Annison & Lewis 1959). This large organ increases the transit time of the dietary material through the alimentary canal, allowing the rather slow microbial fermentation of the plant polymers to proceed. Much information on the physiology and structure of the ruminant alimentary canal can be found in Hungate (1966).

The rumen microorganisms ferment the dietary material in order to obtain energy and carbon for their own growth and, because of the complexity of the diet, a range of fermentation and other reactions occurs. Carbohydrates are fermented, lipids and proteins hydrolysed and some of the products utilised by the microorganisms. Some amino acids including those essential to the host are deaminated. The net result is that little soluble carbohydrate or soluble protein enters the abomasum. However, vitamins synthesised by the microorganisms are made available to the host animal, toxic dietary constituents may be detoxified and volatile fatty acids are generated during the fermentation. The volatile fatty acids are chiefly acetate, propionate and butyrate which are absorbed by the host animal through the rumen epithelium (Pfander & Phillipson 1953) and used as major energy sources. The acetate alone supplies 50 to 60% of the animals energy requirements (Hungate 1966). Propionate is used for gluconeogenesis, accounting for some 50% of the glucose used by the ruminant, and butyrate may also provide energy by conversion to acetyl-CoA in the liver.

Other products of fermentation—carbon dioxide and methane—are lost from the animal by eructation. The microbial cells produced at the expense of the digesta are digested by the host animal as they pass through the abomasum and small intestine.

An important part of the digestion of plant tissues by the ruminant is the effect of chewing and ruminating upon the physical structure of the plant tissue. Virtually all ingested food is subjected to chewing although at high rates of intake little chewing may occur during the actual feeding. However, by ruminating, i.e. regurgitating coarse, less dense material from the rumen (Evans *et al.* 1973) followed by repeated rechewing and swallowing, considerable physical breakdown of the plant tissue can be induced (Reid *et al.* 1979). Chewing releases cell contents from the diet and it assists the microbial breakdown of the tissues in the rumen by providing access for the microorganisms.

Ingested food is mixed with saliva in the mouth before swallowing. The saliva contains high levels of sodium and bicarbonate ions and lower levels of phosphate, chloride, potassium and calcium ions (McDougall 1948; Phillipson & Mangan 1959), urea and mucoproteins. The total salivary excretion in cattle may be as much as $190\ l\ d^{-1}$ (Bailey 1961) and the bicarbonate content may exceed the equivalent of 300 g $NaHCO_3$ d^{-1} entering the bovine rumen. The function of the bicarbonate is to buffer the rumen pH at or slightly below pH 7.0; the phosphate neutralises the acids produced during the microbial fermentation (Counotte *et al.* 1979).

After each bite, the tongue shifts the food to the back of the mouth and, after several bites, the accumulated food is swallowed *en masse* as a bolus which is carried to the rumen via the oesophagus. The average weight of a bolus in cattle is about 100 g, including the saliva (Hungate 1966). The bolus travels down the oesophagus by peristaltic contractions of the oesophageal wall and is ejected into the mass of digesta in the rumen. Rhythmic muscular contractions of the rumen wall mix the bolus with the rumen contents where the plant particles are inoculated with the microorganisms present.

Uniformity of distribution of particles is rarely obtained, and stratification occurs with the lighter, larger particles in the upper layers in the dorsal sac of the rumen and smaller denser particles in the ventral sac and reticulum (Evans *et al.* 1973). When the animal is fed on a low-roughage, high concentrate diet, stratification is at a minimum and may be absent (Balch *et al.* 1955).

The rumen may be considered as a continuous culture microbial system, with regular inputs of nutrients, removal of waste products and an overflow system which passes the digesta to the abomasum.

29.2 The rumen environment

The physical and chemical conditions prevailing in the rumen vary considerably. Parameters affecting these conditions include the quantity and quality of the diet, the frequency of feeding, the time after feeding and the health of the animal. Extensive information regarding the physical and chemical conditions in the rumen may be found in Hungate (1966) and Clarke (1977). The rumen is a warm, anaerobic, chemically reducing environment rich in organic matter but often deficient in readily metabolisable compounds. From the microbiological standpoint it may be considered as three interconnecting environments: the liquid phase (the rumen fluid proper); the solid phase (the digesta) and the rumen epithelium. Microbial movement occurs between all three environments. Overlying the rumen contents is a gaseous phase consisting chiefly of carbon dioxide, nitrogen and methane, but which contains low levels of hydrogen sulphide and oxygen (McArthur & Miltimore 1961).

The liquid phase of the rumen fluid has been examined more extensively both microbiologically and chemically than the solid phase or the rumen epithelium. However, it is in the digesta that the most significant reactions, the digestion of cellulose and hemicellulose, occur. Many rumen bacteria are associated with the digesta (Akin & Amos 1975; Cheng *et al.* 1977) as are phycomycete fungi (Orpin 1977a, b; Bauchop 1979a) and, at times, a significant proportion of the ciliate protozoon population (Orpin 1979a; 1980b). The bacteria associated with the digesta may be more than twice as numerous as in the fluid phase (Warner 1962) and about half of the total bacterial population (Weller *et al.* 1958).

The rumen epithelium supports a population of adherent bacteria (Bauchop *et al.* 1975; McCowan *et al.* 1978; Cheng *et al.* 1979b). Many of the bacterial isolates made from the epithelium have been shown to be ureolytic, facultative anaerobes (Wallace *et al.* 1979). This population appears to be independent of the rumen fluid bacterial population and may be important in controlling the transport of urea into the rumen (Cheng & Wallace 1979). The composition of the normal epithelial bacterial populations varies with their position on the rumen wall (Bauchop *et al.* 1975; McCowan *et al.* 1978) where they are sandwiched between the highly anaerobic rumen fluid and the aerobic epithelial cells.

Although rumen bacteria are in contact with the rumen epithelium, antibody production to rumen bacteria occurs not in the rumen epithelium but in the wall of the caecum and possibly in that of the colon and small intestine (Sharpe *et al.* 1975). There is little evidence that rumen microorganisms *per se* are pathogenic to the host and indeed they may play a significant role in protecting the host against pathogens by the stimulation of the production of cross-reacting antigens.

29.3 Rumen microbiology

The microbial population of the rumen is complex, containing many different types of interacting prokaryotic and eukaryotic microorganisms, including bacteria (Hungate 1966), ciliate protozoa (Clarke 1977), flagellate protozoa (Jensen & Hammond 1964), phycomycete fungi (Orpin 1975, 1976a, 1977b; Bauchop 1979b), amoebae (Braune 1914) and bacteriophages (Adams *et al.* 1966; Hoogenraad *et al.* 1967; Orpin & Munn 1974). The majority of the dominant microorganisms are obligate anaerobes found in no other habitats, but facultatively anaerobic bacteria are present and may play a significant role in reactions on the rumen epithelium (Wallace *et al.* 1979) and may reach transient high levels of population density (Hungate 1966). Possibly their major role in rumen ecology is their ability to scavenge oxygen from the system which might otherwise limit the growth of the obligate anaerobes.

Because of the association of microorganisms with the digesta and rumen epithelium it is difficult to enumerate these organisms, and the majority of estimates have been made using filtered rumen contents. The number of all rumen microorganisms varies from animal to animal and fluctuates with time after feeding (Warner 1966), the diet and the health of the animal (Hungate 1966). The bacteria are normally present at 1×10^{10} to 1×10^{11} ml^{-1}, the ciliate protozoa at 1×10^4 to 1×10^6 ml^{-1} and flagellate protozoa at 1×10^3 to 1×10^5 ml^{-1}. The population densities of other rumen microorganisms (phycomycetes, amoebae, bacteriophages) are still to be ascertained. Because the numbers and the state of nutrition of the microorganisms varies with time, so does their nutritive value to the host (Czerkawski 1976). A review of the physiology of rumen microorganisms can be found in Hungate (1966) and their biochemical activities in Prins (1977).

Many interrelationships exist between rumen microorganisms, particularly relating to the supply of primary nutrients, the supply of growth factors and predation by one species on another. Readily metabolisable plant constituents such as soluble sugars and proteins disappear from the digesta shortly after it enters the rumen (Hungate 1966) leaving the plant cell walls as the major component of the digesta. Only a few rumen organisms are cellulolytic and many of the other organisms rely upon the release of nutrients or fermentation products from plant cell walls by the organisms capable of digesting these tissues. Extracellular hemicellulases are probably responsible for the release of many soluble carbohydrates into the rumen fluid, but the concentration of cellulase in the liquid phase of rumen fluid is low. The cellulolytic rumen bacteria are unlikely to release significant quantities of soluble carbohydrates into the rumen liquor because of their close attachment to the tissues they are digesting but they do produce succinate, acetate, formate and carbon dioxide (*Bacteroides succinogenes*) and hydrogen (*Ruminococcus flavefaciens*) or acetate, ethanol, hydrogen and carbon dioxide (*Ruminococcus albus*) as fermentation products. The succinate can be used for an energy source *in vitro* by *Selenomonas ruminantum* and the formate and carbon dioxide can be utilised for energy production by methanogens (Wolin 1975; Latham & Wolin 1977; Prins 1977). Perhaps the most closely linked system so far identified is the fermentation of benzoate to methane by an inseparable mixture of bacteria (Evans 1977). No specific interrelations of this type have been demonstrated *in vivo* because of the melée of microorganisms but no doubt the principle is operative.

Antagonistic relationships occur, with predation of bacteria (Coleman 1975a, b), and some ciliates (Coleman 1978a, b) and phycomycete zoospores (Orpin 1975) by ciliates. Other interactions which are known to occur, but the significance of which has yet to be determined, include the apparent parasitism of ciliates by chitrid fungi (Lubinsky 1955a, b), the destruction of bacteria by bacteriophages and the dominance of certain ciliate populations over others (Eadie 1962b).

29.4 Fate of plant material in the rumen

Active invasion of freshly ingested plant particles occurs by both ciliate protozoa (Orpin 1979a, 1982) and motile rumen bacteria (Orpin 1980). The holotrich protozoa show the greatest taxis, responding chemotactically to soluble carbohydrates diffusing from the tissues. *Isotricha intestinalis* and *I. prostoma* possess a specialised region on the anterior dorsilateral cell surface which is used to attach the cells to surfaces, provided soluble carbohydrates and soluble protein are present (Orpin & Hall 1977; Orpin & Letcher 1978). These conditions occur in the bolus (Reid *et al.* 1962) and attachment has been demonstrated *in vivo*. Attachment by the isotrichs is flexible and at times transient and may be followed by insinuation of the cells into crevices in the plant tissue and probably functions to maintain the cells near the source of soluble carbohydrates which are their major carbon sources (Hungate 1966).

Many species of oligotrich protozoa also associate with freshly ingested plant tissues (Bauchop & Clarke 1976; Orpin 1979a, 1982) after taxis to diffusing soluble carbohydrates (Orpin 1979b). *Epidinium ecaudatum,* attached itself by its oral cavity to these tissues (Bauchop 1979c) as does *Eudiplodinium maggii* (C. G. Orpin, unpublished observations). Both *Epidinium ecaudatum* and *Eudiplodinium maggii* are capable of digesting cellulose and hemicellulose (Hungate 1942, 1943; Coleman 1979). Rumen ciliates dependant upon starch, such as, *Entodinium* spp., *Ophryoscolex caudatus* and *Polyplastron multivesiculatum* (Coleman 1979), showed less association with dietary material at low particle densities.

Association of the ciliates with plant particles occurs within 5 to 30 min of the material entering the rumen, the speed depending on the protozoon species and the nature of the diet (Orpin 1982). The ciliate population involved in these transfers varies with the composition of the population and the diet but may reach 46% of the population with entodinia, which show the least association of the ciliates examined, and up to 96% with *Isotricha intestinalis* and *Ophryoscolex caudatus* (Orpin 1982). This results in a decrease in the expected contribution of the ciliates to the outflow of microbial nitrogen from the rumen (Weller & Pilgrim 1974), since the ciliates behave as though they are part of the particulate phase of the rumen fluid.

Motile rumen bacteria, such as *Butyrivibrio fibrisolvens*, *Lachnospira multiparus* and *Selenomonas ruminantium*, associate with hay particles *in vitro* with great rapidity (Orpin 1980). In these species, 70 to 95% of the population in the range 1×10^8 to 1×10^{10} ml^{-1} became associated with the particles within 10 min of incubation. *Lachnospira multiparus* alone is capable of causing extensive fragmentation of tissues (Cheng *et al.* 1979a). Association of the non-motile adherent cellulolytic bacteria *Ruminococcus flavefasciens* and *Bacteroides succinogenes* occurs at a slower rate, mostly with cut edges of cell walls (Latham *et al.* 1978b).

Adhesion of mixed rumen bacterial populations to plant tissues has been examined extensively by Akin's group (Akin *et al.* 1973, 1974; Akin & Burdick 1975; Akin 1976) and a useful review may be found in Akin (1979). Adhesion occurs through the mediation of either an extracellular fibrous capsule composed of glycoprotein (Cheng *et al.* 1977) or a thinner electron-dense layer in conjunction with an apparently flexible cell wall (Akin *et al.* 1974). Pure cultures of rumen bacteria showed that adhesion by *Ruminococcus albus* and *R. flavefasciens* is mediated by extracellular fibrous capsules (Patterson *et al.* 1975; Latham *et al.* 1978a); that by *Bacteroides succinogenes* by a thinner external layer and flexible cell wall (Latham *et al.* 1978b) which conforms to the shape of the underlying substratum. Adhesion may be more prominent on some plant cell types than others, for example *R. flavefasciens* adhered strongly to and digested the epidermal, schlerenchyma, phloem and mesophyll cells, but neither adhered to nor digested the cuticle, protoxylem cell walls or chloroplasts (Latham *et al.* 1978a). Attachment was most prominent at damaged regions of the cell walls. Adhesion of rumen large bacteria and *Lampropedia merismopedioides* to plant cuticular surfaces has been demonstrated by Clarke (1979) but the mechanism of adhesion is not known.

The effect of this is that the combined association—both active and passive—of rumen bacteria results in the removal of some 20 to 25% of the bacterial population of rumen liquor within 30 min of the host animal having eaten (Orpin 1980a).

Zoospores of rumen phycomycete fungi also invade freshly ingested plant tissues. These tissues are located chemotactically (Orpin & Bountiff 1978) and peak flagellate invasion occurs 18 to 20 min (*Neocallimastix frontalis*) and about 1 h (*Piromonas communis* and *Sphaeromonas communis*) after feeding (Orpin 1977a). The sites of invasion are the stomata and damaged regions of the particle where the zoospores encyst and germinate. The speed of germination appears to be controlled, at least in part, by the concentration of soluble carbohydrates (Orpin & Bountiff 1978), and the growing rhizoid penetrates the plant particle. Although many zoospores are released into the supernatant fluid soon after the animal eats, zoospore production may occur (at lower rates) at any time during the day.

The rumen phycomycetes are capable of digesting up to 45% of the dry weight of grass hay and wheat straw, digesting cellulose, xylan and hemicelluloses (Orpin & Letcher 1979; Orpin & Hart 1981), and so decreasing the size of the plant particles in the digesta.

Electron microscopy revealed that mesophyll and phloem cells of plant tissues are digested by mixed rumen microorganisms prior to the outer bundle sheath and the epidermis. Whilst adhesion of bacteria often occurs prior to degradation of the underlying surface, cell walls in the vicinity of, but not necessarily in contact with the bacteria, may also be degraded (Akin & Amos 1975).

Rumination and microbial digestion results in the gradual decrease in particle sizes and increase in particle density (Evans *et al.* 1973). The denser particles graduate to the reticulum where they are passed to the omasum and finally the abomasum where digestion of the residual particles and the microorganisms commences. A comparison of digesta particle sizes in the various stomach compartments is given by Becker *et al.* (1963); the chemical composition of rumen particulate material by Czerkawski (1976) and quantitative information on particle size changes in the rumen by Reid *et al.* (1977) and Evans *et al.* (1973).

Experimental

29.5 Sampling

It must be remembered that many rumen microorganisms are strict anaerobes and exposure to air during sampling should be avoided. The ciliates and phycomycetes are sensitive to temperature changes and it is advisable to hold rumen samples at or near 39°C using a vacuum flask or incubator if these organisms are to be examined. Rumen contents deteriorate rapidly after removal from the rumen, and meaningful information can only be obtained if the subsequent manipulations are carried out as soon as possible after removal from the animal. Any surgery that is necessary must comply with pertinent legislation in the country in which it is performed.

If sufficient animals are available, sacrifice and rapid removal of the rumen contents provides a reliable method of obtaining uncontaminated material. Since rumen contents often show stratification of the solid phase (Schalk & Amadon 1928; Evans *et al.* 1973) the entire contents should be mixed in bulk and subsampled to obtain a representative sample. Disadvantages of this method are cost (of animals), limitation to a single sampling and, probably, the distance (if performed at an abattoir) from the laboratory.

Rumen contents may be obtained from live animals using a stomach tube and gag, but samples are not representative and may be contaminated with mucus and saliva. In addition, anaerobic conditions cannot be maintained easily. Stomach tubing, however, is the only economic method to survey a large number of animals but results will be only of qualitative value. The methods are well described by Hungate (1966).

Cud sampling (Davey & Briggs 1959) has been employed but the cud contains chiefly rumen liquor and the larger digesta particles and is therefore not representative and may be used for qualitative microbiological evaluation.

It is common practice for rumen microbiologists

to filter rumen fluid through muslin to remove coarse food particles. The filtrate, although easy to handle, will not contain a representative sample of the rumen microbial population since organisms associated with the food particles, particularly the phycomycete fungi (Bauchop 1979a, b; Orpin & Letcher 1979), will be preferentially removed.

Direct access to the rumen can be obtained by the provision of a permanently cannulated fistula. Hecker (1974) fully describes the construction and fitting of suitable cannulae. Rumen cannulae used for small animals commonly have an internal bore diameter of 2 to 3 cm, sufficient for the insertion of a sampling tube of 1 to 1.5 cm diameter. Those used on cattle may have a bore as large as 15 cm, sufficient for the insertion of the samplers arm. A rubber or plastic cannula is normally fitted with a removable plug.

The provision of a rumen cannula allows uncontaminated samples of rumen contents to be obtained by aspiration from different regions in the rumen. If a representative sample of the contents is required, several samples should be taken from different regions in the rumen, mixed and subsampled. Alternatively, the rumen contents may be emptied via the cannula, subsampled and replaced (Hungate 1966). In cattle, samples can be obtained by hand from known regions (by reference to internal structures) of the reticulo-rumen, and boli may be obtained before they mix with the rumen contents (Reid *et al.* 1962). A permanent cannula is essential for experiments involving continuous sampling (Corbett *et al.* 1976).

The maintenance of fistulated animals is expensive and it is usually not feasible, on economic grounds, to fistulate large numbers of animals. Because of the difficulty in taking representative samples, many measurements will be required in most studies on rumen contents and the measurements analysed statistically, in order to arrive at a meaningful result.

Various workers have devised automatic equipment for obtaining samples of rumen contents at predetermined times (Canaway *et al.* 1965; Farrell *et al.* 1970) including equipment carried on the back of a free-ranging animal (Corbett *et al.* 1976). All these types of equipment suffer from blockages in sampling tubes caused by particulate material, including those which have a filter designed to reduce blocking. The only way to avoid this is to feed the animal on a finely ground diet, but this may render the experiment valueless as the digestibility may be changed by alteration in the particle size.

When examining the contribution of particular species or groups of rumen microorganisms to the nutrition of the host animal, it is often necessary to sample the abomasum contents to obtain information on the outflow of material from the rumen. The abomasum may be sampled in living animals by the use of a surgically prepared cannula and fistula or re-entrant cannula (Hecker 1974). This is the easiest way to obtain regular abomasal samples, but samples may be obtained directly through the abdominal wall using a hypodermic needle (Evans & Spurrell 1967), or via cannulae in the omasum (Willes & Mendel 1964) or duodenum (McLeay & Titchen 1970) by passage of a suitable tube.

Samples of rumen epithelium can be obtained from slaughtered animals. Portions of rumen epithelium may be excised soon after death of the host animal, but post-mortem changes may affect the bacterial population of the tissue. An alternative approach, using sheep infused with nutrient solutions both abomasally and intraruminally (Wallace *et al.* 1979), provided samples of sloughed epithelial cells harbouring microbial populations apparently identical to those on the intact epithelium, and could be used as experimental epithelial populations in the study of the facultative anaerobes of the epithelium.

29.6 Cultural techniques and equipment

Most of the dominant rumen bacteria are obligate anaerobes that will not grow in media prepared conventionally and it is necessary to use highly reduced culture media for their isolation and cultivation. Media of this type contain a reducing agent and an Eh indicator (usually resazurin) and are prepared under an oxygen-free gas phase. The preparation of media and methods of handling cultures are described by Hungate (1969), Latham and Sharpe (1971) and Holdeman and Moore (1972), using butyl rubber stoppered culture tubes for liquid cultures and agar-containing media in roll bottles for the isolation of rumen bacteria. These methods have largely been superseded by the use of culture tubes with a screw cap fitted with a butyl rubber septum (Bellco Glass Inc., Vineland N.J., USA) or culture tubes fitted with serum bottle closures (Bellco Glass Inc.) (Latham & Wolin 1978; Balch *et al.* 1979). Isolations may be made using agar media in rolled serum bottles (Miller & Wolin

1974), prepared in a similar manner to roll tubes (Hungate 1969). The advantages of these culture vessels are that the culture medium can be autoclaved within the vessel and constituents destroyed by heat may be added aseptically after autoclaving by syringe through the rubber septum without the tube being opened. Inoculation and transfer of samples may be effected similarly and estimates of gas production made by syringe (Prins 1971) with little risk of oxygen entry. The gas samples may be analysed quantitatively (Holdeman & Moore 1972).

Recent refinements in anaerobic glove boxes incorporating a palladium catalyst and hydrogen to remove any oxygen present in the gas phase have permitted the use of aerobic techniques in an anaerobic environment (Edwards & McBride 1975). This equipment, together with an ultra-low oxygen chamber, has permitted the isolation of methanogens on agar plates. The author has also performed viable counts using poured agar plates prepared in the anaerobic chamber and isolated many strains of rumen bacteria in a fraction of the time taken using roll-tubes. The manipulations are performed in the chamber atmosphere of 95% nitrogen and 5% hydrogen and incubations performed in anaerobe jars under the appropriate gas phase. If the whole chamber is heated, it is unnecessary to remove the anaerobe jars for incubation.

The facultatively anaerobic bacteria may be cultured using conventional anaerobic techniques (Barnes *et al.* 1978). Ciliate protozoa may be cultured in stoppered glass tubes in media containing cysteine as a reducing agent (Coleman 1978a). Since the cultures contain many facultative anaerobes, stringent anaerobic procedures are unnecessary both in media preparation and handling cultures.

The phycomycete fungi may be cultured using equipment similar to that used for bacteria (Orpin 1975, 1976a, 1977b) but growth in liquid cultures is poor. The incorporation of agar or plant particles provides a substratum and prevents growth from occurring only on the bottom of the culture tubes. Stringent anaerobic precautions are necessary.

29.7 Enumeration, culture and identification of rumen bacteria

Due to the nature of rumen contents and the association of bacteria with the digesta, the rumen epithelium, the surface of ciliates (Coleman 1975b; Imai & Ogimoto 1978) and within the ciliates (White 1969), there is no method available to enumerate the bacteria with accuracy. Methods of determining total numbers of bacteria and the problems involved are discussed by Hobson (1961), Warner (1962) and Hungate (1966).

The easiest method for determining the total number of bacteria in rumen contents is by a direct microscopic count on diluted samples of filtered rumen fluid. This method unfortunately measures both living and dead cells; there is some interference by plant particles (Warner 1962) and cells associated with the digesta are not counted. The latter can be minimised by blending the sample before filtration (Warner 1962). Cells in the dilution may be counted with or without fixation in a standard counting chamber such as a haemocytometer chamber for large bacteria (Warner 1962) or a Helber chamber (Barnes *et al.* 1978) for small bacteria, viewed either with bright field illumination, after staining with gentian violet (Hungate 1966), or phase contrast illumination. Warner (1962) found that the error in counting was only ± 5% if 2000 bacteria were counted by this method. Total counts may be made using this method or different morphological types may be counted separately. It is necessary to ensure that no large particles are present in the samples counted using the Helber chamber since the coverslip will not fit correctly. Ciliates in the sample may burst and obscure the bacteria. The errors associated with counting chambers are discussed by Meynell and Meynell (1970).

The Coulter counter has been used with some success by Hobson and Mann (1970) and Orpin (1977c) but problems arise with the orifice blocking frequently with digesta particles and the method does not differentiate between living and dead cells and small digesta fragments.

An alternative approach can be made using cultural counts obtained by diluting rumen fluid and inoculating a known quantity into an appropriate culture medium and either counting the number of colonies which grow in agar-containing media (colony counts) or diluting until no tubes show growth (dilution counts) (Hungate 1966). Counts obtained in this way are usually much lower than those obtained by direct counts and only reflect numbers of bacteria culturable in that medium, but can give information on relative numbers. Two types of culture media are usually employed, habitat-simulating and niche-simulating media. The former attempts to reproduce rumen conditions, the latter provides a medium designed to encourage the selective growth of organisms metabolising specific chemical components of the rumen ecosystem. Habitat-simulating media are used when counts of the total viable bacterial populations are made and are often based on rumen fluid supplemented with low levels of organic compounds including several soluble carbohydrates (Bryant & Burkey 1953; Kistner 1960; Bryant & Robinson 1961; Hungate 1966; Hobson 1969; Henning & van der Walt 1978). The rumen fluid supplies compounds essential for growth of some rumen bacteria including volatile fatty acids, branched-chain fatty acids, haemin and vitamins. The rumen fluid has been successfully replaced by a mixture of known compounds (Caldwell & Bryant 1966): whilst this may not provide the necessary factors for all rumen bacteria it removes the effect that rumen fluid of variable composition may have on comparative estimations.

Niche-stimulating media are used to enumerate bacteria that ferment a specific compound. The basal medium may or may not contain rumen fluid but contains the specific compound under examination as the major carbon and energy source. Media for the isolation of cellulolytic, amylolytic, proteolytic and lipolytic bacteria are given by Hobson (1969), hemicellulytic bacteria by Henning and van der Walt (1978) and Dehority and Grubb (1976) and lactolytic bacteria by Mackie and Heath (1979). When rumen fluid is used, it may be artificially depleted of fermentable material (Dehority & Grubb 1976) prior to incorporation into the medium.

The methanogenic bacteria can be isolated and cultured using media and equipment described by Hungate (1950) with modifications (Bryant *et al.* 1968; Wolfe 1971), using normal butyl rubber stoppered culture tubes and roll tubes. Under these conditions growth of methanogens is slow but by using tubes fitted with serum bottle closures the atmosphere in the culture vessels may be pressurised (Balch & Wolfe 1976; Balch *et al.* 1979) and the substrate levels increased, resulting in an increased growth rate of the methanogens. Alternatively the gas pressure may be increased using a pressure vessel (Balch & Wolfe 1976).

Other media have been developed to isolate other organisms including *Selenomonas ruminantium* (Tiwari *et al.* 1969) and spirochaetes (Stanton & Canole-Parola 1979). Media for facultatively anaerobic bacteria have been published by Barnes *et al.* (1978).

The rumen large bacteria, Quin's Oval (Quin 1943) or *Magnoovum eadii* (Eadie 1962a; Orpin 1976b) colloquially known as Eadie's Oval, and *Oscillospira guilliermondii* have not been cultured axenically though both Quin's Oval (Orpin 1972b) and Eadie's Oval (Orpin 1972a) have been cultured *in vitro* with other bacteria. Large strains of *Selenomonas ruminantium* have been cultured axenically (Prins 1971). Enumeration of these morphologically distinct species is easy by direct microscopy (Warner 1966; Orpin 1977c) using a haemocytometer. Identification is based on the morphology of each species and light and electron micrographs have been published by Hungate (1966) for Quin's Oval, *Oscillospira* spp. and large selenomonads; Eadie (1962a) and Orpin (1976b) for Eadie's Ovals; and Prins (1971) for large selenomonads.

Identification of other bacteria may be made on a basis of cell morphology, substrate utilisation, fermentation patterns and other biochemical tests. Reference should be made to Hungate (1966), Holdeman and Moore (1972) and Buchanan and Gibbons (1975). Tentative identification of methanogens is possible by their fluorescence under ultraviolet illumination (Mink & Dougan 1977).

29.8 Enumeration, culture and identification of ruman ciliates

The ciliates may be enumerated in rumen contents by microscopically counting the number of cells of each species in a sample of diluted rumen fluid fixed with formaldehyde (Warner 1962) or iodine (Coleman 1978a). Some species may be difficult to identify and an examination of the size, shape and location of the macronucleus, micronucleus or skeletal plates may be necessary. Nuclei may be stained with haematoxyclin (Kudo 1954) or methylene blue (Dehority & Potter 1974). Fixation with iodine not only preserves the cells well but facilitates counting by heavily staining intracellular polysaccharides and aids identification by staining the skeletal plates in a single operation. Formaldehyde fixation followed by staining with methyl green (Clarke 1964) has also been applied but storage of formaldehyde-fixed preparations is inadvisable as the ciliates sometimes deteriorate (Coleman 1978a). If nuclei and skeletal plates are to be examined, samples should be obtained before the animal has been fed because these structures may be obscured by ingested plant particles or storage polysaccharides in replete cells.

Errors in counting may arise due to the association of the ciliates with particulate material, especially soon after the animal has fed. These may be minimised by vigorously shaking the sample before filtration to remove the large plant particles or by allowing the sample to cool to about 30°C, at which temperature no significant attachment of ciliates occurs. Other errors may be incurred during sampling and during pipetting since the large ciliates sediment readily at 1 **g** in diluted suspensions and sediment within the pipette. All operations must be undertaken rapidly but sedimentation can be alleviated by the incorporation of glycerol (Adam 1951).

Identification of rumen oligotrich ciliates is by no means easy. Many papers dealing with the oligotrich ciliates rely on the use of caudal species (Dogeil 1927; Kofoid & MacLennan 1930, 1932, 1933) as taxonomic characters. These have been shown to be very variable in clone cultures of ciliates (Clarke 1963; Coleman 1978a) and of little taxonomic value. Because of this both Hungate (1966) and Coleman (1978a) believe that there are fewer species than those described by earlier authors. Confusion has arisen in the literature because some species have been given several names since they were first recognised. It is recommended that workers identifying ciliates should consult Dogeil (1927), Kofoid and MacLennan (1930, 1932, 1933) and Lubinsky (1957, 1958a, b), followed by a comparison with published micrographs (Hungate 1966; Coleman 1979; Imai *et al.* 1977). Unfortunately no reference collections are available for comparison.

The holotrich ciliates, *Isotricha intestinalis*, *I. prostoma* and *Dasytricha ruminantium* may be identified by the size of the cell and the position of the oral cavity and light micrographs are published in Hungate (1966). The other genera of rumen holotrichs, *Butschlia* and *Charonia*, occur more rarely and line drawings may be found in Hungate (1966) and photomicrographs of *Butschlia parva* in Dehority (1970) and *Charonia vertriculi* in Dehority and Mattos (1978).

Many species of oligotrich protozoa have been cultured *in vitro* but none have been cultured successfully axenically. A good review of cultural methods can be found in Coleman (1978a). Essentially the initial isolation is made by selecting individual cells by micromanipulator from mixed rumen contents or a mixed culture *in vitro* and inoculating into anaerobic buffered salts solution. Subsequent treatment depends upon the species concerned but invariably involves the daily addition of substrate in the form of particulate starch, wholemeal flour or grass particles. Initial isolation may be made by differential centrifugation if there is sufficient difference in mass between the organism to be isolated and the other species present; this has been done with *Entodinium simplex* (Coleman 1978a).

The holotrich ciliates are more difficult to cultivate and only *Dasytricha ruminantium* has been grown *in vitro* for more than 60 d (Clarke & Hungate 1966). In this instance it was necessary to add soluble carbohydrate once daily and remove and replace the supernant fluid 2 to 4 h afterwards.

It is likely that any ciliate species may be grown *in vivo* in the rumen of a defaunated animal as the sole ciliate species and cells harvested by differential centrifugation. This has been done with *Isotricha intestinalis*, *I. prostoma*, *Dasytricha ruminantium*, *Ophryoscolex caudatus*, *Epidinium ecaudatum*, *Eudiplodinium maggii* and *Entodinium caudatum* in the author's laboratory. For the holotrich protozoa, the inoculum was 100 cells of the appropriate species picked by micromanipulator (Coleman 1978a) from rumen contents containing mixed ciliates. The inocula used with the other species were pure cultures grown *in vitro*. This method allows large quantities of cells to be produced with little·labour, provided they can be separated from the rumen contents. It may be necessary to manipulate the diet (or bedding) of the animal until the right combination is found which produces digesta particles of the right size which separate easily from the ciliates.

29.9 Enumeration, culture and identification of rumen flagellate protozoa

Direct microscopic counts may be made on diluted rumen contents using a haemocytometer (Warner 1962). No cultural techniques are yet available for viable counting. The flagellate protozoa are identified by their morphology (Braune 1913; Kudo 1954; Jensen & Hammond 1964) observed under the light microscope.

Jensen and Hammond (1964) isolated and cultured several species of rumen flagellate using media based on that used for enteric trichomonads.

29.10 Enumeration, culture and identification of rumen phycomycetes

No satisfactory method is yet available for enumerating the rumen phycomycete fungi. This is because the vegetative stage of the organisms is intimately associated with the digesta and the population density of the flagellates in the rumen fluid does not necessarily reflect the numbers of vegetative growths on the digesta and only represents those flagellates which have actually swum free from the digesta. The zoospores may be counted microscopically in fresh rumen fluid and can be distinguished from the flagellate protozoa by their characteristic motility, high refractivity and number and location of flagella. Flagellates of *Neocallimastix frontalis* are multiflagellated and larger than the singly flagellated *Piromonas communis* and *Sphaeromonas communis* (Orpin 1975, 1976a, 1977b). In fixed preparations counts may be made in the same way as for the flagellate protozoa.

Three species of rumen phycomycetes (Orpin 1975, 1976a, 1977b) have been cultured *in vitro* in the absence of bacteria. Removal of bacteria can be accomplished using antibiotics.

29.11 Continuous culture and artificial rumens

The rumen itself is a kind of continuous culture system, with a continual inflow of saliva or food mixed with saliva and a continual outflow of digesta to the omasum. Various workers (Czerkawski 1976) have attempted to establish systems *in vitro* capable of duplicating the rumen. These have met with varying degrees of success and the major difficulty has been the supply of particulate plant material. The most successful designs have been those of Weller and Pilgrim (1974) and the 'Rusitec' of Czerkawski and Breckenridge (1977). Both systems rely upon the daily manual addition of plant material contained in a nylon or terylene bag to an established culture initially started by inoculating with rumen digesta (Weller & Pilgrim 1974) or rumen fluid (Czerkawski & Breckenridge 1977). In both methods the bags of plant material are moved vertically in the surrounding fluid, simulating the movement of digesta in the rumen. The easiest to operate is probably the Rusitec, since the digesta and freshly added plant material are in two separate chambers; in other apparatus they are in the same chamber and the removal of old digesta is more difficult. In the Rusitec a stable microbial population develops with volatile fatty acid production (Czerkawski & Breckenridge 1979a, b) closely approximating to that of the rumen.

Continuous culture of rumen bacteria in chemostats has been effected but it is difficult to maintain anaerobic conditions. Equipment has been designed to facilitate the use of soluble and insoluble substrates (Slyter 1975). These chemostat cultures are not representative of the rumen conditions but are useful for the production of cells for metabolic experiments and for experiments designed to study interactions between bacterial species. Further information may be found in Hobson (1965, 1969) and Hobson and Summers (1978).

29.12 Storage of isolates

At present there is no known method for the storage of viable rumen ciliates or phycomycete fungi. Both are killed by exposure to low temperatures and cultures must be growing to remain viable.

The facultatively anaerobic bacteria may be stored at 2 to 4°C for some months but it is advisable to subculture every month. The strictly anaerobic bacteria may be stored for a few weeks at 2 to 4°C but freezing at −20°C may decrease viability (Hobson 1969). Many species of facultatively anaerobic and strictly anaerobic rumen bacteria have been freeze-dried successfully (White *et al.* 1974; Phillips *et al.* 1975) and cultures remain viable for 6 months or more after freezing rapidly to −60°C (Hobson 1969).

29.13 Defaunation

Defaunation of an adult animal of rumen ciliate protozoa may be accomplished using dioctyl sodium sulphosuccinate (manoxol OT) (Abou Akkada *et al.* 1968; Orpin 1977c). Care must be taken that the particular batch of manoxol used does not contain components toxic to the animals and it is recommended that a trial defaunation be performed on a single animal to check this, prior to treating a number of animals. It is by far the easiest method available to date since other methods involve employing toxic copper salts and starvation (Becker 1929; Christiansen *et al.* 1965); or removing the entire rumen contents, killing the protozoa and replacing the fluid; or by a multistage process involving removal of rumen contents, treatment of the rumen with formaldehyde and replacement of rumen contents (Jouany & Senaud 1979).

Young animals, separated soon after birth from their mothers and isolated from other ruminants, develop rumens that are free of ciliates (Eadie & Gill 1971; Williams & Dinusson 1973).

29.14 Study of microbe–microbe interactions

Predation of bacteria by ciliates can be estimated by incubating anaerobically ^{14}C-labelled bacteria with washed suspensions of protozoa (Coleman 1971, 1978b), separating the bacteria and protozoa by differential centrifugation and measuring the [^{14}C]carbon in the protozoa. The technique has many pitfalls such as the adhesion of bacteria to the protozoa and plant particles in the preparation and several controls are necessary (Coleman 1978b). The digestion of bacteria is measured by the release of [^{14}C]carbon into the supernatant fluid or into the supernatant from sonicated protozoa. Predation of ciliates by other ciliates (e.g. *Epidinium ecaudatum* by *Polyplastron multivesiculatum*) can be measured by counting populations using light microscopy (Coleman *et al.* 1972).

Batch and continuous mixed cultures of rumen bacteria have been used to examine competition between specific microorganisms (Hobson & Summers 1978) and nutrient transfer reactions (Wolin 1975). Experiments of this type involve growing the organisms separately and measuring parameters such as growth yields or the flow of carbon or hydrogen in the system, then growing the organisms together and measuring the same parameters. It is difficult to deduce from experiments of this type what would happen *in vivo* since the spatial relationship of the organisms is not known and no doubt other interactions with other species occur.

29.15 Microbe–plant tissue interactions

Interactions of this sort can be studied by electron microscopy both with mixed populations of microorganisms (Akin 1976) and known species (Latham *et al.* 1978a, b; Cheng *et al.* 1979a) using both transmission and scanning electron microscopy. The light microscope can be used to study some interactions of ciliates and phycomycete fungi with particles but resolution is poor compared to scanning electron microscopy which has been employed effectively by Bauchop and Clarke (1976) and Bauchop (1979a).

Chemotaxis to plant tissues by rumen ciliates can be studied using plant particles contained in a basket immersed in rumen fluid or suspension of known composition (Orpin 1979a, 1982). This provides information on the population transfers from the solid to the liquid phase of rumen fluid but gives no information on any attachment mechanisms nor does it allow for the effects of mixing in the rumen. Chemotaxis to specific compounds which are released from plant tissues may be studied either with capillaries (Orpin & Bountiff 1978; Orpin & Letcher 1978) or, for the less motile ciliates, cotton wool impregnated with the compound being examined (Orpin 1979b).

Uptake of bacteria onto plant particles has been estimated quantitatively using ^{35}S- and ^{14}C-labelled bacteria (Orpin 1980). It is likely that capillary techniques as used by Adler (1973) and Palleroni (1976), but employed in an anaerobic atmosphere, could be used to measure rumen bacterial chemotaxis to specific compounds *in vitro*.

Adhesion mechanisms can be elucidated using transmission and scanning electron microscopy (Akin & Amos 1975; Akin 1976; Cheng *et al.* 1977). Light microscopy provides information on attachment where specific organelles (Orpin & Letcher 1978) or the oral cavity of ciliates (Bauchop 1979c) is employed. Chemical analysis of particles after incubation with microorganisms can be achieved using the procedures outlined in Rogers and Perkins (1968). Since the efficiency of different procedures varies and the products analysed by different procedures are rarely identical, it is essential that researchers use a single analytical procedure for comparative work.

It is theoretically possible to identify bacteria attached to plant particles using immunofluorescent techniques. These methods have been applied to rumen bacteria by Hobson and Mann (1957), Hobson *et al.* (1962) and Jarvis *et al.* (1967). The drawback is that the particular serotypes being examined have to be cultured *in vitro* before the fluorescent antiserum can be prepared but the technique is useful for assessing the abundance of particular bacterial isolates in rumen contents.

29.16 Gnotobiotic animals

It is difficult to duplicate the effects of the host animal on rumen metabolism *in vitro*. For experiments where it is crucial that the sheep factor is included, resort must be made to gnotobiotic animals. The preparation and maintenance of gnotobiotic animals is expensive but there is no substitute. Experiments on the establishment of several rumen bacterial species have been performed by Lysons *et al.* (1971, 1976), Mann and Stewart (1974), Lysons and Alexander (1975) and Hobson and Summers (1978) but to date little has been published on this kind of work. Although several bacterial species thought to be capable of performing the essential functions of the rumen have been inoculated into the rumens of these animals, the lambs did not survive for more than 11 to 12 weeks (Hobson & Summers 1978). The successful rearing of animals with defined rumen microfloras will allow experiments on both the ecology of rumen microorganisms and rumen function to be performed.

29.17 Measurement of microbial biomass and activity

Fermentation of substrates by rumen microorganisms results in the generation of ATP, release of waste products into the surrounding fluid and growth of the cells. Measurements of substrate utilisation and waste product accumulation can be used to measure microbial activity as can the measurement of any microbial cellular component which is directly related to growth. ATP is an accepted indicator of microbial biomass but its measurement in rumen contents is difficult (Forsberg & Lam 1977). Fermentation products may be measured and if gas production is estimated in the presence of excess substrate, results of experiments *in vivo* and *in vitro* are comparable (El-Shazly & Hungate 1965). The ability of rumen microbes to ferment plant cell walls can also be estimated in this way (Moir 1976). The concentration of volatile fatty acids and other metabolites in the rumen can be estimated by isotope dilution *in vivo* (Murray *et al.* 1975). This method reduces problems such as unrepresentative sampling, inherent with experiments *in vitro*.

The measurement of cellular components depends upon the separation of the microbial population from contaminating plant particles for quantitative accuracy but acceptable results may be obtained by measuring components not found in plant tissues. Neither diaminopimelic acid (DAPA) nor 2-aminoethylphosphonic acid (AEP) are found in plants so can be used as markers for bacteria and ciliate protozoa, respectively. Unfortunately DAPA is found in cell walls of lysed bacteria, but it can be used to estimate bacterial growth rather than biomass. AEP is a useful ciliate marker but the results of Dawson and Kemp (1967) show that the AEP-containing lipids may vary in concentration with different ciliate populations.

Total microbial nitrogen can be estimated by the incorporation of [^{35}S]sulphur (from $^{35}SO_4^{2-}$) into microbial protein which is then estimated (Walker *et al.* 1975), but estimates may be inaccurate since much of the microbial amino acid sulphur may be derived from sources other than sulphate or sulphide (Prins 1977). Microbial nitrogen can also be estimated from measurement of microbial nucleic acids (Smith & McAllan 1970). Whilst this measures a parameter of living cells, the extraction, analysis and interpretation of results is difficult since the ratio of nucleic acid to protein varies with the nutrition of the organism.

Bucholtz and Bergen (1973) demonstrated a highly significant correlation between the incorporation of [^{32}P]phosphorus into phospholipids and cell growth in rumen microorganisms.

More information on methods *in vivo* and *in vitro* of determining rumen microbial activity and biomass may be found in Clarke (1977).

Recommended reading

CLARKE R. T. J. & BAUCHOP T. (1977) *Microbial ecology of the gut*. Academic Press, London.

HOBSON P. N. (1976) *The Microflora of the Rumen.* Patterns of Progress, Meadowfield Press, Durham.

HOLDEMAN L. V. & MOORE W. E. C. (1972) *Anaerobe Laboratory Manual.* Virginia Polytechnic Institute and State University, Blacksburg, Virginia.

HUNGATE R. E. (1966) *The Rumen and its Microbes.* Academic Press, London.

References

ABOU AKKADA A. R., BARTLEY E. E., BERUBE R., FINA L. R., MEYER R., HENDRICKS D. & JULIUS F. (1968) Simple method to remove completely ciliate protozoa from adult ruminants. *Applied Microbiology* **16**, 1475–7.

ADAMS K. M. G. (1951) The quantity and distribution of the ciliate protozoa in the large intestine of the horse. *Parasitology* **41**, 301–11.

ADAMS J. C., GAZAWAY J. A., BRAILSFORD M. D., HARTMAN P. A. & JACOBSON N. L. (1966) Isolation of bacteriophages from the bovine rumen. *Experientia* **22**, 717–18.

ADLER J. (1973) A method for measuring chemotaxis and use of the method to determine optimum conditions for chemotaxis in *Escherichia coli*. *Journal of General Microbiology* **74**, 77–91.

AKIN D. E. (1976) Ultrastructure of rumen bacterial attachment to forage cell walls. *Applied and Environmental Microbiology* **31**, 562–8.

AKIN D. E. (1979) Microscopic evaluation of forage digestion by rumen microorganisms—a review. *Journal of Animal Science* **48**, 701–10.

AKIN D. E. & AMOS H. E. (1975) Rumen bacterial degradation of forage cell walls investigated by electron microscopy. *Applied Microbiology* **29**, 692–701.

AKIN D. E., AMOS H. E., BARTON F. E. & BURDICK D. (1973) Rumen microbial degradation of grass tissue revealed by scanning electron microscopy. *Agronomy Journal* **65**, 825–8.

AKIN D. E. & BURDICK D. (1975) Percentage of tissue types in tropical and temperate grass leaf blades and degradation of tissues by rumen microorganisms. *Crop Science* **15**, 661–8.

AKIN D. E., BURDICK D. & MICHAELS G. E. (1974) Rumen bacterial interrelationships with plant tissue during degradation revealed by transmission electron microscopy. *Applied Microbiology* **27**, 1149–56.

ANNISON E. F. & LEWIS D. (1959) *Metabolism in the Rumen*. Methuen, London.

BAILEY C. B. (1961) Saliva secretion and its relation to feeding in cattle. 3. The rate of secretion of mixed saliva in the cow during eating, with an estimate of the magnitude of the total daily secretion of mixed saliva. *British Journal of Nutrition* **15**, 443–51.

BALCH C. C., BALCH D. A., BARTLETT S., BARTRAM M. P., JOHNSON V. W., ROWLAND S. J. & TURNER J. (1955) Studies of the secretion of milk of low fat content by cows on diets low in hay and high in concentrates. VI. The effect of the physical and biochemical processes of the reticulo-rumen. *Journal of Dairy Research* **22**, 270–89.

BALCH W. E., FOX G. E., MAGRUM L. J., WOESE C. R. & WOLFE R. S. (1979) Methanogens: re-evaluation of a unique biological group. *Microbiological Reviews* **43**, 260–96.

BALCH W. E. & WOLFE R. S. (1976) New approach to the cultivation of methanogenic bacteria: 2-mercaptoethanesulphonic acid (HS-CoM)-dependent growth of *Methanobacterium ruminantium* in a pressurised atmosphere. *Applied and Environmental Microbiology* **32**, 781–91.

BARNES E. M., MEAD G. C., IMPEY C. S. & ADAMS B. W. (1978) Analysis of the avian intestinal flora. In: *Techniques for the Study of Mixed Populations* (Eds D. W. Lovelock & R. Davies), pp. 89–105. Academic Press, London.

BAUCHOP T. (1977) Foregut fermentation. In: *Microbial Ecology of the Gut* (Eds R. T. J. Clarke & T. Bauchop), pp. 223–50. Academic Press, London.

BAUCHOP T. (1979a) Rumen anaerobic fungi of cattle and sheep. *Applied and Environmental Microbiology* **38**, 148–58.

BAUCHOP T. (1979b) The rumen anaerobic fungi: colonizers of plant fibre. *Annales Recherches Vétérinaires* **10**, 246–8.

BAUCHOP T. (1979c) The rumen ciliate *Epidinium* in primary degradation of plant tissues. *Applied and Environmental Microbiology* **37**, 1217–23.

BAUCHOP T. & CLARKE R. T. J. (1976) Attachment of the ciliate *Epidinium* Crawley to plant fragments in the sheeps rumen. *Applied and Environmental Microbiology* **32**, 417–22.

BAUCHOP T., CLARKE R. T. J. & NEWHOOK J. C. (1975) Scanning electron microscopic study of bacteria associated with the rumen epithelium of sheep. *Applied Microbiology* **30**, 668–75.

BECKER E. R. (1929) Methods of rendering the rumen and reticulum of ruminants free from their normal infusorian fauna. *Proceedings of the National Academy of Sciences, U.S.A.* **15**, 435–9.

BECKER E. R., MARSHALL S. P. & ARNOLD P. T. D. (1963) Anatomy, development and functions of the bovine omasum. *Journal of Dairy Science* **46**, 835–9.

BRAUNE R. (1913) Untersuchungen über die im Wilderkauermagen Vorkommenden Protozoen. *Archiv für Protistenkunde* **32**, 111–70.

BRYANT M. P. & BURKEY L. A. (1953) Cultural methods and some characteristics of some of the more numerous groups of bacteria in the bovine rumen. *Journal of Dairy Science* **36**, 205–17.

BRYANT M. P., MCBRIDE B. C. & WOLFE R. S. (1968) Hydrogen-oxidizing methane bacteria. I. cultivation and methanogenesis. *Journal of Bacteriology* **95**, 1118–23.

BRYANT M. P. & ROBINSON I. M. (1961) An improved, non-selective culture medium for rumen bacteria and its use in determining diurnal variation in numbers of bacteria in the rumen. *Journal of Dairy Science* **44**, 1446–56.

BUCHANAN R. E. & GIBBONS N. E. (1975) *Bergey's Manual of Determinative Bacteriology*, 8e. Williams & Wilkins Co., Baltimore.

BUCHOLTZ H. F. & BERGEN W. G. (1973) Microbial phospholipid synthesis as a marker for microbial protein synthesis in the rumen. *Applied Microbiology* **25**, 504–13.

CALDWELL D. R. & BRYANT M. P. (1966) Medium without rumen fluid for non-selective enumeration and isolation of rumen bacteria. *Applied Microbiology* **14**, 794–801.

CANAWAY R. L., TERRY R. A. & TILLEY J. M. A. (1965) An automatic sampler of fluids from the rumen of fistulated sheep. *Research in Veterinary Science* **6**, 416–32.

Cheng K-J., Akin D. E. & Costerton J. W. (1977) Rumen bacteria: interaction with particulate dietary components and response to dietary variation. *Federation Proceedings* **36**, 193–7.

Cheng K-J., Dinsdale D. & Stewart C. S. (1979a) Maceration of clover and grass leaves by *Lachnospira multiparus*. *Applied and Environmental Microbiology* **38**, 723–9.

Cheng K-J., McCowan R. P. & Costerton J. W. (1979b) Adherent epithelial bacteria in ruminants and their roles in digestive tract function. *American Journal of Clinical Nutrition* **32**, 139–48.

Cheng K-J. & Wallace R. J. (1979) The mechanism of passage of endogenous urea through the rumen wall and the role of ureolytic epithelial bacterial in the urea flux. *British Journal of Nutrition* **42**, 553–7.

Christiansen W. C., Kawashima C. R. & Burroughs W. (1965) Influence of protozoa upon rumen acid production and liveweight gains in lambs. *Journal of Animal Science* **24**, 730–4.

Clarke R. T. J. (1963) The cultivation of some rumen oligotrich protozoa. *Journal of General Microbiology* **33**, 401–8.

Clarke R. T. J. (1964) Ciliates of the rumen of domestic cattle, (*Bos taurus* L.). *New Zealand Journal of Agricultural Research* **8**, 1–9.

Clarke R. T. J. (1977) Methods of studying gut microbes. In: *Microbial Ecology of the Gut* (Eds R. T. J. Clarke & T. Bauchop), pp. 1–33. Academic Press, London.

Clarke R. T. J. (1979) Niche in pasture fed ruminants for the large bacteria *Oscillospira, Lampropedia* and Quin's & Eadie's Ovals. *Applied and Environmental Microbiology* **37**, 654–7.

Clarke R. T. J. & Bauchop T. (1977) *Microbial Ecology of the Gut*. Academic Press, London.

Clarke R. T. J. & Hungate R. E. (1966) Culture of the rumen holotrich ciliate *Dasytricha ruminantium* Schuberg. *Applied Microbiology* **14**, 340–5.

Coleman G. S. (1971) The cultivation of rumen Entodiniomorphid protozoa. In: *Isolation of Anaerobes* (Eds D. A. Shapton & R. E. Board), pp. 159–76. Academic Press, London.

Coleman G. S. (1975a) The interrelationships between rumen ciliate protozoa and bacteria. In: *Digestion and Metabolism in the Ruminant* (Eds I. W. McDonald & A. C. J. Warner). University of New England Publishing Unit, Armidale, New South Wales.

Coleman G. S. (1975b) The role of bacteria in the metabolism of rumen entodiniomorphid protozoa. In: *Symbiosis*, Symposia of the Society of Experimental Biology, vol. 29, pp. 533–58. Cambridge University Press, Cambridge.

Coleman G. S. (1978a) Rumen Entodiniomorphid protozoa. In: *Methods of Cultivating Parasites in vitro* (Eds A. E. R. Taylor & J. R. Baker), pp. 39–54. Academic Press, London.

Coleman G. S. (1978b) Methods for the study of the metabolism of rumen ciliate protozoa and their closely associated bacteria. In: *Techniques for the Study of Mixed Populations* (Eds D. W. Lovelock & R. Reeves), pp. 144–63. Academic Press, London.

Coleman G. S. (1979) Rumen ciliate protozoa. In: *Biochemistry and Physiology of Protozoa* (Eds M. Levandowsky & S. H. Hutner), pp. 381–408. Academic Press, London.

Coleman G. S., Davies J. I. & Cash M. A. (1972) The cultivation of the rumen ciliates *Epidinium ecaudatum caudatum* and *Polyplastron multivesiculatum in vitro*. *Journal of General Microbiology* **73**, 509–21.

Corbett J. L., Lynch J. J., Nicol G. R. & Beeston J. W. V. (1976) A versatile peristaltic pump designed for grazing lambs. *Laboratory Practice* **25**, 458–62.

Counotte G. H. M., van't Klooster A. T., van der Guilen J. & Prins R. A. (1979) An analysis of the buffer system in the rumen of dairy cattle. *Journal of Animal Science* **49**, 1536–44.

Czerkawski J. W. (1976) Chemical composition of microbial matter in the rumen. *Journal of the Science of Food & Agriculture* **27**, 621–32.

Czerkawski J. W. & Breckenridge G. (1977) Design and development of a long-term rumen simulation technique (Rusitec). *British Journal of Nutrition* **38**, 371–4.

Czerkawski J. W. & Breckenridge G. (1979a) Experiments with the long-term rumen simulation technique (Rusitec); response to supplementation of basal rations. *British Journal of Nutrition* **42**, 217–28.

Czerkawski J. W. & Breckenridge G. (1979b) Experiments with the long-term rumen simulation technique (Rusitec); use of soluble food and an inert solid matrix. *British Journal of Nutrition* **42**, 229–45.

Davey L. A. & Briggs C. A. E. (1959) The normal flora of the bovine rumen. Bacteriological evaluation of rumen contents by the examination of cud. *Journal of Agricultural Science* **52**, 187–8.

Dawson R. M. C. & Kemp P. (1967) The aminoethylphosphonate-containing lipids of rumen protozoa. *Biochemical Journal* **105**, 837–42.

Dehority B. A. (1970) Occurrence of the ciliate protozoa *Butschlia parva* Schuberg in the rumen of the ovine. *Applied Microbiology* **19**, 179–81.

Dehority B. A. & Grubb J. A. (1976) Basal medium for the selective enumeration of rumen bacteria utilizing specific energy sources. *Applied and Environmental Microbiology* **32**, 703–11.

Dehority B. A. & Mattos W. R. (1978) Diurnal changes and effect of ration on concentrations of the rumen ciliate *Charon vertriculi*. *Applied Environmental Microbiology* **36**, 953–8.

Dehority B. A. & Potter E. L. (1974) *Diplodinium flabellum*: occurrence and numbers in the rumen of sheep with a description of two new subspecies. *Journal of Protozoology* **21**, 686–93.

DOGEIL A. V. (1927) Monographie de familie Ophryoscolecidae. *Archiv für Protistenkunde* **59**, 1–227.

EADIE J. M. (1962a) The development of rumen microbial populations in lambs and calves under various conditions of management. *Journal of General Microbiology* **29**, 563–78.

EADIE J. M. (1962b) Inter-relationships between certain rumen ciliate protozoa. *Journal of General Microbiology* **29**, 579–88.

EADIE J. M. & GILL J. C. (1971) The effect of the absence of rumen ciliate protozoa on growing lambs fed a roughage-concentrate diet. *British Journal of Nutrition* **26**, 155–67.

EDWARDS T. & MCBRIDE B. C. (1975) New method for the isolation of methanogenic bacteria. *Applied Microbiology* **29**, 540–5.

EL-SHAZLY K. & HUNGATE R. E. (1965) Method of measuring diaminopimelic acid in total rumen contents and its application to the estimation of bacterial growth. *Applied Microbiology* **14**, 27–30.

EVANS E. W., PEARCE G. R., BURNETT J. & PILLINGER S. L. (1973) Changes in some physical characteristics of digesta in the reticulorumen of cows. *Journal of Nutrition* **29**, 357–76.

EVANS L. & SPURRELL F. A. (1967) Technique for the direct injection of materials into the ruminant abomasum. *Journal of Applied Physiology* **22**, 1030–2.

EVANS W. C. (1977) Biochemistry of the bacterial catabolism of aromatic compounds in anaerobic environments. *Nature, London* **270**, 17–22.

FARRELL D. J., CORBETT J. L. & LENG R. A. (1970) Automatic sampling of blood and ruminal fluid of grazing sheep. *Research in Veterinary Science* **11**, 217–20.

FORSBERG C. W. & LAM K. (1977) Use of adenosine 5′-triphosphate as an indicator of the microbial biomass of rumen contents. *Applied and Environmental Microbiology* **33**, 528–37.

HECKER P. A. (1974) Examination of methods for enumerating hemicellulose utilizing bacteria in the rumen. *Applied and Environmental Microbiology* **38**, 13–17.

HENNING P. A. & VAN DER WALT A. E. (1978) Inclusion of xylan in a medium for the enumeration of total culturable rumen bacteria. *Applied and Environmental Microbiology* **35**, 1008–11.

HOBSON P. N. (1961) Techniques of counting rumen organisms. In: *Digestive Physiology and Nutrition in the Ruminant* (Ed. D. Lewis), pp. 107–18. Butterworths, London.

HOBSON P. N. (1965) Continuous culture of some anaerobic and facultatively anaerobic rumen bacteria. *Journal of General Microbiology* **38**, 167–80.

HOBSON P. N. (1969) Rumen bacteria. In: *Methods in Microbiology* (Eds J. R. Norris & D. W. Ribbons), vol. 3B, pp. 133–49. Academic Press, London.

HOBSON P. N. (1976) *The Microflora of the Rumen.* Meadowfield Press, Durham.

HOBSON P. N. & MANN S. O. (1957) Some studies on the identification of rumen bacteria by fluorescent antibodies. *Journal of General Microbiology* **16**, 463–71.

HOBSON P. N. & MANN S. O. (1970) Applications of the coulter counter in Microbiology. In: *Automation, Mechanization and Data Handling in Microbiology* (Eds A. Baillie & R. J. Gilbert), pp. 91–105. Academic Press, London.

HOBSON P. N., MANN S. O. & SMITH W. (1962) Serological tests of a relationship between rumen selenomonads *in vitro* and *in vivo*. *Journal of General Microbiology* **29**, 265–70.

HOBSON P. N. & SUMMERS R. (1978) Anaerobic bacteria in mixed cultures: ecology of the rumen and sewage digesters. In: *Techniques for the Study of Mixed Populations* (Eds D. W. Lovelock & R. Davies), pp. 125–41. Academic Press, London.

HOLDEMAN L. V. & MOORE W. E. C. (1972) *Anaerobe Laboratory Manual.* Virginia Polytechnic Institute and State University, Blacksburg, Virginia.

HOOGENRAAD N. J., HIRD F. J. R., HOLMES I. & MILLIS N. F. (1967) Bacteriophages in rumen contents of sheep. *Journal of General Virology* **1**, 575–6.

HUNGATE R. E. (1942) The culture of *Eudiplodinium neglectum* with experiments on the digestion of cellulose. *Biological Bulletin, Marine Biological Laboratory, Woods Hole, Massachusetts* **83**, 303–19.

HUNGATE R. E. (1943) Further experiments on cellullase digestion by the protozoa in the rumen of cattle. *Biological Bulletin, Marine Biological Laboratory, Woods Hole, Massachusetts* **84**, 157–63.

HUNGATE R. E. (1950) The anaerobic mesophilic cellulolytic bacteria. *Bacteriological Reviews* **14**, 1–49.

HUNGATE R. E. (1957) Microorganisms in the rumen of cattle fed a constant ration. *Canadian Journal of Microbiology* **3**, 289–311.

HUNGATE R. E. (1966) *The Rumen and its Microbes.* Academic Press, London.

HUNGATE R. E. (1969) A roll tube method for cultivation of strict anaerobes. In: *Methods in Microbiology* (Eds J. R. Norris & D. W. Ribbons), vol. 3B, pp. 117–32. Academic Press, London.

IMAI S., KATSUNO M. & TSUNODA K. (1977) Scanning electron microscopy of rumen ciliates in cattle. *Zoological Magazine, Tokyo* **86**, 194–207.

IMAI S. & OGIMOTO K. (1978) Scanning electron and fluorescent microscopic studies on the attachment of spherical bacteria to ciliate protozoa in the ovine rumen. *Japanese Journal of Veterinary Science* **40**, 9–19.

JARVIS B. D. W., WILLIAMS V. J. & ANNISON E. F. (1967) Enumeration of cellulolytic cocci in sheep rumen by using a fluorescent antibody technique. *Journal of General Microbiology* **48**, 161–9.

JENSEN E. A. & HAMMOND D. M. (1964) A morphological study of Trichomonads and related flagellates from the bovine digestive tract. *Journal of Protozoology* **11**, 386–94.

JOUANY J. P. & SENAUD J. (1979) Defaunation du rumen de mouton. *Annales de Biologie Animale, Biochemie et Biophysique* **19**, 89–94.

KISTNER A. (1960) An improved method for viable counts of bacteria of the ovine rumen which fement carbohydrates. *Journal of General Microbiology* **23**, 565–76.

KOFOID C. A. & MACLENNAN R. F. (1930) Ciliates from *Bos indicus* Linn. 1. The genus *Entodinium* Stein. *University of California Publications in Zoology* **33**, 471–544.

KOFOID C. A. & MACLENNAN R. F. (1932) Ciliates from *Bos indicus* Linn. 2. A revision of *Diplodinium* Schuberg. *University of California Publications in Zoology* **37**, 53–152.

KOFOID C. A. & MACLENNAN R. F. (1933) Ciliates from *Bos indicus* Linn. 3. *Epidinium* Crawley, *Epiplastron* gen. nov. and *Ophryoscolex* Stein. *University of California Publications in Zoology* **39**, 1–33.

KUDO R. R. (1954) *Protozoology*, 4e. Charles C. Thomas, Springfield, Illinois.

LATHAM M. J., BROOKER B. E., PETTIPHER G. L. & HARRIS P. J. (1978a) *Ruminococcus flavefasciens* cell coat and adhesion to cotton cellulose and to cell walls in leaves of perennial rye-grass (Lolium perenne). *Applied and Environmental Microbiology* **35**, 156–65.

LATHAM M. J., BROOKER B. E., PETTIPHER G. L. & HARRIS P. J. (1978b) Adhesion of *Bacteroides succinogenes* in pure culture and in the presence of *Ruminococcus flavefasciens* to cell walls of perennial rye-grass (Lolium perenne). *Applied and Environmental Microbiology* **35**, 1166–73.

LATHAM M. J. & SHARPE M. E. (1971) The isolation of anaerobic organisms from the bovine rumen. In: *The Isolation of Anaerobes* (Eds D. A. Shapton & R. G. Board), pp. 133–47. Academic Press, London.

LATHAM M. J. & WOLIN M. J. (1977) Fermentation of cellulose by *Ruminococcus flavefasciens* in the presence and absence of *Methanobacterium ruminantium*. *Applied and Environmental Microbiology* **34**, 297–301.

LATHAM M. J. & WOLIN M. J. (1978) Use of a serum bottle technique to study the interaction between strict anaerobes in mixed culture. In: *Techniques for the Study of Mixed Populations* (Eds D. W. Lovelock & R. Reeves), pp. 113–24. Academic Press, London.

LUBINSKY G. (1955a) On some parasites of parasitic protozoa. I. *Sphaerita hoari* sp.n.-A chytrid parasitizing *Eremoplastron bovis*. *Canadian Journal of Microbiology* **1**, 440–50.

LUBINSKY G. (1955b) On some parasites of parasitic protozoa. II. *Sagittospora cameroni* Gen.n.Sp.-A Phycomycete parasitizing Ophryoscolecidae. *Canadian Journal of Microbiology* **1**, 675–84.

LUBINSKY G. (1957) Studies on the evolution of the Ophryoscolecidae (Ciliata: Oligotricha). III. Phylogeny of the Ophryoscolecidae based on their comparative morphology. *Canadian Journal of Zoology* **35**, 141–59.

LUBINSKY G. (1958a) Ophryoscolecidae (Ciliata: Entodiniomorphida) of the Reindeer (*Rangifer tarandus L.*) from the Canadian Arctic. II. Diplodininae. *Canadian Journal of Zoology* **36**, 937–59.

LUBINSKY G. (1958b) Ophryoscolecidae (Ciliata: Entodiniomorphida) of Reindeer (*Rangifer tarandus* L.) from the Canadian Arctic. I. Entodininae. *Canadian Journal of Zoology* **36**, 819–35.

LYSONS R. J. & ALEXANDER T. J. L. (1975) The gnotobiotic ruminant and *in vivo* studies of defined bacterial populations. In: *Digestion and Metabolism in the Ruminant* (Eds I. W. MacDonald & A. C. I. Warner), pp. 180–92. University of New England Publishing Unit, Armidale, New South Wales.

LYSONS R. J., ALEXANDER T., HOBSON P. N., MANN S. O. & STEWART C. S. (1971) Establishment of a limited rumen microflora in gnotobiotic lambs. *Research in Veterinary Science* **12**, 486–7.

LYSONS R. J., ALEXANDER T. J. L., WELLSTEAD P. D., HOBSON P. N., MANN S. O. & STEWART C. S. (1976) Defined bacterial populations in the rumens of gnotobiotic lambs. *Journal of General Microbiology* **94**, 257–69.

MCARTHUR J. M. & MILTIMORE J. E. (1961) Rumen gas analysis by gas-solid chromatography. *Canadian Journal of Animal Science* **41**, 187–96.

MCBEE R. H. (1977) Fermentation in the hindgut. In: *Microbial Ecology of the Gut*. (Eds R. T. J. Clarke & T. Bauchop), pp. 185–222. Academic Press, London.

MCCOWAN R. P., CHENG K-J., BAILEY C. B. M. & COSTERTON J. W. (1978) Adhesion of bacteria to epithelial cell surfaces within the reticulo-rumen of cattle. *Applied and Environmental Microbiology* **35**, 149–55.

MCDOUGALL E. I. (1948) Studies on ruminant saliva. 1. The composition and output of sheep's saliva. *Biochemical Journal* **43**, 99–109.

MACKIE R. I. & HEATH S. (1979) Enumeration and isolation of lactate-utilizing bacteria from the rumen of sheep. *Applied and Environmental Microbiology* **38**, 416–21.

MCLEAY L. M. & TITCHEN D. A. (1970) Abomasal secretory responses to teasing with food and feeding in the sheep. *Journal of Physiology* **206**, 605–28.

MANN S. O. & STEWART C. S. (1974) Establishment of a limited rumen flora in gnotobiotic lambs fed on a roughage diet. *Journal of General Microbiology* **84**, 379–82.

MEYNELL G. G. & MEYNELL E. (1970) *Theory and Practice in Experimental Bacteriology*, 2e. Cambridge University Press, Cambridge.

MILLER T. L. & WOLIN M. J. (1974) A serum bottle modification of the Hungate technique for cultivating obligate anaerobes. *Applied Microbiology* **27**, 985–7.

MINK R. W. & DOUGAN P. R. (1977) Tentative identification of methanogenic bacteria by fluorescence microscopy. *Applied & Environmental Microbiology* **33**, 713–17.

MOIR K. W. (1976) *In vitro* gas production from plants fermented in rumen fluid: the influence of the plant cell wall. *Laboratory Practice* **25**, 457–8.

MURRAY R. M., BRYANT A. M. & LENG R. A. (1975) In: *Tracer Studies on Non-Protein Nitrogen for Ruminants*, vol. 2, pp. 21–7. International Atomic Energy Agency, Vienna.

ORPIN C. G. (1972a) The culture of the rumen organism Eadie's Oval *in vitro*. *Journal of General Microbiology* **70**, 321–9.

ORPIN C. G. (1972b) The culture *in vitro* of the rumen bacteria Quin's Oval. *Journal of General Microbiology* **73**, 523–30.

ORPIN C. G. (1975) Studies on the rumen flagellate *Neocallimastix frontalis*. *Journal of General Microbiology* **91**, 249–62.

ORPIN C. G. (1976a) Studies on the rumen flagellate *Sphaeromonas communis*. *Journal of General Microbiology* **94**, 270–80.

ORPIN C. G. (1976b) The characterization of the rumen bacterium Eadie's Oval, *Magnoovum* gen.nov.*eadie* sp.nov. *Archives of Microbiology* **111**, 155–9.

ORPIN G. C. (1977a) Invasion of plant tissue in the rumen by the flagellate *Neocallimastix frontalis*. *Journal of General Microbiology* **98**, 423–30.

ORPIN C. G. (1977b) The rumen flagellate *Piromonas communis* its life-history and invasion of plant material in the rumen. *Journal of General Microbiology* **99**, 215–518.

ORPIN C. G. (1977c) Studies on the defaunation of the ovine rumen. *Journal of Applied Bacteriology* **43**, 309–18.

ORPIN C. G. (1979a) Association of rumen ciliate protozoa with plant particles *in vitro*. *Society for General Microbiology Quarterly* **7**, 31–2.

ORPIN C. G. (1979b) Chemotaxis in rumen ciliate protozoa. *Society for General Microbiology Quarterly* **7**, 32.

ORPIN C. G. (1980) Quantitative aspects of the association of rumen bacteria with plant particles *in vitro*. *Society for General Microbiology Quarterly* **7**, 174.

ORPIN C. G. (1982) Migration of rumen ciliate populations to plant material ingested by the host animal. *Journal of General Microbiology*. (In press.)

ORPIN C. G. & BOUNTIFF L. (1978) Zoospore chemotaxis in the rumen phycomycete *Neocallimastix frontalis*. *Journal of General Microbiology* **104**, 113–22.

ORPIN C. G . & HALL F. J. (1977) Attachment of the rumen holotrich protozoon *Isotricha intestinalis* to grass particles. *Proceedings of the Society for General Microbiology* **4**, 82–3.

ORPIN C. G. & HART Y. (1981) Digestion of plant cell wall polymers and disintegration of plant particles *in vitro* by anaerobic rumen phycomycete fungi. *Current Microbiology*. (In press.)

ORPIN C. G. & LETCHER A. J. (1978) Some factors controlling the attachment of the rumen holotrich protozoa. *Isotricha intestinalis* and *I. prostoma* to plant particles *in vitro*. *Journal of General Microbiology* **106**, 33–40.

ORPIN C. G. & LETCHER A. J. (1979) Utilization of cellulose, starch, xylan and other hemicelluloses for growth by the rumen phycomycete *Neocallimastix frontalis*. *Current Microbiology* **3**, 121–4.

ORPIN C. G. & MUNN E. A. (1974) The occurrence of bacteriophages in the rumen and their influence on rumen bacterial populations. *Experientia* **30**, 1018–20.

PALLERONI N. J. (1976) Chamber for bacterial chemotaxis experiments. *Applied and Environmental Microbiology* **32**, 729–30.

PATTERSON H. R., IRVINE R. F., COSTERTON J. W. & CHENG K-J. (1975) Ultrastructure and adhesion properties of *Ruminococcus albus*. *Journal of Bacteriology* **122**, 278–87.

PFANDER W. H. & PHILLIPSON A. T. (1953) The rates of absorption of acetic, propionic and *n*-butyric acids. *Journal of Physiology* **122**, 102–10.

PHILLIPS B. A., LATHAM M. J. & SHARPE M. E. (1975) A method of freeze-drying rumen bacteria and other strict anaerobes. *Journal of Applied Bacteriology* **38**, 319–22.

PHILLIPSON A. T. & MANGAN J. L. (1959) Bloat in cattle. XVI. Bovine saliva: the chemical composition of the parotid, submaxillary and residual secretions. *New Zealand Journal of Agricultural Research* **2**, 990–1001.

PRINS R. A. (1971) Isolation, culture and fermentation characteristics of *Selenomonas ruminantium* var *bryanti* var n. from the rumen of sheep. *Journal of Bacteriology* **105**, 820–5.

PRINS R. A. (1977) Biochemical activities of gut microorganisms. In: *Microbial Ecology of the Gut* (Eds R. T. J. Clarke & T. Bauchap), pp. 73–183. Academic Press, London.

QUIN J. I. (1943) Studies on the alimentary tract of merino sheep in South Africa. VII. Fermentation in the forestomachs of sheep. *Onderstepoort Journal of Veterinary Science and Animal Industry* **18**, 91–112.

REID C. S. W., JOHN A., ULYATT M. J., WAGHORN G. C. & MILLIGAN L. P. (1979) Chewing and the physical breakdown of feed. *Annales Recherches Vétérinaires* **10**, 205–7.

REID C. S. W., LYTTLETON J. W. & MANGAN J. L. (1962) Bloat in cattle. XXIV. A method of measuring the effectiveness of chewing in the release of plant cell contents from ingested feed. *New Zealand Journal of Agricultural Research* **5**, 237–48.

REID C. S. W., ULYATT M. J. & MUNRO J. A. (1977) The physical breakdown of feed during digestion in the rumen. *Proceedings of the New Zealand Society for Animal Production* **37**, 173–5.

ROGERS H. J. & PERKINS H. R. (1968) *Cell Walls and Membranes*. Spon, London.

SCHALK A. F. & AMADON R. S. (1928) Physiology of the ruminant stomach (bovine). *Bulletin of North Dakota Experimental Station* No. **216**.

SHARPE M. E., LATHAM M. J. & REITTER B. (1975) The immune response of the host animal to bacteria in the rumen and caecum. In: *Digestion and Metabolism in the Ruminant* (Eds I. W. McDonald & A. C. I. Warner), pp. 193–204. University of New England Publishing Unit, Armidale, New South Wales.

SLYTER L. L. (1975) Automatic pH control and soluble and insoluble input for continuous culture of rumen microorganisms. *Applied Microbiology* **30**, 330–2.

SMITH R. H. & MCALLAN A. B. (1970) Nucleic acid metabolism in the ruminant. *British Journal of Nutrition* **24**, 545–56.

STANTON T. B. & CANOLE-PAROLA E. (1979) Enumeration and selective isolation of rumen spirochetes. *Applied and Environmental Microbiology* **38**, 956–73.

TIWARI A. D., BRYANT M. P. & WOLFE R. S. (1969) Simple method for isolation of *Selenomonas ruminantium* and some nutritional characteristics of the species. *Journal of Dairy Science* **52**, 2054–6.

WALKER D. J., EGAN A. R., NADER C. J., ULYATT M. J. & STORER G. B. (1975) Rumen microbial protein synthesis and proportions of microbial and non-microbial nitrogen flowing to the intestines of sheep. *Australian Journal of Agricultural Research* **26**, 699–708.

WALLACE R. J., CHENG K-J., DINSDALE D. & ORSKOV E. R. (1979) An independent microbial flora of the epithelium and its role in the ecomicrobiology of the rumen. *Nature, London* **279**, 424–6.

WARNER A. C. I. (1962) Enumeration of rumen microorganisms. *Journal of General Microbiology* **28**, 119–28.

WARNER A. C. I. (1966) Diurnal changes in the concentrations of microorganisms in the rumens of sheep fed limited diets once daily. *Journal of General Microbiology* **45**, 213–35.

WELLER R. A., GRAY F. V. & PILGRIM A. F. (1958) The conversion of plant nitrogen to microbial nitrogen in the rumen of the sheep. *British Journal of Nutrition* **12**, 421–9.

WELLER R. A. & PILGRIM A. F. (1974) Passage of protozoa and volatile fatty acids from the rumen of sheep and from a continuous *in vitro* fermentation system. *British Journal of Nutrition* **32**, 341–51.

WHITE R. W. (1969) Viable bacteria inside the rumen ciliate *Entodinium caudatum*. *Journal of General Microbiology* **56**, 403–8.

WHITE R. W., MACKENZIE A. R. & BOUSFIELD I. J. (1974) The successful freeze drying and retention of biohydrogenation activity of bacteria isolated from the ovine rumen. *Journal of Applied Bacteriology* **37**, vi.

WILLES R. F. & MENDEL V. E. (1964) A permanent omasal fistula for experimental studies in sheep. *American Journal of Veterinary Research* **25**, 1302–6.

WILLIAMS P. P. & DINUSSON W. E. (1973) Ruminal volatile fatty acid concentrations and weight gains of calves reared with and without ruminal ciliated protozoa. *Journal of Animal Science* **36**, 588–99.

WOLFE R. S. (1971) Microbial formation of methane. In: *Advances in Microbial Physiology* (Eds A. H. Rose & J. F. Williamson), vol. 6, pp. 107–46. Academic Press, London.

WOLIN M. J. (1975) Interactions between the bacterial species of the rumen. In: *Digestion and Metabolism in the Ruminant* (Eds I. W. McDonald & A. C. I. Warner), University of New England Publishing Unit, Armidale, New South Wales.

Chapter 30 • Termite Microbial Communities

Betsey Dexter-Dyer Grosovsky
and Lynn Margulis

30.1 Introduction

The hypertrophied hindgut or paunch of *Reticulitermes flavipes* and other rhinotermitids (subterranean termites) harbours a diverse heterotrophic and cellulolytic microbial ecosystem. Many unique genera and species of protists, primarily mastigotes (eukaryotic microorganisms bearing undulipodia, 9+2 cilia or flagella—Margulis 1980), have been identified (Table 1). Some unique bacterial species have also been described. Other members of the microbial community defy identification either because it is impossible to cultivate them or they are morphologically indistinct or variable.

The digestion of cellulose and its utilisation as a carbon source by the host insect requires the presence of hindgut organisms (Breznak 1975). Nitrogen fixed from the air by nitrogen-fixing bacteria may be a major source of this element (Mertins *et al.* 1973; Potrikus & Breznak 1977). On the other hand, in nature, termite carcasses, fungi and faecal waste may provide more important sources of nitrogen (Hungate 1941). Microbes are probably involved in other aspects of termite nutrition as well as absorption of nutrients (Speck *et al.* 1971; Breznak & Pankratz 1977) and metabolism of vitamins (Breznak 1975). These symbioses are generally obligate: many microbes cannot survive more than a few hours outside the termite

Table 1. Classification of microbes of *Reticulitermes flavipes* (after Margulis & Schwartz 1982).

KINGDOM MONERA (Bacteria)
 Phylum Spirochaeta
  Class Pillotaceae
   Pillotina, *Hollandina*, *Clevelandina*, *Diplocalyx* (indistinguishable at light microscope level, scored as pillotina spirochaetes)
  Class Treponemaceae
 Phylum Omnibacteria
   scored together—indistinguishable
 Phylum Clostridia
   Arthromitus spp.

KINGDOM PROTOCTISTA
 Phylum Zoomastigina*
  Class Pyrsonymphida
   Pyrsonympha vertens (Fig. 1)
   Dinenympha gracilis (with or without associated surface rod bacteria) (Figs 2 and 3)
  Class Parabasalia
  Order Hypermastigina
   Spirotrichonympha spp. (Fig. 5)
   Trichonympha agilis (Fig. 4)
  Order Trichomonadia
   Microjoenia fallax
   Holomastigotes elongatum
   tiny indistinguishable forms (*Tricercomitus*-like and other miscellaneous small mastigotes < 10 μm long)

*Since flagellin-containing flagella are restricted to prokaryotes, and cilia and eukaryotic flagella are identical (≡undulipodia or (9+2) organelles) the use of the term 'flagellates' for undulipodiated eukaryotes seems anomalous. Thus we use mastigotes for termite and non-termite eukaryotic microorganisms that bear long 9+2 organelles (Margulis 1980).

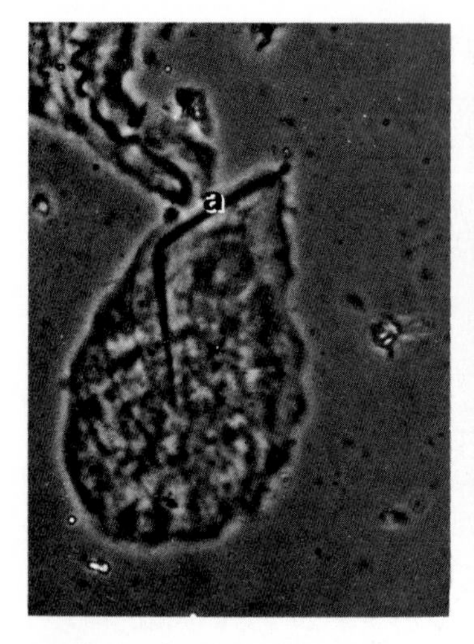

Figure 1. *Pyrsonympha vertens* (× 200). (a), axostyles.

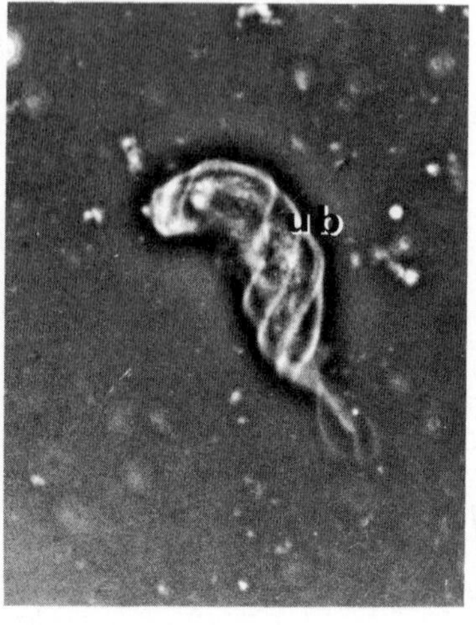

Figure 2. *Dinenympha gracilis*, without rods (× 400). (ub), undulating band.

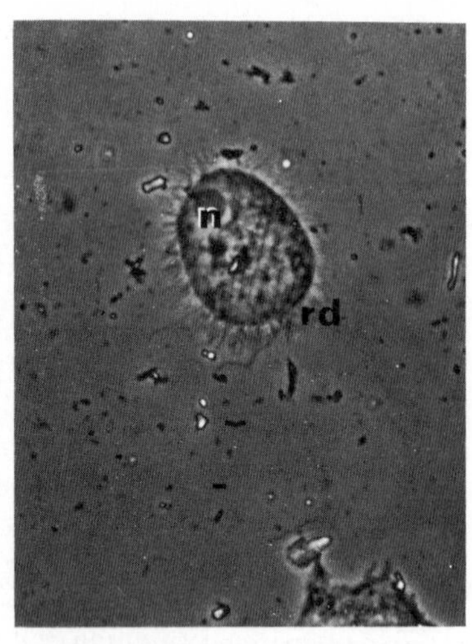

Figure 3. *Dinenympha gracilis*, with rods (× 400). (n), nucleus; (rd), rod-shaped bacteria.

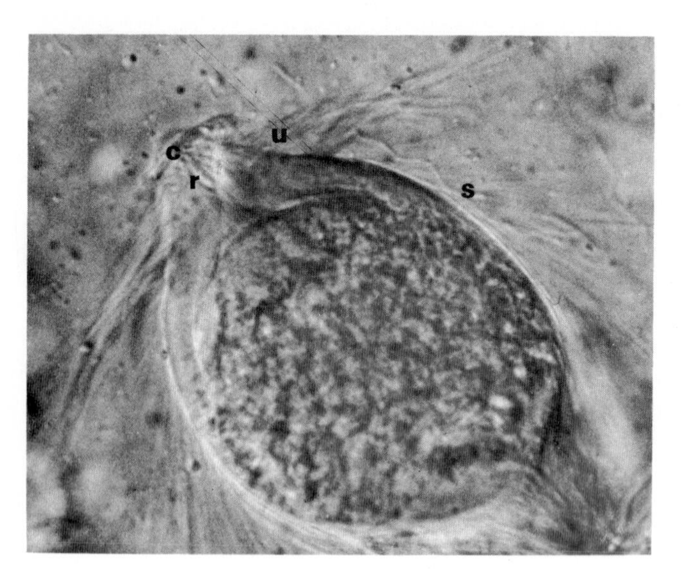

Figure 4. *Trichonympha agilis*. (c), rostral cap; (r), rostrum; (s), spirochaete; (u), undulipodia.

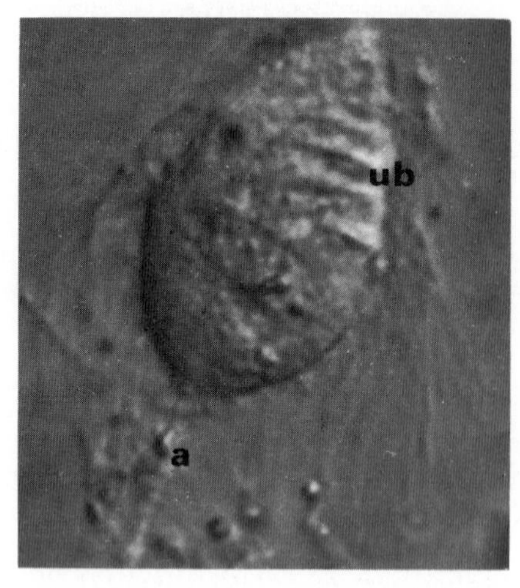

Figure 5. *Spirotrichonympha* sp. (× 400). (a), axostyles; (ub), undulating band.

and the termites cannot survive loss of the microbial community.

The symbiotic microbes enjoy a stable, moist, dark environment. Recent work using redox dyes (Veivers *et al.* 1980) indicated that the hindgut environment is probably anaerobic. This is contrary to previous work (Eutick *et al.* 1976) which suggested that the hindgut was microaerophilic. The hindgut is apparently rich in nutrients and structurally complex so that a large number of microbial species occupy separate niches.

Experimental

30.2 Removal of microbial symbiont

A major method of study of relationships in complex symbioses, in which it is impossible to cultivate the individual components, is to alter the conditions of the associations in order to observe differential survival of members and to identify those functions that correlated with the presence of certain species (Cleveland 1925b, 1926; Breznak 1975; Jeon & Hah 1977).

The termite hindgut microbiota may be removed, either partially or completely, by starvation, elevated temperatures, exposure to high oxygen tension (Cleveland 1925b, 1926) or alteration of the diet (Lund 1930; Cleveland *et al.* 1934; Cleveland 1951; Jakobi & Silva 1959; Bready & Friedman 1963; Speck *et al.* 1971; Lenz & Becker 1972; Breznak *et al.* 1974; Zusková 1974). Treatment with chemicals, such as wood preservatives and antibiotics, may also profoundly affect the structure of termite communities (Table 2). In addition, protists from one species of termite may be transferred to another and the effects of the transfer upon the host termites and their microbial community noted (Cleveland *et al.* 1934; Dropkin 1946).

Table 2. Antibiotics used in termite research.

Class of drug and mechanism of action	Drug	Use in termite research
WOOD PRESERVATIVES	pentachlorophenol	(a) *Kal*, *Ret*, (e) (Lenz & Becker 1972)
	Dieldrin	(a) *Kal*, decreases *Joenia* and Spirochaetes (Lenz & Becker 1972)
	boric acid	(a) *Ret*, variable results
		(a) *Kal*, no strong effect (Lenz & Becker 1972)
	arsenic acid	(a) *Ret*, *Kal*, (e), (f) (Lenz & Becker 1972)
	benzene hexachloride	(a) *Ret*, *Kal*, no effect or somewhat beneficial for symbionts but kills termite sometimes (Lenz & Becker 1972)
	sodium fluoride	(a) *Kal*, (e), (a) *Ret*, variable results (Lenz & Becker 1972)
	copper sulphate	(a) *Ret*, variable results (Lenz & Becker 1972)
SULPHONAMIDES (Huff 1976) inhibit bacterial synthesis of folic acid	triple sulphur	(b) *Ret*, (d) (Bready & Friedman 1963)
	sulphisoxazole	(b) *Ret*, (d) (Bready & Friedman 1963), (a) *Het*, *Ret*, (e), (a) *Kal*, kills *Joenia* and Spirochaetes (Speck *et al.* 1971)
	Sulphtalil	(b) *Euc*, (e), (f) (Jakobi & Silva 1959)
	Aristamic Elkosin	(a) *Het*, (e), (a) *Kal*, kills *Joenia*
		(a) *Ret*, kills spirochaetes, *Trichonympha*, *Holomastigotes* (Speck *et al.* 1971)
	Sulfuno	(a) *Het*, *Ret*, (e) (Speck *et al.* 1971)
	Supronal	(a) *Het*, *Ret*, (e), (a) *Kal*, kills spirochaetes, diminishes *Joenia* (Speck *et al.* 1971)
	Tardamid	(a) *Kal*, no effect, (a) *Ret*, *Het*, *Cop*, (e) (Speck *et al.* 1971)
	Lederkyn	(a) *Het*, *Ret*, (e), (a) *Kal* kills spirochaetes, diminishes *Joenia* (Speck *et al.* 1971)

Class of drug and mechanism of action	Drug	Use in termite research
NITROFURAN MUTAGEN (Ebringer 1978) anti-prokaryote or eukaryote	furazolidone	(b) *Ret*, (c) (Bready & Friedman 1963)
ARSENIC COMPOUNDS (Wilson & Jones 1975)	Acetarson	(b) *Euc*, (e), (f) (Jakobi & Silva 1959)
	3-nitro 4-hydroxy-phenyl arsenic acid	(b) *Ret*, *Cop*, *Neo*, (g) (Zusková 1974)
	arsenic acid	see wood preservatives
INHIBITORS OF ANAEROBES (Edwards 1970; Yoshikawa 1974; Miller 1976) interferes with electron transport involving reduced ferredoxin	metronidazole	(b) *Ret*, *Cop*, *Neo*, (e) (Zusková 1974)
INHIBITOR OF REPLICATION (Huff 1976; Ebringer 1978)	novobiocin	(b) *Ret*, (c) (Bready & Friedman 1963)
DYE	5% (w/v) aqueous acid fuchsin	(b) ter, kills spirochaetes (Cleveland 1928)
PROTEIN SYNTHESIS INHIBITORS:	neomycin sulphate	(b) *Ret*, (c) (Bready & Friedman 1963)
AMINOGLYCOSIDES (Huff 1976; Ebringer 1978)	kanamycin	(a) *Het*, *Ret*, *Cop*, (e), (a) *Kal*, kills *Joenia* and spirochaetes (Speck *et al.* 1971)
anti-30S ribosome unit	dihydrostreptomycin	(a) *Het*, *Ret*, (e) (Speck *et al.* 1971)
	streptomycin	(b) *Ret*, (c) (Bready & Friedman 1963)
		(a) *Het*, *Cop*, (e), (a) *Kal*, kills *Joenia*, *Foania*, spirochaetes
		(a) *Ret*, kills *Trichonympha*, spirochaetes (Speck *et al.* 1971)
PROTEIN SYNTHESIS INHIBITORS:	tetracycline	(b) *Ret*, (d) (Bready & Friedman 1963)
		(a) *Kal*, kills *Joenia*, *Foania*, *Hexamastix*, *Spirochaetes*
MISCELLANEOUS		(a) *Het*, *Ret*, *Cop*, (e) (Speck *et al.* 1971)
(Huff 1976; Ebringer 1978)		
chloramphenicol – 50S	cyclohexamide	(b) *Ret*, (c) (Bready & Friedman 1963)
tetracycline—30S, 40S	chloramphenicol	(b) *Ret*, *Cop*, *Neo*, (g) (Zusková 1974)
cyclohexamide—60S		(b) *Ret*, (c) (Bready & Friedman 1963)
		(a) *Het*, *Ret*, Cop, (e), (a) *Kal*, kills spirochaetes (Speck *et al.* 1971)
tetracycline	antibiotic mixture	(a) *Cop*, stops nitrogen fixation, kills mastigotes (Breznak & Brill 1973)
penicillin		
chloramphenicol		(a) *Cry*, *Ret*, stops nitrogen fixation
streptomycin sulphate		(a) *Ret*, stops methane production (Breznak *et al.* 1974)
PROTEIN SYNTHESIS INHIBITORS:	spiramycin	(a) *Kal*, kills spirochaetes
		(a) *Het*, *Ret*, *Cop*, (e) (Speck *et al.* 1971)
(Huff 1976; Ebringer 1978)		
NON POLYENE MACROLIDES anti-50S ribosome unit	erythromycin	(b) *Ret*, (c) (Bready & Friedman 1963)

Class of drug and mechanism of action	Drug	Use in termite research
CELL WALL INHIBITORS (Huff 1976; Ebringer 1978)	penicillin G	(b) *Ret*, (c) (Bready & Friedman 1963) (b) *Cry*, removes bacteria-like striations from *Urinympha* (Cleveland 1951)
all Gram-positive cell wall inhibitors except Polymyxin which is a Gram-negative inhibitor and increases permeability		(a) *Het*, *Ret*, *Cop*, (e), (a) *Kal*, kills spirochaetes and *Joenia* (Speck *et al.* 1971)
	bacitracin	(b) *Ret*, (c) (Bready & Friedman 1963)
	vancomycin	(b) *Ret*, (c) (Bready & Friedman 1963)
	polymyxin B sulphate	(b) *Ret*, (c) (Bready & Friedman 1963)
TOXIC WOOD EXTRACTS	Lapachon	(a) *Kal*, *Cryp*, *Zoo*, *Ret*, (e) (Lenz & Becker 1972)
	Tectoquinone	(a) *Kal*, kills *Joenia*, diminishes smaller mastigotes (a) *Cryp*, (e) (a) *Zoo*, kills *Trichonympha*, *Streblomastix*, reduces *Trichomonas* (a) *Ret*, kills *Trichonympha*, *Spirotrichonympha* reduces other mastigotes (Lenz & Becker 1972)
	Lapachol	(a) *Kal*, (g), (a) *Zoo*, (e) (a) *Cryp*, kills pentatrichomonads (a) *Ret*, variable, *Trichonympha* most sensitive (Lenz & Becker 1972)
MISCELLANEOUS	tyrothrycin	(b) *Ret*, (c) (Bready & Friedman 1963)
	streptovaricin	(b) *Ret*, (c) (Bready & Friedman 1963)
	colistin sulphate	(a) *Kal*, kills spirochaetes and *Joenia* (a) *Het*, *Ret*, *Cop*, (e) (Speck *et al.* 1971)
	Fe Cl_3	(b) *Euc*, (e), (f) (Jakobi & Silva 1959)

Abbreviations for Table 2

TERMITE GENERA		LOCATION
Cop	*Coptotermes*	World-wide, mainly tropical
Cryp	*Cryptotermes*	World-wide, except temperate Eurasia
Euc	*Eucryptotermes*	S. America
Het	*Heterotermes*	World-wide, mainly tropical
Kal	*Kalotermes*	World-wide, mainly subtropical & tropical
Neo	*Neotermes*	Cosmopolitan
Ret	*Reticulitermes*	N. America; Eurasia (temperate only)
Zoo	*Zootermopsis*	N. America
ter	unnamed termite species	
WOOD-EATING COCKROACH GENUS		
Cry	*Cryptocercus*	N. America, restricted to S.E. mountains, N.W. coast

Key:
(a) quantitative data
(b) non-quantitative data
(c) weakly effective in removing bacteria
(d) moderately effective in removing bacteria
(e) kills all mastigotes and spirochaetes
(f) kills termite
(g) moderately effective in killing mastigotes

30.2.1 Chemical Treatments

Antibiotic treatments have been used to remove selected species of microorganisms from the hindgut. Usually specific quantities of antibiotics are dissolved in water which is used to moisten filter paper. This food source is then administered to the insects.

The earliest work was by Cleveland (1928), who found that 5% (w/v) aqueous acid fuchsin removed the spirochaetes from termite hindguts. Later, Cleveland (1951) studied *Urinympha talea*, a mastigote of the wood-eating cockroach *Cryptocercus punctulatus*. He noted that the surface of the protist was covered with striations embedded beneath the pellicle. He suspected that these striations were bacteria. Treating the cockroaches with penicillin caused the bacteria-like striations to be removed. This was followed, 3 to 5 d later, by the death of the cockroaches perhaps due to the loss of the bacteria. Acetarson, sulphanil or ferric chloride was used to purge the microbes from *Eucryptotermes* sp. termites (Jakobi & Silva 1959). These treatments eventually resulted in the death of the termites.

Bready and Friedman (1963) were not successful in exterminating all bacteria from the hindgut of *Reticulitermes flavipes*. However, they selectively removed certain types of bacteria by using antibiotics either singly or in combinations.

The effects of antibiotics, wood preservatives, and wood-based toxins on termite hindgut ecosystems have been extensively studied (Speck *et al.* 1971). Variable results have been reported, such as large differences in response between individual termites in the same experiment (Speck *et al.* 1971; Lenz & Becker 1972). Mastigotes were removed from various termite species which were fed [^{14}C]-labelled nutrients such as glucose. They concluded that in some species of termites the mastigotes and/or bacteria seemed to be increasing the rate of incorporation of certain amino acids into protein.

Breznak *et al.* (1973) demonstrated that nitrogen fixation in *Coptotermes formosonus* is performed by bacteria in the hindgut. Antibiotic treatment of the termites resulted in the loss of bacteria together with the loss of nitrogen fixation. After 20 d, there was a reduction in the numbers of total protists. Similar experiments in *Cryptocercus* sp., demonstrated that antibiotics reduced the numbers of bacteria and nitrogen-fixing activity (Breznak *et al.* 1974). Antibiotic treatment also reduced emission of methane in *Reticulitermes flavipes* but not in *Cryptocercus* sp. Gut bacteria appear to have a greater responsibility for methane production in *R. flavipes* than in *Cryptocercus* sp.

Several anti-mastigote drugs were administered to species of *Reticulitermes*, *Coptotermes* and *Neotermes* (Zusková 1974). Among others, flagyl (metronidazole) completely eradicated all of the mastigotes in the hindgut. Chloramphenicol and several other antibiotics were effective against some of the mastigotes.

The remarkable efficacy of metronidazole against the termite mastigotes is beginning to be understood: the target of action in the cell is thought to be the hydrogenosome. This newly discovered organelle is found in some anaerobic protists probably including termite mastigotes (Müller 1980). Hydrogenosomes have been characterised as anaerobic equivalents of mitochondria. They contain enzymes which decarboxylate phosphoenolpyruvate to acetate and sometimes produce carbon dioxide and water. They lack cytochromes but contain ferredoxin-type proteins. Apparently they are bounded by a double membrane and some contain a structure resembling circular DNA. These features, which are similar to those in members of the free-living genus of anaerobic bacteria *Clostridium*, have prompted some to suggest that the hydrogenosome may be of symbiotic origin (Müller 1980). Metronidazole, which is an effective antibiotic against anaerobic bacteria, blocks hydrogenosome function and thus acts against anaerobic protists that may contain these organelles.

30.2.2 Natural Loss

Natural loss of microbial species from the termite gut community has been reported (Smythe 1972). Whether the absence of *Pyrsonympha* and *Dinenympha* species from *Reticulitermes* was due to their initial absence in the founding primary reproductive pair or to the age or condition of the colony, or other variable is not known. Some castes (reproductives, soldiers and young instars which receive pre-digested proctodeal or stomodeal food) are deficient in certain protist species which Cleveland (1925c) felt was due to differences in the caste's mode of nutrition. However, in healthy, freshly collected termites (*Pterotermes occidentis*, *Kalotermes schwartzi*) major members of the protist community are represented in the same ratios in all castes, even if the total number of symbionts is lowered (To *et al.* 1978, 1980). Furthermore, in nature the relative proportions of mastigotes and spirochaetes apparently differ with season and/or geographical location since such community differences have been seen in *Reticulitermes hesperus* from Southern California (D. Chase, personal communication). Natural variation of community composition occurs with age of colony, season, type of wood and geographical location (Lenz & Becker 1972). Thus it is wise to begin community studies with the same source of insect hosts from within the same colony.

The patterns of species loss due to moulting, stage of life cycle, or chemical treatment, and the order in which the population was replaced was studied by Clayton (1954). Some protist species are widespread in different termite hosts, others are restricted to a single species (Kirby 1938). Yamin (1979) has compiled literature on the distribution of termite protist species.

30.3 Maintenance of termite stocks and subcultures

Much of the following discussion of experimental techniques is based upon some recent studies of heat-treated or starved termites (B. Grosovsky & L. Margulis, unpublished observations).

Subterranean (Rhinotermitids) and drywood (Mastotermitids, Calotermitids) termites can generally be maintained for at least several months in the wood in which they are found nesting. Sawed wood piled in layers and sealed at the ends to prevent loss of moisture is suitable. The data in Figs 6 to 8 derive from studies of *Reticulitermes flavipes* nesting in a pile of logs and boards. The wood was cut into 60 cm lengths and stored in an incubator (22°C) with a water tray in the bottom. Logs or boards may be wrapped in plastic sheets but in any case the relative humidity must be kept high and the quantity of water maintained at a level appropriate to the termite species. Planks of wood arranged in layers are useful because they can be removed individually with minimal disturbance to the colony.

Termites are very sensitive to vibration and other disruptions. Individuals are best removed from stock colonies by gentle brushing with small flexible-haired paint brushes and kept in jars or Petri dishes with moistened filter paper discs. Small foam rubber discs (1 cm diameter) will maintain the moisture in starvation subcultures. In this technique pieces of wood debris (>2 mm) are removed with forceps, so that only fine debris and sawdust remain. The sample colony controls may be fed on cellulose fibres, filter paper or 'wood soup'—a boiled and dried mixture of nesting wood from the colony inoculated with pieces of wood from the old colony. Fungi and bacteria are introduced into the new cultures on the wood fragments and on the bodies of the termites and may help the termites acclimatise to the new subcolony. For antibiotic or chemical experiments feeding cannot be rigorously controlled and therefore the precise quantity of chemical administered is not known.

Before commencing an experiment the termites should be left undisturbed for at least 24 h. By that time, because they are vibration sensitive, photophobic and shun desiccating conditions, most termites will have gathered under the filter paper. The filter paper with the fine debris can then be removed carefully with forceps leaving the experimental termites in the transparent dish.

In the heat and starvation studies reported here either a clean disc of moistened filter paper was placed over the termites, or 30 to 50 termites were gently tapped into a second clean dish containing new paper. This quantity of termites is both easy to count and sufficient to persist for a time required by the experiments. Larger subcultures survive for the longest times (Becker 1969), apparently because they permit the termites to cluster together and thereby reducing surface area, maintaining optimal humidity and controlling bacterial and fungal growth. The same number of individuals in control cultures are stored at 22°C in incubators and the filter paper re-moistened about every three days. Unavoidable fungal growth, arising from spores carried on the termite bodies, develops in some dishes but generally filter paper cultures survive at least 120 days. In some cases cellulolytic fungi may even enhance the survival time of starved termite colonies (Cleveland 1928) but prolific fungal growth is deleterious.

30.4 Community census

For starvation cultures paper was replaced by a small foam rubber sponge which did not have to be dampened so frequently. For heat treatment studies the cultures were placed in a 36°C incubator for various time intervals: 1, 5, 10, 12, 15, 20 and 25 h, and the water replenished when needed.

The number of termites in each subculture is counted; if the termites are too active to be counted accurately they may be immobilised by refrigerating them at below zero temperatures for a minute or so (Dropkin 1946). To take a community census of symbionts, mature pseudergates or secondary reproductives are sacrificed on microscope slides. The head and thorax are removed with a single-edged razor blade, thus the anterior midgut is severed from the foregut. The abdomen is placed into a small drop of Trager's U solution (Trager 1934). Using two pairs of fine forceps (one pair grips the most posterior abdominal segment and the second holds the anterior abdomen) the gut is removed; the abdomen is discarded and the gut finely macerated. If, in moribund termites, the gut tend to disintegrate it may be necessary to macerate the entire abdomen and remove debris. A cover glass sealed at the edges with Vasoline is placed on the macerated gut preparation which is then observed with phase contrast light microscopy, usually at a magnification of × 400. After checking that the symbiont preparation is evenly distributed, each of ten random microscope fields of vision are counted. The major components of the microbial community of *Reticulitermes flavipes* are shown in Table 3.

Table 3. Symbiont populations in *Reticulitermes flavipes*: ranges in untreated controls.*

Trichonympha agilis	4 to 12
Spirotrichonympha spp.	0 to 4
Pyrsonympha vertens	1 to 15
Dinenympha gracilis	23 to 100
Microjoenia fallax	6 to 20
Holomastigotes elongatum	0 to 2
small mastigotes	present
Arthromitus sp.	present
pillotina spirochaetes	present
rods on *Dinenympha gracilis*	present

*Termites collected from wooded region in south-eastern Massachusetts in September. Counts represent the range of microbial numbers in ten microscope fields (at × 400).

In taking censuses of termite colonies, individuals with dark abdomens (indicating that they have been feeding) ought to be chosen for analysis if possible. If counted too soon after a moult, the quantity of paunch microbes may be anomalously low. Alates, soldiers, and primary reproductives ought not to be included in any census in order to avoid inaccuracies caused by caste development on symbiont community. If the numbers of subcultured termites decreased drastically to fewer than five termites, the individuals are not counted, instead populations are scored according to the total number of days the rest survived.

30.5 Heat treatment and starvation

In the original data reported here the termite viability and mastigote and bacterial diversity at 15 d intervals were plotted for each heat and starvation treatment. Untreated control termites were counted and the number of paunch symbionts in the controls and treated samples compared. A selection of results are shown in Figs 6 to 8.

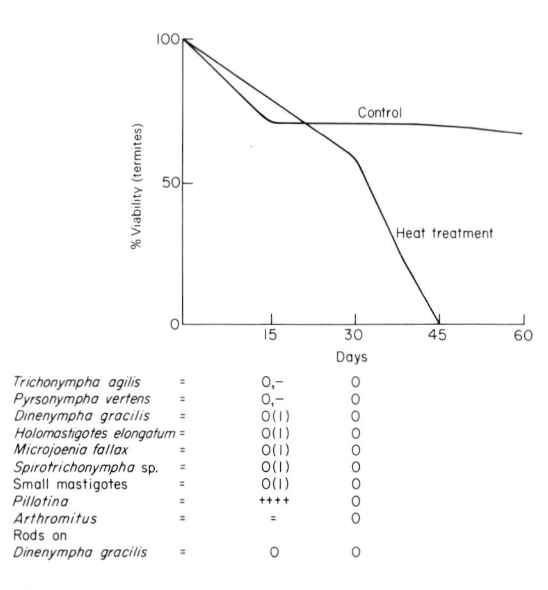

Figure 6. Symbiont populations in *Reticulitermes flavipes*: heat treatment 15 h at 36°C. Measurements taken after 15 d and 30 d.

Key to symbols
0 = complete elimination of the microbe by the treatment;
0 (*number*) = the number indicates the number of days that had elapsed before the microbe was eliminated;
= = the number of microbes falls within the range of values for untreated controls;
— = the number of microbes is fewer than the lowest seen in the untreated controls;

+ = the number of microbes is greater than the highest seen in untreated controls;
+ + + or + + + + = bloom—usually large number of microbes. + + + + is reserved for spectacular blooms such as when thousands of packed coordinated spirochaetes undulate in synchrony.

At least three termites were examined per entry, more than one symbol indicates differences in microbiota in termites treated the same way.

Since *Holomastigotes elongatum* and *Spirotrichonympha* spp. are found with such low frequency in untreated termites, the symbol (–) is not used for them; if they were absent in treated colonies for several consecutive days and then for all subsequent days the symbol (0) is used. The symbol (–) is applied to *Pyrsonympha vertens* only if fewer than five individuals per ten fields were seen for several consecutive days. The community composition is listed for each 15 d interval under the plot of termite viability for each treatment. Each entry represents the averaged results of 3 to 5 Petri dishes, several termites in each.

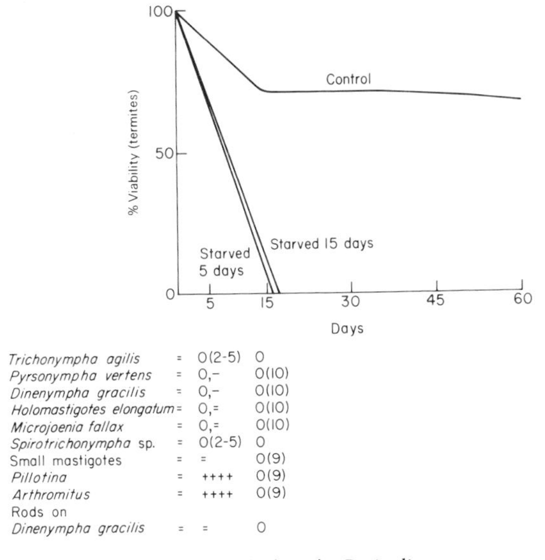

Figure 7. Symbiont populations in *Reticulitermes flavipes* after starvation for 5 and 15 d. Measurements taken after 5 d and 15 d. (For key to symbols see legend to Fig. 6.)

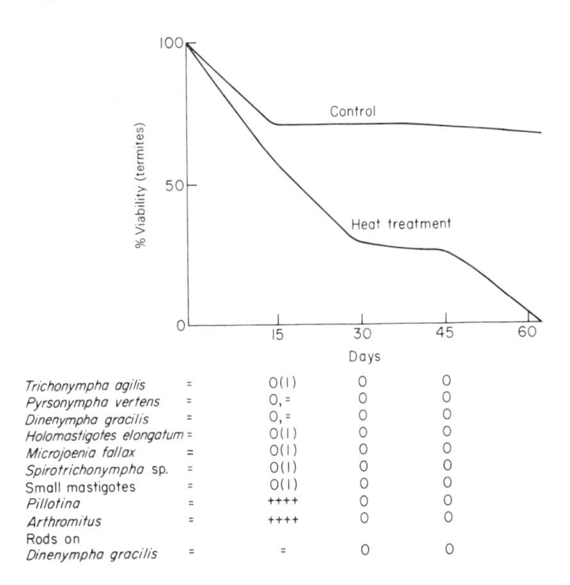

Figure 8. Symbiont populations in *Reticulitermes flavipes* after heat treatment (12 h at 36°C) and starvation (5 d). Measurements taken after 15 d, 30 d and 45 d. (For key to symbols see legend to Fig. 6.)

Five to fifteen hours of 36°C heat exposure (Fig. 6) or starvation (Figs 7 and 8) of termites produced a bloom of pillontins, spirochaetes and *Arthromitus* sp.: microbes which are often closely associated with each other in termites. Heat and starvation may be used as a mechanism to enrich for these microtubule-containing spirochaetes (Margulis *et al.* 1978, 1981).

Some of the milder heat treatments (1 to 10 h) were not detrimental to the termites nor to species of *Trichonympha*, *Pyrsonympha*, *Dinenympha* and *Microjoenia*. However, *Holomastigotes* and *Spirotrichonympha* species were eliminated by 10 h treatment. Prolonged heat treatment (15, 20, 25 h) is detrimental and eventually lethal for both the termites and their microsymbionts (Fig. 6).

Incubation at 36°C for 15 h (Fig. 6) resulted in an initial bloom of spirochaetes. Since it was accompanied by sharp decreases in *Trichonympha agilis* and *Pyrsonympha vertens* populations and the elimination of the other morphologically distinguishable symbionts, it is likely that it represents the best method by which microtobule-containing large spirochaetes may be enriched. Thirty days after the 15 h elevated temperature treatment, the entire symbiotic microbiota was eliminated and within 46 d all the termites had died.

The frequency distribution of mastigotes and bacteria in the hindgut ecosystem was responsive to changes in temperature and starvation regimes. Heat alters many variables affecting the growth of microbes. In addition to altering rates of metabolic reactions, it can affect pH, water activity, ion activity, viscosity, hydration, aggregation of macromolecules and solubility of gases (Farrell & Rose 1967).

The results of starvation of the termites reported in Fig. 7 are similar to those of Clayton (1954) and Lenz and Becker (1972). The two hypermastigotes *Trichonympha agilis* and *Spirotrichonympha* sp. were most sensitive: they were eliminated in two to five days. These species, found only in termite and wood-eating cockroaches, probably depend on cellulose as their sole carbon source. The polymastigotes and conspicuous bacterial morphotypes were not eliminated until the ninth or tenth day of the starvation regime and probably have other sources of nutrition besides cellulose. The spirochaetes and *Arthromitus* sp. thrived during the first week of starvation but died on the ninth day. Presumably an immediate consequence of starvation is a sudden increase of nutrients or other factors favourable for the growth of these two types of associated bacterial symbionts. Thus starvation represents a second method by which microtubule-containing spirochaetes may be enriched.

Termites which are starved for five days and then fed filter paper cannot be rescued. They lose their remaining mastigotes and die at about the same time as those termites starved for the entire experimental period of 15 d. All the mastigotes except *Trichonympha agilis*, *Pyrsonympha vertens*, and *Dinenympha gracilis* are killed by heat treatment at 36°C for 12 h. *T. agilis* is very sensitive to starvation; it dies prior to the fifth day. One expects that only *P. vertens*, *D. gracilis* and bacteria will survive a combination of stresses: heat treatment for 12 h and 5 d starvation. As expected (Fig. 8) *P. vertens* and *D. gracilis* survived for more than a week in some termites, but after the fifteenth day the entire symbiotic microbiota had succumbed; by the sixty-fourth day all the termites had died. Since in general termites starved for 5 d would have died prior to 21 d it seems likely that elevated temperature treatment helped the starved termites survive for more than two months (perhaps by enhancing the growth of cellulolytic fungi).

Although the trends are consistent, clearly there are uncontrolled variables in these experiments. The discrepancies may be due to differences in the original termite populations (they were collected at various locations in south-eastern Massachusetts, from different colonies and at different times of the year). Cellulolytic fungi and associated bacteria unavoidably present in the cultures may also affect the results.

30.6 Microbial ecology: the termite paunch ecosystem

In this section, when a mastigote is referred to as being cellulolytic, the possibility that bacteria associated with the mastigote may actually be the source of the cellulase is not excluded. One can infer that *Trichonympha agilis* is cellulolytic. These hypermastigotes are always present in healthy pseudergates. Furthermore, *T. agilis* is the first, or one of the first, mastigotes to die when the hosts are starved. In healthy termites wood particles are observed within *T. agilis*; they are taken in via a temporary posterior cytostome (Cleveland 1925a). This trichonymphid, like others of the genus, is probably directly dependent upon cellulose digestion.

Certain mastigotes, including some trichonymphids, can be grown axenically on cellulose media (Yamin 1978a,b, 1980a,b; Table 4) although the growth *in vitro* of *Trichonympha agilis* has not been achieved. Most members of the termite paunch community are unculturable, although some can be maintained on defined media (Buhse *et al.* 1975).

Table 4. Axenically grown mastigotes from termites* (courtesy of M. A. Yamin).

Insect	Mastigote	Cellulose digesting
Cryptocercus punctulatus	*Trichonympha chula*	yes
	Trichonympha sp.	yes
Zootermopsis	*Trichonympha sphaerica*	yes
	Tricercomitus termopsidis	no
	Trichomitopsis termopsidis	yes
Pterotermes occidentis	*Metadevescovina polyspira*	yes

*See Yamin (1978a,b; 1979) for details of growth medium.

Although *Trichonympha agilis* may be directly dependent on cellulose, it alone cannot support the termite. In the experiments described here after other mastigotes were eliminated by 15 or 20 h of heat treatment, the *T. agilis* population declined and the termites died. *T. agilis* is regularly associated with pillotina spirochaetes which course among its undulipodia; it appears as if the spirochaetes are grazing on some substance produced by the *Trichonympha* species. Cellulose-digesting mastigotes may be producing metabolites in addition to acetate, carbon dioxide and hydrogen, and possibly even glucose (Yamin 1980a,b).

Pyrsonympha vertens, which is always present in healthy workers and has been maintained in pure culture with cellulose as the sole carbon source, is directly dependent on a cellulose supply (Buhse *et al.* 1975).

Pyrsonympha vertens has been observed to consume wood particles within phagocytotic vesicles (Bloodgood *et al.* 1974) and wood particles may be observed in its phagocytotic vacuoles. Old colonies of termites that had depleted most of their food supply, sometimes lack *P. vertens* (Smythe 1972), a species which is thought to be particularly sensitive to a decrease in cellulose. On the other hand, when *Reticulitermes* sp. was starved in the experiments reported here *P. vertens* seemed less sensitive than other mastigotes to a lack of cellulose (Fig. 7) suggesting it is only facultatively cellulolytic.

Holomastigotes elongatum neither directly depends on cellulose nor is the species essential to termite viability. It is either absent or present in low numbers in healthy workers. The loss of *H. elongatum* after ten hours of elevated temperatures did not affect the viability of the termite. This mastigote which has absorptive nutrition (Kudo 1947) probably requires some cellulolytic product as carbon source. *H. elongatum* persists after the most sensitive mastigotes are eliminated when the host termite is starved (Fig. 7). In fact Buhse *et al.* (1975) reported that *H. elongatum* could not be grown on a cellulose-containing medium.

Spirotrichonympha spp. which were present in low numbers or absent altogether in healthy pseudergate workers are apparently not important for the viability of the termite. When the hyper-

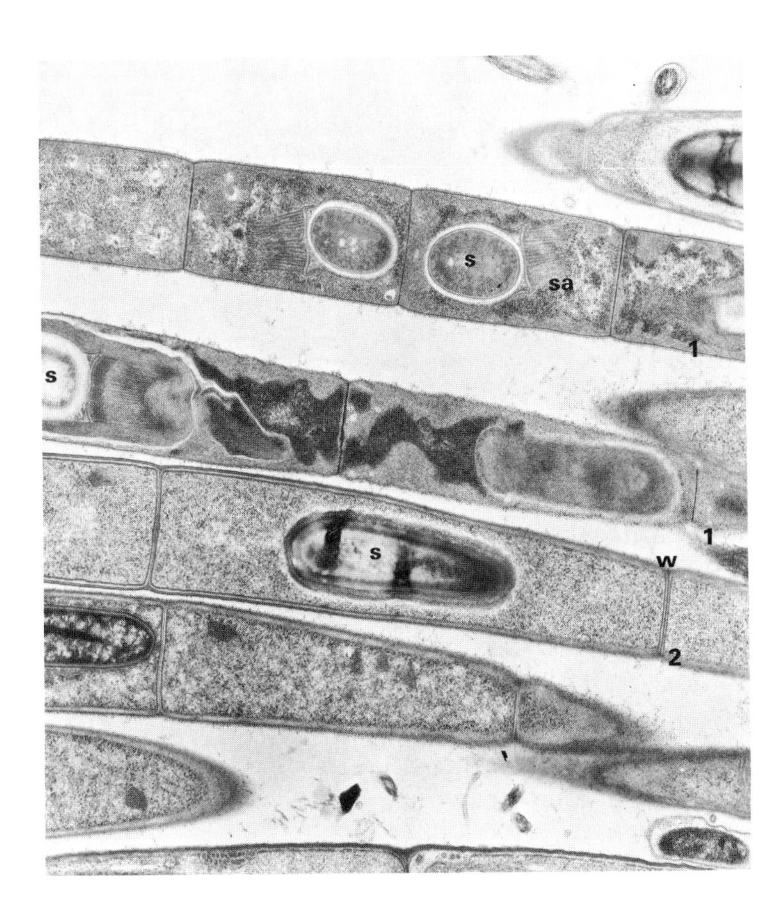

Figure 9. *Arthromitus* spp. from *Reticulitermes tibialis*. Several species are present in the same termite and at least two are represented here. On the basis of the spore (s) morphology, the spore appendage (sa) and the cell wall morphology and the mode of host attachment (w) the first *Arthromitus* (1) is a different species than the second (2). The unlabelled trichome in which a spore is forming at the left is probably a member of the second species. (Transmission electron micrograph × 50 000, courtesy of D. Chase.)

mastigote was eliminated by incubation for ten hours, the termites were not adversely affected. The cellulose medium devised by Buhse *et al.* (1975) could not support *Spirotrichonympha* sp., one of the first mastigotes eliminated by starvation (Fig. 7); it probably has a direct dependence on a cellulose source.

Dinenympha gracilis is closely associated with *Pyrsonympha vertens*. Natural termite populations lacking *P. vertens* often lacked *D. gracilis* as well (Smythe 1972). Both *Dinenympha gracilis* and *P. vertens* react similarly after both starvation and heating; only *D. gracilis* and *P. vertens* remain (Fig. 8). When the host was starved they were lost together on day ten (Fig. 7). Buhse *et al.* (1975) maintained both polymastigotes on a cellulose medium.

Wood particles were not observed in *Dinenympha gracilis* and it has been claimed that it is not cellulolytic (Kirby 1924). However, its close association with the cellulose-dependent species *Pyrsonympha vertens* suggests that *D. gracilis* may utilise the products of cellulose breakdown. Its food supply, whatever it is, must be abundant enough to support quite large populations. The numbers of rod-shaped bacteria associated with the surface of *Dinenympha gracilis* increased after one hour of heat treatment and decreased with starvation. Although the temperature elevation treatment increases the proportion of *Dinenympha vertens* the importance of this symbiont and its epibiota to the termite is unknown. The mechanism of regulation of surface bacteria on *Dinenympha* host is also unknown.

Microjoenia fallax probably does not directly depend on cellulose since it is one of the longest-surviving mastigotes during cellulose starvation (Fig. 7). It is doubtful if the *Arthromitus* sp. or pillotinas depend directly on cellulose either since they are not seen associated with wood and since they also persist after cellulose has been excluded from the diet (Fig. 2). Since they tend to bloom together near the undulipodia as the trichonymphids are dying the latter may yield nutrients for these bacteria and eventually huge numbers of synchronously undulating pillotinas are observed. *Arthromitus*, apparently an anaerobic or facultatively aerobic filamentous spore former (Fig. 9), probably resists starvation and heat by sporulation.

Some 30 to 40 isolates of bacteria have been removed from surface sterilised hindguts of *Pterotermes occidentis* from the Sonoran desert (To *et al.* 1978). Most are facultative anaerobes and many, such as the treponeme spirochaetes and *Arthromitus* sp. do not survive more than a single transfer. The roles of these species as well as those of the small polymastigotes in the hindgut ecosystem are virtually unknown.

Acknowledgements

This work was based in large part on the Masters Thesis of Betsey Dexter-Dyer Grosovsky (Boston University, Department of Biology, 1978). We are grateful to L. P. To, D. Chase, M. A. Yamin and J. Breznak for providing us with comment and unpublished observations. We are especially grateful to M. A. Yamin for critical reading and comment on the manuscript. Support for this research was provided by NASA, NSF and the Guggenheim Foundation to Lynn Margulis and by the Boston University Graduate School.

Recommended reading

BREZNAK J. A. (1975) Symbiotic relationship between termites and their microbiota. *Symposium Society of Experimental Biology* **29**, 559–80. Cambridge University Press, Cambridge.

CLEVELAND L. R. (1926) Symbiosis among animals with special reference to termites and their intestinal flagellates. *Quarterly Review of Biology* **1**, 51–60.

KRISHNA K. & WEESNER F. M. (1969) *Biology of Termites*, vols I, II. Academic Press, New York.

TO L., MARGULIS L., CHASE D. & NUTTING W. (1980) The symbiotic microbial community of the Sonoran Desert termite: *Pterotermes occidentis*. *Biosystems* **13**, 109–37.

YAMIN M. A. (1979) Flagellates of the order Trichomonadida Kirby, Oxymonadida Grassé and Hypermastigida Grassi and Foa reported from lower termites (Isoptern Families Mastotermidae, Kalotermitidae, Hodotermitidae, Termopsidae, Rhinotermitidae and Serritermitida) and from the wood-feeding roach *Cryptocercus* (Dictyoptera: Cryptocercidae). *Sociobiology* **4**, 3–119.

References

BECKER G. (1969) Rearing of termites and testing methods. In: *Biology of Termites* (Eds K. Krishna & F. M. Weesner), vol. 1, pp. 351–85. Academic Press, New York.

BLOODGOOD R. A., MILLER K. R., FITZHARRIS T. P. & MCINTOSH J. R. (1974) The ultrastructure of *Pyrsonympha* and its associated microorganisms. *Journal of Morphology* **143**, 77–106.

BREADY J. K. & FRIEDMAN S. (1963) The nutritional requirements of termites in axenic cultures. *Annals of the Entomological Society of America* **56**, 706–8.

BREZNAK J. A. (1975) Symbiotic relationship between termites and their microbiota. *Symposium Society of Experimental Biology* **29**, 559–80. Cambridge University Press, Cambridge.

BREZNAK J. A., BRILL W. J., MERTINS J. W. & COPPELL H. C. (1973) Nitrogen fixation in termites. *Nature, London* **244**, 577–80.

BREZNAK J. A., MERTINS J. W. & COPPELL H. C. (1974) Nitrogen fixation and methane production in *Cryptocercus*. *University of Wisconsin Forestry Notes*, no. 184.

BREZNAK J. A. & PANKRATZ H. S. (1977) *In situ* morphology of the gut microbiota of termites. *Journal of Applied and Environmental Microbiology* **33**, 406–26.

BUHSE H. E., STAMLER S. & SMITH H. E. (1975) Protracted maintenance of symbiotic polymastigote flagellates outside their termite host. *Journal of Protozoology* **22**, 11a–12a.

CLAYTON T. S. (1954) Defaunation studies on the protozoa of *R. flavipes*. *Journal of Protozoology* **1**, 6.

CLEVELAND L. R. (1925a) Method by which *T. campanula* a protozoon in intestine of termites ingests solid particles of wood for food. *Biological Bulletin* **48**, 282.

CLEVELAND L. R. (1925b) Ability of termites to live perhaps indefinitely on a diet of pure cellulose. *Biological Bulletin* **48**, 289–91.

CLEVELAND L. R. (1925c) The feeding habits of termite castes and its relation to their intestinal flagellates. *Biological Bulletin* **48**, 295.

CLEVELAND L. R. (1926) Symbiosis among animals with special reference to termites and their intestinal flagellates. *Quarterly Review of Biology* **1**, 51–60.

CLEVELAND L. R. (1928) Further observations and experiments on the symbiosis between termites and their intestinal protozoa. *Biological Bulletin* **5**, 231–7.

CLEVELAND L. R. (1951) Hormone induced sexual cycles of flagellates. VII. One division meiosis and ontogeny without cell division in Urinymphas. *Journal of Morphology* **88**, 385–440.

CLEVELAND L. R., HALL S. R., SANDERS E. P. & COLLIER J. (1934) The wood feeding roach *Cryptocercus*, its protozoa, and the symbiosis between protozoa and roach. *Memoirs of the American Academy of Arts and Sciences* **17**, 185–342.

DROPKIN V. H. (1946) The use of mixed colonies of termites in the study of host symbiont relationship. *Journal of Parasitology* **32**, 247.

EBRINGER L. (1978) Effect of drugs on chloroplasts. In: *Progress in Molecular and Subcellular Biology* (Ed. F. E. Hahn), vol. 6, pp. 271–350. Springer-Verlag, Berlin.

EDWARDS J. (1970) The mode of action of metronidazole against *Trichomonas vaginalis*. *Journal of General Microbiology* **63**, 297–302.

EUTICK M. L., O'BRIEN R. W. & SLAYTOR M. (1976) Aerobic state of guts of *Nausitermes* and *Coptotermes*. *Journal of Insect Physiology* **22**, 1377–80.

FARRELL J. & ROSE A. (1967) Temperature effects on microorganisms. *Annual Review of Microbiology* **21**, 101–20.

HUFF E. (1976) *Physicians Desk Reference*. Charles Baker, Jr. Medical Economics Company, Oradell, New Jersey.

HUNGATE R. E. (1941) Experiments on the nitrogen economy of termites. *Entomological Society of America. Annals* **34**, 467–89.

JAKOBI H. & SILVA J. D. L. (1959) Sobre a Defaunacao Bioquemica dos termitas. *Revista brasileira de biologia* **19**, 2.

JEON K. & HAH J. H. (1977) Effects of chloramphenicol on endosymbionts of *Amoeba proteus*. *Journal of Protozoology* **24** (2), 289–93.

KIRBY H. (1924) Morphology and mitosis of *Dinenympha* fimbriata sp. nov. *University of California Publications in Zoology* **26**, 199–220.

KIRBY H. (1938) Host parasite relations in the distribution of protozoa in termites. *University of California Publications in Zoology* **41**, 189–211.

KRISHNA K. & WEESNER F. M. (1969) *Biology of Termites*, vols I and II. Academic Press, New York.

KUDO R. (1947) *Protozoology*. Charles C. Thomas, Springfield, Illinois.

LENZ M. & BECKER G. (1972) Spezifische Empfindlichkeit Symbiontischer Flagellaten von Termiten gegenüber toxischen Stortzen. *Zeitschrift für Angewandte Zoologie* **59**, 205–36.

LUND A. E. (1930) The effect of diet upon the fauna of *Termopsis*. *University of California Publications* **36**, 81–96.

MARGULIS L. (1980) Undulipodia, flagella and cilia. *Biosystems* **12**, 105–8.

MARGULIS L. & SCHWARTZ K. V. (1982) *Five Kingdoms: A Guide to the Phyla of Life on Earth*. W. H. Freeman & Co., San Francisco.

MARGULIS L., TO L. & CHASE D. (1978) Microtubules in prokaryotes. *Science* **200**, 1118–24.

MARGULIS L., TO L. & CHASE D. (1981) Microtubules, undulipodia and *Pillotina* spirochetes. In: *Origins and Evolution of Eukaryotic Intracellular Organelles* (Ed. J. F. Frederick), pp. 356–68. New York Academy of Sciences, New York.

MERTINS J. W., COPPELL H. C., BREZNAK J. A. & BRILL W. J. (1973) Demonstration of nitrogen fixation in termites. *University of Wisconsin Forestry Notes*, No. 168.

MÜLLER M. (1980) The hydrogenosome. In: *The Eukaryotic Microbial Cell* (Eds G. W. Gooday, D. Lloyd &

A. P. J. Trinci), pp. 127–43. Cambridge University Press, Cambridge.

MÜLLER M., LINDMARK D. G. & MCLAUGHLIN T. (1977) Mode of action of metronidazole on anaerobic micro-organisms. In: *Metronidazole* (Eds S. M. Feingold, J. McFadzean & F. J. C. Roe), pp. 12–19. Amsterdam, Excerpta Medica.

POTRIKUS C. J. & BREZNAK J. A. (1977) Nitrogen fixing *Enterobacter agglomerans* isolated from the guts of woodeating termites. *Applied and Environmental Microbiology* **33**, 392–9.

SMYTHE R. V. (1972) Feeding and survival at constant temperatures by normally and abnormally faunated *Reticulitermes virginicus*. *Annals of the Entomological Society of America* **65**, 756–7.

SPECK V., BECKER G. & LENZ M. (1971) Ernährungsphysiologische Untersuchungen an Termiten nach selektiven medikamentaser Ausschaltung der Darmsymbioton. *Zeitschrift für Angewandte Zoologie* **58**, 475–95.

TO L., MARGULIS L., CHASE D. & NUTTING W. (1980) The symbiotic microbial community of the Sonoran Desert termite: *Pterotermes occidentis*. *Biosystems* **13**, 109–37.

TO L., MARGULIS L. & CHEUNG A. T. W. (1978) *Pillotina* and *Hollandina*: the distribution and behavior of large spirochaetes symbiotic in termites. *Microbios* **22**, 103–33.

TRAGER W. (1934) The cultivation of a cellulose digesting flagellate. *Biological Bulletin* **66**, 182–90.

VEIVERS P. C., O'BRIAN R. W. & SLAYTOR M. (1980) The redox state of the gut of termites. *Journal of Insect Physiology* **26**, 75–7.

WILSON C. O. & JONES T. E. (1975) *American Drug Index*. J. P. Lippincott, Philadelphia.

YAMIN M. A. (1978a) Axenic cultivation of the flagellate *Tricercomitus divergens* Kirby from the termite *Cryptotermes cavifrons* (Banks). *Journal of Parasitology* **64**, 1122–3.

YAMIN M. A. (1978b) Axenic cultivation of the cellulolytic flagellate *Trichomitopsis termopsidas* from the termite *Zootermopsis*. *Journal of Protozoology* **25**, 535–8.

YAMIN M. A. & TRAGER W. (1979) Cellulolytic activity of an axenically-cultivated termite flagellate, *Trichomitopsis termopsidis*. *Microbiology* **113**, 417–20.

YAMIN M. A. (1979) Flagellates of the orders Trichomonadida Kirby, Oxymonadida Grassé, and Hypermastigida Grassi and Foà reported from lower termites (Isoptern Families Mastotermidae, Kalotermitidae, Hodotermitidae, Termopsidae, Rhinotermitidae and Serritermitida) and from the wood-feeding roach *Cryptocercus* (Dictyoptera: Cryptocercidae). *Sociobiology* **4**, 3–119.

YAMIN M. A. (1980a) Cellulose metabolism by the termite flagellate *Trichomitopsis termopsidis*. *Applied and Environmental Microbiology* **39**, 859–63.

YAMIN M. A. (1980b) Axenic cultivation and metabolism of mutualistic cellulose-digesting flagellates of termites. Ph.D thesis, Rockefeller University.

YOSHIKAWA T. (1974) *In vitro* resistance of *Neisseria* to Metronidazole. *Antimicrobial Agents and Chemistry* **6**, 327–9.

ZUSKOVÁ Z. (1974) Defaunation of protozoans in termites. *Journal of Protozoology* **21**, 459.

Chapter 31 • Nematode-Destroying Fungi

George L. Barron

31.1 Introduction

More than 150 species of fungi are known to attack nematodes or their eggs (Barron 1977, 1981). Nematode-destroying fungi can be classified as either predatory or endoparasitic. Predators produce an extensive hyphal system in the environment and at intervals along the length of the hyphae trapping devices are produced to catch and hold live nematodes. The victim is then penetrated by fungal hyphae and its entire body contents rapidly consumed.

Organs of capture have been described in detail by Drechsler (1934, 1941a) and reviewed by Barron (1977) but in general nematodes are captured either by adhesive or non-adhesive devices. Adhesive devices include hyphae, knobs, branches and nets, whereas non-adhesive devices are either non-constricting (passive) rings or constricting (active) rings. In the lower (non-septate) fungi nematodes are captured by means of adhesive produced directly on the hyphae. The hyphae are either coated with adhesive along their entire length or are capable of producing adhesive at any point in response to nematode contact. *Stylopage hadra* (Drechsler 1935) is one of the most common predators capturing nematodes in this way. In the higher (septate) fungi the adhesive branch is the simplest of the organs of capture produced. Trapping structures are only a few cells in height and arise as short laterals growing from prostrate hyphae on the substrate. A thin film of adhesive material is secreted over the entire surface of the branch. *Dactylella gephyropaga* (Drechsler 1937) is one of the most common predators capturing nematodes using adhesive branches.

Studies on predatory fungi indicate that adhesive nets are more common than any other trapping device and that *Arthrobotrys oligospora* Fresenius is the most common species capturing nematodes with adhesive nets (Barron 1977). Nets may be prostrate but are more commonly raised above the surface of the substratum. An erect lateral branch grows out from a vegetative hypha, curves round and grows down to eventually anastomose with the parent hypha. A branch from this primary loop or another branch from the hypha repeats this process and, through anastomoses, a complex three-dimensional network is built up. The net is coated with a thin adhesive film.

Some predators capture nematodes using adhesive knobs; a method that has evolved in both the Deuteromycetes and the Basidiomycetes. In the Deuteromycetes the knob is sometimes sessile on the hypha but more often a globose to subglobose adhesive cell is produced at the apex of a slender, non-adhesive stalk. The adhesive material forms a thin film over the surface of the knob. The knob itself is separated from the support stalk by a septum at its point of origin. Nematodes adhere to a knob and eventually are caught by several knobs with subsequent penetration and destruction. *Dactylaria candida* (Drechsler 1937) is the most common species attacking nematodes in this way. Fungi producing adhesive knobs often produce non-constricting rings as an alternative trapping device.

In predatory Basidiomycetes the adhesive knob is very distinctive, consisting of a secretory cell shaped like an hour glass. At maturity the secretory cell is engulfed in a large, spherical ball of viscous, sticky material. The adhesive is much more profuse than in the Deuteromycetes. All predatory Basidiomycetes are characterised by having hyphae with clamp connections and are contained within the genus *Nematoctonus* (Drechsler 1941b).

In predatory fungi with non-constricting (passive) rings, erect lateral branches arise from the prostrate hyphae. The branch is at first very slender but widens in the upper part then curves around following a circular pathway until the tip finally anastomoses with the support stalk to form a three- or four-celled ring. Non-constricting rings are passive in their action and when the nematode enters the ring, its forward motion causes the ring to wedge around the body. Such rings are virtually impossible to dislodge and during the struggle the ring often breaks off at the stalk apex and the host swims off with the ring appearing as a collar around its body or head region. However, the detached ring is still functional and gives rise to hyphae which penetrate and assimilate the nematode.

Dactylaria candida produces non-constricting rings as well as adhesive knobs, although in most strains the adhesive knobs are said to predominate. Although this is true in plate culture conditions, it may not be true in the natural environment.

The constricting ring is produced in the same manner as the non-constricting ring but the support stalk is usually shorter and much stouter. When a nematode moves into the ring, it triggers off a response by the fungus such that the three cells composing the ring swell rapidly inwards with such force that they pinch in the body of the nematode and hold it securely allowing no chance for escape. Swelling is strictly inwards with the outside diameter of a triggered ring being much the same as that of the untriggered ring. It takes about 0.1 s for the ring cells to inflate to their maximum size. Of the predatory fungi capturing nematodes using constricting rings, *Dactylaria brochopaga* (Drechsler 1937) is one of the most frequently encountered.

After capture by a trapping device, the nematode is colonised rapidly by assimilative hyphae and in 12 to 24 h the contents of the host have been completely digested. The protoplasm in the assimilative hyphae is then translocated for further growth and reproduction and the host appears as a shell filled with the now empty hyphae of the predator. Conidiophores are produced singly or in terminal clusters at the apex of the conidiophore or its branches. The tall conidiophores and relatively large conidia can be seen readily with a dissecting microscope.

In contrast to predatory fungi, endoparasites have no extensive hyphal development outside the body of the host. In most species only evacuation tubes (in the case of lower fungi) or conidiophores and conidia (in the case of higher fungi) are produced externally. In a few species fertile hyphae arise from the body of the nematode and trail over the surface of the agar. Such hyphae are not assimilative and are strictly for the purpose of producing conidia.

Conidiophores and conidia of endoparasites are very small. Because of their small size, endoparasites are more difficult to locate under the dissecting microscope than predators.

A number of Deuteromycetes have developed adhesive spores as the primary method of host infection; the best known are species of *Verticillium* and *Meria*.

In *Meria coniospora* (Drechsler 1941b) the spores are tear-drop-shaped and at maturity have a small adhesive bud at the distal end. As with other adhesive devices, adhesion may require a short lag period before the material cements itself to the host wall. The conidia of *Meria* can attack at any point on the body wall but most infections are in the head region in the vicinity of the mouth. Once attached, spores are very difficult to dislodge and no amount of thrashing on the part of the victim seems effective in releasing them. In Petri dishes dozens, or sometimes even hundreds, of conidia may be attached to a single nematode. One spore, however, is sufficient to cause infection and subsequent death. The attached spore germinates, penetrates directly through the cuticle, and forms an infection hypha in the body cavity.

Within a few days the body of the nematode is filled with the hyphae of the parasite. Some break out through the body wall at a number of points. External development is restricted to the development of conidiophores and conidia of the asexual reproductive stage. Conidia gather in clusters or short chains at the mouths of the tubular conidiogenous cell.

Harposporium anguillulae Zopf is the most commonly encountered endoparasite with ingested spores. The spores tend to be more or less crescent-shaped with a very sharp point at one end. The ends of the spore are not precisely in the same plane and in this species and in many other species of *Harposporium* the spores resemble part of the wide helix. The sharp point is usually reflexed back towards the main axis of the spore. During the process of swallowing, the point of the spore apparently penetrates between the muscle fibres of the oesophagus and becomes lodged. A germ tube develops from the centre of the convex side of the spore, penetrates the oesophageal muscle and enters the body cavity of the host. Colonisation of the host and further development is similar to the genus *Meria*.

In lower fungi, flagellate spores are produced which are attracted to the host by secretions from the body orifices. Following chemical gradients, zoospores migrate to and eventually encyst on the host cuticle although in some of the lower fungi the zoospores show no chemotactic response. After a short swimming phase, presumably for dispersal, the zoospore encysts and produces an adhesive apical bud which allows it to attach to passing nematodes. In *Haptoglossa zoospora* zoospores encyst then produce a highly specialised injection cell which shoots an infective sporidium through the host cuticle.

Experimental

31.2 Isolation and maintenance of nematodes

A continuing supply of nematodes is necessary either as bait to recover nematode-destroying fungi or to maintain and purify the fungi once they have been located. Nematode cultures also facilitate study of the details on the biology of nematode–fungus associations which are either not known or not fully understood.

31.2.1 The Baermann Funnel

Nematodes can be recovered from soil or organic debris using the Baermann funnel technique (Hooper 1970). Samples (10 to 100 g) of freshly collected material are placed on two layers of tissue paper supported by nylon mesh glued to the base of a tin-foil pie dish from which the bottom has been removed (Fig. 1). The ends of the tissue are folded over and the sample covered with a large watch glass to prevent evaporation of water. The level of water in the funnel is raised until it touches the bottom of the pie dish. Water is soaked up by the soil and the tissue. The apparatus should be checked after an hour and water added as required to give a broad contact between the water and the soil. Nematodes wriggle through the tissue, sink down the funnel by gravity and are collected at the base in a small (5 ml) collecting tube.

If the sample is variable (containing large particles of debris) a higher nematode yield may be obtained by using the preliminary screening techniques described by Flegg (1967).

(1) A 250 ml aliquot of well-mixed soil or debris is washed through a 20 mesh sieve into a pail and the debris on the screen discarded.

(2) The soil and water in the pail are mixed thoroughly, allowed to stand for about 30 s, decanted through a 275 to 325 mesh sieve and the residue (containing the nematodes) collected in a beaker.

(3) Water is added to the pail and step 2 repeated.

(4) The residues recovered from 275 to 325 mesh screen are placed on a Baermann funnel apparatus as indicated previously.

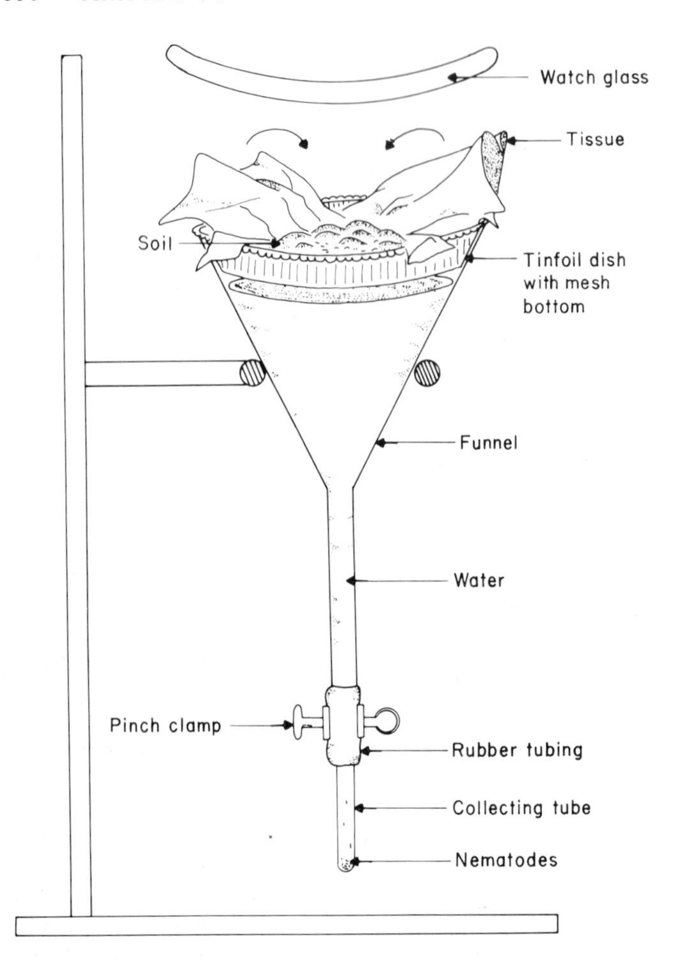

Figure 1. Baermann funnel apparatus (after Barron 1977).

31.2.2 Culturing Nematodes

The population obtained from the Baermann funnel is usually a mixture of nematode species. The nematode suspension (1 ml) is pipetted onto a 7-day-old water agar (WA) plate. The Petri dish environment is unsuitable for the survival of most species of nematodes and, after incubation for 7 days at room temperature (20 to 30°C), only one or a few species will survive as dominant populations. From the surviving population a large, gravid female is removed on a fine needle and transfered to a drop of water on a plate of nutrient medium. Nigon's medium (Dougherty 1960) has proved most useful for culturing nematodes and contains in distilled water (g l^{-1}) $MgSO_4.7H_20$, 0.75; K_2HPO_4, 0.75; NaCl, 2.75; KNO_3, 3.0; peptone, 2.5; lecithin, 1.0; and agar, 15.0. The gravid female should be inspected under the low power of a microscope to confirm that the eggs are present in the body. This female should be able to lay sufficient eggs to start a stock culture containing only a single species of nematode. Nematodes hatching out feed on bacteria which are carried over as natural contaminants or which can be supplemented by the addition of selected bacteria, such as *Escherichia coli*.

Non-sterile cultures of nematodes are ideal for most routine studies such as recovery of predators or endoparasites or to study infection cycles. Nematodes grow fairly well on Nigon's medium but seldom form the dense, writhing masses found on decaying organic matter Useful materials for encouraging nematode populations are either peanut butter (Hooper 1970) or dried green pea soup mix. A little peanut butter in the centre of a Nigon plate is excellent for encouraging rapid development of nematodes. If pea soup is used, it is first autoclaved in its tin foil package and powdered before the package is opened. About 0.5 g is sprinkled over a 2.5 cm diameter area in the centre of the plate and covered with a few drops of nematode suspension. Plates are incubated at 20 to 30°C and large numbers of nematodes are available in a few days, especially at the higher temperatures.

31.2.3 Duddington's Techniques I

Several methods for culturing nematodes have been proposed by Duddington (1955) and are useful alternatives to those described above. Duddington (1955) prefered to use weak maize meal agar. The meal (20 g) was warmed with tap water (1 litre) for one hour at 70°C and filtered through glass wool before adding agar (2% w/v) and autoclaving. According to Duddington (1955) this method permitted the growth of enough bacteria to provide food for protozoa and nematodes but was too weak to encourage vigorous growth of aggressive saprophytes, such as a *Rhizopus* species.

Duddington (1955) suggested rabbit dung agar as another useful medium. This was made by steeping rabbit pellets in tap water for several days, filtering through glass wool, diluting to a pale straw colour, adding agar and autoclaving. As can be seen from the above description preparation of this medium does not seem to be critical and there is a wide margin for error. The disadvantage of dung infusion medium, according to Duddington (1955), was that it often became soft and churned up by nematode activity. He also pointed out the advantage of this medium for studying endoparasites which are obscured on the richer maize-meal agar.

Plates of dung infusion medium are inoculated with fresh horse or sheep dung or other material rich in eelworms. In a few days nematodes cover the plates in enormous numbers. One such plate is now used to inoculate further plates of the same medium by cutting out small pieces of agar where inspection has shown them to be free of contaminating, predaceous fungi. The second series culture is used to inoculate plates of weak maize-meal agar, Nigon's medium or water agar.

The danger in Duddington's technique is the possibility of contamination with nematode parasites. If, however, a succession of transfers is made at short intervals (24 h), the nematodes usually outgrow the parasites.

Culturing nematodes in association with one or several species of bacteria (axenic culture) is satisfactory for most purposes. For more critical or sophisticated studies, however, it is sometimes necessary to have microbe-free nematodes (axenic culture) growing either in a nutrient medium or in association with a pure culture of a fungus in the case of fungal-feeding nematodes.

The development and maintenance of axenic cultures of nematodes requires more demanding techniques and is beyond the scope of this chapter but an excellent and detailed review of methods on all aspects of nematode culture is given by Hooper (1970).

31.2.4 Culture Chamber

Panagrellus redivivus is one of the nematodes most commonly used to study nematode-destroying fungi. Winkler and Pramer (1961) described a simple apparatus to provide a constant supply of large numbers of relatively clean nematodes. The chamber serves for both the culture and collection of the nematodes.

The apparatus consists of a 15 cm diameter funnel with a ground lip and attached stopcock. A glass-rod saddle supports a heavy Petri dish (9 cm) bottom and approximately 25 g oatmeal in 45 ml of water. The stopcock is closed and water added until the level of liquid in the funnel is 3 to 4 mm below the open end of the dish. The moist oatmeal is inoculated with *Panagrellus redivivus* and the unit closed, using a ground-glass plate as a lid. The top of the dish is about 1.0 cm below the lid. Adults move out of oatmeal over the edge of the dish into the water and drop by gravity to the stopcock where they can be run off as required. Winkler and Pramer (1961) estimated that a chamber can produce 1.6×10^7 nematodes in 12 d at 20°C. Although more nematodes are produced at higher temperatures, 20°C is preferred because the population is maintained over a longer period.

31.3 Isolation and maintenance of fungi

Sites which over a long period of time have encouraged the development and maintenance of numerous and diverse species of nematodes are most likely to harbour a wide range of predators and parasites (Barron 1977). Old compost piles, rotting wood, leaf mould, old pasture soils and agricultural soils are particularly good sources of nematodes and their parasites. Duddington (1951) found that soil from under Bryophytes was an especially rich source.

31.3.1 Sprinkling Techniques

Drechsler's method

Drechsler's technique (Drechsler 1941) for discovering nematode-destroying fungi is both simple and effective. It is based on the observation that agar plate cultures prepared to isolate Oomycetes, such as *Pythium* and *Phytophthora*, from discoloured roots or other decaying plant material, often encouraged abundant multiplication of free-living nematodes. Drechsler found that these nematodes were frequently destroyed in large numbers by a variety of predaceous and parasitic fungi.

A pinch of leaf mould or other organic debris is sprinkled over a plate of weak maize meal agar previously overgrown with a *Pythium* or *Phytophthora* species. After the addition of the organic debris to the Petri dish there is a development of saprophytic 'weed' fungi. This is followed by a gradual increase in the natural populations of nematodes introduced in the debris. Often the aggressive predatory fungi decimate these nematode populations. Predatory fungi may disappear with time or there may be a further flush of nematode activity followed by the development of additional predators or endoparasites. Frequently mites or worms introduced with the organic particles cause problems. Observations are made frequently and regularly over a long period of time. The material is well worked by a variety of organisms and nematode-destroying fungi are often difficult to locate.

This method approaches the natural system most closely and encourages the development of a variety of nematodes and subsequently a wide range of parasites or predators.

Baited plates

Adding a suspension of nematodes to a Petri dish prior to the addition of the soil or organic debris has proved one of the simplest and most reliable methods of recovering nematode-destroying fungi (Wyborn *et al.* 1969).

A Petri dish containing water agar (2% w/v) or weak maize meal agar, (0.25 to 0.5 strength, Difco) is baited with 1 ml of a heavy suspension of stock nematodes (5000 ml^{-1}) and sprinkled lightly with 0.5 to 1.0 g of soil or organic debris. Five days after treatment Petri dishes are inspected using a dissecting binocular microscope. The light source should be set at a very low angle to the horizontal to pick out the delicate conidiophores or hyphae growing from dead hosts. Plates are inspected at 1 to 3 d intervals for several weeks.

The presence of numerous nematodes on baited plates exerts a fungistatic effect which prevents profuse development of saprophytic fungi. In many cases the only fungi found are the predators or endoparasites of nematodes which allows a much more rapid inspection of the plates. This method is very good for predators but is not the best method to recover endoparasites. As only one nematode species is normally used as bait, the parasites present in the sample which are incapable of attacking the bait host are not recovered. If aerial growth of saprophytic fungi becomes a problem using the sprinkling method, then the nematodes can be introduced to the Petri dish a day or two before adding the soil or debris.

31.3.2 Duddington's Techniques II

Duddington's method for the recovery of predaceous fungi (Duddington 1955) involves 2 g soil scattered over the bottom of a sterile Petri plate then a layer of weak maize meal agar, cooled to near solidification is poured over the soil. Mixing is kept to a minium as observations are difficult to make if the soil is scattered over the entire plate.

Duddington (1955) also recommended a culture rejuvenation technique, which may be repeated several times, for extending the life of predaceous cultures. Sterile agar, cooled to near solidification, is poured as a thin layer over the surface of an old culture.

31.3.3 Differential Centrifugation

Predatory fungi are aggressive and often sporulate profusely over the agar surface. Petri plates may become littered with dead corpses of captured hosts and endoparasites become increasingly difficult to spot. The differential centrifugation technique (Barron 1969) was designed to separate the relatively small-spored endoparasites from large-spored predators.

About 200 g well-mixed soil is measured out in a beaker, added to 250 ml water in a Mason jar and blended for 30 to 60 s. The soil-water mixture is passed through a soil screen to remove the coarse material. The mixture is passed through a further, finer screen and the sediment discarded. The sample passing through the screen is poured into two 50 ml centrifuge tubes and centrifuged at 750 **g** for 2 min to remove the heavier soil particles and large spores of the predatory fungi. The supernatant is decanted, retained and centrifuged again at 2500 **g** for 1 h. The supernatant is discarded and the residue saved. A few drops of water are added to the residue and stirred with a glass rod. The mixture is poured on to 2% (w/v) water agar and spread over about half the plate. A few drops of a heavy suspension of nematodes are added to each plate and the plates incubated at room temperature and inspected at intervals. Usually within 10 d parasitised nematodes are found on the plates.

The water agar plates should be poured several days before the addition of soil and nematodes. This allows excess water to be rapidly absorbed and makes observations much easier.

It is obvious that only those species of parasites capable of attacking the bait nematode are recovered using this technique. Saprophytic fungi do not grow on the plate and because a large part of the plate is relatively free from soil, the infected nematodes can be spotted and transferred with relative ease.

31.3.4 THE BAERMANN FUNNEL

The use of the Baermann funnel for studying nematode-destroying fungi was first proposed by Giuma and Cooke (1972) and is based upon the reasonable assumption that in any natural population of nematodes, a small percentage will be infected by fungal parasites. During the period immediately following capture, nematodes are still sufficiently mobile to be recovered by this procedure.

Nematodes are recovered from soil or debris using the Baermann funnel technique as outlined previously (see section 32.2.1). After 24 to 48 h the contents of the collecting tube are poured into a 10 ml centrifuge tube and the nematode suspensions concentrated by centrifugation at 1000 r/min for 3 min. The supernatant is discarded and the nematodes resuspended in 3 ml distilled water and incubated at room temperature (18 to 22°C). Examination of suspensions immediately after centrifugation show that very few nematodes have obvious symptoms of fungal attack. After 4 d or more of incubation in distilled water, samples are placed in glass cavity blocks and re-examined. Nematodes with thalli within them or with hyphae emerging from them are removed by means of a fine Pasteur pipette to a sterile 10 ml centrifuge tube containing 5 ml sterile, distilled water and the suspension concentrated as before. The supernatant is discarded and the nematodes resuspended in a further 5 ml sterile, distilled water and again centrifuged. This process is repeated four times. After the final wash nematodes are once more transferred to glass cavity blocks in a little distilled water. Those nematodes with hyphae emerging from them are transferred to plates of 2% (w/w) cornmeal agar while those containing thalli are left in the cavity blocks and examined at daily intervals in water to observe the development of the parasites.

Fertile hyphae usually grow out rapidly from dead nematodes placed on agar and bacterial contamination is usually negligible. Giuma and Cooke (1972) noted that the technique allowed reasonably rapid isolation of endozoic parasites without either the necessity for their maintenance on nematode populations or the use of antibiotics in the agar medium employed.

Barron (1977) recommended an alternative method of handling Baermann extracts. The collecting tube containing the nematodes is removed from the funnel. The supernatant (4 ml) is pipetted off gently with a sterile pipette or eye dropper and the remaining 1 ml (containing the nematodes) is poured onto a water agar Petri dish. Petri dishes are examined at daily intervals for 10 d routinely or for longer periods if preliminary observations warrant it. In some cases the nematode suspension is poured onto fresh water agar plates and in other cases 7 d water agar plates are used. In the former case the 1 ml nematode suspension forms a water layer over the entire surface of the agar and this remains for many days. If necessary, additional water can be added to prevent drying out. This type of treatment favours flagellate endoparasites, such as *Catenaria* and *Myzocytium* species. If 7 d plates are used the 1 ml liquid is quickly taken up by the agar and the plates have no obvious film of free water. This type of treatment favours *Nematoctonus* and *Harposporium* species and the Hyphomycete group in general and eliminates many of the strictly flagellate species. The Baermann funnel favours the recovery of endoparasites but predaceous species are sometimes found.

31.3.5 PURIFICATION AND CULTURE

Non-axenic cultures

Barron (1977) described a method for isolating and maintaining specific endoparasite-nematode cultures. Once an infected nematode is located it is transferred to a Petri dish containing water agar seeded with a drop of nematode suspension. If the host is susceptible, and the environmental conditions suitable, numerous infected nematodes are soon found. The time taken to establish a new infection cycle varies considerably from parasite to parasite but is normally 2 to 7 d.

The process of purification by transferring single infected nematodes can be repeated. The host nematode can also be washed using centrifugation (Giuma & Cooke 1972) in order to eliminate contaminating organisms. Alternatively, individual nematodes can be washed by transferring through a series of sterile saline washes using deep well slides.

With care, non-axenic cultures of an endoparasite and its nematode host can be maintained indefinitely with regular transfers. Studies using non-axenic cultures have permitted life histories and infection cycles to be established with a much greater degree of confidence than previously when observations were made directly on infected hosts recovered from the primary Petri dish on which the parasite was located.

Axenic cultures

Pure cultures of predators are easy to obtain by streaking using the method of Aschner and Kohn (1958). The tip of mounted needle is flamed and cooled in agar. A spore or spore cluster is lightly touched with the cooled needle and the dislodged conidia will adhere to the tip. The needle is streaked lightly over the surface of a malt extract plate. The Petri dish is inverted and inspected under low power magnification ($\times$ 100). The conidia are located and germination and growth checked at daily intervals. Predatory fungi usually produce aerial mycelia and therefore it is not difficult to pick off the spores with a needle and it is not necessary to add antibiotics to the medium.

With the exception of *Catenaria* species the non-septate endoparasites have not yet been obtained in pure culture. Cultures of Hyphomycetes, however, can be obtained using the methods described by Aschner and Kohn (1958). These workers successfully streaked the conidia of *Harposporium anguillulae* on glucose-yeast-extract agar with aureomycin (30 mg ml^{-1}). The conidia of endoparasites, however, do not germinate readily. Also, the conidiophores are short and it is difficult to pick up conidia without touching the surface of the dead host or the substrate below, both of which are heavily contaminated with microorganisms. Aschner and Kohn (1958) recommended an alternative method of purification involving transfer of the infected nematodes as follows.

Nematodes which are infected with the fungus, which is not yet sporulating, are picked up on a fine needle and transferred to a Petri dish containing malt extract agar or potato dextrose agar supplemented with aureomycin (0.1 mg ml^{-1}). Instead of producing conidiophores and conidia, the hyphae from the nematode continue to grow in a vegetative manner into the nutrient medium surrounding the nematode. When sufficient growth is present, removal of hyphal fragments from the edge of the growing colony usually results in bacteria-free cultures on the first transfer. With slower-growing species a second transfer onto agar containing aureomycin is often necessary in order to eliminate the bacteria.

Stimulating production of conidia

Some species, such as *Meria coniospora*, produce numerous conidia in culture. On the other hand, certain species or strains are virtually sterile in culture producing few or no conidia and certainly not enough for experimental purposes. This was found for *Harposporium anguillulae* by Aschner and Kohn (1958). To stimulate sporulation these workers took a small piece of a pure colony and fragmented it into pieces (the size of a pinhead or less) which were then placed on water agar plates. They found that such fragmented mycelium under low nutrient conditions developed hyphae which, with regard to spore formation were identical to those observed in natural infections.

Maintenance of cultures

Fungal cultures frequently lose viability after several months refrigeration. Also, while some species sporulate freely on agar media, others are sterile and reinfection of the nematodes is not readily obtained from the non-sporulating cultures. To maintain these fungi in a condition in which they may be easily and quickly recovered in the active parasitic state, they can be maintained in soil cultures (Barron 1969).

Small screw-topped vials (30 ml) are half-filled with moist, screened soil, and sterilised. A heavy suspension (1.0 ml) of healthy nematodes is added to a plate containing a large number of heavily infected nematodes and the mixture swirled gently for a few seconds. About 0.5 ml of the resultant suspension (containing fresh nematodes, infected nematodes, conidia and bacteria) is added to each of the vials of soil. The vials are allowed to stand at room temperature for several days to allow infection of the fresh nematode population to become well established and are then refrigerated at 5°C.

When required, the vial is shaken to loosen the soil which is scattered over a plate of water agar freshly seeded with nematodes. In the case of aggressive parasites, such as *Meria coniospora* and *Harposporium anguillulae*, numerous infected nematodes can be found within a week of seeding.

31.3.6 Staining Predatory Fungi

Faust and Pramer (1964) discovered that the monoazo dye Janus Green was most useful for staining predatory fungi. The dye is used at a concentration of 0.01% (w/v) in 0.2M sodium acetate-acetic acid buffer pH 4.6 and added drop by drop to the test material.

Faust and Pramer (1964) applied the staining solution directly to surface cultures of *Arthrobotrys conoides, A. dactyloides* and *Dactylella ellipsospora* on maize meal extract agar and on cellophane in association with the nematode *Panagrellus redivivus*. Fungal walls and particulates were stained dark green to blue. Trapping devices of the predatory fungi were stained most intensely. Nematodes, captured and killed, were stained bright yellow, whilst living nematodes were not stained at all. Faust and Pramer (1964) also suggested that this stain might be useful for direct observation of nematode-trapping fungi in soil.

31.3.7 Estimation of Nematode-Destroying Fungi

As pointed out by Wyborn *et al.* (1969) anomalous results in studies of nematode-destroying fungi are often the consequence of non-standardised techniques. These workers took into consideration such factors as soil weight, nutrient medium, temperature and nematode numbers. They found that the most convenient weight of soil was 0.5 g, the most suitable medium was 2% (w/v) water agar and that incubation at room temperature (18 to 30°C) was satisfactory. They concluded that the greatest single factor influencing growth of predatory fungi was the level of stimulation (i.e. the number of nematodes added to each Petri dish as bait). Wyborn *et al.* (1969) also found that the greatest number of nematode-destroying fungi was recovered when 5000 nematodes were added per plate and suggested that a standard level of stimulation should be employed to obtain quantitative and qualitative assessments of fungal species present.

31.4 Egg parasites

Fungal parasites of nematode eggs, such as *Rhopalomyces* and *Helicocephalum* species are recorded occasionally using baited plates and soil sprinkling techniques (Barron 1977). There have been few attempts, however, to develop selective methods designed to recover egg parasites. A notable exception is the research of Stirling *et al.* (1978) who carried out an excellent study on a novel fungus, *Dactylella oviparasitica*, and its ability to control the root-knot nematode *Meloidogyne*.

Stirling *et al.* (1978) noted that 25 years ago Lovell peach was used as a root stock for peaches in California but its susceptibility to root-knot nematodes (*Meloidogyne* spp.) led to its replacement by Nemaguard rootstock in most areas. A recent survey showed that several old San Joaquin Valley orchards on Lovell rootstock had unexpectedly low root-knot nematode populations. It was suspected that the nematode was under natural biological control in these orchards and further investigation revealed root-knot nematode eggs parasitised by a fungus. This is one of the best examples of the biological control of a plant parasitic nematode in the field by a naturally occurring antagonist. Four methods for detecting and evaluating the significance of *Dactylella oviparasitica* were described by Stirling (1979).

31.4.1 Detection of *Dactylella Oviparasitica*

Detection in egg masses

The gelatinous matrix of *Meloidogyne* egg masses collected from host plants in the field is partially dissolved by treatment in 1% (w/v) NaCl for about 2 min and the eggs examined for parasitic fungi. Clumps of parasitised eggs are washed in sterile water and added to cornmeal agar (contains in distilled water (g l^{-1}): cornmeal infusion, 50; agar, 15); glucose-peptone agar (contains in distilled water (g l^{-1}): glucose, 10; peptone, 10; agar, 16) or YPSS agar (contains in distilled water (g l^{-1}): yeast extract, 4; K_2HPO_4, 1; $MgSO_4.7H_2O$, 0.5; soluble starch, 20; agar, 16). Inoculated media are examined daily for *Dactylella oviparasitica*. If estimates of the number of parasitised and unparasitised eggs are desired, egg masses are macerated and eggs counted.

Detection in root samples

About 1 g of roots are spread over the surface of quarter-strength cornmeal agar and the plates incubated at about 24°C for at least 1 month. A distinct succession of organisms colonises roots and agar, similar to that observed with methods used to isolate parasites and predators of soil nematodes. After about 1 month predaceous fungi begin to decline and conidia of *Dactylella oviparasitica* can often be seen protruding from roots.

Detection in field soil

Root pieces containing egg masses from *Meloidogyne*-infected plants grown in the greenhouse are mixed with soil collected from the field. The soil and roots are placed in containers and egg masses are examined 15 to 25 d later for *Dactylella oviparasitica*. Fungal activity *in vivo* can be evaluated by placing soil and roots in porous ceramic tubes, bags of fine nylon screening or other materials which allow free movement of water and gases and burying the containers in the field.

Testing field soil in the greenhouse

Tomato seedlings are planted in field soil in pots, and second-stage *Meloidogyne* larvae are added if this soil contains few root-knot nematodes. Plants are grown at 25 to 27°C for about 40 d then the egg masses removed and examined for parasitism by fungi. Since *Dactylella oviparasitica* is associated with plant roots, incorporation of roots into the soil or the use of rhizosphere soil enhances the chances of detecting the fungus.

31.5 Activity of nematode-destroying fungi in soil

It is difficult to measure accurately the numbers and kinds of nematode-destroying fungi in soil or organic debris. It is equally difficult to estimate the levels of activity of fungi or the effects of various types of soil amendments or treatments upon these activities. Nevertheless, despite the complexities involved some attempts have been made on qualitative and quantitative estimations of nematode-destroying fungi in their natural habitat.

31.5.1 Effect of Soil Amendments

Cooke (1961) developed a method using agar discs in order to obtain evidence of the effects of soil amendments on the predaceous activity of indigenous populations of nematode-trapping fungi.

Discs (1 cm diameter and about 3 mm thick) are cut from plates of weak maize-meal agar and placed on microscope slides, four discs usually being arranged in a row on each slide. Each slide is buried singly in 15 cm Petri dishes containing soil so that the discs lie about 1 cm below the soil surface. The soil is passed through a 7 mm mesh sieve prior to use and the moisture content held constant throughout the period of the experiment. At weekly intervals the slides are removed and replaced by others bearing fresh discs. On removal from the soil the disc surfaces are washed with a strong, fine jet of water from a washbottle. The discs adhere to the slide and can be examined immediately, unfixed and unstained, under the microscope.

Nutrients or organic amendments can be added to the soil before screening to study the effects of supplements on the activity of nematode-destroying fungi. Cooke (1961) found that a diversity of nematode-destroying fungi (both trap-forming predators and endozoic parasites) appeared on the disc surfaces during the decomposition of cabbage leaf tissue, glucose, and sucrose. The discs were, in effect, used as microquadrats recording, to some extent, what happened in that region of the soil in intimate contact with their surfaces.

31.5.2 Competitive Saprophytic Ability

Cooke (1963a) used an agar plate method to test the saprophytic ability of individual predatory fungi against a complete soil flora.

Predatory fungi were grown on sand-maize-meal mixture (contains in distilled water (g 15 ml^{-1}): quartz sand, 100; maize-meal, 3) for 6 weeks at room temperature. The cultures were passed through a 2 mm mesh sieve before being mixed with a medium loam soil which had been passed through a 2 mm mesh sieve and brought to 50% of the water-holding capacity. The sieved, sand-maize-meal culture was mixed with soil to give mixtures containing the following percentages (by weight) of inoculum: 98, 90, 75, 25 and 10. Then 10 g of each mixture was spread in an even layer on the bottom of a sterile Petri dish. Over this was poured 10 ml sterile distilled water agar cooled to 40°C. After the agar had set, discs were cut from the agar-impregnated mixture with a sterile cork borer (4 mm diameter). For every treatment five plates of Difco cornmeal agar were each inoculated with four equally spaced inoculum-soil discs. After growth for 5 d the total fungal colony arising from each disc was assigned a rating of 1, 11/12, 5/6, 3/4 and so on down to 0 according to the proportion of total fungal colony that was occupied. The ratings for each of the 20 discs were added together and a percentage calculated.

This method can only be used effectively if the nematode-trapping fungus under test forms distinctive segments of the total colony or sporulates freely enough for it to be identified and given a rating.

31.5.3 Predaceous Activity

As pointed out by Cooke (1963b) behaviour of predatory fungi in Petri dishes and sterile soil is not necessarily a measure of their activity in non-sterile soil. Cooke (1963a, b) studied predaceous activity nematode-destroying fungi when added to soil and estimated the intensity and duration of any activity on the nematode populations in soil. He prepared a sand-maize-meal inoculum as above and mixed it with sieved soil to give 200 g amounts of mixture with 2, 5, 10 and 15% (w/v) of inoculum. Each 200 g amount was placed in a 15 cm Petri dish and, at weekly intervals, two samples (10 g) were removed from each dish and their nematode population assessed using the Baermann technique. Uninoculated soil is used as a control and the moisture content of all the soils is maintained at 50% of the total water-holding capacity.

31.5.4 Rhizosphere

Predatory fungi are markedly stimulated in soil by the addition of organic supplements and particularly by green manuring (Linford 1937). Peterson and Katznelson (1965) suggested that nematode-destroying fungi might be expected to be abundant in the root zone because of root excretions and decomposing, sloughed-off dead or dying root parts. They suggested that the relative abundance of nematodes in the root zone could encourage a rich flora of nematode-destroying fungi. On the other hand they argued that many microorganisms in the rhizosphere might compete for organic substrates and have antagonistic effects.

Peterson and Katznelson (1965) also studied the incidence of nematode-destroying fungi in the root vicinity. They did this by growing plants in pots and soil beds in either the greenhouse or in a growth chamber for 4 to 6 weeks after which time they were removed for testing. Unplanted soil was used as the control. Rhizosphere soil was obtained by collecting soil particles which remained in close contact with the roots after loosely adhering soil and large clumps had been removed by shaking. The roots were washed in five changes of sterile water and segments of 0.5 to 1.0 cm placed on cornmeal agar containing chlortetracycline (30 mg l^{-1}). The rhizosphere soil and the root-free unplanted soil were distributed on plates of cornmeal agar (10 to 20 mg per plate). Entire rootlets with their adhering rhizosphere soil can be placed on the agar. Fifty plates were prepared for each material, i.e. root segments, rhizosphere soil and root-free soil, and a suspension of nematodes added to each plate. The nematodes were removed from stock plates by means of the Baermann funnel technique, washed in three changes of sterile water, and added to the plates of roots. This procedure was employed to stimulate trap formation which served as an index of the presence of predaceous fungi in the sample. Plates were thoroughly examined after 3 and 6 weeks incubation at 20°C in plastic bags (used to prevent evaporation of moisture).

31.5.5 Enumeration

The enumeration of nematode-destroying fungi may be achieved by using the method of Klemmer and Nakano (1964). Each soil sample was reduced to a slurry of even consistency by blending with an equal volume of water. A 0.5 ml aliquot of each slurry was pipetted into an empty Petri dish and dispersed in 30 ml cooled but still liquid 3% (w/v) water agar. Strips (3 mm × 60 mm) of the hardened soil/agar mixtures were cut from these dishes by using two parallel scalpels. The strips are transferred to the centres of empty Petri dishes and the plates flooded with 30 ml of cornmeal agar adjusted to 10% of the specified concentration of nutrients. Finally, 1 ml saprophytic nematodes suspension is pipetted over the soil/agar strips. After several days incubation, the dishes were examined for the presence of trapping structures. It is normally possible to distinguish between hyphae originating from separate loci within the strips. Such hyphae bearing one or more trapping structures are each given a count of one.

By standardising the procedure described here, counts per dish can be converted to counts per gram soil. They are directly comparable to other microbial counts obtained by conventional soil-dilution plating techniques.

Eren and Pramer (1965) described a simple dilution and plating procedure to enumerate nematode-trapping fungi in soil. A known quantity of soil is suspended and diluted serially in sterile water blanks. Suspensions (0.1 ml) are added to the surface of each of three Petri plates containing water agar poured previously and permitted to solidify. Inoculated plates are incubated at 28°C for 3 d, after which nematodes are added to the surface of the agar in each plate. The plates (35 mm diameter) are returned to the incubator and examined periodically under the microscope at a magnification of × 100 for organelles of capture characteristic of the various species of nematode-trapping fungi. Supplementary additions of nematodes are made when necessary and final observations recorded after three weeks' storage. The most-probable-number of nematode-trapping fungi in soil is determined by reference to probability tables, based on the number of positive results obtained when three replicates are tested at each of six dilutions.

This method was evaluated by Eren and Pramer (1965) using *Arthrobotrys conoides*, a nematode-trapping fungus which produces networks of adhesive hyphal loops for the capture of prey. When spores of *A. conoides* were harvested from agar cultures, washed, counted and added at known concentration to soil, recoveries averaged ±15%. Since organelles of capture can originate from both spores and hyphal fragments, the method measures all predaceous propagules.

A third method of enumerating nematode-trapping fungi was described by Mankau (1975). Replicate 10 g samples of soil are passed through a dry standard sieve, soaked in water for 10 min and suspended in about 500 ml of tap water in beakers. Using this method, up to 25 g samples can be analysed from mineral soils containing little organic matter, but in soil with an abundance of light organic debris, sample size may have to be limited to 5 g. Samples are mixed with a magnetic stirrer. The soil suspension is washed through a graded series of sieves, depending upon the soil type. The most commonly used are: No. 20 or 32, No. 100, and No. 325 or 400. The suspension is decanted through each sieve separately into a flask and allowed to stand briefly so that heavy mineral particles settle out. The sieves can be supported in funnels over flasks which receive the filtrate. Each sieve is rinsed separately with a small stream of water into a small beaker. With practice, the materials from each of three sieves can be pooled into a 25 ml beaker; these are set aside to allow fine suspended material to settle. The suspensions are filtered through a Millipore filter system containing a filter paper disc (fast-type) on a sintered glass base under vacuum. It is important not to clog the filter paper and thus slow the passage of the suspension. Simultaneous filtration of a number of samples can be accomplished on a multi-place vacuum manifold. Filtration is stopped when water leaves the mat. The filter paper disc is removed and placed on the surface of quarter-strength cornmeal agar in the centre of a Petri dish. Plastic grid dishes facilitate precise scanning of the dishes.

As described by Mankau (1975) the microflora and microfauna migrate on and into the agar where they can be observed easily. Fungi and bacteria develop sparsely, followed by microphagous organisms, particularly protozoa and nematodes. In about 6 to 12 d, parasites and predators of nematodes, particularly nematode-trapping fungi, appear. The succession of organisms which develop in the dishes approximates that observed in the colonisation of natural substrates in the soil. Quantitative comparisons between soil samples are made by counting predation events and parasitised nematodes or species of organisms present in replicate dishes after a specific incubation period.

The dishes are kept at room temperature and can be observed for a year or more if suitably stored to prevent desiccation.

31.5.6 FUNGISTASIS

Under conditions which would appear to be favourable, the spores of most fungi fail to germinate when placed in contact with natural soils (Dobbs & Hinson 1953). Using a modified agar-disc technique, Mankau (1962) was able to demonstrate that a water diffusible compound was present in all California soils tested which inhibited the germination of the conidia of the predatory fungi studied, namely, *Arthrobotrys arthrobotryoides* (adhesive nets), *A. dactyloides* (constricting rings) and *Dactylella ellipsospora* (adhesive knobs).

Agar-disc technique

Jackson (1958) used an agar-disc technique to examine fungistasis. Soil samples are passed through a 2 mm sieve, either immediately after collecting or after air drying for 12 h, and thoroughly mixed before testing. Soil (40 to 60 g) is placed in a Petri dish and brought to approximately 60% of water-holding capacity by the addition of distilled water.

Agar discs are prepared by pouring 9 ml molten 2% (w/v) water agar into a Petri dish standing on a level surface, giving a layer of agar just under 1.5 mm in thickness. The discs are cut from the agar with a flamed 7.5 mm diameter cork-borer and removed on the tip of a flamed scalpel. One thickness of Whatman no. 1 filter paper is interposed between the agar disc and the soil. For each test, four pieces of filter paper (1 cm^2) are placed on the surface of the soil in a Petri dish and an agar disc placed on each of the squares. Either immediately, or after 1 to 4 h of incubation to allow diffusion of inhibitory substances from soil into the agar disc, the surface of each disc is inoculated with a drop of a spore suspension in distilled water on the end of a glass rod. Control and treated discs are inoculated with samples of the same spore suspension. Controls are similarly prepared, but the agar discs are placed on filter papers saturated with distilled water in dishes containing no soil. After inoculation the Petri dish is incubated at 25°C for the minimum period required to obtain good germination of the spores on the control discs.

After incubation the agar discs are removed to microscope slides and covered with a drop of lactophenol and a coverslip. Germinated and ungerminated spores in each of four random microscope fields on each agar disc are counted, the magnification used depending on the size and density of distribution of the spores. Spores are recorded as having germinated when they produce a germ tube of one spore diameter or more in length. Results are expressed as a percentage of the total number of spores that germinate.

Fluorescent brightener technique

Eren and Pramer (1968) used a fluorescent brightener to facilitate direct estimations of fungistasis. They found that the ease and precision of measurements on growth and spore germination of *Arthrobotrys conoides* in soil was greatly increased using this method. The brightener is disodium 4,4′-bis (4-anilion 6-bis (2-hydroxyethyl) amino-s-triazin-2-yl-amino) 2-2′-stilbene disulphonate. Tests showed that there is no reduction in germination of spores of *Arthrobotrys conoides* suspended for 5 h in brightener solutions ranging in concentration from 0.001 to 0.1% (w/w).

To evaluate the influence of various factors on germination in soil, spores of *Arthrobotrys conoides* are suspended in a 0.025% (w/v) brightener solution in distilled water for 1 h. Treated spores are harvested by centrifugation at 2000 **g** for 10 min at 4°C, washed three times with distilled water to remove excess brightener and suspended to a concentration of 1×10^4 spores ml^{-1}. Portions of soil (5 g) in small Petri plates (35 mm diameter) are appropriately amended or otherwise treated and inoculated with 0.6 ml spore suspension. All soil samples are brought to 70% of water-holding capacity and stored at 28°C.

A quantity of soil (0.5 g) is removed from each plate and used to measure spore germination after 1 and 3 d. For this purpose, the soil sample is suspended in 9.5 ml water. The suspension is agitated for 1 min, permitted to settle for 30 s and a 0.4 ml aliquot is withdrawn from one cm below the liquid surface. This is spread uniformly on the 16 cm^2 surface area of a glass slide, permitted to air dry and observed microscopically.

Spores appear bright blue on a pink or red background of soil particles and the contrast is sufficiently sharp to distinguish spores from soil. Adequate quantities of brightener are translocated to colour hyphae which are newly emerged.

31.6 Physiological studies on nematode-destroying fungi

31.6.1 Attractants

Monoson and Ranieri (1972) studied the nematode *Aphelenchus avenae* in association with the net-forming predator *Arthrobotrys musiformis* and observed that the movement of nematodes was not random. They suggested that nematodes were attracted to the trapping organs (adhesive nets) of the predator due to the presence of a nematode-attracting substance (NAS) produced by the trapping devices. Monoson *et al.* (1973) claimed that extracts of *Arthrobotrys musiformis* and *Monacrosporium doedycoides*, from hyphae in which traps had been induced, contained a NAS, while extracts from hyphae without such traps did not display any attraction potential. The ability of fungi in general and nematode-destroying fungi in particular to attract nematodes has been the topic of considerable research.

It was shown by Townshend (1964) that the fungal-feeding nematode *Aphelenchus avenae* was attracted to 57 out of 59 fungi tested. Although no predatory fungi were among those tested, there is no reason to suppose that these might not also attract the nematodes.

Attraction of nematodes to fungi

The attractiveness of fungi to *Aphelenchus avenae* is determined by a technique designed for the Chi-squared test and described by Townshend (1964). Two 0.5 cm discs of fungus are cut from a 2-week-old plate culture and placed in opposite quadrants in a sterile Petri dish of peptone-dextrose agar (PDA) not more than 1.5 cm from the rim. Sterile nematodes are pipetted onto 1 cm discs of thick filter paper previously sterilised on germination pads in Petri dishes. The germination pad soaks up the excess moisture that would have trapped the nematodes when the disc was placed face down on the agar. A nematode-bearing disc or nematode suspension is placed in the centre of a dish of PDA midway between the fungal colonies which have been incubated 20 to 24 h at 25°C. Sterile fibrous paper discs are placed in the other quadrants. The test dishes are examined after a further 18 h incubation at 25°C and the nematodes under each disc counted.

Using a modified Townshend technique, Jansson and Nordbring-Hertz (1979) studied attraction of nematodes to predaceous fungi and found that of 14 fungi tested 10 attracted and 1 repelled nematodes and 3 fungi had no effect. It was noted by Jansson and Nordbring-Hertz (1979) that the attraction intensity increased with increasing dependence on nematodes for nutrients.

Attraction intensity

The same assay procedure described above is used for measuring attraction intensity except that the numbers of nematodes accumulating under each disc is counted at 1 h intervals for 6 h (Jansson & Nordbring-Hertz 1979). There is a linear increase in the number of attracted nematodes with time and the slope of the line is used as a measure of the attraction intensity.

Attraction of nematodes to trapping devices

Field and Webster (1977) developed a simple technique to demonstrate that nematodes are attracted to adhesive nets, constricting rings or adhesive knobs.

A marked sterile microscope slide is place into a Petri dish, and covered with 15 ml water agar. Discs (9 mm diameter) cut from 3-week-old cultures of the predaceous fungi are located as shown in Fig. 2. One disc is an untreated control and the other is from a culture which has been treated ('stimulated') 24 h earlier with nematode extract of 10% (v/v) sterile horse serum. The treated areas are checked microscopically for the presence of traps. A nematode suspension containing 40 to 60 nematodes (*c.* 20 μl) carefully applied to the centre point of the agar-covered slide, equidistant from the 2 discs used. After 12 h the slide is systematically scanned and the number of nematodes moving towards each disc scored. The results are subject to the Chi-squared test, against the hypothesis that if the treated discs were no more attractive than untreated discs, then equal numbers of nematodes would move towards each disc.

The possibility that the horse serum or nematode extract is the cause of the attraction can be eliminated in several ways.

(1) Uninoculated cornmeal agar plates are flooded with 10% (v/v) sterile horse serum and left for 24 h. The movement of the *Rhabditis* sp. is then tested towards discs treated in this way compared with untreated discs.

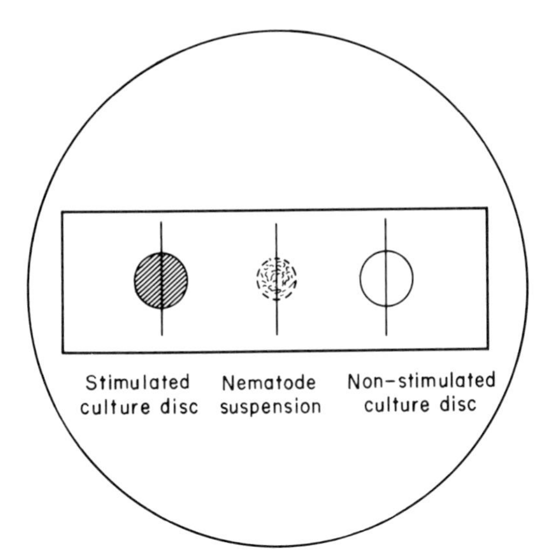

Figure 2. Method for comparing the movement of nematodes. A marked glass slide in a Petri dish is covered with tap-water agar. The two culture discs are placed 4 cm apart on the slide and an aliquot of nematode suspension placed at the midpoint between them (after Field & Webster 1977).

(2) Cornmeal agar discs treated with horse serum are tested against unstimulated discs from a 3-week-old culture of *Arthrobotrys anchonia*, using *Rhabditis* sp.

(3) The movement of *Rhabditis* sp. towards cornmeal agar discs with and without the addition of nematode extract is tested.

(4) The movement of *Aphelenchus* sp. towards cornmeal agar with or without horse serum is tested.

In no case did the presence of horse serum or nematode extract cause more attraction than that of a control not containing these substances.

Attraction of nematodes to fungal extracts

Balan *et al.* (1976) investigated the attraction of nematodes to fungal extracts. 21-day-old cultures of *Monacrosporium rutgeriensis* grown on agar were pressed through a nylon screen (1 mm mesh). Agar rectangles from 15 Petri dishes were transferred into a 500 ml flask to which 100 ml acetone was added and shaken on a rotary shaker for 40 min. After filtration this procedure was repeated and the acetone extracts pooled. The acetone was removed by distillation *in vacuo* at a temperature not exceeding 60°C. The residue was extracted twice with equal volumes of chloroform and separated in a separating funnel: the pooled chloroform extracts were concentrated.

From a Petri dish with a 3 mm layer of sterile 3% (w/v) water agar, blocks 10 mm diameter were cut with a cork-borer and transferred to a glass plate. Filtrate (0.2 ml) of the medium on which predaceous fungi had grown was pipetted onto the surface of the blocks, and allowed to diffuse into the agar for 20 min and any remaining liquid removed. The blocks were inverted in a 10 cm diameter Petri dish containing 10 ml *Panagrellus redivivus* suspension (2000 nematodes ml^{-1}). Up to 14 blocks with various samples were accommodated around the perimeter of the dish. Control blocks are prepared in the same manner but using sterile medium instead of the filtered medium in which the predaceous fungus grew.

To prevent sliding, the blocks were pressed down by brass rings (height 5 mm, outer diameter 10 mm, inner diameter 6 mm) through which the interface between the bottom of the dish and the agar blocks can be observed with an inverted objective lens or with a dissecting microscope. A magnification is used which allows the observation of the whole area within a ring in one field.

The diffusion of the test substance into the water which surrounds the agar blocks is relatively slow because blocks are pressed down to the bottom of the Petri dish and only an undisturbed thin film of water is in contact with that part of the block onto which the sample was applied. The test substance forms a deposit under the agar and diffuses slowly from the edges of the block. Thus, a satisfactory concentration gradient is formed, with the highest concentration in the centre under the blocks.

If attractants are present, nematodes move against this gradient and accumulate under the blocks in great numbers. Only few nematodes collect under blocks lacking attractants or under control blocks. The dishes are left undisturbed in a strictly horizontal position for 1 h after which the number of nematodes within the ring is estimated.

Attraction assay technique

Another method of measuring attraction was described by Balan and Gerber (1972). One end of a metal cylinder, similar to that used in antibiotic assays, is covered with a polypropylene screen held in place by a segment of rubber tubing. The cylinder with the screen bottom is then filled with washed and ignited sand (grain size 0.3 to 0.8 mm).

The tools used and the method of cylinder assembly are illustrated in Fig. 3.

To a layer of 20 g of sand in the bottom of a Petri dish, is added 7 ml of sterile medium; an even distribution of the sand–medium slurry is achieved by gentle swirling of the plates in the horizontal position, after which 2 ml of a suspension of nematodes (10 000 ml^{-1}) in sterile medium are slowly pipetted into the disc and evenly distributed by further swirling. Three sand-filled cylinders are then saturated with an excess of the growth to be assayed and three with sterile medium (controls). This is done in such a way as to remove all air pockets and bubbles from the sand screen. Cylinders of both kinds are finally transferred in an alternating manner to form a circle in the Petri dish containing the nematodes. During the assay, nematodes move from the suspension in the dish and enter the sand in the cylinders through the openings in the screen. If no attractants are present, the number of nematodes in each test cylinder should not be significantly different from that in the control cylinders.

After one hour, the cylinders of the controls and the treatments are transferred into two Erlenmeyer flasks containing appropriate volumes of water to obtain a concentration of 50 to 100 nematodes per ml. The number of nematodes in the two suspensions is then counted in a Scott Counting Slide using eight 1 ml samples for suspension and averaging the results. For each assay, a separate Petri dish and a separate set of controls is used. According to Balan and Gerber (1972) the results of parallel experiments differ by less than 10%.

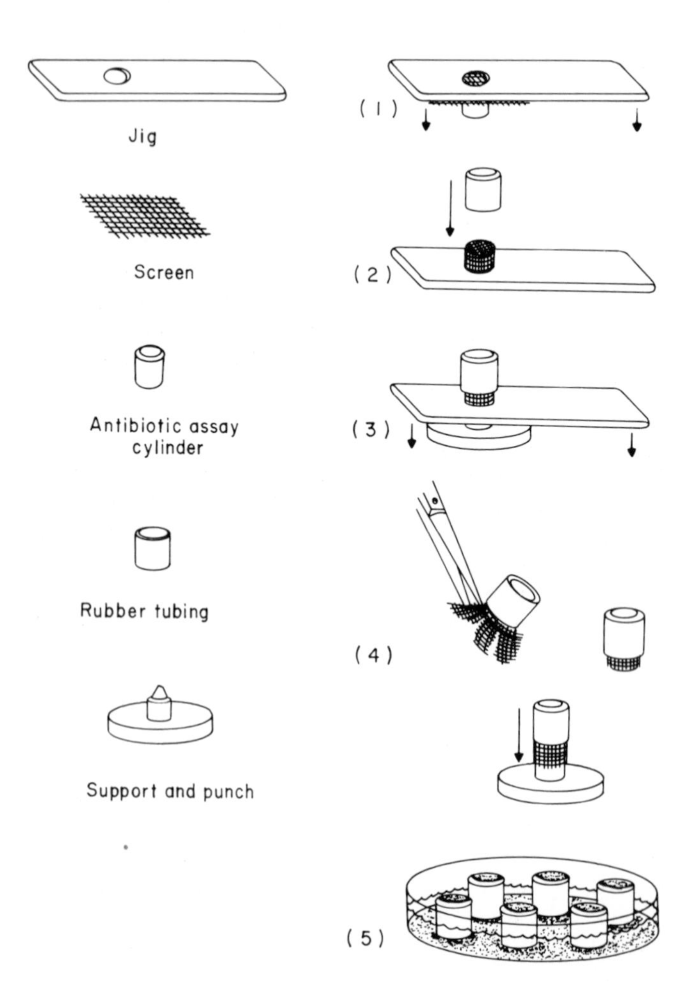

Figure 3. Cylinder assembly for attraction assay. The screen is shaped around the end of the cylinder by pressing down the jig: (1) a segment of rubber tubing is slipped half-way over the cylinder (2) which is pressed out of the jig by the support and punch (3). The excess screen is trimmed and the rubber tubing is pushed all the way down over the cylinder standing on the support (4). A Petri dish with cylinders is arranged for the assay (5) (after Balan & Gerber 1972).

31.6.2 Demonstration of Toxin Production by Conidia

Giuma and Cooke (1971) demonstrated that individuals of *Aphelenchus avenae* became immobile and died when the adhesive conidia of *Nematoctonus haptocladus* and *N. concurrens* became attached to their cuticles. The following method was used to demonstrate this for *N. concurrens*.

Populations of *Aphelenchus avenae* are maintained on colonies of *Pyrenochaeta terrestris* on 2% (w/v) malt agar at 20°C. Within 2 weeks large numbers of nematodes congregate at the colony margins. Discs (1 cm diameter) are removed aseptically from the margin either for transfer to fresh plates or for collection of nematodes. For collection, 8 to 10 discs are placed in a sterile muslin bag, which is then immersed in sterile distilled water in a Baermann funnel. After 24 h a 5 ml aliquot of nematode suspension is run off. The nematodes are concentrated by centrifugation and after counting are resuspended to give a suspension of known density.

The fungus is grown on plates of 2% (w/v) cornmeal agar at room temperature. Conidia are removed at 14 and 21 d by washing with sterile distilled water and, after counting, used at known densities. Stock plates are inoculated with conidial suspensions spread evenly over the surface of the agar.

Experiments are conducted in Petri dishes containing 15 ml of 2% (w/v) distilled water agar. To each is added 0.5 ml distilled water containing 2000 nematodes and 0.5 ml distilled water containing 1.5×10^6 conidia. The suspensions are distributed evenly over the agar surface with a glass spreader. Dishes are incubated at room temperature and observations were made at regular intervals.

Adhesive processes are usually produced by all conidia within 24 h of planting. After 48 to 72 h branches arose from the germ tube and produced hyphae which themselves bore terminal or lateral adhesive processes.

Giuma and Cooke (1971) found that conidia of *Nematoctonus concurrens* attach to nematodes by their adhesive cells relatively soon after germination. The commonest region of attachment is near the anterior end of the nematode and individuals are frequently seen moving with several conidia attached. According to Giuma and Cooke (1971) nematodes invariably become immobile on contact with an adhesive process and die before there is any sign of cuticle penetration by the fungus. Although they found it difficult to determine how soon after contact immobilisation occurs, it appears to take place within 24 h. They report that after death there is some hyphal growth, frequently extensive, over the surface of the body before penetration takes place.

31.6.3 Testing for Nematicidal Activity

Giuma and Cooke (1971) tested solutions for nematicidal activity. About 50 nematodes were suspended in 0.05 ml of water and added to a 10 × 75 mm test tube containing 0.2 ml of the solution to be tested; the tubes were stoppered and kept at 24°C. Nematicidal activity, indicated by an irreversible loss of motility, was evaluated microscopically (× 20) directly in the test tubes. Controls in tap water (or sterile water) were also run and nematode motility was observed for 72 h.

31.6.4 Formation of Trapping Devices

Couch (1937) found that rings are produced sporadically or not at all in pure cultures of *Dactylella bembicodes*. If some water containing nematodes is added to the cultures, however, rings are produced abundantly. Commandon and de Fonbrune (1938) showed that sterile filtrate, in which nematodes had been living, induced ring formation. Apparently nematodes secrete a compound (or compounds) which is responsible for morphogenesis of the trapping devices. Subsequent studies by Roubaud and Deschiens (1939) showed that this trap-inducing agent was not restricted to nematodes and a wide range of materials of animal origin were capable of inducing trap formation; human blood serum and extracts from earthworms were particularly effective. The term nemin was proposed by Pramer and Stoll (1959) for the substance(s) inducing trap formation. Nemin was not inactivated by exposure to 100°C for 10 min. Wootton and Pramer (1966) found yeast extract induced trap formation. Using chromatographic techniques, to purify the compounds in the yeast extract, they established that the active agents were valine, leucine and isoleucine. Of these amino acids, valine was the most active and elicited a response in either the D or L configurations at levels of 0.01 mg ml^{-1}.

Norbring-Hertz (1973) demonstrated the ability of valyl peptides under conditions of low level nutrition to induce formation of capture organs in *Arthrobotrys oligospora*. In general the technique used is simply the addition of crude extracts, at various dilutions, to cultures of nematode-destroying fungi growing on water agar or weak nutrient agar. The number of trapping devices in a known area is then estimated.

In some cases (Norbring-Hertz 1973) more sophisticated techniques (e.g. ion exchange chromatography, gel chromatography) are employed to purify compounds prior to testing their morphogenic effects. These techniques, however, are not peculiar to studies on nematode-destroying fungi and will not be discussed in detail here.

Spontaneous trap formation

It has been shown by many workers (Barron 1977) that formation of trapping devices in nematode destroying fungi is markedly stimulated by nematodes, nematode extract, blood serum, peptides, amino acids and other compounds. Cooke (1963c), however, studied spontaneous trap formation by 13 species of predatory fungi growing on water agar using the following method.

Conidia are washed with sterile distilled water from cultures growing on slopes of Difco cornmeal agar and filtered through sterile muslin. The conidial suspension is spread on plates of sterile distilled water agar so that there are 25 to 30 conidia per plate. Spores usually germinate after 24 h and are examined over a period of five days; about 100 conidia of each species are observed for signs of trap formation. Cooke (1963c) found that traps formed spontaneously and at low levels in only three of the species tested.

31.6.5 Mechanisms of Adhesion

Most predatory fungi capture nematodes by adhesion to specific capture organs (Barron 1977). In *Arthrobotrys oligospora* the trapping device is a three-dimensional adhesive network. Nordbring-Hertz and Mattiasson (1979) have presented evidence to indicate that the firmness of attachment to the trap, despite violent struggling on the part of the nematode, is initially the result of lectin on the traps which binds to a carbohydrate on the nematode's surface. Nordbring-Hertz and Mattiasson (1979) suggested that if a lectin–carbohydrate interaction is involved in adhesion, then pre-exposure of the lectin-carrying surfaces to specific carbohydrates would prevent attachment of the nematodes to the traps.

Arthrobotrys oligospora is grown on strips of dialysis membrane placed on the surface of a low-nutrient mineral salts medium. Trap formation is induced by addition of a small amount of nematodes to 3 to 4 d old cultures. Trap-containing strips (7 to 14 d) are transferred to empty Petri dishes and flooded for at least 20 to 24 h with the appropriate carbohydrate (20 to 200 mM solution in distilled water or in phosphate-buffered saline). After removal of excess carbohydrate, 1 to 2 drops of a nematode suspension in water (*c*. 1000 nematodes ml^{-1}) are added to the fungal culture to test capture ability. Microscopic observations of capture are made after 30 min, 3 h and about 24 h.

A number of carbohydrates were tested and it was found that pretreatment of the fungus with N-acetyl-D-galactosamine at low concentrations (20 mM) completely inhibited the capture of nematodes over a 24 h period.

31.6.6 Spontaneous Trap Induction

Balan and Lechevalier (1972) demonstrated that abundant trap formation in *Arthrobotrys dactyloides* (constricting rings) could be induced by adverse conditions resulting in lack of nutrients and/or water.

The fungus is cultivated on cellophane (300 PD-62, DuPont, Wilmington, Delaware, USA) which is attached by a rubber band to a cylindrical glass ring slightly smaller than the bottom of the Petri dishes used. The cellophane and ring thus resemble a drum with only one side covered. This assembly is sterilised separately and then transferred, cellophane down, to the surface of solidified Difco cornmeal agar (17 g l^{-1}), inoculated in the centre, and incubated in the dark at 28°C. Abundant trap formation is observed after 12 to 16 d and the results are evaluated by counting the number of visible traps in ten microscopic fields (× 100) situated at approximately equal distances from each other on a randomly selected line from the centre of the dish to its periphery.

It was found by Balan and Lechevalier (1972) that a high yield of traps was obtained when the fungus was grown on a double layer of cellophane but scarcely any traps were formed on cornmeal agar without cellophane, on perforated cellophane, or on cellophane superimposed on a richer medium containing yeast extract and commercial glucose. They interpreted the paucity of traps on perforated cellophane and on flat cellophane not attached to a glass ring as indicating that the microfilm of moisture on the surface of the cellophane carried sufficient nutrient and/or water to inhibit trap formation.

Recommended reading

BARRON G. L. (1977) *The Nematode-Destroying Fungi.* Canadian Biological Publications, Guelph, Ontario, Canada.

BARRON G. L. (1981) Parasites and predators of microscopic animals. In: *Biology of Conidial Fungi* (Eds B. Kendrick & G. Cole). pp. 167–200. Academic Press, New York.

DUDDINGTON C. L. (1957) *The Friendly Fungi.* Faber & Faber, London.

DUDDINGTON C. L. (1962) Predaceous Fungi and the Control of Eelworms. In: *Viewpoints in Biology* (Eds C. Duddington & J. Carthy), vol. 1, pp. 151–200. Butterworths, London.

References

ASCHNER M. & KOHN S. (1958) The biology of *Harposporium anguillulae. Journal of General Microbiology* **19**, 182–9.

BALAN J. & GERBER N. (1972) Attraction and killing of the nematode *Panagrellus redivivus* by the predaceous fungus *Arthrobotrys dactyloides. Nematologica* **18**, 163–73.

BALAN J., KRIZKOVA L., NEMEC P. & KOLOZSVARY A. (1976) A qualitative method for detection of nematode attracting substances and proof of production of three different attractants by the fungus *Monacrosporium rutgeriensis. Nematologica* **22**, 306–11.

BALAN J. & LECHEVALIER H. A. (1972) The predaceous fungus *Arthrobotrys dactyloides*: induction of trap formation. *Mycologia* **64**, 919–22.

BARRON G. L. (1969) Isolation and maintenance of endoparasitic nematophagous Hyphomycetes. *Canadian Journal of Botany* **47**, 1899–902.

BARRON G. L. (1977) *The Nematode-Destroying Fungi.* Canadian Biological Publications, Guelph, Ontario, Canada.

BARRON G. L. (1981) Parasites and Predators of Microscopic Animals. In: *Biology of Conidial Fungi* (Eds. B. Kendrick & G. Cole). pp. 167–200. Academic Press, New York.

COMMANDON J. & DE FONBRUNE P. (1938) Recherches expérimentales sur les champignons prédateurs des nématodes du sol. Les pièges garrotteurs. *Comptes Rendue de l'Académie des Sciences, Paris* **129**, 620–5.

COOKE R. C. (1961) Agar disk method for the direct observation of nematode-trapping fungi in the soil. *Nature, London* **191**, 1411–12.

COOKE R. C. (1963a) The predaceous activity of nematode-trapping fungi added to soil. *Annals of Applied Biology* **51**, 295–9.

COOKE R. C. (1963b) Succession of nematophagous fungi during the decomposition of organic matter in the soil. *Nature, London* **197**, 205.

COOKE R. C. (1963c) Ecological characteristics of nematode-trapping Hyphomycetes. I. Preliminary studies. *Annals of Applied Biology* **52**, 431–7.

COUCH J. N. (1937) The formation and operation of the traps in the nematode-catching fungus, *Dactylella bembicodes* Drechsler. *Journal of the Elisha Mitchell Society* **53**, 301–9.

DOBBS C. G. & HINSON W. H. (1953) A widespread fungistasis in soil. *Nature, London* **172**, 197–9.

DOUGHERTY E. C. (1960) Cultivation of Aschelminths, especially Rhabditid nematodes. In: *Nematology* (Eds J. N. Sasser & W. R. Jenkins), pp. 297–318. University of North Carolina Press, Chapel Hill.

DRECHSLER C. (1934) Organs of capture in some fungi preying on nematodes. *Mycologia* **26**, 135–44.

DRECHSLER C. (1935) A new species of conidial phycomycete preying on nematodes. *Mycologia* **27**, 206–15.

DRECHSLER C. (1937) Some Hyphomycetes that prey on free-living terricolous nematodes. *Mycologia* **29**, 447–552.

DRECHSLER C. (1940) Three new hyphomycetes preying on free-living terricolous nematodes. *Mycologia* **32**, 448–470.

DRECHSLER C. (1941a) Predaceous fungi. *Biological Reviews of the Cambridge Philosophical Society* **16**, 265–90.

DRECHSLER C. (1941b) Some Hyphomycetes parasitic on free-living terricolous nematodes. *Phytopathology* **31**, 773–802.

DUDDINGTON C. L. (1951) The ecology of predacious fungi. I. Preliminary survey. *Transactions of the British Mycological Society* **34**, 322–31.

DUDDINGTON C. L. (1955) Notes on the technique of handling predaceous fungi. *Transactions of the British Mycological Society* **38**, 97–103.

DUDDINGTON C. L. (1957) *The Friendly Fungi.* Faber & Faber, London.

DUDDINGTON C. L. (1962) Predaceous Fungi and the Control of Eelworms. In: *Viewpoints in Biology* (Eds C. Duddington & J. Carthy), vol. 1, pp. 151–200. Butterworths, London.

EREN J. & PRAMER D. (1965) The most probable number of nematode-trapping fungi in soil. *Soil Science* **99**, 285.

EREN J. & PRAMER D. (1968) Use of a fluorescent brightener as aid to studies of fungistasis and nematophagous fungi in soil. *Phytopathology* **58**, 644–6.

FAUST M. A. & PRAMER D. (1964) A staining technique for the examination of nematode-trapping fungi. *Nature, London* **204**, 94–5.

FIELD J. I. & WEBSTER J. (1977) Traps of predaceous fungi attract nematodes. *Transactions of the British Mycological Society* **68**, 467–9.

FLEGG J. J. M. (1967) Extraction of *Xiphinema* and *Longidorus* species from soil by a modification of Cobb's decanting and sieving technique. *Annals of Applied Biology* **60**, 429–37.

GIUMA A. Y. & COOKE R. C. (1971) Nematoxin production by *Nematoctonus haptocladus* and *N. concurrens*. *Transactions of the British Mycological Society* **56**, 89–94.

GIUMA A. Y. & COOKE R. C. (1972) Some endozoic parasites on soil nematodes. *Transactions of the British Mycological Society* **59**, 213–18.

HOOPER D. J. (1970) Culturing nematodes. In: *Laboratory Methods for Work with Plant and Soil Nematodes* (Ed. J. F. Southey), pp. 96–113. Technical Bulletin No. 2. Ministry of Agriculture, Fisheries and Food, United Kingdom.

JACKSON R. M. (1958) An investigation of fungistasis in Nigerian soils. *Journal of General Microbiology* **18**, 248–58.

JANSSON H. & NORDBRING-HERTZ B. (1979) Attraction of nematodes to living mycelium of nematophagous fungi. *Journal of General Microbiology* **112**, 89–93.

KLEMMER W. & NAKANO R. Y. (1964) A semi-quantitative method of counting nematode-trapping fungi in soil. *Nature, London* **203**, 1085.

LINFORD M. B. (1937) Stimulated activity of natural enemies of nematodes. *Science* **85**, 123–4.

MANKAU R. (1962) Soil fungistasis and nematophagous fungi. *Phytopathology* **52**, 611–15.

MANKAU R. (1975) A semi-quantitative method for enumerating and observing parasites and predators of soil nematodes. *Journal of Nematology* **7**, 119–22.

MONOSON H. L., GALSKY A. G., GRIFFIN J. A. & MCGRATH E. J. (1973) Evidence for and partial characterization of a nematode-attraction substance. *Mycologia* **65**, 78–86.

MONOSON H. L. & RANIERI G. M. (1972) Nematode attraction by an extract of a predaceous fungus. *Mycologia* **64**, 628–31.

NORDBRING-HERTZ B. (1973) Peptide-induced morphogenesis in the nematode-trapping fungus *Arthrobotrys oligospora*. *Physiologia Plantarum* **29**, 223–33.

NORDBRING-HERTZ B. & MATTIASSON B. (1979) Action of a nematode-trapping fungus shows lectin-mediated host/microorganism interaction. *Nature, London* **281**, 477–9.

OLTHOF TH. H. A. & ESTEY R. H. (1963) A nematoxin produced by the nematophagous fungus *Arthrobotrys oligospora* Fresenius. *Nature, London* **197**, 514–5.

PETERSON E. A. & KATZNELSON H. (1964) Occurrence of nematode-trapping fungi in the rhizosphere. *Nature, London* **204**, 1111–12.

PETERSON E. A. & KATZNELSON H. (1965) Studies on the relationship between nematodes and other soil microorganisms. IV. Incidence of nematode-trapping fungi in the vicinity of plant roots. *Canadian Journal of Microbiology* **11**, 491–5.

PRAMER D. & STOLL N. R. (1959) Nemin: a morphogenic substance causing trap formation by predaceous fungi. *Science* **129**, 966–7.

ROUBAUD M. E. & DESCHIENS R. (1939) Sur les agents de formation des dispositifs de capture chez les Hyphomycetes prédateurs de nématodes. *Comptes Rendue de l'Académie des Sciences, Paris* **209**, 77–9.

STIRLING G. R. (1979) Techniques for detecting *Dactylella oviparasitica* and evaluating its significance in field soils. *Journal of Nematology* **11**, 99–100.

STIRLING G. R., MCKENRY M. V. & MANKAU R. (1978) Biological control of root-knot nematode on peach. *California Agriculture* **32**, 6–7.

TOWNSHEND J. L. (1964) Fungus hosts of *Aphelenchus avenae* Bastian, 1865 and *Bursaphelenchus fungivorus* Franklin and Hooper 1962 and their attractiveness to these nematode species. *Canadian Journal of Microbiology* **10**, 727–37.

WINKLER E. J. & PRAMER D. (1961) A chamber for culturing and collecting the nematode *Panagrellus redivivus*. *Nature, London* **192** 472–3.

WOOTTON L. M. O. & PRAMER D. (1966) Valine-induced morphogenesis in *Arthrobotrys conoides*. *Annual Proceedings of the American Society for Microbiology*, p. 75.

WYBORN C. H. E., PRIEST D. & DUDDINGTON C. L. (1969) Selective technique for the determination of nematophagous fungi in soils. *Soil Biology and Biochemistry* **1**, 101–2.

Part 6
Stressed Environments

Chapter 32 • Extreme Natural Environments

David W. Smith

32.1 Introduction

Extreme natural environments and the effect of stress on microorganisms have been reviewed by many in recent years (Colwell & Morita 1974; Holding *et al.* 1974b; Gray & Postgate 1976; Heinrich 1976; Brock 1978; Kushner 1978a; West & Skujins 1978). The coverage in these volumes is exhaustive and no attempt will be made here to summarise this information. The emphasis in this chapter, in keeping with the topic of the book, will be the experimental procedures which have been used in natural environment studies with brief systematic review where necessary.

There are two basic ways to define an extreme natural environment:

(1) in terms of some specific physical and/or chemical parameter(s);

(2) in terms of the organisms present.

The first approach considers the factor of interest in a given environment in a quantitative way, asking, for example, if the temperature is extreme. The second approach considers the environment and asks if the number of organisms is restricted, that is, if there is lower species diversity. An environment with low diversity may then be judged to be extreme in the factor of interest. Both of these approaches make judgements as to what is extreme by comparison to anthropocentric norms of 'average' environmental conditions. This objection is generally accepted for the first case, but the species diversity definition has been held to be more rigorous. The diversity approach actually reduces to the same base as the factor approach because it is necessary to decide what is low diversity and what is high diversity. It therefore seems reasonable to conclude that there is no absolute definition of extreme (Brock 1969; Alexander 1976; Kushner 1978b). Nonetheless, the concept is useful and there are a number of environments and a number of specific organisms in them which are judged to be extreme by general consensus.

This chapter will examine five factors which are often considered as delimiters of extreme environments: temperature, pH, water availability, radiation and hydrostatic pressure. It is particularly important to realise that these parameters are not independent; they are interrelated both with each other and with other physical and chemical factors. Before considering these five parameters in detail, it is instructive to reflect on some common features of extreme environment studies. Many research projects embrace a study of the overall habitat as well as a specific component; for example, investigations by Brock's group of extremely hot and extremely acid habitats quickly became a study of Yellowstone National Park itself. Such digressions may be unavoidable and even desirable as the physical nature of an area is a relevant consideration in ecological studies. The danger, however, is that certain organisms become too closely identified with the area of research and the general ecological observations become obscured. One of Brock's important contributions is that he showed the wide geographical occurrence of high temperature and low pH environments in New Zealand (Brock & Brock 1970, 1971), Iceland (Brock & Brock 1966; Brock 1967), El Salvador and Italy (Brock 1978). Another example is the study of pressure resistance which becomes a marine microbiology investigation since the areas of high hydrostatic pressure are all deep ocean habitats.

Another problem is that studies of extreme environments tend to become centred on a specific organism or small group of organisms. Therefore rather than a study of the ecological effects of acid environments on microorganisms, for example, we find examinations of *Thiobacillus*, *Cyanidium*, *Sulfolobus*, *Thermoplasma*, and *Bacillus* species.

32.2 Temperature

By far the greatest number of extreme temperature studies are concerned with hot environments, such as those found in Yellowstone National Park because they are fairly accessible when compared to low temperature environments. However, cold environments ($<10°C$) are far more extensive, encompassing the Arctic and Antarctic continents as well as the large majority (*c*. 90%) of ocean waters (Baross & Morita 1978). A detailed survey of organisms capable of growth at high temperature and their ecology was presented by Tansey and Brock (1978). In addition to its primary effects on biological reactivity and cellular stability, temperature provides good examples of the interrelationships discussed above. In fact in any environment, however hot or cold, temperature affects gas solubilities, diffusion rates, viscosity, density and chemical reactivity.

32.3 Acidity and alkalinity

Extensive areas of extremely low pH exist, usually in geothermal areas which have significant hydrogen sulphide emissions. This hydrogen sulphide is rapidly oxidised by microorganisms upon exposure to the air (Schoen & Rye 1970; Mosser *et al.* 1973). There is, therefore, a relationship between the low pH and high temperature. However, this relationship is variable for two reasons:

(1) there are many geothermal habitats which do not have hydrogen sulphide emissions and therefore do not become acid, as seen in many geyser basins in Yellowstone Park (Brock 1978);

(2) acid formed near a hydrogen sulphide source is often carried into cooler bodies of water in sufficient quantities to create extremely low pH (Brock 1978).

An additional type of environment with low pH is the coal refuse pile (slag heap). Although these environments are man-made, many have persisted for decades and this alone makes them extremely interesting subjects for study. The acidity in these piles also arises from oxidation of reduced sulphur compounds, but here the source is usually sulphide minerals, primarily ferrous disulphide (pyrite) (Dugan *et al.* 1970; Singer & Stumm 1970). Sulphide minerals are stable in the absence of oxygen, but coal processing and the disposal of the mineral refuse exposes the pyrites to oxygen which leads to its rapid microbiological oxidation to sulphuric acid. Although most of these piles self-heat and often even ignite, the heat input is much less than in geothermal areas, so that any effect of the acid generally occurs at non-extreme temperatures. These piles also provide a good example of the interaction between one extreme and another. The acid leaching always brings with it high concentrations of heavy metals whose effect cannot be ignored in microbiological studies. In fact, heavy metal leaching by acidophilic bacteria is industrially very important (Murr *et al.* 1978).

Finally, it should be noted that there are few stable high pH environments and, therefore, few microorganisms which grow at high pH (Souza *et al.* 1974). Brock (1971) has measured the pH of a large number of hot springs and shown two consistent peaks in pH distribution at 3.0 and 8.5. These values are maintained by the buffering action of sulphate in the acid areas and bicarbonate and silicate in the alkaline regions. The absence of extremely basic ($pH > 12$) environments probably reflects the lack of an adequate buffer to maintain such a level.

32.4 Water availability

Water is a universal biological requirement. There are two types of environment with extreme water limitation: the hypersaline and the xerophilic. Hypersaline areas are not uncommon, for example the Great Salt Lake in Utah, the Dead Sea in the Middle East, the Salton Sea in California and many smaller bodies of water (Kushner 1978c). These areas receive little run-off and have high evaporation rates, leaving extremely high salt concentrations. There is one other important type of hypersaline environment which is clearly not natural, but may be more common even though transient. This is the saltern, an area where sea water is retained and allowed to evaporate to collect the salt for commercial purposes. Large areas along San Francisco Bay are covered by these enclosures which turn bright red from the growth of *Halobacterium* spp. as the salt concentration approaches saturation.

Very dry environments are abundant on all continents, including the frozen tundra. In these areas water is either totally absent or in the form of ice or very tightly bound to the mineral surroundings (matric binding). It has become clear that the best way to describe water presence is in terms of its availability either as water activity or as water potential (Griffin 1972; Brown 1976). These expressions can be viewed as reflecting the energy costs necessary for an organism to obtain its water. These quantitative measures have been applied to saline studies (Brown 1976; Kushner 1978c) and to several dry areas (Smith & Brock 1973a, b; Brock 1975a, b, c, d).

There are two other areas of study in which water limitation is important: aerobiology and food microbiology. The subject of airborne microorganisms has been reviewed by Dimmick and Akers 1969, Gregory 1973 and Edmonds 1979. Water can be limiting to these organisms, but it is highly variable with the relative humidity. Direct study is enormously difficult and has been largely limited to taxonomic surveys in natural areas and examinations of factors affecting survival in artificial aerosols. Food microbiology was the first area to deal quantitatively with the problems of water availability (Scott 1957), and essentially all the field studies of the past two decades have been based upon the methods created during food research.

32.5 Radiation

In addition to photosynthetically useful light, the biologically significant forms of natural radiation are ultraviolet and ionising radiation (Nasim & James 1978). Radiation is important in our understanding of aerobiology, since high radiation levels occur in the atmosphere, especially at high altitudes. With the obvious exception of those wavelengths suitable for photosynthesis, radiation is generally harmful to microorganisms although there are examples of remarkable radiation resistance, namely *Micrococcus radiodurans*. This organism is several orders of magnitude more resistant to ultraviolet damage than most other bacteria which have been tested. It is now apparent that the ability of *M. radiodurans* to survive excessive radiation is based not on an innate resistance to radiation but rather on an extremely efficient system to repair damage after it occurs (Nasim & James 1978).

32.6 Pressure

The average hydrostatic pressure in the ocean is about 380 atmospheres (Morita 1974) so that the majority of marine environments may be judged to be extreme. Even though the oceans are the most abundant extreme environments in the world, they are among the least studied. The obvious explanation is the nearly insurmountable technical difficulties involved in dealing with great depths. Hydrostatic pressure increases at the rate of about 1 atmosphere for each 10 m of depth, therefore the average pressure of 380 atmospheres equates to an average depth of 3800 m. Conventional oceanographic gear can collect samples from extreme depths, but the process is lengthy—up to 18 h from 10 000 m (ZoBell & Morita 1957). This process causes depressurisation to 1 atmosphere and warming from probably less than 5°C to the ambient temperature at sea level. Physiological damage to microorganisms in the sample is almost certain (Seki & Robinson 1969; ZoBell 1970; Morita 1976) although this point cannot be established without comparison to an undecompressed sample, an experiment made possible by recent developments in sampling devices, but not yet specifically reported.

Experimental

32.7 Introduction

It is important to realise that, with the exception of high pressure investigations, there are relatively few techniques which are unique to the examination of extreme environments. Most studies have considered basic microbiological processes and have used established field procedures with modifications as necessary for the parameter in question. Most of these modifications are quite minor and consist of obvious adjustments, for example incubating samples for thermophiles at elevated temperatures and those for acidophiles at lowered pH. That such modifications need only be minor contains an important point which is usually implicit, but needs explicit statement: a given experiment in an extreme environment will examine a certain physiological type of microorganism or a specific physiological process. For example, our water potential studies in the dry, hot acid soils of Yellowstone Park (Smith & Brock 1973a, b) did not examine water availability directly, but rather the effect of variations in water availability on photosynthetic [^{14}C]carbon dioxide fixation by a soil alga. The distinction is not superfluous, nor does it represent an error in field studies. It is the case that extreme physical and chemical factors are not studied in the abstract; they are considered as effectors of a carefully defined and delineated process. This point will be developed below.

There are three basic aspects of extreme environment studies:

(1) cultural isolations;

(2) physiological experiments;

(3) molecular mechanisms of tolerance and adaptation.

Not all of these features are apparent for each extreme parameter and there is no intention to imply a chronological sequence. Indeed, for some parameters, one or more of these stages is absent, as occurs in high radiation and high pressure environments.

Extreme environments have been investigated sufficiently well such that it is generally accepted that the limits placed on microbial life are quite broad. However, it is much less clear how microorganisms respond as the limit for a given parameter is approached. There is a diversity of responses ranging from tolerance by the indigenous community to the selection of organisms which perform optimally and even become dependent upon the extreme condition (Brock 1978). Sometimes it is difficult to distinguish between microbial survival and microbial activity under extreme conditions. Generally, more is understood about molecular adaptations to extreme environments than ecological functioning. Nevertheless, considerable ecological knowledge exists for thermophiles (and to a lesser degree acidophiles) although there is scant information about the effects of high radiation on microorganisms *in situ*.

32.8 Temperature

32.8.1 High Temperature

Thermophilic microorganisms are active in and have been isolated from an enormous array of natural and man-made habitats (Tansey & Brock 1978). However, few of these areas have been appropriate for detailed investigations of physiological ecology because the habitat was either transient or logistically difficult to examine. The clearest exception to these problems are the stable hot spring areas which have proven amenable to study. This type of habitat and the organisms within it are by far the best studied of all the extreme environments.

Microscopic observations of hot spring waters have long demonstrated bacterial presence in boiling water. However, physical presence cannot be equated with activity or, indeed, even with viability. Culture experiments may seem a logical approach, but the design of a suitable culture medium and the choice of growth conditions for an unknown microorganism are almost impossible tasks (Brock 1978). Brock's group obtained convincing evidence for high temperature growth and activity *in situ* in two rather different ways. First they inserted microscope slides into springs and used them as surfaces on which microorganisms could develop (Bott & Brock 1969). Slides were placed in two groups: the first group was incubated for 24 to 48 h, while the second was removed at intervals less than the presumed organism doubling time of the attached bacteria and irradiated with ultraviolet light. All cells present on slides of the second group at the end of the incubation period were presumed to represent total migration and attachment. The cells on the non-irradiated slides (the first group) arose from random attachment as well as growth after attachment. The difference in cell counts between the two groups is taken as growth *in situ*. Supporting this conclusion was the observation that most of the cells on the non-irradiated slides were clustered, presumably reflecting microcolony development from a pioneer attached cell. It was clear that growth occurred at

91°C, essentially in boiling water at Yellowstone's elevation (mean elevation of 2130 m, theoretical boiling point of pure water is 93°C).

The second experimental approach was the incubation of hot spring samples with radioisotopically labelled substrates such as amino acids (Brock *et al.* 1971). Incubations were performed at several different temperatures for 2 h and incorporation into cellular material was determined by trapping the cells on membrane filters for scintillation counting. The results of these experiments not only showed that microbiological activity occurred at high temperatures, but also demonstrated that the activity was maximal at these temperatures. As Tansey and Brock (1978) noted, it is interesting that, despite this demonstration of activity *in situ*, it has not been possible to obtain cultures of these bacteria. The same conclusion about maximal physiological activity occurring at high temperatures has been made with other processes as well: [^{35}S]sulphur oxidation to [^{35}S]-sulphate by *Sulfolobus* sp. at 85 to 90°C (Mosser *et al.* 1973); [^{14}C]carbon dioxide fixation by soil autotrophs at 60 to 70°C (Fliermans & Brock 1972); and photosynthetic [^{14}C]carbon dioxide fixation by flexibacteria (Bauld & Brock 1973). It must be emphasised that the numerical value of the observed optimum temperature did not always match the environmental temperature very well; for example, Doemel (1970) showed that isolates of *Cyanidium caldarium* fixed [^{14}C]carbon dioxide optimally at 45°C regardless of the temperature of the environment from which it was isolated. Similarly the *Sulfolobus* sp. study of Mosser *et al.* (1973) failed to find temperature-sensitive strains although such strains were clearly identified for the cyanobacterium genus *Synechococcus* by Peary and Castenholz (1964).

A fortuitous circumstance surrounding studies of hot spring bacteria is that thermal gradients frequently occur as the water flows from the spring and rapidly cools (Brock 1978). This means that field incubations over a wide variety of temperatures can be performed quickly and easily by placing incubation vials every few meters along the thermal gradient. The gradient has more than just experimental design value, however, as a large temperature change over a small distance allows other factors to remain constant. It is, therefore, possible to consider the effects of variations in temperature virtually independently of other features (Brock 1978).

Molecular investigations of thermophiles include: membrane characteristics (Ray *et al.* 1971a, b); nucleic acid metabolism (Zeikus & Brock 1971); ribosome structure and stability (Zeikus *et al.* 1970); and protein structure and function (Freeze & Brock 1970; Singleton & Amelunxen 1973; Singleton 1976). As expected, obligate thermophiles are distinct from their more mesophilic relatives. Some changes are extensive (lipid composition affects membrane fluidity responses to temperature) but, interestingly, some of the more significant effects in terms of thermostability and activity seem to be brought about by rather subtle adjustments (Singleton 1976). It is fair to conclude that thermophilic bacteria are different from mesophiles, but that there is no single subcellular mechanism which will easily explain the difference.

32.8.2 Low Temperature

Microbial life at low temperature has received much less attention (see Morita 1975 for definitions of psychrophiles, psychrotrophs and psychrotolerant organisms). Part of the difficulty is that relatively stable, cold environments are less accessible and their study is more difficult. Most of the ocean has a stable temperature of <5°C (Morita 1974), but is at great depth, presenting technological problems as well as obvious interaction with pressure as a complicating factor. Terrestrial cold environments, such as tundra and permanent snow fields, are plentiful but are difficult to sample for technical as well as human comfort reasons. Also, many of the tundra areas are frozen, making water availability an interacting parameter. There is, however, substantial evidence from cold environments not under extreme pressure that microbial activity can occur at low temperature. Baross and Morita (1978) summarised results concerning microbial primary production and heterotrophic consumption at temperatures near 0°C in ice algae, snow algae and microorganisms of cold surface marine waters. Fuhrman and Azam (1980) recently demonstrated significant growth of bacteria in Antarctic waters, using a sensitive technique to measure DNA replication *in situ*. Vishniac and Hempfling (1979) described the existence of an indigenous microbial community in the cold, dry deserts of the Antarctic. Their work centred on yeasts, typified by *Cryptococcus vishniacii*, but some interesting inferences were drawn concerning the general microbial ecology of this extreme environment, especially with regard to physical factors such as water availability and biotic factors such as potential primary productivity by endolithic algae (Friedmann & Ocampo 1976). Vishniac and Hempfling (1979) also noted that many of the Antarctic psychrophiles have temperature ranges for growth of up to 20°C or greater. This finding is in contrast to the well-studied marine psychrophiles, which have a very narrow range. It is probable that the differences reflect the seasonably variable Antarctic tundra temperatures and virtually constant ocean temperatures.

Cellular processes decline in rate as the temperature is lowered. The temperature below which growth ceases may bring about the failure of an enzyme activity, protein synthesis, transport or membrane fluidity (Evison & Rose 1965; Morita 1975; Arthur & Watson 1976; Watson *et al.* 1976; Inniss & Ingraham 1978). Psychrophiles placed at elevated temperatures display altered morphology, leakage of cellular components and occasionally cell lysis (Inniss & Ingraham 1978). Some mesophiles respond to lowered temperature by decreasing the percentage of unsaturated fatty acids in the lipids of their membranes. Inniss and Ingraham (1978) reviewed several aspects of the rather elaborate modifications which occur, while noting that the only organisms for which this change has been well established are two mesophiles. It would seem valuable to survey a number of psychrophiles for their fatty acid composition and any changes that occur as the temperature declines.

32.9 Acidity and alkalinity

A comprehensive treatment of the ecological and biochemical features of organisms capable of growth at extreme pH values is presented by Langworthy (1978).

32.9.1 Acidic Environments

Many microorganisms, both prokaryotic and eukaryotic, grow well in and are adapted to extremely acid (pH < 2) conditions, for example, *Thiobacillus ferrooxidans* (Rao & Berger 1971); *Bacillus acidocaldarius* (Darland & Brock 1971); *Sulfolobus acidocaldarius* (Brock *et al.* 1972); *Thermoplasma acidophilum* (Darland *et al.* 1970); the alga *Cyanidium caldarium* (Doemel & Brock 1971); and several fungi (Langworthy 1978). The obligate nature of the acid requirement has, for the most part, been demonstrated in laboratory cultures. There are two reasons for the limited field data on this point. First, experimental variation of acidity in the field is not only technically awkward; it also brings with it the possibility of sample alterations by changing equilibria, solubilities and precipitations. Second, there are relatively few natural pH gradients in acid areas, unlike the valuable temperature gradients mentioned above (see section 32.8.1) and even the gradients that do exist contain variations in temperature and chemical features. Most field data directly relating to pH effects were obtained by collecting samples and isolating cultures from a variety of environments and correlating environmental pH with growth response. This approach has provided valuable, although somewhat indirect, information about pH effects *in situ* (Doemel 1970; Brock 1973).

A number of reports exist on the internal pH of acidophilic microorganisms, especially bacteria (Langworthy 1978). Direct measurements of cell lysates and sophisticated measurements of internal/external distribution of ionisable compounds have clearly demonstrated that the internal pH values of even the most extreme acidophiles are near neutrality. As Langworthy (1978) points out, many enzymes have been isolated from acidophiles with pH optima near neutrality, providing circumstantial evidence for non-acidic conditions intracellularly. The acidophilic thiobacilli have received much attention and, although some chemical differences from non-acidophiles do exist, there is no striking feature to account for their obligate acidophily. However, extreme acidophiles which are also thermophiles (*Bacillus acidocaldarius*, *Sulfolobus acidocaldarius*, *Thermoplasma acidophilum*) have a number of chemical features which distinguish them. Langworthy (1978) described changes in membrane composition in terms of lipid type and fatty acid type which are unique to these bacteria.

32.9.2 Alkaline Environments

Langworthy (1978) has summarised the few reports which exist on natural alkalinophiles, most of which are based on studies *in vitro*. The scarcity of high pH environments is probably significant in the observation that most microorganisms isolated from environments pH > 11.0 have culture pH optima which are much lower, usually 6.0 to 9.0

(cf. Ohta *et al.* 1975). Presumably there has been insufficient selection for obligately alkalinophilic organisms. The internal pH values of alkalinophiles are also closer to neutrality than to the medium pH (Langworthy 1978). There have been virtually no reports of chemical modifications which are peculiar to these bacteria although such modifications may not exist, since the highest pH areas are only 3 to 4 units away from 7 while the low pH areas are 6 to 7 units away. We must remember the precaution expressed by Kushner (1978c), however: 'the correlation of an unusual cellular feature with the ability of that organism to grow in an extreme environment does not constitute demonstration of cause and effect.'

32.10 Water availability

Water availability is a quantitative concept describing how much effort an organism must expend to obtain the liquid water it needs. This discussion will consider microorganisms which are incapable of direct accumulation of water (bacteria, fungi and algae). These organisms must control their intracellular water content by controlling their intracellular solute content, since water will flow across the membrane in response to concentration gradients. The expressions defining water availability are most simply presented in terms of solutions.

There are two fundamental terms: water activity (a_w) and water potential (ψ). Pure water has the maximum availability, with an a_w of 1.0 and a ψ of 0. When compared to pure water, all solutions have decreased availability with a_w less than 1.0 and a negative value for ψ. a_w is a unitless parameter ranging from 0 to 1.0. ψ is an energy expression, usually expressed in pressure units called bars: 1 bar $= 1 \times 10^6$ dyne cm^{-2} $= 0.987$ atmospheres $= 75$ cm mercury pressure $= 1022$ cm water pressure. All solutions have negative potentials with respect to pure water, and their potentials are in units of bars. These expressions are interconvertible as follows:

$$a_w = \frac{\text{r.h.}}{100}$$

and

$$\psi = \frac{1000\,RT}{\text{Wa}} \ln(a_w)$$

where r.h. is the relative humidity over a given solution; R is the gas constant; T is the absolute temperature; and Wa is the molecular weight of water (Lang 1967). Solutions differ from pure water in three ways termed the colligative properties: increased osmotic pressure, depressed freezing point and elevated boiling point. There are two measurement principles for these parameters: the freezing point depression displayed by a solution and the r.h. over the solution or system in question. The most common expression of decreased water availability, osmotic pressure, is rarely measured directly.

Although these basic relations have been developed using aqueous systems, the concept is successfully applied to areas in which water availability is limited by adsorption to surfaces rather than by solute interactions. The first detailed use was in food microbiology (Scott 1957), although the last 20 years have seen an increasing application to soil and other solid phase systems such as plant leaves (Brown & van Haveren 1972). These substrates, often simply referred to as dry, retain water by tenacious surface interactions. This water is difficult for organisms to obtain, and gives the environment a low water potential. This type of water binding is referred to as matric water potential as opposed to solute interactions, which are referred to as osmotic water potential (Griffin 1969). It may be seen that in all environments water is limited by both matric and osmotic effects and usually one or the other will predominate. Hypersaline environments have almost exclusively osmotic forces at work while dry environments have a mixture of the two with a usual predominance of matric.

32.10.1 Hypersaline Environments

There are many lakes with salinities far in excess of that of sea-water (Nissenbaum 1975; Brown 1976; Post 1977; Kushner 1978c). Man-made areas of hypersalinity also abound, namely salterns which have been used for centuries for the preparation of solar salt from sea-water. Indeed, Darwin (1845) commented on the extreme environment of the brine shrimp in such ponds. An impressive variety of microorganisms, invertebrates such as shrimp, and grazing flies have been observed and examined. Post (1977) presented a thorough analysis of the types of bacteria, algae, fungi and protozoa which can be isolated from Great Salt Lake. Many of the isolates require high levels of salt (typically but not exclusively sodium chloride) for activity and growth in culture. Such organisms are termed halophiles or salt loving. Despite the long history of qualitative analyses, there are no direct measurements of the activity or growth of extreme halophiles in their natural environments. However, the evidence from cultures *in vitro* is overwhelming. For example, the absolute requirement of the halophilic bacteria (*Halobacterium* spp.) in culture almost certainly means that they are active in the areas of isolation, but presence cannot be taken as absolute evidence of growth. This conclusion has recently been underscored by the observations of Horowitz (1976) in the Antarctic: the Don Juan pond is saturated with calcium chloride, has a water activity near 0.45, and freezes at −48°C (Meyer *et al.* 1962). Bacteria can be observed in this water but Cameron *et al.* (cited in Baross & Morita 1978) believed their isolate of *Achromobacter* sp. from this pond was probably inactive there, while Horowitz (1976) has compelling evidence that these bacteria are washed in from the surroundings and do not grow in the pond.

The mechanisms of tolerance of and adaptation to high salt environments have been established in remarkable detail, especially considering the paucity of ecological background. There are two different types of response: adaptation through protein alterations as evidenced by the halophilic bacteria, and tolerance by creation of compatible solutes as occurs with the halophilic algal genus, *Dunaliella*, and osmophilic yeasts.

Protein modification is a simplified term for the multiple ways in which these bacteria differ from non-halophiles. It has been known for some time (Baxter 1959) that the proteins of halophiles are acidic, that is, they have an excess of negative charges. Lanyi (1974) fractionated the cells of halophilic bacteria and found that cytoplasmic proteins, ribosomal proteins, proteins from various portions of the cell envelope and gas vacuole proteins all have an excess of acidic over basic amino acid residues of 13 to 22%. A major role of the cations of the required salt is to combine with these negatively charged sites and thereby prevent disruptive electrostatic repulsion. Kushner (1978c) summarised these observations and points out their likely oversimplification. An additional effect of high salt concentrations is to increase the strength of hydrophobic interactions within the protein by increased ordering of water in the region of the salt (Lanyi 1974; Brown 1976). The final noteworthy feature is that the external proteins of these bacteria are stabilised by sodium chloride while the intracellular proteins are both stabilised and activated by potassium chloride (Kushner 1978c). Consistent with these findings is that halophiles grown in a medium of 3.3M-Na^+ and 0.05M-K^+ can have internal concentrations of 0.80M-Na^+ and 5.3M-K^+ (Matheson *et al.* 1975). However, these values are averages of entire cells and, as Kushner (1978c) points out, there may be substantial differences in concentration in different parts of a cell.

Compatible solutes are synthesised internally by the cell in amounts sufficient to counteract the stress from high external solute concentrations. The word compatible means that the solute can be present in substantial concentrations without disturbing cellular processes. The so-called osmophilic yeasts make a number of polyalcohols when grown in media with extremely high sugar concentrations (Brown & Simpson 1972; Brown 1974). Rose (1976) and others have pointed out that these organisms are more properly referred to either as facultatively osmophilic or as osmotolerant since they grow much more slowly at high solute concentration than at low. The algal genus *Dunaliella* makes large amounts of glycerol in response to salinity stress (Ben-Amotz & Avron 1973; Borowitzka & Brown 1974) and the production of such solutes is an impressive procedure to allow growth in high solute conditions. The energy cost of this synthesis is considerable, but is presumably selected for because it allows growth in an environment from which almost all other organisms are excluded. It must be noted that the ability to function in a high solute environment requires more than osmotic adjustment. The chemical nature of the specific solutes must also be considered; for example, the halophilic bacteria have achieved osmotic and ionic balance by adapting to high salt levels while *Dunaliella* spp. apparently cannot withstand high intracellular salt concentrations and have evolved a rather more elaborate response involving internal solute synthesis.

32.10.2 Arid (Desertic) Environments

Any area, whatever the extent, which has an evaporative potential in excess of precipitation may be characterised as a desert. This definition is clearly met by great expanses in Northern Africa and portions of North America. There are also localised areas which have extraordinary evaporation potentials due to geothermal activity, such as those found in Yellowstone Park in the Western

United States (Brock 1978). A third type of environment with extremely limited water availability may be referred to as a physiological desert. The tundra biome typically has large amounts of its water unavailable because it is frozen. In such areas spring activity of soil bacteria appears to result more from thawing (and presumably increased water availability) than from temperature increases *per se* (Holding *et al.* 1974a).

Many field studies in dry environments have been focused on individual organisms (Lange *et al.* 1970 cited in Rogers 1971; Rogers 1971; Smith & Brock 1973a, b; Friedmann & Galun 1974; Brock 1975a, b), although some more general physiological studies have been carried out (Holding *et al.* 1974a; West & Skujins 1978). Microorganisms may survive the dry conditions, but none have been demonstrated to function optimally at these low water potentials (Smith 1978). The effect of decreasing water potential on natural populations has been examined in two ways.

(1) A number of samples may be collected from different environments or at different times so that a range of water potential values are represented. Determinations of microorganism number or some physiological activity are then made and correlations with the water potential *in situ* are made (Smith & Brock 1973b; Rychert *et al.* 1978);

(2) The second procedure involves collection of samples, alteration of water potential and measurement of viability or physiological activity at the new levels. These experiments have been done routinely with laboratory cultures (Adebayo *et al.* 1971), but much less often *in situ* (Smith & Brock 1973a; Brock 1975b). The results obtained with natural populations manipulated in this way show that these organisms are sensitive to decreasing water potentials and tolerate dry environments rather than thriving in them. In fact, long periods of inactivity due to low water potential may have to be endured before the opportunity arises for active growth (Brock 1975b).

Measurement of soil water potential requires equilibration in an environment of known relative humidity (Harris *et al.* 1970) or with a calibrated material such as filter paper (Griffin 1972) or within a small chamber in a psychrometer (Rawlins 1972). The psychrometer method is by far the most flexible procedure, allowing multiple rapid (5 min) determination of water potential. Control of water potential has been accomplished in soil tensiometers (Griffin 1963a, b) and pressure apparatus (Rose cited in Griffin 1972), but these devices require long equilibration and are rather limited in convenience and the range of water potentials over which they are effective (Griffin 1972). The agar dish method of Harris *et al.* (1970) is relatively rapid and allows good control over the water potential of the sample.

Microorganisms are better able to withstand osmotic water potential stress than they are matric stress (Sommers *et al.* 1970; Adebayo & Harris 1971; Brock 1975b). This conclusion is emphasised by the existence of microorganisms such as *Halobacterium* spp. and *Dunaliella* spp. which grow only in high solute environments, although the effects of saturated salt solutions on the halophilic bacteria are much more complex than simple osmotic events (Kushner 1978c). A microorganism faced with a water potential stress must attain an internal potential lower than the external or else cell water will be lost (Adebayo *et al.* 1971). If this stress is predominantly matric, then the organism is severely limited. The paucity of solutes in such a dry environment means that active accumulation of solutes is insufficient. The alternative would seem to be the internal synthesis of solutes to lower the water potential, either *de novo* as in *Dunaliella* spp. and the osmophilic yeasts, or by degradation of intracellular polymers such as nucleic acids, proteins, and carbohydrates to their more osmotically active constituent monomers. Apparently few organisms have the capability of massive solute synthesis like *Dunaliella* spp., and the degradation approach would necessitate extreme selectivity of which compounds were degraded. The severe limitations imposed by matric stress are therefore quite reasonably more difficult to respond to than those arising from osmotic stress.

A separate example of extreme desiccation is provided by the frozen Antarctic deserts which have been described as the driest environments in the world (Horowitz *et al.* 1972; Vishniac & Mainzer cited in Vishniac & Hempfling 1979). The effect of cold in this environment has already been discussed above, but the dry nature deserves separate consideration. Horowitz *et al.* (1972) considered low water availability the biggest stress in the area, and indeed were unable to isolate microorganisms from several areas, concluding that they were sterile environments. Vishniac and Hempfling (1979) isolated a variety of microorganisms from the Antarctic soils, indicating that there must be some micro-environments with sufficient liquid water (and nutrients) to support growth.

Airborne microorganisms have received little direct ecological attention (Gregory 1971; Gregory 1973; Akers *et al.* 1979). It is clear that all forms of microorganisms are present in the atmosphere, even at altitudes in excess of 27 000 m (Burch 1967). Fulton (1966a, b) found the highest bacterial densities above 3000 m at a temperature of <5°C. Baross and Morita (1978) describe the isolation of viable microorganisms from the Antarctic air at a temperature of −20°C, a remarkable combination of low water availability, low temperature and possibly high radiation. The activity of airborne microorganisms *in situ* is less clear, and a large portion of natural aerobiology investigation has involved taxonomic descriptions of sampled atmospheres. Gregory (1971) summarises the evolution of sampling methods and points out that quite different methods are employed by different investigators, thereby making geographical comparisons very difficult. Gregory (1971) also cautions against excessive extrapolation of laboratory results to the natural environment, citing some unexpectedly rapid losses of viability in *Escherichia coli* populations.

There has been an abundance of investigation of the factors affecting the survival and activity of microorganisms in experimentally created aerosols (Dimmick 1969; Strange & Cox 1976; Edmonds 1979). These studies have made major contributions to the understanding of transmission of diseases through the air. While significant in an epidemiological sense, the majority of this work is not appropriate in this discussion which concentrates on more natural systems. A clear result of the aerosol studies is the expected controlling effect of relative humidity on survival. Airborne organisms are also exposed to high radiation levels. Repair mechanisms are present, but the efficiency of the overall repair process is greatly impaired under dry conditions (Webb & Tai 1968; Webb & Walker 1968).

32.11 Radiation

All terrestrial and surface aquatic organisms live in potentially extreme radiation from sunlight. This conclusion follows from observations of coliform susceptibility to natural sunlight (Bellaire *et al.* 1977) and is supported circumstantially by the variety of systems for repair of radiation-induced damage in most microorganisms (Nasim & James 1978). There are two general forms of harmful radiation: ultraviolet and ionising. Ultraviolet radiation (wavelength <400 nm) is a significant component of sunlight, although Nasim and James (1978) remind us that Luckiesh and Knowles (1948) reported lethal effects of solar wavelengths up to 700 nm. Natural ionising radiation arises from cosmic rays or from terrestrial radioactivity. Man-made ionising radiations have increased enormously in significance in the past 40 years (UNSCEAR 1972). Even though these radiations are still a minority of the planetary total, they are locally highly concentrated. Some bacteria have been reported growing in the waters of 'swimming-pool-type nuclear reactors' (Prince—cited in Vallentyne 1963).

The few ecological studies of extreme radiation have been in relation to tolerances displayed by photosynthetic microorganisms (Goldman *et al.* 1963; Stepanek 1965; Brock & Brock 1969). Many of these observations have been extended in the laboratory (Jorgensen & Steemann Nielsen 1965; Yentsch & Lee 1966; van Baalen 1968). Jorgensen (1964) demonstrated that a *Chlorella* sp. reduced its chlorophyll content in response to high light intensity. The response to high light levels was rather complex, however. Brock and Brock (1969) demonstrated that natural mats of thermophilic cyanobacteria are not affected by the high ultraviolet intensity of their high altitude (*c.* 2100 m) natural environment. A large factor in this observation is that mat-forming organisms exhibit considerable self-shading.

However, the general observation is that high radiation intensities are damaging and therefore most studies are concerned with the mechanism of this damage. As a simplification, we may divide radiation damage into two components:

(1) damage caused directly by the radiation;

(2) damage effected by some intracellular compound which has interacted with the radiation.

32.11.1 Direct Damage

The most directly radiation-sensitive portion of the microbial cell is its nucleic acid, more specifically the DNA. As Bridges (1976) noted, there are two different reasons for this importance of DNA, depending on the type of radiation. Ultraviolet light, especially in the region of 250 to 260 nm, is absorbed by the DNA. The predominant effect is the formation of covalently linked dimers between adjacent pyrimidine residues (usually thymines) on the DNA (Setlow & Carrier 1966). There are additional effects of ultraviolet on DNA, including strand breakage, cross-linking, and binding to protein (Rahn 1972, 1973). Ionising radiation probably affects many cellular components, but is especially harmful to the DNA because there is (relatively) a considerable amount of DNA per cell and because damage to this primary genetic information will be extremely difficult to recover from (Bridges 1976; Hutchinson 1966). The damage is likely to be severe—single- and double-strand breaks which are reparable by few microorganisms (Bridges 1976).

DNA damage is repaired through a variety of enzymic systems (Bridges 1976; Nasim & James 1978). These repair activities can be divided into two groups, according to their need for visible light.

(1) Photoactivation is the enzymic splitting of thymine dimers relatively soon after their formation and before the next cycle of DNA replication. Although the photoreactivating enzyme appears to be widespread and important in the organisms which possess it, there is no clear correlation between overall radiation resistance and the ability to photoreactivate (Nasim & James 1978). Perhaps this specific enzymic activity should be viewed as one weapon in a larger arsenal.

(2) Nasim and James (1978) also describe a light-dependent process termed photo protection. In this action visible light can cause a delay in the occurrence of cell division, thereby allowing more time for repair and delaying the negative effects of the UV damage.

Dark repair activities are of two types.

(1) The enzymic removal of the dimer which is accompanied by the cleavage of several nucleotides. New bases are inserted, using the remaining strand as a template, and the newly synthesised segment is sealed into place with ligase.

(2) The second mechanism is termed post-replication recombinational repair. In this system the dimers are skipped over during replication, leaving gaps on the new strand where the dimers were on the old strand. The gaps are filled by exchange between old (irradiated) and newly synthesised DNA molecules. As Nasim and James (1978) point out, this basic scheme can be much more complicated, involving multiple growth cycles.

Finally, the mechanisms of resistance to ultraviolet and ionising radiations may be somewhat independent, since mutants can be found which repair ultraviolet damage but are sensitive to ionising radiation. This observation may result from inability to repair single-strand breaks (Nasim & James 1978).

32.11.2 Photochemical Damage

A considerable amount of indirect damage is caused by the photochemical creation of singlet oxygen, following reaction of light with an intracellular photosensitiser (Krinsky 1976). Many molecules have been proposed as sensitisers, but porphyrins, widely distributed in the biosphere, are especially effective (Spikes 1975). In many cases the mechanism of damage is unidentified, but the target is clear. Respiratory pigments and cytochromes (Kashket & Brodie 1962; Bragg 1971); membranes (Krinsky 1974); enzymes which are haem proteins such as catalase (Mitchell & Anderson 1965); enzyme cofactors (Krinsky 1976); and membrane transport (D'Aoust *et al.* 1974) have all been shown to be photochemically susceptible (see Krinsky (1976) for a detailed analysis of these multiple effects).

Protection against many photochemical effects is related to the presence of carotenoid pigments. Many organisms display increased photosensitivity when mutants are isolated which lack carotenoids: photosynthetic bacteria (Griffiths *et al.* 1955; Sistrom *et al.* 1956); *Corynebacterium* sp. (Kunisawa & Stanier 1958); cyanobacteria (Krinsky 1968); and *Sarcina* sp. (Mathews & Sistrom 1960). Carotenoid pigments quench either the singlet oxygen or the photoactivated sensitisers (Krinsky 1976). The significance of radiation as an extreme factor may be deduced by the abundance of mechanisms which exist to detoxify its effects, presumably in response to selective pressure. Such conclusions are, of course, rather speculative, since very little has been done in the way of quantifying radiation effects in natural environments.

32.12 Pressure

The basic ecological question is whether or not microorganisms exist which can be called barophiles. That is, has the huge extent of high pressure habitat in the lower layers of the oceans selected for organisms which grow best under the elevated pressures found there? As has been noted previously (see section 32.6), there are extraordinary technological problems associated with answering this question.

One consequence of the technological difficulties is the rather incongruous fact that most of the information we have about putative barophiles has been obtained in experiments with organisms which are clearly harmed by pressures greater than one atmosphere. Some of this work has used bacteria from the deep ocean (Jannasch & Wirsen 1977; Ehrlich 1978), while some has used surface and terrestrial organisms (Pope *et al.* 1976; Marquis & Matsumura 1978). These latter studies with terrestrial organisms have been valuable in providing rather detailed explanations for the harmful effects of high pressure in terms of cellular processes such as transport and protein synthesis and of general reactivity as shown in the molecular volume approach (Marquis & Matsumura 1978). These findings are of great interest, but the emphasis here will be on the ecology of natural high pressure forms and environments.

The term barophile was coined by ZoBell and Oppenheimer (1950) to describe bacteria which 'grow preferentially or exclusively at high hydrostatic pressures'. ZoBell and Morita (1957) characterised an organism which grew slowly at 700 atmospheres but was indeed an obligate barophile. Most attempts at barophile isolation have obtained only barotolerant organisms (Jannasch *et al.* 1976; Schwarz *et al.* 1976; Jannasch & Wirsen 1977). Improved sampling devices have greatly aided the study of barophiles. In recent years two types of sampling device have been developed which allow the collection of undecompressed water samples from depth. The device of Jannasch's group (Jannasch *et al.* 1973, 1976; Jannasch & Wirsen 1977) is deployed by conventional cable methods and has been used to recover undecompressed samples from 3100 m with a stated ability to sample from 6000 m. This system has the advantage of allowing transfers of sample aliquots to a variety of conditions without decompression. The device used by Yayanos and his co-workers (Yayanos 1976, 1977, 1978; Yayanos *et al.* 1979) is fundamentally different in that it is a free vehicle involving no cable equipment. The immediate advantage of this design is that greater depths can be sampled conveniently. At present it is not possible to make undecompressed transfers after recovery, but the device has been successfully used at a depth of 5780 m. Jannasch and Wirsen (1977) have found only barotolerant organisms in their samples, collected from depths up to 3100 m. Yayanos *et al.* (1979) have demonstrated obligate barophily in a *Spirillum* sp. isolated from 5700 m. This bacterium, which has a doubling time of 4.5 h was isolated from a sample which was collected and maintained at 580 bars (*c.* 572 atmospheres) for 5 months. At that time the sampling trap was decompressed and opened and media were inoculated and returned to high pressure (570 atmospheres), low temperature (2 to 4°C) conditions. They isolated the spirillum and found in further work that growth was poor at 1 atmosphere and 825 bars, with a broad optimum around 500 bars.

Even this microbiologically fundamental process of culture isolation is impressive technologically. The key features of the successful effort of Yayanos *et al.* (1979) are:

(1) the sample was obtained undecompressed and with minimum temperature change from great depth (Yayanos 1978);

(2) bacterial enrichment was allowed to proceed for 5 months under pressure;

(3) growth and isolation were facilitated by a new procedure using silica gel-solidified media to minimise depressurisation time (Dietz & Yayanos 1978).

Of major importance were the great sample depth and the long enrichment period under pressure. Yayanos *et al.* (1979) note their inability to cultivate all the different morphological types in their sample, a regret common to many field microbiology efforts, even in non-extreme environments. There are two improvements which may be suggested.

(1) Samples may yield more barophiles if never depressurised. It was necessary for Yayanos *et al.* (1979) to depressurise their samples and cultures for inoculations and transfers.

(2) Only water samples were collected although it is generally believed that microbial activity is greater on particles and in sediments than in adjacent water columns (Brock 1966; Alexander 1971).

The technological problems of collecting and manipulating undecompressed sediment samples aseptically are overwhelming, but their solution should be well rewarded by new information. A word of restraint is warranted here. Fuhrman and Azam (1980) recently presented evidence that the activity of suspended marine bacteria may be comparable to that supposed for sediments. Their results were derived from experiments which very sensitively measured rates of nucleic acid synthesis. These techniques may be ideally suited for examination of barophiles *in situ*: the populations are basically undefined and presumed to be growing slowly (Morita (1976), but see calculated growth rates of Yayanos *et al.* (1979) above).

One of the most exciting discoveries in marine biology history is the Galapagos rift community (Ballard 1977; Corliss *et al.* 1979). This community is apparently driven by the primary productivity of sulphur-oxidising, carbon dioxide-fixing bacteria, probably belonging to the genus *Thiobacillus* (Jannasch & Wirsen 1979). Even though most evidence is observational in the field or based on cultures in the laboratory, significant inferences can be drawn concerning the activity of these bacteria *in situ*. Such indirect conclusions are probably the best evidence available concerning the ecology of deep sea microorganisms. The technology involved in obtaining these results is worth noting. Samples were collected (and decompressed) by a submersible during a large-scale multidisciplinary investigation (Galapagos Biology Expedition Participants 1979). Although there is much to be said for direct sampling, the expense involved restricts most studies to sampling from the sea surface.

32.13 Nutrient limitation

A condition occasionally referred to as extreme is nutrient limitation (Hanson 1976; Brock 1978). Many examinations of natural environments have found them to be nutrient deficient, especially in organic content. As a group, microorganisms have an impressive ability to accumulate dilute nutrients, as shown by the growth of organisms in distilled water in the laboratory (Poindexter 1964). It is difficult to classify low nutrient areas as a distinct type of extreme environment, largely because natural areas of low nutrient concentration are usually extreme in some other parameter; for example, the low nutrient condition of deserts follows from low primary productivity due to water limitation. Indeed in most systems extreme in one or more of the five parameters dealt with in this chapter, productivity in general is greatly restricted. Hot springs, acid drainage, salterns, the atmosphere, and the deep sea are all characterised by general (although irregular) nutrient limitation. The effects of these extreme parameters would greatly confuse any systematic approach to the study of nutrient limitations on microorganisms. Notwithstanding, traditional microbiological culture media have grossly excessive nutrient concentrations and may lead us to inaccurate and unnatural conclusions (Veldkamp 1977). An alternative approach is to grow organisms in nutrient-limited conditions in continuous cultures and many experiments have been conducted in this way, demonstrating responses quite different from those seen in batch cultures (Tempest & Neijssel 1978). Competition between organisms in nutrient-limited continuous cultures is discussed by Harder and Veldkamp (1971) and Kuenen *et al.* (1977). The results of these experiments may allow us to make ecologically relevant conclusions, although they make use of nutrient limitation as a tool more than studying it as an independent parameter.

32.14 Redox

Another parameter not considered here as extreme is redox potential, although it has been suggested as such (Vallentyne 1963; Brock 1978). The reason for exclusion of redox is again the principle of parameter interaction. Wide variations of redox do occur in nature, but it is necessary to consider carefully the controlling factors in a given area. Each environment has a characteristic redox level determined by the identity of redox active compounds present, for example, many marine sediments have very low redox levels because there is an excess of sulphide produced microbiologically during sulphate reduction (Whitfield 1969; Jorgensen 1977). The key points here are that:

(1) redox is a consequence of chemical composition, not a primary parameter;

(2) the chemicals which have a controlling redox activity are highly reactive biologically.

The sulphide example given is perhaps an extreme one, but more subtle redox effects will be difficult to distinguish from the often subtle biological and chemical forces driving them.

32.15 Conclusions

Table 1 presents some clear examples in which interaction of extremes has been found. This table is not inclusive of all known interactions, nor does it just present examples in which an environment is known to be extreme in more than one parameter simultaneously. The purpose is to provide examples in which clear evidence exists for functional interaction of the parameters on microbial activity. Many of the results lead to important ecological and mechanistic information, but often these interactions add complexity to experimental design and interpretation. The importance of interaction can be major and is often apparent in a field investigation. The presentation of the summary in Table 1 is designed to remind us that multiple factors are more often the rule than the exception.

It has been noted (Kushner 1978c) that a major impetus for several extreme environment studies was an interest in extraterrestrial life. Indeed, major support was received to devise instruments and experiments for the supposedly harsh conditions to be encountered on Mars by the Viking Program. We now have results from life detection experiments which 'have been shown to require no biological explanations' (Vishniac & Hempfling 1979), thereby diminishing extraterrestrial incentive and funding.

The possibility of extreme environments serving as natural models of polluted systems also stimulated research, especially in the parameters of high temperature and low pH (Brock 1978; Langworthy 1978). This approach was valid, but it appears now that studies of hot and/or acid environments have given us an increased understanding of the basic biology involved rather than allowing prediction of pollution effects.

Whatever the reasons that persuaded individuals to study extreme environments, much of interest has been learned. In this chapter I have tried to emphasise that, despite technological bottlenecks, the main limitation in studying extreme environments and their microorganisms is not procedural but inspirational: when the interest and commitment are there, the problems can be solved. What has been accomplished, mostly in the last 25 years, is a confirmation of the basic ubiquity of microorganisms, regardless of any anthropocentric definition of 'normal' or 'suitable' environments. Brock (1969) began his first review of microorganisms in extreme environments with an insightful and compelling quote from Darwin. We may do well, in closing, to refer to these remarks again.

> 'Well may we affirm, that every part of the world is habitable! Whether lakes of brine, or . . . warm mineral springs—the wide expanse and depths of the ocean—the upper regions of the atmosphere, and even the surface of perpetual snow—all support organic beings.'
> (CHARLES DARWIN, *Voyage of H.M.S. Beagle*, ch. IV)

Table 1. Interacting parameters in natural environments.

Parameters	Effect	Reference
Pressure and temperature	maximum barotolerance occurs at optimum growth temperature	Marquis & Matsumura (1978)
Pressure and temperature	pressure stimulation of serine deamination by whole cells depends on growth temperature	Albright & Morita (1972)
Salinity and temperature	optimum salinity and optimum temperature are functions of each other	Morita (1975)
Salinity and temperature	lower temperature reduced salt requirement of moderately halophilic bacteria	Kushner (1978c)
Salinity and temperature	lower temperature reduced salt requirement of yeast	Onishi (1963)
Water availability and temperature	decomposition at low a_w required high temperature and at low temperature required high a_w	Flanagan & Veum (1974)
Radiation and temperature	adaptation to brighter illumination by Arctic phytoplankton requires higher (>0°C) temperature	Wetzel (1975)
Water availability and radiation	in airborne microorganisms, radiation repair most efficient at high a_w	Webb & Tai (1968)
Water availability and radiation	alga in hot acid soil adapted to subsurface layer to maximise a_w and illumination	Smith & Brock (1973a)
Temperature and pH	natural hot springs more likely to have bacteria if extreme in only one parameter	Brock & Darland (1970)

Recommended reading

ALEXANDER M. (1976) Natural selection and the ecology of microbial adaptation in a biosphere. In: *Extreme Environments* (Ed. M. R. Heinrich), pp. 3–25. University Park Press, Baltimore.

BAROSS J. A. & MORITA R. Y. (1978) Microbial life at low temperatures. In: *Microbial Life in Extreme Environments* (Ed. D. J. Kushner), pp. 9–71. Academic Press, London.

EHRLICH H. L. (1978) How microbes cope with heavy metals, arsenic and antimony in their environment. In: *Microbial Life in Extreme Environments* (Ed. D. J. Kushner), pp. 381–408. Academic Press, London.

FRIEDMANN E. I. & GALUN M. (1974) Desert algae, lichens, and fungi. In: *Desert Biology* (Ed. G. W. Brown), pp. 165–212. Academic Press, New York.

GRAY T. R. G. & POSTGATE J. R. (1976) *The Survival of Vegetative Microbes. Symposium of the Society of General Microbiology* vol. 26. Cambridge University Press, Cambridge.

GREGORY P. H. (1973) *The Microbiology of the Atmosphere*, 2e. Halsted Press, Wiley, New York.

HEINRICH M. R. (1976) *Extreme Environments*. Academic Press, New York.

KUSHNER D. J. (1978a) *Microbial Life in Extreme Environments*. Academic Press, London.

ZOBELL C. E. (1970) Pressure effects on morphology and life processes. In: *High Pressure Effects on Cellular Processes* (Ed. A. Zimmerman), pp. 85–130. Academic Press, New York.

References

ADEBAYO A. A. & HARRIS R. F. (1971) Fungal growth responses to osmotic as compared to matric water potential. *Soil Science Society of America Proceedings* **35**, 465–9.

ADEBAYO A. A., HARRIS R. F. & GARDNER W. R. (1971) Turgor pressure of fungal mycelia. *Transactions of the British Mycological Society* **57**, 145–51.

AKERS T. G., EDMONDS R. L., KRAMER C. L., LIGHTHART B., MCMANUS M. L., SCHLICHTING H. E. JR., SOLOMON A. M. & SPENDLOVE J. C. (1979) Sources and characteristics of airborne materials. In: *Aerobiology, the Ecological Systems Approach* (Ed. R. L. Edmonds), pp. 11–84. Dowden, Hutchinson and Ross, Stroudsburg, Pennsylvania.

ALBRIGHT L. J. & MORITA R. Y. (1972) Effect of environmental parameters of low temperature and hydrostatic pressure on L-serine deamination by *Vibrio marinus*. *Journal of the Oceanographic Society of Japan* **28**, 63–70.

ALEXANDER M. (1971) *Microbial Ecology*. John Wiley & Sons, New York.

ALEXANDER M. (1976) Natural selection and the ecology of microbial adaptation in a biosphere. In: *Extreme Environments* (Ed. M. R. Heinrich), pp. 3–25. University Park Press, Baltimore.

ARTHUR H. & WATSON K. (1976) Thermal adaptation in yeast: growth temperatures, membrane, lipid, and cytochrome composition of psychrophilic, mesophilic, and thermophilic yeasts. *Journal of Bacteriology* **128**, 56–68.

BALLARD R. D. (1977) Notes on a major oceanographic find. *Oceanus* **20**, 35–44.

BAROSS J. A. & MORITA R. Y. (1978) Microbial life at low temperatures. In: *Microbial Life in Extreme Environments* (Ed. D. J. Kushner), pp. 9–71. Academic Press, London.

BAULD J. & BROCK T. D. (1973) Ecological studies of *Chloroflexis*, a gliding, photosynthetic bacterium. *Archiv für Mikrobiologie* **92**, 267–84.

BAXTER R. M. (1959) An interpretation of the effects of salts on the lactic dehydrogenase of *Halobacterium salinarum*. *Canadian Journal of Microbiology* **5**, 47–57.

BELLAIRE J. T., PARR-SMITH G. A. & WALLIS I. G. (1977) Significance of diurnal variations in fecal coliform die-off rates in the design of ocean outfalls. *Journal of the Water Pollution Control Federation* **49**, 2022–32.

BEN-AMOTZ A. & AVRON M. (1973) The role of glycerol in the osmotic regulation of the halophilic alga *Dunaliella parva*. *Plant Physiology* **51**, 875–8.

BOROWITZKA L. J. & BROWN A. D. (1974) The salt relations of marine and halophilic species of the unicellular green alga, *Dunaliella*. *Archiv für Mikrobiologie* **96**, 37–52.

BOTT T. L. & BROCK T. D. (1969) Bacterial growth rates above 90°C in Yellowstone hot springs. *Science* **164**, 1411–12.

BRAGG P. D. (1971) Effect of near-ultraviolet light on the respiratory chain of *Escherichia coli*. *Canadian Journal of Biochemistry* **49**, 492–5.

BRIDGES B. A. (1976) Survival of bacteria following exposure to ultraviolet and ionizing radiations. In: *The Survival of Vegetative Microbes* (Eds T. R. G. Gray & J. R. Postgate), *Symposium of the Society for General Microbiology* vol. 26, pp. 183–208. Cambridge University Press, Cambridge.

BROCK T. D. (1966) *Principles of Microbial Ecology*. Prentice-Hall, Englewood Cliffs, New Jersey.

BROCK T. D. (1969) Microbial growth under extreme conditions. In: *Microbial Growth* (Eds P. Meadow & S. J. Pirt), *Symposium of the Society for General Microbiology* vol. 19, pp. 15–41. Cambridge University Press, Cambridge.

BROCK T. D. (1971) Bimodal distribution of pH values of thermal springs of the world. *Bulletin of the Geological Society of America* **82**, 1393–4.

BROCK T. D. (1973) Lower pH limit for the existence of blue-green algae: evolutionary and ecological implications. *Science* **179**, 480–3.

BROCK T. D. (1973) Primary colonization of Surtsey, with special reference to the blue-green algae. *Oikos* **24**, 239–43.

BROCK T. D. (1975a) Effect of water potential on a *Microcoleus* (Cyanophyceae) from a desert crust. *Journal of Phycology* **11**, 316–20.

BROCK T. D. (1975b) The effect of water potential on photosynthesis in whole lichens and in their liberated algal components. *Planta* **124**, 13–23.

BROCK T. D. (1975c) Salinity and the ecology of *Dunaliella* from Great Salt Lake. *Journal of General Microbiology* **89**, 285–92.

BROCK T. D. (1975d) Effect of water potential on growth and iron oxidation by *Thiobacillus ferrooxidans*. *Applied Microbiology* **29**, 495–501.

BROCK T. D. (1978) *Thermophilic Microorganisms and Life at High Temperatures*. Springer-Verlag, New York.

BROCK T. D. & BROCK M. L. (1966) Temperature optima for algal development in Yellowstone and Iceland hot springs. *Nature, London* **209**, 733–4.

BROCK T. D. & BROCK M. L. (1969) Effect of light intensity on photosynthesis by thermal algae adapted to natural and reduced sunlight. *Limnology and Oceanography* **14**, 334–41.

BROCK T. D. & BROCK M. L. (1970) The algae of Waimangu Cauldron (New Zealand): distribution in relation to pH. *Journal of Phycology* **6**, 371–5.

BROCK T. D. & BROCK M. L. (1971) Microbiological studies of thermal habitats of the central volcanic region, North Island, New Zealand. *New Zealand Journal of Marine and Freshwater Research* **5**, 233–57.

BROCK T. D., BROCK K. M., BELLY R. T. & WEISS R. L. (1972) *Sulfolobus*: a new genus of sulfur-oxidizing bacteria living at low pH and high temperature. *Archiv für Mikrobiologie* **84**, 54–68.

BROCK T. D., BROCK M. L., BOTT T. L. & EDWARDS M. R. (1971) Microbial life at 90°C: the sulfur bacteria of Boulder Spring. *Journal of Bacteriology* **108**, 303–14.

BROCK T. D. & DARLAND G. K. (1970) Limits of microbial existence: temperature and pH. *Science* **169**, 1316–18.

BROWN A. D. (1974) Microbial water relations: features of the intracellular composition of sugar-tolerant yeasts. *Journal of Bacteriology* **118**, 769–77.

BROWN A. D. (1976) Microbial water stress. *Bacteriological Reviews* **40**, 803–46.

BROWN A. D. & SIMPSON J. R. (1972) Water relations of sugar-tolerant yeasts: the role of intracellular polyols. *Journal of General Microbiology* **72**, 589–91.

BROWN R. W. & VAN HAVEREN B. P. (1972) *Psychrometry in Water Relations Research*. Utah Agricultural Experiment Station, Utah State University.

BURCH C. W. (1967) Microbes in the upper atmosphere and beyond. In: *Airborne Microbes* (Eds P. H. Gregory & J. L. Monteith), *Symposium of the Society for General Microbiology* vol. 17, pp. 354–74. Cambridge University Press, Cambridge.

CAMERON R. E., MORELLI F. A. & RANDALL L. P. (1972) Aerial, aquatic, and soil microbiology of Don Juan Pond, Antarctica. *Antarctic Journal of the United States* **7**, 254–8.

COLWELL R. R. & MORITA R. Y. (1974) *Effect of the Ocean Environment on Microbial Activities*. University Park Press, Baltimore.

CORLISS J. B., DYMOND J., GORDON L. I., EDMONT J. M., VON HERZEN R. P., BALLARD R. D., GREEN K., WILLIAMS D., BAINBRIDGE A., CRANE K. & VAN ANDEL T. H. (1979) Submarine thermal springs on the Galápagos rift. *Science* **203**, 1073–83.

D'AOUST J. Y., GIROUX J., BARRON L. R., SCHNEIDER H. & MARTIN W. G. (1974) Some effects of visible light on *Escherichia coli*. *Journal of Bacteriology* **120**, 799–804.

DARLAND G. & BROCK T. D. (1971) *Bacillus acidocaldarius* sp. nov., an acidophilic thermophilic spore-forming bacterium. *Journal of General Microbiology* **67**, 9–15.

DARLAND G., BROCK T. D., SAMSONOFF W. & CONTI S. F. (1970) A thermophilic, acidophilic mycoplasma isolated from a coal refuse pile. *Science* **170**, 1416–18.

DARWIN C. (1962) *Voyage of H.M.S. Beagle* (Ed. L. Engel). Doubleday, Garden City, New York.

DIETZ A. S. & YAYANOS A. A. (1978) Silica gel media for isolating and studying bacteria under hydrostatic pressure. *Applied and Environmental Microbiology* **36**, 966–8.

DIMMICK R. L. (1969) Production of biological aerosols. In: *An Introduction to Experimental Aerobiology* (Eds R. L. Dimmick & A. B. Akers), pp. 22–45. Wiley-Interscience, New York.

DIMMICK R. L. & AKERS A. B. (1969) *An Introduction to Experimental Aerobiology*. Wiley-Interscience, New York.

DOEMEL W. N. (1970) The physiological ecology of *Cyanidium caldarium*. Ph.D. thesis, Indiana University, U.S.A.

DOEMEL W. N. & BROCK T. D. (1970) The upper temperature limit of *Cyanidium caldarium*. *Archiv für Mikrobiologie* **72**, 326–32.

DOEMEL W. N. & BROCK T. D. (1971) The physiological ecology of *Cyanidium caldarium*. *Journal of General Microbiology* **67**, 17–32.

DUGAN, P. R., MACMILLAN C. D. & PFISTER R. M. (1970) Aerobic heterotrophic bacteria indigenous to pH 2.8 acid mine water: microscopic enumeration of acid streams. *Journal of Bacteriology* **101**, 973–81.

EDMONDS R. L. (1979) *Aerobiology, the Ecological Systems Approach*. Dowden, Hutchinson and Ross, Stroudsburg, Pennsylvania.

EHRLICH H. L. (1978) How microbes cope with heavy metals, arsenic and antimony in their environment. In: *Microbial Life in Extreme Environments* (Ed. D. J. Kushner), pp. 381–408. Academic Press, London.

EVISON L. M. & ROSE A. H. (1965) A comparative study on the biochemical bases of the maximum temperature for growth of three psychrophilic micro-organisms. *Journal of General Microbiology* **40**, 349–64.

FLANAGAN P. W. & VEUM A. K. (1974) Relationships between respiration, weight loss, temperature and moisture in organic residues on tundra. In: *Soil Organisms and Decomposition in Tundra* (Eds A. J. Holding, O. W. Heal, S. F. MacLean Jr. & P. W. Flanagan), pp. 249–77. Tundra Biome Steering Committee, Stockholm.

FLIERMANS C. B. & BROCK T. D. (1972) Ecology of sulfur-oxidizing bacteria in hot acid soils. *Journal of Bacteriology* **111**, 343–50.

FREEZE H. & BROCK T. D. (1970) Thermostable aldolase from *Thermus aquaticus*. *Journal of Bacteriology* **101**, 541–50.

FRIEDMANN E. I. & GALUN M. (1974) Desert algae, lichens, and fungi. In: *Desert Biology* (Ed. G. W. Brown), pp. 165–212. Academic Press, New York.

FRIEDMANN E. I. & OCAMPO R. (1976) Endolithic blue-green algae in the Dry Valleys: primary producers in the Antarctic desert ecosystem. *Science* **193**, 748–74.

FUHRMAN J. A. & AZAM F. (1980) Bacterioplankton secondary production estimates for coastal waters of British Colombia, Antarctica, and California. *Applied and Environmental Microbiology* **39**, 1085–95.

FULTON J. D. (1966a) Microorganisms of the upper atmosphere. III. Relationship between altitude and micropopulation. *Applied Microbiology* **14**, 233–40.

FULTON J. D. (1966b) Microorganisms of the upper atmosphere. IV. Microorganisms of a land air mass as it traverses an ocean. *Applied Microbiology* **14**, 241–4.

Galapagos Biology Expedition Participants (1979) Galapagos '79: initial findings of a biology quest. *Oceanus* **22**, 2–10.

GOLDMAN C. R., MASON D. T. & WOOD B. J. B. (1963) Light injury and inhibition in Antarctic freshwater phytoplankton. *Limnology and Oceanography* **8**, 313–22.

GRAY T. R. G. & POSTGATE J. R. (1976) *The Survival of Vegetative Microbes. Symposium of the Society of General Microbiology* vol. 26. Cambridge University Press, Cambridge.

GREGORY P. H. (1971) Airborne microbes: their significance and distribution. *Proceedings of the Royal Society of London Series B* **177**, 469–83.

GREGORY P. H. (1973) *The Microbiology of the Atmosphere*, 2e. Halsted Press, Wiley, New York.

GRIFFIN D. M. (1963a) Soil physical factors and the ecology of fungi. I. Behavior of *Curvularia ramosa* at small soil water suctions. *Transactions of the British Mycological Society* **46**, 273–80.

GRIFFIN D. M. (1963b) Soil physical factors and the ecology of fungi. II. Behavior of *Pythium ultimum* at small soil water suctions. *Transactions of the British Mycological Society* **46**, 368–72.

GRIFFIN D. M. (1969) Soil water in the ecology of fungi. *Annual Review of Phytopathology* **7**, 289–310.

GRIFFIN D. M. (1972) *Ecology of Soil Fungi*. Chapman and Hall, London.

GRIFFITHS M., SISTROM W. R., COHEN-BAZIRE G. & STANIER R. Y. (1955) Function of carotenoids in photosynthesis. *Nature, London* **176**, 1211–15.

HANSON R. S. (1976) Dormant and resistant stages of procaryotic cells. In: *Chemical Evolution of the Giant Planets* (Ed. C. Ponamperuma), pp. 107–20. Academic Press, New York.

HARDER W. & VELDKAMP H. (1971) Competition of marine psychrophilic bacteria at low temperatures. *Antonie van Leeuwenhoek* **37**, 51–63.

HARRIS R. F., GARDNER W. R., ADEBAYO A. A. & SOMMERS L. E. (1970) Agar dish isopiestic equilibration method for controlling the water potential of solid substrates. *Applied Microbiology* **19**, 536–7.

HEINRICH M. R. (1976) *Extreme Environments*. Academic Press, New York.

HOLDING A. J., COLLINS, V. G., FRENCH D. D., D'SYLVA B. T. & BAKER J. H. (1974a) Relationship between viable bacterial counts and soil characteristics in tundra. In: *Soil Organisms and Decomposition in Tundra* (Eds A. J. Holding, O. W. Heal, S. F. MacLean Jr. & P. W. Flanagan), pp. 49–64. Tundra Biome Steering Committee, Stockholm.

HOLDING A. J., HEAL Q. W., MACLEAN JR. S. F. & FLANAGAN P. W. (1974b) *Soil Organisms and Decomposition in Tundra*. Tundra Biome Steering Committee, Stockholm.

HOROWITZ N. H. (1976) Life in extreme environments: biological water requirements. In: *Chemical Evolution of the Giant Planets* (Ed. C. Ponamperuma), pp. 121–8. Academic Press, New York.

HOROWITZ N. H., CAMERON R. E. & HUBBARD J. S. (1972) Microbiology of the dry valleys of Antarctica. *Science* **176**, 242–5.

HUTCHINSON F. (1966) The molecular basis for radiation effects on cells. *Cancer Research* **26**, 2045–52.

INNISS W. E. & INGRAHAM J. L. (1978) Microbial life at low temperatures: mechanisms and molecular aspects. In: *Microbial Life in Extreme Environments* (Ed. D. J. Kushner), pp. 73–104. Academic Press, London.

JANNASCH H. W. & WIRSEN C. O. (1977) Retrieval of concentrated and undecompressed microbial populations from the deep sea. *Applied and Environmental Microbiology* **33**, 642–6.

JANNASCH H. W. & WIRSEN C. O. (1979) Chemosynthetic primary production at East Pacific sea floor spreading centers. *Bioscience* **29**, 592–8.

JANNASCH H. W., WIRSEN C. O. & TAYLOR C. D. (1976) Undecompressed microbial populations from the deep sea. *Applied and Environmental Microbiology* **32**, 360–7.

JANNASCH H. W., WIRSEN C. O. & WINGET C. L. (1973) A bacteriological pressure-retaining deep-sea sampler and culture vessel. *Deep-Sea Research* **20**, 661–4.

JORGENSEN B. B. (1977) The sulfur cycle of a coastal

marine sediment (Limfjorden, Denmark). *Limnology and Oceanography* **22**, 814–32.

JORGENSEN E. G. (1964) Adaptation to different light intensities in the diatom *Cyclotella Meneghiniana* Kutz. *Physiological Plantarum* **17**, 136–45.

JORGENSEN E. G. & STEEMANN NIELSEN E. (1965) Adaptation in plankton algae. In: *Primary Productivity in Aquatic Environments* (Ed. C. R. Goldman), pp. 37–46. University of California Press, Berkeley.

KASHKET E. R. & BRODIE A. F. (1962) Effects of near-ultraviolet irradiation on growth and oxidative metabolism of bacteria. *Journal of Bacteriology* **83**, 1093–100.

KRINSKY N. I. (1968) The protective functions of carotinoid pigments. In: *Photophysiology* (Ed. A. C. Giese), vol. 3, pp. 123–95. Academic Press, New York.

KRINSKY N. I. (1974) Membrane photochemistry and photobiology. *Photochemistry and Photobiology* **20**, 532–5.

KRINSKY N. I. (1976) Cellular damage initiated by visible light. In: *The Survival of Vegetative Microbes* (Eds T. R. G. Gray & J. R. Postgate), *Symposium of the Society for General Microbiology* vol. 26, pp. 209–39. Cambridge University Press, Cambridge.

KUENEN J. G., BOONSTRA J., SCHRODER H. G. J. & VELDKAMP H. (1977) Competition for inorganic substrates among chemoorganotrophic and chemolithotrophic bacteria. *Microbial Ecology* **3**, 119–30.

KUNISAWA R. & STANIER R. Y. (1958) Studies on the role of carotenoid pigments in a chemoheterotrophic bacterium, *Corynebacterium poinsettiae*. *Archiv für Mikrobiologie* **31**, 146–56.

KUSHNER D. J. (1978a) *Microbial Life in Extreme Environments*. Academic Press, London.

KUSHNER D. J. (1978b) Introduction: a brief overview. In: *Microbial Life in Extreme Environments* (Ed. D. J. Kushner), pp. 1–7. Academic Press, London.

KUSHNER D. J. (1978c) Life in high salt and solute concentrations: halophilic bacteria. In: *Microbial Life in Extreme Environments* (Ed. D. J. Kushner), pp. 317–68. Academic Press, London.

LANG A. R. G. (1967) Osmotic coefficients and water potentials of sodium chloride solutions from 0 to 40°C. *Australian Journal of Chemistry* **20**, 2017–23.

LANGE O. L., SCHULZE E. D. & KOCH W. (1970) Experimentellökologische Untersuchungen an Flechten der Negev-Wuste. II. CO_2-Gaswechsel und Wasserhaushalt von *Ramalina maciformis* (Del.) Bory am natürlichen Standort während der sommerlichen Trocken-periode. *Flora, Jena* **149**, 345–53.

LANGWORTHY T. A. (1978) Microbial life in extreme pH values. In: *Microbial Life in Extreme Environments* (Ed. D. J. Kushner), pp. 279–315. Academic Press, London.

LANYI J. K. (1974) Salt dependent properties of proteins from extremely halophilic bacteria. *Bacteriological Reviews* **38**, 272–90.

LUCKIESH M. & KNOWLES T. (1948) Radiosensitivity of *Escherichia coli* to ultraviolet energy (λ 2537) as affected by irradiation of preceding cultures. *Journal of Bacteriology* **55**, 369.

MARQUIS R. E. & MATSUMURA P. (1978) Microbial life under pressure. In: *Microbial Life in Extreme Environments* (Ed. D. J. Kushner), pp. 105–58. Academic Press, London.

MATHESON A. T., YAGUCHI M. & VISENTIN L. P. (1975) The conservation of amino acids in the N-terminal position of ribosomal and cytosol proteins from *Escherichia coli*, *Bacillus stearothermophilus*, and *Halobacterium cutirubrum*. *Canadian Journal of Biochemistry* **53**, 1323–7.

MATHEWS M. M. & SISTROM W. R. (1960) The function of the carotenoid pigments in *Sarcina lutea*. *Archiv für Mikrobiologie* **35**, 139–46.

MEYER G. M., MORROW M. B., WYSS O., BERG T. E. & LITTLEPAGE J. Q. (1962) Antarctica: the microbiology of an unfrozen saline pond. *Science* **138**, 1103–4.

MITCHELL R. L. & ANDERSON I. C. (1965) Photoinactivation of catalase in carotenoidless tissues. *Crop Science* **5**, 588–91.

MORITA R. Y. (1974) Hydrostatic pressure effects on microorganisms. In: *Effect of the Ocean Environment on Microbial Activities* (Eds R. R. Colwell & R. Y. Morita), pp. 133–8. University Park Press, Baltimore.

MORITA R. Y. (1975) Psychrophilic bacteria. *Bacteriological Reviews* **39**, 144–67.

MORITA R. Y. (1976) Survival of bacteria in cold and moderate hydrostatic pressure environments with special reference to psychrophilic and barophilic bacteria. In: *The Survival of Vegetative Microbes* (Eds T. R. G. Gray & J. R. Postgate) *Symposium of the Society for General Microbiology* vol. 26, pp. 279–98. Cambridge University Press, Cambridge.

MOSSER J. L., MOSSER A. G. & BROCK T. D. (1973) Bacterial origin of sulfuric acid in geothermal habitats. *Science* **179**, 1323–4.

MURR L. E., TORMA A. E. & BRIERLY J. A. (1978) *Metallurgical Applications of Bacterial Leaching and Related Microbiological Phenomena*. Academic Press, New York.

NASIM A. & JAMES A. P. (1978) Life under conditions of high irradiation. In: *Life in Extreme Environments* (Ed. D. J. Kushner), pp. 409–39. Academic Press, London.

NISSENBAUM A. (1975) The microbiology and biogeochemistry of the Dead Sea. *Microbial Ecology* **2**, 139–61.

OHTA K., KIYOMIYA A., KOYAMA N. & NOSOH Y. (1975) The basis of the alkalophilic property of a species of *Bacillus*. *Journal of General Microbiology* **86**, 259–66.

ONISHI H. (1963) Osmophilic yeasts. *Advances in Food Research* **12**, 53–94.

PEARY J. & CASTENHOLZ R. W. (1964) Temperature strains of a thermophilic blue-green alga. *Nature, London* **202**, 720–1.

POINDEXTER J. S. (1964) Biological properties and classification of the *Caulobacter* group. *Bacteriological*

Reviews **28**, 231–95.

POPE D. H., SMITH W. P., ORGRINC M. A. & LANDAU J. V. (1976) Protein synthesis at 680 atm: is it related to environmental origin, physiological, or taxonomic group? *Applied and Environmental Microbiology* **31**, 1001–2.

POST F. J. (1977) The microbial ecology of the Great Salt Lake. *Microbial Ecology* **3**, 143–65.

PRINCE A. E. (1960) Space age microbiology. Introduction. In: *Developments in Industrial Microbiology* vol. 1, pp. 13–14. Plenum Press, New York.

RAHN R. O. (1972) Ultraviolet irradiation of DNA. In: *Concepts in Radiation Cell Biology*, pp. 1–56. Academic Press, New York.

RAHN R. O. (1973) Denaturation in ultraviolet-irradiated DNA. In: *Photophysiology* (Ed. A. C. Giese), vol. 8, pp. 231–55. Academic Press, London.

RAO G. S. & BERGER L. R. (1971) The requirement of low pH for growth of *Thiobacillus thiooxidans*. *Archiv für Mikrobiologie* **79**, 338–44.

RAWLINS S. L. (1972) Theory of thermocouple psychrometers for measuring plant and soil water potential. In: *Psychrometry in Water Relations Research* (Eds R. W. Brown & B. P. van Haveren), pp. 43–50. Utah Agricultural Experiment Station, Utah State University.

RAY P. H., WHITE D. C. & BROCK T. D. (1971a) Effect of growth temperature on the lipid composition of *Thermus aquaticus*. *Journal of Bacteriology* **108**, 227–35.

RAY P. H., WHITE D. C. & BROCK T. D. (1971b) Effect of temperature on the fatty acid composition of *Thermus aquaticus*. *Journal of Bacteriology* **106**, 25–30.

ROGERS R. W. (1971) Distribution of the lichen *Chondropsis semiviridis* in relation to its heat and drought resistance. *New Phytologist* **70**, 1069–77.

ROSE A. H. (1976) Osmotic stress and microbial survival. In: *The Survival of Vegetative Microbes* (Eds T. R. G. Gray & J. R. Postgate), *Symposium of the Society for General Microbiology* vol. 26, pp. 155–82. Cambridge University Press, Cambridge.

ROSE C. W. (1966) *Agricultural Physics*. Pergamon Press, Oxford.

RYCHERT R., SKUJINS J., SORENSEN D. & PROCELLA D. (1978) Nitrogen fixation by lichens and free-living microorganisms in deserts. In: *Nitrogen in Desert Ecosystems* (Eds N. E. West & J. Skujins), pp. 20–30. Dowden, Hutchinson and Ross, Stroudsburg, Pennsylvania.

SCHOEN R. & RYE R. O. (1970) Sulfur isotope distribution in solfataras, Yellowstone National Park. *Science* **170**, 1082–4.

SCHWARZ J. R., YAYANOS A. A. & COLWELL R. R. (1976) Metabolic activities of the intestinal microflora of a deep-sea invertebrate. *Applied and Environmental Microbiology* **31**, 46–8.

SCOTT W. J. (1957) Water relations of food spoilage microorganisms. *Food Research* **7**, 83–127.

SEKI H. & ROBINSON D. B. (1969) Effect of decompression on activity of microorganisms in sea water. *Internationale Revue der Gesamten Hydrobiologie* **54**, 201–5.

SETLOW R. B. & CARRIER W. L. (1966) Pyrimidine dimers in ultraviolet-irradiated DNA's. *Journal of Molecular Biology* **17**, 237–54.

SINGER P. C. & STUMM W. (1970) Acidic mine drainage: the rate determining step. *Science* **167**, 1121–3.

SINGLETON R. (1976) A comparison of the amino acid composition of proteins from thermophilic and non-thermophilic origins. In: *Extreme Environments* (Ed. M. R. Heinrich), pp. 189–200. Academic Press, New York.

SINGLETON R. & AMELUNXEN R. A. (1973) Proteins from thermophilic microorganisms. *Bacteriological Reviews* **37**, 320–42.

SISTROM W. R., GRIFFITHS M. & STANIER R. Y. (1956) The biology of a photosynthetic bacterium which lacks colored carotenoids. *Journal of Cellular and Comparative Physiology* **48**, 473–515.

SMITH D. W. (1978) Water relations of microorganisms in nature. In: *Microbial Life in Extreme Environments* (Ed. D. J. Kushner), pp. 369–80. Academic Press, London.

SMITH D. W. & BROCK T. D. (1973a) The water relations of the alga *Cyanidium caldarium* in soil. *Journal of General Microbiology* **79**, 219–31.

SMITH D. W. & BROCK T. D. (1973b) Water status and the distribution of *Cyanidium caldarium* in soil. *Journal of Phycology* **9**, 330–2.

SOMMERS L. E., HARRIS R. F., DALTON F. N. & GARDNER W. R. (1970) Water potential relations of three root-infecting *Phytophthora* species. *Phytopathology* **60**, 932–4.

SOUZA K. A., DEAL P. H., MACK H. M. & TURNBILL E. E. (1974) Growth and reproduction of microorganisms under extremely alkaline conditions. *Applied Microbiology* **28**, 1066–8.

SPIKES J. D. (1975) Porphyrins and related compounds as photodynamic sensitizers. *Annals of the New York Academy of Sciences* **244**, 496–508.

STEPANEK M. (1965) Numerical aspects of nannoplankton production in reservoirs. In: *Primary Productivity in Aquatic Environments* (Ed. C. R. Goldman), pp. 293–307. University of California Press, Berkeley.

STRANGE R. E. & COX C. S. (1976) Survival of dried and airborne bacteria. In: *The Survival of Vegetative Microbes* (Eds T. R. G. Gray & J. R. Postgate), *Symposium of the Society for General Microbiology* vol. 26, pp. 111–54. Cambridge University Press, Cambridge.

TANSEY M. R. & BROCK T. D. (1978) Microbial life at high temperatures: ecological aspects. In: *Microbial Life in Extreme Environments* (Ed. D. J. Kushner), pp. 159–216. Academic Press, London.

TEMPEST D. W. & NEIJSSEL O. M. (1978) Eco-physiological aspects of microbial growth in aerobic nutrient-limited environments. In: *Advances in Microbial Ecology* (Ed. M. Alexander), vol. 2, pp. 105–53. Plenum Press, New York.

UNSCEAR (United Nations Scientific Committee on the Effects of Atomic Radiation) (1972) *Ionizing Radiation: Levels and Effects* vol. 1. United Nations, New York.

VALLENTYNE J. R. (1963) Environmental biophysics and microbial ubiquity. *Annals of the New York Academy of Sciences* **108**, 342–52.

VAN BAALEN C. (1968) The effects of ultraviolet irradiation on a coccoid blue-green alga: survival, photosynthesis, and photoreactivation. *Plant Physiology* **43**, 1689–95.

VELDKAMP H. (1977) Ecological studies with the chemostat. In: *Advances in Microbial Ecology* (Ed. M. Alexander), vol. 1, pp. 59–94. Plenum Press, New York.

VISHNIAC H. S. & HEMPFLING W. P. (1979) Evidence of an indigenous microbiota (yeast) in the Dry Valleys of Antarctica. *Journal of General Microbiology* **112**, 301–14.

VISHNIAC W. V. & MAINZER S. E. (1972) Soil microbiology studied *in situ* in the Dry Valleys of Antarctica. *Antarctic Journal of the United States* **7**, 88–9.

WATSON K., ARTHUR H. & SHIPTON W. A. (1976) Leucosporidium yeasts: obligate psychrophiles which alter membrane-lipid and cytochrome composition with temperature. *Journal of General Microbiology* **97**, 11–18.

WEBB S. J. & TAI C. C. (1968) Lethal and mutagenic actions of 3200–4000 Å light. *Canadian Journal of Microbiology* **14**, 727–35.

WEBB S. J. & WALKER J. L. (1968) The influence of cell water content on the inactivation of RNA by partial desiccation and ultra-violet light. *Canadian Journal of Microbiology* **14**, 565–72.

WEST N. E. & SKUJINS J. J. (1978) *Nitrogen in Desert Ecosystems.* Dowden, Hutchinson and Ross, Stroudsburg, Pennsylvania.

WETZEL R. G. (1975) *Limnology.* W. B. Saunders, Philadelphia.

WHITFIELD M. (1969) E_h as an operational parameter in estuarine studies. *Limnology and Oceanography* **14**, 547–58.

YAYANOS A. A. (1976) Recovery of amphipods from deep ocean trenches at near *in situ* pressure: description of the instrument and scanning electron microscopy of animal-associated microbes. *Biophysical Journal* **16**, 180a.

YAYANOS A. A. (1977) Simply actuated closure for a pressure vessel: design for use to trap deep-sea animals. *Review of Scientific Instrumentation* **48**, 786–91.

YAYANOS A. A. (1978) Recovery and maintenance of live amphipods at a pressure of 580 bars from an ocean depth of 5700 meters. *Science* **200**, 1056–9.

YAYANOS A. A., DIETZ A. S. & VAN BOXTEL R. (1979) Isolation of a deep-sea bacterium and some of its growth characteristics. *Science* **205**, 808–9.

YENTSCH C. S. & LEE R. W. (1966) A study of photosynthetic light reactions, and a new interpretation of sun and shade phytoplankton. *Journal of Marine Research* **24**, 319–37.

ZEIKUS J. G. & BROCK T. D. (1971) Protein synthesis at high temperatures: aminoacylation of tRNA. *Biochimica et Biophysica Acta* **228**, 736–45.

ZEIKUS J. G., TAYLOR M. W. & BROCK T. D. (1970) Thermal stability of ribosomes and RNA from *Thermus aquaticus*. *Biochimica et Biophysica Acta* **204**, 512–20.

ZOBELL C. E. (1970) Pressure effects on morphology and life processes. In: *High Pressure Effects on Cellular Processes* (Ed. A. Zimmerman), pp. 85–130. Academic Press, New York.

ZOBELL C. E. & MORITA R. Y. (1957) Barophilic bacteria in some deep sea sediments. *Journal of Bacteriology* **73**, 563–8.

ZOBELL C. E. & OPPENHEIMER C. H. (1950) Some effects of hydrostatic pressure on the multiplication and morphology of marine bacteria. *Journal of Bacteriology* **60**, 771–81.

Chapter 33 • Microbial Ecology of Domestic Wastes

Ken P. Flint

33.1 Introduction

In Britain the majority of domestic properties are connected to a sewerage system leading to a sewage treatment plant where at least primary, and generally also secondary, treatment of the wastes is made so that the sewage effluent discharged to the rivers is purified about 95% with respect to its biological oxygen demand (BOD). BOD is an arbitrary measure of the amount of oxygen used by a sample of water in a given time (normally 5 d) at a given temperature (usually 20°C) during the oxidation of organic matter. Discharged effluent is supposed to meet the Royal Commission's requirement of 20 mg BOD l^{-1}, 30 mg suspended solids (SS) l^{-1} and to be diluted at least eightfold by the receiving water. Unfortunately, these criteria are not always met and many rivers, because they have a major town situated on them, receive large amounts of sewage effluents during times of low flow, that is most summers. These rivers can be adversely affected by the large volumes of effluent entering them. The river quality is affected by the removal of oxygen from the water to satisfy the BOD of the effluent and in extreme cases this leads to completely anoxic conditions and fish deaths but generally these rivers are not able to support fish life anyway. The layman often judges river quality by its fish stocks, hence the emphasis that the Thames Water Authority have put on the fact that salmon have now returned to the river after a long absence. Of more concern to the microbiologist is the number of potentially pathogenic bacteria, viruses and parasites that can be discharged to rivers, and their survival in river waters destined for consumption, irrigation or recreational use and in sewage sludges often used as fertilisers. There are a large number of waterborne diseases, for example, dysentry, cholera, typhoid, polio, infectious hepatitis, tularaemia, leptospirosis, ascaris and gardiasis, which are virtually unknown in this country because of the efficiency of our waste treatment works but which are prevalent in many parts of the world.

33.1.1 Composition of Domestic Sewage

Domestic wastes are largely composed of metabolic wastes, waste waters and in some cases the macerated animal and vegetable products of the kitchen waste disposal units. A typical waste would have the following composition approximately: lipids, 33%; protein, 25%; ash, 20%; cellulose, 8%; starch, 8%; and lignin, 6%. The initial BOD would be between 275 and 300 mg l^{-1} and the suspended solids content between 300 and 350 mg l^{-1}. These organic components are mainly degraded to carbon dioxide, methane, water and inorganic salts during the treatment processes by the action of a multitude of microorganisms. The biochemistry of the processes involved was reviewed by Higgins and Burns (1975).

33.1.2 Waste Treatment Processes

The aim of sewage treatment is primarily to enable waste water to be disposed of safely without causing a pollution or environmental problem and, secondly, to conserve natural resources by allowing the water to be recycled to industry, agriculture or the home. This aim is achieved by three methods:

(1) the reduction in BOD;

(2) the reduction in the numbers of pathogenic microorganisms;

(3) the reduction in the levels of inorganic nutrients, such as phosphorus and nitrogen, which can lead to eutrophication problems in standing waters.

The basic treatment process has three stages.

(1) *Primary treatment*. Raw sewage is first passed through screens to remove the large pieces of insoluble material, then through a degritting chamber and on to primary sedimentation tanks where some of the biomass and flocculated organic material settles out and is removed as raw sludge. This can be treated by anaerobic digestion to yield methane and a dried sludge which can be used as a fertiliser as long as the toxic heavy metal content is low.

In many coastal towns this is the only treatment wastes receive prior to the discharge of the effluent into the sea although this situation may change in the future now that the European Economic Commission has brought in regulations governing the levels of enteric bacteria in the sea, especially near bathing beaches. The effluent from the primary sedimentation tanks (settled sewage) passes to the second stage of treatment.

(2) *Secondary treatment*. The settled sewage undergoes oxidation by microorganisms growing either as a film on the surface of stones, slag or corrugated plastics (trickling or percolating filters) or as flocs in suspension (activated sludge). The microbial biomass produced by growth on the organic content of the wastes is removed in a settlement tank and passed to the anaerobic digester or dumped. The settled effluent may be discharged or, if necessary, given a final treatment.

(3) *Tertiary treatment*. This is applied, if it is necessary, to upgrade an effluent because of problems caused in the receiving waters, for example, eutrophication. Inorganic compounds can be removed either by chemical means or by biological activity in sewage lagoons. In some cases the final effluent is sterilised by chlorination particularly if the effluent is passed directly to a water reclamation plant as is often the case in arid countries such as South Africa and Israel.

33.1.3 Microbial Ecology of Waste Treatment

Microorganisms isolated from sewage may be either faecal in origin or typical aquatic bacteria involved in biodegradation processes.

Crude sewage contains a large number of non-pathogenic faecal bacteria, e.g. *Escherichia coli, Proteus* spp., *Serratia* spp., *Bifidobacteria* spp., as well as smaller numbers of pathogens like *Clostridium perfringens, Streptococcus faecalis* and *Salmonella* spp. It is this group of faecal organisms whose numbers are reduced by the treatment processes.

Activated sludge plants are continuously inoculated by the return of the sludge to the incoming settled sewage. Hence the dominant organisms are those which can grow under the prevailing conditions and which either produce a zoogloeal matrix (floc) or which can thrive embedded in the matrix produced by other bacteria. It was believed for a long time that the floc was a product of a single bacterium (*Zoogloea ramigera*) but it has now been established that many bacteria can produce flocs under aerobic conditions and high carbon concentrations (Benedict & Carlson 1971). It has been established that most of the biochemical activity is produced by the non-zoogloeal bacteria embedded in the matrix, i.e. by the mainly Gram-negative bacteria of the sludge, particularly species of *Pseudomonas, Flavobacterium, Nocardia, Achromobacter, Alcaligenes* and *Mycobacterium* (Unz & Dondero 1967). The zoogloeal mass is desirable as this settles out and allows the discharge of a clarified effluent with a low BOD. The free-swimming bacteria in the influent are mainly of faecal origin and many are removed by the activities of protozoa which are attached by stalks to the flocs or which crawl over the floc surfaces preying on bacteria adsorbed to the surface but not embedded in the matrix. A full list of protozoa found in activated sludge plants is given by Curds (1975). Few algae are found in activated sludge plants and fungi play little, if any, role in the removal of BOD although they can cause a problem known as bulking in which the sludge is very difficult to settle due to the presence of filamentous fungi (or filamentous bacteria, such as *Sphaerotilus natans*, or species of *Beggiatoa, Leptothrix, Leucothrix* and *Thiothrix*).

A trickling filter is a specialised habitat which can take many weeks to reach a stable state. The upper areas of the bed are dominated by algae (species of *Stigeoclonium, Ulothrix, Chlorella, Euglena* and *Phormidium*) and by fungi (species of *Geotrichium, Mucor, Aureobasidium, Subbaromyces* and *Fusarium*) which can often clog a filter. The dominant bacteria (species of *Pseudomonas, Achromobacter, Flavobacterium* and *Alcaligenes*) are typical Gram-negative aquatic bacteria, many of which can develop zoogloeal habits and coat the support material to produce a film, generally about 2 mm thick. The lower reaches of the bed are often dominated by chemolithotrophs, e.g. *Nitrosomonas* sp. and *Nitrobacter* sp. (which remove ammonia from the wastes by conversion to nitrate). These are often absent, or present in very low numbers, in activated sludge. Protozoal numbers vary with depth and a list of those species found in trickling filters is given by Curds (1975).

One of the chief concerns of the microbiologist is the survival of potential pathogens through the treatment process. Most studies have concentrated on the survival of *Escherichia coli* which has had a long history as an indicator of organism, in the belief that if *E. coli* numbers are considerably reduced (by 99%) then all pathogens are reduced by a similar amount or more. Factors which are believed to influence the survival of *E. coli* are temperature, BOD, light intensity and predation by protozoa, bacteriophage and *Bdellovibrio* sp. (Pike & Carrington 1979). More recently attention has been given to the survival of viruses, particularly the human pathogens, such as polio, and to the survival of parasitic ova of tapeworms. Many of the ova and viral particles are adsorbed onto the surface of flocs and hence are removed from the sewage stream but are not inactivated. They are generally present in very low numbers in sewage effluent but are present in sewage sludge in much higher numbers and could constitute a health hazard if the sludge is used as a fertiliser without first being subject to anaerobic digestion. Many viral particles have been shown to survive for up to four months after application of the sludge to soil (Butler & Balluz 1979; Pahren *et al.* 1979).

Much of the methodology used in the microbial, chemical and physical analyses of domestic wastes and sewage is standardised so that the results of water authorities in various parts of the country, and even between countries, ought to be directly comparable. The standard recommended techniques are laid down in *Standard Methods for the Examination of Water and Waste Water* (American Public Health Association 1975) and *Analysis of Raw, Potable and Waste Waters* (Department of the Environment 1972) and these texts should be routinely consulted. Here I wish to examine four parameters of major importance to any ecological study of wastes:

(1) the measurement of biochemical oxygen demand (BOD), chemical oxygen demand (COD) and nitrogenous oxygen demand (NOD);

(2) the estimation of microbial and viral numbers in sewage and the techniques for examining pathogen survival;

(3) the estimation of biochemical parameters of sewage;

(4) the estimation of some chemical parameters of sewage responsible for toxic effects.

Experimental

33.2 Measurement of biochemical, chemical and nitrogenous oxygen demands

33.2.1 BOD Determination

BOD is usually determined by the dilution method in which the dissolved oxygen (DO) concentration of a water sample is determined initially and after incubation at 20°C for 5 d, a value which is the standard BOD_5 value. Numerous precautions need to be taken to ensure a significant and accurate result. The samples must be free from preservatives and assayed as soon as possible after collection since it has been shown that up to 40% of the initial BOD can be removed by standing at room temperature for less than 8 h after collection. The assay must be carried out in glass-stoppered, narrow-necked bottles with a capacity of 250 ml. These bottles ought to be acid washed but providing they are kept solely for BOD estimation they only need rinsing between assays. It is often necessary to dilute the water sample to ensure that less than 70% of the DO is used during the 5 d incubation period. As a general guide river water needs a dilution of 0 to 4 volumes, sewage effluent 1 to 9 volumes, settled sewage 14 to 99 volumes and crude sewage 29 to 199 volumes. The diluent would be the river water to which the sewage effluent is to be discharged but this is generally impractical because of collection problems and its intrinsic variance. Deionised water should be used in preference to distilled water (particularly if the latter is from a copper still as this could have a high metal content). It is recommended that the following are added to 1 l of deionised water: 0.125 mg $FeCl_3.6H_20$; 27.5 mg $CaCl_2$; 25 mg $MgSO_4.7H_20$; and 1 ml of buffer pH 7.2 containing 42.5 g KH_2PO_4; 8.8 g NaOH and 2 g, $(NH_4)_2SO_4$ per litre. The amended dilution water is saturated with oxygen by bubbling air through it and used immediately; old dilution water should not be topped up. The sample to be assayed is diluted and mixed thoroughly without vigorous shaking which induces the formation of minute air bubbles. The diluted water is transferred to two BOD bottles and the initial concentration of DO determined, as outlined below, in one of the bottles. A blank of the dilution water is similarly treated. The sample and blank are incubated in a water bath at 20° ± 0.5°C for 5 d and the final DO concentration determined. The BOD_5 of the initial sample can now be calculated (Department of the Environment 1972).

Dissolved oxygen concentration is usually determined by the titrimetric Winkler method or one of its modifications. The basis of the method is the preparation of manganous hydroxide, in a completely full glass-stoppered bottle, which absorbs any oxygen present to form higher hydroxides which on subsequent acidification in the presence of iodide, liberate iodine in an amount chemically equivalent to the original dissolved oxygen content. The iodine can be determined by titration with sodium thiosulphate using starch as an indicator (Department of the Environment 1972).

The details of the method are as follows. To the sample is added 2 ml manganous sulphate solution (500 g $MnSO_4.5H_2O$ l^{-1}) below the liquid surface, and 2 ml alkaline iodine solution (400 g NaOH l^{-1}; 900 g NaI_2 l^{-1}) is added rapidly at the surface. The stopper is replaced, avoiding entrapment of air bubbles, and the contents mixed by inverting the bottle. A precipitate (brown if oxygen is present) forms which is allowed to settle to the bottom of the bottle. Then 4 ml 9M sulphuric acid or 85 to 90% (v/v) phosphoric acid (not recommended if heavy metals are present in the sample) is added and the contents again mixed. The precipitate should dissolve immediately. The sample (100 ml) is removed and titrated with 0.025M sodium thiosulphate solution using 2 ml starch (5 g l^{-1}) as the end point indicator. This method can only be used if the nitrite concentration of the sample is less than 50 μg nitrogen l^{-1}. For higher concentrations it is necessary to add 300 ml of sodium azide (25 g l^{-1}) solution to a litre of alkaline-iodide solution.

Instruments are becoming available for the direct measurement of dissolved oxygen. The sensors recommended for use with sewage samples are membrane covered and are more useful for determining the oxygen concentration *in situ* than in BOD bottles (American Public Health Association 1975).

BOD can be determined manometrically although these techniques are not yet accepted as an alternative to the dilution method. The great advantage of using a manometer is that it is possible to use much smaller sample volumes and the BOD is obtained in a much shorter time period than the 5 d required by the dilution method. The use of the Warburg manometer has been reviewed by Montgomery (1967). Because of the nature of the manometer equipment, it is essential that the shaking speed is sufficient to replenish oxygen in the liquid phase as fast as it is being utilised and that carbon dioxide is rapidly absorbed.

One of the chief advantages of the manometric techniques is in the study of the inhibition of metabolic activity in sewage treatment plants. Many treatment plants are rapidly inactivated by toxic industrial wastes and manometric analysis can rapidly detect inhibitory effects. Also they can give a guide to the dilution necessary to eliminate any toxic effect of the wastes in the normal BOD determination by the dilution method. In general BOD_5 values determined by the dilution method and extrapolated from the manometric method are in good agreement, but as the BOD_5 value determined by the dilution method is such a universally accepted value it seems unlikely that a manometric technique can as yet replace the dilution method (Montgomery 1967).

33.2.2 COD Determination

Chemical oxygen demand is determined by boiling the sample in acid dichromate solution in the presence of a silver catalyst. The dichromate is reduced by the organic matter and the remainder is determined by titration with ferrous sulphate in the presence of 1:10 phenanthroline as an indicator. The test is carried out by the addition of 5 ml 0.25M potassium dichromate, 10 ml concentrated H_2SO_4 and 1 ml of a saturated solution of silver sulphate in 9M sulphuric acid to 5 ml of sample and refluxing for 2 h. After cooling, 45 ml of water are added to the flask and the contents titrated against 0.125M ferrous sulphate (American Public Health Association 1975).

33.2.3 NOD Determination

Nitrogenous oxygen demand is determined from the total nitrogen as estimated by the Kjeldahl method. Here the sample 100 ml is boiled with 1 g catalyst (4 g selenium, 4 g copper sulphate and 250 g sodium sulphate) and 10 ml ethyl alcohol and 10 ml concentrated sulphuric acid. The contents of the flask are diluted with 150 ml water and neutralised using phenolphthalein as an indicator. The ammonia can then be determined by the Nessler procedure (American Public Health Association 1975).

Table 1 shows the typical values expected for the oxygen demands for domestic sewage treated by trickling filters and activated sludge. It can be seen that whereas the trickling filter appears to be slightly less efficient than the activated sludge treatment on a BOD_5 basis, there is a much greater reduction in total oxygen demand in the trickling filter (77% compared to 56%) due to the much higher nitrification activity (NOD reduced by 73% compared to 23%).

33.2.4 Evaluation of the Estimation of Oxygen Demand

Recently the use of BOD_5 has been criticised (Sherrard *et al.* 1979; Stones 1979; Aziz & Tebbutt 1980). Many workers use the BOD_5 value as a control parameter which it obviously is not, for several reasons: the long response time of the test; the fact that it cannot be accurately, and independently, calibrated; and the inability of the test to differentiate between carbonaceous oxygen demand and nitrogenous oxygen demand. The former is caused by the utilisation of oxygen to form carbon dioxide from the carbonaceous constituents of the wastes and the latter by the utilisation of oxygen to produce nitrate from the organic nitrogen compounds and ammonia present in wastes. During the course of a dilution assay there is generally an induction period during which the nitrifying bacteria actively grow but this period is absent if there is a strong nitrifying element to the microbial flora already, i.e. in a sewage effluent but not in the crude sewage. This can lead to a marked underestimation of the total oxygen demand of a waste and could result in the situation where a high total oxygen demand waste required the same treatment area and degree as a lower total oxygen demand waste which had the same carbonaceous oxygen demand but a much reduced nitrogenous oxygen demand. The nitrogenous oxygen demand, particularly from an activated sludge plant, can be substantial but is rarely considered by authorities when investigating the oxygen demand of effluents, and hence the polluting potential of an effluent can be seriously underestimated if the BOD_5 is relied upon to give an adequate estimation of total oxygen demand.

Another major failing of the conventional dilution test is that it does not estimate the total carbonaceous oxygen demand. A substantial portion of the carbonaceous content of a waste,

Table 1. A comparison of oxygen demand methods for two methods of waste treatment. The values are adapted from Stones (1979). TOD is total oxygen demand as determined by the sum of COD and NOD.

	Trickling filters (mg l^{-1})		%
	Influent	Effluent	reduction
BOD_5	256	23	91
COD	630	140	77.8
NOD	211	57	73
TOD	841	197	76.6
	Activated sludge (mg l^{-1})		%
	Influent	Effluent	reduction
BOD_5	258	18	93.0
COD	565	174	69.2
NOD	225	173	23.1
TOD	790	347	56.1

particularly one receiving an industrial component, may not be readily amenable to biochemical breakdown by the bacterial flora of the sewage works. A more meaningful figure for the determination of sewage strength is given by the chemical oxygen demand and the BOD_5 should be kept as a measure of the amenability of sewage to biochemical breakdown. Even so, no account is being taken here of the added load due to nitrogenous waste, which is of particular importance if there is a lot of waste from the dairy industry, and a better estimate of plant loading would be given by the total oxygen demand, determined by the sum of chemical and nitrogenous oxygen demands. The BOD value can be used to determine the ability of a sewage microflora to degrade actively the organic content of the wastes (a high proportion of COD measured as BOD_5 would indicate this).

In conclusion, the importance of NOD to the total loading of a sewage plant and its contribution to the overall oxygen demand of effluents has been overlooked but it is of major concern with the increase in the use of activated sludge as the main means of sewage treatment (the saving in land space in these days of high land prices is a major economic concern to councils). Activated sludge has a very poor nitrifying potential and can, therefore, seriously deplete the oxygen content of river waters in satisfying NOD rather than the BOD_5 which could easily meet consent requirements. Many authorities even include allylthiourea in the dilution water to suppress nitrification when assaying BOD_5 thus obscuring the total oxygen demand of the effluents, but there is a long way to go before TOD receives universal support as the measure of oxygen demand which should be allowed for sewage effluents discharged to waters.

It is probably up to the academic to begin to change attitudes by teaching the limitations rather than the uses of the BOD test.

33.3 Enumeration of bacteria

The characterisation of the types and numbers of microorganisms present in crude sewage or involved in sewage treatment involves their isolation on agar plates. The nature of microbial growth in treatment process, i.e. fixed films in trickling filters and flocs in activated sludge processes, makes the isolation and enumeration of microorganisms very difficult and in many cases non-reproducible (see Chapter 6).

33.3.1 General Culture Techniques for Bacteria

There have been many studies on the use of different methods to isolate bacteria from sewage treatment plants, particularly the aerobic heterotrophs of activated sludge (as these bacteria are the species most actively involved in the biochemical oxidation of organic matter). Gayford and Richards (1970) examined activated sludge microorganisms and found the highest viable counts were obtained using homogenisation in a Silverson mixer (the blades of which were sterilised in 70% (v/v) ethanol prior to use) for 15 min in the presence of 0.01% (w/v) sodium pyrophosphate and 0.01% (w/v) Lubrol W (neither of which are toxic at these concentrations to a *Pseudomonas* species considered typical of sewage bacteria). The Lubrol W, a non-ionic surfactant, is believed to act by aiding floc separation and the sodium pyrophosphate by preventing bacterial reaggregation. There was a 13-fold increase in numbers after this treatment compared with homogenisation alone. Gayford and Richards (1970) determined bacterial numbers by growth on nutrient agar at 19°C for 44 h but observed better growth on sterilised activated sludge solidified with agar (a 5-fold increase compared with nutrient agar). A very similar result was obtained by Prakasam and Dondero (1967) but as the composition of activated sludge varies from batch to batch it is not recommended that it is used as a medium for routine use. The surface spread plate is also recommended over the pour plate technique giving a viable count ratio of 1:0.84 (Gayford & Richards 1970).

Pike *et al.* (1972) recommended the use of casitone-glycerol-yeast extract agar (CGY) for the enumeration of viable heterotrophs in activated sludge. Their homogenisation technique consisted of ultrasonication in a Kerry cleaning bath of a sample diluted 1:10 in a sodium tripolyphosphate solution (5 mg l^{-1}). The plates were incubated at 22°C for 6 d. This agar (5 g casitone, 5 g glycerol, 1 g yeast extract, in 1 l water at pH 7.2) has been found to give consistently higher counts (at least 3-fold more) than nutrient agar for the examination of sewage effluents (Flint & Hopton 1977b).

Irrespective of the homogenisation technique used and the nature of agar medium, it must be borne in mind that not all bacteria will grow on any one medium and only a fraction of the total number of microorganisms present in sewage can be estimated by plate counting techniques. The estimation of numbers of bacteria present in trickling filter films has received very little attention because of the difficulties of removing the film from the surface of stones or plastic media.

33.3.2 MEMBRANE FILTER TECHNIQUES FOR SPECIFIC GROUPS OF ORGANISMS

There are many accounts of the types of microorganisms to be found thriving in activated sludge or trickling filter plants (Allen 1944; Hawkes 1963; Pike & Carrington 1972; Pike 1975) and specific media have been devised for the isolation and, more importantly, for the enumeration of a number of the bacterial species used as biological indicators. These developments have been made in response to recent concern over the fate of pathogens and indicator organisms through the course of sewage treatment with particular reference to the potential use of sewage sludge as a fertiliser and the use of sewage effluent for irrigation purposes, particularly in arid countries (Grabow *et al.* 1974).

Most of these techniques involve the use of membrane filters because the media and techniques have been developed for the microbiological quality control of drinking water and require large samples to be concentrated in order to obtain the necessary counts. Smaller volumes or dilutions are needed if the membrane filter techniques are used with crude or treated sewage. Many of these membrane filter techniques allow the identification of a bacterial colony to the species level without first purifying individual colonies, thus enabling a more rapid identification of specific indicator bacteria. This is of major importance since it is widely believed that if the indicator bacteria are detected in drinking water then there is a possibility that pathogenic microorganisms could also have avoided the purification steps and found their way into drinking water supplies. A major symposium has recently discussed this topic (James & Evison 1979).

There is abundant literature on the fate of *Escherichia coli* and plasmid-carrying *E. coli* during sewage treatment (Grabow *et al.* 1974; van der Drift *et al.* 1977) but little on the fate of other commonly used indicator organisms. The following examples are recently published techniques for the detection of indicator organisms. In all cases they have only been used on river waters receiving sewage effluent but providing sewage samples are adequately diluted and homogenised the same techniques should be applicable.

(1) *Clostridium perfringens (welchii)* (Bisson & Cabelli 1979). The medium contains in distilled water ($g\,l^{-1}$): tryptone, 30; yeast extract, 20; sucrose, 5; cysteine, 1; $MgSO_4.7H_2O$, 0.1; bromocresol purple, 0.4; and agar, 15. The medium is adjusted to pH 7.6 and autoclaved at 15 p.s.i. for 15 min. The medium is allowed to cool to 50°C and the following selective agents added: D-cycloserine, 400 mg; polymyxin B sulphate, 25 mg; indoxyl-β-B-glucoside, 600 mg; 0.5% (w/v) phenolphthalein diphosphate, 2.0 ml; and 4.5% (w/v) ferric chloride, 0.2 ml. After filtering the sample, the membranes are incubated anaerobically for 18 to 24 h at 45°C. Yellow colonies on the filters indicate the ability to utilise sucrose and are counted. The plates are now exposed to ammonia vapour for 30 s and red-purple colonies, previously scored as positive for sucrose utilisation, denote the presence of acid phosphatase activity and are taken to be *C. perfringens*. Bisson and Cabelli (1979) indicate that greater than 90% of these presumptive colonies are in fact *C. perfringens*.

(2) *Aeromonas hydrophilia* (Rippey & Cabelli 1979). This organism has recently been suggested as a useful indicator bacterium (Kaper *et al.* 1979). The medium contains in distilled water ($g\,l^{-1}$): tryptose, 5; trehalose, 5; yeast extract, 2; NaCl, 3; KCl, 2; $MgSO_4.7H_2O$, 0.2; $FeCl_3$, 0.01; bromothymol blue, 0.04; and agar, 14. The medium is adjusted to pH 8.0, autoclaved at 15 p.s.i. for 15 min, cooled to 50° C and the following added: ethanol, 10 ml; ampicillin, 20 mg; and sodium desoxycholate, 100 mg. The filters are incubated overnight at 30° C and circular yellow colonies scored. The filters are transferred to plates of mannitol medium containing in distilled water ($g\,l^{-1}$): tryptose, 5; mannitol, 5; yeast extract, 2; NaCl, 3; KCl, 2; $MgSO_4.7H_2O$, 0.2; $FeCl_3$, 0.1; and bromothymol blue, 0.08. The medium is adjusted to pH 8.0 and autoclaved at 15 p.s.i. for 15 min. 10 mg sodium desoxycholate is added after sterilisation. Only those colonies which remain yellow after further incubation are *Aeromonas* spp. The filter is transferred to a pad of buffered saline (to remove traces of organic acids) and then to a Millipore prefilter pad soaked in Kovacs oxidase reagent. Those yellow colonies which rapidly develop a purple halo are scored as *A. hydrophila*.

(3) *Escherichia coli* (Dufour & Cabelli 1975). The medium contains in distilled water (g l^{-1}): proteose peptone, 5; yeast extract, 3; lactose, 10; NaCl, 7.5; K_2HPO_4, 3.3; KH_2PO_4, 1; sodium lauryl sulphate, 0.2; sodium desoxycholate, 0.1; bromocresol purple, 0.08; bromophenol red, 0.08; and agar, 15. The medium is autoclaved and dispensed. The filter is placed on the surface of the agar and is incubated at 35° C for 2 h and 44.5° C for 18 h. Yellow colonies are scored and the membrane filter transferred to a Millipore prefilter pad saturated with urease reagent for 10 to 15 min. Those colonies which remain yellow are scored as *E. coli* and those which turn pink-red are probably *Klebsiella* spp.

A recent development using this medium but with increased bromocresol purple/bromophenol red indicator concentration (0.15 g l^{-1} of each) and omitting the phosphate buffer, has enabled faecal coliform (all the yellow colonies) to be identified within 7 h, a significant decrease in the time taken to identify positively *Escherichia coli*. The microcolonies are easily visible under a low power binocular microscope within this time (Reasoner *et al.* 1979).

(4) *Enterococci* (faecal streptococci—*Streptococcus faecalis, S. faecium, S. avium*) (Levin M. A. *et al.* 1975). The medium contains in distilled water (g l^{-1}): peptone, 10; yeast extract, 30; NaCl, 15; aesculin, 1; agar, 15; sodium azide, 0.15; actidone, 0.05. The medium is sterilised by autoclaving. It is cooled to 50° C and the following added (g l^{-1}): nalidixic acid, 0.24; triphenyl tetrazolium chloride, 0.15. The pH is adjusted before autoclaving to 7.1. The filter is incubated for 48 h at 41° C and dark red colonies scored. The filter is transferred to aesculin iron agar plates containing in distilled water (g l^{-1}): aesculin, 1; ferric citrate, 0.5; and agar, 15). The plates are incubated at 41° C for 30 min. Colonies with a black precipitate observed from the reverse side of the plates are scored as presumptive faecal streptococci.

33.3.3 Biochemical Cultural Techniques

As well as selective cultural techniques for specific groups of bacterial species, there are also a number of techniques available for the enumeration of bacteria capable of degrading specific organic compounds or possessing certain enzymes. These techniques have been used to investigate changes in species composition and their functions as the nature of the influent to an activated sludge works was changed by the addition of different industrial effluents (Rawlings & Woods 1978). These techniques obviously suffer from the same problems as viable counts of bacteria determined by plating methods as only a fraction of those bacteria possessing the biochemical function to be measured will grow on the medium used. Also the determination of some enzymes on plates is not possible because the enzyme is produced only under certain conditions; for example, alkaline phosphatase in many organisms is a derepressible enzyme produced only in the absence of inorganic phosphate. Thus, bacteria producing this enzyme cannot be determined on plates because the phosphate level needed to obtain an observable colony is higher than the level at which the synthesis of the enzyme is derepressed (Flint & Hopton 1977a).

There are basically two techniques for assaying enzyme activity on plates.

(1) Those that allow colonies to develop first, subsequently the plate is flooded with a substrate plus suitable buffer and observation made of the development of a coloured product which indicates enzyme activity.

(2) Those that include the substrate in the agar plate and produce an indication of enzyme activity during growth of the colonies.

The latter technique is best used for extracellular enzymes. The methods used for the investigation of some enzymes are given below.

(1) *Phosphatase enzymes* (Flint & Hopton 1977a,b). The phosphatase enzyme system is generally determined by flooding incubated plates with 5% (w/v) *p*-nitrophenyl phosphate (pNPP) in an appropriate buffer (that is, pH 5.0 for acid phosphatases, pH 7.0 for neutral phosphatases and pH 9.0 for alkaline phosphatases), leaving at room temperature for 5 min and scoring, as positive, those colonies which develop a yellow coloration. Under acid conditions the product (*p*-nitrophenol) is colourless and the yellow colour can only be determined after neutralisation with ammonia solution or sodium hydroxide. If the plate contains a large number of yellow colonies, then phenolphthalein solution (1% (w/v) in an appropriate buffer) is used in place of pNPP. The hydrolysis product is red in the presence of ammonia. Using either substrate the presence of a halo around the colony denotes the production of extracellular enzyme. This technique has been used successfully to determine the numbers of phosphatase-produc-

ing bacteria in sewage, sewage effluent and a variety of fresh waters.

(2) *Pectolytic enzymes* (Hankin *et al.* 1971). Serial dilutions of homogenised sewage are spread-plated onto nutrient agar containing 1% (w/v) citrus pectin as a substrate. After 6 d incubation the plates are flooded with 1% (w/v) hexadecyltrimethylammonium bromide to precipitate the unused pectin. Positive colonies, i.e. those producing pectolytic enzymes, are denoted by the formation of a clear halo.

(3) *Proteolytic enzymes* (Hankin *et al.* 1971). Serial dilutions of homogenised sewage are spread-plated onto nutrient agar containing 0.4% (w/v) gelatin. After incubation the plates are flooded with a saturated solution of ammonium sulphate to precipitate unused gelatin and clear haloes around colonies denote the possession of protease enzymes.

(4) *Amylase enzymes* (Hankin & Sands 1974). Serial dilutions of homogenised sewage are plated on to nutrient agar containing 0.2% (w/v) starch and after incubation the plates are flooded with Lugol's iodine. Positive colonies are indicated by a clear halo against a dark blue-black background.

(5) *Lipolytic enzymes* (Hankin *et al.* 1971). Serial dilutions are plated onto a medium containing in distilled water (g l^{-1}): Tween 20 (polyoxyethylene sorbitan monolaurate), 10 ml; peptone, 10; NaCl, 5; $CaCl_2.2H_2O$, 0.1; and agar, 20. The medium is adjusted to pH 7.4. The precipitation of calcium laurate denotes the ability of colonies to cleave the lauric acid moiety from Tween 20.

(6) *Cellulolytic enzymes* (Hankin & Anagnostakis 1977). Serial dilutions are plated on to nutrient agar containing 1% (w/v) carboxymethylcellulose. After incubation the plates are flooded with 1% (w/v) hexadecylmethylammonium bromide to precipitate unhydrolysed cellulose and positive colonies are denoted by a clear halo.

Other enzyme systems could be amenable to analysis on plates and the techniques which have in the main been applied so far to soil analysis could be adapted easily to the analysis of sewage waters (see Burns 1978 for a review of techniques for the study of soil enzymes).

33.3.4 Microscopical Techniques

Culture techniques only estimate a small proportion of the total microbial population of a sewage works and to gain an estimate of total microbial population it is necessary to resort to ATP estimation (see section 33.5.1) or microscopical techniques. The wide variety of microscopical methods used for freshwater and sediment analyses have been reviewed by Jones (1979) and most of these techniques are readily adaptable to sewage works. The floc or film needs homogenising prior to observations being made and the total bacterial count can be determined by counting the bacteria in the squares of a standard haemocytometer. The bacterial count obtained in this way is up to three orders of magnitude higher than that obtained by viable counts, but it is difficult to differentiate live from dead organisms using the light microscope and also difficult to differentiate between inert detrital particles and microorganisms at the limits of the microscope resolution (Pike & Curds 1971).

Some cytological staining methods are of use. It is recommended that smears are made on coverslips, air dried and stained using standard methods (Pike & Carrington 1979). Poly-β-hydroxybutyric acid, a polymer associated with floc formation, can be demonstrated using Sudan Black and the numbers of Gram-positive and Gram-negative organisms differentiated using the tannic acid-crystal violet modification recommended by Bissett (1955).

Recently attempts have been made to study the ecology of specific groups of microorganisms *in vivo* using immunofluorescence staining techniques (Belser 1979). Fluorescent antibodies were prepared against the nitrifying bacterium *Nitrosomonas europeae* and used to estimate its population density in activated sludge. This technique has also been used successfully against nitrifying bacteria in soils (Rennie & Schmidt 1977; Gosserand & Cleyet-Marel 1979). The major drawback of the method is that antibodies are very specific and can only be used if the dominant species are known or if the presence or absence of a particular species is to be determined. One of the advantages of such a technique is that it could be used to determine the organisms at the centre of flocs.

The method can also be combined with microautoradiography (the study of uptake of labelled substrates by bacteria attached to glass slides), to study the metabolic activity of organisms in flocs and free swimming (Brock & Brock 1968; Fliermans & Schmidt 1975; Meyer-Reil 1978). The method has also been used to show that *Escherichia coli* species are removed from sewage wastes by predation by protozoa as well as by non-specific attachment to sludge flocs and subsequent sedimentation (van der Drift *et al.* 1977).

The ecology and role of filamentous organisms in sewage works is also more readily studied by microscopy than by cultural methods as these organisms are difficult to grow on plates. In a normal healthy sewage plant, filamentous organisms are present in very low numbers and their isolation requires growth of sludge aliquots in stationary culture in the presence of a high nitrogen to carbon ratio (van Ween 1973). Filaments develop on the flocs and can be isolated by removal using a Pasteur pipette. These filaments have been identified mainly as species of *Sphaerotilus, Beggiatoa, Nocardia, Thiothrix, Leucothrix, Lineola* and *Geotrichum*. These resemble the filamentous organisms responsible for the undesired sewage condition known as bulking and also resemble sewage fungus, often found growing in rivers receiving sewage effluent (Dondero 1975).

33.4 Estimation of virus numbers

One of the areas of domestic waste treatment which is at present under closest investigation is the estimation of viral numbers and the survival of viruses throughout the course of sewage treatment (see Chapter 5). There has been much speculation in recent years about the potential for survival in sewage, rivers and drinking waters of viruses pathogenic to humans which can occur in large numbers (up to 1×10^6 viral particles (g faeces)$^{-1}$), even in humans not manifesting disease symptoms. The presence of these viruses in sewage effluent and hence in waters used for recreation, drinking or irrigation purposes could pose a health hazard (Hoff & Becker 1968; Cabelli 1976, 1979; Mackowiak *et al.* 1976; Shuval 1976; Wood 1979). Human viruses have not been investigated in as much detail as either pathogenic bacteria or bacteriophage because of isolation difficulties from sewage plants and the problems experienced in growing these viruses in tissue culture. The techniques used mainly for the isolation of human viruses from river waters and sewage effluents and the viruses isolated by these methods have been reviewed in recent years (Hill *et al.* 1971; Berg *et al.* 1976; Sobsey 1976; Babiuk *et al.* 1977; Wallis *et al.* 1979). The techniques for the study of bacteriophages have also been reviewed recently (Ayres 1977; Balluz *et al.* 1978; Butler & Balluz 1979; see Chapter 5).

The major problem facing the isolation of viral particles from sewage is that most of the particles are embedded in the complex organic material of the floc or slime layers and these particles cannot readily be separated. Many studies, therefore, have concentrated on the viruses present in sewage effluent, i.e. after the organic material has been removed in the sedimentation process, on the premise that only these particles pose a potential health threat. This does, however, mean that there are few studies on the survival of viruses through the sewage treatment process.

The following methods have been used to isolate the viruses present in sewage and sewage effluents.

(1) *Precipitation on to alum* (Lal & Lund 1975). 75 mg aluminium sulphate is stirred with 100 ml of sample for 2 h and the precipitate pelleted by low-speed centrifugation. The viral particles are eluted into 0.1M-Tris-HCl buffer, pH 9.0 for 2 h, the precipitate removed by centrifugation and the eluate neutralised.

(2) *Two phase concentration* (Lund & Hedstrom 1966). This method is used to recover added enterovirus from sewage. 20 g of 5M-NaCl, 58 g of 30% (w/v) polyethyleneglycol (PEG) (Carbowax 6000) and 2 to 7 g sodium dextran sulphate are added to 200 ml of sample and the pH adjusted to 7.2. The solution is left overnight at 4°C in a separating funnel and the phases collected. The viruses are concentrated in the bottom phase about 2 ml in volume.

(3) *Acid precipitation* (Lydholm & Nielsen 1980). 1M-HCl is added to the sample to give a final pH of 3.5. After stirring for 1 h the precipitate is pelleted at 16 000 ***g*** and taken up in 4 volumes of 10% (w/v) beef extract at pH 7.0. After 2 h the precipitate is removed by centrifugation and the eluate containing the viruses adjusted to pH 7.2.

(4) *Direct elution*. This is by far the easiest method to use and the one recommended for the recovery of viruses from sewage flocs. The solid material is resuspended in either 10% (w/v) beef extract (Nielsen & Lydholm 1980), 1% (w/v) sodium dodecylsulphate (SDS) (Ward & Ashley 1977) or 5% (w/v) glycine (Hurst *et al.* 1978) and homogenised for up to 2 h. The bacteria and sludge are removed by centrifugation and the eluate tested directly or after concentration. Concentration can be achieved by ultrafiltration (Amicon, UK) using PM10 filters (nominal retention of particles greater than 10 000 molecular weight) under 5 p.s.i. nitrogen gas pressure or by acid precipitation.

(5) *Adsorption on to resins.* Ion exchange resins are used extensively for the recovery of viruses from river and drinking water but they do not have much application to sewage works (Goyal *et al.* 1980).

(6) *Direct inoculation.* Some investigations have attempted to isolate viruses by directly inoculating tissue culture cells with sewage. The technique showed a good recovery of added viruses but toxicity effects due to other components of the sewage posed problems for the tissue culture (Subrahmanyan 1977).

None of the above techniques can recover all the viruses present in sewage and, just as with bacteria, there is a cultural problem. The most common tissue culture cell line used for enumeration of viral particles is the HeLa cell line which is only useful for the isolation of polioviruses, echoviruses and some of the adenoviruses and coxsackieviruses. Infectious hepatitis virus requires injection into test animals for its detection and hence is rarely studied.

Adsorption of viral particles onto the organic material in sewage poses, at present, an unsolved problem for those investigating viral numbers in sewage. In fact this process is believed to be the major mechanism by which viruses are removed and eventually inactivated from sewage. Wellings *et al.* (1976) showed that poliovirus was apparently reduced in number by three orders of magnitude during the course of sewage treatment if only the liquor was examined for free virus, but studies of the liquor plus solid phase revealed that the total viable viral numbers remained relatively constant. This constitutes an additional health problem if the residual sludge from sewage treatment is used as a fertiliser without further treatment (Lund & Ronne 1973; Nielsen & Lydholm 1980). In fact soils amended with sludge have been found to harbour coxsackievirus for up to four months (Pahren *et al.* 1979).

There has been much more work carried out on the detection and survival of bacteriophages since they are much easier to cultivate. Butler and Balluz (1979) compared the survival characteristics of phage f2 and poliovirus to determine whether or not the former could be used as an indicator for pathogenic virus survival through the course of sewage and water treatment processes. Phage f2 was found to disappear faster than poliovirus because of its greater adsorption to the floc material rather than actual death.

There has been much work on the factors affecting the survival of bacteriophages, for example, temperature, salinity, agitation, solar radiation, pollutants and clay minerals mainly using bacteriophage added to sewage effluents or crude sewage rather than using naturally occurring viruses (Balluz *et al.* 1977; Babich & Stotzky 1980). Temperature and sunlight have been suggested as the major factors responsible for viral inactivation and clay minerals (montmorillonite, vermiculite, kaolinite and attaplugite) protected the viral particles from inactivation, presumably by adsorption.

The presence and survival of viruses in sewage remains one of the few areas where our knowledge is slight and there are no standard methods yet available despite the wealth of evidence which shows that viruses do survive sewage treatment and have been responsible for waterborne disease outbreaks.

33.5 Estimation of biochemical parameters of waste waters

A considerable amount of attention is paid to the biochemistry of waste waters, particularly with a view to defining the functioning of a sewage works in terms of some parameter, e.g. ATP, which is easier to estimate than microbial numbers.

33.5.1 Microbial ATP

There has been much discussion about the viability of microorganisms in a floc and particularly how much of the dead biomass still retaining some enzymic activity is present (Patterson *et al.* 1970; Levin G. V. *et al.* 1975). ATP is suggested as an indicator of viability as it is a non-conservative component of cells and hence is associated only with growing microorganisms. The solid phase of an activated sludge mixed liquor is composed of active microorganisms, inert or moribund microbial solids and inert debris and the usefulness of being able to quantify the active microbial biomass for treatment plant design and operating characteristic purposes is well recognised.

Bucksteeg (1966) showed that sludge activity was a function of organic load and that it showed a more rapid response to organic load variations than did the total solids content or the bacterial counts. Patterson *et al.* (1969, 1970) used sludge activity measurements (mainly ATP) to detect toxicity problems caused by increased loading of industrial effluents. Recently it has been suggested by Kennicutt (1980) that ATP measurements could be used as an indicator of toxicity. He showed that mercury at 1 mg l^{-1} caused a 100% decrease in sludge ATP levels in less than 1 h whereas non-toxic chloroform at saturating concentrations caused less than a 30% decrease in 6 h.

ATP is measured by using the luciferin-luciferase system after extraction of the ATP using boiling 0.2M-Tris-HCl buffer pH 7.8 for 10 min (Chiu *et al.* 1973; Nelson & Lawrence 1980). The ATP measurement can be converted to microbial dry weight or numbers by using the conversion figures calculated from pure cultures of typical sewage or aquatic bacteria. For activated sludge figures of 1 to 10 μg ATP is equivalent to 1×10^9 cells. For *Escherichia coli* 1.2 μg ATP is equivalent to 1×10^9 cells. A list of these conversion figures is presented in Nelson and Lawrence (1980). The technique, although rapid, has a major drawback in that it is by no means certain that only bacterial ATP is being measured in the system. Protozoal and fungal ATP can cause a substantial measurement variation due to the large size of these organisms relative to a bacterium but, nevertheless, ATP estimation is considered to be a rapid means of assessing the performance of activated sludge plant operations (Levin G. V. *et al.* 1975) although its use with the trickling filter is limited due to the difficulties of removing the film rapidly from different depths of the bed.

33.5.2 Microbial DNA

Estimation of microbial DNA has been discussed by Agardy and Shephard (1967) and Hattingh and Siebert (1967). Like ATP, DNA has been suggested as a parameter which could be used to control digester loading and activated sludge loading. Again the major problem is the significance of DNA measurements and the relationship between protozoal and bacterial DNA levels. DNA is isolated by homogenising the sludge, adding 1 ml cold TCA to 2.5 ml homogenate and centrifuging (5 ml 95% (v/v)). Ethanol is added to the supernatant which is centrifuged and the pellet resuspended in 3 ml 10% (w/v) TCA. The TCA extract is heated for 15 min at 90° C, centrifuged and the extract kept. The extraction is repeated 3 times and the extracts pooled. The amount of DNA is determined by the addition of 10 ml diphenylamine to 5 ml extract and heating at 100° C for 15 min. The absorbance is read at a wavelength of 600 nm.

33.5.3 Enzymes

Recently there have been a number of studies on the activity of various enzymes, in particular in activated sludge tanks receiving different quality wastes. Dolyagin and Zubova (1976) point out that the purification of waste waters is related to the biochemical and physiological activity of the microbes present rather than to the biomass. They studied the levels of catalase and protease enzymes as biochemical indicators of microbial activity in activated sludge plants. Enzyme activity could also give some indication of the health of sewage works, for example, numerous enzymes are inhibited by heavy metals, pesticides and other chemicals which could prove toxic to sewage works. Enzymes, in common with ATP measurement, are easier to measure than biomass and provide an answer to the health of sewage much faster than the standard methods. More information on enzyme assays is given by Flint and Hopton (1977a, b) and Rawlings and Woods (1978) and most of the assays used for soil enzymes can easily be adapted to the assay of sewage enzymes (Burns 1978).

33.6 Estimation of chemical parameters

The chemical composition of sewage is a major factor which determines the composition of the microbial flora and governs the type of treatment processes required to purify the waste. The chemical composition is more important for industrial wastes than domestic wastes because of their potential toxicity to the sewage process. Most chemical analyses are listed in *Standard Methods for the Examination of Water and Waste Water* (American Public Health Association 1975) but there are assays available for some important organic components which are not included.

(1) *Cellulose* (Strickland & Parsons 1972). The amount of cellulose present in wastes, particularly in wastes which undergo an anaerobic digestion stage, is determined by estimation of reducing sugars by the anthrone procedure after hydrolysis using 3M-NaOH.

(2) *Protein* (Sridhar & Pillai 1973). The different types of protein present at various stages of sewage treatment have been determined using a modification of the Lowry method.

(3) *Free sugars* (Whittaker & Vallentyne 1957). The free sugars are extracted in 70% (v/v) ethanol at pH 11.0 for 24 h. The eluate is filtered, acidified to pH 2.0 and passed through an XAD macroreticular resin filter to remove interfering organic compounds. The individual free sugars can be determined by thin layer chromatography against known standards. Total free sugars can be detected by the anthrone method (Strickland & Parsons 1972).

(4) *Chlorinated organic compounds*. The detection of these compounds has become a major concern in recent years since the discovery that the majority of these compounds are carcinogens and that they can be produced during chlorination of either sewage effluent or water in a purification plant. Up to 129 potential carcinogens have been discovered in the drinking water supplies of Cincinatti, Ohio and New Orleans (Harris *et al.* 1977).

Because of the low concentrations of these compounds present, any detection process relies on an efficient concentration procedure. Baird *et al.* (1979) used an XAD-2 macroreticular resin and eluted the organic compounds from this using ether. The ether fraction is concentrated further by evaporation under nitrogen (in total a 20 000-fold concentration). The residues are determined by gas chromatography.

The carcinogenic nature of the concentrated compounds or of the initial drinking water or waste effluent can be shown using direct animal feed studies and by observing changes in mortality, blood sugar, enzyme levels or histopathological changes (Ettinger 1961; Truhat *et al.* 1979), or by the simpler Ames test (Ames *et al.* 1972) which measures the back-mutation rate of a *his*$^-$ *Salmonella typhimurium* in the presence of an activating liver microsome preparation.

A number of other bioassays involving the use of algae, protozoa, fish or molluscs are given in *Standard Methods for the Examination of Water and Waste Water* (American Public Health Association 1975). Recently the cell multiplication inhibition bioassay has been criticised by Bringmann and Kuhn (1980) because some known carcinogens could not be detected using some of the standard test organisms.

Much work remains to be done on the quantities and origins of chlorinated organic compounds in wastes and drinking waters. Some of these compounds, particularly the chlorinated pesticides can reach levels which are toxic to sewage purification processes and many pass unchanged through the sewage works process, to build up in the environment (Jolley 1978).

Recommended reading

AMERICAN PUBLIC HEALTH ASSOCIATION (1975) *Standard Methods for the Examination of Water and Waste Water*, 14e. American Public Health Association, New York.

BELSER L. W. (1979) Population ecology of nitrifying bacteria. *Annual Review of Microbiology* **33**, 309–33.

BERG G., BODILY H. L., LENNETTE E. H., MELNICK J. L. & METCALF T. G. (1976) *Viruses in Water*. American Public Health Association, Washington.

CURDS C. R. & HAWKES H. A. (1975) *Ecological Aspects of Used Water Treatment*. Academic Press, London.

DEPARTMENT OF THE ENVIRONMENT (1972) *Analysis of Raw, Potable and Waste Waters*. HMSO, London.

DONDERO N. C. (1975) *Sphaeotilus-Leptothrix* group. *Annual Review of Microbiology* **29**, 407–28.

HANKIN L. & SANDS D. C. (1974) Bacterial production of enzymes in activated sludge systems. *Water Pollution Control Federation Journal* **46**, 2015–25.

HAWKES H. A. (1963) *The Ecology of Activated Sludge*. Pergamon Press, Oxford.

JAMES A. & EVISON L. E. (1979) *Biological Indicators of Water Quality*. John Wiley & Sons, Chichester.

JOLLEY R. E. (1978) *Water Chlorination; Environmental Impact and Health Effects*, Vols 1 and 2. Ann Arbor Science Publishers, Ann Arbor, Michigan, U.S.A.

JONES J. G. (1979) *A Guide to Methods for Estimating Microbial Numbers and Biomass in Fresh Waters*. Freshwater Biological Association, Publication 39, Ambleside, Cumbria.

SKINNER F. A. & SHEWAN J. M. (1977) *Aquatic Microbiology*. Society for Applied Bacteriology Symposium Series 6. Academic Press, London.

SYKES G. A. & SKINNER F. A. (1971) *Microbial Aspects of Pollution*. Society for Applied Bacteriology Symposium Series 1, Academic Press, London.

References

AGARDY F. J. & SHEPHARD N. C. (1967) DNA—a rational basis for digester loading. *Water Pollution Control Federation Journal* **37**, 1236–42.

ALLEN L. A. (1944) The bacteriology of activated sludge. *Journal of Hygiene, Cambridge* **43**, 424–31.

AMERICAN PUBLIC HEALTH ASSOCIATION (1975) *Standard Methods for the Examination of Water and Waste Water*, 14e. American Public Health Association, New York.

AMES B. N., GURNEY E. G., MILLER J. A. & BARTSCH H. (1972) Carcinogens as frameshift mutagens: Metabolites and derivities of 2-acetylaminofluorene and other aromatic amine carcinogens. *Proceedings of National Academy of Sciences U.S.A.* **69**, 3128–32.

AYRES P. A. (1977) Coliphages in sewage and the marine environment. In: *Aquatic Microbiology* (Eds F. A. Skinner & J. M. Shewan) pp. 275–98. Academic Press, London.

AZIZ J. A. & TEBBUTT T.H.Y. (1980) Significance of COD, BOD and TOD correlation in kinetic models of biological oxidation. *Water Research* **14**, 319–24.

BABICH H. & STOTZKY G. (1980) Reduction in inactivation rates of phage by clay minerals in lake water. *Water Research* **14**, 185–7.

BABIUK L. A., MOHAMMED K., SPENCE L., FAUVEL M. & PETRO R. (1977) Rotavirus isolation and cultivation in the presence of trypin. *Journal of Clinical Microbiology* **6**, 610–17.

BAIRD R., SELNA M., HASKINS J. & CHAPPELLE D. (1979) Analysis of selected trace organics in advanced wastewater treatment. *Water Research* **13**, 493–502.

BALLUZ S. A., JONES H. H. & BUTLER M. (1977) The persistence of poliovirus in activated sludge treatment. *Journal of Hygiene, Cambridge* **78**, 165–73.

BALLUZ S. A., BUTLER M. & JONES H. H. (1978) The behaviour of f2 coliphage in activated sludge treatment. *Journal of Hygiene, Cambridge* **80**, 237–42.

BELSER L. W. (1979) Population ecology of nitrifying bacteria. *Annual Review of Microbiology* **33**, 309–33.

BENEDICT R. G. & CARLSON D. A. (1971) Aerobic heterotrophic bacteria in activated sludge. *Water Research* **5**, 1023–30.

BERG G., BODILY H. L., LENNETTE E. H., MELNICK J. L. & METCALF T. G. (1976) *Viruses in Water*. American Public Health Association, Washington.

BISSETT K. A. (1955) *The Cytology and Life History of Bacteria*. E. & S. Livingstone, Edinburgh.

BISSON J. W. & CABELLI V. J. (1979) MF enumeration method for *Clostridium perfringens*. *Applied and Environmental Microbiology* **37**, 55–66.

BRINGMANN G. & KUHN R. (1980) Comparison of the toxicity thresholds of water pollutants to bacteria, algae and protozoa in the cell multiplication inhibition test. *Water Research* **14**, 231–41.

BROCK M. L. & BROCK T. D. (1968) The application of microautoradiographic techniques to ecological studies. *Mitteilung der internationalen Vereinigung für theoretische und angewandte Limnologie* **15**, 1–29.

BUCKSTEEG W. (1966) Determination of sludge activity—a possible method of controlling activated sludge plants. *Advances in Water Pollution Research* **2**, 83–102.

BURNS R. G. (1978) *Soil Enzymes*. Academic Press, London.

BUTLER M. & BALLUZ S. A. (1979) A comparison of the behaviour of poliovirus and f2 coliphage in activated sludge treatment. In: *Biological Indicators of Water Quality* (Eds A. James & L. Evison), pp. 19–21. John Wiley & Sons, Chichester.

CABELLI V. J. (1976) Indicators of recreational water quality. In: *Bacterial Indicators/Health Hazards Associated with Water* (Eds A. W. Hoadley & B. Dutka), pp. 65–79. American Society for Testing and Materials, Tallahassee, Florida, U.S.A.

CABELLI V. J. (1979) Evaluation of recreational water quality: EPA approach. In: *Biological Indicators of Water Quality* (Eds A. James & L. Evison), pp. (14)1–(14)23. John Wiley & Sons, Chichester.

CHIU S. Y., KAO I. C., ERIKSON L. E. & FAN F. T. (1973) ATP pools in activated sludge. *Water Pollution Control Federation Journal* **45**, 1746–58.

CURDS C. R. (1975) Protozoa. In: *Ecological Aspects of Used Water Treatment* (Eds C. R. Curds & H. A. Hawkes), pp. 203–68. Academic Press, London.

CURDS C. R. & HAWKES H. A. (1975) *Ecological Aspects of Used Water Treatment*. Academic Press, London.

DEPARTMENT OF THE ENVIRONMENT (1972) *Analysis of Raw, Potable and Waste Waters*. HMSO, London.

DOLYAGIN A. B. & ZUBOVA M. Z. (1976) Biochemical characteristics of activity of sludge microorganisms. *Hydrobiologica Journal* **12**, 74–6.

DONDERO N. C. (1975) *Sphaerotilus–Leptothrix* group. *Annual Review of Microbiology* **29**, 407–28.

DUFOUR A. P. & CABELLI V. J. (1975) Membrane filter procedure for enumerating component genera of the coliform group in sea water. *Applied Microbiology* **29**, 826–33.

ETTINGER M. B. (1961) Wastewater treatment and drinking water standards. *Water Pollution Control Federation Journal* **33**, 1290–2.

FLIERMANS C. B. & SCHMIDT E. L. (1975) Autoradiography and immunofluorescence combined for autecological study of single cell activity with *Nitrobacter* as a model system. *Applied Microbiology* **30**, 674–7.

FLINT K. P. & HOPTON J. W. (1977a) Substrate specificity and ion inhibition of bacterial and particle associated alkaline phosphatase of waters and sewage sludge. *European Journal of Applied Microbiology* **4**, 195–204.

FLINT K. P. & HOPTON J. W. (1977b) Seasonal variation in alkaline phosphatase activity of waters and sewage sludges. *European Journal of Applied Microbiology* **4**, 205–15.

GAYFORD C. G. & RICHARDS J. P. (1970) Isolation and enumeration of aerobic heterotrophic bacteria in activated sludge. *Journal of Applied Bacteriology* **33**, 342–50.

GOSSERAND A. & CLEYET-MAREL J. C. (1979) Isolation from soils of nitrobacter: evidence for noval serotypes using immunofluorescence. *Microbial Ecology* **5**, 197–205.

GOYAL S. M., ZERDA K. S. & GERBA C. P. (1980) Concentration of coliphage from large volumes of water and wastewater. *Applied and Environmental Microbiology* **39**, 85–91.

GRABOW W. O. K., PROZESKY O. W. & SMITH L. S. (1974) Drug resistant coliforms call for review of water quality standards. *Water Research* **8**, 1–9.

HANKIN L. & ANAGNOSTAKIS S. L. (1977) Solid media containing carboxymethylcellulose to detect Cx cellulase activity. *Journal of General Microbiology* **98**, 109–15.

HANKIN L. & SANDS D. C. (1974) Bacterial production of enzymes in activated sludge systems. *Water Pollution Control Federation Journal* **46**, 2015–25.

HANKIN L., ZUCKER M. & SANDS D. C. (1971) Improved solid medium for the detection and enumeration of pectolytic bacteria. *Applied Microbiology* **22**, 205–9.

HARRIS R. H., PAGE T. & EPSTEIN S. S. (1977) Drinking water and cancer mortality in Louisiana. *Science* **193**, 55–8.

HATTINGH W. H. J. & SIEBERT M. L. (1967) Determination of the DNA content of anaerobic sludge. *Water Research* **1**, 197–203.

HAWKES H. A. (1963) *The Ecology of Activated Sludge*. Pergamon Press, Oxford.

HIGGINS I. J. & BURNS R. G. (1975) *The Chemistry and Microbiology of Pollution*. Academic Press, London.

HILL W. F., AKIN E. W. & BENTON W. H. (1971) Detection of viruses in water: A review of methods and applications. *Water Research* **5**, 967–95.

HOFF J. C. & BECKER R. C. (1968) The accumulation and elimination of crude and clarified poliovirus suspensions by shellfish. *American Journal of Epidemiology* **90**, 53–61.

HURST C. J., FARRAH S. R., GERBA C. P. & MELNICK J. L. (1978) Development of quantitative methods for the detection of enteroviruses in sewage sludges during inactivation and following land disposal. *Applied and Environmental Microbiology* **36**, 81–9.

JAMES A. & EVISON L. E. (1979) *Biological Indicators of Water Quality*. John Wiley & Sons, Chichester.

JOLLEY R. E. (1978) *Water Chlorination; Environmental Impact and Health Effects*, Vols 1 and 2. Ann Arbor Science Publishers, Ann Arbor, Michigan, U.S.A.

JONES J. G. (1979) *A Guide to Methods for Estimating Microbial Numbers and Biomass in Freshwaters*. Freshwater Biological Association, Publication 39, Ambleside, Cumbria.

KAPER S., SEIDLER R. J., LOCKMAN H. & COLDWELL R. R. (1979) Medium for the presumptive identification of *Aeromonas hydrophila* and Enterobacteriaceae. *Applied and Environmental Microbiology* **38**, 1023–6.

KENNICUTT M. C. (1980) ATP as an indicator of toxicity. *Water Research* **14**, 325–8.

LAL S. M. & LUND E. (1975) Recovery of virus by chemical precipitation followed by elution. *Progress in Water Technology* **7**, 687–93.

LEVIN G. V., SCHROT J. R. & HESS W. L. (1975) Methodology for the application of ATP determination in wastewater treatment. *Environmental Science and Technology* **9**, 961–5.

LEVIN M. A., FISHER J. R. & CABELLI V. J. (1975) Membrane filter technique for enumeration of enterococci in marine waters. *Applied Microbiology* **30**, 66–71.

LUND E. & HEDSTROM C. E. (1966) The use of an aqueous polymer phase system for enterovirus isolations from sewage. *American Journal of Epidemiology* **84**, 283–91.

LUND E. & RONNE V. (1973) Isolation of virus from sewage treatment plant sludges. *Water Research* **7**, 863–71.

LYDHOLM B. & NIELSEN A. L. (1980) Methods for the detection of virus in waste water, applied to samples from small scale treatment systems. *Water Research* **14**, 169–73.

MACKOWIAK P. A., CARAWAY C. I. & PORTNOY B. L. (1976) Oyster associated hepatitis: lessons from the Louisiana experience. *American Journal of Epidemiology* **103**, 181–91.

MEYER-REIL L-A. (1978) Autoradiography and epifluorescence microscopy combined for the determination of number and spectrum of actively metabolising bacteria in natural waters. *Applied and Environmental Microbiology* **36**, 506–12.

MONTGOMERY H. A. C. (1967) Determination biochemical oxygen demand by respirometric methods. *Water Research* **1**, 631–62.

NELSON F. O. & LAWRENCE A. W. (1980) Microbial viability measurements and activated sludge kinetics. *Water Research* **14**, 217–25.

NIELSEN A. L. & LYDHOLM B. (1980) Methods for the isolation of viruses from raw and digested waste water sludges. *Water Research* **14**, 175–8.

PAHREN H. R., LUCAS J. B., RYAN J. A. & DOTSON G. K. (1979) Health risks associated with land application of municipal sludge. *Water Pollution Control Federation Journal* **51**, 2588–601.

PATTERSON J. W., BREZONIK P. L. & PUTNAM H. D. (1969) Sludge activity parameters and their application to toxicity measurements in activated sludge. *Proceedings 24th Industrial Waste Conference, Engineering Bulletin of Purdue University*, pp. 127–54. Purdue University, Lafayette, Indiana, USA.

PATTERSON J. W., BREZONIK P. L. & PUTNAM H. D. (1970) Measurement and significance of ATP in activated sludge. *Environmental Science and Technology* **4**, 569–75.

PIKE E. B. (1975) Aerobic bacteria. In: *Ecological Aspects of Used Water Treatment* (Eds C. R. Curds & H. A. Hawkes), vol. 1, pp. 1–63. Academic Press, London.

PIKE E. B. & CARRINGTON E. G. (1972) Recent developments in the study of bacteria in the activated sludge process. *Water Pollution Control (Great Britain)* **71**, 583–605.

PIKE E. B. & CARRINGTON E. G. (1979) The fate of enteric bacteria and pathogens during sewage treatment. In: *Biological Indicators of Water Quality* (Eds

A. James & L. E. Evison), pp. (20)1–(20)32. John Wiley & Sons, Chichester.

PIKE E. B., CARRINGTON E. G. & ASHBURNER P. A. (1972) An evaluation of procedures for enumerating bacteria in activated sludge. *Journal of Applied Bacteriology* **35**, 309–21.

PIKE E. B. & CURDS C. R. (1971) The microbial ecology of the activated sludge process. In: *Microbial Aspects of Pollution* (Eds G. A. Sykes & F. A. Skinner), pp. 123–47. Academic Press, London.

PRAKASAM T. B. S. & DONDERO N. (1967) Aerobic heterotrophic bacterial population of sewage and activated sludge. I. Enumeration. *Applied Microbiology* **15**, 461–7.

RAWLINGS D. E. & WOODS D. R. (1978) Bacteriology and enzymology of fellmongery activated sludge systems. *Journal of Applied Bacteriology* **44**, 131–9.

REASONER D. J., BLANNON J. C. & GELDRICH C. E. (1979) Rapid faecal coliform test. *Applied and Environmental Microbiology* **38**, 229–36.

RENNIE R. J. & SCHMIDT E. L. (1977) Immunofluorescence studies of *Nitrobacter* populations in soils. *Canadian Journal of Microbiology* **23**, 1011–17.

RIPPEY S. R. & CABELLI V. J. (1979) Membrane filter procedure for enumeration of *Aeromonas hydrophila* in freshwaters. *Applied and Environmental Microbiology* **38**, 108–13.

SHERRARD J. H., FRIEDMAN A. A. & RAND M. C. (1979) BOD_5: are there alternatives available? *Water Pollution Control Federation Journal* **51**, 1799–804.

SHUVAL H. I. (1976) Health consideration in water renovation and reuse. In: *Water Renovation and Reuse* (Ed. H. I. Shuval), pp. 33–72. Academic Press, London.

SKINNER F. A. & SHEWAN J. M. (1977) *Aquatic Microbiology*. Society for Applied Bacteriology Symposium Series 6. Academic Press, London.

SOBSEY M. D. (1976) Methods for detecting enteric viruses in water and wastewater. In: *Viruses in Water* (Eds G. Berg, H. L. Bodily, E. H. Lennette, J. L. Melnick & T. G. Metcalf), pp. 89–127. American Public Health Association, Washington.

SRIDHAR M. K. C. & PILLAI S. C. (1973) Protein in wastewater and wastewater sludges. *Water Pollution Control Federation Journal* **45**, 1595–1600.

STONES T. (1979) A critical examination of the uses of BOD test. *Effluent and Waste Treatment Journal* **19**, 250–4.

STRICKLAND J. D. H. & PARSONS T. R. (1972) *A Practical Handbook of Seawater Analysis*. Fishery Research Board of Canada, Ottawa.

SUBRAHMANYAN T. P. (1977) Persistence of enterovirus in sewage sludge. *Bulletin of World Health Organisation* **55**, 431–4.

SYKES G. A. & SKINNER F. A. (1971) *Microbial Aspects of Pollution*. Society for Applied Bacteriology Symposium Series 1, Academic Press, London.

TRUHAT R., GAK J. C. & GRAILLOT C. (1979) Recherches sur les risques pouvant résulter de la pollution chimique des eaux d'alimentation. I. Etude de la toxicité à long terme chez le rat et la souris des micropollutants organiques chloroforma extractibles à partir d'eau livrés à la consommation humaine. *Water Research* **13**, 689–97.

UNZ R. F. & DONDERO N. C. (1967) The predominant bacteria in natural zoogloeal colonies. *Canadian Journal of Microbiology* **13**, 1671–82.

VAN DER DRIFT C., VAN SEGGEDEN E., STUMM C., HOL W. & TWINTE J. (1977) Removal of *E. coli* in wastewater by activated sludge. *Applied and Environmental Microbiology* **34**, 315–19.

VAN WEEN W. L. (1973) Bacteriology of activated sludge, in particular the filamentous bacteria. *Antonie van Leewenhoek* **39**, 189–205.

WALLIS G., MELNICK J. L. & GERBA C. P. (1979) Concentration of viruses from water by membrane chromatography. *Annual Review of Microbiology* **33**, 413–37.

WARD R. L. & ASHLEY C. S. (1977) Identification of the virucidal agents in wastewater sludge. *Applied and Environmental Microbiology* **33**, 860–4.

WELLINGS F. M., LEWIS A. L. & MOUNTAIN C. W. (1976) Demonstration of solids associated viruses in wastewater and sludge. *Applied and Environmental Microbiology* **31**, 354–8.

WHITTAKER J. R. & VALLENTYNE J. R. (1957) On the occurrence of free sugars in lake sediment extracts. *Limnology and Oceanography* **2**, 98–110.

WOOD P. C. (1979) Public health aspects of shellfish from polluted waters. In: *Biological Indicators of Water Quality* (Eds A. James & L. E. Evison), pp. (13)1–(13)18. John Wiley & Sons, Chichester.

Chapter 34 • Oil Degradation in Hydrocarbon- and Oil-Stressed Environments

Sally Stafford, Paul Berwick, David E. Hughes and David A. Stafford

34.1 Introduction

Most oil discharges at sea, estuaries or inland waterways are brought about by deliberate action, although an increasing proportion are accidental. They come from ships cleaning out bunker oil or oil tankers discharging oily washings (Beastall 1977). Occasionally large spillages occur from stricken tankers, such as the Amoco Cadiz spillage off the French coast or the Torrey Canyon off the Cornish coast. The immediate availability of a wide variety of organic compounds enables the indigenous microbial population to respond rapidly to changing environmental conditions. Much carbon is readily available, but the presence of other nutrients is essential for a high microbial activity to be realised. Such nutrients include nitrogen (present in oil) and phosphate which has to be added extraneously to increase both population size and activity.

The type of oil spilt can influence the rate at which the oil spreads on water surfaces and thus the potential area and volume available to microbial attack. For example, lighter crudes, such as Nigerian Light and Brega (Libya), evaporate and spread more quickly than Venezuelan crudes of lower volatility (Beastall 1977). The spread oils undergo both autoxidation and microbial attack and oxidise at the rate of about 0.2 tonnes km^{-2} d^{-1} for a slick about 0.1 mm or more in thickness. This assumes good dispersion of the oil which often does not occur because thick oil may be wind and wave blown to produce 'chocolate mousse' and tarballs. In this case most microbial activity is on the surface of the emulsion and nutrients diffuse slowly from it to provide for microbial growth. As microbes attach themselves to these emulsions and tars, the density increases and some sedimentation of the smaller particles of microbes plus tar occurs. The activity of these microorganisms can be affected by environmental conditions such as pH, oxygen tension and nutrient availability which will determine the rate at which the microbes degrade hydrocarbons, aromatics and heterocyclics. Maximum hydrocarbon oxidation occurs between 25 to 37°C although some oxidation has been noted at −1°C (Zobell 1973). About 90% of the sea surface is at 4°C or lower and so the activity of psychrophilic microbes is important, although the rate of oil degradation is still slow. This only highlights the difficulties in interpreting rates of oil breakdown in laboratory cultures and relating those activities to sea-water microbial breakdown of oils *in situ*.

The oil–water–microbe phase relationships influence the biodegradation patterns considerably and the state of the phase is important when attempting to determine biodegradability. Difficulties are created when microbial identification, and especially enumeration, is attempted using standard plating techniques. The use of homogenisation does not of itself increase the cell number determinations when using the plate count technique (see section 34.3), but ultrasonics and/or emulsification may separate cells more efficiently for purposes of cell counting (see section 34.3.1).

When determining cell numbers by using selective growth media, in one study in the U.K., the highest numbers of oil-degrading organisms were found near the large ports in the Bristol Channel, where chronic oil pollution frequently occurs locally (Beastall 1977). In the Gulf of Mexico, similar techniques using modified sea-water yeast extract for total viable counts showed that the proportion of oil-degrading and sulphur-oxidising bacteria were found in higher proportions near to oil platforms (Hollaway *et al.* 1980). However, no major differences were noted in taxonomic groups or physiological types between polluted and non-polluted sources. The genera *Pseudomonas*, *Vibrio*, *Moraxella*, *Acinetobacter* and *Aeromonas* were all found in oil-polluted and non-polluted sites with *Pseudomonas* spp. predominating. Others have found that organism diversity increased in the oil-polluted habitats due to the extra organic material made available (Hood *et al.* 1975). The seeding of oil spills with oil-degrading microorganisms will not accelerate the rate of oil biodegradation because of the low availability of nitrogen and phosphorus. Between 1 and 10 mg phosphorus and about 6 to 60 mg nitrogen are required to biodegrade 1 g oil (Beastall 1977). Some oils contain nitrogen in heterocyclic fractions and often only the addition of phosphorus is necessary.

One other important factor governing the rate of oil breakdown using microbes is that of agitation. Whilst severe agitation can cause tarball or mousse formation, adequate aeration at sea enhances microbial activity on oils; for example, it has been shown that the oxidation of octadecane and naphthalene occurred at a faster rate in oxidised sediments (+500 mV) at sea than reduced sediments (−200 mV) (Delaune *et al.* 1980). The sediment conditions were simulated in the laboratory where the redox potential was carefully controlled. From the ^{14}C-labelled substrates [^{14}C]carbon dioxide was collected and counted to quantify the rate of hydrocarbon breakdown by microbes in varying environmental conditions (Lee & Anderson 1974; Ward & Brock 1978). In this way residence times of oil fractions can be calculated for sea, estuarine and sediment environments; for example, [^{14}C]octadecane had an approximate residence time of 50 d in oxidised sediments and over 200 d in the reduced sediment (Delaune *et al.* 1980). After transition of oil deposits from reduced to oxidised states microbial activity increased rapidly.

The chemical analyses cited in the methods section, including UV absorption (see section 34.8.3) and gas liquid chromatography (GLC) (see section 34.8.5) have been used successfully to monitor the spread of oil slicks at sea and the degradation of their constituents (Law 1978). The accuracy of the UV fluorescence method when monitoring aromatic breakdown depends upon the aromatic standards used. Furthermore the presence of two oils of different compositions may affect the accuracy (Ward & Brock 1978). GLC traces are useful at fingerprinting oil sources and often pristane (C_{17}) and phytane (C_{18}) are resistant to microbial attack, and this is very clearly shown by using GLC analyses of water samples.

It is very difficult to relate laboratory-determined breakdown rates to actual rates at sea, and thus laboratory studies using the techniques described in the methods sections must be used with the chemical monitoring; for example, in the laboratory, known mixed microbial cultures can degrade between 26% and 98% of oils within 30 d at 25°C. Oils from Texas, Louisiana, Oklahoma, Pennsylvania and California lost between 0.78% and 98.8% of their major constituents (Hughes & Beastall 1976). Crude oil is a complex mixture of hydrocarbons, aromatic and heterocyclic compounds which also contain sulphur and nitrogen. Using the techniques described below it is possible to measure activities of pure and mixed populations of microbes and to demonstrate their selectivity in degrading compounds sequentially; for example, there often appears to be a preference for *n*-alkanes over *iso*-alkanes (Hughes & Beastall 1976). Individual species are not capable of degrading all oil constituents, although genetic manipulation and incorporation of a wide variety of gene-controlled catabolic breakdown sequences into a single species may move some way towards it. In nature a complex interplay between species and available substrates is essential for the complete mineralisation of oil constituents. Some microbes can degrade a wide variety of compounds, while others are seemingly only capable of degrading one compound. The natural population of microbes may have adaptive enzymes that are fairly easily induced when environmental pressures change, such as when an oil spill occurs. The size of the oil contamination determines whether the microbes can biodegrade the constituents in a reasonable time and as we have seen this may depend on available oxygen, nutrients and the temperature.

Crude oils from which the light hydrocarbons (short chain) and aromatic fractions, which can be inhibitory to microorganisms, have been removed, constitute weathered oils. These may be degraded by microbes in aqueous environments although microbial attack can also occur in terrestrial environments. The rate and extent of degradation of these modified or weathered fractions depends on the nutritional conditions. The main physical conditions influencing the rate of degradation are temperature and the degree of dispersal of the oil in the water. The problems of mousse formation not only influence the microbial activity but also the way in which microbiologists attempt to enumerate the bacteria. Problems of separating the bacteria from the oil are outlined in section 34.3.

Large masses of oil are degraded mainly on the surface and form hard tarball crusts. Eventually these degrade, leaving recalcitrant ashphaltenes (long-chain hydrocarbons) which amount to about 20 to 40% of the weight of the original oil (Beastall 1977). These residues are almost biologically inert in sediment or soil deposits.

The two most difficult problems encountered when attempting to understand what happens during oil degradation are the estimation of the total number of viable microorganisms and the role of individual microorganisms within that population.

The methods developed for determining total biomass include emulsification and plating with or without homogenisation, as well as filtration (see section 34.3). Specific hydrocarbons can be incorporated into liquid and solid media to determine growth of population groups on those fractions as carbon sources. ATP and adenylate charge are also useful parameters for the measurement of viable biomass (see section 34.7.4). Other methods can include protein and DNA estimations (Callely & Stafford 1977): they can be tedious and often the errors do not justify their use. The main problem with them when compared with the ATP method is that of their conservative properties. Non-viable cells lose ATP quickly due to the presence of ATPase activity, and thus this method is much more reliable for determining the presence of living cells (Stafford & Hughes 1976).

The susceptibility to breakdown for various classes of degradable components is generally assumed to be, in order: straight chain aliphatic compounds, branched chain aliphatic compounds, aromatic heterocyclic compounds and polycyclic compounds. However, many components appear to degrade more slowly when tested alone than when associated with oil (Hughes & McKenzie

1975). Cometabolism plays an important role in the breakdown of specific components, and one active group of microbes degrading one hydrocarbon may require the breakdown product of another hydrocarbon to increase their metabolic activity. It is, therefore, important when monitoring the activity of pure and mixed microbes with pure and mixed substrates (oil components), that good analytical tools are available to measure the microbial catabolic processes. Thus in section 34.8 a wide range of analytical techniques, including gravimetric analysis, UV analysis, oil fractionation, infra-red spectroscopy and gas liquid chromatography are described.

As well as the inorganic nutrients which influence the growth and activity of oil-degrading organisms, organic components can also affect microbial degradation of oil; for example, detergents, carbohydrates, sewage and other organic compounds may give an initial stimulation to biological activity, especially when present in low concentrations. At high concentrations they may cause a variety of inhibitory effects (Callely *et al.* 1977). This effect is marked, especially with biodegradable detergents, which oil-degrading microorganisms prefer as substrates to the oil and to the fatty acid intermediates. Thus detergents may act as negative feedback inhibitors of oil degradation pathways (Cain 1977).

Some attempts have been made to determine respiration rates of microbes exposed to oils and oil fractions in order to determine acclimation times, the amount of oxidation and any possible metabolic inhibitors that may be present in oils. The usefulness of this technique is described in section 34.8.7 as a screening procedure for determining the potential biodegradability of oil components.

Every opportunity should be seized to study the microbial degradation on the occasion of oil spillages although this is rarely done in the haste to attempt to deal with the problem of pollution. Natural populations of microbes in and around an oil slick can then be compared in physiological range and type with those immediately outside the polluted area. As the oil is autoxidised and biologically degraded any changes in microbial activity can be monitored by both chemical analytical techniques and by microbial analysis. Thus a time course in changes in the microbial ecology can be better established with respect to oil spills and a time profile of activity may also be monitored.

Experimental

34.2 Sampling methods

The collection of samples of sea-water or bottom deposits for analysis is one of the most expensive and involved procedures in marine ecology. Sampling away from land involves using boats and often trained crew and if the sampling points are many hours from shore, then an on-board laboratory is required. This, in part, explains why less information on marine microbiology is available than in other fields of microbial ecology.

Much of the methodology in this experimental section, if not referenced, is based on the authors' unpublished observations.

34.2.1 Plankton Sampling

There are three methods of collecting plankton involving nets, pumps and traps. The details of these sampling methods are given in Chapters 3, 6 and 15. Niskin samplers for sea-water have a distinct advantage over other systems in that large sterile bags (up to 5 l) can be used and large volumes are often required for filtration methods for microbial enumeration (see section 34.6.2).

34.2.2 Sunken Oil and Oil Bottom Deposits

Divers are sent to depths up to 30 m and use sterile core samplers for sampling oil and oil-contaminated sea-bed material. The core is kept on ice for transport to the laboratory, where a middle section is removed for either oil analysis or microbial counting.

34.2.3 Oil Sampling on Water Surfaces

Samples of water from immediately below the oil layer are obtained aseptically using a suction method. Water is drawn into a thistle funnel and transferred to a U-tube. The large end of the U-tube is marked so that the funnel can be held just below the surface during sampling and a rubber bung is removed from the funnel when it is in the required position. The U-tube is connected to a 50 Sterilin sampling bottle and suction applied.

34.2.4 Dosing and Sampling of Soils

Fresh crude oil is applied to the surface of soil by sprinkling through a watering can. Normally levels of 1 to 5 l m^{-2} are used. Seven days after the addition of oil three replicate soil cores of 10 cm diameter and 10 cm in depth are taken from the amended areas, and similar cores are taken from the control non-oiled areas. The cores are placed in a deep freeze overnight (−18°C) to solidify and allow the soil cores to be cut horizontally into 2 cm slices.

The prepared material (approximately 50 g wet weight) is extracted with a 90% (v/v) benzene: 10% (v/v) methanol mixture in a soxhlet condenser. The extract is evaporated in a rotary evaporator at 35°C and dried under a stream of nitrogen. The weighed extracts are resuspended in 2 ml heptane (IP standard) and a sample containing not more than 20 mg material is eluted through an activated silica gel column. The eluent is evaporated under nitrogen and taken up into 0.1 ml of heptane for analysis of the carbon preference index by gas liquid chromatography.

Soil cores are taken at various times after the addition of the crude oil and sectioned in the laboratory as described above. The top 3 cm and 6 cm of the soils are analysed. These depths are chosen to ensure that the oil is extracted to the maximum depth of penetration. The residual oil is estimated on a composite of five replicate soil samples, randomly taken from locations at each site by the following procedure. A weighed sample (*c.* 25 g) is extracted with a 90% (w/v) benzene: 10% (v/v) methanol mixture in a soxhlet apparatus for a period of not less than 2 h, or until the eluent from the upper flask is clear. The extract is evaporated to 5 ml in a rotary evaporator at 35°C, transferred to a gravimetric bottle and evaporated to dryness under a stream of nitrogen. The weight of solvent extractable material can be determined.

A sample of the extract is taken up in 1 ml *n*-heptane (IP standard) and eluted through a silica-gel column prior to urea adduction of the normal alkanes.

34.2.5 Changes in Microbial Activity in Response to Oil Addition

Soil bacteriostasis is a phenomenon where the components of a soil tend to limit the growth of microorganisms. This may be examined by determining the effects of control and amended soils on the growth of named organisms, previously isolated from soil systems. Standard dilution plate techniques with 1% (w/v) yeast extract–mineral salt agar can be used to isolate the required test organisms. Bacterial colonies are picked from the plates to give isolates for analysis. The colonies are checked for purity and identified.

The method used to assess bacteriostasis is described as follows. Soil samples are sieved (2 mm mesh size) and 40 g placed in Petri dishes. Discs 10 mm in diameter and 4 mm thick are aseptically cut from 1.5% (w/v) sterile, glass distilled water agar adjusted to pH 7.0. The test discs are transferred to strips of washed, sterilised Whatman No. 1 filter paper placed in close contact with the soil. The control discs are transferred aseptically onto strips of sterile, moistened, pre-washed Whatman No. 1 filter paper in sterile Petri dishes.

The discs are left in position for 18 h at 25°C before inoculating them with 0.01 ml bacterial suspension in distilled water. Standardisation of the inocula is not necessary, provided that control and soil discs are inoculated from the same suspension. All the discs are incubated at 25°C for 48 h after which time five replicates are removed, stained with acidic aniline blue and examined microscopically.

The size of the bacterial colonies on soil and control discs are compared. Bacteriostasis is demonstrated if colony development is less on the discs in contact with the soil, compared with the control discs.

34.3 Problems of microbial enumeration

One of the initial difficulties encountered in the study of the microbial degradation of oil is finding a satisfactory method for estimating microbial populations (see Chapter 6). Conventional counting procedures, such as plate count techniques, are often unsuitable without modification. Crude oil is mainly insoluble in water and microorganisms tend to adhere to the oil droplets and yet it is necessary to obtain a stable suspension of microorganisms in water, free from oil droplets before a plate count can be attempted. This can be achieved by either emulsifying alone or emulsifying and homogenising crude oil cultures. Conventional counting methods can then be used to determine the total number of viable organisms present in the sample.

34.3.1 Emulsification and Homogenisation of Samples Containing Crude Oil

The value of procedures involving emulsification and homogenisation can be demonstrated with the following examples for the enumeration of microorganisms in an oil/sinker mixture obtained from contaminated water samples (D. A. Stafford, unpublished observations).

The use of a homogeniser to break up crude oil cultures, previously emulsified with the detergent Corexit 7661 (Esso) did not increase the number of viable microorganisms obtained by plate counts (Table 1). It is possible that at high homogeniser speeds microbial cell walls were disrupted simultaneously as more organisms were released into suspension. However, this is unlikely at the forces used in the experiment described in Table 1. If there had been an increase in cell numbers after homogenisation, it would have been advantageous to investigate the use of ultrasonic dispersal methods (Williams *et al.* 1970). However, the results suggested that emulsification alone is sufficient for the separation of microbial cells from oil.

There was an increase in the number of oil droplets in a given volume with increasing homogenisation speeds (Table 2). Only a slight difference was observed in the size of the oil droplets during homogenisation although it was difficult to count the number of oil droplets which were less than 2 μm diameter.

Table 1. Homogenisation of crude oil cultures and determination of viable counts.

Speed setting on Polytron for 1 min	Number of organisms ($ml^{-1} \times 10^{-7}$)	Average number of microorganisms ($ml^{-1} \times 10^{-7}$)	Log number of microorganisms (ml^{-1})	Average of duplicate samples (log number of microorganisms ml^{-1})
0	49			
	54			
	49	51	8.71	
	53			
	67			
	66	62	8.79	8.75
3	38			
	39			
	58	45	8.65	
	52			
	61			
	57	57	8.76	8.71
5	54			
	59			
	61	58	8.76	
	60			
	44			
	59	54	8.73	8.75
7	75			
	58			
	58	64	8.81	
	43			
	41			
	61	48	8.68	8.78
9	75			
	79			
	75	76	8.88	8.88

Table 2. Homogenisation of crude oil cultures showing the number of droplets, determined by haemocytometer cell counts, as a function of homogenisation speed.

Speed setting on Polytron for 1 min	Number of oil droplets ($ml^{-1} \times 10^{-3}$) (Average of duplicates)	Log number oil droplets (Log number ml^{-1})
0	1.6	
	1.7 (1.65)	8.22
3	4.0	
	2.7 (3.35)	8.53
5	14.1	
	10.3 (12.20)	9.09
7	16.1	
	18.8 (17.45)	9.24
9	30.2	
	20.0 (25.10)	9.40

34.3.2 Estimation of Total Viable Numbers of Microorganisms in Crude Oil Cultures

The use of 50% (w/v) Corexit in emulsification of crude oil/sinker mixtures inoculated with microorganisms appeared to be successful (Table 3). Oil/sinker mixtures are composed of crude oil and siliconised pulverised fly ash mixed in the ratio of 1 to 2 to produce a compact 'cake'. There was little difference between the total viable counts obtained

Table 3. Estimation of total viable numbers of bacteria in sinker/oil mixtures, using Corexit as an emulsifier.

Sample	Organisms stock culture number	Log number bacteria (ml original inoculum)$^{-1}$
Original inoculum	HC 20	6.53
	HC 25	7.79
	HC 26	7.65
	HC 42	5.30
	HC 46	5.72
	Total	8.05
Original inoculum exposed to 50% (w/v) Corexit for 30 s	HC 20	6.42
	HC 25	7.90
	HC 26	7.36
	HC 42	5.60
	HC 46	5.00
	Total	8.03
Inoculum in oil/sinker mixture exposed to 50% (w/v) Corexit for 30 s	HC 20	5.58
	HC 25	6.38
	HC 26	7.88
	HC 42	5.00
	HC 46	5.00
	Total	7.92
Inoculum in oil/sinker mixture exposed to 50% (w/v) Corexit for 30 min	HC 20	5.80
	HC 25	6.66
	HC 26	8.08
	HC 42	5.00
	HC 46	5.00
	Total	8.20

from the original inoculum (11.1×10^7 organisms ml^{-1}) and the sample exposed to 50% (w/v) Corexit for 30 s (10.6×10^7 organisms ml^{-1}). Inoculated oil/sinker mixtures emulsified with Corexit gave similar results. When the sample was exposed to Corexit for 30 s or 30 min, the total counts obtained were 8.3×10^7 and 16.0×10^7 organisms (ml inoculum)$^{-1}$, respectively. The differences in these counts and those of the original inoculum were within the limits of experimental error. The slightly higher total counts obtained after the mixture was exposed to Corexit for 30 min at room temperature may have been due to growth during this period. The differential plate count (Table 3) showed an increase in the numbers of most of the organisms after this time.

Corexit can also be used successively in the emulsification of floating crude oil cultures, where there is normally an increase of nearly 100-fold in the number of viable microorganisms obtained by plate counts compared with the untreated samples.

34.3.3 Estimation of Hydrocarbon-Degrading Organisms

The bacteria capable of utilising crude oil as the sole carbon and energy source can be isolated from samples, such as estuarine mud, using the following enrichment technique. Approximately 1 g mud is inoculated into 100 ml artificial sea-water (Rila Marine Mix, New Jersey, USA) supplemented with 1% (w/v) ammonium sulphate and 1% (w/v) potassium monohydrogen phosphate. To this is added 0.1 g crude oil at the time of inoculation and a further 0.1 g after 48 h. Three successive subcultures are made after 9 d incubation periods at 25°C and from this enrichment culture, hydrocarbon-oxidising bacteria can be isolated and purified by conventional techniques (see Chapters 1 and 6).

34.4 Problems associated with the use of emulsification agents

The previous sections have described the need to use emulsification agents for counting microorganisms in crude oil. However, emulsification agents, such as Corexit 7661, must first be examined for a number of important properties, such as toxicity towards hydrocarbon-oxidising bacteria and ability to emulsify floating crude oil and oil in oil/sinker mixtures.

34.4.1 Toxicity Tests

Oxford cylinder plate technique

The different types of organisms obtained from crude oil enrichment cultures can be tested for toxicity effects by the Oxford cylinder plate procedure. The organisms are grown in nutrient broth for 3 d at 25°C and seeded plates prepared by mixing 0.5 ml culture with 15 ml nutrient agar in a Petri dish. A range of concentrations of Corexit in distilled water are prepared and the inhibitory effect of these can be tested by adding a small volume (1 to 2 ml) to open-ended metal cylinders placed on the surface of the bacterial seeded agar plates. The emulsifier diffuses into the surrounding agar whilst the plates are incubated for 24 h at 25°C. The size of the zones of growth inhibition are a measure of the toxicity of the Corexit solutions.

Time-dilution technique

Various concentrations of emulsifier such as Corexit at 0.1%, 1.0% and 10% (v/v), may be prepared in suitable growth media and tested for toxicity against bacteria isolated from crude oil enrichment culture. The procedure involves incubating the bacteria in the presence of the different emulsifier concentrations for periods up to 1 h at room temperature and then plating a loopful out onto a suitable growth medium, such as nutrient agar.

34.4.2 Emulsification Capability

The volume of Corexit required to completely emulsify crude oil may be determined by mixing artificial sea-water samples containing 1% (v/v) crude oil (unweathered) with different proportions of the detergent Corexit. The samples are agitated for 30 s and left to stand for 10 min. Similarly, different volumes of 50% (w/v) Corexit in nutrient broth are added to known weights of oil/sinker mixtures. Each sample is agitated for 30 s and left to stand for 10 min. The degree of emulsification is recorded.

Samples of the oil/sinker mixture (5 g) are inoculated with 1 ml of the culture by thoroughly mixing with a spatula. From this, samples are taken and Corexit is added in the proportion 1 g sample: 20 ml 50% (w/v) Corexit in nutrient broth. Emulsification is achieved by stirring on a Rotamix for 30 s and the samples are left for a specified time. Tenfold dilutions are prepared in nutrient broth and the number of viable bacteria in each sample counted on nutrient agar by the conventional plate count technique. The bacterial population in the original inoculum is also counted in the same way, without the addition of Corexit. The plates are incubated at 25°C for 5 d. The number of viable bacteria are then determined by a differential plate count.

Enriched artificial sea-water containing 1% (v/v) crude oil is inoculated with a drop of culture from a crude oil enrichment flask. This is incubated at 30°C for 3 d on an orbital shaker. After this time the oil is partially degraded and dispersed as small droplets throughout the medium. A 10 ml sample of this culture is used for the viable count determinations. It is mixed on a Rotamix for 30 s and two 0.5 ml samples are taken and viable counts made by the plating procedure. To the remaining 9 ml are added 1 ml Corexit. This is mixed for 30 s and dilution plated on nutrient agar. The plates are incubated at 30°C for 4 d and the number of viable organisms determined.

(1) *Crude oil cultures.* Enriched artificial sea-water (150 ml) containing 2% (v/v) crude oil in a 250 ml flask is inoculated with 1 ml from a crude oil enrichment culture. This is incubated at 30°C for 5 d on an orbital shaker. There is partial breakdown of the oil after this time. After incubation, 6 ml of Corexit is added to give a final concentration of 4% (v/v) in the medium. The contents of the flask are shaken vigorously until a homogenous suspension is obtained. From this mixture 5 ml samples are taken for microbial counts.

(2) *Homogenising of samples.* Each sample is mixed for 1 min either on a Rotamix or on a homogeniser (e.g. Polyton, Model PT2, Northern Media Supply Ltd., Hull). The samples are kept on ice during the homogenising procedure. After each sample has been treated, the probe on the homogeniser is washed thoroughly: 2 × 10 ml sterile distilled water, 1 × 10 ml 70% (v/v) ethanol, 3 × 10 ml sterile distilled water for 1 min each. The last wash is tested for contamination by spreading 1 ml on a nutrient agar plate.

(3) *Plate counts.* The number of viable microorganisms in each sample is estimated by preparing tenfold dilutions in nutrient broth (Oxoid) and plating out 0.1 ml, in triplicate, from each dilution. The plates are incubated at 30°C for 3 d.

(4) *Examination of oil globules.* After homogenising, the oil droplets in each sample are examined in two ways:

(i) the number of oil globules per ml is estimated using a haemocytometer slide;

(ii) the size of 100 oil globules from each sample is measured microscopically using a double image eyepiece.

34.5 Biodegradation of floating crude oil and crude oil sinker mixtures and the effects of aeration, temperature and nutrients on degradation

34.5.1 Inoculum Preparation

For all studies the inocula are prepared as follows. Bacterial isolates (or a naturally occurring mixed population) capable of utilising crude oil as sole source of carbon and energy are grown. Each organism (or mixture) is grown at 30°C for 5 d in artificial sea-water supplemented with 0.1% (w/v) $(NH_4)_2HPO_4$ and 0.1% (v/v) crude oil. The organisms are harvested in a centrifuge at 10 000 ***g*** washed and resuspended in sea-water. An inoculum containing a mixture of a number of organisms is prepared by mixing equal volumes of the suspensions of the same optical density.

34.5.2 Estimation of Available Phosphorus and Nitrogen in Sea-water

Available phosphorus and nitrogen are estimated by the method of Murphy and Riley (1962) as described by Strickland and Parsons (1968). Before analysis, samples are filtered through Whatman No. 1 paper to remove oil. Estimations are made using the method of Solozano (1969).

34.5.3 Preparation of Oil/Sinker Mixtures

Sterilised crude oils, are mixed with pulverised siliconised fly ash (PFA) in the ratio of 1 part oil to 2 parts PFA. One ml of the inoculum and 10 g of oil/sinker mixture are combined and 1 g amounts placed in an open glass Petri dish. This is lowered to the bottom of a 3 l beaker containing 2 l sea-water enriched with 0.1% (w/v) $(NH_4)_2SO_4$ and 0.1% (w/v) K_2HPO_4 and is incubated at 30°C with aeration. Sterile controls are also prepared. Air can be replaced by nitrogen to simulate anaerobic conditions. Periodically, 1 g samples are removed and the number of viable bacteria present determined. The oil is extracted and analysed by gas liquid chromatography.

Similar systems are prepared with and without the addition of enrichments and, from these, samples are taken periodically for ammonia and phosphorus determinations.

34.5.4 Low-Temperature Studies Using Floating Oil

Two series of 500 ml flasks are prepared containing 200 ml artificial sea-water and 0.1% (v/v) crude oil supplemented with approximately 0.1% (w/v) $(NH_4)_2HPO_4$. To one series is added 1 ml natural mixed microbial population obtained from crude oil enrichment medium at 8°C. These are incubated at 8°C without aeration. To the other series is added 1 ml of natural mixed microbial population obtained from crude oil enrichment medium incubated at 25°C. This series is incubated at 25°C without aeration. Sterile controls are also prepared. One flask at each temperature is used for microbial plate counts. The plates are incubated at either 8° or 25°C. At intervals, one flask from each series is taken and the media analysed for ammonia content; phosphorus content; and alkane degradation. Experiments using these methods have shown good microbial growth and oil breakdown (Byrom & Beastall 1971).

34.6 Measurement of the influence of dispersants on oil biodegradation

34.6.1 Effect of Dispersant on the Microbial Degradation of Crude Oil at Sea

One method used widely in Great Britain for treating oil at sea is emulsification with chemical dispersants. A relatively non-toxic, non-ionic dispersant, BP 1100X, has been tested for its biodegradability, and its effect on the rate of biodegradation of crude oil in the sea (D. A. Stafford, unpublished observations).

Flasks containing 200 ml artificial sea-water enriched with 0.1% (w/v) $(NH_4)_2HPO_4$ and 1% (v/v) BP 1100X are inoculated with 1 ml BP 1100X enrichment culture and incubated at either 8° or 25°C with aeration for 2 weeks. Samples are taken at regular intervals and tested for the presence of dispersant.

The cobaltothiocyanate colorimetric procedure of Greff *et al.* (1965) is used for the determination of the concentration of non-ionic dispersant. This method is based on the formation of a blue complex between ammonium cobaltothiocyanate reagent with an ethoxylated compound.

(1) *Inoculum.* The original inoculum for enrichment cultures is obtained from oil-contaminated muds. Enrichment cultures are prepared by 3 successive subcultures at weekly intervals from enriched sea-water medium containing either 1% (v/v) Kuwait crude oil, 1% (v/v) BP 1100X, or 1% (v/v) of both oil and BP 1100X. All cultures are grown at 25°C with aeration.

(2) *Incubation conditions.* Flasks are prepared containing 200 ml artificial sea-water enriched with 0.1% (w/v) $(NH_4)_2HPO_4$ with added carbon source of either 1% (v/v) Kuwait crude oil or an equal mixture of 1% (v/v) oil and 1% (v/v) BP 1100X. The enrichment cultures are used separately as inocula. The inoculated flasks are incubated at 25°C in an orbital shaker.

(3) *Biodegradation studies.* The rate of oil breakdown was followed by a combination of visual assessment of oil release and dispersal, and by GLC analysis. Using these techniques it was shown that there is a rapid reduction in the concentration of BP 1100X during microbial growth, although breakdown was slower at 8°C than at 25°C. After two weeks at either temperature, no dispersant can usually be detected (Table 4). It would appear that the surface active agent portion of BP 1100X is readily biodegradable.

Table 4. Rate of biodegradation of 1% (v/v) BP 1100X in an enriched sea-water medium.

Incubation time (d)	Concentration of BP 1100X	
	8°C	25°C
0	1.00	1.00
2	0.66	0.54
7	0.04	0.01
14	0.00	0.00

The comparative rates of oil breakdown at 25°C with different enrichment culture inocula are summarised in Table 5.

As expected, the oil alone was degraded most rapidly when inoculated with an oil enrichment

Table 5. Microbial degradation of Kuwait crude oil dispersed with BP 1100X.

Sea-water medium supplemented with 1% (v/v) of each component	Enrichment culture inoculum grown on	Comparative rates of oil biodegradation at 25°C
Oil	Oil	Very fast
Oil	Oil + dispersant	Fast
Oil + dispersant	Oil + dispersant	Fast
Oil + dispersant	Oil	Slow
Oil	Dispersant	Slow
Oil + dispersant	Dispersant	Very slow

culture. However, the emulsified oil was not broken down rapidly by the same inoculum. Both oil and dispersant enrichment were required for rapid breakdown of emulsified oil. Thus it would appear that enrichment of both microbial populations was necessary before the dispersed oil droplets were attacked.

The results of GLC analysis (Figs 1 and 2) show the biodegradation pattern of the heptane-soluble portion of the Kuwait crude and the dispersant. The breakdown of the kerosene-like solvent rather than the surface active agent portion of BP 1100X is shown by this method. The alkane components of the emulsified crude oil were degraded rapidly in the presence of dispersant and oil enrichment inoculum. However, there appeared to be little difference in the chromatograms obtained for emulsified oil after four weeks' incubation with only a dispersant inoculum (Fig. 1). After the same period with both inocula there was almost complete breakdown of the alkanes (Fig. 2).

34.7 Methods of assessing microbial activity

34.7.1 Biological Oxygen Demand

The dissolved oxygen content of the sample is determined before and after incubation for five

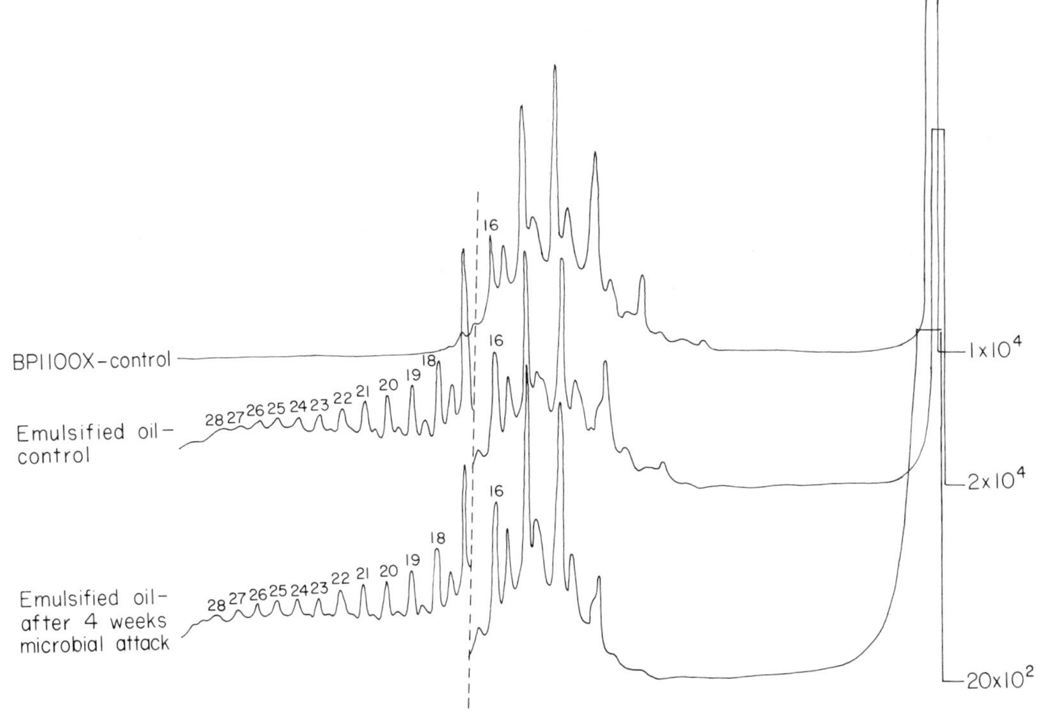

Figure 1. Degradation of crude oil, emulsified with BP1100X, by an oil enrichment culture. Gas liquid chromatogram (carbon chain length denoted by small numbers).

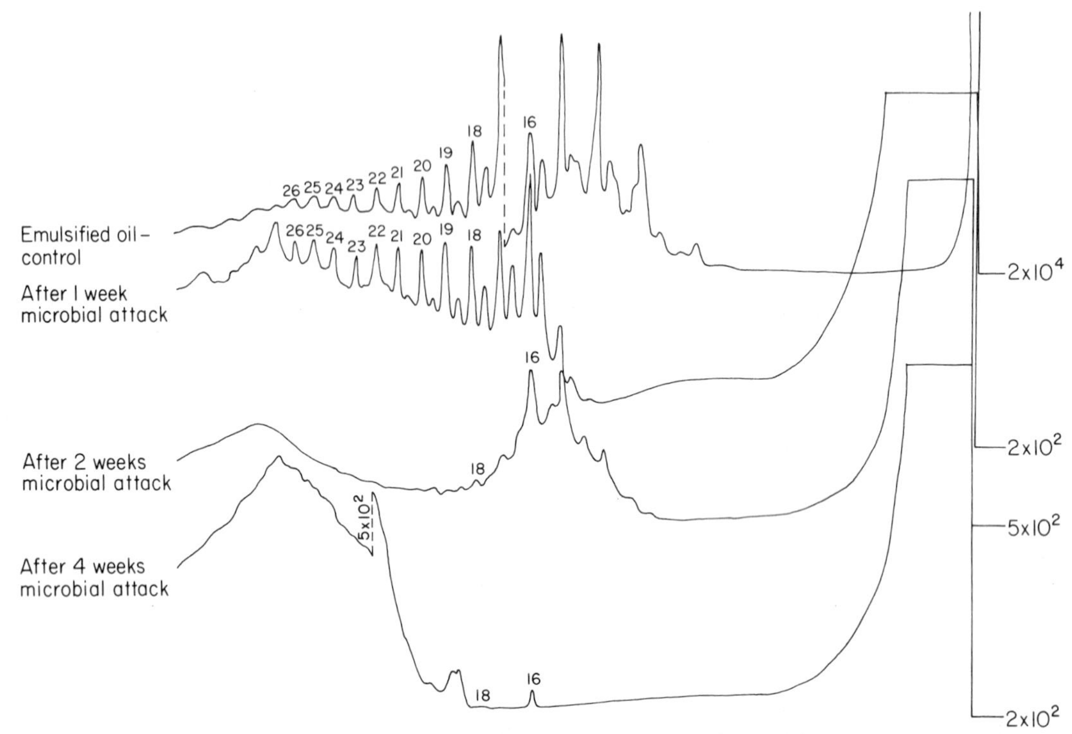

Figure 2. Degradation of crude oil, emulsified with BP1100X, by oil-dispersant enrichment culture. Gas liquid chromatogram (carbon chain length denoted by small numbers).

days at 20°C. This is known as the biological oxygen demand (BOD_5) of a sample and is usually defined as the amount of oxygen required by microorganisms over a period of 5 d while oxidising decomposable organic matter under aerobic conditions.

For each sample to be determined, 1 litre of dilution water is prepared. To this 1 ml each of ferric chloride solution, calcium chloride solution, magnesium sulphate solution and phosphate buffer solution is added. The prepared dilution water is saturated with oxygen by bubbling air through it for a minimum of two hours. The water is used immediately.

The method follows that described in the Department of Environment Handbook (1972). The BOD_5, expressed in mg oxygen l^{-1} of sample is calculated as:

$$BOD_5 = \frac{\text{Difference of filtration in sample}}{\text{Difference of filtration in blank}}.$$

The BOD test is widely used to determine the polluting strength of domestic and industrial wastes in terms of the oxygen that they will require if discharged into natural water courses in which aerobic conditions exist. The test is one of the most important in the assessment and control of water (see Chapter 33).

In the case of domestic and many industrial waste waters, it has been found that the 5-day BOD value is about 70 to 80% of the total BOD which is a large enough percentage of the total to be considered representative.

As has already been mentioned, BOD is defined as the amount of oxygen required by microorganisms to biodegrade organic materials in a sample of waste water. There are many factors which affect biological growth such as temperature, availability of nutrients, oxygen supply, pH and the presence of toxins.

Ruchhoff (1950) summarised the factors which may be expected to affect the extent of substrate oxidation as follows:

(1) the number and complexity of the microflora and microfauna;

(2) the concentration and complexity of the organic constituents in the substrate;

(3) the presence of the necessary mineral nutrients;

(4) the pH and buffering capacity of the substrate;
(5) the presence of toxic or enzyme inhibitory compounds;
(6) the temperature;
(7) the presence of dissolved oxygen or other hydrogen acceptors.

34.7.2 Hydrocarbon-Oxidising Bacteria in Crude Oil, Marine Sediment or Sea-water Samples

The purpose of this section is to discuss methods for determining the percentage of hydrocarbon-oxidising bacteria in samples of crude oil, crude oil/sinker mixtures, sea-water or marine sediments. Four different methods are assessed.

Millipore filter method

The method described by Goetz and Tsuneishi (1959) for enumerating coliform bacteria in water has been modified (Department of Microbiology, University College, Cardiff) for enumerating hydrocarbon oxidisers. Samples are filtered through a 0.45 mm Millipore membrane filter. Oil and oil/sinker mixtures are emulsified with Corexit and diluted before filtration. Sediment samples are also diluted. Microbes in water samples were concentrated by filtering large volumes (500 ml or 1 l). Microorganisms collected on the Millipore filter are then placed on an absorbent disc, previously saturated with enriched artificial sea-water containing 1% (v/v) hexadecane. Controls are also prepared with no hexadecane in the medium. Filters are incubated, in a Petri dish sealed with Parafilm, at 25°C for at least a week. Hydrocarbon-oxidising microorganisms should produce visible, discrete colonies on the membrane filter.

Soaked filter paper technique

A modification of the method described by Jones and Edington (1968) is used. Special agar-Noble (Difco) plates are prepared containing 0.1% (w/v) $(NH_4)_2HPO_4$ with and without the addition of 0.01% (w/v) yeast extract (Oxoid). Estuarine water samples can be diluted with 50% (v/v) artificial sea-water and aliquots spread on the plates, by standard procedures. A sterile filter paper soaked in either dodecane or hexadecane is placed in the lid of each inverted Petri dish. Controls, containing no hydrocarbon are also prepared. The plates are sealed with Parafilm and incubated at 25°C for two weeks. Plate counts are simultaneously made on both nutrient agar (Oxoid) and Zobells agar in order to compare total viable counts with the number of hydrocarbon oxidisers.

Biodegradation of the dispersant, therefore, is rapid under certain conditions. However, dispersed crude oil is only degraded rapidly in the presence of organisms adapted to both growth on the oil and dispersant. It would appear that the dispersant must be degraded before the oil is made available for microbial attack. It is, therefore, possible that unless the oil is dispersed over a large area that the free oil droplets may coalesce before they are degraded.

Liquid hydrocarbon agar media

The method used is similar to that described by Baruah *et al.* (1967). Bacteriological counts are made using this method on the same water samples used in the soaked filter paper technique. Activated silica gel is prepared by drying 1 g amounts silica gel SG 41 thin layer Chromedia, particle size 5 to 10 μm at 160°C overnight. Samples are then stored in a desiccator until required. Dodecane and hexadecane are adsorbed onto separate samples of activated silica gel by mixing, aseptically, 0.7 ml sterile hydrocarbon (sterilised through a Millipore Swinny filter adaptor 0.45 μm) with 1 g silica and 5 ml diethyl ether in a Petri dish. After evaporation of the ether, each silica gel sample is divided equally between 12 Petri dishes. Cooled molten Noble agar (15 ml) enriched with 0.1% (w/v) $(NH_4)_2HPO_4$, with or without the addition of 0.01% (w/v) yeast extract, is added to each Petri dish and mixed thoroughly to distribute the silica gel. The final concentration of hydrocarbon in the agar is approximately 2.5 g l^{-1}. Spread dilution plates, using 50% (v/v) artificial sea-water for the dilutions, are prepared and counts made after two weeks incubation at 25°C. These results are compared with those of the soaked filter paper technique.

Most-probable-number (MPN) technique

The proportion of hydrocarbon-oxidising bacteria in estuarine samples is estimated by the MPN technique as described by Gunkel (1967). Five tubes of each tenfold dilution of the samples are prepared in artificial sea-water enriched with 0.1% (w/v) $(NH_4)_2HPO_4$. To each tube is added crude oil to give a final concentration of 0.1% (v/v). Similar series of dilution tubes are prepared containing either 0.1% (v/v) dodecane or hexadecane. Series of nutrient broth dilution tubes are used as controls. All tubes are incubated at 25°C and results read after two, three and four weeks. Before a final reading is taken, after four weeks, a drop of concentrated HCl is added to each dilution to remove any precipitate which may obscure results. Tubes in which there is definite turbidity are recorded as positive. The number of hydrocarbon oxidisers in the original samples are calculated using most-probable-number tables.

34.7.3 General Conclusions on Microbial Activity Assessment

Although more colonies tend to develop on membranes with hexadecane in the medium than on the controls without hexadecane, this method for assessing the proportion of hydrocarbon oxidisers in a sample has certain disadvantages.

Oil samples, even when emulsified, leave droplets of oil on the membranes which obscure visible growth. Similarly, during filtration of mud samples, particles collect on the membranes which makes enumeration of colonies difficult and inaccurate. The number of hydrocarbon-oxidising bacteria in sea-water is relatively low and thus large volumes are required to obtain significant results. Again, particles in the sea-water concentrate on the membrane, together with the microorganisms, and obscure the results.

Organisms, other than hydrocarbon oxidisers are capable of growing on Noble agar; agar itself can be used by the organisms as a carbon source. The high proportion of agar digesters growing on Noble agar plates compared with non-selective agar plates is indicative of this, and digestion of the agar makes it difficult to count the number of colonies accurately. Difficulty is also encountered in counting colonies on agar containing silica gel. The silica gel is dispersed throughout the medium in the form of white flakes which are not always easily distinguished from the colonies.

There is often little difference in the counts obtained on nutrient agar and Noble agar with or without hydrocarbons in either the agar or the atmosphere above the agar. Media supplemented with 0.01% (w/v) yeast extract support a slightly greater population than media without yeast extract. The low concentration of yeast extract is there to provide a supply of growth factors but it also contains sources of carbon which may be used for microbial growth. Lower counts are obtained on Zobell's agar (a media for marine microbes) than on nutrient agar. A possible explanation is that the microorganisms found in estuarine water may not all be tolerant of high salt concentrations. Neither method is accurate for the estimation of hydrocarbon oxidisers in a population where Noble agar is used as the solidifying agent. It is probable that either washed Noble agar or washed Ion agar No. 2 (Oxoid) may be more suitable (Jones & Edington 1968).

There is a significant difference between counts obtained in water samples diluted in sea-water with both added hydrocarbon and diluted in nutrient broth. Counts in nutrient broth tubes were always higher. There is often growth of the organisms in the sea-water and hydrocarbon dilution tubes after two weeks but there is usually no increase in the number of positive tubes after three weeks incubation.

Dilutions with added 0.1% (v/v) hexadecane give higher counts than either 0.1% (v/v) dodecane or 0.1% (v/v) crude oil dilutions. It is probable that the most volatile fractions and water-soluble components in the crude oil may be toxic to some of the organisms, for example, it would appear that a greater number of organisms are capable of oxidising hexadecane than dodecane.

Results obtained using the MPN technique give quantitative results which can be used in a comparison of the proportion of hydrocarbon oxidisers in different samples. Hexadecane would seem to be the most appropriate hydrocarbon to use as a carbon source as this supports growth of a greater number of organisms than other hydrocarbons tested.

The incubation temperature of 25°C is chosen to encourage growth of estuarine microorganisms which were not obligate psychrophiles. However, in an assessment of hydrocarbon-oxidising marine psychrophiles, lower temperatures should be used. Gunkel (1967) suggests that maximum numbers can be obtained at 18°C.

34.7.4 Adenylate Charge Studies

A small number of determinations of adenylate

charge can be made on control and test soils and sea-waters, during the period of oil degradation. Adenylate charge is the mathematical ratio of adenosine triphosphate (ATP), adenosine diphosphate (ADP) and adenosine monophosphate (AMP). Depending on the expression of the ratio, various numerical scales can be derived to indicate the energy state of the microbial system; for example (Atkinson & Walton 1967):

$$\frac{\mathrm{ATP}+\frac{1}{2}\mathrm{ADP}}{\mathrm{ATP}+\mathrm{ADP}+\mathrm{AMP}}.$$

The ratio of high energy phosphate to the metabolic pool phosphates, allows for a scale running from 0 to 1 where 0 is equivalent to cell death, and 1 is equivalent to maximal activity of the biota within the system.

All the adenosine phosphates can be linked enzymically to the luciferin/luciferase complex (firefly lantern extract) resulting in light emission which can be measured with a suitable photomultiplier, amplified and displayed on a chart recorder. Using such a system it is possible to show adenylate charges. The repeatability of the assay is, however, suspect owing to the difficulty in obtaining reproducible recoveries of the adenosine phosphates from soils or infected waters even on addition of excess ATP. The results achieved so far, however, indicate that the biota of samples tested show a higher activity after oil addition.

34.8 Measurement of rates and extent of degradation of hydrocarbon and oils

34.8.1 Oil Content Determination (Mousse)

The mousse is processed following a method developed in the Microbiology Department, University College, Cardiff. From the analytical point of view, however, there is one chief disadvantage of this method in that the oil and water content of a 'chocolate mousse' emulsion is variable, and certainly heterogeneous in respect of several tonnes of material. It is, therefore, important to determine a mean value for the oil and water content of the material in order to relate 'mousse' by weight to 'water-free oil' by weight. It is anticipated, however, that the variable water content of 'chocolate mousse' would limit the accuracy involved in sampling.

Approximately 10 g of mousse is placed in a pre-weighed 50 ml glass round-bottomed flask (Quickfit) and re-weighed to 5 significant figures. The material is transferred to a separating funnel in stages, by dissolving the oil emulsion in no less than 150 ml carbon tetrachloride. A further 50 ml of water is added to improve separation of the two phases. After shaking for two minutes and consequent partition of the aqueous and non-aqueous phases, the solvent is removed by decanting the solution in stages into a pre-weighed 50 ml round-bottomed flask (Quickfit) supported by an electrothermal heater. The flask neck forms a seal with a cone at one end of a water-cooled condenser.

The heater is disconnected when approximately half the solution remaining is oil. Initially the oil samples are dried with a nitrogen stream in a water bath at 80°C. There is no appreciable loss in weight when samples are dried in an oven at 80°C for 6 h and placed in a desiccator for a further 2 h. The oil content of mousse is given by:

$$\frac{\text{weight of oil}}{\text{weight of mousse}}\ \mathrm{g\,l^{-1}}\ \text{(at room temperature)}.$$

The procedure can be repeated with further samples, and the mean value calculated. An example of oil weight loss during breakdown is given in Fig. 3.

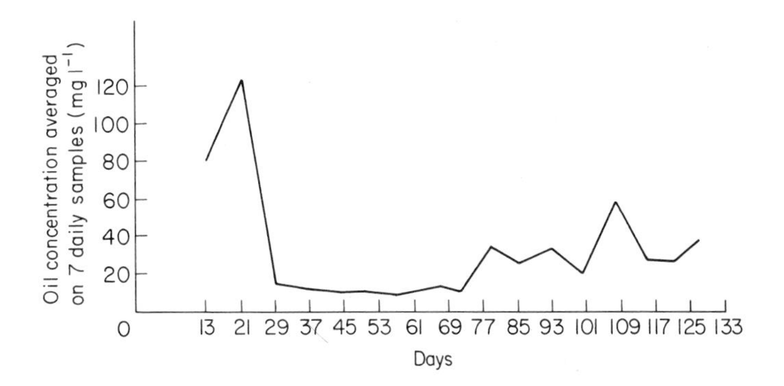

Figure 3. Oil concentration in mousse. Gravimetric (mass weight) determinations were carried out on a daily basis. 500 ml samples of the mid-tank liquor were withdrawn by siphon. A 50 ml aliquot was extracted into 150 ml carbon tetrachloride in a separating funnel and shaken. After separation of the two phases the non-aqueous fraction was collected. The solvent was removed by distillation. The oil was dried and weighed. The weight of the oil in mg l^{-1} was given by: (weight of oil in 50 ml) × 20.

34.8.2 The Calibration of Oil Extracted from Mousse

Approximately 10 g mousse is purified by the extraction of oil from water and suspended solids as described in section 34.8.1.

A stock solution of 1×10^4 mg l^{-1} of oil is prepared using volumetric glassware. Subsequent serial dilutions are prepared to give the samples required for the construction of a calibration curve by ultraviolet absorption as described in section 34.8.3. (Fig. 4.)

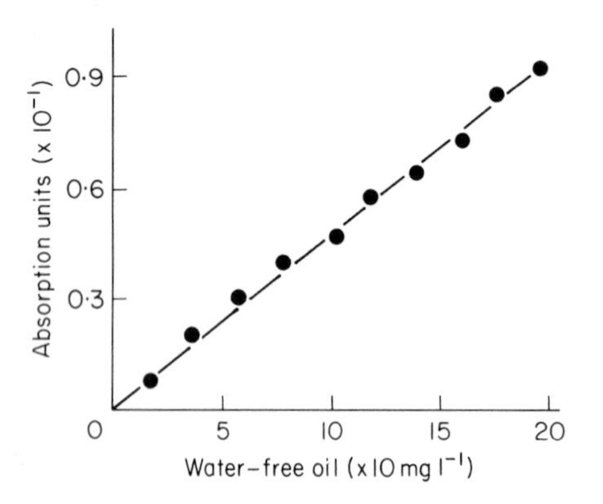

Figure 4. A calibration curve of oil extracted from BP Christos-Bitas mousse using carbon tetrachloride. Standard oil solutions from 200 to 2000 mg l^{-1} were prepared and their absorbence values measured at 263 nm.

34.8.3 Comparison of Ultraviolet Absorption and Gravimetric Determinations for the Measurement of Oil Residues

For each determination a 50 ml aliquot is withdrawn, extracted into carbon tetrachloride and dried as described in section 34.8.1.

The weight of oil in 8 l is given by:

$$\text{Weight of oil (g) in 50 ml} \times \frac{8000}{50}.$$

The concentration of oil is given by:

$$\text{Weight of oil (g) in 50 ml} \times 2 \times 10^4 \text{ mg l}^{-1}.$$

Ultraviolet absorption

For each determination a 50 ml aliquot is withdrawn and extracted into 150 ml carbon tetrachloride as described in section 34.8.2. The solution is read against a carbon tetrachloride blank using a pair of quartz cuvettes of 1 cm path length in a recording spectrophotometer. An absorption maxima is obtained at $UV_{(263)}$. Consequently all readings are taken at this wavelength. The comparative figures are shown in Fig. 5.

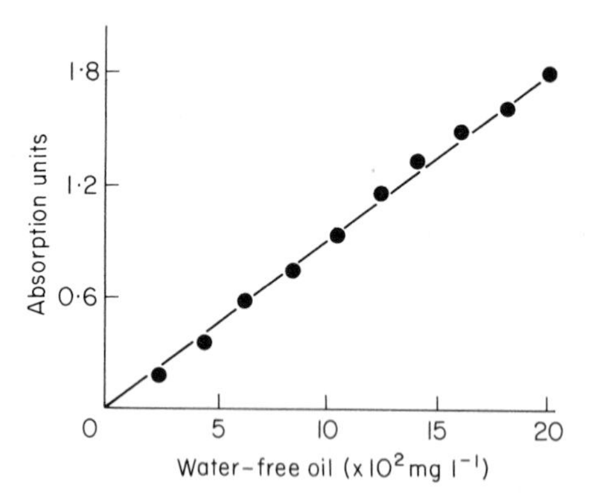

Figure 5. A calibration curve of oil extracted from BP Christos-Bitas mousse using carbon tetrachloride. Standard oil solutions from 20 to 200 mg l^{-1} were prepared and their absorbence values measured at 263 nm.

In comparing Gravimetric (Mass weight) determinations with UV absorption a mean discrepancy of 1.73 is found between these two methods of measurements. In other words the oil concentrations determined by $UV_{(263)}$ were an average 1.73 times higher than gravimetric measurements.

These findings could be accounted for by evaporation of the carbon tetrachloride solvent in routine samples relative to the calibration standards. Such a situation may have occurred whenever there is a delay in reading the samples.

Since UV absorption at 263 nm is due principally to conjugated double bonds, then any decrease in oil fractions rich in the *pi* bonds would account for the additional fall in oil concentration. Later work with column chromatography showed approximately 50% reduction in aromatics measured in terminal residues of mousse degraded bacteriologically (S. Stafford, unpublished observations).

34.8.4 Oil Fractionation

The Institute of Petroleum's Standard Procedure (IP 143/77) is used with slight modifications to the solvent type and volumes used.

The oil residues remaining at the end of degradation experiments can be examined by infra-red spectroscopy.

Infra-red spectroscopy

The oil sample or fraction which is stored in carbon tetrachloride is dried in a ventilated cabinet at 52°C to give a 1:1 ratio of oil residue and solvent. Using a Pasteur pipette a smear of the material is made on a pair of potassium bromide cells measuring 2.5 cm in diameter by 0.5 cm in thickness. The smeared surfaces are separated by a neoprene spacer of 0.05 cm thickness and the cells are pressed together gently but firmly to expel any air in the cell chamber formed (Price 1973). Infra-red scans are run from 400 to 4000 cm^{-1} on each sample.

In contrast to the relatively few absorption peaks observed in the ultraviolet region for most organic compounds, the infra-red spectrum provides a rich array of absorption bands due to electronic excitation, vibrational change and rotational change (Kawahara 1972). Many of the absorption bands cannot be assigned accurately, but in some cases it is possible to specify the molecular configurations present; for example, Kawahara *et al.* (1974) have made extensive use of the discreet wavelengths: 3050 cm^{-1} (aromatic CH absorption); 2925 cm^{-1} (methylene asymmetric stretch vibration); 1600 cm^{-1} (aromatic absorption); 1375 cm^{-1} (C–

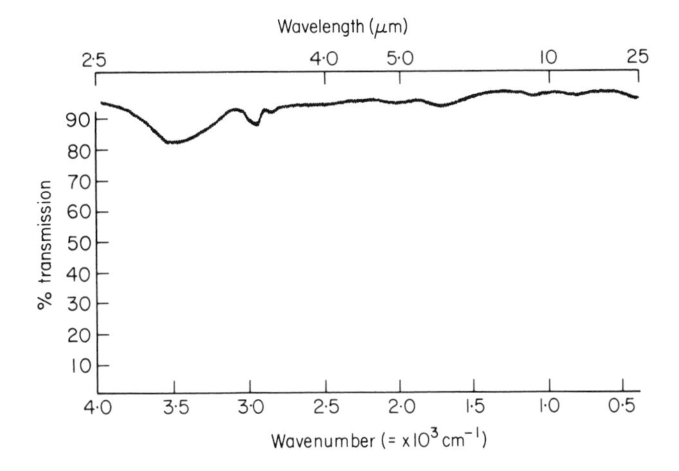

Figure 6. Infra-red spectroscopy. The sample was sedimented residue from the floor of the digester. At the end of the experiment the aeration pump was turned off. After 24 h approximately 5 l of slurry was dredged and allowed to settle. The solids (200 ml) were extracted into 800 ml carbon tetrachloride. The oil residue was separated by distillation and dried in a ventilated cabinet at 52°C to give an approximate 1:1 ratio of oil residue to solvent. The material was smeared on a pair of potassium bromide cells. Residual solvent was removed by further heating at 52°C for 2 min. The cells were pressed together, assembled and placed in the sample holder. An infra-red scan was run from 400 to 4000 cm^{-1}.

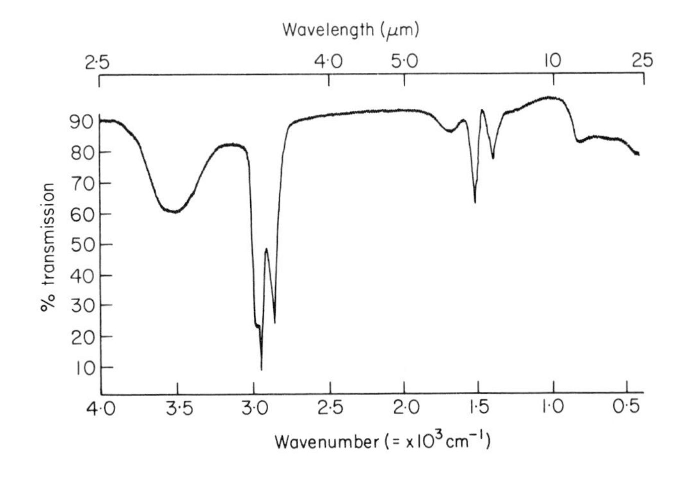

Figure 7. Infra-red spectroscopy. The oil from 5 g Christos-Bitas mousse was extracted into 50 ml carbon tetrachloride and dried in a ventilated cabinet at 52°C to give an approximate 1:1 ratio of oil residue to solvent. The material was smeared on a pair of potassium bromide cells. Residual solvent was removed by further heating at 52°C for 2 min. The cells were pressed together, assembled and placed in the sample holder. An infra-red scan was run from 400 to 4000 cm^{-1}.

CH_3 branching); 810 cm^{-1} (aromatic absorption); and 720 cm^{-1} (chain methylene group) in the fingerprinting of oil and asphalt residues (see Figs 6 and 7).

While there is no doubt that the comparison of oil spectra offers a reliable tool in distinguishing between fractions of a widely different origin available in a concentrated form, the effect of using diluted oil samples has been to reduce the absorption peak heights and to broaden the peak widths such that the spot frequency corresponding to a given absorption maxima is uncertain. These observations have led to disagreement in the literature as to which discreet wavelengths should be used to identify oils. Kawahara *et al.* (1974), Mattson (1977a, b), Lynch and Brown (1973) and Spencer (1975) all specify different spot frequencies for fingerprinting oil residues. In an attempt to improve the reliability of identifying petroleum pollutants, use has been made of linear discriminant function analysis where the number of replicated samples becomes important in the statistical representation of the results. Combination methods have been used where oil samples are examined by gas chromatography as well (Kawahara 1972).

It is concluded that IR spectroscopy is a valuable tool in confirming the presence or absence of specific carbon groups, for instance aromatics, in degraded oil residues. The literature suggested, however, that the derived data should be interpreted only in conjunction with other analytical methods.

34.8.5 Gas Liquid Chromatography

Conventional GLC of oil samples is generally carried out on a non-polar liquid phase such as SE-30 Silicone or Silicone Gum Rubber OV1 or OV17 which effects separation of peaks in approximate order of boiling points. The chromatogram consists usually of an unresolved hump, which varies depending on the nature of the oil, and a series of well resolved peaks due primarily to *n*-alkanes and isoprenes. The unresolved hump contains the aromatic cycloaliphatic and branched aliphatic hydrocarbons. To effect a better resolution, open tubular capillary columns can be employed. Using these columns under optimum conditions it is possible to resolve most of the hump so prevalent in packed columns chromatography (R. H. Clemett, unpublished observations).

Samples of the oil residues remaining at the end of the experiment were analysed by GLC. These samples comprised:

(1) Sample A, before contact with hydrocarbon oxidising bacteria;

(2) Sample B, homogeneous oil in liquor suspension collected from the middle of an oil-enriched tank;

(3) Sample C, surface mousse remaining;

(4) Sample D, sedimented residue remaining from the floor of the tank.

Each sample is extracted with sufficient *n*-pentane, to provide a 10% (w/v) blend. This precipitates any asphaltenes present which would otherwise block the column. The extract is then filtered through Whatman No. 42 filter paper to remove the suspended asphalts, and passed through a column of florisil to remove any inorganic material present. After evaporation to 2 to 5 ml the extract is injected onto a GLC apparatus employing a capillary column and using a flame ionisation detector (Fig. 8).

The normal alkane peaks are identified by spiking the sample with a trace amount of a standard normal alkane. As the samples are in the kerosene boiling range and above, the calculations of the normal alkane/isoprenoid hydrocarbon ratios for nC_{17}/pristane and nC_{18}/phytane are effected. This makes the comparison of carbon peaks on different chromatograms possible. It should be noted that an alternative method for identifying the normal alkane peaks would have been to chromatograph a synthetic mixture of *n*-alkanes of known composition under identical conditions in the comparison exercise.

Urea adduction technique

The urea adduction technique involves the addition of a saturated solution of urea in methanol (0.5 ml) to a solution of sample (5 mg) in *n*-heptane (4 ml) and acetone (2 ml) in a centrifuge tube. The solvent is evaporated in a stream of nitrogen, and the crystals washed (3×5 ml of *n*-heptane, centrifuging when necessary). The crystals are dissolved in distilled water which is pre-extracted with *n*-heptane and brought to pH 11.0 with sodium hydroxide, and the *n*-alkane recovered by extracting with *n*-heptane (3×5 ml). The extracts are washed with distilled water and dried with anhydrous sodium sulphate before evaporating to 0.1 ml for GLC analysis.

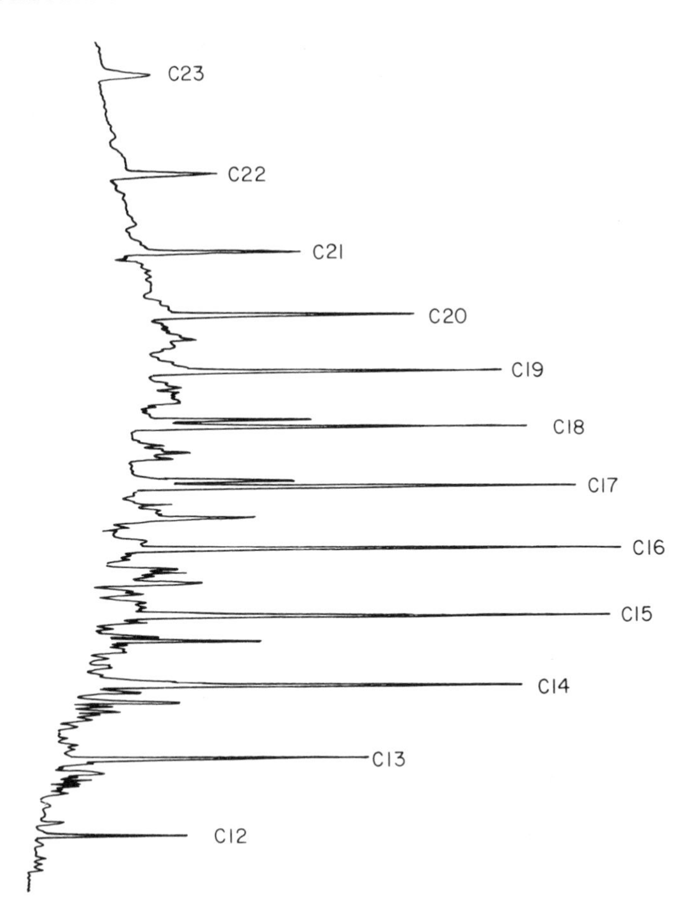

Figure 8. Qualitative gas liquid chromatography. Christos-Bitas mousse (4 g) was extracted into 100 ml *n*-pentane, filtered and passed through a column of florisil. After evaporation to 2 to 5 ml the extract was injected into a gas liquid chromatograph employing a 50 m × 0.5 mm internal diameter stainless steel column coated with silicone fluid OV 101. The carrier gas was nitrogen at a flow rate of 2 ml min^{-1}. The combustion gas was hydrogen at a flow rate of 30 ml min^{-1}. Air to support combustion was adjusted to 200 ml min^{-1}. The sample volume was 2 μl. The injection temperature was 225°C. The oven was programmed from 70 to 220°C at 6°C min^{-1}.

34.8.6 Separation of Saturates, Aromatics and Heterocyclics by Elution Chromatography

The American National Standard ANSI/ASTM:D2549-76 (1976) is followed with slight modifications to the solvent type and volumes used.

Bauxite 20 to 60 mesh is activated by heating at 538°C for 16 h and is transferred to an airtight container while still hot and protected thereafter from moisture. A matched pair of Pyrex chromatographic columns, 76 cm in length having an internal diameter of 1 cm is constructed for the analysis, with a 100 ml bulb fitted at the top of the column to provide a reservoir. A 2 mm opening at the lower end facilitates collection of the fractions.

Each column is cleaned with chromic-sulphuric acid, distilled or demineralised water, acetone and dry air. A small plug of glass wool is introduced into the column, care being taken to press it firmly into the lower end to prevent the flow of silica gel from the column. Small increments of dry silica gel (150 mesh) are added while vibrating the column along its length by means of a Vortex mixer until the tightly packed silica gel extends to the 60 cm height mark on the chromatographic column. The vibration is continued while the bauxite (40 mesh) is added until this layer extends to the 76 cm height mark. The column is vibrated for an additional 3 min after the filling is completed. Then 50 ml of *n*-pentane is added to the top of the column to prewet the absorbent. When the solvent reaches the bottom of the silica gel bed, the column is ready for use.

The filtrate is added to the column and the fraction is collected in a 500 ml beaker. Complete elution of the fraction is effected by the addition of a further 200 ml *n*-pentane to the top of the column. The elutant containing solvent and the saturate fraction is transferred in stages to a pre-weighed, round-bottomed flask. The solvent is removed and the saturate fraction dried. The elution procedure is repeated using 250 ml toluene to obtain the aromatic fraction, and repeated using 250 ml methanol to obtain the heterocyclic fraction. An intermittent nitrogen stream at 1 to 2 p.s.i. is applied via an airtight fitting to the top of the column to facilitate the complete removal of the elutant for each of the three fractions. The round-bottomed flasks are reweighed, and the weight of each fraction is expressed as a percentage of the total.

34.8.7 Respirometry

Substrate oxidation is proportional to oxygen uptake in bacterial systems. Manometric methods have been used to follow oxygen uptake by bacterial cultures grown on hydrocarbons by Liu and Dutka (1973). For any respective hydrocarbon, the rate of oxygen uptake and the total amount of oxygen consumed by the culture are greater than the endogenous level where the carbon source is absent. It may, therefore, be concluded that since the hydrocarbon has stimulated oxygen uptake, it has in fact been utilised, in part or in whole.

In oil investigations, oxygen uptake studies can be followed using Gilson differential respirometry. The initial steps involve calculating the optimum amount of water-free oil that can be extracted from mousse, oils or emulsions and using concentrations in the respirometry experiments which are comparable with these values. The volume of a 1% (v/v) solution of oil in carbon tetrachloride required to be added to the main chamber of the respirometer flasks may be determined and treated as follows: in order to add 20 mg oil to the experimental system, 2.0 ml 1% (v/v) oil in carbon tetrachloride is added to the main chamber of each flask. The solvent is evaporated by passing a stream of nitrogen over the mouth of the flask for about 15 min, a period of time which allows removal of all the solvent and ensures that the oil forms a thin coat over the bottom and walls of the flask. This is considered to be important in providing a maximum surface area for bacterial contact and ensures reliable oxygen uptake measurements.

For normal respirometry, the total volume of cell suspension is maintained at 2.0 ml and usually comprises 1.5 ml buffer and 0.5 ml cell suspension in buffer. Various concentrations of sodium chloride can be added to determine the optimum concentration for maximum oxygen uptake.

Sodium chloride (0.02 g) is weighed and added to the side arm of those flasks requiring 1% (w/v) sodium chloride; 0.5 g sodium chloride is weighed and added to the side arm of those flasks requiring 2.5% (w/v) sodium chloride. In each flask the centre well is filled with corrugated Whatman filter paper 1×0.5 cm^2. The cells suspended in buffer (0.5 ml), then buffer alone (1.5 ml) are added to the side arm and tipped into the main chamber. Oxygen uptake is followed at around 20°C. After pre-incubation for five minutes the cells and buffer are tipped into the main chamber. The carbon dioxide was absorbed in 0.2 ml of 20% (w/v) potassium hydroxide placed in the centre well and absorbed by the filter paper. Oxygen uptake can be measured in the presence of oil residues (Fig. 9).

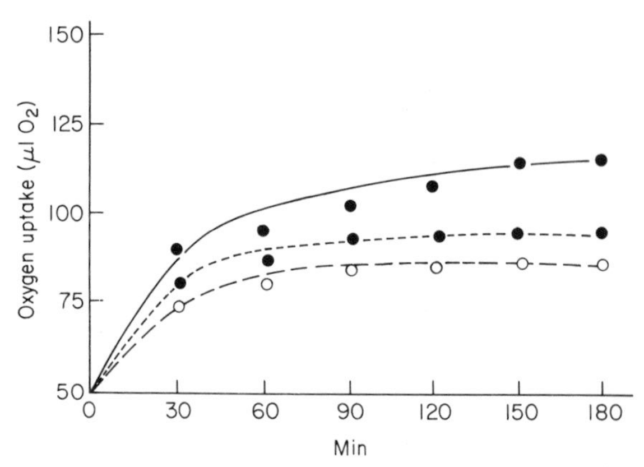

Figure 9. Respirometry: partial oxidation of Christos-Bitas mousse, by a pure culture in the absence of sodium chloride. The oil was extracted from Christos-Bitas mousse using carbon tetrachloride, dried, weighed and dissolved in a known volume of carbon tetrachloride. A given volume of the soil solution was added to the empty Gilson flask to provide concentrations of 1% (v/v) and 2.5% (v/v) oil in 2 ml of cell suspension. The carbon tetrachloride was evaporated with a nitrogen stream. Each Gilson flask was filled with 2 ml of centrifuged cells in buffer (0.2M-$K_2H\ PO_4/KH_2\ PO_4$; pH 7.5). Oxygen uptake was followed for 180 min at 20°C with conventional methodology using the Gilson Differential Respirometer. After pre-incubation for 5 min the cells and buffer were tipped into the main chamber. The carbon dioxide was absorbed in 0.2 ml of 20% (w/v) potassium hydroxide placed in the centre well and absorbed by the filter paper. (●—●), 2.5% (v/v) oil; (●----●), 1% (v/v) oil; (○--○), endogenous.

Recommended reading

CALLELY A. G., FORSTER C. F. & STAFFORD D. A. (1977) *Treatment of Industrial Effluents*. Hodder & Stoughton, London.

FLOODGATE G. D. (1972) Biodegradation of hydrocarbons in the sea. In: *Water Pollution Microbiology* (Ed. R. Mitchell), pp. 153–71. John Wiley & Sons, New York.

References

ANSI/ASTM:D 2549–76 (1976) Standard method of representative aromatics and non-aromatic fractions of high boiling oils by elution chromatography. In: *Annual Book of ASTM Standards* Part 24 ASTM. Washington.

BARUAH J. N., ALROY Y. & MATELEO R. I. (1967) Incorporation of liquid hydrocarbons in to agar media. *Applied Microbiology* **15**, 961–5.

BEASTALL S. (1977) Treatment and natural fate of oil spillages. In: *Treatment of Industrial Effluents* (Eds A. G. Callely, C. F. Forster & D. A. Stafford), pp. 328–45. Hodder & Stoughton, London.

BYROM J. A. &. BEASTALL S. (1971) Microbial degradation of crude oil with particular emphasis on pollution. In: *Microbiology 1971*, pp. 73–86. Applied Science Publishers Ltd., Amsterdam.

CAIN R. B. (1977) Surfactant biodegradation in waste waters. In: *Treatment of Industrial Effluents* (Eds A. G. Callely, C. F. Forster & D. A. Stafford), pp. 283–327. Hodder & Stoughton, London.

CALLELY A. G., FORSTER C. F. & STAFFORD D. A. (1977) *Treatment of Industrial Effluents*. Hodder & Stoughton, London.

CALLELY A. G. & STAFFORD D. A. (1977) Microbiological and biochemical aspects. In: *Treatment of Industrial Effluents*, (Eds A. G. Callely, C. F. Forster & D. A. Stafford), pp. 129–48. Hodder & Stoughton, London.

DELAUNE R. D., HAMBRICK G. A. & PATRICK W. H. (1980) Degradation of hydrocarbons in oxidized and reduced sediments. *Marine Pollution Bulletin* **11**, 103–6.

Department of the Environment Handbook for Raw and Potable Wastes (1972) HMSO London.

EVANS E. D., KENNY G. S., MEINSCHEIN W. D. & BRAY E. E. (1957) Distribution of *n*-paraffins and separation of saturated hydrocarbons from recent marine sediments. *Analytical Chemistry* **29**, 12–14.

GOETZ A. & TSUNEISHI N. (1959) Bacteriological test for air-borne irritants. *Industrial and Engineering Chemistry Product Research and Development* **51**, 772–4.

GREFF R. A., SETZKORN E. A. & LESLIE W. D. (1965) A colourimetric method of determination of parts per million of non-ionic surfactants. *Journal of American Oil Chemists Society* **42**, 180–5.

GUNKEL W. (1967) Experimentell-ökologische Untersuchungen über die limitierenden Faktoren des mikrobiellen Ölabbanes im marinen Milieu. *Helgoländer wissenschaftliche Meeresuntersuchungen* **15**, 210–25.

HOLLAWAY S. L., FAW G. M. & SIZEMORE R. K. (1980) The bacterial community composition of an active oil field in the northwestern Gulf of Mexico. *Marine Pollution Bulletin* **11**, 153–6.

HOOD M. A., BISHOP W. S., BISHOP R. W., MEYERS S. P. & WHELAN T. (1975) Microbial indicators of oil-rich marsh sediments. *Applied Microbiology* **30**, 982–7.

HUGHES D. E. & BEASTALL S. (1976) Bacterial degradation of crude oil. In: *The Control of Oil Pollution* (Ed. J. Wardley-Smith), pp. 39–46. Graham & Trotman Ltd., London.

HUGHES D. E. & MCKENZIE P. (1975) The microbial degradation of oil at sea. *Proceedings of the Royal Society B*, **189**, 375.

JONES J. G. & EDINGTON M. A. (1968) An ecological survey of hydrocarbon oxidizing microorganisms. *Journal of General Microbiology* **52**, 381–90.

KAWAHARA F. K. (1972) Characterization and identification of spilled residual fuel oils by gas chromatography and infrared spectrophotometry. *Journal of Chromatographic Science* **10**, 629–36.

KAWAHARA F. K. & BALLINGER D. G. (1970) Characterization of oil slicks on surface waters. *Industrial and Engineering Chemistry Product Research and Development* **2**, 553–8.

KAWAHARA F. K., SANTNER J. F. & JULIAN E. C. (1974) Characterization of heavy residual fuel oils and asphalts by infrared spectrophotometry using statistical discriminant function analysis. *Analytical Chemistry* **46**, 266–73.

LAW R. J. (1978) Petroleum hydrocarbon analyses conducted following the wreck of the supertanker *Amoco Cadiz*. *Marine Pollution Bulletin* **9**, 293–6.

LEE R. F. & ANDERSON J. W. (1974) Fate and effect of naphthalenes: controlled ecosystem pollution experiment. *Journal of Water Bulletin and Marine Sciences* **27**, (1), 127–37.

LIU D. L. & DUTKA B. J. (1973) Biological oxidation of hydrocarbons in aqueous phase. *Water Pollution Control Federation. Journal* **45**, 232–9.

LYNCH P. F. & BROWN C. W. (1973) Identifying source of petroleum by infrared spectroscopy. *Environmental Science and Technology* **7**, 1123–7.

MATTSON J. S. (1971) Fingerprinting of oil by infrared spectrometry. *Analytical Chemistry* **43**, 1872–3.

MATTSON J. S. (1977a) Multivariate statistical approach to the fingerprinting of oils by infrared spectrometry. *Analytical Chemistry* **49**, 297–302.

MATTSON J. S. (1977b) Classification of petroleum pollutants by linear discriminant function analysis of infrared spectral patterns. *Analytical Chemistry* **49**, 500–2.

MURPHY J. & RILEY J. P. (1962) A modified single solution method for the determination of phosphate in natural waters. *Analytica Chimica Acta* **27**, 31–6.

PRICE W. J. (1973) *The Principles and Practice of Infrared Spectroscopy*. Pye-Unicam Ltd., Cambridge.

RUCHHOFF C. C. (1950) Discussion in simplified method for analysis of BOD data. *Sewage and Industrial Wastes* **22**, 1343.

SOLOZANO L. (1969) Determination of ammonia in natural waters by the phenolhypochlorite method. *Limnology and Oceanography* **14**, 799–801.

SPENCER M. J. (1975) *Oil identification using infrared spectrometry*. Department of Transportation, US Coast Guard, UM-RSMAS No. 7502 (Dot—CG-81-75-1364).

STAFFORD D. A. & HUGHES D. E. (1976) The microbiology of the activated sludge process. *CRC Critical Reviews in Environmental Control* **6**, 233.

STRICKLAND J. D. H. & PARSONS T. R. (1968) Determination of reactive phosphorus. In: *A Practical Handbook of Seawater Analysis*, Ie. Fisheries Research Board of Canada, Bulletin No. 167.

WARD D. M. & BROCK T. D. (1978) Anaerobic metabolism of hexadecane in sediments. *Geomicrobiology Journal* **1**, 1–9.

WILLIAMS A. R., STAFFORD D. A., CALLELY A. G. & HUGHES D. E. (1970) Ultrasonic dispersal of activated sludge flocs. *Journal of Applied Bacteriology* **33**, 656–63.

ZOBELL C. E. (1973) Bacterial degradation of mineral oils at low temperature. In: *The Microbial Degradation of Oil Pollutants* (Eds D. G. Ahearn & S. P. Meyers), pp. 153–61. Centre for Wetland Resources, Publication Number LSU-SG-73-01.

Chapter 35 • Effect of Pesticides on Soil Microorganisms

Michael P. Greaves

35.1 Introduction

35.1.1 The Potential Problem of Pesticides in the Environment

Pesticides are toxic chemicals which act by interfering with biochemical reactions in the target organisms. As yet, complete specificity has not been achieved generally and there is a risk that pesticides will affect non-target organisms. In view of the increasing use of pesticides in agriculture there is concern that the environment will be harmed (Kornberg 1979). Although in recent years there have been few reports of environmental damage, other than those of drift affecting neighbouring crops, the legacy of the harm caused by organochlorine pesticides in the 1950s remains.

The fear expressed most frequently is that 'the soil will be poisoned'. This assumes that microorganisms are important in maintaining soil fertility and are also the most important agents which detoxify pesticides in soil. Thus, any chemical which seriously affects the soil microflora may harm soil fertility and crop production.

Less tangible side-effects may occur in aquatic ecosystems contaminated by pesticides. Effects on aquatic microorganisms may ultimately be reflected in higher organisms as a result of imbalances within food chains. On the other hand, deliberate use of pesticides in water may produce more obvious side-effects. Thus, the control of aquatic weeds with some herbicides has caused the development of blooms, usually involving filamentous algae.

Hislop (1976) has reviewed the effects of pesticides on the microflora of plant surfaces and concluded that 'non-selective toxicants cause major shifts in the microbiological equilibrium but induce resistance in pathogens less often than selective toxicants that cause minimal disturbance to the microflora of the host plant'. Both instances could be classed as undesirable side-effects.

35.1.2 Registration of Pesticides

The annual world usage of toxic agricultural chemicals is about 1×10^6 tonnes comprising approximately 500 products containing some 80 active ingredients, but very few of these are used in quantities over 1×10^5 tonnes (Yeo 1979). They can be regarded as minor global pollutants and any environmental problems associated with their use are likely to be relatively localised.

Even so, the regulation of the manufacture and sale of pesticides is greater than for any other group of chemicals. Registration schemes are operated by many countries to ensure that risks to man, wildlife and the environment are minimised or eliminated. Waitt (1975) has described the development of registration requirements. Amongst the recent extensions of these requirements is the need for manufacturers to submit data from ecological studies for consideration by the registration authorities. At first concern focused on wildlife, such as birds, fish and mammals. Later, the below-ground macroorganisms were included and, most recently, the microorganisms. To some extent the agrochemical industry anticipated these requirements and had already established research programmes to investigate the potential risks (Johnen 1978).

Although all registration requirements have a common aim, they usually differ in detail. Thus, the UK Pesticide Safety Precautions Scheme (PSPS), a voluntary agreement negotiated between government and industry (Bunyan & Stanley 1979), emphasises the importance of dialogue between these two to reach agreements on pesticide safety evaluation. Similarly, dialogue between the manufacturer and the technical secretariat of PSPS is used to reach agreement on the package of tests to be applied. The flexibility of this system allows the use of general guidelines to the manufacturer which, with regard to soil microorganisms, state simply 'relevant work might include . . . measurement in the laboratory, on samples from the field, of major effects on soil microbiological processes such as total respiration, nitrification and rate of organic matter decomposition' (MAFF 1971).

The testing of soil and soil organisms suggested in the booklet *Agricultural Pesticides* (Council of Europe 1973) is less specific, suggesting that ' . . . there may be a need to evaluate the possible effects of pesticides or their metabolites on the soil microflora.'

In contrast, the Environmental Protection Agency (EPA), which is responsible for regulating manufacture and sale of pesticides in the United States, has published a relatively detailed guideline (EPA 1978a). They require that 'data shall be supplied on oxygen consumption, carbon dioxide evolution, nitrogen cycle reactions and measurement of enzyme activity for dehydrogenase or phosphatase'. In certain situations further data on degradation of starch, cellulose, protein and pectin are required. A permitted alternative approach is to provide data on effects on a number of pure or mixed cultures of microorganisms. Some limited recommendations on acceptable methods are also given.

The data requirements for effects on aquatic microorganisms follow a similar pattern. The PSPS and Council of Europe requirements are covered in broad statements similar to those applying to soil organisms. The requirements of the EPA, as yet only in draft form (EPA 1978b), are more detailed and complex than those for effects on soil.

At the time of writing, attempts are being made to harmonise registration requirements of different countries. Also, considerable effort is being devoted to improving the methods used to obtain the necessary data. Additional tests are being considered to ensure that as many aspects of environmental safety as possible are covered. However, it must be remembered that the balance between risk and benefit is precarious. Too great an insistence on safety may increase costs to the point where a potentially efficacious pesticide cannot be profitable and is therefore abandoned. This may be of greater detriment to man's welfare than the environmental risks ensuing from the use of the compound.

35.1.3 Methods of Measuring Pesticide Side-Effects

It is beyond the scope of this chapter to consider in detail the methods available for studying the impact of pesticides on all microbial environments. Accordingly, attention is focused on soil, the environment which is the main recipient of pesticides. The methods described or referred to have been selected in order to form a basis from which investigative systems can be developed. Many of the methods have been used in the author's laboratory and have proved useful. A comprehensive range of methods is presented throughout this book and it must be stressed that any method useful in microbial ecology may also be applied in assessing stress induced by pesticides.

Increasing use of pesticides, especially herbicides, in or near water and increasing demands for water for drinking, industrial and amenity use, emphasise the importance of research into pesticide stress in aquatic environments. Interested readers should refer to the following publications for detailed information on appropriate methods (Vollenweider 1969; Ware & Roan 1970; Sorokin & Kadota 1972; Collins *et al.* 1973; Hurlbert 1975; Butler 1977; Mackereth *et al.* 1978; Jones 1979).

Plant surfaces, especially leaves of agricultural crops, are frequently exposed to relatively high doses of pesticide. As yet, there is little information as to the interaction of these chemicals with the microflora present. Preece and Dickinson (1971), Dickinson and Preece (1976) and Chapter 24 are important sources of methodology.

Recent reviews illustrate the diversity of the methods available for studying pesticide-induced stresses in the soil microflora (Greaves *et al.* 1976; Grossbard 1976; Hissett & Gray 1976; Tu & Miles 1976; Anderson 1978a, b; Johnen 1978; Greaves 1979; Simon-Sylvestre & Fournier 1979; Greaves & Malkomes 1980). Some methods suggested by the Southern Weed Science Society of America for investigating herbicide-microorganism interactions are given by Curl and Rodriguez-Kabana (1972, 1977). Many fundamental concepts of microbial response to stress have been reviewed by Strange (1976).

The major problem facing the researcher is: which of the available methods are the most valid for investigating pesticide effects and, possibly more problematical, how are they applied for the best results? A further difficulty stems from the recognition that none of the methods available at present is ideally suited to the requirements of microbial ecologists. This dilemma is exaggerated by the difficulties of experimentation in the field. The complexities of these issues are dealt with in the excellent, comprehensive book edited by Hill and Wright (1978).

Most authorities responsible for the registration of pesticides state, or imply, a preference for the functional approach in which effects on microbial processes, e.g. respiration and nitrification, are examined. Recently, a series of international workshops has resulted in the publication of recommended tests of pesticide side-effects in soil (Greaves *et al.* 1980). This report also indicates areas requiring further research, recognising that while the recommended tests are the best available, they are not ideal.

Experimental

35.2 Soil sampling and pretreatment

Adequately controlled field experiments are difficult, if not impossible, to design. Thus, pesticide side-effects are usually investigated in laboratory or glasshouse conditions and require that soil is removed from the field and treated (e.g. sieved or air-dried) prior to the experiment. As such operations can severely modify the response of the soil microflora to pesticides (Wingfield *et al.* 1977; Wingfield 1980b), the methods used must be chosen with care and reported fully and accurately.

Detailed descriptions of site selection, sampling methods and soil treatments are given by the Federal Working Group on Pest Management (1974), Parkinson *et al.* (1971) and Williams and Gray (1973). Information specifically related to investigation of pesticide side-effects is given by Greaves *et al.* (1980).

A particular point of contention is the suitability of undisturbed soil cores in research into pesticide effects on soil microflora. This author believes that thay can play an important role in fundamental studies but they are less suitable for monitoring side-effects for registration purposes (see section 35.9.1).

The major requirements of sampling can be summarised as follows.

(1) The site(s) chosen should be representative of the general area(s) in which the pesticide is used. The area sampled should reflect the major features of the main site.

(2) As field error is greater than analytical error, the pattern of sampling should be designed in order to minimise this error.

(3) Most pesticides remain near the soil surface so it is rarely necessary to sample deeper than 8 to 10 cm.

(4) Sample preparation (e.g. sieving and other processes) should be minimal as it can modify the response of the microflora to pesticides.

(5) Full details of the sampling site (e.g. agricultural use and treatment), sampling method and treatments should always be recorded.

35.3 Experimental design

Just as soil sampling must be carefully planned to minimise field errors, so must the subsequent use of the soil. Samples from a well-mixed, bulk sample from the field tend to become highly individual in their microbiological properties. Experimental design must, therefore, pay regard to adequate replication. The actual level of replication which is necessary will vary depending on the measurement being made. Between-sample variation in microbial populations counted by plate dilution methods is usually high and as many as ten replicates may be required. In contrast, variation in nitrification between samples may be low enough to permit as few as four replicates. While the many texts on statistics for biologists can be used to design adequate experiments, it is more satisfactory to discuss the requirements with an experienced statistician.

Some other general points to consider in designing pesticide experiments are the following.

(1) The nature of the pesticide and its formulation will influence experiment design. Thus, persistence affects the duration of the experiment as will formation of a potentially toxic metabolite. Granular formulations or volatile compounds, normally incorporated into soil, may impose different design features than compounds which are foliar- or surface-applied.

(2) Temperature, soil moisture and other incubation conditions will obviously affect the results. In general, it is most satisfactory to attempt to simulate the conditions normally encountered in the field at the time when the pesticide is used. Thus, the standard 25°C incubation temperature, often used by microbiologists, may give different effects to those encountered at the usual spring temperature of approximately 10°C, which applies when most herbicides are used on arable crops in the UK.

(3) The two main criteria for judging the importance of pesticide effects (see section 35.10) are the amplitude and duration of the effect. In order that these may be ascertained with reasonable accuracy, sampling must be frequent throughout the experiment. In practice, weekly samples are normally adequate. Samples taken only at the start and finish of an experiment are hopelessly inadequate.

(4) Pesticide dosage is a most important factor. Many reports in the literature present data showing large effects which result from treatment with unnaturally high doses of pesticide. This may be valuable as a means of identifying ways in which organisms might react to pesticides and pointing to areas which might be investigated more critically. On the other hand, it is not necessarily a valid means of assessing the potential effect of the pesticide on soil fertility. The researcher has an obligation to make this clear and usually this is best done by showing the effect of a dose equivalent to that used in normal agricultural practice.

(5) Application of the pesticide to the soil can be achieved in many ways ranging from simple addition from a pipette to application with a sophisticated sprayer. It may be left as a surface layer or mixed into the soil as uniformly as possible. Obviously, the actual detail will depend on the pesticide used, the equipment available and the objectives of the experiment. However, the microbial ecologist should recognise that pesticide distribution in soil is not uniform. Concentration gradients exist and will result in a range of effects. These may be of greater interest and value to the ecologist than an overall measure of effect in a homogenous sample containing an average concentration of pesticide. This is true particularly if the research concerns the rhizosphere (see section 35.8) as in many situations pesticides will be confined to soil above the rooting zone. Thus, any effect on rhizosphere organisms is likely to be indirect, mainly due to the pesticide's effect on the plant. In non-cultivation systems, such as soft fruit growing, a considerable amount of root material is very near the soil surface and here pesticides may exert direct effects on the rhizosphere microflora.

35.4 Degradation of pesticides

It is not the purpose of this chapter to describe methods of determining the microbial degradation of pesticides. For this the reader is referred to comprehensive reviews by Hill and Wright (1978) and Hance (1980). Nonetheless, degradation must be considered briefly in view of its interaction with the side-effects of the pesticide on the microflora. In particular, knowledge of the degradation rate is essential to allow proper experimental design and interpretation of data on effects observed. The duration of exposure to the pesticide, and production of metabolites which are more toxic than the parent compound are obviously important, as is the utilisation of the pesticide or its products as sources of carbon or energy by the microbial population. It is recommended that, in investigations of pesticide effects on soil microorganisms, measurements of degradation rate are made. Information on metabolites formed can often be obtained from the literature or from the manufacturer.

35.5 Measurement of microbial processes

35.5.1 Respiration

Evolution of carbon dioxide or uptake of oxygen are measured most frequently to assess the activity of the soil microflora and its response to pesticides. A general discussion of the available methods, their advantages and disadvantages, is given by Anderson (1978a). Soil respiration, measured by carbon dioxide evolution, can be determined in the field (Wallis & Wilde 1957; Witkamp 1966; Reiners 1968). The disadvantages of this approach are numerous, the chief one being that it overestimates microbial respiration by including the respiration of the macroflora and fauna.

In general, the impact of pesticides on microbial respiration is determined most conveniently in laboratory experiments. Two basic systems have been developed for this purpose:

(1) jars or flasks containing both soil and a carbon dioxide absorbent;

(2) containers of soil through which carbon dioxide-free air flows before passing through a carbon dioxide absorbent.

Static systems may be affected by oxygen limitation and usually are used for short experiments. Gas-flow systems may not permit complete removal of carbon dioxide from the soil but can be used for long periods. This is necessary in many investigations of pesticide effects, particularly if the chemical is persistent or metabolites exert toxic effects.

Respiration measured as oxygen consumption is not favoured as much as the methods described above. The reasons for this are detailed by Stotzky (1965) and Anderson (1978a). It is not proposed to detail methods here as a comprehensive review is given by Anderson (1978a). However, the Warburg technique of measuring oxygen uptake has found frequent application and, in short-term experiments, can produce useful data. An interesting method which could be used more widely in pesticide studies is described by Howard (1968). This uses undisturbed cores of soil and permits them to be maintained under field conditions for most of the experimental period.

Two points should be borne in mind when using respiration methods. Regardless of the method used, the data can only indicate the gross metabolic activity of the total biological system in soil plus the contributions due to chemical reactions. Thus, it can only detect gross changes in microbial respiration and may not detect even major shifts of microbial balance. Furthermore, consideration must always be given to the potential contribution to carbon dioxide production which can arise from degradation of the pesticide being treated. At normal concentrations this will be of little significance but, at the very high concentrations often used by experimenters, may assume some importance.

Soil used in respiration studies may be amended by the addition of a substrate such as glucose or plant material. Release of carbon dioxide from this system may be regarded as being close to the maximum potential respiration of soil as the mineralisation of the substrate should involve the activity of a large proportion of the microflora. Certainly, the use of ^{14}C-labelled substrate allows the effects of pesticides on the degradation of plant residues to be followed in a more meaningful way than conventional methods. Even so, it still presents only gross measurements and may not detect potentially important shifts of specific microbial balance. Stotzky (1974) has suggested the use of substrates used only by certain species, while Anderson and Domsch (1973a, b, 1974) have used antibiotics to distinguish bacterial and fungal respiration in soils. Both these approaches hold promise for more definitive investigations of effects of pesticides on microbial respiration.

35.5.2 Nitrogen Transformations

One stated aim of assessing effects of pesticides on the soil microflora is to ensure that those microbial functions contributing to soil fertility are not harmed. Transformation of organic nitrogen to inorganic forms available as plant nutrients has, therefore, become a major target for study. Usually, two parts of the overall process, ammonification and nitrification, are studied (see Chapters 9, 11 and 12).

Ammonification, the production of ammonium-nitrogen from organic forms, includes the degradation of proteins into smaller molecules (proteolysis) as well as the deamination of amides and amino acids (Anderson 1978a). Nitrification, the oxidation of ammonium-nitrogen to nitrite and nitrate, is sensitive to pesticides and other toxicants.

There are many methods of studying nitrogen transformations, including simple incubations of soil in containers with or without added nitrogen substrates and the more complex perfusion methods based on that of Lees and Quastel (1946). Each type of method has its advantages but the former finds more frequent application in recent studies. In particular, recently published recommended tests (Greaves *et al.* 1980) describe a protocol which is believed to be one of the most valid ways of providing data for registration purposes. This method involves incubation of soil with 0.5% (w/w) lucerne meal as a source of organic nitrogen and measuring ammonium, nitrite and nitrite release over a period of time. If the pesticide has no effect on the system, nitrate should accumulate as the intermediates, ammonium and nitrite, are usually transitory. If ammonification is affected, ammonium production and, thus, nitrate production decreases below that in control soil. Equally, if nitrification is affected ammonium-nitrogen accumulates. This effect can be checked by incubating the soil with ammonium sulphate, at 100 μg nitrogen (g soil)$^{-1}$.

Denitrification is seldom studied in work with pesticide–soil interactions and, as pointed out by Anderson (1978a), there are many difficulties associated with measurement of gaseous nitrogen production from soils which are also producing carbon dioxide. One of the best methods available is that of Payne (1973) which uses gas chromatography to separate nitrogen, nitrous and nitric oxides.

35.5.3 NITROGEN FIXATION

Accounts of nitrogen fixation and of methods for studying the process in both symbiotic and free-living organisms are given by Quispel (1974), Burns and Hardy (1975), Stewart (1975), Nutman (1976), Newton *et al.* (1977) and Anderson (1978a) (see also Chapters 13 and 25).

Symbiotic nitrogen fixation by *Rhizobium* spp. is an important contributor to world agriculture and so the impact of pesticides on the process is also important. In the relatively restricted literature dealing with pesticide interactions with *Rhizobium* spp., a range of methods has been used (Helling *et al.* 1971; Greaves *et al.* 1976; Anderson 1978a). Generally, it has been shown that pure cultures of *Rhizobium* spp. are resistant to pesticides used at realistic concentrations (Kecskes 1972; Greaves *et al.* 1976; Grossbard 1976) and effects on nodulation are likely to result from phytotoxic effects on the host. Accordingly, examination of pesticide effects is best made using the legume–*Rhizobium* spp. combination and studies with pure cultures are not recommended.

Studies of the symbiosis should determine, in one test, the responses of both plant and the bacterial symbionts to the pesticide (Greaves *et al.* 1980). Suitable methods were described by Johnen *et al.* (1979) who concluded that nodule activity (measured by the acetylene reduction method—Hardy *et al.* 1968; Johnen *et al.* 1979), nodule weight, plant weight or nodule numbers were reliable indicators of pesticide effects. However, they also showed that, because of large standard errors associated with results from acetylene reduction methods, nodule counts and nodule weights, the methods only detected major differences between treatments. Greaves *et al.* (1978) formed similar conclusions and drew attention to lack of reproducibility between repeat experiments. They concluded that the most reliable indications of pesticide effects were obtained, with the greatest ease in practice, from measurements of plant growth and yield. This was reflected in the recommendations by Greaves *et al.* (1980) which suggested that, if effects were found using plant growth and yield measurements, further investigations using the acetylene reduction method should be undertaken. This method is very sensitive but presents several difficulties. One is the need to grow the legumes under strictly controlled environmental conditions, particularly light intensity. In addition, very large numbers of replicates are required to allow detection of small, but significant, effects. It is essential that excised nodules are not used as they give false, high values for nitrogenase activity.

Present knowledge indicates that nitrogen fixation by free-living microorganisms, either in the rhizosphere of crops or in root-free soil, is of little significance in agriculture. As such it need not be included as the subject of general tests in regulatory requirements (Greaves *et al.* 1980). For pesticides used in rice culture, however, effects on the growth and nitrogen-fixing activity of the cyanobacteria may be important, as might that on the *Azolla*-*Anabaena* symbiosis. These effects may be studied conveniently using the acetylene reduction method.

35.5.4 Phosphorus Mineralisation

Effects of pesticides on phosphorus mineralisation in soils have been studied by few workers (Simon-Sylvestre & Fournier 1979). This is due to the lack of precise knowledge of the process itself (Cosgrove 1967) and to the well-known difficulties of measuring phosphate, particularly inorganic forms, in soil.

The studies which have been reported appear to rely on soil incubations, or field sampling, followed by extraction and measurement of inorganic phosphate. Although such methods are simple, they can do no more than indicate gross effects and reveal nothing of the means by which pesticides may affect the phosphorus cycle.

The lack of literature dealing with pesticide effects in this process precludes any assessment of which method(s) might be most suitable. There is no reason, however, why any of the methods for investigating the microbial mineralisation of phosphorus in normal, unstressed, soils should not be used to advantage. Details of these methods are given earlier in this book (see Chapter 33).

35.5.5 Mineralisation of Other Elements

As with phosphorus, there is virtually no literature concerned with the effects of pesticides on the mineralisation of elements such as sulphur and manganese (Simon-Sylvestre & Fournier 1979). This does not, necessarily, reflect a lack of importance to agriculture or of interest to the microbial ecologist, but the absence of basic knowledge of the processes and of the methods for their study. Again, simple soil incubation techniques with extraction of the element of interest may be a convenient, but crude, method of study. The metabolism of sulphur-containing organic compounds in soil has been reviewed by Freney (1967) and Alexander (1977). Alexander (1977) has also reviewed the mineralisation of several other elements. Both these reviews give references to methods which may be applicable in pesticide investigations.

35.5.6 Organic Matter Decomposition

A wide range of techniques is available to quantify organic matter degradation. Perhaps that used most frequently is the measurement of carbon dioxide evolution from soils with added plant material (see section 35.5.1). Alternatively, one of the varied buried substrate techniques can be used. Latter and Howson (1977) and Wingfield (1980a) describe methods of following cellulose degradation using different forms of cloth as substrate. The cloth used by Latter and Howson (1977) had the disadvantage that it required boiling to reduce the starch and tallow content. Wingfield (1980a) used the more satisfactory Shirley Test Cloth (Shirley Institute, Manchester, UK) which is woven specially for burial tests and requires no pretreatment. Latter and Howson (1977) discuss the method in detail giving many useful references and conclude that the method is a valuable indicator of activity on cellulose substrates, a conclusion confirmed by work on assessment of effects of herbicides on organic matter breakdown (Wingfield 1980a, b).

A useful method of following organic matter breakdown in localised sites in soil is described by Webley and Duff (1962). The substrate is ground and mixed with an inert base such as kaolinite. The mixture is pressed into tablets which are buried in soil. The tablets are recovered after the required time and the residual substrate analysed. Obviously, this technique is most suitable for insoluble substrates.

A rapid technique of assessing protein degradation has been described by Cullimore and Ball (1978). Pieces of fixed, unexposed colour film are buried and, after incubation, proteolysis is measured as the degree of light transmission through the film resulting from degradation of the gelatin layers of the film. This is achieved with a simple photocell system (Cullimore & Ball 1978) or an image analysing computer such as the Quantimet (Image Analysing Computers Ltd., Cambridge, UK) (M.P. Greaves & J. M. Johnson, unpublished observations). A simpler assessment can be made by overlaying the etched film with a 100 square grid and scoring the number of squares which show etching (M.P. Greaves & J. M. Johnson, unpublished observations). This technique has been used successfully to assess the potential effects of herbicides on protein degradation in soils.

The most widely used and possibly most productive burial technique is the litter bag method (Heath *et al.* 1966; Parkinson & Lousier 1975). Among the factors to be considered is the size of bag mesh to be used. Heath *et al.* (1966) suggested that 0.003 mm excluded all soil animals, 0.5 mm admitted small soil animals but excluded earthworms and 7.0 mm admitted all soil animals including earthworms. The choice of substrate depends on the pesticide and its principle area of use (forest, orchard, grassland, arable). Thus straw, leaves of grass, apple, beech and others have all been used: they may be entire for small leaves or cut lengths or discs for the larger forms. Assessment of degradation may be gravimetric, photometric or by simple visual assessment. Generally, however, weight loss is the most useful measure. In addition, changes in nutrient content of the buried plant material can be measured.

The simple methods described represent only a proportion of the many available. Other methods which could be applied to pesticide-stressed environments can be obtained from Kilbertus *et al.* (1975), Anderson and Macfadyen (1976) and Lohm and Persson (1977).

35.6 Measurement of enzyme activity

Many reactions involving soil organic matter transformations are catalysed by extracellular enzymes, some of which exist either free in the soil or associated with colloidal clays and organic matter (Burns 1978). Today, emphasis is placed on the role of soil enzymes in soil fertility and the effects of pollution and agrochemicals on these enzymes. However, the methods are recognised as being deficient in many ways and new and improved methods may require the re-evaluation of existing knowledge (Skujins 1978). Nonetheless, enzyme activities may be useful indicators of the effects of pesticides and one regulatory agency (EPA 1978a) specifies an assessment of pesticides on phosphatase or dehydrogenase activities.

Cervelli *et al.* (1978) have reviewed the interactions between agrochemicals and soil enzymes and the methods of measuring enzyme activities have been comprehensively covered by Burns (1978). Thus, the methods will not be detailed here but their value to the study of pesticide stress in microbial environments will be considered.

35.6.1 Phosphatases

Phosphatases are studied frequently in relation to pesticide effects, presumably because of an assumed relationship to phosphorus nutrition of crops. R. G. Burns, in a contribution to a report of recommended tests for measuring pesticide effects (Greaves *et al.* 1980) comments 'Phosphatase is a collection of enzymes, usually measured by using an artifical substrate (e.g. *p*-nitrophenyl phosphate), and whose activity bears little relation to total phosphate availability in soils.' This, in conjunction with difficulties of quantifying the actual contribution of soil phosphatases and with the lack of agreed methods (problems common to all soil enzymes), is sufficient reason not to use this enzyme as an indicator of pesticide stress on microbial communities. It may have some value in studies of pure cultures.

35.6.2 Dehydrogenases

This group of enzymes, like phosphatases, has been recommended as an indicator of pesticide stress (EPA 1978a). Again, R. G. Burns (quoted in Greaves *et al.* 1980) states 'Dehydrogenases reflect a broad range of microbial oxidative activities, do not accumulate to any extent (. . . they are not soil enzymes) and yet they do not consistently correlate to microbial numbers, carbon dioxide evolution or oxygen consumption. Additionally, activity may be dependent upon the nature and concentration of amended carbon substrates and alternative electron acceptors.' Lack of unequivocal methods adds to the inadvisability of using dehydrogenase activity as a measure of pesticide stress on natural environments.

35.6.3 Urease

Urease is unique among soil enzymes due to its role in decomposing an important fertiliser (urea) and has, therefore, been studied more intensively than other soil enzymes (Bremner & Mulvaney 1978). The available literature contains many contradictions, often due to poor techniques, and indicates the need for more research into the nature and properties of this enzyme.

The methods used to measure urease activity in soil are numerous and, because little attempt has been made to evaluate most of them, it is difficult to recommend any one. However, useful methods include those of Zantua and Bremner (1975), Cervelli *et al.* (1976), Gauthier *et al.* (1976) and Lethbridge and Burns (1976).

35.6.4 Other Enzymes

The literature abounds with reports of the effects of pesticides on soil enzymes such as amylase, invertase and cellulase. There are nearly as many methods as there are publications and, consequently, a variety of reported effects. In addition, there are claims that particular enzymes are good indicators of soil biological activity, soil fertility and so on but the evidence to support these claims fully is sparse. Enzyme activity is only one factor which can contribute to soil fertility and so correlations between activity and fertility are likely to be the exception rather than the rule.

Nevertheless, the sensitivity, and especially the accuracy, of enzyme determinations make them an attractive proposition for measuring pesticide effects on environments. The range of enzymes and methods for their measurement are detailed by Burns (1978). It is necessary, however, to stress that the methods used must be carefully chosen to suit the objectives of the work. It should be considered whether there is a real need for a buffered system and attention should be paid to the effects of variables such as soil type, soil environmental conditions and so on. Excellent accounts of the importance of such factors with specific regard to urease are given by Kiss *et al.* (1975) Bremner and Mulvaney (1978) and Burns (1978).

35.7 Microbial populations

As the requirements of the regulatory authorities have evolved and research efforts have intensified, it has become accepted that studies of pesticide effects on microbial populations are unlikely to yield data which are useful in predicting the likely effects of pesticides in natural environments. However, in fundamental studies they can be useful, providing information which can aid the clarification of effects on biological processes in soil. Furthermore, they are often invaluable in studies of effects on pathogens and their activity. Readers are referred to reviews by Altman and Campbell (1977) and Papavizas and Lewis (1979).

Papavizas and Lewis (1979) make a detailed evaluation of pesticide–pathogen interactions. They conclude that present studies 'provide ample data, but what do they mean and how can they be correlated to disease in a natural ecosystem? . . . It would be a disservice to our attempts to elucidate pesticide–plant pathogen interactions in nature if we derive our conclusions from such (laboratory and greenhouse) observations only.' Similar cautions apply to studies of interactions between pesticides and the saprophytic microorganisms in soil or any other environment.

35.7.1 Viable Counts

Many techniques for counting viable microorganisms have been discussed by Parkinson *et al.* (1971) and earlier in this book (see Chapter 6). All these methods, based on dilution plate techniques or other methods using agar plates, suffer from many well-known disadvantages which may be grouped under three headings (Gray & Williams 1971):

(1) the selective nature of growth media;

(2) problems of transferring organisms from their natural environment to growth medium;

(3) the problems arising from the different growth phases of microorganisms in soil.

If these problems are considered carefully, viable counts can give useful data concerning pesticide–microflora interactions. However, they should only be used in conjunction with other methods of measuring microbial functions to obtain their major value, which is to indicate the reasons why microbial populations or processes may have responded to a pesticide.

35.7.2 Total Counts

Direct observation of microbial cells, either in soil extracts or in soils *in situ*, can be achieved in many ways using normal or fluorescent stains, fluorescent antisera and examination by light or electron microscopy (Parkinson *et al.* 1971; see Chapter 6). In addition, these methods allow the direct observation of microbial form and arrangement in the soil. Information of this kind may help in assessing the importance of pesticide effects, for example, observation of algae colonising contact slides in soil (McCann & Cullimore 1979) allowed the effect of herbicides to be determined in relation to depth in soil. Obviously, an effect limited to the top few millimetres of the soil profile is much less severe than one extending to 2 or 3 cm. A major problem is that there is still no really reliable method of distinguishing live from dead cells, or for distinguishing some bacteria from propagules such as actinomycete spores. An interesting method is described by Peterson and Frederick (1979) who claim it gives good estimates of microbial numbers and biomass in soil. It is agreed that viable counts generally underestimate populations and that direct cell counts provide an overestimate.

35.7.3 Biomass Measurement

The counting of microorganisms is often the preliminary stage to calculating biomass. The determination of biomass is necessary in order to compare satisfactorily the population densities of different microorganisms as mere comparisons of numbers can give a distorted view of the relative importance of different microbial groups. Basic methods of determining biomass have been described by Gray and Williams (1971), Parkinson *et al.* (1971) and earlier in this book (see Chapter 6).

More recently, Jenkinson and Powlson (1976) described a respirometric method of biomass determination which gives results that correlate well with those obtained from optical microscopy (Jenkinson *et al.* 1976). Anderson and Domsch (1978) reported a modification of the method of Jenkinson and Powlson (1976). Anderson and Domsch (1978) also suggested that the method can be combined with the selective inhibition method (Anderson & Domsch 1973b) to allow the separate contribution of bacteria and fungi to be determined. Lynch and Panting (1980) also modified the method of Jenkinson and Powlson (1976) and found it suitable for comparative studies but not for providing an absolute measure of biomass. Lynch and Panting (1980) showed clearly the need to consider carefully how the soil should be treated prior to experimentation. They pointed out that sieving the soil, as recommended by Jenkinson and Powlson (1976), can result in smaller biomass estimates and decided that intact soil cores ought to be used. An important feature of this physiological method is that it can be applied to a variety of soils and conditions. Many methods are restricted in their application by soil organic matter content, pH and the biomass concentration itself.

One further method, an example of obtaining biomass estimates from measurement of a specific cellular component, deserves mention. ATP measurements are increasingly used to estimate biomass in natural environments. There are many sources of error in the technique (Wildish *et al.* 1979) and it seems that this method will find greatest use in evaluating the effects of environmental variables on 'the total microbial life flux of the system' (Lee *et al.* 1971) and then only if comparing data obtained using exactly the same technique and if care is taken to avoid microbial dephosphorylation of ATP (Wildish *et al.* 1979).

35.7.4 Growth Rates and Microbial Distribution

In general, the rate of microbial growth in soil is slow. There are periods of activity, e.g. following the incorporation of crop residues, followed by long periods of inactivity. A different situation occurs in soil influenced by plant roots and the rhizosphere (see section 35.8 and Chapter 23) where a continuous or semi-continuous supply of organic materials maintains relatively high growth rates.

At present there are few suitable methods to measure accurately the growth rate of microorganisms in normal soil. It is impossible to observe soil organisms for long enough without amending or altering the soil. Several methods have been developed for observing growth rates in disturbed soils, either in the laboratory or in the field. These are detailed by Parkinson *et al.* (1971) and Brock (1971) and include a variety of approaches such as growth tubes, replica plating, buried film and nylon mesh. Considerable attention is also given to the growth of microorganisms in model systems using sand, vermiculite, glass beads and other materials in place of soil. These methods give useful information on effects of pesticides on microbial growth rates. However, it must be remembered that none indicates more than potential growth rates since the actual microbial growth rates *in situ* are beyond precise measurement at present.

As well as giving data on growth rates, many of these techniques can give valuable information on microbial distribution. This information is critical in evaluating the overall importance of the side-effects of pesticides. A pesticide may be capable of destroying a very large proportion of the microbes in soil but its movement in soil may be limited. Thus, its effect is limited and soil may be reinoculated from the underlying and unaffected horizons when the pesticide concentration has fallen below the toxic level. Clearly, a knowledge of the distribution of microorganisms in the soil profile, in relation to the distribution of the pesticide, will help assessments of the likelihood of a successful recolonisation. Equally, the distribution of microorganisms at the micro-environment level (see

Chapter 7) is important in terms of their interaction with pesticides. Microorganisms are intimately associated with organic matter particles in soils and are likely to be protected to some extent against those pesticides which are preferentially adsorbed by clays. Again, Parkinson *et al.* (1971) detail several valuable methods which can be used to determine microbial distribution and arrangement at this level.

35.7.5 Pure Cultures

Because of their simplicity and the wealth of information yielded, pure culture experiments with pesticides have been used frequently. Greaves *et al.* (1980) have rejected them for regulatory studies as 'isolated organisms may be metabolically atypical of their form in soil. Furthermore, they can change progressively during storage. They are normally stimulated to artificially high metabolic rates by growth in normal laboratory media. They are removed from their normal ecological associations. Interpretation of results is difficult and extrapolation to field situations impossible. Thus, they are not able to reflect, in any meaningful way, the side-effects pesticides may have.'

Nonetheless, pure culture studies have value in fundamental research, for example, studies of the effect of pesticides on pure cultures in different media will quickly reveal nutrient-herbicide interactions which are potentially important in soil. The miniaturised methods of testing the toxicity of pesticides to microorganisms described by Cooper *et al.* (1978) are a valuable aid in rapidly screening large numbers of pesticides and in helping to decide whether further, more complicated investigations are necessary. A further example is the assay of effects of pesticides on growth and nitrogen-fixing ability of free-living microorganisms, such as *Anabaena* sp., which might explain observed side-effects on paddy rice.

The methods for pure culture experiments are numerous, including growth in media such as nutrient solutions, agar, sand and soil, and using systems as divergent as test tubes and continuous culture equipment. The use of chemostats, both batch and continuous, is rightly gaining in popularity (see section 35.9.3) and valuable information can be gained in this way regarding pesticide effects on potential growth rates, community interactions and metabolic processes (see Chapters 16 and 20). It should be stressed that such techniques are not limited to pure cultures, in the sense of single species, but are extremely valuable for examining mixtures of organisms, albeit isolated from their natural environment (Senior *et al.* 1976).

35.8. Rhizosphere studies

All pesticides used in agriculture are applied either directly to a crop at a particular stage of growth or to soil prior to crop emergence. In each case, the pesticides are present in the crop and/or in the soil near crop roots for some period of time, depending on the pesticide's persistence. Thus, if the chemical is toxic to microorganisms, it has the potential to affect directly the rhizosphere or root-region microflora. Chemicals which are not toxic to microorganisms may affect the microflora indirectly through effects on crop physiology or root morphology which may change root exudation patterns.

Greaves (1978, 1980) has shown, for example, that normal application rates of the herbicide mecoprop (($\pm$)-2-(4-chloro-2-methylphenoxy) propionic acid) can be associated with increased invasion of plant roots by bacteria, presumably as a result of root-stunting caused by the herbicide. Other herbicides, e.g. dichlobenil (2,6-dichlorobenzonitrile) and trifluralin (2,6-dinitro-N,N-dipropyl-4-trifluoromethylaniline) are known to be associated with root damage which may also involve microbial responses.

Techniques of studying the rhizosphere microflora are, perhaps, not so numerous as those for organisms in root-free soil. Most studies are based on some form of viable counting of populations, followed by isolation and testing for potential activities (Greaves & Webley 1965).

A variety of techniques applicable to rhizosphere studies are described by Harley and Scott Russell (1979). Rovira (1979) and Lynch (see Chapter 23) have described more recent developments. Bowen and Rovira (1976) have drawn attention to some changes from traditional studies concerning the numbers and types of microorganisms in the root region. An important feature is the release of organic materials from roots, especially if the exudates are changed by or include the pesticide applied. This will affect many aspects of microbial interaction with the root, including colonisation patterns. In all investigations of interactions between foliar-applied pesticides and rhizosphere microflora, it must be remembered that the amount of chemical retained on the plant is far more important than that delivered by the sprayer. It is the amount deposited on the plant that determines what is taken up and translocated and thus the effect, if any, on the root microflora. Rovira (1979) advocates observation of roots by light and electron microscopy and modelling of root region populations. He further suggests areas of research such as interactions between root pathogens and root function, root-associated nitrogen fixation and mycorrhizae as requiring more attention. Rovira (1979) refers to many detailed techniques which can be used, and which are potentially valuable, for research into pesticide interactions with root region microflora.

Root-associated nitrogen fixation, especially symbiotic nitrogen fixation by legumes, deserves special mention as it is included in regulatory requirements (see Chapter 25). As mentioned earlier (see section 35.5.3), there are several problems associated with these techniques which must be resolved before unequivocal assessments of the effects of pesticides on the legume–*Rhizobium* symbiosis can be made.

Mycorrhizae also require intensive research with regard to pesticide effects. In particular, the vesicular-arbuscular forms are important in view of their wide distribution in agricultural crops. Comprehensive reviews, including many methods appropriate for pesticide studies, are given by Sanders *et al.* (1975) and in Chapter 26.

It is particularly difficult to recommend individual methods for use in studies of pesticide–root region microflora interactions. Root region microbiology is ill defined and few studies have included pesticides. The development of methods has, until recently, been slow. The increasing use of light microscopy and transmission and scanning electron microscopy is an exciting development as it permits a clearer understanding of the spatial distribution of organisms around roots and of the development of populations of organisms within root tissues (Fig. 1).

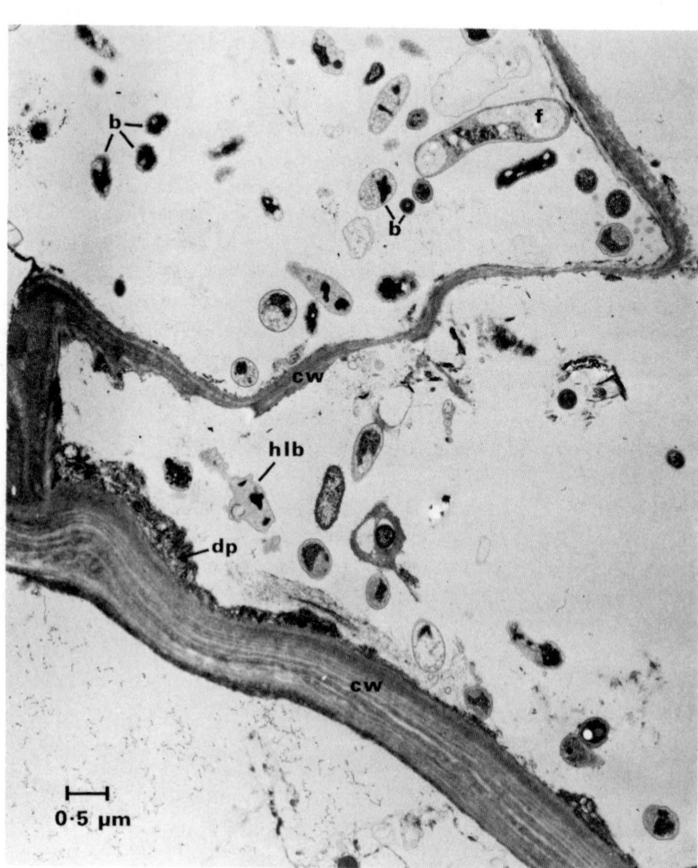

Figure 1. Electron micrograph showing microbial colonisation of wheat root cortex cells: b, bacteria; cw, cell wall; dp, decomposing protoplasm; f, fungus; hlb, helically lobed bacterium (by permission of J. A. Sargent, ARC Weed Research Organization).

35.9 Microcosms

One of the oldest, and least resolved, arguments in microbial ecology is that concerning the relative merits of laboratory versus field studies (Bull 1980). In assessing the effects of pesticides on microbial functions it is accepted that field studies are essential but in the majority of cases it is impossible to carry them out effectively. Consequently, there has been some emphasis in recent years on the development and use of laboratory models of ecosystems to investigate, for example, nutrient cycles and the fate of pollutants or pesticides.

Bull (1980) has reviewed some aspects of model ecosystems, popularly known as microcosms, and points out that 'microcosms should be viewed as analytical tools and not reproductions of ecosystems.' Nevertheless, it is considered by many that microcosms, provided that 'they are made functionally similar to the ecosystem being mimicked . . . are sound investigative systems . . .' (Bull 1980). However, it must be remembered that the data obtained are generally a function of the microcosm design and will only reveal potential activity and effects. Useful reviews of microcosm design and application are given by Witt and Gillett (1977) and Chapter 16.

35.9.1 Soil Cores

Among the simplest microcosm designs is the soil core. This may be taken from bare soil (Wingfield *et al.* 1977) or from a plant community such as grassland (Draggan 1977). Wingfield *et al.* (1977) have shown that effects of the herbicide dalapon on the microflora were less marked in the top 1 cm of soil cores than in sieved soils, even though the dalapon concentration (180 μg (g dry soil)$^{-1}$) in the top 1 cm of the cores was higher than the concentration obtained in the disturbed soils (110 μg (g dry soil)$^{-1}$). This showed that the use of soil cores had several advantages over disturbed soils. In addition, it suggested that the results obtained were more indicative of the likely effects of the herbicide in the field. Conversely, Johnen and Drew (1977) considered that the disadvantages of variation between cores precluded their use for routine investigations. K. H. Domsch (quoted in Greaves *et al.* 1980) suggested that the effects of natural phenomena (i.e. natural variations) can be used to evaluate the effects of pesticides and so variation between cores may be a useful property. Provided proper replication is used, cores can contribute greatly to our knowledge of the interactions between pesticides and the soil microflora.

35.9.2 Complex Systems

Considerable attention has focused on complex microcosms during the past decade. A wide range of systems has been used including the terrestrial/aquatic system described by Metcalf *et al.* (1971) and Metcalf (1973), as well as the more sophisticated microagroecosystems, such as large blocks of cropped soil held in boxes in a glasshouse (Nash & Beall 1977; Witt & Gillett 1977). They are designed, principally, for examining the environmental fate of chemicals rather than for assessing their side-effects on microorganisms. Nonetheless, they should all be useful in such assessments, as they are more realistic models of the actual environment where pesticides are used than are the present 'flasks of sieved soil'.

35.9.3 Chemostats

In his elegant consideration of biodegradation, Bull (1980) details many facets of the use of chemostats for modelling natural environments. His comments apply whether the objective of the study is biodegradation of a pesticide or its effects on a mixed community of microorganisms. An excellent example of change in community structure which may arise from pesticide treatment, and of the usefulness of the chemostat in detecting and evaluating the change, is given by Senior *et al.* (1976). Slater (1979) has described the fundamentals of population and community growth kinetics in both closed and open growth systems. Veldkamp (1976) has surveyed some of the possibilities which the continuous-flow culture technique presents in studying interrelationships between microorganisms and their environment. It takes little imagination to see how the use of chemostats can provide fundamental data essential to the development of methods which allow the prediction of pesticide side-effects with confidence.

35.10 Interpretation and evaluation of data

It is clear from the information presented in this chapter, and elsewhere in this book, that there are sufficient methods available to make data acquisition all too easy. Even if the methods applied are limited to just two or three, as recommended by Greaves *et al.* (1980), the amount of data accumulated can be very large. Consequently, the problem of interpretation and evaluation of this data is equally large.

Until recently, judgement of importance has been a subjective process depending on factors such as the experimenter's training, experience and bias (Greaves 1978). More recently, K. H. Domsch (quoted in Greaves *et al.* 1980 and in Greaves & Malkomes 1981) has described a scheme of evaluation which introduces an objective element. Domsch proposes that the magnitude and duration of side-effects ought to be compared to those of effects of natural stresses, such as drought. Only if the pesticide effects are greater than those of natural stress should they be considered potentially risky and be investigated further. Risk evaluation is a matter of defining what is tolerable. Domsch's scheme fits this pattern in that effects of natural stress often have to be tolerable (as we cannot eliminate them) and thus set a baseline for comparison with pesticide-induced effects. Laskowski (1979) independently proposed a similar approach. As yet, such schemes are at an early stage and considerably more developmental work is required.

35.11 Conclusions

This chapter follows several which describe many methods applicable to pesticide studies. All methods have intrinsic advantages and disadvantages which make them more or less suitable for particular applications and so no attempt has been made to recommend any method as being right for a particular job. Methods must be chosen carefully according to the objectives of the experiments and this chapter is, therefore, confined to a survey of the types of method used commonly in pesticide studies. By presenting background references it is hoped the reader will be able to construct and apply methods successfully for his own experiments and their objectives.

It is worth stating, despite its obvious nature, that a pesticide-stressed environment is little different from any other environment in terms of the restraints and requirements it places on the ecologist. The major difference is the presence of the pesticide. Thus, the physico-chemical nature of the pesticide, its concentration and distribution in the environment and its persistence are important factors to be considered when designing experiments.

Lee and Jones (1978) stated 'there is little point in developing a highly sensitive, accurate and precise analytical procedure when some of the other information needed for hazard assessment is known only to orders of magnitude'. Most pesticide–soil microflora investigations stem from the assumption that toxic effects may affect microbial contributions to soil fertility. Another fear is that environmental quality may be impaired. Neither of these aspects of microbiology can be precisely defined, let alone quantified. Until quantities can be ascribed to contributions to soil fertility, perhaps investigations of pesticide-stressed environments should be limited to simple experimentation, such as that recommended by Greaves *et al.* (1980), and more energy be spent investigating the role of microorganisms in normal environments.

Recommended reading

GREAVES M. P., DAVIES H. A., MARSH J.A.P. & WINGFIELD G. I. (1976) Herbicides and soil microorganisms. *CRC Critical Reviews in Microbiology* **5**, 1–38.

GREAVES M. P. & MALKOMES H-P. (1981) Effects on soil microflora. In: *Interactions Between Herbicides and the Soil* (Ed. R. J. Hance), pp. 223–53. Academic Press, London.

GROSSBARD E. (1976) Effects on the soil microflora. In: *Herbicides: Physiology, Biochemistry, Ecology* (Ed. L. J. Audus), pp. 99–147. Academic Press, London.

HILL I. R. & WRIGHT S. J. L. (1978) *Pesticide Microbiology*. Academic Press, London.

HURLBERT S. H. (1975) Secondary effects of pesticides on aquatic ecosystems. *Residue Reviews* **57**, 81–148.

PAPAVIZAS G. C. & LEWIS J. A. (1979) Side-effects of pesticides on soil-borne plant pathogens. In: *Soil-Borne Plant Pathogens* (Eds B. Schippers & W. Gams), pp. 483–505. Academic Press, London.

SIMON-SYLVESTRE G. & FOURNIER J-C. (1979) Effects of pesticides on the soil microflora. *Advances in Agronomy* **31**, 1–92.

STOTZKY G. (1974) Activity, ecology, and population dynamics of micro-organisms in soil. In: *Microbial Ecology* (Eds A. I. Laskin & H. Lechevalier), pp. 57–135. CRC Press, Cleveland, USA.

TU C. M. & MILES J. R. W. (1976) Interactions between insecticides and soil microbes. *Residue Reviews* **64**, 17–65.

WARE G. W. & ROAN C. C. (1970) Interaction of pesticides with aquatic micro-organisms and plankton. *Residue Reviews* **33**, 15–45.

References

ALEXANDER M. (1977) *Introduction to Soil Microbiology*. John Wiley & Sons, London.

ALTMAN J. & CAMPBELL C. L. (1977) Effect of herbicides on plant diseases. *Annual Review of Phytopathology* **15**, 361–85.

ANDERSON J. M. & MACFADYEN A. (1976) *The Role of Terrestrial and Aquatic Organisms in Decomposition Processes*. Blackwell Scientific Publications, Oxford.

ANDERSON J. P. E. & DOMSCH K. H. (1973a) Selective inhibition as a method for estimation of the relative activities of microbial populations in soils. In: *Modern Methods in the Study of Microbial Ecology, Ecological Bulletin 17* (Ed. T. Rosswall), pp. 281–2. Swedish Natural Science Research Council, Stockholm.

ANDERSON J. P. E. & DOMSCH K. H. (1973b) Quantification of bacterial and fungal contributions to soil respiration. *Archiv für Mikrobiologie* **93**, 113–27.

ANDERSON J. P. E. & DOMSCH K. H. (1974) Use of selective inhibitors in the study of respiratory activities and shifts in bacterial and fungal populations in soil. *Annals of Microbiology* **24**, 189–94.

ANDERSON J. P. E. & DOMSCH K. H. (1978) A physiological method for the quantitative measurement of microbial biomass in soils. *Soil Biology and Biochemistry* **10**, 215–21.

ANDERSON J. R. (1978a) Some methods for assessing pesticide effects on non-target soil micro-organisms and their activities. In: *Pesticide Microbiology* (Eds I. R. Hill & S. J. L. Wright), pp. 247–312. Academic Press, London.

ANDERSON J. R. (1978b) Pesticide effects on non-target soil micro-organisms. In: *Pesticide Microbiology* (Eds I. R. Hill & S. J. L. Wright), pp. 313–53. Academic Press, London.

BOWEN G. D. & ROVIRA A. D. (1976) Microbial colonization of plant roots. *Annual Review of Phytopathology* **14**, 121–44.

BREMNER J. M. & MULVANEY R. L. (1978) Urease activity in soils. In: *Soil Enzymes* (Ed. R. G. Burns), pp. 149–96. Academic Press, London.

BROCK T. D. (1971) Microbial growth rates in nature. *Bacteriological Reviews* **35**, 39–58.

BULL A. T. (1980) Biodegradation: Some attitudes and strategies of micro-organisms and microbiologists. In: *Contemporary Microbial Ecology* (Eds D. C. Ellwood, J. N. Hedger, M. J. Latham, J. M. Lynch & J. H. Slater), pp. 107–36. Academic Press, London.

BUNYAN P. J. & STANLEY P. I. (1979) Assessment of the environmental impact of new pesticides for regulation purposes. *1979 Proceedings British Crop Protection Conference—Pests and Diseases*, pp. 881–91.

BURNS R. C. & HARDY R. W. F. (1975) *Nitrogen-Fixation in Bacteria and Higher Plants*. Springer-Verlag, New York.

BURNS R. G. (1978) *Soil Enzymes*. Academic Press, London.

BUTLER G. L. (1977) Algae and pesticides. *Residue Reviews* **66**, 19–62.

CERVELLI S., NANNIPIERI P., GIOVANNINI G. & PERNA A. (1976) Relationships between substituted urea herbicides and soil urease activity. *Weed Research* **16**, 365–8.

CERVELLI S., NANNIPIERI P. & SEQUI P. (1978) Interactions between agrochemicals and soil enzymes. In: *Soil Enzymes* (Ed. R. G. Burns), pp. 251–93. Academic Press, London.

COLLINS V. G., JONES J. G., HENDRIE M.S., SHEWAN J. M., WYNN-WILLIAMS D. D. & RHODES M. E. (1973) Sampling and estimation of bacterial populations in the aquatic environment. In: *Sampling—Microbiological Monitoring of Environments* (Eds R. G. Board & D. W. Lovelock), pp. 77–110. Academic Press, London.

COOPER S. L., WINGFIELD G. I., LAWLEY R. & GREAVES M. P. (1978) Miniaturized methods for testing the toxicity of pesticides to micro-organisms. *Weed Research* **18**, 105–7.

COSGROVE D. J. (1967) Metabolism of organic phosphates in soil. In: *Soil Biochemistry* (Eds A. D. McLaren & G. H. Peterson), pp. 216–28. Marcel Dekker, New York.

Council of Europe (1973) *Agricultural Pesticides*. Council of Europe, Strasbourg.

CULLIMORE D. R. & BALL L. (1978) New monitoring system for proteolysis in soil as influenced by selected herbicidal applications. *Applied and Environmental Microbiology* **36**, 959–61.

CURL E. A. & RODRIGUEZ-KABANA R. (1972) Microbial interactions. In: *Research Methods in Weed Science* (Ed. R. E. Wilkinson), pp. 161–94. Southern Weed Science Society, Auburn, Alabama.

CURL E. A. & RODRIGUEZ-KABANA R. (1977) Herbicide-plant disease relationships. In: *Research Methods in Weed Science* (Ed. B. Truelove), pp. 173–191. Southern Weed Science Society, Auburn, Alabama.

DICKINSON C. H. & PREECE T. F. (1976) *Microbiology of Aerial Plant Surfaces*. Academic Press, London.

DRAGGAN S. (1977) Effects of substrate type and arsenic dosage level on arsenic behaviour in grassland microcosms. I. Preliminary results on ^{74}As transport. In: *Terrestrial Microcosms and Environmental Chemistry* (Eds J. M. Witt & J. W. Gillett), pp. 102–10. National Science Foundation, Washington.

EPA (1978a) Environmental Protection Agency, Registration of pesticides in the United States, proposed guidelines. *Federal Register* **43**, (132) Part II, 29696–29741.

EPA (1978b) *Guidelines for Registering Pesticides in the United States. Subpart J, Hazard Evaluation: Non-Target Plants and Micro-Organisms*. Criteria and Evaluation Division, Office of Pesticide Programs, U.S. EPA, Washington.

Federal Working Group on Pest Management (1974) *Guidelines on Sampling and Statistical Methodologies for Ambient Pesticide Monitoring*. Federal Working Group on Pest Management, Washington.

FRENEY J. R. (1967) Sulfur-containing organics. In: *Soil Biochemistry* (Eds A. D. McLaren & G. H. Peterson), pp. 229–59. Marcel Dekker, New York.

GAUTHIER S. M., ASHTAKALA S. S. & LENOIR J. A. (1976) Inhibition of soil urease activity and nematocidal action of 3-amino-1,2,4-triazole. *Horticultural Science* **11**, 481–2.

GRAY T. R. G. & WILLIAMS S. T. (1971) *Soil Micro-Organisms*. Oliver & Boyd, Edinburgh.

GREAVES M. P. (1978) Problems and progress in the evaluation of herbicide safety to the soil microflora. *Report of Weed Research Organization 1976–77*, pp. 95–103. Agricultural Research Council, London.

GREAVES M.P. (1979) Measurement and interpretation of side-effects of pesticides on microbial processes. *Proceedings 1979 British Crop Protection Conference—Pests and Diseases*, pp. 469–75.

GREAVES M. P. (1980) Herbicide effects on the root microflora. *Report of Weed Research Organization 1978–79*, pp. 28–30. Agricultural Research Council, London.

GREAVES M. P., DAVIES H. A., MARSH J. A. P. & WINGFIELD G. I. (1976) Herbicides and soil micro-organisms. *CRC Critical Reviews in Microbiology* **5**, 1–38.

GREAVES M. P., LOCKHART L. A. & RICHARDSON W. G. (1978) Measurement of herbicide effects on nitrogen fixation by legumes. *Proceedings 1978 British Crop Protection Conference—Weeds*, pp. 581–5.

GREAVES M. P. & MALKOMES H-P. (1980) Effects on soil microflora. In: *Interactions between Herbicides and the Soil* (Ed. R. J. Hance), pp. 223–53. Academic Press, London.

GREAVES M. P., POOLE N. J., DOMSCH K. H., JAGNOW G. & VERSTRAETE W. (1980) *Recommended Tests for Assessing the Side-Effects of Pesticides on the Soil Microflora. Technical Report, Agricultural Research Council Weed Research Organization*, Number 59.

GREAVES M. P. & WEBLEY D. M. (1965) A study of the breakdown of organic phosphates by micro-organisms from the root region of certain pasture grasses. *Journal of Applied Bacteriology* **28**, 454–65.

GROSSBARD E. (1976) Effects on the soil microflora. In: *Herbicides: Physiology, Biochemistry, Ecology* (Ed. L. J. Audus), pp. 99–147. Academic Press, London.

HANCE R. J. (1980) *Interactions between Herbicides and the Soil*. Academic Press, London.

HARDY R. W. F., HOLSTEN R. D., JACKSON E. K. & BURNS R. C. (1968) The acetylene-ethylene assay for N_2 fixation: laboratory and field evaluation. *Plant Physiology* **43**, 1185–207.

HARLEY J. L. & SCOTT RUSSELL R. (1979) *The Soil-Root Interface*. Academic Press, London.

HEATH G. W., ARNOLD M. K. & EDWARDS C. A. (1966) Studies in leaf litter breakdown. I. Breakdown rates of leaves of different species. *Pedobiologia* **6**, 1–12.

HELLING S., KEARNEY P. C. & ALEXANDER M. (1971) Behaviour of pesticides in soils. *Advances in Agronomy* **23**, 147–229.

HILL I. R. & WRIGHT S. J. L. (1978) *Pesticide Microbiology*. Academic Press, London.

HISLOP E. C. (1976) Some effects of fungicides and other agrochemicals on the microbiology of the aerial surfaces of plants. In: *Microbiology of Aerial Plant Surfaces* (Eds C. H. Dickinson & T. F. Preece), pp. 41–74. Academic Press, London.

HISSETT R. & GRAY T. R. G. (1976) Microsites and time changes in soil microbe ecology. In: *The Role of Terrestrial and Aquatic Organisms in Decomposition Processes* (Eds J. M. Anderson & A. Macfadyen), pp. 23–39. Blackwell Scientific Publications, Oxford.

HOWARD P. J. A. (1968) The use of Dixon and Gibson respirometers in soil and litter respiration studies. *Merlewood Research and Development Paper No. 5*. Nature Conservancy Council, UK.

HURLBERT S. H. (1975) Secondary effects of pesticides on aquatic ecosystems. *Residue Reviews* **57**, 81–148.

JENKINSON D. S. & POWLSON D. S. (1976) The effects of biocidal treatments on metabolism in soil. V. A method for measuring soil biomass. *Soil Biology and Biochemistry* **8**, 209–13.

JENKINSON D. S., POWLSON D. S. & WEDDERBURN R. W. (1976) The effects of biocidal treatments on metabolism in soil. III. The relationship between soil biovolume, measured by optical microscopy and the flush of decomposition caused by fumigation. *Soil Biology and Biochemistry* **8**, 189–202.

JOHNEN B. G. (1978) Recent advances in the study of effects of pesticides on the population dynamics of non-target micro-organisms. *1978 Proceedings British Crop Protection Conference—Weeds*, pp. 1037–46.

JOHNEN B. G. & DREW E. A. (1977) Ecological effects of pesticides on soil micro-organisms. *Soil Science* **123**, 319–24.

JOHNEN B. G., DREW E. A. & CASTLE D. L. (1979) Studies on the effect of pesticides on symbiotic nitrogen fixation. In: *Soil-Borne Plant Pathogens* (Eds B. Schippers & W. Gams), pp. 513–23. Academic Press, London.

JONES J. G. (1979) *A Guide to Methods for Estimating Microbial Numbers and Biomass in Fresh Water. Scientific Publication No. 39*. Freshwater Biological Association, Ambleside, Cumbria, UK.

KECSKES M. (1972) A survey of herbicide sensitivity and resistance of rhizobia. In: *Symposia Biologica Hungarica 11* (Ed. J. Szegi), pp. 405–15. Akademiai Kiado, Budapest.

KILBERTUS G., REISINGER O., MOUREY A. & CONCELA DA FONSECA J. A. (1975) *Biodégradation et Humification. Proceedings of 1st International Colloquium on Biodegra-*

dation and Humification. Pierron, Sarraguemines.

KISS S., DRAGAN-BULARDA M. & RADULESCU D. (1975) Biological significance of enzymes accumulated in soil. *Advances in Agronomy* **27**, 28–87.

KORNBERG H. (1979) *Royal Commission on Environmental Pollution, 7th Report, Agriculture and Pollution*. HMSO, London.

LASKOWSKI D. A. (1979) Effects of herbicides on microorganisms and microbial processes. *Abstracts of 1979 Meeting of the Weed Science Society of America*, p. 118.

LATTER P. M. & HOWSON G. (1977) The use of cotton strips to indicate cellulose decomposition in the field. *Pedobiologia* **17**, 145–55.

LEE C. C., HARRIS R. F., WILLIAMS J. D. H., SYERS J. K. & ARMSTRONG D. E. (1971) Adenosine triphosphate in lake sediments. II. Origin and significance. *Soil Science Society of America Proceedings* **35**, 86–91.

LEE G. F. & JONES R. A. (1978) *ASTM Summary of 1978 Activities, Environmental Chemistry—Fate Modelling. Occasional Paper No. 36*. Environmental Engineering Programme, Colorado State University, Fort Collins.

LEES H. & QUASTEL J. H. (1946) Biochemistry of nitrification in soil. I. Kinetics of, and the effect of poisons on, soil nitrification as studied by a soil perfusion technique. *Biochemical Journal* **40**, 803–15.

LETHBRIDGE G. & BURNS R. G. (1976) Inhibition of soil urease by organophosphorus insecticides. *Soil Biology and Biochemistry* **8**, 99–102.

LOHM U. & PERSSON T. (1977) *Soil Organisms as Components of Ecosystems. Ecological Bulletin No. 25*. Swedish Natural Science Research Council, Stockholm.

LYNCH J. M. & PANTING L. M. (1980) Cultivation and the soil biomass. *Soil Biology and Biochemistry* **12**, 29–34.

MCCANN A. E. & CULLIMORE D. R. (1979) Influence of pesticides on the soil algal flora. *Residue Reviews* **72**, 1–31.

MACKERETH F. J. H., HERON J. & TALLING J. F. (1978) *Water Analysis: Some Revised Methods for Limnologists. Scientific Publication No. 36*. Freshwater Biological Association, Ambleside, Cumbria, UK.

MAFF (Ministry of Agriculture, Fisheries and Food) (1971) *Pesticides Safety Precautions Scheme Agreed between Government Departments and Industry, Appendix D*, p. 7. Ministry of Agriculture, Fisheries and Food, Pesticides Branch, London.

METCALF R. L. (1973) A laboratory model ecosystem evaluation of the chemical and biological behaviour of radio labelled micropollutants. *Proceedings of the FAO/IACA/WHO Symposium on Nuclear Techniques in Comparative Studies of Food and Environmental Contamination*, pp. 49–62. Otaniemi, Finland.

METCALF R. L., SAUGHA G. K. & KAPOOR I. P. (1971) Model ecosystem for the evaluation of pesticide biodegradability and ecological magnification. *Environmental Science and Technology* **5**, 709–13.

NASH R. G. & BEALL M. L. (1977) A microagroecosystem to monitor the environmental fate of pesticides. In: *Terrestrial Microcosms and Environmental Chemistry* (Eds J. M. Witt & J. W. Gillett), pp. 86–94. National Science Foundation, Washington.

NEWTON W. E., POSTGATE J. R. & RODRIGUEZ-BARRUECO C. (1977) *Recent Developments in Nitrogen Fixation*. Academic Press, London.

NUTMAN P. S. (1976) *Symbiotic Nitrogen Fixation in Plants*. Cambridge University Press, Cambridge.

PAPAVIZAS G. C. & LEWIS J. A. (1979) Side-effects of pesticides on soil-borne plant pathogens. In: *Soil-Borne Plant Pathogens* (Eds B. Schippers & W. Gams), pp. 483–505. Academic Press, London.

PARKINSON D., GRAY T. R. G. & WILLIAMS S. T. (1971) *Methods for Studying the Ecology of Soil Micro-Organisms*. Blackwell Scientific Publications, Oxford.

PARKINSON D. & LOUSIER J. D. (1975) Litter decomposition in a cool temperate woodland. In: *Biodégradation et Humification. I. Proceedings of 1st International Colloquium on Biodegradation and Humification* (Eds G. Kilbertus, O. Reisinger, A. Mourey & J. A. Concela da Fonseca), pp. 75–87. Pierron, Sarraguemines.

PAYNE W. J. (1973) The use of gas chromatography for studies of denitrification in ecosystems. In: *Modern Methods in the Study of Microbial Ecology, Ecological Bulletin 17* (Ed. T. Rosswall), pp. 263–8. Swedish Natural Science Research Council, Stockholm.

PETERSON H. L. & FREDERICK L. R. (1979) A direct microscopic ratio-method using polystyrene beads to determine microbial numbers in soil. *Soil Biology and Biochemistry* **11**, 77–83.

PREECE T. F. & DICKINSON C. H. (1971) *Ecology of Leaf Surface Micro-Organisms*. Academic Press, London.

QUISPEL A. (1974) *The Biology of Nitrogen Fixation*. North-Holland Publishing Company, Amsterdam.

REINERS W. A. (1968) Carbon dioxide evolution from the floor of three Minnesota forests. *Ecology* **49**, 471–83.

ROVIRA A. D. (1979) Biology of the soil-root interface. In: *The Soil-Root Interface* (Eds J. L. Harley and R. Scott Russell), pp. 145–60. Academic Press, London.

SANDERS F. E., MOSSE B. & TINKER P. B. (1975) *Endomycorrhizas*. Academic Press, London.

SENIOR E., BULL A. T. & SLATER J. H. (1976) Enzyme evaluation in a microbial community growing on the herbicide dalapon. *Nature, London* **263**, 476–9.

SIMON-SYLVESTRE G. & FOURNIER J-C. (1979) Effects of pesticides on the soil microflora. *Advances in Agronomy* **31**, 1–92.

SKUJINS J. (1978) History of abiontic soil enzyme research. In: *Soil Enzymes* (Ed. R. G. Burns), pp. 1–49. Academic Press, London.

SLATER J. H. (1979) Microbial population and community dynamics. In: *Microbial Ecology: a Conceptual Approach* (Eds J. M. Lynch & N. J. Poole), pp. 45–63. Blackwell Scientific Publications, Oxford.

SOROKIN Y. I. & KADOTA H. (1972) *Techniques for the Assessment of Microbial Production and Decomposition in Fresh Waters. IBP Handbook No. 23.* Blackwell Scientific Publications, Oxford.

STEWART W. D. P. (1975) *Nitrogen-Fixation by Free-Living Micro-Organisms. International Biological Programme 6.* Cambridge University Press, Cambridge.

STOTZKY G. (1965) Microbial respiration. In: *Methods of Soil Analysis. Part 2. Chemical and Microbiological Properties* (Eds C. A. Black, D. D. Evans, J. L. White, L. E. Ensminger & F. E. Clark), pp. 1550–72. American Society of Agronomy, Madison.

STOTZKY G. (1974) Activity, ecology, and population dynamics of micro-organisms in soil. In: *Microbial Ecology* (Eds A. I. Laskin & H. Lechevalier), pp. 57–135. CRC Press, Cleveland, U.S.A.

STRANGE R. E. (1976) *Patterns of Progress. Microbial Response to Mild Stress.* Meadowfield Press Ltd., Shildon, UK.

TU C. M. & MILES J. R. W. (1976) Interactions between insecticides and soil microbes. *Residue Reviews* **64**, 17–65.

VELDKAMP H. (1976) *Patterns of Progress. Continuous Culture in Microbial Physiology and Ecology.* Meadowfield Press Ltd., Shildon, UK.

VOLLENWEIDER R. A. (1969) *A Manual on Methods for Measuring Primary Production in Aquatic Environments. IBP Handbook No. 12.* Blackwell Scientific Publications, Oxford.

WAITT A. W. (1975) Pesticide legislation and industry. *Pesticide Science* **6**, 199–208.

WALLIS G. W. & WILDE S. A. (1957) Rapid determination of carbon dioxide evolved from forest soils. *Ecology* **38**, 359–61.

WARE G. W. & ROAN C. C. (1970) Interaction of pesticides with aquatic micro-organisms and plankton. *Residue Reviews* **33**, 15–45.

WEBLEY D. M. & DUFF R. B. (1962) A technique for investigating localized microbial development in soils. *Nature, London* **194**, 364–5.

WILDISH D. J., POOLE N. J. & JOLES S. J. (1979) Problems in determining soil ATP. *Bulletin of Environmental Contamination and Toxicology* **23**, 192–5.

WILLIAMS S. T. & GRAY T. R. G. (1973) General principles and problems of soil sampling. In: *Sampling—Microbiological Monitoring of Environments* (Eds R. G. Board & D. W. Lovelock), pp. 111–23. Academic Press, London.

WINGFIELD G. I. (1980a) Effect of asulam on cellulose decomposition in three soils. *Bulletin of Environmental Contamination and Toxicology* **24**, 473–6.

WINGFIELD G. I. (1980b) Effects of time of soil collection and storage on microbial decomposition of cellulose in soil. *Bulletin of Environmental Contamination and Toxicology* **24**, 671–5.

WINGFIELD G. I., DAVIES H. A. & GREAVES M. P. (1977) The effect of soil treatment on the response of the soil microflora to the herbicide dalapon. *Journal of Applied Bacteriology* **43**, 39–46.

WITKAMP M. (1966) Decomposition of leaf litter in relation to environment, microflora and microbial respiration. *Ecology* **47**, 194–201.

WITT J. M. & GILLETT J. W. (1977) *Terrestrial Microcosms and Environmental Chemistry.* National Science Foundation, Washington.

YEO D. (1979) A view from the agricultural chemicals industry. In: *Safety of Chemicals in the Environment.* Proceedings Harwell Environmental Seminar, pp. 119–27.

ZANTUA M. I. & BREMNER J. M. (1975) Comparison of methods of assaying urease activity in soils. *Soil Biology and Biochemistry* **7**, 291–5.

Chapter 36 • Gaseous and Heavy Metal Air Pollutants

Harvey Babich and Guenther Stotzky

36.1 Introduction

Industrial and domestic activities have accelerated the rates of transfer of numerous elements through the hydrosphere, lithosphere and atmosphere. This extraneous impact on the natural cycling of these elements has resulted in their increased mobilisation, transport, deposition and subsequent accumulation in the various microbial ecosystems. This chapter will discuss the effects of atmospheric pollutants, specifically gaseous and heavy metal particulates, on microbial ecology, inasmuch as these contaminants, once in the atmosphere, are subjected to passive and turbulent deposition processes which eventually introduce them into terrestrial and aquatic environments. Consequently, any discussion of atmospheric pollutants must include their effects on the activities, ecology and population dynamics of microbes in terrestrial and aquatic environments (Babich & Stotzky 1972, 1974, 1978e).

Microorganisms are directly involved in atmospheric pollution in that they are the source of substantial quantities of various gaseous pollutants, both inorganic and organic, and they serve as sinks for the removal of numerous atmospheric constituents. Microbes are also responsible for the biotransformation of many pollutants, and in this chapter special mention will be made of the methylation of heavy metals; a process which results in organometallic compounds which are usually more toxic than were the original inorganic forms. Microorganisms and viruses in the airborne state may also be considered as atmospheric pollutants (Babich & Stotzky 1978e).

Microbes and viruses are recipients of, and responders to, atmospheric pollutants. Studies performed *in vitro* and *in situ* have demonstrated that atmospheric pollutants may adversely affect: the generation time of bacteria; spore germination, mycelial proliferation, fruiting body formation and spore production by fungi; microbial respiratory activity; photosynthesis of cyanobacteria, algae and lichens; nitrogenase activity of microbes involved in dinitrogen fixation; and viral infectivity. Furthermore, microbial activities (such as nitrification, denitrification, litter decomposition and mineralisation of carbon, nitrogen and phosphorus) and interactions (such as host–parasite, host-saprophyte and mutualism) are often adversely affected by atmospheric pollutants (Babich & Stotzky 1974, 1978e).

The physico-chemical characteristics of the environment into which a pollutant is deposited affect the chemical form and, hence, the availability and toxicity of the pollutant to the microbiota. The environmental factors which may alter the toxicity of both gaseous and heavy metal pollutants include pH, relative humidity, moisture content, aeration, inorganics (both cationic and anionic), soluble and particulate organics, clay minerals, cation exchange capacity, temperature, light, and hydrostatic pressure. Other abiotic factors, such as the concentration of the pollutant, the length of exposure, interactions between pollutants, and the chemical speciation of the pollutant, also influence the responses of microbes to atmospheric pollution (Babich & Stotzky 1980a, b).

36.2 Interrelations between microbial activities and atmospheric pollution

36.2.1 Microbes as Sources of Atmospheric Pollutants

Many of the gaseous constituents in the troposphere arise from biological and abiological processes occurring on or near the surface of the Earth. These atmospheric gases are derived from both natural (biotic and abiotic) and anthropogenic (industrial and domestic) activities. Natural abiotic sources include emissions from volcanoes, forest fires ignited by lightning, chemical reactions in soil and aquatic systems, and chemical and photochemical reactions in the troposphere itself. Natural biotic sources include emissions from living microorganisms, plants and animals, and from decaying organic matter. Industrial (e.g. smelting, mining, sewage treatment) and domestic (e.g. incineration, automobile exhaust) activities contribute significant quantities of gases to the atmosphere (Babich & Stotzky 1972, 1974).

Some atmospheric gases (e.g. peroxyacetyl nitrate—PAN) are derived only from anthropogenic sources, although most originate from both natural and anthropogenic sources. The degree of pollution is defined as the difference between the concentration of the background levels arising from natural sources and the measured concentration of gases (Babich & Stotzky 1974).

On a global basis, the quantities of a gas derived from natural sources may exceed those emitted from anthropogenic activities. For example, 3.6×10^{11} kg yr^{-1} of carbon monoxide is emitted from anthropogenic sources, whereas natural sources are responsible for 3.0×10^{12} kg yr^{-1} (Rasmussen *et al.* 1975). Anthropogenic carbon monoxide results from the incomplete combustion of carbonaceous fuels, such as petroleum, coal and wood, with automobile emissions being the main anthropogenic source. Abiotically, carbon monoxide is produced during atmospheric oxidation of methane, by forest and prairie fires ignited by lightning, in clouds, by photochemical oxidation of organic matter at the surface of oceans, and in volcanoes (Babich & Stotzky 1972, 1974; Nozhevnikova & Yurganov 1978). The oceans are also natural biotic sources of carbon monoxide (Swinnerton *et al.* 1970): marine kelps, diatoms, siphonophores and the bacterial decomposition of marine phytoplankton evolve carbon monoxide. Terrestrial green plants, freshwater algae, soil fungi, seeds during germination, and haem catabolism in animals also produce carbon monoxide (Babich & Stotzky 1972, 1974). Natural emissions of nitrogen dioxide, hydrogen sulphide and ammonia also exceed those from anthropogenic sources (Rasmussen *et al.* 1975).

If natural emissions of carbon monoxide and some other gases exceed anthropogenic emissions, why are these gases considered to be pollutants? The answer to this is that natural emissions occur over a much greater surface area than do anthropogenic emissions and, as they are subject to a greater dilution effect by atmospheric turbulence, their background concentrations are low. Conversely, anthropogenic emissions occur in restricted areas, often near urban complexes, and, therefore, their concentrations are high and more hazardous to the local biosphere (Babich & Stotzky 1974).

Microorganisms, either directly through their metabolism or indirectly through their alteration of organic matter, are the principal sources of biologically derived gases in the atmosphere; for example, atmospheric ammonia is released during microbial deamination of organic matter but also arises from nitrate reduction (Schlegel 1974).

Denitrifying bacteria in soils and aquatic environments account for most of the production of gaseous nitric oxide, nitrous oxide and dinitrogen (see Chapter 12). The denitrifiers are facultative aerobes and include such species as *Pseudomonas denitrificans*, *Paracoccus denitrificans* and *Thiobacillus denitrificans* (Schlegel 1974). Nitrous oxide is also formed by the oxidation of ammonium by the autotrophic nitrifier, *Nitrosomonas europaea*; by the reduction of nitrate by heterotrophic bacteria, such as *Bacillus subtilis*, *Escherichia coli* and *Aerobacter aerogenes* (Yoshida & Alexander 1970); and by the reduction of nitrite by soil fungi, such as *Aspergillus flavus*, *Penicillium atrovenetum* (Yoshida & Alexander 1970), *Fusarium oxysporum* and *Fusarium solani* (Bollag & Tung 1972).

Anaerobic bacterial reduction of nitrate to nitrite in soil is the major source of nitrogen oxides in the atmosphere. In acidic soils, nitrite is converted to nitrous acid which decomposes chemically to nitric oxide and nitrogen dioxide (Allison 1955). Photosynthetic microbes, located in the top layer of soil and capable of reducing nitrate to nitrite, are presumably another source of atmospheric nitric oxide and nitrogen dioxide (Makarov 1969). Similarly, nitrogen dioxide volatilised from silos containing corn and alfalfa has a microbial origin: nitric oxide, initially emitted from microbial denitrification of nitrate in silage, is oxidised in the atmosphere to nitrogen dioxide (Peterson *et al.* 1958; Scaletti *et al.* 1960).

Both aerobic and anaerobic microorganisms in soil and aquatic environments are the main sources of hydrogen sulphide and of many organic sulphur-containing volatiles (e.g. methyl mercaptan, dimethyl sulphide, dimethyl disulphide, carbon disulphide and carbonyl sulphide). In the atmosphere, hydrogen sulphide is rapidly oxidised to sulphur dioxide, which constitutes a portion of the natural background levels of sulphur dioxide but to which are added large quantities from the combustion of coal and petroleum, from the refining of petroleum, and from the smelting of non-ferrous metals (Stotzky & Schenk 1976; Babich & Stotzky 1978b).

The primary biogenic source of hydrogen sulphide in the atmosphere is the reduction of sulphate to sulphide, by bacteria of the genera *Desulfovibrio* and *Desulfotomaculum*, which occurs in anaerobic environments containing organic matter (e.g. swamps, tidal flats, salt marshes, stratified lakes—Schlegel 1974; Babich & Stotzky 1978b). Lesser amounts of hydrogen sulphide are produced by bacterial decomposition of sulphur-containing amino acids in aerobic environments; for example, hydrogen sulphide was evolved by *Proteus vulgaris, Serratia marcescens, Escherichia coli, Clostridium sporogenes, Pseudomonas aeruginosa, Bacillus subtilis* and *Alcaligenes faecalis* when cysteine was used as the sulphur source (Starkey 1950).

Studies using synthetic media supplemented with sulphur-containing organic compounds have demonstrated the formation and evolution of volatile sulphur compounds by algae, bacteria and fungi. During the degradation of methionine, for example, methyl mercaptan was evolved by *Microsporum gypseum, Aspergillus niger* (Starkey 1956), *Aspergillus oryzae, Fusarium culmorum, Streptomyces lavendulae, Pseudomonas fluorescens, Clostridium tetani* (Kadota & Ishida 1972) and *Clostridium tetanomorphum* (Starkey 1956). Methyl mercaptan and dimethyl sulphide were evolved by *Scopulariopsis brevicaulis* (Starkey 1956) and methyl mercaptan and dimethyl disulphide were produced by *Proteus* species (Hayward *et al.* 1977). During the decomposition of dimethyl-β-propiothetin, dimethyl sulphide was evolved by the unicellular marine algae, *Skeletonema costatum, Phaeodactylum tricornutum* and *Syracosphaera cartera* (Kadota & Ishida 1972). When grown on a complex organic medium, strains of *Brevibacterium linens* produced methyl mercaptan, dimethyl disulphide and methyl thioacetate (Cuer *et al.* 1979). *Schizophyllum commune*, a wood-rotting fungus, methylated sulphate to form methyl mercaptan (Challenger & Charlton 1947), while *Pseudomonas fluorescens* and *Pseudomonas aeruginosa* produced dimethyl disulphide from sulphate (Rasmussen 1974).

The oceans are also sources of atmospheric dimethyl sulphide (Lovelock *et al.* 1972) which is probably of biotic origin as emissions of this gas have been detected from marine algae (Kadota & Ishida 1972). Coastal and oceanic waters are also sources of carbon disulphide, but it is not known whether the gas is of biotic or abiotic origin (Lovelock 1974).

There is relatively little information on the evolution of organic, sulphur-containing volatiles from freshwater systems. Pond water containing species of the green algae, *Spirogyra, Gonium, Pandorina* and *Eudorina*, produced dimethyl sulphide and methyl mercaptan, while waters containing large populations of the cyanobacteria genera, *Nostoc* and *Anabaena*, produced dimethyl sulphide as well as some dimethyl disulphide. However, axenic cultures of *Pandorina, Eudorina* and *Gonium* species did not evolve sulphur-containing gases, whereas hydrogen sulphide, dimethyl sulphide, dimethyl disulphide and, occasionally, methyl mercaptan were detected from axenic cultures of *Anacystis, Synechococcus, Plectonema* and *Oscillatoria* species. The pond waters containing green algae were apparently contaminated with bacteria, which were the producers, rather than the algae, of the sulphur-containing organic volatiles (Rasmussen 1974). Cyanobacterial mats, collected from a hot spring effluent and incubated anaerobically, produced hydrogen sulphide, methyl mercaptan and dimethyl sulphide, which were probably the result of the decomposition of the cyanobacteria by other microorganisms (Zinder *et al.* 1977).

Microbial decomposition of methionine in soil, under both aerobic (Banwart & Bremner 1975a; Francis *et al.* 1975) and waterlogged (Banwart & Bremner 1975a) conditions, resulted in the volatilisation of methyl mercaptan, dimethyl sulphide and dimethyl disulphide. Soils, either aerobic or waterlogged, evolved carbon disulphide after amendment with cysteine or cystine, and evolved ethyl mercaptan, ethyl methyl sulphide and diethyl disulphide after amendment with ethionine or sulphur-ethyl cysteine (Banwart & Bremner 1975a). As emissions of hydrogen sulphide were not detected from these soils, it was suggested that organic, sulphur-containing volatiles, and not hydrogen sulphide, are the main sulphur-containing gases emitted from soil (Bremner & Steele 1978).

A range of sulphur-containing organic volatiles is also produced during the microbial degradation of animal manures (Banwart & Bremner 1975b). Soils amended with manures from cattle, poultry, sheep and swine or with sewage sludge and maintained under either aerobic or anaerobic conditions evolved methyl mercaptan, dimethyl sulphide, dimethyl disulphide, carbon disulphide and carbonyl sulphide, but not hydrogen sulphide (Banwart & Bremner 1976).

As mentioned above, carbon monoxide is one of the products of the metabolic activities of microbes and the oceans have been described as a major source of this gas. Marine bacteria (i.e. *Alginomonas* and *Brevibacterium* species—Junge *et al.* 1972) and diatoms, as well as the bacterial decomposition of marine plankton (Babich & Stotzky 1974), produce carbon monoxide. Some freshwater algae and cyanobacteria (Nozhevnikova & Yurganov 1978) produce carbon monoxide during metabolism of photosynthetic pigments. *Streptococcus mitis* and *Bacillus cereus* form carbon monoxide during anaerobic growth in the presence of haem compounds, such as erythrocytes, haemoglobin, myoglobin and cytochrome *c* (Engel *et al.* 1972), and some fungi (e.g. *Aspergillus niger, Fusarium* sp., *Cephalosporium* sp.) produce carbon monoxide during degradation of plant flavonoids (Westlake *et al.* 1961).

On a global basis, atmospheric methane is the result primarily of the anaerobic bacterial decomposition of organic matter in sewage, lakes, ponds, marshes, paddy fields and swamps. The methanogenic bacteria are anaerobes and belong to the genera, *Methanosarcina, Methanobacillus, Methanobacterium, Methanospirillum* and *Methanococcus* (Mah *et al.* 1977). Methane is also formed microbiologically in the digestive tracts of animals, particularly ruminants (Ehhalt 1974), and in the decaying heartwood of elm, cottonwood, poplar and willow trees (Zeikus & Ward 1974). Some methanogenic bacteria, such as *Methanosarcina barkeri* and *Methanobacterium formicicum*, reduce carbon monoxide to methane (Bortner *et al.* 1974), and methanogens present in anaerobic freshwater sediments and sewage sludge metabolise methyl mercaptan and dimethyl sulphide to carbon dioxide and methane (Zinder & Brock 1978). Species of *Pseudomonas*, isolated from lake sediments, degrade methyl mercury to volatile elemental mercury and methane (Spangler *et al.* 1973).

Microorganisms that produce ethylene have been isolated from aerobic and waterlogged soils and from aquatic systems (Primrose 1976; Primrose & Dilworth 1976; Considine *et al.* 1977) and are now considered to be ubiquitous in the environment (Primrose 1979). Fungi that evolve ethylene include *Mucor hiemalis* (Lynch 1974), *Penicillium cyclopium, Penicillium crustosum* (Considine *et al.* 1977), *Penicillium digitatum, Blastomyces dermatitidis, Histoplasma capsulatum, Fusarium oxysporum* (Primrose 1979), *Saccharomyces cerevisiae* (Thomas & Spencer 1977) and *Agaricus bisporus* (Wood & Hammond 1977). Bacteria that evolve ethylene include *Xanthomonas phaseoli, Pseudomonas solanacearum, Erwinia carotovora, Agrobacterium tumefaciens* (Swanson *et al.* 1979), *Enterobacter aerogenes, Escherichia coli* (Primrose & Dilworth 1976), *Enterobacter cloacae, Serratia liquefaciens* (Primrose 1976), *Acinetobacter calcoaceticus, Bacillus mycoides* and *Rhizobium trifolii* (Primrose 1979).

Fungi produce various complex organic volatiles which may be important in intra- and inter-species interactions (Stotzky & Schenck 1976), for example, hexa-1,3,5-triyne is evolved by *Fomes annosus*; trimethyl ethylene, pelargonaldehyde, and trimethyl amine are evolved by *Puccinia graminis* var. *triciti* and dimethyl and trimethyl arsine are evolved by *Penicillium brevicaule* (Hutchinson 1971). *Trichoderma viride* produces 6-pentyl α-pyrone; *Ceratocystis variospora* evolves 6-methyl 5-penten 2-one; *Phellinus igniarius* evolves methyl benzoate (Collins 1976); *Agaricus campestris* evolves 2,3-dimethyl 1-pentene; *Thielaviopsis basicola* evolves ethanol, formaldehyde, and acetaldehyde; and *Penicillium digitatum* evolves acetylene, propylene, ethane and propane (Stotzky & Schenck 1976). Complex organic volatiles are also released from actinomycetes. The earthy, musty odour characteristic of species of *Streptomyces* has been identified as trans-1,10-dimethyl-trans-9-decalol (Collins 1976). Heptaene and pentaene have also been identified as organic volatiles evolved from species of *Streptomyces* (Fries 1973).

36.2.2 Microbes as Sinks for Atmospheric Pollutants

Atmospheric gases are removed from the atmosphere by various scavenging processes, termed sinks. The most important removal mechanisms are:

(1) precipitation, in which the pollutant is removed by rain-out (i.e. absorption of gases in clouds) and by wash-out (i.e. absorption and particle capture by falling raindrops, snow or hail);

(2) chemical reactions which may involve photic energy and result in oxidised products (e.g. photo-oxidation of hydrogen sulphide to sulphur dioxide) or aerosols (by conversion of gaseous sulphur dioxide to sulphate particulates);

(3) dry deposition, which involves removal by gravitation or turbulent diffusion (i.e. eddy diffusion);

(4) direct absorption or adsorption by various substances on the Earth's surface, including water, soil, plants and microorganisms (Rasmussen *et al.* 1975).

As atmospheric pollutants are eventually deposited onto terrestrial and into aquatic ecosystems, there is no clear distinction between atmospheric pollution and terrestrial and aquatic pollution.

Soil is a major sink for the removal of many gaseous pollutants. The removal by soil of sulphur dioxide (Bremner & Banwart 1976; Ferenbaugh *et al.* 1979), nitrogen dioxide (Abeles *et al.* 1971; Ferenbaugh *et al.* 1979), ozone (Turner *et al.* 1973), hydrogen sulphide (Smith *et al.* 1973; Bremner & Banwart 1976), methyl mercaptan (Smith *et al.* 1973) and ammonia (Malo & Purvis 1964) appears to be primarily by abiotic processes. Conversely, microorganisms appear to be responsible for the removal by soil of nitrous oxide (Blackmer & Bremner 1976), carbon monoxide (Inman & Ingersoll 1971; Bartholomew & Alexander 1979), ethylene and acetylene (Smith *et al.* 1973), dimethyl sulphide, dimethyl disulphide, carbonyl sulphide and carbon disulphide (Bremner & Banwart 1976). Microorganisms in soil may also be a minor sink for the removal of atmospheric sulphur dioxide, as some fungi isolated from soil, e.g. species of *Alternaria*, *Penicillium*, *Chaetomium*, *Colletotrichum*, *Trichoderma*, *Rhizopus* and *Fusarium oxysporum*, were able to remove sulphur dioxide from the atmosphere (Craker & Manning 1974). Furthermore, although the removal of atmospheric sulphur dioxide, hydrogen sulphide, nitrogen dioxide and ammonia by soil is predominantly an abiotic process, once these gases come into contact with the soil solution, they form soluble products which may be utilised by the microbiota.

The removal of carbon monoxide by soil has been attributed to the metabolic activities of fungal and bacterial populations. Carbon monoxide is readily oxidised to carbon dioxide by fungi (e.g. *Penicillium digitatum*, *Penicillium restrictum*, *Mucor hiemalis*, *Haplosporangium parvum*, *Mortierella vesiculata* and species of *Aspergillus*—Inman & Ingersoll 1971) and by bacteria (primarily actinomycetes; for example *Nocardia salmonicolor*, *Nocardia autotrophica*, *Nocardia asteroides*, *Streptomyces griseus*, *Actinoplanes humiferus*, *Actinoplanes philippinensis*, *Microbispora rosea*, *Agromyces ramosus* and *Mycobacterium phlei*). However, this oxidation appears to be cometabolic and does not provide energy for the cells (Bartholomew & Alexander 1979). Other bacteria capable of oxidising carbon monoxide include *Hydrogenomonas carboxydovorans*, *Seliberia carboxydohydrogena*, *Pseudomonas carboxydoflava*, *Pseudomonas gazotropha*, *Comamonas compransoris*, *Achromobacter carboxydus* (Nozhevnikova & Yurganov 1978); a species of *Rhodopseudomonas* (Hirsch 1968); the methanogens, *Methanosarcina barkeri* and *Methanobacterium formicicum* (Bortner *et el.* 1974); the methane-oxidising bacteria, *Methylomonas albus* and *Methylosinus trichosporium* (Hubley *et al.* 1974); and the cyanobacterium, *Nostoc* sp. (Chappelle 1962). Carbon monoxide is also oxidised by species of the algae, *Chlorella*, *Euglena*, *Scenedesmus*, *Oshomonas* and *Ankistrodesmus* (Chappelle 1962).

The main sink for methane in the atmosphere appears to be the troposphere, wherein it is chemically destroyed. Although soil does not appear to be a significant sink for methane (Ehhalt 1974), bacteria capable of oxidising methane have been isolated from soils and sediments; for example, *Bacillus hexacarbovorum*, *Mycobacterium flavum*, *Mycobacterium rubrum*, *Mycobacterium methanicum*, *Pseudomonas fluorescens*, *Methanomonas carbonatophila*, *Methanomonas methanooxidans* and *Methylococcus capsulatus* (Coty 1969). An alga of the genus *Chlorella* also oxidised methane (Enebo 1967).

Both fungi and bacteria are capable of utilising more complex gaseous alkanes; for example, species of *Graphium*, *Phialophora* and *Acremonium*, isolated from raw sewage, oxidised ethane, propane and n-butane (Davies *et al.* 1974), and *Penicillium nigricans*, *Allescheria boydii*, *Graphium cumeiferum*, and *Gliocladium* sp., isolated from soils near natural and domestic gas seepages, utilised *n*-butane (McLee *et al.* 1972). Gaseous alkanes, such as *n*-butane, ethane and propane, are also metabolised by bacteria isolated from soil (McLee *et al.* 1972; Babich & Stotzky 1974).

In contrast to the alkanes, there is little information on the microbial utilisation of gaseous alkenes. Microbial metabolism accounts for the removal of atmospheric ethylene by soil (Abeles *et al.* 1971), and ethylene oxidising bacteria of the genus *Mycobacterium* have been identified (de Bont 1976). Nitrifiers are a potential sink for acetylene, as indicated by the success of the acetylene reduction technique, whereby acetylene is reduced to ethylene as an indication of dinitrogen fixation (Babich & Stotzky 1974).

36.2.3 POLLUTANT TRANSFORMATION BY MICROBES

Microorganisms are also involved in another aspect of air pollution, i.e. pollutant transformation. Industrial and domestic activities emit large quantities of non-gaseous, inorganic pollutants, such as mercury, cadmium, lead, tin, selenium, tellurium and arsenic. As a result of various scavenging processes, these contaminants are deposited into aquatic and on to terrestrial environments, where at least one biochemical activity of microbes, i.e. methylation, can change their chemical form.

Methylation results in the conversion of inorganics to organic volatiles (although metallic mercury is also a volatile element) and changes their potential ecological significance. Some of the altered properties of organometal compounds that may affect their ecological impact include their tendency to form complexes; relative and absolute water and lipid solubilities; valence state; wide distribution as a result of their volatility; and greater toxicity to the biota than their comparable inorganic forms (Stotzky & Schenck 1976).

Both fungi (e.g. *Aspergillus niger*, *Scopulariopsis brevicaulis*, *Saccharomyces cerevisiae*) and bacteria (e.g. *Pseudomonas fluorescens*, *Mycobacterium phlei*, *Escherichia coli*, *Aerobacter aerogenes*, *Bacillus megaterium*) are able to methylate mercury (Vonk & Sijpesteijn 1973; Hamdy & Noyes 1975). Although methylation occurs both aerobically and anaerobically, more takes place aerobically (Stotzky & Schenck 1976; Summers & Silver 1978).

Species of *Pseudomonas*, *Alcaligenes*, *Acinetobacter*, *Flavobacterium* and *Aeromonas*, isolated from sediment of Lake Ontario, were able to transform trimethyl lead to tetramethyl lead (Wong *et al.* 1975), and a species of *Pseudomonas*, isolated from sediment of Chesapeake Bay, was able to methylate tin and cadmium (Summers & Silver 1978).

Species of *Aeromonas*, *Flavobacterium* and *Pseudomonas* isolated from lake sediments methylated inorganic selenium to dimethyl selenide and dimethyl diselenide (Chau *et al.* 1976), and a species of *Corynebacterium* isolated from soil methylated selenium to the volatile dimethyl selenide (Doran & Alexander 1977). Dimethyl selenide was volatilised by the fungi, *Schizophyllum commune* (Challenger & Charlton 1947), *Scopulariopsis brevicaulis*, *Penicillium chrysogenum* (Challenger 1945) and species of *Penicillium* (Fleming & Alexander 1972; Janda & Fleming 1978), *Fusarium*, *Acremonium* and *Verticillium* (Janda & Fleming 1978).

Scopulariopsis brevicaulis, *Penicillium chrysogenum* (Challenger 1945) and a species of *Penicillium* isolated from sewage (Fleming & Alexander 1972) converted inorganic tellurium to dimethyl telluride and *Scopulariopsis brevicaulis* methylated inorganic arsenic to volatile trimethyl arsine (Challenger 1945).

Experimental

36.3 Response of microorganisms to atmospheric pollutants

36.3.1 STUDIES *IN VITRO*

General principles

A variety of techniques have been used to study the effects of gaseous pollutants on microbes *in vitro*. Three methods of exposing microorganisms to these pollutants have been employed: gaseous exposure; volatile exposure; and solubility product solutions. In direct gaseous exposures, the gas is bubbled through suspensions of microorganisms or over cultures supported on agar, filter paper discs or cellophane strips. In volatile exposures, the microorganisms are supported above volatile solutions although, in contrast to direct fumigation, it is difficult to control the concentration of the volatiles. In solubility product solutions, the microbial cultures are placed into solutions of a salt of the pollutant; for example, solutions of sodium sulphite will yield, depending on the pH, sulphurous acid, bisulphite and sulphite: all solubility products of gaseous sulphur dioxide. Similarly, solutions of sodium nitrite and sodium nitrate have been used to study the effects of gaseous nitrogen dioxide as, in solution, nitrogen dioxide yields nitrous acid, nitric acid, nitrite ions, nitrate ions and/or hydrogen ions depending on the pH. The rationale of this third approach is that before a gaseous pollutant interacts with a microorganism, it must first penetrate the aqueous film surrounding the cell and, hence, the effect of the pollutant is mediated by its rate of entry into an aqueous external and, subsequently, aqueous internal environment. Any effects of the gas are, therefore, probably not the result of the gaseous form of the pollutant, but of its solubility products. However, in these types of exposures, extraneous ions (e.g. sodium as sodium sulphite, sodium nitrite and sodium nitrate) are used and are introduced concomitantly into the system and their effects must be accounted for (Babich & Stotzky 1974; Stotzky & Schenk 1976).

In studies with heavy metals, solutions of the metal salts (e.g. cadmium chloride, lead nitrate) are added to a liquid or solid medium, which is subsequently inoculated with the test microorganisms. There are primarily two problems with this approach:

(1) the metal salts introduce extraneous anions, such as chloride and nitrate into the medium;

(2) the form of the heavy metal used in the study may not reflect the chemical form that is being emitted as a pollutant, e.g. water-soluble cadmium chloride versus water-insoluble cadmium sulphide; inorganic lead nitrate versus organic tetraethyl lead.

One of the major questions when conducting experiments involving environmental contaminants is how relevant are they to the 'real world' when the concentrations used *in vitro* are considerably above those encountered in the environment. Some studies have demonstrated a reciprocal relationship between pollution concentration, c, and length of exposure time, t; that is, exposure for long periods of time at low (chronic) concentrations elicited responses from the microbiota that were equivalent to short exposures at high (acute) concentrations, suggesting that $c \times t$ was a constant. When, for example, ozone in concentrations from 1.4 to 1400 p.p.m. (μg ml^{-1}) was bubbled through aqueous suspensions of spores of *Sclerotinia fructicola* for 3 h, it was noted that when the exposure time was doubled, the gas concentration required for kill was about 50% (Watson 1942). Reciprocal $c \times t$ relationships were also noted in the germination of sclerotia of *Botrytis* sp., *Rhizoctonia tuliparum* and *Sclerotium delphinii* exposed to sulphur dioxide, hydrogen sulphide or ammonia at concentrations of 1 to 1000 p.p.m. for 1 to 16 h (McCallan & Weedon 1940), and in the toxicity of sulphur dioxide (Couey & Uota 1961; Couey 1965) and ozone (Hibben & Stotzky 1969) to fungal spores. Because of problems of contamination, desiccation and the long delay before results are obtained, it is desirable to reduce the time of exposure to the pollutant. This, however, often results in using concentrations that are considerably higher than those that occur naturally.

The chemical form of the pollutant is an important factor when assessing the effects on the microbiota. Water-soluble and water-insoluble inorganic compounds and lipid-soluble organic compounds of the same pollutant (e.g. lead nitrate, lead oxide and tetraethyl lead, respectively) have different mobilities and affinities for microbes and, thereby, exert differential toxicities (Table 1).

The physiological state of the test microorganism can affect its response to a pollutant; for example, phosphorus-starved cells of *Saccharomyces ellipsoideus* were more susceptible to injury by cadmium than were nitrogen-starved cells, which, in turn, were more susceptible than carbon-starved cells (Nakamura 1961). The aeration status affects the overall physiology of microorganisms, and hence, it can influence the response of microbes to pollutants; for example, nitrite was more toxic to denitrifiers, such as *Pseudomonas* sp., under aerobic conditions (Bollag & Henninger 1978), whereas the toxicity of copper to *Chlorella vulgaris* was more pronounced under anaerobic conditions (McBrien & Hassall 1967).

The morphological stage of the test microorganism can also affect its response to a pollutant; for

Table 1. Examples of the effects of the chemical form of heavy metal pollutants on their toxicity to microorganisms.

Pollutant	Effect
Lead	concentrations of Pb from 0.1 to 5.0 p.p.m. as Pb $(NO_3)_2$ inhibited, but as Pb acetate stimulated, growth of *Chlamydomonas eugametos* (Hutchinson 1973)
	Pb, as Pb acetate, was toxic to *Aspergillus niger*, whereas Pb, as insoluble compounds (i.e. PbO, PbS and $Pb_3(OH)_2(CO_3)_2$ was non-toxic (Zlochevskaya & Rukhadze 1968)
	bacteria and fungi, isolated from Pb-polluted soils, accumulated greater quantities of Pb when in the form of $PbBr_2$, $Pb(NO_3)_2$ and Pb acetate than when in the insoluble forms of PbS, PbO and Pb^0 (Aickin & Dean 1979)
	tetraethyl Pb, but not $Pb(NO_3)_2$, induced formation of giant, multinucleate cells in the chrysophycean flagellate, *Poterioochromonas malhamensis* (Roderer & Schnepf 1977)
	tetraethyl Pb was more toxic than tetraphenyl Pb to *Chlorella vulgaris* (Vallee & Ulmer 1972)
Cadmium	Cd^{2+} was more toxic than $Cd(CN)_4^{2-}$ to respiratory activity of a mixed microbiota in activated sludge containing small amounts of particulates (Cenci & Morozzi 1977); in the presence of greater amounts of particulates, $Cd(CN)_4^{2-}$ was more toxic than Cd^{2+}, possibly due to the greater sorption of Cd^{2+} to the negatively charged particulates (Morozzi & Cenci 1978)
	Cd as $CdCl_2$, but not as $Cd(NO_3)_2$, was mutagenic to *Bacillus subtilis* (Nishioka 1975)
Iron	insoluble forms of Fe, such as $Fe(OH)_3$ and $FeCO_3$, were less toxic than soluble forms, such as Fe citrate, Fe cysteine and $Fe_2(SO_4)_3$, to strains of *Sphaerotilus* (Chang *et al.* 1979)
	Fe^{2+}, but not Fe^{3+}, was mutagenic to *Escherichia coli* (Catlin 1953)
Manganese	Mn as $Mn(NO_3)_2$, $MnCl_2$, $MnSO_4$ and $Mn(CH_3COO)_2$, but not as $KMnO_4$, was weakly mutagenic to *Bacillus subtilis* (Nishioka 1975)
Nickel	Ni^{2+} was more toxic than $Ni(CN)_4^{2-}$ to respiratory activity of a mixed microbiota in activated sludge (Morozzi & Cenci 1978)
Chromium	Cr^{6+} (as CrO_3), but not Cr^{2+} (as $CrCl_2$), reduced the rate of fermentation of a rumen microbiota (Forsberg 1978)
	Cr^{6+} (as K_2CrO_4), but not Cr^{3+} (as $CrK(SO_4)_2$), was mutagenic to *Escherichia coli* (Venitt & Levy 1974)
	Cr^{6+} (as K_2CrO_4 and $K_2Cr_2O_7$), but not Cr^{3+} (as $CrCl_3$), was mutagenic to *Bacillus subtilis* (Nishioka 1975)
	Cr^{6+} (as $Na_2Cr_2O_7$, K_2CrO_4, CrO_3 and K_2CrO_4), but not Cr^{3+} (as $CrK(SO_4)_2$) or Cr^{2+} (as $CrCl_2$), was mutagenic to *Salmonella typhimurium* (Petrilli & deFlora 1977)
Mercury	$HgCl_2$ was more toxic than was $Hg(NO_3)_2$ to spores of *Tilletia triciti* (Bodnar & Terenyi 1932)
	$HgCl_2$ reduced growth of *Phaeodactylum tricornutum, Chaetoceros galvestonensis* and *Cyclotella nana* to a greater extent than did dimethyl mercury (Hannan & Patouillet 1972)
	methyl Hg^+ reduced photosynthesis of a mixed marine phytoplankton to a greater extent than did equivalent concentrations of Hg as $HgCl_2$ (Knauer & Martin 1972)
	phenylmercuric acetate was more toxic than inorganic Hg^{2+} to species of *Pseudomonas* (Nelson & Colwell 1975)
	phenylmercuric acetate was less toxic to growth of *Chlamydomonas* sp., *Chlorella* sp. and *Phaeodactylum tricornutum* than was $HgCl_2$ (Nuzzi 1972)
	photosynthesis of *Nitzschia delicatissima* was inhibited to a greater extent by phenylmercuric acetate than by diphenylmercury (Harriss *et al.* 1970)
	Hg as CH_3HgCl and $CH_3COOHgC_6H_5$, but not as $HgCl_2$, was mutagenic to *Bacillus subtilis* (Nishioka 1975)

Table 1. cont.

Pollutant	Effect
Mercury (cont.)	no differences in toxicity were noted for Hg^{2+} and $Hg(CN)_4^{2-}$ to respiratory activity of a mixed microbiota in activated sludge (Morozzi & Cenci 1978)
Zinc	$Zn(CN)_4^{2-}$ was more toxic than Zn^{2+} to respiratory activity of a mixed microbiota in activated sludge (Morozzi & Cenci 1978)
	$Zn(CN)_4^{2-}$ (Kozloff *et al.* 1957) and mixtures of $ZnCl_3^-/ZnCl_4^{2-}$ (Babich & Stotzky 1978a) were more toxic than Zn^{2+} to bacteriophages of *Escherichia coli*

example, sclerotia of *Rhizoctonia tuliparum, Sclerotium delphinii* and *Botrytis* sp. were more resistant to sulphur dioxide than were the vegetative mycelial stages (McCallan & Weedon 1940). Fruiting body formation is more sensitive to some pollutants than the hyphae involved in vegetative growth; for example, concentrations of cadmium that inhibited the formation of conidia by *Aspergillus niger* and *Trichoderma viride*, of sporangiospores by *Rhizopus stolonifer* (Babich & Stotzky 1977a), and of sclerotia by *Sclerotium rolfsii* (le Tourneau 1978) did not appreciably inhibit the vegetative growth of these fungi. Capsulated strains of *Klebsiella aerogenes* were more resistant to cadmium and copper than were non-capsulated strains, possibly because the extracellular polysaccharides complexed the heavy metals and, thereby, reduced their uptake by the cells (Bitton & Freihofer 1978).

*Specific studies**.

Many studies on the effects of gaseous and heavy metal pollutants on microbes *in vitro* have merely established the concentrations of the pollutants that are lethal to the responding microorganisms; for example, 0.3 mg l^{-1} copper, 0.65 mg l^{-1} cadmium or 0.7 mg l^{-1} zinc was lethal to *Selenastrum capricornutum* (Bartlett *et al.* 1974); 1 mg l^{-1} cadmium was toxic to *Chlamydomonas reinhardtii* (Fennikoh *et al.* 1978); and 50 mg l^{-1} lead was lethal to *Paramecium multimicronucleatum* (Ruthven & Cairns 1973). An irradiated atmosphere containing 0.5 p.p.m. nitrogen dioxide was toxic to *Serratia marcescens* (Jacumin *et al.* 1964); 0.15 μl l^{-1} (p.p.b.) nitrogen dioxide, 20 μl l^{-1} sulphur dioxide and 3.0 μl l^{-1} sulphur dioxide plus 3.0 μl l^{-1} gaseous formaldehyde were toxic to aerosolised *Rhizobium meliloti* (Won & Ross 1969); 5 p.p.m. nitrogen dioxide was toxic to aerosolised Venezuelan equine encephalomyelitis virus (Ehrlich & Miller 1972); and mixtures of ozone and olefins (e.g. cyclohexene, trans 2-butene) were toxic to *Escherichia coli, Staphylococcus aureus* (Druett & Packman 1972), *Micrococcus albus* (Dark & Nash 1970) and ϕX-174 bacteriophage of *Escherichia coli* (deMik *et al.* 1977).

Other investigations have been concerned primarily with determining the concentrations of gaseous and heavy metal pollutants that affect the growth rate of microorganisms. Irradiated mixtures of nitrogen dioxide and butene-1 inhibited growth of, but did not kill, *Escherichia coli* (Estes 1967), and ambient air containing elevated concentrations of nitric oxide was fungistatic to *Penicillium pfefferianum, Neurospora crassa, Alternaria brassicola* and *Aspergillus niger* (Marchesani 1969). In other studies, mycelial growth of *Chaetomium* sp. and *Pestalotiopsis* sp. was inhibited by 110 p.p.m. lead, of *Pleurophomella* sp. and *Gnomonia platani* by 275 p.p.m. lead (Smith 1977), of *Pleurophomella* sp. and *Aureobasidium pullulans* by 57 p.p.m. aluminium (Smith *et al.* 1978), of *Botrytis cinerea, Penicillium vermiculatum* and *Fomes annosus* by 100 p.p.m. cadmium and of *Aspergillus niger, Scopulariopsis brevicaulis* and *Phycomyces blakesleeanus* by 1000 p.p.m. cadmium (Babich & Stotzky 1977a). Growth of *Chlamydomonas eugametos* and *Haematococcus capensis* was inhibited by 0.1 p.p.m. lead (Hutchinson 1973), and 6.1 p.p.m. cadmium inhibited growth of *Scenedesmus quadracauda* (Klass *et al.* 1974). Methylmercury acetate, at a concentration of 1×10^{-8}M, was bacteriostatic to *Rhodopseudomonas capsulata* (Jeffries & Butler 1975).

* The authors have retained the units of concentration used in the original papers. For comparison purposes the reader should regard p.p.m. as μg g^{-1}, μg ml^{-1} or μl g^{-1} (or mg kg^{-1} etc.); p.p.b. are parts per 1×10^9 (or μg l^{-1} etc).

More definitive studies, however, have differentiated between inhibitory effects of pollutants on the generation times of microorganisms; for example, heavy metals have been shown to extend greatly the lag phase of sensitive microorganisms. For instance, concentrations of phenylmercuric acetate ranging from 5×10^{-9} to 2.5×10^{-8}M increased the lag phase of *Chorella pyrenoidosa* (Ben-Bassat & Mayer 1975), and that of *Agrobacterium tumefaciens* was lengthened in the presence of 5 and 10 p.p.m. cadmium (Babich & Stotzky 1977a). The lag phase of *Dunaliella salina* was extended from 3 to 18 d in the presence of 2.5 p.p.m. copper (Pace *et al.* 1977), and in the presence of 80 μg l^{-1} copper, the lag phase of *Selenastrum capricornutum* was extended by 6 d (Bartlett *et al.* 1974). Concentrations of 4 and 6 p.p.m. mercury resulted in lag periods of 4 and 6 d respectively, in *Anabaena inaequalis* (Stratton *et al.* 1979); 2 and 4 μM mercury resulted in lag periods of 3 and 8 d, respectively, in *Chlamydomonas variabilis* (Delcourt & Mestre 1978); 5 p.p.m. mercury induced a lag period of 24 h in *Arthrobacter* sp. and of 72 h in *Vibrio* sp.; 15 p.p.m. mercury induced a lag period of 41 h in *Bacillus* sp.; and 20 p.p.m. mercury induced a lag of 70 h in *Citrobacter* sp. (Vaituzis *et al.* 1975).

The doubling time of *Anabaena inaequalis* was increased from 0.2 to 5 and 8 d in the presence of 4 and 6 p.p.b. mercury, respectively (Stratton *et al.* 1979), and 0.05 mg l^{-1} copper reduced the growth rate of *Spirulina platensis* by 60% (Kallqvist & Meadows 1978). The mycelial growth rates of *Fusarium solani, Cunninghamella echinulata, Aspergillus niger* and *Trichoderma viride* were decreased in the presence of 10 mM zinc (Babich & Stotzky 1978a); 50 p.p.m. lead reduced those of *Rhizoctonia solani* and *Aspergillus giganteus*; 100 p.p.m. lead reduced those of *Fusarium solani, Trichoderma viride* and *Cunninghamella echinulata*; and 500 p.p.m. lead decreased those of *Botrytis cinerea* and *Penicillium brefeldianum* (Babich & Stotzky 1979a).

In addition to reducing the growth of microbes, gaseous and heavy metal pollutants have been shown experimentally to influence adversely the formation and germination of fungal spores, bacterial conjugation, DNA-mediated bacterial transformation and viral infectivity (Table 2). Alterations in microbial growth and reproductive potentials may influence several aspects of microbial ecology; for instance, inhibition of spore germination of a sensitive species could result in increased proliferation of more resistant competitors, with the subsequent loss of the sensitive species from the community.

Gaseous and heavy metal pollutants have also been shown to induce morphological abnormalities in sensitive microorganisms (Table 3). These abnormalities may be permanent or transient; an example of the latter is the conversion of aerial to subterranean hyphae during fumigation with ozone and the reversion to aerial growth when ozone is removed (Hibben & Stotzky 1969).

Biochemical activities of microorganisms, such as photosynthesis, cellular respiration, dinitrogen fixation, denitrification and bioluminescence, are also adversely affected by gaseous and heavy metal pollutants (Table 4). Furthermore, within the same microorganism, each biochemical process may have a differential sensitivity to the same pollutant, for example, photosynthesis of *Chlorella vulgaris* (McBrien & Hassall 1967) and of *Chlamydomonas reinhardtii* (Gross & Dugger 1969) was more sensitive to copper and PAN, respectively, than was respiration, and 5 p.p.m. sulphur dioxide decreased the rate of photosynthesis but increased the rate of respiration in *Euglena gracilis* (deKoning & Jegier 1968b). Dinitrogen fixation by *Anabaena cylindrica* was more sensitive to bisulphite than was photosynthesis (Hallgren & Huss 1975). Overall growth of *Anabaena inaequalis* was inhibited by 2 p.p.b. mercury whereas 100 p.p.b. mercury was required to inhibit both photosynthesis and dinitrogen fixation (Stratton *et al.* 1979), and growth of *Spirulina platensis* was more sensitive to copper than was photosynthesis (Kallqvist & Meadows 1978).

Although most studies have shown that increasing the concentration of a toxic pollutant increases the extent of the adverse microbial response, some exceptions have been reported, for example, 25 but not 100 p.p.m. cadmium inhibited photosynthesis of a *Chlorella* sp. (Mills & Colwell 1977), and although 0.1 p.p.m. lead inhibited growth of *Haematococcus capensis*, the inhibitory effects were reduced or eliminated at higher concentrations (Hutchinson 1973). Furthermore, several studies have shown that atmospheric pollutants may elicit stimulatory responses from microorganisms (Table 5), although these stimulatory responses, upon closer examination, may not be as beneficial as initially evaluated; for example, although ozone stimulated spore production by *Alternaria solani,*

Table 2. Examples of the adverse effects of some gaseous and heavy metal pollutants on the reproductive potential of microorganisms.

Inhibited microbial response	Pollutant	Microorganism
Formation of fungal spores	ozone	*Botrytis cinerea* (Krause & Weidensaul 1978), *Colletotrichum lindemuthianum* (Treshow *et al.* 1969)
	cadmium	*Aspergillus niger, Trichoderma viride, Rhizopus stolonifer* (Babich & Stotzky 1977a), *Sclerotium rolfsii* (le Tourneau 1978)
Germination of fungal spores	ozone	*Verticillium albo-atrum, Verticillium dahliae, Colletotrichum lagenarium, Fusarium oxysporum, Botrytis allii, Trichoderma viride, Aspergillus niger, Penicillium egyptiacum* (Hibben & Stotzky 1969), *Botrytis cinerea* (Krause & Weidensaul 1978)
	sulphur dioxide	*Phytophthora infestans* (Saunders 1970), *Diplocarpon rosae* (Saunders 1966), *Botrytis cinerea, Alternaria* sp. (Couey & Uota 1961; Couey 1965)
Conjugation	zinc	*Escherichia coli* (Ou & Anderson 1972)
DNA-mediated transformation	mercury, phenyl mercury, methyl mercury, cadmium, peroxyacetyl nitrate, bisulphite	*Bacillus subtilis* (Peak & Belser 1969; Groves *et al.* 1974; Shapiro 1977)
Viral infectivity	mercury	bacteriophages of *Staphylococcus aureus* (Babich & Stotzky 1979b)
	zinc	foot-and-mouth disease virus (Polatnick & Bachrach 1978)
	peroxyacetyl nitrate	bacteriophage of *Serratia marcescens* (Peak & Belser 1969)
	bisulphite	bacteriophages of *Escherichia coli* (Babich & Stotzky 1978c)

Alternaria oleraceae, Fusarium dianthe, Fusarium oxysporum, Glomerella cingulata and *Helminthosporium* sp., subsequent germination of these spores was greatly reduced (Kuss 1950). Exposure of *Trichoderma viride* and *Aspergillus terreus* to 1 p.p.m. ozone for 7 d stimulated mycelial growth but inhibited the subsequent formation of conidia (Stotzky, unpublished data).

Inhibition of growth, abnormalities in morphology, decreases in reproductive potentials, and alterations in biochemical activities of microbes as a result of their exposure to gaseous and heavy metal pollutants are only the manifestations of events occurring on the cellular and subcellular levels. The mechanisms whereby pollutants exert their direct effect include the inhibition of specific enzymatic reactions, disruption of the integrity of the cell envelope, inhibition of intracellular transport of materials, degradation of pigments involved in photosynthesis and adverse effects on the genome (Table 6).

Levels of a pollutant that are lethal to a majority of microbes may cause mutation in some and, thereby, increase the selection of cells that can tolerate the elevated concentration of the pollutant. In addition, microbial cells have developed other mechanisms of resistance to high levels of pollutants, especially to the heavy metals. These mechanisms include the following.

(1) Detoxication is achieved by production of extracellular organic material. Examples of this mode of detoxication are the chelation of ambient copper by oxalic acid secreted by *Penicillium ochrochloron* (Stokes & Lindsay 1979), of lead by citric

Table 3. Examples of the effects of gaseous and heavy metal pollutants on the morphology of microorganisms.

Pollutant	Microorganism	Effect
Sulphur dioxide	*Xanthoria fallax, Xanthoria parietina, Parmelia caperata, Physcia millegrana*	bleaching of chlorophyll, permanent plasmolysis, brown dots on chloroplasts (Rao & le Blanc 1966)
	Serratia marcescens, Sarcina lutea	pigmentation delayed (Babich & Stotzky 1974)
Ozone	*Escherichia coli*	formation of mucoid cells with excessive capsular polysaccharides (Hamelin & Chung 1974a)
	Escherichia coli	formation of long, aseptate filamentous cells (Hamelin & Chung 1976)
	Alternaria solani	conidia germinate while still attached to conidiophore (Rich & Tomlinson 1968)
	Colletotrichum lindemuthianum	loss of pigmentation; light refractive globules in hyphae (Treshow *et al.* 1969)
	Trichoderma viride, Botrytis allii, Penicillium egyptiacum, Phytophthora cactorium, Penicillium expansum, Sclerotinia fructicola, Colletotrichum lindemuthianum, Alternaria oleracea, Helminthosporium sativum	suppression of aerial hyphae (Watson 1942; Hibben & Stotzky 1969; Treshow *et al.* 1969)
Irradiated mixture of hexene-1 and nitrogen dioxide	*Serratia marcescens*	increased pigmentation (Jacumin *et al.* 1964)
Cadmium	*Serratia marcescens, Aspergillus niger*	pigmentation delayed (H. Babich & G. Stotzky, unpublished observations)
	Anabaena inaequalis	increase in filament length, heterocyst frequency, and loss of cellular contents from filament apical cells (Stratton & Corke 1979)
Lead	*Bacillus subtilis*	induction of coccoid-shaped cells; abnormal cross-wall formation; formation of protoplasts (Barrow & Tornabene 1979)
Zinc	*Schroederella schroederi Thalassiosira rotula*	inhibition of cell division; formation of long, curved cells (Kayser 1977)
Mercury	*Tetrahymena pyriformis*	swollen and distorted mitochondria; numerous lysosomes; highly condensed chromatin (Tingle *et al.* 1973)
	Vibrio sp.	giant, pleomorphic cells; elongated cells; formation of spheroplasts (Vaituzis *et al.* 1975)
	Arthrobacter sp.	irregular septum formation (Vaituzis *et al.* 1975)
	Bacillus sp.	irregular and multiple septa (Vaituzis *et al.* 1975)
	Enterobacter sp.	pleomorphic cells; outgrowth of cell wall, plasmolysis (Vaituzis *et al.* 1975)
	Flavobacterium sp.	convoluted cell wall; plasmolysis; irregular mesosomes (Vaituzis *et al.* 1975)

Table 4. Examples of the adverse effects of gaseous and heavy metal pollutants on biochemical activities of microorganisms.

Inhibited microbial response	Pollutant and responding microorganism
Photosynthesis	5 p.p.m. Pb and 17 p.p.m. Pb reduced the rate of photosynthesis of *Cosmarium botrytis* and *Navicula pelliculosa*, respectively (Malanchuk & Gruendling 1973)
	10^{-10}M-Cu inhibited photosynthesis of *Oscillatoria theibauti* (Reuter *et al.* 1979)
	100 p.p.b. Hg or 1 p.p.m. Cd reduced the rate of photosynthesis of *Anabaena inaequalis* (Stratton & Corke 1979; Stratton *et al.* 1979)
	0.1 mM-HSO_3^- inhibited photosynthesis of *Anabaena flos-aquae, Anacystis nidulans, Calothrix anomala* and *Fischerella muscicola* (Wodzinski *et al.* 1978)
	1 p.p.m. SO_2 inhibited photosynthesis of *Anabaena flos-aquae* and *Chlamydomonas reinhardtii* (Wodzinski & Alexander 1978)
	5 p.p.m. SO_2 or 0.2 to 1.0 p.p.m. O_3 reduced photosynthesis of *Euglena gracilis* (deKoning & Jegier 1968a, b)
	1 mM-NO_2^- inhibited photosynthesis of *Calothrix anomala* and *Fischerella muscicola* (Wodzinski *et al.* 1978)
Aerobic respiration	6 p.p.m. Cd inhibited respiration of *Escherichia coli* (Zwarun 1973)
	2 p.p.m. SO_2 inhibited respiration of the mycobionts of the lichens, *Cladonia cristatella* and *Caloplaca holocarpa*, whereas 6 p.p.m. SO_2 was required to inhibit respiration of the phycobionts of these lichens (Showman 1972)
	0.075 p.p.m. gaseous formaldehyde or propionaldehyde inhibited respiration rates of *Euglena gracilis* (deKoning & Jegier 1970a)
Nitrogen fixation	100 p.p.b. Hg inhibited nitrogenase activity of *Anabaena inaequalis* (Stratton *et al.* 1979)
	concentrations of 0.025 to 0.125 p.p.m. Cd, Pb or Zn inhibited nitrogen fixation in *Westiellopsis* sp. (Henriksson & DaSilva 1978)
	simulated acid rain (i.e. solutions of H_2SO_4) inhibited nitrogen fixation in *Lobaria pulmonaria* (Denison *et al.* 1977)
	HSO_3^- inhibited nitrogenase activity of *Bejerinckia indica* (Wodzinski *et al.* 1978), *Anabaena cylindrica* (Hallgren & Huss 1975) and *Collema tenax* (Sheridan 1979)
	nitrogen fixation in *Nostoc muscorum, Azotobacter vinelandii* and *Clostridium pasteurianum* was inhibited by CO (Lind & Wilson 1942; Bradbeer & Wilson 1963)
Denitrification	50 p.p.m. Cd inhibited denitrification of *Pseudomonas aeruginosa* and *Pseudomonas* sp. (Bollag & Bababasz 1979)
Bioluminescence	irradiated mixtures of cis 2-butene and NO (Serat *et al.* 1965) and of cis 2-butene and NO_2 (Serat *et al.* 1969) reduced luminescence of *Photobacterium phosphoreum*
	20 p.p.m. SO_2 reduced luminescence of *Photobacterium phosphoreum* (Smith & Sie 1969)

acid secreted by *Aspergillus niger* (Gadd & Griffiths 1978), and the production of red pigments (1,4,5, 8-tetrahydroxyanthraquinones) by *Pyrenophora avenae* that chelate phenyl-mercury and thus reduce its uptake (Greenaway 1971).

(2) Intracellular uptake and accumulation are prevented. Strains of *Staphylococcus aureus* that carry a plasmid conferring resistance to cadmium have cell membranes that are impermeable to cadmium (Chopra 1971, 1975; Kondo *et al.* 1974; Tynecka *et al.* 1975), and *Scenedesmus acutiformis*, an alga resistant to copper, has a cell membrane that is impermeable to that metal (Mierle & Stokes 1976). *Micrococcus luteus* and *Azotobacter* sp. rapidly accumulate lead, but this is immobilised within the cell wall (Tornabene & Edwards 1972).

Table 5. Examples of the stimulatory effects of gaseous and heavy metal pollutants on microorganisms.

Stimulated microbial response	Pollutant	Microorganism
Growth	5 to 10 p.p.m. cadmium	*Escherichia coli, Streptococcus faecalis* (Doyle *et al.* 1975)
	0.5 to 5 p.p.m. lead	*Chlorella vulgaris* (Hutchinson 1973)
	275 p.p.m. lead	*Aureobasidium pullulans, Epicoccum* sp. (Smith 1977)
	7 to 71 p.p.m. lead	*Gnomonia platani* (Smith *et al.* 1978)
	8.3 p.p.m. manganese	*Aureobasidium pullulans* (Smith *et al.* 1978)
	92 p.p.m. iron	*Chaetomium* sp. (Smith *et al.* 1978)
Spore production	10 p.p.m. ozone for 70 h	*Colletotrichum lagenarium* (Hibben & Stotzky 1969)
	10 to 60 p.p.m. ozone for 4 h d^{-1} for 3 d	*Alternaria oleraceae* (Treshow *et al.* 1969)
Spore germination	10 p.p.m. ozone for 1 or 2 h	*Penicillium egyptiacum, Aspergillus terreus* (Hibben & Stotzky 1969)
	10 to 30 p.p.m. sulphur dioxide for 3 h	*Diplocarpon rosae* (Saunders 1966)
Photosynthesis	25 p.p.b. mercury	*Anabaena inaequalis* (Stratton *et al.* 1979)
Respiration	5 p.p.m. sulphur dioxide	*Euglena gracilis* (deKoning & Jegier 1968b)
Nitrogen fixation	0.005 to 0.125 p.p.m. nickel; 0.005 to 0.025 p.p.m. lead	*Nostoc muscorum* (Henriksson & DaSilva 1978)
	25 p.p.b. mercury	*Anabaena inaequalis* (Stratton *et al.* 1979)
Bioluminescence	500 p.p.m. ozone for 3 h	*Armillaria mellea* (Berliner 1963)

Strains of *Saccharomyces cerevisiae* that are tolerant of copper produce large amounts of hydrogen sulphide, which precipitates the copper as copper sulphide in and around the cell wall (Gadd & Griffiths 1978).

(3) The toxicant is removed from the cell and the cellular environment. Strains of *Escherichia coli, Staphylococcus aureus* and *Pseudomonas aeruginosa*, which carry plasmids conferring resistance to mercury (Summers & Lewis 1973; Clark *et al.* 1977), and *Chlorella pyrenoidosa* (Ben-Bassat & Mayer 1975) volatilise Hg^{2+} as Hg^0.

(4) Detoxication may be achieved intracellularly. The resistance to mercury of strains of *Staphylococcus aureus* which carry the plasmid for mercury resistance was correlated with the uptake of the toxicant, followed by an intracellular mechanism which changed the mercury to an unidentified innocuous form (Kondo *et al.* 1974). In the presence of toxic levels of cadmium, cells of *Escherichia coli* developed abnormal morphologies and ceased to divide. During this extended lag phase, the cells repaired the damage caused by cadmium, redistributed the intracellular cadmium from the cell membrane to the cell wall, resumed normal morphology and then divided (Mitra *et al.* 1975).

Studies performed *in vitro* have shown that gaseous and heavy metal pollutants exert a variety of overt and covert influences on the microbiota. These influences presumably affect the activity, ecology and population dynamics of microorganisms in natural habitats.

36.3.2 STUDIES *IN SITU**

General principles

When investigating the effects of atmospheric pollutants on the microbiota, studies must be

* *In situ* is distinguished in this chapter from *in vitro* in that samples from natural microbial habitats were used in the *in situ* studies, even though the natural samples were disturbed and altered.

Table 6. Some biochemical and molecular mechanisms accounting for the toxic effects of gaseous and heavy metal pollutants on microorganisms.

Mechanism of toxicity	Comments
Biochemical activity	inhibition of photosynthesis in *Ankistrodesmus falcatus* by Cu was correlated with an adverse effect on the electron transport reactions of photosystem II (Shioi *et al.* 1978)
	inhibition of Fe-oxidising activity of *Thiobacillus ferrooxidans* by Hg was correlated to an adverse effect on cytochrome oxidase (Imai *et al.* 1975)
	growth inhibition of *Escherichia coli* by Cd, Zn, Cu, Hg and Ni was due to decreases in RNA and protein syntheses (Blundell & Wild 1969)
	irradiated mixtures of NO_2^+ plus butene-1 inhibited the activity of glutamate dehydrogenase of *Escherichia coli* (Estes 1967)
	inhibition of photosynthesis in *Euglena gracilis* by O_3 was due to diminished rate of NADH formation (deKoning & Jegier 1969)
Cell membrane/cell wall integrity	SO_2 caused leakage of K^+ from *Umbilicaria muhlenbergii* (Puckett *et al.* 1977)
	O_3 caused leakage of K^+ from *Chlorella sorokiniana* (Chimiklis & Heath 1975)
	O_3 caused leakage of the cell contents from *Escherichia coli* (Scott & Lesher 1963)
	O_3 destroyed the protein coat of ϕX-174 bacteriophage of *Escherichia coli*, resulting in reduced infectivity (deMik & deGroot 1977)
Intracellular transport	NO_2^- inhibited uptake of leucine and glucose by *Escherichia coli* (Nagy *et al.* 1977)
	NO_2^- inhibited active transport uptake of glucose by *Pseudomonas aeruginosa* (Rowe *et al.* 1979)
Pigmentation	HSO_3^- inhibited photosynthesis of *Umbilicaria muhlenbergii* by irreversibly oxidizing chlorophyll *a* (Puckett *et al.* 1973)
	PAN inhibited photosynthesis of *Chlamydomonas reinhardtii* by destroying chlorophyll *a* and *b* and the carotenoids (Gross & Dugger 1969)
Cell division	formation of giant, multinucleated cells of *Poterioochromonas malhamensis* by tetraethyl Pb was correlated to adverse effects on cytokinesis (Roderer 1979)
DNA[1]	growth inhibition of *Escherichia coli* by Cd was correlated to breakages in the DNA (Mitra & Bernstein 1978)
	mixtures of O_3 and cyclohexene inactivated aerosolized ϕX-174 bacteriophage of *Escherichia coli*, with inactivation being correlated to breakages in the DNA (deMik & deGroot 1978)
	HSO_3^- inhibited growth and cell division of *Chlorella pyrenoidosa* by impairing DNA synthesis (Das & Runeckles 1974)
	O_3 induced nutritional mutants in *Escherichia coli* (Hamelin & Chung 1974a, b)
	PAN was mutagenic to *Serratia marcescens* (Peak & Belser 1969)
	SO_2 (i.e. HSO_3^-) was mutagenic to *Escherichia coli, Micrococcus aureus, Saccharomyces cerevisiae*, and to T4, ϕX-174 and λ bacteriophages of *Escherichia coli* (Fishbein 1976; Shapiro 1977)
	NO_2 (i.e. HNO_2) was mutagenic to T2, T4 and ϕX-174 bacteriophages of *Escherichia coli*, tobacco mosaic virus, tobacco rattle virus, polyoma virus, Newcastle disease virus, *Escherichia coli, Salmonella typhimurium, Bacillus subtilis, Bacillus megaterium, Saccharomyces cerevisiae, Schizosaccharomyces pombe, Aspergillus nidulans* and *Neurospora crassa* (Babich & Stotzky 1974; Fishbein 1976)

Table 6. cont.

Mechanism of toxicity	Comments
DNA (cont.)	Mn was mutagenic to T4 bacteriophage of *Escherichia coli* (Orgel & Orgel 1965), *Escherichia coli* (Demerec & Hanson 1951), *Bacillus subtilis* (Nishioka 1975) and *Plectonema boryanum* (Singh & Kashyap 1978)
	Cd and Cu were mutagenic to *Escherichia coli* (pol A^+/pol A^-) (H. Babich & G. Stotzky, unpublished observations)

[1]See Table 1 for other examples of heavy metals that are mutagenic to microorganisms

conducted in the natural habitat of the microbes as well as *in vitro*, as the responses to the same pollutant may be different under these conditions; for example, 500 p.p.m. zinc did not inhibit denitrification by *Pseudomonas* sp. in broth, whereas the same concentration of zinc added to soil inoculated with these bacteria was inhibitory (Bollag & Bababasz 1979). Conversely, fungi tolerated higher concentrations of cadmium in soil than in synthetic medium (Babich & Stotzky 1977c).

The data derived from studies performed in one natural microbial environment are often used to predict the effects of both acute and chronic exposures to pollutants on the microbiota in other natural habitats. However, as the deposition of the same pollutant into different environments may result in different responses from the indigenous microbiota, it is necessary to perform studies in a variety of environments; for example, *Penicillium vermiculatum, Penicillium asperum, Aspergillus niger, Aspergillus fischeri* and *Cunninghamella echinulata* tolerated higher concentrations of cadmium in an acidic soil than in an alkaline soil (Babich & Stotzky 1980a). *Aeromonas* sp., *Agrobacterium tumefaciens* and ϕ11M15 bacteriophage of *Staphylococcus aureus* tolerated mercury better in sea-water than in lake water (Babich & Stotzky 1979b), and *Tetrahymena pyriformis* tolerated lead better in soft than in hard water, whereas mercury was more toxic in hard water (Carter & Cameron 1973).

Furthermore, studies conducted *in situ* must recognise, as was previously emphasised for studies *in vitro*, that different forms of a pollutant may exert different toxicities; for example, elevated concentrations of vanadium in soil, derived primarily from fuel combustion, decreased the acid phosphatase activity in soil, but different chemical forms of vanadium exerted different degrees of inhibition in the order sodium orthovanadate (Na_3VO_4) > sodium metavanadate ($NaVO_3$), vanadyl sulphate ($VOSO_4$) > vanidic oxide (V_2O_5) (Tyler 1976a). Synthesis of amylase was inhibited in soils amended with different forms of lead, in the order lead acetate > lead chloride ($PbCl_2$) > lead sulphide (PbS) > lead sulphate ($PbSO_4$) > lead monoxide (PbO), and the inhibition was correlated with reductions in the number of amylase-producing bacteria (Cole 1977).

The methodologies used to study the effects of gaseous and heavy metal pollutants on microorganisms in natural environments include analysing either specific microbial populations or their biochemical activities. Studies of specific microbial populations may evaluate alterations in the numbers and diversities of species, differences in the extent of tolerance of different populations to pollutants and changes in various microbial interactions—such as symbiotic, host–saprophyte and host–parasite interactions—that occur in polluted environments. Studies on effects of pollution on the biochemical activities of microbes have evaluated changes in denitrification, nitrification, dinitrogen fixation, mineralisation of defined organic forms of carbon, nitrogen and phosphorus, litter decomposition, respiration and enzyme synthesis and activity.

Specific studies

Gaseous and heavy metal pollutants adversely affect the survival, growth, species diversity and interactions of microbes in natural environments, and many studies have concentrated on recording these changes using standard isolation, enumeration and identification methods (see Chapters 1 to 6). For instance, the growth rates of *Penicillium vermiculatum* and *Aspergillus flavipes* were reduced in an acidic soil amended with 100 p.p.m. cadmium, whereas 250 p.p.m. cadmium were necessary to reduce the growth rates of *Aspergillus fischeri* and *Aspergillus niger* and 1000 p.p.m. cadmium were required to decrease the growth rates of *Aspergillus janus, Penicillium asperum, Trichoderma viride* and *Cunninghamella echinulata*. Furthermore, in soils amended with 100 p.p.m. cadmium, the inhibitory effects of *Serratia marcescens* to *Aspergillus niger* were enhanced, whereas those of *Agrobacterium radiobacter* to *Aspergillus niger* were reduced (Babich & Stotzky 1977c). Survival of the marine bacterium, *Aeromonas* sp., in filtered seawater and of *Agrobacterium tumefaciens* and ϕ11M15 bacteriophage of *Staphylococcus aureus* in filtered lake water was reduced in the presence of 1 p.p.m. mercury (Babich & Stotzky 1979b).

Lower numbers of bacteria were recovered from an acidic forest soil fumigated with 1.0 $\mu l\ l^{-1}$ sulphur dioxide for 48 h than from non-exposed soils (Grant *et al.* 1979), and soil acidified by sulphur dioxide pollution had greater numbers of fungi but decreased bacterial populations compared to non-polluted soils (Wainwright 1978, 1979). Sulphur-oxidising microbes (i.e. *Thiobacillus thioparus, Thiobacillus novellus, Alternaria tenuis, Aureobasidium pullulans, Epicoccum nigrum* and *Cephalosporium* sp.) were isolated in greater numbers from leaves, litter and soils polluted with sulphur dioxide than from similar substrates obtained from non-polluted sites (Wainwright 1978, 1979). Gaseous nitrogen dioxide added to soil at a concentration equivalent to 51 $\mu g\ (g\ soil)^{-1}$ nitrogen increased the number of nitrite-oxidising microbes (Ghiorse & Alexander 1977), but growth of aerobic and anaerobic asymbiotic dinitrogen-fixing bacteria was inhibited in soil exposed to 100 μg nitrogen dioxide $(ml\ air)^{-1}$ (Smith & Mayfield 1978).

In the area surrounding a zinc factory, the species diversities of corticolous, saxicolous, lignicolous and terricolous lichen communities were reduced as compared to non-polluted areas (Nash 1972). Decreases in the number of microbial species were also noted in soils from sites which were heavily contaminated with cadmium, lead, zinc and copper as compared to soils obtained from non-contaminated sites (Hartman 1974). Soils within 2 km of a smelter in Pennsylvania and contaminated with zinc had lower numbers of bacteria, including actinomycetes, and fungi as compared to soils from control sites. Furthermore, zinc-tolerant microorganisms, especially non-spore-forming eubacteria and actinomycetes, were readily isolated from the zinc-contaminated but not from the control soils (Jordan & Lechevalier 1975). The phylloplane microbiota (particularly bacteria and pigmented yeasts) of cabbages and pine saplings in an area near a smelter showed reductions in both numbers and species diversity as compared to the microbiota of plants from control sites. Microorganisms isolated from leaves contaminated with heavy metals were also more tolerant of higher concentrations of mixtures of zinc, lead and cadmium than were those isolated from leaves from non-polluted areas (Gingell *et al.* 1976).

Indeed, the presence of microbes tolerant of heavy metals in environments contaminated with heavy metals is a common observation. Bacteria isolated from activated sewage sludge had high tolerances to heavy metals (Horitsu & Tomoyeda 1975; Horitsu *et al.* 1978); water and sediment samples collected from a heavily polluted site in the Chesapeake Bay contained larger populations of heavy metal-resistant bacteria, principally of the genera *Bacillus, Erwinia, Mycobacterium* and *Pseudomonas*, than samples from non-polluted sites (Austin *et al.* 1977); and large populations of species of *Bacillus* which were tolerant of mercury have been isolated from polluted sediments in the New York Bight (Timoney *et al.* 1978). *Chlorella* sp. and *Scenedesmus* sp. isolated from lake waters containing high levels of copper and nickel were more tolerant of these metals than were laboratory strains (Stokes *et al.* 1973). Numerous metal-resistant fungi were isolated from soil contaminated by mine draining, and when cultured in laboratory media, *Penicillium lilacinum, Paecilomyces* sp. and *Synnematium* sp. tolerated 10 000 p.p.m. cadmium, whilst *Pencillium waskmani* and a *Trichoderma* sp. tolerated 8000 p.p.m. cadmium (Tatsuyama *et al.* 1975). Almost all of the bacteria isolated from a Douglas fir needle litter amended with 2 mM cadmium and incubated for two weeks were resistant to 5 mM cadmium in synthetic media, whereas only 1% of the bacteria isolated from unamended litter was resistant to 5 mM cadmium (Lighthart 1979). Mercury-resistant strains of *Pseudomonas, Bacillus, Penicillium* and *Aspergillus* have been isolated from soils treated with mercury-containing fungicides (Antonovics *et al.* 1971).

Deposition of gaseous and heavy metal pollutants into aquatic and onto terrestrial ecosystems influences biochemical activities of microorganisms in these environments. Denitrification by *Pseudomonas aeruginosa, Pseudomonas denitrificans* and

Pseudomonas sp. was inhibited in autoclaved soils amended with 50 μg g^{-1} cadmium or 500 μg g^{-1} zinc; 50 μg g^{-1} copper inhibited denitrification by *Pseudomonas aeruginosa* and an unidentified *Pseudomonas* species, but not by *Pseudomonas denitrificans*. Denitrification by the indigenous microbiota in soil was inhibited by amendments of 250 μg g^{-1} cadmium, 250 μg g^{-1} copper, 500 μg g^{-1} zinc or 1000 μg g^{-1} lead (Bollag & Bababasz 1979).

Rates of nitrification were reduced in a soil (pH 7.2) fumigated with 5 p.p.m. nitrogen dioxide (Labeda & Alexander 1978) and in acidic soils (pH 5.0—Bozian & Stotzky 1976; Labeda & Alexander 1978), but not in a neutral soil (pH 7.2—Labeda & Alexander 1978), fumigated with sulphur dioxide. Reductions in nitrification were also noted in soils treated with solutions of simulated acid rain (i.e. solutions of sulphuric acid—Tamm 1976), with 5 μmol (g soil)$^{-1}$ of nickel, mercury, chromium or cadmium (Liang & Tabatabai 1978) or with 1000 μg g^{-1} zinc (Premi & Cornfield 1969; Wilson 1977). Nitrification in water samples of the Chesapeake Bay was inhibited by additions of 10 p.p.m. mercury or 100 p.p.m. copper, cadmium or lead (Mills & Colwell 1977).

The addition of up to 1000 p.p.m. nickel to soil inhibited nitrification to a greater extent than mineralisation of carbon and nitrogen, suggesting a greater sensitivity to nickel of the autotrophic nitrifiers than of the heterotrophic microbiota (Giashuddin & Cornfield 1978). Conversely, soil amended with 9 to 18 μmol g^{-1} cadmium chloride, 9 to 22 μmol g^{-1} cadmium acetate or 121 μmol g^{-1} lead acetate exhibited enhanced nitrification, giving rise to the suggestion that the general microbiota may be more sensitive to heavy metals than the nitrifiers (Tyler *et al.* 1974).

Dinitrogen fixation in red clover plants inoculated with *Rhizobium trifolii* was inhibited by fumigation with 0.01% (v/v) carbon monoxide (Lind & Wilson 1941), whilst the nitrogenase activity of soil aggregates was reduced after exposure to 3.5 to 35 μg nitrogen dioxide ml^{-1} air (Smith & Mayfield 1978), and 300 μM lead and 18 μM cadmium inhibited the nitrogenase activity of soybean nodules containing *Rhizobium japonicum* (Huang *et al.* 1974).

Mineralisation of carbon (determined by the evolution of carbon dioxide), nitrogen (determined by the conversion of organic nitrogen to ammonium) and phosphorus (determined by the conversion of organic phosphorus to phosphate) is also adversely influenced by pollutants. Carbon mineralisation was reduced in soils amended with 40 μg g^{-1} mercury (Landa & Fang 1978), and both carbon and nitrogen mineralisation were decreased in soils amended with nickel, either 1000 p.p.m. as nickel sulphate or 5000 p.p.m. as nickel oxide (Giashuddin & Cornfield 1978, 1979). Soils obtained from sites surrounding a brass mill in Sweden and contaminated with copper and zinc had lower rates of phosphorus mineralisation than soils obtained from non-contaminated sites (Tyler 1976a). The addition to aerobic soils of 100 p.p.m. copper (as copper sulphate) increased, at 1000 p.p.m. had no effect on, and at 10 000 p.p.m. decreased nitrogen mineralization. However, nitrogen mineralisation was not reduced if the soils amended with 10 000 p.p.m. copper were maintained anaerobically or when 10 000 p.p.m. copper were added as copper carbonate to aerobic soils (Premi & Cornfield 1969). Soil respiration was decreased after fumigation with sulphur dioxide, addition of bisulphite (Grant *et al.* 1979) or treatment with simulated acid rain (Tamm 1976), and, when measured after 23 d incubation, it was stimulated by the addition of a mixture of 10 p.p.m. cadmium and 100 p.p.m. zinc but inhibited by a mixture of 10 p.p.m. cadmium and 1000 p.p.m. zinc (Chaney *et al.* 1978).

As pollutants adversely affect microbial mineralisation processes, the accumulation of organic matter (e.g. litter) often occurs in polluted environments. For example, the accumulation of organic matter in the surface soil surrounding an aluminium factory in India that was emitting gaseous and particulate fluoride compounds was correlated with the high concentrations of fluoride in the litter and soil, which presumably were toxic to the microbes involved in the decomposition processes (Rao & Pal 1978). The rates of decomposition of a spruce needle litter, obtained from sites around metal-processing industries in Sweden emitting copper, zinc, cadmium and nickel, were reduced in comparison to litter obtained from non-polluted sites (Tyler 1972; Ruhling & Tyler 1973). Treatment of Douglas fir needle litter with 1000 μg g^{-1} cadmium, mercury, nickel, zinc or cadmium decreased the rates of decomposition (Spalding 1979). The decomposition of leaf material placed in an artificial aquatic microcosm supplemented with 5 or 10 μl l^{-1} cadmium was decreased as compared to decomposition in a microcosm not treated with cadmium. Scanning electron microscopy showed that the leaf material in the cadmium-treated system had less microbial colonisation than leaf material obtained from the control microcosm (Giesy 1978).

The principal steps in litter decomposition and humus formation are primarily controlled by extracellular enzymes of microbial origin which may accumulate in the environment (Burns 1978). Pollutants can influence the rates and extent of organic matter decomposition through either the inhibition of enzyme activity or the decreased production of enzymes. Acid phosphatase activity was reduced both in an organic soil by the addition of 100 mg kg^{-1} vanadium (Tyler 1976b) and in soils contaminated with various heavy metals surrounding a brass mill in Sweden (Tyler 1976a). The synthesis of cellulase and xylanase was reduced in soils treated with either 1000 μg g^{-1} cadmium or mercury (Spalding 1979), and lead acetate, lead chloride and lead sulphide reduced the synthesis of amylase at lead concentrations of 0.45 mg g^{-1} soil or higher (Cole 1977).

The reduction of the photosynthetic rates of microbes or toxicities to photosynthetic microbes by pollutants not only decreases the biomass and, thereby, adversely affects food webs, but it also decreases the energy content of the recipient environment. Concentrations of 25 p.p.m. cadmium or mercury inhibited photosynthesis of phytoplankton in water samples from Chesapeake Bay (Mills & Colwell 1977). The disappearance of lichens from cities and from areas surrounding industrial complexes has been attributed to atmospheric pollutants, principally sulphur dioxide. Transplantation of bark plugs bearing lichens from trees in non-polluted areas to trees in polluted areas often resulted in decreased growth, reductions in photosynthesis, morphological abnormalities and, subsequently, death (Babich & Stotzky 1974). Heavy metals may also adversely affect lichens; for example, the growth rate of *Pseudoparmelia baltimorensis* was reduced in areas contaminated with high concentrations of atmospheric lead (Lawrey & Hale 1979).

Delicately balanced microbial interactions, such as mutualism and parasitism, may be affected by atmospheric pollution. In mutualism, each component is beneficial to the other, and any adverse effect on one of the components will subsequently affect the entire community. Fumigation of seedlings of 'Troyer' citrange with ozone reduced mycorrhizal infection and chlamydospore production by *Glomus fasciculatus* (McCool *et al.* 1979). *Rhizobium* nodules were absent from pinto bean plants exposed to ozone (0.1 to 0.15 μl l^{-1} for 8 h d^{-1} for 28 d), whereas nodules were present on plants exposed to charcoal-filtered air (Manning *et al.* 1971), and nodulation was reduced on kidney bean and soybean plants treated with simulated acid rain (Shriner 1977).

Plants injured by pollutants may be more susceptible to invasion by facultative saprophytic and parasitic fungi than are healthy plants; for example, the infectivity of potato (Manning *et al.* 1969), geranium (Manning *et al.* 1970) and broadbean (Manning 1975) leaves by *Botrytis cinerea* was increased after fumigation of the plants with ozone, and replication of tobacco mosaic virus was stimulated in pinto bean fumigated with fluoride (Treshow *et al.* 1967) and in tobacco treated with ozone (Brennan & Leone 1970). Conversely, infection of strawberry by *Xanthomonas fragariae* (Laurence & Wood 1978a) and of soybean by *Pseudomonas glycinea* (Laurence & Wood 1978b) was reduced by exposures to ozone. Obligate fungal parasitism may also be retarded by exposure of the host plants to pollutants that damage the host tissue; for example, infectivity of oat by *Puccinia coronata*, of wheat by *Puccinia graminis* and of barley by *Erysiphe graminis* was reduced by exposures to ozone (Heagle 1975), and the infectivity of *Uromyces phaseoli* on kidney bean was reduced by treatment with simulated acid rain (Shriner 1977).

Field studies have reported both increases and decreases of plant diseases in polluted areas. In regions of high sulphur dioxide pollution, various plant diseases caused by fungi were either reduced or eliminated; namely, *Rhytisma acerinum* infection of Sycamore leaves (Bevan & Greenhalgh 1976); blister rust (*Cronartium ribicola*) of eastern white pine; mildew (*Microsphaera alni*) of oak and lilac; rose infection by *Diplocarpon rosae* and *Sphaerotheca pannisa*; foliar disease caused by *Hysterium pulicare*, *Hypodermella laricis*, *Lophodermium pinastri* and *Venturia inaequalis*; and rust disease of wheat by *Puccinia graminis* (Saunders 1971; Heagle 1973; Babich & Stotzky 1974). Blackspot disease of roses was eliminated in areas containing high concentrations of ammonia of anthropogenic origin (Saunders 1970). Conversely, proliferation of plant pathogenic fungi has been noted in areas polluted with sulphur dioxide; e.g. trees damaged by sulphur dioxide had a higher incidence of infection by *Armillaria mellea*, *Glocophyllum abietinum*, *Trametes heteromorpha* and *Trametes serialis*;

spruce needles injured by sulphur dioxide showed a higher incidence of infection by *Lophodermium piceae*; and pine needles injured by sulphur dioxide had increased infectivity by *Rhizosphaera kalkhoffi* (Heagle 1973).

36.4 Abiotic factors affecting pollutant toxicity

Various biotic and abiotic factors appear to influence the sensitivity of microorganisms to pollutants. The physiological and morphological states of the responding microbes and the chemical form (speciation) of the pollutant have already been discussed as factors that influence pollutant toxicity. The influence of other abiotic factors, however, has not received sufficient attention. Among these factors are the possible synergistic and antagonistic interactions between pollutants. Atmospheric pollutants, whether particulate or gaseous, are seldom, if ever, present in the environment as individual constituents; more usually, domestic and industrial activities simultaneously emit several pollutants that may be deposited into a common microbial habitat. Interactions, whether synergistic or antagonistic, between pollutants may evoke responses from the indigenous microbiota that are different from those evoked by exposure to each pollutant individually.

No synergistic effects were noted between sulphur dioxide and ozone on the germination of conidia of *Microsphaera alni* (Hibben & Taylor 1975) or on the photosynthetic activity of *Euglena gracilis* (deKoning & Jegier 1970b). However, sulphur dioxide reduced the toxicity of gaseous formaldehyde to *Rhizobium meliloti* in the airborne state, presumably because the bisulphite formed in the presence of moisture reacted with the aldehyde moiety of formaldehyde to form a product that was less toxic (Won & Ross 1969).

Air containing ozone together with different olefin vapours (e.g. pentene, cyclohexene) was more toxic to aerosolised *Escherichia coli, Micrococcus albus* and *Staphylococcus aureus* than either ozone or an olefin alone, presumably because ozone cleaved the double bond of the olefins to yield a highly toxic mixture containing aldehydes or ketones and a peroxide zwitterion (Dark & Nash 1970; Druett & Packman 1972). A combination of 40 p.p.b. ozone and 1000 p.p.b. cyclohexene was highly toxic to and induced many breaks in the DNA of aerosolised *Escherichia coli*, whereas neither ozone nor cyclohexane alone appreciably affected survival (deMik & deGroot 1978). Similarly, mixtures of ozone and trans-2-butene or of ozone and cyclohexene were toxic to aerosolised ϕX-174 coliphage, causing many breaks in DNA (deMik & deGroot 1977; deMik *et al.* 1977).

Interactions also occur between heavy metal pollutants (Table 7). Antagonistic interactions between heavy metal cations can result from competition between the cations for common sites on the surface of the target cell, with the more efficient competitor preventing the uptake of the other cation. Synergistic effects between heavy metal cations on microbes may result from the adsorption of both cations on the surface of the cell, with the adsorption of one metal increasing the permeability to the second heavy metal (Babich & Stotzky 1978d, 1980a).

Although not studied extensively and seldom emphasised in environmental studies of the effects of pollution on microorganisms, the abiotic physico-chemical characteristics of the recipient environment influence the toxicity of pollutants to microbes in that environment. This aspect has been reviewed recently (Babich & Stotzky 1978d, 1979c, 1980; Stotzky & Babich 1980), and, therefore, only a few examples will be presented in addition to a summary of the environmental characteristics that have been shown to influence the toxicity of both gaseous and heavy metal pollutants to microbes (Tables 8 and 9).

When sulphur dioxide reacts with water, sulphurous acid, hydrogen, bisulphite, and/or sulphite ions are formed, depending on the pH and buffering capacity of the environment. In alkaline environments, sulphite is the principal anionic product of sulphur dioxide dissolution, whereas in environments with a pH between 2.0 and 7.5, bisulphite is the dominant species. As substantial quantities of undissociated sulphurous acid are present only below pH 2.0 (Puckett *et al.* 1973), this potential toxicant is of limited environmental concern. The detrimental effects of sulphur dioxide on the microbiota appear to be due primarily to bisulphite and the excessive hydrogen ions produced, with the toxicity of the latter being dependent on the buffering capacity of the recipient environment; sulphite is non-toxic. In alkaline environments with strong buffering capacities, such as sea-water, dissolution of sulphur dioxide yields sulphite, which is oxidized to sulphate, a nutrient, and hydrogen ions which are complexed in the carbonate-bicarbonate buffer system. In ecosystems with

Table 7. Examples of antagonistic and synergistic interactions between heavy metal pollutants[1].

Pollutant	Antagonist	Microorganism
Nickel	cadmium	*Physarum polycephalum* (Chin *et al.* 1978)
	zinc	*Chlorella* sp. (Upitis *et al.* 1973)
Cadmium	manganese, iron	*Chlorella pyrenoidosa* (Hart & Scaife 1977; Hart *et al.* 1979)
	mercury, nickel	*Physarum polycephalum* (Chin *et al.* 1978)
	zinc	*Aspergillus niger* (Laborey & Lavollay 1967, 1973)
	iron, tin, lead, chromium	*Escherichia coli* (Ohta & Udaka 1977)
	zinc	*Euglena gracilis* (Nakano *et al.* 1978)
	manganese, zinc, iron	*Chlorella* sp. (Upitis *et al.* 1973)
Chromium	iron	*Chlorella* sp. (Upitis *et al.* 1973)
Zinc	manganese	*Lactobacillus pentosus, Lactobacillus arabinosus* (MacLeod & Snell 1950)
Copper	cadmium	*Selenastrum capricornutum* (Bartlett *et al.* 1974)
Mercury	cadmium	*Physarum polycephalum* (Chin *et al.* 1978)
Lead	manganese	*Selenastrum capricornutum, Chlorella stigmatophora* (Christensen & Scherfig 1979)

Pollutant	Synergist	Microorganism
Lead	cadmium	*Physarum polycephalum* (Chin *et al.* 1978)
Zinc	cadmium	*Physarum polycephalum* Chin *et al.* 1978), *Klebsiella aerogenes* (Pickett & Dean, 1976)
Manganese	copper	*Selenastrum capricornutum* (Christensen & Scherfig 1979)
Copper	cadmium	*Physarum polycephalum* (Chin *et al.* 1978)
	nickel	*Chlorella vulgaris, Haematococcus capensis* (Hutchinson 1973)

[1] In most studies, the microbial response studied was growth. Antagonists decreased the inhibitory effects of the indicated pollutant, whereas synergists enhanced the toxicity of the pollutant.

poor buffering capacity, such as most lakes, the excessive hydrogen ions generated by dissolution of sulphur dioxide will lower the pH of these environments, and the toxicity of sulphur dioxide to the indigenous microbiota is due to both bisulphite and hydrogen ions (Babich & Stotzky 1980a).

There is little information on the effects of organic matter on the toxicity of gaseous pollutants to microbes. However, the toxicity of ozone to viruses and microbes decreased in the presence of soluble organics, presumably as a result of the 'ozone demand' of organic matter (Table 8).

Several studies have shown that the toxicity of heavy metals to viruses and microbes is dependent on the pH of the medium (Table 9), but it is difficult to determine exactly the mechanism whereby pH influences heavy metal toxicity. One of the effects of pH may be its influence on the chemical speciation and mobility of heavy metals; for example, in the presence of increasing OH^- concentrations, Zn^{2+} forms $ZnOH^+$, $Zn(OH)_2$, $Zn(OH)_3^-$ and $Zn(OH)_4^{2-}$, and in sea-water, with an average pH of 8.2, zinc apparently exists primarily as $Zn(OH)_2$ (Hahne & Kroontje 1973). However, other abiotic and biotic parameters are also affected

Table 8. Examples of the influence of abiotic environmental factors on the toxicity of some gaseous pollutants and their solubility products to microorganisms and viruses.

Abiotic factor	Pollutant	Microbial response
pH	sulphur dioxide	SO_2 inhibited nitrification in acidic soil (pH 5.0) but not in a neutral soil (pH 7.2) (Labeda & Alexander 1978)
		mineralisation of a protein hydrolysate was inhibited by HSO_3^- to a greater extent in soil at pH 3.89 than at pH 4.01 (Grant *et al.* 1979)
		inactivation of poliomyelitis virus, coxsackievirus and echovirus by HSO_3^- was greater at pH 5.0 than at pH 7.0 (Salo & Cliver 1978)
		inactivation of T1 bacteriophage of *Escherichia coli* by SO_3^{2-} and HSO_3^- increased as the pH was decreased from 7.0 to 5.0 (Babich & Stotzky 1978c)
		photosynthesis by *Cladonia alpestris, Cladonia deformis, Umbilicaria muhlenbergii* and *Stereocaulon paschale* was inhibited by HSO_3^- at pH 3.2 but not at pH 4.4 or 6.6 (Puckett *et al.* 1973; Nieboer *et al.* 1976)
		HSO_3^- was more inhibitory to photosynthesis by *Fischerella muscicola, Calothrix anomala* and *Anabaena flos-aquae* at pH 6.0 than at pH 7.7 (Wodzinski *et al.* 1978)
		HSO_3^- was more inhibitory to nitrogen fixation and photosynthesis by *Sterocaulon paschale* and *Anabaena cylindrica* at pH 5.8 than at pH 6.5 or 7.3, respectively (Hallgren & Huss 1975)
		the toxicity of HSO_3^- to photosynthesis by *Chlorococcum* sp. increased as the pH was decreased (Sheridan 1978)
		HSO_3^- was more toxic than SO_3^{2-} to growth of *Botrytis cinerea, Cunninghamella echinulata, Phycomyces blakesleeanus, Aspergillus niger, Trichoderma viride* (Babich & Stotzky 1978c), *Saccharomyces cerevisiae* (Rahn & Conn 1944), *Escherichia coli* (Rahn & Conn 1944; Babich & Stotzky 1978c), *Pseudomonas aeruginosa, Bacillus cereus* and *Serratia marcescens* (Babich & Stotzky 1978c), and decreasing the pH increased the toxicity of both anions
	nitrogen dioxide	the bacteriostatic effect of NO_2^- to *Staphylococcus aureus* increased as the pH was decreased (Castellani & Niven 1955)
		inhibition of photosynthesis by *Fisherella muscicola, Ankistrodesmus falcatus, Ulothrix fimbriata, Calothrix anomala* and *Anabaena flos-aquae* by NO_2^- was greater at pH 6.0 than at pH 7.7 (Wodzinski *et al.* 1978)
pH	ammonia	inactivation of poliomyelitis virus by NH_3 was evident only at pH values above 7.5 (Ward & Ashley 1978)
		growth of *Aspergillus niger* in soils amended with NH_4NO_3 was inhibited when the pH was increased above 6.0 with $CaCO_3$ (Rosenzweig & Stotzky 1980)
	ozone	mutagenicity of O_3 to *Escherichia coli* decreased as the pH was increased from 5.0 to 9.0 (Hamelin & Chung 1974b)

Table 8. cont.

Abiotic factor	Pollutant	Microbial response
Relative humidity (r.h.) and moisture	sulphur dioxide	2.5 and 5 mg m^{-3} SO_2 were most toxic to aerosolized *Serratia marcescens* at mid-range r.h. levels (i.e. 60%) (Lighthart *et al.* 1971)
		20 μl l^{-1} SO_2 was most toxic to aerosolised *Rhizobium meliloti* at high r.h. levels (i.e. 95%) (Won & Ross 1969)
		3.6 p.p.m. SO_2 was more toxic to aerosolised Venezuelan equine encephalomyelitis virus at 30 than at 60% r.h. (Berendt *et al.* 1972)
		spores of *Alternaria* sp. were more sensitive to SO_2 at 90 than at 80% r.h., and wet spores were more sensitive than dry spores (Couey 1965)
		increasing the r.h. increased the toxicity of SO_2 to spores of *Botrytis cinerea* (Couey & Uota 1961)
	nitrogen dioxide	0.15 μl l^{-1} NO_2 was most toxic to aerosolised *Rhizobium meliloti* at high r.h. levels (i.e. 95%) (Won & Ross 1969)
	ozone	aerosolised *Streptococcus salivarius* was sensitive to 0.025 to 2.4 p.p.m. O_3 only at r.h. values above 50% (Elford & van den Ende 1942)
		wet spores of *Botrytis cinerea, Aspergillus terreus* and *Fusarium oxysporum* were more sensitive to O_3 than were dry spores (Hibben & Stotzky 1969)
	carbon monoxide	85 p.p.m. CO was toxic to aerosolised *Serratia marcescens* only at r.h. values below 75% (Lighthart 1973)
Soluble organics	ozone	O_3 was more toxic to growth of *Penicillium jenseni* in a mineral salts medium than in a similar medium amended with yeast extract (G. Stotzky, unpublished observations)
		O_3 was more toxic to *Staphylococcus aureus* in an inorganic than in an organic broth (Ingram & Haines 1949)
		Streptococcus salivarius aerosolised from an organic broth was more tolerant to O_3 than when aerosolised from water suspensions (Elford & van den Ende 1942)
		Staphylococcus aureus, Salmonella typhimurium, Escherichia coli, Shigella flexneri, Pseudomonas fluorescens and *Vibrio cholera* were more sensitive to O_3 when exposed in saline than in a secondary effluent containing dissolved organics (Burleson *et al.* 1975)
		unwashed cells of *Bacillus cereus* and *Escherichia coli* harvested from an organic medium were more resistant to O_3 than were cells washed in saline (Broadwater *et al.* 1973)
Clay minerals	sulphur dioxide	the extent of inhibition by SO_2 of nitrification in soil was reduced by the addition of montmorillonite but not of kaolinite (Bozian & Stotzky 1976)
Light	sulphur dioxide	at 30 and 60% r.h. and a light intensity of 308 mcal cm^{-2} min^{-1} and at 60% r.h. and a light intensity of 40 mcal cm^{-2} min^{-1}, there was a synergistic interaction between light and 3.6 p.p.m. SO_2 to survival of aerosolised Venezuelan equine encephalomyelitis virus; at 30% r.h. and a light intensity of

Table 8. cont.

Abiotic factor	Pollutant	Microbial response
		40 mcal cm^{-2} min^{-1}, there was an antagonistic interaction between light and SO_2 on survival of the virus in the airborne state (Berendt *et al.* 1972)
Temperature	sulphur dioxide	toxicity of SO_2 to spores of *Botrytis cinerea* increased 1.5 times for each 10°C rise between 0 and 30°C (Couey & Uota 1961)
		the toxicity of HSO_3^- to photosynthesis by *Chlorococcum* sp. increased as the temperature was increased from 0 to 30°C (Sheridan 1978)
	ozone	species of *Achromobacter, Pseudomonas, Proteus, Mucor, Botrytis, Thamnidium* and *Penicillium* were more sensitive to O_3 at 0 than at 20°C (Ingram & Haines 1949)

by pH. Lowering the pH increases the quantity of H^+, which may compete with heavy metal cations for sites on the cell surface. The ambient pH also influences the charge of the ionogenic groups on the surface of microbial cells and, thus, affects the affinity of the cell surface for the metal ions. The physiological state and the biochemical activities of microbes are also affected by changes in pH, and this can affect their sensitivity to heavy metals.

Other ambient cations, in addition to hydrogen ions, influence heavy metal toxicity to microbes. The decreased toxicity of heavy metals to microbes in the presence of increasing concentrations of non-toxic cations (e.g. K^+, Ca^{2+}, Mg^{2+}—Table 9) probably reflects competition between the non-toxic cations and the heavy metal cations for surface ionogenic groups of the target microbial cell.

Heavy metal cations form coordination complexes not only with hydroxyl ions but also with other inorganic anionic ligands, such as chloride, the dominant inorganic anion in the oceans. The various coordination complexes have different stabilities, and in the oceans, with an average pH of 8.2 and an average chloride concentration of 20 000 p.p.m., zinc occurs predominantly as $Zn(OH)_2$, lead as $PbOH^+$, cadmium as a mixture of $CdCl_2$, and $CdCl_3^-$, and mercury as a mixture of $HgCl_3^-$ and $HgCl_4^{2-}$ (Hahne & Kroontje 1973). The different chemical forms of the same heavy metal may exert different toxicities; e.g. bacteria and bacteriophages were both more sensitive to mercury as Hg^{2+} than as mixtures of $HgCl_3^-$ and $HgCl_4^{2-}$ and tolerated mercury better in sea-water than in lake water (Babich & Stotzky 1979b).

Other inorganic anions, notably sulphide, carbonate and phosphate, react with heavy metals to form insoluble complexes. Deposition of a heavy metal into an environment containing high concentrations of these ligands results in its precipitation as an insoluble salt and thereby in a decrease in the availability and, hence, toxicity of the pollutant to the indigenous microbiota (Table 9).

Insoluble inorganics, such as hydrous aluminosilicate clay minerals (e.g. kaolinite, attapulgite, montmorillonite), also affect heavy metal toxicities. Most crystalline clay minerals possess surfaces which are predominantly negatively charged and to which charge-compensating cations (e.g. H^+, K^+, NH_4^+, Na^+, Ca^{2+}, Mg^{2+}) are adsorbed. These cations are not permanent components of the clays and are constantly being exchanged by other cations in the environment (Stotzky 1972). Heavy metal cations introduced into an environment may be exchanged for non-toxic cations on the exchange complex of the clays and, thereby, removed, at least temporarily, from solution and uptake by the microbiota. Studies with cadmium in both synthetic media (Babich & Stotzky 1977b, 1978f) and soil (Babich & Stotzky 1977c, 1978f) and with lead in synthetic media (Babich & Stotzky 1979a) have shown that the toxicity of these heavy metals to microbes could be reduced or totally eliminated by the incorporation of clay minerals into the media or soils (Table 9). The protective effects of the clay minerals increased as the cation exchange capacity of the clays increased.

The studies presented in Tables 7, 8 and 9, although conducted primarily in synthetic media and indicating that more studies are needed in

Table 9. Examples of the influence of some abiotic environmental factors on the toxicity of some heavy metal pollutants to microorganisms and viruses.

Abiotic factor	Pollutant	Microbial response
pH	copper	*Penicillium nigricans* grew on media supplemented with 4 × 10^{-2}M-Cu at pH 2.8 but not pH 5.6 (Singh 1977)
		Scytalidium sp. grew on media containing 1M-Cu at pH 0.3 to 2.0 but was sensitive to 4 × 10^{-5}M-Cu on media at pH 6.7 (Starkey 1973)
		the toxicity of Cu to spores and mycelium of *Fusarium lycopersici* decreased as the pH was decreased (Horsfall 1956)
		growth and photosynthesis of *Chlorella pyrenoidosa* was inhibited by Cu to a greater extent at pH 8.0 than at pH 5.0 (Steemann Nielsen *et al.* 1969)
		Cu was more inhibitory to growth of *Chlorella pyrenoidosa sorokiniana* at pH 7.0 than at pH 5.0 (Hannan & Patouillet 1972)
	mercury	the toxicity of Hg to spores and mycelium of *Fusarium lycopersici* decreased as the pH was decreased (Horsfall 1956)
		Hg was more toxic to *Phaeodactylum tricornutum* in sea-water at pH 8.13 than at pH 8.45 (Hannan & Patouillet 1979)
	aluminium	*Chlorella pyrenoidosa* tolerated Al better at pH levels between 8.0 and 9.0 than at pH 4.6 (Foy & Gerloff 1972)
	cadmium	growth inhibition of *Alcaligenes faecalis, Bacillus cereus, Agrobacterium tumefaciens, Nocardia corallina, Rhizopus stolonifer, Aspergillus niger* and *Trichoderma viride* by Cd was greater at alkaline (i.e. pH 8.0 and 9.0) than at neutral or acidic pH levels (Babich & Stotzky 1977a)
		Penicillium asperum, Penicillium vermiculatum, Aspergillus niger and *Cunninghamella echinulata* were more sensitive to Cd in an acidic soil than in an alkaline soil (Babich & Stotzky 1980a)
		Cd extended the generation times of *Chlorella pyrenoidosa* to a greater extent at pH 7.0 than at pH 8.0 (Hart & Scaife 1977)
	iron	Fe citrate was more inhibitory to respiration of strains of *Sphaerotilus* at pH 6.0 than at pH 7.0 (Chang *et al.* 1979)
	lead	the toxicity of Pb to *Trichoderma viride* and *Aspergillus niger* increased as the pH was increased from acidic to alkaline levels (Babich & Stotzky 1979a)
Inorganic cations	copper	K reduced the toxicity of Cu to growth of *Chlorella pyrenoidosa* (Steemann Nielsen *et al.* 1969)
	cadmium	Mg and Ca reduced the toxicity of Cd to growth of *Aspergillus niger* (Laborey & Lavollay 1973, 1977)
		Mg (Abelson & Aldous 1950) and Ca (Ohta & Udaka 1977) reduced the toxicity of Cd to growth of *Escherichia coli*
	zinc	Mg reduced the toxicity of Zn to growth of *Aspergillus niger* (Adiga *et al.* 1962; Laborey & Lavollay 1973)
	nickel	Mg reduced the toxicity of Ni to growth of *Escherichia coli, Aerobacter aerogenes, Torulopsis utilis* (Abelson & Aldous 1950) and *Aspergillus niger* (Abelson & Aldous 1950; Adiga *et al.* 1962)

Table 9. cont.

Abiotic factor	Pollutant	Microbial response
Inorganic anions	copper	PO_4^{3-} reduced the toxicity of Cu to *Aerobacter aerogenes*, presumably due to the precipitation of Cu as $Cu_3(PO_4)_2$ (MacLeod *et al.* 1967)
	mercury	the toxicity of Hg to survival of *Aeromonas* sp., *Acinetobacter* sp., *Agrobacterium tumefaciens, Erwinia herbicola,* P1 coliphage and φ11M15 bacteriophage of *Staphylococcus aureus* was reduced by Cl^-, due to the formation of $HgCl_3^-/HgCl_4^{2-}$ mixtures which exerted less toxicity than Hg^{2+} (Babich & Stotzky 1979b)
	lead	PO_4^{3-} reduced the toxicity of Pb to *Chlamydomonas reinhardtii* (Schulze & Brand 1978), *Aspergillus giganteus* and *Fusarium solani* (Babich & Stotzky 1979a), presumably due to the precipitation of $Pb_3(PO_4)_2$
		CO_3^{2-} reduced the toxicity of Pb to *Aspergillus giganteus* and *Fusarium solani*, presumably due to the precipitation of Pb as $PbCO_3$ (Babich & Stotzky 1979a)
	zinc	the toxicity of Zn to photosynthesis by *Selenastrum capricornutum* was decreased by additions of S^{2-}, presumably due to the formation of ZnS (Hendricks 1978)
		the toxicity of Zn to the survival of T1, T7, P1 and φ80 coliphages and to growth of *Aspergillus niger* was increased by Cl^-, due to the formation of $ZnCl_3^-/ZnCl_4^{2-}$ mixtures which exerted greater toxicity than Zn^{2+} (Babich & Stotzky 1978a)
Aeration	zinc	Zn at 10 000 p.p.m., reduced nitrogen mineralisation in soil to a greater extent under anaerobic than aerobic conditions (Premi & Cornfield 1969), possibly due to the precipitation of Zn as ZnS under anaerobiosis
Soluble organics	copper	cysteine reduced the toxicity of Cu to *Escherichia coli* (Hirsch 1961)
		cysteine and yeast extract reduced the toxicity of Cu to *Aerobacter aerogenes* (MacLeod *et al.* 1967)
		Cu chelated with EDTA was less toxic than free Cu to *Phaeodactylum tricornutum* (Bentley-Mowat & Reid 1977), *Chlorella* sp. and *Scenedesmus* sp. (Stokes & Hutchinson 1975)
		Cu complexed with malic, oxalic and tartaric acids was less toxic than free Cu to *Candida albicans* (Avakyan & Rabotnova 1971)
		Cu complexed with amino acids (e.g. glycine, aspartic acid, glutamic acid, arginine) was less toxic than free Cu to *Candida utilis* (Avakyan 1971)
	mercury	tryptone decreased the toxicity of Hg to the anaerobic bacteria, *Bacteroides* sp. and *Clostridium* sp. (Hamdy & Wheeler 1978)
		increasing the concentration of proteose peptone decreased the toxicity of CH_3Hg^+ to *Tetrahymena pyriformis* (Hartig 1971)
		cysteine and yeast extract protected *Rhodopseudomonas capsulata* against bacteriostatic concentrations of CH_3Hg^+ (Jeffries & Butler 1975)
		glutathionine reduced the toxicity of phenyl Hg^+ to *Phaeodactylum tricornutum* (Nuzzi 1972)

Table 9. cont.

Abiotic factor	Pollutant	Microbial response
	cadmium	cysteine reduced the toxicity of Cd to *Staphylococcus aureus* (Tynecka & Zylinska 1974)
		EDTA protected *Escherichia coli* against Cd toxicity (Ohta & Udaka 1977)
		pyruvic acid, gluconic acid, citric acid and aspartic acid reduced the toxicity of Cd to *Klebsiella aerogenes* (Pickett & Dean 1976)
	lead	Pb chelated with nitrilotriacetic acid was not toxic to *Chlamydomonas reinhardtii*, whereas equivalent concentrations of free Pb were toxic (Schulze & Brand 1978)
		EDTA reduced the toxicity of Pb to *Phaeodactylum tricornutum* (Schulz-Baldes & Lewin 1976)
		tryptone, yeast extract, cysteine, succinic acid and increasing concentrations of neopeptone reduced the toxicity of Pb to *Fusarium solani, Aspergillus giganteus, Rhizoctonia solani* and *Cunninghamella echinulata* (Babich & Stotzky 1979a)
	zinc	pyruvic acid, gluconic acid, citric acid and aspartic acid reduced the toxicity of Zn to *Klebsiella aerogenes* (Pickett & Dean 1976)
Particulate organics	mercury	addition of organic-rich sediment to media protected *Bacteroides* sp. and *Clostridium* sp. against Hg toxicity (Hamdy & Wheeler 1978)
	lead	humic acid protected *Rhizoctonia solani, Fusarium solani, Aspergillus giganteus* and *Cunninghamella echinulata* against inhibitory concentrations of Pb (Babich & Stotzky 1979a)
Clay minerals	cadmium	montmorillonite and, to a lesser extent, kaolinite protected *Bacillus megaterium, Agrobacterium tumefaciens, Nocardia corallina, Botrytis cinerea, Aspergillus niger, Phycomyces blakesleeanus, Trichoderma viride, Thielaviopsis paradoxa, Pholiota marginata, Scopulariopsis brevicaulis, Schizophyllum* sp. and *Chaetomium* sp. against inhibitory concentrations of Cd in synthetic media (Babich & Stotzky 1977b)
		Aspergillus fischeri, Aspergillus niger, Penicillium asperum, Penicillium vermiculatum and *Trichoderma viride* tolerated Cd better in soils amended with montmorillonite than in unamended soils or soils amended with kaolinite (Babich & Stotzky 1977c)
	lead	montmorillonite, attapulgite and kaolinite protected *Aspergillus giganteus, Rhizoctonia solani, Fusarium solani* and *Cunninghamella echinulata* against inhibitory concentrations of Pb in synthetic media (Babich & Stotzky 1979a)
Cation exchange capacity	cadmium	the greater protection afforded by montmorillonite, as compared to equivalent concentrations of kaolinite, against the toxicity of Cd to bacteria and fungi in synthetic media was correlated with the higher cation exchange capacity of montmorillonite (Babich & Stotzky 1977b)

Table 9. cont.

Abiotic factor	Pollutant	Microbial response
	lead	the sequence of protection of the clay minerals, montmorillonite > attapulgite > kaolinite, against Pb toxicity to fungal growth was correlated with the cation exchange capacities of the clays (Babich & Stotzky 1979a)
	zinc	100 μg g^{-1}Zn reduced nitrification in soil with a low cation exchange capacity (i.e. 3.13 mEq 100 g^{-1}) to a greater extent than in soil with a higher cation exchange capacity (i.e. 12.43 mEq 100 g^{-1}) (Wilson 1977)
Temperature	copper	the tolerance of *Paramecium tetraurelia* to Cu decreased as the temperature was increased from 12 to 34°C (Szeto & Nyberg 1979)
	mercury	0.05 μg l^{-1} Hg inhibited growth of *Nitzschia* sp. at 15 but not at 25°C (Knowles & Zingmark 1978)
	zinc	the toxicity of Zn to growth of *Nitzschia linearis* in soft water increased as the temperature was increased from 22 to 30°C (Cairns *et al.* 1972)
	chromium	the toxicity of Cr^{3+} to growth of *Nitzschia linearis* in soft water and to *Navicula seminulum* in hard water decreased as the temperature was increased from 22 to 30°C (Cairns *et al.* 1972)
		inhibition of *Cyclotella meneghiniana* by 0.5 to 4.0 mg l^{-1} Cr^{3+} increased as the temperature was increased from 5 to 25°C (Cairns *et al.* 1978)
Hydrostatic pressure	copper	0.1 and 1.0 mg l^{-1} Cu depressed growth of the marine bacterium, BIII39 (a Mn^{2+} oxidiser), at 1 atm; at 272 atm, the inhibitory effects of Cu were not evident, but at 340 atm, Cu was again inhibitory to growth. At 1 atm, 0.1 mg l^{-1} Cu did not affect cell growth, but 1 mg l^{-1} Cu was slightly toxic to bacterium BIII88 (a Mn^{4+} reducer); at 340 atm, 0.1 mg l^{-1} Cu slightly enhanced cell yield, but 1 mg l^{-1} Cu was slightly toxic. At 1 atm, 0.1 mg l^{-1}Cu was toxic to bacterium BIII32 (a Mn^{4+} reducer), and increasing the concentration to 1 mg l^{-1} Cu increased the toxicity; at 340 atm, 0.1 mg l^{-1} Cu was not toxic, although 1 mg l^{-1} Cu reduced cell yield (Arcuri & Ehrlich 1977)
	nickel	increasing the hydrostatic pressure from 1 to 340 atm increased the toxicity of Ni to growth of bacterium BIII39; at 1 atm, 1 mg l^{-1} Ni was not toxic to bacterium BIII88 and bacterium BIII32, but at 340 atm, this concentration reduced cell yields of both bacteria (Arcuri & Ehrlich 1977)

natural habitats, clearly show that the responses of microbes and viruses to heavy metal and gaseous pollutants are influenced by the physico-chemical characteristics of the recipient environment. Furthermore, these studies also emphasise the importance of selecting the proper media for the isolation and enumeration of microbial populations resistant to heavy metals from natural environments. The use of a rich nutrient medium, amended with heavy metals for maximum isolation of microbes presumably tolerant to heavy metals, may yield an overestimation of tolerant populations as a result of the complexing and detoxication of the heavy metals by the inorganic and organic constituents of the medium. Conversely, the use of a nutrient-poor medium to avoid this masking of the effects of the added heavy metals by the constituents of the medium may not be adequate to support the growth of fastidious, but heavy metal-tolerant, microbes, thus yielding an underestimation of microbes tolerant to heavy metals (Babich & Stotzky 1980a, b); for example, a concentration of 1000 p.p.m. lead inhibited growth of *Fusarium solani* and *Cunninghamella echinulata* on an acidic minimal nutrient agar (AMNA), but the same concentration of lead was not inhibitory when these fungi were grown on tomato juice agar (TJA), Sabouraud dextrose agar (SDA) or Czapek solution agar (CSA). Furthermore, *Rhizoctonia solani* was not only more tolerant of 1000 p.p.m. lead on TJA, SDA or CSA than on AMNA, but this fungus tolerated lead better on TJA than on SDA, the more acidic medium with less organic matter, and best on CSA, an alkaline agar containing phosphate amendments (Babich & Stotzky 1980b). One method for resolving this dilemma between the detoxication effects of the isolation medium and the need to isolate as many organisms as possible from natural habitats might be to use a nutrient-rich medium for the initial isolation and then replica-plating from this medium to plates containing a less nutrient-rich medium into which the heavy metal of interest has been incorporated.

The composition of the growth medium also influences the response of microbes to gaseous pollutants; for example, various bacteria, including *Mycobacterium smegmatis, Bacillus megaterium, Arthrobacter globiformis, Agrobacterium tumefaciens* (G. Stotzky, unpublished observations), *Sarcina lutea* and *Serratia marcescens* (H. Babich & G. Stotzky, unpublished observations), tolerated sulphur dioxide better when exposed in media buffered to pH 7.0 than in non-buffered media initially adjusted to pH 7.0. A concentration of 100 p.p.h.m. ozone inhibited germination of fungal spores on agar media but did not adversely affect germination of spores fumigated in a liquid broth or dry on cellophane strips (Hibben & Stotzky 1969).

36.5 Conclusions

Although often neglected in ecological and environmental studies of air pollution, microorganisms are intimately involved in many aspects of such pollution. Microorganisms are producers of many inorganic and organic gases, which essentially constitutes the natural background levels of these gases. The microbiota, in general, has apparently attained some tolerance, either through morphological or physiological modifications, to the potentially toxic effects of those gases that are direct products of microbial metabolism. For example, of the most prevalent gaseous atmospheric pollutants (hydrogen sulphide, nitric oxide, ammonia, carbon monoxide, methane, nitrogen dioxide, sulphur dioxide and ozone), nitrogen dioxide, sulphur dioxide and ozone, which are not directly evolved by microorganisms, are the most toxic to them. Microbes are also sinks for atmospheric pollutants and, therefore, are important natural, detoxication systems; are involved in the biogeochemical cycling of various elements; and may eventually utilise these pollutants as nutrient or energy sources. Microbes are also involved in the biotransformation of one pollutant (e.g. a heavy metal) into another potentially more toxic pollutant (e.g. the methylated form of that heavy metal). In many cases, the final effect of a pollutant on the biosphere may be determined by the microbiota; e.g. mercury enters food chains and webs as lipid-soluble methyl mercury.

Microorganisms and viruses are influenced, usually adversely, as is the macrobiota, by gaseous and heavy metal pollutants. However, microbial growth and viral replication may also be stimulated or not affected by these pollutants. The final response of an organism is dependent on the concentration of the pollutant, the length of exposure, the chemical speciation of the pollutant and the physiological state of the organism. Furthermore, the abiotic physico-chemical factors of the synthetic media in which the experiments are conducted and of the natural environment of the microbes influence the chemical form and mobility of a pollutant, thereby either potentiating or attenuating its toxicity to the microbiota. Consequently, the final response of microbes to a pollutant is influenced by various abiotic and biotic factors, which must be recognised when evaluating or predicting the long-term effects of air pollution on the microbiotic component of the biosphere.

Recommended reading

BABICH H. & STOTZKY G. (1974) Air pollution and microbial ecology. *Critical Reviews in Environmental Control* **4**, 353–421.

BABICH H. & STOTZKY G. (1978) Effects of cadmium on the biota: influence of environmental factors. *Advances in Applied Microbiology* **23**, 55–117.

BABICH H. & STOTZKY G. (1978) Atmospheric sulfur compounds and microbes. *Environmental Research* **15**, 513–31.

BABICH H. & STOTZKY G. (1980) Environmental factors that influence the toxicity of heavy metal and gaseous pollutants to microorganisms. *Critical Reviews in Microbiology* **8**, 99–145.

STOTZKY G. & BABICH H. (1980) Mediation of the toxicity of pollutants to microbes by the physicochemical composition of the recipient environment. In: *Microbiology—1980* (Ed. D. Schlessinger), pp. 352–4. American Society for Microbiology, Washington, D.C.

STOTZKY G. & SCHENCK S. (1976) Volatile organic compounds and microorganisms. *Critical Reviews in Microbiology* **4**, 333–82.

SUMMERS A. O. & SILVER S. (1978) Microbial transformation of metals. *Annual Review of Microbiology* **32**, 637–72.

References

ABELES F. B., CRAKER L. E., FORRENCE L. E. & LEATHER G. R. (1971) Fate of air pollutants: removal of ethylene, sulfur dioxide, and nitrogen dioxide by soil. *Science* **173**, 914–16.

ABELSON P. H. & ALDOUS E. (1950) Ion antagonisms in microorganisms: interference of normal magnesium metabolism by nickel, cobalt, cadmium, zinc, and manganese. *Journal of Bacteriology* **60**, 401–13.

ADIGA P. R., SASTRY K. S. & SARMA P. S. (1962) The influence of iron and magnesium on the uptake of heavy metals in metal toxicities in *Aspergillus niger*. *Biochimica et Biophysica Acta* **64**, 546–8.

AICKIN R. M. & DEAN A. C. R. (1979) Lead accumulation by micro-organisms. *Microbios Letters* **5**, 129–33.

ALLISON F. E. (1955) The enigma of soil nitrogen balance sheets. *Advances in Agronomy* **7**, 213–50.

ANTONOVICS J., BRADSHAW A. D. & TURNER R. G. (1971) Heavy metal tolerance in plants. *Advances in Ecologic Research* **1**, 1–85.

ARCURI E. J. & EHRLICH H. L. (1977) Influence of hydrostatic pressure on the effects of the heavy metal cations of manganese, copper, cobalt, and nickel on the growth of three deep-sea bacterial isolates. *Applied and Environmental Microbiology* **33**, 282–8.

AUSTIN B., ALLEN D. A., MILLS A. L. & COLWELL R. R. (1977) Numerical taxonomy of heavy metal-tolerant bacteria isolated from an estuary. *Canadian Journal of Microbiology* **23**, 1433–47.

AVAKYAN Z. A. (1971) Comparative toxicity of free ions and complexes of copper and amino acids to *Candida utilis*. *Microbiology* **40**, 363–8.

AVAKYAN Z. A. & RABOTNOVA I. L. (1971) Comparative toxicity of free ions and complexes of copper with organic acids for *Candida albicans*. *Microbiology* **40**, 262–6.

BABICH H. & STOTZKY G. (1972) Ecologic ramifications of air pollution. *Transactions of the Society of Automotive Engineers* **81**, 1955–71.

BABICH H. & STOTZKY G. (1974) Air pollution and microbial ecology. *Critical Reviews in Environmental Control* **4**, 353–421.

BABICH H. & STOTZKY G. (1977a) Sensitivity of various bacteria, including actinomycetes, and fungi to cadmium and the influence of pH on sensitivity. *Applied and Environmental Microbiology* **33**, 681–95.

BABICH H. & STOTZKY G. (1977b) Reductions in the toxicity of cadmium to microorganisms by clay minerals. *Applied and Environmental Microbiology* **33**, 696–705.

BABICH H. & STOTZKY G. (1977c) Effect of cadmium on fungi and on interactions between fungi and bacteria in soil: influence of clay minerals and pH. *Applied and Environmental Microbiology* **33**, 1059–66.

BABICH H. & STOTZKY G. (1978a) Toxicity of zinc to fungi, bacteria, and coliphages: influence of chloride ions. *Applied and Environmental Microbiology* **36**, 906–14.

BABICH H. & STOTZKY G. (1978b) Atmospheric sulfur compounds and microbes. *Environmental Research* **15**, 513–31.

BABICH H. & STOTZKY G. (1978c) Influence of pH on inhibition of bacteria, fungi, and coliphages by bisulfite and sulfite. *Environmental Research* **15**, 405–17.

BABICH H. & STOTZKY G. (1978d) Effects of cadmium on the biota: influence of environmental factors. *Advances in Applied Microbiology* **23**, 55–117.

BABICH H. & STOTZKY G. (1978e) Atmospheric pollution: impacts on and interactions with microbial ecology. In: *Microbial Ecology, Proceedings in Life Sciences* (Eds M. W. Loutit & J. A. R. Miles), pp. 13–17. Springer-Verlag, New York.

BABICH H. & STOTZKY G. (1978f) Effect of cadmium on microbes *in vitro* and *in vivo*: influence of clay minerals. In: *Microbial Ecology, Proceedings in Life Sciences* (Eds M. W. Loutit & J. A. R. Miles), pp. 412–15. Springer-Verlag, New York.

BABICH H. & STOTZKY G. (1979a) Abiotic factors affecting the toxicity of lead to fungi. *Applied and Environmental Microbiology* **38**, 506–13.

BABICH H. & STOTZKY G. (1979b) Differential toxicities of mercury to bacteria and bacteriophages in sea and lake water. *Canadian Journal of Microbiology* **25**, 1252–7.

BABICH H. & STOTZKY G. (1980a) Environmental factors that influence the toxicity of heavy metal and gaseous pollutants to microorganisms. *Critical Reviews in Microbiology* **8**, 99–145.

BABICH H. & STOTZKY G. (1980b) Physicochemical factors that affect the toxicity of heavy metals to microbes in aquatic habitats. In: *Proceedings of the American Society for Microbiology Conference on Aquatic Microbial Ecology*. (Eds R. R. Colwell & J. Foster), pp. 181–203. University of Maryland Sea Grant Publication, College Park, Maryland.

BANWART W. L. & BREMNER J. M. (1975a) Formation of volatile sulfur compounds by microbial decomposition of sulfur-containing amino acids in soils. *Soil Biology and Biochemistry* **7**, 359–64.

BANWART W. L. & BREMNER J. M. (1975b) Identification of sulfur gases evolved from animal manures. *Journal of Environmental Quality* **4**, 363–6.

BANWART W. L. & BREMNER J. M. (1976) Evolution of volatile sulfur compounds from soils treated with sulfur-containing organic materials. *Soil Biology and Biochemistry* **8**, 439–43.

BARROW W. & TORNABENE T. G. (1979) Chemical and ultrastructural examination of lead-induced morphological convertants of *Bacillus subtilis*. *Chemico-Biological Interactions* **26**, 207–22.

BARTHOLOMEW G. W. & ALEXANDER M. (1979) Microbial metabolism of carbon monoxide in culture and in soil. *Applied and Environmental Microbiology* **37**, 932–7.

BARTLETT L., RABE F. W. & FUNK W. H. (1974) Effects of copper, zinc, and cadmium on *Selenastrum capricornutum*. *Water Research* **8**, 179–85.

BEN-BASSAT D. & MAYER A. M. (1975) Volatilization of mercury by algae. *Physiologia Plantarum* **33**, 128–32.

BENTLEY-MOWAT J. A. & REID S. M. (1977) Survival of marine phytoplankton in high concentrations of heavy metals, and uptake of copper. *Journal of Experimental Marine and Biological Ecology* **26**, 249–64.

BERENDT R. F., DORSEY E. L. & HEARN H. J. (1972) Virucidal properties of light and SO_2. I. Effect on aerosolized Venezuelan equine encephalomyelitis virus. *Proceedings of the Society of Experimental Biology and Medicine* **139**, 1–5.

BERLINER M. D. (1963) *Armillaria mellea*: an ozonophilic basidiomycete. *Nature, London* **197**, 309–10.

BEVAN R. J. & GREENHALGH G. N. (1976) *Rhytisma acerinum* as a biological indicator of pollution. *Environmental Pollution* **10**, 271–85.

BITTON G. & FREIHOFER V. (1978) Influence of extracellular polysaccharides on the toxicity of copper and cadmium toward *Klebsiella aerogenes*. *Microbial Ecology* **4**, 119–25.

BLACKMER A. M. & BREMNER J. M. (1976) Potential of soil as a sink for atmospheric nitrous oxide. *Geophysical Research Letters* **3**, 739–42.

BLUNDELL M. R. & WILD D. G. (1969) Inhibition of bacterial growth by metal salts. *Biochemistry Journal* **115**, 207–11.

BODNAR J. & TERENYI A. (1932) Biochemistry of the smut diseases of cereals. *Hoppe-Seyler's Zeitschrift für Physiologische Chemie* **207**, 78–92 (cited in: Horsfall J. G. (1956), *Principles of Fungicidal Action*, p. 145. Chronica Botanica Co., Waltham, Massachusetts).

BOLLAG J-M. & BABABASZ W. (1979) Effect of heavy metals on the denitrification process in soil. *Journal of Environmental Quality* **8**, 196–201.

BOLLAG J-M. & HENNINGER N. M. (1978) Effects of nitrite toxicity on soil bacteria under aerobic and anaerobic conditions. *Soil Biology and Biochemistry* **10**, 377–81.

BOLLAG J-M. & TUNG G. (1972) Nitrous oxide release by soil fungi. *Soil Biology and Biochemistry* **4**, 271–6.

BORTNER M. H., KUMMLER R. H. & JAFFE L. S. (1974) Carbon monoxide in the earth's atmosphere. *Water, Air, and Soil Pollution* **3**, 17–52.

BOZIAN R. H. & STOTZKY G. (1976) Inhibition of nitrification in soil by SO_2, and the effect of kaolinite and montmorillonite on inhibition. *Agronomy Abstracts* p. 135.

BRADBEER C. & WILSON P. W. (1963) Inhibitors of nitrogen fixation. In: *Metabolic Inhibitors* (Eds R. M. Hochster & J. H. Quastel), vol. 2, pp. 595–614. Academic Press, New York.

BREMNER J. M. & BANWART W. L. (1976) Sorption of sulfur gases by soils. *Soil Biology and Biochemistry* **8**, 79–83.

BREMNER J. M. & STEELE C. G. (1978) Role of microorganisms in the atmospheric sulfur cycle. *Advances in Microbial Ecology* **2**, 155–201.

BRENNAN E. & LEONE I. A. (1970) Interaction of tobacco mosaic virus and ozone in *Nicotiana sylvestris*. *Journal of the Air Pollution Control Association* **20**, 470.

BROADWATER W. T., HOEHN R. C. & KING P. H. (1973) Sensitivity of three selected bacterial species to ozone. *Applied Microbiology* **26**, 391–3.

BURLESON G. R., MURRAY T. M. & POLLARD M. (1975) Inactivation of viruses and bacteria by ozone, with and without sonication. *Applied Microbiology* **29**, 340–4.

BURNS R. G. (1978) *Soil Enzymes*. Academic Press, London.

CAIRNS J., BUIKEMA A. L., HEATH A. G. & PARKER B. C. (1978) Effects of temperature on aquatic organism sensitivity to selected chemicals. Virginia Water Resources Research Center, Bulletin 106. pp. 88. Virginia Polytechnic Institute, Blacksburg, Virginia.

CAIRNS J., LANZA G. R. & PARKER B. C. (1972) Pollution related structural and functional changes in aquatic communities with emphasis on freshwater algae and protozoa. *Proceedings of the Academy of Natural Science, Philadelphia* **124**, 79–127.

CARTER J. W. & CAMERON I. L. (1973) Toxicity bioassay of heavy metals in water using *Tetrahymena pyriformis*, *Water Research* **7**, 951–61.

CASTELLANI A. G. & NIVEN C. F. (1955) Factors affecting the bacteriostatic action of sodium nitrite. *Applied Microbiology* **3**, 154–9.

CATLIN B. W. (1953) Response of *Escherichia coli* to

ferrous ions. I. Influence of temperature on the mutagenic action of Fe^{++} for a streptomycin-dependent strain. *Journal of Bacteriology* **65**, 413–21.

CENCI G. & MOROZZI G. (1977) Evaluation of the toxic effects of Cd^{2+} and $Cd(CN)_4^{2-}$ ions on the growth of mixed microbial population of activated sludges. *Science of the Total Environment* **7**, 131–43.

CHALLENGER F. (1945) Biological methylation. *Chemical Reviews* **36**, 315–61.

CHALLENGER F. & CHARLTON P. T. (1947) Studies on biological methylation. Part X. The fission of the mono- and di-sulphide links by moulds. *Journal of the Chemical Society* 424–9.

CHANEY W. R., KELLY J. M. & STRICKLAND R. C. (1978) Influence of cadmium and zinc on carbon dioxide evolution from litter and soil from a black oak forest. *Journal of Environmental Quality* **7**, 115–19.

CHANG Y., PFEFFER J. T. & CHIAN E. S. K. (1979) Comparative study of different iron compounds in inhibition of *Sphaerotilus* growth. *Applied and Environmental Microbiology* **38**, 385–9.

CHAPPELLE E. W. (1962) Carbon monoxide metabolism. *Developmental and Industrial Microbiology* **3**, 99–122.

CHAU Y. K., WONG P. T. S., SILVERBERG B. A., LUXON P. L. & BENGERT G. A. (1976) Methylation of selenium in the aquatic environment. *Science* **192**, 1130–1.

CHIMIKLIS P. E. & HEATH R. L. (1975) Ozone-induced loss of intracellular potassium ion from *Chlorella sorokiniana*. *Plant Physiology* **56**, 723–7.

CHIN B., LESOWITZ G. S. & BERNSTEIN I. A. (1978) A cellular model for studying accommodation to environmental stressors: protection and potentiation by cadmium and other metals. *Environmental Research* **16**, 432–42.

CHOPRA I. (1971) Decreased uptake of cadmium by a resistant strain of *Staphylococcus aureus*. *Journal of General Microbiology* **63**, 265–7.

CHOPRA I. (1975) Mechanisms of plasmid-mediated resistance to cadmium in *Staphylococcus aureus*. *Antimicrobial Agents and Chemotherapy* **7**, 8–14.

CHRISTENSEN E. R. & SCHERFIG J. (1979) Effects of manganese, copper, and lead on *Selenastrum capricornutum* and *Chlorella stigmatophora*. *Water Research* **13**, 79–92.

CLARK D. L., WEISS A. A. & SILVER S. (1977) Mercury and organomercurial resistances determined by plasmids in *Pseudomonas*. *Journal of Bacteriology* **132**, 186–96.

COLE M. A. (1977) Lead inhibition of enzyme synthesis in soil. *Applied and Environmental Microbiology* **33**, 262–8.

COLLINS R. P. (1976) Terpenes and odoriferous materials from microorganisms. *Lloydia* **39**, 20–4.

CONSIDINE P. J., FLYNN N. & PATCHING J. W. (1977) Ethylene production by soil microorganisms. *Applied and Environmental Microbiology* **33**, 977–9.

COTY V. F. (1969) A critical review of the utilization of methane. In: *Biotechnology and Bioengineering Symposium No. 1*, (Ed. E. L. Gaden), pp. 105–17. Interscience Publishers, Inc., New York.

COUEY H. M. (1965) Inhibition of germination of *Alternaria* spores by sulfur dioxide under various moisture conditions. *Phytopathology* **55**, 525–7.

COUEY H. M. & UOTA M. (1961) Effect of concentration, exposure time, temperature, and relative humidity on the toxicity of sulfur dioxide to the spores of *Botrytis cinerea*. *Phytopathology* **51**, 815–19.

CRAKER L. E. & MANNING W. J. (1974) SO_2 uptake by soil fungi. *Environmental Pollution* **6**, 309–11.

CRESPI H. L., HUFF D., DABOLL H. F. & KATZ J. J. (1972) Carbon monoxide in the biosphere: CO emission by fresh-water algae, Final Report to: Coordinating Research Council, 30 Rockefeller Plaza, New York.

CUER A., DAUPHIN G., KERGOMARD A., DUMONT J. P. & ADDA J. (1979) Production of S-methylthioacetate by *Brevibacterium linens*. *Applied and Environmental Microbiology* **38**, 332–4.

DARK F. A. & NASH T. (1970) Comparative toxicity of various ozonized olefins to bacteria suspended in air. *Journal of Hygiene* **68**, 245–53.

DAS G. & RUNECKLES V. C. (1974) Effects of bisulfite on metabolic development in synchronous *Chlorella pyrenoidosa*. *Environmental Research* **7**, 353–62.

DAVIES J. S., ZAJIC J. E. & WELLMAN A. M. (1974) Fungal oxidation of gaseous alkanes. *Developmental and Industrial Microbiology* **15**, 256–62.

DE BONT J. A. M. (1976) Oxidation of ethylene by soil bacteria. *Antonie van Leeuwenhoek* **42**, 59–71.

DEKONING H. W. & JEGIER Z. (1968a) Quantitative relation between ozone concentration and reduction of photosynthesis of *Euglena gracilis*. *Atmospheric Environment* **2**, 615–16.

DEKONING H. W. & JEGIER Z. (1968b) A study of the effects of ozone and sulfur dioxide on the photosynthesis and respiration of *Euglena gracilis*. *Atmospheric Environment* **2**, 321–6.

DEKONING H. W. & JEGIER Z. (1969) Effect of ozone on pyridine nucleotide reduction and phosphorylation of *Euglena gracilis*. *Archives of Environmental Health* **18**, 913–16.

DEKONING H. W. & JEGIER Z. (1970a) Effect of aldehydes on photosynthesis and respiration of *Euglena gracilis*. *Archives of Environmental Health* **20**, 720–2.

DEKONING H. W. & JEGIER Z. (1970b) Effects of sulfur dioxide and ozone on *Euglena gracilis*. *Atmospheric Environment* **4**, 357–61.

DELCOURT A. & MESTRE J. C. (1978) The effects of phenylmercuric acetate on the growth of *Chlamydomonas variabilis* Dang. *Bulletin of Environmental Contamination and Toxicology* **20**, 145–8.

DEMEREC M. & HANSON J. (1951) Mutagenic action of manganous chloride. *Cold Spring Harbor Symposia on Quantitative Biology* **16**, 215–28.

DEMIK G. & DEGROOT I. (1977) Mechanisms of inactivation of bacteriophage φX174 and its DNA in aerosols by ozone and ozonized cyclohexene. *Journal of Hygiene* **78**, 199–211.

DEMIK G. & DEGROOT I. (1978) Breaks induced in the deoxyribonucleic acid of aerosolized *Escherichia coli* by ozonized cyclohexene. *Applied and Environmental Microbiology* **35**, 6–10.

DEMIK G., DEGROOT I. & GERBRANDY J. L. F. (1977) Survival of aerosolized bacteriophage φX174 in air containing ozone-olefin mixtures. *Journal of Hygiene* **78**, 189–98.

DENISON R., CALDWELL B., BORMANN B., ELDRED L., SWANBERG C. & ANDERSON S. (1977) The effects of acid rain on nitrogen fixation in western Washington coniferous forests. *Water, Air, and Soil Pollution* **8**, 21–34.

DORAN J. W. & ALEXANDER M. (1977) Microbial transformations of selenium. *Applied and Environmental Microbiology* **33**, 31–7.

DOYLE J. J., MARSHALL R. T. & PFANDER W. H. (1975) Effects of cadmium on the growth and uptake of cadmium by microorganisms. *Applied Microbiology* **29**, 562–4.

DRUETT H. A. & PACKMAN L. P. (1972) The germicidal properties of ozone–olefin mixtures. *Journal of Applied Bacteriology* **35**, 323–9.

EHHALT D. H. (1974) The atmospheric cycle of methane. *Tellus* **26**, 58–70.

EHRLICH R. & MILLER S. (1972) Effect of NO_2 on airborne Venezuelan equine encephalomyelitis virus. *Applied Microbiology* **23**, 481–4.

ELFORD W. J. & VAN DEN ENDE J. (1942) An investigation of the merits of ozone as an aerial disinfectant. *Journal of Hygiene* **42**, 240–65.

ENEBO L. (1967) A methane-consuming green alga. *Acta Chemica Scandinavica* **21**, 625–32.

ENGEL R. R., MATESEN J. M., CHAPMAN S. S. & SCHWARTZ S. (1972) Carbon monoxide production from heme compounds by bacteria. *Journal of Bacteriology* **112**, 1310–15.

ESTES F. L. (1967) The effect of initial concentration of reactants on the biological effectiveness of photochemical reaction products. *Atmospheric Environment* **1**, 159–71.

FENNIKOH K. B., HIRSHFIELD H. I. & KNEIP T. J. (1978) Cadmium toxicity in planktonic organisms of a freshwater food web. *Environmental Research* **15**, 357–67.

FERENBAUGH R. W., GAUD W. S. & STATES J. S. (1979) Pollutant sorption by desert soils. *Bulletin of Environmental Contamination and Toxicology* **22**, 681–7.

FISHBEIN L. (1976) Atmospheric mutagens. I. Sulfur oxides and nitrogen oxides. *Mutation Research* **32**, 309–330.

FLEMING R. W. & ALEXANDER M. (1972) Dimethylselenide and dimethyltelluride formation by a strain of *Penicillium*. *Applied Microbiology* **24**, 424–9.

FORSBERG C. W. (1978) Effects of heavy metals and other trace elements on the fermentative activity of the rumen microflora and growth of functionally important rumen bacteria. *Canadian Journal of Microbiology* **24**, 298–306.

FOY C. D. & GERLOFF G. C. (1972) Response of *Chlorella pyrenoidosa* to aluminum and low pH. *Journal of Phycology* **8**, 268–71.

FRANCIS A. J., DUXBURY J. M. & ALEXANDER M. (1974) Evolution of dimethylselenide from soils. *Applied Microbiology* **28**, 248–50.

FRANCIS A. J., DUXBURY J. M. & ALEXANDER M. (1975) Formation of volatile organic products in soils under anaerobiosis. II. Metabolism of amino acids. *Soil Biology and Biochemistry* **7**, 51–6.

FRIES N. (1973) Effects of volatile organic compounds on the growth and development of fungi. *Transactions of the British Mycological Society* **60**, 1–21.

GADD G. M. & GRIFFITHS A. J. (1978) Microorganisms and heavy metal toxicity. *Microbial Ecology* **4**, 303–17.

GHIORSE W. C. & ALEXANDER M. (1977) Effect of nitrogen dioxide on nitrite oxidation and nitrite-oxidizing populations in soil. *Soil Biology and Biochemistry* **9**, 353–5.

GIASHUDDIN M. & CORNFIELD A. H. (1978) Incubation study on effects of adding varying levels of nickel (as sulphate) on nitrogen and carbon mineralization in soil. *Environmental Pollution* **15**, 231–4.

GIASHUDDIN M. & CORNFIELD A. H. (1979) Effects of adding nickel (as oxide) to soil on nitrogen and carbon mineralization at different pH levels. *Environmental Pollution* **19**, 67–70.

GIESY J. P. (1978) Cadmium inhibition of leaf decomposition in an aquatic microcosm. *Chemosphere* **7**, 467–75.

GINGELL S. M., CAMPBELL R. & MARTIN M. H. (1976) The effect of zinc, lead, and cadmium pollution on the leaf surface microflora. *Environmental Pollution* **11**, 25–37.

GRANT I. F., BANCROFT K. & ALEXANDER M. (1979) Effect of SO_2 and bisulfite on heterotrophic activity in an acid soil. *Applied and Environmental Microbiology* **38**, 78–83.

GREENAWAY W. (1971) Relationship between mercury resistance and pigment production in *Pyrenophora avenae*. *Transactions of the British Mycological Society* **56**, 37–44.

GROSS R. E. & DUGGER W. M. (1969) Responses of *Chlamydomonas reinhardtii* to peroxyacetyl nitrate. *Environmental Research* **2**, 256–66.

GROVES D. J., WILSON G. A. & YOUNG F. E. (1974) Inhibition of transformation of *Bacillus subtilis* by heavy metals. *Journal of Bacteriology* **120**, 219–26.

HAHNE H. C. H. & KROONTJE W. (1973) Significance of pH and chloride concentration on behavior of heavy metal pollutants: mercury (II), cadmium (II), zinc (II), and lead (II). *Journal of Environmental Quality* **2**, 444–50.

HALLGREN J-E. & HUSS K. (1975) Effects of SO_2 on photosynthesis and nitrogen fixation. *Physiologia Plantarum* **34**, 171–6.

HAMDY M. K. & NOYES O. R. (1975) Formation of methyl mercury by bacteria. *Applied Microbiology* **30**, 424–32.

HAMDY M. K. & WHEELER S. R. (1978) Inhibition of bacterial growth by mercury and the effects of protective agents. *Bulletin of Environmental Contamination and Toxicology* **20**, 378–86.

HAMELIN C. & CHUNG Y. S. (1974a) The effect of low concentrations of ozone on *Escherichia coli* chromosome. *Mutation Research* **28**, 131–2.

HAMELIN C. & CHUNG Y. S. (1974b) Optimal conditions for mutagenesis by ozone in *Escherichia coli* K12. *Mutation Research* **24**, 271–9.

HAMELIN C. & CHUNG Y. S. (1976) Rapid test for assay of ozone sensitivity in *Escherichia coli. Molecular and General Genetics* **145**, 191–4.

HANNAN P. J. & PATOUILLET C. (1972) Effects of pollutants on growth of algae. Report of NRL Progress, pp. 1–8.

HANNAN P. J. & PATOUILLET C. (1979) An algal toxicity test and evaluation of adsorption effect. *Journal of the Water Pollution Control Federation* **51**, 834–40.

HARRISS R. C., WHITE D. B. & MACFARLANE R. B. (1970) Mercury compounds reduce photosynthesis by plankton. *Nature, London* **170**, 736–7.

HART B. A., BERTRAM P. E. & SCAIFE B. D. (1979) Cadmium transport by *Chlorella pyrenoidosa*. *Environmental Research* **18**, 327–35.

HART B. A. & SCAIFE B. D. (1977) Toxicity and bioaccumulation of cadmium in *Chlorella pyrenoidosa*. *Environmental Research* **14**, 401–13.

HARTIG W. J. (1971) Studies on mercury toxicity in *Tetrahymena pyriformis*. *Journal of Protozoology* Supplement **18**, 26.

HARTMAN L. M. (1974) A preliminary report: fungal flora of the soil as conditioned by varying concentrations of heavy metals. *American Journal of Botany* Supplement **61**, 23.

HAYWARD N. J., JEAVONS T. H., NICHOLSON A. J. C. & THORNTON A. G. (1977) Methyl mercaptan and dimethyl disulfide production from methionine by *Proteus* species detected by head-space gas-liquid chromatography. *Journal of Clinical Microbiology* **6**, 187–94.

HEAGLE A. S. (1973) Interactions between air pollutants and plant parasites. *Annual Review of Phytopathology* **11**, 365–88.

HEAGLE A. S. (1975) Response of three obligate parasites to ozone. *Environmental Pollution* **9**, 91–5.

HENDRICKS A. C. (1978) Response of *Selenastrum capricornutum* to zinc sulfides. *Journal of the Water Pollution Control Federation* **50**, 163–8.

HENRIKSSON L. E. & DASILVA E. J. (1978) Effects of some inorganic elements on nitrogen-fixation in blue-green algae and some ecological aspects of pollution. *Zeitschrift für Allgemeine Mikrobiologie* **18**, 487–94.

HIBBEN C. R. & STOTZKY G. (1969) Effects of ozone on the germination of fungus spores. *Canadian Journal of Microbiology* **15**, 1187–96.

HIBBEN C. R. & TAYLOR M. P. (1975) Ozone and sulfur dioxide effects on the lilac powdery mildew fungus. *Environmental Pollution* **9**, 107–14.

HIRSCH H. M. (1961) Small colony variants of *Escherichia coli. Journal of Bacteriology* **81**, 448–58.

HIRSCH P. (1968) Photosynthetic bacterium growing under carbon monoxide. *Nature, London* **217**, 555–6.

HORITSU H., TAKAGI M. & TOMOYEDA M. (1978) Isolation of a mercuric chloride-tolerant bacterium and uptake of mercury by the bacterium. *Journal of Applied Microbiology and Biotechnology* **5**, 279–90.

HORITSU H. & TOMOYEDA M. (1975) A comparative study on characters of heavy metallic ion-tolerant microorganisms. In: *Proceedings of the First Intersectional Congress of the International Association of Microbiological Societies* (Ed. T. Hasegawa), vol. 2, pp. 522–6. Science Council of Japan.

HORSFALL J. G. (1956) *Principles of Fungicidal Action*. Chronica Botanica Co., Waltham, Mass.

HUANG C-Y., BAZZAZ F. A. & VANDERHOEF L. N. (1974) The inhibition of soybean metabolism by cadmium and lead. *Plant Physiology* **54**, 122–4.

HUBLEY J. H., MITTEN J. R. & WILKINSON J. F. (1974) The oxidation of carbon monoxide by methane-oxidizing bacteria. *Archiv für Mikrobiologie* **95**, 365–8.

HUTCHINSON S. A. (1971) Biological activity of volatile fungal metabolites. *Transactions of the British Mycological Society* **57**, 185–200.

HUTCHINSON T. C. (1973) Comparative studies of the toxicity of heavy metals to phytoplankton and their synergistic interactions. *Water Pollution Research, Canada* **8**, 68–90.

IMAI K., SUGIO T., TSUCHIDA T. & TANO T. (1975) Effect of heavy metal ions on the growth and iron-oxidizing activity of *Thiobacillus ferroxidans. Agricultural and Biological Chemistry* **39**, 1349–54.

INGRAM M. & HAINES R. B. (1949) Inhibition of bacterial growth by pure ozone in the presence of nutrients. *Journal of Hygiene* **47**, 146–58.

INMAN R. E. & INGERSOLL R. B. (1971) Uptake of carbon monoxide by soil fungi. *Journal of the Air Pollution Control Association* **21**, 646–7.

INMAN R. E., INGERSOLL R. B. & LEVY F. A. (1971) Soil: a natural sink for carbon monoxide. *Science* **172**, 1229–31.

JACUMIN W. J., JOHNSTON D. R. & RIPPERTON L. A. (1964) Exposure of microorganisms to low concentrations of various pollutants. *American Industrial Hygiene Association Journal* **25**, 595–600.

JANDA J. M. & FLEMING R. W. (1978) Effect of selenate

toxicity on soil microflora. *Journal of Environmental Science and Health* **A13**, 697–706.

JEFFRIES T. W. & BUTLER R. G. (1975) Growth inhibition of *Rhodopseudomonas capsulata* by methylmercury acetate. *Applied Microbiology* **30**, 156–8.

JORDAN M. J. & LECHEVALIER M. P. (1975) Effects of zinc-smelter emissions on forest soil microflora. *Canadian Journal of Microbiology* **21**, 1855–65.

JUNGE C. E., SEILER W., SCHMIDT U., BOCK R., GREESE K. D., RADLER R. & RUGER H. J. (1972) Kohlenoxyd und Wasserstoffproduktion mariner Mikroorganismen in Nahrmedien mit syntätischem Seewasser. *Naturwissenschaften* **59**, 514–15.

KADOTA H. & ISHIDA Y. (1972) Production of volatile sulfur compounds by microorganisms. *Annual Review of Microbiology* **26**, 127–38.

KALLQVIST T. & MEADOWS B. S. (1978) The toxic effect of copper on algae and rotifers from a soda lake (Lake Nakuru, East Africa). *Water Research* **12**, 771–5.

KAYSER H. (1977) Effect of zinc sulphate on the growth of mono- and multi-species cultures of some marine plankton algae. *Helgolander wissenschafter Meeresuntersuchung* **30**, 682–96.

KLASS E., ROWE D. W. & MASSARO E. J. (1974) The effect of cadmium on population growth of the green alga, *Scenedesmus quadracauda. Bulletin of Environmental Contamination and Toxicology* **12**, 442–5.

KNAUER G. A. & MARTIN J. J. (1972) Mercury in a marine pelagic food chain. *Limnology and Oceanography* **17**, 868–76.

KNOWLES S. C. & ZINGMARK R. G. (1978) Mercury and temperature interactions on the growth rates of three species of freshwater phytoplankton. *Journal of Phycology* **14**, 104–9.

KONDO I., ISHIKAWA T. & NAKAHARA H. (1974) Mercury and cadmium resistances mediated by the penicillinase plasmids in *Staphylococcus aureus. Journal of Bacteriology* **117**, 1–7.

KOZLOFF L. M., LUTE M. & HENDERSON K. (1957) Viral invasion. I. Rupture of thiol ester bonds in the bacteriophage tail. *Journal of Biological Chemistry* **228**, 511–28.

KRAUSE C. R. & WEIDENSAUL T. C. (1978) Effects of ozone on the sporulation, germination, and pathogenicity of *Botrytis cinerea. Phytopathology* **68**, 195–8.

KUSS F. R. (1950) The effect of ozone on fungus sporulation. M.S. thesis, University of New Hampshire, Durham, New Hampshire, U.S.A.

LABEDA D. P. & ALEXANDER M. (1978) Effects of SO_2 and NO_2 on nitrification in soil. *Journal of Environmental Quality* **7**, 523–6.

LABOREY F. & LAVOLLAY J. (1967) Sur la toxicité exercée par Zn^{++} et Cd^{++} dans la croissance d'*Aspergillus niger*, l'antagonisme de ces ions et l'interaction Mg^{++}–Zn^{++}–Cd^{++}. *Comptes Rendus de l'Academie des Sciences*, Series D **264**, 2937–40.

LABOREY F. & LAVOLLAY J. (1973) Sur la nature des antagonismes responsables de l'interaction des ions Mg^{++}, Cd^{++}, et Zn^{++} dans croissance d'*Aspergillus niger. Comptes Rendus de l'Academie des Sciences*, Series D **276**, 529–32.

LABOREY F. & LAVOLLAY J. (1977) Sur l'antitoxicité du calcium et du magnesium a l'egard du cadmium dans la croissance d'*Aspergillus niger. Comptes Rendus de l'Academie des Sciences*, Series D **284**, 639–42.

LANDA E. R. & FANG S. C. (1978) Effect of mercuric chloride on carbon mineralization in soil. *Plant and Soil* **49**, 179–83.

LAURENCE J. A. & WOOD F. A. (1978a) Effects of ozone on infection of wild strawberry by *Xanthomonas fragariae. Phytopathology* **68**, 689–92.

LAURENCE J. A. & WOOD F. A. (1978b) Effects of ozone on infection of soybean by *Pseudomonas glycinea. Phytopathology* **68**, 441–5.

LAWREY J. D. & HALE M. E. JR. (1979) Lichen growth responses to stress induced by automobile exhaust pollution. *Science* **204**, 423–4.

LE TOURNEAU D. (1978) Inhibition of sclerotial formation of *Sclerotium rolfsii* by cadmium. *Mycologia* **70**, 849–52.

LIANG C. N. & TABATABAI M. A. (1978) Effects of trace elements on nitrification in soils. *Journal of Environmental Quality* **7**, 291–3.

LIGHTHART B. (1973) Survival of airborne bacteria in a high urban concentration of carbon monoxide. *Applied Microbiology* **25**, 86–91.

LIGHTHART B. (1979) Enrichment of cadmium-mediated antibiotic resistant bacteria in a Douglas-fir (*Pseudotsuga menziesii*) litter microcosm. *Applied and Environmental Microbiology* **37**, 859–61.

LIGHTHART B., HIATT V. E. & ROSSANO A. T. (1971) The survival of airborne *Serratia marcescens* in urban concentrations of sulfur dioxide. *Journal of the Air Pollution Control Association* **21**, 639–42.

LIND C. J. & WILSON P. W. (1941) mechanisms of biological nitrogen fixation; VIII. Carbon monoxide as an inhibitor of nitrogen fixation by red clover. *Journal of the American Chemical Society* **63**, 3511–14.

LIND C. J. & WILSON P. W. (1942) Carbon monoxide inhibition of nitrogen fixation by *Azotobacter. Archives of Biochemistry* **1**, 59–72.

LOVELOCK J. E. (1974) CS_2 and the natural sulphur cycle. *Nature, London* **248**, 625–6.

LOVELOCK J. E., MAGGS R. J. & RASMUSSEN R. A. (1972) Atmospheric dimethyl sulphide and the natural sulphur cycle. *Nature, London* **237**, 452–3.

LYNCH J. M. (1974) Mode of ethylene formation by *Mucor hiemalis. Journal of General Microbiology* **83**, 407–11.

MCBRIEN D. C. H. & HASSALL K. A. (1967) The effect of toxic doses of copper upon respiration, photosynthesis, and growth of *Chlorella vulgaris. Physiologia Plantarum* **20**, 114–17.

MCCALLAN S. E. A. & WEEDON F. R. (1940) Toxicity of

ammonia, chlorine, hydrogen cyanide, hydrogen sulfide, and sulfur dioxide gases. II. Fungi and bacteria. *Contributions of the Boyce Thompson Institute for Plant Research* **11**, 331–42.

McCool P. M., Menge J. A. & Taylor O. C. (1979) Effects of ozone and HCl gas on the development of the mycorrhizal fungus *Glomus fasciculatus* and growth of "Troyer" citrange. *Journal of the American Society for Horticultural Science* **104**, 151–4.

McLee A. G., Kormendy A. C. & Wayman M. (1972) Isolation and characterization of n-butane-utilizing microorganisms. *Canadian Journal of Microbiology* **18**, 1191–5.

MacLeod R. A., Kuo S. C. & Gelinas R. (1967) Metabolic injury to bacteria. II. Metabolic injury induced by distilled water or Cu^{++} in the plating diluent. *Journal of Bacteriology* **93**, 961–9.

MacLeod R. A. & Snell E. E. (1950) The relation of ion antagonism to the inorganic nutrition of lactic acid bacteria. *Journal of Bacteriology* **59**, 783–92.

Mah R. A., Ward D. M., Baresi L. & Glass T. L. (1977) Biogenesis of methane. *Annual Review of Microbiology* **31**, 309–41.

Makarov B. N. (1969) Liberation of nitrogen dioxide from soil. *Soviet Soil Science* **1**, 20–5.

Malanchuk J. L. & Gruendling G. K. (1973) Toxicity of lead nitrate to algae. *Water, Air, and Soil Pollution* **2**, 181–90.

Malo B. A. & Purvis E. R. (1964) Soil absorption of atmospheric ammonia. *Soil Science* **97**, 242–7.

Manning W. J. (1975) Interactions between air pollutants and fungal, bacterial, and viral plant pathogens. *Environmental Pollution* **9**, 87–90.

Manning W. J., Feder W. A., Papia P. M. & Perkins I. (1971) Influence of foliar ozone injury on root development and root surface fungi on pinto bean plants. *Environmental Pollution* **1**, 305–12.

Manning W. J., Feder W. A. & Perkins I. (1970) Ozone injury increases infection of geranium leaves by *Botrytis cinerea*. *Phytopathology* **60**, 669–70.

Manning W. J., Feder W. A., Perkins I. & Glickman M. (1969) Ozone injury and infection of potato leaves by *Botrytis cinerea*. *Plant Disease Reporter* **53**, 691–3.

Marchesani W. (1969) A study of the effect of air pollution on fungus. *Atmospheric Environment* **3**, 685.

Mierle G. & Stokes P. M. (1976) Heavy metal tolerance and metal accumulation by planktonic algae, In: *Trace Substances in Environmental Health—X* (Ed. D. D. Hemphill), pp. 113–22. University of Missouri, Columbia, Missouri.

Mills A. L. & Colwell R. R. (1977) Microbiological effects of metal ions in Chesapeake Bay water and sediment. *Bulletin of Environmental Contamination and Toxicology* **18**, 99–103.

Mitra R. S. & Bernstein I. A. (1978) Single-strand breakage in DNA of *Escherichia coli* exposed to Cd^{2+}. *Journal of Bacteriology* **133**, 75–80.

Mitra R. S., Gray R. H., Chin B. & Bernstein I. A. (1975) Molecular mechanisms of accommodation in *Escherichia coli* to toxic levels of Cd^{2+}. *Journal of Bacteriology* **121**, 1180–8.

Morozzi G. & Cenci G. (1978) Comparison of the toxicity of some metals and their tetracyanide complexes on the respiration of nonacclimated activated sludges. *Zentralblatt für Bakteriologie, Parasitenkunde, Infektionskrankheiten und Hygiene 1. Abteilung Originale, Reihe B* **167**, 478–88.

Nagy Z., Solymossy M. & Antoni F. (1977) Lethal effect of nitrous acid on *Escherichia coli*. *Mutation Research* **42**, 191–204.

Nakamura H. (1961) Adaptation of yeast to cadmium. II. Significance of phosphorus in cadmium resistance. *Memoirs of the Konan University, Science Series* **5**, 89–98.

Nakano Y., Okamoto K., Toda S. & Fuwa F. (1978) Toxic effects of cadmium on *Euglena gracilis* grown in zinc deficient and zinc sufficient media. *Agricultural and Biological Chemistry* **42**, 901–7.

Nash T. H., III (1972) Simplication of the Blue Mountain lichen communities near a zinc factory. *Bryologist* **75**, 315–24.

Nelson J. D. & Colwell R. R. (1975) The ecology of mercury-resistant bacteria in Chesapeake Bay. *Microbial Ecology* **1**, 191–218.

Nieboer E., Richardson D. H. S., Puckett K. J. & Tomassini F. D. (1976) The phytotoxicity of sulphur dioxide in relation to measurable responses in lichens, In: *Effects of Air Pollutants on Plants* (Ed. T. A. Mansfield), pp. 61–85. Cambridge University Press, Cambridge.

Nishioka H. (1975) Mutagenic activities of metal compounds in bacteria. *Mutation Research* **31**, 185–9.

Nozhevnikova A. N. & Yurganov L. N. (1978) Microbiological aspects of regulating the carbon monoxide content in the earth's atmosphere. *Advances in Microbial Ecology* **2**, 203–44.

Nuzzi R. (1972) Toxicity of mercury to phytoplankton. *Nature, London* **237**, 38–40.

Ohta T. & Udaka S. (1977) Isolation of cadmium- and mercury-sensitive mutants of *Escherichia coli* and some factors influencing their sensitivities. *Agricultural and Biological Chemistry* **41**, 461–6.

Orgel A. & Orgel L. E. (1965) Induction of mutations in bacteriophage T4 with divalent manganese. *Journal of Molecular Biology* **14**, 453–7.

Ou J. T. & Anderson T. F. (1972) Effect of Zn^{2+} on bacterial conjugation: inhibition of mating pair formation. *Journal of Bacteriology* **111**, 177–85.

Pace F., Ferrara R. & del Carratore G. (1977) Effects of sub-lethal doses of copper sulphate and lead nitrate on growth and pigment composition of *Dunaliella salina* Teod. *Bulletin of Environmental Contamination and Toxicology* **17**, 679–85.

Peak M. J. & Belser W. L. (1969) Some effects of the air

pollutant, peroxyacetyl nitrate, upon deoxyribonucleic acid and nucleic acid bases. *Atmospheric Environment* **3**, 385–97.

PETERSON W. H., BURRIS R. H., SANT R. & LITTLE H. N. (1958) Production of toxic gases (nitrogen oxides) in silage making. *Journal of Agricultural and Food Chemistry* **6**, 121–6.

PETRILLI F. L. & DEFLORA S. (1977) Toxicity and mutagenicity of hexavalent chromium on *Salmonella typhimurium*. *Applied and Environmental Microbiology* **33**, 805–9.

PICKETT A. W. & DEAN A. C. R. (1976) Cadmium and zinc sensitivity and tolerance in *Klebsiella (Aerobacter) aerogenes*. *Microbios* **15**, 79–91.

POLATNICK J. & BACHRACH H. L. (1978) Effect of zinc and other chemical agents on foot-and-mouth disease virus replication. *Antimicrobial Agents and Chemotherapy* **13**, 731–4.

PREMI P. R. & CORNFIELD A. H. (1969) Effects of addition of copper, manganese, zinc, and chromium compounds on ammonification and nitrification during incubation of soil. *Plant and Soil* **31**, 345–52.

PRIMROSE S. B. (1976) Ethylene-forming bacteria from soil and water. *Journal of General Microbiology* **97**, 343–6.

PRIMROSE S. B. (1979) Ethylene and agriculture: the role of the microbe. *Journal of Applied Bacteriology* **46**, 1–25.

PRIMROSE S. B. & DILWORTH M. J. (1976) Ethylene production by bacteria. *Journal of General Microbiology* **93**, 177–81.

PUCKETT K. J., NIEBOER E., FLORA W. P. & RICHARDSON D. H. S. (1973) Sulphur dioxide: its effect on photosynthetic ^{14}C fixation in lichens and suggested mechanisms of phytotoxicity. *New Phytologist* **72**, 141–54.

PUCKETT K. J., TOMASSINI F. D., NIEBOER E. & RICHARDSON D. H. S. (1977) Potassium efflux by lichen thalli following exposure to aqueous sulphur dioxide. *New Phytologist* **79**, 135–45.

RAHN O. & CONN J. E. (1944) Effect of increase in acidity on antiseptic efficiency. *Industrial and Engineering Chemistry* **36**, 185–7.

RAO D. N. & LEBLANC F. (1966) Effects of sulphur dioxide on the lichen algae, with special reference to chlorophyll. *Bryologist* **69**, 69–75.

RAO D. N. & PAL D. (1978) Effect of fluoride pollution on the organic matter content of soil. *Plant and Soil* **49**, 653–6.

RASMUSSEN R. A. (1974) Emission of biogenic hydrogen sulfide. *Tellus* **26**, 254–60.

RASMUSSEN K. H., TAHERI M. & KABEL R. L. (1975) Global emissions and natural processes for removal of gaseous pollutants. *Water, Air, and Soil Pollution* **4**, 33–64.

RICH S. & TOMLINSON H. (1968) Effects of ozone on conidiophore and conidia of *Alternaria solani*. *Phytopathology* **58**, 444–6.

RODERER G. (1979) Hemmung der Cytokinese und Bildung von Riesenzellen bei *Poterioochromonas malhamensis* durch organische Bleiverbindungen und andere Agenzien. *Protoplasma* **99**, 39–51.

RODERER G. & SCHNEPF E. (1977) Tetraethyl lead and triethyl lead inhibit cytokinesis of the chrysophycean flagellate *Poterioochromonas*. *Naturwissennschaften* **64**, 588.

ROSENZWEIG W. D. & STOTZKY G. (1980) Influence of environmental factors on antagonism of fungi by bacteria in soil: nutrient levels. *Applied and Environmental Microbiology* **39**, 354–60.

ROWE J. J., YARBROUGH J. M., RAKE J. B. & EAGON R. G. (1979) Nitrite inhibition of aerobic bacteria. *Current Microbiology* **2**, 51–4.

RUETER J. G., MCCARTHY J. J. & CARPENTER E. J. (1979) The toxic effect of copper on *Oscillatoria (Trichondesmium) theibautii*. *Limnology and Oceanography* **24**, 558–62.

RUHLING A. & TYLER G. (1973) Heavy metal pollution and decomposition of spruce needle litter. *Oikos* **24**, 402–16.

RUTHYEN J. A. & CAIRNS J. (1973) Response of freshwater protozoan artificial communities to metals. *Journal of Protozoology* **20**, 127–35.

SALO R. J. & CLIVER D. O. (1978) Inactivation of enteroviruses by ascorbic acid and sodium bisulfite. *Applied and Environmental Microbiology* **36**, 68–75.

SAUNDERS P. J. W. (1966) The toxicity of sulphur dioxide to *Diplocarpon rosae* Wolf causing blackspot of roses. *Annals of Applied Biology* **58**, 103–14.

SAUNDERS P. J. W. (1970) Air pollution in relation to lichens and fungi. *Lichenologist* **4**, 337–49.

SAUNDERS P. J. W. (1971) Modification of the leaf surface and its environment by pollution, In: *Ecology of Leaf Surface Micro-organisms* (Eds T. F. Preece & C. H. Dickinson), pp. 81–9. Academic Press, New York.

SCALETTI J. V., GATES C. E., BRIGGS R. A. & SCHUMAN L. M. (1960) Nitrogen dioxide production from silage. I. Field survey. *Agronomy Journal* **52**, 369–72.

SCHLEGEL H. G. (1974) Production, modification, and consumption of atmospheric trace gases by microorganisms. *Tellus* **26**, 11–20.

SCHULZ-BALDES M. & LEWIN R. A. (1976) Lead uptake in two marine phytoplankton organisms. *Biological Bulletin* **150**, 118–27.

SCHULZE H. & BRAND J. J. (1978) Lead toxicity and phosphate deficiency in *Chlamydomonas*. *Plant Physiology* **62**, 727–30.

SCOTT D. B. M. & LESHER E. C. (1963) Effect of ozone on survival and permeability of *Escherichia coli*. *Journal of Bacteriology* **85**, 567–76.

SERAT W. F., BUDINGER F. E. & MUELLER P. K. (1965) Evaluation of biological effects of air pollutants by use of luminescent bacteria. *Journal of Bacteriology* **90**, 832–3.

SERAT W. F., KYONO J. & MUELLER P. K. (1969)

Measuring the effect of air pollutants on bacterial luminescence: a simplified procedure. *Atmospheric Environment* **3**, 303–9.

SHAPIRO R. (1977) Genetic effects of bisulfite (sulfur dioxide). *Mutation Research* **39**, 149–76.

SHERIDAN R. P. (1978) Toxicity of bisulfite to photosynthesis and respiration. *Journal of Phycology* **14**, 279–81.

SHERIDAN R. P. (1979) Impact of emissions from coal-fired electricity generating facilities on N_2-fixing lichens. *Bryologist* **82**, 54–8.

SHIOI Y., TAMAI H. & SASA T. (1978) Inhibition of photosystem II in the green alga *Ankistrodesmus falcatus* by copper. *Physiologia Plantarum* **44**, 434–8.

SHOWMAN R. E. (1972) Residual effects of sulfur dioxide on the net photosynthetic and respiratory rates of lichen thalli and cultured lichen symbionts. *Bryologist* **75**, 335–41.

SHRINER D. S. (1977) Effects of simulated rain acidified with sulfuric acid on host–parasite interactions. *Water, Air, and Soil Pollution* **8**, 9–14.

SINGH N. (1977) Effect of pH on the tolerance of *Penicillium nigricans* to copper and other heavy metals. *Mycologia* **69**, 750–5.

SINGH S. P. & KASHYAP A. K. (1978) Manganese toxicity and mutagenesis in two blue-green algae. *Environmental and Experimental Botany* **18**, 47–53.

SMITH E. A. & MAYFIELD C. I. (1978) Effects of nitrogen oxide on selected soil processes. *Water, Air, and Soil Pollution* **9**, 33–43.

SMITH K. A., BREMNER J. M. & TABATABAI M. A. (1973) Sorption of gaseous atmospheric pollutants by soils. *Soil Science* **116**, 313–19.

SMITH L. JR. & SIE E. H. C. (1969) Response of luminescent bacteria to common atmospheric pollutants. *Proceedings of the Institute of Environmental Sciences, 1969 Annual Technical Meeting*, pp. 154–7. Institute of Environmental Sciences, Mt. Prospect, Illinois.

SMITH W. H. (1977) Influence of heavy metal leaf contaminants on the *in vitro* growth of urban-tree phylloplane-fungi. *Microbial Ecology* **3**, 231–9.

SMITH W. H., STASKAWICZ B. J. & HARKOV R. S. (1978) Trace-metal pollutants and urban-tree leaf pathogens. *Transactions of the British Mycological Society* **70**, 29–33.

SPALDING B. (1979) Effects of divalent metal chlorides on respiration and extractable enzymatic activities of Douglas-fir needle litter. *Journal of Environmental Quality* **8**, 105–9.

SPANGLER W. J., SPIGARELLI J. L., ROSE J. M. & MILLER H. M. (1973) Methyl mercury: bacterial degradation in lake sediments. *Science* **180**, 192–3.

STARKEY R. L. (1950) Relations of microorganisms to transformations of sulfur in soils. *Soil Science* **70**, 55–65.

STARKEY R. L. (1956) Transformations of sulfur by microorganisms. *Industrial and Engineering Chemistry* **48**, 1429–37.

STARKEY R. L. (1973) Effect of pH on toxicity of copper to *Scytalidium* sp., a copper-tolerant fungus, and some other fungi. *Journal of General Microbiology* **78**, 217–25.

STEEMANN NIELSEN E., KAMP-NIELSEN L. & WIUM-ANDERSEN W. (1969) The effect of deleterious concentrations of copper on the photosynthesis of *Chlorella pyrenoidosa*. *Physiologia Plantarum* **22**, 1121–33.

STOKES P. M. & HUTCHINSON T. C. (1975) Copper toxicity to phytoplankton, as affected by organic ligands, other cations, and inherent tolerance of algae to copper. In: *Symposium of the International Joint Commission on the Great Lakes* (Eds R. W. Andrew, P. V. Wodson & D. E. Konaslevich), pp. 159–85. Windsor, Canada.

STOKES P. M., HUTCHINSON T. C. & KRAUTER K. (1973) Heavy-metal tolerance in algae isolated from contaminated lakes near Sudbury, Ontario. *Canadian Journal of Botany* **51**, 2155–68.

STOKES P. M. & LINDSAY J. E. (1979) Copper tolerance and accumulation in *Penicillium ochro-chloron* isolated from copper-plating solution. *Mycologia* 71, 796–806.

STOTZKY G. (1972) Activity, ecology, and population dynamics of microorganisms in soil. *Critical Reviews in Microbiology* **2**, 59–137.

STOTZKY G. & BABICH H. (1980) Mediation of the toxicity of pollutants to microbes by the physicochemical composition of the recipient environment. In: *Microbiology—1980* (Ed. D. Schlessinger), pp. 352–4. American Society for Microbiology, Washington, D.C.

STOTZKY G. & SCHENCK S. (1976) Volatile organic compounds and microorganisms. *Critical Reviews in Microbiology* **4**, 333–82.

STRATTON G. W. & CORKE C. T. (1979) The effect of cadmium ion on the growth, photosynthesis, and nitrogenase activity of *Anabaena inaequalis*. *Chemosphere* **8**, 277–82.

STRATTON G. W., HUBER A. L. & CORKE C. T. (1979) Effect of mercuric ion on the growth, photosynthesis, and nitrogenase activity of *Anabaena inaequalis*. *Applied and Environmental Microbiology* **38**, 537–43.

SUMMERS A. O. & LEWIS E. (1973) Volatilization of mercuric chloride by mercury-resistant plasmid-bearing strains of *Escherichia coli*, *Staphylococcus aureus*, and *Pseudomonas aureus*. *Journal of Bacteriology* **113**, 1070–3.

SUMMERS A. O. & SILVER S. (1978) Microbial transformation of metals. *Annual Review of Microbiology* **32**, 637–72.

SWANSON B. T., WILKINS H. F. & KENNEDY B. W. (1979) Factors affecting ethylene production by some plant pathogenic bacteria. *Plant and Soil* **51**, 19–26.

SWINNERTON J. W., LINNENBOM V. J. & LAMONTAGNE R. A. (1970) The ocean: a natural source of carbon monoxide. *Science* **167**, 984–6.

SZETO C. & NYBERG D. (1979) The effect of temperature on copper tolerance of *Paramecium*. *Bulletin Environmental Contamination and Toxicology* **21**, 131–5.

TAMM C. O. (1976) Acid precipitation: biological effects in soil and on forest vegetation. *Ambio* **5**, 235–8.

TATSUYAMA K., EGAWA H., SENMARU H., YAMAMOTO H., ISHIOKA S., TAMATSUKURI T. & SAITO K. (1975) *Penicillium lilacinum*: its tolerance to cadmium. *Experientia* **31**, 1037.

THOMAS K. C. & SPENCER M. (1977) L-methionine as an ethylene precursor in *Saccharomyces cerevisiae*. *Canadian Journal of Microbiology* **23**, 1669–74.

TIMONEY J. F., PORT J., GILES J. & SPANIER J. (1978) Heavy-metal and antibiotic resistance in the bacterial flora of sediments of New York Bight. *Applied and Environmental Microbiology* **36**, 465–72.

TINGLE L. E., PAVLAT W. A. & CAMERON I. L. (1973) Sublethal cytoxic effects of mercuric chloride on the ciliate *Tetrahymena pyriformis*. *Journal of Protozoology* **20**, 301–4.

TORNABENE T. G. & EDWARDS H. W. (1972) Microbial uptake of lead. *Science* **176**, 1334–5.

TRESHOW M., DEAN G. & HARNER F. M. (1967) Stimulation of tobacco mosaic virus-induced lesions on bean by fluoride. *Phytopathology* **57**, 756–8.

TRESHOW M., HARNER F. M., PRICE H. E. & KORMELINK J. R. (1969) Effects of ozone on growth, lipid metabolism, and sporulation of fungi. *Phytopathology* **59**, 1223–5.

TURNER N. C., RICH S. & WAGGONER P. E. (1973) Removal of ozone by soil. *Journal of Environmental Quality* **2**, 259–64.

TYLER G. (1972) Heavy metals pollute nature, may reduce productivity. *Ambio* **1**, 53–9.

TYLER G. (1976a) Heavy metal pollution, phosphatase activity, and mineralization of organic phosphorus in forest soils. *Soil Biology and Biochemistry* **8**, 327–32.

TYLER G. (1976b) Influence of vanadium on soil phosphatase activity. *Journal of Environmental Quality* **5**, 216–17.

TYLER G., MORNSJO B. & NILSSON B. (1974) Effects of cadmium, lead, and sodium salts on nitrification in a mull soil. *Plant and Soil* **40**, 237–42.

TYNECKA Z., ZAJAC J. & GOS Z. (1975) Plasmid dependent impermeability barrier to cadmium ions in *Staphylococcus aureus*. *Acta Microbiologica Polonica Series A*. **7**, 11–20.

TYNECKA Z. & ZYLINSKA W. (1974) Plasmid borne resistance to some inorganic ions in *Staphylococcus aureus*. *Acta Microbiologica Polonica Series A* **6**, 83–92.

UPITIS V. V., PAKALNE D. S. & NOLLENDORF A. F. (1973) The dosage of trace elements in the nutrient medium as a factor in increasing the resistance of *Chlorella* to unfavorable conditions of culturing. *Microbiology* **42**, 758–62.

VAITUZIS Z., NELSON J. D., WAN L. W. & COLWELL R. R. (1975) Effects of mercuric chloride on growth and morphology of selected strains of mercury-resistant bacteria. *Applied Microbiology* **29**, 275–86.

VALLEE B. L. & ULMER D. D. (1972) Biochemical effects of mercury, cadmium, and lead. *Annual Review of Biochemistry* **41**, 91–128.

VENITT S. & LEVY L. S. (1974) Mutagenicity of chromates in bacteria and its relevance to chromate carcinogenesis. *Nature, London* **250**, 493–5.

VONK J. W. & SIJPESTEIJN A. K. (1973) Studies on the methylation of mercuric chloride by pure cultures of bacteria and fungi. *Antonie van Leeuwenhoek* **39**, 505–13.

WAINWRIGHT M. (1978) Sulphur-oxidising microorganisms on vegetation and in soils exposed to atmospheric pollution. *Environmental Pollution* **17**, 167–74.

WAINWRIGHT M. (1979) S-oxidation in soils exposed to heavy atmospheric pollution. *Soil Biology and Biochemistry* **11**, 95–8.

WARD R. L. & ASHLEY C. S. (1978) Identification of the virucidal agent in wastewater sludge. *Applied and Environmental Microbiology* **33**, 860–4.

WATSON R. D. (1942) Ozone as a fungicide. Ph.D. thesis, Cornell University, Ithaca, New York.

WESTLAKE D. W. S., ROXBURGH J. M. & TALBOT G. (1961) Microbial production of carbon monoxide from flavonoids. *Nature, London* **189**, 510–11.

WILSON D. O. (1977) Nitrification in three soils amended with zinc sulfate. *Soil Biology and Biochemistry* **9**, 277–80.

WODZINSKI R. S. & ALEXANDER M. (1978) Effect of sulfur dioxide on algae. *Journal of Environmental Quality* **7**, 358–60.

WODZINSKI R. S., LABEDA D. P. & ALEXANDER M. (1978) Effects of low concentrations of bisulfite-sulfite and nitrite on microorganisms. *Applied and Environmental Microbiology* **35**, 718–23.

WON W. D. & ROSS H. (1969) Reaction of airborne *Rhizobium meliloti* to some environmental factors. *Applied Microbiology* **18**, 555–7.

WONG P. T. S., CHAU Y. K. & LUXON P. L. (1975) Methylation of lead in the environment. *Nature, London* **253**, 263–4.

WOOD D. A. & HAMMOND J. B. W. (1977) Ethylene production by axenic fruiting cultures of *Agaricus bisporus*. *Applied and Environmental Microbiology* **34**, 228–9.

YOSHIDA T. & ALEXANDER M. (1979) Nitrous oxide formation by *Nitrosomonas europaea* and heterotrophic microorganisms. *Soil Science Society of America Journal* **34**, 880–2.

ZEIKUS J. G. & WARD J. C. (1974) Methane formation in living trees: a microbial origin. *Science* **184**, 1181–3.

ZINDER S. H. & BROCK T. D. (1978) Production of methane and carbon dioxide from methane thiol and dimethyl sulphide by anaerobic lake sediments. *Nature, London* **273**, 226–8.

ZINDER S. H., DOEMEL W. N. & BROCK T. D. (1977) Production of volatile sulfur compounds during the decomposition of algal mats. *Applied and Environmental Microbiology* **34**, 859–60.

ZLOCHEVSKAYA I. V. & RUKHADZE E. G. (1968) A study of the toxic action of some lead compounds. *Microbiology* **37**, 951–5.

ZWARUN A. A. (1973) Tolerance of *Escherichia coli* to cadmium. *Journal of Environmental Quality* **2**, 353–5.

Index